BIOLOGY

concepts and applications

FIFTH EDITION

CECIE STARR

LISA STARR
Biology Illustrator

BROOKS/COLE
™
THOMSON LEARNING

*Australia • Canada • Mexico • Singapore • Spain
United Kingdom • United States*

BROOKS/COLE

THOMSON LEARNING

BIOLOGY PUBLISHER: Jack C. Carey

DEVELOPMENT EDITOR: Mary Arbogast

ASSISTANT EDITOR: Mark Andrews, Suzannah Alexander

EDITORIAL ASSISTANT: Karen Hansten

MEDIA PROJECT MANAGER: Pat Waldo

TECHNOLOGY PROJECT MANAGER: Keli Amann

MARKETING MANAGER: Tom Ziolkowski

MARKETING ASSISTANT: Maureen Griffin

ADVERTISING PROJECT MANAGER: Linda Yip

SENIOR PROJECT MANAGER, EDITORIAL/PRODUCTION: Teri Hyde

PRINT/MEDIA BUYER: Karen Hunt

PERMISSIONS EDITOR: Bob Kauser

PRODUCTION SERVICE: Lachina Publishing Services, Inc.

TEXT AND COVER DESIGN: Gary Head, Gary Head Design

ART EDITOR AND PHOTO RESEARCHER: Myrna Engler

ILLUSTRATORS: Lisa Starr, Gary Head

COVER IMAGE: From the African savannah, lioness and cub at sunrise. ©Minden Pictures

COVER PRINTER: Phoenix Color Corp (MD)

COMPOSITOR: Angela Harris, John Becker, Preface, Inc.

FILM HOUSE: Tom Anderson, H&S Graphics

PRINTER: Quebecor/World, Versailles

BOOKS IN THE BROOKS/COLE BIOLOGY SERIES

Biology: The Unity and Diversity of Life, Ninth, Starr/Taggart
Biology: Concepts and Applications, Fifth, Starr
Laboratory Manual for Biology, Second, Perry/Morton/Perry
Human Biology, Fourth, Starr/McMillan
Laboratory Manual for Human Biology, Morton/Perry/Perry
Perspectives in Human Biology, Knapp
Human Physiology, Fourth, Sherwood
Fundamentals of Physiology, Second, Sherwood
Psychobiology: The Neuron and Behavior, Hoyenga/Hoyenga

Introduction to Cell and Molecular Biology, Wolfe
Molecular and Cellular Biology, Wolfe

Introduction to Microbiology, Second, Ingraham/Ingraham
Microbiology: An Introduction, Batzing

Genetics: The Continuity of Life, Fairbanks
Human Heredity, Fifth, Cummings
Introduction to Biotechnology, Barnum
Sex, Evolution, and Behavior, Second, Daly/Wilson
Gene Discovery Lab, Benfey

Plant Biology, Rost et al.
Plant Physiology, Fourth, Salisbury/Ross
Plant Physiology Laboratory Manual, Ross

General Ecology, Second, Krohne
Living in the Environment, Twelfth, Miller
Environmental Science, Ninth, Miller
Sustaining the Earth, Fifth, Miller

Oceanography: An Invitation to Marine Science, Third, Garrison
Essentials of Oceanography, Second, Garrison
Marine Life and the Sea, Milne

Brooks/Cole—Thomson Learning
10 Davis Drive
Belmont, CA 94002
USA

Asia
Thomson Learning
60 Albert Street, #15-01
Albert Complex
Singapore 189969

Australia
Nelson Thomson Learning
102 Dodds Street
South Melbourne, Victoria 3205
Australia

Canada
Nelson Thomson Learning
1120 Birchmount Road
Toronto, Ontario M1K 5G4
Canada

Europe/Middle East/Africa
Thomson Learning
Berkshire House
168-173 High Holborn
London WC1 V7AA
United Kingdom

For more information about our products, contact us at:
Thomson Learning Academic Resource Center
1-800-423-0563
For permission to use material from this text, contact us by:
Phone: 1-800-730-2214 **Fax:** 1-800-730-2215
Web: http://www.thomsonrights.com

Library of Congress Cataloging-in-Publication Data
Starr, Cecie.
  Biology : concepts and applications / Cecie Starr; Lisa Starr, biology
    illustrator.—5th ed. p. cm.- -(Brooks/Cole biology series)
  Includes bibliographical references (p.).
  ISBN 0-534-38549-4 (hardback)
  1. Biology I. Title. II. Series.
  QH307.2 .S73 2002
  570—dc21

2001 0 49 945

Student Paperback Edition: 0-534-38558-3
Annotated Instructor's Edition: 0-534-38555-9

CONTENTS IN BRIEF

INTRODUCTION

1 Concepts and Methods in Biology 2

I PRINCIPLES OF CELLULAR LIFE

2 Chemical Foundations for Cells 20
3 Carbon Compounds in Cells 34
4 Cell Structure and Function 50
5 Ground Rules of Metabolism 74
6 How Cells Acquire Energy 92
7 How Cells Release Stored Energy 108

II PRINCIPLES OF INHERITANCE

8 Cell Division and Mitosis 126
9 Meiosis 138
10 Observable Patterns of Inheritance 152
11 Chromosomes and Human Genetics 170
12 DNA Structure and Function 190
13 From DNA to Proteins 200
14 Controls Over Genes 212
15 Recombinant DNA and Genetic Engineering 222

III PRINCIPLES OF EVOLUTION

16 Microevolution 238
17 Speciation 260
18 The Macroevolutionary Puzzle 272

IV EVOLUTION AND BIODIVERSITY

19 The Origin and Evolution of Life 292
20 Prokaryotes, Viruses, and Protistans 310
21 Fungi 332
22 Plants 340
23 Animals: The Invertebrates 356
24 Animals: The Vertebrates 382
25 Biodiversity in Perspective 406

V PLANT STRUCTURE AND FUNCTION

26 Plant Tissues 420
27 Plant Nutrition and Transport 436
28 Plant Reproduction and Development 450

VI ANIMAL STRUCTURE AND FUNCTION

29 Tissues, Organ Systems, and Homeostasis 472
30 Integration and Control: Nervous Systems 486
31 Sensory Reception 508
32 Endocrine Control 520
33 Protection, Support, and Movement 538
34 Circulation 554
35 Immunity 576
36 Respiration 596
37 Digestion and Human Nutrition 612
38 The Internal Environment 632
39 Animal Reproduction and Development 644

VII ECOLOGY AND BEHAVIOR

40 Population Ecology 684
41 Social Interactions 702
42 Community Interactions 718
43 Ecosystems 736
44 The Biosphere 756
45 Human Impact on the Biosphere 782

APPENDIX I   CLASSIFICATION SCHEME
APPENDIX II   UNITS OF MEASURE
APPENDIX III ANSWERS TO SELF-QUIZZES
APPENDIX IV   ANSWERS TO GENETICS PROBLEMS
APPENDIX V   A CLOSER LOOK AT SOME MAJOR
              METABOLIC PATHWAYS
APPENDIX VI   PERIODIC TABLE OF THE ELEMENTS

DETAILED CONTENTS

INTRODUCTION

---

1 CONCEPTS AND METHODS IN BIOLOGY

Why Biology? 2

1.1 **DNA, Energy, and Life** *4*
Nothing Lives Without DNA *4*
Nothing Lives Without Energy *5*

1.2 **Energy and Life's Organization** *6*
Levels of Biological Organization *6*
Interdependencies Among Organisms *7*

1.3 **If So Much Unity, Why So Many Species?** *8*

1.4 **An Evolutionary View of Diversity** *10*
Mutation—Original Source of Variation *10*
Evolution Defined *10*
Natural Selection Defined *10*

1.5 **The Nature of Biological Inquiry** *12*
Observations, Hypotheses, and Tests *12*
An Assumption of Cause and Effect *12*
Experimental Design *13*
Sampling Error *13*
About the Word "Theory" *13*

1.6 *Focus on Science: The Power of Experimental Tests 14*
Biological Therapy Experiments 14
Identifying Important Variables 15
Bias in Reporting Results 15

1.7 **The Limits of Science** *15*

I PRINCIPLES OF CELLULAR LIFE

---

2 CHEMICAL FOUNDATIONS FOR CELLS

Checking Out Leafy Clean-Up Crews 20

2.1 **Regarding the Atoms** *22*
Structure of Atoms *22*
Isotopes—Variant Forms of Atoms *22*

2.2 *Focus on Science: Using Radioisotopes To Track Chemicals and Save Lives 23*

2.3 **What Happens When Atom Bonds With Atom?** *24*
Electrons and Energy Levels *24*
Think "Shells" *24*
From Atoms to Molecules *24*

2.4 **Important Bonds in Biological Molecules** *26*
Ion Formation and Ionic Bonding *26*
Covalent Bonding *26*
Hydrogen Bonding *27*

2.5 **Properties of Water** *28*
Polarity of the Water Molecule *28*
Water's Temperature-Stabilizing Effects *29*
Water's Cohesion *29*
Water's Solvent Properties *29*

2.6 **Acids, Bases, and Buffers** *30*
The pH Scale *30*
How Do Acids Differ From Bases? *30*
Buffers Against Shifts in pH *31*
Salts *31*

---

3 CARBON COMPOUNDS IN CELLS

Carbon, Carbon, In the Sky—Are You Swinging Low and High? 34

3.1 **Properties of Organic Compounds** *36*
The Molecules of Life *36*
Carbon's Bonding Behavior *36*
Functional Groups *36*
How Do Cells Build Organic Compounds? *36*

3.2 **Carbohydrates** *38*
The Simple Sugars *38*
Short-Chain Carbohydrates *38*
Complex Carbohydrates *38*

3.3 **Lipids** *40*
Fats and Fatty Acids *40*
Phospholipids *41*
Sterols and Their Derivatives *41*
Waxes *41*

3.4 **Amino Acids and the Primary Structure of Proteins** *42*
Amino Acid Structure *42*
What Is a Protein's Primary Structure? *42*

3.5 **How Does a Three-Dimensional Protein Emerge?** *44*
Second Level of Protein Structure *44*
Third Level of Protein Structure *44*
Fourth Level of Protein Structure *44*
Glycoproteins and Lipoproteins *45*
Structural Changes by Denaturation *45*

3.6 **Nucleotides and Nucleic Acids** *46*

3.7 *Focus on the Environment: Food Production and a Chemical Arms Race 47*

---

4 CELL STRUCTURE AND FUNCTION

Animalcules and Cells Fill'd With Juices 50

4.1 **Basic Aspects of Cell Structure and Function** *52*
Structural Organization of Cells *52*
Fluid Mosaic Model of Cell Membranes *52*
Overview of Membrane Proteins *53*

4.2 **Cell Size and Cell Shape** 54

4.3 *Focus on Science: Microscopes—Gateways to Cells* 54

4.4 **Defining Features of Eukaryotic Cells** 56
Major Cellular Components 56
Which Organelles Are Typical of Plants? 56
Which Organelles Are Typical of Animals? 56

4.5 **Micrographs of Plant and Animal Cells** 58

4.6 **The Nucleus** 60
Nuclear Envelope 60
Nucleolus 61
Chromosomes 61
What Happens to the Proteins Specified by DNA? 61

4.7 **The Cytomembrane System** 62
Endoplasmic Reticulum 62
Golgi Bodies 62
A Variety of Vesicles 63

4.8 **Mitochondria** 64

4.9 **Specialized Plant Organelles** 65
Chloroplasts and Other Plastids 65
Central Vacuole 65

4.10 **The Cytoskeleton** 66
Mechanisms of Cell Movements 66
Flagella and Cilia 67

4.11 **Cell Surface Specializations** 68
Eukaryotic Cell Walls 68
Matrixes Between Animal Cells 69
Cell-to-Cell Junctions 69

4.12 **Prokaryotic Cells** 70

5   **GROUND RULES OF METABOLISM**

Growing Old With Molecular Mayhem 74

5.1 **Energy and the Underlying Organization of Life** 76
Defining Energy 76
What Can Cells Do With Energy? 76
How Much Energy Is Available? 76
The One-Way Flow of Energy 76

5.2 **Energy Changes and Cellular Work** 78
Energy Inputs, Energy Outputs 78
The Role of ATP 78
The Role of Electron Transfers 79

5.3 **The Directional Nature of Metabolism** 80
Concentration Gradients and Diffusion 80
Which Way Will a Reaction Run? 80
Metabolic Pathways 81

5.4 **Enzyme Structure and Function** 82
Four Features of Enzymes 82
Enzyme–Substrate Interactions 82

5.5 **Factors Affecting Enzyme Action** 84
How Is Enzyme Action Controlled? 84
Helpers in Enzyme Action 84
Does the Environment Influence Enzymes? 84

5.6 **Metabolism and Cell Membranes** 85

5.7 **Protein-Mediated Transport** 86
Passive Transport 86
Active Transport 86

5.8 **Bulk Transport Across Membranes** 87

5.9 **Movement of Water Across Membranes** 88
Osmosis 88
Effects of Tonicity 88
Effects of Fluid Pressure 89

6   **HOW CELLS ACQUIRE ENERGY**

Sunlight and Survival 92

6.1 **Photosynthesis—An Overview** 94
Where the Reactions Take Place 94
But Things Don't *Really* End With Glucose 94

6.2 **Sunlight As an Energy Source** 96
Properties of Light 96
Pigments—The Rainbow Catchers 96
Where Are Photosynthetic Pigments Located? 97

6.3 **The Light-Dependent Reactions** 98
What Happens to the Absorbed Energy? 98
Cyclic and Noncyclic Electron Flow 98
The Legacy—A New Atmosphere 99

6.4 **Case Study: A Controlled Release of Energy** 100

6.5 **The Light-Independent Reactions** 101
How Do Plants Capture Carbon? 101
How Do Plants Build Glucose? 101

6.6 **Fixing Carbon—So Near, Yet So Far** 102
C4 Plants 102
CAM Plants 103

6.7 *Focus on Science: Light in the Deep Dark Sea* 103

6.8 *Focus on the Environment: Autotrophs, Humans, and the Biosphere* 104

7   **HOW CELLS RELEASE STORED ENERGY**

The Killers Are Coming! The Killers Are Coming! 108

7.1 **How Do Cells Make ATP?** 110
Comparison of the Main Types of Energy-Releasing Pathways 110
Overview of Aerobic Respiration 110

7.2 **Glycolysis: First Stage of Energy-Releasing Pathways** 112

7.3 **Second Stage of the Aerobic Pathway** 114
Preparatory Steps and the Krebs Cycle 114
Functions of the Second Stage 114

7.4 **Third Stage of the Aerobic Pathway** *116*
Electron Transport Phosphorylation *116*
Summary of the Energy Harvest *116*

7.5 **Anaerobic Routes of ATP Formation** *118*
Fermentation Pathways *118*
Anaerobic Electron Transport *119*

7.6 **Alternative Energy Sources in the Human Body** *120*
Carbohydrate Breakdown in Perspective *120*
Energy From Fats *120*
Energy From Proteins *120*

7.7 *Commentary: Perspective on Life 122*

II **PRINCIPLES OF INHERITANCE**

8 **CELL DIVISION AND MITOSIS**
Silver In the Stream of Time 126

8.1 **Dividing Cells: The Bridge Between Generations** *128*
Overview of Division Mechanisms *128*
Some Key Points About Chromosomes *128*
Mitosis and the Chromosome Number *128*

8.2 **The Cell Cycle** *130*
The Wonder of Interphase *130*
Mitosis Proceeds Through Four Stages *130*

8.3 **Mitosis** *132*
Prophase: Mitosis Begins *132*
Transition to Metaphase *132*
From Anaphase Through Telophase *133*

8.4 **Division of the Cytoplasm** *134*
Cell Plate Formation in Plants *134*
Cytoplasmic Division of Animal Cells *135*
Perspective on Mitotic Cell Division *135*

8.5 *Focus on Science: Henrietta's Immortal Cells 136*

9 **MEIOSIS**
Octopus Sex and Other Stories 138

9.1 **Comparing Sexual With Asexual Reproduction** *140*

9.2 **How Meiosis Halves the Chromosome Number** *140*
Think "Homologues" *140*
Two Divisions, Not One *141*

9.3 **Visual Tour of the Stages of Meiosis** *142*

9.4 **A Closer Look at Key Events of Meiosis I** *144*
Crossing Over in Prophase I *144*
Metaphase I Alignments *145*

9.5 **From Gametes to Offspring** *146*
Gamete Formation in Plants *146*
Gamete Formation in Animals *146*
More Shufflings at Fertilization *146*

9.6 **Meiosis and Mitosis Compared** *148*

10 **OBSERVABLE PATTERNS OF INHERITANCE**
A Smorgasbord of Ears and Other Traits 152

10.1 **Mendel's Insight Into Inheritance Patterns** *154*
Mendel's Experimental Approach *154*
Some Terms Used in Genetics *155*

10.2 **Mendel's Theory of Segregation** *156*
Predicting Outcomes of Monohybrid Crosses *156*
Testcrosses *157*

10.3 **Independent Assortment** *158*
Predicting Outcomes of Dihybrid Crosses *158*
The Theory in Modern Form *159*

10.4 **Dominance Relations** *160*
Incomplete Dominance *160*
ABO Blood Types: A Case of Codominance *160*

10.5 **Multiple Effects of Single Genes** *161*

10.6 **Interactions Between Gene Pairs** *162*
Hair Color in Mammals *162*
Comb Shape in Poultry *163*

10.7 **How Can We Explain Less Predictable Variations?** *164*
Regarding the Unexpected Phenotype *164*
Continuous Variation in Populations *164*

10.8 **Environmental Effects on Phenotype** *166*

11 **CHROMOSOMES AND HUMAN GENETICS**
The Philadelphia Story 170

11.1 **Chromosomes and Inheritance** *172*
Genes and Their Chromosome Locations *172*
Autosomes and Sex Chromosomes *172*

11.2 *Focus on Science: Karyotyping Made Easy 173*

11.3 **Sex Determination in Humans** *174*

11.4 **Early Questions About Gene Locations** *176*
Linked Genes—Clues to Inheritance Patterns *176*
Crossing Over and Genetic Recombination *177*

11.5 **Human Genetic Analysis** *178*
Constructing Pedigrees *178*
Regarding Human Genetic Disorders *179*

11.6 **Examples of Inheritance Patterns** *180*
Autosomal Recessive Inheritance *180*
Autosomal Dominant Inheritance *180*
X-Linked Recessive Inheritance *181*

11.7 *Focus on Health: Progeria—Too Young to Be Old 182*

11.8 **Changes in Chromosome Structure** *182*
Major Categories of Structural Change *182*
Does Chromosome Structure Evolve? *183*

11.9 Changes in Chromosome Number *184*
Categories and Mechanisms of Change *184*
Change in the Number of Autosomes *184*
Change in the Number of Sex Chromosomes *184*

11.10 *Focus on Bioethics: Prospects in Human Genetics 186*
Phenotypic Treatments 186
Genetic Screening 186
Genetic Counseling 186
Prenatal Diagnosis 186
Regarding Abortion 187
Preimplantation Diagnosis 187

---

12   DNA STRUCTURE AND FUNCTION

Cardboard Atoms and Bent Wire Bonds 190

12.1 Discovery of DNA Function *192*
Early and Puzzling Clues *192*
Confirmation of DNA Function *192*

12.2 DNA Structure *194*
What Are the Components of DNA? *194*
Patterns of Base Pairing *195*

12.3 *Focus on Bioethics: Rosalind's Story 196*

12.4 DNA Replication and Repair *196*
How Is a DNA Molecule Duplicated? *196*
Monitoring and Fixing the DNA *197*

12.5 *Focus on Science: Cloning Mammals—A Question of Reprogramming DNA 198*

---

13   FROM DNA TO PROTEINS

Beyond Byssus 200

13.1 How Is DNA Transcribed Into RNA? *202*
The Three Classes of RNA *202*
The Nature of Transcription *202*
Finishing Touches on mRNA Transcripts *202*

13.2 Deciphering the mRNA Transcripts *204*
What Is the Genetic Code? *204*
Structure and Function of tRNA and rRNA *204*

13.3 How Is mRNA Translated? *206*
Stages of Translation *206*
What Happens to the New Polypeptides? *207*

13.4 Do Mutations Affect Protein Synthesis? *208*
Common Gene Mutations and Their Sources *208*
Causes of Gene Mutations *209*
The Proof Is in the Protein *209*

---

14   CONTROLS OVER GENES

When DNA Can't Be Fixed 212

14.1 Types of Control Mechanisms *214*

14.2 Bacterial Control of Transcription *214*
Negative Control of the Lactose Operon *214*
Positive Control of the Lactose Operon *215*

14.3 Controls in Eukaryotic Cells *216*
Case Study: X Chromosome Inactivation *216*
Case Study: Hormones as Control Agents *216*
Case Study: Sunlight and Phytochrome *217*

14.4 *Focus on Science: Lost Controls and Cancer 218*
The Cell Cycle Revisited 218
Oncogenes 218
Characteristics of Cancer 218
Here's to Suicidal Cells! 219

---

15   RECOMBINANT DNA AND GENETIC ENGINEERING

Mom, Dad, and Clogged Arteries 222

15.1 A Toolkit for Making Recombinant DNA *224*
Restriction Enzymes *224*
Modification Enzymes *224*
Cloning Vectors for Amplifying DNA *224*
Reverse Transcriptase To Make cDNA *225*

15.2 PCR—A Faster Way To Amplify DNA *226*
What Are Primers? *226*
What Are the Reaction Steps? *226*

15.3 *Focus on Bioethics: DNA Fingerprints 227*

15.4 How Is DNA Sequenced? *228*

15.5 From Haystacks to Needles—Isolating Genes of Interest *229*
What Are Probes? *229*
Screening for Genes *229*

15.6 Using the Genetic Scripts *230*

15.7 Designer Plants *230*
Regenerating Plants From Cultured Cells *230*
How Are Genes Transferred Into Plants? *230*

15.8 Gene Transfers in Animals *232*
Supermice and Biotech Barnyards *232*
Mapping and Using the Human Genome *232*

15.9 *Focus on Bioethics: Who Gets Enhanced? 233*

15.10 Safety Issues

III   PRINCIPLES OF EVOLUTION

---

16   MICROEVOLUTION

Designer Dogs 238

16.1 Early Beliefs, Confounding Discoveries *240*
The Great Chain of Being *240*
Questions From Biogeography *240*
Questions From Comparative Morphology *240*
Questions About Fossils *241*

16.2 **A Flurry of New Theories** 242
Squeezing New Evidence Into Old Beliefs 242
Voyage of the *Beagle* 242

16.3 **Darwin's Theory Takes Form** 244
Old Bones and Armadillos 244
A Key Insight—Variation in Traits 244

16.4 **Individuals Don't Evolve—
Populations Do** 246
Examples of Variation in Populations 246
The "Gene Pool" 246
Stability and Change in Allele Frequencies 246
Mutations Revisited 247

16.5 *Focus on Science: When Is a Population
Not Evolving?* 248

16.6 **Natural Selection Revisited** 249

16.7 **Directional Change in the
Range of Variation** 250
What Is Directional Selection? 250
The Case of the Peppered Moths 250
Pesticide Resistance 250
Antibiotic Resistance 251

16.8 **Selection Against or in Favor of
Extreme Phenotypes** 252
Stabilizing Selection 252
Disruptive Selection 253

16.9 **Special Types of Selection** 254
Sexual Selection 254
Maintaining Two or More Alleles 254
Sickle-Cell Anemia—Lesser of Two Evils? 254

16.10 **Gene Flow** 255

16.11 **Genetic Drift** 256
Chance Events and Population Size 256
Bottlenecks and the Founder Effect 256
Genetic Drift and Inbred Populations 257

17 SPECIATION

The Case of the Road-Killed Snails 260

17.1 **On the Road to Speciation** 262
What Is a Species? 262
Reproductive Isolating Mechanisms 262

17.2 **Speciation in Geographically Isolated
Populations** 264
Allopatric Speciation Defined 264
The Pace of Geographic Isolation 264
Allopatric Speciation on Archipelagos 265

17.3 **Models for Other Speciation Routes** 266
Sympatric Speciation 266
Parapatric Speciation 267

17.4 **Patterns of Speciation** 268
Branching and Unbranched Evolution 268
Evolutionary Trees and Rates of Change 268
Adaptive Radiations 268
Extinctions—End of the Line 269

18 THE MACROEVOLUTIONARY PUZZLE

Legends of the Flood 272

18.1 **Fossils—Evidence of Ancient Life** 274
Fossilization 274
Fossils in Sedimentary Rock Layers 274
Interpreting the Fossil Record 275

18.2 *Focus on Science: Dating Pieces of the
Macroevolutionary Puzzle* 276
Origin of the Geologic Time Scale 276
Radiometric Dating 277

18.3 **Evidence From Biogeography** 278
An Outrageous Hypothesis 278
Drifting Continents, Changing Seas 278

18.4 **Evidence From Comparative Morphology** 280
Morphological Divergence 280
Morphological Convergence 281

18.5 **Evidence From Patterns of Development** 282
Developmental Program of Larkspurs 282
Developmental Program of Vertebrates 282

18.6 **Evidence From Comparative Biochemistry** 284
Protein Comparisons 284
Nucleic Acid Comparisons 284
Molecular Clocks 284

18.7 **How Do We Interpret the Evidence?** 286
Identifying, Naming, and Classifying Species 286
Groupings of Species—The Higher Taxa 286
Phylogenetic Classification Schemes 287

18.8 *Focus on Science: Case Study—Interpreting
and Misinterpreting the Evidence* 288

IV EVOLUTION AND BIODIVERSITY

19 THE ORIGIN AND EVOLUTION OF LIFE

In the Beginning... 292

19.1 **Conditions on the Early Earth** 294
Origin of the Earth 294
The First Atmosphere 294
Synthesis of Organic Compounds 294

19.2 **Emergence of the First Living Cells** 296
Origin of Agents of Metabolism 296
Origin of Self-Replicating Systems 296
Origin of the First Plasma Membranes 297

19.3 **Origin of Prokaryotic and Eukaryotic Cells** 298

19.4 *Focus on Science: Where Did Organelles Come From?* 300

19.5 **Life in the Paleozoic Era** *302*

19.6 **Life in the Mesozoic Era** *304*
Speciation on a Grand Scale *304*
Rise of the Ruling Reptiles *304*

19.7 *Focus on Science: Horrendous End to Dominance* *306*

19.8 **Life in the Cenozoic Era** *307*

---

20 PROKARYOTES, VIRUSES, AND PROTISTANS

The Mostly Microscopic Multitudes *310*

20.1 **Characteristics of Prokaryotic Cells** *312*
Two Prokaryotic Kingdoms *312*
Splendid Metabolic Diversity *312*
Cell Sizes and Shapes *313*
Structural Features *313*

20.2 **Major Bacterial Groups** *314*
Archaebacteria *314*
Eubacteria *314*
Regarding the "Simple" Bacteria *315*

20.3 **Bacterial Growth and Reproduction** *316*

20.4 **Defining Characteristics of Viruses** *317*

20.5 **Viral Multiplication Cycles** *318*
What Happens During Viral Infections? *318*
What About Viroids and Prions? *319*

20.6 **Kingdom at the Crossroads** *320*

20.7 **Parasitic and Predatory Molds** *320*

20.8 **The Animal-Like Protistans** *322*
Amoeboid Protozoans *322*
Ciliated Protozoans *322*
Animal-Like Flagellates *323*

20.9 **The Notorious Sporozoans** *324*

20.10 *Focus on Health: The Nature of Infectious Diseases* *325*

20.11 **The (Mostly) Single-Celled Algae** *326*
A Sampling of Single-Celled Diversity *326*
Algal Blooms and the Cell From Hell *327*

20.12 **Red Algae** *327*

20.13 **Brown Algae** *328*

20.14 **Green Algae** *328*

---

21 FUNGI

Ode to the Fungus Among Us *332*

21.1 **Characteristics of Fungi** *334*
Mode of Nutrition *334*
Major Groups *334*
Key Features of Fungal Life Cycles *334*

21.2 **Consider the Club Fungi** *334*
A Sampling of Spectacular Diversity *334*
Example of a Fungal Life Cycle *335*

21.3 *Focus on Health: A Look at the Unloved Few* *336*

21.4 **Spores and More Spores** *336*

21.5 **The Symbionts Revisited** *338*

---

22 PLANTS

Pioneers in a New World *340*

22.1 **Evolutionary Trends Among Plants** *342*
Overview of the Plant Kingdom *342*
Evolution of Roots, Stems, and Leaves *342*
From Haploid to Diploid Dominance *342*
Evolution of Pollen and Seeds *343*

22.2 *Focus on the Environment: Ancient Carbon Treasures* *344*

22.3 **The Bryophytes** *344*

22.4 **Existing Seedless Vascular Plants** *346*
Whisk Ferns *346*
Lycophytes *346*
Horsetails *346*
Ferns *347*

22.5 **The Rise of the Seed-Bearing Plants** *348*

22.6 *Focus on the Environment: Good-bye, Forests* *349*

22.7 **Gymnosperm Diversity** *350*

22.8 **Angiosperms—The Flowering, Seed-Bearing Plants** *351*

22.9 **Visual Overview of Flowering Plant Life Cycles** *352*

22.10 *Commentary: Seed Plants and People* *353*

---

23 ANIMALS: THE INVERTEBRATES

Madeleine's Limbs *356*

23.1 **Overview of the Animal Kingdom** *358*
General Characteristics of Animals *358*
Diversity in Body Plans *358*

23.2 **Puzzles About Origins** *360*

23.3 **Sponges—Success in Simplicity** *360*

23.4 **Cnidarians—Tissues Emerge** *362*
Cnidarian Body Plans and Life Cycles *362*
Examples of Cnidarian Diversity *363*

23.5 **Acoelomate Animals—And the Simplest Organ Systems** *364*
Characteristics of Flatworms *364*
Classes of Flatworms *364*

23.6 **Roundworms** *365*

23.7 *Focus on Health: A Rogue's Gallery of Worms 366*

23.8 **Two Major Divergences** *367*

23.9 **A Sampling of Mollusks** *368*
Molluscan Diversity *368*
Evolutionary Experiments With Body Plans *369*

23.10 **Annelids—Segments Galore** *370*
Advantages of Segmentation *370*
Annelid Adaptations—A Case Study *370*

23.11 **Arthropods—The Most Successful Organisms on Earth** *372*
Arthropod Diversity *372*
Adaptations of Insects and Other Arthropods *372*

23.12 **A Look at Spiders and Their Kin** *373*

23.13 **A Look at the Crustaceans** *374*

23.14 *How Many Legs?* *375*

23.15 **A Look at Insect Diversity** *376*

23.16 **The Puzzling Echinoderms** *378*

---

24 ANIMALS: THE VERTEBRATES

Making Do (Rather Well) With What You've Got 382

24.1 **The Chordate Heritage** *384*
Characteristics of Chordates *384*
Chordate Classification *384*

24.2 **Invertebrate Chordates** *384*
Tunicates *384*
Lancelets *384*

24.3 **Evolutionary Trends Among the Vertebrates** *386*
Puzzling Origins, Portentous Trends *386*
The First Vertebrates *387*

24.4 **Existing Jawless Fishes** *387*

24.5 **Existing Jawed Fishes** *388*
Cartilaginous Fishes *388*
Bony Fishes *388*

24.6 **Amphibians** *390*
Origin of Amphibians *390*
Frogs and Toads *390*
Salamanders *391*
Caecilians *391*

24.7 **The Rise of Reptiles** *392*

24.8 **Birds** *394*

24.9 **The Rise of Mammals** *396*

24.10 **From Early Primates to Hominids** *398*

24.11 **Emergence of Early Humans** *400*

24.12 *Focus on Science: Out of Africa—Once, Twice, or . . . 402*

---

25 BIODIVERSITY IN PERSPECTIVE

The Human Touch 406

25.1 **On Mass Extinctions and Slow Recoveries** *408*

25.2 **The Newly Endangered Species** *410*
Habitat Loss and Fragmentation *410*
Other Contributing Factors *411*

25.3 *Focus on the Environment: Case Study— The Once and Future Reefs 412*

25.4 *Focus on Science: Rachel's Warning 414*

25.5 **Conservation Biology** *414*
The Role of Systematics *414*
Bioeconomic Analysis *415*
Sustainable Development *415*

25.6 **Reconciling Biodiversity With Human Demands** *416*
A Case for Strip Logging *416*
Ranching and the Riparian Zones *416*

V PLANT STRUCTURE AND FUNCTION

---

26 PLANT TISSUES

Plants Versus the Volcano 420

26.1 **Overview of the Plant Body** *422*
Shoots and Roots *422*
Three Plant Tissue Systems *422*
Meristems—Where Tissues Originate *423*

26.2 **Types of Plant Tissues** *424*
Simple Tissues *424*
Complex Tissues *424*
Dicots and Monocots—Same Tissues, Different Features *425*

26.3 **Primary Structure of Shoots** *426*
How Stems and Leaves Form *426*
Internal Structure of Stems *426*

26.4 **A Closer Look at Leaves** *428*
Similarities and Differences Among Leaves *428*
Leaf Fine Structure *428*

26.5 **Primary Structure of Roots** *430*
Taproot and Fibrous Root Systems *430*
Internal Structure of Roots *430*
Regarding the Sidewalk-Buckling, Record-Breaking Root Systems *431*

26.6 **Accumulated Secondary Growth— The Woody Plants** *432*

27 PLANT NUTRITION AND TRANSPORT

Flies for Dinner 436

27.1 **Soil and Its Nutrients** *438*
Properties of Soil *438*
Which Nutrients Are Essential for Plant Growth? *438*
Leaching and Erosion *439*

27.2 **How Do Roots Absorb Water and Mineral Ions?** *440*
Absorption Routes *440*
Specialized Absorptive Structures *440*

27.3 **How Is Water Transported Through Plants?** *442*
Transpiration Defined *442*
Cohesion–Tension Theory of Water Transport *442*

27.4 **How Do Stems and Leaves Conserve Water?** *444*
The Water-Conserving Cuticle *444*
Controlled Water Loss at Stomata *444*

27.5 **How Are Organic Compounds Distributed Through Plants?** *446*
Translocation *446*
Pressure Flow Theory *446*

28 PLANT REPRODUCTION AND DEVELOPMENT

A Coevolutionary Tale 450

28.1 **Reproductive Structures of Flowering Plants** *452*
Think Sporophyte and Gametophyte *452*
Components of Flowers *452*
Where Pollen and Eggs Develop *453*

28.2 *Focus on Health: Pollen Sets Me Sneezing 453*

28.3 **A New Generation Begins** *454*
From Microspores to Pollen Grains *454*
From Megaspores to Eggs *454*
From Pollination to Fertilization *454*

28.4 **From Zygotes to Seeds and Fruits** *456*
Formation of the Embryo Sporophyte *456*
Seeds and Fruit Formation and Dispersal *456*

28.5 **Asexual Reproduction of Flowering Plants** *458*
Asexual Reproduction in Nature *458*
Induced Propagation *458*

28.6 **Patterns of Early Growth and Development— An Overview** *460*
Seed Germination *460*
Genetic Programs, Environmental Cues *460*

28.7 **Effects of Plant Hormones** *462*

28.8 *Focus on Science: Foolish Seedlings! 463*

28.9 **Adjustments in the Rate and Direction of Growth** *464*
What Are Tropisms? *464*
Responses to Mechanical Stress *465*

28.10 **Biological Clocks and Their Effects** *466*
Rhythmic Leaf Movements *466*
Flowering—A Case of Photoperiodism *466*

28.11 **Life Cycles End, And Turn Again** *468*
Senescence *468*
Entering and Breaking Dormancy *468*

VI ANIMAL STRUCTURE AND FUNCTION

29 TISSUES, ORGAN SYSTEMS, AND HOMEOSTASIS

Meerkats, Humans, It's All the Same 472

29.1 **Epithelial Tissue** *474*
General Characteristics *474*
Cell-to-Cell Contacts *474*
Glandular Epithelium and Glands *475*

29.2 **Connective Tissue** *476*
Soft Connecting Tissues *476*
Specialized Connective Tissues *476*

29.3 **Muscle Tissue** *478*

29.4 **Nervous Tissue** *479*

29.5 *Focus on Science: Frontiers in Tissue Research 479*

29.6 **Organ Systems** *480*
Overview of Major Organ Systems *480*
Tissue and Organ Formation *481*

29.7 **Homeostasis and Systems Control** *482*
Concerning the Internal Environment *482*
Mechanisms of Homeostasis *482*

30 INTEGRATION AND CONTROL: NERVOUS SYSTEMS

Why Crack the System? 486

30.1 **Neurons—The Communication Specialists** *488*
Functional Zones of a Neuron *488*
A Neuron at Rest, Then Moved to Action *488*
Gradients Required for Action Potentials *488*

30.2 **Triggering and Propagating Action Potentials** *490*
Approaching Threshold *490*
An All-or-Nothing Spike *490*
The Direction of Propagation *491*

30.3 **Chemical Synapses** *492*
A Smorgasbord of Signals *492*
Synaptic Integration *492*
How Is Neurotransmitter Removed From the Synaptic Cleft? *493*

30.4 **Paths of Information Flow** 494
Blocks and Cables of Neurons 494
Reflex Arcs 494

30.5 *Focus on Health: Skewed Information Flow* 496

30.6 **Invertebrate Nervous Systems** 496
Regarding the Nerve Net 496
On the Importance of Having a Head 496

30.7 **Vertebrate Nervous Systems—An Overview** 498
Evolution of the Spinal Cord and Brain 498
The System's Functional Divisions 498

30.8 **The Major Expressways** 500
Peripheral Nervous System 500
The Spinal Cord 501

30.9 **The Vertebrate Brain** 502
Evolutionary High Points 502
Functional Divisions of the Human Brain 502
Protection at the Blood–Brain Barrier 503

30.10 **Memory** 504

30.11 *Focus on Health: Drugging the Brain* 504

31 SENSORY RECEPTION

Different Strokes for Different Folks 508

31.1 **Overview of Sensory Pathways** 510

31.2 **Somatic Sensations** 511

31.3 **Senses of Hearing and Balance** 512
Evolution of the Vertebrate Ear 512
A Closer Look at Inner Ear Functions 513

31.4 **Sense of Vision** 514
A Sampling of Invertebrate Eyes 514
Vertebrate Eyes 514

31.5 **Visual Perception** 516

31.6 **Senses of Taste and Smell** 517

32 ENDOCRINE CONTROL

Hormone Jamboree 520

32.1 **The Endocrine System** 522
Hormones and Other Signaling Molecules 522
Discovery of Hormones and Their Sources 522

32.2 **Signaling Mechanisms** 524
The Nature of Hormonal Action 524
Characteristics of Steroid Hormones 524
Characteristics of Peptide Hormones 525

32.3 **The Hypothalamus and Pituitary Gland** 526
Posterior Lobe Secretions 526
Anterior Lobe Secretions 526

32.4 **Examples of Abnormal Pituitary Output** 528

32.5 **Sources and Effects of Other Hormones** 529

32.6 **Feedback Control of Hormonal Secretions** 530
Negative Feedback From the Adrenal Cortex 530
Local Feedback in the Adrenal Medulla 530
Cases of Skewed Feedback From the Thyroid 530
Feedback Control of the Gonads 531

32.7 **Responses to Local Chemical Changes** 532
Secretions From Parathyroid Glands 532
Effects of Local Signaling Molecules 532
Secretions From Pancreatic Islets 532

32.8 **Hormones and the Environment** 534
Daylength and the Pineal Gland 534
Thyroid Function and Frog Habitats 534
Comparative Look at a Few Invertebrates 535

33 PROTECTION, SUPPORT, AND MOVEMENT

Of Men, Women, and Polar Huskies 538

33.1 **Vertebrate Skin** 540

33.2 *Focus on Health: Sunlight and Skin* 541
The Vitamin Connection 541
Suntans and Shoe-Leather Skin 541
Sunlight and the Front Line of Defense 541

33.3 **Skeletal Systems** 542

33.4 **A Closer Look at Bones and Joints** 544
Bone Structure and Function 544
Bone Formation and Remodeling 544
Bones and the Blood Level of Calcium 545
Where Bones Meet—Skeletal Joints 545

33.5 **Skeletal–Muscular Systems** 546
How Do Muscles and Bones Interact? 546
Human Skeletal–Muscular System 547

33.6 **A Closer Look at Muscles** 548
Skeletal Muscle Structure and Function 548
Sources of Energy for Contraction 549

33.7 **Control of Muscle Contraction** 550

33.8 **Properties of Whole Muscles** 551
Muscle Tension and Muscle Fatigue 551
Effects of Exercise and Aging 551

34 CIRCULATION

Heartworks 554

34.1 **Circulatory Systems—An Overview** 556
General Characteristics 556
Evolution of Vertebrate Circulatory Systems 557
Links With the Lymphatic System 557

34.2 **Characteristics of Blood** 558
Functions of Blood 558
Blood Volume and Composition 558

34.3 *Focus on Health: Blood Disorders* 560

34.4 **Blood Transfusion and Typing** 560
Concerning Agglutination 560
ABO Blood Typing 561
Rh Blood Typing 561

34.5 **Human Cardiovascular System** 562

34.6 **The Heart Is a Lonely Pumper** 564
Heart Structure 564
Cardiac Cycle 564
Cardiac Muscle Contraction 565

34.7 **Blood Pressure in the Cardiovascular System** 566
Arterial Blood Pressure 566
How Can Arterioles Resist Flow? 567
Controlling Mean Arterial Blood Pressure 567

34.8 **From Capillary Beds Back to the Heart** 568
Capillary Function 568
Venous Pressure 569

34.9 *Focus on Health: Cardiovascular Disorders* 570

34.10 **Hemostasis** 572

34.11 **Lymphatic System** 572
Lymph Vascular System 572
Lymphoid Organs and Tissues 573

35 IMMUNITY

Russian Roulette, Immunological Style 576

35.1 **Three Lines of Defense** 578
Surface Barriers to Invasion 578
Nonspecific and Specific Responses 578

35.2 **Complement Proteins** 579

35.3 **Inflammation** 580
The Roles of Phagocytes and Their Kin 580
The Inflammatory Response 580

35.4 **Overview of the Immune System** 582
Defining Features 582
The Key Defenders 582
The Main Targets 582
Control of Immune Responses 583

35.5 **How Lymphocytes Form and Do Battle** 584
B and T Cell Formation 584
Lymphocyte Battlegrounds 585

35.6 **Antibody-Mediated Response** 586
The Roles of Antibodies 586
Classes of Immunoglobulins 586

35.7 **Cell-Mediated Response** 588

35.8 *Focus on Science: Cancer and Immunotherapy* 589

35.9 **Defenses Enhanced, Misdirected, or Compromised** 590
Immunization 590
Allergies 590
Autoimmune Disorders 591
Deficient Immune Responses 591

35.10 *Focus on Health: AIDS—The Immune System Compromised* 592

36 RESPIRATION

Conquering Chomolungma 596

36.1 **The Nature of Respiration** 598
The Basis of Gas Exchange 598
Which Factors Influence Gas Exchange? 598

36.2 **Invertebrate Respiration** 599

36.3 **Vertebrate Respiration** 600
Gills of Fishes and Amphibians 600
Lungs 600

36.4 **Human Respiratory System** 602
What Are the System's Functions? 602
From Airways Into the Lungs 603

36.5 **Cyclic Reversals in Air Pressure Gradients** 604
The Respiratory Cycle 604
Breathing at High Altitudes 605

36.6 **Gas Exchange and Transport** 606
Exchanges at the Respiratory Membrane 606
Oxygen Transport 606
Carbon Dioxide Transport 606
Matching Air Flow With Blood Flow 607

36.7 *Focus on Health: When the Lungs Break Down* 608

37 DIGESTION AND HUMAN NUTRITION

Lose It—And It Finds Its Way Back 612

37.1 **The Nature of Digestive Systems** 614
Incomplete and Complete Systems 614
Correlations With Feeding Behavior 614

37.2 **Visual Overview of the Human Digestive System** 616

37.3 **Into the Mouth, Down the Tube** *617*

37.4 **Digestion in the Stomach and Small Intestine** *618*
The Stomach *618*
The Small Intestine *618*
The Role of Bile in Fat Digestion *619*
Controls Over Digestion *619*

37.5 **Absorption in the Small Intestine** *620*
Structure Speaks Volumes About Function *620*
What Are the Absorption Mechanisms? *621*

37.6 **Disposition of Absorbed Organic Compounds** *622*

37.7 **The Large Intestine** *623*
Colon Functioning *623*
Colon Malfunctioning *623*

37.8 **Human Nutritional Requirements** *624*
New, Improved Food Pyramids *624*
Carbohydrates *624*
Lipids *625*
Proteins *625*

37.9 **Vitamins and Minerals** *626*

37.10 *Focus on Science: Weighty Questions, Tantalizing Answers* *628*

38 THE INTERNAL ENVIRONMENT
Tale of the Desert Rat *632*

38.1 **Urinary System of Mammals** *634*
The Challenge—Shifts in Extracellular Fluid *634*
Components of the Urinary System *635*
Nephrons—Functional Units of Kidneys *635*

38.2 **Urine Formation** *636*
Factors Influencing Filtration *636*
Reabsorption of Water and Sodium *637*

38.3 *Focus on Health: When Kidneys Break Down* *638*

38.4 **The Body's Acid–Base Balance** *638*

38.5 **On Fish, Frogs, and Kangaroo Rats** *639*

38.6 **Maintaining Body Temperature** *640*
Heat Gains and Heat Losses *640*
Responses to Stressful Temperatures *640*

39 ANIMAL REPRODUCTION AND DEVELOPMENT
From Frog to Frog and Other Mysteries *644*

39.1 **The Beginning: Reproductive Modes** *646*
Sexual Versus Asexual Reproduction *646*
Costs and Benefits of Sexual Reproduction *646*

39.2 **Stages of Development—An Overview** *648*

39.3 **Early Marching Orders** *650*
Information in the Egg Cytoplasm *650*
Cleavage—The Start of Multicellularity *650*
Cleavage Patterns *651*

39.4 **How Do Specialized Tissues and Organs Form?** *652*
Cell Differentiation *652*
Morphogenesis *653*

39.5 **Pattern Formation** *654*
Embryonic Signals in Pattern Formation *654*
A Theory of Pattern Formation *654*
Evolutionary Constraints on Development *655*

39.6 **Reproductive System of Human Males** *656*
Where Sperm Form *656*
Where Semen Forms *656*
Cancers of the Prostate and Testis *657*

39.7 **Male Reproductive Function** *658*
Sperm Formation *658*
Hormonal Controls *658*

39.8 **Reproductive System of Human Females** *660*
The Reproductive Organs *660*
Overview of the Menstrual Cycle *660*

39.9 **Female Reproductive Function** *662*
Cyclic Changes in the Ovary *662*
Cyclic Changes in the Uterus *663*

39.10 **Visual Summary of the Menstrual Cycle** *664*

39.11 **Pregnancy Happens** *665*
Sexual Intercourse *665*
Fertilization *665*

39.12 **Formation of the Early Embryo** *666*
Cleavage and Implantation *666*
Extraembryonic Membranes *667*

39.13 **Emergence of the Vertebrate Body Plan** *668*

39.14 **Why Is the Placenta So Important?** *669*

39.15 **Emergence of Distinctly Human Features** *670*

39.16 *Focus on Health: Mother as Provider, Protector, Potential Threat* *672*

39.17 **Giving Birth** *673*

39.18 **From Birth Onward** *674*
Nourishing the Newborn *674*
Regarding Breast Cancer *674*
Postnatal Development *674*
Why Do Animals Age? *674*

39.19 Control of Human Fertility *676*
Some Ethical Considerations *676*
Birth Control Options *676*

39.20 *Focus on Health: Sexually Transmitted Diseases* *678*

39.21 *Focus on Bioethics: To Seek or End Pregnancy* *680*

VII ECOLOGY AND BEHAVIOR

40 POPULATION ECOLOGY

Tales of Nightmare Numbers *684*

40.1 Characteristics of Populations *686*

40.2 *Focus on Science: Elusive Heads to Count* *687*

40.3 Population Size and Exponential Growth *688*
Gains and Losses in Population Size *688*
From Ground Zero to Exponential Growth *688*
What Is the Biotic Potential? *689*

40.4 Limits on the Growth of Populations *690*
What Are the Limiting Factors? *690*
Carrying Capacity and Logistic Growth *690*
Density-Dependent Controls *691*
Density-Independent Factors *691*

40.5 Life History Patterns *692*
Life Tables *692*
Patterns of Survival and Reproduction *692*

40.6 *Focus on Science: Natural Selection Among the Guppies* *693*

40.7 Human Population Growth *694*
How We Began Sidestepping Controls *694*
Present and Future Growth *695*

40.8 Control Through Family Planning *696*

40.9 Population Growth and Economic Development *698*
Demographic Transition Model *698*
A Question of Resource Consumption *699*

40.10 Social Impact of No Growth *699*

41 SOCIAL INTERACTIONS

Deck the Nest With Sprigs of Green Stuff *702*

41.1 Behavior's Heritable Basis *704*
Genes and Behavior *704*
Hormones and Behavior *704*
Instinctive Behavior Defined *705*

41.2 Learned Behavior *706*

41.3 The Adaptive Value of Behavior *707*

41.4 Communication Signals *708*
The Nature of Communication Signals *708*
Examples of Communication Displays *708*
Illegitimate Signalers and Receivers *709*

41.5 Mates, Parents, and Individual Reproductive Success *710*
Sexual Selection Theory and Mating Behavior *710*
Costs and Benefits of Parenting *711*

41.6 Costs and Benefits of Living in Social Groups *712*
Cooperative Predator Avoidance *712*
The Selfish Herd *712*
Dominance Hierarchies *712*
Regarding the Costs *713*

41.7 *Focus on Science: Why Sacrifice Yourself?* *714*

41.8 An Evolutionary View of Human Social Behavior *715*

42 COMMUNITY INTERACTIONS

No Pigeon Is an Island *718*

42.1 Which Factors Shape Community Structure? *720*
The Niche *720*
Categories of Species Interactions *720*

42.2 Mutualism *721*

42.3 Competitive Interactions *722*
Competitive Exclusion *722*
Resource Partitioning *723*

42.4 Predation and Parasitism *724*
Dynamics of Predator–Prey Interactions *724*
Dynamics of Parasite–Host Interactions *725*

42.5 *Commentary: A Coevolutionary Arms Race* *726*
Camouflage *726*
Warning Coloration *726*
Mimicry *726*
Moment-of-Truth Defenses *726*
Predator Responses to Prey *727*

42.6 Forces Contributing to Community Stability *728*
A Successional Model *728*
The Climax-Pattern Model *728*
Cyclic, Nondirectional Changes *728*
Restoration Ecology *729*

42.7 Community Instability *730*

42.8 *Focus on the Environment: Exotic and Endangered Species* *730*
The Plants That Ate Georgia *730*
The Rabbits That Ate Australia *731*

42.9 Patterns of Biodiversity *732*
What Causes Mainland and Marine Patterns? *732*
What Causes Island Patterns? *732*

---

43 ECOSYSTEMS

Crêpes for Breakfast, Pancake Ice for Dessert *736*

43.1 The Nature of Ecosystems *738*
Overview of the Participants *738*
Structure of Ecosystems *739*

43.2 The Nature of Food Webs *740*
How Many Energy Transfers? *740*
Two Categories of Food Webs *741*

*43.3 Focus on Science: Biological Magnification
in Food Webs* *742*

43.4 Studying Energy Flow Through Ecosystems *743*
What Is Primary Productivity? *743*
What Are Ecological Pyramids? *743*

43.5 Focus on Science: Energy Flow at Silver Springs *744*

43.6 Biogeochemical Cycles—An Overview *745*

43.7 Hydrologic Cycle *746*

43.8 Sedimentary Cycles *747*
A Look at the Phosphorus Cycle *747*
Eutrophication *747*

43.9 Carbon Cycle *748*

*43.10 Focus on Science: From Greenhouse
Gases to a Warmer Planet?* *750*
Greenhouse Effect *750*
Global Warming Defined *750*
Evidence of an Intensified Greenhouse Effect *750*
Causes and Effects: Some Predictions *750*

43.11 Nitrogen Cycle *752*
Cycling Processes *752*
Nitrogen Scarcity *753*
Human Impact on the Nitrogen Cycle *753*

---

44 THE BIOSPHERE

Does a Cactus Grow in Brooklyn? *756*

44.1 Air Circulation Patterns and Regional Climates *758*

44.2 The Ocean, Landforms, and Regional Climates *760*
Ocean Currents and Their Effects *760*
Regarding Rain Shadows and Monsoons *761*

44.3 Realms of Biodiversity *762*

44.4 Soils of Major Biomes *764*

44.5 Deserts *765*

**44.6 Dry Shrublands, Dry Woodlands,
and Grasslands** *766*

**44.7 Tropical Rain Forests and Other
Broadleaf Forests** *768*

44.8 Coniferous Forests *770*

44.9 Arctic and Alpine Tundra *771*

44.10 Freshwater Provinces *772*
Lake Ecosystems *772*
Stream Ecosystems *773*

44.11 The Ocean Provinces *774*
Primary Productivity in the Ocean *774*
Hydrothermal Vents *774*

44.12 Wetlands and the Intertidal Zone *776*
Mangrove Wetlands and Estuaries *776*
Rocky and Sandy Coastlines *776*
Upwelling Along the Coasts *777*

44.13 Focus on Science: Rita in the Time of Cholera *778*
El Niño Southern Oscillation *778*
A Cholera Connection *779*

---

45 HUMAN IMPACT ON THE BIOSPHERE

An Indifference of Mythic Proportions *782*

45.1 Air Pollution—Prime Examples *784*
Smog *784*
Acid Deposition *784*

**45.2 Ozone Thinning—Global Legacy
of Air Pollution** *786*

**45.3 Where to Put Solid Wastes? Where To
Produce Food?** *787*
Oh Bury Me Not, In Our Own Garbage *787*
Converting Marginal Lands for Agriculture *787*

**45.4 Deforestation—An Assault on
Finite Resources** *788*

*45.5 Focus on Bioethics: You and the
Tropical Rain Forest* *790*

45.6 Who Trades Grasslands for Deserts? *791*

45.7 A Global Water Crisis *792*
Consequences of Heavy Irrigation *792*
Water Pollution *792*
The Coming Water Wars *793*

45.8 A Question of Energy Inputs *794*
Fossil Fuels *794*
Nuclear Energy *794*

45.9 Alternative Energy Sources *796*
Solar–Hydrogen Energy *796*
Wind Energy *796*
What About Fusion Power? *796*

*45.10 Focus on Bioethics: Biological Principles
and the Human Imperative* *797*

Preface

Teachers of introductory biology know all about the Red Queen effect, whereby one runs as fast as one can to stay in the same place. New and refined information from hundreds of fields of inquiry piles up daily, and somehow teachers are expected to distill it into Biology Lite, a zip through the high points, and help students deepen their understanding of a world of unbelievable richness.

Restricting textbook content runs the risk of splintering understanding of that world, as when an emphasis on human biology inadvertently reinforces archaic notions that everything on Earth is here in the service of Us; or when a molecular focus excludes knowledge of whole organisms. Here I am reminded of the well-intentioned but lethal blanketing of habitats with DDT.

We offer this book as a coherent account of the sweep of life's diversity and its underlying unity. Through its examples of problem solving and experiments, it shows the power of thinking critically about the natural world. It highlights key concepts, current understandings, and research trends for major fields of biological inquiry. It explains the structure and function of a broad sampling of organisms in enough detail so that students can develop a working vocabulary about life's parts and processes.

We start with an overview of the basic concepts and scientific methods. Next are three units on the principles of biochemistry, inheritance, and evolution—a conceptual framework for exploring life's unity and diversity. Units on comparative anatomy and physiology of plants, then animals, follow. The last unit focuses on the patterns and consequences of organisms interacting with one another and with the environment. This conceptual organization parallels the levels of biological organization.

CONCEPT SPREADS Feedback from teachers of more than 3 million students helped us refine our approach. We keep the story line in focus for students by subscribing to the question "How do you eat an elephant?" and its answer, "One bite at a time." We organize descriptions, art, and supporting evidence for each concept on two facing pages at most. Each *concept spread* starts with a numbered tab and concludes with boldfaced summaries of key points (*see below*). On-page summaries allow students to check their understanding of one concept before turning to another.

The clear conceptual organization within chapters also offers teachers flexibility in assigning text material to fit course requirements. For example, those who spend little time on photosynthesis may choose to bypass the spreads on properties of light and the chemiosmotic theory of ATP formation. They may or may not assign Focus essays (one on the global impact of photosynthesis, another on how photosynthesis may have started in the first place). All of the concept spreads flow as parts of the same story, but some clearly offer depth that can be treated as optional.

Concept spreads are not gimmicks. Ongoing feedback guides decisions about when to add depth and when to loosen core material with applications. Within spreads,

Numbered tabs indicate the start of a new concept as the chapter's story unfolds. The gold tabs identify basic chapter concepts. Blue tabs identify Focus essays that enrich the basics with examples of experiments (to demonstrate the power of critical thinking), of the nature of life, and of applying the basics to issues of human interest.

This icon signifies that our interactive CD-ROM further explores the concept being illustrated.

Topic section ends with a summary of key concepts.

Website expands on the section's topic.

headings and subheadings help students keep track of the hierarchy of information. Transitions between spreads help them keep the greater story in focus and discourage memorization for its own sake. To avoid disrupting the basic story line while still attending to interested students, we include some details in optional, enriching illustrations.

Our organization helps students find assigned topics easily and translates into a tangible outcome—improved test scores. Its underlying logic also helps them develop enough confidence to dig deeper into biological science.

BALANCING CONCEPTS WITH APPLICATIONS Chapters start with a lively or sobering application and an advance organizer—a list of key concepts. Essays parallel the core material to help maintain student interest in the basics. The essays afford more depth on medical, environmental, and social issues without interrupting conceptual flows. Briefer applications are integrated in the text. The final pages index all applications separately for fast reference.

FOUNDATIONS FOR CRITICAL THINKING To help students sharpen their capacity for critical thinking, we walk them through experiments that yielded evidence in favor of or against hypotheses. The main index lists all of the experiments we selected for the book (see the entries *Experiment, examples*, and *Test, observational*).

We use certain chapter introductions as well as whole chapters to show students some productive results of critical thinking. The introductions to Mendelian genetics (Chapter 10), DNA structure and function (12), speciation (17), immunology (35), and behavior (41) are examples. Also, each chapter has a set of *Critical Thinking* questions, most from Katherine Denniston. Daniel Fairbanks created many of the *Genetics Problems*, which help students grasp the principles of inheritance (Chapters 10 and 11).

VISUAL OVERVIEWS OF CONCEPTS We simultaneously develop text and art as inseparable parts of the same story. We give visual learners a means to work their way through a visual overview of major processes before reading the corresponding text. Students repeatedly let us know how much they appreciate this art. The overview illustrations include step-by-step descriptions of biological parts and processes. Instead of "wordless" diagrams, we break down information into a series of illustrated callouts. In Figure 13.10, for example, callouts integrated with the art walk students through the stages by which a mature mRNA transcript becomes translated.

Many anatomical drawings are integrated overviews of structure/function. Students need not jump back and forth from text, to tables, to illustrations, and back again to see how an organ system is put together and what its parts do. We hierarchically arrange descriptions of parts to reflect a system's structural and functional organization.

ZOOM SEQUENCES Many illustrations progress from macroscopic to microscopic views of a system or process. Figure 6.3, for example starts with a plant leaf and ends with reaction sites in the chloroplast. Figures 33.15 and 33.16 start with a ballerina's biceps and move down through levels of skeletal muscle contraction.

COLOR CODES Consistent use of colors for molecules, cell structures, and processes helps students track what is going on. We use these colors throughout the book:

ICONS Small icons next to an illustration help students relate a topic to the big picture. For instance, a cell icon reminds them of the location of the plasma membrane relative to the cytoplasm. Some icons relate reactions and processes to cell locations and to one another. Others are reminders of evolutionary relationships among groups, as in Chapters 23 and 24. A multimedia icon directs students to art in the CD-ROM packaged in its own envelope at the back of their book. Others direct them to supplemental material on the Web and to InfoTrac® College Edition, an online database of full-length articles from hundreds of top academic journals and popular sources.

MAJOR CONTENT REVISIONS New to this edition is a chapter on biodiversity and mass extinctions (25). The chapters on principles of metabolism (5), photosynthesis (6), gene control (14), macroevolution (18), immunity (35), principles of animal reproduction/development (39), and ecosystems (43) are significantly reorganized. Like the other chapters, they are now more succinct and accessible owing to line-by-line snipping, as indirectly evidenced by the more open page layouts. Users will find major new art, crisper definitions, new end-of-chapter material, and new or updated material throughout, as highlighted here:

Unit I. Principles of Cellular Life Chapter 2 has new atomic and molecular models. Chapter 3 shows the non-ionized and ionized forms of carboxyl and amino groups (a point of confusion for students, who see both in the literature); its hemoglobin model is new. Chapter 4 has new text and art on the cytoskeleton and, as elsewhere, distinguishes between prokaryotic cell and bacterial cell. Chapter 5 has a new introduction on enzymes and aging, and weaves in fascinating experiments with *Caenorhabditis elegans*. The chapter has a new Critical Thinking question on cyanide poisoning. Revised and reorganized Chapter 6 has livelier writing, as in Section 6.1, and a condensed Summary. It has new text and art on photosystems; the function of electron transport systems is stated more clearly. Art comparing carbon-fixing strategies has accurate cross-sections of C3 and C4 leaves. There are new Critical Thinking questions on the evolution of photosynthesis and on bioluminescence. Chapter 7, a tough topic, has more user-friendly art.

Unit II. Principles of Inheritance In Chapter 8, we moved the overview of chromosome structure forward, added a bit on the cell cycle, refined the mitosis art, and added photographs of Henrietta Lacks and HeLa cells. Micrographs accompany the meiosis diagram in Chapter 9; we finally found some uncomplicated ones that won't confuse students. Chapter 10 has refined definitions of monohybrid and dihybrid crosses, and new text and art on experiments that revealed environmental effects on *Achillea millefolium* phenotype. Chapter 11 has a CML update, the crossing over sketch is moved to the overview section, its karyotyping essay is simplified, and it has a new human example of an outcome of crossing over and genetic recombination. More human genetic disorders are covered (see Table 11.1). We expanded the Summary and added a Critical Thinking question on perceptions of genetic disorders. Chapter 12 includes a new model for DNA and mentions new evidence of the function of DNA's structure (page 195). The updated DNA cloning essay includes concerns about random mistakes in cloning processes. We added a Critical Thinking question on cloning extinct mammals. Updated Chapter 13 includes a recently deduced molecular model for the eukaryotic ribosome; text and art on translation are modified accordingly and show the rRNA catalytic site. The transposon material is updated here and elsewhere; the distinction between ionizing and non-ionizing radiation is clarified. The summary is tighter.

Chapter 14 has a shorter introductory essay on cancer and the nature of gene control; details are in an updated essay on the cell cycle and cancer. A new section gives an overview of control mechanisms. The lactose operon section also describes lactose intolerance. The point of X chromosome inactivation (dosage compensation) is clearly stated. A new invertebrate example (ecdysone control) is included. Chapter 15 includes a recent genetic engineering experiment on aspens plus updates on the mammalian cloning issue, the Human Genome Initiative, and gene therapy for bubble boys. We have an extended Critical Thinking question on genetically engineered food, and another on minimal organisms. We moved description and illustration of bacterial conjugation to Section 20.3.

Unit III. Principles of Evolution *Archaeopteryx* is now in Chapter 18. We refined the text on stabilizing and disruptive selection and balanced polymorphism and added Self-Quiz questions plus a Critical Thinking question on the rise of TB cases. Chapter 17 has new photographs for mechanical isolation and sympatric speciation. Chapter 18 has a new introduction based on insights into the biblical flood story as a way to put catastrophism and evolutionary theories in a cultural context. Next is a new section on the fossil record, a new focus essay on radiometric dating and the geologic time scale, and revised sections on biogeography and comparative morphology with new art. A revised comparative embryology section connects transposons to primate evolution; the comparative biochemistry section is updated and has a better explanation of how molecular clocks are calibrated. A new section on taxonomy and classification follows; it has a simple description of how to construct a cladogram, and considers the pros and cons of the three-domain scheme. A new essay uses *Archaeopteryx* to show how evidence of evolution can be interpreted and misinterpreted. The Summary is revised.

Unit IV. Evolution and Biodiversity Chapter 19 has updates on the RNA world, the selective advantage of DNA structure, the Cambrian "explosion," Paleozoic and Mesozoic mass extinctions, and new reconstructions of early life (Figures

19.14 and 19.16). We moved the visual summary of Earth and life history to Chapter 18, but a concise version also is at the chapter's end. Chapter 20 has stunning new micrographs and art, more on archaebacteria and Lyme disease, a bit on bacterial conjugation, an update on prions and BSE, a bit on toxoplasmosis, a new table on the eight deadliest infectious diseases, and a look at food poisoning. The section on single-celled algae is revised, with a nod to The Cell From Hell. Two new Critical Thinking questions consider unsterilized needles and foot-and-mouth disease. Chapter 21 has new ascomycetes micrographs, a bit on household molds (page 336), an update on some imperfect fungi, and a better section on symbionts. Chapter 22 has new opening photos and more concise sections on evolutionary trends and the rise of seed-bearing plants. Chapter 23 has revised bits on cnidarians, roundworms, spiders and scorpions, and insects, and new Critical Thinking questions. Lobe-finned fishes are back with the fishes in Chapter 24, which also has an example of frog deformities, a cutaway turtle shell and skeleton, and more on birds and mammals, as in new Critical Thinking questions with art. Its sections on primates and human evolution have updated text and art, including new skull reconstructions. The summary is extensively revised.

New Chapter 25 puts biodiversity and extinction crises in evolutionary perspective. It starts with the Easter Island story and the nature of mass extinctions, slow recoveries, and individual species losses. It looks at threatened and endangered species, including their global distribution. It has a case study of coral reefs, an essay on Rachel Carson, and a section on conservation biology, including the role of systematics and bioeconomic analysis. It has two positive examples of recent attempts to reconcile biodiversity with human population growth and economic pressures (strip logging in the tropics and preservation of riparian zones).

Unit V. Plant Structure and Function Line-by-line text changes for clarity and a number of art refinements have improved this unit, already recognized as exceptionally accurate and clear thanks to Nancy Dengler, John Jackson, Cleon Ross, Thomas Rost, and Frank Salisbury.

Unit VI. Animal Structure and Function Chapter 29 has new micrographs and art, and an update on lab-grown tissues. Chapter 30 has a tighter introduction, some new art, sections rewritten for clarity (e.g., 30.1, 30.2, and 30.9), more on multiple sclerosis, and updates on drug addiction. The Summary is revised; we added new Critical Thinking questions. Chapter 31 has revised sections on hearing and vision; Section 31.6 has brief updates (umami, and human pheromones). Chapter 32 has a tighter introduction, an update on signaling molecules, and refined passages on cortisol, thyroid hormone, and prostaglandin secretions. A new section correlates frog deformities, hot-spot habitats, and thyroid malfunctioning. Chapter 33 has updates on vitamin D, a comparison of vertebrate skeletons, a rewrite on calcium metabolism, and new art on muscle contraction.

Chapter 34 has important new art on blood cell lineages, cardiac muscle cells, veins, hemostasis, and lymph nodes, and refined text on blood pressure. Chapter 35, a significant revision, has updates on blood cell lineages and immune functions, autoimmunity, immune deficiencies, and immunoglobulins; new molecular models; easier-to-follow text and art; and an AIDS update. Chapter 36 has new art and text on the respiratory cycle and the respiratory membrane. Chapter 37 has new text on food pyramids and protein-energy malnutrition and dietary excess, plus new art on villus fine structure. Chapter 38 has crisper definitions. Chapter 39 has major updates as well as text rewrites, especially on pattern formation and aging.

Unit VII. Ecology and Behavior　Chapter 40 has updates on human population demographics, TFRs, age structure art, and California's rolling blackouts. The behavior chapter (41) is earlier in the unit to better correspond with levels of biological organization. The text is tighter; some overlaps between sections are eliminated. The example on altruism among termites is now a Critical Thinking question.

Chapter 42 has tie-ins to Chapter 25 on biodiversity. It has updated material on the Canadian lynx/snowshoe hare puzzle, a new table on some species introductions, and a Criticial Thinking question on frog deformities.

Chapter 43 (ecosystems) has major new sections with stunning art on the nature of food chains and food webs, using tallgrass prairie as the key example, followed by an essay on biological magnification based on the original DDT study. It has new art on seasonal shifts in an omnivore's diet, rewritten text for the carbon cycle, and updated text and art for the essay on global warming. It describes the effect of fertilizers on soil ion exchange.

Chapter 44 has new art on environmental gradients and on climate zones, new icon maps for biomes, new text and photographs of tundra, and a revised wetlands section. The coral reef section is now in Chapter 25. Chapter 45 has updates on ozone thinning and nuclear energy, and some changes to the essay on tropical rain forests.

SUPPLEMENTS　The Instructors' Examination copy lists the comprehensive package of print and multimedia supplements to this book, including online resources.

A COMMUNITY EFFORT　This book is the current version of an educational effort that started twenty-six years ago. About 2,000 teachers and researchers have contributed to its ongoing refinement, and along the way we have helped more than 3 million students gain insights into the nature of biology. Much of our art and text, which is accurate and pedagogically sound as an outcome of years of research, brainstorming, and classroom testing, is now standard fare in many textbooks. The individuals listed at right deserve recognition for their commitment to quality in education.

We entrusted this edition to Gary Head, Lisa Starr, Teri Hyde, Grace Davidson, Myrna Engler, and Angela Harris —an exceptionally dedicated publishing team. Each is a superior talent in publishing. Heidi Marschner, Diane Kimmel, and Karen Hunt helped out. Pat Waldo, Chris Evers, and Steve Bolinger created a terrific multimedia package. Susan Badger and Jack Carey especially, and Kathie Head and Sean Wakely, thank you so much for your creative, enlightened management.

Cecie Starr, June 2001

General Advisors/Contributors

JOHN ALCOCK　*Arizona State University*
GEORGE COX　*San Diego State University*
KATHERINE DENNISTON　*Towson State University*
MELANIE DEVORE　*Georgia College and State University*
DANIEL FAIRBANKS　*Brigham Young University*
TOM GARRISON　*Orange Coast College*
PAUL HERTZ　*Barnard College*
JOHN JACKSON　*North Hennipin Community College*
EUGENE KOZLOFF　*University of Washington*
KAREN MESSLEY　*Rock Valley College*
CLEON ROSS　*Colorado State University*
LAURALEE SHERWOOD　*West Virginia University*
STEPHEN WOLFE　*University of California, Davis*

Fifth Edition Reviewers

BENIVENGA, STEPHEN　*University of Wisconsin, Oshkosh*
BORGESON, CHARLOTTE　*University of Nevada*
BUTTON, JERRY　*Portland Community College*
CLARK, DEBORAH C.　*Middle Tennessee University*
DABNEY, MICHAEL W.　*Hawaii Pacific University*
DEGROOTE, DAVID K.　*St. Cloud University*
FENSTER, EUGENE J.　*Longview Community College*
HEIM, WERNER G.　*The Colorado College*
JOHNSTON, TIMOTHY　*Murray State University*
JUILLERAT, FLORENCE　*Indiana University – Purdue University*
KILBURN, KERRY S.　*Old Dominion University*
LYNG, R. DOUGLAS　*Indiana University – Purdue University*
MASON, ROY B.　*Mt. San Jacinto College*
MENDELSON, JOSEPH R.　*Utah State University*
MORENO, JORGE A.　*University of Colorado, Boulder*
NADLER, KENNETH D.　*Michigan State University*
NAPLES, VIRGINIA　*Northern Illinois University*
POLCYN, DAVID M.　*California State University, San Bernardino*
RUETER, JOHN　*Portland State University*

Contributors of Influential Reviews

ALDRIDGE, DAVID　*North Carolina Agricultural/Technical State University*
ARMSTRONG, PETER　*University of California at Davis*
BAJER, ANDREW　*University of Oregon*
BAKKEN, AIMEE　*University of Washington*
BARBOUR, MICHAEL　*University of California, Davis*
BARHAM, LINDA　*Meridian Community College*
BARKWORTH, MARY　*Utah State University*
BELL, ROBERT A.　*University of Wisconsin, Stevens Point*
BENDER, KRISTEN　*California State University, Long Beach*
BINKLEY, DAN　*Colorado State University*
BLEEKMAN, GEORGE　*American River College*
BRENGELMANN, GEORGE　*University of Washington*
BRINSON, MARK　*East Carolina University*
BROWN, ARTHUR　*University of Arkansas*
BUCKNER, VIRGINIA　*Johnson County Community College*
CALVIN, CLYDE　*Portland State University*
CASE, CHRISTINE　*Skyline College*
CASE, TED　*University of California, San Diego*
CHRISTIANSEN, A. KENT　*University of Michigan*
COLAVITO, MARY　*Santa Monica College*
CONKEY, JIM　*Truckee Meadows Community College*
CROWCROFT, PETER　*University of Texas at Austin*
DANIELS, JUDY　*Washtenau Community College*
DAVIS, JERRY　*University of Wisconsin, La Crosse*
DELCOMYN, FRED　*University of Illinois, Urbana*
DEMMANS, DANA　*Finger Lakes Community College*
DEMPSEY, JEROME　*University of Wisconsin*

DENETTE, PHIL *Delgado Community College*
DENGLER, NANCY *University of California, Davis*
DESAIX, JEAN *University of North Carolina*
DETHIER, MEGAN *University of Washington*
DEWALT, R. EDWARD *Louisiana State University*
DIBARTOLOMEIS, SUSAN *Millersville University of Pennsylvania*
DIEHL, FRED *University of Virginia*
DLUZEN, DEAN *Northeastern Ohio Universities College of Medicine*
DONALD-WHITNEY, CATHY *Collin County Community College*
DOYLE, PATRICK *Middle Tennessee State University*
DUKE, STANLEY H. *University of Wisconsin, Madison*
DYER, BETSEY *Wheaton College*
EDLIN, GORDON *University of Hawaii, Manoa*
EDWARDS, JOAN *Williams College*
ELMORE, HAROLD W. *Marshall University*
ENDLER, JOHN *University of California, Santa Barbara*
ENGLISH, DARREL *Northern Arizona University*
ERWIN, CINDY *City College of San Francisco*
EWALD, PAUL *Amherst College*
FALK, RICHARD *University of California, Davis*
FISHER, DAVID *University of Hawaii, Manoa*
FISHER, DONALD *Washington State University*
FLESSA, KARL *University of Arizona*
FONDACARO, JOSEPH *Hoechst Marion Roussel, Inc.*
FRAILEY, CARL *Johnson County Community College*
FRISBIE, MALCOLM *Eastern Kentucky University*
FROEHLICH, JEFFREY *University of New Mexico*
FULCHER, THERESA *Pellissippi State Technical Community College*
GAGLIARDI, GRACE S. *Bucks County Community College*
GENUTH, SAUL M. *Mt. Sinai Medical Center*
GHOLZ, HENRY *University of Florida*
GIBSON, THOMAS *San Diego State University*
GOODMAN, H. MAURICE *University of Massachusetts Medical School*
GOSZ, JAMES *University of New Mexico*
GREGG, KATHERINE *West Virginia Wesleyan College*
GUTSCHICK, VINCENT *New Mexico State University*
HARRIS, JAMES *Utah Valley Community College*
HARTNEY, KRISTINE BEHRENTS *California State University, Fullerton*
HASSAN, ASLAM *Univeristy of Illinois College of Veterinary Medicine*
HELGESON, JEAN *Collin County Community College*
HESS, WILFORD M. *Brigham Young University*
HUFFMAN, DAVID *Southwest Texas State University*
INEICHER, GEORGIA *Hinds Community College*
INGRAHAM, JOHN L. *University of California, Davis*
JENSEN, STEVEN *Southwest Missouri State University*
JOHNSON, LEONARD R. *University of Tennessee College of Medicine*
JUILLERAT, FLORENCE *Indiana University, Purdue University*
KAREIVA, PETER *University of Washington*
KAUFMAN, JUDY *Monroe Community College*
KAYE, GORDON I. *Albany Medical College*
KAYNE, MARLENE *Trenton State College*
KELLNER, CHRIS *Arkansas Tech University*
KENDRICK, BRYCE *University of Waterloo*
KEYES, JACK L. *Linfield College, Portland Campus*
KILLIAN, JOELLA C. *Mary Washington College*
KIRKPATRICK, LEE A. *Glendale Community College*
KREBS, CHARLES *University of British Columbia*
KREBS, JULIA E. *Francis Marion University*
KUTCHAI, HOWARD *University of Virginia Medical School*
LANZA, JANET *University of Arkansas, Little Rock*
LASSITER, WILLIAM *University of North Carolina*
LEVY, MATTHEW *Mt. Sinai Medical Center*
LEWIS, LARRY *Salem State College*
LITTLE, ROBERT *Medical College of Georgia*
LOHMEIER, LYNNE *Mississippi Gulf Coast Community College*
LOPO, ALINA C. *University of California-Los Angeles Medical Center*
LUMSDEN, ANN *Florida State University*
MACKLIN, MONICA *Northeastern State University*
MANN, ALAN *University of Pennsylvania*
MARTIN, JAMES *Reynolds Community College*
MARTIN, TERRY *Kishwaukee College*
MATSON, RONALD *Kennesaw State College*
MATTHEWS, ROBERT *University of Georgia*
MAXWELL, JOYCE *California State University, Northridge*
MCCLURE, JERRY *Miami University*

MCEDWARD, LARRY *University of Florida*
MCKEAN, HEATHER *Eastern Washington State University*
MCKEE, DOROTHY *Auburn University, Montgomery*
MCNABB, ANN *Virginia Polytechnic Institute and State University*
MICKLE, JAMES *North Carolina State University*
MILLER, G. TYLER *Wilmington, North Carolina*
MINORKSY, PETER V. *Western Connecticut State University*
MOISES, HYLAN C. *University of Michigan Medical School*
MOORE-LANDECKER, ELIZABETH *Glassboro State University*
MORRISON-SHETLER, ALLISON *Georgia State University*
MORTON, DAVID *Frostburg State University*
MURPHY, RICHARD *University of Virginia Medical School*
MYRES, BRIAN *Cypress College*
NAGARKATTI, PRAKASH *Virginia Polytechnic Institute & State University*
NELSON, RILEY *University of Texas at Austin*
NORRIS, DAVID *University of Colorado*
PEARCE, FRANK *West Valley College*
PECHENIK, JAN *Tufts University*
PECK, JAMES H. *University of Arkansas, Little Rock*
PERRY, JAMES *University of Wisconsin, Center-Fox Valley*
PETERSON, GARY *South Dakota State University*
PIPERBERG, JOEL *Millersville University*
PLETT, HAROLD *Fullerton College*
REESE, R. NEIL *South Dakota State University*
REEVE, MARIAN *Emeritus, Merritt Community College*
REID, BRUCE *Kean College of New Jersey*
RENFROE, MICHAEL *James Madison University*
REZNICK, DAVID *University of California, Riverside*
RICKETT, JOHN *University of Arkansas, Little Rock*
ROBBINS, ROBERT *National Science Foundation*
ROIG, MATTIE *Broward Community College*
ROSE, GREIG *West Valley College*
ROST, THOMAS *University of California, Davis*
RUIBAL, RODOLFO *University of California, Riverside*
SALISBURY, FRANK *Utah State University*
SCHAPIRO, HARRIET *San Diego State University*
SCHLESINGER, WILLIAM *Duke University*
SCHNEIDEWENT, JUDY *Milwaukee Area Technical College*
SCHNERMANN, JURGEN *University of Michigan School of Medicine*
SCHREIBER, FRED *California State University, Fresno*
SELLERS, LARRY *Louisiana Tech University*
SHEGAL, PREM *East Carolina University*
SHONTZ, NANCY *Grand Valley State University*
SHOPPER, MARILYN *Johnson County Community College*
SLOBODA, ROGER *Dartmouth College*
SMITH, JERRY *St. Petersburg Junior College, Clearwater Campus*
SMITH, MICHAEL E. *Valdosta State College*
SMITH, ROBERT L. *West Virginia University*
SOLOMON, NANCY *Miami University*
STEARNS, DONALD *Rutgers University*
STEELE, KELLY P. *Appalachian State University*
STEINERT, KATHLEEN *Bellevue Community College*
SUMMERS, GERALD *University of Missouri*
SUNDBERG, MARSHALL D. *Louisiana State University*
SWANSON, ROBERT *North Hennepin Community College*
SWEET, SAMUEL *University of California, Santa Barbara*
TAYLOR, JANE *Northern Virginia Community College*
TERHUNE, JERRY *Jefferson Community College, University of Kentucky*
TIZARD, IAN *Texas A&M University*
TROUT, RICHARD E. *Oklahoma City Community College*
TYSER, ROBIN *University of Wisconsin, LaCrosse*
WAALAND, ROBERT *University of Washington*
WAHLERT, JOHN *City University of New York, Baruch College*
WALSH, BRUCE *University of Arizona*
WARING, RICHARD *Oregon State University*
WARNER, MARGARET R. *Purdue University*
WEBB, JACQUELINE F. *Villanova University*
WEIGL, ANN *Winston-Salem State University*
WEISS, MARK *Wayne State University*
WELKIE, GEORGE W. *Utah State University*
WENDEROTH, MARY PAT *University of Washington*
WHITE, EVELYN *Alabama State University*
WHITENBERG, DAVID *Southwest Texas State University*
WINICUR, SANDRA *Indiana University, South Bend*
YONENAKA, SHANNA *San Francisco State University*

Introduction

Current configurations of the Earth's oceans and land masses—the geologic stage upon which life's drama continues to unfold. This composite satellite image reveals global energy use at night by the human population. Just as biological science does, it invites you to think more deeply about the world of life—and about our impact upon it.

CONCEPTS AND METHODS IN BIOLOGY

Why Biology?

Leaf through a newspaper on any given Sunday and you might get an uneasy feeling that the natural world is spinning out of control. Northern forests are burning fiercely (Figure 1.1). The Arctic ice cap is melting and Greenland's glaciers are becoming lakes. Heat waves and droughts are breaking records everywhere, and maybe the whole atmosphere is warming up. The last individual of an ancient primate lineage bit the dust just as the human population neared the 6.1 billion mark. Geneticists decoded yet another garbled message in human DNA that causes another horrible disease. People are still arguing about when, exactly, the mass of human embryonic cells busily dividing inside a woman's womb is "alive." In a bizarre application of genetic engineering, somebody decided to make an "artistic statement" by producing a glow-in-the-dark bunny. Every bit of the live bunny, eyes included, turns fluorescent green in black light.

It's enough to make you throw down the paper and go sit in a park. It's enough to make you wish you lived in the good old days when things were so much simpler.

Of course, read up on the good old days and you'll discover they weren't so good. For example, in 1918, maybe when your grandparents were alive, the Spanish flu swept around the world. Infected people often died within hours of the first symptoms. Before the pandemic subsided, *30 to 40 million* people died. Unlike most flu viruses, which mainly endanger infants, the elderly,

Figure 1.1 Biology starts with the premise that any aspect of nature—including this forest fire in Montana—has one or more underlying causes. By giving us a way of thinking critically about life, biology helps us understand how to deal with nature, and with our place in it.

and the very ill, the Spanish flu virus tended to take down healthy, young adults, the ones who actually support most of the social infrastructure. Then, as today, many people thought the world was coming apart at the seams. They felt helpless and hopeless before a force of nature on the rampage.

What it boils down to is this: For a couple of million years at least, humans and their immediate ancestors have been trying to make sense of the natural world. We observe, we come up with ideas about it, we test the ideas. However, the more pieces of the puzzle we fit together, the bigger the puzzle gets. We are now smart enough to know that it's almost overwhelmingly big.

You might choose to walk away from the challenge and simply let others tell you what to think. Or you might choose to develop your own understanding of the puzzle. Maybe you're interested in the pieces that affect your health, your home, the food you eat, and your offspring, should you choose to reproduce. Maybe you just find the connections among organisms and their environment fascinating. Regardless of the focus, **biology**—the scientific study of life—can help you deepen your perspective on the nature of life.

Start with a question that seems simple enough: *What is life?* Offhandedly, you might say you know it when you see it. Yet the question opens up a story that began at least 3.8 billion years ago and has been unfolding in countless directions ever since.

From a biological perspective, "life" is an outcome of ancient events by which lifeless matter—atoms and molecules—became organized into the first living cells. "Life" is a way of capturing and using energy and raw materials. "Life" is a way of sensing and responding to the environment. "Life" is a capacity to reproduce. And "life" evolves, which means the traits that characterize individuals of a population change from one generation to the next.

Throughout this book, you will come across many examples of how organisms are constructed, how they function, where they live, what they do. The examples support concepts which, when taken together, convey what "life" is. This chapter introduces you to the basic concepts. It sets the stage for forthcoming descriptions of scientific observations, experiments, and tests that help show how you can develop, modify, and refine your views of life. As you continue your reading, you may find it useful to return to this simple overview as a way to reinforce your grasp of the details.

Key Concepts

1. Unity underlies the world of life, for all organisms are alike in key respects. They consist of one or more cells made of the same kinds of substances, put together in the same basic ways. Their activities require inputs of energy, which they must get from their surroundings. All organisms sense and respond to changing conditions in their environment. They all have a capacity to grow and reproduce, based on instructions contained in DNA.

2. The world of life shows immense diversity. Many millions of different kinds of organisms, or species, now inhabit the Earth. Many millions more lived in the past. Each species is unique in some of its traits—that is, in some aspects of its body plan, body functioning, and behavior.

3. Theories of evolution, especially a theory of evolution by natural selection as formulated by Charles Darwin, help explain life's diversity. The theories unite all fields of biological inquiry into a single, coherent whole.

4. Biology, like other branches of science, is based on systematic observations, hypotheses, predictions, and observational and experimental tests. The external world, not internal conviction, is the testing ground for scientific theories.

DNA, ENERGY, AND LIFE

Nothing Lives Without DNA

DNA AND THE MOLECULES OF LIFE   Picture a frog on a rock, busily croaking. Even without thinking about it, you know the frog is alive and the rock is not. Could you explain why? After all, both consist only of protons, electrons, and neutrons. But these units of matter are building blocks for diverse atoms—the building blocks for the larger and more diverse bits of matter called molecules. It is at the molecular level that differences between living and nonliving things start to emerge.

You will never, ever find a rock made of nucleic acids, proteins, carbohydrates, and lipids. In nature, only *cells* build these molecules, which you will read about later. All living things are composed of one or more cells, the smallest units of matter having a capacity for life. And the signature molecule of cells is a nucleic acid known as **DNA**. No chunk of granite or quartz has it.

Encoded in DNA's structure are the instructions for assembling a dazzling array of proteins from a limited number of smaller building blocks, the amino acids. By analogy, if you follow suitable instructions and invest energy in the task, you might organize a heap of a few kinds of ceramic tiles (representing amino acids) into diverse patterns (representing proteins), as in Figure 1.2.

Enzymes are among the proteins. When these worker molecules get an energy boost, they swiftly build, split, and rearrange the molecules of life. As described later, some work with nucleic acids called RNAs to carry out DNA's protein-building instructions. Think of this as a flow of information, from *DNA to RNA to protein*. This molecular trinity is central to our understanding of life.

Figure 1.2 Examples of objects built from the same materials by different assembly instructions.

THE HERITABILITY OF DNA   We humans tend to think we all enter the world abruptly and leave it the same way. But we are much more than this. *We and all other organisms are part of a journey that began at least 3.8 billion years ago, with the origin of the first living cells.*

Under present-day conditions on Earth, new cells arise only from cells that already exist. They do so by **inheritance**, which is an acquisition of traits by way of transmission of DNA from parents to offspring. Why do baby storks look like storks and not pelicans? Because they inherited stork DNA, which isn't exactly the same as pelican DNA in its molecular details.

Reproduction refers to actual mechanisms by which parents produce offspring. Often it starts when a sperm fertilizes an egg. However, a fertilized egg would never form if the sperm and egg had not formed earlier. And both formed from DNA instructions passed on countless times—from cell to cell—according to the principles of inheritance and reproduction.

For frogs and humans and other large organisms, DNA also guides **development**, a series of stages that transform a single cell into a multicelled adult, typically with tissues and organs specialized for certain tasks. For instance, a moth is the adult stage of a winged insect (Figure 1.3). It starts as a fertilized egg that develops

Figure 1.3   "The insect"—a series of stages of development. Different adaptive properties emerge at each stage. Shown here, a silkworm moth, from the egg (**a**), to a larval stage (**b**), to a pupal stage (**c**), and on to the winged form of the adult (**d**,**e**).

Figure 1.4 Response to signals from pain receptors, activated by a lion cub flirting with disaster.

into an immature larval stage. That stage, a caterpillar, eats leaves and grows rapidly until an internal alarm clock goes off. Then, its tissues become remodeled into a different stage—a pupa. In time an adult emerges. An adult form is adapted to reproduce. Adult moths make sperm or eggs, and the color, pattern, and fluttering frequency of its wings are adapted to attract a mate.

And so "the insect" is a series of organized stages, each of which must be completed before the next begins. Instructions for completing each stage were written into moth DNA long before each moment of reproduction— and so the ancient moth story continues.

Nothing Lives Without Energy

ENERGY DEFINED Moths and everything else in the universe have some amount of **energy**, which we define as the capacity to do work. And nothing—absolutely nothing—happens in the universe without a *transfer* of energy. For instance, suppose some undisturbed atom in a cell absorbs enough energy from the sun to get its electrons jumping. A molecule nearby grabs an excited electron and transfers it to machinery that converts the extra energy into a form the cell can use. In such ways, atoms give up, share, and grab electrons; and molecules form, are rearranged, and are split apart. Such molecular work allows cells to stay alive, grow, and reproduce.

METABOLISM DEFINED Each living cell has the capacity to (1) obtain and convert energy from its surroundings and (2) use energy to maintain itself, grow, and make more cells. We call this capacity **metabolism**. Think of one cell in a leaf that makes food by photosynthesis. It intercepts sunlight energy and converts it to chemical energy, in the form of ATP molecules. ATP helps drive hundreds of events by transferring energy to metabolic workers—in this case, enzymes that put together sugar molecules. ATP also forms by aerobic respiration. This process can release energy that cells have tucked away in sugars and other kinds of molecules.

SENSING AND RESPONDING TO ENERGY It's often said that only organisms respond to the environment. Yet even a rock shows responsiveness, as when it yields to the force of gravity and tumbles down a hill or changes its shape slowly under the repeated battering of wind, rain, or tides. The difference is this: *Organisms can sense changes in their surroundings and then make compensatory, controlled responses to the changes.* How? Each organism has **receptors**, which are molecules and structures that detect stimuli. A **stimulus** is a specific form of energy that a receptor can detect. Examples are sunlight energy,

heat energy, a hormone molecule's chemical energy, and the mechanical energy of a bite (Figure 1.4).

Cells adjust metabolic activities in response to signals from receptors. Each cell (and organism) can withstand only so much heat or cold. It must rid itself of harmful substances. It requires certain foods, in certain amounts. Yet temperatures do shift, harmful substances might be encountered, and food is sometimes plentiful or scarce.

Finish a snack, and simple sugars leave your gut and enter your blood. Blood, along with the tissue fluid that bathes your cells, makes up your *internal* environment. Unchecked, too much or too little sugar in the blood can cause diabetes and other problems. When the level rises, the pancreas (a glandular organ) secretes more insulin. Most of your body cells have receptors for this hormone, which stimulates cells to take up sugar. When enough cells do so, the blood sugar level returns to normal.

Organisms respond so exquisitely to energy changes that their internal operating conditions remain within tolerable limits. We call this a state of **homeostasis**. It is one of the key defining features of life.

All organisms consist of one or more cells, the smallest units of life. Under present-day conditions, new cells form only when cells that are already alive reproduce.

DNA, the molecule of inheritance, encodes protein-building instructions, which RNAs help carry out. Many proteins are enzymes that speed up cellular work, which includes building all the complex molecules characteristic of life.

Cells live only for as long as they engage in metabolism. They acquire and transfer energy that is used to assemble, break down, stockpile, and dispose of materials in ways that promote survival and reproduction.

Single cells and multicelled organisms sense and respond to environmental conditions in ways that help maintain their internal operating conditions.

ENERGY AND LIFE'S ORGANIZATION

Levels of Biological Organization

Taken as a whole, the metabolic activities of single cells and multicelled organisms maintain a great pattern of organization in nature, as in Figure 1.5. Consider the hierarchy. Life's properties emerge when DNA and other molecules are organized into cells. A **cell** is the smallest unit of organization having a capacity to survive and reproduce on its own, given DNA instructions, suitable conditions, building blocks, and energy inputs. This is obvious for amoebas and other free-living cells. Yet the definition also fits **multicelled organisms** composed of many specialized, interdependent cells that are typically organized in tissues and organs. Do you question this? After all, your own cells could never live all alone in nature, because body fluids must continuously bathe them. Yet even isolated human cells remain alive under controlled conditions in laboratories around the world. Researchers routinely maintain isolated human cells for important experiments, as in cancer studies.

Cells and multicelled organisms usually are part of a **population**: a group of organisms of the same kind. A zebra herd is an example. The next level of organization is the **community**: all populations of all species living in the same area, such as the African savanna's bacteria, grasses, trees, zebras, lions, and so on. The next level is the **ecosystem**: a community *together with* its physical and chemical environment. The **biosphere**, the highest level, includes all parts of the Earth's crust, waters, and atmosphere in which organisms live. Astoundingly, *this globe-spanning organization begins with the convergence of energy, certain materials, and DNA in tiny, individual cells.*

BIOSPHERE
All regions of the Earth's crust, waters, and atmosphere that sustain life

ECOSYSTEM
Community and its physical environment

COMMUNITY
Populations of all species occupying the same area

POPULATION
Group of individuals of the same kind (that is, the same species) occupying the same area

MULTICELLED ORGANISM
Individual consisting of interdependent cells typically organized in tissues, organs, and organ systems

ORGAN SYSTEM
Two or more organs interacting chemically, physically, or both in ways that contribute to organism's survival

ORGAN
Structural unit in which tissues, combined in specific amounts and patterns, perform a common task

TISSUE
Organized aggregation of cells and substances functioning together in a specialized activity

CELL
Smallest unit with the capacity to live and reproduce, independently or as part of multicelled organism

ORGANELLE
Membrane-bound internal compartment for specialized reactions (most prokaryotic cells have none)

MOLECULE
Unit of two or more bonded-together atoms of the same element or different elements

ATOM
Smallest unit of an element (a fundamental substance) that still retains the properties of that element

SUBATOMIC PARTICLE
Electron, proton, neutron, or some other fundamental unit of matter

populations of shrubs and trees

community (all of the populations living in the same area)

populations of grasses

population of zebras

ecosystem (community together with its physical environment)

multicelled individual

once-live multicelled individual, about to revert to molecules and atoms

Figure 1.5 Levels of organization in nature.

beetle larva

Figure 1.6 Example of the one-way flow of energy and the cycling of materials through the biosphere.

Producers capture, convert, and use or store some energy from the sun.

PRODUCERS

NUTRIENT CYCLING

CONSUMERS, DECOMPOSERS

ONE-WAY FLOW OF ENERGY

Energy gets transferred from one organism to another; in time, all flows back to the environment.

(**a**) Plants of a warm, dry grassland called the African savanna capture energy from the sun and use it to build plant parts. Some of the energy ends up inside plant-eating organisms, such as this adult male elephant. He eats huge quantities of plants to maintain his eight-ton self. He produces piles of solid wastes—dung—that still hold some unused nutrients and energy. Although most organisms might not recognize it as such, elephant dung is an exploitable food source.

(**b**) And so we next have little dung beetles scrambling to the scene almost simultaneously with the uplifting of an elephant tail. Working rapidly, they carve fragments of moist dung into round balls, which they roll off and bury in burrows. In those balls the beetles lay eggs—a reproductive behavior that will help assure their forthcoming offspring (**c**) of a compact food supply.

Thanks to beetles, dung does not pile up and dry out into rock-hard mounds in the intense heat of the day. Instead, the surface of the land is tidied up, the beetle offspring get fed, and the leftover dung accumulates in beetle burrows—there to enrich the soil that nourishes the plants that feed (among others) the elephants.

Interdependencies Among Organisms

A great flow of energy into the world of life starts with **producers**, which are plants and other organisms that make their own food. Animals are **consumers**. Directly or indirectly, they depend on food energy that's stored in tissues of producers. For example, some energy gets transferred to zebras after they've browsed on plants. It gets transferred again when a lion devours a zebra, as in Figure 1.5. And it is transferred again when fungal and bacterial decomposers extract energy from tissues and remains of lions, elephants, or any other organism. **Decomposers** break down sugars and other biological molecules to simpler materials, some of which is cycled back to the producers. In time, all of the energy that the plants originally captured from the sun's rays returns to the environment, but that's another story.

For now, keep in mind that organisms connect with one another by a one-way flow of energy *through* them and a cycling of materials *among* them, as in Figure 1.6. Their interconnectedness affects the structure, size, and composition of populations and communities. It affects ecosystems, even the biosphere. Understand the extent of their interactions and you will gain insight into food shortages, cholera epidemics, acid rain, global warming, biodiversity losses, and other modern-day problems.

Nature shows levels of organization. The characteristics of life emerge at the level of single cells and extend through populations, communities, ecosystems, and the biosphere.

A one-way flow of energy through organisms and a cycling of materials among them organizes life in the biosphere. In nearly all cases, energy flow starts with energy from the sun.

IF SO MUCH UNITY, WHY SO MANY SPECIES?

So far, we have focused on life's unity, on characteristics that all living things have in common. Think of it! They consist of the same "lifeless" materials. They can stay alive by metabolism—by ongoing energy transfers at the cellular level. They interact in their requirements for energy and raw materials. They have a capacity to sense and respond to their environment in highly specific ways. They all have the capacity to reproduce based on instructions encoded in DNA. They all inherited DNA from individuals of a preceding generation.

Superimposed on the common heritage is immense diversity. You share the planet with many millions of different kinds of organisms, or **species**. Many millions more preceded us during the past 3.8 billion years, but their lineages vanished; they are extinct. For centuries, scholars have tried to make sense of the confounding diversity. One of them, Carolus Linneaus, came up with a classification scheme that assigns a two-part name to each newly identified species. The first part designates the **genus** (plural, genera). Each genus encompasses all species that seem closely related by way of their form, functioning, and ancestry. The second part of the name designates a particular species within a genus.

For instance, *Quercus alba* is the white oak's scientific name, and *Q. rubra* the red oak's name. (Once you spell out the genus name in a document, you can abbreviate it thereafter.) Use such names as passports when you visit the web, and enter worlds of information.

Biologists also classify life's diversity by assigning species to groups at more encompassing levels. Among other things, they group genera that apparently share a common ancestor into the same *family*, related families into the same *order*, related orders into the same *class*, and related classes into the same *phylum* (plural, phyla). At a higher level, they assign related phyla to the same *kingdom*. They are still refining the groups. For instance, scholars once recognized only two kingdoms—animals and plants. Later, biologists recognized five kingdoms. Today, most agree that there should be at least six: the **Archaebacteria**, **Eubacteria**, **Protista**, **Fungi**, **Plantae**, and **Animalia** (Figure 1.7). We adhere to a six-kingdom scheme in this book. At this point in your reading, it is enough simply to become acquainted with a few of the defining features of their members.

Archaebacteria and eubacteria are all single cells, of a type called *prokaryotic*. The word means they don't have a nucleus. (In other cell types, this membrane-bound sac keeps DNA separate from the rest of the cell's interior.)

Figure 1.7 A few representatives of life's diversity.

KINGDOM ARCHAEBACTERIA. (**a**) From the muck of an anaerobic (oxygen-free) habitat, a colony of cells (*Methanosarcina*).

KINGDOM EUBACTERIA. (**b**) A "true" bacterium sporting a number of bacterial flagella. Flagella are used for motility.

KINGDOM PROTISTA. (**c**) A trichomonad that lives in a termite's gut. Compared to bacteria, most protistans are much larger and have greater internal complexity. They range from microscopically small single cells, such as this one, to kelps as tall as trees.

KINGDOM FUNGI. (**d**) A stinkhorn fungus. Some species of fungi are parasites and some cause diseases, but the vast majority are decomposers. Without decomposers, communities would gradually become buried in their own wastes.

KINGDOM PLANTAE. (**e**) Trunk of a redwood growing near the coast of California. Like nearly all plants, redwoods produce their own food by photosynthesis. (**f**) Flower of a plant from the family of composites. Its colors and patterning guide bees to nectar. The bees get food. The plant gets help reproducing. Like many other organisms, they interact in a mutually beneficial way.

KINGDOM ANIMALIA. (**g**,**h**) Male bighorn sheep competing for females and so displaying a characteristic of the kingdom— they actively move about in their environment.

Different kinds are producers, consumers, and decomposers. Of all kingdoms, theirs show the most metabolic diversity.

Diverse archaebacteria live in hot springs, salt lakes, and other habitats as harsh as the ones that prevailed when cells first evolved. The eubacteria are more common and widespread than archaebacteria. Their name means "true bacteria," so in this book we'll just refer to them as bacteria (singular, bacterium).

Characterizing protistans is not easy. Like the plants, fungi, and animals, they are all *eukaryotic*, meaning their DNA is inside a nucleus. Most are larger and show far more internal complexity than prokaryotic cells. Beyond this generalization are confoundingly diverse single-celled producers, consumers, and decomposers, as well as giant multicelled "seaweeds." The food webs of most aquatic provinces start with photoautotrophic protistans.

Most fungi, including the common field mushrooms sold in grocery stores, are multicelled. These eukaryotic decomposers and consumers feed in a distinctive way. They secrete enzymes that digest food outside of the fungal body, then their cells absorb the digested bits.

Both plants and animals are multicelled, eukaryotic organisms. Nearly all plants are photosynthetic, with cellular pipelines for water and solutes that extend through roots, stems, and leaves. Most live on land and are the major producers for communities. Animals are consumers that ingest other organisms or their tissues. Among their ranks are herbivores (which graze on the photosynthesizers), carnivores (meat eaters), parasites, and scavengers. Unlike plants, all animals actively move about during at least some stage of their life.

Pulling this together, you start to get a sense of what it means when someone says life has unity *and* diversity.

Unity threads through the world of life, for all organisms are alike in important ways. They are composed of the same substances, which are assembled in the same basic ways. They engage in metabolism, and they sense and respond to their environment. They all have a capacity to reproduce, based on heritable instructions encoded in their DNA.

Immense diversity threads through the world of life, for organisms differ enormously in body form, in the functions of their body parts, and in their behavior.

To make the study of life's diversity more manageable, we group organisms into six kingdoms—the archaebacteria, eubacteria, protistans, fungi, plants, and animals.

AN EVOLUTIONARY VIEW OF DIVERSITY

Given that organisms are so much alike, what could account for their great diversity? One key explanation is called evolution by means of natural selection. A few simple examples will be enough to introduce you to its premises, which build on the simple observation that individuals of a population vary in their traits.

Mutation—Original Source of Variation

DNA has two striking qualities. Its instructions work to ensure that offspring will resemble their parents, yet they also permit variations in the details of most traits. As an example, having five fingers on each hand is a human trait. Yet some humans are born with six fingers on each hand instead of five. This is an outcome of a **mutation**—a heritable change in DNA. Mutations are the original source of variations in heritable traits.

Many mutations are harmful. A change in even a bit of DNA may be enough to sabotage the body's growth, development, or functioning. Yet some variations are harmless or beneficial. A classic case is a mutation that results in dark-colored moths in a population of light-colored moths. Moths fly at night and rest in the day, when birds that eat them are active. Birds tend to miss light-colored moths resting on light-colored tree trunks. These moths are camouflaged; they "hide in the open."

Suppose people build coal-burning factories nearby. In time, soot-laden smoke darkens all the tree trunks. Dark moths on darkened trunks aren't as conspicuous to bird predators (Figure 1.8), so they are more likely to survive and reproduce. Where soot accumulates, the variant (dark) form of this trait is more adaptive than the light form. An **adaptive trait** is any form of a trait that helps an individual survive and reproduce under prevailing environmental conditions.

Figure 1.8 How different forms of a trait (body surface coloration) are adaptive to two different conditions. (**a**) Light-colored moths (*Biston betularia*) on a nonsooty tree trunk are well hidden from predators, but dark ones stand out. (**b**) The dark color is more adaptive in places where soot darkens the tree trunks.

Figure 1.9 *Facing page:* One outcome of artificial selection practices: just a few of the 300+ varieties of domesticated pigeons. Breeders started with variant forms of traits in captive populations of wild rock doves.

WILD ROCK DOVE

Evolution Defined

Think about that population of light-colored moths in a sooty forest. At some point, a DNA mutation arose and resulted in a moth of a darker color. When the mutated individual reproduced, some of its offspring inherited the trait. Birds saw and ate many light-colored moths, but most of the dark ones escaped detection and lived long enough to reproduce. So did their dark offspring. So did *their* offspring, and so on. Said another way, the frequency of one form of the trait increased relative to the frequency of the other. In time, the dark form may even become the most common form. People may end up referring to "the population of dark-colored moths."

Evolution is under way. In biology, the word means genetically based change in a line of descent over time. Like moths, individuals of most populations typically show different forms of many (or most) of their traits. The relative frequencies of the different forms typically change over successive generations.

Natural Selection Defined

Long ago, the naturalist Charles Darwin used pigeons to explain a conceptual connection between evolution and variation in traits. Domesticated pigeons differ in their size, feather color, and other traits (Figure 1.9). As Darwin knew, pigeon breeders select certain forms of traits. For example, if they are after black, curly-edged tail feathers, they will let only individual pigeons with the most black and most curl in their tail feathers mate and reproduce. So "black" and "curly" will be the most common forms of tail feathers in the captive pigeon population. Different forms of these traits will become less common or will be eliminated.

Pigeon breeding is a case of **artificial selection**, for the selection among different forms of a trait is taking place in an "artificial" environment—under contrived, manipulated conditions. Yet Darwin saw the practice as a simple model for *natural* selection, a favoring of some forms of traits over others in nature.

Whereas breeders are "selective agents" promoting the reproduction of some individuals but not others in captive populations, different selective agents operate across the range of variation among pigeons in the wild. Pigeon-eating peregrine falcons are such agents. Swifter or more effectively camouflaged individuals of a pigeon population have a far better chance of escaping from falcons and living long enough to reproduce, compared to the not-so-swift or too-conspicuous individuals among them. What Darwin identified as **natural selection** is simply a difference in *which* individuals of a population survive and reproduce in a given generation. The difference is an outcome of some individuals having more adaptive forms of traits.

Unless you happen to be a pigeon breeder, variation among pigeons may not be a burning issue in your life. So think about something closer to home. **Antibiotics** are toxic secretions of certain soil-dwelling bacteria and fungi; they work against bacterial and fungal competitors for nutrients. In the 1940s, we learned to use antibiotics to kill disease-causing bacteria. Doctors prescribed these "wonder drugs" even for mild infections. In some cases antibiotics were added to toothpaste and chewing gum.

As it turned out, antibiotics are powerful agents of natural selection. Consider streptomycin, an antibiotic that shuts down some essential proteins in bacteria that are its targets. However, certain strains of bacteria have mutated in ways that help them resist streptomycin. The mutations slightly altered the molecular form of the proteins, so streptomycin can't bind to them. Unlike nonmutated cells, the mutants survive and reproduce. Said another way, an antibiotic acts against the bacteria susceptible to it but *favors* variant strains that resist it!

Antibiotic-resistant strains are making it difficult to treat bacterial diseases, including tuberculosis, typhoid, gonorrhea, and staph infections. A few superbugs can no longer be eliminated from the most susceptible people.

When resistance to antibiotics evolves by selection processes, our strategies for using antibiotics must also evolve to overcome the new defenses. That's why drug companies try to modify parts of antibiotic molecules. Such molecular changes in the laboratory might yield more effective antibiotics. But these may work only until new generations of ever more resistant superbugs enter the deadly evolutionary competition for nutrients.

Later in the book, we will consider mechanisms by which populations of moths, pigeons, bacteria, and all other organisms evolve. Meanwhile, keep in mind the following points about natural selection. They are central to biological inquiry, for they have consistently proved useful in explaining a great deal about nature.

1. Individuals of a population vary in form, function, and behavior. Much of this variation is heritable; it can be transmitted from parents to offspring.

2. Some forms of heritable traits are more adaptive to prevailing conditions. They improve an individual's chance of surviving and reproducing, as by helping it secure food, a mate, hiding places, and so on.

3. Natural selection is the outcome of differences in survival and reproduction among variant individuals in a given generation.

4. Natural selection leads to a better fit with prevailing environments. The adaptive forms of traits tend to become more common than other forms. Thus the population's characteristics change; it evolves.

In the evolutionary view, then, *life's diversity is the sum total of variations in traits that have accumulated in different lines of descent generation after generation, as by natural selection and other processes of change.*

Mutations in DNA introduce variations in heritable traits.

Although many mutations are harmful, some give rise to variations in form, function, or behavior that are adaptive under prevailing environmental conditions.

Natural selection is a result of differences in survival and reproduction among individuals of a population that vary in one or more heritable traits. The process helps explain evolution—changes in lines of descent over the generations.

THE NATURE OF BIOLOGICAL INQUIRY

The preceding sections sketched out major concepts in biology. Now consider approaching this or any other collection of "facts" with a critical attitude. *"Why should I accept that they have merit?"* The answer requires insight into how biologists make inferences about observations and test the predictive power of the inferences against actual experience in nature or the laboratory.

Observations, Hypotheses, and Tests

To get a sense of "how to do science," start by following some practices that pervade scientific research:

1. Observe some aspect of nature, carefully check what others have found out about it, then frame a question or identify a problem related to your observation.

2. Develop **hypotheses**, or educated guesses, about possible answers to questions or solutions to problems.

3. Using hypotheses as a guide, make a **prediction**—that is, a statement of what you should observe in the natural world if you were to go looking for it. This is often called the "if–then" process. (*If* gravity does not pull objects toward Earth, *then* it should be possible to observe apples falling up, not down, from a tree.)

4. Devise ways to **test** the accuracy of your predictions, as by making systematic observations, building models, and conducting experiments. **Models** are theoretical, detailed descriptions or analogies that help us visualize something that has not yet been directly observed.

5. If tests do not confirm the prediction, check to see what might have gone wrong. For example, maybe you overlooked some factor that influenced the test results. Or maybe the hypothesis is not a good one.

6. Repeat the tests or devise new ones—the more the better, for hypotheses that withstand many tests are likely to have a higher probability of being useful.

7. Objectively analyze and report the test results and the conclusions drawn from them.

You might hear someone refer to these practices as "the scientific method," as if all scientists march to the drumbeat of an absolute, fixed procedure. They do not. Many observe, describe, and report on some subject, then leave it to others to hypothesize about it. Some are lucky; they stumble onto information they aren't even looking for. (Of course, chance does favor a mind that's prepared to know what the information means.) It is not one single method scientists have in common. Rather, it is a critical attitude about being shown rather than told and taking a logical approach to problem solving.

Logic encompasses thought patterns by which an individual draws a conclusion that does not contradict evidence used to support it. Lick a cut lemon and you

Figure 1.10 Generalized sequence of steps involved in an experimental test of a prediction based on a hypothesis.

notice it's mouth-puckeringly sour. Lick ten more. Each time you notice the same thing, so you conclude that all lemons are mouth-puckeringly sour. You correlated one specific (lemon) with another (sour). By this pattern of thinking, called **inductive logic**, an individual derives a general statement from specific observations.

Express the generalization in "if–then" terms, and you have a hypothesis: "If you lick any lemon, then you will get an extremely sour taste in your mouth." By this pattern of thinking, called **deductive logic**, an individual makes inferences about specific consequences or specific predictions that must follow from a hypothesis.

You decide to test the hypothesis by sampling lemons in your area. One variety, the Meyer lemon, is mellow for a lemon. You also find that some people can't taste anything. So you modify the hypothesis: "If most people lick any lemon *except* the Meyer lemon, they will get a very sour taste in their mouth." Suppose you sample all known lemon varieties in the world and conclude the modified hypothesis is a good one. You'll never prove it beyond all shadow of a doubt, for there may be lemon trees in places we don't even know about. You *can* say the hypothesis has a high probability of not being wrong.

Comprehensive observations are a logical means to test the predictions that flow from hypotheses. So are **experiments**. These tests simplify observation in nature or the laboratory by manipulating and controlling the conditions under which observations are made. When suitably designed, observational and experimental tests allow you to predict that something will happen if a hypothesis isn't wrong (or won't happen if it *is* wrong). Figure 1.10 gives a general idea of the steps involved.

AN ASSUMPTION OF CAUSE AND EFFECT Experiments start from the premise that any aspect of nature has one or more underlying causes. With this premise, science is distinct from faith in the supernatural (meaning "beyond nature"). Experiments deal with potentially falsifiable hypotheses. Hypotheses of this sort can be tested in the natural world in ways that might disprove them.

a Natalie, blindfolded, randomly plucks a jellybean from a jar of 120 green and 280 black jellybeans. That's a ratio of 30 to 70 percent.

b The jar is hidden before she removes her blindfold. She observes a single green jellybean in her hand and assumes the jar holds only green jellybeans.

c Still blindfolded, Natalie randomly picks 50 jellybeans from the jar and ends up with 10 green and 40 black ones.

d The larger sample leads her to assume one-fifth of the jar's jellybeans are green and four-fifths are black (a ratio of 20 to 80). Her larger sample more closely approximates the jar's green-to-black ratio. The more times Natalie repeats the sampling, the greater the chance she will come close to knowing the actual ratio.

Figure 1.11 A simple demonstration of sampling error.

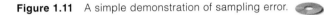

EXPERIMENTAL DESIGN To get conclusive test results, experimenters rely on certain practices. They refine test designs by searching the literature for information that may relate to their inquiry. They design experiments to test one prediction of a hypothesis at a time. Each time, they set up a **control group**: a standard for comparison with one or more experimental groups. Ideally, a control group is identical with the experimental group *except* for the one variable being investigated. **Variables** are specific aspects of objects or events that may differ or change over time and among individuals.

Section 1.6 shows the design of a recent experiment. As you will see, the experimenters directly manipulated a single variable in an attempt to support or disprove a prediction. They also tried to hold constant any other variables that could influence the results.

SAMPLING ERROR Rarely can experimenters observe *all* individuals of a group. Rather, they must use subsets or samples of populations, events, and other aspects of nature. They must avoid **sampling error**, or risking tests with groups that by chance are not representative of the whole. Generally, distortion in test results is less likely when the samplings are large and when they are repeated (Figure 1.11).

About the Word "Theory"

Suppose no one has disproved a hypothesis after years of rigorous tests. Suppose scientists use it to interpret more data or observations, which could involve more hypotheses. When a hypothesis meets these criteria, it may become accepted as a **scientific theory**.

You may hear someone apply the word "theory" to a speculative idea, as in the expression "It's only a theory." But a scientific theory differs from speculation for this reason: *Researchers have tested its predictive power many times and in many ways in the natural world and have yet to find evidence that disproves it.* This is why the theory of natural selection is respected. We use it successfully to explain diverse issues, such as how life originated, how plant toxins relate to plant-eating animals, what the sexual advantage of feather color might be, why certain cancers run in families, or why antibiotics are losing effectiveness. Together with the record of Earth history, the theory has even helped explain how life evolved.

Maybe an exhaustively tested theory is as close to the truth as scientists can get with the evidence available. For example, Darwin's theory stands, with only minor modification, after more than a century's worth of many thousands of different scientific tests. We can't prove it holds under all possible conditions; an infinite number of tests would be required to do so. As for any theory, we can only say it has a *high or low probability* of being a good one. So far, biologists haven't found evidence that refutes Darwin's theory. Yet they still keep their eyes open for any new information and new ways of testing that might disprove its premises.

This last point gets us back to the value of thinking critically. Scientists must keep asking themselves: *Will observations or experiments show that a hypothesis is false?* They expect one another to put aside pride or bias by testing ideas even in ways that may prove them wrong. Even if an individual doesn't or won't do this, others will—for science proceeds as a community that is both cooperative and competitive. Ideally, its practitioners share ideas, knowing it's just as useful to expose errors as to applaud insights. Individuals can and do change their mind when presented with contradictory evidence. This is a strength of science, not a weakness.

A scientific approach to studying nature is based on asking questions, formulating hypotheses, making predictions, devising tests, and objectively reporting the results.

A scientific theory is a testable explanation about the cause or causes of a broad range of related phenomena. It remains open to tests, revision, and tentative acceptance or rejection.

The Power of Experimental Tests

BIOLOGICAL THERAPY EXPERIMENTS Worry about the new strains of pathogenic (disease-causing) bacteria that resist antibiotics, and you won't be alone. Bruce Levin of Emory University and Jim Bull of the University of Texas have been looking for possible alternatives to antibiotic therapy. They studied literature on a *biological therapy* that enlists bacteriophages—"bacteria eaters"—to fight infections. Bacteriophages, a class of viruses, attack a narrow range of bacterial strains. When they contact a target cell, they inject enzymes and genetic material into it. What happens next is a hostile takeover of the cell's metabolic machinery. The cell makes many new viral particles, then dies after viral enzymes rupture its outer membrane. Virus particles escape and infect new cells.

The biologists wondered, as others had decades ago, whether injections of bacteriophages could help people fight bacterial infections. The discovery of antibiotics in the 1940s had diverted attention away from that idea, at least in the West. Now the bacteria eaters were starting to look good again. Levin and Bull focused on earlier experiments conducted by two British researchers,

Figure 1.12 *Escherichia coli* cell dividing in two. This is a harmless strain.

H. Williams Smith and Michael Huggins. Would they be able to duplicate the results of those researchers?

Levin and Bull selected 018:K1:H7, which is a strain of *Escherichia coli*. Like a harmless *E. coli* strain that lives in the intestines of humans and other mammals (Figure 1.12), it leaves the body in feces. The bacteriophages used in laboratory work usually ignore 018:K1:H7. So the biologists looked for some *E. coli* killers in samples from a sewage treatment plant in Atlanta. They isolated two kinds of bacteriophages and named them *H* (Hero, an effective killer) and *W* (the less effective Wimp).

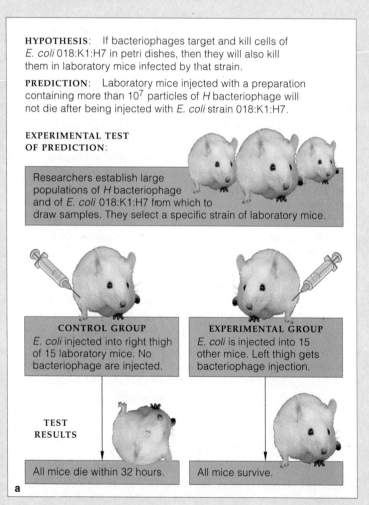

HYPOTHESIS: If bacteriophages target and kill cells of *E. coli* 018:K1:H7 in petri dishes, then they will also kill them in laboratory mice infected by that strain.

PREDICTION: Laboratory mice injected with a preparation containing more than 10^7 particles of *H* bacteriophage will not die after being injected with *E. coli* strain 018:K1:H7.

EXPERIMENTAL TEST OF PREDICTION:

Researchers establish large populations of *H* bacteriophage and of *E. coli* 018:K1:H7 from which to draw samples. They select a specific strain of laboratory mice.

CONTROL GROUP
E. coli injected into right thigh of 15 laboratory mice. No bacteriophage are injected.

EXPERIMENTAL GROUP
E. coli is injected into 15 other mice. Left thigh gets bacteriophage injection.

TEST RESULTS

All mice die within 32 hours.

All mice survive.

a

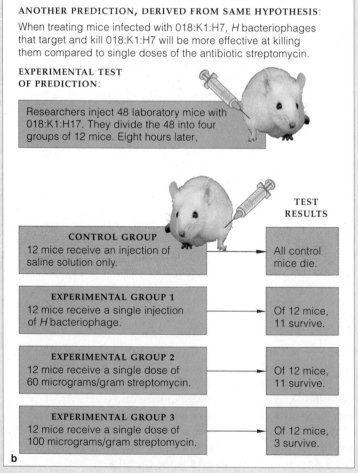

ANOTHER PREDICTION, DERIVED FROM SAME HYPOTHESIS:

When treating mice infected with 018:K1:H7, *H* bacteriophages that target and kill 018:K1:H7 will be more effective at killing them compared to single doses of the antibiotic streptomycin.

EXPERIMENTAL TEST OF PREDICTION:

Researchers inject 48 laboratory mice with 018:K1:H17. They divide the 48 into four groups of 12 mice. Eight hours later,

TEST RESULTS

CONTROL GROUP
12 mice receive an injection of saline solution only.

All control mice die.

EXPERIMENTAL GROUP 1
12 mice receive a single injection of *H* bacteriophage.

Of 12 mice, 11 survive.

EXPERIMENTAL GROUP 2
12 mice receive a single dose of 60 micrograms/gram streptomycin.

Of 12 mice, 11 survive.

EXPERIMENTAL GROUP 3
12 mice receive a single dose of 100 micrograms/gram streptomycin.

Of 12 mice, 3 survive.

b

Figure 1.13 Two experiments to compare the effectiveness of bacteriophage injections against antibiotics for treating a bacterial infection.

With graduate student Terry DeRouin and laboratory technician Nina Moore Walker, the biologists grew a large population of 018:K1:H7 in a culture flask. They selected a specific strain of laboratory mice, all females of the same age. Fifteen mice of one experimental group were each injected with 018:K1:H7 in one thigh and more than 10^7 particles of H bacteriophage in the other. All mice survived. Fifteen mice of a control group were injected only with 018:K1:H7. None survived. Figure 1.13a outlines this experimental test.

Figure 1.13b outlines a second experimental test of another prediction derived from the same hypothesis. The results reinforced Smith and Huggins's conclusion that certain bacteriophages are as good as or better than antibiotics at stopping specific bacterial infections.

Each type of bacteriophage targets specific strains of one or at most a few bacterial species. It now takes too long to discover which bacterium is infecting a patient, so the right bacteriophage might not be enlisted until it's too late. New procedures may speed up identification.

IDENTIFYING IMPORTANT VARIABLES   In nature, many factors can influence the outcome of an infection. For example, genetic differences among infected individuals lead to differences in immune responses to an infection. Age, nutrition, and health at the time of infection also may affect the outcome. Some pathogens are deadlier than others, and so on.

That's why researchers work to simplify and control variables in experiments. Variables, recall, are specific aspects of objects or events that may differ over time and among individuals. All *E. coli* cells in Levin and Bull's experiments were descended from the same parent cell, and they were raised on the same nutrients at the same temperature in a flask. They could be expected to respond the same way to the bacteriophage attack. The mice were the same age and sex, and they were raised under identical laboratory conditions. Each mouse in each experimental group received the same amount of bacteriophages or streptomycin. Each mouse in a control group received the same injection of saline solution. The focus was on *one variable*—a specific bacteriophage versus a specific antibiotic—in a simple, controlled, artificial situation.

BIAS IN REPORTING RESULTS   Whether intentionally or not, experimenters run a risk of interpreting their data in terms of what they want to prove or dismiss. A few have even been known to fake measurements or nudge findings to reinforce their bias. That is why science emphasizes presenting test results in *quantitative* terms—that is, with actual counts or some other precise form. Doing so will allow other experimenters to check or test the results readily and systematically, as Levin and Bull did. At this writing, they are assembling a detailed report of their own experiments for publication in a science journal.

THE LIMITS OF SCIENCE

The call for objective testing strengthens the theories that emerge from scientific studies. It also puts limits on the kinds of studies that can be carried out. Beyond the realm of science, some events remain unexplained. Why do we exist, for what purpose? Why does any one of us have to die at a particular moment? Such questions lead to *subjective* answers. These come from within, as an outcome of all the experiences and mental connections that shape human consciousness. Because people differ vastly in this regard, subjective answers do not readily lend themselves to scientific analysis and experiments.

This is not to say subjective answers are without value. No human society can function for long unless its members share a commitment to certain standards for making judgments, even subjective ones. Moral, aesthetic, philosophical, and economic standards vary from one society to the next. But they all guide people in deciding what is important and good, and what is not. All attempt to give meaning to what we do.

Every so often, scientists stir up controversy when they happen to explain some part of the world that was considered to be beyond natural explanation—that is, as belonging to the supernatural. This is commonly the case when a society's moral codes are interwoven with religious narratives. Exploring a longstanding view of the natural world from a scientific perspective might be misinterpreted as questioning morality, even though the two are not the same thing.

As one example, centuries ago in Europe, Nicolaus Copernicus studied the planets and concluded the Earth circles the sun. Today this seems obvious enough. Back then, it was heresy. The prevailing belief was that the Creator made the Earth—and, by extension, humans—the immovable center of the universe. Later a respected scholar, Galileo Galilei, studied the Copernican model of the solar system, thought it was a good one, and said so. He was forced to retract his statement publicly, on his knees, and put the Earth back as the fixed center of things. (Word has it that when he stood up he muttered, "Even so, it *does* move.") Later still, Darwin's theory of evolution ran up against the same prevailing belief.

Today, as then, society has sets of standards. Those standards might be questioned when some new, natural explanation runs counter to supernatural beliefs. This doesn't mean that scientists who raise the questions are less moral, less lawful, less sensitive, or less caring than anyone else. It simply means one more standard guides their work: *The external world, not internal conviction, must be the testing ground for scientific beliefs.*

Systematic observations, hypotheses, predictions, tests— in all these ways, science differs from systems of belief that are based on faith, force, or simple consensus.

SUMMARY   *Gold* indicates text section

1. There is unity in the living world, for all organisms share these characteristics: They consist of one or more cells. They are assembled from the same kinds of atoms and molecules according to the same laws of energy. They engage in metabolism, and they also sense and respond to specific aspects of the environment. Their DNA gives them a capacity to survive and reproduce. Table 1.1 lists these characteristics. *1.1*

2. The hierarchy of biological organization includes cells, multicelled organisms, populations, communities, ecosystems, and the biosphere. *1.2*

3. Many millions of species (kinds of organisms) exist. Many millions more existed in the past but are extinct. Classification schemes are ever more inclusive groupings of species, starting with genus and proceeding through family, order, class, and phylum, to kingdom. *1.3*

4. Life's diversity arises through mutations: heritable changes in the structure of DNA molecules. The changes are the basis for variation in heritable traits. These are traits that parents bestow on their offspring, including most details of body form and function. *1.4*

5. Darwin's theory of evolution by natural selection is central to biological inquiry. Its key premises are: *1.4*

   a. Individuals of a population differ in the details of their shared heritable traits. Variant forms of traits may affect the ability to survive and reproduce.

   b. Natural selection is the outcome of differences in survival and reproduction among individuals that differ in one or more traits. Adaptive forms of a trait tend to become more common; less adaptive ones become less common or vanish. Thus the defining traits may change over successive generations; the population may evolve.

6. The methods of scientific inquiry are diverse. These terms are important aspects of the methods: *1.5*

   a. Theory: Explanation of a broad range of related phenomena supported by numerous tests. An example is Darwin's theory of evolution by natural selection.

   b. Hypothesis: A proposed explanation of a specific phenomenon. Sometimes called an educated guess.

   c. Prediction: A claim about what can be expected in nature, based on premises of a theory or hypothesis.

   d. Test: An attempt to produce actual observations that match predicted or expected observations.

   e. Conclusion: A statement about whether a theory or hypothesis should be accepted, modified, or rejected, based on tests of predictions derived from it.

7. Logic is a pattern of thought by which an individual draws a conclusion that does not contradict evidence used to support the conclusion. Inductive logic means individuals derive a general statement from specific observations. Deductive logic means individuals make inferences about particular consequences or predictions

Table 1.1 *Summary of Life's Key Characteristics*

SHARED CHARACTERISTICS THAT REFLECT LIFE'S UNITY

1. Organisms consist of one or more cells.

2. Organisms are constructed of the same kinds of atoms and molecules according to the same laws of energy.

3. Organisms engage in metabolism; they acquire and use energy and materials to survive and reproduce.

4. Organisms sense and make controlled responses to their internal and external environments.

5. Heritable instructions encoded in DNA give organisms their capacity to grow and reproduce. DNA instructions also guide the development of complex multicelled organisms.

6. Characteristics that define a population of organisms can change through the generations; the population can evolve.

FOUNDATIONS FOR LIFE'S DIVERSITY

1. Mutations (heritable changes in the structure of DNA) give rise to variation in heritable traits, including most details of body form, functioning, and behavior.

2. Diversity is the sum total of variations that accumulated in different lines of descent over the past 3.8 billion years, as by natural selection and other processes of evolution.

that must follow from a hypothesis. Such a pattern of thinking is often expressed in "if–then" terms. *1.5*

8. Predictions that flow from hypotheses can be tested by comprehensive observations or by experiments in nature or the laboratory. *1.5*

9. Experimental tests simplify observations in nature or the laboratory because conditions under which the observations are made can be precisely manipulated and controlled. They are based on a premise that any aspect of nature has one or more underlying causes, whether obvious or not. With such an assumption of cause and effect, only those hypotheses that can be tested in ways that might disprove them are scientific. *1.5, 1.6*

10. A control group is a standard against which one or more experimental groups (test groups) are compared. Ideally, it is the same as each experimental group in all variables except the one being investigated. *1.5, 1.6*

11. A variable is a specific aspect of an object or event that might differ over time and between individuals. Experimenters directly manipulate the variable they are studying to support or disprove a prediction. *1.5, 1.6*

12. Test results may be distorted by sampling error—chance differences between a population, event, or some other aspect of nature and the samples that were chosen to represent it. Distortion is less likely when samplings are large and when they are repeated. *1.5, 1.6*

13. Scientific theories arise from systematic observations, hypotheses, predictions, and experimental testing. The external world is the testing ground for theories. *1.7*

Review Questions

For this chapter and subsequent chapters, *italics* after a review question identify the section where you can find answers. They include section numbers and *CI* (for Chapter Introduction).

1. Why is it difficult to formulate a simple definition of life? *CI*

2. Name the molecule of inheritance in cells. *1.1*

3. Write out simple definitions of the following terms: *1.1*
   a. cell   b. metabolism   c. energy   d. ATP

4. How do organisms sense changes in their surroundings? *1.1*

5. Study Figure 1.5. Then, on your own, arrange and define the levels of biological organization. *1.2*

6. Study Figure 1.6. Then, on your own, make a sketch of the one-way flow of energy and the cycling of materials through the biosphere. To the side of the sketch, write out definitions of producers, consumers, and decomposers. *1.2*

7. List the shared characteristics of life. *CI, 1.3*

8. What are the two parts of the scientific name for each kind of organism? *1.3*

9. List the six kingdoms of species as outlined in this chapter. Also list some of their general characteristics. *1.3*

10. Define mutation and adaptive trait. Explain the connection between mutation and the diversity of life. *1.4*

11. Write brief definitions of evolution, artificial selection, and natural selection. *1.4*

12. Define and distinguish between: *1.5*
   a. hypothesis and prediction
   b. observational test and experimental test
   c. inductive and deductive logic
   d. speculation and scientific theory

13. With respect to experimental tests, define variable, control group, and experimental group. *1.5*

14. What does sampling error mean? *1.5*

Self-Quiz   ANSWERS IN APPENDIX III

1. _____ is the capacity of cells to extract energy from sources in their environment, and to transform and use energy to grow, maintain themselves, and reproduce.

2. _____ is a state in which the internal environment is being maintained within tolerable limits.

3. The _____ is the smallest unit of life.

4. If a form of a trait improves chances for surviving and reproducing in a given environment, it is a(n) _____ trait.

5. The capacity to evolve is based on variations in heritable traits, which originally arise through _____ .

6. You have some number of traits that also were present in your great-great-great-great-grandmothers and great-great-great-great-grandfathers. This is an example of _____ .
   a. metabolism   c. a control group
   b. homeostasis   d. inheritance

7. DNA molecules _____ .
   a. contain instructions for traits
   b. undergo mutation
   c. are transmitted from parents to offspring
   d. all of the above

8. For many years in a row, a dairy farmer allowed his best milk-producing cows but not the poor producers to mate. Over many generations, milk production increased. This outcome is an example of _____ .
   a. natural selection   c. evolution
   b. artificial selection   d. both b and c

9. A control group is _____ .
   a. a standard against which experimental groups are compared
   b. identical to experimental groups except for one variable
   c. a standard with several variables against which an experimental group is compared
   d. both a and b

10. A specific aspect of an object or event that may change over time or change among individuals is a _____ .
   a. control group   c. variable
   b. experimental group   d. sampling error

11. The fewer the number of individuals from a population that are chosen at random for an experimental group, _____ .
   a. the greater the chance of sampling error
   b. the smaller the chance of sampling error
   c. the less likely that differences among them will distort the test results

12. Match the terms with the most suitable descriptions.
   _____ adaptive trait   a. statement of what you should find in nature if you were to go looking for it
   _____ natural selection   b. proposed explanation; educated guess
   _____ scientific theory   c. improves chances to survive and reproduce in prevailing environment
   _____ hypothesis   d. related set of hypotheses that form a broadly useful, testable explanation
   _____ prediction   e. outcome of differences in survival and reproduction among individuals that differ in details of one or more traits

Critical Thinking

1. Some spiders (*Dolomedes*) that feed on insects around ponds occasionally capture tadpoles and small fishes, as in Figure 1.14, and isn't that fun to think about? While they are immature, the female spiders confine themselves to a small patch of vegetation next to the pond. When sexually mature, they mate and store sperm that fertilize the eggs. Only then do they move out and occupy larger areas around the pond. Develop hypotheses to explain what might cause the spiders to live in different places at different times. Design an experiment to test each hypothesis.

Figure 1.14 A spider (*Dolomedes*) and its prey: a tiny minnow. The spider delivered paralyzing venom as well as digestive enzymes into its captive and is now sucking predigested juices from it. Such spiders can move about below the water's surface. Hairs on their body trap oxygen (for aerobic respiration) during hunting expeditions.

2. A scientific theory about some aspect of nature rests upon inductive logic. The assumption is that, because an outcome of some event has been observed to happen with great regularity, it will happen again. However, we can't know this for certain, because there is no way to account for all possible variables that may affect the outcome. To illustrate this point, Garvin McCain and Erwin Segal offer a parable:

> Once there was a highly intelligent turkey. It lived in a pen, attended by a kind, thoughtful master, and had nothing to do but reflect on the world's wonders and regularities. It observed some major regularities. Morning always began with the sky getting light, followed by the clop, clop, clop of its master's friendly footsteps, then by the appearance of delicious food. Other things varied—sometimes the morning was warm and sometimes cold—but food always followed footsteps. The sequence of events was so predictable that it became the basis of the turkey's theory about the goodness of the world. One morning, after more than one hundred confirmations of the theory, the turkey listened for the clop, clop, clop, heard it, and had its head chopped off.

The turkey learned the hard way that explanations about the world only have a high or low probability of not being wrong. Today, some people take this uncertainty to mean that "facts are irrelevant—facts change." If that is so, should we just stop doing scientific research? Why or why not?

3. Witnesses in a court of law are asked to "swear to tell the truth, the whole truth, and nothing but the truth." What are some of the problems inherent in the question? Can you think of a better alternative?

4. Many popular magazines publish an astounding number of articles on diet, exercise, and other health-related topics. Some authors recommend a specific diet or dietary supplement. What kinds of evidence do you think the articles should include so that you can decide whether to accept the recommendations?

5. Although scientific information is often used when making a decision, it can't tell an individual what's "right" or "wrong." Give an example of this from your own experience.

6. As the old saying goes, Everybody complains about the weather, but nobody does anything about it. Maybe you can start thinking about how we humans might change at least one aspect of local weather conditions. Two researchers at Arizona State University found that, at least for their investigation of the northeast coast of North America, it really does rain more on weekends, when we'd rather be out having fun!

As R. Cerveny and R. Balling Jr. reasoned, if the amount of rain falling each day of the week is a random event, then rules of probability should apply. (*Probability* means the chance that each possible outcome of an event will occur is proportional to the number of ways it can be reached.) Thus, each day of the week should get one-seventh (14.3 percent) of the total rainfall for the week. But it turns out Monday is driest (13.1 percent). Days of the week are wetter, and Saturday is the wettest of all (with 16 percent of the total). Sunday is a bit above average.

What causes this effect? As the researchers hypothesized, if *air pollution* is greater on weekdays than weekends, then human activities might influence regional patterns of rainfall. Monday through Friday, coal-burning factories and gas-powered vehicles release quantities of tiny particles into the air. The particles can promote updrafts and act as "platforms" for the formation of water droplets. Air pollution must build up during the week and extend into Saturday. Over the weekend, fewer factories operate and fewer commuters are on the road. Therefore, by Monday, the air must be cleaner—and drier.

What kind of evidence do you suppose Cerveny and Balling gathered to test the hypothesis? Jot down a few ideas and check them against the researchers' article in the 6 August 1998 issue of *Nature*. Whether or not you continue in biology, this exercise will be good practice in searching through scientific literature for actual data that you can evaluate on topics of interest to you.

7. Scientists devised experiments to shed light on whether different fish species of the same genus compete in their natural habitat. They constructed twelve ponds, identical in chemical composition and physical characteristics. Then they released the following individuals in each pond:

Ponds 1, 2, 3:	Species A	300 individuals each pond
Ponds 4, 5, 6:	Species B	300 individuals each pond
Ponds 7, 8, 9:	Species C	300 individuals each pond
Ponds 10, 11, 12:	Species A, B, C	300 individuals of each species in each pond

Does this experimental design take into consideration all factors that can affect the outcome? If not, how would you modify it?

Selected Key Terms

For this chapter and subsequent chapters, these are the **boldface** terms that appear in the text, in the sections indicated here by *italic* numbers (or *CI*, short for Chapter Introduction). As a study aid, make a list of the terms, write a definition for each, then check it against the one in the text. You will use these terms later on. Becoming familiar with each one will help give you a foundation for understanding the material in later chapters.

adaptive trait *1.4*	ecosystem *1.2*	mutation *1.4*
Animalia *1.3*	energy *1.1*	natural selection *1.4*
antibiotic *1.4*	Eubacteria *1.3*	Plantae *1.3*
Archaebacteria *1.3*	evolution *1.4*	population *1.2*
artificial selection *1.4*	experiment *1.5*	prediction *1.5*
biology *CI*	Fungi *1.3*	producer *1.2*
biosphere *1.2*	genus *1.3*	Protista *1.3*
cell *1.2*	homeostasis *1.1*	receptor *1.1*
community *1.2*	hypothesis *1.5*	reproduction *1.1*
consumer *1.2*	inductive logic *1.5*	sampling error *1.5*
control group *1.5*	inheritance *1.1*	species *1.3*
decomposer *1.2*	metabolism *1.1*	stimulus *1.1*
deductive logic *1.5*	model *1.5*	test, scientific *1.5*
development *1.1*	multicelled	theory, scientific *1.5*
DNA *1.1*	organism *1.2*	variable *1.5*

Readings

Carey, S. 1994. *A Beginner's Guide to the Scientific Method.* Belmont, California: Wadsworth. Paperback.

Committee on the Conduct of Science. 1989. *On Being a Scientist.* Washington, D.C.: National Academy of Sciences. Paperback.

McCain, G., and E. Segal. 1988. *The Game of Science.* Fifth edition. Pacific Grove, California: Brooks/Cole. Paperback.

Moore, J. 1993. *Science as a Way of Knowing.* Cambridge, Massachusetts: Harvard University Press.

On-Line readings at Student Guide for InfoTrac: www.brookscole.com/biology

I Principles of Cellular Life

Living cells of one kind of plant (Elodea), observed with the aid of a microscope. Each of these rectangular cells contains efficient chemical factories called chloroplasts (the round, bright green parts inside).

CHEMICAL FOUNDATIONS FOR CELLS

Checking Out Leafy Clean-Up Crews

Right now you are breathing in oxygen. You would die without it. Two centuries ago, no one had a clue to what oxygen is, where it comes from, and how it helps keep people alive. Then researchers started to unlock the secrets of this chemical substance and others. As their knowledge of chemistry deepened, they conjured up such amazing things as synthetic fertilizers, nuclear power, nylon stockings, lipsticks, aspirin, antibiotics, and plastic parts of refrigerators, computers, television sets, jet planes, and cars.

Today, chemical "magic" brings us benefits *and* problems. For instance, feeding the human population would be impossible without widespread applications of synthetic fertilizers that boost crop yields. But crop plants don't take up every last bit of fertilizer. Nitrogen, phosphorus, and other nutrients in runoff from fields enter lakes, rivers, and seas—where they "feed" the protistans that cause huge fish kills (Section 20.11).

What about diverting the runoff to holding ponds, where water can evaporate? Farmers commonly do this. However, evaporation ponds are like magnets that pull selenium from the soil. High levels of selenium are toxic to grazing animals and waterfowl.

In 1996, Norman Terry and others at the University of California, Berkeley, grew cattails and other grasses in ten experimental plots contaminated by agricultural runoff (Figure 2.1). As they knew, plants can take up

selenium and convert some of it to a far less toxic gas. They measured how much selenium settled in mud, entered plant tissues, and escaped into the air. As they predicted, before runoff trickled away from the plots, the plants significantly reduced the selenium levels.

Terry's research is an example of **bioremediation**: the use of living organisms to withdraw harmful substances from the environment. Another example is the use of sunflowers to clean a pond at Chernobyl, in the Ukraine. Following a total meltdown at a nuclear power plant, strontium 90, cesium 137, and other extremely nasty radioactive elements contaminated the surroundings, including the pond (Section 45.8).

Fertilizers, wetlands, people, pumpkins, the air you breathe—*every solid, liquid, or gaseous substance in and around you is a collection of one or more elements.* Think of **elements** as fundamental forms of matter that have mass and take up space. Here on Earth, we can't break an element apart into something else except in physics laboratories. Suppose you break a chunk of copper into smaller and smaller bits. The smallest bit you end up with, even a single atom, is still copper.

Ninety-two elements occur naturally on Earth. Four of these—oxygen, hydrogen, carbon, and nitrogen— are the most abundant kinds in your body, as they are in all other living things (Figure 2.2). Organisms also incorporate phosphorus, potassium, sulfur, calcium,

Figure 2.1 At large experimental wetlands in Corcoran, California, cattails, bulrushes, *Spartina*, and other grasses help clean up the water from agricultural runoff by taking up selenium.

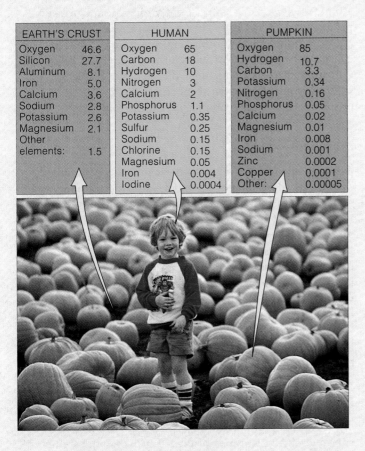

EARTH'S CRUST		HUMAN		PUMPKIN	
Oxygen	46.6	Oxygen	65	Oxygen	85
Silicon	27.7	Carbon	18	Hydrogen	10.7
Aluminum	8.1	Hydrogen	10	Carbon	3.3
Iron	5.0	Nitrogen	3	Potassium	0.34
Calcium	3.6	Calcium	2	Nitrogen	0.16
Sodium	2.8	Phosphorus	1.1	Phosphorus	0.05
Potassium	2.6	Potassium	0.35	Calcium	0.02
Magnesium	2.1	Sulfur	0.25	Magnesium	0.01
Other		Sodium	0.15	Iron	0.008
elements:	1.5	Chlorine	0.15	Sodium	0.001
		Magnesium	0.05	Zinc	0.0002
		Iron	0.004	Copper	0.0001
		Iodine	0.0004	Other:	0.00005

Figure 2.2 Proportions of elements in a human body and the fruit of pumpkin plants, compared to proportions of elements in the materials of the Earth's crust. How are the proportions similar? How do they differ?

sodium, and chlorine, as well as trace elements such as iodine. A *trace* element simply is one that represents less than 0.01 percent of body weight.

Each organism's structure and functions depend on availability of certain kinds and amounts of elements. For example, your muscles will become sore and weak, and your brain simply will not work properly, without daily intakes of magnesium. Similarly, older leaves of plants droop, turn yellow, and die without magnesium.

As you might deduce from this story, safeguarding the environment, our food supplies, and our health starts with knowledge of chemistry. So do efforts to minimize side effects of its applications. Learn about the structure and behavior of chemical substances, and you'll be better equipped to assess risks as well as benefits associated with them.

Key Concepts

1. All substances consist of one or more elements, such as hydrogen, oxygen, and carbon. Each element is composed of atoms, which are the smallest units of matter that still display the element's properties. The atoms of elements are composed of protons, electrons and, except for the hydrogen atom, neutrons.

2. The atoms that make up each kind of element have the same number of protons and electrons. They may differ slightly from one another in their number of neutrons. Variant forms of an element's atoms are called isotopes.

3. Atoms have no overall electric charge unless they become ionized—that is, unless they lose electrons or acquire more of them. An ion is an atom or molecule that has lost or gained one or more electrons. With an extra electron, the charge becomes positive; with an electron loss, the charge becomes negative.

4. Whether a given atom will interact with other atoms depends on how many electrons it has and how they are arranged. When energetic interactions unite two or more atoms, this is a chemical bond.

5. The molecular organization and activities of living things arise largely from ionic, covalent, and hydrogen bonds between atoms.

6. Life probably originated in water and is exquisitely adapted to its properties. Foremost among its properties are temperature-stabilizing effects, cohesiveness, and a capacity to dissolve or repel a variety of substances.

7. All organisms depend on the controlled formation, use, and disposal of hydrogen ions (H^+). The pH scale is a measure of the concentration of H^+ ions in solutions.

REGARDING THE ATOMS

Structure of Atoms

What are the smallest particles that retain the properties of an element? **Atoms**. A line of about a million of them would fit in the period ending this sentence. Small as atoms are, physicists have split them into more than a hundred kinds of smaller particles. The only subatomic particles you will need to consider in this book are the ones called **protons**, **electrons**, and **neutrons**.

All atoms have one or more protons, which carry a positive electric charge (p^+). Except for hydrogen, atoms also have one or more neutrons, which are uncharged. Protons and neutrons make up the atom's core region, or atomic nucleus (Figure 2.3). Zipping about the nucleus and occupying most of the atom's volume are one or more electrons, which carry a negative charge (e^-). Each atom has just as many electrons as protons. This means an atom carries no net charge, overall.

Each element has a unique *atomic* number, which is the number of protons in its atoms. The atomic number

for the hydrogen atom, with its one proton, is 1. For the the carbon atom, with six protons, it is 6 (Table 2.1).

Each element also is assigned a *mass* number, which is the combined number of protons and neutrons in the atomic nucleus. For example, the carbon atom, with six protons and six neutrons, has a mass number of 12.

Why bother with atomic and mass numbers? They can give you a sense of whether and how substances will interact. *This information can help you predict how substances might behave in individual cells, in multicelled organisms, and in the environment, under many conditions.*

Isotopes—Variant Forms of Atoms

Athough all atoms of an element have the same number of protons, they do not all have the same number of neutrons. Atoms that vary in neutron number are called **isotopes**. Most naturally occurring elements have two or more isotopes. Carbon has three, nitrogen has two, and so on. A superscript number to the left of the symbol for an element signifies the isotope. Thus carbon's three isotopes are ^{12}C (or carbon 12, the most common form, with six protons, six neutrons), ^{13}C (six protons, seven neutrons), and ^{14}C (six protons, eight neutrons).

All isotopes of an element interact with other atoms in the same way. This means cells can use any isotope of an element for metabolic activities.

What about radioactive isotopes (radioisotopes)? A physicist, Henri Becquerel, discovered them in 1896 after he put a heavily wrapped rock on top of an unexposed photographic plate inside a desk drawer. The rock held energy-emitting isotopes of uranium. A few days after the plate was exposed to the emissions, a faint image of the rock appeared on it. Marie Curie, a coworker, gave the name "radioactivity" to this chemical behavior.

As we now know, a **radioisotope** is an isotope that has an unstable nucleus and becomes more stabilized by spontaneously emitting energy and particles. (These particles are much smaller than protons, electrons, and neutrons.) The process, radioactive decay, transforms a radioisotope into an atom of a different element at a known rate. For instance, over a predictable time span, carbon 14 becomes nitrogen 14, as Section 18.2 describes.

Elements are forms of matter that occupy space, have mass, and cannot be degraded to something else by ordinary means.

Atoms, the smallest particles unique to each element, have one or more positively charged protons, negatively charged electrons, and (except for hydrogen) neutrons.

Most elements have two or more isotopes, which are atoms that differ in the number of neutrons. A radioisotope has an unstable nucleus and stabilizes itself by spontaneously emitting particles and energy at a known rate.

---

electron
proton
neutron

HYDROGEN   HELIUM

Figure 2.3 The hydrogen atom and helium atom according to one model for atomic structure. The model is highly simplified; the nucleus of these two representative atoms really would be an invisible speck at the scale employed here.

Table 2.1	*Atomic Number and Most Common Mass Number of Selected Elements*		
Element	Symbol	Atomic Number	Mass Number
Hydrogen	H	1	1
Carbon	C	6	12
Nitrogen	N	7	14
Oxygen	O	8	16
Sodium	Na	11	23
Magnesium	Mg	12	24
Phosphorus	P	15	31
Sulfur	S	16	32
Chlorine	Cl	17	35
Potassium	K	19	39
Calcium	Ca	20	40
Iron	Fe	26	56
Iodine	I	53	127

Using Radioisotopes to Track Chemicals and Save Lives

Radioisotopes make splendid tracers. Like a shipping label, a **tracer** is any substance that has a radioisotope attached to it. Researchers track tracers after delivering them into a cell, a multicelled body, an ecosystem, or some other system. Devices detect emissions from a tracer and follow it through a pathway or pinpoint its destination.

For example, Melvin Calvin and other botanists used tracers to track specific steps of photosynthesis. As they knew, all isotopes of an element have the same number of electrons, so all must interact with other atoms the same way. Plants, they hypothesized, should be able to use any carbon isotope when they build carbohydrates. By putting plant cells in a medium enriched with a tracer (^{14}C rather than ^{12}C), they tracked the uptake of carbon through each reaction step in the formation of sugars and starches.

Botanists also use a radioisotope of phosphorus (^{32}P) as a tracer to identify how plants take up and use soil nutrients and synthetic fertilizers. They use such findings in work to improve crop yields.

Besides being research tools, radioisotopes also are medical diagnostic tools. Clinicians use kinds that swiftly decay into harmless elements. Consider how they might analyze a human thyroid. This is the only gland of ours that uses iodine to build thyroid hormones, which greatly influence our body's growth and functions. If a patient's symptoms point to abnormal output of thyroid hormones, clinicians may inject a trace amount of a radioisotope of iodine (^{123}I) into the patient's blood. Then they may use a photographic imaging device to scan the gland. Figure 2.4 shows examples of the type of images they get.

PET (for *Positron-Emission Tomography*) also uses radioisotopes to form images of body tissues. Suppose clinicians attach a tracer to glucose or another molecule. They inject the labeled glucose into the patient, who is moved into a PET scanner (Figure 2.5a). Because all cells must use glucose, cells throughout the body take up the labeled glucose. Uptake is greater in some tissues than others, depending on their metabolic activity at the time the patient is being examined. Laboratory devices detect the radioactive emissions and use them to form an image. Such images may reveal variations or abnormalities in metabolic activity, as in Figure 2.5.

Radioisotopes energetic enough to kill cells have uses, also. In *radiation therapy*, radioisotopes kill or impair the activity of their targets: abnormal body cells. For example, emissions from a source of radium 226 or cobalt 60 may destroy small, localized cancers. As another example, the emissions from plutonium 238 drive artificial pacemakers that smooth out irregular heartbeats. The radioisotope is sealed in a case so its emissions won't damage body cells.

normal thyroid

enlarged

cancerous

Figure 2.4 Location of the human thyroid relative to the trachea (windpipe), and scans of the thyroid gland from three patients.

detector ring inside PET scanner   only body section inside ring

The ring intercepts emissions from the labeled molecules

a   b   c

d PET scan (*red* indicates areas of greatest activity; *blue*, of least activity)

Figure 2.5 (**a**) Patient moving into a PET scanner. (**b,c**) Inside, a ring of detectors intercepts radioactive emissions from labeled molecules that were injected into the patient. Computers analyze and color-code the number of emissions from each location in the scanned body region.

(**d**) Different colors in a brain scan signify differences in metabolic activity. Cells of this brain's left half absorbed and used the labeled molecules at expected rates. However, cells in the right half showed little activity. The patient was diagnosed as having a neurological disorder.

WHAT HAPPENS WHEN ATOM BONDS WITH ATOM?

Electrons and Energy Levels

Cells stay alive because energy inherent in all electrons makes things happen (Figure 2.6). Countless atoms in cells acquire, share, and donate electrons. Atoms of some elements do this easily; others don't. What determines whether an atom will interact with others? *The outcome depends on the number and arrangement of its electrons.*

Tinker with magnets and you can get a sense of the attractive force between unlike charges (+ −) as well as the repulsive force between like charges (++ or − −). Electrons carry a negative charge. In atoms, the positive charge of protons attracts them but they repel each other. They respond to the pulls and pushes by moving about in different orbitals (Figure 2.7). Orbitals are volumes of space around the atomic nucleus in which electrons are likely to be at any instant. Visualize three preschoolers circling a cookie jar, not yet expert in the art of sharing. Each is drawn to the cookies but dreads being shoved away by the others. Two might bob about on opposite sides of the jar to avoid a direct hit. But all three never, ever occupy the same space at the same time.

Each orbital can house one or at most two electrons. Atoms differ in how many occupied orbitals they have because they differ in their number of electrons.

Hydrogen is the simplest atom. It has a lone electron in a spherical orbital, closest to the nucleus. The orbital corresponds to the *lowest available energy level*. In other atoms, two electrons fill the first orbital, and two more electrons occupy a second spherical orbital around the first. Larger atoms have even more electrons occupying orbitals farther from the nucleus, at *higher energy levels*.

An electron at any energy level has a specific amount of energy. On its own, it cannot move to a different level. But suppose an atom absorbs extra energy when sunlight hits it.

Its first orbital has one electron.

Its atomic nucleus has one proton.

a Shell model of a hydrogen atom

Figure 2.6 How even a lone electron can make things happen. Physicists confined electrons inside a bubble of liquid helium, then used fluctuating sound waves to pop the bubble. When an electron escaped, it made the white-centered red flash shown here.

If the energy input is just the right amount, the electron will become excited enough to reach a higher energy level. Left to itself, it will quickly return to the lower energy level by emitting the extra energy. Most often, its emission is in the form of light. You will read about this event in the sections on photosynthesis.

Think "Shells"

The **shell model** is a simple way to show how electrons are distributed in an atom. By this model, a number of "shells" enclose all orbitals available to electrons, as in Figure 2.8. The first, spherical orbital is inside the first shell. Enclosing the first shell is a second shell at a higher energy level. The second shell has four more available orbitals. More orbitals fit in a third shell, a fourth shell, and so on through large, complex atoms of the heavier elements. Appendix VI lists all of these elements.

From Atoms to Molecules

How can you predict whether two atoms will interact? Check for electron vacancies in their outermost shell. Vacancies mean an atom might give up, gain, or share electrons under suitable conditions, which of course will change the distribution or number of its electrons. You can use the shell model to picture what goes on here.

Figure 2.8 shows the electron distribution for some atoms. Electrons are assigned to an energy level (a shell). Count the vacancies—one or more unfilled orbitals in

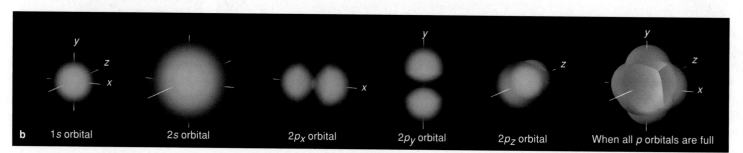

1s orbital	2s orbital	$2p_x$ orbital	$2p_y$ orbital	$2p_z$ orbital	When all p orbitals are full

b

Figure 2.7 Electron arrangements in atoms, which are viewed in terms of three axes. Each axis (x, y, or z) is perpendicular to the other two. (**a**) A hydrogen atom's electron occupies a spherical 1s orbital, at the lowest energy level. (**b**) Every atom has a 1s orbital occupied by one or two electrons. Other orbitals include the 2s and p orbitals at the second energy level. Each can hold two electrons.

Figure 2.8 Examples of the shell model for the distribution of electrons in atoms. Successive shells correspond to higher energy levels from the nucleus. Hydrogen, carbon, and other atoms with electron vacancies (unfilled orbitals) inside their outermost shell tend to give up, accept, or share electrons. Helium and other atoms that have no electron vacancies in their outermost shell are inert; they show little if any tendency to interact with other atoms.

HYDROGEN
$1p^+$, $1e^-$

HELIUM
$2p^+$, $2e^-$

FIRST SHELL. This first energy level corresponds to the $1s$ orbital. It holds one or at most two electrons.

CARBON
$6p^+$, $6e^-$

NITROGEN
$7p^+$, $7e^-$

OXYGEN
$8p^+$, $8e^-$

SECOND SHELL. The $2s$ orbital and the three p orbitals fit in this shell, which corresponds to the second energy level. One or at most two electrons can occupy each of these orbitals.

SODIUM
$11p^+$, $11e^-$

CHLORINE
$17p^+$, $17e^-$

CALCIUM
$20p^+$ $20e^-$

THIRD AND FOURTH SHELLS. More electrons can occupy as many as eight orbitals inside the third shell, which corresponds to the third energy level. Sodium and chlorine are examples. Still more electrons can occupy orbitals inside the fourth shell (as in calcium), and so on.

Figure 2.9 Chemical bookkeeping. We use symbols for elements when writing *formulas*, which identify the composition of compounds. For instance, water has the formula H_2O. The subscript number tells you two hydrogen (H) atoms are present for each oxygen (O) atom. Use such symbols and formulas when writing *chemical equations:* representations of the reactions among atoms and molecules. Substances entering a reaction (reactants) are to the left of the reaction arrow, and products are to the right, as shown by this chemical equation for photosynthesis.

REACTANTS:

$12H_2O$ + $6CO_2$

WATER   CARBON DIOXIDE

24 hydrogens   6 carbons
12 oxygens   12 oxygens

sunlight energy

PRODUCTS:

$6O_2$ + $C_6H_{12}O_6$ + $6H_2O$

OXYGEN   GLUCOSE   WATER

12 oxygens   6 carbons   12 hydrogens
             12 hydrogens   6 oxygens
             6 oxygens

each atom's outermost shell. As you can see, helium is one of the atoms with no vacancies. It is an *inert* atom; it shows little tendency to enter chemical reactions.

Now look again at Figure 2.2, which lists the most abundant of the elements making up a typical organism (a human). These elements include hydrogen, oxygen, carbon, and nitrogen. Atoms of all four elements have electron vacancies. Because of this, *they tend to fill the vacancies by forming bonds with other atoms.*

Each **chemical bond** is a union between the electron structures of atoms. Take a moment to study Figure 2.9. It summarizes a few conventions we use when talking about metabolic reactions. A **molecule** forms when two or more atoms join up. Some molecules consist of only one element. Molecular nitrogen (N_2), with two nitrogen atoms, is like this. So is molecular oxygen (O_2).

The molecules of **compounds** consist of two or more different elements in proportions that never vary. Water is an example. In each molecule of water you find one oxygen atom bonded to two hydrogen atoms. Water

molecules in rainclouds, the ocean, a Siberian lake, your bathtub, the petals of a leaf or flower, or anywhere else always have twice as many hydrogen as oxygen atoms.

In a **mixture**, two or more elements intermingle in proportions that can vary (and usually do). For instance, you make a mixture by swirling water with the sugar sucrose, a compound of carbon, hydrogen, and oxygen.

Electrons occupy orbitals, or volumes of space around an atom's nucleus. By a simplified model, orbitals are arranged as a series of shells enclosing the nucleus. Successive shells correspond to levels of energy, which become higher with distance from the nucleus.

One or at most two electrons occupy any orbital. Atoms with unfilled orbitals in their outermost shell tend to interact with other atoms; those with no vacancies do not.

In molecules of an element, all atoms are the same kind. In molecules of a compound, atoms of two or more elements are bonded together, in unvarying proportions.

IMPORTANT BONDS IN BIOLOGICAL MOLECULES

Eat your peas! Drink your milk! For as long as you've been alive, you've been eating complex carbohydrates, proteins, and other "biological molecules." Only living cells make and use these molecules, which consist of a few kinds of atoms joined by a few kinds of bonds—mainly ionic, covalent, and hydrogen bonds.

Ion Formation and Ionic Bonding

An atom, recall, has just as many electrons as protons, so it carries no net charge. That balance can change for atoms having a vacancy—an unfilled orbital—in their outermost shell. For example, a chlorine atom has such a vacancy and can acquire another electron. A sodium atom has a lone electron in its outermost shell, and that electron can be pulled or knocked out. An atom that has either lost or gained one or more electrons is an **ion**. The balance between its protons and its electrons has shifted, so now the atom is ionized; it has become positively or negatively charged (Figure 2.10*a*).

In living cells, neighboring atoms commonly accept or donate electrons among themselves. When one loses an electron and another grabs it, both become ionized. Depending on cellular conditions, the two ions might not separate; they might stay together as a result of the mutual attraction of opposite charges. An association of two ions that have opposing charges is known as an **ionic bond**. You see one outcome of ionic bonding in Figure 2.10*b*, which shows a portion of a crystal of table

salt, or NaCl. In such crystals, sodium ions (Na^+) and chloride ions (Cl^-) interact through ionic bonds.

Covalent Bonding

Suppose two atoms, each with an unpaired electron in its outermost shell, meet up. Each exerts an attractive force on the other's unpaired electron but not enough to yank it away. Each becomes more stable by *sharing* its unpaired electron with the other. A sharing of a pair of electrons is a **covalent bond**. Example: A hydrogen atom partly fills the electron vacancy in its outermost shell when it's covalently bonded to another hydrogen atom, as shown at right.

In structural formulas, a single line between two atoms represents a *single* covalent bond. Molecular hydrogen (H_2) has such a bond (H—H). Also, two

Two hydrogen atoms, each with a single proton, share two electrons

MOLECULAR HYDROGEN

atoms share two electron pairs of electrons in a *double* covalent bond, as in molecular oxygen (O=O). Two atoms share three pairs of electrons in a *triple* covalent bond, as in molecular nitrogen (N≡N). These examples happen to be gaseous molecules. Each time you breathe in, a stupendous number of H_2, O_2, and N_2 molecules flow through your nose and down toward your lungs.

Covalent bonds are nonpolar or polar. In a *nonpolar* covalent bond, the atoms are exerting the same pull on

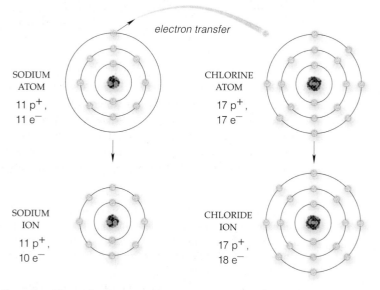

electron transfer

SODIUM ATOM
11 p⁺,
11 e⁻

CHLORINE ATOM
17 p⁺,
17 e⁻

SODIUM ION
11 p⁺,
10 e⁻

CHLORIDE ION
17 p⁺,
18 e⁻

1 mm

a Formation of a sodium ion and a chloride ion

Figure 2.10 **(a)** Ionization by way of an electron transfer. In this case, a sodium atom donates the single electron in its outermost shell to a chlorine atom, which has an unfilled orbital in *its* outermost shell. A sodium ion (Na^+) and a chloride ion (Cl^-) are the outcome of this interaction. **(b)** In each crystal of table salt, or NaCl, many sodium and chloride ions remain together owing to the mutual attraction of opposite charges. Their interaction is a case of ionic bonding.

b Crystals of sodium chloride (NaCl)

Figure 2.18 Sulfur dioxide emissions from a coal-burning power plant. Airborne pollutants such as sulfur dioxide dissolve in water vapor to form acidic solutions. They are a major component of acid rain.

batteries. Fossil fuel burning—as by power plants and cars—and nitrogen fertilizers release strong acid that alter the pH of rain (Figure 2.18). Certain regions are sensitive to the pH of this *acid rain* because of their soil type and vegetation cover. The modified chemistry of habitats in such regions can harm organisms. We will return to this topic in Section 45.1.

Buffers Against Shifts in pH

Metabolic reactions are sensitive to even slight shifts in pH, for H^+ and OH^- can combine with many different molecules and alter their functions. Normally, control mechanisms minimize unsuitable shifts in pH, as they do when HCl enters the stomach in response to a meal. Many of the controls involve buffer systems.

A **buffer system** is a partnership between a weak acid and the base that forms when the acid dissolves in water. The two work like a pair to counter slight shifts in pH. Remember, the OH^- level rises when any strong base enters a fluid. But a weak acid can neutralize part of the incoming OH^- by combining with it. The weak acid's partner (a base) forms as a result of its action. Later, if a strong acid floods in, the base will accept H^+ from it, thus becoming the buffer system's acid.

Bear in mind, the action of a buffer system cannot make new hydrogen ions or eliminate ones that already are present. It can only bind or release them.

In all complex, multicelled organisms, diverse buffer systems operate in the internal environment—in blood and tissue fluids. For example, metabolic reactions in the vertebrate lungs and kidneys help control the acid–base balance of this environment, at levels suitable for life (Chapters 37 and 38). For now, simply think of what happens when the blood level of H^+ decreases and the blood is not as acidic as it should be. At such times, carbonic acid that is dissolved in blood releases H^+ and so becomes the partner base, bicarbonate:

$$H_2CO_3 \longrightarrow HCO_3^- + H^+$$
CARBONIC ACID BICARBONATE

When blood becomes more acidic, more H^+ becomes bound to the base, thus forming the partner acid:

$$HCO_3^- + H^+ \longrightarrow H_2CO_3$$
BICARBONATE CARBONIC ACID

Uncontrolled shifts in pH have drastic outcomes. If blood's pH (7.3–7.5) declines even to 7, this invites a *coma*, a sometimes irreversible state of unconsciousness. An increase to 7.8 can lead to *tetany*, a potentially lethal condition in which skeletal muscles start contracting uncontrollably. In *acidosis*, carbon dioxide builds up in blood, too much carbonic acid forms, and blood pH plummets. *Alkalosis* is an uncorrected increase in blood pH. Both conditions can be lethal.

Salts

Salts are compounds that release ions *other than* H^+ and OH^- in solutions. Salts and water often form when a strong acid and strong base interact. Depending on a solution's pH value, salts can form and dissolve easily. Consider how sodium chloride forms, then dissolves:

$$\text{HCl (acid)} + \text{NaOH (base)} \longrightarrow \text{NaCl (salt)} + H_2O$$
HYDROCHLORIC   SODIUM   SODIUM CHLORIDE
ACID   HYDROXIDE

$$Na^+ \quad Cl^- \text{ (ionization)}$$

Many salts dissolve into ions that play key roles in cells. For example, nerve cell activity depends on ions of sodium, potassium, and calcium. Your muscles contract with the help of calcium ions. And water absorption by plant cells depends largely on potassium ions.

Hydrogen ions (H^+) and other ions dissolved in the fluids inside and outside cells affect cell structure and function.

When dissolved in water, acidic substances release H^+, and basic (alkaline) substances accept them. Certain acid–base interactions help maintain the pH value of a fluid—that is, its H^+ concentration—by a buffering action.

A buffer system counters slight shifts in pH by releasing hydrogen ions when their concentration is too low or by combining with them when the concentration is too high.

Salts are compounds that release ions other than H^+ and OH^-, and many of those ions have key roles in cell functions.

SUMMARY   *Gold* indicates text section

1. Chemistry helps us understand the nature of all the substances that make up cells, organisms, and the Earth, its waters, and the atmosphere. Each substance consists of one or more elements. Of the ninety-two naturally occurring elements, the most common in organisms are oxygen, carbon, hydrogen, and nitrogen. Organisms also have lesser amounts of many other elements, such as calcium, phosphorus, potassium, and sulfur. *CI*

2. Table 2.2 summarizes some key chemical terms that you will encounter throughout this book. *CI, 2.1–2.6*

3. Elements consist of atoms. An atom has one or more positively charged protons plus an equal number of negatively charged electrons and (except for hydrogen atoms) one or more uncharged neutrons. The protons and neutrons occupy the core region, or the atomic nucleus. Most elements have isotopes, which are two or more forms of atoms that have the same number of protons but different numbers of neutrons. *2.1*

4. Whether an atom interacts with others depends on the number and arrangement of its electrons, which occupy orbitals (volumes of space) in a series of shells around the atomic nucleus. When an atom has unfilled orbitals in its outermost shell, it tends to bond with atoms of other elements. *2.3*

5. An atom has no net charge. But it may lose or gain one or more electrons and so become an ion, which has an overall positive or negative charge. *2.4*

6. Generally, a chemical bond is a union between the electron structures of atoms. *2.3*

   a. In an ionic bond, a positive ion and negative ion stay together because of the mutual attraction of their opposite charges. *2.4*

   b. Atoms often share one or more pairs of electrons in single, double, and triple covalent bonds. Electron sharing is equal in nonpolar covalent bonds, and it is unequal in polar covalent bonds. The interacting atoms have no net charge, but the bond is slightly negative at one end and slightly positive at the other, overall. *2.4*

   c. In a hydrogen bond, one covalently bonded atom (such as oxygen and hydrogen) with a slight negative charge is weakly attracted to the slight positive charge of a hydrogen atom that is taking part in a different polar covalent bond. *2.4*

7. Polar covalent bonds join together three atoms in a water molecule (two hydrogens and one oxygen). The polarity of a water molecule invites hydrogen bonding between water molecules. Such hydrogen bonding is the basis of liquid water's ability to resist temperature changes more than other fluids do, display internal cohesion, and easily dissolve polar or ionic substances. The properties greatly influence the metabolic activity, structure, shape, and internal organization of cells. *2.5*

8. By the pH scale, a solution is assigned a number that reflects its H^+ concentration. This ranges from 0 (highest concentration) to 14 (lowest). At pH 7, the H^+ and OH^- concentrations are equal. Acids release H^+ in water, and bases combine with them. Buffer systems help maintain pH values of blood, tissue fluids, and the fluid inside cells. *2.6*

Table 2.2 *Summary of Important Players in the Chemical Basis of Life*

ELEMENT	Fundamental form of matter that occupies space, has mass, and cannot be broken apart into a different form of matter by ordinary physical or chemical means.
ATOM	Smallest unit of an element that still retains the characteristic properties of that element.
Proton (p^+)	Positively charged particle of the atomic nucleus. All atoms of an element have the same number of protons, which is the atomic number. A proton without an electron zipping around it is a hydrogen ion (H^+).
Electron (e^-)	Negatively charged particle that can occupy a volume of space (orbital) around an atomic nucleus. All atoms of an element have the same number of electrons. Electrons can be shared or transferred among atoms.
Neutron	Uncharged particle of the nucleus of all atoms except hydrogen. For a given element, the mass number is the number of protons and neutrons in the nucleus.
MOLECULE	Unit of matter in which two or more atoms of the same element, or different ones, are bonded together.
Compound	Molecule composed of two or more different elements in unvarying proportions. Water is an example.
Mixture	Intermingling of two or more elements in proportions that can and usually do vary.
ISOTOPE	One of two or more forms of atoms of an element that differ in their number of neutrons.
Radioisotope	Unstable isotope, having an unbalanced number of protons and neutrons, that emits particles and energy.
Tracer	Molecule of a substance to which a radioisotope is attached. In conjunction with tracking devices, it can be used to follow the movement or destination of that substance in a metabolic pathway, the body, etc.
ION	Atom that has gained or lost one or more electrons, thus becoming positively or negatively charged.
SOLUTE	Any molecule or ion dissolved in some solvent.
Hydrophilic substance	Polar molecule or molecular region that can readily dissolve in water.
Hydrophobic substance	Nonpolar molecule or molecular region that strongly resists dissolving in water.
ACID	Substance that donates H^+ when dissolved in water.
BASE	Substance that accepts H^+ when dissolved in water; OH^- forms directly or indirectly afterward.
SALT	Compound that releases ions other than H^+ or OH^- when dissolved in water.

Review Questions

1. What is an element? Name four elements (and their symbols) that make up more than 95 percent of the body weight of all living organisms. *CI*

2. Define atom, isotope, and radioisotope. *2.1*

3. How many electrons can occupy each orbital around an atomic nucleus? Using the shell model, explain how the orbitals available to electrons are distributed in an atom. *2.3*

4. Define molecule, compound, and mixture. *2.3*

5. Distinguish between:
   a. ionic and hydrogen bonds *2.4*
   b. polar and nonpolar covalent bonds *2.4*
   c. hydrophilic and hydrophobic interactions *2.5*

6. If a water molecule has no net charge, then why does it attract polar molecules and repel nonpolar ones? *2.5*

7. Label the atoms in each water molecule in the sketch below. Indicate which parts of each molecule carry a slight positive charge (+) and which carry a slight negative charge (−). *2.5*

8. Define acid and base. Then describe the behavior of a weak acid in solutions having a high or low pH value. *2.6*

Self-Quiz   ANSWERS IN APPENDIX III

1. Electrons carry (a) _____ charge.
   a. positive   b. negative   c. zero

2. Atoms share electrons unequally in a(n) _____ bond.
   a. ionic   c. polar covalent
   b. nonpolar covalent d. hydrogen

3. An individual water molecule shows _____ .
   a. polarity   d. spherelike hydration
   b. hydrogen-bonding capacity   e. a and b
   c. notable heat resistance   f. all of the above

4. In liquid water, spheres of hydration form around _____ .
   a. nonpolar molecules   d. solvents
   b. polar molecules   e. b and c
   c. ions   f. all of the above

5. Hydrogen ions (H^+) are _____ .
   a. the basis of pH values   d. dissolved in blood
   b. unbound protons   e. both a and b
   c. targets of certain buffers f. all of the above

6. When dissolved in water, a(n) _____ donates H^+; however, a(n) _____ accepts H^+.

7. Match the terms with the most suitable descriptions.
   ____ trace element
   ____ buffer system
   ____ chemical bond
   ____ temperature
   a. weak acid and its partner base work as a pair to counter pH shifts
   b. union between electron structures of two atoms
   c. less than 0.01% of body weight
   d. measure of molecular motion in some defined region

Critical Thinking

1. An ionic compound forms when calcium combines with chlorine. Referring to Figure 2.8, give the compound's formula. (*Hint:* Be sure the outermost shell of each atom is filled.)

2. David, an inquisitive three-year-old, poked his fingers into warm water inside a metal pan on the stove, then touched the pan and got a nasty burn. Explain why water in a metal pan heats up far more slowly than the pan itself.

3. When molecules absorb microwaves, which are a form of electromagnetic radiation, they move more rapidly. Explain why a microwave oven can heat foods.

4. From what you know about cohesion, devise a hypothesis to explain why water forms droplets.

5. Edward is trying to study a chemical reaction that an enzyme catalyzes (speeds up). H^+ forms in the reaction, but the enzyme is destroyed at low pH. What can he include in his reaction mix to protect the enzyme while he studies the reaction?

6. Through interactions with water and ions, a soluble protein becomes dispersed in cellular fluid. An electrically charged cushion around it makes this happen. Using the sketch at right as a guide, explain the chemical interactions through which this cushion forms. Start with the major bonds in the protein molecule.

7. Reactants and products often are expressed in moles. A *mole* is a certain number of atoms or molecules of any substance, just as "a dozen" can refer to any twelve cats, roses, and so forth. Molar weight, in grams, equals the total atomic weight of all atoms making up that substance. For example, the atomic weight of carbon is 12, so one mole of carbon weighs 12 grams. A mole of oxygen (atomic weight 16) weighs 16 grams.
   State why a mole of water (H_2O) weighs 18 grams, and why a mole of glucose ($C_6H_{12}O_6$) weighs 180 grams.

Selected Key Terms

acid *2.6*	evaporation *2.5*	mixture *2.3*
atom *2.1*	hydrogen bond *2.4*	molecule *2.3*
base *2.6*	hydrogen ion (H^+) *2.6*	neutron *2.1*
bioremediation *CI*	hydrophilic	pH scale *2.6*
buffer system *2.6*	substance *2.5*	proton *2.1*
chemical bond *2.3*	hydrophobic	radioisotope *2.1*
cohesion *2.5*	substance *2.5*	salt *2.6*
compound *2.3*	hydroxide ion (OH^-) *2.6*	shell model *2.3*
covalent bond *2.4*	ion *2.4*	solute *2.5*
electron *2.1*	ionic bond *2.4*	temperature *2.5*
element *CI*	isotope *2.1*	tracer *2.2*

Readings

Ritter, P. 1996. *Biochemistry: A Foundation.* Pacific Grove, California: Brooks/Cole. Good, easy-to-read introduction to biochemistry, with plenty of human-interest applications.

On-Line readings at Student Guide for InfoTrac: www.brookscole.com/biology

CARBON COMPOUNDS IN CELLS

Carbon, Carbon, In the Sky—Are You Swinging Low and High?

High in the mountains of the Pacific Northwest, vast forests of conifers have endured another murderously cold winter (Figure 3.1). Like other organisms, these evergreen, cone-bearing trees can't grow or reproduce without liquid water. Yet through the winter months, water in their habitat is locked away from them in the form of snow and ice.

The trees get by anyway. Their needle-shaped leaves have an epidermis, a layer of interconnected cells with a thick, waxy wall that restricts water loss. About the only way precious water escapes is through small gaps across the epidermis that close or open in response to changing conditions. During autumn's cool, dry days, the trees go dormant; their metabolic activities idle and growth ceases. Even so, water conserved inside them is enough to keep their leaf cells alive.

When spring arrives, rising temperatures and the water from melting snow invite renewed growth. Tree roots soak up the mineral-laden water. Carbon dioxide, a gaseous molecule consisting of one carbon and two oxygen atoms, moves from the air into each leaf. And now the trees engage in photosynthetic magic, turning water and carbon dioxide into sugars, starches, and other carbon-based compounds. They are the premier producers of the northern forests.

Producers, recall, are organisms that use self-made organic compounds as structural materials and packets of energy. One way or another, compounds made by producers of forests and other ecosystems all over the world nourish every consumer and decomposer.

Today, plants of the northern forests and plains of Canada, the United States, and elsewhere are breaking dormancy earlier than they did just two decades ago. Why? No one knows for sure, but the change in their life cycles may be one outcome of a long-term change in the global climate.

Figure 3.1 A forest of conifers beneath the first snows of winter on Silver Star Mountain. The form, function, and very survival of these trees—and of people, bears, and all other organisms—start with the carbon atom and its diverse molecular partners in organic compounds.

For example, researchers studying carbon dioxide in the air have long known that its concentration shifts seasonally. It declines in the spring and summer, when photosynthesizers use up huge quantities of carbon dioxide. It rises at other times of year, when population sizes of decomposers burgeon. (Decomposers release this gas as a metabolic by-product.) The researchers discovered that the spring decline starts a full week earlier compared to the mid-1970s. Also, the seasonal swings are more dramatic—up to 20 percent greater in Hawaii and a whopping 40 percent in Alaska.

Swings in the atmospheric concentration of carbon dioxide were greatest in 1981 and 1990. Intriguingly, atmospheric *temperatures* were uncommonly high in 1981 and 1990. Is this possible evidence of a long-term rise in the world's atmospheric temperature, a *global warming*? If so, could global warming be promoting the longer growing season and wider seasonal swings?

The picture gets more intricate. We humans burn gasoline, coal, and other fossil fuels for energy (Figure 3.1). Fossil fuels are rich in carbon. As you will read in later chapters, burning them may be releasing enough carbon—as carbon dioxide—to add to global warming.

For now, the point to keep in mind is this: *Carbon permeates the world of life—from the energy-requiring activities and structural organization of cells, to physical and chemical conditions that span the globe and influence ecosystems everywhere.*

With this chapter, we turn to life-giving properties that emerge from the molecular structure of carbon-rich compounds. Study the chapter well, including the summary in Section 3.8. It will serve as a foundation for understanding how different organisms put such compounds together, how they use them, and how the effects of those uses ripple through the biosphere.

Key Concepts

1. Organic compounds have a backbone of one or more carbon atoms to which hydrogen, oxygen, nitrogen, and other atoms are attached. We define cells partly by their capacity to assemble the organic compounds known as carbohydrates, lipids, proteins, and nucleic acids.

2. Cells assemble large biological molecules from their pools of smaller organic compounds, including simple sugars, fatty acids, amino acids, and nucleotides.

3. Glucose and other simple sugars are carbohydrates. So are organic compounds of two or more covalently bonded sugar units. The most complex carbohydrates that cells assemble are polysaccharides, many of which consist of hundreds or thousands of sugar units.

4. Lipids are greasy or oily compounds that show little tendency to dissolve in water but that easily dissolve in nonpolar compounds, such as other lipids. Neutral fats (triglycerides), phospholipids, waxes, and sterols are important lipids.

5. Cells use carbohydrates and lipids as building blocks and as major sources of energy.

6. Proteins have diverse roles. Many serve as structural materials. Many are enzymes, a type of molecule that greatly increases the rate of specific metabolic reactions. Other kinds transport cell substances, contribute to cell movements, trigger changes in cell activities, and defend the body against injury and disease.

7. ATP and other kinds of nucleotides have essential roles in metabolism. DNA and RNA, strandlike nucleic acids assembled from nucleotide units, are the basis of inheritance and reproduction.

PROPERTIES OF ORGANIC COMPOUNDS

The Molecules of Life

Under present-day conditions in nature, *only living cells synthesize carbohydrates, lipids, proteins, and nucleic acids.* These are the molecules characteristic of life. Different classes of biological molecules serve as instant energy sources, structural materials, metabolic workers, cell-to-cell signals, and libraries of hereditary information.

The molecules of life are **organic compounds**, with hydrogen and often other elements covalently bonded to carbon atoms. This term is a holdover from a time when chemists thought "organic" substances were the ones they got from animals and vegetables, as opposed to "inorganic" substances they got from minerals. The term persists even though scientists now make organic compounds in laboratories. And it persists even though we now have reason to believe that organic compounds were present on Earth before organisms were.

Carbon's Bonding Behavior

By far, organisms consist mainly of oxygen, hydrogen, and carbon (Figure 2.2). The oxygen and hydrogen are mainly in the form of water. Put aside the water, and carbon makes up more than half of what's left.

Carbon's importance in life arises from its versatile bonding behavior. As shown in the sketch below, *each carbon atom can share pairs of electrons with as many as four other atoms.* Each covalent bond formed this way is relatively stable. Such bonds join carbon atoms together in chains. The chains are a backbone to which many other elements, such as hydrogen, oxygen, and nitrogen, become attached.

The bonding arrangement is the start of wonderful three-dimensional shapes of organic compounds. A chain of carbon atoms, bonded covalently one after another, forms a backbone from which other atoms can project:

Many carbon backbones coil back on themselves in a ring structure, which we can diagram in such ways as:

Figure 3.2 Examples of functional groups.

Functional Groups

A carbon backbone with only hydrogen atoms bonded covalently to it is a **hydrocarbon**, which is a very stable structure. Besides hydrogen, biological molecules have **functional groups**: various atoms or clusters of them covalently bonded to the backbone (Figure 3.2).

To get a sense of their importance, think of **alcohols**: sugars and other organic compounds with one or more hydroxyl groups (—OH). Alcohols dissolve fast because water molecules hydrogen-bond with —OH groups. Or think of a protein. Its shape starts at bonding patterns between amino and carboxyl groups of its backbone. Amino groups also serve as buffers against declines in pH. Other functional groups contribute to differences between males and females of many species (Figure 3.3).

How Do Cells Build Organic Compounds?

Using carbon (from carbon dioxide), water, and sunlight (for energy), the photosynthetic cells described earlier make sugars. The simple sugars are one of four major families of small organic compounds you'll find in every cell; the other three are fatty acids, amino acids, and nucleotides. Cells use these to build all of the biological molecules they require for their structure and functions.

HO

AN ESTROGEN

OH

O

TESTOSTERONE

Figure 3.3 Observable differences in traits between male and female wood ducks (*Aix sponsa*). Two sex hormones have key roles in the development of feather color and other traits that help males and females recognize each other (and therefore influence reproductive success). Both hormones—testosterone and one of the estrogens—have the same carbon ring structure. As you can see at *left*, however, they differ in a single functional group (—OH) attached to the ring.

Figure 3.4
Examples of metabolic reactions by which most biological molecules are put together, rearranged, and broken apart.

enzyme action at functional groups

a Two condensation reactions. Enzymes remove an —OH group and H atom from two molecules, which covalently bond as a larger molecule. Two water molecules form.

enzyme action at functional groups

b Hydrolysis, a water-requiring cleavage reaction. Enzyme action splits a molecule into three parts, then attaches an —OH group and an H atom derived from a water molecule to each exposed site.

How do they do it? It will take more than one chapter to sketch out answers (and best guesses) to the question. At this point, simply be aware that reactions by which cells assemble, rearrange, and split apart organic compounds require more than energy inputs. They require **enzymes**, a class of proteins that make specific metabolic reactions proceed faster than they would on their own. Different kinds mediate different reactions. In chapters to come, you will read about five categories of reactions:

1. *Functional-group transfer*. One molecule gives up a functional group, which another molecule accepts.

2. *Electron transfer*. One or more electrons stripped from one molecule are donated to another molecule.

3. *Rearrangement*. A juggling of its internal bonds converts one type of organic compound into another.

4. *Condensation*. Through covalent bonding, two molecules combine to form a larger molecule.

5. *Cleavage*. A molecule splits into two smaller ones.

To sense what goes on, think about this **condensation reaction**: Enzymes split off an —OH group from one molecule and an H atom from another, a covalent bond forms between the two at their exposed sites, and the discarded atoms may form water, or H_2O (Figure 3.4*a*).

Starch and other polymers form by many condensation reactions. **Polymers** are molecules of three to millions of subunits that may or may not be identical. Individual subunits of larger molecules are known as **monomers**.

Another example: A type of cleavage reaction called **hydrolysis** is like condensation in reverse (Figure 3.4*b*). Enzymes split molecules at specific groups, then attach one —OH group and a hydrogen atom derived from a water molecule to the exposed sites. With hydrolysis, cells can cleave polymers into smaller molecules when these are required for building blocks or energy.

Organic compounds have diverse, three-dimensional shapes and functions that start with their carbon backbone and the bonding arrangements that arise from it. Functional groups covalently bonded to such backbones add enormously to the structural and functional diversity.

Carbohydrates, lipids, proteins, and nucleic acids are the main biological molecules, the organic compounds that only living cells can assemble under conditions that now occur in nature.

Cells assemble, rearrange, and degrade organic compounds mainly through enzyme-mediated reactions involving the transfer of functional groups or electrons, rearrangement of internal bonds, and a combining or splitting of molecules.

CARBOHYDRATES

Which biological molecules are most abundant? The **carbohydrates**. Most consist of carbon, hydrogen, and oxygen in a 1:2:1 ratio $(CH_2O)_n$. Their functional groups easily dissolve in water, and cells use them as structural materials, transportable forms of energy, or storage forms of energy. The three main classes are **monosaccharides**, **oligosaccharides**, and **polysaccharides**.

The Simple Sugars

"Saccharide" comes from a Greek word meaning sugar. A *mono*saccharide (meaning one sugar monomer) is the simplest carbohydrate. It has at least two —OH groups bonded to the carbon backbone plus an aldehyde or a ketone group. Most readily dissolve in water. Common types have a backbone of five or six carbon atoms that tends to form a ring structure when dissolved in cells or body fluids. Ribose and deoxyribose, the sugar unit of RNA and DNA, respectively, have five carbon atoms. Glucose has six (Figure 3.5a). Most organisms use glucose as the main energy source. Glucose also is a precursor (parent molecule) of many compounds and a building block for larger carbohydrates. Vitamin C (a well-known sugar acid) and glycerol (an alcohol that has three —OH groups) are other compounds with sugar monomers.

Short-Chain Carbohydrates

Unlike the simple sugars, an *oligo*saccharide is a short chain of covalently bonded sugar monomers. (*Oligo–* means "a few.") The ones called *di*saccharides consist of only two sugar units. Lactose, sucrose, and maltose are examples. Lactose, a sugar in milk, has one glucose and one galactose unit. Sucrose, the most plentiful sugar in nature, has a glucose and a fructose unit (Figure 3.5c). Plants convert larger carbohydrates to sucrose, which is easily transported through leaves, stems, and roots. Table sugar is sucrose crystallized from sugarcane and sugar beets. The backbone of some proteins and other large molecules often bears oligosaccharide side chains.

Complex Carbohydrates

The "complex" carbohydrates, or *poly*saccharides, are straight or branched chains of many sugar monomers (often hundreds or thousands) of the same or different types. Cellulose, starch, and glycogen, the most common polysaccharides, consist only of glucose. Cellulose is a structural component of plant cell walls (Figure 3.6a). Like steel rods in reinforced concrete, fibers of cellulose are tough, insoluble, and resistant to weight loads and mechanical stress. Plants store their photosynthetically produced sugars as large starch molecules (Figure 3.6b). Enzymes can readily hydrolyze starch to glucose units.

a Structure of glucose

b Structure of fructose

c Formation of a sucrose molecule from two simple sugars

glucose   fructose

sucrose   + H_2O

Figure 3.5 Straight-chain and ring forms of (**a**) glucose and (**b**) fructose. For reference purposes, carbon atoms of simple sugars are commonly numbered in sequence, starting at the end closest to the molecule's aldehyde or ketone group. (**c**) Condensation of two monosaccharides into a disaccharide.

If both cellulose and starch consist of glucose, why are their properties so different? The answer starts with differences in covalent bonding patterns between their glucose units, which are joined together in chains.

In cellulose, many glucose chains stretch out side by side and hydrogen-bond to one another at —OH groups (Figure 3.7a). This bonding arrangement stabilizes the chains into a tightly bundled pattern that resists being digested, at least by most enzymes.

In starch, the pattern of covalent bonding puts each glucose unit at an angle relative to the next unit in line. The chain ends up coiling like a spiral staircase (Figure 3.7b). The coils are not especially stable. In starches that have branched chains, they are even less stable. Many —OH groups project outward from the coiled chains, and this makes them readily accessible to enzymes.

In animals, glycogen is the sugar-storage equivalent of starch in plants. Muscle and liver cells store a lot of it. When the level of sugar in blood declines, liver cells degrade glycogen, so glucose is released and enters the

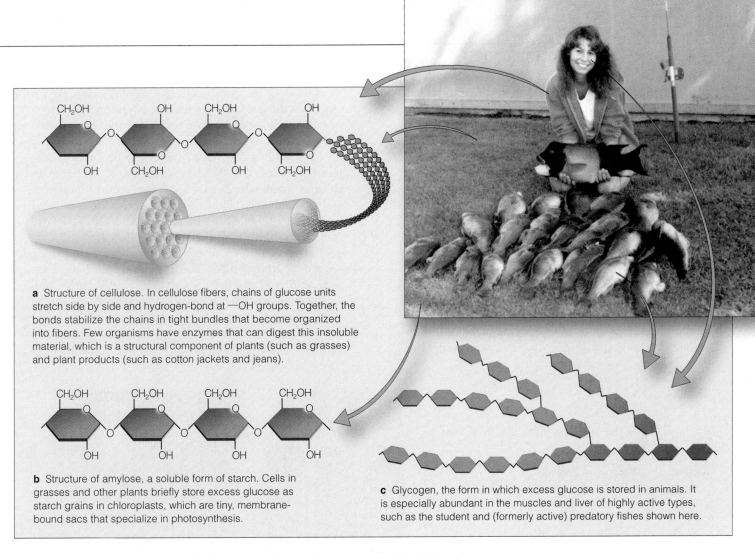

a Structure of cellulose. In cellulose fibers, chains of glucose units stretch side by side and hydrogen-bond at —OH groups. Together, the bonds stabilize the chains in tight bundles that become organized into fibers. Few organisms have enzymes that can digest this insoluble material, which is a structural component of plants (such as grasses) and plant products (such as cotton jackets and jeans).

b Structure of amylose, a soluble form of starch. Cells in grasses and other plants briefly store excess glucose as starch grains in chloroplasts, which are tiny, membrane-bound sacs that specialize in photosynthesis.

c Glycogen, the form in which excess glucose is stored in animals. It is especially abundant in the muscles and liver of highly active types, such as the student and (formerly active) predatory fishes shown here.

Figure 3.6 Molecular structure of cellulose, starch, and glycogen, and their typical locations in a few organisms. All three carbohydrates consist only of glucose units.

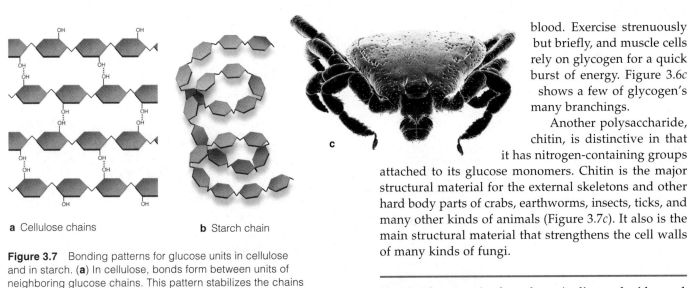

a Cellulose chains   **b** Starch chain

Figure 3.7 Bonding patterns for glucose units in cellulose and in starch. (**a**) In cellulose, bonds form between units of neighboring glucose chains. This pattern stabilizes the chains and lets them become tightly bundled. (**b**) In amylose, a form of starch, bonding between the units makes the chains coil, which orients the bonds in a way that is accessible to enzymes. (**c**) Scanning electron micrograph of a tick. Its body covering is a protective cuticle reinforced with chitin.

blood. Exercise strenuously but briefly, and muscle cells rely on glycogen for a quick burst of energy. Figure 3.6c shows a few of glycogen's many branchings.

Another polysaccharide, chitin, is distinctive in that it has nitrogen-containing groups attached to its glucose monomers. Chitin is the major structural material for the external skeletons and other hard body parts of crabs, earthworms, insects, ticks, and many other kinds of animals (Figure 3.7c). It also is the main structural material that strengthens the cell walls of many kinds of fungi.

The simple sugars (such as glucose), oligosaccharides, and polysaccharides (such as starch) are carbohydrates. Every cell requires carbohydrates as structural materials, stored forms of energy, and transportable packets of energy.

LIPIDS

If something is greasy or oily to the touch, you can bet it is a lipid or has lipid components. **Lipids** are nonpolar hydrocarbons. They resist dissolving in water but easily dissolve in nonpolar substances (think butter into warm cream sauce). Cells use different lipids as energy stores, as structural materials (for example, in cell membranes and surface coatings), and as signaling molecules. Let's look first at the kinds having fatty acid components—fats, phospholipids, and waxes. Then we'll consider the sterols, each with a backbone of four carbon rings.

Fats and Fatty Acids

Lipids known as **fats** have one, two, or three fatty acids attached to glycerol. Each **fatty acid** has a backbone of as many as thirty-six carbon atoms, a carboxyl group (—COOH) at one end, and hydrogen atoms occupying

most or all of the remaining bonding sites. Most stretch out like a flexible tail. Tails that are *unsaturated* have one or more double bonds. *Saturated* tails have single bonds only. Figure 3.8a shows examples.

Most animal fats have many saturated fatty acids, which pack together by weak interactions. They remain solid at temperatures that keep most plant fats liquid, as "vegetable oils." Similar packing interactions in plant fats aren't as stable because of rigid kinks in their fatty acid tails. That is why vegetable oils flow freely.

Butter, lard, vegetable oils, and other natural fats are mostly **triglycerides**. These "neutral" fats have three fatty acid tails attached to a glycerol unit (Figure 3.8b). Triglycerides are the body's most abundant lipids and its richest energy source. Gram for gram, they yield more than twice as much energy when degraded, compared to complex carbohydrates such as starches. Quantities of

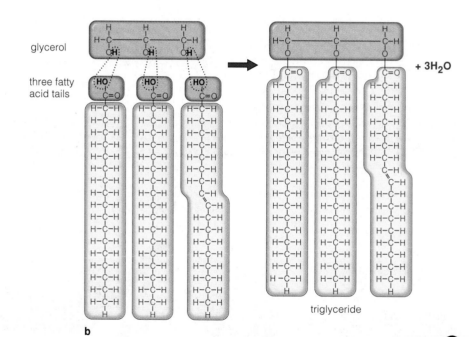

glycerol

three fatty acid tails

+ 3H₂O

triglyceride

b

stearic acid   oleic acid   linolenic acid

a

Figure 3.8 *Above:* (**a**) Structural formulas for three fatty acids. In stearic acid, the carbon backbone is fully saturated with hydrogen atoms. Oleic acid, with a double bond in its backbone, is an unsaturated fatty acid. Linolenic acid, with its three double bonds, is a polyunsaturated fatty acid. (**b**) Condensation of fatty acids and a glycerol molecule into a triglyceride.

Figure 3.9 Triglyceride-protected penguins taking the plunge.

a One of the phospholipids

hydrophilic head (*orange*)

two hydrophobic tails

b Cholesterol

one hydrophobic tail

coronary artery

atherosclerotic plaques

heart muscle tissue

c Cholesterol-rich plaques in a coronary artery

d Water-repelling, waxy coat on cherries

e Honeycomb, a structure made of beeswax

Figure 3.10 (**a**) Structural formula of a typical phospholipid in animal and plant cell membranes. Are its hydrophobic tails saturated or unsaturated? (**b**) Structural formula of cholesterol, the major sterol of animal tissues. (**c**) Your liver synthesizes enough cholesterol for the body. A fat-rich diet may result in excessively high cholesterol levels in blood and the formation of abnormal masses of material in certain blood vessels (arteries). Such *atherosclerotic plaques* may clog the arteries that deliver blood to the heart. (**d**) Demonstration of the water-repelling attribute of a cherry cuticle. (**e**) Honeycomb, a structural material constructed of a firm, water-repellent, waxy secretion called beeswax.

triglycerides are stored as droplets in the cells of body fat (adipose tissue) in every vertebrate. A thick layer of triglycerides under the skin helps penguins and some other animals survive (Figure 3.9). It insulates the body against the extreme temperatures of icy habitats.

One fatty acid is a precursor of eicosanoids. These local signaling molecules include prostaglandins. As you will see, they have roles in contraction, defense, message flow in the nervous system, and other key processes.

Phospholipids

A **phospholipid** has a glycerol backbone, two fatty acid tails, and a hydrophilic "head" with a phosphate group and another polar group (Figure 3.10*a*). Phospholipids are the main materials of cell membranes, which have two layers of lipids. Heads of one layer are dissolved in the cell's fluid interior, and heads of the other layer are dissolved in the surroundings. Sandwiched between the two are all the fatty acid tails, which are hydrophobic.

Sterols and Their Derivatives

Sterols are among the many lipids with no fatty acids. Sterols differ in the number, position, and type of their functional groups, but all have a rigid backbone of four fused-together carbon rings, as below. Eukaryotic cells have sterols in their membranes. Cholesterol is the most common type in tissues of animals. Figure 3.10*b* shows its structure. Also, cholesterol gets remodeled into

sterol backbone

compounds such as vitamin D (necessary for good bones and teeth), steroids, and bile salts. The steroids include sex hormones of the sort shown in Figure 3.3. These govern gamete formation and secondary sexual traits, such as feather color and hair distribution. Bile salts have roles in the digestion of fats in the small intestine.

Waxes

Waxes have long-chain fatty acids tightly packed and linked to long-chain alcohols or carbon rings. All have a firm consistency; all repel water. The cuticle covering aboveground plant parts consists mostly of waxes and another lipid, cutin (Figure 3.10*d*). It restricts water loss and fends off certain parasites. Waxy secretions protect, lubricate, and impart pliability to skin and to hair. Birds secrete waxes, fatty acids, and fats from preen glands to waterproof feathers. Bees use beeswax to construct honeycomb, which houses not only honey but new bee generations (Figure 3.10*e*). Tiny organisms drifting in the seas even use waxes as their main energy source.

Being largely hydrocarbon, lipids can dissolve in nonpolar substances, but they resist dissolving in water.

Triglycerides, or neutral fats, have a glycerol head and three fatty acid tails. They are the body's major energy reservoirs. Phospholipids are the main components of cell membranes.

Sterols such as cholesterol serve as membrane components and precursors of steroid hormones and other compounds. Waxes are firm yet pliable components of water-repelling and lubricating substances.

AMINO ACIDS AND THE PRIMARY STRUCTURE OF PROTEINS

Have you ever wondered how a permanent wave works (Figure 3.11)? By the end of this section and the next, you'll have an idea of how the process alters individual hairs. You will see how the structure of hair and many other body parts begins with the structure of proteins.

Of all large biological molecules, **proteins** are the most diverse. Proteins called enzymes make reactions proceed faster than they would on their own. Structural proteins are the stuff of spider webs, butterfly wings, feathers, bone, and many other body parts and products. Some proteins transport substances across membranes and through fluids. Nutritious proteins abound in eggs and seeds. Protein hormones are signals for change in cell activities. Many proteins act as weapons against pathogens. Amazingly, cells build thousands of diverse proteins from only twenty kinds of amino acids!

Figure 3.11 Appearance of the hair of actress Nicole Kidman (**a**) before and (**b**) after a permanent wave. The difference, as you will see from this section and the next, starts with strings of amino acids in polypeptide chains.

Figure 3.12 Structural formulas for eight of twenty common amino acids. The *green* boxes highlight their R groups, which are side chains that have functional groups. Each type of side chain contributes in a major way to the distinctive properties of each amino acid.

Amino Acid Structure

Every **amino acid** is a small organic compound that consists of an amino group, a carboxyl group (which is an acid), a hydrogen atom, and one or more atoms called its R group. Figure 3.12 shows the structure of some amino acids you'll be considering later on in the book. As you can see, their parts generally are bonded covalently to the same carbon atom, as in Figure 3.13.

Figure 3.13 Generalized structural formula for amino acids.

What Is a Protein's Primary Structure?

When a cell synthesizes a protein, enzymes link amino acids, one after the other, by peptide bonds. This type of covalent bond forms between the amino group ($-NH_3^+$) of one amino acid and the carboxyl group ($-COO^-$) of the next amino acid in line, as Figure 3.14 shows.

FIVE OF THE NONPOLAR AMINO ACIDS

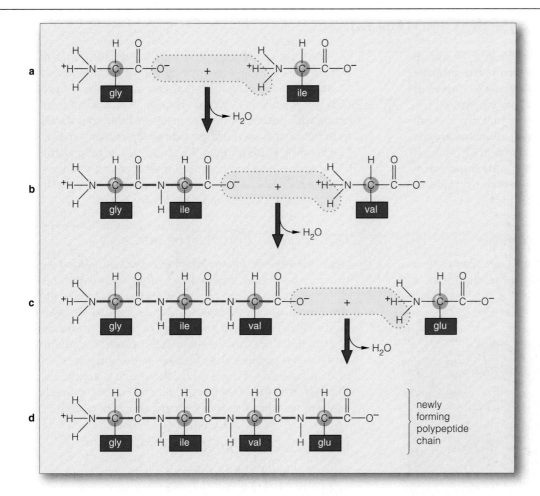

Figure 3.14 Peptide bond formation during protein synthesis.

(**a**) The first two amino acids shown are glycine (gly) and isoleucine (ile). They are at the start of the sequence for a polypeptide chain.

(**b**) A peptide bond forms between glycine and isoleucine during a condensation reaction. A water molecule forms as a by-product.

(**c**) A peptide bond forms between isoleucine and another amino acid, valine (val). Water again forms.

(**d**) Remember, DNA specifies the order in which different kinds of amino acids follow one another in a growing polypeptide chain. In this case, glutamate (glu) is the fourth amino acid specified.

In this case, the outcome will be one of the two polypeptide chains of an insulin molecule, as shown in Figure 3.15. Chapter 14 details steps by which DNA instructions are translated into proteins.

Figure 3.15 Diagram of the amino acid sequence for an insulin molecule found in cattle. This protein serves as a hormone. It consists of two polypeptide chains linked at two bridges (—S—S—).

When peptide bonds join three or more amino acids, we have a **polypeptide chain**. In such chains, the carbon backbone has nitrogen atoms positioned in this regular pattern: —N—C—C—N—C—C—.

For each kind of protein, different amino acids are selected one at a time from the twenty kinds available. Instructions encoded in the cell's DNA determine the selection. Overall, the protein's amino acid sequence is unique and is its *primary* structure (Figures 3.14 and 3.15).

Different cells make thousands of different proteins. Many are fibrous, with polypeptide chains organized as strands or sheets. Collectively, these fibers contribute to a cell's shape and organization. Globular proteins have one or more chains folded in compact, rounded shapes. Most enzymes are globular proteins. So are actin and a few other proteins that help cells, and cell parts, move.

Regardless of the type of protein, its shape and its function arise from the primary structure—that is, from information encoded in its amino acid sequence. That sequence influences which parts of a polypeptide chain will coil, bend, or interact with other chains. The type and arrangement of atoms in the coiled, stretched-out, and folded parts dictate whether the protein will function as, say, an enzyme, a transporter, a receptor, or even an unlucky target for a bacterium or virus.

Each protein consists of one or more polypeptide chains of amino acids. The amino acid sequence (which kind of amino acid follows another in the chain) is unique for each kind of protein and gives rise to its unique structure, chemical behavior, and function.

HOW DOES A THREE-DIMENSIONAL PROTEIN EMERGE?

The preceding section gave you a sense of how amino acids form a polypeptide chain, which is the primary structure of proteins. Now consider a few examples of how that structure dictates the final protein shape.

For the most part, the primary structure gives rise to a protein's shape in two ways. First, it allows hydrogen bonds to form between different amino acids along a polypeptide chain's length. Second, it puts R groups in positions that make them interact. The interactions force the chain to bend and twist in specific ways.

Second Level of Protein Structure

Hydrogen bonds form at regular, short intervals along a new polypeptide chain. They give rise to the protein's

secondary structure—a coiled or an extended pattern. Picture the chain as a set of rigid playing cards joined by sticks (covalent bonds) that can swivel a bit. Each card is a peptide group, and some of its atoms can form bonds with nearby atoms. For instance, hydrogen bonds form between every third peptide group in many chains. The bonding pattern forces the groups to swivel and coil like a spiral staircase (Figure 3.16a). In other proteins, hydrogen bonds hold two or more chains side by side in an extended, sheetlike array (Figure 3.16b).

Third Level of Protein Structure

Most polypeptide chains that have a coiled secondary structure bend even more. Why? Along their length are amino acids, including the bulky proline, that make the chain bend and loop outward. Some R groups far apart on the chain interact to hold loops in specific positions. A chain that folded as a result of bend-producing amino acids and R-group interactions is at the third level of protein structure. Figure 3.17a shows one example: the compact structure of the protein globin. Hydrogen bonds between functional groups brought about its folding.

Fourth Level of Protein Structure

Now picture many bonds forming between *four* globin molecules. Then put an iron-containing functional group called heme close to the center of each one. You end up with **hemoglobin**, an oxygen-transporting, respiratory protein (Figure 3.17b). As you read this, each mature red blood cell inside your body is busily transporting a billion molecules of oxygen. All that oxygen is bound to 250 million hemoglobin molecules.

one peptide group

b

Figure 3.16 Bonding patterns between peptide groups of polypeptide chains. Extensive hydrogen bonding (*dotted* lines) can bring about (**a**) coiling of one chain or (**b**) sheetlike arrays of two or more chains.

a

heme group

alpha chain

beta chain

coiled and twisted polypeptide chain of one globin molecule

a

b

beta chain

alpha chain

Figure 3.17 (**a**) Globin molecule. This coiled polypeptide chain associates with a heme group, an iron-containing functional group that strongly attracts oxygen. There is a whole family of globins, identical in most parts of their amino acid sequences but unique in other parts.

(**b**) Hemoglobin, which transports oxygen in blood. This protein consists of four polypeptide chains (globin molecules) and four heme groups. Two chains, designated *alpha*, differ slightly from the other two (*beta* chains) in their amino acid sequence. We sketched the coiled parts of each chain as if they were enclosed in a transparent tube to help you visualize all four globin molecules.

a   dead, flattened cells
of a shaft of hair

b   one hair cell

hair's cuticle

keratin macrofibril

microfibril (three chains coiled into one strand)

coiled keratin polypeptide chain

disulfide bridges between two keratin chains

bridges broken

hair wrapped around curlers

different bridges form

c

Hemoglobin is a classic example of the fourth level of protein structure. In all proteins at this level, hydrogen bonds and sometimes covalent bonds between R groups join two or more polypeptide chains into one molecule.

Hemoglobin and most enzymes are examples of the **globular proteins**; their chains are folded in a rounded shape. **Fibrous proteins** are long strands or sheets of polypeptide chains. Examples are keratin (Figure 3.18) and collagen, the most common animal protein. Skin, bone, corneas, arteries, and many other animal body parts depend on collagen's strength.

Glycoproteins and Lipoproteins

Some proteins have other organic compounds attached to their polypeptide chains. For example, **lipoproteins** form when certain proteins circulating in blood join up with cholesterol, triglycerides, and phospholipids that were absorbed from the gut after a meal. Similarly, most **glycoproteins** are proteins that have linear or branched oligosaccharides attached. Almost every protein at the surface of animal cells are glycoproteins. So are most proteins secreted from cells and many proteins in blood.

Structural Changes by Denaturation

Breaking weak bonds of a protein or any other large molecule disrupts its three-dimensional shape, an event called **denaturation**. For example, weak hydrogen bonds are sensitive to increases and decreases in temperature and pH. If the temperature or pH exceeds a protein's range of tolerance, its polypeptide chains will unwind or change shape, and the protein will lose its function. Consider the protein albumin, concentrated in the "egg white" of uncooked chicken eggs. When you cook eggs,

Figure 3.18   Structure of hair. (**a**,**b**) Hair cells develop from modified skin cells. They synthesize polypeptide chains of the protein keratin. Disulfide bridges link three chains into fine fibers, which become bundled into larger, cable-like fibers. The fibers nearly fill the cells, which eventually die off. (**a**) Dead, flattened cells form a tubelike cuticle around the developing hair shaft.

(**c**) For a permanent wave, hair is exposed to chemicals that break disulfide bridges. When hairs wrap around curlers, their polypeptide chains are held in new positions. Exposure to a different chemical causes disulfide bridges to form again. But, compared to before, the new ones form between different sulfur-bearing amino acids. The displaced bonding locks the hair in curled positions (compare Figure 3.11).

the heat does not disrupt the strong covalent bonds of albumin's primary structure. But it destroys weaker bonds contributing to the three-dimensional shape. For some proteins, denaturation might be reversed when normal conditions are restored—but albumin isn't one of them. There is no way to uncook a cooked egg.

Proteins have a primary structure, which is the sequence of different kinds of amino acids along a polypeptide chain. An individual's DNA specifies that sequence.

Proteins have a secondary structure: a coiled pattern or an extended, sheetlike pattern that arises by hydrogen bonding at short, regular intervals along a polypeptide chain.

At the third level of protein structure, bonding at certain amino acids makes a coiled chain bend and loop at certain angles, in certain directions. Interactions among R groups in the chain hold the loops in characteristic positions.

At the fourth level of protein structure, many hydrogen bonds and other interactions join two or more polypeptide chains. Many proteins are this structurally complex.

3.6

NUCLEOTIDES AND NUCLEIC ACIDS

Nucleotides are small organic compounds with a sugar, one or more phosphate groups, and a base. Their sugar, unit, either ribose or deoxyribose, has a five-carbon ring structure. All the bases have a single or double carbon ring structure that incorporates nitrogen.

One nucleotide, **ATP** (adenosine triphosphate), has a string of three phosphate groups attached to its sugar:

ATP easily transfers a phosphate group to many other molecules, thereby energizing them to enter a reaction. That's why ATP is central to metabolism (Section 5.2).

Some of the nucleotides are part of **coenzymes**. These enzyme helpers transfer hydrogen atoms plus electrons stripped from molecules to other reaction sites. NAD+ is an example (Sections 5.2 and 7.1). Other coenzymes are chemical messengers between cells and inside cells.

Nucleic acids are composed of nucleotides. **DNA** (deoxyribonucleic acid) is a double-stranded molecule of four kinds of nucleotides that differ in their bases. The bases are adenine, guanine, thymine, or cytosine (Figure 3.19). Their sequential order in a DNA strand encodes information on making all proteins that cells need to survive and reproduce. Bases in some parts of

Figure 3.19 Four kinds of nucleotides that function as components of DNA. Two nucleotide bases, adenine and guanine, have a double ring structure. Ttwo other bases, thymine and cytosine, have a single ring structure.

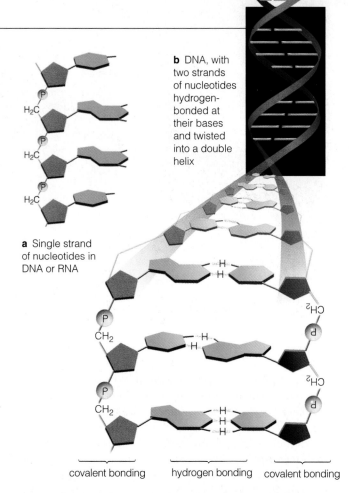

b DNA, with two strands of nucleotides hydrogen-bonded at their bases and twisted into a double helix

a Single strand of nucleotides in DNA or RNA

covalent bonding   hydrogen bonding   covalent bonding

Figure 3.20 (**a**) A series of covalently bonded nucleotides in a single strand of DNA or RNA. (**b**) DNA double helix.

the sequence are unique to each species. As Figure 3.20 shows, hydrogen bonds between bases join the two DNA strands, which twist spirally into a double helix.

Like DNA, **RNAs** (ribonucleic acids) are assembled from four kinds of nucleotides. Unlike DNA, the bases are adenine, guanine, cytosine, and *uracil*. Also unlike DNA, most RNA molecules are single nucleotide strands. RNAs do not encode protein-building instructions, but different classes carry out these instructions in cells. One kind even catalyzes peptide bond formation. Ever since the first cells emerged, RNAs have served in processes by which genetic information is used to build proteins. We return to DNA and RNA in Chapters 12, 13, and 19.

Some nucleotides function as energy carriers or chemical messengers. Others are subunits of coenzymes and nucleic acids. The nucleic acids are single or double strands of covalently bonded nucleotides.

DNA has two helically coiled, hydrogen-bonded strands of nucleotides. Its sequence of bases encodes information about all proteins a cell requires to survive and reproduce. Different classes of single-stranded nucleic acids called RNAs carry out DNA's protein-building instructions.

Food Production and a Chemical Arms Race

Next time you shop for groceries, reflect on what it takes to give you a daily supply of organic compounds. For example, lettuce typically grows in fertilized cropland. Weeds typically compete with lettuce for the fertilizer's nutrients. Each year, these food pirates and others ruin or gobble up nearly half of what people try to grow.

Most plants aren't entirely defenseless. They evolved under intense selection pressure of attacks by insects and other organisms. In nature, they often repel the attackers with toxins. A **toxin** is an organic compound, a normal metabolic product of one species, but its chemical effects harm or kill individuals of a different species that contact it. Humans, too, encounter traces of natural toxins, even in such familiar edibles as hot peppers, potatoes, figs, celery, rhubarb, and alfalfa sprouts. Still, we do not die in droves, so we must have chemical defenses against these toxins.

A few thousand years ago, farmers used sulfur, lead, arsenic, and mercury to kill insects. They freely dispensed these exceedingly toxic metals until the late 1920s, when someone figured out they were poisoning people. Traces of the toxic metals still turn up in contaminated lands.

Also, early farmers took a cue from plants and used *organic* compounds extracted from leaves, flowers, and roots as natural pesticides. In 1945, scientists started to develop synthetic toxins and to understand mechanisms by which toxins attack their targets. *Herbicides* kill weeds by disrupting metabolism and growth. Most *insecticides* clog a target insect's airways, disrupt nerves and muscles, or block reproduction. *Fungicides* work against harmful fungi, including a mold that makes aflatoxin, one of the deadliest poisons. By 1995, people in the United States were spraying or spreading more than 1.25 billion pounds of toxins through fields, gardens, homes, and industrial and commercial sites (Figure 3.21).

Some pesticides also kill birds and other predators that help control pest population sizes. And targeted pests are developing resistance to the chemical arsenal, for reasons given earlier (Chapter 1). Pesticides can't be haphazardly released, for they are easily inhaled, ingested with food, or absorbed by skin. Certain types are active for weeks or years. Some trigger rashes, headaches, hives, asthma, and joint pain in millions of people or trigger life-threatening allergic reactions in abnormally sensitive individuals.

DDT and other long-lived pesticides are now banned in the United States (Section 43.3). Even rapidly degradable types, such as malathion and other organophosphates (Figure 3.21*d*), must meet rigorous application standards and safety tests. Still, many farmers argue that limiting pesticide applications will invite devastating crop losses, and then how will we feed the hungry multitudes?

b Atrazine. Farmers spray this herbicide directly on leaves of corn and other crop plants. It kills weeds within a few days but doesn't hurt crop plants. Other herbicides that don't persist long are glyphosate (Roundup), daminozide (Alar), and alachlor (Lasso).

c Dichlorodiphenyltrichloroethane, or DDT. This nerve cell poison breaks down two to fifteen years after its application. Chlordane is another type of insecticide that persists for a long time in the environment.

d Malathion. Like other organophosphates, this nerve cell poison lasts one to twelve weeks. Organophosphates break down faster than chlorinated hydrocarbons but are far more toxic. About four dozen kinds are cheap and effective. They account for half of all insecticide uses in the United States. The Environmental Protection Agency wants to minimize pesticide residues on food. At this writing methyl parathion is banned for food crops; applications of azenphos methyl must end at least three weeks before harvest time. Farmers are contesting these actions and want the EPA to consider economic and trade issues as well as human health.

Figure 3.21   (**a**) Lettuce, a common crop plant, and a low-flying crop duster with its rain of pesticides. (**b–d**) A few of the major pesticides, some more toxic than others.

SUMMARY

1. Organic compounds consist of carbon, hydrogen, and often other elements. The carbon atoms are often bonded covalently to form a linear or ring-shaped backbone, and functional groups are attached to the backbone. *3.1*

2. In nature, only living cells assemble large organic compounds called biological molecules (Table 3.1). Cells build these complex carbohydrates, lipids, proteins, and nucleic acids from simple sugars, fatty acids, amino acids, nucleotides, and other small organic molecules. *3.1*

3. Cells assemble, rearrange, and break down organic compounds by different enzyme-mediated reactions, as when they transfer functional groups or electrons. *3.1*

4. Cells use the simple sugars (such as glucose) and the oligosaccharides for energy and as building blocks. They use complex carbohydrates (such as starch, cellulose, and glycogen) as structural materials and as energy storage forms. They use lipids (especially triglycerides) as energy storage forms, structural components of cell membranes, and precursors of other compounds. *3.2–3.3*

5. Proteins consist of polypeptide chains of amino acids. Proteins, the most diverse biological molecules, serve in enzyme activity, structural support, transport, changes in cell shape and activities, movements, and defense against disease. Nucleotides function as energy carriers (especially ATP), coenzymes, and subunits for strands of the nucleic acids DNA and RNA, which interact and form the basis of inheritance and reproduction. *3.4–3.6*

Table 3.1 *Summary of the Main Organic Compounds in Living Things*

Category	Main Subcategories	Some Examples and Their Functions	
CARBOHYDRATES . . . contain an aldehyde or a ketone group, and one or more hydroxyl groups	**Monosaccharides** (simple sugars)	Glucose	Energy source
	Oligosaccharides	Sucrose (a disaccharide)	Most common form of sugar transported through plants
	Polysaccharides (complex carbohydrates)	Starch, glycogen Cellulose	Energy storage Structural roles
LIPIDS . . . are largely hydrocarbon; generally do not dissolve in water but dissolve in nonpolar substances, such as other lipids	**Lipids with fatty acids**		
	Glycerides: One, two, or three fatty acid tails attached to glycerol backbone	Fats (e.g., butter), oils (e.g., corn oil)	Energy storage
	Phospholipids: Phosphate group, one other polar group, and (often) two fatty acids attached to glycerol backbone	Phosphatidylcholine	Key component of cell membranes
	Waxes: Long-chain fatty acid tails attached to alcohol	Waxes in cutin	Conservation of water in plants
	Lipids with no fatty acids		
	Sterols: four carbon rings; the number, position, and type of functional groups differ among various sterols	Cholesterol	Component of animal cell membranes; precursor of many steroids and of vitamin D
PROTEINS . . . are one or more polypeptide chains, each with as many as several thousand covalently linked amino acids	**Fibrous proteins** Long strands or sheets of polypeptide chains; often tough, water-insoluble	Keratin Collagen	Structural component of hair, nails Structural component of bone
	Globular proteins One or more polypeptide chains folded and linked into globular shapes; many roles in cell activities	Enzymes Hemoglobin Insulin Antibodies	Great increase in rates of reactions Oxygen transport Control of glucose metabolism Tissue defense
NUCLEIC ACIDS (AND NUCLEOTIDES) . . . are chains of units (or individual units) that each consist of a five-carbon sugar, phosphate, and a nitrogen-containing base	**Adenosine phosphates**	ATP cAMP (Section 32.2)	Energy carrier Messenger in hormone regulation
	Nucleotide coenzymes	NAD^+, $NADP^+$, FAD	Transport of electrons, protons (H^+), from one reaction site to another
	Nucleic acids Chains of thousands to millions of nucleotides	DNA, RNAs	Storage, transmission, translation of genetic information

Review Questions

1. Define organic compound. Name the type of chemical bond that predominates in the backbone of such a compound. *3.1*

2. Define hydrocarbon. Also define functional group and give an example. *3.1*

3. Name the molecules of life and the families of small organic compounds from which they are built. Do they break apart most easily at their hydrocarbon portion or at functional groups? *3.1*

4. Describe the difference between a condensation reaction and hydrolysis. *3.1*

5. Select one of the carbohydrates, lipids, proteins, or nucleic acids described in this chapter. Speculate on how its functional groups and bonds between the carbon atoms in its backbone contribute to its final shape and function. *3.2 through 3.8*

6. Which item listed includes all of the other items listed? *3.3*
   a. triglyceride   c. wax   e. lipid
   b. fatty acid   d. sterol   f. phospholipid

7. Explain how hemoglobin's three-dimensional shape arises, starting with the primary structure of its four chains. *3.4, 3.5*

Self-Quiz   ANSWERS IN APPENDIX III

1. Each carbon atom can share pairs of electrons with as many as _____ other atoms.
   a. one   b. two   c. three   d. four

2. Hydrolysis is a (an) _____ reaction.
   a. functional group transfer   d. condensation
   b. electron transfer   e. cleavage
   c. rearrangement   f. both b and d

3. _____ is a simple sugar (monosaccharide).
   a. Glucose   c. Ribose   e. both a and b
   b. Sucrose   d. Chitin   f. both a and c

4. In unsaturated fats, fatty acid tails have one or more _____ .
   a. single covalent bonds   b. double covalent bonds

5. _____ are to proteins as _____ are to nucleic acids.
   a. Sugars; lipids   c. Amino acids; hydrogen bonds
   b. Sugars; proteins   d. Amino acids; nucleotides

6. A denatured protein or DNA molecule has lost its _____ .
   a. hydrogen bonds   c. function
   b. shape   d. all of the above

7. _____ are nucleotides.
   a. ATP   c. RNA subunits
   b. DNA subunits   d. all of the above

8. Match each molecule with the most suitable description.
   _____ long sequence of amino acids   a. carbohydrate
   _____ the main energy carrier   b. phospholipid
   _____ glycerol, fatty acids, phosphate   c. protein
   _____ two strands of nucleotides   d. DNA
   _____ one or more sugar monomers   e. ATP

Critical Thinking

1. Jack decided to celebrate summer by making a crab salad and a peach pie for some friends. He had such a good time he forgot to put the cap on the bottle of olive oil after making the salad. He also didn't tighten the lid on the shortening can after making the pie crust.

   A few weeks later, the oil was rancid but the shortening still smelled okay. Both substances are fats. Both were stored in a dark cupboard at room temperature. Yet one substance spoiled and the other stayed fresh. Speculate on which molecular bonds were the starting point for the difference.

2. In the following list, identify which is the carbohydrate, the fatty acid, the amino acid, and the polypeptide:
   a. $^+NH_3$—CHR—COO$^-$ c. (glycine)$_{20}$
   b. $C_6H_{12}O_6$ d. $CH_3(CH_2)_{16}COOH$

3. A clerk in a health-food store tells you that certain "natural" vitamin C tablets extracted from rose hips are better for you than synthetic vitamin C tablets. Given your understanding of the structure of organic compounds, what would be your response? Design an experiment to test whether the vitamins differ.

4. Xanthophyll is a yellow pigment. Rabbits that eat green, leafy plants with xanthophyll accumulate this yellow molecule in body fat but not in muscles. What chemical properties of the molecule might cause this selective accumulation?

5. Cows can digest cellulose in grasses, but people can't. A cow's four-chambered stomach holds populations of microbes that are not normal residents of the human stomach. What kind of metabolic reactions do you think the microbes carry out? What do you think might happen to a cow being treated with an antibiotic that kills the microbes?

6. A Gary Larson cartoon states that sheep with steel wool have no natural enemies. You might laugh at the thought without knowing what wool is. Wool is the keratin-rich, soft undercoat of sheep, Angora goats, llamas, and other hairy mammals. Keratin is a strandlike, fibrous protein. Many hydrogen bonds hold it in a helical shape. Visualize three such chains coiled together, with many disulfide bridges in their end regions that stabilize them into tight, ropelike fibers. Each hair has linear groups of the fibers, which resist stretching (Figure 3.18). Given keratin's structure, speculate why, when you run a wool sweater through the hot cycle of a dryer, it shrinks pathetically and permanently.

7. Reflect on the component atoms of the different classes of organic compounds. Now look at the structural formulas for the pesticides shown in Figure 3.21. Which one is a glycoprotein? An organophosphate? A chlorinated hydrocarbon?

8. Many farmers believe crop losses will be disastrous if pesticides are banned without sufficient research and effective alternatives. Many environmental activists are pushing for rapid bans to protect people and the environment. What do you think?

Selected Key Terms

alcohol *3.1*	fibrous protein *3.5*	oligosaccharide *3.2*
amino acid *3.4*	functional group *3.1*	organic compound *3.1*
ATP *3.6*	globular protein *3.5*	phospholipid *3.3*
carbohydrate *3.2*	glycoprotein *3.5*	polymer *3.1*
coenzyme *3.6*	hemoglobin *3.5*	polypeptide chain *3.4*
condensation	hydrocarbon *3.1*	polysaccharide *3.2*
reaction *3.1*	hydrolysis *3.1*	protein *3.4*
denaturation *3.5*	lipid *3.3*	RNA *3.6*
DNA *3.6*	lipoprotein *3.5*	sterol *3.3*
enzyme *3.1*	monomer *3.1*	toxin *3.7*
fat *3.3*	monosaccharide *3.2*	triglyceride *3.3*
fatty acid *3.3*	nucleic acid *3.6*	wax *3.3*
	nucleotide *3.6*	

Readings

Ritter, P. 1996. *Biochemistry: An Introduction*. Pacific Grove, California: Brooks/Cole. Great human applications.

On-Line readings at Student Guide for InfoTrac:
www.brookscole.com/biology   49

CELL STRUCTURE AND FUNCTION

Animalcules and Cells Fill'd With Juices

Early in the seventeenth century, a scholar by the name of Galileo Galilei put two glass lenses inside a cylinder. With this instrument he happened to look at an insect, and later he described the stunning geometric patterns of its tiny eyes. Thus Galileo, who wasn't a biologist, was one of the first to record a biological observation made through a microscope. The study of the cellular basis of life was about to begin. First in Italy, then in France and England, scholars set out to explore a world whose existence had not even been suspected.

At midcentury Robert Hooke, Curator of Instruments for the Royal Society of England, was at the forefront of those studies. When Hooke first turned a microscope to thinly sliced cork from a mature tree, he observed tiny compartments (Figure 4.1*c*). He gave them the Latin name *cellulae*, meaning small rooms—hence the origin of the biological term "cell." They

actually were the interconnecting walls of dead plant cells, which is what cork is made of, but Hooke didn't think of them as being dead because neither he nor anyone else at the time knew cells could be alive. In other plant tissues, he observed cells "fill'd with juices" but didn't have a clue to what they represented.

Given the simplicity of their instruments, it's just amazing that the pioneers in microscopy observed as much as they did. Antony van Leeuwenhoek, a Dutch shopkeeper, had exceptional skill in constructing lenses and possibly the keenest vision of all (Figure 4.1*a*). By the late 1600s, he was discovering natural wonders everywhere, including "many very small animalcules, the motions of which were very pleasing to behold," in scrapings of tartar from his teeth. Elsewhere he observed protistans, sperm, and even a bacterium—an organism so small it would not be seen again for another two centuries!

a

b

c

d

Figure 4.1 Early glimpses into the world of cells. (**a**) Antony van Leeuwenhoek, with microscope in hand. (**b**) Robert Hooke's compound microscope and (**c**) his drawing of cell walls from cork tissue. (**d**) One of van Leeuwenhoek's early sketches of sperm cells. (**e**) Cartoon evidence of the startling impact of microscopic observations on nineteenth-century London.

In the 1820s, improvements in lenses brought cells into sharper focus. Robert Brown, a botanist, noticed an opaque spot in a variety of cells and called it a nucleus. In 1838 another botanist, Matthias Schleiden, wondered if the nucleus had something to do with development. As he hypothesized, each plant cell might develop as an independent unit even though it's part of the plant.

By 1839, after years of studying animal tissues, the zoologist Theodor Schwann had this to say: Animals as well as plants consist of cells and cell products—and even though the cells are part of a whole organism, to some extent they have an individual life of their own.

A decade later a question remained: Where do cells come from? Rudolf Virchow, a physiologist, completed his own studies of a cell's growth and reproduction— that is, its division into two daughter cells. Every cell, he reasoned, comes from a cell that already exists.

And so, by the middle of the nineteenth century, microscopic analysis had yielded three generalizations, which together constitute the **cell theory**. *First, every organism is composed of one or more cells. Second, the cell is the smallest unit having the properties of life. Third, the continuity of life arises directly from the growth and division of single cells.* All three insights still hold true.

This chapter is not meant for memorization. Read it simply to gain an overview of current understandings of cell structure and function. In later chapters, you can refer back to it as a road map through the details. With its images from microscopy, this chapter and others in the book can transport you into spectacular worlds of juice-fill'd cells and animalcules.

Key Concepts

1. All organisms consist of one or more cells. The cell is the smallest unit that still retains the characteristics of life. Since the origin of life, each new cell has arisen from a preexisting cell. These are the three generalizations of the cell theory.

2. All cells have a plasma membrane. This outermost, double-layered membrane separates a cell's interior from its surroundings. Also, all cells contain cytoplasm, an organized internal region where energy conversions, protein synthesis, movements of cell parts, and other required activities proceed. In eukaryotic cells only, the DNA is enclosed within a membrane-bound nucleus. In prokaryotic cells (archaebacteria and eubacteria), the DNA simply is concentrated in part of the cell interior.

3. The plasma membrane and internal cell membranes consist of lipids and proteins. The lipids are organized as two adjacent layers—that is, as a bilayer. They give a membrane its basic structure and prevent water-soluble substances from freely crossing it. Diverse proteins are embedded in the bilayer or positioned at its surfaces. They carry out many functions, such as transporting water-soluble substances across the membrane.

4. Organelles such as the nucleus are membrane-bound compartments within eukaryotic cells. They physically separate metabolic reactions and allow them to proceed in orderly ways. Prokaryotic cells do not have comparable organelles.

5. When cells are growing, they increase faster in volume than in surface area. This physical constraint on growth influences their size and shape.

6. Different microscopes modify light rays or accelerated beams of electrons in ways that allow us to form images of incredibly small specimens. They are the foundation for our current understanding of cell structure and function.

e

BASIC ASPECTS OF CELL STRUCTURE AND FUNCTION

Inside your body and at its moist surfaces, trillions of cells live in interdependency. In northern forests, pollen grains—four-celled structures—escape from pine trees. In scummy pondwater, a single-celled amoeba moves on its own. For humans, pines, amoebas, and all other organisms, the **cell** is the smallest unit that lives on its own or has the potential to do so (Figure 4.2). Each cell is structurally organized for metabolism. It senses and responds to its environment. And heritable instructions in its DNA give it the potential to reproduce.

Figure 4.2 (**a**) A row of cells at the surface of a tiny tube inside a human kidney. Each cell has a nucleus (the sphere stained darker red). (**b**) A winged pollen grain. One of its cells will give rise to a sperm that may meet up with an egg and help start a new pine tree.

Structural Organization of Cells

Cells differ enormously in size, shape, and activities, as you might gather by comparing a tiny bacterium with one of your relatively giant liver cells. Yet they are alike in three respects: They all start out life with a plasma membrane, a region of DNA, and a region of cytoplasm:

1. **Plasma membrane.** This thin, outermost membrane maintains the cell as a distinct entity. By doing so, it allows metabolic events to proceed apart from random events in the environment. A plasma membrane does not isolate the cell interior. Substances and signals continually move across it in highly controlled ways.

2. **Nucleus** or **nucleoid.** Depending on the species, the DNA occupies a membrane-bound sac (nucleus) in the cell or simply a region of the cell interior (nucleoid).

3. **Cytoplasm.** Cytoplasm is everything between the plasma membrane and region of DNA. It consists of a semifluid matrix and other components, such as **ribosomes** (structures on which proteins are built).

This chapter introduces two fundamentally different kinds of cells. **Eukaryotic cells** have organelles, tiny sacs having one or more outer membranes, in the cytoplasm. One, the nucleus, is their defining feature. **Prokaryotic cells** don't have a nucleus; nothing intervenes between their DNA and cytoplasm. The only prokaryotic cells are the archaebacteria and eubacteria. All other organisms—from amoebas to peach trees to puffball mushrooms to whales and zebras—consist of eukaryotic cells.

Fluid Mosaic Model of Cell Membranes

Fluid bathes the two surfaces of a cell membrane and is vital for its functions. The membrane itself has a fluid quality; it isn't a rigid barrier between cytoplasm and extracellular fluid. Why is this so? The answer starts with phospholipids. A **phospholipid** has a hydrophilic head (a phosphate group, another polar group, and a glycerol backbone) and two hydrophobic tails (fatty acids). The head dissolves in water; the tails don't. Immerse many phospholipids in water. They may spontaneously cluster

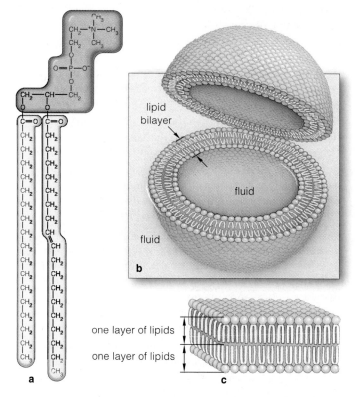

one layer of lipids

one layer of lipids

Figure 4.3 Lipid bilayer organization of cell membranes. (**a**) One of the phospholipids, the most abundant molecules of membranes. (**b,c**) These and other lipids are arranged into two layers. Their hydrophobic tails are sandwiched between the hydrophilic heads. In cells, the heads are dissolved in cytoplasm on one side of the bilayer and in extracellular fluid on the other side.

as a sheet at the surface. Or they may form two-layered spheres beneath the surface, with all tails sandwiched between all the heads. This **lipid bilayer** arrangement is the structural basis of cell membranes (Figure 4.3).

Figure 4.4 shows a bit of membrane corresponding to the **fluid mosaic model**. By this model, cell membranes have a mixed composition—a *mosaic*—of many diverse phospholipids, sterols, and proteins. Phospholipids are the most abundant components. Cholesterol is the most

Golgi body

vesicle

microfilaments
(components of
cytoskeleton)

mitochondrion

chloroplast

microtubules
(components of
cytoskeleton)

central vacuole

rough endoplasmic reticulum
(rough ER)

ribosomes (attached to rough ER)

ribosomes (free in cytoplasm)

smooth endoplasmic reticulum
(smooth ER)

DNA + nucleoplasm
nucleolus   } NUCLEUS
nuclear envelope

plasma
membrane

cell wall

a Components of a typical plant cell.

nuclear envelope
nucleolus   } NUCLEUS
DNA + nucleoplasm

vesicle

lysosome

rough ER

ribosomes
(attached to
rough ER
and free in
cytoplasm)

smooth ER

vesicle

Golgi body

pair of
centrioles

components of { microfilaments
cytoskeleton   { microtubules

plasma
membrane

mitochondrion

b Components of a typical animal cell.

MICROGRAPHS OF PLANT AND ANIMAL CELLS

The sketches in the preceding section cannot do justice to eukaryotic cell structures. Here are specific examples to give you a better sense of their exquisite details.

Figure 4.9 Transmission electron micrograph of a photosynthetic cell from a blade of Timothy grass (*Phleum pratense*), in cross-section. The sketches highlight key organelles.

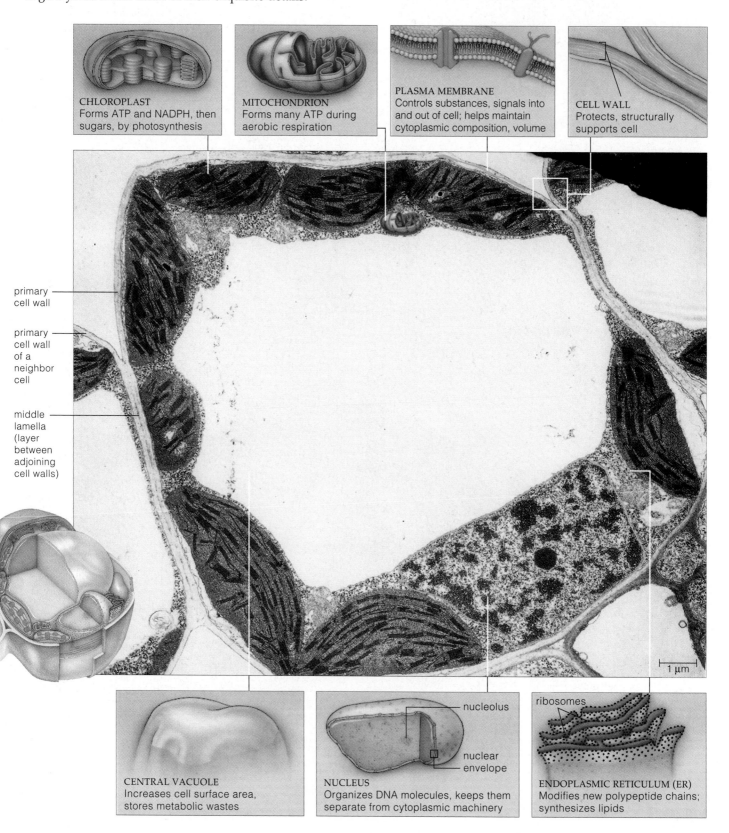

CHLOROPLAST
Forms ATP and NADPH, then sugars, by photosynthesis

MITOCHONDRION
Forms many ATP during aerobic respiration

PLASMA MEMBRANE
Controls substances, signals into and out of cell; helps maintain cytoplasmic composition, volume

CELL WALL
Protects, structurally supports cell

primary cell wall

primary cell wall of a neighbor cell

middle lamella (layer between adjoining cell walls)

CENTRAL VACUOLE
Increases cell surface area, stores metabolic wastes

NUCLEUS
Organizes DNA molecules, keeps them separate from cytoplasmic machinery

nucleolus

nuclear envelope

ribosomes

ENDOPLASMIC RETICULUM (ER)
Modifies new polypeptide chains; synthesizes lipids

1 µm

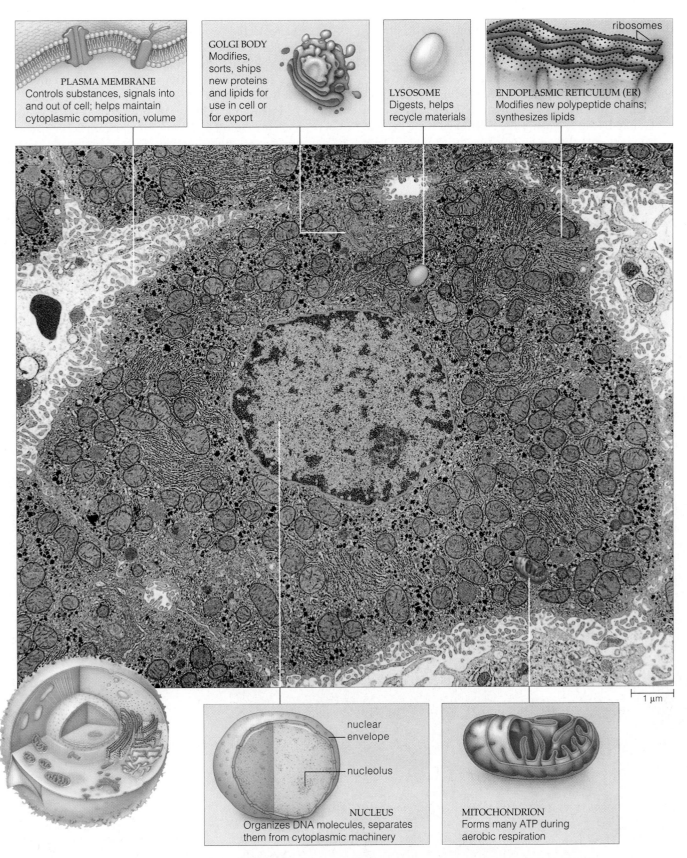

PLASMA MEMBRANE
Controls substances, signals into and out of cell; helps maintain cytoplasmic composition, volume

GOLGI BODY
Modifies, sorts, ships new proteins and lipids for use in cell or for export

LYSOSOME
Digests, helps recycle materials

ribosomes

ENDOPLASMIC RETICULUM (ER)
Modifies new polypeptide chains; synthesizes lipids

1 µm

nuclear
envelope

nucleolus

NUCLEUS
Organizes DNA molecules, separates them from cytoplasmic machinery

MITOCHONDRION
Forms many ATP during aerobic respiration

Figure 4.10 Transmission electron micrograph of an animal cell, cross-section. The sketches highlight key organelles. The specimen is a cell from the liver of a rat.

THE NUCLEUS

Constructing, operating, and reproducing cells simply cannot be done without carbohydrates, lipids, proteins, and nucleic acids. It takes a class of proteins—enzymes —to build and use these molecules. Said another way, a cell's structure and function begin with proteins. *And instructions for building proteins are located in DNA.*

Unlike bacteria, eukaryotic cells have their genetic instructions distributed among several to many DNA molecules of different lengths. Each human body cell, for instance, has forty-six DNA molecules. End to end,

the molecules would extend outward for a meter or so. Similarly, stretch a frog cell's twenty-six DNA molecules end to end and they would extend ten meters. We have very little idea what frogs are doing with so much of it, but that's a lot of DNA!

Eukaryotic cells, recall, protect their DNA inside the nucleus. Figure 4.11 shows this organelle's structure. A nucleus has two functions. *First*, it physically separates all of the DNA molecules from the cytoplasm's complex metabolic machinery. Localizing DNA this way makes it easier for a parent cell to copy its hereditary instructions before the time comes for it to divide. Why? DNA molecules can be sorted into parcels —one for each of the two daughter cells that form later. *Second*, the outer membranes of the nucleus serve as a boundary where cells can control the passage of substances and signals to and from the cytoplasm. Table 4.2 lists the components of the nucleus.

Nuclear Envelope

Unlike cells, a nucleus has two outer membranes, one wrapped around the other. Its double-membrane system is the **nuclear envelope**. It has not one but two lipid bilayers in which many molecules of proteins are embedded (Figure 4.12). The nuclear envelope surrounds the nucleoplasm, the fluid portion of a nucleus. The innermost surface contains attachment sites for strandlike proteins that anchor DNA to the envelope and also help keep it organized. On the outer surface is a profusion of ribosomes. Proteins are built on membrane-bound ribosomes or on other ribosomes positioned in the cytoplasm.

cytoplasm
nucleus
plasma membrane
nuclear envelope
nucleoplasm
nucleolus
chromatin

Table 4.2 *Components of the Nucleus*

Nuclear envelope	Pore-riddled double-membrane system that selectively controls the passage of various substances into and out of the nucleus
Nucleoplasm	Fluid interior portion of the nucleus
Nucleolus	Dense cluster of RNA and proteins that will be assembled into subunits of ribosomes
Chromosome	One DNA molecule and many proteins that are intimately associated with it
Chromatin	Total collection of all DNA molecules and their associated proteins in the nucleus

Figure 4.11 Pancreatic cell nucleus. The small arrows on this transmission electron micrograph point to pores where control systems operate to restrict or permit the passage of specific substances across the nuclear envelope.

nuclear pore (cluster of proteins that spans both bilayers)

one of two lipid bilayers (this one faces the cytoplasm)

NUCLEAR ENVELOPE

one of two lipid bilayers (this one faces the nucleoplasm)

a   1 μm   200 nm   b

Figure 4.12 (**a**) Part of the outer surface of a nuclear envelope. *Left:* This specimen was fractured to show the intimate layering of its two lipid bilayers. *Right:* Nuclear pores. Each pore across the envelope is an organized array of membrane proteins. The pore permits the selective transport of substances into and out of the nucleus. (**b**) Sketch of the nuclear envelope's structure.

Like all cell membranes, a nuclear envelope's lipid bilayers prevent water-soluble substances from moving freely into and out of a nucleus. However, many pores, each a clustering of proteins, span both bilayers. Small, water-soluble molecules and ions can cross the nuclear envelope at these pores. As you will see later in the book, large molecules (including ribosomal subunits) move across the bilayers at pores in highly controlled ways.

Nucleolus

Take another look at the nucleus in Figure 4.11. Inside is a somewhat spherical, dense mass of material. What is it? When eukaryotic cells grow, one or more of these masses form inside the nucleus. Each is a **nucleolus** (plural, nucleoli), a construction site for many molecules of structural proteins and of RNAs in charge of peptide bond formation during protein synthesis. Some RNAs and proteins combine to form ribosomal subunits, then pass through nuclear pores into the cytoplasm. There, two subunits converge as an intact ribosome.

Chromosomes

When eukaryotic cells are not busy dividing, their DNA looks like thin threads. Many proteins are attached to the threads, like beads on a string. The beaded threads have a grainy appearance except at high magnification (Figure 4.11). When a cell prepares to divide, however, it duplicates its DNA molecules so each daughter cell will get all required hereditary instructions. Then each molecule folds into a compact structure, proteins and all.

Early microscopists bestowed the name *chromatin* on the seemingly grainy substance and *chromosomes* on the condensed structures. We now define **chromatin** as the cell's collection of DNA, together with all proteins associated with it (Table 4.2). Each **chromosome** is one DNA molecule and its associated proteins, regardless of whether it is in threadlike or condensed form:

one chromosome (one threadlike DNA molecule + proteins; not duplicated)

one chromosome (threadlike but now duplicated; two DNA molecules + proteins)

one chromosome (duplicated and also condensed)

In other words, the appearance of "the chromosome" changes over the life of a eukaryotic cell! In chapters to come, you will look at different aspects of chromosomes, so you may find it useful to remember this point.

What Happens to the Proteins Specified by DNA?

Outside of the cell nucleus, new polypeptide chains for proteins are assembled on ribosomes. What happens to them? Many are stockpiled in the cytoplasm or get used at once. Many others enter a cytomembrane system. As you will read shortly, this system includes endoplasmic reticulum, Golgi bodies, and a variety of vesicles.

Thanks to DNA's instructions, many proteins take on particular, final forms in the cytomembrane system. Lipids also are packaged and assembled in the system by enzymes and other proteins (which also were built according to DNA's instructions). As you will see next, vesicles deliver the proteins and lipids to specific sites within the cell or to the plasma membrane, for export.

The nucleus, an organelle with two outer membranes, keeps the DNA molecules of eukaryotic cells separated from the metabolic machinery of the cytoplasm.

The localization makes it easier to organize the DNA and to copy it before a parent cell divides into daughter cells.

Pores across the nuclear envelope help control the passage of many substances between the nucleus and cytoplasm.

THE CYTOMEMBRANE SYSTEM

The **cytomembrane system** is a series of organelles in which lipids are assembled and new polypeptide chains are modified into final proteins. Its products are sorted and shipped to different destinations. Figure 4.13 shows how its organelles—the ER, Golgi bodies, and various vesicles—functionally interconnect with one another.

Endoplasmic Reticulum

The functions of the cytomembrane system begin with **endoplasmic reticulum**, or **ER**. In animal cells, the ER is continuous with the nuclear envelope, and it extends through cytoplasm. Its membrane regions appear rough or smooth, depending mainly on whether ribosomes are attached to the membrane facing the cytoplasm.

We typically observe *rough* ER arranged into stacks of flattened sacs with many ribosomes attached (Figure 4.14*a*). Every new polypeptide chain is synthesized on ribosomes. But only the newly forming chains having a built-in signal can enter the space within rough ER or become incorporated into ER membranes. (That signal is a sequence of fifteen to twenty specific amino acids.) Once the chains are in rough ER, enzymes may attach oligosaccharides and other side chains to them. Many specialized cells secrete the final proteins. Rough ER is abundant in such cells. For example, in your pancreas, ER-rich gland cells make and secrete enzymes that end up in the small intestine and help digest your meals.

Smooth ER is free of ribosomes and curves through cytoplasm like connecting pipes (Figure 4.14*b*). Many cells assemble most lipids inside the pipes. Smooth ER is well developed in seeds. In liver cells, certain drugs as well as toxic metabolic wastes are inactivated in it. Sarcoplasmic reticulum, a type of smooth ER in skeletal muscle cells, functions in contraction.

Golgi Bodies

In **Golgi bodies**, enzymes put the finishing touches on proteins and lipids, sort them out, then package them inside vesicles for shipment to specific locations. For example, an enzyme in one Golgi region might attach a phosphate group to a new protein, thereby giving it a mailing tag to its proper destination.

Commonly, a Golgi body looks rather like a stack of pancakes; it is composed of a series of flattened, membrane-bound sacs (Figure 4.15). In functional terms, the last portion of a Golgi body corresponds to the top pancake. Here, vesicles form as patches of the membrane bulge out, then break away into the cytoplasm.

5 Vesicles budding from the Golgi membrane transport finished products to the plasma membrane. The products are released by exocytosis.

4 Proteins and lipids take on final form in the space inside the Golgi body. Different modifications allow them to be sorted out and shipped to their proper destinations.

3 Vesicles bud from the ER membrane and then transport unfinished proteins and lipids to a Golgi body.

2 In the membrane of smooth ER, lipids are assembled from building blocks delivered earlier.

1 Some polypeptide chains enter the space inside rough ER. Modifications begin that will shape them into the final protein form.

SECRETORY PATHWAY

assorted vesicles

Golgi body

smooth ER

rough ER

DNA instructions for building polypeptide chains leave the nucleus and enter the cytoplasm.

The chains (*green*) are assembled on ribosomes in the cytoplasm.

Some vesicles form at the plasma membrane, then move into the cytoplasm. These *endocytic* vesicles might fuse with the membrane of other organelles or remain intact, as storage vesicles.

Other vesicles bud from ER and Golgi membranes, then fuse with the plasma membrane. The contents of these *exocytic* vesicles are thereby released from the cell.

Figure 4.13 Cytomembrane system, a membrane system in the cytoplasm that synthesizes, modifies, packages, and ships proteins and lipids. *Green* arrows highlight a secretory pathway by which some proteins and lipids are packaged and released from many types of cells, including gland cells that secrete mucus, sweat, and digestive enzymes.

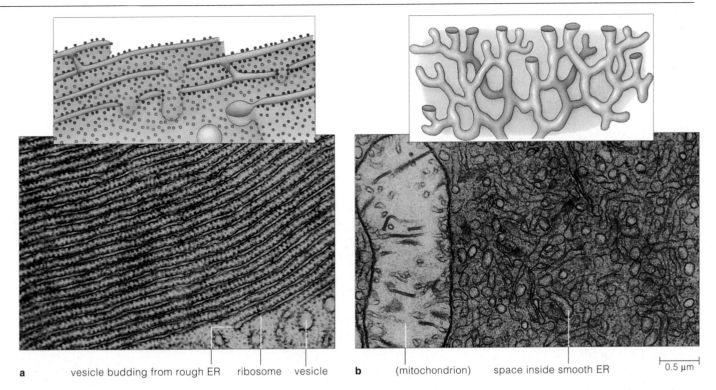

a   vesicle budding from rough ER   ribosome   vesicle   **b**   (mitochondrion)   space inside smooth ER   0.5 μm

Figure 4.14 Transmission electron micrographs and sketches of endoplasmic reticulum. (**a**) Many ribosomes dot the flattened surfaces of rough ER that face the cytoplasm. (**b**) This section reveals the diameters of many interconnected, pipelike regions of smooth ER.

ROUGH ER   SMOOTH ER

internal space   budding vesicle

GOLGI BODY

0.25 μm

Figure 4.15 Sketch and micrograph of a Golgi body from an animal cell.

A Variety of Vesicles

Vesicles are tiny, membranous sacs that move through the cytoplasm or take up positions in it. A common type, the lysosome, buds from Golgi membranes of animal cells and some fungal cells. Lysosomes are organelles of intracellular digestion. They hold enzymes that digest complex carbohydrates, proteins, nucleic acids, and some lipids. Often they fuse with vesicles that formed from patches of plasma membrane that surrounded bacteria, molecules, and other items that docked at membrane receptors. Lysosomes even digest whole cells or cell parts. For example, when a tadpole is developing into an adult frog, its tail gradually disappears. Lysosomal enzymes are responding to developmental signals and are helping to destroy cells that make up the tail.

Peroxisomes, another type, are sacs of enzymes that break down fatty acids and amino acids. A product of the reactions, hydrogen peroxide, is toxic, as the Chapter 5 introduction describes. But enzyme action converts it to water and oxygen or delivers it to reaction sites that break down alcohol. Drink alcohol, and peroxisomes of liver and kidney cells will degrade nearly half of it.

Many proteins take on final form and lipids are synthesized in the ER and Golgi bodies of the cytomembrane system.

Lipids, proteins (such as enzymes), and other items become packaged in vesicles destined for export, storage, membrane building, intracellular digestion, and other cell activities.

MITOCHONDRIA

Recall, from Section 3.6, that ATP is an energy carrier. At many different reaction sites, it delivers energy—in the form of phosphate groups—that drives nearly all cell activities. Many ATP molecules form when organic compounds are fully broken down to carbon dioxide and water in a **mitochondrion** (plural, mitochondria).

Only eukaryotic cells contain these organelles. Figure 4.16 gives an idea of their structure. ATP-forming reactions that proceed in mitochondria extract more energy from organic compounds than can be done by any other means. They cannot run to completion without plenty of oxygen. Like all other land-dwelling vertebrates, every time you breathe in, you take in oxygen mainly for the mitochondria in cells—in your case, many trillions of individual cells.

Each mitochondrion has a double-membrane system. The outer membrane faces the cytoplasm. Most commonly, the inner one is repeatedly folded.

What's the point of such an intricate system? It creates two compartments. Enzymes and other proteins of electron transport systems stockpile hydrogen ions in the outer compartment. Then the ions flow out from the compartment in controlled ways. Energy inherent in the flow drives ATP formation, as Chapter 7 describes. Oxygen binds to the spent electrons to form water, an end product of the reactions. Its removal keeps the transport systems clear for operation.

All eukaryotic cells contain one or more mitochondria. You may find only one in a single-celled yeast. You might find one thousand or more in cells that demand a lot of energy, such as those of muscles. Reflect on the profusion of mitochondria in Figure 4.10, which is a micrograph of merely one thin section from a liver cell. It alone tells you that the liver is a notably active, energy-demanding organ.

In size and biochemistry, mitochondria are a lot like bacteria. They even contain their own DNA and some ribosomes. They divide on their own. Did they evolve from ancient prokaryotic cells? Maybe. One idea is that a predatory, amoebalike cell engulfed other cells, which managed to avoid being digested. Perhaps the engulfed cells and their descendants were still able to reproduce. Suppose they became permanent, protected residents. If so, then some structures and functions they formerly needed to maintain independent life would no longer

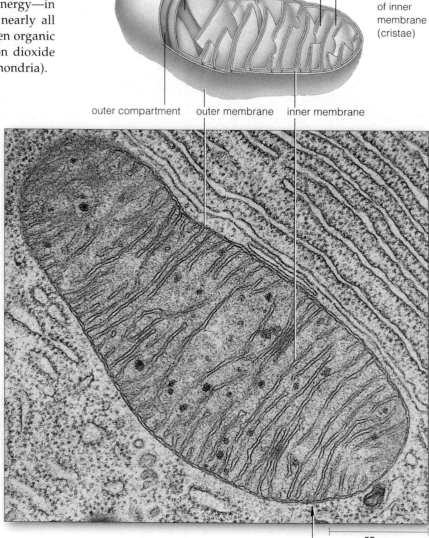

Figure 4.16 Sketch and transmission electron micrograph, thin section, of a typical mitochondrion. This organelle specializes in the production of large quantities of ATP, the main energy-carrying molecule between different reaction sites in cells. The process that produces ATP in mitochondria can't proceed without free oxygen.

be necessary and might have been lost. In time the descendants may have evolved into mitochondria. We return to this topic in Section 19.4.

The organelles called mitochondria are the ATP-producing powerhouses of all eukaryotic cells.

Energy-releasing reactions proceed at the compartmented, internal membrane system of mitochondria. The reactions, which require oxygen, produce far more ATP than can be made by any other cellular reactions.

4. Label the parts of this bacterial cell. *4.12*

5. Describe three features that all cells have in common. After reviewing Table 4.3, write a paragraph on the key differences between eukaryotic and prokaryotic cells. *4.1, 4.4, 4.11, 4.12*

6. Briefly characterize the structure and function of the cell nucleus, nuclear envelope, and nucleolus. *4.6*

7. Define chromosome and chromatin. Do chromosomes always have the same appearance during a cell's life? *4.6*

8. Which organelles are part of the cytomembrane system? *4.7*

9. Is this statement true or false: Plant cells have chloroplasts, but not mitochondria. Explain your answer. *4.8, 4.9*

10. What are the functions of the central vacuole? *4.9*

11. Define cytoskeleton. How does it aid in cell functions? *4.10*

12. What gives rise to the 9 + 2 array of cilia and flagella? *4.10*

13. Cell walls are typical of which organisms: bacteria, protistans, fungi, plants, animals? Are the walls impermeable? *4.11*

14. In certain plant cells, is a secondary wall deposited inside or outside the surface of the primary wall? *4.11*

15. In multicelled organisms, coordinated interactions depend on linkages and communications between cells. What types of junctions occur between adjacent animal cells? Plant cells? *4.11*

Self-Quiz   ANSWERS IN APPENDIX III

1. Cell membranes consist mainly of a _____ .
   a. carbohydrate bilayer and proteins
   b. protein bilayer and phospholipids
   c. lipid bilayer and proteins
   d. none of the above

2. Organelles _____ .
   a. are membrane-bound compartments
   b. are typical of eukaryotic cells, not prokaryotic cells
   c. separate chemical reactions in time and space
   d. all of the above are features of the organelles

3. Cells of many protistans, plants, and fungi, but not animals, commonly have _____ .
   a. mitochondria   c. ribosomes
   b. a plasma membrane   d. a cell wall

4. Is this statement true or false: The plasma membrane is the outermost component of all cells. Explain your answer.

5. Unlike eukaryotic cells, prokaryotic cells _____ .
   a. lack a plasma membrane   c. do not have a nucleus
   b. have RNA, not DNA   d. all of the above

6. Match each cell component with its function.
   ____ mitochondrion   a. synthesis of polypeptide chains
   ____ chloroplast   b. initial modification of new
   ____ ribosome   polypeptide chains
   ____ rough ER   c. final modification of proteins;
   ____ Golgi body   sorting, shipping tasks
                         d. photosynthesis
                         e. formation of many ATP

Critical Thinking

1. Why is it likely that you will never encounter a predatory two-ton living cell on the sidewalk?

2. In compound light microscopes with blue filters, lenses transmit only blue light. Look at the spectrum of visible light in Figure 6.5. Why is blue light so efficient for viewing objects at high magnification?

3. Your professor shows you an electron micrograph of a cell with large numbers of mitochondria and Golgi bodies. You notice that this particular cell also contains a great deal of rough endoplasmic reticulum. What kinds of cellular activities would require such an abundance of the three kinds of organelles?

4. Each year in the United States alone, there are about 40,000 reported cases of *cystic fibrosis* (CF). This is a fatal genetic disorder that, at this writing, is incurable. On average, CF individuals can expect to live no longer than thirty years.

Affected glands secrete more than they should, which has far-reaching effects. In time, after digestive enzymes clog a duct between the pancreas and small intestine, food can't be digested properly. Even when food intake increases, malnutrition results. Cysts form in the pancreas, which degenerates and becomes fibrous (hence the disorder's name). Thick mucus builds up in the respiratory tract. CF individuals have trouble expelling airborne bacteria and particles that enter the lungs (Figure 4.26).

The disorder may arise from a defective protein in the plasma membrane of gland cells that secrete mucus, digestive enzymes, and sweat. Review Section 4.7, then name the organelles involved in the secretory pathway in those cells.

Selected Key Terms

basal body *4.10*	ER (endoplasmic	nucleoid *4.1*
cell *4.1*	reticulum) *4.7*	nucleolus *4.6*
cell cortex *4.10*	eukaryotic cell *4.1*	nucleus *4.1, 4.6*
cell theory *CI*	flagellum *4.10*	organelle *4.4*
cell wall *4.11*	fluid mosaic	phospholipid *4.1*
central vacuole *4.9*	model *4.1*	plasma membrane *4.1*
centriole *4.10*	Golgi body *4.7*	primary wall *4.11*
chloroplast *4.9*	intermediate	prokaryotic cell *4.1*
chromatin *4.6*	filament *4.10*	pseudopod *4.10*
chromosome *4.6*	lipid bilayer *4.1*	ribosome *4.1*
cilium *4.10*	microfilament *4.10*	secondary wall *4.11*
cytomembrane	microtubule *4.10*	surface-to-volume
system *4.7*	micrograph *4.3*	ratio *4.2*
cytoplasm *4.1*	mitochondrion *4.8*	vesicle *4.7*
cytoskeleton *4.10*	nuclear envelope *4.6*	wavelength *4.3*

Readings

Alberts, B., et al. 1994. *Molecular Biology of the Cell*. Third edition. New York: Garland. See Chapter 16: The Cytoskeleton.

On-Line readings at Student Guide for InfoTrac: www.brookscole.com/biology

5

GROUND RULES OF METABOLISM

Growing Old With Molecular Mayhem

Somewhere in those slender strands of DNA inside your cells are snippets of instructions for making two kinds of enzymes: superoxide dismutase and catalase (Figure 5.1). Both are *antioxidants*. Like vitamins E and C, they help keep you from growing old before your time by cleaning house, so to speak. Working as a team, they neutralize many potentially toxic wastes by using oxygen (O_2) to strip hydrogen atoms from molecules. Many other kinds of organisms, ranging from bacteria to roundworms to plants, also depend on them.

Stripping hydrogen atoms from molecules happens to release electrons, and O_2 normally picks them up. Sometimes it picks up only one. That one electron isn't enough to complete the reaction. But it's enough to give the oxygen a negative charge (O_2^-).

Like other unbound molecular fragments with the wrong number of electrons, O_2^- is a **free radical**. Free radicals slip away from a variety of enzyme-catalyzed reactions, including the digestion of fats and amino acids. They also form when x rays and other forms of ionizing radiation bombard water and other molecules. They even escape from electron transport chains.

Free radicals are highly reactive. When they dock with a molecule, they alter its structure and function.

They even attack molecules that aren't open to just any reaction. Such molecules include DNA, cell membrane lipids, and membrane receptors which, when activated, tell cells it's time to die (Section 14.4).

Enter superoxide dismutase. Under its biochemical prodding, two rogue oxygen molecules combine with hydrogen ions. Hydrogen peroxide (H_2O_2) and O_2 are the outcome. Hydrogen peroxide, a normal by-product of some reactions, is lethal to cells when it accumulates. Enter catalase. Under its prodding, two molecules of hydrogen peroxide react and split into water and oxygen: $2H_2O_2 \longrightarrow 2H_2O + O_2$. This crucial reaction normally occurs before hydrogen peroxide does major damage.

As we age, our cells make copies of enzymes in ever diminishing numbers, in crippled form, or both. When this happens to superoxide dismutase and catalase, free radicals and hydrogen peroxide accumulate. Like loose cannons, they careen through cells and blast away at the structural integrity of proteins, DNA, lipids, and other molecules. Cells suffer or die.

Those brown "age spots" on an older person's skin are evidence of free radical assaults (Figure 5.2a). Each spot is a mass of brownish-black pigments that build up in cells when free radicals take over—all for the want

Figure 5.1 Computer models of the molecular structure of (**a**) superoxide dismutase and (**b**) catalase. Both enzymes help keep free radicals inside the body in check.

Figure 5.2 (**a**) Owner of skin with a spattering of age spots, evidence of free radicals on the loose. At one time he, like the boy in (**b**), had many smoothly functioning molecules of superoxide dismutase and catalase. (**c**) The worm of choice for aging studies, *Caenorhabditis elegans*.

of specific enzymes. Worse yet, free radicals also may usher in heart disease and other disorders.

Knowing this, Simon Melov, Bernard Malfroy, and their colleagues did experiments with the roundworm *Caenorhabditis elegans* (Figure 5.2*c*). These tiny worms don't live more than a month. However, when synthetic superoxide dismutase and catalase were added to their diet, their life span increased greatly. In some cases it actually doubled! The researchers also fed enzymes to worms that were genetically engineered to succumb prematurely to free radicals. Those worms lived just as long as the unaltered ones.

With this example, we turn to **metabolism**, the cell's capacity to acquire energy and use it to build, degrade, store, and release substances in controlled ways. Each extra day in a worm's life reminds us that cells must continually take in energy-rich solutes, store things, build membranes, replenish enzymes, and check out their DNA. A constant supply of energy drives every one of their activities. At times, the activities may seem remote from your interests. But they help define who you are and who you will become, age spots and all.

Key Concepts

1. Cells engage in metabolism, or chemical work. They use energy to stockpile, build, rearrange, and break apart substances. Cells use energy for mechanical work, as when they move flagella. They also channel energy into electrochemical work, as when they move ions into or out of an organelle compartment.

2. All organisms secure energy from outside sources. Sunlight energy is the primary source for the web of life. Another source is chemical bond energy of molecules in the physical environment or in other, edible organisms.

3. ATP is the main carrier of energy in cells. It couples reactions that release energy with reactions that require it. ATP activates molecules—primes them for chemical change—by transferring a phosphate group to them.

4. Many metabolic reactions involve oxidation–reduction reactions, or electron transfers among substances. Such transfers often are made in electron transport systems.

5. Chemical reactions proceed far too slowly on their own to sustain life. In cells, specific enzymes greatly increase the rate of specific reactions. Coenzymes, such as NAD^+, assist enzymes in the reactions or transfer electrons and hydrogen released during the reactions to other sites.

6. Many enzymes mediate reactions in orderly sequences called metabolic pathways. The coordinated operation of these pathways maintains, increases, and decreases the relative amounts of substances in cells.

7. Much of metabolism relies on concentration gradients. These are differences in the number of molecules or ions of a substance between two regions. Molecules tend to move to the region of lower concentration, a behavior called diffusion.

8. Cells work with and against concentration gradients across the plasma membrane and internal membranes. Specific solutes cross membranes by passive and active transport. Water crosses by osmosis. Materials also cross the plasma membrane by endocytosis and exocytosis.

ENERGY AND THE UNDERLYING ORGANIZATION OF LIFE

Defining Energy

If you've ever watched a house cat stalk a mouse, you know it "freezes" its position to avoid detection before springing at its unsuspecting prey. Like everything else in the universe that's stationary, the cat has a store of **potential energy**—a capacity to do work, simply owing to its position in space and the arrangement of its parts. When that cat springs, some of its potential energy is transformed into **kinetic energy**, the energy of motion.

Energy on the move does work when it imparts motion to other things. In skeletal muscle cells within the cat, ATP gave up some of its potential energy to molecules of contractile units and set them in motion. The combined motions in many muscle cells resulted in the movement of whole muscles. The transfer of energy from ATP also resulted in the release of another form of energy called **heat**, or *thermal* energy.

The potential energy of molecules has its own name: **chemical energy**. It is measurable, as in kilocalories. A kilocalorie is the same thing as 1,000 calories, which is the amount of energy it takes to heat 1,000 grams of water from 14.5°C to 15.5°C at standardized pressure.

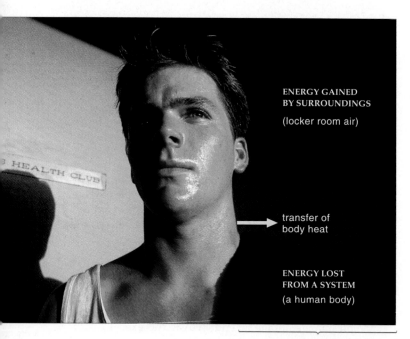

ENERGY GAINED
BY SURROUNDINGS

(locker room air)

transfer of
body heat

ENERGY LOST
FROM A SYSTEM

(a human body)

NET ENERGY CHANGE = 0

Figure 5.3 Example of how the total energy content of any system, *together with its surroundings*, remains constant.

"System" means all matter in a specific region, such as a human body, a plant, a DNA molecule, or a galaxy. "Surroundings" can be a small region in contact with the system or as vast as the whole universe. The system shown (a human male) is giving off heat to the surroundings (a locker room) by evaporative water loss from sweat. What one region loses, the other region gains, so the total energy content of both does not change.

What Can Cells Do With Energy?

All organisms have specific adaptations for securing energy from their environment. Some capture energy from the sun, and others extract energy from inorganic or organic substances in the environment. Regardless of the source, energy inputs become coupled to thousands of energy-requiring processes in cells. Cells use energy for *chemical* work—to stockpile, build, rearrange, and break apart substances. They channel it into *mechanical* work—to move flagella and other cell structures and (in multicelled species) the whole body or portions of it. They channel it into *electrochemical* work—to move charged substances into or out of the cytoplasm or an organelle compartment.

How Much Energy Is Available?

Like single cells, we cannot create energy from scratch; we must get it from someplace else. Why? According to the **first law of thermodynamics**, the total amount of energy in the universe remains constant. More energy cannot be created; existing energy cannot vanish. It can only be converted from one form to some other form.

Think about what the first law means. The universe has only so much energy, distributed in various forms. One form may be converted to another, as when corn plants absorb energy from the sun and convert it to the chemical energy of starch. After you eat and digest corn, your cells extract energy from starch and convert it to other forms, such as mechanical energy for movement.

With each metabolic conversion, a bit of the energy escapes to the surroundings, as heat. Even when you "do nothing," your body gives off about as much heat as a 100-watt lightbulb because of conversions in your cells. The energy being released is transferred to atoms and molecules making up the air, and the conversion of thermal to kinetic energy "heats up" the surroundings (Figure 5.3). The kinetic energy increases the number of ongoing, random collisions among molecules in the air. And with each collision, a bit more energy is released as heat. However, none of the energy ever vanishes.

The One-Way Flow of Energy

Energy available for conversions in cells resides mainly in covalent bonds. Glucose, glycogen, starches, fatty acids, and other organic compounds have organized arrangements of many of these bonds and are said to have a high energy content. When the compounds enter metabolic reactions, bonds break or become rearranged. During the molecular commotion, some amount of heat energy is lost to the surroundings. In general, cells can't recapture energy lost as heat.

Figure 5.4 Example of the one-way flow of energy into the world of life that compensates for the one-way flow of energy out of it. The sun continually loses energy, much of it in the form of wavelengths of light (Section 6.2). Living cells intercept some of the energy and convert it to useful forms of energy, stored in bonds of organic compounds. Each time a metabolic reaction proceeds in cells, stored energy is released—and some inevitably is lost to the surroundings, mostly as heat.

The lower photograph shows green, water-dwelling, photosynthetic cells (*Volvox*). They live in tiny, spherical colonies. The orange cells serve in reproduction. They form new colonies inside the parent sphere.

ENERGY LOST
one-way flow of energy from sun to Earth's environment

ENERGY GAINED
one-way flow of energy from environment to organisms

Producer organisms harness sun's energy, use it to build organic compounds from simple raw materials available in their environment.

All organisms tap potential energy stored in organic compounds to drive energy conversions that keep them alive. Some energy is lost with each conversion.

ENERGY LOST
one-way flow of energy from organisms back to the environment

For example, your cells release usable energy from glucose when they break all of its covalent bonds. After many steps, six molecules of carbon dioxide and six of water remain. Compared with glucose, these leftovers have more stable arrangements of atoms, but chemical energy in their bonds is much less than the overall chemical energy of glucose. Why? *Some energy was lost at each step leading to their formation.* Said another way, glucose is a better source of usable energy.

What about the heat that was transferred from cells to their surroundings when carbon dioxide formed? That heat just isn't useful. Cells can't convert it to other forms, so they can't use it to do work.

Bad news for cells of the remote future: The amount of "low-quality" energy in the universe is increasing. No energy conversion can ever be 100 percent efficient; even highly efficient ones lose heat. So the total amount of energy in the universe is spontaneously flowing from forms rich in energy to forms having less of it. Billions of years from now, energy may not be available for conversions; all of it may be dissipated in space.

Without energy inputs to maintain it, any organized system tends to become more and more disorganized over time. **Entropy** is a measure of the degree of a system's disorder. Think of the Egyptian pyramids—once highly organized, presently crumbling, and many thousands of years from now, dust. It seems the ultimate destination of pyramids and everything else in the universe is a state of maximum entropy. And that, basically, is the point to remember about the **second law of thermodynamics**.

Can life be one glorious pocket of resistance to the depressing flow toward maximum entropy? After all, in every new organism, new bonds form and hold atoms together in precise arrays. So molecules become more organized and have a richer store of energy, not poorer!

Yet a simple example will show that the second law does indeed apply to life on Earth. The primary energy source for life is the sun—which has been releasing energy since it first formed. Plants can capture sunlight energy, convert it to other forms, then lose energy to other organisms that feed, directly or indirectly, on the plants. At each energy transfer from the sun onward, some energy is lost as heat and joins the universal pool. *Overall, energy still flows in one direction.* The world of life maintains its amazing degree of organization only because it is being resupplied with energy that's being lost from someplace else (Figure 5.4).

The amount of energy in the universe remains constant. Energy can undergo conversion from one form to another, but it cannot be created out of nothing or destroyed.

The total amount of energy in the universe is spontaneously flowing from usable to nonusable forms.

A steady flow of sunlight energy into the interconnected web of life compensates for the steady flow of energy leaving it.

ENERGY CHANGES AND CELLULAR WORK

Energy Inputs, Energy Outputs

When cells convert one form of energy to another, the amount of potential energy available to them changes. The greater the amount a cell taps into, the larger the energy change, and the more work can be done.

Imagine a Martian watching the NASA Rover scoot around her planet. She decides to push it to the top of a rocky hill, which requires tapping into potential energy stored in her muscle cells (Figure 5.5a). Once the Rover is precariously perched on top, it has enough potential energy (owing to its position) to roll to the base on its own. The higher the Rover is relative to its final position at the hill's base, the greater the energy change (Figure 5.5b). This same rule operates in the cellular world, in which temperature and pressure remain fairly constant: *Energy changes in living cells tend to proceed spontaneously in the direction that results in a decrease in usable energy.*

Recall that glucose ($C_6H_{12}O_6$) is built from carbon dioxide ($6CO_2$) and water ($12H_2O$). Each substance has potential energy in chemical bonds, but the bond energy of glucose surpasses those of the other two substances combined. Thus, the assembly of glucose from carbon dioxide and water is not something that will proceed spontaneously. Picture carbon dioxide and water at the base of an energy hill. On their own, they simply don't have enough energy for an uphill run to make glucose.

In photosynthetic cells, *energy inputs* from the sun drive reactions by which glucose is built. The result is a net increase in usable energy for the cell (Figure 5.5c). Said another way, the reaction sequence that produces glucose is **endergonic** (meaning "energy in").

Now picture the reactions running in reverse, from glucose (at the top of the energy hill) to carbon dioxide and water (at the base). Energetically, a downhill run is favorable. The reaction sequence proceeds spontaneously and ends with a net loss in energy; it is **exergonic**, which means "energy out" (Figure 5.5d). All cells release usable energy from glucose and other large organic molecules.

The Role of ATP

Cells survive by *coupling* energy inputs to energy outputs. **ATP** (adenosine triphosphate) is the main coupling agent. This nucleotide consists of a five-carbon sugar (ribose), a nucleotide base (adenine), and a tail of three phosphate groups. Hundreds of different enzymes can split off the tail's outer phosphate group and attach it to another molecule, which thereby is primed to enter a reaction. **Phosphorylation** is the name for this type of phosphate-

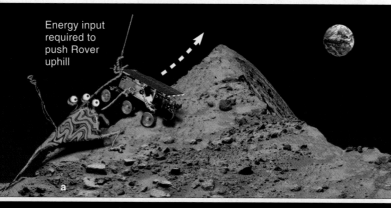

Energy input required to push Rover uphill

a

Potential energy released by the downhill run (but not captured to do useful work)

b

Figure 5.5 Simple examples of energy changes involved in (**a**,**b**) mechanical work and (**c**,**d**) chemical work.

product with more energy (plus by-products $6O_2$ and $6H_2O$)

ENERGY IN

6   12

energy-poor starting substances

c

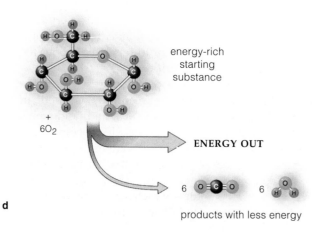

energy-rich starting substance

+ $6O_2$

ENERGY OUT

6   6

products with less energy

d

Figure 5.13 Model of the enzyme hexokinase. Hexokinase attaches a phosphate group to glucose. (**a**) A glucose molecule (color-coded *red*) is heading for the active site, which is a cleft in the enzyme (*green*). (**b**) When glucose enters the site, parts of the enzyme briefly close in around it and prod the molecule to enter into the reaction.

(**c**) Induced-fit model of enzyme–substrate interactions. Only when a substrate is bound in place is an enzyme's active site complementary to it. The fit is most precise during the transition state of a reaction. However, the enzyme–substrate complex is short-lived. Why? Usually, the attractive forces holding it together are weak.

two substrate molecules

substrates contacting active site of enzyme

active site

TRANSITION STATE (tightest binding but least stable)

end product

enzyme unchanged by the reaction

c

or enzymes may promote it; this makes no difference. The activation energy is an energy barrier—something like a hill. And that barrier must be surmounted *one way or another* before a reaction will proceed.

An enzyme makes the energy barrier smaller, so to speak. How? Every enzyme contains one or more **active sites**. These are crevices in the molecule's surface where the enzyme interacts with its substrates and catalyzes a reaction. Figure 5.13 shows the active site of a specific enzyme that primes a glucose molecule for reaction by attaching a phosphate group to it.

According to Daniel Koshland's **induced-fit model**, part of each substrate's surface has chemical groups that are almost but not quite complementary to the chemical groups in an active site. When substrates first settle into the site, the contact strains some of their bonds. Strained bonds are easier to break, so they promote formation of new bonds (in products). Also, charged or polar groups in the active site often cause the electric charge to shift in substrates, which primes them for conversion.

When substrates fit most precisely in an active site, they are in an activated, *transition* state and will react (Figure 5.13c). Now the reaction must proceed, just as the NASA Rover must roll down the Martian hill when something pushes it over the crest (Figure 5.12).

What induces the transition state or gets substrates over the energy barrier once that state is reached? The following are examples of the mechanisms involved.

1. *Helping substrates get together.* Substrate molecules rarely collide when there aren't many of them around. Snag them at an active site, and it's like boosting their local concentration. The reaction rate increases by 10,000 to 10,000,000 times, depending on the reaction.

2. *Orienting substrates in positions that favor reaction.* Without enzymes, substrates collide from random directions. Weak but extensive bonding at an active site puts reactive chemical groups on precise collision courses much more often.

3. *Promoting acid–base reactions.* In many cases, acidic or basic side groups of amino acids that are part of the enzyme's active site are poised to donate or accept hydrogen atoms from substrates. The loss or addition destabilizes a substrate's covalent bonds and makes them easier to break. Hydrolysis works this way.

Depending on the type of enzyme, mechanisms of the sort just described work alone or in combination with one another to bring about the transition state.

Enzymes catalyze (speed) the rate at which specific reactions reach equilibrium. They do so by lowering the amount of activation energy necessary to make substrates react.

Enzymes change the rate, not the outcome, of a reaction. They act only on specific substrates. They may catalyze the same reaction repeatedly, as long as substrates are available.

FACTORS AFFECTING ENZYME ACTION

You probably don't get much done when you feel too hot or cold or out of sorts because you ate too many sour plums or salty potato chips. When the cupboard is bare, you focus on food. Maybe you call a friend to go shopping with you, and when you drive too fast to the grocery store, police tend to slow you down. In such respects, you have a lot in common with enzymes. They, too, respond to shifts in temperature, pH, and salinity, and to the relative abundances of particular substances. Many depend on helpers for specific tasks. And all normal enzymes respond to metabolic police.

How Is Enzyme Action Controlled?

Each cell controls its enzyme activity. By coordinating control mechanisms, it maintains, lowers, or increases the concentrations of substances. Controls adjust how fast enzyme molecules are built and become available, or they work on enzymes that are already synthesized.

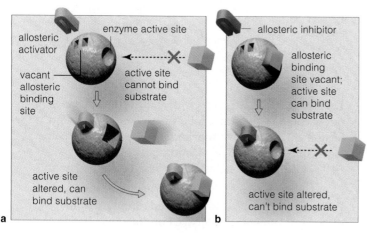

Figure 5.14 Allosteric control of enzymes. (**a**) An active site is unblocked when an activator protein binds to a vacant allosteric site. (**b**) An active site is blocked when an inhibitor protein binds to a vacant allosteric site.

Figure 5.15 One example of feedback inhibition of a metabolic pathway. Five kinds of enzymes act in sequence. They convert a substrate to the end product, tryptophan.

For example, with *allosteric* control, some molecule binds to an enzyme at a site other than the active site. (*Allo–* means different, *steric* means structure, or state.) The active site changes shape in a way that allows or prevents enzyme action. Figure 5.14 shows models for the binding, which is reversible. Study these models, then picture a bacterium synthesizing tryptophan (and other amino acids) used to build proteins. Soon, enough proteins are built. But tryptophan synthesis won't stop until the concentration of tryptophan molecules triggers **feedback inhibition**. By this feedback mechanism, a change caused by an activity *shuts down that activity*. In this pathway, a feedback loop starts and ends with an allosteric enzyme. When tryptophan accumulates, the unused molecules bind with the allosteric site. Binding shuts down the enzyme, hence the rest of the pathway.

What if tryptophan is scarce when the demand for it steps up? The enzyme molecules are not shut down, so tryptophan will be built. Such feedback loops quickly adjust concentrations of many substances (Figure 5.15).

In humans and other multicelled organisms, control of enzyme activity is just amazing. Cells not only work to stay alive, they work with other cells in ways that benefit the whole body! For example, this vast cellular work relies on hormones, a type of signaling molecule. Specialized cells release hormones into tissues. Any cell having receptors for a given hormone responds to it, then its program for building a protein or some other activity changes. The hormone trips cell controls into action—and the activities of certain enzymes change.

Helpers in Enzyme Action

Enzyme action often requires coenzymes or metal ions. These help the reaction or transfer electrons, atoms, or functional groups to a different reaction site. You'll be reading about NAD^+, $NADP^+$, and FAD, all coenzymes derived from vitamin precursors. One of the metal ions, ferrous iron (Fe^{++}), is a helper in cytochrome action. Cytochromes are proteins in cell membranes, including the inner membrane of chloroplasts and mitochondria. They are part of major electron transport systems.

Does the Environment Influence Enzymes?

Temperature, recall, is a measure of molecular motion. A rise in temperature boosts reaction rates by making substrates collide more often with active sites. Yet past some temperature (which differs among enzymes), the increased motion disrupts the weak bonds holding the enzyme in its three-dimensional shape. Substrates no longer bind to the active site, so reaction rates decline sharply, as Figure 5.16*a* shows. When temperatures rise above or fall below the range of tolerance, metabolism

Figure 5.16 How the action of one type of enzyme shifts when (a) temperatures and (b) pH values fall outside the range of tolerance. (c) Siamese cats show observable effects of change in temperature. Fur on the ears and paws has more of a dark brown pigment, melanin, than the rest of the body does. A heat-sensitive enzyme controlling melanin production is less active in warmer parts of the body, which end up with lighter fur.

is disrupted. That's what happens with extremely high fevers. People usually do not survive when the body's internal temperature reaches 44°C (112°F).

pH values that extend beyond an enzyme's range of tolerance also take a toll (Figure 5.16b). Most enzymes work best when the pH is between 6 and 8. Trypsin, for example, is active in the small intestine, where pH is 8 or so. Pepsin, a protein-digesting enzyme, is one of the exceptions. It functions in gastric fluid, a highly acidic liquid (pH of about 1–2) that denatures most enzymes.

In addition, most enzymes do not work well when fluids are saltier than usual. High salinity (extremely high ion concentrations) disrupts the interactions that help hold enzymes in their three-dimensional shapes.

Control mechanisms govern the synthesis of new enzymes and stimulate or inhibit the activity of existing enzymes. By controlling enzymes, cells control the concentrations and kinds of substances available to them.

Coenzymes and metal ions help enzymes catalyze reactions or transfer electrons and atoms to other reaction sites.

Enzymes function best when the cellular environment stays within limited ranges of temperature, pH, and salinity. The actual ranges differ from one type of enzyme to the next.

METABOLISM AND CELL MEMBRANES

All molecules contain energy in the form of chemical bonds. Cell membranes help make that energy available for cell functioning. How? *They selectively concentrate reactants and end products in amounts most favorable for specific reactions.* Think about the water bathing a cell membrane. Plenty of substances are dissolved in it. But the kinds and amounts are not the same next to the two membrane surfaces. The membrane itself established and is maintaining these differences, which are vital for metabolism. How? Each cell membrane shows **selective permeability**: Because of its molecular structure, a cell membrane permits some substances but not others to cross it in certain ways, at certain times (Figure 5.17).

For instance, being largely nonpolar, the hydrocarbon chains of the lipid bilayer passively let carbon dioxide, molecular oxygen, and other small, nonpolar molecules cross the membrane. Some water molecules cross, also. They are polar, but they slip through gaps that open up when the chains flex and bend.

By contrast, large, polar molecules such as glucose seldom move freely across the bilayer. Neither do ions. Such water-soluble substances can cross it passively or actively through the interior of transport proteins. Water molecules in which they are dissolved cross with them.

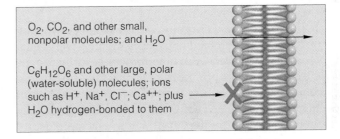

O_2, CO_2, and other small, nonpolar molecules; and H_2O

$C_6H_{12}O_6$ and other large, polar (water-soluble) molecules; ions such as H^+, Na^+, Cl^-; Ca^{++}; plus H_2O hydrogen-bonded to them

Figure 5.17 Selective permeability of cell membranes. Small, nonpolar molecules and some water molecules cross the lipid bilayer. Ions and large, polar, water-soluble molecules and the water dissolving them cannot cross the bilayer; transport proteins spanning the bilayer must help them across.

Membrane structure also allows cells to import and export substances in bulk across the plasma membrane. The processes are called endocytosis and exocytosis.

At any time in a cell's life, all of these mechanisms are operating simultaneously. Let's take a look at how they work, because they will occupy our attention in many chapters to follow.

Metabolic reactions depend on the chemical energy inherent in concentrated amounts of molecules and ions. Cells have mechanisms for increasing or decreasing those concentrations across the plasma membrane and internal cell membranes.

PROTEIN-MEDIATED TRANSPORT

Different kinds of proteins span the lipid bilayer of cell membranes. They let water-soluble molecules and ions diffuse across the membrane, through their tunnel-like interior (Figure 5.18). When a solute enters the tunnel's opening on one side of the membrane, it weakly binds to the protein. The protein's folded shape changes, so that the opening closes behind the bound solute and opens in front of it. The solute is exposed to the fluid bathing the other side of the membrane. The binding site reverts to its former state and the solute is released.

Passive Transport

Many of the proteins simply let solutes diffuse freely across the membrane. **Passive transport** is the name for an unassisted flow of solutes through the interior of a transport protein, down their concentration gradients. Energetically, the flow costs only what the cell already spent to produce and maintain the gradients. Passive transport itself adds nothing more to the energy cost. The *net* direction of movement during a given interval depends on how many molecules or ions of the solute are randomly contacting the transport proteins (Figure 5.18). Encounters are more frequent on the side of the membrane where the solute concentration is greatest. Therefore, the solute's *net* movement is in the direction where it is less concentrated.

If nothing else were going on, the passive, two-way transport would proceed until concentrations were equal across the membrane. But other processes influence the outcome. For instance, nearly all cells take up glucose from the blood, then use it for biosynthesis and instant energy. When the level of glucose in blood is high, they can maintain the gradient even if uptake is rapid. How? As fast as glucose molecules diffuse into the cells, others are entering metabolic reactions. In other words, when the cells use glucose, they help maintain a concentration gradient that favors the uptake of *more* glucose.

Active Transport

Only in a dead cell have solute concentrations become equal on both sides of membranes. Living cells never stop expending energy to pump potassium and other solutes to and from their interior. In **active transport**, energy-driven mechanisms called "membrane pumps" make solutes cross cell membranes *against* concentration gradients, so that their net movement is in the direction where they are more concentrated.

In most cases, ATP molecules deliver the energy for active transport. When it gives up one of its phosphate groups to a transport protein, the chemical fit between the solute and the protein binding site improves on one side of the membrane (Figure 5.19*a,b*).

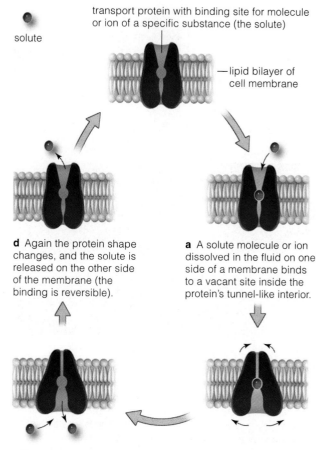

transport protein with binding site for molecule or ion of a specific substance (the solute)

solute

—lipid bilayer of cell membrane

d Again the protein shape changes, and the solute is released on the other side of the membrane (the binding is reversible).

a A solute molecule or ion dissolved in the fluid on one side of a membrane binds to a vacant site inside the protein's tunnel-like interior.

c Once it is exposed to fluid on the other side of the membrane, the binding site releases the solute. The newly vacated binding site attracts another molecule or ion of the solute.

b Binding alters the shape of the protein. The once-open part of the protein molecule closes behind the solute and another part opens in front of it.

Figure 5.18 Passive transport across a cell membrane. Solutes are able to move in both directions through transport proteins. In passive transport, *net* movement will be down the concentration gradient (from higher to lower) until concentrations are the same on both sides of the membrane.

One type of active transport system, the calcium pump, helps keep the calcium concentration inside a cell at least a thousand times lower than outside. Another, the sodium–potassium pump, moves potassium ions (K^+) across the plasma membrane. Activation of this protein helps bind a sodium ion (Na^+) on one side of a cell membrane. After the Na^+ makes the crossing, it is released. This release facilitates the binding of K^+ at a different binding site on the protein, which reverts to its original shape after K^+ is delivered to the other side.

Through operation of such active transport systems, concentration and electric gradients can be maintained across membranes. The gradients are vital to many cell activities and physiological processes, including muscle contraction and nerve cell (neuron) function. We return

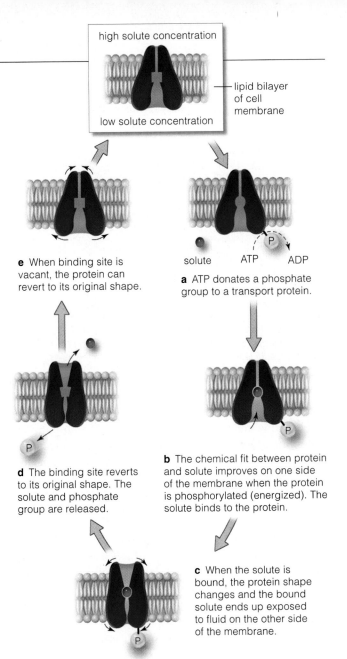

e When binding site is vacant, the protein can revert to its original shape.

a ATP donates a phosphate group to a transport protein.

solute   ATP   ADP

d The binding site reverts to its original shape. The solute and phosphate group are released.

b The chemical fit between protein and solute improves on one side of the membrane when the protein is phosphorylated (energized). The solute binds to the protein.

c When the solute is bound, the protein shape changes and the bound solute ends up exposed to fluid on the other side of the membrane.

Figure 5.19 Active transport across a cell membrane. ATP transfers a phosphate group to a transport protein. The transfer sets in motion reversible changes in the protein's shape that result in a greater net movement of solute particles *against* the concentration gradient.

to the mechanisms of active transport in later chapters. For now, keep these points in mind:

Transport proteins span the lipid bilayer of cell membranes. They bind solutes on one side of the bilayer and reversibly change shape. This shunts the solute through their interior.

With passive transport, a solute simply diffuses through the protein; its net movement is down its concentration gradient.

With active transport, the net diffusion of a solute is uphill, against its concentration gradient. The transporting protein must be activated, as by ATP energy, to counter the energy inherent in the gradient.

BULK TRANSPORT ACROSS MEMBRANES

Transport proteins are fine for moving a few small ions and molecules at a time. But cells use vesicles to expel or take in big items or lots of them. By **exocytosis**, a vesicle fuses with the plasma membrane, and its contents are released outside (Figure 5.20*a*). By **endocytosis**, the cell secures items at its outer surface. An indentation forms at the plasma membrane, balloons inward, and pinches off. The resulting endocytic vesicle transports or stores its contents in the cytoplasm (Figure 5.20*b*).

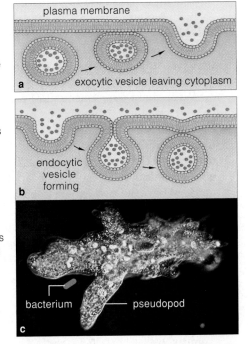

Figure 5.20 (**a**) Exocytosis. Cells release substances when an exocytic vesicle's membrane and the plasma membrane fuse. (**b**) Endocytosis. A patch of plasma membrane balloons inward beneath water and solutes outside. It pinches off as an endocytic vesicle that moves into the cytoplasm. (**c**) A phagocyte, *Amoeba proteus*. Its pseudopods (lobes of cytoplasm) are about to form an endocytic vesicle around a target bacterium.

plasma membrane

a exocytic vesicle leaving cytoplasm

b endocytic vesicle forming

bacterium   pseudopod

c

In all cells, endocytic vesicles form around extracellular fluid regardless of which substances are dissolved in it. Sometimes cell receptors chemically recognize and bind specific substances, such as hormones and lipoproteins. **Phagocytosis** (meaning "cell eating") is an active form of endocytosis. Some white blood cells engulf viruses, bacteria, and other pathogens. Amoebas and some other protistans engulf food this way. These phagocytes form pseudopods, which flow over a target and converge to make a vesicle (Figure 5.20*c*). The vesicle then fuses with lysosomes, the enzymes of which digest the contents.

Whereas transport proteins in a cell membrane deal only with ions and small molecules, exocytosis and endocytosis move larger packets of material across the plasma membrane.

By exocytosis, a cytoplasmic vesicle fuses with the plasma membrane, so that its contents are released outside the cell. By endocytosis, a small patch of the plasma membrane sinks inward and seals back on itself, forming a vesicle inside the cytoplasm. Membrane receptors often mediate this process.

MOVEMENT OF WATER ACROSS MEMBRANES

By far, more water diffuses across cell membranes than any other substance, so the key factors that influence its directional movement deserve special attention.

Osmosis

Turn on a faucet or watch a waterfall. The moving water demonstrates **bulk flow**—the mass movement of one or more substances in response to pressure, gravity, or some other external force. Bulk flow accounts for some movement of water through plants and animals. With each beat, your heart creates fluid pressure that drives a large volume of blood, which is mainly water, through interconnected blood vessels. Sap flows in conducting tissues that thread through maple trees, and this, too, is an example of bulk flow.

What about the movement of water into and out of cells or organelles? A membrane intervening between two regions permits the small, polar water molecules to cross but restricts the passage of ions and large polar molecules. **Osmosis** is the name for the diffusion of water molecules in response to a water concentration gradient between two regions separated by a selectively permeable membrane.

Concentrations of solutes in water on the two sides of a membrane influence osmosis, as you can see from Figure

HYPOTONIC CONDITIONS	HYPERTONIC CONDITIONS	ISOTONIC CONDITIONS

Water diffuses into red blood cells, which swell up

Water diffuses out of the cells, which shrink

No net movement of water, no change in cell size or shape

1 liter of distilled water — 1 liter of 10% sucrose solution — 1 liter of 2% sucrose solution

2% sucrose solution

a

b

Figure 5.21 Demonstration of how a solute concentration gradient influences osmotic movement. Start with a container divided by a membrane that water but not proteins can cross. Pour water into the left compartment. Pour the same volume of a protein-rich solution into the right compartment. There, proteins occupy some of the space. The net diffusion of water in this example is from left to right (large *gray* arrow).

selectively permeable membrane between two compartments

water molecule

protein molecule

5.21. *The side with more solute particles has a lower water concentration.* Why is this so? Imagine dissolving some glucose in water. Compared to an equivalent volume of water, the glucose solution has fewer water molecules, because each molecule of glucose occupies some of the space formerly occupied by molecules of water.

It's mainly the *total number* of molecules or ions, not the type of solute, that dictates water's concentration. Dissolve some amount of an amino acid or urea in one liter of water, and the concentration of water changes about as much as it did in the glucose solution. Add some sodium chloride (NaCl) to one liter of water, and it dissociates into equal numbers of sodium ions and chloride ions. There are now twice as many particles of solute as there were in the glucose solution—and so the water concentration has decreased proportionately.

Effects of Tonicity

Given that water molecules tend to move osmotically to a region where water is *less* concentrated, the direction of their movement tends to be toward a region where solutes are *more* concentrated. Figure 5.22 illustrates this tendency. Suppose you decide to make a simple observational test of this statement. You construct three sacs from a membrane that water but not sucrose can cross. You fill each sac with a solution that's 2 percent

Figure 5.22 Effect of tonicity on water movement. (**a**) Arrow widths show the direction and relative amounts of water movement. (**b**) The micrographs correspond to the sketches. They show shapes that human red blood cells assume when you put them in fluids of higher, lower, and equal solute concentrations. Normally, solutions inside and outside red blood cells are in balance. This type of cell does not have any built-in mechanisms that would help it adjust to drastic changes in solute levels in its fluid surroundings.

Figure 5.24 (**a**) Young plant cells. (**b**) This is what plasmolysis (an osmotically induced loss of internal fluid pressure) does to such cells. Their cytoplasm and central vacuole shrink and the plasma membrane moves away from the cell wall.

Figure 5.23 Increase in fluid volume owing to osmosis. In time, the net diffusion across a membrane separating two compartments is equal, but the fluid volume in compartment 2 is greater because the membrane is impermeable to solutes.

sucrose. Next you immerse one sac in a liter of distilled water (which has no solutes), one in a solution that's 10 percent sucrose. You immerse the third sac in a solution that's 2 percent sucrose. Each time, tonicity dictates the extent and direction of water movement (Figure 5.22).

Tonicity refers to the relative solute concentrations of two fluids. When two fluids on opposing sides of a membrane differ in solute concentration, the one having fewer solutes is the **hypotonic solution**. The one with more is the **hypertonic solution**. Water tends to diffuse from hypotonic to hypertonic fluids. **Isotonic solutions** have the same solute concentrations, so water shows no net osmotic movement from one to the other.

Normally, fluids on the inside and outside of your cells are isotonic. If a tissue fluid becomes drastically hypotonic, too much water would diffuse into the cells and they'd burst. If the fluid became too hypertonic, an outward diffusion of water would shrivel the cells. Most cells have built-in mechanisms that adjust to shifts in tonicity. Red blood cells don't. (Figure 5.22 shows what happened to them in an observational test of the effects of tonicity differences.) That is why severely dehydrated patients are given infusions of a solution isotonic with blood. Such solutions move by bulk flow from a bottle mounted on a rack above the patient, through a tube, and directly into an incised vein.

Effects of Fluid Pressure

Animal cells generally avoid bursting by engaging in ongoing selective transport of solutes across the plasma membrane. Cells of plants and many protistans, fungi, and bacteria also avoid that unpleasant prospect with the help of pressure exerted on their cell walls.

Pressure differences as well as solute concentrations influence the osmotic movement of water. Take a look at Figure 5.23. It shows how water continues to diffuse across a membrane between a hypotonic solution and a hypertonic solution until the water concentration is the same on both sides. As you can see, the *volume* of the

formerly hypertonic solution has increased (because its solutes cannot diffuse out). Any volume of fluid exerts **hydrostatic pressure**, or a force directed against a wall, membrane, or some other structure enclosing the fluid. The greater the fluid's solute concentration, the greater will be the hydrostatic pressure it exerts.

Living cells cannot increase in volume indefinitely (Section 4.2). At some point, hydrostatic pressure that builds up in the cell counters water's inward diffusion. That point is the osmotic pressure, the amount of force preventing further increase in a solution's volume.

Think of a young plant cell, with its pliable primary wall. As it grows, water diffuses into it and hydrostatic pressure increases against the wall. The wall expands, so the cell volume can increase. The thin walls are strong enough for the cell's internal fluid pressure to develop to the point where it counterbalances water uptake.

Plant cells are vulnerable to water losses, which can happen when the soil dries or becomes too salty. Water stops diffusing in and starts diffusing out, so internal fluid pressure falls. Such osmotically induced shrinkage of cytoplasm is called plasmolysis (Figure 5.24). Plants adjust somewhat to the loss of pressure, as when they actively take up potassium ions against a concentration gradient by mechanisms outlined in Section 5.7.

As you will read in Chapters 34 and 36, hydrostatic and osmotic pressure also influence the distribution of water in the blood, tissue fluid, and cells of animals.

Osmosis is the net diffusion of water between two solutions that differ in water concentration and that are separated by a selectively permeable membrane. The greater the number of molecules and ions dissolved in a solution, the lower its water concentration will be.

Water tends to move osmotically to regions of greater solute concentration (from hypotonic to hypertonic solutions). There is no net diffusion between isotonic solutions.

The fluid pressure that a solution exerts against a membrane or wall also influences the osmotic movement of water.

SUMMARY

1. Cells store, break down, and dispose of substances by acquiring and using energy and raw materials from outside sources. Metabolism, the sum of these energy-driven activities, underlies their survival. *CI*

2. Two laws of thermodynamics affect life. *First*, energy undergoes conversion from one form to another, but its total amount neither increases nor decreases; the total amount in the universe holds constant. *Second*, energy spontaneously flows in one direction, from usable forms to forms that are less and less usable. *5.1*

3. All matter has potential energy (as measured by the capacity to do work) owing to its position in space and the arrangement of its parts. Cells use potential energy to do mechanical, chemical, and electrical work. *5.1*

4. Without energy, a cell (like all organized systems) gets disorganized. It loses chemical potential energy in each metabolic reaction, mainly as heat. It stays alive (organized) by balancing energy outputs and inputs. The sun is life's primary energy source. Plants and other photosynthesizers convert sunlight energy to chemical bond energy of organic compounds. Plants, as well as organisms that feed on plants and each other, use energy in organic compounds to do cellular work. *5.1*

5. Cells conserve energy by coupling energy-releasing reactions with energy-requiring ones. ATP is the main coupling agent; it primes molecules to react by giving up a phosphate group to them. *5.2*

6. In aerobic respiration, coenzymes (NAD^+ and FAD) transfer electrons and hydrogen from glucose to transport systems, where electron transfers yield energy to drive ATP formation. In photosynthesis, electron transfers also drive ATP formation. The coenzyme $NADP^+$ transfers electrons and hydrogen to sites where they will combine with carbon and oxygen to form glucose. *5.2*

7. Much of metabolism is directional; it works with and against concentration gradients. Movement of the ions or molecules of a substance from a region of higher to lower concentration is called diffusion. Diffusion rates are influenced by concentration, electric, and pressure gradients, and by temperature and molecular size. *5.3*

8. Metabolic pathways are orderly sequences of enzyme-mediated reactions (Table 5.1). Photosynthesis and other biosynthetic pathways build energy-rich molecules from small molecules having less energy. Aerobic respiration and other degradative pathways break down energy-rich molecules to small ones of lower energy content. *5.3*

9. Enzymes are catalysts; they greatly enhance reaction rates for specific substrates by lowering the activation energy needed to start the reaction. They make bonds easier to break by binding substrates at their active site. Nearly all are proteins; some RNAs are catalytic. *5.4*

Table 5.1 *Participants in Metabolic Pathways*	
SUBSTRATE	Substance that enters a metabolic reaction or metabolic pathway; also called a reactant
INTERMEDIATE	Substance formed between reactants and end products of a pathway
END PRODUCT	Substance left at end of reaction or pathway
ENZYME	Usually a protein that enhances reaction rates; some RNAs also do this
COFACTOR	Coenzyme (such as NAD^+) or metal ion; assist enzymes or taxis electrons, hydrogen, or functional groups between reaction sites
ATP	Main energy carrier in cells; couples energy-releasing reactions with energy-requiring ones
TRANSPORT PROTEIN	Protein that passively assists substances across a cell membrane or actively pumps them across

10. Different controls over enzyme action influence the kinds and amounts of substances available at any time in the cell. Enzyme action is best within a limited range of temperature, pH, and salinity. *5.5*

11. Oxygen, carbon dioxide, and other small nonpolar molecules diffuse across a membrane's lipid bilayer. Ions and large, polar molecules such as glucose cross it with the passive or active help of transport proteins. Water moves through proteins and across the bilayer. *5.6*

12. Transport proteins move water-soluble substances across membranes by reversibly changing shape. Passive transport doesn't require energy inputs; solutes simply diffuse through a protein's interior. Active transport uses energy boosts, as from ATP. The protein pumps a solute against its concentration gradient (Figure 5.25). *5.7*

13. By exocytosis, the membrane of a vesicle fuses with the plasma membrane; the vesicle contents are released outside (Figure 5.25). By endocytosis, a patch of plasma membrane forms a vesicle by way of receptor recognition of specific solutes, by indiscriminate uptake of solutes dissolved in extracellular fluid, or by phagocytosis ("cell eating," an engulfment of cells and substances). *5.8*

14. Osmosis is a special case of diffusion by which water moves across a selectively permeable membrane. *5.9*

Review Questions

1. State the first and second laws of thermodynamics. Does life violate the second law? *5.1*

2. Define and give a few examples of potential energy. *5.1*

3. How does ATP prime a molecule to enter a reaction? *5.2*

4. Define and describe four key features of enzymes. *5.4*

5. If all transport proteins work by changing shape, then how do passive transporters differ from active transporters? *5.7*

6. Define diffusion in terms of concentration gradients, electric gradients, and pressure gradients. What is osmosis? *5.3, 5.9*

7. Distinguish between exocytosis and endocytosis. *5.8*

Concentration gradient

low

high

DIFFUSION ACROSS LIPID BILAYER

Lipid-soluble molecules and water molecules diffuse across

PASSIVE TRANSPORT (facilitated diffusion)

Water-soluble molecules and ions diffuse through interior of transport proteins. No energy boost required.

ACTIVE TRANSPORT

Specific solutes pumped across, through transport proteins against gradient; requires energy boost

energy input

EXOCYTOSIS

Vesicle moves to plasma membrane, fuses with it; contents released outside

ENDOCYTOSIS

Vesicle forms from patch of plasma membrane sinking into cytoplasm

Figure 5.25 Summary of membrane crossing mechanisms.

Self-Quiz   ANSWERS IN APPENDIX III

1. Electron transport systems involve _____ .
   a. enzymes, coenzymes
   b. electron transfers
   c. cell membranes
   d. all of the above

2. Immerse a living red blood cell in a hypotonic solution, and water will _____ .
   a. move into the cell
   b. move out of the cell
   c. show no net movement
   d. move in by endocytosis

3. _____ can readily diffuse across a lipid bilayer.
   a. Glucose   b. Oxygen   c. Carbon dioxide   d. b and c

4. Sodium ions cross a membrane through transport proteins that receive an energy boost. This is an example of _____ .
   a. passive transport
   b. active transport
   c. facilitated diffusion
   d. a and c

5. Match each substance with the most suitable description.
   ____ coenzyme or metal ion
   ____ selective membrane crossing
   ____ substance entering a reaction
   ____ formed during a reaction
   ____ substance at end of reaction
   ____ enhances reaction rate
   ____ mainly ATP

   a. reactant or substrate
   b. enzyme
   c. enzyme helper
   d. intermediate
   e. end product
   f. energy carrier
   g. transport protein

Critical Thinking

1. *Cyanide poisoning* results after cyanide, a toxic compound, binds to an enzyme of electron transport systems. Binding stops the enzyme from donating electrons in the system. What effect will this have on ATP production? From what you know of ATP function, what effect will this have on a person's health?

2. *Pyrococcus furiosus*, a bacterium that grows in volcanic vents, thrives at 100°C, the boiling point of water. Enzymes isolated from *P. furiosus* cells don't function well below 100°C. Which aspect of enzyme structure keeps them stable and active at such extreme temperatures? (*Hint:* Review Section 3.5's summary of interactions that maintain protein structure.)

empty contractile vacuole

filled contractile vacuole

Figure 5.26 Contractile vacuoles of *Paramecium.*

3. Water moves osmotically into *Paramecium*, a protistan of aquatic habitats. If unchecked, the influx will bloat the cell and then rupture its plasma membrane. A mechanism that requires energy involves contractile vacuoles to expel the excess (Figure 5.26). Water enters tubelike extensions of this organelle and collects in a central space in the vacuole. A filled vacuole contracts and squirts excess water out of a pore that opens to the outside. Are the fluid surroundings hypotonic, hypertonic, or isotonic relative to this cell's cytoplasm?

4. *Vibrio cholerae*, a bacterium, causes *cholera*, which causes severe diarrhea and losses of up to twenty liters of fluid a day. Drink contaminated water, and *V. cholerae* cells adhere to the intestinal lining and secrete a toxic metabolic product. In turn, cells of the lining secrete chloride ions (Cl^-). Sodium ions (Na^+) follow Cl^- into the intestinal fluid. Correlate the massive fluid loss with these ion movements.

5. Many cultivated fields in California are heavily irrigated. Over the years, most of the water has evaporated from the soil, leaving behind all of the solutes in the irrigation water. What kinds of problems will the altered soil conditions cause plants?

6. Imagine you're a juvenile shrimp living in an *estuary*, where fresh water draining from the land mixes with saltwater from the sea. Many people own homes around a lake and want boat access to the sea. They ask their city for permission to build a canal to your estuary. If they succeed, what may happen to you?

Selected Key Terms

active site 5.4
active transport 5.7
ATP 5.2
bulk flow 5.9
chemical energy 5.1
concentration gradient 5.3
diffusion 5.3
electron transport system 5.2
endergonic reaction 5.2
endocytosis 5.8
entropy 5.1
enzyme 5.4
exergonic reaction 5.2
exocytosis 5.8
feedback inhibition 5.5
first law of thermodynamics 5.1
free radical CI
heat 5.1

hydrostatic pressure 5.9
hypertonic solution 5.9
hypotonic solution 5.9
induced-fit model 5.4
isotonic solution 5.9
kinetic energy 5.1
metabolic pathway 5.3
metabolism CI
osmosis 5.9
oxidation–reduction (redox) reaction 5.2
passive transport 5.7
phagocytosis 5.8
phosphorylation 5.2
potential energy 5.1
second law of thermodynamics 5.1
selective permeability 5.6

Readings

Adams, S. 22 October 1994. "No Way Back." *New Scientist.* A refreshing look at the second law of thermodynamics.

Ritter, P. 1996. *Biochemistry: A Foundation.* Brooks/Cole: Pacific Grove, California.

On-Line readings at Student Guide for InfoTrac: www.brookscole.com/biology

HOW CELLS ACQUIRE ENERGY

Sunlight and Survival

Think about the last time you were hungry and craved a bit of apple, lettuce, chicken, pizza, or any other food. Where did it come from? For the answers, look past the refrigerator, the market or restaurant, or even the farm. Look to individual plants, the starting point for nearly all edibles you put into your mouth. Plants use environmental sources of energy and raw materials to make glucose and other organic compounds. Organic compounds, recall, are built on a framework of carbon atoms. So the questions become these:

1. *Where does the carbon come from in the first place?*

2. *Where does the energy come from to drive the synthesis of carbon-based compounds?*

Answers vary, because they depend on an organism's mode of nutrition.

Plants generally are "self-nourishing" organisms, or **autotrophs**. Their carbon source is carbon dioxide (CO_2), a gaseous compound in the air and dissolved in water. Plants, some bacteria, and many protistans are *photo*autotrophs, which means they use sunlight for **photosynthesis**. By this process, energy from the sun drives the formation of ATP and a coenzyme, NADPH. ATP delivers energy to sites where glucose and other carbohydrates are built. NADPH delivers electrons and hydrogen building blocks to the same sites.

Many other organisms are **heterotrophs**, meaning they must feed on autotrophs, one another, and organic wastes. (*Hetero*– means other, as in "being nourished by other organisms.") That is how most bacteria, many protistans, and all fungi and animals stay alive. Unlike plants, heterotrophs can't feed themselves with energy and raw materials from the physical environment.

How do we know such things? We didn't have a clue until the mid-seventeenth century. Before then, most people assumed that plants got raw materials to make food from soil. By 1882 a few chemists had an inkling that plants use sunlight, water, and something in the air to make food. A botanist, T. Englemann, wondered: What parts of sunlight do plants favor? As he knew, plants and algae release oxygen during photosynthesis. He also knew that, like many other organisms, some free-living bacteria use oxygen for aerobic respiration.

Englemann hypothesized: If bacterial cells need oxygen, then they will gather in places where the most photosynthesis is going on. He put bacterial cells in a water droplet, then put them on a microscope slide with a green alga (*Spirogyra*). He used a crystal prism to break up a beam of sunlight and cast a spectrum of colors across the slide. Bacteria congregated mostly where violet and red light fell on the green alga. The algal cells released more oxygen in the part illuminated by light of those colors, which happens to be the best light for photosynthesis (Figure 6.1).

Such observations showed us how photosynthesis works. They also helped reveal one of nature's great patterns, *for photosynthesis is the main pathway by which carbon and energy enter the web of life.* Once a cell makes or takes up organic compounds, it uses or stores them. *All* autotrophs and heterotrophs store energy in organic compounds and release it by other processes. Aerobic respiration, the main energy-releasing process, requires oxygen. Figure 6.2 previews the chemical links between photosynthesis and aerobic respiration—the focus of this chapter and the next.

Such processes might seem far removed from your daily interests. But remember, the food that nourishes you and most other organisms can't be produced or used without them. We will return to this point in later chapters. It provides perspective on many issues, such as nutrition, dieting, agriculture, human population growth, genetic engineering, and the impact of pollution on our sources of food—hence on our survival.

A crystal prism breaks up a beam of light into a spectrum of colors, which are cast across a droplet of water on a microscope slide.

bacteria (*white*)

part of an algal strand stretched out across a microscope slide

400   450   500   550   600   650   700

Colors associated with wavelengths of light (nanometers)

Figure 6.1 Results from T. Englemann's observational test that correlated portions of visible light with photosynthesis in *Spirogyra*, a strandlike green alga. Many oxygen-requiring bacterial cells moved to the colors where algal cells released the most oxygen, which is a by-product of their photosynthetic activity.

Plants capture sunlight energy, which drives photosynthesis.

PHOTOSYNTHESIS

1. H_2O is split by light energy. Its oxygen escapes; its electrons, hydrogen are used at transport systems that release energy for ATP formation. Coenzymes pick up the electrons, hydrogen.

2. ATP energy drives synthesis of glucose from hydrogen and electrons (delivered by the coenzymes), plus carbon and oxygen (from carbon dioxide).

water, carbon dioxide

oxygen

AEROBIC RESPIRATION

1. Glucose is degraded to carbon dioxide, water; coenzymes pick up its electrons, hydrogen.

2. Coenzymes give up electrons, hydrogen to oxygen-requiring transport systems that release energy to drive ATP formation.

ATP available to drive many cellular tasks

Figure 6.2 Links between photosynthesis—the main energy-requiring process in the world of life—and aerobic respiration, the main energy-releasing process.

Key Concepts

1. Plants, some bacteria, and many protistans use sunlight energy, carbon dioxide, and water to make glucose and other carbohydrates, which have a backbone of carbon atoms. The metabolic process by which they do this is called photosynthesis.

2. Photosynthesis is the main process by which carbon and energy become available to the web of life.

3. In the first stage of photosynthesis, sunlight energy is trapped and converted to chemical bond energy in ATP molecules. Typically, water molecules are split, their electrons and hydrogen atoms are picked up by the coenzyme $NADP^+$ to form NADPH, and their oxygen atoms are released as a by-product.

4. In the second stage of photosynthesis, ATP delivers energy for biosynthesis. NADPH delivers electrons and hydrogen, which combine with carbon and oxygen (both from carbon dioxide) to form glucose.

5. In the photosynthetic cells of plants and in many protistans, both stages of the reactions proceed inside organelles called chloroplasts.

6. Often we summarize photosynthesis this way:

$$12H_2O + 6CO_2 \xrightarrow{\text{LIGHT ENERGY}} 6O_2 + C_6H_{12}O_6 + 6H_2O$$

WATER   CARBON DIOXIDE   OXYGEN   GLUCOSE   WATER

PHOTOSYNTHESIS—AN OVERVIEW

Sit outdoors on a warm, cloudless day and you'll get hot but you'll never build your own food. How do plants do it? Think of it! They have to plug into the sun's energy, but not so much that it might cook them. They have to do magic and turn the sun's energy into ATP's chemical bond energy. They have to get carbon, oxygen, and hydrogen from somewhere and combine them into carbohydrates. As you might imagine, these multiple tasks can't all be done in the same place. Different functions, different structures.

leaf's upper surface   photosynthetic cells

a Sow thistle (*Sonchus*) leaf

b Cutaway of a small section from the leaf. Its upper and lower surfaces enclose many photosynthetic cells.

c One of the photosynthetic cells, with green chloroplasts.

Where the Reactions Take Place

Plants and protistans called algae carry out the tasks in **chloroplasts**: organelles specialized for photosynthesis (Section 4.7). Figure 6.3 shows the structure of one type. All chloroplasts have two outermost membranes that surround a mostly fluid interior, the **stroma**. Another membrane, in the stroma, is highly folded into flattened sacs. The sacs, **thylakoids**, interconnect by channels. It makes no difference how many sacs a chloroplast has or whether they are stacked as in Figure 6.3*d,e*. The inner membrane always forms one functional compartment.

The capture and conversion of sunlight energy to chemical energy takes place at thylakoids. With these *light-dependent* reactions, plants get the electrons and hydrogen for making food. Amazingly, sunlight splits water molecules. Water's oxygen diffuses away. But its electrons flow through transport systems embedded in the thylakoid membrane. (Remember Section 5.2?) As electrons flow through the systems, hydrogen moves inside the thylakoid compartment. ATP forms when it flows out. Waste not, want not: the helpful coenzyme NADP⁺ picks up the spent electrons and hydrogen.

Making glucose doesn't require light. These *light-independent* reactions proceed elsewhere, in the stroma. This is where ATP gives up energy, the coenzyme gives up electrons and hydrogen, and carbon dioxide (CO_2) is dismantled for its carbon and oxygen atoms.

Overall, you can think about all the reactants and products just described in terms of a simple equation:

LIGHT ENERGY

$$12H_2O + 6CO_2 \longrightarrow 6O_2 + C_6H_{12}O_6 + 6H_2O$$

two outer layers of membrane

inner membrane system (thylakoids, all interconnecting by channels)

d Cutaway view of one chloroplast

Figure 6.3 Zooming in on sites of photosynthesis inside a leaf from a typical plant.

In case you're wondering how we know what we know: Remember the description of tracers in Section 2.2? By attaching radioisotopes to atoms of the reactants shown in the summary equation, researchers found out where each atom of the reactants ends up in the products:

REACTANTS: $12H_2O$ $6CO_2$

PRODUCTS: $6O_2$ $C_6H_{12}O_6$ $6H_2O$

But Things Don't <u>Really</u> End With Glucose

Notice how we show *glucose* as the end product of the reactions. We do so to keep the chemical bookkeeping simple. But you'll find little glucose in a chloroplast.

a

Figure 6.11 (**a**) Filling in details for Figure 6.3—ATP and NADPH formation in the noncyclic pathway of photosynthesis. The *yellow* arrows show electron flow. Electrons released from water molecules split by photolysis move through two photosystems (*light green*) and two transport systems (*dark green*). The joint operation of the two photosystems boosts electrons to an energy level high enough to drive NADPH formation. Also, as you will read shortly, hydrogen ions move across the thylakoid membrane in ways that lead to ATP formation at proteins called ATP synthases.

(**b**) Diagram the energy changes associated with the noncyclic pathway, and you will end up with this "Z scheme."

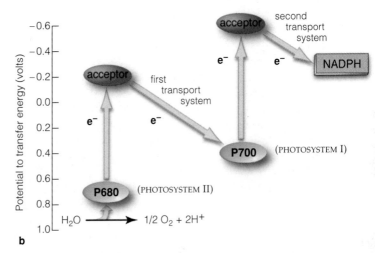

b

simply note that when electrons leave the first transport system, they still have a bit of extra energy. They get even more upon their arrival at photosystem I, which sunlight also happens to be bombarding. They are able to enter the second transport system only because the energy boost puts them at a higher energy level.

Today, both the cyclic and noncyclic pathways of electron flow operate in all photosynthetic organisms. Which one dominates at a given time depends on the organism's metabolic demands for ATP and NADPH.

The Legacy—A New Atmosphere

On sunny days, you can see bubbles of oxygen escaping from aquatic plants (Figure 6.12). When the noncyclic pathway first evolved, the oxygen it released dissolved in seawater, wet mud, and similar prokaryotic habitats. By about 1.5 billion years ago, enormous quantities of

Figure 6.12 Visible evidence of photosynthesis—bubbles of oxygen escaping from the leaves of *Elodea*, a plant that lives in freshwater habitats.

dissolved oxygen were escaping to what had been an oxygen-free atmosphere. The atmosphere now changed forever. And so did the world of life. The vast, global increase in free oxygen favored the evolution of aerobic respiration. This became the most efficient pathway for releasing usable energy from organic compounds. And so the emergence of the noncyclic pathway allowed you and every other animal to be around today, breathing the oxygen that helps keep your cells alive.

In the light-dependent reactions, sunlight energy drives the release of electrons from photosystems in thylakoid membranes. The electron flow through nearby transport systems results in the formation of ATP, NADPH, or both.

ATP alone forms in a cyclic pathway that uses one type of photosystem. Both ATP and NADPH form by a noncyclic pathway in which electrons flow from water, through two types of photosystems, and finally to NADP$^+$.

All photosynthetic species employ one or both pathways in response to the cell's changing needs for ATP and NADPH.

Oxygen, a by-product of the noncyclic pathway, changed the early atmosphere and made aerobic respiration possible.

CASE STUDY: A CONTROLLED RELEASE OF ENERGY

Spill some sugar into a campfire and you'll release its energy all at once, but not in a form that does any good. Events in the chloroplast control the release of energy from incoming sunlight—and they direct it into doing useful cellular work.

Sunlight prods enzymes to repeatedly split water molecules into oxygen, hydrogen, and electrons. The oxygen combines to form O_2. In this form it diffuses out

from the chloroplast and out of the cell. The free hydrogen (H^+) does not leave. It accumulates in the thylakoid, as shown in Figure 6.13a.

electrons.) Certain components of the transport system shunt the ions across the membrane, as in Figure 6.13b.

Therefore, *by a combination of photolysis and electron transport, H^+ has now become more concentrated in the thylakoid compartment than in the stroma.* In addition, the unequal distribution of these positively charged ions means there is a difference in electric charge across the membrane. An H^+ concentration gradient and an electric gradient have become established.

The combined force of the two gradients propels H^+ into the stroma, through transport proteins called ATP synthases. This type of protein spans the thylakoid membrane (Figure 6.13c). ATP synthases have built-in machinery that responds to the ion flow. They catalyze

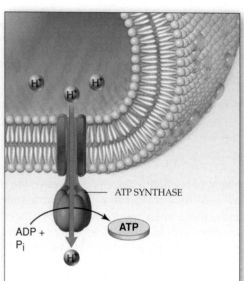

a Hydrogen ions released by photolysis (splitting of water molecules) accumulate inside the thylakoid compartment. (O_2 forms when two oxygen atoms combine. It diffuses out of the chloroplast, then out of the photosynthetic cell.)

b More hydrogen ions accumulate in the compartment as components of electron transport systems accept excited electrons (from photolysis). Components of the system also pick up hydrogen ions in the stroma and shunt them across the membrane.

c Ions follow the concentration and electric gradients across the thylakoid membrane. They flow to the stroma, through the interior of ATP synthases that span the membrane. The flow drives the formation of ATP from the cell's pool of ADP and P_i.

Figure 6.13 From photosynthesis, an example of how energy is released and directed to cellular tasks. Sunlight energy sets in motion events that build concentration and electric gradients, which are tapped to make ATP. It also leads to the formation of NADPH.

At the same time, an acceptor molecule picks up the electrons and transfers them to an electron transport system that is positioned next to it in the thylakoid membrane. The membrane has many, many transport systems. While all the systems are accepting electrons, they're also picking up hydrogen ions from the stroma. (These ions are attracted to the negative charge of the

the attachment of unbound phosphate to a molecule of ADP in the stroma, thus forming an ATP molecule.

The sequence of events just described is called the chemiosmotic model for ATP formation in chloroplasts. As you will read in the next chapter, the same model applies to ATP formation in mitochondria.

In chloroplasts, ongoing photolysis and transfers of electrons in transport systems set up H^+ concentration and electric gradients across the thylakoid membrane. Ion flow from the thylakoid compartment to the stroma drives ATP formation.

THE LIGHT-INDEPENDENT REACTIONS

The **light-independent reactions** are the "synthesis" part of photosynthesis. They occur as a cyclic pathway called the **Calvin–Benson cycle**. The reactions require energy from ATP, hydrogen and electrons from NADPH, and carbon and oxygen from carbon dioxide (CO_2), which is present in the air (or water) that surrounds photosynthetic cells. We say these reactions are light-independent because they do not depend directly on sunlight. They proceed just as well in the dark, as long as ATP and NADPH are available.

How Do Plants Capture Carbon?

Let's track a CO_2 molecule that diffuses into air spaces inside a leaf and ends up next to a photosynthetic cell. It diffuses into the cell, then into a chloroplast's stroma. An enzyme attaches the carbon atom of CO_2 to **RuBP** (ribulose bisphosphate), a compound with a backbone of five carbon atoms. **Rubisco** (for RuBP carboxylase) is the enzyme's name. Its action produces an unstable six-carbon intermediate that splits into two molecules of **PGA** (phosphoglycerate). PGA is a stable molecule with a three-carbon backbone. Incorporating a carbon atom from CO_2 into a stable organic compound is called **carbon fixation**. Quite simply, food can't be produced without this first step of the Calvin–Benson cycle.

The cycle yields phosphorylated glucose, and it also regenerates the RuBP. For our purposes, we can focus just on the carbon atoms of the substrates, intermediates, and end products, as in Figure 6.14.

How Do Plants Build Glucose?

Each PGA accepts a phosphate group from ATP, plus hydrogen and electrons from NADPH. The resulting intermediate is called **PGAL** (phosphoglyceraldehyde). To build *one* six-carbon sugar phosphate, carbon atoms from six CO_2 must be fixed, and twelve PGAL must form. Most of the PGAL becomes rearranged to form new RuBP, which can be used to fix more carbon. But two of the PGAL combine, thus forming glucose with a phosphate group attached to its six-carbon backbone.

When phosphorylated this way, glucose is primed to enter other reactions. Plants use it as a building block for their main carbohydrates—sucrose, cellulose, and starch. *The synthesis of these organic compounds by other pathways marks the end of the light-independent reactions.*

With six turns of the cycle, enough RuBP molecules form to replace the ones used in carbon fixation. The ADP, NADP$^+$, and phosphate remaining diffuse through the stroma, back to the light-dependent reaction sites. There they may be converted again to NADPH and ATP.

Photosynthetic cells convert phosphorylated glucose to sucrose or starch during daylight hours. Sucrose is

Figure 6.14 Light-independent reactions of photosynthesis. *Red* circles signify carbon atoms of key molecules. All intermediates have one or two phosphate groups attached; for clarity, we show only the phosphate group on glucose. Many water molecules that formed in the light-dependent reactions enter this pathway; six remain when it's over. Appendix V (Figure C) shows details.

the most readily transportable carbohydrate in plants. Starch is the most common storage form. The cells also convert excess PGAL to starch. They briefly store starch (as starch grains) inside the stroma. After the sun goes down, they convert starch to sucrose for export to other living cells in leaves, stems, and roots. Ultimately, the products and intermediates of photosynthesis end up as energy sources and as building blocks for all of the lipids, amino acids, and other organic compounds that plants require for growth, survival, and reproduction.

In the light-independent reactions (the Calvin–Benson cycle), carbon is "captured" from carbon dioxide, glucose forms during reactions that require ATP and NADPH, and RuBP (necessary to capture the carbon) is regenerated.

FIXING CARBON—SO NEAR, YET SO FAR

If sunlight intensity, air temperature, rainfall, and soil composition never varied, photosynthesis might be the same in all plants. But environments differ—so details of photosynthesis differ. Comparing a few carbon-fixing adaptations to stressful environments makes this point.

C4 Plants

All plants must take up CO_2 for growth. But CO_2 isn't always plentiful inside leaves, which have a waxy cover that helps plants conserve water. The CO_2 diffuses in and O_2 diffuses out mainly at **stomata** (singular, stoma). These microscopic openings across the leaf surface close on hot, dry days. Closure helps a plant conserve water,

but then CO_2 can't get into leaves. If photosynthetic cells are busy, oxygen accumulates. A high O_2 concentration in a leaf triggers *photorespiration*. This process wastes fixed CO_2 and lowers the plant's sugar-making capacity. Remember rubisco, the enzyme that attaches carbon to RuBP in the Calvin–Benson cycle? Well, rubisco also can attach *oxygen* to it. When CO_2 gets used up and O_2 builds up in the leaf because it can't diffuse out, only one (not two) PGA forms. So does one glycolate molecule, which is later degraded.

The many **C3 plants** include sunflowers and beans. "C3" refers to *three*-carbon PGA, the first intermediate of its carbon-fixing pathway (Figure 6.15*a*). The first one to form when corn and many other plants fix carbon is *four*-carbon oxaloacetate. Hence *their* name, **C4 plants**.

When C4 plants close stomata on hot, dry days, they still maintain enough CO_2 in leaves by *fixing carbon twice*, in two types of photosynthetic cells. Mesophyll cells, the first type, use CO_2 to form oxaloacetate—which is transferred to bundle-sheath cells at leaf veins (Figure

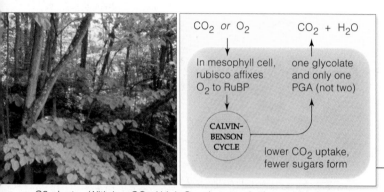

a C3 plants. With low CO_2 / high O_2, photorespiration predominates.

b C4 plants. With low CO_2 / high O_2, Calvin–Benson cycle predominates.

c CAM plants. With low CO_2 / high O_2, Calvin–Benson cycle predominates.

Leaf of basswood (*Tilia*), a typical C3 plant, cross-section

Leaf of corn (*Zea mays*), a typical C4 plant, cross-section

Figure 6.15 Three ways of fixing carbon in hot, dry weather, when there is too little CO_2 and too much O_2 in leaves. (**a**) The C3 pathway is common in evergreens and many nonwoody plants of temperate zones, such as basswood and Kentucky bluegrass. (**b**) The C4 pathway is common in grasses and other plants that evolved in the tropics and that fix CO_2 twice. Corn, sugarcane, and crabgrass are examples. (**c**) CAM plants open stomata and fix carbon at night. They include orchids, pineapples, and cacti and many other succulents.

6.15*b*). There, CO_2 is released and its local concentration rises, so rubisco doesn't use oxygen. In mesophyll cells, CO_2 gets fixed again in the Calvin–Benson cycle. With this pathway, C4 plants can get by with small stomata, lose less water, and make more glucose than C3 plants can when conditions are hot, bright, and dry.

Experiments show that photorespiration lowers the photosynthetic efficiency of many C3 plants. Example: Grow hothouse tomatoes at CO_2 levels high enough to block photorespiration and you see a dramatic rise in growth rates. So why hasn't natural selection eliminated this wasteful process? Look to rubisco, the enzyme that starts the process. This enzyme evolved long ago, when the atmosphere had little O_2 and abundant CO_2. Maybe some gene coding for its structure can't mutate without having bad effects on carbon-fixing activity. Or maybe the pathway that degrades glycolate to carbon dioxide has proved to be so adaptive that it can't be eliminated. After all, glycolate is toxic at high concentrations.

The C4 pathway evolved separately in many lineages of flowering plants over the past 50 to 60 million years. Before then, when atmospheric CO_2 levels were higher, C3 plants had the selective advantage. Which pathway will be most adaptive in the future? Atmospheric CO_2 levels have been increasing for decades and may double in the next fifty years. If so, photorespiration will again decline—and many vital crop plants may benefit.

CAM Plants

We see a carbon-fixing adaptation to desert conditions in cacti. A cactus is one of the succulents; it has juicy, water-storing tissues and thick surface layers that restrict water loss. It cannot open stomata on hot days without losing precious water. It opens them and fixes CO_2 *at night*. Its cells store the resulting intermediate in their central vacuoles, then use it for photosynthesis the next day when stomata are closed. Many plants are adapted this way. They are **CAM plants** (short for *Crassulacean Acid Metabolism*). Unlike C4 species, CAM plants do not fix carbon twice in different types of cells. They fix it in the same cells but at different times (Figure 6.15*c*).

Many plants die during prolonged droughts, but some CAM plants survive by closing stomata even at night. They keep fixing the CO_2 from aerobic respiration. Not much forms, but it's enough to maintain low metabolic rates. CAM plants do grow slowly. Try growing a cactus plant in Seattle or other places with a mild climate, and it will compete poorly with C3 and C4 plants.

C4 plants and CAM plants both have modified ways of fixing carbon for photosynthesis. The modifications counter stresses imposed by hot, dry conditions in their environments.

Light in the Deep Dark Sea

During the late 1980s a graduate student, Cindy Lee Van Dover, was puzzling over an eyeless shrimp. Divers had discovered it at a hydrothermal vent deep in the Atlantic Ocean. **Hydrothermal vents** are fissures in the seafloor where molten rock rises and mixes with cold seawater. Mineral-rich water, superheated by 650 degrees, spews into the perpetually dark surroundings. The sides of the vents teem with life, including shrimps and bacteria.

When Van Dover saw videotapes of shrimps clinging to the vent, she noticed a pair of bright strips on their back. By studying specimens in the lab, she found that the strips are connected to a nerve. Could eyeless shrimps have a novel sensory organ?

EYELESS SHRIMP

Steven Chamberlain, a neuroscientist, confirmed the strips are a sensory organ with light-absorbing pigmented cells (photoreceptors). Later, Ete Szuts, a pigment expert, isolated the shrimp pigment. Its absorption spectrum is the same as that for rhodopsin, a visual pigment in eyes as complex as yours.

Van Dover had a hunch: If the strips serve as an eye, then we should find *light* on the seafloor. On later dives, special cameras confirmed her hypothesis. Like coils in a toaster oven, vents release heat (infrared radiation). They also release faint radiation at the low end of the visible spectrum—light up to nineteen times brighter than expected for infrared. A billion billion photons per square inch per second strike typical sunlit trees. Studies of photoautotrophic bacteria living 72 meters (240 feet) below the Black Sea's surface show that they encounter a trillion photons per square inch per second. *Just as many photons are available to hydrothermal vent organisms*.

About this time Van Dover casually asked a colleague, "Hey, what if there's enough light for photosynthesis?" The response was, "What a stupid idea."

Euan Nisbet, who studies ancient environments, thought about this. The first cells arose around 3.8 billion years ago. As Nisbet and Van Dover hypothesized, What if they arose at hydrothermal vents? They could have used inorganic compounds such as hydrogen sulfide (for their hydrogen and electrons) and carbon dioxide (for carbon). Survival probably depended on being able to move away from light at vents and thereby avoid being boiled alive. Millions of years later, some bacterial descendants were evolving in hot springs near the ocean's surface. In time they used their light-detecting machinery to absorb light from the sun. They were adapted to a new energy source; they had become photosynthetic.

Some observations support the intriguing hypothesis that light-sensing machinery of deep-sea bacteria became modified for shallow-water photosynthesis. A few clues: Absorption spectra for the chlorophyll in evolutionarily ancient photosynthetic bacteria correspond to the light measured at hydrothermal vents. Also, photosynthetic machinery contains iron, sulfur, manganese, and other minerals—which are abundant at hydrothermal vents.

Autotrophs, Humans, and the Biosphere

We conclude this chapter with a story that reinforces how photosynthesizers and other autotrophs fit in the world of living things. It is a story of mind-boggling numbers of single-celled and multicelled species on land and in the sunlit waters of the Earth.

Each spring, you sense renewed growth of autotrophs on land as leaves unfurl and lawns and fields turn green. You might not be aware that uncountable numbers of single-celled autotrophs also drift through the surface waters of the world ocean. You can't see them without a microscope; a row of 7 million cells of one aquatic species would be less than a quarter-inch long. In some regions, a cupful of seawater might hold 24 million cells of one species. And that wouldn't include cells of all the other aquatic species.

Most of the drifters are photoautotrophic prokaryotic cells and protistans. Together they are the "pastures of the seas," producers that ultimately feed most other marine organisms. The pastures "bloom" in the spring, when nutrient inputs sustain rapid reproduction. At that time seawater becomes warmer and enriched with nutrients that winter currents churn up from the deep.

Until NASA gathered data from satellites in space, we had no idea of their numbers and distribution. Figure 6.16*a* shows visual evidence of their activities one winter in the surface waters of the North Atlantic Ocean. Figure 6.16*b* shows a springtime bloom stretching from North Carolina all the way past Spain!

Collectively, the cells help shape the global climate, for they deal with staggering numbers of reactant and product molecules. For instance, they sponge up nearly half the carbon dioxide that we humans release each year, as when we burn fossil fuels or burn vast tracts of forests to clear land for farming. Without the aquatic photoautotrophs, atmospheric carbon dioxide would accumulate more rapidly and possibly contribute to global warming, as described in Section 43.10. If the atmosphere warms by only a few degrees, then sea levels will rise and all lowlands near the coasts of continents and islands will become submerged.

Although such global change is a real possibility, tons of industrial wastes, raw sewage, and fertilizers in runoff from croplands still drain into the ocean each day. The pollutants seriously alter the chemical composition of seawater. How long will the marine photoautotrophs be able to function in this chemical brew? The answer will have impact on your own life in more ways than one.

Other autotrophs also influence your life in ways you might not expect. At hydrothermal vents, in hot springs, in waste heaps of coal mines are diverse prokaryotic cells classified as *chemo*autotrophs. They, too, use carbon dioxide as a carbon source. But they use inorganic compounds in their environment as energy sources.

For example, some archaebacteria living on the sides of hydrothermal vents strip hydrogen and electrons from hydrogen sulfide. Other kinds of chemoautotrophs living in soil obtain energy from nitrogen-containing wastes and remains of animals and other organisms.

Although they, too, are microscopically small, the chemoautotrophs also exist in monumental numbers. They influence the cycling of nitrogen, phosphorus, and other elements through the biosphere. We will consider their impact on the environment in later chapters. In this unit, we turn next to pathways by which cells can release the chemical bond energy of glucose and other biological molecules—the chemical legacy of autotrophs everywhere.

a Photosynthetic activity in winter.

b Photosynthetic activity in spring.

Figure 6.16   Two satellite images that help convey the magnitude of photosynthetic activity during springtime in the North Atlantic portion of the world ocean. In these color-enhanced images, *red-orange* shows where chlorophyll is most concentrated.

SUMMARY   *Gold* indicates text section

1. Cell structure and functions are based on organic compounds, the synthesis of which depends on sources of carbon and energy. Plants and other autotrophs get carbon from carbon dioxide and energy from the sun or inorganic compounds. Animals and other heterotrophs aren't self-nourishing; they get carbon and energy from organic compounds synthesized by autotrophs. *CI*

2. Photosynthesis is the main process by which carbon and energy enter the web of life. Its two stages are the light-dependent and light-independent reactions. This equation and Figure 6.17 summarize the process: *6.1*

$$12H_2O + 6CO_2 \xrightarrow{\text{LIGHT ENERGY}} 6O_2 + C_6H_{12}O_6 + 6H_2O$$

WATER   CARBON   OXYGEN   GLUCOSE   WATER
        DIOXIDE

3. Chloroplasts are organelles of photosynthesis in algae and plants. Two outer membranes enclose its semifluid interior (stroma). A third membrane in the stroma forms a system of connected sacs (thylakoids) and channels. The light-dependent reactions occur at thylakoids. The light-independent reactions occur in the stroma. *6.1*

4. The light-dependent reactions start at photosystems that each include 200 to 300 pigments. *6.2–6.4*

a. Chlorophyll *a*, the main photosynthetic pigment, absorbs all wavelengths of visible light but only a little of the yellow-green and green ones. Accessory pigments (e.g., carotenoids) absorb other wavelengths.

b. A cyclic pathway forms ATP only. Photosystem I gives up excited electrons to a transport system, which sends the "spent" ones back to the photosystem.

c. The noncyclic pathway forms ATP and NADPH. Water molecules are split into hydrogen, electrons, and oxygen. Excited electrons enter photosystem II, then a transport system, photosystem I, and another transport system. $NADP^+$ picks up the electrons and hydrogen to form NADPH. By 1.5 billion years ago, oxygen released by this pathway had accumulated in the atmosphere. It ultimately made possible aerobic respiration.

5. ATP energy drives the light-independent reactions (by phosphate-group transfers). Hydrogen and electrons from NADPH, plus carbon and oxygen (from carbon dioxide), are used as building blocks for glucose. Most glucose is used at once to synthesize starch, cellulose, and other end products of photosynthesis. *6.5*

a. Cyclic reactions (Calvin–Benson cycle) start when rubisco, an enzyme, affixes carbon from CO_2 to the five-carbon RuBP. The resulting intermediate splits into two PGA that ATP phosphorylates. Hydrogen and electrons from NADPH help form two PGAL.

b. For every six carbon atoms that enter the cycle, twelve PGAL form. Two are used to form a six-carbon sugar phosphate. The rest are used to regenerate RuBP.

light-independent reactions in stroma of chloroplast

CO_2 into leaf O_2 out

light-dependent reactions at thylakoids of chloroplast

Figure 6.17 Summary of photosynthesis. In the light-dependent reactions, sunlight energy is converted to chemical bond energy of ATP. In a noncyclic pathway, water molecules are split by photolysis. $NADP^+$ conserves the released electrons and hydrogen (as NADPH). Oxygen is released as a by-product.

ATP energy drives the light-independent reactions. Hydrogen and electrons from NADPH, plus carbon and oxygen from carbon dioxide, are used to form glucose. In the Calvin–Benson cycle, the RuBP on which the cycle turns is regenerated, and new water molecules form. *Each turn* of the cycle requires one CO_2, three ATP, and two NADPH. Because each glucose molecule has a backbone of six carbon atoms, its formation requires six turns of the cycle.

6. Rubisco, a carbon-fixing enzyme, evolved when air had far more CO_2 and less O_2. Today, when CO_2 is used up and O_2 builds up in leaves, rubisco attaches oxygen (not carbon) to RuBP. This process, photorespiration, is wasteful: one PGA and glycolate form. Glycolate can't be used to form sugars and is later degraded. *6.6*

7. Photorespiration predominates in C3 plants such as sunflowers under hot, dry conditions. Stomata close, so O_2 from photosynthesis builds up in leaves to levels higher than CO_2 levels. Corn and other C4 plants raise the CO_2 level by fixing carbon twice, in two cell types. CAM plants such as cacti raise the CO_2 level by fixing carbon at night, when stomata are open. *6.6*

Review Questions

1. A cat eats a bird, which earlier ate a caterpillar that chewed on a weed. Which organisms are autotrophs? Which are the heterotrophs? *CI*

2. Summarize the photosynthesis reactions as an equation. Name where each stage takes place inside a chloroplast. *6.1*

3. What is the function of ATP in photosynthesis? What is the function of NADPH? *6.1*

4. Which of the following pigments are most visible in a maple leaf in summer? Which become the most visible in autumn? *6.3*
   a. chlorophylls   c. anthocyanins
   b. phycobilins   d. carotenoids

5. How does chlorophyll *a* differ in function from accessory pigments during the light-dependent reactions? *6.2*

6. Fill in Figure 6.18 for the light-dependent reactions. *6.3*

7. With respect to the light-dependent reactions, how do the cyclic and noncyclic pathways of electron flow differ? *6.3*

8. What substance does *not* take part in the Calvin–Benson cycle: ATP, NADPH, RuBP, carotenoids, O_2, CO_2, or enzymes? *6.5*

9. Fill in the blanks for Figure 6.19. Which substances are the original sources of carbon atoms and hydrogen atoms used in the synthesis of glucose in the Calvin–Benson cycle? *6.1, 6.5*

10. On hot, dry days, oxygen from photosynthesis accumulates in leaves. Explain why, and explain what happens in C3 plants versus C4 plants. *6.6*

11. Are photoautotrophs the only "self feeders"? If not, give examples and where you might find them. *6.7, 6.8*

Self-Quiz   ANSWERS IN APPENDIX III

1. Photosynthetic autotrophs use _____ from the air as a carbon source and _____ as their energy source.

2. In plants, light-dependent reactions proceed at the _____ .
   a. cytoplasm   c. stroma
   b. plasma membrane   d. thylakoid membrane

3. In the light-*dependent* reactions, _____ .
   a. carbon dioxide is fixed   c. CO_2 accepts electrons
   b. ATP and NADPH form   d. sugar phosphates form

Figure 6.19 Review Figure 6.17. Then, on your own, fill in the blanks (*red* lines) with the names of the key reactants, intermediates, and products of photosynthesis.

4. The light-*independent* reactions proceed in the _____ .
   a. cytoplasm
   b. plasma membrane
   c. stroma

5. When a photosystem absorbs light, _____ .
   a. sugar phosphates are produced
   b. electrons are transferred to ATP
   c. RuBP accepts electrons
   d. light-dependent reactions begin

6. Identify which of the following substances accumulates inside the thylakoid compartment of chloroplasts during the light-dependent reactions:
   a. glucose   c. chlorophyll   e. hydrogen ions
   b. carotenoids   d. fatty acids

Figure 6.18
Review Figure 6.11, which shows the light-dependent reactions of photosynthesis. Then, on your own, fill in the blanks (*red* lines) with the names of key components and activities.

7. The Calvin–Benson cycle starts when _____ .
   a. light is available
   b. light is not available
   c. carbon dioxide is attached to RuBP
   d. electrons leave a photosystem

8. ATP phosphorylates _____ in the light-independent reactions.
   a. RuBP   b. NADP$^+$ c. PGA   d. PGAL

9. Match each event with its most suitable description.
   ____ photon absorption
   ____ NADPH formation
   ____ CO$_2$ fixation
   ____ PGAL formation
   ____ ATP formation only

   a. rubisco required
   b. ATP, NADPH required
   c. electrons cycled back to photosystem I
   d. energy matches amount required to excite electrons
   e. photolysis required

Critical Thinking

1. Suppose a garden in your neighborhood is filled with red, white, and blue petunias. Explain each of the three floral colors in terms of which wavelengths of light it is absorbing and which wavelengths it is transmitting or reflecting.

2. About 200 years ago, Jan Baptista van Helmont performed an experiment on the nature of photosynthesis. He wanted to know where growing plants acquire the raw materials necessary to increase in size. For his experiment, he planted a tree seedling weighing 5 pounds in a barrel filled with 200 pounds of soil. He watered the tree regularly.

   Five years passed. Van Helmont again weighed the tree and the soil. The tree weighed 169 pounds, 3 ounces. The soil weighed 199 pounds, 14 ounces. Because the tree weight had increased so much and soil weight had decreased so little, he concluded the tree had gained weight after absorbing the water he had added to the barrel.

   Given what you know about the composition of biological molecules, why was van Helmont's conclusion misguided? Reflect on the current model of photosynthesis and then give a more plausible explanation of his results.

3. Melanie is using radioactively labeled carbon atoms ($^{14}CO_2$) in the laboratory. The photosynthesizing plants she is studying take them up. Identify the compound in which the labeled carbon will appear first: NADPH, PGAL, pyruvate, or PGA.

4. The green alga (*Codium*) in Figure 6.20a grows in sunlit seawater. Its main pigments absorb red light, and its accessory pigments help harvest photon energy in the sunlit waters. Some of its pigments even function as shields against ultraviolet radiation. Other species of green algae live in freshwater ponds and lakes, just beneath the soil surface, and on rocks. Some survive in snow.

   The red alga in Figure 6.20b was found on a tropical reef in clear, warm water. Its phycobilins, which mask the color of chlorophyll, are good at absorbing wavelengths of green and blue-green light. Those wavelengths can penetrate deep water. Some of its relatives live in very deep, dimly lit waters, and their pigments are nearly black.

   The variation in pigments invites a question: If trapping energy is so vital to life, then why doesn't each photosynthetic pigment go after the whole rainbow? In other words, why aren't all pigments black?

   The earliest photoautotrophs evolved in the seas, so their pigments evolved in the seas, also. Speculate on how natural selection may have favored the evolution of different pigments at different depths, starting with the first kinds that may have evolved at hydrothermal vents (Section 6.7).

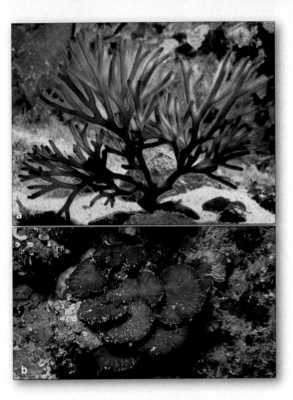

Figure 6.20 (**a**) A green alga (*Codium*) that grows in coastal waters. (**b**) A red alga from a tropical reef.

5. Fireflies emit light when enzymes called luciferases convert chemical energy to light energy. The reactions start when ATP transfers a phosphate group to a substance called luciferin. Luciferases, with a little help from oxygen, convert activated luciferin to a different molecule. Their action boosts electrons of the molecule to a higher energy level. The excited electrons quickly return to a lower energy level, and they release energy as fluorescent light (Section 6.2). When organisms flash with it, this is called *bioluminescence*. Electrons also get an energy boost in the light-dependent reactions of photosynthesis. So why aren't plants naturally bioluminescent?

Selected Key Terms

autotroph *CI*
C3 plant *6.6*
C4 plant *6.6*
Calvin–Benson cycle *6.5*
CAM plant *6.6*
carbon fixation *6.5*
carotenoid *6.2*
chlorophyll *6.2*
chloroplast *6.1*
electromagnetic spectrum *6.2*
electron transport system *6.3*
heterotroph *CI*
hydrothermal vent *6.7*
light-dependent reactions *6.3*
light-independent reactions *6.5*

PGA *6.5*
PGAL *6.5*
photolysis *6.3*
photon *6.2*
photosynthesis *CI*
photosystem *6.2*
pigment *6.2*
reaction center *6.3*
rubisco *6.5*
RuBP *6.5*
stoma (stomata) *6.6*
stroma *6.1*
thylakoid *6.1*
wavelength *6.2*

Readings

Bazzaz, F., and E. Jajer. January 1992. "Plant Life in a CO$_2$-Rich World." *Scientific American*, 266:68–74.

Zimmer, C. November 1996. "The Light at the Bottom of the Sea." *Discover*, 63–73.

On-Line readings at Student Guide for InfoTrac:
www.brookscole.com/biology

7

HOW CELLS RELEASE STORED ENERGY

The Killers Are Coming! The Killers Are Coming!

Thanks to selective breeding experiments gone wrong, descendants of "killer" bees that flew out of South America a few decades ago buzzed across the border between Mexico and Texas. By 1995, they had invaded 13,287 square kilometers of southern California and were busily setting up colonies. By 1998, when nectar-rich flowers bloomed profusely in the deserts after heavy El Niño storms, they moved even farther west and north than biologists had predicted.

When provoked, the bees are terrifying. To give an example, when a construction worker started a tractor a few hundred yards from a hive, thousands of agitated bees flew into action. They entered a nearby subway station, where they stung passengers on the platform and in the trains. They killed one person and injured a hundred more.

Where did the bees come from? In the 1950s, some queen bees were shipped over from Africa to Brazil for breeding experiments. Why? Honeybees are big business. They are the source of nutritious honey. Also, they are rented to commercial orchards, where their collective pollinating activities may significantly enhance fruit production. Imagine yourself enclosing an orchard tree in a cage that keeps out all pollinators. Without pollinators to spread pollen about, less than 1 percent of that tree's flowers will set fruit. Now imagine putting a hive of honeybees inside the cage with the tree. As many as 40 percent of the flowers will set fruit.

Compared to their relatives in Africa, the bees of Brazil are sluggish pollinators and honey producers. By cross-breeding the two strains, researchers thought they might come up with a strain of mild-mannered but zippier bees. So they put some local bees and imported ones together inside netted enclosures, complete with artificial hives. Then they let nature take its course.

Figure 7.1 One of the relatively mild-mannered honeybees buzzing in for a landing on a flower, wings beating with energy provided by ATP. If this were one of its Africanized relatives protecting a hive, possibly you would not stay around to watch the landing. The two kinds of bees look alike. How can we tell them apart? From our own biased perspective, Africanized bees are the ones with an attitude problem.

Twenty-six African queen bees escaped. That was bad enough. Then beekeepers got wind of preliminary experimental results. After learning that the first few generations of offspring were more energetic but not overly aggressive, they imported hundreds of African queens and encouraged them to mate with the locals. And they set off a genetic time bomb.

Before long, African bees became established in commercial hives—and in wild bee populations. Their traits became dominant. The "Africanized" bees do everything other bees do, but they do more of it faster. Their eggs develop into adults more quickly. Adults fly more rapidly, outcompete other bees for nectar, and even die sooner.

When something disturbs their hives or swarms, Africanized bees become extremely agitated. They can remain that way for as long as eight hours. Whereas a mild-mannered honeybee might chase an intruding animal fifty yards or so, a squadron of Africanized bees will chase it a quarter of a mile. If they catch up with it, they collectively can sting it to death.

Doing things faster means having a continuous supply of energy and efficient ways of using it. An Africanized bee's stomach can hold thirty milligrams of sugar-rich nectar—which is enough fuel to fly sixty kilometers. That's more than thirty-five miles! Besides this, compared to other kinds of bees, an Africanized bee's flight muscle cells have larger mitochondria. These organelles specialize in releasing a great deal of energy from sugars and other organic compounds, then converting it to the energy of ATP.

Whenever they tap into the stored energy of organic compounds, Africanized bees reveal their biochemical connection with other organisms. Study a primrose or puppy, a mold growing on stale bread, an amoeba in pondwater, or a bacterium living on your skin, and you will discover that their energy-releasing pathways differ in some details. But all of the pathways require characteristic starting materials. They yield predictable products and by-products. And they yield the universal energy currency of life—ATP.

In fact, throughout the biosphere, organisms put energy and raw materials to use in amazingly similar ways. *At the biochemical level, we find undeniable unity among all forms of life.* We will return to this idea in the concluding section of the chapter.

Key Concepts

1. The cells of all organisms can release energy stored in glucose and other organic compounds, then use it in ATP production. The energy-releasing pathways differ from one another. But the main types all start with the breakdown of glucose to pyruvate.

2. The initial breakdown reactions, known as glycolysis, can proceed in the presence of oxygen or in its absence. Said another way, glycolysis can be the first stage of either aerobic or anaerobic pathways.

3. Two kinds of energy-releasing pathways are entirely anaerobic, from start to finish. We call them fermentation and anaerobic electron transport. They proceed only in the cytoplasm, and none yields more than a small amount of ATP for each glucose molecule metabolized.

4. Another pathway, aerobic respiration, also starts in the cytoplasm. But this one runs to completion in organelles called mitochondria. Compared to the other pathways, aerobic respiration releases far more energy from glucose.

5. Aerobic respiration has three stages. First, pyruvate forms from glucose (during glycolysis). Second, different reactions break down the pyruvate to carbon dioxide. These reactions liberate electrons and hydrogen, which coenzymes deliver to electron transport systems. Third, stepwise electron transfers through the systems help set up the conditions that favor ATP formation. Free oxygen accepts the electrons at the end of the line and combines with hydrogen, thereby forming water.

6. Over evolutionary time, photosynthesis and aerobic respiration became linked on a global scale. The oxygen-rich atmosphere, a long-term outcome of photosynthetic activity, sustains aerobic respiration, which has become the dominant energy-releasing pathway. And most kinds of photosynthesizers use carbon dioxide and water from aerobic respiration as raw materials when they synthesize organic compounds:

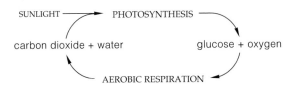

HOW DO CELLS MAKE ATP?

Organisms stay alive by taking in energy. Plants and all other organisms that engage in photosynthesis get energy from the sun. Animals get energy secondhand, thirdhand, and so on, by eating plants and one another. Regardless of its source, energy must be in a form that can drive thousands of life-sustaining reactions. Energy that becomes converted into the chemical bond energy of adenosine triphosphate—ATP—serves that function.

Plants make ATP during photosynthesis and use it to make glucose and other carbohydrates. *But they and every other organism also produce ATP by breaking down glucose and other carbohydrates, lipids, and proteins.* The breakdown reactions strip electrons and hydrogen from intermediates. Transfers of electrons through transport systems, of the sort described in Section 5.2, are central to the energy-releasing pathways.

Comparison of the Main Types of Energy-Releasing Pathways

The first energy-releasing pathways evolved 3.8 billion years ago, when conditions were different on Earth. The atmosphere had little free oxygen, so the pathways must have been *anaerobic;* they could proceed without using oxygen. Many prokaryotic cells and protistans still live in places where oxygen is absent or not always available. To make ATP, they use anaerobic pathways (fermentation or anaerobic electron transport). Some cells in your body use an anaerobic route for short periods, but only when they aren't getting enough oxygen. Your cells, like most others, rely more on **aerobic respiration**, an oxygen-dependent pathway of ATP formation. Each breath you take provides your actively respiring cells with a fresh supply of oxygen.

Make note of this point: *The main energy-releasing pathways all start with the same reactions in the cytoplasm.* During this initial stage of reactions, called **glycolysis**, enzymes cleave and rearrange a glucose molecule into two molecules of **pyruvate**, an organic compound with a backbone of three carbon atoms. Once glycolysis is over, the energy-releasing pathways differ. Most importantly, only the aerobic pathway continues in a mitochondrion (Figure 7.2). There, oxygen serves as the final acceptor

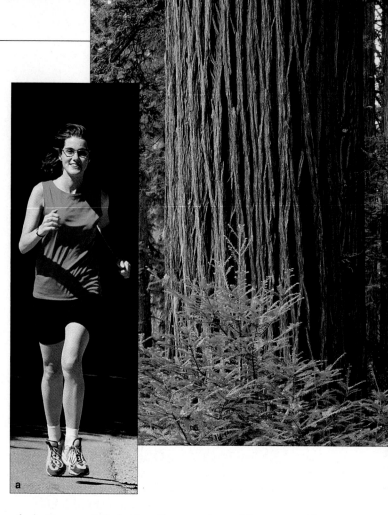

of electrons used during the reactions. The anaerobic pathways start and end in the cytoplasm. A substance other than oxygen is the final electron acceptor.

As you examine the energy-releasing pathways in sections to follow, keep in mind that the reaction steps do not proceed by themselves. Enzymes catalyze each step, and the intermediate molecules formed at one step serve as substrates for the next enzyme in the pathway.

Overview of Aerobic Respiration

Of all energy-releasing pathways, aerobic respiration gets the most ATP for each glucose molecule. Whereas anaerobic routes typically have a net yield of two ATP molecules, the aerobic route commonly yields thirty-six or more. If you were a bacterium, you wouldn't require much ATP. Being far larger, more complex, and highly active, you rely absolutely on the aerobic route's high yield. When a glucose molecule is the starting material, aerobic respiration can be summarized this way:

$$C_6H_{12}O_6 \ + \ 6O_2 \ \longrightarrow \ 6CO_2 \ + \ 6H_2O$$

GLUCOSE   OXYGEN   CARBON   WATER
                               DIOXIDE

However, as you can see, the summary equation only tells us what the substances are at the start and finish of the pathway. In between are three reaction stages.

Figure 7.2 Where the aerobic and anaerobic pathways of ATP formation start and finish up.

AEROBIC RESPIRATION

start (glycolysis) in cytoplasm

completed in mitochondrion

ANAEROBIC ENERGY-RELEASING PATHWAYS

start (glycolysis) in cytoplasm

completed in cytoplasm

Figure 7.3 Overview of aerobic respiration, which has three stages. From start to finish, the typical net energy yield from each glucose molecule is thirty-six ATP. (**a,b**) Only this pathway delivers enough ATP to construct and maintain redwoods and all other large, multicelled organisms. It alone delivers enough ATP for highly active animals, including bees, humans, and kingfishers and other birds.

(**c**) In the first stage (glycolysis), enzymes partially break down glucose to pyruvate. In the second stage, which is mainly the Krebs cycle, enzymes completely break down pyruvate to carbon dioxide. NAD^+ and FAD pick up electrons and hydrogen stripped from intermediates at both stages. In the final stage (electron transport phosphorylation), the reduced coenzymes (NADH and $FADH_2$) give up electrons and hydrogen to transport systems. Energy released during the flow of electrons through the systems drives ATP formation. Oxygen accepts the electrons at the end of the third stage.

Use Figure 7.3 while we outline the reactions. Again, glycolysis is the first stage. The second stage is mainly a cyclic pathway, the **Krebs cycle**. Enzymes break down pyruvate to carbon dioxide and water, thereby releasing electrons and hydrogen.

NAD^+ (for nicotinamide adenine dinucleotide) and **FAD** (flavin adenine dinucleotide) function in glycolysis and the Krebs cycle. Remember, these coenzymes assist enzymes by accepting electrons and hydrogen stripped from intermediates. The unbound hydrogen atoms are naked protons, or hydrogen ions (H^+). They tag along with the oppositely charged electrons. In reduced form, the coenzymes are abbreviated NADH and $FADH_2$.

Little ATP forms in glycolysis or the Krebs cycle. The large energy harvest comes in the third stage, after the coenzymes give up the electrons and hydrogen to many electron transport systems. These systems are the machinery for **electron transport phosphorylation**. It sets up H^+ concentration and electric gradients, which drive ATP formation at nearby membrane proteins. It is during this final stage that so many ATP molecules are produced. As it ends, oxygen inside the mitochondrion accepts the "spent" electrons from the last component of each transport system. Oxygen picks up hydrogen at the same time and thereby forms water.

Cells drive nearly all metabolic activities by releasing energy from glucose and other organic compounds and converting it to chemical bond energy of ATP. All of the main energy-releasing pathways start inside the cytoplasm with glycolysis, a stage of reactions that break down glucose to pyruvate.

The most common anaerobic pathways, which include the fermentation routes, end in the cytoplasm. Each has a net energy yield of two ATP.

Aerobic respiration, an oxygen-dependent pathway, runs to completion in the mitochondrion. From start (glycolysis) to finish, it commonly has a net energy yield of thirty-six ATP.

7.2

GLYCOLYSIS: FIRST STAGE OF ENERGY-RELEASING PATHWAYS

Let's track what happens to a glucose molecule in the first stage of aerobic respiration. Remember, the same things happen to glucose in the anaerobic pathways.

GLUCOSE

As you know, glucose is one of the simple sugars (Section 3.2). Each molecule consists of six carbon, twelve hydrogen, and six oxygen atoms, all covalently bonded together. Its carbons are the backbone. In the cytoplasm, glucose or some other simple sugar is partly broken down during glycolysis to two molecules of pyruvate, a three-carbon compound:

glucose ⟶ glucose–6–phosphate ⟶ 2 pyruvate

The first steps of glycolysis are *energy-requiring*. As Figure 7.4 indicates, they proceed only when two ATP molecules each transfer a phosphate group to glucose. Such phosphate-group transfers, remember, are known as phosphorylations. In this case, they raise the energy content of glucose to a level high enough to allow entry into the *energy-releasing* steps of glycolysis.

The first energy-releasing step cleaves the activated glucose into two molecules. We can call each of them PGAL (phosphoglyceraldehyde). Each PGAL becomes converted to an unstable intermediate that gives up a phosphate group to ADP, and so ATP forms. The next intermediate in the sequence does the same thing.

Thus, a total of four ATP form by **substrate-level phosphorylation**. We define this metabolic event as the direct transfer of a phosphate group from a substrate of a reaction to some other molecule—in this case, ADP. Remember, though, two ATP were invested to jump-start the reactions. So the *net* energy yield is two ATP.

Meanwhile, the coenzyme NAD$^+$ picks up electrons and hydrogen atoms liberated from each PGAL, thus becoming NADH. When NADH delivers its cargo to a different reaction site, it reverts to NAD$^+$. Said another way, NAD$^+$—like other coenzymes—is reusable.

In sum, glycolysis converts energy stored in glucose to a transportable form of energy, in ATP. NAD$^+$ picks up electrons and hydrogen stripped from glucose. These have roles in the next stage of reactions. So do the end products of glycolysis—the two molecules of pyruvate.

Glycolysis is a stage of reactions that partially break down glucose or some other carbohydrate to two molecules of pyruvate, thereby releasing energy.

Two NADH and four ATP form. However, when we subtract the two ATP required to start the reactions, the *net* energy yield of glycolysis is two ATP per glucose molecule.

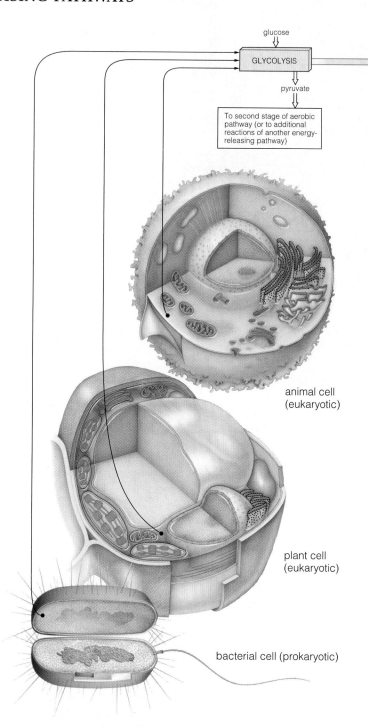

Figure 7.4 Glycolysis, the first stage of the main energy-releasing pathways. All prokaryotic and eukaryotic cells use glycolysis, which occurs in the cytoplasm. In this example, glucose is the starting material. Two pyruvate, two NADH, and four ATP form. Because cells invest two ATP to start glycolysis, the *net* energy yield is two ATP. Appendix V (Figure A) gives more details.

Depending on the type of cell and environmental conditions, the pyruvate may enter the second set of reactions of the aerobic pathway, including the Krebs cycle. Or it may be used in other reactions, such as those of fermentation.

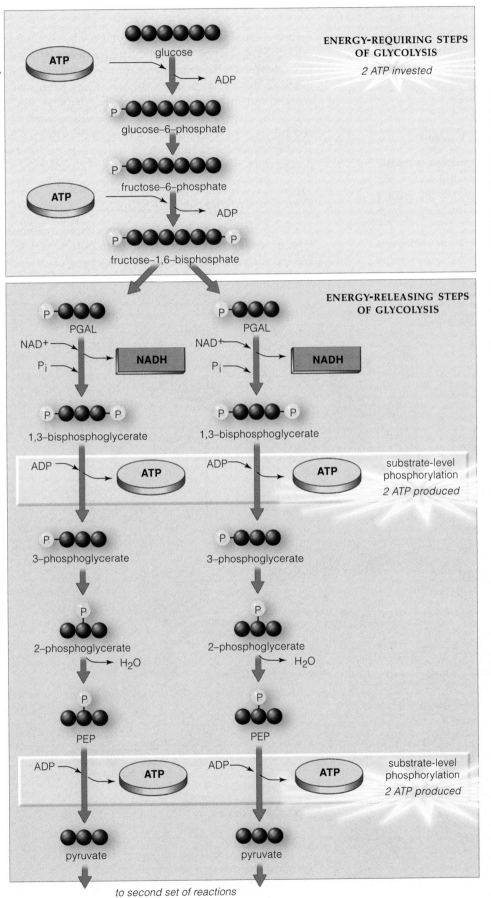

ENERGY-REQUIRING STEPS OF GLYCOLYSIS

2 ATP invested

glucose

glucose–6–phosphate

fructose–6–phosphate

fructose–1,6–bisphosphate

ENERGY-RELEASING STEPS OF GLYCOLYSIS

PGAL

1,3–bisphosphoglycerate

substrate-level phosphorylation

2 ATP produced

3–phosphoglycerate

2–phosphoglycerate

PEP

substrate-level phosphorylation

2 ATP produced

pyruvate

to second set of reactions

a This diagram tracks the six carbon atoms (*red* spheres) of the glucose molecule shown above. Glycolysis starts with an energy investment of two ATP.

b One ATP makes a phosphate-group transfer to glucose, and some atoms in glucose are rearranged in response.

c A phosphate-group transfer from the second ATP causes rearrangements that form fructose–1,6–bisphosphate. This intermediate can be split easily.

d It splits at once into two molecules, each with a three-carbon backbone. We can call these two PGAL.

e Two NADH form when each PGAL gives up two electrons and a hydrogen atom to NAD+.

f One ATP forms as each PGAL also combines with inorganic phosphate (P_i) and transfers a phosphate group to ADP.

g *Thus, two ATP formed by the direct transfer of a phosphate group from two intermediates of the reactions.* The original energy investment of two ATP has been paid off.

h In the next two reactions, the two intermediates each release a hydrogen atom and an —OH group, which then combine to form water.

i Two 3–phosphoenolpyruvate (PEP) molecules result. Each PEP transfers a phosphate group to ADP.

j *Once again, two ATP have formed by substrate-level phosphorylation.*

k In sum, the net energy yield from glycolysis is two ATP for each glucose molecule entering the reactions. Two molecules of pyruvate, the end product, may enter the next set of reactions in an energy-releasing pathway.

SECOND STAGE OF THE AEROBIC PATHWAY

Suppose two pyruvate molecules formed by glycolysis leave the cytoplasm and enter a **mitochondrion** (plural, mitochondria). In this organelle, the second and third stages of the aerobic pathway are completed. Figure 7.5 shows its structure and functional zones.

Preparatory Steps and the Krebs Cycle

It is during the second stage that the glucose is finally broken down completely to carbon dioxide and water. Two ATP form, but the most striking part of the second stage is the transfer of electrons and hydrogen from intermediates to many coenzymes.

In a few preparatory reactions, an enzyme removes a carbon atom from each pyruvate molecule. An enzyme

helper, coenzyme A, becomes **acetyl–CoA** by combining with the two-carbon fragment. The fragment is passed to **oxaloacetate**, the entry point for the Krebs cycle. The name of this cyclic pathway honors Hans Krebs, who worked out many of its details in the 1930s. It also is called the citric acid cycle. *Six* carbon atoms enter the second stage of reactions (three in each pyruvate). And *six* depart, in six carbon dioxide molecules, during the preparatory steps and the cycle proper (Figure 7.6).

Functions of the Second Stage

Think of the second stage as having three functions. First, it loads electrons and hydrogen onto NAD+ and FAD, resulting in NADH and FADH$_2$. Second, it forms

Figure 7.5 (**a**) Transmission electron micrograph and sketch of a mitochondrion, thin section.

(**b,c**) Functional zones in the mitochondrion. An inner membrane system divides this organelle's interior into two compartments. The second and third stages of aerobic respiration occur here. Coenzymes pick up electrons and hydrogen from second-stage reactions, then deliver them to electron transport systems at the inner membrane. Operation of the transport systems in the third stage drives ATP formation at ATP synthases, special proteins that span the membrane.

inner compartment   outer compartment   cytoplasm

outer mitochondrial membrane

inner mitochondrial membrane

1 Pyruvate from cytoplasm enters inner mitochondrial compartment.

OUTER COMPARTMENT

acetyl-CoA

Krebs Cycle

NADH

NADH

FADH$_2$

ATP

3 NADH and FADH$_2$ give up electrons and H+ to membrane-bound electron transport systems

4 As electrons move through the transport system, H+ is pumped to outer compartment.

2 Krebs cycle and preparatory steps: NAD+ and FADH$_2$ accept electrons and hydrogen stripped from the pyruvate. ATP forms. Carbon dioxide forms.

ATP   ATP   ATP   ATP

5 Oxygen accepts electrons, joins with H+ to form water

free oxygen

ADP + P$_i$

INNER COMPARTMENT

6 Following its gradients, H+ flows back into inner compartment, through ATP synthases. The flow drives ATP formation

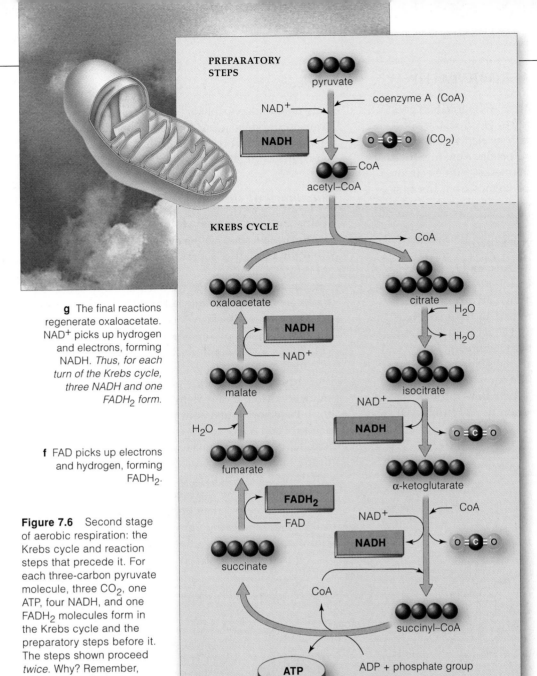

PREPARATORY STEPS

pyruvate

NAD^+ → coenzyme A (CoA)

NADH

$O=C=O$ (CO_2)

—CoA

acetyl–CoA

KREBS CYCLE

CoA

oxaloacetate

citrate

H_2O

H_2O

NADH

NAD^+

malate

isocitrate

NAD^+

H_2O

NADH

$O=C=O$

fumarate

α-ketoglutarate

FADH$_2$

FAD

NAD^+ → CoA

NADH

$O=C=O$

succinate

CoA

succinyl–CoA

ATP

ADP + phosphate group

a Pyruvate from glucose enters a mitochondrion. A carbon atom is released, as CO_2. A coenzyme binds with the two-carbon fragment, becoming acetyl–CoA. NADH forms as NAD^+ picks up hydrogen and electrons. *Thus, one NADH forms during a few steps preceding the Krebs cycle.*

b The four-carbon oxaloacetate is the entry point into the Krebs cycle. Acetyl–CoA transfers two carbons to it, forming citrate, a six-carbon compound. Adding and removing water changes citrate to another intermediate.

c Another carbon atom is released, as CO_2. NADH forms when NAD^+ picks up hydrogen and electrons.

d Another carbon atom is released as CO_2, another NADH forms, and a coenzyme A molecule binds to the intermediate. *At this point, for each turn of the cycle, three carbon atoms have been released.* This balances out the three carbons that entered the mitochondrion (in pyruvate).

e A phosphate group replaces the coenzyme A and is attached to ADP. *Thus, for each turn of the Krebs cycle, one ATP forms by substrate-level phosphorylation.*

g The final reactions regenerate oxaloacetate. NAD^+ picks up hydrogen and electrons, forming NADH. *Thus, for each turn of the Krebs cycle, three NADH and one FADH$_2$ form.*

f FAD picks up electrons and hydrogen, forming FADH$_2$.

Figure 7.6 Second stage of aerobic respiration: the Krebs cycle and reaction steps that precede it. For each three-carbon pyruvate molecule, three CO_2, one ATP, four NADH, and one FADH$_2$ molecules form in the Krebs cycle and the preparatory steps before it. The steps shown proceed *twice*. Why? Remember, glucose was degraded earlier to *two* pyruvate molecules.

two ATP by substrate-level phosphorylations. Third, it rearranges Krebs cycle intermediates into oxaloacetate. This is important; cells have only so much oxaloacetate, which must be regenerated to keep the reactions going.

The two ATP do not add much to the small yield from glycolysis. But the reactions load ten coenzymes with electrons and hydrogen. So far, then, the final stage of the aerobic pathway will get all of these coenzymes:

Glycolysis:	2 NADH
Pyruvate conversion before Krebs cycle:	2 NADH
Krebs cycle:	2 FADH$_2$ + 6 NADH
Coenzymes sent to third stage:	2 FADH$_2$ + 10 NADH

Overall, these are the key points to remember about the second stage of aerobic respiration:

During the second stage of aerobic respiration, two pyruvate molecules from glycolysis enter a mitochondrion.

All of pyruvate's carbon atoms (originally from glucose) are released, in the form of carbon dioxide. Two ATP form. The oxaloacetate that is the entry point for the second-stage cyclic reactions is regenerated.

The breakdown of pyruvate releases enough electrons and hydrogen to reduce ten coenzymes. With the two coenzymes that formed in glycolysis, they will deliver electrons and hydrogen to sites of the final stage of the aerobic pathway.

THIRD STAGE OF THE AEROBIC PATHWAY

In the aerobic pathway's third stage, ATP formation goes into high gear. This stage requires electron transport systems and ATP synthases, both located in the inner membrane that divides the mitochondrion into two compartments (Figure 7.7). Both interact with electrons and hydrogen, delivered by coenzymes from the first two stages of the aerobic pathway.

Electron Transport Phosphorylation

The electrons enter transport systems, and hydrogen is released and so becomes ionized (H$^+$). As the electrons are transferred through the systems, hydrogen ions in the inner compartment are shuttled to the outer one. These repeated shuttlings set up H$^+$ concentration and electric gradients across the inner membrane.

The only way ions can follow the gradient and flow back to the inner compartment is through the interior of ATP synthases (Figure 7.7). Their flow through the transport proteins drives formation of ATP from ADP and unbound phosphate. Free oxygen helps keep the transport systems clear for operation. It withdraws the spent electrons at the end of the transport systems and combines with H$^+$ to form water, a by-product.

Summary of the Energy Harvest

Thirty-two ATP typically form during the third stage of aerobic respiration. When you add these to the yield from the preceding stages, the net harvest is thirty-six ATP from one glucose molecule (Figure 7.8). That's a lot! Anaerobic pathways may use up eighteen glucose molecules to get the same net yield.

Think of thirty-six ATP as a typical yield only. The amount depends on the type of cell and the prevailing conditions. For example, it is less when an intermediate is pulled away from the reactions and used elsewhere.

Also, NADH from the cytoplasm can't even enter a mitochondrion. It only delivers electrons and hydrogen to it; transport proteins in the membrane shuttle these across. Then NAD$^+$ or FAD already inside accept them and deliver the electrons to transport systems. But FAD drops them off at a *lower* step in a transport system "staircase," so its deliveries produce less ATP.

One final point. Glucose, recall, has more energy (stored in more covalent bonds) than the products of its full breakdown (carbon dioxide and water). About 686 kilocalories of energy are released when each mole of

Figure 7.7 Electron transport phosphorylation, the third and final stage of aerobic respiration. The reactions occur at electron transport systems and at ATP synthases, a type of transport protein, in the inner mitochondrial membrane. Each electron transport system consists of specific enzymes, cytochromes, and other proteins. The membrane divides the mitochondrion's interior into two compartments.

NADH and FADH$_2$ give up electrons and hydrogen to the transport systems. As electrons are transferred *through* the systems, the unbound hydrogen (H$^+$) is shuttled across the membrane, to the outer compartment:

The H$^+$ concentration is now greater in the outer compartment. Concentration and electric gradients across the membrane have been set up. H$^+$ follows these gradients, through the interior of ATP synthases. Energy associated with the flow drives the formation of ATP from ADP and unbound phosphate (P$_i$). Hence the name, electron transport *phosphorylation*:

Do these events sound familiar? They should. ATP forms in much the same way in chloroplasts. By the *chemiosmotic* model, H$^+$ concentration and electric gradients across a cell membrane drive ATP formation. In this case, H$^+$ flows in the opposite direction compared to the flow in chloroplasts.

a In glycolysis, 2 ATP used; 4 ATP form by *substrate-level phosphorylation.* So *net* yield is 2 ATP.

b In Krebs cycle of second stage, 2 ATP form by *substrate-level phosphorylation.*

c In third stage, NADH from glycolysis used to form 4 ATP by *electron transport phosphorylation.*

d In third stage, NADH and FADH$_2$ from second stage used to make 28 ATP by *electron transport phosphorylation.*

TYPICAL NET ENERGY YIELD: **36 ATP**

Figure 7.8 Summary of the harvest from the energy-releasing pathway of aerobic respiration. Commonly, thirty-six ATP form for each glucose molecule. The net yield varies according to shifting concentrations of reactants, intermediates, and end products. It also varies among different types of cells.

Cells differ in how they use the NADH from glycolysis, which can't enter mitochondria. These NADH give up electrons and hydrogen to transport proteins in the outer mitochondrial membrane, which shuttle them across. NAD$^+$ or FAD already inside accept them, forming NADH or FADH$_2$.

Any NADH inside a mitochondrion delivers electrons to the highest entry point into a transport system. When it does, enough H$^+$ is pumped across the inner membrane to make *three* ATP. FADH$_2$ delivers them to a lower entry point. Fewer hydrogen ions are pumped, so only *two* ATP can form.

In liver, heart, and kidney cells, for example, electrons and hydrogen from glycolysis enter the highest entry point of transport systems, so the energy harvest is thirty-eight ATP. More commonly, as in skeletal muscle and brain cells, they are transferred to FAD, so the harvest is thirty-six ATP.

glucose is degraded to these more stable end products. (Section 2.7, *Critical Thinking* question 7, gives a simple definition of molar weight.) Much of the freed energy escapes as heat, but about 39 percent is stored in ATP.

In the final stage of the aerobic pathway, many coenzymes deliver electrons to transport systems in a mitochondrion's inner membrane. The membrane forms two compartments.

Coenzymes give up hydrogen and electrons to the transport systems. Electron transfers through these systems cause the unbound hydrogen (H$^+$) to be shuttled into the inner compartment. The resulting H$^+$ concentration and electric gradients drive ATP formation as H$^+$ flows back across the membrane. Oxygen is the final electron acceptor.

Again, from start (glycolysis in the cytoplasm) to finish (in mitochondria), the pathway commonly has a net yield of thirty-six ATP for every glucose molecule metabolized.

ANAEROBIC ROUTES OF ATP FORMATION

So far, we have tracked the fate of a glucose molecule through the pathway of aerobic respiration. We turn now to its use as a substrate for fermentation pathways. Remember, these are anaerobic pathways; they do *not* use oxygen as the final acceptor of the electrons that ultimately drive the ATP-forming machinery.

Fermentation Pathways

Diverse kinds of organisms use fermentation pathways. Many are prokaryotic cells and protistans that reside in marshes, bogs, mud, deep-sea sediments, the animal gut, canned foods, sewage treatment ponds, and other oxygen-free settings. Some fermenters actually will die if they are exposed to oxygen. Bacteria responsible for many diseases, including botulism and tetanus, are like this. Other kinds of fermenters, including the bacterial "employees" of yogurt manufacturers, are indifferent to the presence of oxygen. Still others can use oxygen, but they also might use a fermentation pathway when oxygen becomes scarce. Even your muscle cells do this.

Glycolysis serves as the first stage of fermentation pathways, just as it does for aerobic respiration. In these pathways also, enzymes split glucose and rearrange the fragments into two pyruvate molecules. Here again, two NADH form, and the net energy yield is two ATP.

However, fermentation reactions do not completely break down glucose to carbon dioxide and water. Also, they produce no more ATP beyond the tiny yield from glycolysis. *Fermentation's final steps only regenerate NAD$^+$, the coenzyme with a key role in the breakdown reactions.*

Fermentation yields enough energy to sustain many single-celled anaerobic organisms. It even helps some aerobic cells through times of stress. But it isn't enough

to sustain large, multicelled organisms. And that is one reason why you will never see an anaerobic elephant.

LACTATE FERMENTATION   With these points in mind, take a look at Figure 7.9, which tracks the main steps of **lactate fermentation**. During this anaerobic pathway, the *pyruvate* molecules from glycolysis—the first stage of its reactions—accept hydrogen and electrons from NADH. This transfer regenerates NAD$^+$. At the same time, it converts each pyruvate to lactate, a three-carbon compound. You may hear people call this compound "lactic acid," which is the non-ionized form. Lactate, its ionized form, is far more common in cellular fluids.

Some bacteria, such as *Lactobacillus*, rely exclusively on this anaerobic pathway. Their activities often spoil food. Even so, some fermenters have commercial uses, as when they break down glucose in huge vats where cheeses, yogurt, and sauerkraut are produced.

Also, in humans, rabbits, and many other animals, some types of cells use lactate fermentation for a quick fix of ATP. When your own demands for energy are intense but brief—say, during a short race—muscle cells use this pathway. They can't do so for long; they would throw away too much of glucose's energy for too little ATP. Muscles fatigue and lose their ability to contract when these cells deplete their glucose supply.

ALCOHOLIC FERMENTATION   In the anaerobic route called **alcoholic fermentation**, enzymes convert each pyruvate molecule that formed during glycolysis to an intermediate form: acetaldehyde. The NADH transfers electrons and hydrogen to this form and thus converts it to an alcoholic end product—ethanol (Figure 7.10).

Some single-celled fungi called yeasts are famous for their use of this pathway. One type, *Saccharomyces cerevisiae,* makes bread dough rise. Bakers mix the yeast with sugar, then blend both into dough. Fermenting yeast cells release carbon dioxide. Bubbles of this gas expand the dough (make it rise). Oven heat forces the gas out of the dough, and a porous product remains.

Beer and wine producers use yeasts on a large scale. Vintners use wild yeasts living on grapes and cultivated strains of *S. ellipsoideus,* which remain active until the alcohol concentration in wine vats exceeds 14 percent. (Wild yeast cells die when the concentration exceeds 4 percent.) That is why some birds get drunk on naturally fermenting berries. That is why landscapers don't plant prodigious berry-producing shrubbery near highways;

GLYCOLYSIS

$C_6H_{12}O_6$

2 ATP
energy input
2 ADP

2 NAD$^+$

2 NADH

4 ATP
energy output

2 pyruvate

2 ATP net

LACTATE FORMATION

electrons, hydrogen
from NADH

2 lactate

Figure 7.9 Lactate fermentation. In this anaerobic pathway, electrons end up in lactate, the reaction product.

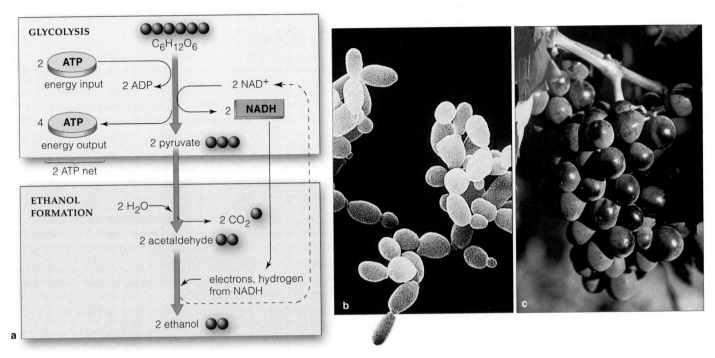

Figure 7.10 (**a**) Alcoholic fermentation, one of the anaerobic pathways. Acetaldehyde, an intermediate of the reactions, is the final acceptor of electrons. Ethanol is the end product. Yeasts, single-celled organisms, use this pathway. (**b**) Activity of one species of *Saccharomyces* makes bread dough rise. (**c**) Another species lives on sugar-rich tissues of ripe grapes.

drunk birds doodle into windshields (Figure 7.11). Wild turkeys similarly have been known to get tipsy when they gobble fermenting apples in untended orchards.

Anaerobic Electron Transport

Especially among prokaryotic cells, we see less common energy-releasing pathways, some of which are topics of later chapters in the book. For example, many species have key roles in the global cycling of sulfur, nitrogen, and other crucial elements. Collectively, their metabolic activities influence nutrient availability for organisms in ecosystems everywhere.

For example, certain archaebacteria and eubacteria use **anaerobic electron transport**. Electrons from some organic compound flow through transport systems in the plasma membrane. Often an inorganic compound in the environment serves as the final electron acceptor. The net energy yield varies, but it is always small.

Even as you read this, some anaerobic species that live in waterlogged soil are stripping electrons from a variety of compounds. They dump electrons on sulfate. Hydrogen sulfide, a putrid-smelling gas, is the result. Sulfate-reducing species also thrive in aquatic habitats that are enriched with decomposed organic material. They even live on the ocean floor, near hydrothermal vents. As described in Section 44.11, they form the food production base for unique communities.

In fermentation pathways, an organic substance that forms during the reactions serves as the final acceptor of electrons from glycolysis. The reactions regenerate NAD$^+$, which is required to keep the pathway operational.

In anaerobic electron transport, an inorganic substance (but not oxygen) usually serves as the final electron acceptor.

For each glucose molecule metabolized, the anaerobic pathways typically have a net energy yield of two ATP molecules, which only form during glycolysis.

Figure 7.11 Robin feasting on the fermented berries of a pyracantha bush.

ALTERNATIVE ENERGY SOURCES IN THE HUMAN BODY

So far, you have looked at what happens after a lone glucose molecule enters an energy-releasing pathway. Now you can start thinking about what cells do when they have too many or too few glucose molecules.

Carbohydrate Breakdown in Perspective

THE FATE OF GLUCOSE AT MEALTIME   While you or any other mammal is eating, glucose and other small organic molecules are being absorbed across the gut lining, then blood transports them through the body. A rise in the glucose level in blood prompts the pancreas to release insulin. This hormone stimulates cells to take up glucose faster. Cells convert the incoming glucose to glucose–6–phosphate and "trap" it in the cytoplasm. (When phosphorylated, glucose cannot be transported back out, across the plasma membrane.) Look again at Figure 7.4, and you see that glucose–6–phosphate is the first activated intermediate of glycolysis.

If your glucose intake exceeds cellular demands for energy, ATP-producing machinery goes into high gear. Unless a cell is rapidly using its ATP, the cytoplasmic concentration of ATP can rise to high levels. When that happens, the glucose–6–phosphate gets diverted into a biosynthesis pathway. This pathway assembles glucose units into glycogen, a storage polysaccharide (Section 3.3). The pathway is especially favored in muscle and liver cells, which maintain the largest glycogen stores.

THE FATE OF GLUCOSE BETWEEN MEALS   When you are not eating, glucose is not entering the bloodstream, and its level in the blood declines. If the decline were not countered, that would be bad news for the brain, your body's glucose hog. At any time, your brain is taking up more than two-thirds of the freely circulating glucose, because its many hundreds of millions of cells use this sugar alone as their preferred energy source.

The pancreas responds to the decline by secreting glucagon. This hormone prompts liver cells to convert glycogen back to glucose and send it back to the blood. Only liver cells do this; muscle cells won't give it up. The blood glucose level rises, and brain cells keep on functioning. Thus, *hormones control whether your body's cells use free glucose as an energy source or tuck it away.*

A word of caution: Don't let the preceding examples lead you to believe cells squirrel away large amounts of glycogen. In adult humans, glycogen makes up merely 1 percent or so of the body's total energy reserves, the energy equivalent of two cups of cooked pasta. Unless you eat on a regular basis, you will deplete the liver's small glycogen stores in less than twelve hours.

Of the total energy reserves in, say, a typical adult who eats well, 78 percent (about 10,000 kilocalories) is concentrated in body fat and 21 percent in proteins.

Energy From Fats

How does the body access its huge reservoir of fats? A fat molecule, recall, has a glycerol "head" and one, two, or three fatty acid "tails." Most fats get stored in your body as triglycerides, with three tails each. Triglycerides accumulate in fat cells of adipose tissue, which forms at buttocks and other strategic places beneath the skin.

When blood glucose levels decline, triglycerides are tapped as an energy alternative. Enzymes in fat cells cleave the bonds between the glycerol and fatty acids, which enter the blood. Enzymes in the liver convert the glycerol to PGAL, an intermediate of glycolysis. Nearly all cells take up the circulating fatty acids. Enzymes cleave the backbone of the fatty acids. The fragments are converted to acetyl–CoA, which can enter the Krebs cycle (Figures 7.6 and 7.12).

Compared to glucose, a fatty acid tail has far more carbon-bound hydrogen atoms, so it yields far more ATP. Between meals or during sustained exercise, fatty acid conversions supply about half of the ATP that muscle, liver, and kidney cells require.

What happens if you eat too many carbohydrates? Exceed the glycogen-storing capacity of your liver and muscle cells, and the excess gets converted to fats. *Too much glucose ends up as excess fat.* In the United States, 25 percent of the people have a combination of genes that lets them eat as much as they like without gaining weight. A diet too rich in carbohydrates keeps the other 75 percent fat. Their insulin levels stay elevated, which "tells" the body to store fat, not use it for energy. We return to this topic in Section 32.7 and Chapter 37.

Energy From Proteins

Eat more proteins than your body requires to grow and maintain itself, and its cells won't store them. Enzymes split dietary proteins into amino acid units. Then they remove the amino group ($-NH_3^+$) from each unit, and ammonia (NH_3) forms. What happens to the leftover carbon backbones? The outcome varies, depending on conditions of the moment. Maybe they'll get converted to fats or carbohydrates. Or maybe they'll enter the Krebs cycle (Figure 7.12), where coenzymes pick up the hydrogen and electrons stripped from carbon atoms. The ammonia that forms undergoes conversions and becomes urea. This nitrogen-containing waste product would be toxic if it accumulated to high concentrations. Normally your body excretes urea, in urine.

As this brief discussion makes clear, maintaining and accessing the body's energy reserves is complicated business. Hormonal controls over the disposition of glucose are special only because glucose is the fuel of choice for the all-important brain. However, as you will

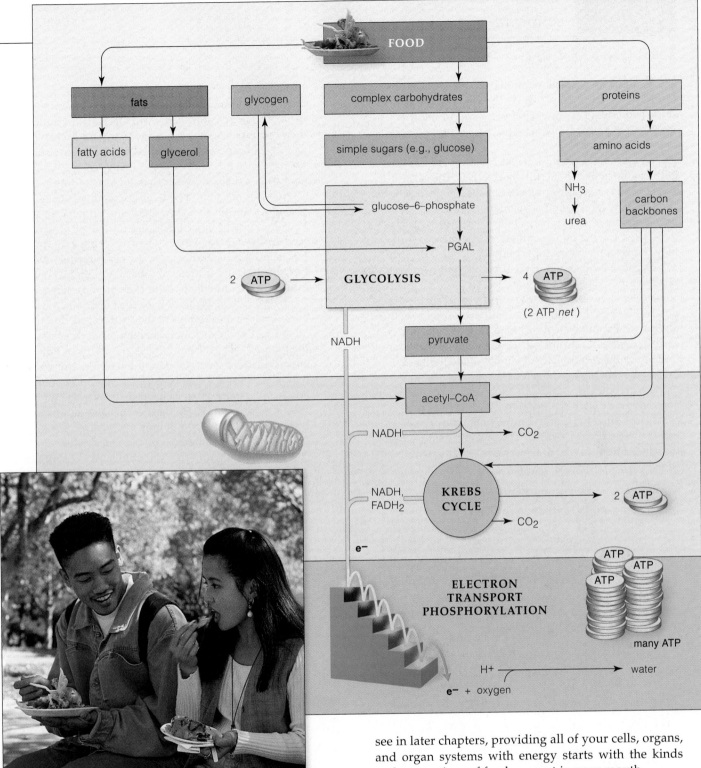

FOOD

fats → fatty acids, glycerol

glycogen

complex carbohydrates → simple sugars (e.g., glucose)

proteins → amino acids → NH_3 → urea; carbon backbones

glucose–6–phosphate → PGAL

2 ATP → **GLYCOLYSIS** → 4 ATP (2 ATP *net*)

NADH

pyruvate

acetyl–CoA

NADH → CO_2

KREBS CYCLE → 2 ATP

NADH, $FADH_2$ → CO_2

e^-

ELECTRON TRANSPORT PHOSPHORYLATION

ATP ATP ATP ATP — many ATP

$H+$ → water

e^- + oxygen

Figure 7.12 Alternative energy sources in the human body. The diagram shows reaction sites where a variety of organic compounds can enter the stages of aerobic respiration.

Complex carbohydrates, fats, and proteins cannot enter the aerobic pathway directly. In humans and other mammals, the digestive system, and individual cells, must first break apart these molecules into simpler, degradable subunits.

see in later chapters, providing all of your cells, organs, and organ systems with energy starts with the kinds and proportions of food you put in your mouth.

This concludes our look at aerobic respiration and other energy-releasing pathways. The section to follow may help you get a sense of how they fit into the larger picture of life's evolution and interconnectedness.

In humans and other mammals, the entrance of glucose or other organic compounds into an energy-releasing pathway depends on the kinds and proportions of carbohydrates, fats, and proteins in the diet as well as on the type of cell.

Perspective on Life

In this unit, you read about photosynthesis and aerobic respiration—the main pathways by which cells trap, store, and release energy. What you may not know is that the two pathways became linked, on a grand scale, over evolutionary time.

When life originated, some 3.8 billion years ago, the Earth's atmosphere had little free oxygen. The earliest single-celled organisms probably made ATP by reactions similar to glycolysis, and fermentation pathways must have dominated. More than a billion years passed. Then oxygen-producing photosynthetic cells appeared—and they changed the course of evolution.

Oxygen, a by-product of the noncyclic pathway of photosynthesis, began to accumulate in the atmosphere. Some cells were able to use the oxygen to accept electrons, possibly as a chance outcome of mutated proteins in their electron transport systems. In time, descendants of those early aerobic cells abandoned photosynthesis. Among them were the forerunners of animals and all other organisms that engage in aerobic respiration.

With aerobic respiration, the flow of carbon, hydrogen, and oxygen through the metabolic pathways of living organisms came full circle. For the final products of this aerobic pathway—carbon dioxide and water—are the same materials that are necessary to build organic compounds in photosynthesis:

Perhaps you have difficulty fathoming the connection between yourself—a highly intelligent being—and such remote-sounding events as energy flow and the cycling of carbon, hydrogen, and oxygen. Is this really the stuff of humanity?

Think back, for a moment, on the structure of a water molecule. Two hydrogen atoms sharing electrons with an oxygen atom may not seem close to your daily life. Yet, through that sharing, water molecules show polarity—and they hydrogen-bond with one another. Their chemical behavior is a beginning for the organization of lifeless matter that leads to the organization of all living things.

For now you can imagine other molecules interspersed through water. The nonpolar kinds resist interaction with water, and the polar kinds dissolve in it. On their own, the phospholipids among them assemble into a two-layered film. Such lipid bilayers, remember, serve as the framework of all cell membranes, hence all cells. From the beginning, the cell has been the basic *living* unit.

The essence of life is not some mysterious force. It is metabolic control. With a cell membrane to contain them, reactions *can* be controlled. With mechanisms built into their membranes, cells can respond to energy changes and shifting concentrations of substances in the environment. The response mechanisms operate by "telling" proteins—enzymes—when and what to build or tear down.

And it is not some mysterious force that creates the proteins themselves. DNA, the slender double-stranded treasurehouse of inheritance, has the chemical structure—*the chemical message*—that allows molecule to reproduce molecule, one generation after the next. In your own body, DNA strands tell trillions of cells how countless molecules must be built or torn apart for their stored energy.

So yes, carbon, hydrogen, oxygen, and other atoms of organic molecules represent the stuff of you, and us, and all of life. But it takes more than molecules to complete the picture. Life exists as long as an unbroken flow of energy sustains its organization. Molecules are assembled into cells, cells into organisms, organisms into communities, and so on up through the biosphere. It takes energy inputs from the sun to maintain these levels of organization. And energy flows through time in one direction—from organized to less organized forms. Only as long as energy continues to flow into the web of life can life continue in all its rich expressions.

In short, life is no more *and no less* than a marvelously complex system of prolonging order. Sustained by energy transfusions from the sun, life continues through its capacity for self-reproduction. For with the hereditary instructions contained in DNA, energy and materials can be organized, generation after generation. Even with the death of individuals, life elsewhere is prolonged. With each death, molecules are released and may be cycled once more, as raw materials for new generations.

In this flow of energy and cycling of material through time, each birth is affirmation of our ongoing capacity for organization, each death a renewal.

SUMMARY

1. Phosphate-group transfers from ATP are central to metabolism. Autotrophic cells alone tap energy from the environment to make ATP for building carbohydrates. *All* cells make ATP by pathways that release chemical energy from organic compounds, such as glucose. *7.1*

2. After glucose enters an energy-releasing pathway, enzymes strip electrons and hydrogen from reaction intermediates formed along the way. Coenzymes pick these up and deliver them to other reaction sites, where the pathway ends. NAD^+ is the main coenzyme; FAD is also used in the aerobic route. Reduced forms of these coenzymes are designated NADH and $FADH_2$. *7.1–7.3*

3. All of the main energy-releasing pathways start with glycolysis. This stage of reactions begins and ends in the cytoplasm. It can be completed either in the presence of oxygen or in its absence. *7.1, 7.2*

   a. In glycolysis, enzymes break down glucose to two pyruvate molecules. Two NADH and four ATP form.

   b. The net energy yield is two ATP (because two ATP had to be invested up front to get the reactions going).

4. Aerobic respiration continues on through two more stages: (1) the Krebs cycle and a few steps preceding it, and (2) electron transport phosphorylation. The stages occur only in mitochondria of eukaryotic cells. *7.1, 7.3*

5. The second stage of the aerobic pathway starts when an enzyme strips a carbon atom from each pyruvate. Coenzyme A binds the two-carbon fragment to form acetyl–CoA, and transfers this to oxaloacetate, the entry point of the Krebs cycle. These cyclic reactions and a few steps before them load ten coenzymes with electrons and hydrogen (eight NADH, plus two $FADH_2$). Two ATP form. Three carbon dioxide molecules are released for each pyruvate that entered this second stage. *7.3*

6. The third stage of the aerobic pathway proceeds at a membrane dividing the mitochondrion's interior into two compartments. Electron transport systems and ATP synthases are embedded in this inner membrane. *7.4*

   a. Coenzymes deliver electrons and hydrogen from the first two stages to transport systems. These systems function to establish an H^+ concentration gradient and electric gradient across the membrane.

   b. H^+ follows the gradients and flows from the outer to the inner compartment. It does so through the interior of ATP synthases. Energy released by the ion flow drives the formation of ATP from ADP and unbound phosphate.

   c. Oxygen picks up spent electrons at the end of the transport system and combines with hydrogen ions to form water. That is, oxygen is the final acceptor of the electrons that initially resided in glucose.

7. Aerobic respiration has a typical net energy yield of thirty-six ATP for each glucose molecule metabolized. Yields vary by cell type and cellular conditions. *7.4*

8. The fermentation pathways and anaerobic electron transport also start with glycolysis, but they do not use oxygen. They are anaerobic, start to finish. *7.5*

   a. Lactate fermentation has a net energy yield of two ATP, which form in glycolysis. The remaining reaction regenerates NAD^+. The two NADH from glycolysis transfer electrons and hydrogen to two pyruvate from glycolysis. Two lactate molecules are the end products.

   b. Alcoholic fermentation has a net energy yield of two ATP from glycolysis, and its remaining reactions serve to regenerate NAD^+. Enzymes convert pyruvate from glycolysis to acetaldehyde, and carbon dioxide is released. The NADH from glycolysis transfer electrons and hydrogen to the two acetaldehyde molecules, thus forming two ethanol molecules, the end products.

   c. Certain bacteria use anaerobic electron transport. Electrons are stripped from various organic compounds and travel through transport systems in the bacterial cell's plasma membrane. An inorganic compound in the environment often serves as the final electron acceptor.

9. In humans and other mammals, simple sugars such as glucose from carbohydrates, glycerol and fatty acids from fats, and carbon backbones of amino acids from proteins can enter ATP-producing pathways. *7.6*

Review Questions

1. Is this true or false: Aerobic respiration occurs in animals but not plants, which make ATP only by photosynthesis. *7.1*

2. Using the diagram below of the aerobic pathway, fill in the blanks with the number of molecules of pyruvate, coenzymes, and end products. Write in the net ATP formed in each stage, then the net ATP formed from start (glycolysis) to finish. *7.1*

3. Is glycolysis energy-*requiring* or energy-*releasing*? Or do both kinds of reactions occur during glycolysis? *7.2*

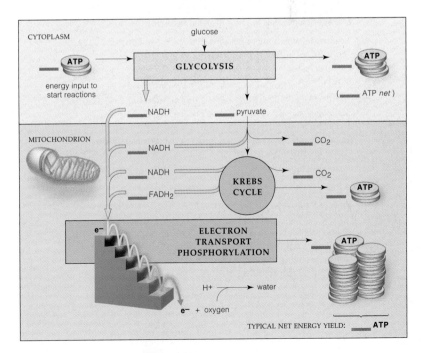

4. In what respect does *electron transport* phosphorylation differ from *substrate-level* phosphorylation? *7.2, Figure 7.7*

5. Sketch the double-membrane system of the mitochondrion and show where electron transport systems and ATP synthases are located. *7.3*

6. Name the compound that is the entry point for the Krebs cycle, and state whether it directly accepts the pyruvate from glycolysis. For each glucose molecule, how many carbon atoms enter the Krebs cycle? How many depart from it, and in what form? *7.3*

7. Is this statement true or false: Muscle cells cannot contract at all when deprived of oxygen. If true, explain why. If false, name the alternative(s) available to them. *7.5*

Self-Quiz   ANSWERS IN APPENDIX III

1. Glycolysis starts and ends in the _____ .
   a. nucleus   c. plasma membrane
   b. mitochondrion   d. cytoplasm

2. Which of the following does *not* form during glycolysis?
   a. NADH   c. $FADH_2$
   b. pyruvate   d. ATP

3. Aerobic respiration is completed in the _____ .
   a. nucleus   c. plasma membrane
   b. mitochondrion   d. cytoplasm

4. In the last stage of aerobic respiration, _____ is the final acceptor of electrons that originally resided in glucose.
   a. water   c. oxygen
   b. hydrogen   d. NADH

5. _____ engage in lactate fermentation.
   a. *Lactobacillus* cells   c. Sulfate-reducing bacteria
   b. Muscle cells   d. a and b

6. In alcoholic fermentation, _____ is the final acceptor of the electrons stripped from glucose.
   a. oxygen   c. acetaldehyde
   b. pyruvate   d. sulfate

7. The fermentation pathways produce no more ATP beyond the small yield from glycolysis, but the remaining reactions _____ .
   a. regenerate ADP   c. dump electrons on an inorganic
   b. regenerate NAD^+ substance (not oxygen)

8. In certain organisms and under certain conditions, _____ can be used as an energy alternative to glucose.
   a. fatty acids   c. amino acids
   b. glycerol   d. all of the above

9. Match the event with its most suitable metabolic description.
   ____ glycolysis   a. ATP, NADH, $FADH_2$, CO_2,
   ____ fermentation   and water form
   ____ Krebs cycle   b. glucose to two pyruvate
   ____ electron transport   c. NAD^+ regenerated, two ATP net
        phosphorylation   d. H^+ flows through ATP synthases

Critical Thinking

1. Diana suspects that a visit to her family doctor is in order. After eating carbohydrate-rich food, she always experiences sensations of being intoxicated and becomes nearly incapacitated, as if she had been drinking alcohol. She even wakes up with a hangover the next day. Having completed a course in freshman biology, Diana has an idea that something is affecting the way her body is metabolizing glucose. Explain why.

2. The cells of your body absolutely do not use nucleic acids as alternative energy sources. Suggest why.

3. The human body's energy needs and its programs for growth depend on balancing the levels of amino acids in blood with proteins in cells. Cells of the liver, kidneys, and intestinal lining are especially important in this balancing act. When the levels of amino acids in blood decline, lysozymes in cells can rapidly digest some of their proteins (structural and contractile proteins are spared, except in cases of malnutrition). The amino acids released this way enter the blood and thereby help maintain the required levels.

   Suppose you embark on a body-building program. You already eat plenty of carbohydrates. However, a nutritionist recommends that you follow a protein-rich diet that includes protein supplements. Speculate on how extra dietary proteins will be put to use, and in which tissues.

4. Each year, Canada geese lift off in precise formation from their northern breeding grounds. They head south to spend the winter months in warmer climates, then make the return trip in spring. As is the case for other migratory birds, their flight muscle cells are efficient at using fatty acids as an energy source. (Remember, the carbon backbone of fatty acids can be cleaved into fragments that can be converted to acetyl–CoA for entry into the Krebs cycle.)

   Suppose a lesser Canada goose from Alaska's Point Barrow has been steadily flapping along for three thousand kilometers and is nearing Klamath Falls, Oregon. It looks down and notices a rabbit sprinting like the wind from a coyote with a taste for rabbit. With a stunning burst of speed, the rabbit reaches the safety of its burrow.

   Which energy-releasing pathway predominated in muscle cells in the rabbit's legs? Why was the Canada goose relying on a different pathway for most of its journey? And why wouldn't the pathway of choice in goose flight muscle cells be much good for a rabbit making a mad dash from its enemy?

5. Reflect on this chapter's introduction, then question 4. Now speculate on which energy-releasing pathway is predominating in agitated Africanized bees that are chasing a farmer through a cornfield.

Selected Key Terms

acetyl–CoA 7.3
aerobic respiration 7.1
alcoholic fermentation 7.5
anaerobic electron transport 7.5
electron transport
   phosphorylation 7.1
FAD 7.1
glycolysis 7.1

Krebs cycle 7.1
lactate fermentation 7.5
mitochondrion 7.3
NAD^+ 7.1
oxaloacetate 7.3
pyruvate 7.1
substrate-level
   phosphorylation 7.2

Readings

DeMauro, S. October 1998. "Hooked on Mitochondria." *Quest.* Volume 5, Number 5. A brief look at mitochondrial disorders.

Wolfe, S. 1995. *An Introduction to Molecular and Cellular Biology.* Belmont, California: Wadsworth. Exceptional reference text.

II Principles of Inheritance

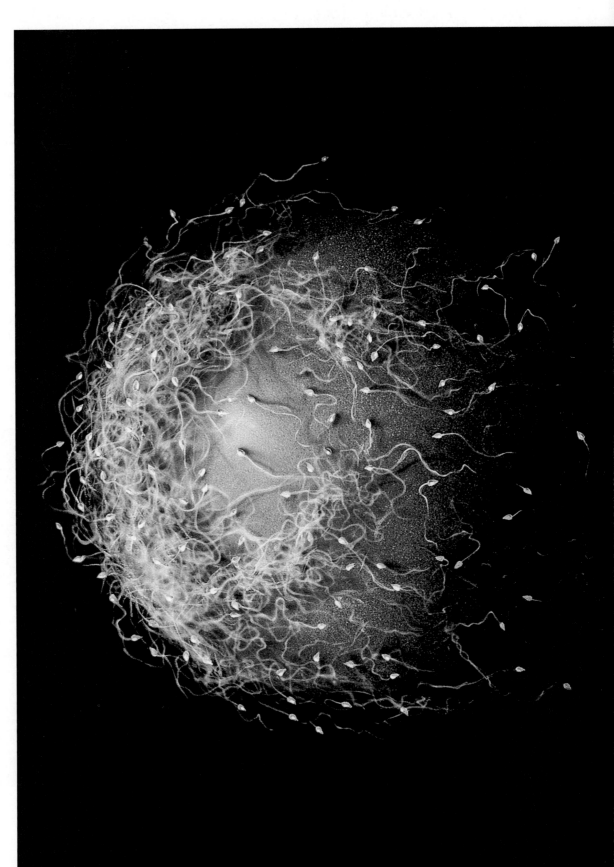

Human sperm, one of which will penetrate this mature egg and so set the stage for the development of a new individual in the image of its parents. This exquisite art is based on a scanning electron micrograph.

CELL DIVISION AND MITOSIS

Silver In the Stream of Time

Five o'clock, and the first rays from the sun dance over the wild Alagnak River of the Alaskan tundra. It is September, and life is ending and beginning in the clear, cold waters. By the thousands, mature silver salmon have returned from the open ocean to spawn in their shallow home waters. The females are tinged with red, the color of spawners, and they are dying.

This morning you watch a female salmon releasing translucent eggs into a shallow depression, which her fins made in the gravel riverbed (Figure 8.1). Moments later a male sheds a cloud of sperm, and the eggs are fertilized. Trout and other predators eat most of them. But wait around and you'll see that some eggs survive and give rise to a new generation.

Within three years the pea-sized eggs have become streamlined salmon, fashioned from billions of cells. A few of their cells will develop into eggs or sperm. In time, on some September morning, they will take part in an ongoing story of birth, growth, death, and rebirth.

Your body, too, emerged through *cell divisions*, as it did for salmon and all other multicelled species. Inside your mother, a fertilized egg divided in two, then the two became four, and so on until billions of cells were growing, developing in specialized ways, and dividing at different times to form different parts. Your body now has about 65 trillion living cells. Many are still dividing. Every five days, for instance, some divisions replace the tissue that lines your small intestine.

Understanding cell division—and, ultimately, how new individuals are put together in the image of their parents—begins with answers to three questions. *First*, what instructions are necessary for inheritance? *Second*, how are those instructions duplicated for distribution into daughter cells? *Third*, by what mechanisms are the duplicated instructions parceled out to daughter cells? We'll need more than one chapter to consider the nature of cell reproduction and other mechanisms of inheritance. Even so, the points made early in this chapter can help you keep the overall picture in focus.

Begin with the word **reproduction**. In biology, this means that parents produce a new generation of cells or multicelled individuals like themselves. The process starts in cells that are programmed to divide. And the ground rule for division is this: *Parent cells must provide*

sexually mature
female salmon

Figure 8.1　The last of one generation and the first of the next in Alaska's Alagnak River.

future arm and hand of embryo, five weeks old

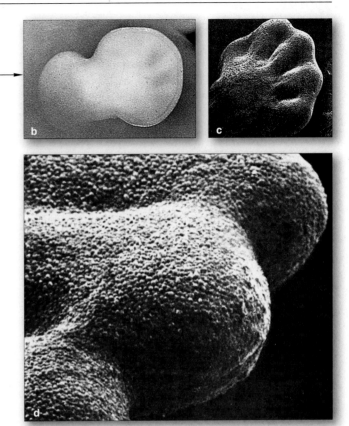

a crosswall. That new crosswall bridges the cytoplasm and divides the parent cell into two daughter cells.

Cytoplasmic Division of Animal Cells

Unlike plant cells, an animal cell isn't confined within a cell wall, and its cytoplasm typically "pinches in two." Look at Figure 8.9, a surface view of a newly fertilized animal egg. An indentation is forming about midway between the egg's two poles ("ends"). The indentation in its plasma membrane is a **cleavage furrow**. It's the first visible sign that the cytoplasm of an animal cell is dividing. The furrow will extend all around the cell and continue to deepen along the plane of the former spindle's midpoint until the cell is cut in two.

A band of microfilaments beneath the cell's plasma membrane generates force for a cut. These cytoskeletal elements are organized to interact and slide past one another (Section 4.10). As they do, they pull the plasma membrane inward until there are two daughter cells. Each has a nucleus, cytoplasm, and plasma membrane.

Perspective on Mitotic Cell Division

This concludes our introduction to mitotic cell division. Look now at your hands and try to visualize the cells making up your palms, thumbs, and fingers. Imagine the divisions that produced all the generations of cells that preceded them as you were developing early on, inside your mother (Figure 8.10). And be grateful for the astonishing precision of the mechanisms that led to their formation at certain times, in certain numbers, for the alternatives can be terrible indeed. Why? Good

Figure 8.10 Transformation of a paddlelike structure into a human hand by mitosis, cytoplasmic divisions, and other processes that are necessary for embryonic development. (**d**) You can see many individual cells at this high magnification. (**e**) Fingers at a later stage of development.

health—and survival itself—depends absolutely on the proper timing and the completion of cell cycle events, including mitosis. Certain genetic disorders arise from mistakes in the duplication or distribution of even one chromosome. Also, when normal controls that prevent cells from dividing are lost, unchecked cell divisions may destroy surrounding tissues and, ultimately, the organism. Section 8.5 describes a landmark case of such losses, which we will explore further in Section 14.4.

After mitosis, a separate mechanism cuts the cytoplasm into two daughter cells, each with a daughter nucleus.

Cytoplasmic division in plants often involves the formation of a cell plate and a crosswall between the adjoining, new plasma membranes of daughter cells.

Cleavage is a form of cytoplasmic division in animals. Rings of microfilaments around a parent cell's midsection slide past one another in a way that pinches the cytoplasm in two.

Henrietta's Immortal Cells

Each human starts out as a single fertilized egg. By the time of birth, mitotic cell divisions and other processes have resulted in a human body of about a trillion cells. Even in adults, billions of cells still divide. For example, cells of the stomach's lining divide every day. Liver cells usually do not divide, but if part of the liver becomes injured or diseased, repeated cell divisions will yield more new cells until the damaged part is finally replaced.

In 1951, George and Margaret Gey of Johns Hopkins University were trying to develop a way to keep human cells dividing *outside* the body. With such isolated cells, these researchers and others could investigate basic life processes. They also could conduct studies of cancer and other diseases without having to experiment directly on patients and gamble with human lives. The Geys used normal and diseased human cells, which local physicians had sent them. But they couldn't stop the descendants of those precious cells from dying out within a few weeks.

Mary Kubicek, a laboratory assistant, worked with the Geys in their efforts to start a self-perpetuating lineage of cultured human cells. She was about to give up after dozens of failed attempts. Even so, in 1951 she decided to prepare one more sample of cancer cells for culture. She gave the sample the code name **HeLa cells**, for the first two letters of the patient's first and last names.

The HeLa cells began to divide. And divide. And divide again! By the fourth day there were so many cells that Kubicek subdivided them into more culture tubes. Unfortunately, tumor cells in the patient were just as vigorous. Six months after she was diagnosed as having cancer, tumor cells had spread to tissues throughout her body. Only eight months after the diagnosis, Henrietta Lacks, a young woman from Baltimore, was dead.

Although Henrietta passed away, some of her cells lived on in the Geys' laboratory as the first successful human cell culture. In time, HeLa cells were shipped to research laboratories all over the world. Some even traveled into space for experiments on the *Discoverer XVII* satellite. Each year hundreds of research papers describe work based on HeLa cells.

Henrietta was thirty-one when runaway cell divisions killed her. And now, decades later, her legacy is helping humans everywhere—in cellular descendants that are still dividing, day after day.

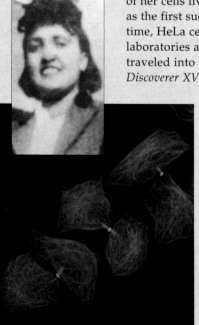

Figure 8.11 *Above:* Henrietta Lacks, a casualty of cancer. Her cellular contribution to science is still helping others. *Below:* Dividing HeLa cells.

SUMMARY
Gold indicates text section

1. Through specific division mechanisms, a parent cell provides each of its daughter cells with the hereditary instructions (DNA) and cytoplasmic machinery required to start up its own operation. *CI, 8.1*

   a. In eukaryotic cells, the nucleus divides by mitosis or meiosis. Cytoplasmic division typically follows.

   b. Prokaryotic cells divide by prokaryotic fission.

2. Each eukaryotic chromosome is one DNA molecule with numerous proteins attached. The chromosomes in a given cell differ in length, shape, and which portion of the hereditary instructions they carry. *8.1*

   a. The "chromosome number" is to the sum total of chromosomes in cells of a given type. Cells having a diploid chromosome number ($2n$) contain two of each kind of chromosome.

   b. Mitosis divides the nucleus into two equivalent nuclei, each with the same chromosome number as the parent cell. It maintains the chromosome number from one cell generation to the next.

   c. Mitosis is the basis of growth, tissue repair, and cell replacements among multicelled eukaryotes. It is the basis of asexual reproduction in many single-celled eukaryotes. (Meiosis occurs only in germ cells.)

3. A duplicated chromosome has two DNA molecules attached at the centromere. For as long as the two stay connected to each other, they are sister chromatids. *8.1*

4. A cell cycle starts when a new cell forms. It proceeds through interphase and ends when the cell reproduces by mitosis and cytoplasmic division. In interphase, a cell carries out its functions. When it is to divide again, the cell increases in mass, roughly doubles the number of its cytoplasmic components, then duplicates each of its chromosomes in preparation for division. *8.2*

5. Mitosis has four continuous stages: *8.3*

   a. Prophase. The duplicated, threadlike chromosomes start to condense. A spindle starts to form. The nuclear envelope starts to break up; its remnants form vesicles during the *transition* to metaphase (or prometaphase). Some microtubules from both poles of the developing spindle push the poles apart. Others directly attach to one of two sister chromatids of each chromosome.

   b. Metaphase. *At* metaphase, all chromosomes have become aligned at the spindle equator.

   c. Anaphase. Microtubules pull sister chromatids of each chromosome away from each other, to opposite spindle poles. Now each type of parental chromosome is represented by a daughter chromosome at both poles.

   d. Telophase. Chromosomes decondense to threadlike form. A new nuclear envelope forms around them. Each nucleus has the parental chromosome number.

6. Separate mechanisms divide the cytoplasm near the end of nuclear division or afterward (in plants, by cell plate formation; in animals, by cleavage). *8.4*

_____ _____ _____ _____ _____

Review Questions

1. Define mitosis and meiosis, two mechanisms that operate in eukaryotic cells. Does either one divide the cytoplasm? *8.1*

2. Define somatic cell and germ cell. *8.1*

3. What are chromosomal proteins? Give an example, and briefly state how it interacts with a DNA molecule. *8.1*

4. What is a chromosome called when it is in the unduplicated state? In the duplicated state (with two sister chromatids)? *8.1*

5. Describe the microtubular spindle and its functions. What role do motor proteins play in its operation? *8.3*

6. Using the diagram above as a guide, name and describe the key features of the stages of mitosis. *8.3*

7. Briefly explain how cytoplasmic division differs between a typical plant cell and typical animal cell. *8.4*

Self-Quiz   ANSWERS IN APPENDIX III

1. Mitosis and cytoplasmic division function in _____ .
   a. asexual reproduction of single-celled eukaryotes
   b. growth, tissue repair, and sometimes asexual reproduction in many multicelled eukaryotes
   c. gamete formation in prokaryotes
   d. both a and b

2. A duplicated chromosome has _____ chromatid(s).
   a. one   b. two   c. three   d. four

3. In a chromosome, a _____ is a constricted region with attachment sites for microtubules.
   a. chromatid b. cell plate   c. centromere   d. cleavage

4. A somatic cell having two of each type of chromosome has a(n) _____ chromosome number.
   a. diploid   b. haploid   c. tetraploid   d. abnormal

5. Interphase is the part of the cell cycle when _____ .
   a. a cell ceases to function
   b. a germ cell forms its spindle apparatus
   c. a cell grows and duplicates its DNA
   d. mitosis proceeds

6. After mitosis, the chromosome number of a daughter cell is _____ the parent cell's.
   a. the same as   c. rearranged compared to
   b. one-half   d. doubled compared to

7. Only _____ is not a stage of mitosis.
   a. prophase   b. interphase   c. metaphase   d. anaphase

8. Match each stage with the events listed.
   _____ metaphase   a. sister chromatids move apart
   _____ prophase   b. chromosomes start to condense
   _____ telophase   c. chromosomes decondense and
   _____ anaphase   daughter nuclei form
                      d. all duplicated chromosomes are
                         aligned at the spindle equator

Critical Thinking

1. Suppose you have a way to measure the amount of DNA in a single cell during the cell cycle. You first measure the amount at the G1 phase. At what points during the rest of the cell cycle would you predict changes in the amount of DNA per cell?

2. The cervix is part of the uterus, a chamber in which embryos develop. The *Pap smear* is a screening procedure that can detect *cervical cancer* in its earliest stages. Treatments range from freezing precancerous cells or killing them with a laser beam to removal of the uterus (a hysterectomy). Treatment is 90+ percent if this cancer is detected early. Survival chances plummet to less than 9 percent after the cancer spreads.

   Most cervical cancers develop slowly. Unsafe sex increases the risk. A key risk factor is infection by human papillomaviruses that cause genital warts (Section 39.19). In 93 percent of all cases, viral genes coding for the tumor-inducing proteins had become inserted into the DNA of previously normal cervical cells.

   Not all women request Pap smears. Many wrongly believe the procedure is costly. Many don't recognize the importance of abstinence or "safe" sex. Others simply don't want to think about whether they have cancer. Knowing what you've learned so far about the cell cycle and cancer, what would you say to a woman who falls into one or more of these groups?

3. Pacific yews (*Taxus brevifolius*) face extinction. People started stripping its bark and killing trees when they heard that *taxol*, a chemical extract from the bark, may fight breast and ovarian cancer. (Synthesizing taxol in laboratories may save the species.) Taxol is a poison that prevents microtubule disassembly. What does this tell you about its potential as an anticancer drug?

4. X rays and gamma rays emitted from some radioisotopes chemically damage DNA, especially in cells engaged in DNA replication. High-level exposure can result in *radiation poisoning*. Hair loss and damage to the lining of the stomach and intestines are two early symptoms. Speculate why. Also speculate on why highly focused radiation therapy is used against some cancers.

Selected Key Terms

anaphase *8.2*	cytoplasmic	mitosis *8.1*
cell cycle *8.2*	division *8.4*	nucleosome *8.1*
cell plate	diploid (chromosome	prophase *8.2*
formation *8.4*	number) *8.1*	reproduction *CI*
centriole *8.3*	germ cell *8.1*	sister
centromere *8.1*	HeLa cell *8.5*	chromatid *8.1*
chromosome *8.1*	histone *8.1*	somatic cell *8.1*
chromosome	interphase *8.2*	spindle
number *8.1*	meiosis *8.1*	apparatus *8.2*
cleavage furrow *8.4*	metaphase *8.2*	telophase *8.2*

Readings

Murray, A., and M. Kirschner. March 1991. "What Controls the Cell Cycle?" *Scientific American* 264(3): 56–63.

On-Line readings at Student Guide for InfoTrac:

MEIOSIS

Octopus Sex and Other Stories

The couple clearly are interested in each other. First he caresses her with one tentacle, then with another—and another and another. She reciprocates with a hug here, a squeeze there. This goes on for hours. Finally the male reaches under his mantle, a fold of tissue that drapes around most of his body. He removes a packet of sperm from a reproductive organ and inserts it into an egg chamber beneath the female's mantle. For every sperm that fertilizes an egg, a new octopus may develop.

Unlike the one-to-one coupling between a male and a female octopus, sex for the slipper limpet is a group enterprise. Slipper limpets are marine animals, relatives of the familiar land snails. Before becoming transformed into a sexually mature adult, a slipper limpet must go through a free-living stage of development called a larva. When a limpet larva is about to undergo its programmed transformation, it settles on a rock or pebble or shell. If it settles down all by itself, it will become a female. If another larva settles on the first limpet and proceeds to develop, that second limpet will function right off as a male. However, the second one will switch gears and develop into a female if a third limpet develops into a male on top of *it*. Later, that third limpet will also become a female if still another limpet develops into a male on top of it—and so on amongst ten or more limpets!

Slipper limpets typically live in such piles, with the bottom one always being the oldest female and the uppermost one being the youngest male (Figure 9.1*a*). Until they switch their sex, male limpets release sperm, which

Figure 9.1 Examples of variations in reproductive modes of eukaryotic organisms. (**a**) Slipper limpets, busily perpetuating the species by group participation in sexual reproduction. The tiny crab in the foreground is merely a passerby. (**b**) Live birth of an aphid, a type of insect that reproduces sexually in autumn but can switch to an asexual mode in summer.

fertilize a female's eggs, which grow up to become males but then, most likely, females—and so it goes, from one sexually flexible generation to the next.

Limpets are not alone in having unusual variations in their mode of reproduction. For example, sexual reproduction is common in many life cycles, but so are asexual episodes based on mitotic cell divisions. Orchids, dandelions, and many other plants reproduce very well with or without sex. Aquatic animals called flatworms can engage in sex or split their small body into two roughly equivalent parts, which each grow into a new flatworm.

And what about those aphids! In summer, nearly all aphids are females, which produce more females from *unfertilized* egg cells (Figure 9.1b). Only when autumn approaches do male aphids develop and do their part in the sexual phase of the life cycle. Even then, females that manage to survive over the winter can do without the opposite sex. Come summer, they begin another round of producing offspring all by themselves.

These examples only hint at the immense variation in reproductive modes among eukaryotic organisms. And yet, despite the variation, *sexual* reproduction dominates nearly all of the life cycles. And it always involves the same events. Briefly, cells set aside for sexual reproduction duplicate their chromosomes before they divide. **Germ cells**, immature reproductive cells that develop in male and female animals, are a fine example. They undergo meiosis and cytoplasmic division. Cellular descendants of germ cells mature and become **gametes**, or sex cells. When a male and female gamete manage to get together, they form the first cell of a new individual by way of fertilization.

With this chapter, we turn to the kinds of cells that serve as the bridge between generations of organisms. Specialized phases of reproduction and development, including asexual episodes, loop out from the basic life cycle of many eukaryotic species. Regardless of the specialized details, all of the life cycles turn on three events: *meiosis, the formation of gametes, and fertilization*. These three interconnected events are the hallmarks of sexual reproduction. As you will see in many chapters throughout the book, they have contributed to the diversity of life.

Key Concepts

1. Sexual reproduction proceeds through three events: meiosis, gamete formation, and fertilization. Sperm and eggs are familiar gametes.

2. Meiosis, a nuclear division mechanism, occurs only in cells that are set aside for sexual reproduction. The immature germ cells of male and female animals are examples. Meiosis sorts out a germ cell's chromosomes into four new nuclei. After meiosis, gametes form by way of cytoplasmic division and other events.

3. Cells with a diploid chromosome number contain two of each type of chromosome characteristic of the species. The two function as a pair during meiosis. Commonly, one chromosome of the pair is maternal, with hereditary instructions from a female parent. The other is paternal, with the same categories of hereditary instructions from a male parent.

4. Meiosis divides the chromosome number by half for each forthcoming gamete. Thus, if both parents have a diploid chromosome number ($2n$), the gametes that form will be haploid (n). Later, a union of two gametes at fertilization will restore the diploid number in the new individual ($n + n = 2n$).

5. Each pair of chromosomes typically swap segments during meiosis and so exchange hereditary information about certain traits. Also, meiosis randomly assigns one of each pair of chromosomes to a forthcoming gamete; *which* one of the pair ends up in a given gamete is a matter of chance. Hereditary instructions are further shuffled at fertilization. All three reproductive events lead to variations in traits among offspring.

6. In most plants, spore formation and other events intervene between meiosis and gamete formation.

COMPARING SEXUAL WITH ASEXUAL REPRODUCTION

When an orchid, flatworm, or aphid reproduces all by itself, what sort of offspring does it get? By the process of **asexual reproduction**, one parent alone produces offspring, and each offspring inherits the same number and kinds of genes as its parent. **Genes** are particular stretches of chromosomes—that is, of DNA molecules. Taken together, the genes for each species contain all the heritable bits of information necessary to make new individuals. Rare mutations aside, this means asexually produced individuals can only be *clones*, or genetically identical copies of the parent.

Inheritance gets much more interesting with **sexual reproduction**. This process involves meiosis, formation of gametes, and fertilization (union of the nuclei of two gametes). In most sexual reproducers, such as humans, the first cell of a new individual contains *pairs of genes* on pairs of homologous chromosomes. Typically, one of each pair is maternal and the other paternal in origin.

If instructions encoded in every pair of genes were identical down to the last detail, sexual reproduction would produce clones, also. Just imagine—you, every single person you know, the entire human population might be a clone, with everybody looking alike. But the two genes of a pair might *not* be identical. Why not? A gene's molecular structure can change; that is what we mean by mutation. Depending on their structure, two genes that happen to be paired in a person's cells may "say" slightly different things about a trait. Each unique molecular form of the same gene is called an **allele**.

Such tiny differences affect thousands of traits. For example, whether your chin has a dimple depends on which pair of alleles you inherited at one chromosome location. One kind of allele at that location says "put a dimple in the chin." Another kind says "no dimple." This leads us to a key reason why members of sexually reproducing species don't all look alike. *Through sexual reproduction, offspring inherit new combinations of alleles, which lead to variations in the details of their traits.*

This chapter gets into the cellular basis of sexual reproduction. More importantly, it starts you thinking about far-reaching effects of gene shufflings at different stages of the process. The process introduces variations in traits among offspring that are typically acted upon by agents of natural selection. Thus, *variation in traits is a foundation for evolutionary change.*

Asexual reproduction produces genetically identical copies of the parent. Sexual reproduction introduces variations in the details of traits among offspring.

Sexual reproduction dominates the life cycles of eukaryotic species. Meiosis, formation of gametes, and fertilization are the basic events of this process.

HOW MEIOSIS HALVES THE CHROMOSOME NUMBER

Think "Homologues"

Think back to the preceding chapter and its focus on mitotic cell division. Unlike mitosis, **meiosis** divides chromosomes into separate parcels not once but *twice* prior to cell division. Unlike mitosis, it is the first step leading to the formation of gametes.

Gametes, recall, are sex cells such as sperm or eggs. In nearly all multicelled eukaryotic organisms, gametes develop from cells that arise in specialized reproductive structures or organs. Figure 9.2 shows a few examples of where gametes form.

anther (where male gametes originate)

ovules inside the ovary (where female gametes originate)

a Flowering plant

(penis)

testis (where sperm originate)

(vagina)

ovary (where eggs develop)

b Human male   **c** Human female

Figure 9.2 Examples of gamete-producing structures.

As you know, the **chromosome number** is the sum total of chromosomes in cells of a given type (Section 8.1). Germ cells start out with the same chromosome number as somatic cells (the rest of the body's cells). If a cell has a **diploid number** (2n), it has a *pair* of each type of chromosome, often from two parents. Except for a pairing of nonidentical sex chromosomes, each pair has the same length, shape, and assortment of genes. And they line up with each other at meiosis. We call them **homologous chromosomes** (*hom*– means alike).

Figure 9.3 From a diploid cell of a human male, twenty-three pairs of homologous chromosomes. The last two are a pair of sex chromosomes (XY or XX). An XY pair does differ in length, shape, and which genes they carry, but the two still pair up as homologues during meiosis.

one pair of homologous chromosomes

As you can probably deduce from Figure 9.3, your own germ cells have 23 + 23 homologous chromosomes. After meiosis, 23 chromosomes—one of each type—end up in gametes. That is, meiosis halves the chromosome number, so gametes have a **haploid number** (*n*).

Two Divisions, Not One

Meiosis is like mitosis in some ways, but the outcome is different. As in mitosis, a germ cell duplicates its DNA in interphase. The two DNA molecules and their associated proteins remain attached at the centromere, the notably constricted region along their length. For as long as they remain attached, we call them **sister chromatids**:

one chromosome in the duplicated state

As in mitosis, the microtubules of a spindle apparatus move the chromosomes in prescribed directions.

With meiosis alone, however, *chromosomes go through two consecutive divisions that end with the formation of four haploid nuclei*. There is no interphase between the nuclear divisions, which we call meiosis I and meiosis II:

interphase (*DNA replication before meiosis I*)	MEIOSIS I	no interphase (*no DNA replication before meiosis II*)	MEIOSIS II
	PROPHASE I		PROPHASE II
	METAPHASE I		METAPHASE II
	ANAPHASE I		ANAPHASE II
	TELOPHASE I		TELOPHASE II

Each duplicated chromosome becomes aligned with its partner—*homologue to homologue*—during meiosis I.

After the two chromosomes of each pair have lined up with each other, they are moved apart:

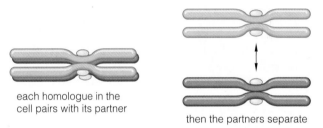

each homologue in the cell pairs with its partner

then the partners separate

The cytoplasm typically starts to divide at some point after each homologue is separated from its partner. The two daughter cells formed this way are haploid; each has *one* of each type of chromosome. But don't forget, the chromosomes are still in the duplicated state.

Next, during meiosis II, *the two sister chromatids of each chromosome are separated from each other*:

one chromosome (duplicated)

two chromosomes (unduplicated)

Each sister chromatid is now a chromosome in its own right. Four nuclei now form. Often the cytoplasm divides once more. The final outcome is four haploid cells.

Figure 9.4, on the next two pages, gives you a closer look at key events of meiosis and their consequences.

Meiosis is a type of nuclear division mechanism. It reduces the chromosome number of a parental cell by half—to the haploid number (*n*)—in daughter cells.

Meiosis proceeds only in immature cells set aside for sexual reproduction, such as animal germ cells. It is the first step leading to the formation of gametes.

VISUAL TOUR OF THE STAGES OF MEIOSIS

MEIOSIS I

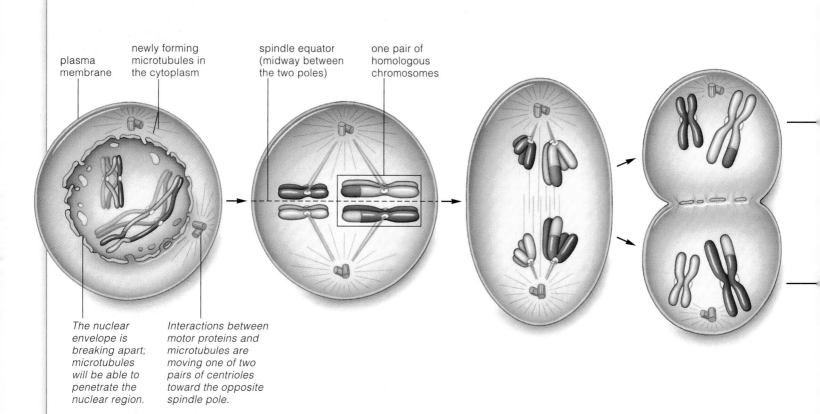

plasma membrane

newly forming microtubules in the cytoplasm

spindle equator (midway between the two poles)

one pair of homologous chromosomes

The nuclear envelope is breaking apart; microtubules will be able to penetrate the nuclear region.

Interactions between motor proteins and microtubules are moving one of two pairs of centrioles toward the opposite spindle pole.

PROPHASE I

Each duplicated chromosome is in threadlike form, but now starts to condense. It pairs with its homologue, and the two typically swap segments. The swapping, called crossing over, is indicated by the break in color on the pair of larger chromosomes. Some of the microtubules of a newly forming spindle become attached to each chromosome's centromere.

METAPHASE I

As in mitosis, motor proteins attached to microtubules move the chromosomes and move the spindle poles apart. They tug the chromosomes into position midway between the spindle poles. Thus the spindle becomes fully formed, owing to dynamic interactions among the motor proteins, microtubules, and the chromosomes themselves.

ANAPHASE I

Microtubules extending from the poles and overlapping at the spindle equator *lengthen* and push the poles apart. Other microtubules extending from the poles to the chromosomes *shorten*, thereby pulling each chromosome away from its homologous partner. These motions move the homologous partners to opposite poles.

TELOPHASE I

The cytoplasm of the germ cell divides at some point. There are now two haploid (*n*) cells. Each cell has one of each type of chromosome that was present in the parent (2*n*) cell. *However, all chromosomes are still in the duplicated state.*

Figure 9.4 Sketches of meiosis in a generalized animal cell. This nuclear division mechanism reduces the chromosome number in immature reproductive cells by half (to the haploid number) for forthcoming gametes. To keep things simple, we track only two pairs of homologous chromosomes. Maternal chromosomes are shaded *purple* and paternal chromosomes *blue*. The light micrographs above show corresponding stages in the formation of pollen grains in a lily (*Lilium regale*).

MEIOSIS II

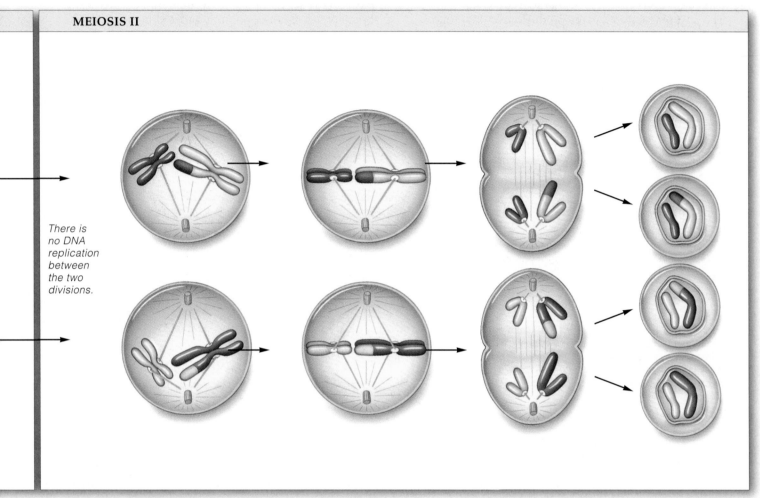

There is no DNA replication between the two divisions.

PROPHASE II
Microtubules have already moved one member of the centriole pair to the opposite pole of the spindle in each of the two daughter cells. Now, during prophase II, microtubules attach to the chromosomes, and motor proteins drive the movement of chromosomes toward the spindle's equator.

METAPHASE II
In each daughter cell, interactions among motor proteins, spindle microtubules, and each duplicated chromosome have moved all of the chromosomes so that they are positioned at the spindle equator, midway between the two poles.

ANAPHASE II
The attachment between the two chromatids of each chromosome breaks. Each of the former "sister chromatids" is now a chromosome in its own right. Motor proteins drive the movement of the newly separated chromosomes to opposite poles of the spindle.

TELOPHASE II
By the time telophase II is over, there will be four daughter nuclei. When cytoplasmic division is completed, each new, daughter cell will have a haploid chromosome number (*n*). All of the chromosomes will now be in the unduplicated state.

Of the four haploid cells that form by way of meiosis and cytoplasmic divisions, one or all may develop into gametes and function in sexual reproduction. In plants, cells that form after meiosis is over may develop into spores, which take part in a stage of the life cycle that precedes gamete formation. (The telophase II micrograph shows spores that will develop into pollen grains.)

A CLOSER LOOK AT KEY EVENTS OF MEIOSIS I

Study the overview in Sections 9.2 and 9.3, and you can sense the basic function of meiosis: *the reduction of the chromosome number by half for the forthcoming gametes.* However, two other major events occur during meiosis: crossing over at prophase I and the random alignment of homologues at metaphase I. Both contribute greatly to the adaptive advantage of sexual reproduction.

The advantage, recall, is the production of offspring with new combinations of alleles. Those combinations are translated into a new generation of individuals that differ in the details of some number of traits.

Crossing Over in Prophase I

Prophase I of meiosis is a time of major gene shufflings. Reflect on Figure 9.5a, which shows two chromosomes condensed to threadlike form. All chromosomes in a germ cell condense this way. As they do, each is drawn close to its homologue. Molecular interactions stitch homologues together point by point along their length, with little space between. The intimate, parallel array favors **crossing over**, a molecular interaction between two of the *non*sister chromatids of a pair of homologous

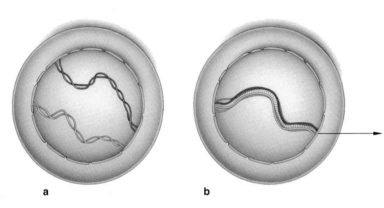

a   b

Figure 9.5 Key events of prophase I, the first stage of meiosis. For clarity, this diagram shows only one pair of homologous chromosomes and one crossover event. (Typically, more than one crossover occurs.) *Blue* signifies the paternal chromosome; *purple* signifies its maternal homologue.

(**a**) Both chromosomes were duplicated earlier, in interphase. Early in prophase I, the sister chromatids of each duplicated chromosome are in thin, threadlike form. They are positioned so closely together, they look like a single thread.

(**b**) Each chromosome becomes zippered to its homologue, so all four chromatids are intimately aligned. When the two sex chromosomes have different forms (such as X paired with Y) they still get zippered together, although only in a small region along their length.

(**c**,**d**) We show the pair of chromosomes as if they were already condensed, then teased apart to give you a sense of what goes on. Bear in mind, their double-stranded DNA molecules are still tightly aligned at this stage. The intimate contact allows one crossover (and usually more) to happen at intervals along the length of nonsister chromatids.

(**e**) Nonsister chromatids exchange segments. As prophase I ends, the chromosomes continue to condense into thicker, rodlike forms. Then they unzipper from each other except at places where they physically cross each other. Such places are called chiasmata (singular, chiasma, meaning "cross"). Each chiasma does not last long, but it's indirect evidence of a crossover at some place in the chromosomes.

(**f**) What is the function of crossing over? It breaks up old combinations of alleles and puts new ones together in pairs of homologous chromosomes.

c Diagram of a pair of duplicated homologous chromosomes

d Crossover between nonsister chromatids of the two chromosomes

chiasma

e Nonsister chromatids exchange segments

f Homologues have new combinations of alleles

chromosomes. Nonsister chromatids break at the same places along their length. At these break points, they exchange corresponding segments—that is, genes.

Gene swapping would be pointless if each type of gene never varied. But remember, a gene can come in slightly different forms: alleles. You can bet that some number of the alleles on one chromosome will *not* be identical to their partner alleles on the homologue. Each crossover is a chance to swap slightly different versions of hereditary instructions for particular traits.

We will look at the mechanism of crossing over in later chapters. For now, it is enough to remember this: *Crossing over leads to genetic recombination, which in turn leads to variation in traits among offspring.*

Metaphase I Alignments

Major shufflings of whole chromosomes begin during the transition from prophase I to metaphase I, the second stage of meiosis. Suppose the shufflings are happening right now in one of your germ cells. Crossovers have already made genetic mosaics of the chromosomes, but put this aside in order to simplify tracking. Just call the twenty-three chromosomes you inherited from your mother the *maternal* chromosomes and the twenty-three homologues from your father the *paternal* chromosomes.

Microtubules already oriented one chromosome of each pair toward one spindle pole and its homologue toward the other pole (refer to Section 8.3). They are busily moving all of the chromosomes, which soon will become positioned at the spindle's equator.

Will all maternal chromosomes be directed to one spindle pole and all paternal chromosomes directed to the other? Maybe, but probably not. Remember, initial contacts between microtubules and chromosomes are random. Because of the random grabs, the positioning of maternal or paternal chromosomes at the spindle's equator at metaphase I follows no particular pattern. Carry this thought one step further. *Either one* of each pair of homologous chromosomes can end up at either spindle pole after they move apart at anaphase I.

Think about the possibilities when you are tracking merely three pairs of homologues. As you can see from Figure 9.6, by metaphase I, three pairs of homologues may be arranged in any one of four possible positions. Here, eight combinations (2^3) of maternal and paternal chromosomes are possible for forthcoming gametes.

Of course, a human germ cell has twenty-three pairs of homologous chromosomes, not just three. So every time a human germ cell gives rise to sperm or eggs, we can expect a grand total of *8,388,608* (or 2^{23}) possible combinations of maternal and paternal chromosomes!

Moreover, in each sperm or egg, many hundreds of alleles inherited from the mother might not "say" the

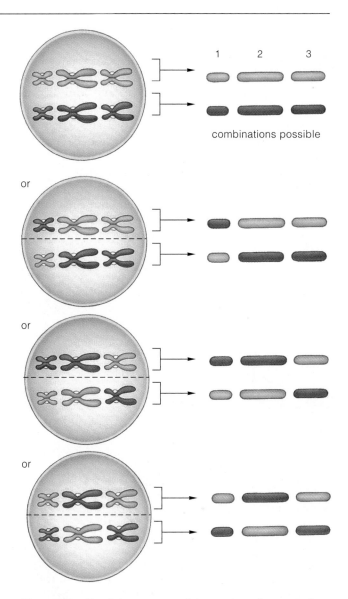

Figure 9.6 Possible outcomes of the random alignment of three pairs of homologous chromosomes at metaphase I. We label three types of chromosomes 1, 2, and 3. The maternal chromosomes are *purple;* paternal ones are *blue.* With merely four possible alignments, eight combinations of maternal and paternal chromosomes are possible in gametes.

exact same thing about hundreds of different traits as the alleles inherited from the father. Are you beginning to get an idea of why such fascinating combinations of traits show up even in the same family?

Crossing over is an interaction between a pair of homologous chromosomes. It breaks up old combinations of alleles and puts new ones together during prophase I of meiosis.

The random attachment and subsequent positioning of each pair of maternal and paternal chromosomes during metaphase I lead to different combinations of maternal and paternal traits in each generation of offspring.

FROM GAMETES TO OFFSPRING

The gametes that form following meiosis are not all the same in their details. For example, human sperm have one tail, opossum sperm have two, and roundworm sperm have none. Crayfish sperm look like pinwheels. Most eggs are microscopic in size, yet an ostrich egg tucked inside its shell is as large as a baseball. From its appearance alone, you might not believe that a plant's gamete is even remotely like an animal's.

Later chapters explain how gametes form in the life cycles of specific organisms, including humans. Figure 9.7 and the rest of this section may help you keep the details in perspective.

a Generalized life cycle for most kinds of plants

b Generalized life cycle for animals

Figure 9.7 Generalized life cycles for (**a**) most plants and (**b**) animals. The zygote is the first cell that forms when the nuclei of two gametes fuse at fertilization.

For plants, a sporophyte (spore-producing body) develops, by way of mitotic cell divisions, from the zygote. After meiosis, gametophytes (gamete-producing bodies) form. A lily plant is a sporophyte. Gametophytes form in parts of its flowers.

Chapters 20 through 24, 28, and 39 have specific examples of life cycles for representative organisms.

Gamete Formation in Plants

For pine trees, apple trees, roses, dandelions, corn, and nearly all other familiar plants, certain events intervene between meiosis and the time that gametes develop and mature. Among other things, spores form.

Spores are haploid resting cells, often walled, that are good at resisting drought, cold, and other adverse environmental conditions. When favorable conditions return, spores germinate (resume growth) and develop into a haploid body or structure that produces gametes. So *gamete*-producing bodies and *spore*-producing bodies develop during the life cycle of most kinds of plants. Figure 9.7*a* is a generalized diagram of these events.

Gamete Formation in Animals

In male animals, gametes form by a process known as spermatogenesis. A diploid germ cell grows in size in the male's reproductive system. It becomes a primary spermatocyte. This large immature cell enters meiosis and cytoplasmic divisions. Four haploid cells result and develop into spermatids (Figure 9.8). These immature cells change in form and develop a tail. Each becomes a **sperm**, a common type of mature male gamete.

In female animals, gametes form by a process called oogenesis. In human females, for instance, a diploid germ becomes an **oocyte**, or immature egg. Unlike sperm, an oocyte stockpiles many cytoplasmic components, and its daughter cells differ in size and function (Figure 9.9). When an oocyte divides after meiosis I, one daughter cell (a secondary oocyte) gets nearly all the cytoplasm. The other cell, the first polar body, is small. Later, both cells undergo meiosis II and cytoplasmic division. One daughter cell of the secondary oocyte develops into a second polar body. The other gets most of the cytoplasm and develops into a gamete. The mature female gamete is called an ovum (plural, ova) or, more often, an **egg**.

And so we have one egg and three polar bodies. The polar bodies don't function as gametes and aren't rich in nutrients or cytoplasm. In time they degenerate. But the very fact that they formed means the egg has a suitable (haploid) chromosome number. Also, by getting most of the cytoplasm, the egg has enough start-up machinery to support the new individual right after fertilization.

More Shufflings at Fertilization

The chromosome number characteristic of the parents is restored at **fertilization**, a time when a female and male gamete unite and their haploid nuclei fuse. Fertilization would double the chromosome number for each new generation if meiosis didn't precede it. Such doublings would disrupt the hereditary instructions, usually for

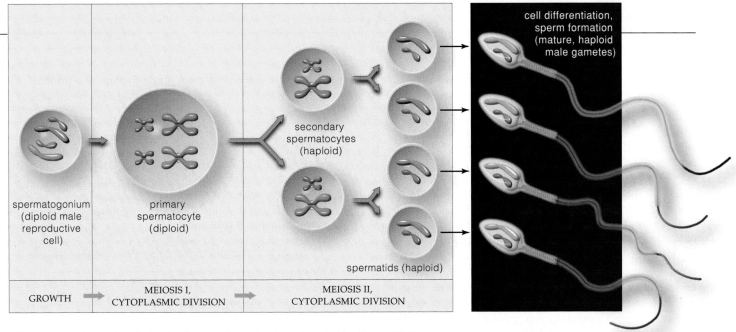

Figure 9.8 Generalized sketch of sperm formation in male animals. Figure 39.7 shows a specific example (how sperm form in human males).

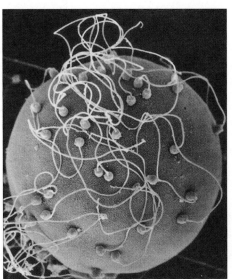

Figure 9.9 Egg formation in female animals. The eggs are far larger than sperm, as the micrograph of sea urchin gametes (*above*) suggests. Also, the three polar bodies are far smaller than the egg. Figure 39.9 gives a specific example (how human eggs form).

the worse. Why? The instructions work as an intricate, fine-tuned package in each individual.

Fertilization contributes to variation in offspring. Reflect on the possibilities for humans alone. During prophase I, an average of two or three crossovers takes place in each human chromosome. Even without these crossovers, the random positioning of pairs of paternal and maternal chromosomes at metaphase I results in one of millions of possible chromosome combinations in each gamete. And of all the male and female gametes that are produced, *which* two actually get together is a matter of chance. The sheer number of combinations that can exist at fertilization is staggering!

Cumulatively, crossing over, the distribution of random mixes of homologous chromosomes into gametes, and fertilization contribute to variation in the traits of offspring.

MEIOSIS AND MITOSIS COMPARED

In this unit our focus has been on two nuclear division mechanisms. Single-celled eukaryotic species reproduce asexually by way of mitosis, followed by cytoplasmic division. Many multicelled eukaryotic species depend on mitosis and cytoplasmic division during episodes of asexual reproduction in their life cycle. All depend on it for growth and tissue repair. By contrast, meiosis occurs only in reproductive cells, such as germ cells that give rise to the gametes used in sexual reproduction. Figure 9.10 summarizes the main similarities and differences between the two nuclear division mechanisms.

The end results of the two mechanisms differ in a crucial way. *Mitotic cell division only produces clones—genetically identical copies of a parent cell. But meiotic cell division, in conjunction with fertilization, promotes variation in traits among offspring.* First, crossing over at prophase I of meiosis puts new combinations of alleles in chromosomes. Second, the random assignment of either member of a pair of homologous chromosomes to either pole of the spindle at metaphase I affects gametes, which end up with mixes of maternal and paternal alleles. And third, different combinations of alleles are brought together simply by chance during fertilization. In later chapters, you will be reading about the ways in which both meiosis and fertilization contribute to the truly stunning diversity and evolution of sexually reproducing organisms.

A *somatic cell* with a diploid chromosome number (2n) is at interphase. Before mitotic division begins, its DNA is replicated (all chromosomes are duplicated).

Figure 9.10 Summary of mitosis and meiosis. Both diagrams use a diploid (2n) animal cell as the example. They are arranged to help you compare similarities and differences between the division mechanisms. The maternal chromosomes are coded *purple* and the paternal chromosomes *blue.*

MEIOSIS I

A *germ cell* with a diploid chromosome number (2n) is at interphase. Before mitotic division begins, its DNA is replicated (all chromosomes are duplicated).

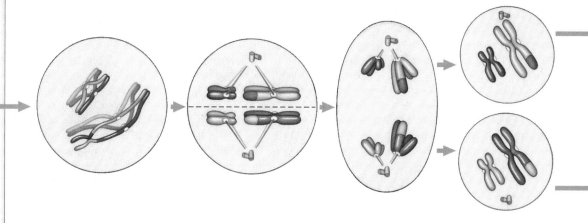

PROPHASE I	METAPHASE I	ANAPHASE I	TELOPHASE I
Each duplicated chromosome (consisting of two sister chromatids) condenses to threadlike form, then rodlike form. *Crossing over* occurs. Each chromosome unzips from its homologue. Each gets attached to the spindle in transition to metaphase.	All chromosomes are now positioned at the spindle's equator.	*Each chromosome is separated from its homologue.* They are moved to opposite poles of the spindle.	When the cytoplasm divides, there are two cells. Each has a haploid (n) number of chromosomes, but these are still in the duplicated state.

MITOSIS

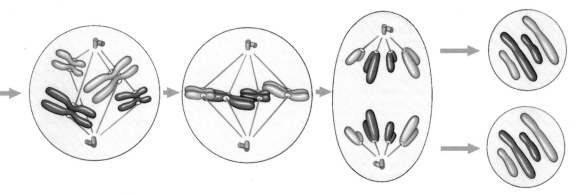

PROPHASE

Each duplicated chromosome (consisting of two sister chromatids) condenses from threadlike form to rodlike form. Each gets attached to the spindle during the transition to metaphase.

METAPHASE

All chromosomes are now positioned at the spindle's equator.

ANAPHASE

Sister chromatids of each chromosome are separated from each other. These new, daughter chromosomes are moved to opposite poles of the spindle.

TELOPHASE

When the cytoplasm divides, there are two cells. Each is diploid (2*n*)—*it has the same chromosome number as the parent cell.*

MEIOSIS II

There is no DNA replication between the two divisions.

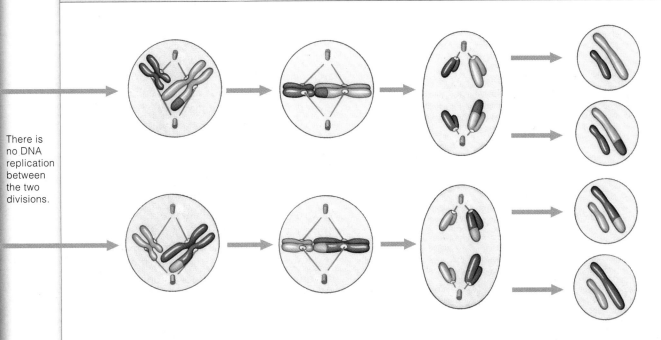

PROPHASE II

Before prophase II, the two centrioles in each new cell were moved apart and a new spindle formed. Now, each chromosome becomes attached to the spindle and starts moving toward its equator.

METAPHASE II

All chromosomes are now positioned at the spindle's equator.

ANAPHASE II

Sister chromatids of each chromosome are separated from each other. These new, daughter chromosomes are moved to opposite poles of the spindle.

TELOPHASE II

Four daughter nuclei form. When the cytoplasm divides, each new cell is haploid (*n*). *The original chromosome number has been reduced by half.* One or all of these cells may become gametes.

1. Eukaryotic life cycles often have asexual as well as sexual phases. Asexual reproduction results in clones. Sexual reproduction (by meiosis, gamete formation, and fertilization) leads to variation in traits. *CI, 9.1, 9.2*

a. Meiosis, a nuclear division mechanism, reduces the chromosome number of a parent germ cell by half. It precedes the formation of haploid gametes, such as sperm in males and eggs in females.

b. At fertilization, a sperm and egg nuclei fuse. This event restores the chromosome number (Figure 9.11).

2. A germ cell with a diploid chromosome number (2*n*) has *two* of each type of chromosome characteristic of its species. Commonly, one of each pair of chromosomes is maternal and the other is paternal. *9.2*

3. Each pair of maternal and paternal chromosomes has homology; the two are alike. Except for a pairing of nonidentical sex chromosomes (e.g., X with Y), the two have the same length, shape, and gene sequence along their length. They interact during meiosis. *9.2*

4. Chromosomes become duplicated during interphase. Each consists of two DNA molecules that stay attached (as sister chromatids) during mitosis. *9.2*

5. Meiosis consists of two consecutive divisions, which both require a microtubular spindle apparatus. *9.1–9.4*

a. During meiosis I, spindle microtubules attach to the centromere region of each duplicated chromosome. Motor proteins attached to the microtubules separate it from its partner, the homologous chromosome.

b. In meiosis II, similar interactions move the sister chromatids of each chromosome away from each other.

6. Meiosis I, the first nuclear division, is characterized by the following events and outcomes: *9.2–9.4*

a. Crossing over occurs in prophase I. Two nonsister chromatids of each pair of homologous chromosomes break at corresponding sites and exchange segments. This puts new combinations of alleles together. Alleles (slightly different molecular forms of the same gene) specify different versions of the same trait.

b. Different combinations of alleles lead to variation in the details of a given trait among offspring.

c. Also in prophase I, a microtubular spindle forms outside the nucleus, and the nuclear envelope starts to break up. In cells with duplicated pairs of centrioles, one pair starts moving to the opposite spindle pole.

d. All of the pairs of homologous chromosomes have become positioned at the spindle equator at metaphase I. Both the maternal chromosome and its homologue have become oriented at random, toward either pole.

e. In anaphase I, spindle microtubules interact with each duplicated chromosome and move it away from its homologue, to opposite spindle poles.

7. Meiosis II (second nuclear division) is characterized by these events and outcomes: *9.2–9.4*

a. At metaphase II, all the duplicated chromosomes are positioned at the spindle equator.

b. Sister chromatids are moved apart in anaphase II. Each is now a separate, unduplicated chromosome.

c. By the end of telophase II, four nuclei—each with a haploid chromosome number (*n*)—have been formed.

8. When the cytoplasm divides, there are four haploid cells. One or all may serve as gametes (or as plant spores that give rise to gamete-producing bodies). *9.4, 9.5*

9. Crossing over, the chance allocation of different mixes of pairs of maternal and paternal chromosomes to different gametes, and the chance of any two gametes meeting at fertilization all contribute to the immense variation in details of traits among offspring. *9.6*

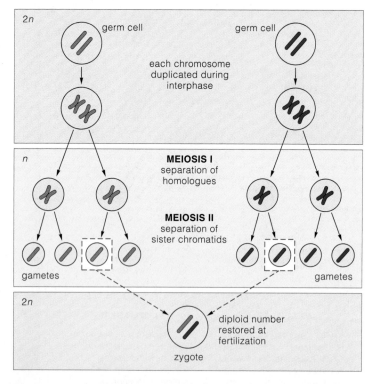

Figure 9.11 Summary of changes in chromosome number at different stages of sexual reproduction, using two diploid (2*n*) germ cells as the example. Meiosis reduces the chromosome number by half (*n*). Union of haploid nuclei of two gametes at fertilization restores the diploid number.

Review Questions

1. The diploid chromosome numbers for the somatic cells of a few organisms are listed at *right*. How many chromosomes will end up in the gametes of each organism? *9.2*

Fruit fly, *Drosophila melanogaster*	8
Garden pea, *Pisum sativum*	14
Corn, *Zea mays*	20
Frog, *Rana pipiens*	26
Earthworm, *Lumbricus terrestris*	36
Human, *Homo sapiens*	46
Chimpanzee, *Pan troglodytes*	48
Amoeba, *Amoeba*	50
Horsetail, *Equisetum*	216

2. A diploid germ cell has four pairs of homologous chromosomes, designated AA, BB, CC, and DD. How would the chromosomes of the gametes be designated? 9.2

3. Look at the chromosomes in the germ cell in the diagram at *right*. Is this cell at anaphase I or anaphase II? 9.3, 9.6

The cell is at anaphase ____ rather than anaphase ____ I know this because:

_____
_____
_____
_____

4. Define meiosis and describe its stages. In what respects is meiosis *not* like mitosis? 9.2, 9.3, 9.6

5. Actor Michael Douglas (Figure 9.12a) inherited a gene from each parent that influences the chin dimple trait. One form of the gene called for a dimple and the other didn't, but one is all it takes for this particular trait. Figure 9.12b shows what the chin of Mr. Douglas might have looked like if he had inherited two ordinary forms of the gene instead. What is the name for the alternative forms of the same gene? 9.1, 9.4

6. Outline the main steps by which gametes form in plants. Do the same for gamete formation in animals. 9.5

7. Genetically speaking, what is the key difference between the outcomes of sexual and asexual reproduction? 9.1, 9.6

Figure 9.12 Example of the chin dimple trait (actually a fissure in the chin surface).

Self-Quiz   ANSWERS IN APPENDIX III

1. Sexual reproduction requires _____ .
   a. meiosis   c. fertilization
   b. gamete formation   d. all of the above

2. An animal cell having two rather than one of each type of chromosome has a _____ chromosome number.
   a. diploid   c. normal gamete
   b. haploid   d. both b and c

3. Generally, a pair of homologous chromosomes _____ .
   a. carry the same genes   c. interact at meiosis
   b. are the same length, shape d. all of the above

4. Meiosis _____ the parental chromosome number.
   a. doubles   c. maintains
   b. reduces   d. corrupts

5. Meiosis is a division mechanism that produces _____ .
   a. two cells   c. eight cells
   b. two nuclei   d. four nuclei

6. Before the onset of meiosis, all chromosomes are _____ .
   a. condensed   c. duplicated
   b. released from protein d. both b and c

7. Duplicated chromosomes move away from their homologue and end up at the opposite spindle pole during _____ .
   a. prophase I   c. anaphase I
   b. prophase II   d. anaphase II

8. Sister chromatids of each duplicated chromosome move apart and end up at opposite spindle poles during _____ .
   a. prophase I   c. anaphase I
   b. prophase II   d. anaphase II

9. Match each term with its description.
   ____ chromosome   a. different molecular forms
        number   of the same gene
   ____ alleles   b. none between meiosis I and II
   ____ metaphase I   c. pairs of homologous
   ____ interphase   chromosomes are now aligned
                           at the spindle equator
                        d. sum total of chromosomes
                           in all cells of a given type

Critical Thinking

1. Assume you can measure the amount of DNA in a primary oocyte, then in a primary spermatocyte, which gives you a mass *m*. What mass of DNA would you expect to find in each mature gamete (egg and sperm) that forms after meiosis? What mass of DNA would you expect to find (1) in an egg fertilized by one of the sperm and (2) in that egg after the first DNA duplication?

2. Adam has a pair of alleles that influence whether a person is right- or left-handed. One allele says "left," and its partner says "right." Visualize one of his germ cells, in which chromosomes are being duplicated prior to meiosis. Visualize what happens to the chromosomes during anaphase I and II. (It might help to use index cards as models of the sister chromatids of each chromosome.) What fraction of Adam's sperm will carry the gene for right-handedness? For left-handedness?

3. Adam also has one allele for long eyelashes, and a partner allele (on the homologous chromosome) for short eyelashes. What fraction of his sperm will have these gene combinations:
   right-handed, long eyelashes   left-handed, long eyelashes
   right-handed, short eyelashes left-handed, short eyelashes

Selected Key Terms

allele 9.1	germ cell CI
asexual reproduction 9.1	haploid number 9.2
chromosome number 9.2	homologous chromosome 9.2
crossing over 9.4	meiosis 9.2
diploid number 9.2	oocyte 9.5
egg (ovum) 9.5	sexual reproduction 9.1
fertilization 9.5	sister chromatid 9.2
gamete CI	sperm 9.5
gene 9.1	spore 9.5

Readings

Klug, W., and M. Cummings. 1994. *Concepts of Genetics.* Fourth edition. New York: Macmillan.

Wolfe, S. 1995. *Introduction to Molecular and Cellular Biology.* Belmont, California: Wadsworth.

On-Line readings at Student Guide for InfoTrac: www.brookscole.com/biology

OBSERVABLE PATTERNS OF INHERITANCE

A Smorgasbord of Ears and Other Traits

Basketball ace Charles Barkley has them. So does actor Tom Cruise. Actress Joan Chen doesn't, and neither did a monk named Gregor Mendel. To see how you fit in with these folks, use a mirror to check out your ears. Is the fleshy lobe at the base of each ear attached to the side of your head? If so, you and Barkley and Cruise have something in common. Or is the fleshy lobe not attached, so that you can flap it back and forth? If so, you are like Chen and Mendel (Figure 10.1).

Charles Barkley

Tom Cruise

Whether a person is born with attached or detached earlobes depends on a single kind of gene. That gene comes in slightly different molecular forms—alleles. Only one form has information about detached lobes. The information is put to use while a human body is developing inside the mother. It calls for a death signal, which is sent to all cells positioned between the newly forming lobes and the head. Without that signal, the cells don't die and earlobes don't detach.

We all have genes for thousands of traits, including earlobes, cheeks, lashes, and eyeballs. Most traits vary in their details from one person to the next. Remember, we inherit pairs of genes on pairs of chromosomes. In some pairs, one allele has strong effects and overwhelms the other allele's contribution. The outgunned allele is said to be recessive to the dominant one. If you have detached earlobes, dimpled cheeks, long lashes, or large

Figure 10.1 Attached and detached earlobes of a few representative humans. This sampling provides observable evidence of a trait governed by a certain gene, which exists in different molecular forms in the human population. Do you have one or the other version of the trait? It depends on which molecular forms of the gene you inherited from your mother and father. As Gregor Mendel perceived, such easily observable traits can be used to identify patterns of inheritance from one generation to the next.

eyeballs, you have at least one and quite possibly two dominant alleles that affect the trait in a predictable way.

When both alleles of a pair are recessive, nothing masks their effect on a trait. You get *attached* earlobes with one pair of recessive alleles, *flat* feet with another pair, a *straight* nose with another pair, and so on.

How did we discover such remarkable things about our genes? It started with Mendel. By analyzing garden

Gregor Mendel

Joan Chen

pea plants generation after generation, Mendel found indirect but *observable* evidence of how parents bestow units of hereditary information—genes—on offspring. This chapter focuses on both the methods and the results of Mendel's experiments. They remain a classic example of how a scientific approach can pry open important secrets about the natural world. And to this day, they serve as the foundation for modern genetics.

Key Concepts

1. Genes are units of information about heritable traits. Alleles, which are slightly different molecular forms of a gene, specify different versions of the same trait.

2. Genes have specific locations on the chromosomes of a species. Humans, pea plants, and other organisms with a diploid chromosome number inherit *pairs* of genes, at equivalent locations on pairs of homologous chromosomes.

3. When the two members of each pair of homologous chromosomes are moved apart from each other during meiosis, their pairs of genes are moved apart, also, and end up in different gametes. Gregor Mendel found indirect evidence of this gene segregation when he crossbred pea plants showing different versions of the same trait, such as purple versus white flowers.

4. Each pair of homologous chromosomes in a germ cell is sorted out for distribution into one gamete or another independently of how the other pairs of homologous chromosomes are assorted. Mendel discovered indirect evidence of this when he tracked many plants having observable differences in two traits, such as flower color and plant height.

5. The contrasting forms of traits that Mendel happened to study were specified by nonidentical alleles. One allele was dominant, in that its effect on a trait masked the effect of a recessive allele paired with it.

6. Not all traits have such clearly dominant or recessive forms. One allele of a pair may be fully or partially dominant over its partner or codominant with it. Two or more gene pairs often influence the same trait, and some single genes influence many traits.

7. The environment introduces variation in traits.

MENDEL'S INSIGHT INTO INHERITANCE PATTERNS

More than a century ago, people wondered about the basis of inheritance. It was common knowledge that sperm and eggs both transmit information about traits to offspring. But few suspected that the information is organized in units (genes). Instead, the idea was that a father's blob of information "blended" with the mother's blob, like cream into coffee, at fertilization.

However, carried to its logical conclusion, blending would slowly dilute a population's shared pool of hereditary information until there was only a single version left of each trait. If that were so, then why did, say, freckles keep showing up among the children of nonfreckled parents through the generations? Why weren't all the descendants of a herd of white stallions and black mares gray? The blending theory could scarcely explain the obvious variation in traits that people could observe with their own eyes. Nevertheless, few disputed the theory.

Charles Darwin was among the scholarly dissidents. According to the key premise of his theory of natural selection, individuals of a population show variation in heritable traits. Through the generations, the variations that improve the chance of surviving and reproducing show up with greater frequency than those that do not. Less advantageous variations may persist among fewer individuals or they may even disappear. It is not that separate versions of a given trait are "blended out" of the population. Rather, *each version of a trait may persist in a population, at frequencies that can change over time.*

Just before Darwin presented his theory, someone was gathering evidence that eventually would support his key premise. A monk, Gregor Mendel, had already guessed that sperm and eggs carry distinct "units" of information about heritable traits. By carefully analyzing traits of pea plants generation after generation, Mendel found indirect but *observable* evidence of how parents transmit genes to offspring.

Mendel's Experimental Approach

Mendel spent most of his adult life in a monastery in Brno, a city near Vienna that has since become part of the Czech Republic. However, Mendel was not a man of narrow interests who accidentally stumbled onto principles of great import. The monastery of St. Thomas was close to European capitals that were the centers of scientific inquiry.

Having been raised on a farm, Mendel was aware of agricultural principles and their applications. He kept abreast of the breeding experiments and developments described in the available literature. He was a member of the regional agricultural society. He won awards for developing improved varieties of vegetables and fruits. Shortly after entering the monastery, he took courses in mathematics, physics, and botany at the University of Vienna. Few scholars of his time showed interest in both plant breeding and mathematics.

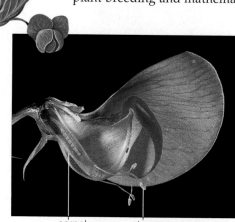

carpel   stamen

Figure 10.2 Garden pea plant (*Pisum sativum*), the organism Mendel chose for experimental tests of his ideas about inheritance.

a Garden pea flower. The section shows the location of its stamens and carpel. Sperm-producing pollen grains form in stamens. Eggs develop, fertilization takes place, and seeds mature inside carpels.

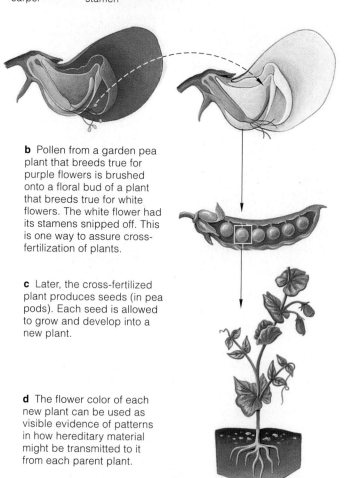

b Pollen from a garden pea plant that breeds true for purple flowers is brushed onto a floral bud of a plant that breeds true for white flowers. The white flower had its stamens snipped off. This is one way to assure cross-fertilization of plants.

c Later, the cross-fertilized plant produces seeds (in pea pods). Each seed is allowed to grow and develop into a new plant.

d The flower color of each new plant can be used as visible evidence of patterns in how hereditary material might be transmitted to it from each parent plant.

Figure 10.14 A rare albino rattlesnake. Like other animals that can't produce melanin, its body surface is white, overall, and its eyes are pink. In birds and mammals, surface coloration arises largely from pigments in feathers, fur, or skin. In fishes, amphibians, and reptiles, it depends on color-bearing cells. Some of these cells contain melanin or yellow-to-red pigments. Others contain crystals that reflect light and thus alter the surface coloration.

The mutation affecting melanin production in the snake shown here had no effect on the production of its yellow-to-red pigments and light-reflecting crystals. That's why the snake's skin appears to be iridescent yellow as well as white. Its eyes look pink because melanin is absent from a tissue layer in each eyeball. Without melanin to absorb it, red light is reflected from blood vessels in the eyes.

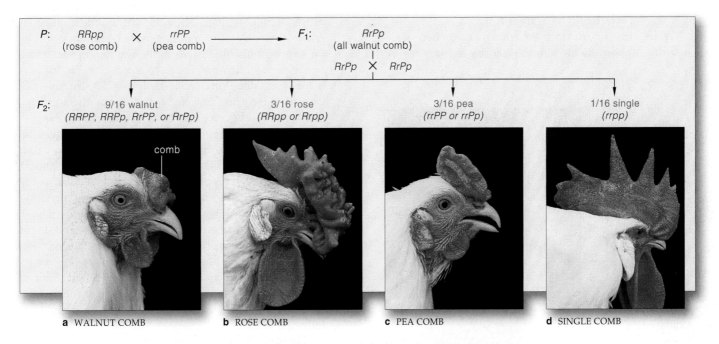

Figure 10.15 Interaction between two genes that affect the same trait in domestic chicken breeds. The initial cross is between a Wyandotte (with a rose comb, **b**, on the crest of its head) and brahma (pea comb, **c**). With complete dominance at the locus for pea comb and at the locus for rose comb, products of the two gene pairs interact and give rise to a walnut comb (**a**). With full recessiveness at both gene loci, products of the genes interact and give rise to a single comb (**d**).

Comb Shape in Poultry

In some cases, interaction between two gene pairs results in a phenotype that neither pair can produce by itself. Geneticists W. Bateson and R. Punnett identified two interacting gene pairs (R and P) that affect comb shape in chickens. Allelic combinations of rr at one gene locus and pp at the other locus result in the least common phenotype, the single comb. The presence of dominant alleles (R, P, or both) results in varied phenotypes.

Take a look at Figure 10.15. The diagram shows the combinations of alleles that interact to specify the rose, pea, and walnut combs shown in the photographs.

Genes often interact, as when alleles of one gene mask the expression of another gene, and when some expected phenotypes may not appear at all.

HOW CAN WE EXPLAIN LESS PREDICTABLE VARIATIONS?

Regarding the Unexpected Phenotype

As Mendel demonstrated, the phenotypic effects of one or two pairs of certain genes show up in predictable ratios when you track them from one generation to the next. Besides this, interactions among two or more gene pairs also can produce phenotypes in predictable ratios, as the example of Labrador coat color demonstrated in Section 10.6.

However, even if you were to track a single gene over the generations, you might find that the resulting phenotypes were not quite what you had expected.

Consider *camptodactyly*, a rare genetic abnormality that affects the shape and the movement of fingers. People who carry the mutant allele for this heritable trait have immobile, bent fingers on both hands. Other people have immobile, bent fingers on the left or right hand only. The fingers of others who carry the mutant allele aren't affected in any obvious way.

What causes such odd variation? Remember, most organic compounds are synthesized by a sequence of metabolic steps. *And different enzymes, each the product of a gene, regulate different steps.* Maybe one gene mutated in one of a number of possible ways. Maybe the gene product blocks the pathway or causes it to run nonstop or not long enough. Maybe poor nutrition or another factor that is variable in the individual's environment affects a crucial enzyme in the pathway. These are the kinds of variable factors that often introduce far less predictable variations in the phenotypes resulting from gene expression.

Continuous Variation in Populations

Generally, individuals of a population display a range of small differences in most traits. This characteristic of populations is called **continuous variation**. It's mainly an outcome of the number of genes affecting a trait and the number of environmental factors influencing their expression. Usually, the greater the number of genes and environmental factors, the more continuous will be the expected distribution of all the versions of that trait.

Look in a mirror at your eye color. The colored part is the iris, a doughnut-shaped, pigmented structure just beneath the cornea. Its color is the cumulative outcome of a number of gene products. Some products take part in the stepwise production and distribution of melanin, the same light-absorbing pigment that influences coat color in mammals. Dark eyes that seem to be almost black have heavy deposits of melanin molecules inside the iris. These molecules absorb most of the incoming light. Melanin deposits aren't as extensive in brown eyes, and some unabsorbed light is being reflected out. Light brown or hazel eyes have even less (Figure 10.16).

Green, gray, or blue eyes don't have green, gray, or blue pigments. Their iris incorporates some amount of melanin, but not much. As a result, many or most of the blue wavelengths of light that do enter the eyeball are reflected out.

How might you describe the continuous variation of some trait within a group, such as the students in Figure 10.17a? The students range from very short to very tall, with average heights much more common than either extreme. You might start out by dividing the full range of different phenotypes into measurable categories. Next, you can count all of the individual students in each category. Doing so will give you the relative frequencies of all the phenotypes, distributed across the range of measurable values.

The bar chart in Figure 10.17c plots the proportion of students in each category against the range of the measured phenotypes. Here, the shortest vertical bars represent categories with the least number of students. The tallest bar represents the category with the greatest number of students. Finally, draw a graph line around

Figure 10.16
A small sampling from the range of continuous variation in the color of human eyes. Products of different pairs of genes interact to produce and then deposit the pigment melanin. Among other things, melanin helps color the eye's iris. Different combinations of alleles result in small differences in eye color. So the frequency distribution for the eye-color trait appears to be continuous over a range from black to light blue.

1	4	8	10	16	16	15	15	14	13	13	11	9	8	8	5	1	2

Number of individuals

60	61	62	63	64	65	66	67	68	69	70	71	72	73	74	75	76	77

Height (inches)

a Students organized according to height, as an example of continuous variation

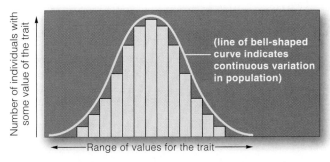

(line of bell-shaped curve indicates continuous variation in population)

Range of values for the trait

b Idealized bell-shaped curve for a population that displays continuous variation in some trait

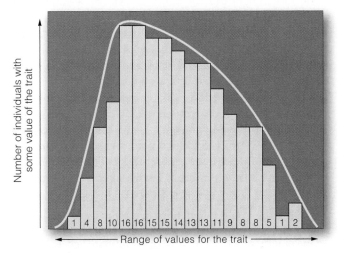

1 4 8 10 16 16 15 15 14 13 13 11 9 8 8 5 1 2

Range of values for the trait

c Specific bell-shaped curve corresponding to the distribution of the trait (height) illustrated by the photograph in (**a**)

Figure 10.17 Continuous variation in body height, a trait that is one of the characteristics of the human population.

(**a**) Suppose you wish to know the frequency distribution for height in a group of 169 biology students at Brigham Young University. You decide on how finely the range of possible heights should be divided. Then you measure each student and assign her or him to the appropriate category. Finally, you divide the number in each category by the total number of all students in all categories.

(**b**) Often, a bar graph is used to depict continuous variation in a population. In such graphs, the proportion of individuals in each category is plotted against the range of measured phenotypes. The curved line above this particular set of bars is an idealized example of the kind of "bell-shaped" curve that emerges for populations showing continuous variation in a trait. The bell-shaped curve in (**c**) is a specific example of this type of diagram.

all the bars and you end up with a bell-shaped curve. Such "bell curves" are typical of populations that show continuous variation in a trait.

Enzymes and other products of genes regulate each step of most metabolic pathways. Mutations, gene interactions, and environmental conditions may affect one or more of the steps. The outcome is variation in phenotypes.

For most traits, individuals of a population show continuous variation—that is, a range of small differences.

The greater the number of genes and environmental factors that can influence a trait, the more continuous will be the expected distribution of all versions of that trait.

ENVIRONMENTAL EFFECTS ON PHENOTYPE

We have mentioned, in passing, that the environment often contributes to variable gene expression among a population's individuals. Before leaving this chapter, consider a few examples of phenotypic variations that arise from environmental effects on genotype.

Possibly you've seen a Himalayan rabbit or Siamese cat. Both furry mammals carry an allele that specifies a heat-sensitive form of one of the enzymes for melanin production. The enzyme is less active at the surface of warm body parts. There, the fur color is lighter than in cooler parts, such as ears and limbs projecting from the main body mass. Figure 10.18 gives an experiment that demonstrated an environmental effect on this allele.

We also can identify environmental effects on genes that govern plant phenotypes. Plant a hydrangea in a garden and you may get pink instead of the expected blue blossoms (Figure 10.19). Genes for its floral color are affected by soil acidity.

Also consider an experiment with yarrow plants. Yarrow can grow from cuttings, which makes it a fine experimental organism; cuttings from the same plant all have the same genotype. Three cuttings were grown at three elevations. The two plants grown at the lowest and highest elevation fared best; the one at the medium elevation grew poorly, as Figure 10.20 shows.

Figure 10.19 Environmental effect on gene expression in *Hydrangea macrophylla*, a favorite garden plant. Even different plants that carry the same alleles have floral colors ranging from pink to blue. The color variation arises because of differences in the acidity of soil in which they happen to be growing.

a Mature cutting at high elevation (3,050 meters)

b Mature cutting at medium elevation (1,400 meters)

c Mature cutting at low elevation (30 meters above sea level)

Figure 10.20 Experiment demonstrating the effect of environmental conditions on gene expression in a yarrow plant (*Achillea millefolium*). Cuttings from the same parent plant were grown in the same soil medium but at three different elevations.

Icepack is strapped onto a hair-free patch.

New hair growing in patch exposed to cold is black.

Figure 10.18 Observable effect of environmental conditions on gene expression in animals. A Himalayan rabbit normally has black hair only on its long ears, nose, tail, and lower leg limbs. For one experiment, a patch of a rabbit's white fur was plucked clean, then an icepack was secured over the hairless patch. Where the colder temperature had been maintained, the hairs that grew back were black.

Himalayan rabbits are homozygous for the *ch* allele of the gene for tyrosinase, an enzyme required to make melanin. This allele specifies a heat-sensitive form of the enzyme, which functions only when the air temperature is below about 33°C. When cells that give rise to hairs grow under warmer conditions, they cannot make melanin, so hairs appear light. This happens in body regions massive enough to conserve a fair amount of metabolic heat. Ears and other slender extremities are cooler because they tend to lose metabolic heat faster.

But remember the sampling error trap (Chapter 1)? The researchers did the same growth experiments for *many* yarrow plants. Phenotypic variation was so great, they couldn't see a consistent pattern in it. For instance, a cutting from one plant developed best at the medium elevation. The conclusion? Different yarrow genotypes react differently across a range of environments.

And so we conclude this chapter, which has dealt with heritable and environmental factors that give rise to variations in phenotype. What is the take-home lesson? Simply this: An individual's phenotype is an outcome of complex interactions among its genes, enzymes and other gene products, and of environmental factors.

Individuals of most populations or species show complex variation for many traits. The variation arises not only from gene mutations and cumulative gene interactions. It arises also as responses to variations in environmental conditions.

1. A gene is a unit of information about a heritable trait. Alleles of a gene are different molecular versions of that information. Through experimental crosses with pea plants, Mendel gathered indirect evidence that diploid organisms have two genes for each trait and that genes keep their identity when transmitted to offspring. *10.1*

2. A homozygous dominant individual has inherited two dominant alleles (*AA*) for the trait being studied. A homozygous recessive has two recessive alleles (*aa*). A heterozygote has two nonidentical alleles (*Aa*). *10.1*

3. For monohybrid crosses, two individuals that breed true for different versions of the same trait produce F_1 offspring that are all heterozygous (*Aa*) for the trait. The experiment is an intercross between two of these F_1 heterozygotes—that is, the monohybrids (*Aa* × *Aa*). Mendel's monohybrid crosses with garden pea plants provided indirect evidence that some forms of a gene may be dominant over other, recessive forms. *10.2*

4. All F_1 offspring of the parental cross *AA* × *aa* were *Aa*. Mendel's crosses between F_1 monohybrids resulted in these combinations of alleles in F_2 offspring: *10.2*

	A	a
A	AA	Aa
a	Aa	aa

AA (dominant)
Aa (dominant)
Aa (dominant)
aa (recessive)
⎫
the expected phenotypic ratio of 3:1

5. Results from Mendel's monohybrid crosses led to the formulation of a theory of segregation. In modern terms, diploid organisms have pairs of genes, on pairs of homologous chromosomes. The genes of each pair segregate from each other at meiosis, so each gamete formed ends up with one or the other gene. *10.2*

6. For dihybrid crosses, individuals that breed true for different versions of *two* traits produce F_1 offspring that are all identical heterozygotes for both traits. The experiment is an intercross between two F_1 dihybrids. Phenotypes of the F_2 offspring from Mendel's dihybrid crosses were close to a 9:3:3:1 ratio: *10.3*

9 dominant for both traits
3 dominant for *A*, recessive for *b*
3 dominant for *B*, recessive for *a*
1 recessive for both traits

7. Mendel's dihybrid crosses led to the formulation of a theory of independent assortment. In modern terms, by the end of meiosis, the gene pairs of two homologous chromosomes have been sorted out for distribution into one gamete or another, independently of how the gene pairs of other chromosomes were sorted out. *10.3*

8. Certain factors influence gene expression. *10.4–10.8*
  a. In some cases, one allele of a pair is incompletely dominant or codominant. *10.4*

  b. Products of pairs of genes may interact in ways that influence the same trait. *10.4, 10.6, 10.7*

  c. One gene may have positive or negative effects on two or more traits, a condition called pleiotropy. *10.5*

  d. Environmental conditions to which an individual is subjected may affect gene expression. *10.8*

Review Questions

1. Distinguish between these terms: *10.1*
  a. gene and allele
  b. dominant allele and recessive allele
  c. homozygote and heterozygote
  d. genotype and phenotype

2. Define a true-breeding lineage. What is a hybrid? *10.1*

3. Distinguish between monohybrid cross, dihybrid cross, and testcross. *10.2, 10.3*

4. Do segregation and independent assortment occur in mitosis, meiosis, or both? *10.2, 10.3*

5. What do the vertical and horizontal arrows of the diagram at right represent? What do the bars and the curved line represent? *10.7*

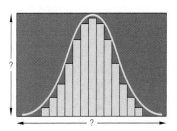

Self-Quiz   ANSWERS IN APPENDIX III

1. Alleles are _____.
  a. different molecular forms of a gene
  b. different phenotypes
  c. self-fertilizing, true-breeding homozygotes

2. A heterozygote has a _____ for the trait being studied.
  a. pair of identical alleles
  b. pair of nonidentical alleles
  c. haploid condition, in genetic terms
  d. a and c

3. The observable traits of an organism are its _____.
  a. phenotype   c. genotype
  b. sociobiology   d. pedigree

4. Second-generation offspring from a cross are the _____.
  a. F_1 generation   c. hybrid generation
  b. F_2 generation   d. none of the above

5. F_1 offspring of the monohybrid cross *AA* × *aa* are _____.
  a. all *AA*   c. all *Aa*
  b. all *aa*   d. 1/2 *AA* and 1/2 *aa*

6. Assuming complete dominance, the F_2 generation following the cross *Aa* × *Aa* will show a phenotypic ratio of _____.
  a. 3:1   c. 1:2:1
  b. 9:1   d. 9:3:3:1

7. Crosses between F_1 pea plants resulting from the cross *AABB* × *aabb* lead to F_2 phenotypic ratios close to _____.
  a. 1:2:1   c. 1:1:1:1
  b. 3:1   d. 9:3:3:1

8. Match each example with the most suitable description.
  _____ dihybrid cross   a. *bb*
  _____ monohybrid cross   b. *AaBb* × *AaBb*
  _____ homozygous condition   c. *Aa*
  _____ heterozygous condition d. *Aa* × *Aa*

1. One gene has alleles *A* and *a*. Another has alleles *B* and *b*. For each genotype listed, what type(s) of gametes will be produced? Assume independent assortment occurs before gametes form.

   a. *AABB*   c. *Aabb*
   b. *AaBB*   d. *AaBb*

2. Refer to Problem 1. What will be the genotypes of offspring from the following matings? Indicate the frequencies of each genotype among them.

   a. *AABB × aaBB*   c. *AaBb × aabb*
   b. *AaBB × AABB*   d. *AaBb × AaBb*

3. In one experiment, Mendel crossed a pea plant that bred true for green pods with one that bred true for yellow pods. All the F_1 plants had green pods. Which form of the trait (green or yellow pods) is recessive? Explain how you arrived at your conclusion.

4. Return to Problem 1. Assume you now study a third gene having alleles *C* and *c*. For each genotype listed, what type(s) of gametes will be produced?

   a. *AABB CC*   c. *Aa BB Cc*
   b. *Aa BB cc*   d. *Aa Bb Cc*

5. Mendel crossed a true-breeding tall, purple-flowered pea plant with a true-breeding dwarf, white-flowered plant. All F_1 plants were tall and had purple flowers. If an F_1 plant self-fertilizes, then what is the probability that a randomly selected F_2 offspring will be heterozygous for the genes specifying height and flower color?

6. At a gene location on a human chromosome, a dominant allele controls *tongue rolling*, an ability to curl up the two sides of the tongue (Figure 10.21). People who are homozygous for a recessive allele at that locus can't roll their tongue. At another gene locus, a dominant allele controls whether the earlobes will be attached or detached (refer to Figure 10.1). These two pairs of genes assort independently. Suppose a tongue-rolling, detached-earlobe woman marries a man who has attached earlobes and can't roll his tongue. Their first child has the phenotype of the father. Given this outcome,

   a. What are the genotypes of the mother, father, and child?

   b. What is the probability that a second child of theirs will have detached earlobes and won't be a tongue roller?

7. Bill and his wife, Marie, are hoping to have children. Both have notably flat feet and long eyelashes, and they tend to sneeze a lot (hence the name of the *achoo syndrome*). Dominant alleles give rise to these traits: *A* (foot arch), *E* (eyelash length), and *S* (chronic sneezing). Bill is heterozygous for all three dominant alleles and Marie is homozygous.

   a. What is Bill's genotype? What is Marie's genotype?

   b. Marie becomes pregnant four times. What is the probability each child will have the dominant phenotypes for all three traits?

   c. What is the probability that each child will have short lashes, high arches, and no chronic tendency to sneeze?

8. *DNA fingerprinting* is a method of identifying individuals by locating unique base sequences in their DNA molecules (Section 15.3). Before researchers refined the method, attorneys often relied on the ABO blood-typing system to settle disputes over paternity. Suppose that you, as a geneticist, are asked to testify during a paternity case in which the mother has type A blood, the child has type O blood, and the alleged father has type B blood. How would you respond to the following statements?

   a. Attorney of the alleged father: "The mother's blood is type A, so the child's type O blood must have come from the father. My client has type B blood; he could not be the father."

   b. Mother's attorney: "Because further tests prove this man is heterozygous, he must be the father."

9. Suppose you identify a new gene in mice. One of its alleles specifies white fur. A second allele specifies brown fur. You want to determine whether the relationship between the two alleles is one of simple dominance or incomplete dominance. What sorts of genetic crosses would give you the answer? On what types of observations would you base your conclusions?

10. Your sister moves away and gives you her purebred Labrador retriever, a female named Dandelion. Suppose you decide to breed Dandelion and sell puppies to help pay for your college tuition. Then you discover that two of her four brothers and sisters show *hip dysplasia*, a heritable disorder arising from a number of gene interactions. If Dandelion mates with a male Labrador known to be free of the harmful genes, can you guarantee to a buyer that puppies will not develop the disorder? Explain your answer.

11. A dominant allele *W* confers black fur on guinea pigs. A guinea pig that is homozygous recessive (*ww*) has white fur. Fred would like to know whether his pet black-furred guinea pig is homozygous dominant (*WW*) or heterozygous (*Ww*). How might he determine his pet's genotype?

12. Red-flowering snapdragons are homozygous for allele R^1. White-flowering snapdragons are homozygous for a different allele (R^2). Heterozygous plants (R^1R^2) bear pink flowers. What phenotypes should appear among F_1 offspring of the crosses listed? What are the expected proportions for each phenotype?

   a. $R^1R^1 \times R^1R^2$ c. $R^1R^2 \times R^1R^2$
   b. $R^1R^1 \times R^2R^2$ d. $R^1R^2 \times R^2R^2$

Notice, in Problem 12, that in cases of incomplete dominance it is inappropriate to refer to either allele of a pair as dominant or recessive. When the phenotype of a heterozygous individual is halfway between those of the two homozygotes, then there is no dominance. Such alleles are usually designated by superscript numerals, as shown here, rather than by uppercase letters for dominance and lowercase letters for recessiveness.

Figure 10.21 A student at San Diego State University exhibiting the tongue-rolling trait for the benefit of a tongue-roll-challenged student.

13. Two pairs of genes affect comb type in chickens (Figure 10.15). When both genes are recessive, a chicken has a single comb. A dominant allele of one gene, P, gives rise to a pea comb. Yet a dominant allele of the other (R) gives rise to a rose comb. An epistatic interaction occurs when a chicken has at least one of both dominants, $P\!-\!R\!-\!$, which gives rise to a walnut comb. Predict the F_1 ratios resulting from a cross between two walnut-combed chickens that are heterozygous for both genes ($PpRr$).

14. As described in Section 10.5, a single mutant allele gives rise to an abnormal form of hemoglobin (Hb^S instead of Hb^A). Homozygotes (Hb^SHb^S) develop sickle-cell anemia. But the heterozygotes (Hb^AHb^S) show few outward symptoms.

Suppose a woman's mother is homozygous for the Hb^A allele and her father is homozygous for the Hb^S allele. She marries a male who is heterozygous for the allele, and they plan to have children. For *each* of her pregnancies, state the probability that this couple will have a child who is:

    a. homozygous for the Hb^S allele

    b. homozygous for the Hb^A allele

    c. heterozygous Hb^AHb^S

15. Certain dominant alleles are so vital for normal development that an individual who is homozygous recessive for a mutant recessive form of the allele is unable to survive. Such recessive, *lethal alleles* can be perpetuated by heterozygotes.

Consider the Manx allele (M^L) in cats. Homozygous cats (M^LM^L) die when they are still embryos inside the mother cat. In heterozygotes (M^LM), the spine develops abnormally, and the cats end up with no tail whatsoever (Figure 10.22).

Two M^LM cats mate. Among their *surviving* progeny, what is the probability that any one kitten will be heterozygous?

16. A recessive allele *a* is responsible for *albinism*, an inability to produce or deposit melanin in tissues. Humans and some other organisms can have this phenotype (Figure 10.23). In each of the following cases, what are the possible genotypes of the father, of the mother, and of their children?

    a. Both parents have normal phenotypes; some of their children are albino and others are unaffected.

    b. Both parents are albino and have only albino children.

    c. The woman is unaffected, the man is albino, and they have one albino child and three unaffected children.

Figure 10.22 Manx cat.

Figure 10.23 An albino male in India.

17. Kernel color in wheat plants is determined by two pairs of genes. Alleles of one pair show incomplete dominance over alleles of the other pair.

For the gene pair at one locus on the chromosome, allele A^1 imparts one dose of red color to the kernel, whereas allele A^2 does not. At the second locus, allele B^1 gives one dose of red color to the kernel, whereas allele B^2 does not. One kernel with genotype $A^1A^1B^1B^1$ is dark red. A different kernel with genotype $A^2A^2B^2B^2$ is white. All other genotypes have kernel colors in between the two extremes.

    a. Suppose you cross a plant grown from a dark red kernel with a plant grown from a white kernel. What genotypes and what phenotypes would you expect among the offspring?

    b. If a plant with genotype $A^1A^1B^1B^2$ self-fertilizes, what genotypes and what phenotypes would be expected among the offspring? In what proportions?

Selected Key Terms

allele *10.1*	hybrid offspring *10.1*
codominance *10.4*	incomplete dominance *10.4*
continuous variation *10.7*	independent assortment *10.3*
dihybrid cross *10.3*	monohybrid cross *10.2*
epistasis *10.6*	multiple allele system *10.4*
F_1 *10.1*	phenotype *10.1*
F_2 *10.1*	pleiotropy *10.5*
gene *10.1*	probability *10.2*
genotype *10.1*	Punnett-square method *10.2*
heterozygous *10.1*	segregation *10.2*
homozygous dominant *10.1*	testcross *10.2*
homozygous recessive *10.1*	true-breeding lineage *10.1*

Readings

Fairbanks, D. J., and W. R. Andersen. 1999. *Genetics: The Continuity of Life*. Monterey, California: Brooks-Cole.

Orel, V. 1996. *Gregor Mendel: The First Geneticist*. New York: Oxford University Press.

On-Line readings at Student Guide for InfoTrac:
www.brookscole.com/biology

11

CHROMOSOMES AND HUMAN GENETICS

The Philadelphia Story

Positioned at strategic locations in chromosomes are genes concerned with cell growth and division. Some specify enzymes and other proteins that perform these tasks. Neighboring genes control when, whether, and how fast the tasks proceed. If something disturbs the neighborhood, cell growth and division can spiral out of control and lead to cancer.

The first abnormal chromosome tied to cancer was named the *Philadelphia chromosome* after the city where someone discovered it. This chromosome shows up in cells of humans affected by a type of leukemia.

The disorder starts with stem cells in bone marrow. All stem cells are unspecialized and retain the capacity for mitotic cell division. A portion of their descendants divide and become specialized. With leukemias, there are too many descendants—far too many of the white blood cells in charge of the body's housekeeping and defense. Leukemic cells may crowd out the stem cells that give rise to red blood cells and platelets. Anemia and internal bleeding follow. Leukemic cells also enter blood and the liver, lymph nodes, spleen, and other organs, where they disrupt basic functions. Left to itself, the cancerous transformation kills the patient.

No one knew about the Philadelphia chromosome until microscopists learned how to identify its physical appearance. Chromosomes, recall, are most condensed at metaphase of mitosis. Then, their size, length, and centromere location are easiest to identify. A **karyotype** is a preparation of metaphase chromosomes based on their defining features. In this chapter, you will learn how to make a karyotype diagram from a photograph of metaphase chromosomes, just like microscopists do. The idea is to line up all of a cell's chromosomes in a certain way and to compare them with a standard karyotype for the species.

It hasn't been easy to identify the Philadelphia chromosome in karyotypes. But *spectral* karyotyping changed things. This newer research and diagnostic tool artificially colors chromosomes, and the colors are clues to their structure (Figure 11.1).

As it turns out, the Philadelphia chromosome is physically longer than its normal counterpart, human chromosome 9. The extra length is actually a piece of chromosome 22! What happened? By chance, both chromosomes broke inside a stem cell. Each broken piece was reattached, but on the wrong chromosome. At the end of chromosome 9, a gene with a role in cell division fused with a gene in chromosome 22. The altered gene region specifies an abnormal protein, and uncontrolled divisions of white blood cells follow.

The Philadelphia story is a glimpse into the world of modern genetics research. Reading it invites you to think about how far you have come in this unit of the book. You first looked at cell division, the starting point of inheritance. You looked at how chromosomes and the genes they carry are shuffled during meiosis, then at fertilization. You mulled over Mendel's insights into inheritance and some exceptions to his conclusions. Here, you have a current example of what we know about the chromosomal basis of inheritance. *How did we get to that understanding?* The answer requires a bit of history, which picks up from where we left Mendel.

In 1884 Mendel had just passed away, and his paper on pea plants had been gathering dust in a hundred

Figure 11.1 Revealing the origin of a killer—how reciprocal translocation between human chromosome 9 and chromosome 22 might appear with the help of spectral karyotyping, a new imaging technique. This particular translocation, the Philadelphia chromosome, results in a gene abnormality that gives rise to chronic myelogenous leukemia (CML), a type of cancer. Gleevec, a new oral drug, inactivates the abnormal protein product of the gene, thus arresting the dangerous cell proliferation. In preliminary tests, the drug put fifty of fifty-three CML patients into remission. So far, its side effects have been mild, compared to severe side effects of chemotherapy.

Figure 11.2 The DNA icon for an Internet site, *Rare Genetic Diseases in Children*, which allows interested individuals to seek information about specific genetic diseases and disorders. At this writing, the Philadelphia chromosome is on the site's topic board.

libraries for nearly two decades. Then the resolving power of microscopes improved and rekindled interest in the hereditary material. Walther Flemming had seen threadlike bodies—chromosomes—in dividing cells. Could chromosomes be the hereditary material?

Microscopists soon realized that each gamete has half the number of chromosomes of a fertilized egg. In 1887, August Weismann hypothesized that a special division process must reduce the chromosome number by half before gametes form. Sure enough, in that same year meiosis was discovered. Weismann promoted another hypothesis: If the chromosome number is halved at meiosis and restored at fertilization, then half of the chromosomes in our cells must come from the father and half from the mother. This view was hotly debated. It prompted a flurry of experimental crosses—just like the ones Mendel had carried out.

Finally, in 1900, researchers came across Mendel's paper while checking literature related to their own genetic crosses. To their surprise, their experimental results confirmed what Mendel's results had already suggested: Diploid cells have two units (genes) for each heritable trait, and the two units are segregated from each other before gametes form.

In the decades to follow, researchers learned more about chromosomes. You'll come across a few high points of their work as we continue with our look at inheritance. As the Philadelphia story tells you, the methods of analysis are not remote from your interests. An inherited collection of information in chromosomal DNA gives rise to traits that, for better or worse, define each organism, young and old alike (Figure 11.2).

Key Concepts

1. Cells of humans and many other sexually reproducing species contain pairs of homologous chromosomes that interact during meiosis. Typically, one chromosome of each pair is maternal in origin, and its homologue is paternal in origin.

2. Each kind of gene has its own specific location, or locus, in a particular type of chromosome.

3. The molecular form of a gene occupying a given locus may be slightly different from one chromosome to the next. All of the different molecular forms of a gene are called alleles.

4. The combination of alleles along the length of a chromosome does not necessarily remain intact through meiosis and gamete formation. By the event called crossing over, some alleles along its length swap places with their partner on the homologous chromosome. Alleles that swap places may or may not be identical.

5. Allelic recombinations contribute to variations in the phenotypes of offspring.

6. A chromosome may change structurally, as when a segment of it is deleted, duplicated, inverted, or moved to a new location. Also, the chromosome number of an individual's cells may change as a result of an improper separation of duplicated chromosomes during meiosis or mitosis.

7. Chromosome structure and the parental chromosome number rarely change. When changes do occur, they may result in genetic abnormalities or genetic disorders.

CHROMOSOMES AND INHERITANCE

Genes and Their Chromosome Locations

Earlier chapters described the structure of chromosomes and what happens to them during meiosis. To refresh your memory and get a general sense of where you are going from here, take a moment to read this list:

1. **Genes** are units of information about heritable traits. The genes of eukaryotic cells are distributed among a number of chromosomes. Each gene has its own location—a gene locus—in one type of chromosome.

2. Any cell with a diploid chromosome number (2*n*) has inherited pairs of **homologous chromosomes**. All but one pair are identical in length, shape, and gene sequence. The single exception is a pairing of nonidentical sex chromosomes, such as X with Y. The two members of a pair of homologous chromosomes interact and segregate from each other during meiosis.

3. A gene at one locus may have the same form or a slightly different one compared to its partner gene on the homologous chromosome. When considering a population as a whole, *which* forms are inherited usually varies from one individual to the next.

4. All the different molecular forms of a gene that are possible at a given locus are called **alleles**. New alleles arise only through mutation.

5. A *wild-type* allele is the most common form of a gene, either in a natural population or in a standard, laboratory-bred strain of a species. Any one of the less common forms of a gene is a *mutant* allele.

6. Genes on the same chromosome are physically linked together. The farther apart two linked genes are, the more vulnerable they are to **crossing over**. By this event, homologous chromosomes exchange corresponding segments (Figure 11.3*a,b*).

7. Crossing over results in **genetic recombination**. The term refers to nonparental combinations of alleles in gametes, then in offspring (Figure 11.3*c*).

8. **Independent assortment** refers to the random alignment of each pair of homologous chromosomes at metaphase I of meiosis. It results in nonparental combinations of alleles in gametes and offspring.

9. On rare occasions, the structure of chromosomes changes abnormally during mitosis or meiosis. So does the parental chromosome number.

Autosomes and Sex Chromosomes

In all but one case, a pair of homologous chromosomes are exactly alike in length, shape, and gene sequence. Microscopists discovered the exception in the late 1800s, using analytical methods such as karyotyping (Section 11.2). A distinctive chromosome is present in females *or* males of many species, but not in both. As an example,

a A pair of duplicated homologous chromosomes (two sister chromatids each). This example has nonidentical alleles at three gene loci (*A* with *a*, *B* with *b*, and *C* with *c*).

b In prophase I of meiosis, a crossover event occurs: two nonsister chromatids exchange corresponding segments.

c This is the outcome of the crossover: genetic recombination between nonsister chromatids (which are shown after meiosis as unduplicated, separate chromosomes).

Figure 11.3 Review of crossing over. As shown earlier in Section 9.4, this event occurs in prophase I of meiosis.

a human male diploid cell has one **X chromosome** and one **Y chromosome** (XY). A human female diploid cell has two X chromosomes (XX). This inheritance pattern is typical of many species, including all mammals and fruit flies. We see a different pattern in butterflies, moths, some birds, and certain fishes. For them, inheriting two identical sex chromosomes results in a male. Inheriting two nonidentical sex chromosomes results in a female.

The human X and Y chromosomes differ physically; one is shorter than the other (Figure 11.4*f*). They also differ in which genes they carry. Even so, they are able to synapse (become zippered together briefly) in a small region along their length. The bit of zippering lets them function as homologues during meiosis.

The human X and Y chromosomes fall into the more general category of **sex chromosomes**. The term refers to types of chromosomes that, in certain combinations, determine a new individual's sex—whether a male or a female will develop. All other chromosomes in cells are the same in both sexes; they are **autosomes**.

Diploid cells have pairs of genes, on pairs of homologous chromosomes. At each gene locus, the alleles (alternative forms of a gene) may be identical or nonidentical.

Crossing over and other events during meiosis give offspring new combinations of alleles and parental chromosomes.

Abnormal events during meiosis or mitosis can change the structure and number of chromosomes.

Autosomes are the pairs of chromosomes that are the same in males and females of a species. One other pair, the sex chromosomes, govern the sex of a new individual.

Karyotyping Made Easy

Karyotype diagrams help answer questions about an individual's chromosomes. Chromosomes are the most condensed and easiest to identify in cells going through metaphase of mitosis. Technicians don't count on finding a cell that happens to be dividing in the body when they go looking for it. They culture cells in vitro (literally, "in glass"). They put a sample of cells, usually from blood, into a glass container. The container holds a solution that stimulates cell growth and mitotic cell divisions.

Adding colchicine to a culture medium arrests cell division at metaphase. This extract of *Colchicum* plants also blocks spindle formation (Section 4.10). If a spindle can't form, sister chromatids of duplicated chromosomes won't be able to separate at anaphase. By using suitable colchicine concentrations and exposure times, technicians can stockpile metaphase cells. This improves the odds of finding candidates for karyotype diagrams.

Following colchicine treatment, the culture medium is transferred to tubes of a centrifuge (Figure 11.4a). Cells have greater mass and density than the solution bathing them. The spinning force moves them farthest from the center of rotation, to the bottom of the attached tubes.

The cells are transferred to a saline solution. When immersed in this hypotonic fluid, they swell (by osmosis) and move apart. The metaphase chromosomes also move apart. The cells are ready to be mounted on a microscope slide, fixed as by air-drying, and stained.

Chromosomes take up some stains uniformly along their length, which allows identification of chromosome size and shape. By other staining procedures, horizontal bands show up along the length of the chromosomes of certain species. If researchers direct a ray of ultraviolet light at the chromosomes, the bands will fluoresce. You can see an example of this in Figure 9.3.

The chromosomes are photographed through the microscope, and the image is enlarged. The photograph is cut apart, one chromosome at a time. Individual cutouts are arranged according to size, shape, and length of the arms. All pairs of homologous chromosomes are horizontally aligned by their centromeres. Figure 11.4f shows a karyotype diagram prepared this way.

a Add cells from a small blood sample to a medium that has a chemical stimulator for mitosis. Add colchicine to arrest mitosis at metaphase. Transfer culture to a centrifuge. (This motor-driven rotary device spins test tubes at high speed. Tube contents respond to the centrifugal force according to their mass, density, and shape.)

b Centrifugation forces cells to bottom of tube. Draw off culture medium. Add a dilute saline solution to tube. Add a fixative.

c Prepare and stain cells for microscopy.

d Put cells on a microscope slide. Observe.

e Photograph one cell through microscope. Enlarge image of its chromosomes. Cut the image apart. Arrange chromosomes as a set.

Figure 11.4 (a–e) How to prepare karyotypes. (f) Human karyotype. Human somatic cells hold 22 pairs of autosomes and 1 pair of sex chromosomes (XX or XY). That is a diploid number of 46. These are metaphase chromosomes; each is in the duplicated state, consisting of two sister chromatids joined at the centromere.

SEX DETERMINATION IN HUMANS

Analyzing human cells, as by karyotyping, has yielded evidence that every normal egg produced by a female has one X chromosome. Half the sperm cells produced by a male carry an X chromosome and half carry a Y.

If an X-bearing sperm fertilizes an X-bearing egg, the new individual will develop into a female. If the sperm happens to carry a Y chromosome, the individual will develop into a male (Figure 11.5).

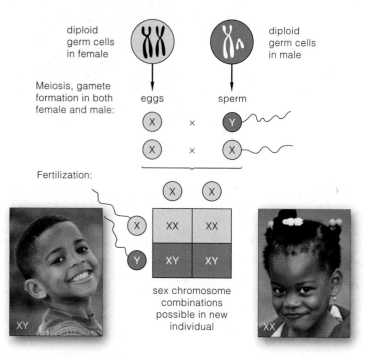

Figure 11.5 Pattern of sex determination in humans.

At this writing, fewer than two dozen genes have been identified on the human Y chromosome. One of them is the master gene for male sex determination. Its expression leads to the formation of testes, the primary male reproductive organs (Figure 11.6). In that gene's absence, ovaries form automatically. Ovaries are the primary female reproductive organs. Testes and ovaries both produce important sex hormones that influence the development of particular sexual traits.

The human X chromosome carries more than 2,300 genes. Like other chromosomes, it carries some genes associated with sexual traits, such as the distribution of body fat and hair. However, most of its genes deal with *nonsexual* traits, such as blood-clotting functions. These genes can be expressed in males as well as in females (males, remember, also carry one X chromosome).

A certain gene on the human Y chromosome dictates that a new individual will develop into a male. In the absence of the Y chromosome (and the gene), a female develops.

---

a A human embryo, eight weeks old and about an inch long. The mass at left is part of the placenta, an organ that forms from maternal and embryonic tissues.

Figure 11.6 Boys, girls, and the Y chromosome.

For about the first four weeks of its existence, a human embryo has neither male nor female traits, regardless of whether it is XY or XX. Then ducts and other internal structures that can develop either way start forming.

(a–c) In an XX embryo, ovaries (the primary female reproductive organs) start to form automatically—*in the absence of a Y chromosome*. By contrast, in an XY embryo, testes (the primary male reproductive organs) start to form during the next four to six weeks. Apparently, a gene region on the Y chromosome governs a fork in the developmental road that can lead to maleness.

The newly forming testes start to produce testosterone and other sex hormones. These hormones are crucial for the development of a male reproductive system. By contrast, in an XX embryo, newly forming ovaries start to produce different kinds of sex hormones, particularly estrogens. Estrogens are crucial for the development of a female reproductive system.

The master gene for male sex determination is named *SRY* (short for the *S*ex-determining *R*egion of the *Y* chromosome). The same gene has been identified in DNA from male humans, chimpanzees, mice, rabbits, pigs, horses, cattle, and tigers, among others. None of the females tested had the gene. Tests with mice indicate that the gene region becomes active about the time that testes are starting to develop.

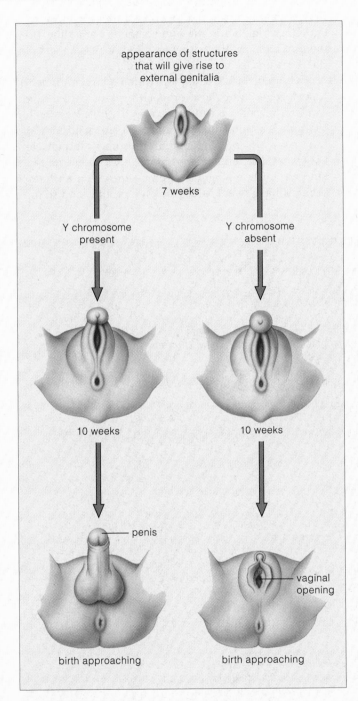

appearance of structures
that will give rise to
external genitalia

7 weeks

Y chromosome
present

Y chromosome
absent

10 weeks

10 weeks

penis

vaginal
opening

birth approaching

birth approaching

b External appearance of reproductive organs
forming in human embryos.

The *SRY* gene resembles DNA regions that specify regulatory
proteins. As described in Chapter 14, regulatory proteins
influence the activity of genes and their products. The *SRY*
gene product apparently regulates a cascade of reactions
that are necessary for male sex determination.

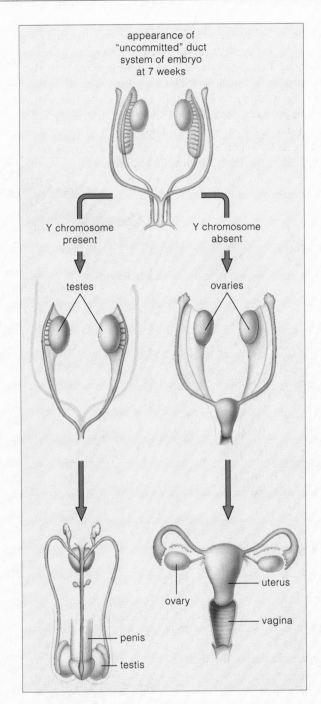

appearance of
"uncommitted" duct
system of embryo
at 7 weeks

Y chromosome
present

Y chromosome
absent

testes

ovaries

ovary

uterus

penis

vagina

testis

c A duct system that forms early in the human embryo. It
may develop into male *or* female primary reproductive organs,
depending on whether the *SRY* gene is present or absent.

EARLY QUESTIONS ABOUT GENE LOCATIONS

Linked Genes—Clues to Inheritance Patterns

By the early 1900s, researchers were suspecting that each gene has a specific location on a chromosome. Through hybridization experiments involving mutant fruit flies (*Drosophila melanogaster*), Thomas Hunt Morgan and his coworkers helped confirm this. For example, they found evidence that a gene for eye color and another gene for wing size are located on the *Drosophila* X chromosome. Figure 11.7 describes one series of their experiments.

During the early *Drosophila* experiments, it seemed as if two mutant genes on the X chromosome (*w* for white eyes and *m* for miniature wings) were "linked." That is, they apparently traveled together through meiosis and ended up together in the same gamete. For a time, they were called "sex-linked genes." Today, researchers use the more precise terms *X-linked* and *Y-linked* genes.

Eventually, researchers identified a large number of linked genes, and those on each type of chromosome came to be called a **linkage group**. *D. melanogaster*, for example, has four linkage groups, corresponding to its four pairs of homologous chromosomes. Indian corn (*Zea mays*) has ten linkage groups, corresponding to its ten pairs of homologous chromosomes. You and other humans have twenty-three linkage groups, and so on.

Figure 11.7 X-linked genes as clues to inheritance patterns.

In the early 1900s Thomas Morgan, an embryologist, was studying patterns of inheritance. He and his coworkers discovered a genetic basis for a relationship between sex determination and some nonsexual traits. For example, human males and females both have blood-clotting factors. Yet blood-clotting disorders (hemophilias) show up most often in males of a family lineage. This sex-specific outcome wasn't like anything Mendel saw; in his hybrid crosses of pea plants, it made no difference which parent carried a recessive allele.

Morgan studied eye color and other nonsexual traits of *Drosophila melanogaster*. This fruit fly can live in small bottles on agar, cornmeal, molasses, and yeast. A female lays hundreds of eggs in a few days and offspring reproduce in less than two weeks. In a single year, Morgan could track traits through nearly thirty generations of thousands of flies.

At first, all flies were wild type for eye color; they had brick-red eyes. Then Morgan got lucky; a mutation in some gene controlling eye color presented him with a white-eyed male.

Morgan established true-breeding strains of white-eyed males and females for **reciprocal crosses**. (In the first of such paired crosses, one parent displays the trait of interest. In the second cross, the other parent displays it.) First he let white-eyed males mate with homozygous red-eyed females. All F_1 offspring had red eyes, and some F_2 males had white eyes (*see diagram*). Then he let true-breeding red-eyed males mate with white-eyed females. The results gave Morgan an eyeful, so to speak. Half the F_1 offspring were red-eyed females, and half were white-eyed males. And of the F_2 offspring, 1/4 were red-eyed females, 1/4 white-eyed females, 1/4 red-eyed males, and 1/4 white-eyed males.

The odd results implied a relationship between an eye-color gene and sex determination. Probably the gene locus was on a sex chromosome. Which one? Because females (XX) could be white-eyed, the recessive allele had to be on one of their X chromosomes. What if the white-eyed males (XY) had the recessive allele on their X chromosome and their Y chromosome had no corresponding eye-color allele? In that case, they had white eyes because they had no dominant allele to mask the effect of the recessive one, as the diagram shows.

And so Morgan's idea of an X-linked gene dovetailed with Mendel's concept of segregation. By proposing that a specific gene is located on an X chromosome but not on the Y, Morgan explained his reciprocal crosses. His experimental results matched predicted outcomes.

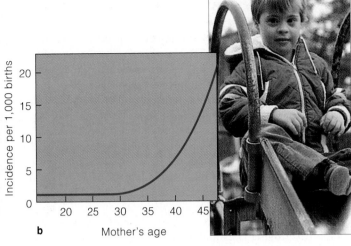

Figure 11.18 Down syndrome. (**a**) Karyotype revealing a trisomic 21 condition (*arrows*). (**b**) Relationship between the frequency of Down syndrome and mother's age at the time of childbirth. Results are from a study of 1,119 affected children who were born in Victoria, Australia, between 1942 and 1957. The young lady above was a lively participant in the Special Olympics, held annually in San Mateo, California.

TURNER SYNDROME Inheriting one X chromosome with no corresponding X or Y chromosome gives rise to *Turner syndrome*, which affects 1 in 2,500 to 10,000 or so newborn girls. Nondisjunction affecting sperm accounts for 75 percent of the cases. Compared to people having other sex chromosome abnormalities, there aren't many who have Turner syndrome. At least 98 percent of all X0 zygotes spontaneously abort early in pregnancy. About 20 percent of all spontaneously aborted embryos with detected chromosome abnormalities have been X0.

Despite the near-lethality, X0 survivors are not as disadvantaged as other aneuploids. They grow up well proportioned but only 4 feet, 8 inches tall, on average. Generally, their behavior is normal during childhood. But most Turner females are infertile. They do not have functional ovaries and so cannot produce eggs or sex hormones. Without the sex hormones, the development of secondary sexual traits, including breasts, is reduced. Possibly as a result of the arrested sexual development and small size, X0 females often are passive and easily intimidated by peers in their teens. Some benefit from hormone therapy and corrective surgery.

KLINEFELTER SYNDROME Of every 500 to 2,000 liveborn males, one inherits one Y and *two* X chromosomes. This XXY condition results mainly from nondisjunction in mothers (about 67 percent of the time, compared to 33 percent in fathers). Symptoms of the resulting *Klinefelter syndrome* develop after the onset of puberty. XXY males are taller than average and sterile, or nearly so. Testes usually are much smaller than average; the penis and

scrotum are not. Often facial hair is sparse, and breasts may be somewhat enlarged. Injections of testosterone, a male sex hormone, can reverse the feminized traits but not the low fertility. Some XXY males display mild mental impairment, but many fall in the normal range for intelligence. Except for low fertility, many affected individuals show no outward symptoms at all.

XYY CONDITION About 1 in 1,000 males has one X and two Y chromosomes. *XYY* males tend to be taller than average. Some may be mildly retarded, but most are phenotypically normal. At one time, all XYY males were thought to be genetically predisposed to become criminals. The erroneous conclusion was based on small numbers of cases in narrowly selected groups, such as prison inmates. Investigators often knew who the XYY males were, and this may have biased their evaluations. There were no **double-blind studies**, by which different investigators gather data independently of one another and then match them up only after both sets of data are completed. In this case, the same investigators gathered the karyotypes and the personal histories. Fanning the stereotype was a sensationalized report in 1968 that a mass-murderer of young nurses was XYY. He wasn't.

In 1976, one Danish geneticist issued a report on a large-scale study based on records of 4,139 tall males, twenty-six years old, who had reported to their draft board. Besides giving results of physical examinations and intelligence testing, the records provided clues to social and economic status, educational history, and any criminal convictions. Only twelve of the males were XYY, which left more than 4,000 for the control group. The only finding of significance was that tall, mentally impaired males who engage in criminal activity are just more likely to get caught—irrespective of karyotype.

Most changes in chromosome number arise as an outcome of nondisjunction during meiosis and gamete formation.

Prospects in Human Genetics

With the arrival of their newborn, parents typically ask, "Is our baby normal?" Quite naturally, they want their baby to be free of genetic disorders, and most of the time it is. But what are the options when it is not?

We do not approach heritable disorders and diseases the same way. We attack diseases with antibiotics, surgery, and other weapons. But how do we attack a heritable "enemy" that can be transmitted to offspring? Should we institute regional, national, or global programs to identify people who might be carrying harmful alleles? Do we tell them they are "defective" and run a risk of bestowing a disorder on their children? Who decides which alleles are harmful? Should society bear the cost of treating genetic disorders before and after birth? If so, should society also have a say in whether an affected embryo will be born at all, or whether it should be aborted? An **abortion** is the expulsion of a pre-term embryo or fetus from the uterus.

Such questions are only the tip of an ethical iceberg. We don't have anwers that are universally accepted.

PHENOTYPIC TREATMENTS Often, symptoms of genetic disorders can be minimized or suppressed by dietary controls, adjustments to environmental conditions, and surgical intervention or hormone replacement therapy.

For example, dietary control works in *phenylketonuria*, or PKU. A certain gene specifies an enzyme that converts one amino acid to another—phenylalanine to tyrosine. If an individual is homozygous recessive for a mutated form of the gene, the first of these amino acids accumulates inside the body. If excess amounts are diverted into other pathways, then phenylpyruvate and other compounds may form. High levels of phenylpyruvate in the blood can impair the functioning of the brain. When affected people restrict their intake of phenylalanine, they are not required to dispose of excess amounts, so they can lead normal lives. Among other things, they can avoid soft drinks and other food products that are sweetened with aspartame, a compound that contains phenylalanine.

Environmental adjustments can counter or minimize symptoms of some disorders, as when albinos avoid direct sunlight. Surgery can repair conditions such as a *cleft lip*, an opening in the upper lip that did not seal, as it should have, during embryonic development. This condition usually arises by interactions among multiple genes from both parents and environmental factors.

GENETIC SCREENING Through large-scale screening programs in the general population, affected persons or carriers of a harmful allele often can be detected early enough to start preventive measures before symptoms develop. For example, most hospitals in the United States routinely screen newborns for PKU, so today it is less common to see people with symptoms of this disorder.

GENETIC COUNSELING If a first child or close relative has a severe heritable problem, prospective parents may worry about their next child. They may request help in evaluating their options from a qualified professional counseler. *Genetic counseling* often includes diagnosis of parental genotypes, detailed pedigrees, and genetic testing for hundreds of known metabolic disorders. Geneticists, too, may be contacted to help predict risks for genetic disorders. During genetic counseling, the prospective parents must be reminded that the same risk usually applies to each pregnancy.

PRENATAL DIAGNOSIS Methods of *prenatal diagnosis* can be used to determine the sex of embryos or fetuses and more than a hundred genetic conditions. (Prenatal means before birth. The term embryo applies until eight weeks after fertilization, after which the term fetus is appropriate.)

Suppose a woman who is forty-five years old becomes pregnant and worries about Down syndrome. She might request prenatal diagnosis by *amniocentesis* (Figure 11.19). With this diagnostic procedure, a clinician withdraws a tiny sample of the fluid inside the amnion, a membranous sac that surrounds the fetus. Some cells that the fetus has sloughed off are suspended in the sample. The cells are cultured and analyzed.

Chorionic villi sampling (CVS) is a different diagnostic procedure. A clinician withdraws cells from the chorion, a fluid-filled, membranous sac that surrounds the amnion.

Removal of about 20 ml of amniotic fluid containing suspended cells that were sloughed off from the fetus

A few biochemical analyses with some of the amniotic fluid

Centrifugation

Quick determination of fetal sex and analysis of purified DNA

Fetal cells

Biochemical analysis for the presence of alleles that cause many different metabolic disorders

Growth for weeks in culture medium

Karyotype analysis

Figure 11.19 Amniocentesis, a prenatal diagnostic tool.

CVS can be perfomed weeks before amniocentesis. It can yield results as early as the ninth week of pregnancy.

Direct visualization of a developing fetus is possible with *fetoscopy*. An endoscope, a fiber-optic device, uses pulsed sound waves to scan the uterus and visually locate particular parts of the fetus, umbilical cord, or placenta (Figure 11.20). Fetoscopy has been used to diagnose blood cell disorders, such as sickle-cell anemia and hemophilia.

All three procedures may accidentally cause infection or puncture the fetus. Occasionally the punctured amnion does not reseal itself quickly, and an excessive amount of amniotic fluid may leak out and cause problems for the fetus. A mother-to-be who requests amniocentesis runs a 1 to 2 percent greater risk of miscarriage. For CVS, she runs a 0.3 percent risk that her future child will have missing or underdeveloped fingers and toes. Fetoscopy increases the risk of a miscarriage by 2 to 10 percent.

Parents-to-be probably should seek counseling from a doctor to help them weigh the risks and benefits of such procedures in terms of their own circumstances. They may wish to ask about the small overall risk of 3 percent that any child will have some kind of birth defect. They may ask about the severity of a genetic disorder that a child might be at risk of developing. And they might consider how old the woman is at the time of pregnancy.

REGARDING ABORTION What happens when prenatal diagnosis does reveal a serious problem? Do prospective parents opt for induced abortion? We can only say here

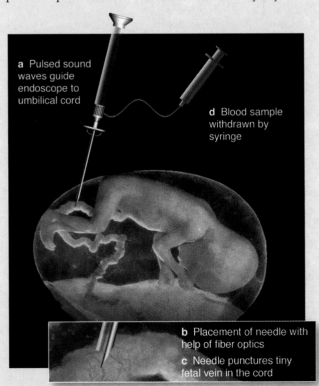

a Pulsed sound waves guide endoscope to umbilical cord

d Blood sample withdrawn by syringe

b Placement of needle with help of fiber optics

c Needle punctures tiny fetal vein in the cord

Figure 11.20 Fetoscopy for prenatal diagnosis.

Figure 11.21 (**a**) Micrograph of the eight-cell stage of human development. (**b**) Appearance of a human fetus at full term. By then, it is about 50 centimeters (20 inches) long.

that they must weigh their awareness of the severity of the genetic disorder against ethical and religious beliefs. Worse, they must play out their personal tragedy on a larger stage, dominated by a nationwide battle between fiercely vocal "pro-life" and "pro-choice" factions. We return to this volatile issue in Sections 39.18 and 39.20.

PREIMPLANTATION DIAGNOSIS Another procedure, *preimplantation diagnosis*, relies on **in-vitro fertilization**. In vitro, recall, means "in glass." Sperm and eggs from prospective parents are placed in an enriched medium in a glass or plastic petri dish. One or more eggs may get fertilized. In two days, mitotic cell divisions may convert one of these into a ball of eight cells (Figure 11.21*a*).

According to one view, the tiny, free-floating ball is a *pre*-pregnancy stage. Like the unfertilized eggs discarded monthly from a woman, it isn't attached to the uterus. Its cells all have the same genes and are not yet committed to giving rise to specialized cells of a heart, lungs, and other organs. Doctors take one of the undifferentiated cells and analyze its genes. If that cell has no detectable genetic defects, the ball is inserted into the uterus.

Some couples who are at risk of passing on muscular dystrophy, cystic fibrosis, and other disorders have opted for the procedure. Many *"test-tube" babies* have been born in good health and are free of the mutant alleles.

In 2000, genetic screening of a dozen fertilized eggs revealed an embryo free of a certain mutated gene. Both prospective parents carried the gene, and their first-born child, Molly, had developed *Fanconi anemia*. She had no thumbs, deformed arms, an incomplete brain, and the prospect of dying soon from leukemia. When her new brother was born, blood stem cells from the umbilical cord were transfused into her. The cells took hold. At this time Molly's bone marrow is producing red blood cells and platelets that are countering the anemia. The first embryo genetically selected to save a life is doing just that.

11.11

SUMMARY   *Gold* indicates text section

1. Genes, the units of instruction for heritable traits, are arranged one after the other along chromosomes. Each gene has its own specific location, or locus, on one type of chromosome. Different molecular forms of that gene, called alleles, may occupy the locus. *11.1*

2. Human somatic cells are diploid (2*n*). They contain twenty-three pairs of homologous chromosomes that interact during meiosis. Each pair has the same length, shape, and genes (except for an XY pairing). *11.1*

3. An allele on one chromosome may or may not be identical to the allele at the equivalent locus on the homologous chromosome. *11.1*

4. Human females have two X chromosomes. Males have one X paired with one Y. All other chromosomes are autosomes (the same in both females and males). A gene on the Y chromosome determines sex. *11.1, 11.3*

5. Karyotypes, used in genetic analysis, are constructed to compare an individual's chromosomes (based upon their defining structural features) against standardized preparations of metaphase chromosomes. *CI, 11.2*

6. Genes on the same chromosome represent a linkage group. However, crossing over (breakage and exchange of segments between homologues) disrupts linkages. The farther apart two gene loci are along the length of a single chromosome, the greater will be the frequency of crossovers between them. *11.4*

7. Pedigrees are charts of genetic connections through lines of descent. Pedigrees provide clues to inheritance of a trait. *11.5*

8. Mendelian patterns of inheritance are characteristic of certain dominant or recessive alleles on autosomes or on the X chromosome. *11.5–11.7*

9. A chromosome's structure may be altered on rare occasions. A segment may be deleted, inverted, moved to a new location (translocated), or duplicated. *11.8*

10. The chromosome number can change. Gametes and offspring may get one more or one less chromosome than the parents (aneuploidy). They may get three or more of each type of chromosome (polyploidy). Nondisjunction during meiosis accounts for most of these changes in chromosome number. *11.9*

11. Changes in chromosome structure or number often result in genetic abnormalities or disorders. Phenotypic treatments, genetic screening, genetic counseling, and prenatal diagnosis are social responses. *11.8–11.10*

12. Crossing over adds to potentially adaptive variation in traits in a population. Most but not all changes in chromosome number or structure are harmful or lethal. Over evolutionary time, some have become established in chromosomes of all species. *11.1, 11.8*

Review Questions

1. What is a gene? What are alleles? *11.1*

2. Distinguish between: *CI, 11.1*
   a. homologous and nonhomologous chromosomes
   b. sex chromosomes and autosomes
   c. karyotype and karyotype diagram

3. Define genetic recombination, and describe how crossing over can bring it about. *11.1, 11.4*

4. Define pedigree. Explain the difference between genetic abnormality and genetic disorder, using examples. *11.5*

5. Contrast a typical pattern of autosomal recessive inheritance with that of autosomal dominant inheritance. *11.6*

6. Describe two clues that often show up when a recessive allele on an X chromosome causes a genetic disorder. *11.6*

7. Distinguish among a chromosomal deletion, duplication, inversion, and translocation. *11.8*

8. Define aneuploidy and polyploidy. Make a simple sketch of an example of nondisjunction. *11.9*

Self-Quiz   ANSWERS IN APPENDIX III

1. _____ segregate during _____ .
   a. Homologues; mitosis
   b. Genes on nonhomologous chromosomes; meiosis
   c. Homologues; meiosis
   d. Genes on one chromosome; mitosis

2. The probability of a crossover occurring between two genes on the same chromosome is _____ .
   a. unrelated to the distance between them
   b. increased if they are closer together on the chromosome
   c. increased if they are farther apart on the chromosome

3. Genetic disorders are caused by _____ .
   a. altered chromosome number   c. mutation
   b. altered chromosome structure   d. all of the above

4. A recognized set of symptoms that characterize a specific disorder is a _____ .
   a. syndrome   b. disease   c. pedigree

5. Chromosome structure can be altered by a _____ .
   a. deletion   c. inversion   e. all of the above
   b. duplication   d. translocation

6. Nondisjunction can be caused by _____ .
   a. crossing over in mitosis
   b. segregation in meiosis
   c. failure of chromosomes to separate during meiosis
   d. multiple independent assortments

7. A gamete affected by nondisjunction would have _____ .
   a. a change from the normal chromosome number
   b. one extra or one missing chromosome
   c. the potential for a genetic disorder
   d. all of the above

8. Match the chromosome terms appropriately.
   ____ crossing over   a. number and defining features of an
   ____ deletion   individual's metaphase chromosomes
   ____ nondisjunction   b. chromosome segment moves to a
   ____ translocation   nonhomologous chromosome
   ____ karyotype   c. disrupts gene linkages at meiosis
   ____ linkage group   d. causes gametes to have abnormal
                              chromosome numbers
                          e. loss of a chromosome segment
                          f. all genes on a given chromosome

Figure 11.22 Mutant fruit fly with vestigial wings.

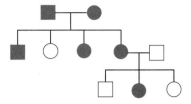

Figure 11.23 Go ahead, identify the mystery pedigree.

Figure 11.24 Two of your linked genes, on a pair of homologous chromosomes.

Critical Thinking—Genetics Problems

ANSWERS IN APPENDIX IV

1. Human females are XX and males are XY.

a. Does a male inherit the X from his mother or father?

b. With respect to X-linked alleles, how many different types of gametes can a male produce?

c. If a female is homozygous for an X-linked allele, how many types of gametes can she produce with respect to that allele?

d. If a female is heterozygous for an X-linked allele, how many types of gametes can she produce with respect to that allele?

2. Expression of a dominant allele at a gene locus governing wing length in *D. melanogaster* results in long wings. Figure 11.22 shows how homozygosity for a recessive allele results in formation of vestigial (short) wings. You cross a homozygous dominant, long-winged fly with a homozygous recessive, vestigial-winged fly. Then you ask a technician to expose the fertilized eggs to a level of x-rays known to induce mutation and deletions. Later, when the irradiated eggs develop into adults, most flies are heterozygous and have long wings. A few have vestigial wings. What might explain these results?

3. *Marfan syndrome* is a genetic disorder with pleiotropic effects. Expression of a mutant allele leads to the absence of a connective tissue (fibrillin) or its abnormal formation in many organs. Affected people are often tall and thin, with double-jointed fingers. The spine curves abnormally. Arms, fingers, and lower limbs are disproportionately long. Eye lenses dislocate easily. Thin tissue flaps in the heart that act like valves to direct blood flow may billow the wrong way when the heart contracts. The heart beats irregularly. The main artery from the heart is fragile. Its diameter widens; its wall may tear.

An estimated 40,000 people in the United States are affected, including a few prominent athletes. Spontaneous mutation gives rise to the allele. Genetic analysis shows it follows a pattern of autosomal dominant inheritance. What is the chance any child will inherit the allele if one parent is heterozygous for it?

4. In the Figure 11.23 pedigree, does the phenotype indicated by *red* circles and squares follow a Mendelian inheritance pattern that is autosomal dominant, autosomal recessive, or X-linked?

5. One kind of *muscular dystrophy*, a genetic disorder, is due to a recessive X-linked allele. Usually, symptoms start in childhood. Over time, a slowly progressing loss of muscle function leads to death, usually by age twenty or so. Unlike color blindness, this disorder is nearly always restricted to males. Suggest why.

6. Suppose you carry two linked genes with alleles *Aa* and *Bb*, respectively, as in Figure 11.24. If the crossover frequency between these two genes is zero, what genotypes would be expected among the gametes you produce, and with what frequencies?

7. Say you have alleles for left-handedness and straight hair on one chromosome and alleles for right-handedness and curly hair on the homologous chromosome. If the two loci for the alleles are very close together along the chromosome, how likely is it that crossing over will occur between them? If they are distant from each other, is a crossover more or less likely to occur?

8. Individuals showing *Down syndrome* usually have an extra chromosome 21, so their body cells contain 47 chromosomes.

a. At which stages of meiosis I and II could a mistake occur that could result in the altered chromosome number?

b. In a few cases, 46 chromosomes are present, including two normal-appearing chromosomes 21 and a longer-than-normal chromosome 14. Explain how this situation can arise.

9. In the human population, mutation of two different genes on the X chromosome causes two types of X-linked *hemophilia* (types A and B). In a few known cases, a woman is heterozygous for both mutant alleles (one on each of her two X chromosomes). All her sons should have either hemophilia A or B. Yet, on very rare occasions, such a woman gives birth to a son who does not have hemophilia, and his one X chromosome does not have either mutant allele. Explain how such an X chromosome could arise.

10. Think back on the chapter definitions of genetic disorder and genetic abnormality. Then think about how subjective we are in terms of our acceptance of a condition that is out of the ordinary. Example: Would you find a cleft lip as acceptable, even attractive, on someone not as gorgeous as Joaquin Phoenix, shown at right? You will not find an answer to this question in Appendix IV. We present it simply as a way to invite critical thinking about what we as a society consider to be "ideal" phenotypes.

Selected Key Terms

abortion *11.10*	genetic abnormality *11.5*	karyotype *CI*
allele *11.1*	genetic disorder *11.5*	linkage group *11.4*
aneuploidy *11.9*	genetic	nondisjunction *11.9*
autosome *11.1*	recombination *11.1*	pedigree *11.5*
crossing over *11.1*	homologous	polyploidy *11.9*
deletion *11.8*	chromosome *11.1*	reciprocal cross *11.4*
disease *11.5*	independent	sex chromosome *11.1*
double-blind	assortment *11.1*	syndrome *11.5*
study *11.9*	inversion *11.8*	translocation *11.8*
duplication *11.8*	in-vitro	X chromosome *11.1*
gene *11.1*	fertilization *11.10*	Y chromosome *11.1*

Readings

Fairbanks, D., and W. R. Andersen. 1999. *Genetics: The Continuity of Life*. Monterey, California: Brooks-Cole.

On-Line readings at Student Guide for InfoTrac:
www.brookscole.com/biology

DNA STRUCTURE AND FUNCTION

Cardboard Atoms and Bent Wire Bonds

One might have wondered, in the spring of 1868, why Johann Friedrich Miescher was collecting cells from the pus of open wounds and, later, from the sperm of a fish. Miescher, a physician, wanted to identify the chemical composition of the nucleus. These particular cells have very little cytoplasm, which makes it easier to isolate the nuclear material for analysis.

Miescher finally succeeded in isolating an organic compound having the properties of an acid. Unlike other substances in cells, it incorporated a notable amount of phosphorus. Miescher called the substance nuclein. He had discovered what came to be known many years later as **deoxyribonucleic acid**, or **DNA**.

The discovery did not cause even a ripple through the scientific community. At the time, no one really knew much about the physical basis of inheritance— that is, *which chemical substance encodes the instructions for reproducing parental traits in offspring*. Few even suspected that the cell nucleus might hold the answer. For a time, researchers generally believed hereditary instructions had to be encoded in the structure of some unknown class of proteins. After all, heritable traits are spectacularly diverse. Surely the molecules encoding information about those traits were structurally diverse also. Proteins are put together from potentially limitless combinations of twenty different amino acids, so the thinking was that they could function as the sentences (genes) in each cell's book of inheritance.

By the early 1950s, however, the results of many ingenious experiments clearly indicated that DNA is the substance of inheritance. Moreover, in 1951, Linus Pauling did something no one had done before. Through his training in biochemistry, a talent for model building, and a few great educated guesses, Pauling deduced the three-dimensional structure of the protein collagen. His discovery was truly electrifying. If someone could pry open the secrets of proteins, then why not assume the same might be done for DNA? And if the structural details of the DNA molecule were worked out, then wouldn't they provide clues to its biological functions? *Who would go down in history as having discovered the very secrets of inheritance?*

Scientists around the world started scrambling after that ultimate prize. Among them were James Watson, a young postdoctoral student from Indiana University, and Francis Crick, an unflappably exuberant researcher at Cambridge University. Exactly how could DNA, a molecule that consists of only four kinds of subunits, hold genetic information? Watson and Crick spent long hours arguing over everything they had read about the size, shape, and bonding requirements of the subunits of DNA. They fiddled with cardboard cutouts of the subunits. They badgered chemists to help them identify any potential bonds they might have overlooked. Then they assembled models from bits of metal connected by wire "bonds" bent at seemingly suitable angles.

In 1953, Watson and Crick put together a model that fit all the pertinent biochemical rules and all the facts about DNA they had gleaned from other sources. They had discovered the structure of DNA (Figures 12.1 and 12.2). And the breathtaking simplicity of that structure enabled them to solve another long-standing riddle— *how the world of life can show such unity at the molecular level and yet show such spectacular diversity at the level of whole organisms.*

Figure 12.1 James Watson and Francis Crick posing in 1953 by their newly unveiled structural model of DNA.

Figure 12.2 A recent computer-generated model of DNA, corresponding to the prototype that Watson and Crick put together decades ago.

Key Concepts

1. In all living cells, DNA molecules are the storehouses of information about heritable traits.

2. In a DNA molecule, two strands of nucleotides twist together, like a spiral stairway. Each strand consists of four kinds of nucleotides that are the same except for one component—a nitrogen-containing base. The four bases are adenine, guanine, thymine, and cytosine.

3. Great numbers of nucleotides are arranged one after another in each strand of the DNA molecule. In at least some regions, the order in which one kind of nucleotide follows another is unique for each species. Hereditary information is encoded in the particular sequence of nucleotide bases.

4. Hydrogen bonds connect the bases of one strand of the DNA molecule to bases of the other strand. As a rule, adenine pairs (hydrogen-bonds) with thymine, and guanine with cytosine.

5. Before a cell divides, its DNA is replicated with the assistance of enzymes and other proteins. Each double-stranded DNA molecule starts unwinding. As it does so, a new, complementary strand is assembled bit by bit on the exposed bases of each parent strand according to the base-pairing rule stated above.

With this chapter, we turn to investigations that led to our current understanding of DNA. The story is more than a march through details of its structure and function. *It also is revealing of how ideas are generated in science.* On the one hand, having a shot at fame and fortune quickens the pulse of men and women in any profession, and scientists are no exception. On the other hand, science proceeds as a community effort, with individuals sharing not only what they can explain but also what they do not understand. Even when an experiment fails to produce the anticipated results, it may turn up information that others can use or lead to questions that others can answer. Unexpected results, too, may be clues to something important about the natural world.

DISCOVERY OF DNA FUNCTION

Early and Puzzling Clues

The year was 1928. Frederick Griffith, an army medical officer, was attempting to develop a vaccine against *Streptococcus pneumoniae*, a bacterium that is one cause of the lung disease pneumonia. (When introduced into the body, vaccines mobilize internal defenses against a real attack. Many vaccines are preparations of killed or weakened bacterial cells.) Griffith never did develop a vaccine. But his work unexpectedly opened a door to the molecular world of heredity.

Griffith isolated and cultured two different strains of the bacterium. He noticed that colonies of one strain had a rough surface appearance, but those of the other strain appeared smooth. He designated the two strains R and S and used them in a series of four experiments:

1. Laboratory mice were injected with live R cells. The mice did not develop pneumonia, as Figure 12.3 shows. *The R strain was harmless.*

2. Other mice were injected with live S cells. The mice died. Blood samples taken from them teemed with live S cells. *The S strain was pathogenic* (disease-causing).

3. S cells were killed by exposure to high temperature. Mice injected with these cells did not die.

4. Live R cells were mixed with heat-killed S cells and injected into mice. The mice died—and blood samples from them teemed with *live* S cells!

What was going on in the fourth experiment? Maybe heat-killed S cells in the mixture weren't really dead. But if that were true, then mice injected with heat-killed S cells alone (experiment 3) would have died. Maybe harmless R cells in the mixture had mutated into a killer form. But if that were true, then mice injected with the R cells alone (experiment 1) would have died.

The simplest explanation was as follows: *Heat killed the S cells but did not destroy their hereditary material— including the part that specified "how to cause infection."* Somehow, that material had been transferred from the dead S cells to living R cells, which put it to use.

Further experiments showed the harmless cells had indeed picked up information on causing infections and were permanently transformed into pathogens. After a few hundreds of generations, descendants of those transformed bacterial cells were still infectious!

The unexpected results of Griffith's experiments intrigued Oswald Avery and his fellow biochemists. Later, they also transformed harmless bacterial cells with *extracts* of killed pathogenic cells. Finally in 1944, after rigorous chemical analyses, they felt confident in reporting that the hereditary substance in their extracts probably was DNA—not proteins, as was then widely believed. To give experimental evidence for their conclusion, they reported that they had added certain protein-digesting enzymes to some extracts, but cells exposed to those extracts were transformed anyway. To other extracts, they had added an enzyme that digests DNA but not proteins. Doing so blocked hereditary transformation.

Despite these impressive experimental results, many biochemists refused to give up on the proteins. Avery's findings, they said, probably applied only to bacteria.

Confirmation of DNA Function

By the early 1950s molecular detectives, including Max Delbrück, Alfred Hershey, Martha Chase, and Salvador Luria, were using viruses as experimental subjects. The viruses they had selected, called **bacteriophages**, infect *Escherichia coli* and other bacteria.

Viruses are biochemically simple infectious agents. They aren't alive, but they hold hereditary information about building more new virus particles. At some point after a virus infects a host cell, viral enzymes take over the cell's metabolic machinery, which starts churning out substances necessary to make new virus particles.

genetic material
viral coat
sheath
base plate
tail fiber
a

---

1 Mice injected with live cells of harmless strain R.

2 Mice injected with live cells of killer strain S.

3 Mice injected with heat-killed S cells.

4 Mice injected with live R cells *plus* heat-killed S cells.

Figure 12.3 Results of Griffith's experiments with a harmless and a pathogenic strain of *Streptococcus pneumoniae*, as described in the text above.

Mice do not die. No live R cells in their blood.

Mice die. Live S cells in their blood.

Mice do not die. No live S cells in their blood.

Mice die. Live S cells in their blood.

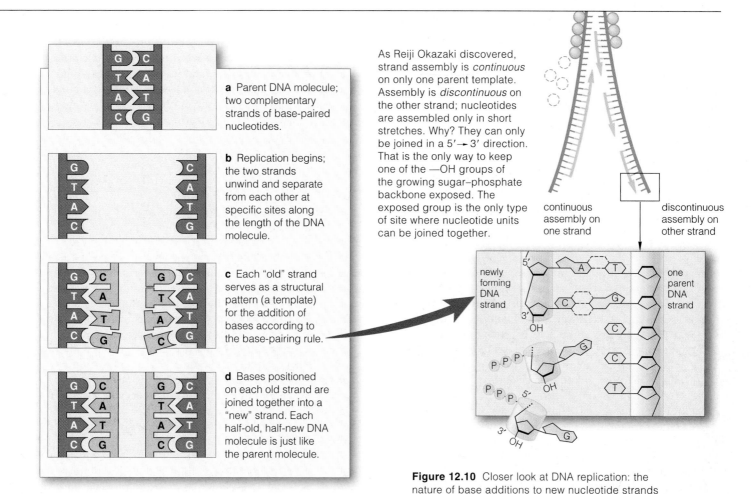

a Parent DNA molecule; two complementary strands of base-paired nucleotides.

b Replication begins; the two strands unwind and separate from each other at specific sites along the length of the DNA molecule.

c Each "old" strand serves as a structural pattern (a template) for the addition of bases according to the base-pairing rule.

d Bases positioned on each old strand are joined together into a "new" strand. Each half-old, half-new DNA molecule is just like the parent molecule.

As Reiji Okazaki discovered, strand assembly is *continuous* on only one parent template. Assembly is *discontinuous* on the other strand; nucleotides are assembled only in short stretches. Why? They can only be joined in a 5'→3' direction. That is the only way to keep one of the —OH groups of the growing sugar–phosphate backbone exposed. The exposed group is the only type of site where nucleotide units can be joined together.

continuous assembly on one strand

discontinuous assembly on other strand

newly forming DNA strand

one parent DNA strand

Figure 12.10 Closer look at DNA replication: the nature of base additions to new nucleotide strands being assembled on a parent template.

of every double-stranded DNA molecule is "old" and half is "new" (Figure 12.9). That is why biologists refer to the process as *semiconservative* replication.

DNA replication uses a team of molecular workers. In response to cellular signals, the replication enzymes become active along the length of the DNA molecule. Together with other proteins, some enzymes unwind the strands in both directions and prevent them from rewinding. Enzyme action jump-starts the unwinding but isn't necessary to unzip hydrogen bonds between the strands; hydrogen bonds are individually weak.

Now enzymes called **DNA polymerases** attach short stretches of free nucleotides to the unwound portions of a parent template (Figure 12.10). The free nucleotides themselves actually drive the strand assembly. Each has three phosphate groups. DNA polymerase splits off two of them, releasing energy that drives the attachments.

DNA ligases fill in tiny gaps between the new short stretches to form a continuous strand. Then enzymes wind the template strand and complementary strand together to form a DNA double helix.

As you will read in Section 15.1, some replication enzymes have uses in recombinant DNA technology.

Monitoring and Fixing the DNA

DNA polymerases, DNA ligases, and other enzymes also engage in **DNA repair**. By this process, enzymes excise and fix altered parts of the base sequence in one strand of the double helix. Suppose a few bases are lost from a strand during replication. DNA polymerases "read" the complementary sequence on the other strand. They and other repair enzymes restore the original sequence. Also, remember crossing over, in which homologous chromosomes exchange segments? The same enzymes of that recombination event repair double-strand breaks. They initiate strand exchange with homologous DNA. Section 13.4 shows what happens when mutation or some other factor compromises the excision–repair function.

DNA is replicated prior to cell division. Enzymes unwind its two strands. Each strand remains intact throughout the process—it is conserved—and enzymes assemble a new, complementary strand on each one.

Enzymes involved in replication also repair the DNA where base-pairing errors have crept into the nucleotide sequence.

Cloning Mammals—A Question of Reprogramming DNA

Imagine the possibility of **cloning**—making a genetically identical copy—of yourself. Is the image that far-fetched? Consider this: Researchers have been cloning complex animals for more than a decade. For example, some use in vitro fertilization methods to grow cattle embryos in petri dishes. After a fertilized egg first starts dividing, they split the early cluster of cells. The two clusters develop as two identical-twin cattle embryos, get implanted in surrogate mothers, and are born as cloned calves (Figure 12.11*a*).

A researcher who clones farm animals derived from embryonic cells has to wait for the clones to grow up to see if they display a desired trait. Using a differentiated cell from an adult would be faster, for a prized genotype would already be known. But at one time, tricking a differentiated cell into reprogramming its DNA to direct the development of a whole embryo seemed impossible.

What does "differentiated" mean? When an embryo first grows from a fertilized egg, all of its cells have the same DNA and are pretty much alike. Then different embryonic cells start using different parts of their DNA. Their unique selections commit them to being liver cells, heart cells, brain cells, and other specialists in structure, composition, and function (Section 15.3).

In 1997 in Scotland, Ian Wilmut coaxed a differentiated sheep cell to become the "uncommitted" first cell of an embryo. His group had slipped nuclei from differentiated cells into unfertilized eggs from which the nucleus had been surgically removed. Of hundreds of modifed eggs, one developed into a whole animal. The cloned lamb, named Dolly, grew into a healthy adult and gave birth to a lamb of her own (Figure 12.11*b*).

In science, extraordinary claims call for extraordinary proof—in this case, successful repeats of the experiment at a time when no one thought mammals could ever be cloned. Researchers around the world have since cloned sheep and mice, cows, pigs, and goats. Some of the mice have been cloned through six generations. Researchers also cloned an endangered species, a guar. (The guar, a

wild ox, died two days after birth owing to a bacterial infection that is typically fatal for young calves.) Pigs are physiologically very much like us, so there is talk of using cloned pigs as sources of a variety of organs that can be harvested and transplanted into humans.

All of which raises some far-out questions. Is human cloning next? For example, will a human female be able to reproduce copies of herself without involving men? Will men be able to pay to have their DNA inserted into a cell stripped of its nucleus, have the cell implanted in a surrogate mother, and make a little repeat of themselves?

Aside from the ethical issues, of all cloning efforts, fewer than 3 percent result in healthy animals. And many of the "successful" clones develop severe medical problems, such as heart and lung abnormalities, sudden and gross obesity, and a compromised immune system.

It now appears that cloning invites random mistakes in gene expression. The cytoplasm of a mammalian egg contains proteins, mRNAs, and other components that have roles in guiding gene expression, starting with the early embryo (Sections 39.3 and 39.4). It takes months or years for an egg that is maturing inside an ovary to stockpile the required components in specific locations in the cytoplasm. Cloning processes use differentiated eggs that must reprogram new DNA within minutes or hours after it is inserted into them.

Even so, Lee Silver, a biologist at Princeton University, suggested that anyone who thinks cloning will move slowly is naive. One couple has already pledged up to 500 million dollars to a controversial company to clone their dead infant. The company says that it has lots of potential customers and has surrogate mothers lined up in various countries.

At this writing, several countries have banned human cloning, but the bans leave room for cloning technology with uses in basic research. Should human cloning be done at all? Both nonscientists and scientists alike are actively debating the issue.

Figure 12.11 (a) Clone of Holsteins, prized milk producers, that resulted from in vitro fertilization. (b) Dolly, a cloned sheep. She started her life as a differentiated cell that was extracted from an adult ewe, then induced to undergo mitotic cell divisions. She is shown with her first lamb. She clearly is able to breed normally and reproduce the old-fashioned way.

SUMMARY   *Gold* indicates text section

1. For all living cells, hereditary information is encoded in DNA (deoxyribonucleic acid). *CI, 12.1*

2. DNA consists of nucleotide subunits. Each of these has a five-carbon sugar (deoxyribose), one phosphate group, and one of four kinds of nitrogen-containing bases (adenine, thymine, guanine, or cytosine). *12.2*

3. A DNA molecule consists of two nucleotide strands twisted together as a double helix. Bases of one strand pair (hydrogen-bond) with bases of the other. *12.2*

4. The bases of the two strands in a DNA double helix pair in constant fashion. Adenine pairs with thymine (**A** to **T**), and guanine with cytosine (**G** to **C**). *Which* base pair follows the next (**A–T, T–A, G–C,** or **C–G**) varies along the length of the strands. *12.2*

5. Overall, the DNA of one species includes a number of unique stretches of base pairs that set it apart from the DNA of all other species. *12.2*

6. During DNA replication, enzymes unwind the two strands of a double helix and assemble a new strand of complementary sequence on each parent strand. Two double-stranded molecules result. One strand of each molecule is old (is conserved); the other is new. *12.4*

7. Some of the enzymes involved in DNA replication also repair DNA where base-pairing errors have been introduced into the nucleotide sequence. *12.4*

Review Questions

1. Name the three molecular parts of a nucleotide in DNA. Also name the four different bases in these nucleotides. *12.2*

2. What kind of bond joins two DNA strands in a double helix? Which nucleotide base-pairs with adenine? With guanine? *12.2*

3. Explain how DNA molecules can show both constancy and variation from one species to the next. *12.2*

Self-Quiz   ANSWERS IN APPENDIX III

1. Which is *not* a nucleotide base in DNA?
   a. adenine   c. uracil   e. cytosine
   b. guanine   d. thymine

2. What are the base-pairing rules for DNA?
   a. A–G, T–C   c. A–U, C–G
   b. A–C, T–G   d. A–T, G–C

3. A DNA strand having the sequence C–G–A–T–T–G would be complementary to the sequence _____ .
   a. C–G–A–T–T–G   c. T–A–G–C–C–T
   b. G–C–T–A–A–G   d. G–C–T–A–A–C

4. One species' DNA differs from others in its _____ .
   a. sugars   b. phosphates   c. base sequence

5. When DNA replication begins, _____ .
   a. the two DNA strands unwind from each other
   b. the two DNA strands condense for base transfers
   c. two DNA molecules bond
   d. old strands move to find new strands

6. DNA replication requires _____ .
   a. free nucleotides   c. many enzymes
   b. new hydrogen bonds   d. all of the above

7. Match the DNA terms appropriately.
   _____ DNA polymerase   a. two nucleotide strands that
   _____ constancy in   are twisted together
         base pairing   b. A with T, G with C
   _____ replication   c. hereditary material duplicated
   _____ DNA double helix   d. replication enzyme

Critical Thinking

1. Chargaff's data suggested that adenine pairs with thymine, and guanine pairs with cytosine. What other data available to Watson and Crick suggested that adenine–guanine and cytosine–thymine pairs normally do not form?

2. One of Matthew Meselson and Frank Stahl's experiments supported the semiconservative model of DNA replication. The researchers made "heavy" DNA by growing *Escherichia coli* in a medium enriched with ^{15}N, a heavy isotope of nitrogen. They prepared "light" DNA by growing *E. coli* in the presence of ^{14}N, the more common isotope. An available technique helped them identify which replicated molecules were heavy, light, or hybrid (one heavy strand, one light). Use two pencils of different colors, one for heavy strands and one for light. Starting with a DNA molecule having two heavy strands, sketch daughter molecules that would form after replication in a ^{14}N-containing medium. Sketch the four DNA molecules that would form if the daughter molecules were replicated a second time in the ^{14}N medium.

3. Mutations (permanent changes in base sequences of genes) are the original source of genetic variation. This variation is the raw material of evolution. Yet how can both statements be true, given that cells have efficient mechanisms to repair DNA before mutations can become established?

4. As Section 4.12 indicates, a pathogenic strain of *E. coli* has acquired an ability to produce a dangerous toxin that causes medical problems and fatalities. This is especially the case for young children who have ingested undercooked, contaminated beef. Develop hypotheses to explain how a normally harmless bacterium such as *E. coli* can become a pathogen.

5. In 1999, scientists discovered a woolly mammoth that had been frozen in glacial ice for the past 20,000 years. They thawed it very carefully so they could use its DNA to clone a woolly mammoth. It turns out there wasn't enough material to work with. But they plan to try again the next time a frozen woolly mammoth comes along. Consider Section 12.5, then speculate on the pros and cons of cloning an extinct animal.

Selected Key Terms

adenine (A) *12.2*	DNA repair *12.4*
bacteriophage *12.1*	DNA replication *12.4*
cloning *12.5*	guanine (G) *12.2*
cytosine (C) *12.2*	nucleotide *12.2*
deoxyribonucleic acid (DNA) *CI*	thymine (T) *12.2*
DNA ligase *12.4*	x-ray diffraction
DNA polymerase *12.4*	image *12.2*

Readings

Watson, J. 1978. *The Double Helix.* New York: Atheneum. Highly personal view of scientists and their methods, interwoven into an account of how DNA structure was discovered.

On-Line readings at Student Guide for InfoTrac:
www.brookscole.com/biology

13

FROM DNA TO PROTEINS

Beyond Byssus

Picture a mussel, of the sort shown in Figure 13.1. Hard-shelled but soft of body, it is using its muscular foot to probe a wave-scoured rock. At any moment, pounding waves can whack the mussel into the water, hurl it repeatedly against the rock with shell-shattering force, and so offer up a gooey lunch for gulls.

By chance, the mussel's foot comes across a crevice in the rock. The foot moves, broomlike, and sweeps the crevice clean. It presses down, forcing air out from underneath it, then arches up. The result is a vacuum-sealed chamber, much like the one that forms when a plumber's rubber plunger is being squished down and up to unclog a drain. Into this vacuum chamber the mussel spews a fluid that's made of keratin and other

proteins. The fluid bubbles into a sticky foam. Now, by curling its foot into a tubelike shape and pumping the foam through it, the mussel produces sticky threads about as wide as a human whisker. It varnishes these threads with another type of protein and ends up with an adhesive. With this adhesive, which we call byssus, the mussel anchors itself to the rock.

Byssus is the world's premier underwater adhesive. Nothing humans have manufactured comes close to it; water degrades or deforms synthetic adhesives. Byssus fascinates biochemists, dentists, and surgeons looking for better ways to do tissue grafts and rejoin severed nerves. Genetic engineers insert mussel DNA into yeast cells. These cells, which reproduce in huge numbers, are

Figure 13.1 Mussels (*Mytilus californianus*) busily demonstrating the importance of proteins for survival. When mussels come across a suitable anchoring site, they use their muscular foot like a plumber's plunger and create a vacuum chamber. In this chamber they manufacture the world's best underwater adhesive from a mix of proteins. The adhesive anchors them to rocks in their wave-swept habitat.

"factories" for translating mussel genes into useful quantities of proteins. This exciting work, like the mussel's own byssus-building efforts, starts with one of life's universal precepts: *Every protein is synthesized in accordance with instructions in DNA.*

You are about to trace the steps leading from DNA to proteins. Many enzymes are players in this pathway. So is another kind of nucleic acid besides DNA. The same steps produce *all* proteins, from mussel-inspired adhesives to the keratin in your hair and fingernails to the insect-digesting enzymes of a Venus flytrap.

Start out by thinking of each cell's DNA as a book of protein-building instructions. The alphabet used to create the book is simple enough: A, T, G, and C (for the nucleotide bases adenine, thymine, guanine, and cytosine). How do you get from that alphabet to a protein? The answer starts with DNA's structure.

DNA, recall, is a double-stranded molecule. Which kind of nucleotide base follows the next along the length of a strand—the **base sequence**—differs from one kind of organism to the next. The two strands unwind entirely from each other when DNA is being replicated. However, at other times in a cell's life, the two strands unwind only in certain regions to expose certain base sequences—genes. Most genes contain instructions for building proteins.

It takes two steps, **transcription** and **translation**, to carry out a gene's protein-building instructions. In eukaryotic cells, transcription proceeds in the nucleus. A newly exposed base sequence in DNA serves as a structural pattern—a template—for assembling a strand of **ribonucleic acid** (RNA) from the cell's pool of free nucleotides. Sooner or later the RNA moves into the cytoplasm, where translation proceeds. At this second step, RNA directs the assembly of amino acids into polypeptide chains. The newly formed chains become folded into the three-dimensional shapes of proteins.

In short, DNA guides the synthesis of RNA, then RNA guides the synthesis of proteins:

$$\text{DNA} \xrightarrow{\textit{transcription}} \text{RNA} \xrightarrow{\textit{translation}} \text{PROTEIN}$$

The newly synthesized proteins will play structural and functional roles in cells. Some even will have roles in synthesizing more DNA, RNA, and proteins.

Key Concepts

1. Life cannot exist without enzymes and other proteins. Proteins consist of polypeptide chains, which consist of amino acids. The sequence of amino acids corresponds to a gene, which is a sequence of nucleotide bases in a DNA molecule.

2. The path leading from genes to proteins consists of two steps, called transcription and translation.

3. In transcription, the double-stranded DNA molecule is unwound at a gene region, then an RNA molecule is assembled on the exposed bases of one of the strands.

4. Translation requires three classes of RNA molecules: messenger RNA, transfer RNA, and ribosomal RNA.

5. In translation, amino acids are joined together to form a polypeptide chain, in a sequence specified by messenger RNA. Transfer RNA delivers the amino acids one at a time to the construction site. Ribosomal RNA catalyzes the chain-building reaction.

6. With few exceptions, the genetic "code words" by which DNA's instructions are translated into proteins are the same in all species of organisms.

7. A mutation is a permanent alteration in a gene's base sequence. Mutations are the original source of genetic variation in populations.

8. Mutations cause changes in protein structure, protein function, or both. The changes may lead to small or large differences in traits among individuals of a population.

HOW IS DNA TRANSCRIBED INTO RNA?

The Three Classes of RNA

Before turning to the details of protein synthesis, be clear on one point. The chapter introduction may have left you with the impression that synthesis of proteins requires only one class of RNA molecules. Actually, it requires three. Transcription of most genes produces **messenger RNA**, or **mRNA**—the only class of RNA that carries *protein-building* instructions. Transcription of some other genes produces **ribosomal RNA**, or **rRNA**, a major component of ribosomes. Ribosomes, recall, are the structural units upon which polypeptide chains are assembled. Transcription of still other genes produces **transfer RNA**, or **tRNA**, which delivers amino acids one by one to a ribosome in the order specified by mRNA.

The Nature of Transcription

An RNA molecule is almost but not quite like a single strand of DNA. RNA, too, consists of only four types of nucleotides. Each nucleotide has a five-carbon sugar, ribose (not DNA's deoxyribose), a phosphate group, and a base. Three types of bases—adenine, cytosine, and guanine—are the same in RNA and DNA. But in RNA, the fourth type of base is **uracil**, not thymine (Figure 13.2). Like thymine, uracil can pair with adenine. This means a new RNA strand can be put together on a DNA region according to base-pairing rules (Figure 13.3).

sugar-phosphate backbone of one strand of nucleotides in a DNA double helix

sugar-phosphate backbone of the other strand of nucleotides

part of the sequence of base pairs in DNA

Figure 13.4 The process of gene transcription, by which an RNA molecule is assembled on a DNA template. The sketch in (**a**) shows a gene region in part of a DNA double helix. In this region, the base sequence of one of the two nucleotide strands (not both) is about to be transcribed into an RNA molecule, in the manner shown in (**b**) through (**e**).

URACIL (U)

base with a single-ring structure

phosphate group

sugar (ribose)

Figure 13.2 Structural formula for one of the four types of RNA nucleotides. The three others have a different base (adenine, guanine, or cytosine instead of the uracil shown here). Compare Section 12.2, which shows DNA's four nucleotides. Notice how the sugars of DNA and RNA differ at one group only (*yellow*).

DNA   G C A T
DNA   C G T A
base pairing during DNA replication

RNA   G C A U
DNA   C G T A
base pairing during transcription

Figure 13.3 An example of base pairing of RNA with DNA during transcription, as compared to base pairing during DNA replication.

Transcription is like DNA replication in another way. Enzymes add nucleotides to a growing RNA strand one at a time, in the 5′ → 3′ direction. Section 12.4 has a simple explanation of strand assembly.

But transcription *differs* from DNA replication in three key respects. First, only a selected stretch of one DNA strand, not the whole molecule, is used as the template. Second, instead of DNA polymerase, the enzyme **RNA polymerase** catalyzes nucleotide additions to a growing RNA strand. Third, at the end of transcription there is a single, free strand of RNA nucleotides, not a double helix.

Transcription is initiated at a **promoter**, a base sequence in DNA that signals the start of a gene. Proteins position an RNA polymerase on DNA and help it bind to the promoter. The enzyme moves along the DNA strand, joining one nucleotide after another (Figure 13.4). When it reaches a certain base, the new molecule is released as a free transcript.

Finishing Touches on mRNA Transcripts

In eukaryotic cells alone, a new mRNA molecule is unfinished. This "pre-mRNA" must be modified before its protein-building instructions can be put to use. Just as a dressmaker might snip off some threads or add bows on a dress before it leaves the shop, so do eukaryotic cells tailor their pre-mRNA.

For example, enzymes attach a cap to the 5′ end of a pre-mRNA molecule. The cap is a nucleotide

RNA polymerase

b A molecule of RNA polymerase binds with a promoter region in the DNA. It will recognize the base sequence located downstream from that site as a template for linking together the nucleotides adenine, cytosine, guanine, and uracil into a strand of RNA.

transcribed
DNA winds
up again

DNA to be
transcribed
unwinds

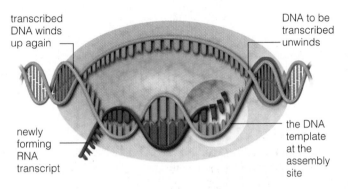

newly
forming
RNA
transcript

the DNA
template
at the
assembly
site

c All through transcription, the DNA double helix becomes unwound in front of the RNA polymerase. Short lengths of the newly forming RNA strand briefly wind up with its DNA template strand. New stretches of RNA unwind from the template (and the two DNA strands wind up again).

growing RNA transcript

direction of transcription ⟶

d What happened at the assembly site? RNA polymerase catalyzed the base-pairing of RNA nucleotides, one after another, with exposed bases on the DNA template strand.

5' A U G G U G U U A G G G C A C G U A U 3'

e At the end of the gene region, the last stretch of the new mRNA transcript is unwound and released from the DNA.

that has a methyl group and phosphate groups bonded to it. Also, enzymes attach a tail of about 100 to 200 nucleotides to pre-mRNA transcripts. The tail becomes wound up with proteins. Later, in the cytoplasm, the cap helps bind the mRNA to a ribosome. Also, enzymes will slowly destroy the wound-up tail from the tip on back. Such tails "pace" the access of enzymes to mRNA.

unit of transcription in DNA strand

exon intron exon intron exon

3' 5'

transcription into pre-mRNA

cap poly-A tail

5' 3'

snipped out *snipped out*

3'

mature mRNA transcript

Figure 13.5 Transcription and modification of new mRNA in the nucleus of eukaryotic cells. Its cap is a nucleotide with functional groups attached. Its tail is a string of adenine nucleotides (hence the name, poly-A tail).

That controlled access determines how long an mRNA molecule will last. It helps keeps the protein-building messages intact for as long as the cell requires them.

Besides these alterations, the pre-mRNA itself gets modified. Most eukaryotic genes contain one or more **introns**, base sequences that must be removed before a pre-mRNA molecule can be translated. These introns intervene between **exons**, the parts that are still in the mRNA when it's translated into protein. As Figure 13.5 shows, introns are transcribed along with the exons but are snipped out before the mRNA leaves the nucleus in mature form. (Think of it this way: *Ex*ons are *ex*ported from the nucleus and *in*trons stay *in* the nucleus.)

It could be that some introns are evolutionary junk, the leftovers of past mutations that led nowhere. Yet many are sites where instructions for building one type of protein can be snipped apart and spliced together in various ways. Alternative splicing allows different cells in your body to use the same gene to make different versions of a pre-mRNA transcript. As a result, the cell has different versions of the protein at its disposal. We return to this topic in Section 14.3.

During gene transcription, a sequence of exposed bases in one of the two strands of a DNA molecule serves as the template for assembling a single strand of RNA. The assembly follows these base-pairing rules: adenine only with uracil, and cytosine only with guanine.

Before leaving the nucleus, each new mRNA transcript, or pre-mRNA, undergoes modification into final form.

DECIPHERING THE mRNA TRANSCRIPTS

What Is the Genetic Code?

Like a strand of DNA, an mRNA molecule is a linear sequence of nucleotides. What are the protein-building "words" encoded in that sequence? Gobind Khorana, Marshall Nirenberg, and other investigators came up with the answer. They showed that ribosomes "read" nucleotide bases *three at a time*, as triplets. Base triplets in an mRNA strand were given this name: **codons**.

Figure 13.6 will give you an idea of how the order of different codons in an mRNA strand dictates the order in which particular amino acids will be added to a growing polypeptide chain.

Count the codons listed in Figure 13.7, and you see that there are sixty-four kinds. Notice how most of the twenty kinds of amino acids correspond to more than one codon. Glutamate corresponds to the code words GAA *and* GAG, for example. AUG codes for the amino acid methionine and is the start point for translation of all mRNA transcripts. Said another way, the "three-bases-at-a-time" selections start at a particular AUG in the transcript's nucleotide sequence. Methionine is the first amino acid in all new polypeptide chains. Codons UAA, UAG,

and UGA do not correspond to amino acids. They are STOP signals that prevent further additions of amino acids to a new polypeptide chain.

The set of sixty-four different codons is the **genetic code**. It is the basis of protein synthesis in all organisms.

Structure and Function of tRNA and rRNA

In a cell's cytoplasm are pools of free amino acids and free tRNA molecules. The tRNAs each have a molecular "hook," an attachment site for amino acids. They also have an **anticodon**, a nucleotide triplet that can base-pair with a codon (Figure 13.8). When tRNAs bind to the codons, they automatically position their attached amino acids in the order specified by mRNA.

A cell has a cytoplasmic pool of sixty-four kinds of codons, but it is able to utilize fewer kinds of tRNAs.

a Part of the amino acid sequence that was put together when mRNA was translated into a polypeptide chain

b Part of the mRNA strand that was transcribed from DNA

c The base sequence of a gene region in DNA

Figure 13.6 The correspondence between genes and proteins. (**a**) A sequence of amino acids, part of a protein's polypeptide chain. (**b**) An mRNA transcript. Every three nucleotide bases, equaling one codon, called for one of the chain's amino acids. (**c**) Exposed bases on one strand of a DNA double helix that was unwound during transcription. It was the template for assembling the mRNA strand. Referring to Figure 13.7, can you fill in the blank codon for tryptophan in the mRNA strand?

first base	second base				third base
	U	C	A	G	
U	phenylalanine	serine	tyrosine	cysteine	U
	phenylalanine	serine	tyrosine	cysteine	C
	leucine	serine	STOP	STOP	A
	leucine	serine	STOP	tryptophan	G
C	leucine	proline	histidine	arginine	U
	leucine	proline	histidine	arginine	C
	leucine	proline	glutamine	arginine	A
	leucine	proline	glutamine	arginine	G
A	isoleucine	threonine	asparagine	serine	U
	isoleucine	threonine	asparagine	serine	C
	isoleucine	threonine	lysine	arginine	A
	methionine (or START)	threonine	lysine	arginine	G
G	valine	alanine	aspartate	glycine	U
	valine	alanine	aspartate	glycine	C
	valine	alanine	glutamate	glycine	A
	valine	alanine	glutamate	glycine	G

Figure 13.7 The genetic code. Codons in mRNA are nucleotide bases "read" in blocks of three. Sixty-one of the base triplets correspond to specific amino acids. Three others are signals that stop translation. The vertical column at *left* lists choices for the first of the three nucleotides in mRNA codons. The top horizontal row lists choices for the second codon. The vertical column at *right* lists choices for the third. For example, reading from left to right, the triplet UGG corresponds to tryptophan. Both UUU and UUC correspond to phenylalanine.

— codon in mRNA

— anticodon in tRNA

tRNA molecule's attachment site for amino acid

amino acid

OH

b   c

Figure 13.8 (**a**) Three-dimensional model for one type of tRNA molecule. (**b**) Simplified model for tRNA that you'll come across in illustrations to follow. The "hook" sketched at the lower end of this model represents the site where a specific amino acid can be attached. (**c**) Structural features common to all tRNAs.

tunnel

b   small ribosomal subunit + large ribosomal subunit ⟶ intact ribosome

Figure 13.9 (**a**) Three-dimensional model for the large subunit of a bacterial ribosome. It consists of two rRNA molecules (*light gray*) and thirty-one structural proteins (*gold*), which stabilize the structure. At one end of a tunnel inside this subunit, rRNA catalyzes polypeptide chain assembly. This is an ancient, highly conserved molecular structure. Its role in life is so vital that the corresponding large subunit of eukaryotic ribosomes, which is bigger, is probably very similar in structure and function. (**b**) Model for the small and large subunits of a ribosome.

How do tRNAs match up with more than one type of codon? According to base-pairing rules, adenine must pair with uracil, and cytosine with guanine. For codon–anticodon interactions, however, the rules loosen up for the first and third bases. To give one example, AUU, AUC, and AUA specify isoleucine. All three of these codons can pair with a single type of tRNA that carries isoleucine. Such freedom in codon–anticodon pairing at a base is known as the "wobble effect."

Before anticodons interact with codons of an mRNA strand, that strand must bind to a ribosome. As shown in Figure 13.9, each ribosome has two subunits. These are assembled in the nucleus from rRNA and structural proteins, which stabilize the ribosome structure. rRNA's enzyme action drives protein synthesis.

At some point the subunits are shipped separately to the cytoplasm. There, intact, functional ribosomes are put together, each from two subunits—but only when messages encoded in mRNA are to be translated.

The nucleotide sequence of both DNA and mRNA encodes protein-building instructions. The genetic code is a set of sixty-four base triplets, which are nucleotide bases read in blocks of three. A codon is a base triplet in mRNA.

Different combinations of codons specify the amino acid sequence of different polypeptide chains, start to finish.

mRNAs are the only molecules that carry protein-building instructions from DNA into the cytoplasm.

tRNAs deliver amino acids to ribosomes, where they base-pair with codons in the order specified by mRNA.

Ribosomes, composed of rRNA and proteins, are structures upon which amino acids are assembled into polypeptide chains. An rRNA catalyzes chain assembly.

HOW IS mRNA TRANSLATED?

Stages of Translation

The protein-building code built into mRNA transcripts of DNA becomes translated at intact ribosomes in the cytoplasm. Translation proceeds through three stages: initiation, elongation, and termination.

During the stage called *initiation*, an initiator tRNA (the only one that can start transcription) and an mRNA transcript are both loaded onto a ribosome. First, the initiator tRNA binds with the small ribosomal subunit.

binding site for mRNA

ELONGATION

P (first binding site for tRNA)   *A* (second binding site for tRNA)

amino acid 1

amino acid 2

d Simplified model for binding sites at one end of the tunnel through the large ribosomal subunit, as shown in Figure 13.9. One site is for an mRNA transcript. Two others are for tRNAs that deliver amino acids to the intact ribosome.

e The initiator tRNA has become positioned in the first tRNA binding site (called *P*) on the ribosomal platform. Its anticodon matches up with the START codon (AUG) of the mRNA, which also has become positioned in *its* binding site. Another tRNA is about to move into the platform's second tRNA binding site (called *A*). It is one that can bind with the codon following the START codon.

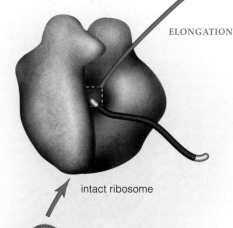

c As the final step of the initiation stage, a large ribosomal subunit joins with the small one. Once this initiation complex has formed, chain *elongation*— the second stage of translation—can get under way.

intact ribosome

INITIATION

b *Initiation*, the first stage of translating the mRNA transcript, is about to begin. An initiator tRNA (one that can start this stage) is loaded onto a platform of a small ribosomal subunit. The small subunit/tRNA complex attaches to the 5' end of the mRNA. It moves along the mRNA and "scans" it for an AUG START codon.

a A mature mRNA transcript leaves the nucleus by passing through pores across the nuclear envelope. Thus it enters the cytoplasm, which contains pools of many free amino acids, tRNAs, and ribosomal subunits.

mRNA transcript

The START codon for the transcript, AUG, matches up with that tRNA's anticodon. At the same time, the AUG binds with the small subunit. Second, a large ribosomal subunit binds with the small subunit. When joined this way, the ribosome, mRNA, and tRNA are an initiation complex (Figure 13.10*b*). The next stage can begin.

In the *elongation* stage of translation, a polypeptide chain is assembled as the mRNA passes between the two ribosomal subunits, a bit like a thread being moved through the eye of a needle. Part of the rRNA molecule located at the center of the large ribosomal subunit has unusual acidity. This region functions as an enzyme. It catalyzes the joining of individual amino acids in the sequence dictated by the codon sequence in the mRNA molecule. Figure 13.10*f–i* shows how a peptide bond forms between the most recently attached amino acid and the next one delivered to the intact ribosome while a polypeptide chain is growing. Here, you might wish to refer to the description and sketch of peptide bond formation in Section 3.4 (Figure 3.15).

During the last stage of translation, *termination*, a STOP codon in the mRNA moves onto the platform. No tRNA has a corresponding anticodon. Proteins known as release factors bind to the ribosome. They trigger enzyme activity that detaches the mRNA *and* the chain from the ribosome (Figure 13.10*j–l*).

Figure 13.10 Translation, the second step of protein synthesis.

Review Questions

1. Are the polypeptide chains of proteins assembled on DNA? If so, state how. If not, state how they are assembled, and on which molecules. *CI, 13.1*

2. Name the three classes of RNA and state the function of each class in protein synthesis. *13.1, 13.2*

3. The pre-mRNA transcripts of eukaryotic cells contain both introns and exons. Are the introns or exons snipped out before the transcript leaves the nucleus? *13.1*

4. Distinguish between codon and anticodon. *13.2*

5. Name the three stages of translation. Briefly describe the key events of each stage. *13.3*

6. Define gene mutation. Give three examples of agents that cause mutations. *13.4*

7. Do all mutations arise spontaneously? Are environmental agents always the trigger for mutation? *13.4*

8. Define and then state the possible outcomes of the following types of mutation: base-pair substitution, base insertion, and insertion of a transposon at a new location in the DNA. *13.4*

Self-Quiz   ANSWERS IN APPENDIX III

1. DNA contains many different genes that are transcribed into different _____ .
    a. proteins
    b. mRNAs only
    c. mRNAs, tRNAs, and rRNAs
    d. all are correct

2. An RNA molecule is _____ .
    a. a double helix
    b. usually single-stranded
    c. always double-stranded
    d. usually double-stranded

3. An mRNA molecule is produced by _____ .
    a. replication
    b. duplication
    c. transcription
    d. translation

4. Each codon calls for a specific _____ .
    a. protein
    b. polypeptide
    c. amino acid
    d. carbohydrate

5. Referring to Figure 13.7, use the genetic code to translate the mRNA sequence UAUCGCACCUCAGGAGACUAG. Notice that the first codon in the frame is UAU. Which amino acid sequence is being specified?

    a. TYR — ARG — THR — SER — GLY — ASP

    b. TYR — ARG — THR — SER — GLY

    c. TYR — ARG — TYR — SER — GLY — ASP

6. Anticodons pair with _____ .
    a. mRNA codons
    b. DNA codons
    c. tRNA anticodons
    d. amino acids

7. Match the terms with the most suitable description.
    _____ alkylating agent
    _____ chain elongation
    _____ exons
    _____ genetic code
    _____ anticodon
    _____ intron
    _____ codon

    a. parts of mature mRNA transcript
    b. base triplet coding for amino acid
    c. second stage of translation
    d. base triplet that pairs with codon
    e. one environmental agent that induces mutation in DNA
    f. set of sixty-four codons for mRNA
    g. parts removed from a pre-mRNA transcript

Figure 13.15 Soft skin tumors on a person with neurofibromatosis, an autosomal dominant disorder.

Critical Thinking

1. Sandra discovered a tRNA with a mutation in DNA that encodes the anticodon 3′-AAU instead of 3′-AUU. In cells with the mutated tRNA, what will be the effect on protein synthesis?

2. A DNA polymerase made an error during the replication of an important gene region of DNA. None of the DNA repair enzymes detected or repaired the damage. A portion of the DNA strand with the error is shown here:

    ```
    ...AATTCC G ACTCCTATGG
       |||||| | ||||||||||
    ...TTAAGG T TGAGGATACC
    ```

    After the DNA molecule is replicated and two daughter cells have formed, one cell is carrying a mutation and the other cell is normal. Develop a hypothesis to explain this observation.

3. *Neurofibromatosis* is a human autosomal dominant disorder caused by mutations in the *NF1* gene. It is characterized by soft, fibrous tumors in the peripheral nervous system and skin as well as abnormalities in muscles, bones, and internal organs (Figure 13.15).
    Because the gene is dominant, an affected child usually has an affected parent. Yet in 1991, scientists reported on a boy who had neurofibromatosis yet whose parents did not. When they examined both copies of his *NF1* gene, they found the copy he had inherited from his father contained a transposon. Neither the father nor the mother had a transposon in any of the copies of their own *NF1* genes. Explain the cause of neurofibromatosis in the boy and how it arose.

Selected Key Terms

alkylating agent *13.4*
anticodon *13.2*
base sequence *CI*
base-pair substitution *13.4*
carcinogen *13.4*
codon *13.2*
deletion (of base) *13.4*
exon *13.1*
gene mutation *13.4*
genetic code *13.2*
insertion (of base) *13.4*
intron *13.1*
ionizing radiation *13.4*

mRNA (messenger RNA) *13.1*
mutation rate *13.4*
nonionizing radiation *13.4*
promoter *13.1*
ribonucleic acid (RNA) *CI*
RNA polymerase *13.1*
rRNA (ribosomal RNA) *13.1*
transcription *CI*
translation *CI*
transposon *13.4*
tRNA (transfer RNA) *13.1*
uracil *13.1*

Readings

Crick, F. 1988. *What Mad Pursuit: A Personal View of Scientific Discovery.* New York: HarperCollins. Crick's autobiography.

Fairbanks, F., and W. R. Anderson. 1999. *Genetics: The Continuity of Life.* Monterey, California: Brooks-Cole.

On-Line readings at Student Guide for InfoTrac:
www.brookscole.com/biology

14

CONTROLS OVER GENES

When DNA Can't Be Fixed

Not long ago, Laurie Campbell turned eighteen and happened to notice a black mole on her skin. It was an odd, encrusted lump with a ragged border. No fool, she quickly made an appointment with her family doctor, who ordered a biopsy. The mole turned out to be a *malignant melanoma*, the deadliest form of skin cancer.

Moles and other tumors are tissue masses that arise after mutations in genes that control cell growth and division. Their cells don't respond to normal controls; they go on dividing as long as conditions for growth remain favorable. All **cancers** are malignant forms of tumors. Laurie was lucky. She detected a cancer in its earliest stage before it could spread through her body. Now she regularly checks out other skin moles. She is aware of having become a statistic. In 2000, there were more than a million cases of skin cancer in the United States and 7,700 deaths from malignant melanoma.

Laurie is smart. She plotted the position of each mole on her body. Once a month, she uses that body map as a guide for a quick, thorough self-examination. Figure 14.1 shows examples of what she looks for. Laurie also schedules a medical examination every six months.

Ultraviolet wavelengths in the sun's rays, tanning lamps, and other sources of nonionizing radiation can cause skin cancer. For instance, they promote covalent bonding between two neighboring thymine bases in a DNA strand. The result is an abnormal, bulky structure in a DNA molecule—a thymine dimer.

thymine dimer

Normally, at least seven gene products interact as a DNA repair mechanism to remove the bulky lesion. But mutation in one or more of those genes can disrupt the mechanism. When that happens, thymine dimers can accumulate in skin cells. They may trigger cancers and other lesions.

You are at risk if you habitually irritate moles, as by shaving or wearing abrasive clothing. You are at risk if your skin is chronically chapped, cracked, or sore. You are at risk if there is a history of cancer in your family or if you have endured radiation therapy. Like Laurie, you are at risk simply if you burn easily. Tanning or otherwise staying out in the sun without protection is ill advised; damaged DNA is a reality for all of us.

Why start a chapter with such awful prospects? Doing so might help focus your attention on how lucky individuals are when gene controls operate as they should. **Gene controls** are molecular mechanisms that govern when and how fast specific genes will be transcribed and translated, and whether gene products will be switched on or silenced.

Figure 14.1 Examples of what can happen when repair enzymes can't fix changes in the nucleotide sequence of DNA. (**a**) *Basal cell carcinoma*, the most common skin cancer. This slow-growing, raised lump may be uncolored, reddish-brown, or black. (**b**) *Squamous cell carcinoma*, the second most common skin cancer. The pink growths, firm to the touch, grow rapidly under the surface of skin exposed to the sun. (**c**) *Malignant melanoma* spreads fastest. The malignant cells form dark, encrusted lumps. They may itch like an insect bite or bleed easily. *Right:* Laurie Campbell avoiding the sun—and melanoma.

In all prokaryotic and eukaryotic cells, many of the controls make transcription rates rise or fall in response to concentrations of nutrients and other substances. For example, genes that specify the enzymes required to metabolize sugars aren't turned on and off; they are transcribed on an ongoing basis. But controls adjust the actual rates of transcription according to how much sugar is available. Controls activate, inactivate, and degrade gene products, including the sugar-digesting enzyme molecules that have finished their task.

In eukaryotic cells only, other controls guide mRNA transcript processing, transport of mature RNAs from the nucleus, and rates of translation in the cytoplasm. (Life is generally short for prokaryotic cells, and their mRNA gets translated as soon as it starts peeling off the DNA.) Other controls guide the modification of new polypeptide chains in the cytomembrane system.

Controls also guide the contribution that eukaryotic cells will make to a long-term program of growth and development. This is especially so for large, complex, multicelled species. As part of the program, different cell lineages activate and suppress fractions of their genes in unique ways. Certain genes are expressed once, some of the time, or not at all. The result? Most cells become specialized in structure, composition, and function. We call this process **cell differentiation.**

For example, all of your cells inherited the same genes, because every one of them descended from a single fertilized egg. They all use genes that specify the enzymes of glycolysis on an ongoing basis. Yet only the immature red blood cells use the genes that specify hemoglobin. Your liver cells access the genes required to synthesize enzymes that neutralize certain toxins, but they are the only ones that do. When your eyes first formed, only certain cells accessed the genes necessary to synthesize crystallin. No other cells used the genes for this type of protein, which helped make transparent fibers of the lens in each eye.

Explaining control of gene activity is like explaining a full symphony orchestra to someone who has never seen one or heard it perform. You have to identify its many separate parts before their interactions start to make sense! Gene controls weave through the story line of many chapters to follow. For that reason, take some time now to become acquainted with just a few of the molecular players and their amazing functions.

Key Concepts

1. In cells, a variety of controls govern when, how, and to what extent genes are expressed. The control elements operate in response to changing chemical conditions and to reception of external signals.

2. Control is exerted by way of regulatory proteins and other molecules that operate before, during, or after gene transcription. The control elements interact with DNA, with RNA that has been transcribed from DNA, or with gene products—that is, with the resulting polypeptide chains or final proteins.

3. Prokaryotic cells depend on rapid control over short-term shifts in nutrient availability and other aspects of their surrounding environment. They commonly rely on regulatory proteins that help make quick adjustments in rates of gene transcription.

4. Eukaryotic cells also rely on controls over short-term shifts in diet and levels of activity. In complex multicelled species, they rely as well on long-term controls over an intricate program of growth and development.

5. Controls over eukaryotic cells come into play when new cells contact one another in developing tissues. They also come into play when these cells start interacting with their neighbors by way of hormones and other signaling molecules.

6. Although all cells of a multicelled organism inherit the same genes, different cell types activate or suppress many of the genes in different ways. Their controlled, selective use of their store of genes leads to the synthesis of the proteins that give each type of cell its distinctive structure, function, and products.

TYPES OF CONTROL MECHANISMS

Different mechanisms control gene expression through interactions with DNA, RNA, and gene products—that is, with new polypeptide chains or final proteins. Some respond to rising or falling concentrations of a nutrient or some other substance. Other mechanisms respond to external signals for change.

The control agents include **regulatory proteins** that intervene before, during, or after gene transcription or translation. They also include signaling molecules such as **hormones**, which initiate change in some cell activity when they dock at suitable receptors.

With **negative control**, certain regulatory proteins trigger a slowdown in some activity. Hence their name, *repressor* proteins. With **positive control**, the regulatory proteins promote or enhance some activity. Hence their name, *activator* proteins. With respect to transcription, control mechanisms involve noncoding base sequences in DNA (which do not specify proteins). Examples are **promoters**, which signal the beginning of a gene; and **enhancers**, which are binding sites for activator proteins.

Figure 14.2 How DNA–histone packing in nucleosomes may be loosened to make gene regions available for transcription. Attaching an acetyl group to a histone makes it loosen its grip on DNA wound around it. Enzymes that attach or detach acetyl groups are known to be associated with transcription.

In eukaryotes, control also is exerted by chemical modification. For instance, regions of newly replicated DNA can be shut down by **methylation**: attachment of methyl groups ($-CH_3$) to nucleotide bases. Inactivated genes are often heavily methylated. And certain genes are activated by demethylation. Also, access to genes is partly controlled through **acetylation**: attaching acetyl groups from the histones that structurally organize the DNA (Section 8.1 and Figure 14.2).

When, how, and to what extent a gene is expressed depends on the type of cell and its functions, on the cell's chemical environment, and on signals for change.

Gene expression is controlled by regulatory proteins that interact with one another, with control elements built into the DNA, with RNA, and with gene products.

Control also is exerted through chemical modifications that inactivate or activate specific gene regions of DNA.

BACTERIAL CONTROL OF TRANSCRIPTION

Think about the dot of the letter "i." About a thousand bacterial cells would stretch side by side across the dot. Just imagine, each of those microscopic specks depends as much on gene controls as you do! When nutrients are plentiful and other environmental conditions also favor growth, the cells swiftly grow and divide. Gene controls promote the rapid synthesis of enzymes that catalyze nutrient digestion and other growth-related activities. Translation starts even before the RNA transcripts have been completed. Remember, bacteria have no nucleus to keep DNA away from ribosomes in the cytoplasm.

Often, prokaryotic genes for enzymes of a metabolic pathway are clustered together in the same sequence as the reaction steps. All genes in such a sequence may be transcribed as one continuous mRNA transcript.

With this bit of background, consider how one kind of prokaryote adjusts transcription rates downward or upward, depending on the availability of nutrients.

Negative Control of the Lactose Operon

Escherichia coli is an enteric bacterium (gut dweller) in all mammals. It survives on glucose, milk sugars such as lactose, and other ingested nutrients. Like other adult mammals, you probably don't drink milk around the clock. When you do, *E. coli* cells in your large intestine (colon) swiftly transcribe three genes for enzymes with roles in sugar-digesting reactions, starting with lactose.

A promoter precedes the three genes, which are next to one another. A different control element intervenes between the promoter and the bacterial genes. It is an **operator**, a binding site for a repressor that can block transcription (Figure 14.3). An arrangement in which a promoter and operator service more than one bacterial gene is an **operon**. Elsewhere in *E. coli* DNA, a different gene codes for this particular repressor protein, which can bind with the operator *or* with a lactose molecule.

When lactose concentration is low, a repressor binds with the operator (Figure 14.4). Being a large molecule, it extends over the promoter and blocks transcription. *So lactose-degrading enzymes aren't built when they aren't required.* When many lactose molecules are moving past in the gut, the chances are greater that one will enter a

Figure 14.3 Computer model for one type of repressor protein. This repressor is bound to an operator in a prokaryotic DNA molecule.

a A repressor protein exerts negative control over three genes of the lactose operon by binding to the operator and inhibiting transcription.

b When the concentration of lactose is low, the repressor is free to block transcription. Being bulky, it overlaps the promoter and prevents binding by RNA polymerase. The enzymes (not needed) are not produced.

c At high concentration, lactose is an inducer of transcription. It binds to and distorts the shape of the repressor—which now cannot bind to the operator. The promoter is exposed and the genes can be transcribed.

Figure 14.4 Negative control of the lactose operon. The operon's first gene codes for an enzyme that splits lactose (a disaccharide) into glucose and galactose. The second enzyme codes for an enzyme that transports lactose into cells. The third helps metabolize certain sugars.

bacterium and bind with the repressor. Binding alters the repressor's shape. The alteration prevents it from binding with the operator, so nothing is stopping RNA polymerase from transcribing the genes. Thus, *lactose-degrading enzymes are synthesized only when required.*

Positive Control of the Lactose Operon

E. coli cells pay far more attention to glucose than to lactose. They transcribe genes for glucose breakdown faster, and continuously. Even if lactose is in the gut, the lactose operon isn't used much *unless there is no glucose.* With no glucose, an activator protein called CAP goes to work. It happens that the lactose operon's promoter is not good at binding RNA polymerase. It does better when CAP adheres to it first. But CAP won't bind to the promoter without first binding to a chemical messenger, cAMP (cyclic adenosine monophosphate, which is like ATP with two phosphate groups removed). When the two adhere to a promoter, they are like an invitation to RNA polymerase to join up and start transcription.

When glucose is plentiful, glycolysis runs full bore, ATP forms, and an enzyme required for cAMP formation

is inhibited. Transcription of the lactose operon genes slows almost to a standstill. When glucose is scarce and lactose is available, though, the enzyme makes cAMP. Molecules of cAMP accumulate, CAP–cAMP complexes form, and lactose operon genes are transcribed fast. So the cells use lactose as an alternative energy source.

We humans are born with lactase, a lactose-digesting enzyme in cells of the intestinal lining. Before three years have passed, lactase levels start to decline in genetically predisposed individuals, and *lactose intolerance* follows. Lactose can't be degraded. It moves into the colon and fans the growth of resident bacterial populations. One gaseous by-product of their metabolism accumulates and distends the colon, causing pain. Short fatty acid chains released by the reactions lead to diarrhea, which is often severe. People avoid the symptoms by drinking milk that incorporates predigested lactose or by taking lactase pills when drinking or eating dairy products.

Transcription rates of bacterial genes for nutrient-digesting enzymes are quickly adjusted downward and upward by control systems that respond to nutrient availability.

CONTROLS IN EUKARYOTIC CELLS

By some estimates, the cells of all complex organisms rarely use more than 5 to 10 percent of their genes at a given time. One way or another, control mechanisms are keeping most of the genes repressed. Which genes are active depends on the type of organism and on the stage of growth and development it's passing through.

Case Study: X Chromosome Inactivation

We introduce a classic case of eukaryotic gene control with this question: At the cellular level, what do female humans and female calico cats have in common? Both have two X chromosomes in their diploid cells, and one is in its threadlike form, but the other is scrunched up even at interphase. The scrunching isn't a chromosome abnormality. It is a programmed shutdown of all but about three dozen genes on *one* of two homologous X chromosomes. This **X chromosome inactivation** occurs in the diploid cells of all female placental mammals.

One X chromosome is inactivated when the females are early embryos, no more than a tiny ball of dividing cells. The outcome is random; *either* chromosome may be condensed this way. One cell might shut down the maternal X chromosome, another cell right next to it might shut down the paternal X chromosome or the maternal X chromosome, and so on. The condensed one looks like a dense spot in an interphase nucleus (Figure 14.5*a*). Researchers named a condensed X chromosome a **Barr body** after Murray Barr, its discoverer.

Once that first random selection is made in a cell, however, all the descendant cells make the exact same selection as they go on dividing to form tissues. When fully developed, then, *each female mammal bears patches of tissue where genes of the maternal X chromosome are being expressed—and patches where genes of the paternal X chromosome are being expressed.* She is a "mosaic" for the X chromosomes! As you know, a pair of alleles on two homologous chromosomes might or might not be identical. When they are not, skin and other tissues of female mammals may have different features from one patch to the next. Mary Lyon, a geneticist, was the discoverer of this **mosaic tissue effect** of random X chromosome inactivation.

We see the effect in human females who are heterozygous for a recessive allele that blocks sweat gland formation. Their skin is a mosaic of tissues with and without sweat glands. This is one symptom of *anhidrotic ectodermal dysplasia.*

Figure 14.5 (a) Light micrograph revealing a Barr body (*arrow*) in a cell's nucleus. A Barr body is an inactivated X chromosome. (b) A mosaic tissue effect that shows up in a human female with anhidrotic ectodermal dysplasia.

unaffected skin (X chromosome with recessive allele was condensed; its allele is inactivated. The dominant allele on other X chromosome is being expressed in this tissue.)

affected skin with no normal sweat glands (In this tissue, the X chromosome with dominant allele has been condensed. The recessive allele on the other X chromosome is being transcribed.)

Where sweat glands are missing, the mutant allele is on an active X chromosome; the dominant allele is on the condensed one (Figure 14.5*b*). We see the same effect in female calico cats, which are heterozygous for a coat color allele on their X chromosomes (Figure 14.6).

What's the point of the shutdown? Remember, male and female mammals differ in their sex chromosomes (XY versus XX). However, a subset of genes, primarily on the X chromosome's long arm, must be expressed at the *same* levels in males and females. Otherwise, early embryos will not develop properly. Inactivating one X chromosome in XX embryos is a **dosage compensation** mechanism; it balances gene expression between sexes during critical early stages of development. The control mechanism works by activating the *Xist* gene on one X chromosome. The gene product interacts with control elements in DNA to condense only that chromosome.

Case Study: Hormones as Control Agents

Hormones, a type of signaling molecule, promote or inhibit gene expression in target cells. Any cell that has receptors for a given hormone is its target. In animals,

Figure 14.6 Why is this female calico cat "calico"? She is heterozygous for an allele on the X chromosome that deals with melanin, a dark pigment. Her coat color depends on which allele was randomly inactivated in the cells making up a given tissue region. In regions where the *active* X chromosome bears the dominant allele, you see dark fur. In regions where the *condensed* X chromosome bears the dominant allele, nothing is masking the recessive allele, so you see yellow fur. The white patches have a different basis. They arise from a gene interaction involving the spotting gene, which blocks melanin synthesis entirely.

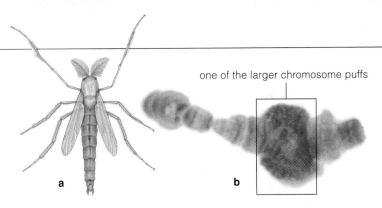

one of the larger chromosome puffs

a   b

Figure 14.7 (**a**) Larva of a midge (*Chironomus*). Midges are flies. The tiny, short-lived, winged adults often swarm together, which helps them find mates fast. They look like mosquitoes but don't bite. Most midge larvae develop and feed rapidly. They continuously transcribe genes for saliva's protein components. Ecdysone is a hormone that binds to a transcription-promoting regulatory protein. The chromosomes loosen and puff out in regions being transcribed. (**b**) Puffs are largest and most diffuse where transcription is most intense. Staining techniques reveal the banding patterns.

Figure 14.8 Experiment that reveals phytochrome's effect on corn seedlings. Two seedlings (the control group) grew in a greenhouse, in sunlight. Two seedlings (the experimental group) grew in the dark for eight days. They could not activate phytochrome, which helps control transcription of genes for the proteins that activate chlorophyll and make rubisco. Leaves of the experimental group never did green up.

control group   experimental group

hormone molecules secreted from cells enter the blood, then are transported to cells that may be quite distant.

Some hormones bind to membrane receptors at the surface of target cells. Others enter cells and bind to a regulatory protein involved in transcription, as when one type activates an enhancer. RNA polymerase will bind avidly to promoter–enhancer complexes.

Consider **ecdysone**, a hormone with roles in many insect life cycles. Insect larvae grow rapidly during the cycles. They feed voraciously on plant parts and other organic matter, and use up lots of saliva to prepare the food for digestion. Their salivary gland cells have made many copies of DNA. These copies remain together in parallel to form a **polytene chromosome**. The cells have ecdysone receptors. When ecdysone docks at a receptor, it triggers rapid transcription of genes in the multiple copies of DNA. Gene regions affected by the hormone puff out during transcription (Figure 14.7). After this, translation of the mRNA transcripts results in protein components of saliva.

Some vertebrate hormones have widespread effects on genes because many types of cells have receptors for them. For instance, somatotropin (growth hormone) promotes the synthesis of the proteins required for cell division, hence for the body's growth. Most cells have somatotropin receptors. Other hormones signal certain cells at certain times. Prolactin is like this. A few days after a female mammal gives birth, prolactin meets up with mammary gland cells and stimulates genes with roles in milk production. Liver and heart cells have the same genes, but they don't have prolactin receptors.

Case Study: Sunlight and Phytochrome

Plant a few seeds from a corn plant in a pot filled with moist, nutrient-rich soil. Let the seeds germinate in full darkness. Eight days later, you will have pale, spindly seedlings that have no chlorophyll (Figure 14.8). If you then expose the seedlings to a single burst of dim light from a flashlight, in less than ten minutes they will be producing functional chlorophyll molecules.

Phytochrome, a blue-green pigment, is a regulatory protein that influences gene transcription in ways that help plants adapt to changes in light conditions.

Far-red wavelengths inactivate phytochrome—and red wavelengths activate it. Far-red prevails at sunset, at night, or in the shade; red prevails at sunrise. The presence of these wavelengths varies from day to night and as the seasons change. By controlling phytochrome activity, the variation controls transcription at certain times of day and year. The genes under control specify enzymes and other proteins that help seeds germinate, stems lengthen and branch, and leaves grow. The gene products also help flowers, fruits, and seeds form.

Experiments by Elaine Tobin and her coworkers at the University of California, Los Angeles, give evidence of phytochrome control. Duckweed (*Lemna*) seedlings grown in the dark markedly increased the number of certain mRNA transcripts after a one-minute exposure to red light. What was transcribed? Genes that specify chlorophyll-binding proteins and rubisco, an enzyme that mediates carbon fixation (Sections 6.5 and 6.6). In the absence of these proteins, chloroplasts don't develop properly and they don't turn green.

As multicelled eukaryotic organisms develop, new cells inherit the same genes, but each type selectively activates and suppresses them. Selective gene expression leads to cell differentiation, whereby cells become specialized in composition, structure, and function.

X chromosome inactivation in XX mammalian embryos is a control mechanism that balances gene expression between sexes. It involves interaction between a gene product and control elements built into an X chromosome.

Hormones and other signals from the external environment help control selective gene expression in eukaryotic cells.

Lost Controls and Cancer

THE CELL CYCLE REVISITED Every second, millions of cells in your skin, bone marrow, gut lining, liver, and elsewhere divide and replace worn-out, dead, and dying predecessors. They don't divide willy-nilly. Mechanisms control the expression of genes that specify enzymes and other proteins required for cell growth, DNA replication, chromosome movements, and cytoplasmic division. And they control when the division machinery is put to rest.

The cell cycle has built-in checkpoints, where proteins monitor chromosome structure and other aspects of the preceding phase of the cycle. They also sense whether or not conditions favor division. The proteins are products of **checkpoint genes**, which are the basis of mechanisms that advance, delay, or block the cell cycle.

For example, the protein products called **kinases** are a class of enzymes that phosphorylate molecules. Some signal that DNA replication is finished and call on the cell to make the transition from interphase to mitosis (Section 8.2). The products called **growth factors** invite transcription of genes with roles in the body's growth. Example: one of these, epidermal growth factor (EGF), activates tyrosine kinase when it binds to target cells in epithelial tissues. Binding is a signal to start mitosis. The product of checkpoint gene *p53* can put the brakes on division when chromosomes are damaged.

ONCOGENES Many cancers are known to arise through mutations in one or more checkpoint genes. A mutated gene that has the potential to induce cancer falls into the broader category of **oncogenes**.

Oncogenes can form following insertion of viral DNA into a cell's DNA. They also can form after mutagens alter the structure of DNA. As you read in Section 13.4, ultraviolet radiation and nonionizing radiation, such as x-rays, are mutagens. So are many natural and synthetic

a Cancer cells break away from their home tissue.

b The metastasizing cells attach to the wall of a blood vessel or lymph vessel and secrete digestive enzymes onto it. They cross the breached wall.

c Cancer cells creep or tumble along inside blood vessels, then leave the bloodstream the same way they got in. They start new tumors in new tissues.

Figure 14.10 Steps in metastasis.

compounds, including asbestos and certain substances in tobacco smoke.

And remember those chromosome alterations and gene mutations described in Sections 11.8 and 13.4? Some cancers arise when base substitutions or deletions alter a gene or one of the control elements that deal with its transcription. Other cancers may arise by translocations or by destabilizing transposons.

CHARACTERISTICS OF CANCER What happens to a cell when it undergoes cancerous transformation? At the very least, all cancer cells display four characteristics. First, *the plasma membrane and cytoplasm change profoundly.* That membrane becomes more permeable, its proteins become lost or altered, and abnormal proteins form. The cytoskeleton shrinks drastically, becomes disorganized, or both (Figure 14.9). Enzyme action shifts, as in amplified reliance on glycolysis. Second, *cancer cells grow and divide abnormally.* Controls against overcrowding in tissues are lost; cell populations reach high densities. The number of small blood vessels that service the growing cell mass increases abnormally. Third, *cancer cells have a weakened capacity for adhesion.* Recognition proteins become lost or altered, so cells can't stay anchored in proper tissues. They break away and may establish colonies in other, distant tissues. **Metastasis** (meh-TA-stuh-SIS) is the name for the process of abnormal cell migration and tissue invasion (Figure 14.10). Fourth, *cancer cells are lethal.*

Figure 14.9 (**a**) A patrol of white blood cells surrounding a cancer cell. (**b**) Benign tumor. (**c**) Malignant tumor.

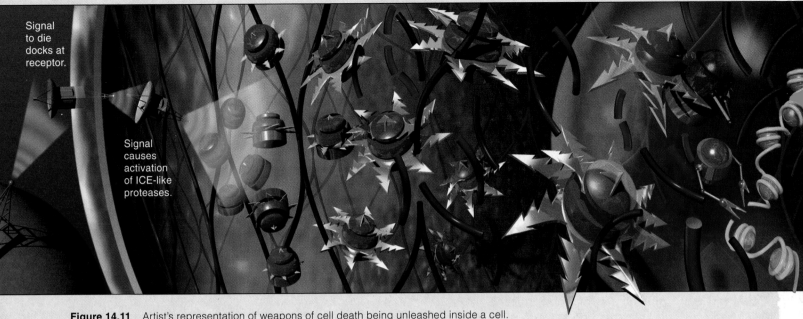

Signal to die docks at receptor.

Signal causes activation of ICE-like proteases.

Figure 14.11 Artist's representation of weapons of cell death being unleashed inside a cell. Normally, cells self-destruct when they finish their functions or become altered in ways that could threaten the body as a whole. When controls over programmed cell death are lost, an altered cell may start dividing abnormally and give rise to cancer.

Unless cancer cells are eradicated, their uncontrollable divisions put the individual on a painful road to death.

Cells of common skin moles and other noncancerous, *benign* tumors grow abnormally but slowly, and they retain the surface recognition proteins that hold them in their home tissue. Unless benign tumors become too large or irritating, doctors usually leave them alone.

Abnormal cells of a *malignant* tumor grow and divide rapidly, with destructive physical and metabolic effects on the surrounding tissues. These are grossly disfigured, metastasizing cancer cells (Figure 14.9c). They break loose from their home tissue, enter lymph or blood vessels, then slip out and invade other tissues where they do not belong. There, they may start growing as new tumors.

Each year in the developed countries alone, 15 to 20 percent of all deaths result from cancer. Cancer isn't just a human problem. Researchers have observed cancers in most of the animal species they have studied to date.

HERE'S TO SUICIDAL CELLS! We conclude with a case study of a gene and its cancerous transformation.

The first cell of a new multicelled individual contains marching orders that will guide its descendants along a program of growth, development, and reproduction, then on to death. As part of that program, many cells heed calls to self-destruct when they finish their prescribed function. If they become altered in ways that might pose a threat to the body as a whole, as by infection or cancer, they can execute themselves. **Apoptosis** (app-uh-TOE-sis) is the name for this form of cell death. It starts with molecular signals that activate and unleash lethal weapons of self-destruction, which were stockpiled earlier in the cell.

Protein-cleaving enzymes called **ICE-like proteases** are such weapons. Think of them as folded pocketknives or lethal Ninja weapons. When popped open, they chop apart structural proteins, including the building blocks of cytoskeletal elements and nucleosomes that organize the DNA (Figure 14.11).

A body cell in the act of suicide shrinks away from its neighbors. Its cytoplasm seems to roil, and its surface repeatedly bubbles outward and inward. No longer are its chromosomes extended through the nucleoplasm; they bunch up near the nuclear envelope. The nucleus, then the cell, breaks apart. Suicidal cells or remnants of them are swiftly engulfed by phagocytic white blood cells, which patrol and protect the body's tissues.

The timing of cell death is predictable in some cells, such as pigment-packed keratinocytes. These form the densely packed sheets of dead cells that are continually sloughed off and replaced at the skin surface. They have a three-week life span, more or less. Keratinocytes and other body cells, even kinds that are supposed to last for a lifetime, can be induced to die ahead of schedule. All it takes is sensitivity to signals that can activate ICE-like proteases and other enzymes of death.

Control genes suppress or trigger programmed cell death. One of these, *bcl-2*, helps keep normal body cells from dying before their time. The knives stay sheathed in cancer cells, which are supposed to—but don't—commit suicide on cue. Maybe they lost normal receptors that would permit contact with their signaling neighbors. Maybe they are receiving abnormal signals about when to grow, divide, or cease dividing. In some cancers, for instance, the suppressor gene *bcl-2* has been tampered with or shut down. Apoptosis is no longer in the cards.

Cancer is a multistep process involving more than one oncogene. But researchers have already identified many of the mutated forms of checkpoint genes that contribute to it. Remember how one form of leukemia, described in Chapter 11's introduction, is responding to the molecular targeting of one gene's product? More anticancer drugs and gene therapies should soon follow.

SUMMARY   **Gold** indicates text section

1. Cells are equipped with controls that govern gene expression; that is, which gene products appear, when, and in what amounts. When control mechanisms come into play depends on cell type, on prevailing chemical conditions, and on signals from other cell types that can change a target cell's activities. *CI, 14.1*

 a. In all prokaryotic and eukaryotic cells, many of the controls adjust transcription rates in response to concentrations of nutrients and other substances.

 b. In eukaryotic cells of complex organisms, controls contribute to a program of growth and development. They allow a fraction of the same genes to be expressed selectively in different lineages of cells. Selective gene expression is the beginning of diverse specializations in cell structure, composition, and function. We call this process cell differentiation.

2. Regulatory proteins (e.g., activators and inhibitors of transcription), hormones, and some DNA sequences are typical control agents. *14.1–14.3*

 a. Diverse control agents interact with one another, with operators and other control elements built into DNA, with RNA, and with gene products.

 b. Gene expression also is controlled with chemical modifications that inactivate and activate gene regions. X chromosome inactivation is an example.

 c. Control elements called promoters are the binding sites in DNA that signal the start of a gene. Enhancers are binding sites in DNA for some activator proteins.

3. In all cell types, two of the most common types of control systems inhibit or enhance gene transcription. In negative control systems, a regulatory protein binds at a specific DNA sequence to block transcription of one or more genes. In positive control systems, a regulatory protein binds to DNA to promote transcription. *14.1*

4. Unlike eukaryotic cells, most prokaryotic cells do not require many genes to live, grow, and reproduce. Most of their control systems adjust transcription rates in response to nutrient availability. Control of operons (groupings of functionally related prokaryotic genes and control elements) is an example. *14.2*

5. Compared to prokaryotic cells, eukaryotic cells use more complex gene controls. Gene expression changes in response to external conditions and also is subject to long-term controls over growth and development. *14.3*

 a. X chromosome inactivation in XX embryos of all female mammals is an outcome of interactions between a product of a gene (*Xist*) and control elements in one X chromosome. This interaction is an example of dosage compensation, a control mechanism that maintains a crucial balance in gene expression between the sexes.

 b. Hormones and phytochromes are two examples of signaling molecules that have roles in selective gene expression. Both can affect transcription.

6. Checkpoint gene products advance, delay, or block the cell cycle. They include growth factors and protein kinases. Some checkpoint gene mutations lead to cancer. A mutated form of such a gene is an oncogene. *14.4*

7. Cancer involves loss of normal controls over the cell cycle and mechanisms of programmed cell death. In all cancer cells, the structure and function of the plasma membrane and cytoplasm are severely compromised. Cancer cells grow and divide abnormally. Mechanisms for self-destruction are suppressed. Cancer cells have a weakened capacity for adhesion to their home tissue. Unless eradicated, they are lethal. *14.4*

Review Questions

1. A plant, fungus, and animal consist of diverse cell types. How might the diversity arise, given that body cells in each of these organisms inherit the same set of genetic instructions? As part of your answer, define cell differentiation and the general way that selective gene expression brings it about. *CI, 14.3*

2. In what fundamental way do negative and positive controls of transcription differ? Is the effect of one or the other form of control (or both) reversible? *14.1*

3. Distinguish between: *14.1, 14.2*
 a. repressor protein and activator protein
 b. promoter and operator
 c. repressor and enhancer

4. Describe one type of control of transcription for the lactose operon in *E. coli*, a prokaryotic cell. *14.2*

5. Using the sketch on the facing page, define three types of gene controls and note the levels at which they take effect. Does the sketch work for both prokaryotic and eukaryotic cells? Why or why not? *14.1–14.3*

6. What is a Barr body? Does it appear in the cells of human males, human females, or both? Explain your answer. *14.3*

7. Briefly describe the general characteristics of normal cells. Then explain the difference between a benign tumor and a malignant tumor. *CI, 14.4*

Self-Quiz   ANSWERS IN APPENDIX III

1. Cell differentiation _____ .
 a. occurs in all complex multicelled organisms
 b. requires different genes in different cells
 c. involves selective gene expression
 d. both a and c
 e. all of the above

2. The expression of a given gene depends on the _____ .
 a. type of cell and its functions   c. environmental signals
 b. chemical conditions   d. all of the above

3. Regulatory proteins interact with _____ .
 a. DNA   c. gene products
 b. RNA   d. all of the above

4. A base sequence signaling the start of a gene is a(n) _____ .
 a. promoter   c. enhancer
 b. operator   d. activator protein

5. In prokaryotic cells but not eukaryotic cells, a(n) _____ precedes the genes of an operon.
 a. lactose molecule   c. operator
 b. promoter   d. both b and c

TRANSCRIPTION *Assembly of RNA on unwound gene regions of DNA molecule*

Pre-mRNA transcript processing — **mRNA** — **rRNA** — **tRNA**

mature mRNA transcripts — protein subunits — ribosomal subunits — mature tRNA

TRANSLATION

Convergence of RNAs

cytoplasmic pools of amino acids, ribosomal subunits, and tRNAs

At an intact ribosome, synthesis of a polypeptide chain at the binding sites for mRNA and tRNAs

ile gly met

FINAL PROTEIN

For use in cell or for export

6. An operon most typically governs _____ .
   a. bacterial genes   c. genes of all types
   b. a eukaryotic gene   d. DNA replication

7. Eukaryotic genes guide _____ .
   a. fast short-term activities   c. development
   b. overall growth   d. all of the above

8. X chromosome inactivation is an example of _____ .
   a. a chromosome abnormality   c. dysfunctional calico cats
   b. dosage compensation   d. dysfunctional women

9. Hormones may _____ gene transcription in target cells.
   a. promote   c. participate in
   b. inhibit   d. both a and b

10. Apoptosis is _____ .
   a. cell division after severe tissue damage
   b. programmed cell death by suicide
   c. a popping sound in mutated toes

11. ICE-like proteases are _____ .
   a. structural proteins   c. environmental signals
   b. lethal weapons   d. low-temperature enzymes

12. Match the terms with their most suitable descriptions.
   _____ phytochrome   a. inhibits gene transcription
   _____ Barr body   b. mutated form of any gene that
   _____ oncogene   can induce cancer
   _____ repressor   c. hormone with key role in
   _____ ecdysone   insect life cycles
                        d. helps plants adapt to daily
                           and seasonal changes in light
                        e. inactivated X chromosome

Critical Thinking

1. Geraldo isolated a strain of *E. coli* in which a mutation has affected the capacity of CAP to bind to a region of the lactose operon, as it would do normally. State how the mutation affects transcription of the lactose operon when cells of this bacterial strain are subjected to these conditions:
   a. Lactose and glucose are available.
   b. Lactose is available but glucose is not.
   c. Both lactose and glucose are absent.

2. *Duchenne muscular dystrophy*, a genetic disorder, affects boys almost exclusively. Early in childhood, muscles begin to atrophy (waste away) in affected individuals, who typically die in their teens or early twenties as a result of respiratory failure. Muscle biopsies of women who carry an allele associated with the disorder reveal some body regions with atrophied muscle tissue. Yet muscles adjacent to a region of atrophy is normal or even larger and more chemically active, as if to compensate for the weakness of the adjoining region. Give a brief explanation of these observations.

3. Unlike most rodents, guinea pigs are well developed at the time of birth. Within a few days, they can eat grass, vegetables, and other plant material. Suppose a breeder decides to separate the baby guinea pigs from their mothers after three weeks. He wants to keep the males and females in different cages. But it is difficult to identify the sex of guinea pigs when they are so young. Suggest a simple test that the breeder can perform to identify their sex.

4. Individuals affected by *pituitary dwarfism* cannot synthesize somatotropin (also called growth hormone). Children with this genetic abnormality will be below the range of normal heights. Develop a hypothesis to explain why therapy that is based on somatotropin injections may be effective.

5. The closer a mammalian species is to humans in its genetic makeup, the more useful information it may yield in laboratory studies of the mechanisms of cancer. Do you support the use of any mammal for cancer research? Why or why not?

Selected Key Terms

acetylation *14.1*
apoptosis *14.4*
Barr body *14.3*
cancer *CI*
cell differentiation *CI*
checkpoint gene *14.4*
dosage compensation *14.3*
ecdysone *14.3*
enhancer *14.1*
gene control *CI*
growth factor *14.4*
hormone *14.1*
ICE-like protease *14.4*
kinase *14.4*

metastasis *14.4*
methylation *14.1*
mosaic tissue effect *14.3*
negative control system *14.1*
oncogene *14.4*
operator *14.2*
operon *14.2*
phytochrome *14.3*
polytene chromosome *14.3*
positive control system *14.1*
promoter *14.1*
regulatory protein *14.1*
X chromosome
    inactivation *14.3*

Readings

Duke, R., D. Ojcius, and J. Ding-E Young. December 1996. "Cell Suicide in Health and Disease." *Scientific American*, 80–87.

Murray, A., and M. Kirschner. March 1991. "What Controls the Cell Cycle?" *Scientific American*, 264(3): 56–63.

Tijan, R. February 1995. "Molecular Machines That Control Genes." *Scientific American*, 54–61.

Travis, J. 5 August 2000. "Silence of the Xs." *Science News* 158: 92–94. Survey of recent investigations into the mechanism of X chromosome inactivation. Raises interesting questions about what has been called "junk DNA" (base sequences that have accumulated in DNA yet have no known function).

RECOMBINANT DNA AND GENETIC ENGINEERING

Mom, Dad, and Clogged Arteries

Butter! Bacon! Eggs! Ice cream! Cheesecake! Possibly you think of such foods as enticing, off-limits, or both. After all, who among us doesn't know about animal fats and the dreaded cholesterol?

Soon after you feast on these fatty foods, cholesterol enters the bloodstream. Cholesterol is important. It is a structural component of animal cell membranes, and without membranes, there would be no cells. Cells also remodel cholesterol into a variety of molecules, such as the vitamin D necessary for the development of good bones and teeth. Normally, however, the liver itself synthesizes enough cholesterol for your cells.

Some proteins circulating in blood combine with cholesterol and other substances to form lipoprotein particles. *High*-density lipoproteins, or *HDLs*, transport cholesterol to the liver, where it's metabolized. Usually, *low*-density lipoproteins, or *LDLs*, end up inside cells that store or use cholesterol. But sometimes too many LDLs form. The excess infiltrates the elastic walls of arteries and helps form atherosclerotic plaques (Figure 15.1). These abnormal masses interfere with blood flow. If they clog one of the tiny coronary arteries that deliver blood to the heart, a heart attack may result.

How well you handle dietary cholesterol depends on the genes you got from your parents. One of the genes specifies a protein receptor for LDLs. Inherit two "good" alleles and your blood level of cholesterol will stay low. Your arteries may never clog up, even with a high-fat diet. Inherit two copies of a certain mutated allele, and

you will develop *familial cholesterolemia*, a rare genetic disorder. Cholesterol levels get so high, many of those affected die of heart attacks in childhood or their teens.

In 1992 a woman from Quebec, Canada, became a milestone in the history of genetics. She was thirty years old. Like two brothers felled in their early twenties by heart attacks, she had inherited the mutant allele for the LDL receptor. She survived a heart attack at sixteen. At twenty-six, she had coronary bypass surgery.

At the time, people were hotly debating the risks and the promises of **gene therapy**—the transfer of one or more normal or modified genes into an individual's body cells to correct a genetic defect or boost resistance to disease. The woman opted for an untried procedure to give her body working copies of the good gene.

Surgeons removed about 15 percent of her liver. Researchers put cells from it in a nutrient-rich medium to promote cell growth and division. *And they spliced the functional allele for the LDL receptor into the genetic material of a harmless virus.* The modified virus infected the cultured liver cells, thereby inserting copies of the good gene into them. When the liver cells went on to reproduce, they made copies of the good gene, also.

About a billion modified cells were infused into the woman's portal vein, a blood vessel that leads directly to the liver. Some cells took up residence in the liver and started to make the missing cholesterol receptor. Two years later, a fraction of the woman's liver cells were behaving normally and sponging up cholesterol from blood. Her LDL blood levels declined nearly 20 percent. There was no sign of the arterial clogging that nearly killed her. Her cholesterol levels are still higher than normal. It's too soon to know whether her life will be prolonged. Yet the intervention gives strong evidence that gene therapy may be able to help some patients.

As you might gather from this pioneering clinical work, recombinant DNA technology has huge potential for medicine, agriculture, and industry. Think of it! For thousands of years, we humans have been changing

Figure 15.1 Life-threatening plaques (*bright yellow, white*) in one of the coronary arteries. These small arteries deliver blood to the heart. Abnormally high levels of cholesterol contribute to the formation of plaques that may clog them. Gene therapies based on recombinant DNA technology have the potential to counter many such threats to health.

Figure 15.2 One outcome of genetic manipulation by way of artificial selection practices: a large kernel from a modern strain of corn next to the tiny kernels of an ancestral species, which was recovered from a prehistoric cave in Mexico.

Key Concepts

1. Genetic experiments have been proceeding for billions of years, through gene mutations, crossing over, genetic recombination, and other natural events.

2. Humans are now purposefully bringing about genetic changes by way of recombinant DNA technology. Such manipulations are called genetic engineering.

3. With this technology, researchers isolate, cut, and splice together gene regions from different species. Then they greatly amplify the number of copies of genes that interest them. The genes, and sometimes the proteins they specify, are produced in quantities that are large enough to use for research and for practical applications.

4. Three activities are at the heart of recombinant DNA technology. First, procedures based on specific types of enzymes are used to cut DNA molecules into fragments. Second, the fragments are inserted into cloning tools, such as plasmids. Third, the fragments containing the genes of interest are identified, then copied rapidly and repeatedly.

5. Genetic engineering involves isolating, modifying, and inserting genes back into the same organism or into a different one. The goal is to beneficially modify traits that the genes influence. Human gene therapy, which focuses on controlling or curing genetic disorders, is an example.

6. The new technology raises social, legal, ecological, and ethical questions regarding its benefits and risks.

genetically based traits. By artificial selection practices, we produced new plants and new breeds of cattle, cats, dogs, and birds from wild ancestral stocks. We were selective agents for meatier turkeys, sweeter oranges, seedless watermelons, spectacular ornamental roses, and big juicy corn kernels (Figure 15.2). We conjured up hybrids such as the mule (horse × donkey) and plants that bear tangelos (tangerine × grapefruit).

Of course, we have to remember we're newcomers on the evolutionary stage. During the 3.8 billion years before we even made our entrance, nature conducted uncountable numbers of genetic experiments. Nature's tools have included mutation, crossing over, and other events that introduce changes in genetic messages. The countless changes gave rise to life's rich diversity.

But the striking thing about human-directed change is that the pace has picked up. Researchers analyze genes with **recombinant DNA technology**. They cut and recombine DNA from different species and insert it into bacterial, yeast, or mammalian cells. The cells replicate their DNA and divide rapidly. They copy the foreign DNA as if it were their own and churn out useful quantities of recombinant DNA molecules. The technology also is the basis of **genetic engineering**. By this process, genes are isolated, modified, and inserted into the same organism or into a different one. Protein products of the modified genes may cover the function of their missing or malfunctioning counterparts.

The new technology does not come without risks. With this chapter, we consider its basic aspects. At the chapter's end, we also address some ecological, social, and ethical questions related to its application.

A TOOLKIT FOR MAKING RECOMBINANT DNA

Restriction Enzymes

In the 1950s, the scientific community was agog over the discovery of DNA's structure. The excitement gave way to frustration. No one could figure out the sequence of nucleotides, the order of genes and gene regions along a chromosome. Robert Holley and his colleagues did sequence a small tRNA. They used digestive enzymes that broke the molecule into fragments small enough to be chemically characterized. But DNA molecules? These are far longer than RNAs. No one knew how to cut one into fragments long enough to have unique, and thus analyzable, sequences. Digestive enzymes can cut DNA, but not in any particular order. There was no telling how the fragments had been arranged in the molecule.

Then, by accident, Hamilton Smith discovered that *Haemophilus influenzae* chops up foreign DNA inserted into it by a bacteriophage. Extracts from *H. influenzae* cells included an enzyme that restricts itself to a specific kind of site in DNA. It was named a **restriction enzyme**.

In time, several hundred strains of bacteria offered up a toolkit of restriction enzymes that recognize and cut specific sequences of four to eight bases in DNA.

Table 15.1 lists a few. They are among many that make *staggered* cuts, which leave single-stranded "tails" on the end of a DNA fragment. Tails made by *Taq*I are two bases long (CG). *Eco*RI makes tails four bases long (AATT).

How many cuts do restriction enzymes make? That depends in part on the molecule. Example: The eight-base sequence recognized by *Not*I is rare in the DNA of mammals, so most fragments are tens of thousands of base pairs long. That's long enough for studying the organization of a **genome**—all of the DNA in a haploid number of chromosomes for a species. For the human species, the genome is about 3.2 billion base pairs long.

Modification Enzymes

DNA fragments with staggered cuts have sticky ends. "Sticky" means a restriction fragment's single-stranded tail can base-pair with a complementary tail of any other DNA fragment or molecule cut by the same restriction enzyme. *Mix DNA fragments cut by the same restriction enzyme, and the sticky ends of any two fragments having complementary base sequences will base-pair and form a recombinant DNA molecule:*

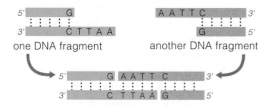

one DNA fragment   another DNA fragment

Notice the nicks where such fragments base-pair. **DNA ligase**, a modification enzyme, can seal these nicks:

Cloning Vectors for Amplifying DNA

Restriction and modification enzymes make it possible to insert foreign DNA into bacterial cells. As you know, each bacterial cell has only one chromosome, a circular DNA molecule. Many also inherit plasmids. A **plasmid** is a very small circle of extra prokaryotic DNA that has just a few genes and that gets replicated along with the bacterial chromosome (Figure 15.3). Bacteria usually can survive without plasmids, but some of the genes offer benefits, as when they confer resistance to antibiotics.

Table 15.1	Examples of Restriction Enzymes	
Bacterial Source	**Enzyme's Abbreviation**	**Its Specific Cut**
Thermus aquaticus	*Taq*I	5′ T C G A 3′ 3′ A G C T 5′
Escherichia coli	*Eco*RI	5′ G A A T T C 3′ 3′ C T T A A G 5′
Nocardia otitidus caviarum	*Not*I	5′ G C G G C C G C 3′ 3′ C G C C G G C G 5′

Figure 15.3 (a) A single plasmid. (b) Plasmids (*arrows*) from a ruptured *Escherichia coli* cell.

FROM HAYSTACKS TO NEEDLES—ISOLATING GENES OF INTEREST

Any genome consists of thousands of genes. *E. coli* has about 4,000, for instance; humans apparently have about 30,000. What if you wanted to learn about or modify the structure of any one of those genes? First you'd have to isolate that gene from all others in the genome.

There are several ways to do this. If part of the gene sequence is already known, you can design primers to amplify the entire gene or part of it by PCR. Often, though, researchers must isolate and clone a gene. First they make a **gene library**. This is a mixed collection of bacteria that contain different cloned DNA fragments. One is the gene of interest. A *genomic* library contains cloned DNA fragments from an entire genome. A *cDNA* library contains DNA derived from mRNA. Often it is the most useful because it is free of introns. But the gene is still hidden like a needle in a haystack. How can you isolate it? One way is to use a nucleic acid probe.

What Are Probes?

A **probe** is a very short stretch of DNA labeled with a radioisotope so that it can be distinguished from other DNA molecules in a given sample. Part of the probe must be able to base-pair with some portion of the gene of interest. Any base-pairing that takes place between sequences of DNA (or RNA) from different sources is called **nucleic acid hybridization**.

How do you acquire a suitable probe? Sometimes part of the gene or a closely related gene has already been cloned, in which case it can be used as the probe. If the gene's structure is a mystery, you still may be able to work backward from the amino acid sequence of its protein product, assuming that protein is already available. By using the genetic code as a guide (Section 13.2), you could build a DNA probe that is more or less similar to the gene of interest.

Screening for Genes

Once you have a gene library and a suitable probe, you are ready to hunt down the gene. Figure 15.9 shows the steps of one isolation method. The first step is to take bacterial cells of the library and spread them apart on the surface of a gelled growth medium in a petri plate. When spread out sufficiently, individual cells undergo division. Each cell starts a colony of genetically identical cells. The bacterial colonies appear as hundreds of tiny white spots on the surface of a culture medium.

After colonies appear, you lay a nylon filter on top of the colonies. Some cells stick to the filter at locations that mirror the locations of the original colonies. You use solutions to rupture the cells, and the DNA so released sticks to the filter. You make the DNA unwind to single strands and then add the probes. The probes hybridize

a Bacterial colonies, each derived from a single cell, grow on a culture plate. Each colony is about 1 millimeter across.

b A nitrocellulose or nylon filter is placed on the plate. Some cells of each colony adhere to it. The filter mirrors how the colonies are distributed on the culture plate.

c The filter is lifted off and put into a solution. The cells stuck to it lyse, but the cellular DNA sticks to the filter.

d Also, the DNA is denatured to single strands at each site. A radioactively labeled probe is added to the filter. It binds to DNA fragments that have a complementary base sequence.

e The probe's location is identified by exposing the filter to x-ray film. The image that forms on the film reveals the colony that has the gene of interest.

Figure 15.9 Use of a probe to identify bacterial colonies that have taken up a gene library.

only with DNA from the colony that took up the gene of interest. If you expose the probe-hybridized DNA to x-ray film, the pattern formed by the radioactivity will identify that colony. With this information you can now culture cells from that colony alone, knowing that it will be the only one that can replicate the cloned gene.

Probes may be used to identify one particular gene among many in gene libraries. Bacterial colonies that have taken up the library can be cultured to isolate the gene.

USING THE GENETIC SCRIPTS

As researchers decoded the genetic scripts of species, they opened the door to astonishing possibilities. For example, genetically engineered bacteria now produce medically valued proteins. Huge bacterial populations produce useful quantities of the desired gene products in stainless steel vats. *Diabetics* who must receive insulin injections every day are among the beneficiaries. At one time, medical supplies of this pancreatic hormone were extracted only from pigs and cattle. Later on, synthetic genes for human insulin were transferred into *E. coli* cells. The cells were the start of populations that became the first large-scale, cost-effective bacterial factory for proteins. Besides insulin, human somatotropin, blood-clotting factors, hemoglobin, interferon, and a variety of drugs and vaccines are also manufactured with the help of genetically engineered bacteria.

Other kinds of modified bacteria hold potential for industry and for cleaning up environmental messes. For instance, as you know, many microorganisms can break down organic wastes and help cycle nutrients through ecosystems. Certain modified bacteria can break down crude oil into less harmful compounds. When sprayed onto oil spills, as from a shipwrecked supertanker, they might help avert an environmental disaster. Others are genetically engineered to sponge up excess phosphates or heavy metals from the environment.

In addition, bacterial species that contain plasmids offer benefits for basic research, agriculture, and gene therapy. For instance, deciphering the messages encoded in bacterial genes helps us reconstruct the evolutionary history of life. Sections 18.6 and 18.7 will give you an idea of how comparisons of DNA or RNA from different organisms can reveal evolutionary secrets. The next two sections give a few more examples of benefits.

What about the "bad" bunch—the pathogenic fungi, bacteria, and viruses? Natural selection favors mutated genes that improve a pathogen's chances of evading a host organism's natural defenses. Mutation is frequent among pathogens that reproduce rapidly. So designing new antibiotics and other defenses against new gene products is a constant challenge. But if we learn about the genes, we may have advance warning of the "plan of attack." The story of HIV, a rapidly mutating virus that causes AIDS, gives insight into the magnitude of the problem (Section 35.10).

Knowing about genes allows us to genetically engineer beneficial microorganisms for uses in medicine, industry, agriculture, and environmental remediation.

Knowing about genes can help us reconstruct evolutionary histories of species.

Knowing about genes may help us design effective, timely counterattacks against rapidly mutating pathogens.

DESIGNER PLANTS

Regenerating Plants From Cultured Cells

Many years ago, Frederick Steward and his coworkers cultured cells from carrot plants. They induced some of them to develop into small embryos. As you will read in Section 28.5, some embryos grew into whole plants. Today, researchers routinely regenerate crop plants and many other plant species from cultured cells. They use ingenious methods to pinpoint a gene in a culture that may contain even millions of cells. Suppose they add a pathogen's toxin to a culture medium. Suppose a few cells happen to carry a gene that confers resistance to the toxin. Those cells will be the only ones to survive.

Whole plants regenerated from preselected cultured cells can be hybridized with other varieties of plants. In this way, desired genes might be transferred to a plant lineage to improve herbicide resistance, pest resistance, and other traits that can benefit crops and gardens.

Consider this larger aspect of the research: The food supply for most of the human population is vulnerable. Generally, farmers prefer genetically similar varieties of high-yield plants and have discarded the more diverse, older varieties. But genetic uniformity makes food crops vulnerable to many kinds of pathogenic fungi, viruses, and bacteria. That is why botanists comb the world for seeds of the wild ancestors of potatoes, corn, and other crop plants. They send their prizes—seeds with genes of a plant's lineage—to **seed banks**. These safe storage facilities are designed to preserve genetic diversity.

The problem is huge. Example: In 1970 a new fungal strain of *Southern corn leaf blight* destroyed much of the United States corn crop. All of the plants carried the same gene that conferred susceptibility to the disease. Ever since that devastating epidemic, seed companies have been much more attentive to offering genetically diverse corn seeds. And where does the diversity come from? Plant breeders call on the seed banks.

How Are Genes Transferred Into Plants?

The Ti (Tumor-inducing) plasmid from *Agrobacterium tumefaciens* is one vehicle for inserting new or modified genes into plants. *A. tumefaciens* infects many kinds of flowering plants (Figure 15.10). Some Ti plasmid genes

crown gall tumor

Figure 15.10
Crown gall tumor on a woody plant, an abnormal tissue growth triggered by a gene in the Ti plasmid.

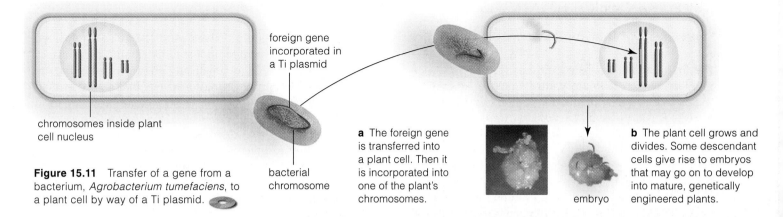

chromosomes inside plant cell nucleus

Figure 15.11 Transfer of a gene from a bacterium, *Agrobacterium tumefaciens*, to a plant cell by way of a Ti plasmid.

foreign gene incorporated in a Ti plasmid

bacterial chromosome

a The foreign gene is transferred into a plant cell. Then it is incorporated into one of the plant's chromosomes.

embryo

b The plant cell grows and divides. Some descendant cells give rise to embryos that may go on to develop into mature, genetically engineered plants.

invade a plant's DNA, then induce the formation of abnormal tissue masses. We call the masses crown gall tumors. Before introducing the plasmid into plant cells, researchers remove the tumor-inducing genes and then insert a desired gene into it. They place plant cells in a culture of the modified bacteria. When some cells take up the gene, whole plants may be regenerated from the cellular descendants, as in Figure 15.11. The expression of foreign genes in plants can sometimes be dramatic, as in Figure 15.12.

In nature, *A. tumefaciens* infects plants called dicots, which include beans, peas, potatoes and other vital crops. Geneticists are working to modify the bacterium so that it might also deliver genes into monocots that are vital food crops, including wheat, corn, and rice. Some use electric shocks or chemicals to deliver modified genes into plant cells. Some also blast microscopic particles coated with DNA into them.

Despite many obstacles, improved varieties of crop plants have been engineered or are in the works. For example, genetically engineered cotton plants display resistance to a herbicide (Figure 15.13). Farmers spray the herbicide in fields of modified cotton plants to kill weeds. The plants are not affected.

Also on the horizon are engineered plants that can be factories for pharmaceuticals. A few years ago, for example, tobacco plants that were engineered to produce hemoglobin and some other proteins were planted in a test field in North Carolina. Afterward, ecologists found

Figure 15.12 A modified plant that glows in the dark. A firefly gene was inserted into the plant's DNA and is being expressed. (Refer to Section 6.9, *Critical Thinking* question 4). The gene's product is luciferase, an enzyme with a role in bioluminescence.

Figure 15.13 (**a**) Control plant (*left*) and three genetically engineered aspen seedlings. Vincent Chiang and coworkers suppressed a regulatory gene involved in a lignin biosynthetic pathway. The modified plants synthesized normal lignin, but not as much. Lignin production decreased by as much as 45 percent—yet cellulose production increased 15 percent. Root, stem, and leaf growth were greatly enhanced. Plant structure did not suffer. Wood harvested from such trees might make it easier to manufacture paper and some clean-burning fuels, such as ethanol. (Lignin, a tough polymer, strengthens the secondary cell walls of plants. Before paper can be made from wood, the lignin must be chemically extracted.)

(**b**) *Left:* Cotton plant, used as the control. *Right:* Genetically engineered cotton plant with a gene for herbicide resistance. Both plants were sprayed with a weedkiller that now has widespread application in cotton fields.

no trace of foreign genes or proteins in the soil or in other plants or animals in the neighborhood. Another example: In a Stanford University laboratory, mustard plants synthesized biodegradable plastic beads that are suitable for use in manufacturing plastics.

Genetically diverse plant species are essential to protecting our vulnerable food supply. Genetic engineers can design plants with new beneficial traits.

GENE TRANSFERS IN ANIMALS

Supermice and Biotech Barnyards

The first mammals enlisted for experiments in genetic engineering were laboratory mice. Consider an example of this work. R. Hammer, R. Palmiter, and R. Brinster corrected a hormone deficiency that leads to dwarfism in mice. Insufficient levels of somatotropin (or growth hormone) cause the abnormality. The researchers used a microneedle to inject the gene for rat somatotropin into fertilized mouse eggs, which they implanted in an adult female. The gene was successfully integrated into mouse DNA, and the eggs developed into mice. Young mice in which that foreign gene was expressed grew 1–1/2 times larger than their dwarf littermates. In other experiments, researchers transferred the gene for human somatotropin into a mouse embryo. The gene became integrated into mouse DNA, and the modified embryo grew up to be a "supermouse" (Figure 15.14).

Today, human genes are being inserted into mouse embryos as part of ongoing research into the molecular basis of genetic disorders, such as Alzheimer's disease. In addition to microneedles, microscopic laser beams are used to penetrate cultured cells.

Animals of "biotech barnyards" are competing with bacterial factories as genetically engineered sources of proteins. For example, goats produce CFTR protein (for treating cystic fibrosis) and TPA (which may lessen the severity of a heart attack). Cattle may produce human collagen for repairing cartilage, bone, and skin.

In 1997 researchers reported the first case of cloning a mammal from an adult cell. They extracted a nucleus from a ewe's mammary gland cell and inserted it into an egg (from another ewe) from which the nucleus had been removed. Signals from the egg cytoplasm sparked the development of an embryo, which was implanted into a surrogate mother. As Section 12.5 describes, this resulted in Dolly, a clone that was herself cloned later on.

Researchers had been cloning animals from embryonic tissue for years, but not adults that already displayed some sought-after trait. Now, if they could refine cloning processes, they might be able to maintain a genotype indefinitely.

Not long after the experiment that resulted in Dolly, genetically engineered clones of mice, cattle, and other animals were being produced. As an example, Steve Stice and his colleagues employed similar cloning methods to make designer cattle. They started with a culture of cells from cattle. They induced the development of six genetically identical calves. And they also engineered specific, targeted alterations in the cloning cell lineage.

Stice's goal is to genetically engineer cattle that are resistant to bovine spongiform encephalopathy, or mad cow disease (Section 20.5). People who eat infected meat also are vulnerable to the disease. Stice's technique also is used to put a human serum albumin gene into dairy cow chromosomes. We use albumin to control blood pressure. At present, this protein must be separated from large quantities of donated human blood. Imagine how much easier it would be to obtain large quantities of the protein from bountiful supplies of milk.

Given how rapidly the technologies are developing, is human cloning not far behind? And what about the possibility of genetically engineering the clones? Here you may wish to reflect once more on Section 12.5.

Mapping and Using the Human Genome

In the early 1980s, biologists were in an uproar over a costly proposal before the National Institutes of Health (NIH) to map the entire human genome. Many argued that benefits would be incalculable—not only for pure research, but also for efforts to cure genetic disorders and combat diseases. Others saw it as a boondoggle; they said such mapping couldn't be done and would divert NIH funding from worthy endeavors. But then Leroy Hood, a molecular biologist, invented automated DNA sequencing (Section 15.4), and the race was on.

By 1990, the Human Genome Initiative was under way. The international effort came with a projected price tag of 3 billion dollars—about 92 cents for each base pair in the heritable script for human life. For the first phase, biologists were to pinpoint the location of the stretches of nucleotides that were already sequenced. Once those "reference markers" were assigned to the chromosomes, biologists could fill in the long spans between them.

By 1991, not even 2,000 genes had been sequenced. But the pace picked up after Craig Venter realized that even a tiny bit of cDNA could be used as a molecular hook to drag the whole sequence of its parent gene out of a

Figure 15.14 Evidence of a successful gene transfer. Two ten-week-old mouse littermates. *Left:* This one weighs 29 grams. *Right:* This one weighs 44 grams. It grew from a fertilized egg into which a gene for human somatotropin had been inserted.

cDNA library. He named the hooks ESTs (for *Expressed Sequence Tags*). By tapping commercial sources, Ventner and his colleague, Mark Adams, put together a cDNA library that was produced from various neurons, the communication cells of nervous systems. Within a few months, the group had sequenced 2,000 more genes. By 1994, many thousands of genes had been decoded.

Later, Ventner and Hamilton Smith used a software program, the TIGR Assembler, to work out the first full genome of an organism—the bacterium *H. influenzae* you read about in Section 15.1. Through application of advanced technologies, a rough draft of the human genome was completed ahead of schedule. Noncoding portions, which are extensive, are still being analyzed.

Even when the whole human genome is sequenced, it will not be easy to manipulate it to advantage. Put yourself in the position of some genetic researcher. You have modified a human gene and you have to figure out how to get it into a host cell of a particular tissue. That's not all. You have to be sure the modified gene gets inserted at a suitable site—say, at or very near the locus for its abnormal counterpart in a chromosome. You must also make sure the cell will synthesize the specified protein at suitable times, in suitable amounts.

Today, most experimenters employ stripped-down viruses as vectors to put genes into cultured human cells. They know that cells can incorporate foreign genes into their DNA. But the viral genetic material can undergo rearrangements, deletions, and other alterations, which can shut down or disrupt gene expression. What about developing synthetic, streamlined versions of human chromosomes? Maybe the cellular machinery for DNA replication and protein synthesis will work on them.

Certain gene therapies simply deliver genetically modified cells into a tissue. This may help, even if the cells only make 10 to 15 percent of a required protein. The chapter introduction is a case in point. But no one can yet predict where the new genes will end up. The danger is that a gene insertion will disrupt the function of some other gene, including the genes controlling cell growth and cell division. One-for-one gene swaps by a process called homologous recombination are possible. Oliver Smithies, one of the best at this, currently puts genes right where they are supposed to go, but only once every 100,000 or so tries. The body may be a bit forgiving of a small load of cells bearing poor swaps. But mistakes in germ cells—which give rise to all cells of a new individual—probably would have unexpected and disastrous effects on the body's development.

Although the technical details are still being worked out, modified genes are being transferred into cells of humans and other mammals during experimental and clinical trials.

Who Gets Enhanced?

This chapter opened with a historic, inspiring case, the first proof that gene therapy may help save human lives. It closes with questions that invite you to consider some social and ethical issues surrounding the application of recombinant DNA technology to the human genome.

To most of us, human gene therapy to correct genetic abnormalities seems like a socially acceptable goal. Let's take this idea one step further. Is it also socially desirable or acceptable to change certain genes of a normal human individual (or sperm or egg) to alter or enhance traits?

The idea of selecting desirable human traits is called *eugenic engineering*. Yet who decides which forms of a trait are most "desirable"? For example, would it be okay to engineer taller or blue-eyed or fair-skinned boys and girls? Would it be okay to engineer "superhumans" with amazing strength or intelligence? Are there some people narcissistic enough to commission a clone of themselves, one with a few genetically engineered "enhancements"?

Recently, more than 40 percent of Americans surveyed say gene therapy would be okay to make their offspring smarter or better looking. A poll of British parents found 18 percent willing to use genetic enhancement to prevent children from being aggressive and 10 percent to keep them from growing up to be homosexual.

There are those who say the DNA of any organism must never be altered. Put aside the fact that nature itself alters DNA much of the time and has done so for nearly all of life's history. The concern is that *we* do not have the wisdom to bring about beneficial changes without causing great harm to ourselves or to the environment. To be sure, when it comes to altering human genes, one is reminded of our very human tendency to leap before we look. And yet, when it comes to restricting genetic modifications of any sort, one also is reminded of another old saying: "If God had wanted us to fly, he would have given us wings." Something about the human experience did give us a capacity to imagine wings of our own making— and that capacity carried us to the frontiers of space.

Where are we going from here? To gain perspective on the question, spend some time reading the history of the human species. It is a history of survival in the face of challenges, threats, bumblings, and sometimes disasters on a grand scale. It is also a story of our connectedness with the environment and with one another.

The basic questions confronting you are these: Should we be more cautious, believing the risk takers may go too far? And what do we as a species stand to lose if risks are not taken? At this writing, a young man with a genetic disorder died after gene therapy that used an adenovirus as a vector. Yet two *SCID-X1* infants with a mutant gene that disabled their immune system may be cured. A viral vector put copies of the nonmutated gene into stem cells taken from their bone marrow. Months after the modified cells were reinfused into their bone marrow, both "bubble boys" left their isolation tents. Their immune system is working normally; they are healthy and living at home.

SAFETY ISSUES

Many years have passed since foreign DNA was first transferred into a plasmid. That gene transfer ignited a debate that will continue well into the next century. The issue is this: *Do potential benefits of gene modifications and gene transfers outweigh potential dangers?*

Genetically engineered bacteria are "designed" so they cannot survive except in the laboratory. As added precautions, "fail-safe" genes are built into the foreign DNA in case they escape. These genes are silent unless the captives are exposed to environmental conditions— whereupon the genes get activated, with lethal results for their owner. Say the package includes a *hok* gene next to a promoter of the lactose operon (Section 14.2). Sugars are plentiful in the environment. If they were to activate the *hok* gene, the protein product of that gene would destroy membrane function and the wayward cell.

What about a worst-case scenario? Remember how retroviruses are used to insert genes into cultured cells? If they escape from laboratory isolation, what might be the consequences? Example: Check out Section 42.8.

And what about genetically engineered plants and animals released into the environment? For example, Steven Lindlow thought about how frost destroys many crops. Knowing that a surface protein of a bacterium promotes formation of ice crystals, he excised the "ice-forming" gene from bacterial cells. As he hypothesized, spraying "ice-minus bacteria" on strawberry plants in an isolated field prior to a frost would help plants resist freezing. He actually had deleted a *harmful* gene from a species, yet a bitter legal battle ensued. The courts ruled in his favor. His coworkers sprayed a strawberry patch; nothing bad happened.

Then there was the potato plant designed to kill the insects that attack it. It also was too toxic for people to eat. Or think of how crop plants compete poorly with weeds for nutrients. Many have been designed to resist weedkillers so farmers can spray for weeds and not worry about killing their crops. Herbicides do have toxic effects on more than their targets (Section 3.7). If crop plants offer herbicide resistance, will farmers be less or more apt to spread them about?

And what if engineered plants or animals transfer modified genes to organisms in the wild? Think of how the advantage in acquiring resources would tilt toward vigorous weeds blessed with herbicide-resistant genes. Such possibilities are why standards for rigorous and extended safety tests must be in place *before* modified organisms enter the environment.

For more on safety issues, turn to *Critical Thinking* question 7 at the chapter's end.

Rigorous safety tests are carried out before genetically modified organisms are released into the environment.

SUMMARY   *Gold* indicates text section

1. Uncountable numbers of gene mutations and other forms of genetic "experiments" have been proceeding in nature for at least 3 billion years. *CI*

2. Through artificial selection practices, humans have manipulated the genetic character of many species for many thousands of years. Currently, recombinant DNA technology has enormously expanded our capacity to genetically modify organisms. *CI*

3. In genetic engineering, specific genes are modified and inserted into the same organism or a different one. In gene therapy, copies of normal or modified genes are inserted into individuals in order to correct a genetic defect or boost resistance to disease. *CI*

4. A genome is all the DNA in the haploid chromosome number for a species. By recombinant DNA technology, a genome can be cut into fragments, then the fragments can be amplified (copied over and over again) to make useful quantities that permit analysis of the nucleotide sequence of the genome or a specific portion of it. *15.1*

5. Researchers work with chromosomal DNA or with cDNA (a DNA strand transcribed by the enzyme reverse transcriptase from a mature mRNA transcript). The first choice is better for questions about DNA regions that control gene expression or that contain introns. cDNA is a better choice for questions about the amino acid sequence of a protein. Either way, researchers use restriction enzymes to cut DNA into fragments. *15.1*

6. The use of plasmids or other cloning vectors is one way to amplify DNA fragments. Many bacteria contain plasmids: small circles of DNA with a few genes in addition to those of the bacterial chromosome. *15.1*

   a. Certain restriction enzymes make staggered cuts that leave the fragments with single-stranded tails. Such tails base-pair with complementary tails of any other DNA cut by the same enzyme, such as plasmid DNA.

   b. DNA ligase, a modification enzyme, seals base-pairing sites between plasmid DNA and foreign DNA. A plasmid modified to accept foreign DNA is a cloning vector; it can deliver foreign DNA into a bacterium or some other cell that can start a population of rapidly dividing descendant cells. All of the descendants have identical copies of the foreign DNA. Collectively, all of the identical copies are a DNA clone.

7. Currently, the polymerase chain reaction (PCR) is the fastest way to amplify fragments of chromosomal DNA or cDNA. Bacterial factories are not needed; the reactions occur in test tubes. The reactions use a supply of nucleotide building blocks and primers: synthetic, short nucleotide sequences that will base-pair with any complementary DNA sequence and that enzymes (DNA polymerases) recognize to start replication. Each round of replication doubles the number of fragments. *15.2*

8. For sexually reproducing species, no two individuals have exactly the same DNA base sequence (except for identical twins). When restriction enzymes are used to cut an individual's DNA, the result is a unique array of restriction fragments called a DNA fingerprint. *15.3*

    a. The DNA in all humans is more than 99 percent identical except at tandem repeats (short stretches of repeated base sequences, such as TTTC). The number and combination of tandem repeats is unique in each individual and can be detected by gel electrophoresis.

    b. DNA fingerprinting has uses in forensic science and in resolving paternity suits.

9. Automated DNA sequencing can rapidly reveal the base sequence of cloned DNA or PCR-amplified DNA fragments. It labels fragments with one of four modifed nucleotides, then separates them according to length by gel electrophoresis. Each label fluoresces a certain color under a laser beam. A machine reads each as it peels off the gel and assembles the whole sequence. *15.4*

10. A gene library is a mixed collection of bacterial cells that took up different cloned DNA or cDNA fragments. A gene may be isolated from the library with use of a probe, a very short stretch of radioactively labeled DNA that is known or suspected to be similar to or identical with part of that gene and can base-pair with it. A base pairing between nucleotide sequences from different sources is called nucleic acid hybridization. *15.5*

11. Generally, recombinant DNA technology and genetic engineering have potential for research and applications in medicine and agriculture, in the home and industry. As with any new technology, the potential benefits must be weighed against potential risks, including ecological and social repercussions. *CI, 15.6–15.10*

Review Questions

1. Distinguish between recombinant DNA technology and genetic engineering. *CI*

2. Distinguish these terms from one another: *15.1*
    a. chromosomal (genomic) DNA and cDNA
    b. cloning vector and DNA clone

3. Define PCR. Can fragments of chromosomal DNA, cDNA, or both be amplified by PCR? *15.2*

4. Define DNA fingerprinting. Briefly describe which portions of the DNA are used in DNA fingerprinting. *15.3*

5. Outline the steps of automated DNA sequencing. *15.4*

6. Define cDNA library, then briefly explain how a gene can be isolated from it. Define probe and nucleic acid hybridization as part of your answer. *15.5*

7. Give three examples of applications that can be derived from knowledge of an organism's genome. *CI, 15.6–15.8*

8. Name one of the ways in which modified genes have been inserted into mammalian cells. *15.8*

9. Define gene therapy. Once the human genome has been fully sequenced, why will it be difficult to manipulate its genes to advantage? *CI, 15.8*

Self-Quiz   ANSWERS IN APPENDIX III

1. _____ is the transfer of normal genes into body cells to correct a genetic defect.
    a. Reverse transcription   c. Gene mutation
    b. Nucleic acid hybridization   d. Gene therapy

2. DNA fragments result when _____ cut DNA molecules at specific sites.
    a. DNA polymerases   c. restriction enzymes
    b. DNA probes   d. RFLPs

3. _____ are small circles of bacterial DNA that are separate from the circular bacterial chromosome.

4. Foreign DNA that was inserted into a plasmid and then replicated many times in a population of bacteria is a _____ .
    a. DNA clone   c. DNA probe
    b. gene library   d. gene map

5. By reverse transcription, _____ is assembled on _____ .
    a. mRNA; DNA   c. DNA; enzymes
    b. cDNA; mRNA   d. DNA; agar

6. PCR stands for _____ .
    a. polymerase chain reaction
    b. polyploid chromosome restrictions
    c. polygraphed criminal rating
    d. politically correct research

7. By gel electrophoresis, fragments of a gene library can be separated according to _____ .
    a. shape   b. length   c. species

8. Automated DNA sequencing relies on _____ .
    a. supplies of standard and labeled nucleotides
    b. primers and DNA polymerases
    c. gel electrophoresis and a laser beam
    d. all of the above

9. Match the terms with the most suitable description.
    ____ DNA fingerprint   a. selecting "desirable" traits
    ____ Ti plasmid   b. deciphering 3.2 billion base pairs
    ____ nature's genetic   of 23 human chromosomes
          experiments   c. used in some gene transfers
    ____ nucleic acid   d. unique array of DNA fragments
          hybridization   inherited in Mendelian pattern
    ____ Human Genome   from each of two parents
          Initiative   e. base pairing of nucleotide
    ____ eugenic   sequences from different
          engineering   DNA or RNA sources
                                 f. mutations, crossovers

Critical Thinking

1. In the following diagram, which restriction enzyme made the cuts (indicated in *red*) in part of a DNA molecule from two different organisms?

Which enzyme has sealed the sticky ends of both together?

2. Ryan, a forensic scientist, obtained a very small bit of DNA from material at a crime scene. In order to examine the sample by DNA fingerprinting, he must amplify the sample by PCR, the polymerase chain reaction. By his estimate, there are 50,000 copies of the DNA in his sample. Calculate the number of copies Ryan will have after fifteen cycles of PCR.

3. A game warden in Africa confiscated eight ivory tusks from elephants. Some tissue is still attached to the tusks. Now she must determine whether the tusks were taken illegally from northern populations of endangered elephants or from other populations of elephants to the south that can be hunted legally. How can she use DNA fingerprinting to find the answer?

4. The Human Genome Initiative is undergoing completion, and knowledge about a number of the newly discovered genes is already being used to detect genetic disorders. Ask yourself: What will be done with genetic information about individuals? Will insurance companies and potential employers request it? At this writing, many women have already refused to take advantage of genetic screening for a gene associated with the development of breast cancer. Should medical records about people participating in genetic research and genetic clinical services be made available to other individuals? If not, how could such information be protected?

5. Lunardi's Market put out a bin of tomatoes having splendid vine-ripened redness, flavor, and texture. The sign posted above the bin identified them as genetically engineered produce. Most shoppers selected unmodified tomatoes in the adjacent bin even though those tomatoes were pale pink, mealy-textured, and tasteless. Which ones would you pick? Why?

6. Avoiding genetically engineered food is probably impossible in the United States. At least 45 percent of cotton, 38 percent of soybean, and 25 percent of corn crops have been engineered to withstand weedkillers or to make their own pesticides. For years, modified corn and soybeans have found their way into breakfast cereals, tofu, soy sauce, vegetable oils, beer, soft drinks, and other food products. They are fed to farm animals.

By contrast, public resistance to genetically engineered food is high in Europe, especially Britain. Many people, including the Prince of Wales, speak out against what the tabloids call "Frankenfood." Protesters routinely vandalize crops (Figure 15.15). Worries abound that such foods may be more toxic, have lower nutritional value, and promote natural selection for antibiotic resistance. Some people worry that designer plants will cross-pollinate with wild plants to produce "superweeds."

Biotechnologists envision a new Green Revolution. They argue that designer plants can hold down food production costs, reduce dependence on pesticides and herbicides, enhance crop yields, and offer improved flavor, nutritional value, even salt tolerance and drought tolerance. Yet the chorus of critics in Europe may provoke a trade war with the United States. The issue is not small potatoes, so to speak. In 1998, the value of

agricultural exports reached about 50 billion dollars. Flattery, threats, and bullying are rampant on both sides of the Atlantic. Restrictions on genetic engineering will profoundly impact United States agriculture, and inevitably the impact will trickle down to what you eat and how much you pay for it.

All of which invites you to read up on scientific research related to this issue and form your own opinions. The alternatives are to be swayed either by media hype (the term Frankenfood, for instance) or by biased reports from groups (such as chemical manufacturers) with their own agendas.

Possibly start with Christopher Bond's article "Politics, Misinformation, and Biotechnology" in the 18 February 2000 *Science* (287:1201). Bond argues that biotechnology attempts to solve real-world problems of sickness, hunger, and dwindling resources; and that "hysteria and unworkable propositions advanced by those who can afford to take their next meal for granted have little currency among those who are hungry."

7. What if it were possible to create life in test tubes? This is the question behind attempts to model and eventually create *minimal organisms*, which we define as living cells having the smallest set of genes necessary to survive and reproduce.

As recent experiments by Craig Ventner and Claire Fraser revealed, *Mycoplasma genitalium*, a bacterium with 517 genes (and 2,209 transposons) might be a candidate. By disabling its genes one at a time in the laboratory, they discovered that it may have only 265–350 essential protein-coding genes.

What if those genes were to be synthesized one at a time and inserted into an engineered cell consisting only of a plasma membrane and cytoplasm? Would the cell come to life? The possibility that it might prompted Ventner and Fraser to seek advice from a panel of bioethicists and theologians. As Arthur Caplan, a bioethicist at the University of Pennsylvania reported, no one on the panel objected to synthetic life research. They felt that much good might come of it, provided scientists didn't claim to have found "the secret of life." The 10 December 1999 issue of *Science* includes an essay from the panel and an article on *M. genitalium* research. Read both, then write down your thoughts about creating life in a test tube.

Selected Key Terms

automated DNA sequencing 15.4	genome 15.1
	nucleic acid hybridization 15.5
cDNA 15.1	PCR 15.2
cloning vector 15.1	plasmid 15.1
DNA clone 15.1	primer 15.2
DNA fingerprint 15.3	probe (nucleic acid) 15.5
DNA ligase 15.1	recombinant DNA technology CI
gel electrophoresis 15.3	restriction enzyme 15.1
gene library 15.5	reverse transcriptase 15.1
gene therapy CI	seed bank 15.7
genetic engineering CI	tandem repeats 15.3

Readings

Cho, M., et al. 10 December 1999. "Ethical Considerations in Synthesizing a Minimal Genome." *Science* 286: 2087–2090.

Pennisi, E. June 2000. "Finally, The Book of Life and Instructions for Navigating It." *Science* 288: 5475.

Watson, J. D., et al. 1992. *Recombinant DNA*. Second edition. New York: Scientific American Books.

"The Human Genome." *Science* 291: 5507. The entire 16 February 2001 issue consolidates current understandings and questions.

Figure 15.15
Activists ripping some genetically modified crop plants from a field in Great Britain.

On-Line readings at Student Guide for InfoTrac:
www.brookscolecom/biology

III Principles of Evolution

Millions of years ago, a bony fish died, and sediments gradually buried it. Today its fossilized remains are studied as one more piece of the evolutionary puzzle.

MICROEVOLUTION

Designer Dogs

We humans have tinkered rather ruthlessly with the modern descendants of a long and distinguished lineage. The lineage originated some 40 million years ago with the appearance of tree-dwelling carnivores that looked rather like small weasels. Their descendants evolved along separate branchings that now include weasels, badgers, otters, skunks, bears, pandas, raccoons —and dogs.

About 50,000 years ago, we began domesticating wild dogs. No doubt the advantages of doing so were important. Times were tough in the days before police protection and supermarkets. Dogs welcomed to the hearth could guard people and their possessions. They could corner, kill, eat, and thus dispose of big rats and other unwelcome vermin.

By 14,000 years ago, we started to develop different varieties (breeds) through artificial selection. Individual dogs having desired forms of traits were selected from each new litter and, later, encouraged to breed. Those having undesired forms of traits were passed over.

After favoring the pick of the litter over hundreds or thousands of generations, we ended up with sheep-herding border collies, badger-hunting dachshunds, wily retrievers, and sled-pulling huskies. And at some point we began to delight in the odd, extraordinary dog. Imagine! In practically no time at all, evolutionarily speaking, we picked our way through the pool of variant dog alleles and came up with such extremes as Great Danes and chihuahuas (Figure 16.1).

Sometimes our canine designs exceeded the limits of biological common sense. For example, how long would a tiny, finicky-eating, nearly hairless, nearly defenseless chihuahua last in the wild? Not long. What about the English bulldog, bred for a very short snout and a compressed face? Long ago, breeders thought these particular traits would allow the dogs to get a better grip on the nose of a bull. (Why they wanted dogs to bite bulls is a story in itself.) So now the roof of the bulldog mouth is ridiculously wide and often flabby, so bulldogs have trouble breathing. Sometimes they get so short of air they pass out.

Through our centuries-old fascination with artificial selection, we produced thousands of varieties of crop plants, ornamental plants, cats, cattle, and birds as well as dogs. With the currently available technologies of genetic engineering, we are now mixing the genes of many different species and producing incredible new varieties, including tobacco plants that produce useful quantities of hemoglobin, and mustard plants that produce plastic.

So, when you hear someone wonder about whether "evolution" takes place, remind yourself that **evolution** simply means genetic change in a line of descent over the generations. Selective breeding practices provide abundant, tangible evidence that heritable changes do, indeed, occur. The actual mechanisms that bring about change are the focus of this chapter. Later chapters will explain their role in the evolution of new species from parent species.

Figure 16.1 Two designer dogs. About 50,000 years ago, humans began domesticating wild dogs. From that ancestral stock, artificial selection resulted in startling diversity among rather closely related breeds, such as the Great Dane (*legs, left*) and the chihuahua (*possibly fearful of being stepped on, right*).

Key Concepts

1. From the fifteenth century onward, explorers found puzzling differences in the world distribution of species. Anatomists identified similarities and differences in the body structure and patterning among the embryos and adult forms of major groups of animals. In sequences of sedimentary rocks, geologists found sequences of fossils. The findings were suggestive of biological evolution, or heritable changes in lines of descent over time.

2. Charles Darwin came of age at a time when scholars were attempting to reconcile the new findings from global explorations, comparative morphology, and geology with prevailing cultural beliefs. He, and Alfred Wallace, came up with a theory of evolution by natural selection to explain the findings. Their evolutionary theory and the theories of others start with variation in traits.

3. All individuals of a population generally have the same number and kinds of genes. Those genes give rise to an array of traits that characterize the population.

4. Mutations may result in two or more slightly different molecular forms of a gene—alleles—that influence a trait in different ways. Individuals of a population vary in the details of a trait when they inherit different combinations of alleles.

5. Any allele may become more or less common in the population relative to other kinds at a gene locus, or it may disappear. *Microevolution* refers to changes in the allele frequencies of a population over time.

6. Allele frequencies change as an outcome of mutation, gene flow, genetic drift, and natural selection. Mutation alone produces new alleles. Gene flow, genetic drift, and natural selection shuffle existing alleles into, through, or out of populations.

EARLY BELIEFS, CONFOUNDING DISCOVERIES

The Great Chain of Being

Our story begins more than two thousand years ago, when the seeds of biological inquiry were starting to take hold among the ancient Greeks. At the time, popular belief held that supernatural beings intervened directly and often in human affairs. As an example, people "knew" that angry gods inflicted a common ailment known as the sacred disease. And yet, from one of the physicians of the school of Hippocrates, these thoughts come down to us:

It seems to me the disease called sacred . . . has a natural cause, as other diseases have. Men think it divine merely because they do not understand it. But if they called everything divine that they did not understand, there would be no end of divine things! . . . If you watch these fellows treating the disease, you see them use all kinds of incantations and magic—but they are also careful in regulating diet. Now if food makes the disease better or worse, how can they say it is the gods who do this? . . . It does not really matter whether you call such things divine or not. In Nature, all things are alike in this, in that they can be traced to preceding causes.

— ON THE SACRED DISEASE (400 B.C.)

He perceived that all aspects of nature have underlying causes—a key premise of modern science (Section 1.5).

Aristotle was foremost among the early naturalists, and he described the world around him in great detail. He had no reference books or instruments to guide him, for biological science in the Western world began with the great thinkers of this age. Yet here was a man who was no mere collector of random tidbits of information. In his thoughtful descriptions of nature we see evidence of a mind perceiving connections between observations and attempting to explain the order of things.

Aristotle believed (as did others) that each kind of organism was distinct from all the rest. Nevertheless, he wondered about organisms with traits that seemed to blur such distinctions. For example, some sponges look very much like plants but do not make their own food, as plants do. They capture and digest it, as animals do. In time, Aristotle came to view nature as a continuum of organization, from lifeless matter through complex forms of plant and animal life.

By the fourteenth century, Aristotle's idea had been transformed into a rigid view of life. A Chain of Being was seen to extend from the "lowest" forms of life to humans, then on up to spiritual beings. Each kind of being, or "species" as it was called, was a separate link in the great chain. All links were designed and forged at the same time, at the same center of creation, and had not changed since then. Scholars thought that once they discovered, named, and described all of the links, the meaning of life would be revealed to them.

a

Figure 16.2 Examples of three species native to three separate geographic realms. (**a**) Ostrich of Africa. *Facing page:* (**b**) Rhea of South America and (**c**) emu of Australia. All three species resemble one another and are unlike most birds in several ways, including their inability to get airborne.

Questions From Biogeography

Until the fifteenth century, naturalists were not aware that the world is a great deal bigger than Europe, so the task of locating and describing all species seemed manageable. Then globe-spanning explorations began. Naturalists were soon overwhelmed by descriptions of tens of thousands of plants and animals that explorers were discovering in Asia, Africa, the Pacific Islands, and the New World.

In 1590 the naturalist Thomas Moufet attempted to sort through the bewildering array. He simply gave up and wrote such gems as this description of locusts and grasshoppers: "Some are green, some black, some blue. Some fly with one pair of wings; others with more; those that have no wings they leap; those that cannot fly or leap they walk . . . Some there are that sing, others are silent." It was not a work of subtle distinctions.

Even so, a few scholars began to examine the world distribution of organisms, a discipline now known as **biogeography** (Section 18.3 and Chapter 44). They soon realized that many plants and animals are unique to isolated places, such as remote oceanic islands. They also were aware that certain species separated by great distances resemble one another (Figure 16.2).

How did so many species get from one center of creation to oceanic islands and other isolated, remote locations? And what do the similarities and differences among them mean?

Questions From Comparative Morphology

By the eighteenth century, many scholars were engaged in **comparative morphology**, the systematic study of similarities and differences in the body plans between major groups, such as different kinds of vertebrates. Think of the bones of a human arm, whale flipper, and

You could do similar calculations for other wing colors. You could go on with your calculations until you run out of paper (or patience). As long as the five assumptions continue to hold, however, the allele frequencies and the range of values for the wing-color trait will not change, as you can see from Figure 16.9.

Therefore, when genotypes and phenotypes do *not* show up in the proportions you predicted on the basis of the Hardy–Weinberg rule, this tells you that one or more conditions of the rule are being violated. And the hunt can begin for the specific evolutionary force, or forces, driving the change.

STARTING POPULATION

490 *AA* butterflies
Dark-blue wings

420 *Aa* butterflies
Medium-blue wings

90 *aa* butterflies
White wings

THE NEXT GENERATION

490 *AA* butterflies

420 *Aa* butterflies

90 *aa* butterflies

NO CHANGE

THE NEXT GENERATION

490 *AA* butterflies

420 *Aa* butterflies

90 *aa* butterflies

NO CHANGE

Figure 16.9 A hypothetical population of butterflies at genetic equilibrium.

NATURAL SELECTION REVISITED

We now turn from our idealized population that never changes to the real-world processes of change. Of these processes, natural selection might account for most of the morphological and physiological changes that have accrued throughout the history of life.

Darwin, recall, was able to explain natural selection after correlating his understanding of inheritance with certain features of populations and their environments. Before we consider the modes of natural selection, let's restate his correlations in modern terms:

1. *Observation:* All populations in nature have the reproductive capacity to increase in numbers through the generations.

2. *Observation:* No population is able to increase indefinitely, for its individuals will run out of food, living space, and other resources that sustain it.

3. *Inference:* Sooner or later, the individuals of a population will end up competing for resources.

4. *Observation:* All of the individuals generally have the same genes, which specify the same assortment of traits. Collectively, their genes represent a pool of heritable information.

5. *Observation:* Most, if not all, kinds of genes occur in slightly different molecular forms (alleles), which give rise to differences in phenotypic details.

6. *Inferences:* Some phenotypes are better than others at helping the individual compete for resources, hence to survive and reproduce. Alleles for those phenotypes therefore increase in the population, and other alleles decrease. In time the genetic change leads to increased **fitness**—an increase in adaptation to the environment.

7. *Conclusions:* **Natural selection** is the outcome of variations in shared traits that affect which individuals of a population survive and reproduce each generation. This microevolutionary process results in adaptation to the environment.

Evolutionary biologists have been documenting the results of natural selection in thousands of field studies of populations of all kinds of organisms. They find that this microevolutionary process has different results. As you will see in sections to follow, sometimes the result is a shift in the range of values for a given trait in some direction. At other times, the result may be stabilization or disruption of an existing range of values.

As Darwin perceived, natural selection is the outcome of variations in shared traits that influence which individuals of a population survive and reproduce in each generation. Natural selection can lead to increased fitness—that is, to an increase in adaptation to the environment.

DIRECTIONAL CHANGE IN THE RANGE OF VARIATION

What Is Directional Selection?

In cases of **directional selection**, allele frequencies that give rise to a range of variation in phenotype tend to shift in a consistent direction. The shift is a response to directional change in the environment or to one or more novel environmental conditions. A new mutation that proves adaptive also sets it in motion. Either way, the forms at one end of the phenotypic range become more common than the midrange forms (Figure 16.10). Let's look at a few documented cases of this outcome.

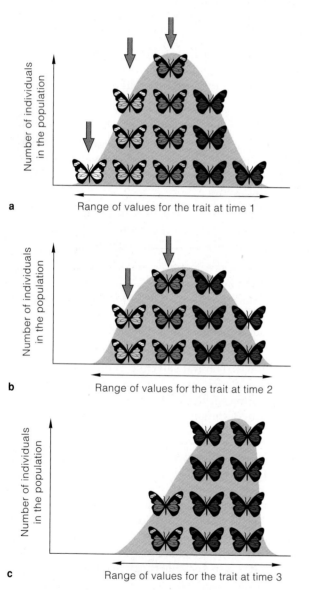

a Range of values for the trait at time 1

b Range of values for the trait at time 2

c Range of values for the trait at time 3

Number of individuals in the population

Figure 16.10 Directional selection, using phenotypic variation within a population of butterflies as the example. A bell-shaped curve (*green*) represents the range of continuous variation in wing color. The most common form (*medium blue*) is between extreme forms of the trait (*white* at one end of the curve, *deep purple* at the other). *Orange* arrows signify which forms are being selected against over time.

The Case of the Peppered Moths

In England, biologists tracked directional selection in about a hundred moth species, including the peppered moth (*Biston betularia*). Peppered moths feed and mate at night. During the day, they rest motionless on birches and other trees. Their wings and body have a mottled pattern in shades that range from light gray to nearly black. Their behavior, coloration, and wing patterning camouflage them from moth-eating birds. Like all birds, the ones that prey on moths hunt during the day.

The industrial revolution started in England in the 1850s. Outpourings of sooty smoke altered conditions in many parts of the surrounding countryside. Before then, light moths were the most common form, and a dark form was rare. Also before conditions changed, light-gray speckled lichens grew thickly on tree trunks. Lichens can camouflage light moths that rest on them, but not dark ones (Figure 16.11*a*).

Lichens are sensitive to air pollution. Between 1848 and 1898, soot and other factory pollutants started to kill the lichens and darken tree trunks. In the modified environment, the less common moth form was better camouflaged (Figure 16.11*b*). Moth collectors came up with a hypothesis: If original conditions favored light moths, then the *changed* conditions favored dark ones.

In the 1950s, H. B. Kettlewell used a *mark–release–recapture method* to test that prediction. He bred both moth forms in captivity and marked hundreds of them so they could be easily identified after being released in the wild. He released them near highly industrialized areas around Birmingham and in an unpolluted area of Dorset. Later on, more dark moths were recaptured in the polluted area and more light ones in the pollution-free area (Table 16.1). Observers also were stationed in blinds near groups of moths that had been tethered to trees. They directly observed birds capturing more of the light moths around Birmingham and more of the dark moths around Dorset. Here was strong evidence that directional selection was operating.

Strict pollution controls went into effect in 1952. Lichens made a comeback. Tree trunks became free of soot, for the most part. As you might have predicted, phenotypes shifted in the reverse direction as a result of natural selection. Where pollution has decreased, the frequency of dark moths has been decreasing as well.

Pesticide Resistance

Widespread use of chemical pesticides in agriculture also results in directional selection. Initial applications kill most of the insects, worms, or other pests, but some individuals usually manage to survive. Some aspect of their structure, physiology, or behavior helps them resist

Figure 16.11 Peppered moths (*Biston betularia*). The light form of the moth is less visible to birds on lichen-covered tree trunks (**a**). The dark form is less visible after pollution kills the lichens and darkens the trunks (**b**).

Table 16.1 *Marked Peppered Moths Recaptured in Polluted and Nonpolluted Areas*

	Near Birmingham (pollution high)	Near Dorset (pollution low)
LIGHT-GRAY MOTHS		
Released	64	393
Recaptured	16 (25%)	54 (13.7%)
DARK-GRAY MOTHS		
Released	154	406
Recaptured	82 (53%)	19 (4.7%)

Data after H. B. Kettlewell.

the chemical effects. When the resistance has a heritable basis, it becomes more common in each new generation. These chemicals are agents of selection that favor the most resistant forms! Today, 450 different species resist one or more pesticides. Worse, the pesticides also kill natural predators of the pests. When freed from natural constraints, the populations of resistant pests burgeon, and crop damage is greater than ever. This outcome of directional selection is called *pest resurgence*.

Maybe pesticide use will decline in fields of plants that are genetically engineered to resist pests. Even modified plants will not win the coevolutionary arms race, but they might help keep food supplies one step ahead of the pests. This is a contentious issue; many consumers are leery of genetically engineered food, as you read in Section 15.11 (*Critical Thinking* question 5).

What about using *biological controls*? By this practice, natural enemies of pests, including parasitic wasps and predatory beetles, are raised in commercial insectaries. Large populations are released at preselected sites. The practice has an advantage in that a control species can coevolve with the pests. Farmers must replace the ones that migrate from the fields or are destroyed at harvest time. They pass on the cost to consumers.

Antibiotic Resistance

When your grandparents were young, tuberculosis, pneumonia, and scarlet fever caused one-fourth of the annual deaths in the United States. Since the 1940s, we have been relying on natural and synthetic antibiotics to fight such bacterial diseases. **Antibiotics**, remember, are toxins that some microorganisms in soil release to destroy bacterial competitors for nutrients (Section 1.4). Streptomycins, for example, prevent protein synthesis in target cells. Penicillins disrupt formation of covalent bonds that hold bacterial cell walls together. Penicillin derivatives weaken the wall until it ruptures.

Antibiotics must be prescribed with restraint and care. Why? Besides executing their intended function, they often disrupt resident bacterial populations that compete for nutrients in the intestines. They can disrupt bacterial and yeast populations in the vagina, as well. Imbalances follow and lead to secondary infections.

Also, antibiotics have been overprescribed in the human population. All too frequently, people demand them for simple infections that they often can fight on their own. As a consequence, antibiotics have lost their punch. Over time, they killed the most susceptible cells of target populations. However, they also favored their replacement by much more resistant cells. Each year, millions of people around the world die from cholera, tuberculosis, and other bacterial diseases. Vancomycin, once held in reserve as "the antibiotic of last resort," is no longer effective against some pathogenic strains of enteric (gut-dwelling) bacteria. In 1996, the World Health Organization announced that, in the dangerous race for supremacy, the pathogens are sprinting ahead.

With directional selection, allele frequencies underlying a range of variation tend to shift in a consistent direction in response to directional change in the environment.

16.8 SELECTION AGAINST OR IN FAVOR OF EXTREME PHENOTYPES

As you have seen, natural selection can bring about a directional shift in a population's range of phenotypic variation. Depending on the prevailing environmental conditions, the process also may favor either the most common or the most extreme phenotypes in that range.

Stabilizing Selection

In **stabilizing selection**, intermediate forms of a trait in a population are favored, so alleles for extreme forms are eliminated (Figure 16.12). This mode of selection tends to counter mutation, gene flow, and genetic drift, and it preserves the most common phenotypes.

The gallmaking fly *Eurosta solidaginis* is evidence of stabilizing selection. Its larvae bore into stems of a tall goldenrod (*Solidago altissima*) and feed on tissues until they metamorphose into adults. In response, the tissue cells multiply rapidly and encase the invaders in a gall, a tumorous mass. We know from genetic analyses that flies with different phenotypes cause galls of different sizes—large, small, and in between—to form.

A wasp parasitizes the larvae (Figure 16.13). It can penetrate small galls, so flies that cause small galls to form are at risk. Also, the downy woodpecker and some other birds eat the larvae. They can peck through large galls, so the flies that cause large galls to form are at risk. Think it through, and you see that the flies that cause *intermediate-sized* galls to form are favored. Their larvae are less vulnerable to wasps *and* to birds.

And so parasitic wasps work against one extreme phenotype and predatory birds work against the other. In this fly population, the intermediate phenotype has the highest survival rate and fitness.

Figure 16.12 Stabilizing selection, using phenotypic variation within a population of butterflies as the example.

a Gall-making fly (producer of tasty larvae)

b Gall on a goldenrod stem that houses fly larvae

c Parasitic wasp (an agent of selection)

d Downy woodpecker (an agent of selection)

Figure 16.13 Example of stabilizing selection. Larvae of the fly *Eurosta solidaginis* (**a**) induce formation of a type of tumor called a gall (**b**) on goldenrod stems. (**c**) A parasitic wasp (*Eurytoma gigantea*) has an egg-laying device that penetrates only the thin wall of small galls. Its eggs develop into larvae, then the wasp larvae eat fly larvae. (**d**) Downy woodpeckers (*Dendrocopus pubescens*) and other birds can use their bill to chisel into large-size galls. They, too, eat the larvae.

Warren Abrahamson and his coworkers monitored twenty *Eurosta* populations in Pennsylvania. They found that larvae inside small and large galls have low relative fitnesses. They also found that larvae in intermediate-size galls have relatively high fitnesses. The fly's natural enemies target larvae in small and large galls; they stabilize the fly's range of phenotypic variation.

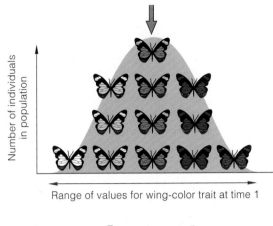

a   Range of values for wing-color trait at time 1

b   Range of values for wing-color trait at time 2

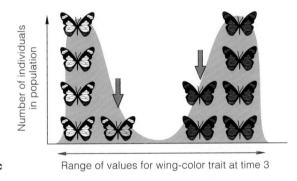

c   Range of values for wing-color trait at time 3

Figure 16.14 Disruptive selection, using phenotypic variation within a population of butterflies as the example.

Disruptive Selection

In **disruptive selection**, forms at both ends of a range of variation are favored and intermediate forms are selected against (Figure 16.14). Thomas Smith found an example of this in a rain forest in Cameroon, West Africa. Smith had read about unusual variation in populations of the black-bellied seedcracker (*Pyrenestes ostrinus*). These African finches have large or small bills—but no sizes in between. The pattern holds for females and males through the geographic range. (This is just remarkable; imagine every person in Texas being four feet *or* six feet tall, with no intermediates.) If the bill pattern is unrelated to gender or geography, what causes it?

Figure 16.15 Disruptive selection among African finches. Feeding trials showed that birds with large bills are better users of hard seeds. Small-billed birds are better at using soft seeds, not hard ones. (**a,b**) Two individuals displaying small and large bill sizes.

(**c**) Graph of survival of juvenile birds during the dry season, when competition for resources is most intense. Individuals with very small, very large, or intermediate-size bills can't feed efficiently on either type of seed; they survive poorly. For this graph, *tan* bars indicate the number of nestlings; *orange* bars indicate the survivors among them. The findings are based on measurements of 2,700 netted individuals.

Smith hypothesized: Seed-cracking ability directly affects finch survival. If only two bill sizes persist, then disruptive selection may be eliminating birds that have intermediate-size bills. What selection pressures could be at work on feeding performance? Cameroon's swamp forests flood during the wet season; lightning-sparked fires burn during the dry season. Two species of sedge (fire-resistant, grasslike plants) dominate these forests. One sedge has hard seeds and the other has soft. When finches reproduce, hard *and* soft seeds are abundant.

All birds prefer soft seeds for as long as they can get them. However, the birds with small bills are better at cracking soft seeds; birds with large bills are better at cracking hard ones. Soft seeds and other food supplies dwindle as the dry season peaks. That is when small-billed birds, and the youngest ones, are at a competitive disadvantage. Many don't survive (Figure 16.15).

Smith also performed experimental crosses between large- and small-billed birds. All offspring had large *or* small bills. Along with other data, the crosses suggest that two alleles at one autosomal gene locus control bill size and feeding performance.

With stabilizing selection, intermediate phenotypes are favored and extreme phenotypes at both ends of the range of variation are eliminated.

With disruptive selection, intermediate forms are selected against; extreme forms in the range of variation are favored.

SPECIAL TYPES OF SELECTION

Sexual Selection

Individuals of most sexually reproducing species have a distinctively male or female phenotype. We call this sexual dimorphism (*dimorphos*, "having two forms"). How does this condition arise, and what maintains it? Natural selection is taking the form of **sexual selection**. The traits being favored are advantageous, with respect to survival and reproduction, simply because males or females prefer them. Through nonrandom mating, the alleles for preferred traits prevail over the generations.

Sexual dimorphism is particularly striking among many mammals and birds, as in Figure 16.16. The males of many species are larger and have flashier coloration and patterning. They are often more aggressive than the females. Remember those male bighorn sheep butting

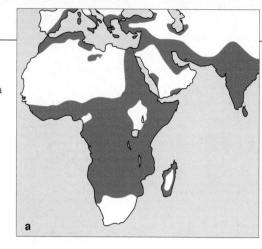

Figure 16.17 (**a**) Distribution of malaria cases in Africa, Asia, and the Middle East in the 1920s, before the start of mosquito control programs. (**b**) The distribution and frequency of people with the sickle-cell trait. Note the close correlation between the maps.

Figure 16.16 An outcome of sexual selection. This ▓▓ male bird of paradise (*Paradisaea raggiana*) is engaged in a flashy courtship display. He caught the eye (and, perhaps, the sexual interest) of the smaller, less colorful female. Males of this species compete fiercely for females, the selective agents. (Why do you suppose drab-colored females have been favored?)

heads (Section 1.3)? Fighting wastes time and energy, and it may cause serious injuries. Why, then, do alleles that contribute to aggressive behavior still persist in a population? An increased chance of mating may offset the costs. Male bighorn sheep fight only to control areas where receptive females gather during a winter rutting season. Winners mate often, with a number of females. Losers won't challenge a stronger, larger male.

Often, females are the agents of selection. They exert direct control over reproductive success by choosing their mates. We return to this topic in Chapter 41.

Maintaining Two or More Alleles

Balancing selection refers to all forms of selection that maintain two or more alleles for a trait in a population. When this type of genetic variation persists over time, we call it **balanced polymorphism** (after *polymorphos*, "having many forms"). A population shows balanced polymorphism when nonidentical alleles for some trait are being maintained at frequencies above 1 percent. Allele frequencies might shift slightly, but over time, they often return to the same values. Smith's African finches are a good example of the effect of balancing selection through the generations.

Sickle-Cell Anemia—Lesser of Two Evils?

Balanced polymorphism may emerge when conditions favor heterozygotes, which carry nonidentical alleles for the trait. Compared to homozygotes, which carry identical alleles for the trait, they have higher fitness.

Let's look at the environmental pressures that favor an Hb^A/Hb^S pairing in humans. Hb^S specifies a mutant form of hemoglobin, an oxygen-transporting protein in blood. Homozygotes (Hb^S/Hb^S) develop *sickle-cell anemia*, a genetic disorder that has serious pleiotropic effects on phenotype (Section 10.8). The frequency of Hb^S is high in tropical and subtropical regions of Asia and Africa. Often, Hb^S/Hb^S homozygotes die in their early teens or early twenties. Yet in those same regions, the heterozygotes (Hb^A/Hb^S) make up nearly one-third

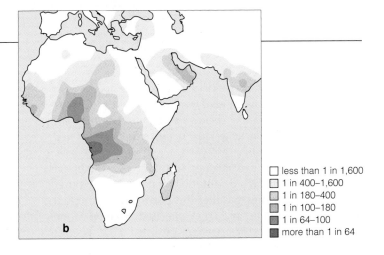

less than 1 in 1,600
1 in 400–1,600
1 in 180–400
1 in 100–180
1 in 64–100
more than 1 in 64

b

of the human population! Why is this combination of alleles maintained at such high frequency?

The balancing act, an outcome of natural selection, is most pronounced in areas with the highest incidence of *malaria* (Figure 16.17 and Section 20.9). A mosquito transmits *Plasmodium*, the parasitic agent of disease, to humans. The parasite multiplies in the liver, and later in red blood cells. The cells rupture and so release new parasites during severe, recurring bouts of infection.

Who is more likely to survive recurring infections? Hb^A/Hb^S heterozygotes. The allelic combination gives them two forms of hemoglobin, with interesting results. They produce enough *nonmutated* molecules to support body functions—and the *altered* molecules distort red blood cells in a way that slows down circulation. The slowdown in blood flow is a factor in getting through recurrences because it hampers the parasite's ability to travel and rapidly infect new cells. The distorted cells are later disposed of as they trickle through the spleen.

So the persistence of the "harmful" Hb^S allele is a matter of relative evils. Natural selection has favored one allelic combination, Hb^A/Hb^S, because its bearers show greater fitness in places where malaria is most prevalent. In these environments, the combination has a higher fitness than either Hb^S/Hb^S or Hb^A/Hb^A.

Malaria has been a selective force for over 2,000 years in tropical and subtropical habitats of Asia and the Middle East. Although sickle-cell anemia occurs at high frequencies in all these regions, its symptoms are far less severe than they are in Central Africa—where the Hb^S allele became established much later in time. It seems likely that other gene products are mediating some of the Hb^S allele's widespread effects; in some way they minimize symptoms of the disorder.

With sexual selection, some version of a trait simply gives the individual an advantage in reproductive success. Sexual dimorphism is one outcome of sexual selection.

Balanced polymorphism is a state in which natural selection is maintaining two or more alleles through the generations at frequencies greater than 1 percent.

GENE FLOW

Individuals of the same species don't necessarily stay put. A population loses alleles whenever an individual permanently leaves it, an act called *emigration*. It gains alleles when new individuals permanently move in, an act called *immigration*. This **gene flow**, a physical flow of alleles between populations, tends to counter genetic differences that we expect to arise through mutation, natural selection, and genetic drift. And it helps keep separated populations genetically similar.

Think of the acorns that blue jays disperse when they store nuts for the winter. Each fall the jays may make hundreds of round trips from acorn-bearing oak trees to bury acorns in the soil of their home territories, which may be up to a mile away (Figure 16.18). The alleles flowing in with "immigrant acorns" help reduce genetic differences that might otherwise arise among neighboring stands of oak trees. Gene flow apparently is operating among peppered moths, also, for the form that doesn't match the prevailing background color is being maintained at frequencies higher than expected.

Figure 16.18 Gene flow among oak populations, courtesy of feathered travel agents. Blue jays hoard acorns in their home territory, but they might shop at nut-bearing trees up to a mile away. Some acorns contribute to the allele pool of an oak population some distance away from the parent tree.

Or think of the millions of people from politically explosive, economically bankrupt countries who seek a more stable homeland. The scale of their movements is unprecedented, but not unique. Throughout human history, immigrations may have minimized many of the genetic differences that otherwise would have built up among geographically separated groups of people.

Gene flow is the physical movement of alleles into and out of a population, through immigration and emigration.

GENETIC DRIFT

Chance Events and Population Size

Genetic drift is a random change in allele frequencies over the generations, brought about by chance alone. The magnitude of its effect on genetic diversity and on the range of phenotypes relates to population size. Its impact tends to be minor or insignificant in very large populations but significant in small ones.

Sampling error, a rule of probability, helps explain the difference. By this rule, the fewer times a chance event occurs, the greater will be the variance from the expected outcome of that occurrence. You saw a simple demonstration of this in Section 1.5.

Also think back on the coin-flipping example given earlier, in Section 10.2. Each time you flip a coin, there is a 50 percent chance it will turn up heads or tails. With, say, only ten flips, the odds are great that the coin will turn up heads 7 times and tails 3 times. But with a thousand flips, the odds it will turn up heads 700 times and tails 300 times are almost nil. Similarly, sampling error applies each time random mating and fertilization take place in a population.

Bear in mind, genetic drift has nothing to do with how a population got small in the first place. *Genetic drift simply increases the chance that an allele will become more or less prevalent when the number of individuals in a population is small.*

Figure 16.19 is a computer simulation of the effect of genetic drift in two populations: one large, the other small. The outcomes parallel the findings of an actual experiment with beetles (*Tribolium*) that mate with one another at random. Researchers grouped 1,320 beetles into twelve populations of 10 beetles and twelve of 100. Which beetles ended up in which group was a matter of chance. At the outset, the frequency of a wild-type allele (call it *A*) was 0.5. The researchers tracked allele *A* for twenty generations. Every time, they randomly removed some offspring to maintain each population's original size. At the end of the experiment, *A* was not the only allele left in the large groups, but it was fixed in seven of the small groups. **Fixation** means that only one kind of allele remains at a particular locus in the population; all individuals are homozygous for it.

Thus, *in the absence of other forces, random change in allele frequencies leads to the homozygous condition and a loss of genetic diversity over the generations.* This happens in all populations; it just happens faster in small ones. Once alleles inherited from an original population are fixed, their frequencies will not change again unless mutation or gene flow introduce new alleles.

Bottlenecks and the Founder Effect

Genetic drift is pronounced when very few individuals rebuild a population or found a new one. This happens after a **bottleneck**, a severe reduction in population size brought about by intense selection pressure or some calamity. Say contagious disease, habitat loss, hunting, or an erupting volcano nearly wipes out a population. Even if a moderate number do survive the bottleneck, allele frequencies will have been altered at random.

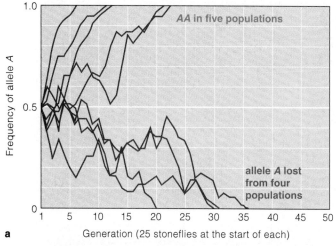

a Generation (25 stoneflies at the start of each)

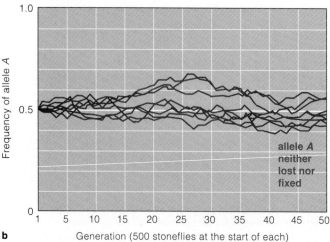

b Generation (500 stoneflies at the start of each)

Figure 16.19 Computer simulation of the effect of genetic drift on one allele's frequency in small populations and in large populations. Equal fitness is assumed for three simulations (*AA* = 1, *Aa* = 1, and *aa* = 1).

(**a**) The size of nine populations of a species (stoneflies, in this case) was maintained at 25 breeding individuals in each generation, through fifty generations. The five graph lines reaching the top of the diagram tell you that allele *A* became fixed in five of the small populations. The four lines plummeting off the bottom of the diagram tell you it was lost from four of them. As you can see, *alleles can be fixed or lost even in the absence of selection.*

(**b**) The size of nine other populations was maintained at 500 individuals in each generation, through fifty generations. Allele *A* did not become fixed in any of these large populations. The magnitude of genetic drift was much less in every generation than in the small populations.

phenotypes of original population

phenotype of
island population

Figure 16.20 Example of the founder effect. A seabird carries a few seeds, stuck to its feathers, from the mainland to a remote oceanic island. By chance, most of the seeds carry an allele for orange flowers that was not common in the original population. In the absence of further gene flow or selection for flower color, genetic drift will fix the allele in the island population.

Example: In the 1890s hunters killed all but twenty of a large population of northern elephant seals. The population eventually recovered. After its size reached 30,000, electrophoretic analysis of a sample of twenty-four genes revealed no variation in the population. A number of alleles had been lost during the bottleneck.

The genetic outcome can be similarly dicey after a few individuals leave a population and establish a new one elsewhere. This form of bottlenecking is called a **founder effect**. By chance, the allele frequencies of the founders may not be the same as those of the original population. In the absence of further gene flow, natural selection will influence those frequencies in drastically different ways because of its interaction with genetic drift. As you might deduce, the founder effect is quite pronounced on isolated islands (Figure 16.20).

Genetic Drift and Inbred Populations

Inbreeding refers to nonrandom mating among closely related individuals, which have many identical alleles in common. Inbreeding is a form of genetic drift in a small population—that is, within the group of relatives that are preferentially interbreeding. Like genetic drift, inbreeding leads to the homozygous condition. It also can lower fitness when the alleles that are increasing in frequency are recessive and have harmful effects.

Most human societies forbid or discourage incest (inbreeding between parents and children or between siblings). But inbreeding among other close relatives is common in small communities that are geographically or culturally isolated from a larger population. The Old

Order Amish of Pennsylvania, for instance, are a highly inbred group having distinct genotypes. One outcome of their inbreeding is the high frequency of a recessive allele that causes *Ellis–van Creveld syndrome*. Affected individuals have extra fingers or toes and short limbs. Section 11.5 shows one affected individual. The allele probably was rare when the small number of founders immigrated to Pennsylvania. Today, though, about 1 in 8 are heterozygous and 1 in 200 are homozygous for it.

Bottlenecks and inbreeding are a bad combination for endangered species, the populations of which have become smaller and vulnerable to extinction. Cheetahs, for instance, apparently survived a drastic bottleneck in the nineteenth century. Survivors mated with their own offspring when no other options were available.

Inbreeding among survivors and their descendants resulted in strikingly similar alleles among the 20,000 existing cheetahs (Figure 16.21). One, a mutated allele, affects fertility. Most male cheetahs have a low sperm count, and 70 percent of the sperm are abnormal. Other shared alleles result in far lower resistance to disease. Therefore, infections that are seldom life-threatening to other cat species can be devastating to cheetahs. In one outbreak of *feline infectious peritonitis* in a wild animal park, the viral pathogen had very little effect on captive lions, but it killed many cheetahs. For them, infection triggered uncontrollable inflammation. Fluid filled the body cavity that houses the heart and other organs, and the cats died in agony. There is no vaccine.

The Florida panther also is highly inbred. Pressure from hunting and urban sprawl has helped put this cat on the endangered species list. Only seventy remain.

Figure 16.21 A few of the remaining cheetahs, which have some bad alleles that made it through a severe bottleneck.

Genetic drift is the random change in allele frequencies over the generations, brought about by chance alone. The magnitude of its effect is greatest in small populations, such as the ones that make it through a bottleneck.

Barring mutation, selection, and gene flow, the chance losses and increases of the various alleles at a given locus lead to the homozygous condition and a loss of genetic diversity.

SUMMARY

1. Awareness of evolution—changes in lines of descent over time—emerged through comparisons of the body structure and patterning for major groups of animals (comparative morphology), questions about the world distribution of plants and animals (biogeography), and observations of fossils in sedimentary rock layers. As first proposed by Charles Darwin and Alfred Wallace, evolution can occur by natural selection. *16.1–16.3*

2. Individuals of a population generally have the same number and kind of genes. But genes come in different allelic forms, and this leads to variations in traits. *16.4*

3. A population is evolving when some forms of a trait (and the underlying alleles) are becoming more or less common relative to the other kinds one generation after the next. This happens through mutation, gene flow, genetic drift, and natural selection (Table 16.2). *16.4*

4. Gene mutations (heritable changes in DNA) are the only source of *new* alleles. New combinations of existing alleles arise by crossing over, independent assortment at meiosis, and mixing of alleles at fertilization. *16.4*

5. Genetic equilibrium, a state in which a population is not evolving, is used as a baseline to measure change. The Hardy–Weinberg rule is this: Genetic equilibrium occurs only if there is no mutation, the population is very large and isolated from other populations of the same species, there is no selection, mating is random, and all members survive and reproduce equally. *16.5*

6. Natural selection is the difference in survival and reproduction among individuals of a population that differ in details of heritable traits. It leads to increased fitness (increased adaptation to the environment). *16.6*

   a. Pressures may shift a range of variation for a trait in one direction (*directional* selection), favor extremes (*disruptive* selection), or eliminate extremes and favor intermediate forms (*stabilizing* selection). *16.7, 16.8*

   b. Selection can result in balanced polymorphism. A population is in this state when nonidentical alleles for a given trait are being maintained over the generations at frequencies greater than 1 percent. *16.9*

   c. Sexual selection, by females or males, can lead to forms of traits that favor reproductive success. Sexual dimorphism (persistence in the phenotypic differences between males and females) is one outcome. *16.9*

7. Gene flow is a change in allele frequencies brought about by the physical movement of alleles into and out of a population (by immigration and emigration). *16.10*

8. Genetic drift is a change in allele frequencies over the generations due to chance alone. Its effect is greater in small populations than in large ones. *16.11*

9. Drift has greatest impact following a bottleneck (a severe reduction, then a recovery, in population size). Alleles making it through a bottleneck may give rise to

Table 16.2 *Summary of Microevolutionary Processes*

MUTATION	A heritable change in DNA
NATURAL SELECTION	Change or stabilization of allele frequencies owing to differences in survival and reproduction among variant individuals of a population
GENETIC DRIFT	Random fluctuation in allele frequencies over time due to chance occurrences alone
GENE FLOW	Change in allele frequencies as individuals leave or enter a population

differences in the range of phenotypes compared to the original population. In bottlenecking, one type of founder effect, a few emigrants from one population establish a small subpopulation in a new environment. *16.11*

Review Questions

1. Define biogeography and comparative anatomy. How did studies in both disciplines contradict the idea that species have remained unchanged since the time of their creation? *16.1*

2. Define evolution. Define evolution by natural selection. Can an individual evolve? *16.1, 16.4, 16.5*

3. Name three broad categories of heritable traits that help characterize a population. *16.4*

4. Explain the difference between continuous variation and polymorphism. *16.4*

5. How do lethal mutations and neutral mutations differ? *16.4*

6. Define genetic equilibrium. Which occurrences can drive allele frequencies away from genetic equilibrium? *16.4, 16.5*

7. Define fitness, with respect to phenotype. *16.6*

8. Identify the mode of selection (stabilizing, directional, or disruptive) for the diagrams on the facing page. *16.7, 16.8*

9. Define bottleneck and the founder effect. Are these cases of genetic drift, or do they merely set the stage for it? *16.11*

Self-Quiz   ANSWERS IN APPENDIX III

1. Biologists define evolution as _____ .
   a. the origin of a species
   b. heritable change in a line of descent over generations
   c. inheritance of characteristics acquired by the individual

2. Darwin saw that populations of Galápagos finches _____ .
   a. show variation in traits
   b. resemble birds in South America
   c. are adapted to different island habitats
   d. all of the above

3. Individuals don't evolve; _____ do.

4. Genetic variation gives rise to variation in _____ traits.
   a. morphological   c. behavioral
   b. physiological   d. all of the above

5. Sickle-cell anemia first appeared in Asia, the Middle East, and Africa. The causative allele entered the U.S. population when people were forcibly brought over from Africa prior to the Civil War. In microevolutionary terms, this is a case of _____ .
   a. mutation   c. gene flow
   b. genetic drift   d. natural selection

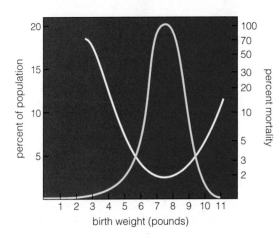

Figure 16.22
The weight distribution for 13,730 human newborns (the *yellow* curve) correlated with mortality rate (*white* curve).

6. Natural selection may occur when there are _____ .
   a. differences in the adaptiveness of forms of traits to prevailing environmental conditions
   b. differences in survival and reproduction among individuals that differ in one or more traits
   c. both a and b

7. Directional selection _____ .
   a. eliminates uncommon forms of alleles
   b. shifts allele frequencies in a steady, consistent direction
   c. favors intermediate forms of a trait
   d. works against adaptive traits

8. Disruptive selection _____ .
   a. eliminates uncommon forms of alleles
   b. shifts allele frequencies in a steady, consistent direction
   c. doesn't favor intermediate forms of a trait
   d. both b and c

9. Match the evolution concepts.
   ____ gene flow
   ____ natural selection
   ____ mutation
   ____ genetic drift

   a. source of new alleles
   b. changes in a population's allele frequencies due to chance alone
   c. allele frequencies change owing to immigration, emigration, or both
   d. differences in survival and reproduction among variant individuals

Critical Thinking

1. Prospects for human newborns of very high or very low birth weight are not good. Being born too small increases the risk of stillbirth or early infant death (Figure 16.22). Also, pre-term instead of full-term pregnancies increase the risks. Are these outcomes of directional, disruptive, or stabilizing selection?

2. For centuries *tuberculosis*, or TB, killed people throughout the world. A bacterium (*Mycobacterium tuberculosis*) causes this contagious lung disease. TB declined steadily in the United States over the years, but caseloads are now rising. The AIDS epidemic, an influx of immigrants from regions where TB is still common, and severe overcrowding in tenements and homeless shelters contribute to the increase. Antibiotics once cured TB in six to nine months. But antibiotic-resistant strains of *M. tuberculosis* have evolved. Two or even three different antibiotics are used simultaneously to block different metabolic pathways of the bacterium. How might this more aggressive approach help get around the problem of antibiotic resistance?

3. A few families in a remote region of Kentucky show a high frequency of *blue offspring*, an autosomal recessive disorder.

Skin of affected individuals appears bright blue. Homozygous recessives lack the enzyme diaphorase. The enzyme catalyzes reactions that maintain hemoglobin in its normal molecular form. Without it, a blue form of hemoglobin accumulates in blood. Skin and its blood capillaries are transparent, so the pigment colors in blood show through and contribute to skin coloration. This gives the skin of nonmutated individuals a pinkish cast—and blue skin its color.

Formulate a hypothesis to explain why the blue-offspring trait is rather common among a cluster of families but rare in the human population at large.

4. In 1996, after reflecting on evidence of evolution that has been accumulating for more than a hundred years, Pope John Paul II acknowledged that the theory of evolution "has been progressively accepted by researchers, following a series of discoveries in various fields of knowledge. The convergence, neither sought nor fabricated, of the results of work that was conducted independently is in itself a significant argument in favor of this theory." His words infuriated individuals who believe in a strict, literal interpretation of the biblical account of creation. Some of these individuals have been demanding that biology instructors treat evolution as "just a theory" and give equal time to the biblical interpretation.

Refer to Section 1.5 on the nature of scientific inquiry. Do you equate scientific and religious explanations of life's origin and history as alternative theories? Why or why not?

Selected Key Terms

allele 16.4
allele frequency 16.4
antibiotic 16.7
balanced polymorphism 16.9
biogeography 16.1
bottleneck 16.11
comparative morphology 16.1
directional selection 16.7
disruptive selection 16.8
evolution CI
fitness 16.6
fixation 16.11
fossil 16.1
founder effect 16.11
gene flow 16.10
gene pool 16.4

genetic drift 16.11
genetic equilibrium 16.4
Hardy–Weinberg rule 16.5
inbreeding 16.11
lethal mutation 16.4
microevolution 16.4
mutation rate 16.4
natural selection 16.6
neutral mutation 16.4
polymorphism 16.4
population 16.4
sampling error 16.11
sexual selection 16.9
stabilizing selection 16.8
uniformity (theory of) 16.2

Readings

Darwin, C. 1957. *Voyage of the Beagle*. New York: Dutton.

Smith, T. B. January 1991. "A Double-Billed Dilemma." *Natural History*, 14–21.

On-Line readings at Student Guide for InfoTrac:
www.brookscole.com/biology

17

SPECIATION

The Case of the Road-Killed Snails

If you happen to be a snail living in a garden in Bryan, Texas, it doesn't take much to keep your genes away from snails in a backyard across the street. By day, the sunbaked asphalt would be about as inviting to a snail as a desert would be to a catfish. Besides, day or night, a street-traversing snail is vulnerable to cars, trucks, skateboards, and bicycles (Figure 17.1a). That strip of asphalt is a formidable barrier to the flow of genes between populations. For snails, that is.

Whether any physical barrier deters gene flow depends in large part on the organism's mode of locomotion or dispersal. It also depends on how fast and how long an organism *can* move in response to environmental factors or its own hormonal signals.

A snail is not swift, and it does not roam very far from its home population. Compare it with the wild black duck, leg-banded in Virginia in 1969, that turned up eight years later

in Korea. Compare it to the wandering albatross, one of the supreme barrier busters. After lifting off from Kerguelen Island in the Indian Ocean, one of these birds soared westward past the southern tip of Africa, across the Atlantic, and around South America's Cape Horn. After traveling thirteen thousand kilometers, it landed in Chile. With a wingspan of 3.65 meters (12 feet) and a lightweight body, that bird was able to exploit the great prevailing winds of the Southern Hemisphere.

And yet, in 1859, some snails did cross an ocean. Humans had transported garden-variety snails (*Helix aspersa*) from France, then released them in California. The idea was that the snails would multiply and meet culinary demands for that French delicacy *escargots aux fines herbes*. It was bad enough that the importers brought over a less tasty species by mistake. Worse, the exotic species exceeded expectations and became an absolute nuisance in

Figure 17.1 (**a**) A snail (*Helix aspersa*) encountering a major barrier to gene flow. It only appears to be reading the warning label. (**b**) Results from a study of neighboring populations of snails, all descended from founders that ended up in a small town in Texas in the 1930s. For each population, a circle represents relative abundances of three alleles (color-coded *green*, *orange*, and *blue*) for an enzyme, leucine aminopeptidase. Genetic variation is greater between populations living on opposite sides of Twenty-Second Street. For example, notice the higher frequencies of the allele coded *orange* in the block to the west of the street.

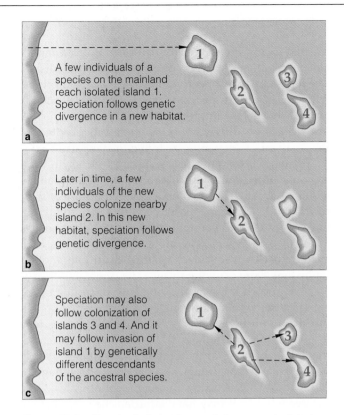

a A few individuals of a species on the mainland reach isolated island 1. Speciation follows genetic divergence in a new habitat.

b Later in time, a few individuals of the new species colonize nearby island 2. In this new habitat, speciation follows genetic divergence.

c Speciation may also follow colonization of islands 3 and 4. And it may follow invasion of island 1 by genetically different descendants of the ancestral species.

Figure 17.6 Sketch of allopatric speciation on an isolated archipelago. (Can you envision other possibilities?)

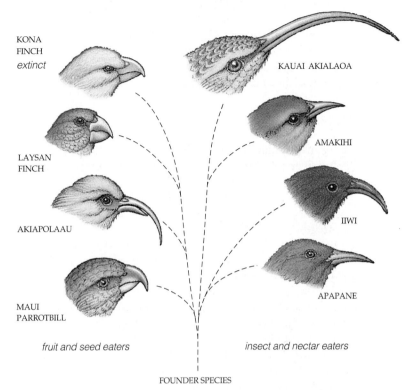

KONA FINCH *extinct*

LAYSAN FINCH

AKIAPOLAAU

MAUI PARROTBILL

KAUAI AKIALAOA

AMAKIHI

IIWI

APAPANE

fruit and seed eaters   *insect and nectar eaters*

FOUNDER SPECIES

Figure 17.7 A few Hawaiian honeycreepers, a fine example of how a new arrival in species-poor habitats on an isolated archipelago can be the start of a flurry of allopatric speciation.

Allopatric Speciation on Archipelagos

An **archipelago** is an island chain some distance away from a continent. Some archipelagos are so close to the mainland that gene flow is more or less unimpeded, and there is little if any speciation. The Florida Keys are like this. Other archipelagos are isolated enough to serve as laboratories for the study of evolution (Figure 17.6). Foremost among these are the Hawaiian Archipelago, nearly 4,000 kilometers from the California coast, and the Galápagos Islands, about 900 kilometers from the Ecuadorian coast. The islands of both chains are merely the tops of great volcanoes, some still active, that rise from the deep seafloor. When each volcano first broke the surface of the sea, its fiery surface was devoid of life.

Remember those Galápagos finches (Section 16.3)? A few finches from the mainland apparently evolved in isolation on one Galápagos island. Later, some of their descendants reached other islands in the chain. In those new and unoccupied habitats—even in different parts of the same habitats—conditions varied, and the island hoppers were subject to a variety of selection pressures. In time, genetic divergences within and between the islands paved the way for new episodes of allopatric speciation. Later on, some island-hopping new species even invaded the island of their ancestors. The distances between islands are enough to foster divergences, but not enough to stop occasional invasions.

Bursts of speciation in the Hawaiian Archipelago have been quite dramatic. They are a premier example of adaptive radiation in a species-poor environment, as described in Section 17.4. The youngest island, Hawaii, formed less than a million years ago. Here alone we see diverse habitats, ranging from lava beds to rain forests to alpine grasslands and high, snow-capped volcanoes. When the ancestors of Hawaiian honeycreepers arrived, they found a veritable buffet of fruits, seeds, nectars, and tasty insects—and not many competitors for them. The near absence of competition from other bird species fanned allopatric speciations. Figure 17.7 only hints at the resulting variation that arose among the different species of Hawaiian honeycreepers. Today, these and thousands of other species of animals and plants that evolved in the archipelago are found nowhere else. As another example of the potential for speciation, the Hawaiian Islands make up less than 2 percent of the world's land mass. Yet they are home to 40 percent of all species of fruit flies (*Drosophila*).

ALLOPATRIC SPECIATION. Some type of physical barrier intervenes between populations or subpopulations of a species and prevents gene flow among them. By doing so, it favors genetic divergence and speciation.

MODELS FOR OTHER SPECIATION ROUTES

Geographic isolation may not always be required for reproductive isolation. In the models for sympatric and parapatric speciation, populations diverge genetically even while they are in contact with one another.

Sympatric Speciation

By the model for **sympatric speciation**, a species may form *within* the home range of an existing species, in the absence of a physical barrier. (*Sym–* means together with, as in "together with others in the homeland.")

EVIDENCE FROM CICHLIDS IN AFRICA   In 1994, Ulrich Schliewen and his coworkers found evidence for the sympatric speciation model. They studied fishes called cichlids that live together in two lakes in Cameroon, West Africa. Each lake basin is the collapsed cone of a small volcano (Figure 17.8). Yet eleven cichlid species coexist in one lake. Nine coexist in the other.

The researchers analyzed mitochondrial DNA from all species in one lake. They did the same for all species in the other lake as the basis for comparison. In their nucleotide sequences, the species of each lake are like each other and *not* like related species in nearby lakes and rivers. For example, nine species in one lake are the only cichlids having a unique base-pair substitution in the gene that specifies the protein cytochrome *b*.

Physical and chemical conditions are too uniform in the lakes to foster geographic separation. For instance, shorelines are so uniform that even small topographic barriers are absent. Thus, allopatric populations could not have evolved, even on a small scale. The lakes are isolated except for tiny creeks trickling in from higher elevations; there has been no gene flow from outside. The lakes must have been colonized before connections with a nearby river system were severed. Also, cichlids are not sluggish swimmers. Besides being very mobile, individuals of different species often make contact in the small lakes; they must be living in sympatry.

Because clusters of species are more closely related to one another than to species anywhere else, we may assume they share a common ancestor. Because of their restricted distribution, we may assume they formed in the same lake. And yet nothing about the lakes isolates different populations. Therefore, the ancestors of each cichlid species were never isolated from one another while they were evolving in separate directions.

Within each lake, species do show small degrees of *ecological* separation; differences in feeding preferences put them in different places. Some cichlid species feed in open waters and others at the lake bottom. Even so, they all *breed* close to the bottom, in sympatry. It may be that small-scale ecological separation was enough to influence sexual selection among potential mates. Over the generations, it may have afforded the reproductive isolation that can lead to speciation.

SPECIATION BY WAY OF POLYPLOIDY   Quite possibly, sympatric speciation has been a prevalent evolutionary event among flowering plants. Consider that about half of all known species of flowering plants are polyploid. **Polyploidy**, remember, is a change in the chromosome number; offspring inherit three or more of each type of chromosome characteristic of the parental stock. Such changes arise when chromosomes separate improperly during meiosis or mitosis. They also arise when a germ cell replicates its DNA but fails to divide, then goes on to function as a gamete (Section 11.9).

Speciation may have been rapid for many flowering plants that engage in self-fertilization or some asexual reproductive mode. Suppose they produced polyploid offspring. If the extra chromosomes paired with each other at meiosis, maybe the extra set of genes did no harm. Common bread wheat is one of the species that might have arisen by polyploidy. In this case, though, cross-fertilization also was involved (Figure 17.9).

Allen Orr has evidence that polyploid animals are rare because of a failed **dosage compensation**. By this normal event, genes on sex chromosomes are expressed at the same levels in females *and* males. For instance, X chromosome inactivation in female mammals means the cells of both males and females have a single active X chromosome (Section 14.3). Polyploidy skews dosage compensation, with bad or fatal effects. The mechanism is absent in plants, so their chromosome doublings are not as problematic as they are in animals.

Figure 17.8 Lake Barombi Mbo in Cameroon, West Africa. Nine kinds of cichlids apparently evolved by way of sympatric speciation in this small, isolated crater lake. There is not even microgeographic isolation. Cichlid populations separate by feeding preferences, but they all breed near the lake bottom, in sympatry.

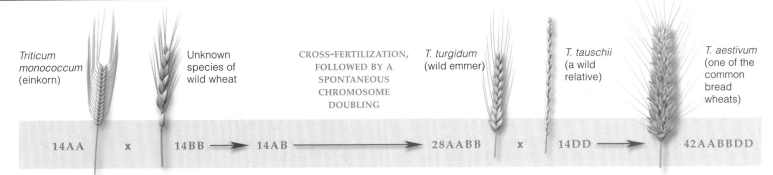

Triticum monococcum (einkorn)

Unknown species of wild wheat

CROSS-FERTILIZATION, FOLLOWED BY A SPONTANEOUS CHROMOSOME DOUBLING

T. turgidum (wild emmer)

T. tauschii (a wild relative)

T. aestivum (one of the common bread wheats)

14AA   x   14BB → 14AB ——————→ 28AABB   x   14DD ——→ 42AABBDD

a By 11,000 years ago, humans had started to cultivate wild wheats. Einkorn (*Triticum monococcum*) is still around. It has a diploid chromosome number of 14 (two sets of 7 chromosomes, shown above as 14AA). Long ago, einkorn wheat probably hybridized with another wild wheat species having the same chromosome number.

b If the AB hybrid offspring were sterile but self-fertilizing, an interbreeding population of AB plants could have arisen by asexual reproduction. About 8,000 years ago, polyploidy must have arisen in such a population. Wild emmer (*T. turgidum*) plants are tetraploid (AABB), with a chromosome number of 28 (two sets of 14). They are fertile; at meiosis, the A chromosomes pair with each other, and the B chromosomes pair with each other.

c Later, an AABB plant probably hybridized with *T. tauschii*, a wild relative of the wild emmer. Its diploid chromosome number must have been 14 (two sets of 7 DD). Populations of the hybrid descendants now include common bread wheats. One of these, *T. aestivum*, has a chromosome number of 42 (six sets of 7 AABBDD).

Figure 17.9 Presumed sympatric speciation in wheat by polyploidy and hybridizations. Wheat grains 11,000 years old have been found in the Near East. Diploid wild wheats still grow there.

BULLOCK'S ORIOLE

BALTIMORE ORIOLE

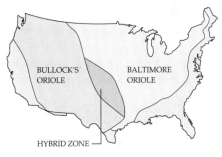

BULLOCK'S ORIOLE

BALTIMORE ORIOLE

HYBRID ZONE

Figure 17.10 A possible setting for parapatric speciation? Hybrids arise along a common border of the geographic ranges of two species of orioles. To the east are Baltimore orioles; to the west are Bullock's orioles. Hybridization was once so common that these two birds were considered subspecies. In 1997, the American Ornithologist Union decided they are separate species.

Parapatric Speciation

By the model for **parapatric speciation**, neighboring populations become distinct species while maintaining contact along a common border. (*Para*– means near, as in "near another homeland.") Interbreeding individuals produce hybrid offspring in the region, which is called a **hybrid zone**. Evidence that might support parapatric speciation is sketchy. Why? It's often difficult to know whether or not geographically separated populations are only subspecies. *Subspecies* refers to geographically distinct populations of the same species.

For example, Bullock's orioles differ from Baltimore orioles in the color patterning of most of their feathers.

Male orioles have different territorial songs and mate with their own kind, usually. As K. Corbin determined, the birds also differ in allele frequencies for several enzymes. Their geographic ranges differ but overlap in the American Midwest (Figure 17.10). Interbreeding was once common in the hybrid zone but is becoming less frequent. Is it an example of parapatric speciation in progress? Maybe. But it might also be an example of "secondary contact." Previously isolated subspecies that diverged recently from a common ancestor may have been getting together again.

SYMPATRIC SPECIATION. **Daughter species arise from a group of individuals within an existing population. This seems to have been a common speciation event among the polyploid species of flowering plants.**

PARAPATRIC SPECIATION. **Adjacent populations evolve into distinct species even while maintaining contact along their common border.**

PATTERNS OF SPECIATION

Branching and Unbranched Evolution

All species—past and present—are related by descent. They share genetic connections through lineages that extend back in time to the molecular origin of the first prototypic cells, some 3.8 billion years ago. Subsequent chapters focus on evidence that supports this view. In anticipation of those chapters, let's start thinking about ways to interpret the large-scale histories of species.

The fossil record provides evidence of two patterns of evolutionary change in lineages. One is branched, the other unbranched. The first is called **cladogenesis** (from the Greek *klados*, meaning branch; and *genesis*, meaning origin). This is the pattern by which a lineage splits, with populations becoming genetically isolated and then diverging in different evolutionary directions. This is the pattern of speciation described earlier.

By the second pattern, **anagenesis**, changes in allele frequencies and in morphology accumulate within an unbranched line of descent. (In this context, *ana–* means renewed.) Directional changes are confined within one lineage, and gene flow never does cease among all of its populations. At some point in time, allele frequencies and morphology have changed so much that we assign a separate name to the descendants. Such changes are occurring in the moth populations you read about earlier.

Evolutionary Trees and Rates of Change

Evolutionary trees summarize information about the continuity of relationship among species. Figure 17.11 is a simple way to start thinking about how tree diagrams are constructed. Each *branch* represents a single line of descent from a common ancestor. And each *branch point* represents a time of genetic divergence and speciation, as brought about by microevolutionary processes.

Figure 17.11 also shows how an evolutionary tree diagram conveys *rates of change*, the estimated length of time between speciation events. Branches having slight angles imply that the species emerged through many small morphological changes over long spans of time (Figure 17.11a). This is the main premise of the **gradual model of speciation**, which actually fits well with many fossil sequences. For example, in many sedimentary rock layers we find sequences of intricately perforated shells of foraminiferans, a common type of protistan. The observable characteristics of foraminiferan fossils in the sequences offer evidence of slow change.

Alternatively, evolutionary tree diagrams for some lineages are constructed with short horizontal branches that make an abrupt 90-degree turn, as in Figure 17.11b. The **punctuation model of speciation** is consistent with such diagrams. By this model, most of the changes in morphology are compressed into a brief period when populations are starting to diverge—say, within only hundreds or thousands of years. Bottlenecks, founder effects, strong directional selection, or a combination of these foster rapid speciation. The daughter species recover fast from the adaptive wrenching, then change little during the next 2 million to 6 million years or so. The proponents of this model argue that reproductive cohesion has indeed prevailed for about 99 percent of the history of most lineages. They cite strong evidence of abrupt change in many parts of the fossil record.

Changes in lines of descent apparently have been gradual, abrupt, or both. Species originated at different times and they differ in how long they have persisted. Remember, some lineages have endured without much change, producing a species here and losing a species there, often over millions of years. Other lineages have branched bushily and sometimes spectacularly during times of adaptive radiation.

Adaptive Radiations

An **adaptive radiation** is a burst of divergences from a single lineage that give rise to many new species, each adapted to an unoccupied or a new habitat or to using a novel resource. Figure 17.12 provides an example. In the past, lineages often diversified in such a way when the member species found themselves in vacant **adaptive zones**. Think of

a Branching at a slight angle; a new species formed through gradual changes in traits over geologic time.

b Horizontal branching; very rapid change in traits around the time of speciation. A long vertical branch means the traits of the new species did not change much thereafter.

c Many branchings of the same lineage at or near the same point in geologic time; an adaptive radiation has taken place.

Figure 17.11 How to read evolutionary tree diagrams.

Figure 17.12 Example of adaptive radiation, as shown by an evolutionary tree diagram. This one started about 65 million years ago, at the start of a geologic era called the Cenozoic, and it led to mammals as different as the opossum and walrus. Variations in the branch widths correspond to the number of groups in the lineage, as represented at different points in time. The wider the branch, the greater the species diversity.

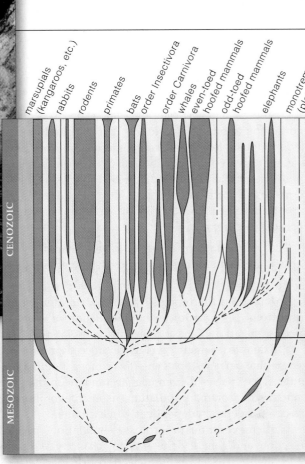

adaptive zones as ways of life, such as "burrowing in seafloor sediments" or "catching winged insects in the air at night." A lineage may radiate into such zones if it has physical, evolutionary, or ecological access to them.

Physical access means a lineage happens to be there when adaptive zones open up. For example, mammals were once distributed through uniform tropical regions of a huge continent. The continent broke up into several land masses that gradually drifted apart. Habitats and resources changed in many different ways on different masses and set the stage for independent radiations.

Evolutionary access means that a modification to an existing structure or function will permit a lineage to exploit the environment in new or more efficient ways. Such modifications are **key innovations**. Example: The forelimbs of certain five-toed vertebrates evolved into wings. And that innovation opened up novel adaptive zones for the ancestors of birds and bats (Section 18.4).

Ecological access means that the lineage can enter vacated adaptive zones or displace the resident species. Sections 25.2, 42.7, and 42.8 offer examples.

Extinctions—End of the Line

Extinction is the irrevocable loss of a species. The fossil record shows twenty or more **mass extinctions**—large, catastrophic events in which entire families and other major groups were abruptly wiped out. For example, 95 percent of all known species disappeared 250 million years ago. In other times, only certain groups vanished.

By studying the statistical distribution of extinctions, David Raup found that most were clustered in time even though they differed in size. Like George Simpson before him, he realized that times of reduced diversity follow extinction events, then new species arise and fill vacant adaptive zones. Apparently, luck has a lot to do

with it. For example, in the past, asteroids repeatedly hit the Earth. Some survivors of one impact radiated into the adaptive zones that dinosaurs had occupied. Among them were mammals—some of which gave rise to your early ancestors. As another example, species not widely distributed tend to be hit hard by pronounced declines in global temperature. This is usually the case for species adapted to warm regions. Unlike species in cold regions, they have nowhere else to go. Asteroids, drifting continents, climatic change—as you will see in later chapters, these are some of the major factors that have contributed to a general pattern of extinctions. In Chapter 25, you also will read about a mass extinction that is currently under way, thanks largely to us.

Lineages have changed gradually, abruptly, or both. Their member species originated at different times and have differed in how long they have persisted.

Adaptive radiations are bursts of divergences from a single lineage that gave rise to many new species, each adapted to a vacant or new habitat or to using a novel resource.

Repeated and often large extinctions occurred in the past. After times of reduced diversity, new species formed and occupied new or vacated adaptive zones.

Taken together, the persistence, branchings, and extinctions of species account for the full range of biological diversity at any point in geologic time.

SUMMARY   *Gold* indicates text section

1. A species is a single kind of organism, recognized partly in terms of its morphology. *17.1*

2. By the biological species concept, a species is one or more populations of individuals that are interbreeding and producing fertile offspring under natural conditions and that are reproductively isolated from other such populations. The concept identifies a species mainly in terms of the portion of alleles that promote or maintain reproductive isolation. Also, it applies only to sexually reproducing organisms. *17.1*

3. The populations of a species have a shared genetic history, are maintaining genetic contact over time, and are evolving independently of other species. *17.1*

4. Speciation is the process by which daughter species form from a population or subpopulation of a parent species; the attainment of reproductive isolation. *17.1*

   a. Example: Genetic divergence proceeds after some geographic barrier stops gene flow among populations of the same species. Differences accumulate in the gene pools of the isolated populations as microevolutionary processes operate independently in each one.

   b. Microevolutionary processes—mutation, natural selection, and genetic drift—may by chance give rise to reproductive isolating mechanisms. Such mechanisms prevent interbreeding between populations and thus promote irreversible genetic differences between them.

   c. Speciation also can occur instantaneously, as by polyploidy or other changes in chromosome number.

5. Prezygotic isolating mechanisms prevent mating or pollination between individuals of populations. They include differences in reproductive timing or behavior, incompatibilities in reproductive structures or gametes, and occupation of different microenvironments in the same area. Postzygotic mechanisms lead to early death, sterility, or unfit hybrid offspring. They take effect after fertilization, while the embryo develops. *17.1*

6. Here are three current models for speciation:

   a. Allopatric speciation. Geographic barriers prevent gene flow between the populations of a species, which genetically diverge in such a way that interbreeding will not occur in nature even if their individuals make contact with each other later on. Allopatric speciation might be the most prevalent speciation route. *17.2*

   b. Sympatric speciation. Reproductive isolation of individuals in the same home range leads to speciation. Speciation by polyploidy is an example. *17.3*

   c. Parapatric speciation. Adjacent populations give rise to a new species while maintaining contact along a border between their home ranges. *17.3*

7. Lineages differ in their time, rates, and direction of speciation. Speciation can proceed gradually, rapidly, or both. Extensive branching of a single lineage (a burst of divergences from it) in the same geologic time span is an adaptive radiation. The lineage may be physically in place when new adaptive zones (ways of life) open up. It may access zones by key innovations. It may enter vacated zones or displace resident species. *17.4*

8. Evolutionary tree diagrams indicate the relationships among groups of species. Each tree branch represents a line of descent (lineage). Branch points are speciation events, as brought about by natural selection, genetic drift, and other microevolutionary processes. *17.4*

9. Extinction is the irrevocable loss of a species. The fossil record shows twenty or more mass extinctions: large, catastrophic events in which entire families or other major groups were abruptly lost. *17.4*

10. Persistences, branchings, and extinctions of species account for the range of biodiversity (Table 17.1). *17.4*

Table 17.1 *Summary of Processes and Patterns of Evolution*

MICROEVOLUTIONARY PROCESSES

Mutation	Original source of alleles	Stability or change in a species is the outcome of balances or imbalances among all of these processes, the effects of which are influenced by population size and by the prevailing environmental conditions.
Gene flow	Preserves species cohesion	
Genetic drift	Erodes species cohesion	
Natural selection	Preserves or erodes species cohesion, depending on environmental pressures	

MACROEVOLUTIONARY PROCESSES

Genetic persistence	Basis of the unity of life. The biochemical and molecular basis of inheritance extends from the origin of first cells through all subsequent lines of descent.
Genetic divergence	Basis of life's diversity, as brought about by adaptive shifts, branchings, and radiations. Rates and times of change varied within and between lineages.
Genetic disconnect	Extinction. End of the line for a species. Mass extinctions are catastrophic events in which major groups are lost abruptly and simultaneously.

Review Questions

1. How does the biological species concept differ from a definition of species based on morphological traits alone? *17.1*

2. Define speciation and describe a speciation model. *CI, 17.2, 17.3*

3. Give examples of reproductive isolating mechanisms. *17.1*

4. With respect to evolutionary tree diagrams, describe what each of the following features represents: *17.4*
   a. a single line
   b. soft-angled branching of line
   c. horizontal branching
   d. vertical continuation of branch
   e. many branchings of a line
   f. dashed line
   g. branch ending before present

5. Define and give an example of an adaptive radiation. *17.4*

6. Define extinction. What kinds of events might bring about extinctions? *17.4*

Self-Quiz   ANSWERS IN APPENDIX III

1. Sexually reproducing individuals of a species _____ .
   a. can interbreed under natural conditions
   b. can produce fertile offspring
   c. have a shared genetic history
   d. all of the above

2. Reproductive isolating mechanisms _____ .
   a. prevent interbreeding   c. reinforce genetic divergence
   b. prevent gene flow   d. all of the above

3. When potential mates occupy overlapping ranges but reproduce at different times, this is a case of _____ isolation.
   a. postzygotic   c. temporal
   b. mechanical   d. gametic

4. In an evolutionary tree diagram, a branch point represents a _____ , and a branch that ends represents _____ .
   a. single species; incomplete data on lineage
   b. single species; extinction
   c. time of divergence; extinction
   d. time of divergence; speciation complete

5. An evolutionary tree diagram with horizontal branches that abruptly become vertical is consistent with the _____ .
   a. gradual model of speciation
   b. punctuation model of speciation
   c. idea of small changes in form over long spans of time
   d. both a and c

6. Match each term with its most suitable description.
   ____ cladogenesis   a. burst of genetic divergences
   ____ anagenesis   from a single lineage
   ____ adaptive   b. catastrophic disappearance
        radiation   of major groups of organisms
   ____ extinction event c. branching lineages
   ____ mass extinction   d. a species lost from a lineage
                          e. genetic and morphological
                             change in unbranched lineage

Critical Thinking

1. You notice several duck species in the same lake habitat. All the females of the various species look quite similar to one another. But the males differ in the patterns and colors of their feathers. Speculate on which forms of reproductive isolation may be keeping each species distinct. How does the appearance of the male ducks provide a clue to the answer?

2. One family of mammals includes true horses (*Equus*), zebras (*Hippotigris* and *Dolichohippus*), and donkeys and asses (*Asinus*). Zebroids are hybrid offspring of wild zebras and domesticated horses that were confined to the same pasture (Figure 17.13). The unnatural confinement breached the reproductive barriers between the two lineages. The barriers have been in place since a genetic divergence got under way more than 3 million years ago. What does the breach suggest about the genetic changes required to attain irreversible reproductive isolation in nature?

3. Suppose a small population that was founded by only a few individuals gets cut off from gene exchange with the main population of their species. By Mayr's view of the *founder effect*, allele frequencies for a few gene products change as a result of genetic drift; then the change triggers a number of changes in other genes that are affected by those alleles. If that is so, then

Figure 17.13   A mixed herd of zebroids and horses.

genetic changes in newly founded populations may be so great that speciation is rapid—so rapid that we will seldom be able to document it through the fossil record. Does this view correspond to the gradual model or the punctuated model of speciation?

4. A key innovation, recall, is some modification in structure or function that permits a species to exploit the environment in a more efficient or novel way, compared to the ancestral species. Identify a key innovation of an existing species, such as humans. Then describe how that innovation might be the basis of an adaptive radiation in environments of the distant future.

5. Richard Lenski uses bacterial populations in culture tubes to develop model systems for studying evolution. He is fond of saying that such a population is the equivalent of the entire human population, that he can replicate it several times over, that bacteria produce several generations in a day, and that he can store them in the deep freeze, then bring them back to active form, unaltered, to directly compare ancestors and descendants. Are bacterial models relevant to evolutionary studies of sexually reproducing organisms? Before you answer, read a short article by P. Raine and M. Travisano entitled "Adaptive Radiation in a Heterogenous Environment" (*Nature*, 2 July 1998, 69–72).

Selected Key Terms

adaptive radiation *17.4*
adaptive zone *17.4*
allopatric speciation *17.2*
anagenesis *17.4*
archipelago *17.2*
biological species concept *17.1*
cladogenesis *17.4*
dosage compensation *17.3*
evolutionary tree *17.4*
extinction *17.4*
gene flow *17.1*
genetic divergence *17.1*
gradual model, speciation *17.4*

hybrid zone *17.3*
key innovation *17.4*
mass extinction *17.4*
parapatric speciation *17.3*
polyploidy *17.3*
punctuation model,
  speciation *17.4*
reproductive isolating
  mechanism *17.1*
speciation *CI*
species *17.1*
sympatric speciation *17.3*

Readings

Futuyma, D. 1998. *Evolutionary Biology*. Third edition. Sunderland, Massachusetts: Sinauer.

Mayr, E. 1976. *Evolution and the Diversity of Life*. Cambridge, Massachusetts: Belknap Press of Harvard University Press.

18

THE MACROEVOLUTIONARY PUZZLE

Legends of the Flood

So far in this unit, you have been considering some evolutionary theories that biologists routinely use to interpret the past and present, even to predict possible futures for the natural world. Today the theories are widely accepted in society at large. This was not the case when early evolutionists in Europe started to work out the details. They rankled more than a few people with their astounding theories, which revealed connections between the geologic record, the fossil record, and the sweep of biological diversity.

As evidence accumulated in support of the theories, many zoologists, geologists, and others came to accept them as guides for research. Others couldn't reconcile them with the Bible, the premier book of the Western world. That book had helped many generations cope with the uncertainties of life, with death, and with natural disasters, including a stupendous flood. Such a flood did indeed occur, but it was assigned meaning at a time when people didn't know much about geologic processes.

For example, William Ryan and Walter Pitman drew from marine geology, genetics, archaeology, botany, and linguistics to discover the location of the biblical flood. They zeroed in on the Black Sea. This landlocked sea is 1,800 meters (6,000 feet) deep. Rainfall and large rivers freshen it, and excess water flows through the Bosporus and Dardanelles straits to the Aegean Sea (Figure 18.1).

What Ryan and Pitman discovered coincides with the biblical account of a great flood. It coincides with stone tablets inscribed in Babylon long before the Bible was written. It coincides with the epic of Gilgamesh and other myths from Greece, Mesopotamia, and the Levant. What follows is the picture they pieced together.

Starting about 120,000 years ago, the climate cooled and evaporation from the sea increased dramatically. So much seawater became airborne, then accumulated on land as snow, that the sea level fell 120 meters (400 feet) below what it is today. Snow compacted into thick ice sheets that advanced over much of Europe, North America, and Asia. Then, about 20,000 years ago, the global climate warmed and a colossal meltdown began. Meltwater drained mainly to the sea, but some collected on land as immense freshwater lakes. One lake was the start of the Black Sea. It rose high enough to connect with a narrow outlet to the Aegean Sea (Figure 18.2a).

Glacial conditions returned, then the climate warmed again. This time nearly all meltwater was diverted to the sea. Little rain fell on land, and evaporation gradually lowered the lake's surface. By 7,600 years ago, it was 150 meters (500 feet) below its outlet to the sea.

When vast ice sheets prevailed, cool, dry deserts and grasslands formed to the south. Modern humans (*Homo sapiens*) witnessed the change. Hunter–gatherers moved into fertile valleys and river deltas in Europe and Asia. Where they encountered water, wild game, and edible plants, they founded settlements—only to abandon them when the climate changed and resources vanished. The lake beckoned the displaced more than once. Being so far below sea level, the lake edge was warmer than the surrounding arid mountains and plateaus. It welcomed the thirsty when springs, ponds, and other lakes dried up. In the moist, fertile soil of that Edenlike oasis, wild wheat and barley took root. So did trading cultures and early attempts at farming.

All the while, the sea was rising. By 7,600 years ago it was high enough to lap over the crest of the Bosporus valley, which plunged to the lake below. The first trickles cut into soft earth. But the cut deepened and water thundered through at 200 times the flow volume of Niagara Falls. For the next ten months, rapidly advancing water submerged valleys and villages, driving humans and wildlife before it.

Figure 18.1 The geologic region where global climate changes led to a catastrophe that altered the course of human history.

a Direction of meltwater flows from Eurasian ice sheet that started about 20,000 years ago

b Freshwater lake (future Black Sea) connects with Mediterranean 14,500 years ago

c The flow of seawater in the same region during flood 7,600 years ago

d Present-day exchange of seawater and freshwater between the Mediterranean and Black seas

Figure 18.2 Historical connection between the Mediterranean and the ice age freshwater lake that became the Black Sea.

These were farmers and fishermen, not nomads who would simply pack up and move on. They built houses and boats; they traded obsidian, leather, herbs, and finely incised pottery. They could not comprehend the force behind their mass dispersal. In time, small bands settled in the Levant, Mesopotamia, Egypt, the Persian Gulf, central Asia, and western Europe. They brought with them precious seeds, farm animals, and fragments of languages, ideas, and technologies. And they carried with them a terrifying story of a monstrous flood.

The story endured. It influenced investigations into processes of nature, as when sixteenth-century biblical scholars argued that the Earth was about 6,000 years old. To them, that was plenty of time for man's creation, fall from grace, and salvation after a punishing flood. The geologic changes that made it into oral and written traditions were such extraordinary catastrophes that people assumed they were divinely invoked. Otherwise the Earth was just an unchanging stage for the human drama. This was a core hypothesis of what came to be known as theories of **catastrophism**.

Biological science emerged within the framework of such influential beliefs, and so did awareness of changes in the Earth and its creatures. Understand this, and you will gain insight into why acceptance of the very idea of biological evolution was so long in coming.

Key Concepts

1. In the evolutionary view, all species that have ever lived are related by descent—some closely and others remotely so.

2. Macroevolution is the name for the major patterns, trends, and rates of change among lines of descent, or lineages, during the history of life. The fossil record, the geologic record, and radiometric dating of rocks yield evidence of macroevolution. So do morphological comparisons of existing groups of species.

3. The fossil record has enough detail to reveal major patterns of descent with modification. The record will never be complete in all of its details, but gaps are to be expected. We know that parts were lost owing to major movements of the Earth's crust and other geologic events. We also know there is a low probability of fossilization in the first place, and of even finding a fossil.

4. Biogeography is the study of the distribution of species through time and through the environment. A theory about crustal movements helps explain some puzzling aspects of the global distribution of life in the past.

5. Comparisons of body forms among major groups may reveal evolutionary relationship. Adults and embryos of different lineages often show similarities in one or more body parts that reflect descent from a common ancestor.

6. Biochemical comparisons within and between major lineages provide strong evidence of macroevolution.

7. Biologists use taxonomy, phylogenetic reconstruction, and classification to study patterns in the diversity of life. Taxonomy is concerned with identifying and naming new species. Phylogenetic reconstruction infers evolutionary connections by certain analytical methods. Classification is concerned with organizing information about species into retrieval systems.

FOSSILS—EVIDENCE OF ANCIENT LIFE

About 500 years ago, Leonardo da Vinci was brooding about seashells entombed in rocks of northern Italy's high mountains, hundreds of kilometers from the sea. How did they get there? By the biblical explanation, floodwaters had deposited the shells in the mountains (Figure 18.3). But many shells were thin, fragile, and intact. Surely they would have been battered to bits if they had been swept across such distances and all the way up into the mountains.

Leonardo also brooded about the rocks. They were stacked like cake layers, and some contained shells but others had none. Then he remembered how large rivers swell with spring floodwaters and deposit silt in the sea. Were the layers deposited long ago, in separate intervals? If so, then shells in the mountains could be evidence of layered communities of organisms that once lived in the seas! Leonardo didn't announce his novel idea. Perhaps he suspected it would be considered heresy and invite deafening silence, imprisonment, or worse.

By the 1700s, **fossils** were accepted as the buried remains and impressions of organisms that had lived in the past. (*Fossil* comes from a Latin word for something that was "dug up.") They were still being interpreted through the prism of prevailing cultural beliefs, as when a Swiss naturalist excitedly unveiled the remains of a giant salamander and announced that they were the skeleton of a man who had drowned in the great flood. By midcentury, however, scholars began to question the interpretations.

Extensive mining, quarrying, and canal excavations were under way. The diggers were finding similar rock layers and similar fossil sequences in distant places, such as the cliffs on both sides of the English Channel. The layers appeared to record the passing of geologic time. Thus, the ordered array of fossils within the rock layers might be a historical record of life—*a fossil record*.

Fossilization

Figure 18.4 shows a few fossils. Most fossils discovered so far are bones, teeth, shells, seeds, spores, and other hard parts. Unless organisms are buried quickly, their soft parts decompose or scavengers make short work of them. Fossilized feces (coprolites) hold residues of species that were eaten in ancient environments. Also, imprints of leaves, stems, tracks, burrows, and other *trace* fossils give indirect evidence of past life.

Fossilization is a very slow process that starts when an organism, or traces of it, becomes buried in volcanic ash or sediments. Sooner or later, water infiltrates the organic remains, which become infused with dissolved metal ions and other inorganic compounds. More and more sediments gradually accumulate above the burial site. They exert ever increasing pressure on the remains. Over great spans of time, the pressure and the chemical changes transform those remains to stony hardness.

Preservation is favored when organisms are buried rapidly in the absence of oxygen. Gentle entombment by volcanic ash or anaerobic mud is best. Preservation also is favored when a burial site remains undisturbed. Most often, however, erosion and other geologic insults crush, deform, break, or scatter the fossils. This is one reason why fossils are relatively rare.

Fossils in Sedimentary Rock Layers

Stratified layers of sedimentary rock, stacked one on top of the other, are a rich source of fossils. Long ago, each layer formed by the gradual deposition of volcanic ash, silt, and other materials. Example: Sand and silt piled up when ancient rivers transported them from land to the sea, just as Leonardo thought. Over time, the sand was compressed into sandstone, the silt into shale. Deposits were not continuous. For instance, as you read in the introduction, sea levels changed when a long-term cooling trend ushered in ice ages. Deposition ceased in some places but resumed after the climate warmed and the sea level rose again.

We call the formation of sedimentary rock layers **stratification**. We can expect that the deepest layers were the first to form, and layers closest to the surface, the last. We can expect that most of the layers formed horizontally, because particles tend to settle in response to gravity. You may

Figure 18.3 From the Sistine Chapel in Italy, Michelangelo's painting of the onset of the biblical flood.

Figure 19.3 (**a**) Representation of the Earth during its formation, when the moon's orbit was much closer than it is today. If the Earth had condensed into a planet of smaller diameter, its gravitational mass would not have been great enough to hold on to an atmosphere. If it had settled into an orbit closer to the sun, water would have evaporated from its hot surface. If the Earth's orbit had been more distant from the sun, its surface would have been far colder, and water would have been locked up as ice. And without liquid water, life as we know it never would have originated on Earth.

(**b**) Stanley Miller's experimental apparatus, used to study the synthesis of organic compounds under conditions that presumably existed on the early Earth. The condenser cools circulating steam so that water droplets form.

kept the mixture circulating and exposed it to a spark discharge to simulate lightning. Amino acids and other small organic compounds formed in less than a week.

In other experiments that simulated conditions on the early Earth, glucose, ribose, deoxyribose, and other sugars formed spontaneously from formaldehyde, and adenine from hydrogen cyanide. Ribose and adenine are part of ATP, NAD$^+$, and other vital nucleotides.

However, if *complex* organic compounds had formed directly in seawater, could they have lasted long? The spontaneous direction of the reactions would have been toward hydrolysis, not condensation, in water. So how did more lasting bonds form?

By one hypothesis, clay in the rhythmically drained muck of tidal flats and estuaries served as templates (structural patterns) for the spontaneous assembly of proteins and other complex compounds. Clay consists of thin, stacked layers of aluminosilicates, and its metal ions attract amino acids. Expose clay to amino acids, warm it with the sun's rays, then alternately moisten it and dry it out. Condensation reactions will occur and form proteins and other complex organic compounds.

By another hypothesis, complex organic compounds formed spontaneously near hydrothermal vents on the seafloor, where species of archaebacteria are thriving today (Section 6.7). Experimental tests by Sidney Fox and others show that when amino acids are heated and placed in water, they spontaneously order themselves into small protein-like molecules he calls proteinoids.

However the first proteins formed, their molecular structure dictated how they could interact with other compounds. Suppose some proteins functioned as weak enzymes and hastened bond formation between amino acids. An enzyme-directed synthesis of proteins would have had selective advantage. Protein configurations that promoted such reactions would win the chemical competition for available amino acids. They would have been favored in another way—for proteins have the capacity to bind metal ions and other metabolic agents.

As you will see, the evolution of metabolism was based on such chemical modifications. For now, simply reflect on the possibility that selection was at work before cells appeared, favoring the chemical evolution of enzymes and other complex organic compounds.

Many different experiments offer indirect evidence that the complex organic molecules characteristic of life could have formed under the physical and chemical conditions that prevailed on the early Earth.

EMERGENCE OF THE FIRST LIVING CELLS

Origin of Agents of Metabolism

A defining characteristic of life is *metabolism*. The word refers to all the reactions by which cells harness energy and use it to drive their activities, such as biosynthesis. During the first 600 million years or so of Earth history, enzymes, ATP, and other organic compounds may have assembled spontaneously, perhaps in the same places. If so, the close association naturally promoted chemical interactions and the start of metabolic pathways.

Imagine an ancient estuary rich in clay deposits. It is a coastal region where seawater mixes with mineral-rich water being drained from the land. There, beneath the sun's rays, countless aggregations of organic molecules stick to the clay. At first there are quantities of an amino acid; call it *D*. All over the estuary, *D* molecules get incorporated into new proteins—until the supply of *D* dwindles. However, suppose a weakly catalytic protein is in the estuary. It can promote the formation of *D* by acting on a plentiful, simpler substance—*C*. By chance, clumps of organic molecules include that enzyme-like protein. The clumps have an advantage in the chemical competition for starting materials.

In time, *C* molecules become scarce. At that point, the advantage tilts to molecular clumps that promote formation of *C* from even simpler substances *B* and *A*. Assume *B* and *A* are carbon dioxide and water. As you know, the atmosphere and the seas have essentially unlimited amounts of carbon dioxide and water. Thus chemical selection has favored a synthetic pathway:

$$A + B \longrightarrow C \longrightarrow D$$

Finally, suppose some clumps are better than others at absorbing and using energy. Which molecular parts could bestow such an advantage? Think of the energy-trapping pathway that now dominates the world of life: photosynthesis. It starts at pigments called chlorophylls. The part of the chlorophyll molecule that absorbs light and gives up electrons is a porphyrin ring structure. Porphyrins also are present in cytochromes, which are part of electron transport systems in all photosynthetic and aerobically respiring cells. Porphyrin molecules are spontaneously assembled from formaldehyde—one of the molecular legacies of cosmic clouds (Figure 19.4). Was porphyrin a major electron transporter of some of the early metabolic pathways? Perhaps.

Origin of Self-Replicating Systems

Another defining characteristic of life is a capacity for reproduction, which now starts with protein-building instructions in DNA. Molecules of DNA are fairly stable and are easily replicated before each cell division. As you know from earlier chapters, molecules of RNA,

four formaldehyde molecules with four pyrrole rings

Figure 19.4 One hypothetical sequence by which formaldehyde, an organic compound, underwent chemical evolution into porphyrin. Formaldehyde was present when the Earth formed. Porphyrin is the light-trapping and electron-donating component of chlorophyll molecules. It also is a component of cytochrome, which is a protein component of electron transport systems that are part of many metabolic pathways.

6H

porphyrin ring system

chlorophyll a

enzymes, and other factors are necessary to translate the DNA instructions into proteins.

Most existing enzymes get assistance from small organic molecules or metal ions called coenzymes. Intriguingly, some categories of coenzymes have a structure identical to that of RNA nucleotides. Another clue: Mix together and then heat up precursors of ribonucleotides and short chains of phosphate groups, and they self-assemble into single strands of RNA. On the early Earth, energy from the sun's rays or from geothermal events might have driven the spontaneous formation of RNA from such starting materials.

Very simple self-replicating systems of RNA, enzymes, and coenzymes have been produced in some laboratories. Did RNA become the information-storing templates on which simple proteins were synthesized? Maybe. We've learned that rRNA catalyzes the translation of mRNA into proteins (Section 13.2). A ribosome is like irregularly shaped pieces of a puzzle joined together. The pieces include many proteins that stabilize the rRNA just about everywhere except at an active site where the large ribosomal subunit contacts the smaller subunit. Ribosomes, remember, are highly conserved structures. The ones in eukaryotic cells are not that different from those in prokaryotic cells. We might

Figure 19.5 Microscopic spheres of (**a**) proteins and (**b**) lipids that self-assembled under abiotic conditions.

take this to mean that rRNA's catalytic behavior is a metabolic function that emerged early in Earth history.

It is possible that an **RNA world** in which RNA was a template for protein synthesis preceded the origin of DNA. Remember, DNA and RNA nucleotides are not that different. Their sugar component differs by a single functional group. Three of their four bases are identical, and RNA's uracil differs from DNA's thymine by only a single functional group.

Then what's the point of having DNA? Compared to RNA, this double-stranded molecule can form long nucleotide chains in far more stable fashion. Computer analyses also reveal that helical coiling of a strandlike molecule is the best way to pack the most subunits in the smallest space. DNA is just a better way to *package* more and more information about building proteins.

Until we identify the chemical ancestors of RNA and DNA, the story of life's origin will be incomplete. Filling in the details will require imaginative sleuthing. For instance, researchers ran a computer program that incorporated information about natural energy sources and simple inorganic compounds of the sort thought to have been present on the early Earth. They asked their advanced computer to subject the chosen compounds to random chemical competition and natural selection as might have occurred untold billions of times in the past. They ran their program again and again. Always the outcome was the same: *Simple precursors invariably evolved—and they spontaneously organized themselves as interacting systems of large, complex molecules.*

Origin of the First Plasma Membranes

Experimental tests are more revealing of the origin of the plasma membrane—the outermost membrane of all living cells. This cellular component consists of a lipid bilayer, studded with diverse proteins that carry out different functions. Its main role is to control which substances move into and out of the cell. Without such control, cells can neither exist nor reproduce.

Figure 19.6 One possible sequence of events that led to the first self-replicating systems, then to the first living cells.

Probably, one avenue of molecular evolution led to **proto-cells**: simple membrane sacs that surrounded and protected information-storing templates and metabolic agents from the environment. We do know that simple membrane sacs can form spontaneously. For example, in one experiment, S. Fox heated amino acids until they formed protein-like chains, which he then placed in hot water. After cooling, the chains assembled into small, stable spheres (Figure 19.5*a*). Like membranes of cells, these proteinoid spheres were selectively permeable to various substances. In other experiments, the spheres picked up free lipid molecules from their surroundings, and a lipid–protein film formed at their surface.

In still other experiments, by David Deamer and his coworkers, fatty acids and glycerol combined to form long-tail lipid molecules under laboratory conditions that simulated conditions in evaporating tidepools. The lipids self-assembled into tiny, water-filled sacs, many of which were like cell membranes (Figure 19.5*b*).

In short, there are gaps in the story of life's origins. But there also is experimental evidence that chemical evolution led to the organic molecules and structures characteristic of life. Figure 19.6 summarizes likely steps in the chemical evolution that preceded the first cells.

Although the story is not yet complete, many laboratory experiments and computer simulations indirectly show that chemical and molecular evolution could have given rise to proto-cells.

ORIGIN OF PROKARYOTIC AND EUKARYOTIC CELLS

The first living cells originated in the **Archean** eon, which lasted from 3.9 billion to 2.5 billion years ago. Those cells emerged as molecular extensions of the evolving universe, of our solar system and the Earth. Maybe they originated at tidal flats or hydrothermal vents (Section 6.7). Fossils show they were like existing bacteria. Specifically, they all were **prokaryotic cells**, with no nucleus. Maybe they were only membrane-bound, self-replicating sacs of RNA, DNA, and other organic molecules. Free oxygen was absent, so they secured energy through anaerobic pathways—fermentation, probably. Energy was plentiful, for geologic processes had already enriched the seas with organic compounds. So "food" was available, there were no predators, and cell structures were free from oxygen attacks.

Hydrogen-Rich, Anaerobic Atmosphere   **Oxygen in Atmosphere: 10%**

ARCHAEBACTERIAL LINEAGE

In a second major divergence, the ancestors of archaebacteria and of eukaryotic cells start down their separate evolutionary roads.

The first major divergence gives rise to eubacteria and to the common ancestor of archaebacteria and eukaryotic cells.

ANCESTORS OF EUKARYOTES

The amount of genetic information increases; cell size increases; the cytomembrane system and the nuclear envelope evolve through modification of cell membranes.

Cyclic pathway of photosynthesis evolves in some anaerobic bacteria.

Noncyclic pathway of photosynthesis (oxygen-producing) evolves in some bacterial lineages.

chemical and molecular evolution, first into self-replicating systems, then into membranes of proto-cells by 3.8 billion years ago

ORIGIN OF PROKARYOTES

EUBACTERIAL LINEAGE

Aerobic respiration evolves in many bacterial groups.

3.8 billion years ago   3.2 billion years ago   2.5 billion years ago

Figure 19.7 Evolutionary tree of life showing possible connections among major lineages and the origins of some eukaryotic organelles.

Some populations of those first prokaryotic cells apparently diverged in two major directions shortly after the time of origin. One lineage gave rise to the **eubacteria**. The other lineage gave rise to the common ancestor of **archaebacteria** and **eukaryotic cells** (Figure 19.7).

Between 3.5 and 3.2 billion years ago, light-trapping pigments, electron transport systems, and other metabolic machinery evolved in some anaerobic eubacteria. With these innovations, the cells became the earliest practitioners of the cyclic pathway of photosynthesis. (You read about this ATP-forming pathway in Section 6.3). Sunlight, an unlimited source of energy, had been tapped. For nearly 2 billion years, photosynthetic descendants of those cells dominated the living world. Their tiny but numerous populations formed large mats in which sediments collected. The mats built up, one atop the other. Calcium deposits hardened and preserved the mats, which came to be called **stromatolites** (Figure 19.8).

Figure 19.8 Stromatolites exposed at low tide in western Australia's Shark Bay. These mounds started forming 2,000 years ago. They are structurally identical with stromatolites that formed more than 3 billion years ago.

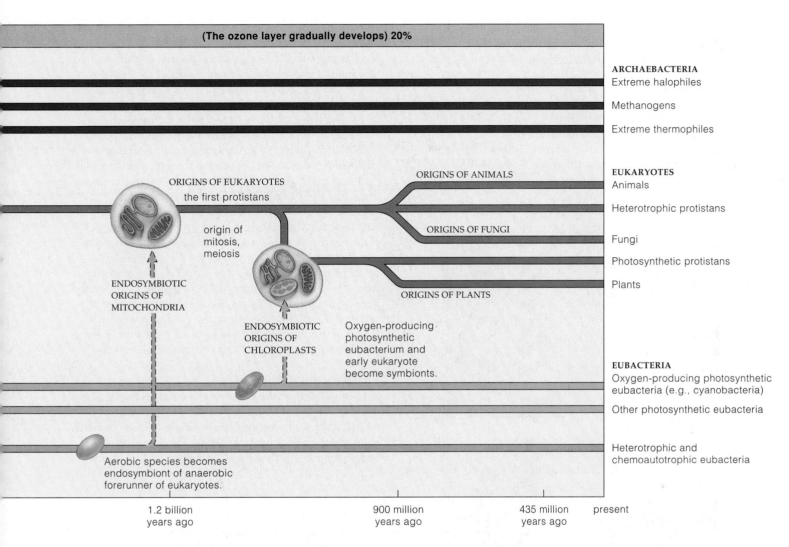

(The ozone layer gradually develops) 20%

ARCHAEBACTERIA
Extreme halophiles

Methanogens

Extreme thermophiles

EUKARYOTES
Animals

ORIGINS OF EUKARYOTES
the first protistans

ORIGINS OF ANIMALS

Heterotrophic protistans

origin of
mitosis,
meiosis

ORIGINS OF FUNGI

Fungi

Photosynthetic protistans

ENDOSYMBIOTIC
ORIGINS OF
MITOCHONDRIA

Plants

ENDOSYMBIOTIC
ORIGINS OF
CHLOROPLASTS

ORIGINS OF PLANTS

Oxygen-producing
photosynthetic
eubacterium and
early eukaryote
become symbionts.

EUBACTERIA
Oxygen-producing photosynthetic
eubacteria (e.g., cyanobacteria)

Other photosynthetic eubacteria

Heterotrophic and
chemoautotrophic eubacteria

Aerobic species becomes
endosymbiont of anaerobic
forerunner of eukaryotes.

1.2 billion
years ago

900 million
years ago

435 million
years ago

present

By the dawn of the **Proterozoic** eon, 2.5 billion years ago, photosynthetic machinery had become altered in some eubacterial species, and the noncyclic pathway of photosynthesis emerged. Oxygen, one of the pathway's by-products, started to accumulate. In time, this had two irreversible effects. First, *an oxygen-rich atmosphere stopped the further chemical origin of living cells.* Except in a few anaerobic habitats, complex organic compounds could no longer form spontaneously and resist attack. Second, *aerobic respiration became the dominant energy-releasing pathway.* In many prokaryotic lineages, selection favored metabolic equipment that neutralized oxygen by using it as an electron acceptor! This key innovation contributed to the rise of multicelled eukaryotes and their invasion of far-flung environments (Section 7.7).

Eukaryotic cells evolved in the Proterozoic, possibly before 1.2 billion years ago. We have fossils, 900 million years old, of complex algae, fungi, and plant spores. As you know, organelles are the premier defining features of eukaryotic cells. *Where did they come from?* The next section describes a few plausible hypotheses.

Rodinia, the first supercontinent, formed 1.1 billion years ago. By about 800 million years ago, stromatolites along its shorelines were declining dramatically. Were newly evolved, bacteria-eating *animals* using them as a concentrated source of food? Fossilized embryos—cell clusters no wider than a pin—give hints that the first animals were soft-bodied forerunners of the modern sponges and marine worms. Before 570 million years ago, in precambrian times, some of their descendants started the first adaptive radiation of animals.

The first living cells evolved by about 3.8 billion years ago, during the Archean eon. All were prokaryotic. Most, if not all, probably made ATP by fermentation routes.

Early on, ancestors of archaebacteria and eukaryotic cells diverged from the lineage that led to modern eubacteria.

Oxygen-releasing photosynthetic bacteria evolved. In time, the oxygen-enriched atmosphere put an end to the further spontaneous chemical evolution of life. That atmosphere was a key selection pressure in the evolution of eukaryotic cells.

Where Did Organelles Come From?

Thanks to Andrew Knoll, William Schopf, Jr., and other globe-hopping microfossil hunters, we have tantalizing evidence of early life. One fossil treasure is a strand of bacterial cells 3.5 billion years old; it was around not long after life originated. Others, from the Proterozoic, were eukaryotic cells with a few membrane-bound organelles in the cytoplasm (Figure 19.9). Most of their descendants have a profusion of organelles (Figure 19.10).

Where did eukaryotic organelles come from? Speculations abound. Some probably evolved by gene mutations and natural selection. For others, researchers make a case for evolution by way of endosymbiosis.

ORIGIN OF THE NUCLEUS AND ER Prokaryotic cells do not have an abundance of organelles. But some species have interesting infoldings of the plasma membrane that incorporate enzymes and other agents of metabolism (Figure 19.11). In early forerunners of eukaryotic cells, similar infoldings into the cytoplasm may have served as passageways to the surface. They may have evolved into ER channels and into an envelope around the DNA.

What would be the advantage of such membranous enclosures? Maybe they protected the genes and protein products from "foreigners." Remember, bacterial species often transfer plasmid DNA among themselves. So do very simple eukaryotic cells called yeasts. A nuclear envelope may have been favored because it got the cell's genes, replication enzymes, and transcription enzymes out of the cytoplasm. It meant vital genetic messages could be copied and read, free of metabolic competition from a potentially unmanageable hodgepodge of foreign genes. Similarly, ER channels might have protected the cell's vital proteins and other organic compounds from metabolically hungry "guests"—foreign cells that became permanent residents in a host cell, as described next.

A THEORY OF ENDOSYMBIOSIS Accidental partnerships between a variety of prokaryotic species probably arose countless times in lineages that gave rise to eukaryotic cells. Maybe some resulted in the origin of mitochondria, chloroplasts, and other organelles. This is a story of endosymbiosis, as developed in greatest detail by Lynn

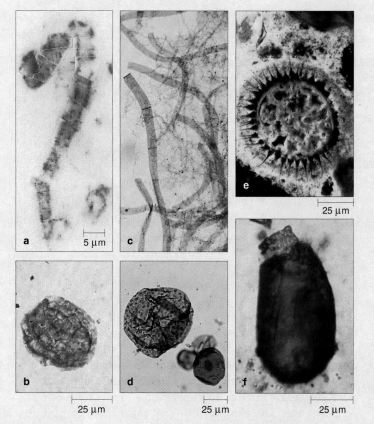

Figure 19.9 From Australia, (**a**) a strand of walled prokaryotic cells 3.5 billion years old and (**b**) one of the oldest known eukaryotes— a protistan 1.4 billion years old. From Siberia, (**c**) a multicelled alga 900 million to 1 billion years old and (**d**) eukaryotic microplankton 850 million years old. (**e**) From China, a eukaryotic cell that lived about 560 million years ago. (**f**) From Spitsbergen, Norway, a protistan that was alive 50 million years before the dawn of the Cambrian.

Figure 19.10
A fine example of the profusion of diverse organelles that are hallmarks of eukaryotic cells: *Euglena*, a single-celled protistan, sliced lengthwise for this micrograph. It also has a long flagellum, which could not fit in the image area at this magnification.

Figure 19.11 (**a**) Model for the origin of endoplasmic reticulum and the nuclear envelope. In the prokaryotic ancestors of eukaryotic cells, infoldings of the plasma membrane gave rise to both cell components. (**b**) Such infoldings are present in the cytoplasm of many kinds of existing bacteria, including *Nitrobacter*, sketched here in cutaway view.

DNA

infolding of plasma membrane

a

b

Margulis. *Endo*– means within; *symbiosis* means living together. In cases of **endosymbiosis**, one species (the guest) spends its entire life cycle *inside* another species (the host). The interaction benefits one or both.

According to one hypothesis, eukaryotic cells evolved by endosymbiosis long after the noncyclic pathway of photosynthesis emerged and oxygen had accumulated to significant levels in the atmosphere. In some groups of bacteria, one type of electron transport system in the plasma membrane had evolved and now incorporated "extra" cytochromes—which donated electrons to oxygen. With this key innovation, such bacteria started extracting energy from organic compounds by aerobic respiration.

Possibly before 1.2 billion years ago, the forerunners of eukaryotes were engulfing aerobic bacteria. Maybe they resembled existing soft-bodied, amoebalike cells that weakly tolerate free oxygen. Maybe they trapped food by sending out cytoplasmic extensions and used endocytic vesicles to get food into the cytoplasm for digestion.

A key point of the theory is that some aerobic bacteria resisted digestion and thrived in the protected, nutrient-rich environment. In time, they were releasing extra ATP. Their hosts came to depend on that ATP for their growth, increased activity, and assembly of hard parts and other structures. The guests benefited, too. They no longer had to seek food or build components for metabolic work the hosts did for them. In time the anaerobic and aerobic cells were incapable of independent life. The guests had evolved into mitochondria, supreme suppliers of ATP.

EVIDENCE OF ENDOSYMBIOSIS Strong evidence favors Margulis's theory. There are plenty of examples of nature continuing to tinker with endosymbionts, including the cell in Figure 19.12. Its mitochondria are like bacteria in size and structure. The inner mitochondrial membrane is like a bacterial plasma membrane. Each mitochondrion replicates its own DNA and divides independently of the host cell's division. A few genetic code words in its DNA and mRNA have unique meanings. They are translated into a few proteins required for special mitochondrial tasks. Compared to the genetic code of living cells, the "mitochondrial code" has a few distinct differences.

Chloroplasts, too, may be stripped-down descendants of oxygen-evolving, photosynthetic bacteria. Perhaps predatory aerobic bacteria engulfed photosynthetic cells. Maybe the cells escaped digestion, absorbed nutrients from the host's cytoplasm, and continued to function. By providing aerobically respiring hosts with oxygen, they would have promoted endosymbiosis.

In their metabolism and overall nucleic acid sequence, chloroplasts resemble some eubacteria. Their DNA self-replicates, and they divide independently of the cell's division. Chloroplasts vary in shape and light-absorbing pigments, just as photosynthetic eubacteria do. Did they originate a number of times in different lineages? Maybe.

Adding to the intrigue are ciliated protistans and marine slugs that "enslave" chloroplasts! The slugs eat algae but retain algal chloroplasts in their gut cells. The chloroplasts draw nutrients from host tissues. They photosynthesize and release oxygen for weeks.

However they arose, new kinds of cells did appear on the evolutionary stage. They had cytomembranes, a nucleus, and mitochondria, chloroplasts, or both. They were eukaryotic cells, the first **protistans**. And with their efficient metabolic strategies, early protistans underwent rapid divergences and adaptive radiations. In no time at all, evolutionarily speaking, some of their descendants gave rise to the great kingdoms of plants, fungi, and animals, as sketched out earlier in Figure 19.7.

cyanobacterium-like structure

mitochondrion

nucleus

Figure 19.12 *Cyanophora paradoxa*, a protistan. Its mitochondria resemble bacteria. Its photosynthetic structures look like spherical cyanobacteria (which are photosynthetic) without the cell wall.

LIFE IN THE PALEOZOIC ERA

We divide the **Paleozoic** into the Cambrian, Ordovician, Silurian, Devonian, Carboniferous, and Permian periods. The supercontinent Rodinia had split apart before this era. Some of its fragments now straddled the equator, and warm, shallow seas lapped their margins. Global conditions restricted any pronounced seasonal changes in winds, ocean currents, and the upward churning of nutrient deposits on the seafloor. Nutrient supplies in equatorial seas were stable but limited. Polar regions were inhospitably icy; and to the south, other fragments had collided to form a new supercontinent, Gondwana:

GONDWANA

Most major animal phyla had evolved earlier in the precambrian seas. Possibly some of their ancestors were among the **Ediacarans**, peculiar organisms shaped like fronds, disks, and blobs that nearly defy classification (Figure 19.13a,b). Like Ediacarans, the early Cambrian animals had highly flattened bodies with a fine surface-to-volume ratio for taking up nutrients (Figure 19.13c). Most thrived on or in seafloor sediments, where dead organisms and organic debris settled. They ranged from sponges to simple vertebrates, and they were diverse.

This was a time when new predators and prey with armorlike shells, spines, mouths, and amazing feeding structures rapidly evolved. We find fossilized animals with punctures, missing chunks, and healed wounds in sedimentary beds. Things were starting to get lively!

What caused the "Cambrian explosion" of diversity? Were asteroids an impetus? After tapering off about 3 billion years ago, the impact rate apparently increased abruptly around this time. Were genes governing early growth and development less intertwined than they are today? If so, selection against novel traits might have been less severe. What about the warm, shallow waters of the equatorial seas? They must have been adaptive zones, with novel opportunities for getting food.

Early in the Ordovician, widespread flooding of the land masses gave rise to broad, tropical seaways. The emergence of the vast marine environments promoted adaptive radiations. Many new reef organisms, such as the swift, shelled predators called nautiloids, evolved. Among their descendants are the chambered nautiluses (Section 25.3). Trilobites, one of the most common but unswift invertebrates, nearly vanished (Figure 19.13d).

Huge glaciers formed across the southern portion of Gondwana at the end of the Ordovician—one of the coldest times in Earth history. Enormous volumes of water were locked up in ice sheets, and this drained shallow seas everywhere. More than 70 percent of all marine groups—including the reef builders, trilobites, placoderms, and jawless fishes—became extinct at the Ordovician–Silurian boundary.

Gondwana drifted north during Silurian times. The sea level rose as great ice sheets melted, and adaptive radiations occurred in the seas. Large coral reefs formed, algae flourished, and the first jawed fishes appeared. Small plants took hold in wet lowlands (Figure 19.14b,c).

Figure 19.13 Representatives from the precambrian and Cambrian seas. Two Ediacarans, about 600 million years old: (**a**) *Spriggina* and (**b**) *Dickensonia*. The oldest known Ediacarans lived 610 million years ago and the most recent in Cambrian times, 510 million years ago. (**c**) From British Columbia's Burgess Shale, a fossilized marine worm. (**d**) A beautifully preserved fossil of one of the earliest trilobites.

Figure 19.14 (**a**) Life in the Silurian seas. Some of these shelled animals (nautiloids) were as long as a kayak. (**b**) At the close of the Silurian and on into the Devonian, forerunners of modern ferns and club mosses dominated swamps and marshes. (**c**) Fossils of a Devonian plant (*Psilophyton*) that may have been one of the earliest ancestors of the conifers and other seed-bearing plants. (**d**) Reptiles (*Dimetrodon*) of a hotter, drier time, the Permian. Some fossils of these carnivores have been found in Texas. Giant club mosses and horsetails had declined. Conifers, cycads, and other gymnosperms replaced them during the Permian.

5 cm

So did fungi and invertebrates, such as spiders, insects, and segmented worms. The Devonian was a pivotal time in vertebrate evolution. Armor-plated fishes with massive jaws diversified in the seas and in fresh water. Fishes that gave rise to amphibians invaded the land. They had lobed fins—the forerunners of legs and other limbs—and simple lungs. Lobed fins and lungs were key innovations that would prove to be advantageous for life out of water, in dry land habitats (Section 24.5).

Warm-water species especially were hit hard by a Devonian mass extinction. Global cooling, a decline in sea level, a meteorite impact, or some combination of these might have been the cause. Afterward, plants and insects embarked on adaptive radiations on land.

In the Carboniferous, land masses were submerged and drained many times. Organic debris accumulated, became compacted, and was converted to coal (Section 22.2). Insects, amphibians, and early reptiles flourished in vast swamp forests of the Permian (Figure 19.14*d*). The early ancestors of cycads, ginkgos, conifers, and other seed-producing plants dominated the forests.

Late in the Permian, 95 percent of all known species were lost in the greatest of all mass extinctions. New evidence suggests that "The Great Dying" followed a direct hit by comet. Whatever the trigger, there was a colossal outpouring of lava, a severe disruption of the global carbon cycle, and a major glaciation.

Pangea had already been forming long before then. By the time the Permian drew to a close, it extended from pole to pole. One world ocean, including the vast Tethys Sea, lapped its margins:

PANGEA ———

——— TETHYS SEA

And just as the Paleozoic mass extinctions had done, the new distribution of oceans, land masses, and land elevations redirected the course of life's evolution.

Early in the Paleozoic era, organisms of all six kingdoms were flourishing in the seas. By the end of the era, many lineages had successfully invaded the land, including the wet lowlands of the new supercontinent Pangea.

LIFE IN THE MESOZOIC ERA

Speciation on a Grand Scale

We divide the **Mesozoic** into the Triassic, Jurassic, and Cretaceous periods. It lasted about 175 million years. In the middle of the Jurassic, the supercontinent Pangea started to break up. Its fragments slowly drifted apart. The fossil record tells us those geographically isolated fragments favored divergences and speciation:

This was an era of spectacular expansion in the range of global diversity. Invertebrates and fishes underwent adaptive radiations in the seas. Conifers and other seed-bearing **gymnosperms,** insects, and reptiles became the visibly dominant lineages on land. Flowering plants—**angiosperms**—emerged during the late Jurassic or the early Cretaceous. In less than 40 million years, they would displace the conifers and related plants in most environments (Figure 19.15 and Section 22.8).

Rise of the Ruling Reptiles

Early in the Triassic, the first **dinosaurs** evolved from a reptilian lineage. They were not much bigger than a turkey. Possibly most species had high metabolic rates, and maybe they were warm-blooded. Many sprinted about on two legs. But the Triassic dinosaurs weren't the dominant land animals. The rulers were *Lystrosaurus* and other plant-eating, mammal-like reptiles that were too big to be bothered by most predators (Figure 9.16).

Adaptive zones opened up for the dinosaurs after a catastrophe that left a string of five craters in what is now France, Quebec, Manitoba, and North Dakota. One crater is as big as Rhode Island. About 214 million years ago, fragments from an asteroid or a comet fell in a row as the Earth spun beneath them, just as a string of huge fragments from comet Shoemaker–Levy 9 hit Jupiter in 1994. The blast waves, global firestorm, lava flows, and earthquakes must have been stupendous. Most animals lucky enough to survive this time of mass extinction were smaller, had higher rates of metabolism, and were less vulnerable than others to drastic changes in outside temperatures. These same attributes also characterized survivors of later mass extinctions.

Descendants of the surviving dinosaurs became the ruling reptiles; they endured for 140 million years. Some species reached monstrous proportions. Among them were the ultrasaurs, taller than a four-story building.

Figure 19.15 Range of diversity among vascular plants of the Jurassic and Cretaceous. Conifers and other gymnosperms were dominant. They started to decline even before flowering plants began a great adaptive radiation that continued into the Cenozoic. *Upper left:* From the Cretaceous, a floral shoot of *Archaeanthus linnenbergeri*. In many traits, this now-extinct flowering plant resembled living magnolias.

Many dinosaurs perished in another mass extinction at the end of the Jurassic, then in a pulse of extinctions during the Cretaceous. Huge outpourings of molten rock may have triggered changes in the global climate. Or maybe asteroids or comets inflicted the blows. Not all lineages survived. Yet some did recover, and new ones evolved. Duckbilled dinosaurs appeared in forests and swamps. Tanklike *Triceratops* and other plant eaters flourished in open habitats. They were prey for the agile and swift *Velociraptor* of motion picture fame.

About 120 million years ago, global temperatures skyrocketed 25 degrees. By one theory, plumes spread out beneath the crust and "greased" the crustal plates into moving twice as fast. A superplume or a rash of them broke through the crust. In what is now the South Atlantic, the crust opened like a zipper. Basalt and lava poured from fissures; volcanoes spewed nutrient-rich ashes. The plumes released great quantities of carbon dioxide, one of the key "greenhouse" gases that absorb some heat radiating from the Earth before it escapes to outer space. The nutrient-enriched planet warmed, and

Figure 19.16 Life in the Mesozoic. (**a**) *Lystrosaurus*, a therapsid (mammal-like reptile) of the Triassic. This tusked herbivore, about 1 meter (three feet) long, fed on fibrous plants in dry floodplains. As Section 18.3 describes, its fossils are distributed on widely separated continents and are evidence in favor of the plate tectonics theory.

(**b**) *Temnodontosaurus*, an ichthyosaur that hunted large squid, ammonites, and other prey in the warm, shallow seaways of the early Jurassic. Fossils meauring 9 meters (30 feet) long have been found in England and Germany. (**c**) From the Cretaceous, cycads, ferns, *A. linnenbergeri* and other flowering plants, and *Deinonychus*, an agile dinosaur with keen eyesight, serrated teeth, and powerful jaws.The short forelimbs of this bipedal carnivore had sharp claws, including a large one shaped like a sickle; hence its name, which means "terrible claw." Fossils 3 meters (nine feet) long have been found in Texas.

it remained warm for 20 million years. Photosynthetic organisms flourished on land and in the shallow seas. Their remains were slowly buried and converted into the world's vast oil reserves.

About 65 million years ago, the last dinosaurs and many marine organisms vanished in a mass extinction. As described in the next section, their disappearance apparently coincided with a direct hit by an asteroid the size of Mount Everest. Imperceptibly over time, the impact site slowly drifted into a position that we now call the northern Yucatán peninsula.

The Mesozoic was a time of major adaptive radiations and of a mass extinction in which the last dinosaurs and many marine organisms disappeared.

Horrendous End to Dominance

It's only been about 55,000 years since the first modern humans walked the Earth. Since then, how many times have people, puffed up with self-importance, set out to conquer neighbors, lands, and seas? Think about it. Then think about the dinosaurs. Were they good at reigning supreme? No question about it; their lineage dominated the land for 140 million years. In the end, did it matter? Not a bit. Sixty-five million years ago, at the Cretaceous–Tertiary (K–T) boundary, nearly all members of their most excellent lineage perished. Why? Bad luck.

A thin layer of iridium-rich rock distributed around the world dates precisely to the K–T boundary. Iridium is rare here but not in asteroids. **Asteroids** are rocky, metallic bodies, 1,000 kilometers to a few meters across, that are hurtling through space. When planets of our solar system were forming, their gravitational pull swept most asteroids from the sky. At least 6,000 still orbit the sun in a belt between Mars and Jupiter (Figure 19.17). The orbits of dozens of others take them across Earth's orbit, like Russian roulette on a cosmic scale.

By analyzing iridium levels in soils, gravity maps, and other evidence, Walter Alvarez and Luis Alvarez hypothesized that an asteroid impact caused the K–T mass extinction. Researchers later identified the impact site. Massive movements in the crust transported it to what is now the northern Yucatán peninsula of Mexico (Figure 19.18). The impact crater is 9.6 kilometers deep and 300 kilometers across—wider than Connecticut. This crater as well as other evidence strongly supports what is now called the **K–T asteroid impact theory**.

To make a crater that big, the asteroid had to hit the Earth at 160,000 kilometers per hour. It blasted at least 200,000 cubic kilometers of dense gases and debris into the sky. The crust heaved violently. Monstrous waves 120 meters high raced across the ocean. They obliterated life on islands, then slammed into coasts of continents. Researchers hypothesized that the atmospheric debris kept out sunlight for months. If so, plants and other

Figure 19.18 Artist's interpretation of what might have happened in the last few minutes of the Cretaceous.

producers would have withered and died. Animals and other consumers would have starved to death.

The hypothesis had problems. By some calculations, the volume of debris blasted aloft would not have been large enough to have had such consequences.

Then comet Shoemaker–Levy 9 slammed into Jupiter. Particles blasted into the Jovian atmosphere triggered an intense heating of an area larger than the Earth. That event supports a **global broiling hypothesis**, proposed first by H. J. Melosh and his colleagues. Briefly stated, energy released at the K–T impact site was equivalent to detonating 100 million nuclear bombs. Trillions of tons of vaporized debris rose in a colossal fireball, then rapidly condensed into particles the size of sand grains. Seconds later, a cooler fireball made of steam, carbon dioxide, and unmelted rock formed. When the debris fell to the Earth, it raised the atmosphere's temperature by thousands of degrees. The sky must have been ten times hotter than it gets above Death Valley in summer. In one horrific glowing hour, nearly all plants erupted in flames and every animal out in the open was broiled alive.

Things haven't settled down much since dinosaurs disappeared. For instance, about 2.3 million years ago, a huge object from outer space hit the Pacific Ocean. At about the same time, vast ice sheets started to form abruptly in the Northern Hemisphere. Long-term shifts in climate may have been ushering in that most recent ice age, but the global impact would have accelerated it. Water vaporized by the impact could have contributed to a global cloud cover that limited the sunlight reaching the surface. Ancestors of humans were around when this happened. The extreme shift in climate surely put their adaptability to the test.

Almost certainly, severe environmental tests await all existing lineages. When we become too smitten with our importance in the world of life, we might do well to step back, from time to time, and reflect on what is going on above and beneath the Earth's surface. Asteroids and superplumes do have a way of leveling the playing field, as they did for gigantic dinosaurs and tiny mammals.

Figure 19.17 What one of the asteroids looks like. This is a small asteroid; it would extend halfway between Baltimore and Washington, D.C.

LIFE IN THE CENOZOIC ERA

The breakup of Pangea triggered events that continued into the present era, the **Cenozoic**. At the dawn of the Cenozoic, major land masses were on collision courses:

Coastlines fractured. The Cascades, Andes, Himalayas, and Alps rose through volcanic activity, uplifting, and other events at crustal rifts and plate boundaries. These geologic changes brought about major shifts in climate that influenced the further evolution of life.

During the Paleocene epoch, climates were warmer and wetter. Tropical and subtropical forests extended farther north and south than they do today. Woodlands and forests spread even into polar regions. Although their key traits evolved before dinosaurs left the scene, mammals now began a major adaptive radiation.

The climate warmed even more during the Eocene epoch. Subtropical forests extended north into polar regions. A variety of mammals, including primates, bats, rhinos, hippos, elephants, horses, and carnivores, emerged. By the late Eocene, the climate grew more cool and dry, and the seasons became pronounced. Woodlands and dry grasslands dominated vast land regions. The patterns of vegetational growth changed so greatly that many mammals became extinct.

From the Oligocene through the Pliocene, the woodlands and grasslands were home to many grazing and browsing animals. Among them were camels and the "giraffe rhinoceros," along with some fearsome carnivores that stalked them (Figure 19.19).

Currently, the distribution of land masses favors biodiversity. The richest ecosystems are tropical forests of South America, Madagascar, and Southeast Asia, and the tropical marine ecosystems of the Pacific archipelagos. Yet we are in the midst of a new mass extinction. About 50,000 years ago, nomadic humans followed the migrating herds of wild animals around the Northern Hemisphere. By a few thousand years later, major groups of mammals were extinct. The pace of extinction has since accelerated as humans hunt

Figure 19.19 Cenozoic species. (**a**) During the early Paleocene, in what is now Wyoming, mammals lived in dense forests of sequoia and other plants. On the ground, raccoonlike *Chriacus* faces a tree-climbing rodent (*Ptilodus*). Higher up is *Peradectes*, a marsupial. (**b**) *Indricotherium*, the "giraffe rhinoceros." At 15 tons and 5.5 meters tall at the shoulder, it was the largest mammal known. It browsed in late Eocene and early Miocene woodlands. (**c**) Saber-tooth cats (*Smilodon*) and small horses from the Pleistocene. Fossils have been recovered from pitch pools in Rancho LaBrea, California.

for food, fur, feathers, or fun, and as natural habitats are converted for farm animals and crops. We'll revisit the current extinction crisis in Chapters 25 and 45.

Crustal movements and mountain building during the Cenozoic triggered major shifts in climate. Conditions favored a great adaptive radiation of mammals, first in tropical forests, then in woodlands and grasslands.

fragments of Rodinia, first supercontinent

Gondwana

Cambrian. Explosion of diversity in equatorial seas; icy polar regions

Silurian. Sea level rises, diverse marine life; plants, invertebrates invade land

SUMMARY

Gold indicates text section

1. The evolutionary story of life begins with the "big bang," a model for the origin of the universe. By this model, all matter and space were once compressed in a fleeting state of incredible heat and density. Time began with the near-instantaneous distribution of matter and energy through the known universe, which has been expanding ever since. Helium and other light elements formed just after the big bang. Heavier elements were created by the formation, evolution, and death of stars. All elements of the solar system, the Earth, and life are products of this physical and chemical evolution. *CI*

2. Four billion years ago, the Earth had a high-density core, an intermediate-density mantle, and a thin, highly unstable crust of low-density rocks. Probably gaseous hydrogen, nitrogen, carbon monoxide, and carbon dioxide made up most of the first atmosphere. Under prevailing conditions, neither free oxygen nor liquid water could accumulate at the Earth's surface. *19.1*

3. For hundreds of millions of years after the crust cooled, runoff from rains put dissolved mineral salts and other compounds in crustal depressions, where the early seas formed. Life as we know it would not have originated without that salty liquid water. *19.1*

4. Many studies and experiments have yielded indirect evidence that life originated under the conditions that existed on the early Earth. *19.1, 19.2*

   a. The composition of cosmic clouds and rocks from planets and the Earth's moon indicate that precursors of complex molecules associated with life were present. In laboratory tests simulating the primordial conditions, including the absence of free oxygen, these precursors spontaneously assembled into sugars (such as glucose), amino acids, and other organic compounds. *19.1*

   b. Chemical principles and computer simulations suggest that metabolic pathways evolved by chemical competition for limited supplies of organic molecules that had accumulated by natural geologic processes in the seas. In the laboratory, self-replicating systems of RNA, enzymes, and coenzymes have been synthesized. Double-stranded, helically coiled DNA probably evolved as the most efficient way of packing the most protein-building instructions in the smallest space. Also, its long nucleotide chains are more stable than RNA. *19.2*

   c. In laboratory simulations of conditions thought to have prevailed on the early Earth, lipids as well as lipid–protein membranes having some of the properties of cell membranes have formed spontaneously. *19.2*

5. Since its origin about 3.8 billion years ago, life has been influenced by changes in the Earth's atmosphere, crust, and oceans. Forces of change included asteroid impacts, volcanism, and crustal plate movements. The forces altered the distribution of land masses and the seas (Figure 19.20) as well as global and local climates.

Another major force of change: activities of organisms, especially the oxygen-releasing photosynthesizers and, far more recently, the human species. *19.2–19.8*

6. The history of life spans the following great intervals (here you may wish to review Section 18.2): *19.2–19.8*
   a. Archean: 3.9 billion to 2.5 billion years ago
   b. Proterozoic: 2.5 billion to 570 million years ago
   c. Paleozoic: 570 to 240 million years ago
   d. Mesozoic: 240 to 65 million years ago
   e. Cenozoic: 65 million years ago to the present

7. The first living cells arose in the Archean and were prokaryotic. The first divergence led to eubacteria and to the ancestor of archaebacteria and eukaryotes. Some eubacteria used a cyclic photosynthesis pathway. *19.3*

8. A noncyclic photosynthetic pathway evolved in the Proterozoic in some eubacterial lineages. Oxygen, the pathway's by-product, accumulated in the atmosphere; it blocked the further spontaneous formation of organic molecules. The spontaneous origin of life was over. The abundance of free oxygen was a selective pressure that favored aerobic respiration, a key step in the origin of eukaryotes. It helped form an ozone layer. This shield against ultraviolet radiation allowed some lineages to move from the seas into low wetlands. *19.3, 19.5*

9. Mitochondria and chloroplasts probably evolved by endosymbiosis between aerobic bacteria and anaerobic bacteria that were the forerunners of eukaryotes. *19.4*

10. By the early Paleozoic, diverse organisms of all six lineages were established in the seas. Before it ended, land was invaded. Ever since, there have been pulses of mass extinctions and adaptive radiations. *19.5–19.8*

11. Dinosaurs were the dominant land animals for 140 million years. Their lineage ended at the Cretaceous–Tertiary (K–T) boundary when an asteroid hit the Earth. In the Cenozoic, mammals began a great radiation into adaptive zones vacated by the dinosaurs. *19.7–19.8*

Review Questions

1. Compare chemical and physical conditions that are thought to have prevailed on the Earth 4 billion years ago with current conditions. *19.1*

2. Describe some experimental evidence for the (a) spontaneous origin of large organic molecules, (b) self-assembly of proteins, and (c) formation of organic membranes and spheres. *19.1, 19.2*

3. Summarize the key points of the theory of endosymbiotic origins for mitochondria and chloroplasts. Cite evidence that favors this theory. *19.4*

Devonian. Jawed fishes evolve, diversify; ancestors of amphibians invade land

Permian into Triassic. Swamp forests (coal source); seed plants evolve

Cretaceous into Tertiary. Extinction of dinosaurs; rise of mammals

Middle Miocene, polar regions again iced over, as in Cambrian

4. Describe the prevailing conditions that probably favored the Cambrian "explosion" of diversity among marine animals, as evidenced by the fossil record. *19.5*

5. When did plants, fungi, and insects invade the land? What kind of vertebrates first invaded the land, and when? *19.5*

6. What were global conditions like when the gymnosperms and dinosaurs originated? *19.6*

7. Briefly explain how an asteroid impact and "global broiling" may have caused the mass extinction at the K–T boundary. *19.7*

8. Would you expect the Paleozoic, Mesozoic, or Cenozoic to be called "the age of mammals"? As part of your answer, explain the differences between global conditions in each era. *19.8*

Self-Quiz   ANSWERS IN APPENDIX III

1. Through study of the geologic record, we know that the evolution of life has been profoundly influenced by _____ .
   a. tectonic movements of the Earth's crust
   b. bombardment of the Earth by celestial objects
   c. profound shifts in land masses, shorelines, and oceans
   d. physical and chemical evolution of the Earth
   e. all of the above

2. _____ was the first to obtain indirect evidence that organic molecules could have been formed on the early Earth.
   a. Darwin   b. Miller   c. Fox   d. Margulis

3. An abundance of _____ was conspicuously absent from the Earth's atmosphere 4 billion years ago.
   a. hydrogen   c. carbon monoxide
   b. nitrogen   d. free oxygen

4. Life as we know it originated by _____ .
   a. 4.6 billion years ago   c. 3.8 billion years ago
   b. 2.8 million years ago   d. 3.8 million years ago

5. Which of the following statements is false?
   a. The first living cells were prokaryotic.
   b. The cyclic pathway of photosynthesis first appeared in some eubacterial species.
   c. Oxygen began accumulating in the atmosphere after the noncyclic pathway of photosynthesis evolved.
   d. In the Proterozoic, the rise in atmospheric oxygen enhanced the spontaneous formation of organic molecules.
   e. All are correct.

6. The first eukaryotic cells emerged during the _____ .
   a. Paleozoic   c. Archean   e. Cenozoic
   b. Mesozoic   d. Proterozoic

7. Match the geologic time interval with the events listed.
   ____ Archean   a. major radiations of dinosaurs, origin
   ____ Proterozoic   of flowering plants and mammals
   ____ Paleozoic   b. chemical evolution, origin of life
   ____ Mesozoic   c. radiations of flowering plants, insects,
   ____ Cenozoic   birds, mammals; humans emerge
                      d. free oxygen present; origin of aerobic
                         metabolism, protistans, fungi, animals
                      e. rise of early plants on land, origin
                         of amphibians, and origin of reptiles

Figure 19.20 Summary of changes in the distribution of land masses and the seas that influenced life's evolution.

Critical Thinking

1. Briefly explain, in terms of hydrophilic and hydrophobic interactions, how proto-cells might have formed in water from aggregations of lipids, proteins, and nucleic acids.

2. Reflect on Figure 19.20 and on this: The Atlantic Ocean is widening, and the Pacific Ocean and Indian Ocean are closing in on each other. Many millions of years from now, continents will collide and form a second Pangea. Write a short essay on what conditions might be like on that future supercontinent and what types of species might survive on it. (*Hint:* Check out the Paleomap Project animations on the Internet.)

3. According to one estimate, there is a chance that about 10^{20} planets have formed in the universe that are capable of sustaining life—but there is only one chance at intelligent life per planet. Given your knowledge of molecular biology and evolutionary processes, do you find this estimate plausible? If so, speculate on why the odds are so low.

4. We know of a number of large asteroids that may intersect Earth's orbit in the distant future. There probably are a number we don't know about. Would you use this as an excuse not to worry about polluting the environment, not to take care of your physical health (as by avoiding drugs), and not to care about our cultural evolution? Why or why not?

Selected Key Terms

angiosperm *19.6*	global broiling hypothesis *19.7*
archaebacterium *19.3*	gymnosperm *19.6*
Archean *19.3*	K–T asteroid impact theory *19.7*
asteroid *19.7*	mantle, of Earth *19.1*
big bang *CI*	Mesozoic *19.6*
Cenozoic *19.8*	Paleozoic *19.5*
crust, of Earth *19.1*	prokaryotic cell *19.3*
dinosaur *19.6*	Proterozoic *19.3*
Ediacaran *19.5*	protistan *19.4*
endosymbiosis theory *19.4*	proto-cell *19.2*
eubacterium *19.3*	RNA world *19.2*
eukaryotic cell *19.3*	stromatolite *19.3*

Readings

de Duve, C. April 1996. "The Birth of Complex Cells." *Scientific American* 274(4): 50–57.

Dott, R., Jr., and R. Batten. 1994. *Evolution of the Earth.* Fifth edition. New York: McGraw-Hill.

Impey, C., and W. Hartman. 2000. *The Universe Revealed.* Belmont, California: Wadsworth. Paperback.

Wright, K. March 1997. "When Life Was Odd." *Discover* 18(3): 52–61. Update on the puzzling Ediacarans.

On-Line readings at Student Guide for InfoTrac: www.brookscole.com/biology

PROKARYOTES, VIRUSES, AND PROTISTANS

The Mostly Microscopic Multitudes

Did a friend ever mention that you are nearly 1/1,000 of a mile tall? Probably not. What would be the point of measuring people in units as big as miles? Even so, we think that way, in reverse, whenever we measure microorganisms. For the most part, **microorganisms** are prokaryotic cells and single-celled protistans too small to be seen without the aid of a microscope.

The bacterial cells in Figure 20.1 are a case in point. To measure them, you would have to divide one meter into a thousand units, or millimeters. Next, you would have to divide one of the millimeters into a thousand smaller units, or micrometers. To give you a sense of how small that is, a single millimeter would be about as small as the dot of this "i." And a *thousand* bacteria would fit side by side on top of the dot!

To be sure, viruses are smaller still, as you might deduce after looking at Figure 20.1. We measure them in nanometers (billionths of a meter). But viruses are not alive. We consider them in this chapter because one kind or another infects just about every living thing.

Prokaryotic cells are the smallest organisms, but they vastly outnumber the individuals in all other kingdoms combined. Their reproductive potential is staggering.

Under ideal conditions, cells of some species divide about every twenty minutes. If that reproduction rate were to hold constant, a single bacterium would have nearly a billion descendants in ten hours! Similarly, huge numbers of single-celled protistans reproduce rapidly in aquatic habitats. Among them are a variety of producers of food webs, helpful symbionts, herbivores and other predators, and parasites (Figure 20.2).

So why don't the unseen multitudes take over the world? Sooner or later, their burgeoning populations use up available nutrients and pollute the surroundings with their own metabolic wastes. They alter the very conditions that initially favored their reproduction. Besides, certain kinds of viruses attack just about every type of bacterium and help keep their population sizes in check. So do seasonal changes in living conditions.

Of course, you probably do not find much comfort in this when you serve as host for one of the pathogenic types. **Pathogens** are infectious, disease-causing agents that invade target organisms and multiply inside or on them. Disease follows when the metabolic activities of pathogenic cells damage body tissues and interfere with their normal functioning.

100 μm · 20 μm · bacteriophage (virus particle) · 225 nm

new virus particles leaving ruptured cell

bacterial cell wall

Figure 20.1 (**a**, **b**) How small are prokaryotic cells? Shown here, *Bacillus* cells on the tip of a pin. (**c**) How small are viruses? Bacteriophage particles are one example. They attacked this bacterial cell, which ruptured open.

Figure 20.2 How diverse are protistans? Shown here, (**a**) *Trypanosoma brucei*, a parasitic, animal-like flagellate that causes African sleeping sickness and (**b**) a predatory foraminiferan that has golden algae living on its spines. Both are single-celled protistans. (**c**) Kelp (*Macrocystis*). Populations of this plantlike, multicelled species form underwater forests as tall as those on land.

Certain pathogens can indeed make you suffer, but they shouldn't give all prokaryotes and protistans a bad name. Think of how single-celled photoautotrophs in the seas provide food and oxygen for entire communities and even affect the global cycling of carbon (Section 6.8). Think of kelp forests, the basis of marine ecosystems (Figure 20.2c). Or think of bacterial species that feed on organic debris. Together with other decomposers, they help cycle the nutrients that sustain entire communities.

From the human perspective, bacteria and protistans are good or bad, or even dangerous. Basically, however, they are simply surviving and reproducing like the rest of us, in ways that are the topics of this chapter.

Key Concepts

1. In structural terms, the simplest forms of life are the prokaryotic cells called archaebacteria and eubacteria. Most are microscopically small. Viruses are smaller still.

2. Prokaryotic cells do not have a profusion of internal, membrane-bound organelles, as eukaryotic cells do. As a group, they show great metabolic diversity, and many kinds display complex behavior. Most reproduce by prokaryotic fission.

3. A virus, a noncellular infectious particle, consists of nucleic acid (either DNA or RNA), a protein coat, and sometimes an outer envelope. It cannot replicate without pirating the metabolic machinery of a specific host cell.

4. Nearly all viral multiplication cycles proceed through five steps: Attachment to a host cell, penetration of its plasma membrane, replication of viral DNA or RNA and synthesis of viral proteins, assembly of new viral particles, and then release from the infected cell.

5. Protistans are easily distinguishable from prokaryotes but difficult to classify with respect to other eukaryotes. Protistans differ enormously in size, morphology, and life-styles.

6. Fungus-like protistans include parasitic and predatory molds that produce spores. Animal-like protistans include nonphotosynthetic flagellated protozoans, sporozoans, and ciliates. Plant-like protistans include photosynthetic flagellates and the red, brown, and green algae.

7. Most of us tend to judge microorganisms through the prism of human interests. Yet their lineages are the most ancient, their adaptations are diverse, and they are simply surviving and reproducing like the rest of us.

CHARACTERISTICS OF PROKARYOTIC CELLS

Prokaryotic cells—bacteria—are the most far-flung and abundant organisms. Thousands of species live nearly everywhere, including deserts, hot springs, glaciers, and the seas. Some have persisted 2,780 meters underground for millions of years! A handful of rich soil may hold billions of prokaryotic cells. Trace any lineage back far enough and you will find prokaryotic ancestors. From *Escherichia coli* to redwoods and elephants, all organisms on Earth interconnect, regardless of size, numbers, and evolutionary distance.

Two Prokaryotic Kingdoms

Just a few decades ago, reconstructing the evolutionary history of prokaryotic cells seemed to be an impossible task. Except for the stromatolites, shown in Section 19.3, the oldest groups are not well represented in the fossil record. Most are not represented at all.

Given their elusive histories, the many thousands of known species are traditionally classified by **numerical taxonomy**. By this practice, the traits of an unidentified prokaryotic cell are compared with those of a known group. Such traits include cell shape, cell wall staining attributes, motility, nutritional requirements, metabolic patterns, and endospores, if any. The greater the total number of traits a cell has in common with the known group, the closer is their inferred relatedness.

Comparative biochemistry, especially new methods of gene sequencing, are yielding insight into prokaryotic phylogenies (Section 18.7 and Appendix I). Remember, rRNAs are essential for protein synthesis, so they can't undergo drastic changes in their overall base sequence without loss of function. The numerous small changes that accumulated in the rRNAs of different lineages can be directly measured. Surprisingly, the measurements are uniting some groups that did not even appear to be related on the basis of other tests. For example, it now seems a key divergence began shortly after prokaryotic cells first appeared on Earth (Section 19.3). One branch led to **eubacteria**, the most common prokaryotic cells. (Here, the *eu–* signifies "typical.") The other branch led to the common ancestor of both the **archaebacteria** and all eukaryotic cells. Table 20.1 and Figure 20.3 show the characteristics both kingdoms still share.

Splendid Metabolic Diversity

All organisms take in energy and carbon to meet their nutritional requirements. But prokaryotic cells display the most diversity in the way they secure resources.

Like plants, *photoautotrophic* species build organic compounds by photosynthesis; they are "self-feeders." They tap sunlight for energy and use carbon dioxide as their carbon source. Their plasma membrane contains

a

Figure 20.3 (a) General body plan of prokaryotic cells. (b) Example of pili, filamentous structures that project from the surface of many types of prokaryotic cells. This *Escherichia coli* cell is dividing in two.

the photosynthetic machinery. Like plants, the oxygen-releasing photosynthesizers get electrons and hydrogen from water molecules. Others are strict anaerobes; they die when exposed to oxygen. They obtain electrons and hydrogen from inorganic compounds, such as hydrogen sulfide and gaseous hydrogen, or H_2. *Chemoautotrophic* archaebacteria and eubacteria are self-feeders, too. Most types use carbon dioxide as a carbon source. Depending on the species, they strip electrons and hydrogen from certain organic compounds or inorganic substances in the environment. Gaseous hydrogen, sulfur, iron, and nitrogen compounds are their kinds of "food."

Photoheterotrophic species are not self-feeders. They use sunlight for energy but get carbon from fatty acids, complex carbohydrates, and other compounds already produced by other organisms. *Chemoheterotrophic* species

Table 20.1 *Characteristics of Prokaryotic Cells*
1. No membrane-bound nucleus.
2. Generally a single chromosome (a circular DNA molecule); many species also contain plasmids.
3. Cell wall present in most species.
4. Reproduction mainly by prokaryotic fission.
5. Great metabolic diversity among species.

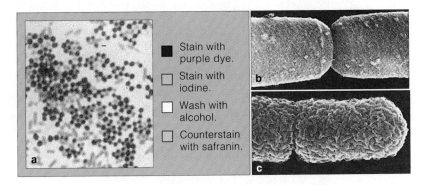

Figure 20.4 Prokaryotic cell walls. (**a**) Gram staining. Cocci and rods smeared on a slide are stained with a purple dye (such as crystal violet), washed off, then stained with iodine. All cells are purple. The slide is washed with alcohol to strip Gram-negative cells of color. The slide is counterstained (with safranin), washed, and dried. The Gram-positive cells (here, *Staphylococcus aureus*) remain purple. The counterstain colors Gram-negative cells (*E. coli*) pink. (**b**) Dividing *Bacillus subtilis* cell. Its wall has a smooth surface texture. (**c**) Dividing *E. coli* cell. Its wrinkled cell wall is characteristic of Gram-negative bacteria.

Figure 20.5 *Helicobacter pylori* cell, with its tuft of flagella. This pathogen can colonize the stomach lining. Untreated infection may lead to gastritis, peptic ulcers, and possibly stomach cancer. *H. pylori* can contaminate water and food, especially unpasteurized milk. Treatment that combines antibiotics with an antacid gets rid of the pathogen, after which ulcers heal.

aren't self-feeders, either. Parasitic types get glucose and other nutrients from living hosts. The saprobes get them from organic products or remains of other organisms.

Cell Sizes and Shapes

The chapter introduction started you thinking about the size of prokaryotic cells. Few species have a width or length greater than 10 micrometers. Three shapes are common among archaebacteria and eubacteria. A round shape is called a coccus (plural, cocci). A rod shape is a bacillus (plural, bacilli). A cell body that is twisted one or more times is a spirillum (plural, spirilla):

coccus   bacillus   spirillum

Don't let the simple categories fool you. Cocci may also be oval or flattened. Bacilli may be skinny like straws or tapered like cigars. Some cells have starlike surface extensions; some spiral types look like corkscrews or commas. Besides this, cells often remain stuck together after cell division, and together they form chains, sheets, and other aggregations.

Structural Features

Only archaebacteria and eubacteria are prokaryotic cells. They originated long before nucleated cells (*pro–*, before; *karyon*, nucleus). None has the profusion of organelles typical of eukaryotic cells. They have only one circular DNA molecule and maybe one or more plasmids. This doesn't mean prokaryotic cells are somehow inferior to eukaryotic cells. Being small-bodied, fast reproducers, they get along well without great internal complexity.

A cell wall usually surrounds the plasma membrane. Remember, this type of semirigid, permeable structure helps a cell maintain shape and resist rupturing when internal fluid pressure rises (Section 4.11). Eubacterial cell walls are made of peptidoglycan (many crosslinked strands of polysaccharides). Clinicians identify many species by wall composition and structure. Figure 20.4 describes a common staining reaction, the Gram stain.

A sticky mesh (glycocalyx) often encloses the wall. It consists of polysaccharides, polypeptides, or both. When highly organized and attached firmly to the wall, we call the mesh a capsule. When less organized and loosely attached to the wall, we call it a slime layer. The mesh helps a cell attach to teeth, mucous membranes, rocks in streams, and other interesting surfaces. It also helps some encapsulated types resist being engulfed by phagocytic, infection-fighting cells of host species.

Some prokaryotic cells have one or more bacterial flagella (singular, flagellum), as in Figure 20.5. These motile structures differ structurally from the eukaryotic flagellum and don't work the same way; they rotate like a propeller. Many types have pili (singular, pilus). These short, filamentous proteins project above the wall, as in Figure 20.3. Some help the cell adhere to surfaces or to one another as a prelude to conjugation (Section 20.3).

Archaebacteria and Eubacteria are kingdoms of prokaryotic cells. These cells have a circular bacterial chromosome and often extra DNA in the form of plasmids. Nearly all species have a cell wall around the plasma membrane. A capsule or slime layer often surrounds the wall.

Prokaryotic cells are classified by biochemical comparisons and numerical taxonomy (the total percentage of observable traits they have in common with a known group).

MAJOR BACTERIAL GROUPS

Archaebacteria

The major groups of archaebacteria are methanogens, extreme halophiles, and extreme thermophiles. All are unique in some aspects of their chemical composition, metabolism, structure, and nucleic acid sequences. They all differ as much from other prokaryotes as from eukaryotes. They are adapted to conditions as hostile as those confronting the first cells on Earth. Hence the name of the kingdom (*archae*– means beginning).

Methanogens ("methane makers") are at home in swamps, sewage, stockyards, animal guts, and other oxygen-free habitats (Figure 20.6*a,b*). Their anaerobic ATP-producing pathway ends in methane (CH_4). Each year, the methanogens release 2 billion tons of methane from termite guts, ruminant guts, wetlands, rice paddies, and landfills. As Section 42.3 describes, the tremendous quantities of this by-product influence carbon dioxide levels in the atmosphere and the global carbon cycle.

Extreme halophiles ("salt lovers") live in very salty water, as in brackish ponds and salt lakes, and near hydrothermal vents (Figure 20.6*c*). Halophiles also spoil salted fish, animal hides, and commercial sea salt. Most make ATP by aerobic pathways. Some also switch to a photosynthetic pathway when oxygen is low.

Extreme thermophiles ("heat lovers") live in highly acidic soils, hot springs, even coal mine wastes (Figure 20.6*d*). Some are the start of food webs at hydrothermal vents, where water reaches 110°C. They get electrons for ATP production from hydrogen sulfide. They are cited as evidence that life originated deep in the oceans.

Eubacteria

Nearly all of the 400 known genera of prokaryotes are eubacteria. Unlike archaebacteria, they have fatty acids in their plasma membrane. In most cases, their cell wall incorporates peptidoglycan. Until the phylogenies are worked out, we must classify them mainly by numerical taxonomy. Here, we focus on their modes of nutrition.

Photoautotrophs. Cyanobacteria, or blue-green algae, are common oxygen-releasing, photosynthetic types. You may have seen them in ponds and lakes, where mucus-sheathed chains of cells form slimy mats in nutrient-enriched water (Figure 20.7). *Anabaena* and other types can convert nitrogen gas to ammonia for biosynthesis. If nitrogen compounds dwindle, modified cells called **heterocysts** synthesize a nitrogen-fixing enzyme. They produce and share nitrogen compounds with other cells in the chains and get carbohydrates in return.

Anaerobic photosynthesizers such as green bacteria get electrons from hydrogen sulfide or hydrogen gas, not water. They might resemble anaerobic bacteria in which the cyclic pathway of photosynthesis evolved.

Figure 20.6 Archaebacteria. (**a**) *Methanococcus jannaschii*. Its DNA is fully sequenced (Section 18.7). (**b**) The stomachs of cattle and other ruminants house methanogens. Cattle release methane as they belch. (**c**) Commercial seawater evaporating ponds, San Francisco Bay, a home of *Halobacterium*. This extreme halophile thrives in places where salt levels greatly exceed that of seawater —for example, 300 versus 35 grams per liter. (**d**) Hot, sulfur-rich water spilling from Emerald Pool in Yellowstone National Park. The spring's yellow-orange rim is evidence of the abundance of carotenoids in extreme thermophiles. See also Figure 1.7*a*.

Chemoautotrophs. Although small, eubacteria in this category have mighty roles in the cycling of nitrogen and other nutrients. For instance, without nitrogen, there are no amino acids. Without amino acids, there is no life. Nitrifying bacteria get electrons from ammonia. Plants use the end product, nitrate, as a nitrogen source.

Chemoheterotrophs. Most eubacteria are in this group. Many have roles as decomposers and as human helpers. We use *Lactobacillus* to help make pickles, buttermilk, and yogurt; we use actinomycetes to make antibiotics. *E. coli* in our gut produces vitamin K and compounds that help us digest fat. It also helps newborns digest milk, and it prevents many food-borne pathogens from colonizing the human gut. Sugarcane and corn rely on the nitrogen-fixing spirochete *Azospirillum*. They take up some nitrogen fixed by this symbiont and give some sugars to it. Legumes such as beans, peas, alfalfa, and clover benefit from *Rhizobium*, a symbiont in their roots.

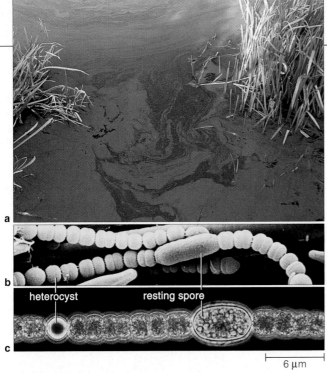

Figure 20.7 A common eubacterium. (**a**) Huge populations of cyanobacteria form dense mats near the surface of a nutrient-enriched pond. (**b,c**) Some cells become resting spores when conditions restrict growth. A nitrogen-fixing heterocyst is also shown. Section 4.12 gives other examples of eubacteria.

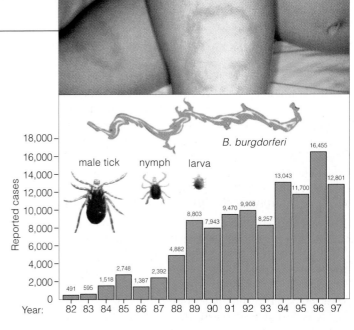

Figure 20.8 Deer ticks (*Ixodes*), the most common vector between all stages of the spirochete *Borrelia burgdorferi* and human hosts. The large photograph shows a rash that developed after infection by *B. burgdorferi*. It is a symptom of the most common tick-borne disease in the United States: Lyme disease. Tick bites deliver the spirochete to new hosts. The graph shows the number of Lyme disease cases in the United States between 1982 and 1997.

Most of the pathogenic prokaryotes are eubacteria. Examples: Pseudomonads are decomposers in soil, but they transfer plasmids with antibiotic-resistance genes. Some *E. coli* infections trigger diarrhea, a leading cause of infant deaths in developing countries. If *Clostridium botulinum* toxins taint fermented grain and improperly sterilized canned foods, you can get *botulism*. This form of poisoning disrupts breathing and can lead to death. *C. tetani*, a relative, causes *tetanus* (Section 30.5).

Like many other soil-dwelling bacteria, *C. tetani* can make endospores. This type of resting structure forms around a copy of the bacterial chromosome and some cytoplasm inside the cell. It forms when the supplies of nitrogen or some other nutrient dwindle and growth is arrested. An endospore is released as a spore when the plasma membrane ruptures. It resists heat, drying out, irradiation, acids, disinfectants, and boiling. It develops into a cell when favorable conditions return.

Many of the chemoautotrophs use insects to taxi to new hosts. Example: Bites from infected blood-sucking ticks transmit *Borrelia burgdorferi* from deer and some other wild animals to humans, and *Lyme disease* is the result. A "bull's-eye" rash, as in Figure 20.8, may form around the bite. Severe headaches, backaches, chills, and fatigue follow. Treatment must be prompt.

Regarding the "Simple" Bacteria

Bacteria are small. Their insides are not elaborate. *But bacteria aren't simple.* Their behavior reinforces this point.

Bacteria move into nutrient-rich regions. Aerobes move toward oxygen; anaerobes avoid it. Photosynthetic types move into light or away from intense light. Many will tumble away from toxins. Such behaviors often start with membrane receptors that change shape when they are stimulated by light or chemical compounds. When a bacterium moves in a different direction, the receptors are stimulated in a different way. This triggers a fleeting "memory," a change in some biochemical condition that can be compared against that of the immediate past.

Some species even show collective behavior, as when millions of *Myxococcus xanthus* cells form a "predatory" colony. The cells secrete enzymes that digest "prey"—cyanobacteria and other microorganisms that get stuck to the colony. Then they absorb breakdown products. What's more, these cells migrate, change direction, and move as a single unit toward what may be food!

Many myxobacteria produce fruiting bodies (spore-bearing structures). Under suitable conditions, cells in a colony differentiate and form a slime stalk, branches, and clusters of spores. The spores are dispersed when a cluster bursts open, and each may form a new colony. Certain eukaryotic species also form such structures.

Archaebacteria and Eubacteria are kingdoms of prokaryotic cells. Archaebacterial species are adapted to habitats that are anaerobic, extremely salty, or acidic. They are thought to resemble the first cells. Eubacterial species are found in most habitats and are the most common prokaryotic cells.

BACTERIAL GROWTH AND REPRODUCTION

Between divisions, prokaryotic cells grow by increasing their component parts. We measure growth of a large, multicelled organism in terms of increases in size, but doing so for a microscopically small bacterium is a bit pointless. Instead, we measure bacterial growth as an increase in the number of cells in a population. Under ideal conditions, each cell divides in two, the division of two cells results in four, four result in eight, and so forth. Many types divide every half hour. Such rates of increase lead to large population sizes in short order.

In nature, conditions that promote growth of some populations might not sustain growth of others. Most species could not survive extreme conditions, as in very hot deserts, Antarctic ice, or deep in the Earth. The few species found in such places reproduce only rarely.

A cell that has nearly doubled in size divides in two. Each daughter cell gets a bacterial chromosome: a circle of double-stranded DNA with very few proteins

Figure 20.10
Sketch of bacterial conjugation. Like many organisms in other kingdoms, many bacteria transfer genes. Recombinations occur between the same or different species. For this sketch, the plasmid's size is hugely exaggerated. The bacterial chromosome is far larger, as you saw in Figure 15.3.

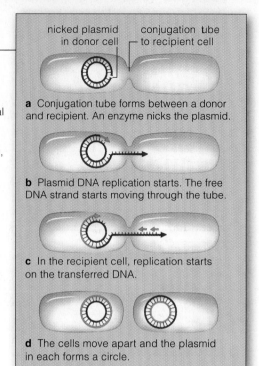

a Conjugation tube forms between a donor and recipient. An enzyme nicks the plasmid.

b Plasmid DNA replication starts. The free DNA strand starts moving through the tube.

c In the recipient cell, replication starts on the transferred DNA.

d The cells move apart and the plasmid in each forms a circle.

Figure 20.9
Prokaryotic fission. This cell division mechanism is used only by prokaryotic cells.

a The bacterial chromosome is attached to the plasma membrane before DNA replication.

b Replication starts and proceeds in two directions from some point in the bacterial chromosome.

c The DNA copy is attached at a membrane site near the attachment site of the parent DNA molecule.

d The two DNA molecules are moved apart by membrane growth between the two attachment sites.

e New membrane and new wall material are added transversely, through the cell's midsection.

f The ongoing, orderly deposition of membrane and wall material at the midsection cuts the cell in two.

attached to it. In some species, a daughter cell buds off of its parent. Most often, however, bacteria reproduce by a division mechanism called **prokaryotic fission.**

As prokaryotic fission gets under way, a parent cell replicates its DNA (Figure 20.9). Both DNA molecules are attached to the plasma membrane at adjacent sites. The cell makes lipid and protein molecules, which are added to the plasma membrane between the two sites. The membrane additions make the two molecules of DNA move apart. Wall material is deposited above the growing membrane. In time, new membrane and new wall material at the cell midsection cut the cytoplasm. Two genetically equivalent daughter cells result.

Daughter cells often inherit one or more plasmids (Section 15.1). These are small circles of extra DNA, replicated independently of the bacterial chromosome.

Genes on the F (Fertility) plasmid confer the means for **bacterial conjugation.** By this mechanism, a donor cell transfers plasmid DNA to a recipient cell. *Salmonella, Streptococcus, E. coli,* and many other species do this. The F plasmid carries genetic instructions for making a sex pilus. Sex pili at the donor cell surface hook onto a recipient cell and pull it next to the donor. After the two cells make contact, a conjugation tube develops between them. Then the plasmid DNA is transferred through the tube in the manner shown in Figure 20.10.

Nearly all bacteria reproduce by prokaryotic fission, a cell division mechanism that follows DNA replication. Each daughter cell inherits a single bacterial chromosome (one DNA molecule). Many species also transfer plasmid DNA.

Figure 20.16 A cellular slime mold, *Dictyostelium discoideum*. (**a**) Its life cycle includes a spore-producing stage. Spores give rise to free-living amoebas, which grow and divide until food (soil bacteria) dwindles. Then, amoebas secrete cyclic AMP, a chemical signal that makes them stream toward one another. Their plasma membrane becomes sticky, so they adhere to one another. A cellulose sheath forms around the cell mass, which starts crawling like a slug (**b–d**). Some slugs may incorporate 100,000 amoebas.

In a migrating slug, the amoebas differentiate into prestalk (*red*), prespore (*white*), and anteriorlike cells (*brown dots*). Prestalk cells secrete ammonia in amounts that vary with temperature and light intensity. The slug moves fastest in response to intermediate levels of ammonia (not too little, not too much). Those levels correspond to warm, moist conditions (not too cold or hot, not soppy or dry). Prestalk and prespore cells differentiate and form a stalked, spore-bearing structure (**e–g**). Anteriorlike cells, which sort into two groups, may help elevate nonmotile spores for dispersal from the top of the spore-bearing structure.

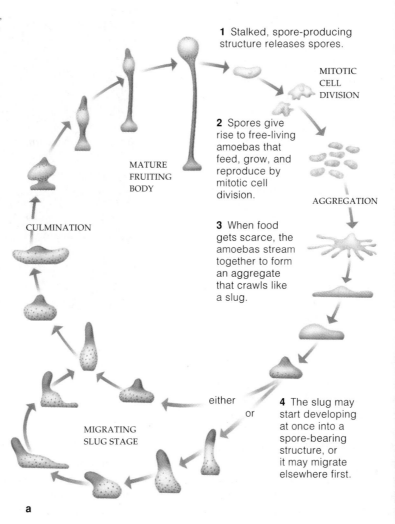

1 Stalked, spore-producing structure releases spores.

MITOTIC CELL DIVISION

2 Spores give rise to free-living amoebas that feed, grow, and reproduce by mitotic cell division.

AGGREGATION

3 When food gets scarce, the amoebas stream together to form an aggregate that crawls like a slug.

either or

4 The slug may start developing at once into a spore-bearing structure, or it may migrate elsewhere first.

MATURE FRUITING BODY

CULMINATION

MIGRATING SLUG STAGE

a

molds (Acrasiomycota) and plasmodial slime molds (Myxomycota). Like true amoebas, they are predatory. Their cells crawl on rotting plant parts and eat bacteria, yeasts, spores, and organic compounds. If nutrients are scarce, starving amoebas gather into a slimy mass that may migrate to a better place and form a spore-bearing structure (Figure 20.16*a*). Slime molds also reproduce sexually.

One of the 500 species of plasmodial slime molds is shown in Figure 20.17. The plasma membrane of these amoebalike cells breaks down when they get together. Cytoplasm can flow freely from cell to cell and thereby distributes nutrients and oxygen throughout the mass. A streaming mass of this sort (the plasmodium) often will occupy several square meters and migrate if food runs out. You might have seen one crossing a lawn or a road, or even climbing a tree.

Most chytrids and water molds are saprobic decomposers of aquatic habitats. Some are single cells. Like slime molds, many are free-living predators some of the time and simple experiments in multicellularity at other times.

During a slime mold life cycle, amoeboid cells aggregate to form a migrating mass. Cells in the mass differentiate, then form reproductive structures and spores or gametes.

Figure 20.17 *Physarum*, a plasmodial slime mold. This aggregation of cells is migrating along a rotting log.

THE ANIMAL-LIKE PROTISTANS

About 65,000 protistan species are informally called the **protozoans** ("first animals"). Why? They may resemble the single-celled, heterotrophic protistans that gave rise to animals. Among these amoeboid protozoans, ciliated protozoans, and animal-like flagellates are predators, parasites, and grazers that actively move about. You'll find free-living species in the seas, damp soil, ponds, lakes, and other aquatic habitats. The symbionts and parasites live in or on moist tissues of other organisms. Some protozoans are notorious pathogens.

Protozoans reproduce either asexually or sexually. Many alternate between reproductive modes when the environment changes. Most often, they reproduce by **binary fission**, an asexual process that divides the body in two parts. The division plane is random for amoebas, longitudinal for flagellates, and transverse for ciliates. Some species also bud from a parent. The life cycles of many parasitic types have encysted stages. Their **cysts** are resistant body coverings made of secreted material; they help the parasite survive stressful conditions.

Amoeboid Protozoans

Amoeboid protozoans (Sarcodina) are soft-bodied cells, some with hard parts. Naked amoebas, foraminiferans, heliozoans, and radiolarians are in this group. All use pseudopods for moving and for capturing prey (Figure 20.18a). Some are gut-dwelling species. Free-living types live in damp soil, fresh water, and saltwater. Most are phagocytes that engulf other protozoans and bacteria. A few parasitize certain invertebrates or vertebrates.

Entamoeba histolytica, a pathogen, completes its life cycle in human intestines. It causes *amoebic dysentery*, an infectious disease most prevalent in subtropical and tropical areas where sanitation is poor. Worldwide, this is a leading cause of death among infants and toddlers, whose immune systems are not yet well developed.

Unlike naked amoebas, foraminiferans have a highly perforated external skeleton of secretions hardened with calcium carbonate (Figures 20.2 and 20.18c). Thin, mucus-covered pseudopods extend through the perforations to trap prey. Like their ancestors, most species live on the seafloor. Of the species we know about, 99 percent are extinct. Over time, fossilized foraminiferan shells were compressed to form a distinctive type of sedimentary rock (Figure 20.18d). We mine the calcified remains for the manufacture of cement and blackboard chalk.

The radiolarians and heliozoans have many slender, reinforced pseudopods radiating from their body. They are well represented in the fossil record, for their silica-hardened parts resist breakdown (Figure 20.18b). Most species drift with ocean currents, as part of plankton. **Plankton** refers to any aquatic community of passive drifters and motile organisms, mostly microscopic.

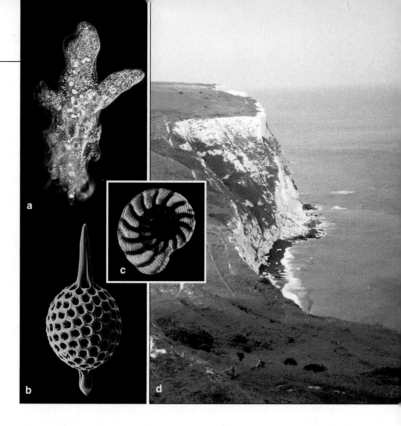

Figure 20.18 Amoeboid protozoans. (**a**) *Amoeba proteus*. (**b**) Radiolarian shell. (**c**) Foraminiferan shell. (**d**) The white cliffs of Dover, England, formed more than 200 million years ago by fossilization and compression of countless foraminiferan shells.

Ciliated Protozoans

Ciliated protozoans (Ciliophora) typically have profuse arrays of cilia at their surface. You read about these fine motile structures in Section 4.10. Most species use them to swim in fresh water or seawater, where they prey on bacteria, algae, and one another. All of the cilia beat in a synchronized pattern over the body surface, calling to mind a soft wind through a field of tall grasses.

Paramecium is typical of the group (Figure 20.19). A mature cell is usually 150 to 200 micrometers long. Its outer body cover, a pellicle, consists of rigid or flexible membranes. A gullet starts as an oral depression at the body surface; there, cilia sweep food-laden water into the cell. Food gets enclosed in enzyme-filled vesicles and is digested. Like amoeboid protozoans, a ciliate's internal solute concentrations are greater than those of external fluids. The cell must counter water's tendency to diffuse inward by osmosis. Like amoebas, it depends on **contractile vacuoles**. Tubes extend from the center of these organelles and collect excess water that moves osmotically into the cell. A filled vacuole contracts and forces water through a pore to the outside (Section 5.10).

Like other protistan groups, ciliates show diversity in life-styles. About 65 percent of the known species are free-living, motile cells. Others attach to substrates, often by stalks. Some form colonies. Many interact with other organisms as symbionts or parasites.

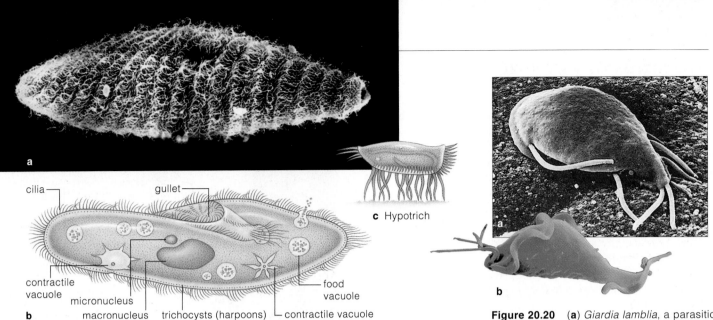

cilia
gullet

c Hypotrich

contractile
vacuole
micronucleus
macronucleus
trichocysts (harpoons)
contractile vacuole
food
vacuole

b

Figure 20.19 (**a,b**) Body plan and scanning electron micrograph of *Paramecium*, a ciliated protozoan. (**c**) Sketch of one of the most animal-like ciliates, which run around on leglike tufts of cilia. Some have a "head" end with modified sensory cilia.

a

b

Figure 20.20 (**a**) *Giardia lamblia*, a parasitic protozoan, adhering to human epithelium by a sucking disk. (**b**) *Trichomonas vaginalis*. This pathogenic species causes trichomoniasis, a highly prevalent, sexually transmitted disease.

Figure 20.21 Diagram of *Trypanosoma brucei*. This parasitic protozoan causes African sleeping sickness. Figure 20.2a is a micrograph of this animal-like flagellate.

undulating membrane (fold of a pellicle, attached to part of flagellum)

mitochondrion

Golgi body

free flagellum

endoplasmic reticulum
nucleus
vacuole
basal body of flagellum

Ciliates reproduce sexually and asexually. Things get interesting because one cell often has two types of nuclei. When it reproduces asexually by binary fission, its small, diploid *micro*nucleus divides by mitosis. Its large *macro*nucleus splits in two a bit sloppily; some DNA may spill out. Most often, sexual reproduction is by a form of conjugation. Cells repeatedly divide their micronuclei, swap two of these with a partner, and let two micronuclei fuse and form a diploid macronucleus to replace the one that disappears. And you probably thought sex among single-celled critters was simple!

Animal-Like Flagellates

Animal-like flagellates (Mastigophora) are free-living predators and parasites bearing one to several flagella. Free-living types live in fresh water and seawater. The parasites live in moist tissues of plants and animals, including humans. *Giardia lamblia*, an internal parasite, causes diarrhea (Figure 20.20a). It contaminates surface waters of aquatic habitats and reservoirs. Cattle, wild animals, and people often harbor the parasite and its cysts, which leave the host in feces. The parasite infects people who drink water or eat food tainted with feces. Many trichomonads are parasitic flagellates. *Trichomonas vaginalis*, a worldwide nuisance, can infect new human hosts during sexual intercourse (Figure 20.20b). Unless a trichomonad infection is treated, it may damage the urinary and reproductive tracts.

Trypanosomes, too, are parasites. *Trypanosoma brucei* causes *African sleeping sickness* (Figures 20.2a and 20.21). Tsetse fly bites transmit it to hosts; it can damage the nervous system. In South America and Mexico, *T. cruzi* causes *Chagas disease*. Bugs pick it up when they feed on infected humans and other animals. *T. cruzi* multiplies in the insect gut and is excreted onto a host. Scratches in skin invite infection. The liver, spleen, eyelids, and the face swell. The infection severely damages the brain and heart. There is no cure for Chagas disease.

Protozoans are single-celled protistans. In some respects they are animal-like, as in the way they actively move about to secure food. They include predators, parasites, and grazers. Some pathogenic species cause dangerous diseases in humans and other animals.

THE NOTORIOUS SPOROZOANS

Sporozoan is an informal name for parasitic protistans that complete part of the life cycle inside specific cells of hosts. These parasites form a motile infective stage (sporozoites); some have encysted stages. One end of the body has a complex of structures to penetrate hosts.

Many sporozoans, such as *Toxoplasma*, cause human diseases (Figure 20.22). *Cryptosporidium* contaminates most freshwater supplies. One strain has put thousands of people in hospitals. Intestinal disorders and other symptoms usually subside after ten days. People with weakened immune systems suffer for months or years.

Four parasitic species of *Plasmodium* cause *malaria*. This long-lasting disease affects more than 100 million people and kills 2.7 million of them every year. Only female *Anopheles* mosquitoes transmit it to humans. A host's bloodstream transports a motile, infective stage

Figure 20.22 *Toxoplasma* is a sporozoan that uses domestic and wild cats as its definitive hosts. Its intermediate hosts are other animals, including humans. Cysts may be in undercooked or raw meat. Flies and cockroaches also move these cysts from cat feces onto food. The disease toxoplasmosis has flu-like symptoms. It isn't common, but it is dangerous for immune-compromised people and during pregnancies. If it infects an unborn child, miscarriage or birth defects result. Any cat, no matter how coddled, can pass cysts. Pregnant women should avoid stray cats and never empty litterboxes, clean up housecat "accidents," or clean out children's sandboxes.

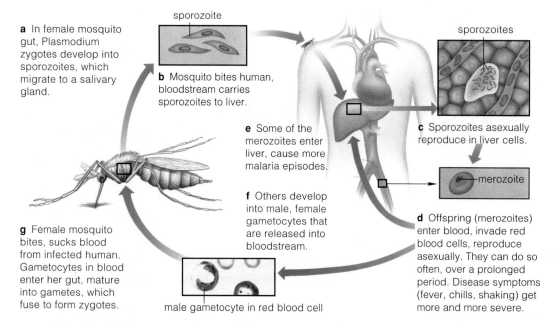

a In female mosquito gut, Plasmodium zygotes develop into sporozoites, which migrate to a salivary gland.

sporozoite

b Mosquito bites human, bloodstream carries sporozoites to liver.

e Some of the merozoites enter liver, cause more malaria episodes.

f Others develop into male, female gametocytes that are released into bloodstream.

g Female mosquito bites, sucks blood from infected human. Gametocytes in blood enter her gut, mature into gametes, which fuse to form zygotes.

male gametocyte in red blood cell

sporozoites

c Sporozoites asexually reproduce in liver cells.

merozoite

d Offspring (merozoites) enter blood, invade red blood cells, reproduce asexually. They can do so often, over a prolonged period. Disease symptoms (fever, chills, shaking) get more and more severe.

Figure 20.23 Life cycle of a sporozoan that causes malaria, a serious disease.

are warm-bodied and have little free oxygen in blood (hemoglobin binds most of it). That stage matures in mosquitoes, which have a lower body temperature and slurp oxygen from the air, along with infected blood.

Historically, malaria has been most prevalent in both the tropical and subtropical regions of Africa. Reported cases in North America and elsewhere are dramatically increasing. This is a result of unprecedented levels of human immigration and globe-hopping travels. Worse, older anti-malarial drugs no longer work against some *Plasmodium* strains. Developing vaccines (preparations that induce the body to build up resistance to a specific pathogen) is difficult. Experimental vaccines just aren't equally effective against all of the stages that develop during the *Plasmodium* life cycle. This is generally the case for developing vaccines against any parasite that has a complex life cycle.

to the liver, where the stage divides repeatedly. Some of the resulting cells (merozoites) penetrate and asexually reproduce inside red blood cells, which they destroy. Others will mature into gametes (Figure 20.23).

Disease symptoms start after infected cells rupture and release merozoites, metabolic wastes, and cellular debris into the blood. They include shaking, chills, a burning fever, and drenching sweat. After one episode, they subside for a few weeks or months, and infected people may feel okay. Relapses recur. In time, anemia may develop and the liver may be grossly enlarged.

The *Plasmodium* life cycle is sensitive to human and mosquito body temperatures and oxygen levels. The gamete-producing stage cannot mature in humans, who

Sporozoans, parasites with a complex of host-penetrating structures, complete their life cycle in specific host cells. Motile infective stages and often encysted stages form.

The Nature of Infectious Diseases

CATEGORIES OF DISEASES   Just by being human, you are a potential host for many pathogenic bacteria, viruses, fungi, protozoans, and parasitic worms. When a pathogen invades your body, it multiplies in cells and tissues; we call this an **infection**. Its outcome, **disease**, results when body defenses cannot be mobilized fast enough to keep the pathogen's activities from interfering with normal functions. With *contagious* diseases, the pathogen can be transmitted by direct contact with body secretions from infected individuals, as in explosive wet sneezes.

During an **epidemic**, a disease spreads rapidly through part of a population for a limited time, then subsides. If an epidemic breaks out in several countries at the same time, this is a pandemic. The AIDS pandemic is like this (Section 35.10). *Sporadic* diseases such as whooping cough break out irregularly and affect very few people. *Endemic* diseases pop up more or less continually but don't spread far in large populations. Tuberculosis is like this.

AN EVOLUTIONARY VIEW   Now look at disease in terms of the pathogen's prospects for survival. *A pathogen stays around only for as long as it has access to outside sources of energy and raw materials.* To a microscopic organism or virus, a human is a treasurehouse of both. With such bountiful resources, it can multiply or replicate itself to amazing population sizes. Evolutionarily speaking, the ones that leave the most descendants win.

Two barriers prevent world dominance by pathogens. First, any species with a history of being attacked by a specific pathogen has coevolved with it and has built-in defenses against it. A premier example is the vertebrate immune system, as described in Chapter 35. Second, if a pathogen is too good at rapidly killing an individual, it may vanish before having time to infect a new one. This is one reason why most have less-than-fatal effects on a host. After all, infected people who live longer can spread more germs and contribute to a pathogen's reproductive or replicative success. Usually, an individual dies only if it becomes host to overwhelming numbers of a pathogen, if it is a novel host (one that has no coevolved defenses against the pathogen), or if a mutant pathogenic strain emerges and can surmount the host's current defenses.

Being equipped with an evolutionary perspective, you probably can work out the numbers on your own. To wit: *The greater the population density of host individuals, the greater the kinds and frequencies of infectious diseases transmitted among them.* This brings us to the bad news.

DRUG-RESISTANT STRAINS   There is an old saying that, when you attack nature, it will come back at you with a pitchfork. At this point in the book, we have looked at antibiotic resistance in several contexts. Here we reinforce the point that infectious diseases are on the upsurge.

Example: Because of economic pressures, the number of working mothers has skyrocketed in the United States. So has the number of preschoolers enrolled in day-care centers. Each center is a crowded population of hosts

intestinal epithelium   *Cryptosporidium* cyst   emerging sporozoite

Table 20.2 *The Eight Deadliest Infectious Diseases*

Disease	Main Agents	Estimated New Cases per Year	Estimated Deaths per Year
Acute respiratory infections*	Bacteria, viruses	1 billion	4.7 million
Diarrheas**	Bacteria, viruses, protozoans	1.8 billion	3.1 million
Tuberculosis	Bacteria	9 million	3.1 million
Malaria	Sporozoans	110 million	2.5–2.7 million
AIDS	HIV	5.6 million	2.6 million
Measles	Viruses	200 million	1 million
Hepatitis B	Virus	200 million	1 million
Tetanus	Bacteria	1 million	500,000

* Includes pneumonia, influenza, and whooping cough.
** Includes amoebic dysentery, cryptosporidiosis, and gastroenteritis.

whose developing immune systems are vulnerable to disease agents, including *Streptococcus pneumoniae*. This bacterium causes pneumonia, meningitis, and middle-ear infections. In people of all ages, *S. pneumoniae* infections result in 40,000 to 50,000 deaths each year. As recently as 1988, drug-resistant strains were almost unheard of. Now they are becoming the rule, not the exception.

Most of our 6.1 billion selves live in crowded cities. During any specified interval, as many as 50 million of us are on the move within and between countries in search of a better life. Is it any wonder the pathogenic agents of cholera, tuberculosis, and other diseases are striking with a vengeance (Table 20.2)?

PATHOGENS IN FOOD SUPPLIES   Food poisoned by infectious agents hits more than 80 million people in the United States each year. Treating food poisoning costs 5 billion to 22 billion dollars. About 9,000 people a year die from it, mainly the very young, the very old, or those with compromised immune systems.

People often dismiss their misery as "the 24-hour flu," not infection by *Salmonella enteriditis* and pathogenic *E. coli* strains. The agents can lurk in undercooked beef or poultry, water, unpasteurized milk or cider, and vegetables grown in manure-fertilized fields. Example: *S. enteriditis* once found its way into some ingredients for a popular ice cream. Almost 225,000 ice cream eaters got stomach cramps, diarrhea, and other symptoms of *salmonella*.

Pathogens in the kitchen may cause at least half of all food poisonings. Carlos Enriquez and his colleagues at the University of Arizona sampled 75 dishrags and 325 sponges in several homes. They found *Salmonella*, *E. coli*, *Pseudomonas*, and *Staphylococcus*, which can live up to two weeks in a wet sponge. You can sanitize sponges simply by putting them through a dishwasher cycle.

THE (MOSTLY) SINGLE-CELLED ALGAE

The rest of this chapter surveys "the algae," the largely photosynthetic protistans. The majority of euglenoids, chrysophytes, and dinoflagellates are single cells, which isn't the case for red, brown, and green algae. Most are components of *phyto*plankton—aquatic communities of mainly microscopic photoautotrophs that are the start of nearly all food webs in ponds, lakes, streams, rivers, the seas, and all other water provinces. Under favorable conditions, certain species undergo an **algal bloom**, a stupendous increase in population size.

A Sampling of Single-Celled Diversity

Fresh water or stagnant ponds and lakes are home to **euglenoids** (Euglenophyta). Most of the 1,000 species of these flagellated, free-living cells are photosynthetic. But all are descended from phagocytic ancestors and still retain a heterotrophic feeding apparatus, now used to take in organic compounds, bacteria, or eukaryotic prey. *Euglena* affords hints of an endosymbiotic past with green algae; its chloroplasts, too, contain chlorophylls *a* and *b* plus carotenoids (Figure 20.24*a* and Section 19.4). Like some protozoans, *Euglena* has a pellicle, a flexible, translucent cover of spiral strips of a protein-rich material. It has a long flagellum and a contractile vacuole that squirts excess water out of the

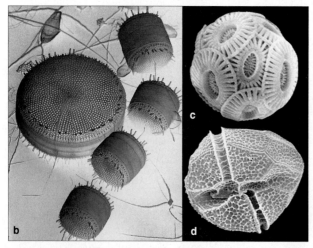

long flagellum

contractile vacuole   chloroplast

eyespot shielding a light-sensitive receptor

a   nucleus   Golgi body   mitochondrion   pellicle

ER

Figure 20.24 Single-celled algae. (**a**) Body plan for *Euglena*. Compare this with the electron micrograph in Section 19.4. (**b**) Diatom shells. (**c**) Coccolithophore shell. (**d**) Dinoflagellate shell.

Figure 20.25 (**a**) Two of the more than twenty-four stages in the life cycle of *Pfiesteria piscicida*. (**b**) The kind of damage inflicted by this cell from hell.

encysted stage   flagellated stage

cell body. An eyespot made of pigments partly shields a light-sensitive receptor, which helps guide the cell to places where light is most suitable for its activities.

Chrysophytes (Chrysophyta) are mostly free-living, photosynthetic cells that contain a distinctive array of pigments: chlorophylls *a*, c_1, and c_2. The golden algae, yellow-green algae, diatoms, and coccolithophores are members of this large group.

Many of the 500 species of golden algae have silica scales or other hard parts and an abundance of a golden-brown pigment, fucoxanthin. Some types resemble true amoebas. We find yellow-green algae in many aquatic habitats. Most of these species are nonmotile, although they all produce flagellated gametes.

The 5,600 existing species of diatoms have a silica "shell." It consists of two perforated parts that overlap like a pillbox (Figure 20.24*b*). For 100 million years or so, finely crushed shells of at least 35,000 extinct species of these mostly photosynthetic cells accumulated at the bottom of lakes and seas. Many sediments contain the deposits; we quarry them for use in filters, insulation, and abrasives. Each year, for instance, 270,000+ metric tons are quarried near Lompoc, California.

Calcium carbonate plates protect coccolithophores (Figure 20.24*c*). Most of the 500 or so existing species are photosynthesizers of marine habitats, especially in the tropics. In the past, accumulations of the plates formed sea sediments, chalk deposits, and limestone deposits, such as Dover's white cliffs. During blooms, these algae secrete mucus that can clog fish gills and a substance so noxious it makes migrating fish change their routes.

Most **dinoflagellates** (Pyrrhophyta) are producers in marine phytoplankton; a few of the 1,200 species are symbionts with corals. Bioluminescent ones make warm seawater shimmer at night. They all have flagella and cellulose plates (Figure 20.24*d*). When they reproduce, the cell and its plates divide into two daughter cells.

Dinoflagellates are yellow-green, green, blue, brown, or red, depending on the pigment array. *Red tides* form near coasts when certain species undergo blooms that turn the water rust-red or brown. Some produce toxins that cause massive kills of fish, birds, and other animals.

Algal Blooms and the Cell From Hell

Since 1991, algal blooms have killed more than a billion fish near the coastlines of North Carolina, Virginia, and Maryland. The agent of death, *Pfiesteria piscicida*, is a dinoflagellate known as "the cell from hell."

The life cycle of *P. piscicida* requires at least twenty-four encysted, flagellated, and amoeboid stages (Figure 20.25). *P. piscicida* waits out harsh conditions as cysts in sediments of estuaries. Its flagellated stages attach to prey and suck out juices. When a large school of oily menhaden or other fish lingers to feed, their excretions incite the encysted cells to emerge and release toxins. The toxins stop fishes from escaping (by slowing them down), then eat away patches of fish skin. *P. piscicida* feeds on sloughed epidermis, blood, and tissue juices oozing from the open sores. And then the flagellates give rise to amoeboids, which feast on the dead fish. All of these changes may unfold within a matter of hours.

Nitrogen, phosphorus, and other nutrients in runoff from fields and in raw sewage fan algal blooms all over the world. Blooms usually develop in warm, shallow water that has become enriched with nutrients. JoAnn Burkholder, a North Carolina State University botanist, correlated *Pfiesteria* blooms with hundreds of millions of gallons of raw sewage from hog and chicken farms. More than 16 million hogs live in 3,500 industrial-scale hog farms in eastern North Carolina. Untreated wastes in holding lagoons often spill into rivers that drain into the poorly flushed Chesapeake Bay estuaries and into rivers in Alabama, Delaware, Florida, and Virginia. For instance, after one heavy rain, a spill from a hog farm exceeded by three times the volume of oil spilled from the tanker *Valdez* (Section 45.8). In 1997 alone, *Pfiesteria* inflicted a 60 million dollar loss on fishing and tourism industries. The magnitude of the loss finally prodded health, agricultural, and environmental agencies into coordinating research to address the problem.

Single-celled photosynthetic protistans include diverse species of euglenoids, chrysophytes, and dinoflagellates. Most species are components of phytoplankton, the food-producing base of most aquatic habitats.

RED ALGAE

Of the 4,100 known species of **red algae** (Rhodophyta), nearly all are marine; only 200 live in freshwater. The life cycle for most species includes multicelled stages, but these do not have tissues or organs. Reproductive modes are diverse, with complex asexual and sexual phases. Figure 20.26 shows one example of a life cycle.

Not all "red algae" are red. Diverse photosynthetic species are green, purple, or almost black. Phycobilins and other accessory pigments mask their chlorophyll *a* (Section 6.2). Parasitic species have little pigmentation.

Red algae have a flexible, slippery texture owing to mucous material in their cell walls. Agar is made from extracts of wall material of several species. This inert, gelatinous substance has uses as a moisture-preserving agent in baked goods and cosmetics, as a setting agent for jellies, and as a culture gel. Carrageenan, extracted from *Eucheuma*, is a stabilizer in paints, dairy products, and many other emulsions.

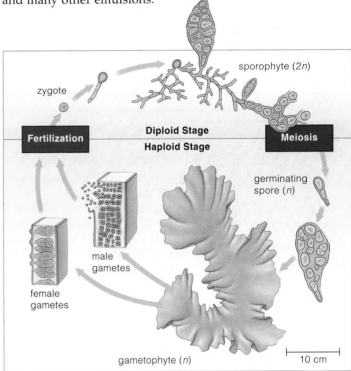

Figure 20.26 Life cycle of *Porphyra*.

Humans find different species of *Porphyra* tasty as well as nutritious. You may know it from sushi bars as *nori*, an algal wrapping for rice and fish.

Red algae are photosynthetic protistans with phycobilins and other accessory pigments that mask chlorophyll *a*. Multicelled stages usually form during their life cycles. Most species are aquatic and show great diversity in size, morphology, life-styles, reproductive modes, and habitats.

BROWN ALGAE

You find most of the 1,500 or so species of **brown algae** (Phaeophyta) living in cool or temperate marine waters, from the intertidal zone into the open ocean. Different kinds are brown, olive-green, or golden, depending on the array of pigments. Brown algae have carotenoids, such as fucoxanthin, and the chlorophylls a, c_1, and c_2. They range in size from the giant kelps—thirty meters long—to the microscopic, threadlike *Ectocarpus*.

Macrocystis, *Laminaria*, and the other giant kelps are the largest, most complex of all protistans. The multicelled sporophytes have stipes (stem-like parts), blades (leaf-like parts), and holdfasts (anchoring structures). Gas-filled bladders on the stipes assist in buoyancy and in keeping blades upright (Figures 20.2c and 20.27a). Each stipe holds tubelike arrays of elongated cells. Through these tubes, sugars and other dissolved products of photosynthesis can quickly move to all living cells in the body.

Giant kelp beds function as productive ecosystems. Think of them as underwater forests in which great numbers of diverse bacteria and protistans, as well as fishes and other animals, carry out their lives.

Humans commercially harvest *Macrocystis* and other kinds of brown algae. We use extracts from them when making ice cream, jelly beans, puddings, salad dressings, canned foods, frozen foods, beer, cough syrups, toothpastes, floor polishes, a number of cosmetics, and paper. Alginic acid in the cell walls of some species is used to make algins, which are added to various products as thickening, emulsifying, and suspension agents. Many people, especially in the Far East, harvest kelps as sources of food and mineral salts, and as a fertilizer for crops.

Figure 20.27 Multicelled brown algae. (**a**) Body plan of *Macrocystis*. (**b**) *Postelsia palmaeformis*, which grows along rocky shorelines.

bladder
blade
stipe
holdfast

The brown algae range in size from the microscopic to the largest of all of the multicelled protistans. Nearly all species live in marine habitats, ranging from the intertidal zone to surface waters of the open ocean.

GREEN ALGAE

Of all protistans, the **green algae** (Chlorophyta) are structurally and biochemically most like the plants and are their closest relatives. Nearly all are photosynthetic. Like plants, they have chlorophylls a and b. They, too, temporarily store carbohydrates in starch grains inside their chloroplasts. The cell wall of some species consists of cellulose, pectins, and other polysaccharides typical of plants.

Figure 20.28 is a sampling of the more than 7,000 known species of green algae. They include single-celled, colonial, filamentous, sheetlike, and tubular forms. You won't be able to see many species without the aid of a microscope. Most, including the *Micrasterias* cell shown in Figure 20.28b, live in freshwater. *Micrasterias* is one of thousands of desmid species, which are important food producers in nutrient-poor ponds and peat bogs. Green algae also grow at the ocean surface, in marine sediments, just below the surface of soil, and on rocks,

Figure 20.28 Representative green algae. (**a**) *Volvox*, a colony of interdependent green algal cells with resemblances to free-living, flagellated cells of the genus *Chlamydomonas*. (**b**) A *Micrasterias* cell reproducing itself. (**c**) From a marine habitat, *Acetabularia*, fancifully called the mermaid's wineglass. Each individual in the cluster is a multinucleate cell mass with a rootlike structure, a stalk, and a cap in which gametes form. (**d**) Sea lettuce (*Ulva*) grows in estuaries and attaches to kelps in the seas. Reproductive cells form inside and are released from the margins of the sporophyte and gametophyte, both of which have the same form shown in the diagram.

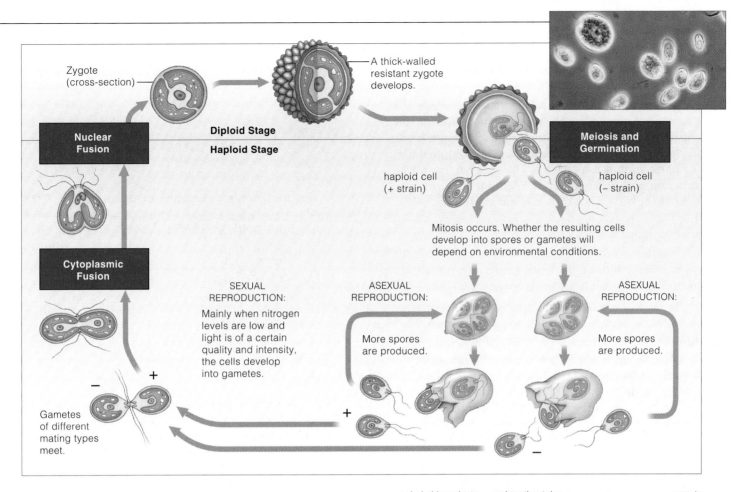

Zygote
(cross-section)

A thick-walled
resistant zygote
develops.

Diploid Stage

Haploid Stage

**Nuclear
Fusion**

**Meiosis and
Germination**

haploid cell
(+ strain)

haploid cell
(– strain)

**Cytoplasmic
Fusion**

Mitosis occurs. Whether the resulting cells
develop into spores or gametes will
depend on environmental conditions.

SEXUAL
REPRODUCTION:

Mainly when nitrogen
levels are low and
light is of a certain
quality and intensity,
the cells develop
into gametes.

ASEXUAL
REPRODUCTION:

More spores
are produced.

ASEXUAL
REPRODUCTION:

More spores
are produced.

+

+

−

Gametes
of different
mating types
meet.

−

Figure 20.29 Life cycle of a species of *Chlamydomonas*, one of
the most common green algae of freshwater habitats. This single-
celled species reproduces asexually, most often. It also reproduces
sexually under certain environmental conditions.

tree bark, other organisms, and snow. Some types are
symbionts with fungi, protozoans, and a few marine
animals. A colonial form (*Volvox*) is a hollow, whirling
sphere of 500 to 60,000 flagellated cells. Those beautiful
white, powdery beaches in tropical regions are largely
the work of uncountable numbers of green algal cells
(*Halimeda*) that formed calcified cell walls, then died
and disintegrated. Some day, green algae might even
accompany astronauts in space. They can grow in very
small spaces, give off vital oxygen, and take up carbon
dioxide exhaled by an aerobically respiring crew.

Green algae employ diverse modes of reproduction.
Chlamydomonas provides a classic example. This single-
celled alga is five to ten micrometers wide, on average.
It is able to reproduce sexually, but most of the time
it engages in asexual reproduction, with typically 16 to
32 daughter cells forming by mitotic cell division in the
confines of the parent cell wall. Daughter cells may live
at home for a while. But sooner or later they depart
by secreting enzymes, which digest what's left of their
parent. Figure 20.29 shows the life cycle of one species.

spiral chloroplasts   conjugation tube   zygote

a   b   c   d

Figure 20.30 One mode of sexual reproduction in *Spirogyra*, or
watersilk. (**a**) This green alga has spiral, ribbonlike chloroplasts.
(**b**) A conjugation tube forms between cells of adjacent haploid
filaments of different mating strains. (**c,d**) The cellular contents
of one strain pass through the tubes into cells of the other strain,
where zygotes form. Zygotes develop thick walls. They undergo
meiosis when they germinate and give rise to haploid filaments.

To give a final example, Figure 20.30 shows *Spirogyra*,
a filamentous green alga, reproducing sexually.

**The green algae show great diversity in size, morphology,
life-styles, and habitats. The structure and biochemistry
of some groups suggest that they have close evolutionary
ties to the plant kingdom.**

SUMMARY   *Gold* indicates text section

1. The microscopic world teems with prokaryotic cells, single-celled protistans, and viruses. *CI*

2. Bacteria are prokaryotic cells without a nucleus or a profusion of organelles. The most common types are the eubacteria. The archaebacteria (methanogens, extreme halophiles, and extreme thermophiles) live in habitats too harsh for most organisms. They are unique in wall structure and other features. Evolutionarily, they are closer to eukaryotic cells than to eubacteria. *20.1, 20.2*

3. Nearly all bacteria have a cell wall that protects the cell and helps it resist rupturing. A capsule or a slime layer may surround it. Some cells use bacterial flagella for motility. Pili (thin filaments) help many cells adhere to surfaces or assist in bacterial conjugation. *20.1*

4. Bacteria as a group show great metabolic diversity, as in their modes of acquiring energy and carbon. They also make behavioral responses to stimuli. *20.1, 20.2*

5. Bacteria reproduce by prokaryotic fission. After the bacterial chromosome is replicated, a parent cell divides into genetically equivalent daughter cells. If plasmids are present, they are replicated independently of the chromosome and may be transmitted to daughter cells. They also may be transferred to cells of the same species or a different species by bacterial conjugation. *20.3*

6. Viruses are nonliving, noncellular agents that infect particular species of nearly all organisms. *20.4*

   a. A virus particle has a core of DNA or RNA and a protein coat that may be enclosed in a lipid envelope from which viral proteins project. Coats of complex viruses have tail fibers and other accessory structures.

   b. A virus particle cannot reproduce on its own. Its genetic material must direct a host cell to synthesize materials necessary to produce new virus particles.

7. Nearly all viral multiplication cycles include five steps: Attachment to a suitable host cell, penetration of it, viral DNA or RNA replication and protein synthesis, assembly of new viral particles, and release. *20.5*

8. Unlike bacteria, protistans and other eukaryotic cells have a nucleus, mitochondria, endoplasmic reticulum with distinct ribosomes, microtubules, and at least two chromosomes (each with DNA and numerous proteins). Many contain chloroplasts. Like other eukaryotic cells, protistans engage in mitosis and meiosis. *20.6*

9. Many protistans are heterotrophs. They include water molds, chytrids, slime molds (cellular and plasmodial types), protozoans (amoebas, animal-like flagellates, and ciliated species), and sporozoans (Table 20.3). *20.6*

10. Like fungi, chytrids and water molds are decomposers and parasites. Their enzymes digest organic matter, and they absorb the breakdown products. Slime molds are phagocytes that form spore-producing structures. *20.7*

Table 20.3 *Summary of Major Protistan Groups*		
HETEROTROPHS *Decomposers, predators, grazers, parasites*		
Chytrids		
Water molds		
Cellular slime molds		
Plasmodial slime molds		
Protozoans:	Amoeboid protozoans (naked amoebas, foraminiferans, heliozoans, radiolarians)	
	Ciliated protozoans	
	Animal-like flagellated protozoans	
	Sporozoans	
AUTOTROPHS, HETEROTROPHS *Photosynthesizers, predators*		
Euglenoids		
Dinoflagellates		
MOSTLY AUTOTROPHS *Photosynthesizers, some parasites*		
Chrysophytes	Golden algae, yellow-green algae, coccolithophores, diatoms	
Red algae		
Brown algae		
Green algae		

11. Like animals, protozoans are predators, grazers, or parasites; some cause dangerous human diseases. The amoebas, foraminiferans, heliozoans, and radiolarians are amoeboid protozoans with or without hard parts. Ciliated protozoans use cilia for motility or for feeding. Animal-like protozoans live freely in aquatic habitats. Like sporozoans, some are parasites. *20.8, 20.9*

12. The euglenoids, chrysophytes, and dinoflagellates are mostly single celled (Table 20.3). Many are part of phytoplankton and undergo algal blooms. *20.11*

13. Red, brown, and green algae differ hugely in body plan, size, reproductive modes, and habitats. Most are multicelled photosynthesizers (Table 20.3). Some species of brown algae are the largest protistans. Plants may be descended from green algal ancestors. *20.12–20.14*

14. It has taken us more than one chapter to consider prokaryotes and protistans—the simplest eukaryotes. Table 20.4 summarizes the similarities and differences among the prokaryotes and eukaryotes.

Review Questions

1. Describe key metabolic and structural features of bacteria. Make sketches of the three basic shapes of bacterial cells. *20.1*

2. Describe one or two eubacteria that benefit humans. Name one or two that cause human diseases. *20.1, 20.2*

3. How is bacterial growth measured? *20.3*

4. Define and describe the characteristics of a virus. *20.4*

5. Review Table 20.4. Then cover it with a sheet of paper and list the major categories of protistans. *20.6, Table 20.4*

6. Select three protistan species, then briefly explain how they affect our affairs, such as human health. *20.8–20.11*

1. _____ are confined to extreme environments, probably much like those that prevailed on the early Earth.
   a. Cyanobacteria   c. Archaebacteria
   b. Eubacteria   d. Protozoans

2. Bacteria reproduce by _____ .
   a. mitosis   c. prokaryotic fission
   b. meiosis   d. longitudinal fission

3. A nondividing bacterial cell has _____ chromosome(s).
   a. one   b. two   c. four   d. many

4. Viruses have a _____ and a _____ .
   a. DNA core; carbohydrate coat
   b. DNA or RNA core; plasma membrane
   c. DNA or RNA core; protein coat

5. The multiplication cycle of all viruses requires _____ .
   a. penetration   b. budding   c. latency

6. During a _____ life cycle, amoeboid cells aggregate and form a migrating mass.
   a. slime mold   b. water mold   c. chytrid   d. sporozoan

7. Amoebas, foraminiferans, and radiolarians are _____ .
   a. protozoans   b. ciliates   c. algae   d. sporozoans

8. Trypanosomes cause which disease(s)?
   a. toxoplasmosis   d. African sleeping sickness
   b. Chagas disease   e. malaria
   c. amoebic dysentery   f. both b and d

9. Euglenoids and chrysophytes are mostly _____ .
   a. photoautotrophs   c. heterotrophs
   b. chemoautotrophs   d. omnivorous

10. Single-celled photosynthetic protistans, which include most euglenoids, chrysophytes, and dinoflagellates, are members of _____ , the "pastures" of most aquatic habitats.
    a. zooplankton   c. brown algae
    b. red algae   d. phytoplankton

11. Match the terms with the most suitable descriptions.
    ____ archaebacteria   a. bacterial cell division mechanism
    ____ eubacteria   b. nonliving infectious particle
    ____ virus   c. agent of malaria
    ____ *Plasmodium*   d. methanogens, extreme halophiles, and extreme thermophiles
    ____ prokaryotic fission   e. most common prokaryotic cells

Critical Thinking

1. Think about Figure 20.1. What does it tell you about the risks that drug abusers take when they share unsterilized needles? What does it tell you about medical uses of needles in developing countries where hygiene is marginal?

2. *Salmonella* bacteria cause food poisoning. Often they live in poultry and eggs. Newly hatched chicks acquire them by eating bacteria-laden feces of healthy adult chickens. Harmless bacteria ingested the same way colonize intestinal cells, leaving no place for *Salmonella* to take hold. Some farmers raise thousands of chicks in confined quarters with no adult chickens. Should they feed chicks a known mixture of bacteria from a lab or a mixture of unknown bacteria from healthy adult chickens? Devise an experiment to test which might stop *Salmonella* colonization.

3. Some strains of picornaviruses cause the highly contagious *foot and mouth disease*. Their hosts include cattle, water buffalo, sheep, and pigs. Symptoms include extensive lesions and major tissue erosion, lameness, weight loss, milk loss, and often death.

Table 20.4   Comparison of Prokaryotes With Eukaryotes

	PROKARYOTES	EUKARYOTES
Organism	Bacteria only	Protistans, fungi, plants, and animals
Level of organization	Single-celled	Single-celled (protistans, mostly), or multicelled, usually tissues, organs
Cell size	Small (1–10 μm)	Large (10–100 μm)
Cell wall	Nearly all are walled	Cellulose or chitin; none in animals
Organelles	Very rarely	Typically profuse
Metabolism	Anaerobic, aerobic	Aerobic modes predominate
Genetic material	Bacterial chromosome, sometimes plasmids	Complex chromosomes (DNA, many associated proteins) within a nucleus
Mode of division	Prokaryotic fission, mostly; also budding	Nuclear division (mitosis, meiosis, or both), and then cytoplasmic division

The viral strain causing a recent epidemic was first identified in Asia, then in Europe and elsewhere. It devastated farmers still recovering from an outbreak of BSE. There is no vaccine; at this writing, more than a million animals have been destroyed to stop the spread of the disease. How is the virus influencing travel and the global economy? *Hint:* Search for foot and mouth disease on the Internet, starting with http://www.cdc.gov/.

4. Suppose you vacation in a developing country. Sanitation and standards of personal hygiene are poor. Having read about parasitic protozoans in water and damp soil, what would you consider safe to drink? Which foods may be best to avoid or which food preparation methods might make them safe to eat?

5. As you read in this chapter, input from heavily fertilized cropland or raw sewage fans algal blooms. Think about it. Do you accept massive kills of aquatic species, birds, and other forms of wildlife as an unfortunate but necessary side effect of human life? If not, how would you stop the pollution? If you could stop it, what sorts of measures would you take to feed the enormous human population, which is dependent on high-yield (and very heavily fertilized) crops? How do you suggest we dispose of or prevent the accumulation of fecal matter and other wastes of more than 6.1 billion people?

Selected Key Terms

algal bloom *20.11*
archaebacterium *20.1*
bacterial conjugation *20.3*
bacteriophage *20.4*
binary fission *20.8*
brown alga *20.13*
chrysophyte *20.11*
chytrid *20.7*
contractile vacuole *20.8*
cyst *20.8*
dinoflagellate *20.11*
disease *20.10*

epidemic *20.10*
eubacterium *20.1*
euglenoid *20.11*
extreme halophile *20.2*
extreme thermophile *20.2*
green alga *20.14*
heterocyst *20.2*
infection *20.10*
lysis *20.5*
methanogen *20.2*
microorganism *CI*
numerical taxonomy *20.1*

pathogen *CI*
plankton *20.8*
prion *20.5*
prokaryotic fission *20.3*
protistan *20.6*
protozoan *20.8*
red alga *20.12*
slime mold *20.7*
sporozoan *20.9*
viroid *20.5*
virus *20.4*
water mold *20.7*

Readings

Margulis, L. 1993. *Symbiosis in Cell Evolution.* Second edition. New York: Freeman. Paperback.

Satchell, M. 28 July 1997. "The Cell From Hell." *U.S. News and World Report,* 26–28.

On-Line readings at Student Guide for InfoTrac:
www.brookscole.com/biology

21

FUNGI

Ode to the Fungus Among Us

When push comes to shove, plants need certain fungi more than they need us. (Actually, they don't need us at all.) Fungi were there, as symbionts, when plants first invaded the land. **Symbiosis**, recall, refers to species that live together and closely interact. In cases of **mutualism**, the interaction benefits both partners or does one of them no harm. Lichens and mycorrhizae are like this.

A **lichen** is a vegetative body in which a fungus is intertwined with one or more photosynthetic organisms. From the Antarctic to the Arctic, lichens live in habitats that are just too hostile to support most organisms.

Lichens absorb minerals from substrates. They make antibiotics against bacteria that can decompose them. They make toxins against invertebrate larvae that graze upon them. Coincidentally, their metabolic products enrich soil or help form new soil from bedrock. This is what happens when lichens colonize barren sites, such as bedrock exposed by retreating glaciers. Conditions improve so much, other species move in and replace the pioneers. This probably happened when plants invaded the land. Cyanobacteria-containing lichens even help maintain ecosystems. They capture nitrogen and convert it to a form that plants use. *Lobaria* alone dishes up 20 percent of the nitrogen needed by old-growth forests in the Pacific Northwest (Figure 21.1).

Lichens also give early warning of deteriorating environmental conditions. How? They absorb toxins but can't get rid of them. From studies in New York City and in England, we know that when lichens die around human habitats, air pollution is getting bad.

a SULFUR SHELF FUNGUS *Polyporus*

Figure 21.2 Fungal species from southeastern Virginia. This sampling hints at the rich diversity within the kingdom Fungi.

Some fungi and young tree roots also are mutualists. Their interaction is a **mycorrhiza** (plural, mycorrhizae), which means "fungus-root." Underground parts of the fungus grow through the soil and afford a huge surface area for absorption. The fungus swiftly takes up many ions of phosphorus and other minerals when these are abundant, and it releases ions to the plant when they are scarce. And what does the plant give the fungus in return? It gives up some sugars. The loss is a trade-off; many plants can't grow well without mycorrhizae.

Many fungi also help plants by being **decomposers**. Figure 21.2 shows just a few of these beneficial species. Like all other decomposers, they break down organic compounds in their surroundings. But few organisms besides fungi digest dinner while it's still on the table. As fungi grow into or on organic matter, their enzyme secretions digest it into bits that fungal cells absorb. We call this mode of nutrition **extracellular digestion and absorption**. Plants benefit when they take up some of the released nutrients. And plants, remember, are the primary producers of nearly all ecosystems.

Keep this global perspective in mind. Why? As you will see, some fungi do cause diseases in humans, pets

Figure 21.1 *Lobaria oregana*, one of the premier lichens.

Figure 21.7 Life cycle of the black bread mold *Rhizopus stolonifer*. Asexual phases are the most common. Mating strains (+ and −) also can reproduce sexually. Either way, haploid spores and mycelia form. A + hypha and − hypha fuse. Two gamete-producing structures form; each has a few haploid nuclei. Their nuclei fuse to form a zygote. The zygote develops a thick wall; it is now a zygospore. It may be dormant for months. Meiosis occurs when the zygospore germinates. Asexual spores form in spore sacs.

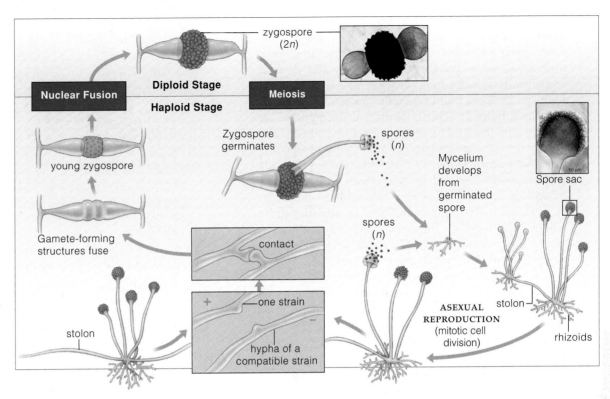

what *Rhizopus stolonifer*, the black bread mold, does to bread. A thick-walled sexual spore—a diploid zygote—forms when it reproduces sexually. A **zygosporangium**, a thin, clear covering, encloses the zygote (Figure 21.7). The zygote proceeds through meiosis. Then it gives rise to a specialized hypha, which bears a spore sac. Some number of spores form in the sporangium; each may give rise to a mycelium. Stalked hyphae grow out from such mycelia. Asexual spores form inside the spore sac on top of each stalk (Figure 21.7).

Or consider the 30,000+ kinds of sac fungi. The vast majority are multicelled. **Ascospores**, the sexual spores of sac fungi, form in sac-shaped cells. In multicelled species, reproductive structures of tightly intertwined hyphae enclose the cells, which we call asci. They look like flasks, globes, and cups (Figures 21.2*b* and 21.6).

The vast majority of sac fungi are multicelled. They include high-priced truffles and morels (Figure 21.6*d*). Truffles are symbionts with roots of hazelnut and oak trees. Pigs and dogs are trained to snuffle out truffles. Other multicelled sac fungi include certain species of *Penicillium* that "flavor" Camembert and Roquefort cheeses and make penicillins, used as antibiotics. We use *Aspergillus* to make citric acid for candies and soft drinks, and to ferment soybeans for soy sauce. Most red, bluish-green, and brown fungal molds that spoil stored foods also are multicelled. The salmon-colored *Neurospora sitophila* can run amok in bakeries and in research laboratories. It produces so many spores that it

is extremely difficult to eradicate. One of its relatives, *N. crassa*, is an important organism in genetic research.

Sac fungi also include about 500 species of single-celled yeasts (although still other yeasts are classified as club fungi). Yeasts reproduce sexually when two cells fuse and become a spore-producing sac. Some types live in the nectar of flowers and on fruits and leaves. Bakers and vintners put fermenting by-products of vast populations of yeasts to use. For example, the carbon dioxide by-products of *Saccharomyces cerevisiae* leaven bread. The commercial production of wine and beer depends on its ethanol end product. Many yeast strains with desired properties have been developed through artificial selection and genetic engineering. Then again, *Candida albicans*, a notorious relative of "good" yeasts, causes vexing infections in humans (Figure 21.6*f*).

What about imperfect fungi? Taxonomically, they are put in a holding station not because they are somehow defective but mainly because no one yet knows what kind of sexual spores they produce (if any). Previously orphaned *Aspergillus*, *Candida*, and *Penicillium* species were recently reunited with their kin—other sac fungi.

Through their rapid production of asexual and sexual spores, fungi take quick advantage of available organic matter, whether it has been discarded or is part of a living or dead organism. Their penchant for spore production is central to their success as decomposers and parasites.

THE SYMBIONTS REVISITED

Recall, from the introduction, that symbiosis refers to species that live together in close ecological association. Often one is a parasite's victim, not a partner. In cases of mutualism, interaction benefits both partners or does one of them no harm. In lichens, a fungus is intertwined with one or more photosynthetic species. Of 13,500 or so known types of lichens, nearly half incorporate sac fungi. Only about 100 species are photosynthetic, and most often these are green algae and cyanobacteria.

A lichen forms after the tip of a fungal hypha binds with a suitable host cell. Both lose their wall, and their cytoplasm fuses or the hypha induces the host cell to cup around it. The fungus and photosynthesizer grow and multiply together. The lichen commonly has distinct layers. The overall pattern of growth is leaflike, erect, pendulous, or flattened (Figures 21.1 and 21.8).

Lichens typically colonize sites that are hostile for most organisms, including sunbaked or frozen rocks, gravestones, even the tops of giant Douglas firs. Almost always, the fungus is the largest portion. Cyanobacteria typically reside in a separate structure inside or outside the main body. The fungus gets a long-term source of nutrients, which it absorbs from photosynthetic cells. Nutrient withdrawals affect the growth of the cells, but the lichen may help shelter them. When more than one fungus is present, it might be a mycobiont, a parasite, or even an opportunist using the lichen as a substrate.

What about those fungal mutualists with young tree roots? Plant growth improves with mycorrhizae, as you can see from Figure 21.9a. The hyphae of some species form a dense net around living cells in roots but do not penetrate them (Figure 21.9b). Others form a velvety

dispersal fragment
(cells of fungus and of
photosynthetic species)

outer layer
of fungal cells

photosynthetic
species

inner layer of
loosely woven
hyphae

outer layer
of fungal cells

a

b

Figure 21.8 (**a**) Structure of one kind of lichen, cross-section. (**b**) Encrusting, flattened lichens growing on a rock.

hyphal
strands

small, young
tree root

a

b

Figure 21.9 (**a**) Mycorrhiza of a hemlock tree. (**b**) Effects of the presence or absence of mycorrhizae on plant growth. The juniper seedlings at left are six months old. They were grown in sterilized, phosphorus-poor soil with a mycorrhizal fungus. The seedlings at right were grown without the fungus.

wrapping around tree roots as the mycelium radiates through soil. These are common in temperate forests. They help trees survive seasonal shifts in temperature and rainfall. About 5,000 fungal species, mostly club fungi, enter into such associations.

Different mycorrhizae develop in about 80 percent of all vascular plants. Fungal hyphae penetrate plant cells, as they do in lichens. Zygomycetes are the fungal partner. Their hyphae branch extensively, forming tree-shaped absorptive structures in plant cells. The hyphae also extend for several centimeters into the soil. Section 27.2 offers a closer look at these beneficial species.

A final note: Since the early 1900s, collectors have recorded data on wild mushroom populations in the forests of Europe. They find the number and kinds of fungi are severely declining. Mushroom gatherers can't be the cause, because inedible as well as edible species are vanishing. However, the decline does correlate with rising air pollution. Vehicle exhaust, smoke from coal burning, and emissions from nitrogen fertilizers pump ozone, nitrogen oxides, and sulfur oxides into the air. Normally, as a tree ages, one species of mycorrhizal fungus gives way to another, in predictable patterns. When fungi die, trees lose this vital support system, and they become vulnerable to severe frost and drought. Are the North American forests at risk, also? Conditions there are deteriorating in comparable ways.

Lichens and mycorrhizae are both symbiotic associations between fungi and other organisms, with mutual benefits.

SUMMARY

Gold indicates text section

1. Fungi are heterotrophs and major decomposers. The saprobes feed on nonliving organic matter; parasites feed on tissues of living organisms. Some fungal species are symbiotic partners with other organisms. Cells of all species secrete digestive enzymes that break down food to small molecules, which the cells absorb. *21.1*

2. Nearly all fungi are multicelled. The food-absorbing part, the mycelium, is a mesh of filaments (hyphae). The aboveground reproductive parts (e.g., mushrooms) form from tightly interwoven hyphae. *21.1, 21.2*

3. Major groups are zygomycetes, ascomycetes (sac fungi), and basidiomycetes (club fungi). Each produces distinctive sexual and asexual spores. When a sexual phase can't be detected or is absent from the life cycle, a fungus is assigned to an informal category called the imperfect fungi. *21.2, 21.4*

4. Lichens are mutualistic associations of fungi with photosynthesizers (e.g., green algae and cyanobacteria). Mycorrhizae are mutualistic associations of a fungus and young roots. Fungal hyphae give up nutrients to the plant and get back carbohydrates. *CI, 21.5*

Review Questions

1. Describe the fungal mode of nutrition, and explain how the structure of mycelia facilitates this mode. *21.1*

2. How does a lichen differ from a mycorrhiza? *CI, 21.5*

3. List conditions that influence fungal spore formation. *21.4*

4. What makes fungi such successful decomposers? *21.3*

Self-Quiz   ANSWERS IN APPENDIX III

1. A mycorrhiza is a _____ .
   a. fungal disease of the foot   c. parasitic water mold
   b. fungus–plant relationship   d. fungus of barnyards

2. Parasitic fungi obtain nutrients from _____ .
   a. tissues of living hosts   c. only living animals
   b. nonliving organic matter   d. none of the above

3. Saprobic fungi derive nutrients from _____ .
   a. nonliving organic matter   c. living animals
   b. living plants   d. both b and c

4. New mycelia form after _____ germinate.
   a. hyphae   b. mycelia   c. spores   d. mushrooms

5. A mushroom is _____ .
   a. the food-absorbing part of a fungal body
   b. the part of the fungal body not constructed of hyphae
   c. a reproductive structure
   d. a nonessential part of the fungus

6. Match the terms appropriately.
   ____ zygomycetes   a. mushrooms, shelf fungi
   ____ conidia   b. type of sexual spore
   ____ hypha   c. *Penicillium's* chains of asexual spores
   ____ club fungi   d. each filament in a mycelium
   ____ ascospore   e. black bread mold
   ____ sac fungi   f. truffles, morels, some yeasts

Figure 21.10 Reproductive structures of *Pilobolus*, a name from a Greek word for "hat-thrower." The "hats" actually are spore sacs.

Critical Thinking

1. *Pilobolus* is a type of fungus that commonly dines on horse dung. Each morning, stalked reproductive hyphae emerge from irregularly spaced piles of dung. By early afternoon, they have dispersed spores to sunlit grasses where horses feed. The spores pass through the horse gut unharmed and exit with their own pile of dung. At the tip of each stalked hypha is a dark-walled, spore-containing sac (Figure 21.10). Just below the sac, the stalk is differentiated into a vesicle, swollen with a fluid-filled central vacuole. At the base of the vesicle is a ring of light-sensitive, pigmented cytoplasm. The stalk bends as it grows until its wall is parallel with the sun's rays and light strikes all of the ring. When that happens, turgor pressure builds up inside the central vacuole until the vesicle ruptures. The forceful blast can propel spore sacs two meters away—which is amazing, considering that the stalk is less than ten millimeters tall. Reflect on the examples of fungi discussed in this chapter. Would you say *Pilobolus* is a zygomycete, club fungus, or sac fungus?

2. Renee sees in the laboratory that the fungus *Trichoderma* grows well in distilled water. It continues to do so even after she rigorously treats the water and glassware to remove all traces of organic carbon. This fungus is not a photoautotroph. Suggest a metabolic life-style that lets it grow under these conditions.

3. *Trichoderma* is being tested as a natural pest control agent. Laboratory experiments demonstrated that some strains of this fungus combat other fungi that cause plant diseases. Some even promote seed germination and plant growth. During one set of twenty trials, workers increased lettuce yields by 54 percent. What concerns must be addressed before *Trichoderma* can be released into the environment for commercial applications?

Selected Key Terms

ascospore *21.4*	fungus *21.1*	sac fungus
basidiospore *21.2*	hypha *21.1*	(ascomycetes) *21.1*
club fungus	lichen *CI*	saprobe *21.1*
(basidiomycetes) *21.1*	mushroom *21.2*	spore (fungal) *21.1*
decomposer *CI*	mutualism *CI*	symbiosis *CI*
extracellular digestion	mycelium *21.1*	zygomycetes *21.1*
and absorption *CI*	mycorrhiza *CI*	zygosporangium *21.4*
	parasite *21.1*	

Readings

Moore-Landecker, E. 1996. *Fundamentals of the Fungi.* Fourth edition. Englewood Cliffs, New Jersey: Prentice-Hall.

On-Line readings at Student Guide for InfoTrac:
www.brookscole.com/biology

PLANTS

Pioneers in a New World

Seven hundred million years ago, no shorebirds stirred and noisily announced the dawn of a new day. There were no crabs to clack their claws together and skitter off to burrows. The only sounds were the rhythmic, muffled thuds of waves in the distance, at the outer limits of another low tide. More than 3 billion years before, life had originated in the waters of the Earth. Now, quietly, the invasion of the land was under way.

Why did it happen then? Astronomical numbers of photosynthetic cells had come and gone, and oxygen-producing types had changed the atmosphere. High above the Earth, the sun's energy had converted much of the oxygen into a dense ozone layer. That layer was a shield against lethal doses of ultraviolet radiation, which had kept life beneath the surface of water and mud.

Were cyanobacteria the first to adapt to intertidal zones, where mud dried out with each retreating tide? Were they the first in the shallow, freshwater streams flowing down to the coasts? Probably. Fossils suggest that later in time, some green algae and fungi made the journey together. Ancient species of green algae that lived near the water's edge or made it onto land gave rise to plants. And through millions of years thereafter, descendant plants became the basis of communities in coastal lowlands, near the snow line of high mountains, and in just about all places in between (Figure 22.1).

We have tantalizing fossils of the first pioneers. We are learning more about them through comparative studies of many modern species. Today, as in the late precambrian, cyanobacteria and green algae form mats in nearshore waters and along freshwater streams of the sort shown in Figure 22.1a. After a volcano erupts or a glacier retreats, cyanobacteria are the first to colonize the barren rocks. Symbiotic interactions between green algae and fungi follow. Their organic by-products and remains accumulate and create pockets of soil. Mosses and other plants soon become established in the newly forming soil and further enrich it.

With this new chapter we turn to the story of plants. Nearly all species are multicelled photoautotrophs that absorb energy from the sun, carbon dioxide from the air, and minerals dissolved in water to synthesize organic compounds. These metabolic wizards also split water molecules. In doing so, they get stupendous numbers of the electrons and hydrogen atoms required for growth into multicellular forms as tall as redwoods, as vast as an aspen forest that is actually one interconnected clone.

We know of at least 295,000 kinds of existing plants. Be glad their ancient ancestors left the water. Without those pioneers in a new world, we humans and other land-dwelling animals never would have made it onto the evolutionary stage.

Figure 22.1 (**a**) Filaments of a green alga massed in a shallow stream. More than 400 million years ago, green algal species that may have been ancestral to all plants lived in similar streams that meandered down to the shores of ancient continents. Land-dwelling descendants of ancestral forms: (**b**) Moss growing on rocks. (**c**) Ponderosa pine high above the floor of Yosemite Valley in the Sierra Nevada of California. (**d**) Flowers of one of the most highly prized of all flowering plants—orchids—growing on a branch of a living tree in a tropical rain forest. With this chapter, we turn to the beginning—and end of the line—of some ancient lineages.

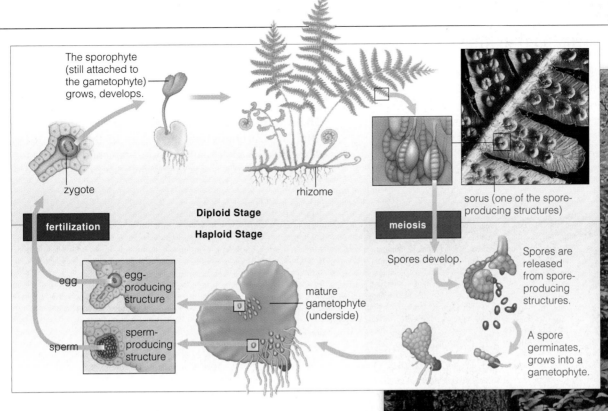

Figure 22.9 Life cycle of a fern. The photograph shows ferns growing in a moist habitat in Indiana. Ferns with finely divided fronds are in the foreground.

The following labels appear within the figure:

The sporophyte (still attached to the gametophyte) grows, develops.

zygote

rhizome

Diploid Stage

fertilization

Haploid Stage

meiosis

sorus (one of the spore-producing structures)

Spores develop.

Spores are released from spore-producing structures.

egg

egg-producing structure

mature gametophyte (underside)

A spore germinates, grows into a gametophyte.

sperm

sperm-producing structure

stems of one species. Its spores give rise to free-living gametophytes 1 millimeter to 1 centimeter across. The sporophytes of most horsetails have rhizomes, hollow photosynthetic stems, and scale-shaped leaves. Stems contain a ringlike array of xylem and phloem strands. Ribs reinforced with silica structurally support the stems and give them a texture like sandpaper. Pioneers of the American West, who did not have many places to store and wash towels, gathered horsetails on their westward journeys and used them as pot scrubbers.

Ferns

With 12,000 or so species, the ferns (Pterophyta) are the largest and the most diverse group of seedless vascular plants. All but about 380 are native to the tropics, but you find them in gardens and in homes throughout the world. Their size range is stunning. The leaves of some floating types are less than 1 centimeter across; some of the tropical tree ferns are 25 meters (82 feet) tall. One climbing fern has a modified leaf stalk 30 meters long.

Most ferns have vascularized rhizomes that give rise to roots and leaves. Exceptions include tropical tree ferns and epiphytes. (*Epiphyte* refers to any aerial plant that grows attached to tree trunks or branches.) While they develop, young fern leaves (fronds) are coiled in a way that resembles a fiddlehead. When mature, these fronds commonly are divided into leaflets.

You may have noticed rust-colored patches on the lower surface of many fern fronds. Each patch, a cluster of spore-producing structures, is a sorus (plural, sori). At dispersal time, the spore-producing structures snap open with such force that the released spores catapult through the air. Each spore that germinates develops into a small gametophyte. You can see an example, a green, heart-shaped gametophyte, in Figure 22.9.

Seedless vascular plants (whisk ferns, lycophytes, horsetails, and ferns) have sporophytes adapted to conditions on land. Yet they have not entirely escaped their aquatic ancestry. When reproducing sexually, they must have water clinging to them so their flagellated sperm can reach the eggs.

THE RISE OF THE SEED-BEARING PLANTS

About 360 million years ago, as the Devonian gave way to the Carboniferous, the first seed-bearing plants arose. In terms of diversity, numbers, and distribution, they would become the most successful groups of the plant kingdom. Seed ferns, gymnosperms, and (much later) angiosperms were the dominant groups. They differed from seedless vascular plants in three crucial respects.

First, compared with seedless vascular plants, the gymnosperms had thickened cuticles, stomata recessed below leaf surfaces, and other water-conserving traits. These traits gave gymnosperms the competitive edge after the Carboniferous gave way to the cooler, drier climates of the Permian. Swamps vanished; and cycads, conifers, and other gymnosperms rose to dominance.

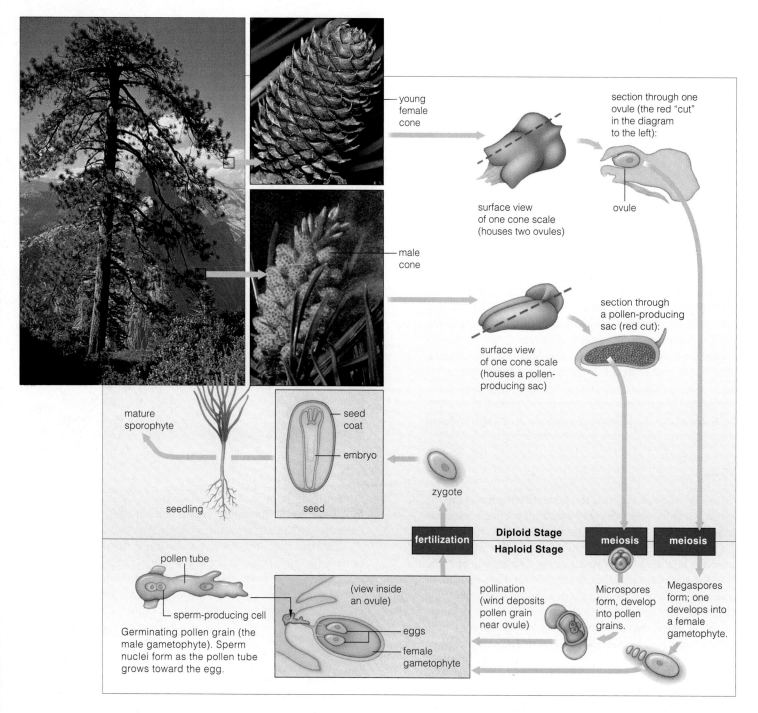

Figure 22.10 Life cycle of one of the conifers, the ponderosa pine (*Pinus ponderosa*).

	CHORDATES
	ECHINODERMS
	ARTHROPODS
	ANNELIDS
COELOMATE ANCESTRY	MOLLUSKS
	ROTIFERS
	ROUNDWORMS
BILATERAL ANCESTRY	FLATWORMS
RADIAL ANCESTRY	CNIDARIANS
MULTICELLED ANCESTRY	SPONGES

SINGLE-CELLED, PROTISTAN-LIKE ANCESTORS

Figure 23.2 A simple evolutionary tree diagram that shows presumed relationships among major groups of animals. Take a moment to study this diagram. We will use it repeatedly as a road map through our discussions of each group, starting with the invertebrates and ending, in the chapter to follow, with chordates —including young Madeleine (*above*).

As you poke through the branches of the animal family tree, keep the greater evolutionary story in mind. At each branch point, microevolutionary processes gave rise to workable changes in body plans. Madeleine's uniquely human traits, and yours, emerged through modification of certain traits that had evolved earlier in countless generations of vertebrates and, before them, in ancient invertebrate forms.

Key Concepts

1. All animals are multicelled, aerobic heterotrophs that ingest or parasitize other organisms. Nearly all kinds have tissues, organs, and organ systems, and most are motile during at least part of their life cycle. Animals reproduce sexually and often asexually, and their embryos develop through a series of continuous stages.

2. Animals originated late in the precambrian. More than 2 million modern species have been identified. Of these, more than 1,950,000 are invertebrates—animals without a backbone. Fewer than 50,000 species are vertebrates—animals with a backbone.

3. Comparisons of the body plans of existing animals, in conjunction with the fossil record, reveal that there were several trends in the evolution of certain lineages. The most revealing aspects of an animal's body plan are its type of symmetry, gut, and cavity (if any) between the gut and body wall; whether it has a distinct head end; and whether it is divided into a series of segments.

4. The placozoans and sponges are structurally simple animals with no body symmetry. Both are at the cellular level of body construction. Cnidarians and comb jellies show radial symmetry, and they are at the tissue level of body construction.

5. Flatworms, roundworms, rotifers, and nearly all other animals that are more complex than the cnidarians show bilateral symmetry. Complex animals have tissues, organs, and organ systems.

6. Not long after the flatworms evolved, divergences began that were the start of two major lineages. One evolutionary branching gave rise to mollusks, annelids, and arthropods. The other gave rise to echinoderms and chordates.

7. By biological measures, including diversity, sheer numbers, and distribution, the arthropods—especially insects—have been the most successful animal group.

OVERVIEW OF THE ANIMAL KINGDOM

General Characteristics of Animals

What, exactly, are **animals**? We can only define them by a list of characteristics, not with a sentence or two. *First*, animals are multicelled. In most cases their body cells form tissues that become arranged as organs and organ systems. The body cells of nearly all species have a diploid chromosome number. *Second*, all animals are heterotrophs that get carbon and energy by ingesting other organisms or by absorbing nutrients from them. *Third*, animals require oxygen for aerobic respiration. *Fourth*, animals reproduce sexually and, in many cases, asexually. *Fifth*, most animals are motile during at least part of the life cycle. *Sixth*, the life cycle includes stages of embryonic development. Briefly, mitotic cell divisions transform an animal zygote into a multicelled embryo. The embryonic cells give rise to primary tissue layers: **ectoderm**, **endoderm**, and, in most species, **mesoderm**. The layers in turn give rise to all tissues and organs of the adult, as described in Section 29.6 and Chapter 39.

Diversity in Body Plans

Mammals, birds, reptiles, amphibians, and fishes are the most familiar animals. All are **vertebrates**, the only animals with a "backbone." And yet, of probably more than 2 million species of animals, fewer than 50,000 are vertebrates! What we call the **invertebrates** are animals with plenty of diverse features, but not a backbone.

Animals are grouped into more than thirty phyla. Table 23.1 lists the groups we describe in this book. The characteristics they share with one another arose early in time, before divergences from a common ancestor gave rise to separate lineages. Later, as morphological differences accumulated among them, the lineages took off in amazingly diverse directions. How might we get a conceptual handle on their modern-day descendants— on animals as different as flatworms, hummingbirds, spiders, toads, humans, and giraffes? We can compare their similarities and differences with respect to five basic features. These are body symmetry, cephalization, type of gut, type of body cavity, and segmentation.

BODY SYMMETRY AND CEPHALIZATION With very few exceptions, animals are radial or bilateral. Those with **radial symmetry** have body parts arranged regularly around a central axis, like spokes of a bike wheel. Thus a cut down the center of a hydra (Figure 23.3a) divides it into equal halves; another cut at right angles to the first divides it into equal quarters. Radial animals live

Table 23.1 *Animal Phyla Described in This Book*

Phylum	Some Representatives	Existing Species
PLACOZOA *Trichoplax*	Simplest animal; plate-like but with layers of cells	1
PORIFERA Poriferans	Sponges	8,000
CNIDARIA Cnidarians	Hydrozoans, jellyfishes, corals, sea anemones	11,000
PLATYHELMINTHES Flatworms	Turbellarians, flukes, tapeworms	15,000
NEMATODA Roundworms	Pinworms, hookworms	20,000
ROTIFERA Rotifers	Tiny body with crown of cilia, great internal complexity	2,000
MOLLUSCA Mollusks	Snails, slugs, clams, squids, octopuses	110,000
ANNELIDA Segmented worms	Leeches, earthworms, polychaetes	15,000
ARTHROPODA Arthropods	Crustaceans, spiders, insects	1,000,000+
ECHINODERMATA Echinoderms	Sea stars, sea urchins	6,000
CHORDATA Chordates	Invertebrate chordates: Tunicates, lancelets	2,100
	Vertebrates: Fishes	21,000
	Amphibians	4,900
	Reptiles	7,000
	Birds	8,600
	Mammals	4,500

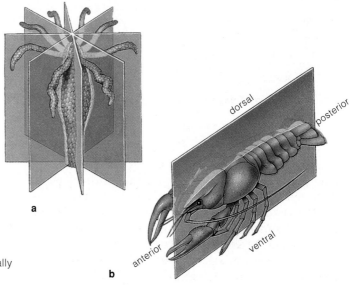

Figure 23.3 Examples of body symmetry. (**a**) A hydra, which is radially symmetrical, and (**b**) a bilaterally symmetrical crayfish.

Figure 23.4 Type of body cavity (if any) in animals.

a No coelom (*acoelomate* animals)

b Pseudocoel (*pseudocoelomate* animals)

c Coelom (*coelomate* animals)

in water. Their body plan is adapted to intercepting food that is coming toward them from any direction.

Animals with **bilateral symmetry** have a right half and left half that are mirror images of each other. Most have an *anterior* end (head) and an opposite, *posterior* end. They have a *dorsal* surface (a back) and an opposite, *ventral* surface (Figure 23.3*b*). As fossils show, this body plan evolved among the first forward-creeping species. Their forward end encountered food and other stimuli first, so there was selection for **cephalization**. By this evolutionary process, sensory structures and nerve cells became more concentrated in the head of many animals. Bilateral body plans and cephalization evolved jointly, in that heritable changes in one affected changes in the other. Over time, pairs of muscles and pairs of sensory structures, nerves, and brain regions were the outcome.

TYPE OF GUT The **gut**, a sac projecting into the body or part of a tube through it, is where food is digested, then absorbed into the internal environment. A saclike gut has a single opening (mouth) for taking in food and disposing of residues. A tubular gut is part of a digestive system that has openings at two ends (mouth and anus). Such "complete" systems have specialized regions that break down, digest, absorb, store, and get rid of material. As more efficient digestive systems evolved, they helped make possible increases in body size and activity.

BODY CAVITIES In between the gut and the body wall of most bilateral animals is a body cavity (Figure 23.4). One type of cavity, a **coelom**, has a unique tissue lining called a peritoneum. This lining also encloses organs in the coelom and helps hold them in place. For example, your body has a coelom, and a sheetlike muscle called a diaphragm divides it into two smaller cavities. Your heart and lungs are positioned in the upper (thoracic) cavity, and your stomach, intestines, and other organs occupy the lower (abdominal) cavity.

The coelom turned out to be a key innovation in the evolution of larger, more complex animals from small ancestral forms. It favored increases in size and activity by cushioning and protecting internal organs.

Some of the less complex invertebrates don't even have a body cavity; tissues fill the region between their gut and body wall. Others have a pseudocoel ("false coelom")—a body cavity, but no peritoneum.

SEGMENTATION Segmented animals have a repeating series of body units that may or may not be similar to one another. The many segments of earthworms have a similar outward appearance. Insect segments are fused into three units (head, thorax, and abdomen) and differ greatly from one another. Especially among the insects, diverse head parts, legs, wings, and other appendages evolved from less specialized segments.

Animals are multicelled, aerobically respiring heterotrophs that ingest other organisms or absorb nutrients from them. Nearly all animals have tissues, organs, and organ systems. Most have a diploid chromosome number.

Animals reproduce sexually and, in many cases, asexually. They go through a period of embryonic development, and most are motile during at least part of the life cycle.

Animal body plans differ in body symmetry, cephalization, type of gut, type of body cavity, and segmentation.

PUZZLES ABOUT ORIGINS

Judging from recent genetic evidence and radiometric dating of fossilized tracks, burrows, and microscopic embryos, animals originated between 1.2 billion and 670 million years ago, during precambrian times. *Where did they come from?* They probably evolved from protistan lineages, but we don't know which ones (Section 19.4).

By one hypothesis, the forerunners of animals were ciliates, like *Paramecium*, and had multiple nuclei in a one-celled body. As they evolved, each nucleus became compartmentalized in individual cells of a multicelled body. However, we don't know of any existing animal that develops by compartmentalization.

By another hypothesis, multicelled animals arose from spherical colonies of a number of flagellated cells, maybe like the *Volvox* colony shown in Section 20.14. In time, as a result of mutation, some cells in the colony became modified in ways that enhanced reproduction and other specialized tasks. And so began the division of labor that characterizes multicellularity.

Suppose colonies became flattened over time and started creeping over the seafloor. A creeping life-style could have favored the evolution of layers of cells, such as those of *Trichoplax adhaerens*. This is the only known **placozoan** (Placozoa, after *plax*, meaning plate; and *zoon*, meaning animal). *T. adhaerens* is a soft-bodied marine animal shaped a bit like pita bread. It has no symmetry and no mouth. It has several thousand cells organized in two layers. It briefly humps up when its body glides over food (Figure 23.5). Gland cells in the lower layer secrete digestive enzymes onto the food, and individual cells absorb breakdown products. Reproduction may be asexual (by budding or fission) or sexual, by means not yet understood. In sum, structurally and functionally, *Trichoplax* is as simple as animals get.

Figure 23.5 Cutaway views of *Trichoplax adhaerens*, an animal with a two-layer body measuring about three millimeters across.

Possibly the question of origins requires more than one answer. It may be that different lineages descended from more than one group of protistan-like ancestors.

Multicelled animals arose from protistan-like ancestors that may have resembled ciliates, colonial flagellates, or both.

SPONGES—SUCCESS IN SIMPLICITY

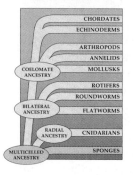

Sponges (Porifera) are animals with no symmetry, no tissues, and no organs. Even so, they are one of nature's success stories. Sponges have been abundant ever since the precambrian, particularly in coral reefs and coastal waters. Of 8,000 or so known species, only about 100 live in fresh water. Sponges are home to tiny worms, shrimps, and many other animals. Some species are big enough for you to sit in; others are as small as a fingernail. Their shapes are flattened, sprawling, lobed, compact, tubelike, and vaselike (Figures 23.6 and 23.7).

The sponge body has no symmetry. Flattened cells line its outer surface and inner cavities. But the linings aren't much more specialized than those cell layers of *Trichoplax*, and they also differ from the tissues of other animals. Amoeboid cells live in a gelatin-like substance between the two linings (Figure 23.7*b*).

Typically, tough protein fibers and sharp, glasslike spicules of calcium carbonate or silica stiffen the body. These skeletal elements may be a reason why sponges as a group have endured so long. Cleveland Hickman put it this way: Most potential predators discover that sampling any sponge is about as pleasant as eating a mouthful of glass splinters embedded in fibrous gelatin. Besides, chemically speaking, many sponges stink.

Water flows into a sponge's body through many microscopic pores and chambers, then out through one or more large openings. It does so when thousands or millions of phagocytic **collar cells** beat their flagella. The cells are part of the inner lining (Figure 23.7*c*). The "collars" are food-trapping structures. Bacteria and other edibles get trapped in them, then are engulfed. Some of the food is transferred to the amoebalike cells for further breakdown, storage, and distribution.

Figure 23.6 A sprawling, red-orange sponge, one of many types that encrust underwater ledges in temperate seas.

Figure 23.7 (**a,b**) Body plan of a simple sponge. The outer lining consists of flattened cells. It also has some contractile cells arranged around the large opening at the top. Their slow contraction directs water flow through the body. In the gelatin-like matrix between the inner and outer linings, amoeba-like cells secrete the materials for spicules and fibers. Other amoeboid cells digest and transport food. They don't lose the capacity to divide, and their cellular descendants can differentiate into any other type of sponge cell. They also have roles in asexual processes, such as gemmule formation.

(**c**) Many flagellated, phagocytic cells line the body's inner canals and chambers. Each has a collar of food-trapping structures called microvilli. Fine filaments connect microvilli to each other to form a "sieve" that strains food particles from the water. Cells at the base of the collar engulf the trapped food.

(**d**) A marine sponge, Venus's flower basket (*Euplectella*). Its silica spicules are fused in a rigid network. A thin layer of cells stretches over them. At the base of the body is an anchoring tuft of spicules. (**e**) Basket sponge releasing a cloud of sperm into the water.

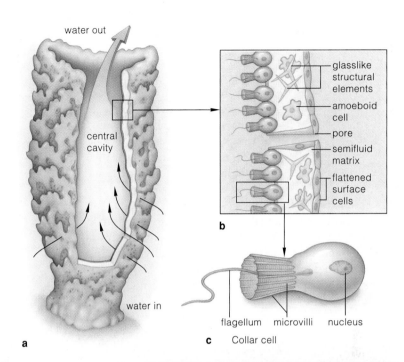

glasslike structural elements

amoeboid cell

pore

semifluid matrix

flattened surface cells

b

water out

central cavity

water in

a

flagellum   microvilli   nucleus

c   Collar cell

d

e

When most sponges reproduce sexually, they simply release sperm into the water (Figure 23.7*e*). But they retain the fertilized eggs and the embryos as they start developing. Each new, young sponge proceeds through a microscopic, swimming larval stage. A **larva** (plural, larvae) is a sexually immature stage that grows and develops into an adult, the sexually mature form of the species. As you will see, the life cycle of many animal species includes larval stages.

Some kinds of sponges also reproduce asexually by fragmentation; small fragments break away from the parent and grow into new sponges. Most freshwater species also reproduce asexually by way of gemmules. These are clusters of sponge cells, some of which form a hard covering around others. The clusters inside are protected from extreme cold or drying out. Later, when favorable conditions return, the gemmules germinate and establish a new colony of sponges.

Sponges have no symmetry, tissues, or organs; they are at the cellular level of construction. Yet they have successfully endured through time, possibly because most predators find their spicule-rich and often stinky bodies unappetizing.

CNIDARIANS—TISSUES EMERGE

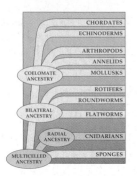

Cnidarians (Cnidaria) are radial, tentacled animals, most of which live in the seas. Jellyfishes belong to the group called scyphozoans. The anthozoans (including sea anemones) and hydrozoans (such as *Hydra*) also are in the phylum. Of roughly 11,000 species, fewer than 50 are found in fresh water. All cnidarians are at the tissue level of organization. They alone make **nematocysts**, capsules with tubular threads that can be discharged. Many threads entangle prey or trap it in a sticky substance oozing from their tip. Some have prey-piercing barbs; others have toxin-delivering tips that sometimes inflict painful stings (Figure 23.8). Hence the name Cnidaria, after the Greek word for nettle.

nematocyst (capsule at free surface of epidermal cell)

Figure 23.8 One type of nematocyst before and after a prey organism (not shown) touched its trigger. The contact made the capsule more "leaky" to water. Water diffused inward. Turgor pressure built up in the capsule and forced the thread to turn inside out. The thread tip swiftly pierced the prey's body.

- outer epithelium (epidermis)
- mesoglea (matrix)
- inner epithelium (gastrodermis)

a   **b**

Cnidarian Body Plans and Life Cycles

The **medusa** (plural, medusae) and **polyp** are the most common body forms of cnidarians. Both have a saclike gut, as in Figure 23.9. Medusae float and look like bells or saucers. The mouth, centered under the bell, often has tentacles and other extensions used in prey capture and feeding. The tubelike polyps have a tentacle-ringed mouth at one end; the other end attaches to a substrate.

Trichoplax draped over and digesting food is like a gut on the run. But the cnidarian gut is a permanent food-processing chamber. Its sheetlike inner lining, the gastrodermis, has glandular cells that secrete digestive enzymes. Epidermis lines the body surfaces (Figure 23.9). Each lining is an **epithelium** (plural, epithelia), a tissue with one free surface bathed in some body fluid or exposed to the outside. All animals more complex than sponges have this tissue. In cnidarians, it incorporates a "nerve net" of interacting **nerve cells**. These receive signals from receptors, then signal **contractile cells** to carry out responses. Such cells can shorten and go back to their original length. The nerve net, a simple nervous system, only controls movement and shape changes.

Between the outer and inner linings is a layer of gelatinous secreted material called mesoglea ("middle jelly"). The volume of mesoglea in jellyfishes imparts buoyancy and serves as a firm yet deformable skeleton

mesoglea-filled bell

tentacles

c   **d**

Figure 23.9 The common cnidarian body plans: (**a**) medusa and (**b**) polyp, midsection. (**c**) Sea wasp (*Chironix*), a jellyfish with tentacles up to fifteen meters long. Its toxin can kill a human within minutes. (**d**) Hydrozoan polyp (*Hydra*) capturing and digesting its prey. (**e**) Sea anemone using its hydrostatic skeleton to escape from a sea star. It closes its mouth, and contractile cells in epithelial tissues generate force against water inside the gut. The body changes shape and thrashes about, generally away from the predator.

part of the water-vascular system

sieve plate

ring canal

ampulla

b

c

d

tube feet on the ventral surface of a sea star arm

upper stomach
lower stomach
anus
gonad
ossicle (stiffening, support structure)
spine
coelom
digestive gland
eyespot
spine
ossicle
a

Figure 23.33 (**a**) Key aspects of the radial body plan of a sea star. Its water-vascular system, combined with great numbers of tube feet, is the basis of locomotion. (**b–d**) Five-armed sea star, with closer views of tube feet.

can become the leader, directing the rest of the body to move in a direction suitable for capturing its prey.

Figures 23.32 and 23.33 show examples of tube feet. These fluid-filled, muscular structures have suckerlike adhesive disks. Sea stars use their tube feet for walking, burrowing, clinging to rocks, or gripping a clam or snail about to become a meal. Tube feet are components of a **water-vascular system** unique to echinoderms. In sea stars, that system includes a main canal in each arm. Short side canals extend from them and deliver water to the tube feet. Each tube foot contains a fluid-filled, muscular structure shaped rather like the rubber bulb on a medicine dropper. As the bulb contracts, it forces fluid into the foot and causes it to lengthen.

Tube feet change shape constantly as muscle action redistributes fluid through the water-vascular system. Hundreds of tube feet may move at a time. After being released, each one swings forward, reattaches to the substrate, then swings backward and is released before swinging forward again. Their motions are splendidly coordinated, so sea stars glide rather than lurch along.

On their ventral surface, sea stars have a formidable feeding apparatus, such as the one shown here:

Some eager sea stars simply swallow their prey whole. Others push part of their stomach outside the mouth and around their prey, then start digesting their meal even before swallowing it. Sea stars get rid of coarse, undigested residues through the mouth. They do have a small anus, but this is of no help in getting rid of an empty clam shell or snail shell.

With their curious traits, echinoderms are a suitable point of departure for this chapter. Even though broad trends have been identified in animal evolution, keep in mind that there are confounding exceptions to the perceived macroevolutionary patterns.

Echinoderms are coelomate animals with spines, spicules, or plates in the body wall. From the evolutionary perspective, they are a puzzling mix of bilateral and radial features.

SUMMARY   **Gold** indicates text section

1. Animals are multicelled, aerobic heterotrophs that ingest or parasitize other organisms. Most kinds have diploid body cells organized into tissues, organs, and organ systems. Animals reproduce sexually and often asexually. They undergo embryonic development. Most are motile during at least part of the life cycle. *23.1*

2. Animals range from structurally simple placozoans and sponges to vertebrates. *23.1*

    a. By comparing major animal phyla and integrating the information with the fossil record, biologists have identified major evolutionary trends among them.

    b. Revealing aspects of body plans are the type of symmetry, gut, and cavity (if any) between the gut and body wall; whether there is a head end; and whether the body is divided into segments (Figure 23.34).

3. *Trichoplax*, the only known placozoan, is the simplest animal. It is composed of little more than two layers of cells with a fluid matrix in between. *23.2*

4. A sponge body has no symmetry. It is at the cellular level of construction. Although it consists of several kinds of cells, these are not organized like the epithelia and other tissues seen in complex animals. *23.3*

5. Cnidarians include the jellyfishes, sea anemones, and hydras. They have radial symmetry, and they are at the tissue level of construction. Cnidarians alone produce nematocysts, capsules with dischargeable threads that are used mainly in prey capture. *23.4*

6. Most animals more complex than cnidarians have bilateral symmetry, tissues, organs, and organ systems. Their gut is saclike, as in flatworms, or complete, with anus and mouth. Those more complex than flatworms have a cavity between the gut and body wall (a coelom, with its peritoneum, or a false coelom). *23.5, 23.6*

7. Two major lineages diverged shortly after flatworms evolved. One (protostomes) gave rise to the mollusks, annelids, and arthropods. The other (deuterostomes) gave rise to echinoderms and chordates. *23.8*

8. All mollusks have a fleshy, soft body and a mantle. Most have either a shell or a remnant of one. Mollusks vary greatly in size, body details, and life-styles. *23.9*

9. Annelids (earthworms, polychaetes, leeches) have complex organs in repeats of coelomic chambers. Their segmented body works as a hydrostatic skeleton. *23.10*

10. Collectively, arthropods are the most successful of all groups in diversity, numbers, distribution, defenses, and capacity to exploit food resources. *23.11–23.15*

    a. All arthropods, including arachnids, crustaceans, and insects, have hardened, jointed exoskeletons. They also have modified segments, specialized appendages, specialized respiratory, nervous, and sensory organs, and (in insects only) wings.

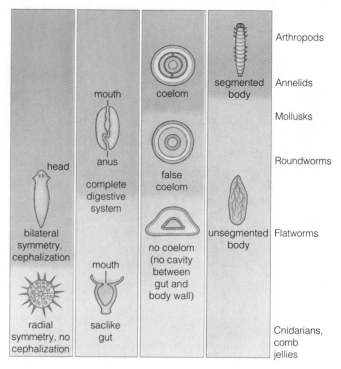

Figure 23.34 Summary of key trends in the evolution of animals, as identified by comparing body plans of major phyla. Not all of these features appeared in every group.

    b. Arthropods develop by growth in size and molting or develop through a series of immature stages, such as larvae and nymphs. Many metamorphose; the tissues of immature forms undergo major reorganization, and body parts are remodeled before the adult emerges.

11. Echinoderms have spines, spicules, or plates in their body wall. Evolutionarily, the phylum is puzzling. The larvae of most species form bilateral features, but they go on to develop into basically radial adults. *23.16*

Review Questions

1. List the six main features that characterize animals. *23.1*

2. When attempting to discern evolutionary relationships among major groups of animals, which aspects of their body plans provide the most useful clues? *23.1*

3. What is a coelom? Why was it important in the evolution of certain animal lineages? *23.1*

4. Name some animals with a saclike gut. Evolutionarily, what advantages does a complete gut afford? *23.1, 23.4–23.6*

5. Choose a species of insect that lives in your neighborhood and describe some of the observable adaptations that underlie its success. *23.15*

Self-Quiz   ANSWERS IN APPENDIX III

1. Which is *not* a general characteristic of the animal kingdom?
    a. multicellularity; most have tissues, many form organs
    b. exclusive reliance on sexual reproduction
    c. motility at some stage of the life cycle
    d. embryonic development during the life cycle

Figure 24.15 (**a**) Galápagos tortoise. Like most turtles, it pulls its head and limbs into its shell when threatened. (**b**) Frilled lizard, flaring a large ruff of neck skin as defensive behavior. (**c**) A rattlesnake in mid-strike. Snakes usually aren't aggressive toward people. Even so, rattlesnakes and other poisonous types bite 8,000 or so Americans annually and kill about 12 of them. King cobras and other snakes bite 200,000 or so people in India and kill about 9,000. (**d**) Tuatara (*Sphenodon*). The two existing species live on small, windswept islands near New Zealand.

powerful jaws, a lengthy snout, and sharp teeth (Figure 24.13). Large crocodiles yank mammals and birds into water, rip them apart by violently spinning their body, then gulp down the chunks. Like the other reptiles and birds, crocodilians adjust their body temperature with behavioral and physiological mechanisms. Like birds, they show complex social behaviors, as when parents guard nests and assist hatchlings into water. Such traits suggest that they share a common ancestor with birds.

The 250 existing species of turtles live inside a shell attached to its skeleton (Figure 24.15*a*). Only sea turtles and other highly mobile types have a reduced shell. The body plan works well; it's been around ever since the Triassic. Turtles use tough, horny plates and strong jaws to chew food. Often they have a fierce disposition. All lay eggs on land, then leave them. Predators eat most of the unattended eggs, so few new turtles ever hatch. Sea turtles have been hunted to the brink of extinction.

About 95 percent of modern-day reptiles, the lizards and snakes, are distantly related to dinosaurs. Most are small. But the Komodo monitor lizard is large enough to prey on young water buffalo, and the longest snake would stretch across ten yards of a football field.

Most lizards, including iguanas and geckos, live in deserts and tropical forests. Most grab insect prey with small, peglike teeth; chameleons capture them with flicks of a tongue that is longer than the body (Section 29.4). Some lizards startle predators and intimidate rivals by flaring a throat fan (Figure 24.15*b*). Many give up their tail when grabbed. The detached tail wriggles and may distract a predator enough to permit a getaway.

In the late Cretaceous, elongated, limbless snakes evolved from short-legged, long-bodied lizards. Some have bony remnants of ancestral hindlimbs (Section 16.1). Snakes are carnivores with flexible skull bones and notably moveable jaws (Figures 24.15*c* and 37.4). Pythons and boas coil around prey, suffocate it into submission, and gulp it down. Fanged types, including coral snakes, rattlesnakes, cottonmouths, copperheads, and king cobras, subdue prey with deadly venom.

Even the most feared snakes are vulnerable during their life cycle; birds and other predators relish snake eggs. Female snakes will store sperm and lay several clutches of fertilized eggs at intervals after they mate, which improves the odds that some eggs will hatch.

Tuataras are like amphibians in some brain features and in how they walk. Their body plan hasn't changed much for 140 million years (Figure 24.15*d*). Like lizards, they have a third "eye," with retina, lens, and nerves to the brain. Being covered with skin, it can only register changes in daylength and light intensity. Does it have roles in reproduction? Maybe (compare Section 32.8). Tuataras, like turtles, can live longer than sixty years. They engage in sex only after they are twenty years old.

Reptiles, with their tough, scaly skin, reliance on internal fertilization, water-conserving kidneys, and amniote eggs, were the first vertebrates to escape dependency on free-standing water in their habitats. The move onto land also involved major modifications in the nervous, circulatory, and respiratory systems.

24.8

BIRDS

Four groups—insects, pterosaurs (now extinct), birds, and bats—are masters of flight. And of those groups, only **birds** grow feathers. **Feathers** are lightweight structures derived from skin that are used for flight and insulation. There are about 28 orders of birds and close to 9,000 named species. They vary enormously in size, proportions, coloration, and capacity for flight. One of the smallest known birds barely tips the scales at 2.25 grams (0.08 ounce). The ostrich, the largest living bird, weighs as much as 150 kilograms (330 pounds). It can't fly but is an impressive sprinter (Figure 16.1). Many birds, such as warblers, parrots, and other perching types, differ markedly in feather coloration and territorial behavior. Bird song and other social behaviors are topics of later chapters. Figures 24.16 through 24.18 highlight just a few basic features of this diverse group of vertebrates.

The first birds evolved during an adaptive radiation of reptiles in the Mesozoic. They apparently diverged from a lineage of small reptiles that ran around on two legs. Feathers evolved as highly modifed reptilian scales.

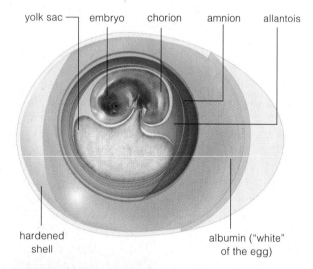

yolk sac — embryo — chorion — amnion — allantois

hardened shell

albumin ("white" of the egg)

Figure 24.16 Generalized sketch of an amniote egg. All birds lay these eggs. The amnion, chorion, allantois, and yolk sac are its four extraembryonic membranes with roles in development. Chapter 39 has an example of how an amniote egg develops. *Left:* Speckled, hard-shelled eggs of a magpie, unhatched in the nest.

Figure 24.17 (**a**) Flight. Of all living vertebrates, only birds and bats fly by flapping their wings. (**b**) Feathers, the key defining characteristic of birds. The flamboyant plumage of this male pheasant is an outcome of sexual selection. This bird is a native of the Himalaya Mountains of India, and it is an endangered species. As is the case for many other kinds of birds, its bright, jewel-colored feathers end up adorning humans—in this case, on the caps of native tribespeople.

(**c**) Many birds, including these Canada geese, display migratory behavior. *Animal migration* is a recurring pattern of movement between two or more regions in response to environmental rhythms. For instance, seasonal change in daylength is a cue that acts on internal timing mechanisms, or biological clocks. The cues trigger physiological and behavioral changes in individuals. Such changes induce migratory birds to make round trips between distant regions that differ in climate. Canada geese spend the summer nesting in marshes and lakes in the northern United States and Canada. Their wintering grounds are in New Mexico and other parts of the southern United States.

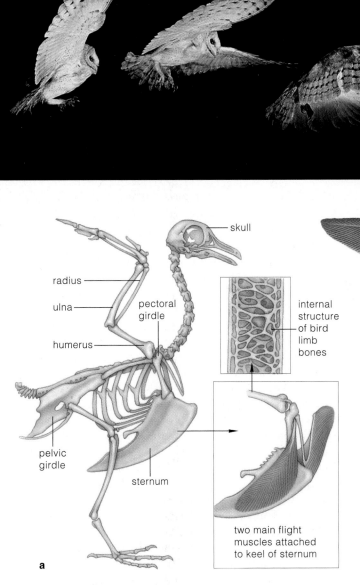

Figure 24.18 (**a**) Body plan of a typical bird. Flight muscles attach to the large, keeled breastbone (sternum). A bird wing is a complicated system of lightweight bones and feathers. (**b**) Feathers gain strength from a hollow central shaft and from tiny barbules that are interlocked in a latticelike array. See Section 18.4 for a look at the evolution of bird forelimbs.

Labels in figure (a): skull, radius, ulna, pectoral girdle, humerus, internal structure of bird limb bones, pelvic girdle, sternum, two main flight muscles attached to keel of sternum

Labels in figure (b): barbule, barb, shaft

Archaeopteryx was in or near that lineage. As you read in Section 18.8, it had reptilian *and* avian traits, including feathers. Birds still share some traits with their closest relatives, the dinosaurs and crocodilians. For example, they still have scales on their legs and some of the same internal structures. They, too, lay amniote eggs. Most show parenting behavior, such as nest guarding.

Bird flight involves more than feathers. It involves bones with a lightweight, honeycombed structure and efficient modes of respiration and circulation. High rates of metabolism sustain flight, and those rates depend on a strong flow of oxygen through the body. Bird lungs connect to elastic ventilating sacs that greatly enhance oxygen uptake (Section 36.3). Like mammals, birds also have a large, durable, four-chambered heart. The heart pumps oxygen-enriched blood to the lungs and to the rest of the body along separate routes, as it probably did in the reptilian ancestors of birds.

Flight also demands an airstream, low weight, and a powerful downstroke that can provide lift (a force at right angles to the airstream). A bird wing, a modified forelimb, is composed of feathers and lightweight bones attached to powerful muscles. Its bones do not weigh much, owing to profuse air cavities in the bone tissue.

For example, the skeleton of a frigate bird, with its seven-foot wingspan, weighs a mere four ounces. That is less than the feathers weigh! A bird's flight muscles attach to an enlarged breastbone (sternum) and to the upper limb bones adjacent to it (Figure 24.18*a*). Muscle contraction creates the powerful downstroke required for flight. Wings, with their long flight feathers, serve as airfoils. Usually, a bird spreads out long feathers on a downstroke and thus increases the size of the surface pushing against air. On the upstroke, a bird folds its feathers somewhat, so that each wing presents the least possible resistance to air.

Like mammals, birds rely on metabolic heat to help maintain a warm body temperature even when outside temperatures decline. Feathers help conserve that heat, and the elastic air sacs help dissipate excess heat.

Of all animals, birds alone have feathers, which they use in flight, in heat conservation, and in socially significant communication displays.

THE RISE OF MAMMALS

Mammals (Mammalia) are the only organisms with hair and mammary glands. Females feed their young with milk, a nutritious fluid that mammary glands secrete into ducts that open onto the body's surface (Figure 24.19*a*). Mammals also are distinctive in that most care for the young for extended periods, with adults serving as models for their behavior. Young mammals are born with a capacity to learn and to repeat behaviors with survival value. It is among mammals that we see the most stunning shows of **behavioral flexibility**, a capacity to expand on basic activities with novel forms of behavior and learning. This trait is more notable in some species than others. It coevolved with the expansion of brain regions, and it became greatest among the mammals called primates.

Unlike reptiles, which generally swallow prey whole, most mammals secure, cut, and in some cases chew food before swallowing it. They also differ in **dentition**, the type, number, and size of teeth. (Figure 24.19*d*). Their incisors (flat chisels or cones) nip or cut food and are large in grazing mammals. Canines have piercing points and are longest in meat eaters. Premolars and molars (cheek teeth) have a bumpy (cusped) platform used to crush, grind, and shear food. If a mammal has large, flat-surfaced cheek teeth, then its early ancestors relied on fibrous plants as the mainstay of their diet.

More than 200 million years ago, a divergence from small reptiles (synapsids) gave rise to therapsids, the ancestors of mammals. By the Jurassic, plant-eating and meat-eating mammals—therians—had evolved. Most were mouse-sized. They had hair and modifications in jaws, teeth, and body form. For instance, their four limbs were positioned upright under their trunk. This skeletal arrangement made it easier to walk erect, but a trunk higher from the ground was not as stable. At that time

Figure 24.20 Location of the placenta in a pregnant human female.

the cerebellum, a brain region dealing with the body's balance and spatial positioning, started expanding.

Therians and dinosaurs coexisted in the Cretaceous. Then the last dinosaurs vanished at the K–T boundary (Section 19.7). With their extinction, new adaptive zones opened up for mammals. *Critical Thinking* question 4 at the chapter's end invites you to take a look at what happened. For now, it is enough to recognize that three major lineages became geographically isolated from one another on distant land masses. Even so, many of their members evolved in similar ways in similar habitats, and in time they came to resemble one another in their form and function. These species are a prime example of **convergent evolution** (Section 18.4 and Table 24.1).

We call the three lineages the egg-laying mammals (monotremes), the pouched mammals (marsupials), and the placental mammals (eutherians). Females of all three lineages lay amniote eggs. Placental mammals do have the competitive edge. They have higher metabolic rates, more precise ways to control body temperature, and a **placenta**. This spongy, blood-filled tissue of maternal and fetal membranes forms in a uterus, a chamber in which embryos develop in relative freedom from harsh external conditions (Figure 24.20). The placenta delivers nutrients and oxygen to the embryo and disposes of its metabolic wastes. Embryos of placental mammals also grow faster in a uterus than marsupials do in a pouch. Many species are fully formed—and less vulnerable to predators—at birth.

Figure 24.19 Three distinctly mammalian traits. (**a**) A human baby busily demonstrating a key defining feature: it derives nourishment from mammary glands. (**b**) Juvenile raccoons, with a thick, furry coat. Unlike reptiles, such as speckled caimans (**c**), mammals (**d**) have an upper row of four kinds of teeth that match up with four kinds on the lower jaw.

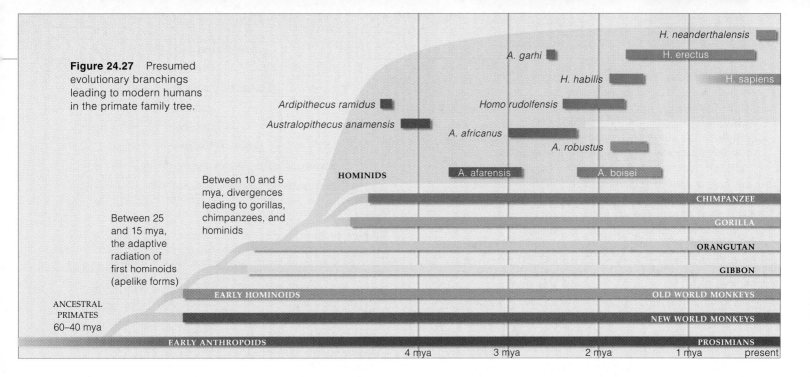

Figure 24.27 Presumed evolutionary branchings leading to modern humans in the primate family tree.

H. neanderthalensis

A. garhi

H. erectus

H. habilis

H. sapiens

Ardipithecus ramidus

Homo rudolfensis

Australopithecus anamensis

A. africanus

A. robustus

HOMINIDS

A. afarensis

A. boisei

Between 10 and 5 mya, divergences leading to gorillas, chimpanzees, and hominids

CHIMPANZEE

GORILLA

Between 25 and 15 mya, the adaptive radiation of first hominoids (apelike forms)

ORANGUTAN

GIBBON

EARLY HOMINOIDS

OLD WORLD MONKEYS

ANCESTRAL PRIMATES 60–40 mya

NEW WORLD MONKEYS

EARLY ANTHROPOIDS

PROSIMIANS

4 mya   3 mya   2 mya   1 mya   present

out of Africa, turned left into Europe, and right into Asia. Some even walked to China; fossils from Southeast Asia and the former Soviet republic of Georgia are 1.8 million and 1.6 million years old. *H. erectus* survived as ice sheets advanced—more than once—into northern Europe, southern Asia, and North America.

Whatever pressures triggered the far-flung travels, this was a time of physical changes, as in skull size and leg length. It also was a time of cultural lift-off for the human lineage. *H. erectus* had a larger brain and was a more creative toolmaker. Its social organization and communication skills must have been well developed. How else can we explain its successful dispersal? From southern Africa to England, different populations used the same kinds of hand axes and other tools to pound, scrape, shred, chop, cut, and whittle material. *H. erectus* withstood environmental challenges by building fires and using furs for clothing. Clear evidence of fire use dates from an ice age in the early Pleistocene.

As Middle Eastern fossils show, *Homo sapiens* evolved by 100,000 years ago. Early *H. sapiens* had smaller teeth and jaws than *H. erectus* (Figure 24.26). Many had a novel feature: a chin. Facial bones were smaller, the skull higher and rounder, the brain larger. *H. erectus* might have developed complex language. But their origin and geographic dispersal are hotly debated (Section 24.12).

One group of early humans, the Neandertals, lived in Europe and in the Near East from 200,000 to 30,000 years ago. Massively built and large brained, some were the first to adapt to the coldest regions (Figure 24.28). Their disappearance coincided with the appearance of anatomically modern humans in the same areas about 40,000 to 30,000 years ago. We have no evidence that Neandertals warred or interbred with the later arrivals. We still do not know what happened to them. We do know that Neandertal DNA has unique sequences, so

Figure 24.28 Human body build correlated with climate. The body of early and modern humans that became adapted to cold climates are heat-conserving (for example, stocky, with short legs) compared to the body of humans adapted to hot climates.

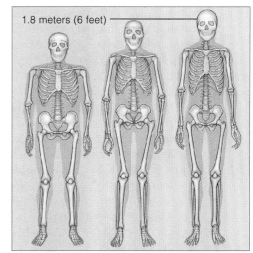

1.8 meters (6 feet)

Neandertal   *Homo erectus*   modern Masai

they might not have contributed to the gene pools of modern European populations.

From 40,000 years ago to today, human evolution has been almost entirely cultural, not biological. And so we leave our story with these conclusions: Humans spread rapidly through the world by devising *cultural* means to deal with a broader range of environments. Compared to their predecessors, they developed rich, varied cultures. Although hunters and gatherers persist in parts of the world, others moved from "stone-age" technology to the age of "high tech," attesting to the great plasticity and depth of human adaptations.

Cultural evolution has outpaced the biological evolution of the only remaining human species, *H. sapiens*. Today, humans everywhere rely on cultural innovation to adapt rapidly to a broad range of environmental challenges.

Out of Africa—Once, Twice, Or . . .

If researchers are interpreting the fossil record of human evolution correctly, then it would seem Africa was the cradle for us all. At this writing, at least, no one has any fossils of humans older than 2 million years *except* in Africa. *H. erectus* coexisted for a time with earlier humans (*H. habilis*) before dispersing from Africa to the cooler grasslands, forests, and mountains of Europe and Asia. This form apparently left Africa in waves between about 2 million and 500,000 years ago. Judging from *H. erectus* fossils from Java, isolated populations may have endured until some time between 53,000 and 37,000 years ago.

Where, on the larger geologic stage, do we place the origin of *H. sapiens*? *Here we find a good example of how the same body of evidence can be interpreted in different ways.* The interpretations are called the **multiregional model** and **African emergence model** for modern human origins. Both attempt to explain the world distribution of fossils of particular ages. Both are based on measurements of genetic distances among existing human populations.

For example, biochemical and immunological studies both indicate that the greatest genetic distance separates *H. sapiens* populations native to Africa from all other human populations. The next greatest distance separates Southeast Asian and Australian populations from others (Figure 24.29). Figure 24.30 summarizes another line of evidence; it correlates *H. sapiens* fossils to specific times.

MULTIREGIONAL MODEL By this model, *H. erectus* populations spread through much of the world by about 1 million years ago. The geographically separated groups faced different selection pressures, so their traits evolved in regionally distinctive ways. Then the subpopulations ("races") of *H. sapiens* evolved from them in different places. They differed phenotypically but did not evolve into separate species because gene flow continued among them, even to the present. Thus, for example, while the armies of Alexander the Great were sweeping eastward, they also contributed "blue-eye genes" from Greeks to the allele pool of generally brown-eyed subpopulations in Africa, the Near East, and Asia.

Figure 24.29 One proposed family tree for populations of modern humans (*Homo sapiens*) that are native to different regions. Branch points show presumed genetic divergences. The tree is based on nucleic acid hybridization studies of many genes (including those for mitochondrial DNA and the ABO blood group) and immunological comparisons.

AFRICAN EMERGENCE MODEL This model does not dispute fossil evidence that populations of *H. erectus* evolved in different ways in different regions. However, it holds that *H. sapiens*—modern humans—originated in sub-Saharan Africa somewhere between 200,000 and 100,000 years ago. Only later did *H. sapiens* populations move out of Africa, into regions along the routes shown in the Figure 24.30 map. In each region that populations of *H. sapiens* settled, they replaced the archaic *H. erectus* populations that had preceded them. Only then did regional phenotypic differences become superimposed on the original *H. sapiens* body plan.

In support of this model, the oldest known *H. sapiens* fossils are from Africa. Also, in Zaire, a finely wrought barbed-bone harpoon and other exquisite tools suggest that African populations were as skilled at making tools as *Homo* populations known earlier from Europe. Also, in 1998, researchers at the University of Texas and in China announced findings from the Chinese Human Genome Diversity Project. Detailed analysis of gene patterns from forty-three ethnic groups in Asia suggest that modern humans moved from Central Asia, along the coast of India, then on into Southeast Asia and southern China. Populations later moved north and northwest into China, then into Siberia, and then on down into the Americas.

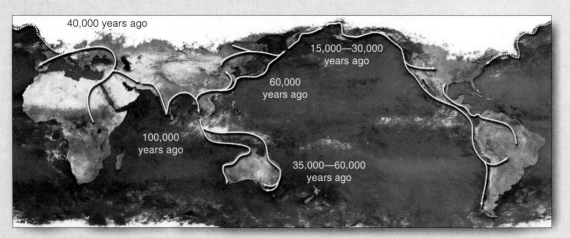

Figure 24.30
Estimated times when populations of early *H. sapiens* were colonizing different regions of the world, based on radiometric dating of fossils. The presumed dispersal routes (*white* lines) seem to support the African emergence model.

SUMMARY   *Gold* indicates text section

1. Generally, chordate embryos have a notochord, a dorsal hollow nerve cord, a pharynx with gill slits (or hints of these), and a tail extending past the anus. Some or all of these traits persist in adults. Chordates with a backbone are vertebrates. Tunicates (e.g., sea squirts) and lancelets are invertebrate chordates. *24.1, 24.2*

2. The eight vertebrate classes are jawless fishes, jawed armored fishes (now extinct), cartilaginous fishes, bony fishes, amphibians, reptiles, birds, and mammals. *24.1*

3. The earliest vertebrates, the jawless fishes that arose in the Cambrian, included ostracoderms. Their modern descendants are lampreys and hagfishes. Jawed fishes also arose in Cambrian times. A once-dominant lineage, the placoderms, became extinct. Other lineages gave rise to the cartilaginous fishes and bony fishes. *24.3–24.5*

4. Four evolutionary trends are associated with certain vertebrate lineages: *24.3, 24.5*

   a. A vertebral column supplanted the notochord as a structural element against which muscles can act. This development led to fast-moving predatory animals.

   b. Jaws evolved from gill-supporting elements. This led to increased predator–prey competition; it favored more efficient nervous systems and sensory organs.

   c. In one lineage of bony fishes, paired fins evolved into fleshy lobes having internal structural elements. The lobes were the forerunners of paired limbs.

   d. In some fish lineages, respiration by gills came to be supplemented by paired lungs, which proved to be adaptive during the invasion of the land. In a related development, the circulatory system became far more efficient at distributing oxygen through the body.

5. The amphibian body plan and reproductive mode are between those of fishes and reptiles. Amphibians were the first vertebrates to invade the land, but they never fully escaped the water. Their skin dries out, and aquatic stages persist in most life cycles. *24.6*

6. Unlike amphibians, reptiles escaped dependence on free-standing water. Key adaptations that permitted them to do so are a toughened, scaly skin that restricts water loss; a reliance upon internal fertilization; more efficient, water-conserving kidneys; and amniote eggs, often leathery or shelled, that protect and metabolically support the embryos. *24.7*

7. Like reptiles, the birds and mammals (both of which descended from reptiles) have efficient circulatory and respiratory systems. They also have a highly developed nervous system and sensory organs. *24.8, 24.9*

   a. Of all organisms, birds alone have feathers, used in flight, heat conservation, and social displays.

   b. Of all organisms, only mammals have mammary glands that produce milk. They have hair or thick skin that functions in insulation. They also have distinctive dentition and a highly developed cerebral cortex, the brain's most complex region. Adults of most species nurture their offspring through an extended period of dependency and learning.

   c. The first mammals evolved from small, hairless reptiles in the Triassic. After the extinction of the last dinosaurs and geologic isolation following the breakup of Pangea, three lineages radiated into new adaptive zones. All three have persisted to the present. They are egg-laying mammals (monotremes), pouched mammals (marsupials), and placental mammals (eutherians).

   d. Placental mammals have higher metabolic rates, more precise control of body temperature, and a more efficient way to nourish and protect embryos. In many places, they outcompeted monotremes and marsupials, which retain many archaic traits. In similar habitats in different parts of the world, some members of the three lineages also show morphological convergence.

8. Primates include the prosimians (lemurs and related forms), tarsioids, and anthropoids (monkeys, apes, and humans). Apes and humans alone are hominoids. Only australopiths and anatomically modern humans (*Homo sapiens*) and others of their lineage, including *H. habilis* to *H. erectus*, are currently classified as hominids. *24.10*

   a. The first primates were tiny, rodentlike mammals that evolved by 60 million years ago in tropical forests.

   b. Hominoids (apelike forms) evolved in the Miocene between 25 to 5 million years ago in Africa. Some gave rise to hominids by at least 5 million years ago.

   c. The origin and early evolution of hominids might coincide with selection pressures that emerged during a long-term change in climate, brought about by shifts in land masses and ocean circulation patterns. Tropical forest gave way to dense woodlands, then grasslands. The hominids were bipedal (upright walkers), they had modified teeth that handled diverse foods. They also had a more complex brain that gave them a capacity to plan for securing scarcer and seasonally available food.

9. Modern humans emerged by modification of traits in ancestral primate forms. Their tree-dwelling ancestors relied less on the sense of smell and more on enhanced daytime vision. Manipulative skills increased when the hands began to be freed from load-bearing functions. Starting with Miocene apes, there were shifts from four-legged climbing to bipedalism, from specialized to omnivorous eating habits, and to remarkable increases in brain complexity and behavior. *24.10*

10. *H. habilis*, the earliest known species of the genus *Homo*, evolved by 2.5 million years ago. It was one of the first stone toolmakers. *Homo erectus*, ancestral to modern humans, evolved by 2 million years ago. *H. erectus* populations radiated out of Africa, into Asia and Europe. The oldest known fossils of early modern humans (*H. sapiens*) date from about 100,000 years ago. About 40,000 years ago, cultural evolution outstripped biological evolution of the human form. *24.10, 24.11*

Review Questions

1. Which features distinguish chordates from other groups of animals? *24.1*

2. List four major trends that apparently developed during the evolution of at least some vertebrate lineages. *24.3*

3. Describe some of the defining characteristics of jawless fishes, cartilaginous fishes, and bony fishes. Which group is the most abundant? *24.3–24.5*

4. Which traits distinguish reptiles from amphibians? *24.6, 24.7*

5. Birds, crocodilians, and dinosaurs share what traits? *24.7, 24.8*

6. List some characteristics that distinguish each of the three mammalian lineages from the reptiles. *24.7, 24.9*

7. What are the key differences between a "hominoid" and a "hominid"? *24.10*

8. Briefly describe some of the conserved physical traits that connect anatomically modern humans with their mammalian ancestors, then their primate ancestors. *24.10*

Self-Quiz   ANSWERS IN APPENDIX III

1. Only _____ have a notochord, a tubular dorsal nerve cord, a pharynx with slits in the wall, and a tail extending past the anus.
   a. echinoderms   d. both b and c
   b. tunicates and lancelets   e. all of the above
   c. vertebrates

2. Gill slits function in _____ .
   a. respiration   d. water regulation
   b. circulation   e. both a and c
   c. food trapping

3. A shift from a reliance on _____ to reliance on _____ was pivotal in the evolution of all vertebrates.
   a. the notochord; a backbone c. gills; lungs
   b. filter feeding; jaws   d. all of the above

4. The first vertebrates included _____ .
   a. bony fishes   c. lobe-finned fishes
   b. jawless fishes   d. both a and b

5. Of all existing vertebrates, _____ are the most diverse.
   a. cartilaginous fishes   d. reptiles
   b. bony fishes   e. birds
   c. amphibians   f. mammals

6. Generally, the only amphibian groups that entirely escaped dependency on free-standing water are _____ .
   a. salamanders   c. caecilians
   b. toads   d. all require it

7. Reptiles moved fully onto land owing to _____ .
   a. tough skin   d. amniote eggs
   b. internal fertilization   e. none of the above
   c. good kidneys   f. all of the above

8. A four-chambered heart evolved first in _____ .
   a. bony fishes   d. mammals
   b. amphibians   e. crocodilians
   c. birds   f. both c and d

9. Amniote eggs form during the life cycle of most _____ .
   a. reptiles   c. mammals
   b. birds   d. all of the above

10. Birds use feathers in _____ .
    a. flight   c. social functions
    b. heat conservation d. all of the above

11. Various mammals _____ .
    a. hatch
    b. finish embryonic development in a pouch
    c. finish embryonic development in a uterus
    d. both b and c
    e. all of the above

12. Early humans _____ .
    a. became adapted to a wide range of environments
    b. became adapted to a narrow range of environments
    c. had flexible bones that cracked easily
    d. were limber enough to swing through the trees

13. Match the organisms with the appropriate features.
    _____ jawless fishes   a. milk glands, behavioral
    _____ cartilaginous   flexibility, thick skin or hair
          fishes   b. respiration by skin and lungs
    _____ bony fishes   c. include coelacanths
    _____ amphibians   d. include hagfishes
    _____ reptiles   e. include sharks and rays
    _____ birds   f. complex social behavior, feathers
    _____ mammals   g. first with amniote eggs

Critical Thinking

1. List factors that could contribute to the collapse of native fish populations in a lake after the accidental introduction of lampreys to it.

2. Think about the flight muscles of birds and their demands for oxygen and ATP energy. What type of organelle would you expect to be profuse in such muscles? Explain your reasoning.

3. Fossil hunters Kathie and Gary unearthed the complete fossilized remains of an early mammal. How can they determine whether the mammal was a herbivore, carnivore, or omnivore?

4. Select one example of morphological convergence listed in Table 24.1. Then use the family names to search the literature and the web for information on their similarities in body form, behavior, and habitats. Write up what you find, and include appropriate illustrations.

5. As you read in this chapter, three mammalian lineages made it to the present: the monotremes (egg-laying mammals), marsupials (pouched mammals), and eutherians (placental mammals). Again, monotremes and marsupials retain many archaic traits. Placental mammals have higher metabolic rates, more precise ways of regulating body temperature, and an efficient way to nourish embryos inside the mother's body.

   Now take a look at Figure 24.31a on the facing page. The fossil record indicates the ancestors of the monotremes and marsupials had entered the southern part of the supercontinent Pangea by the late Jurassic. After the breakup of Pangea, the species on the land mass that would become Australia were already isolated from ancestors of placental mammals, which were evolving elsewhere (Figure 24.19b).

   In the future South America, monotremes were replaced by marsupials and early placental mammals that had evolved independently on other continents. A land bridge rejoined South and North America in the Pliocene. Highly evolved placental mammals radiated southward and rapidly drove many of the previously isolated mammals to extinction. Only opossums and a few other species successfully invaded land to the north.

   Today in Australia, many marsupial species compete with recently introduced placental mammals for resources but are not winning. You can read about one example of this in Section 42.8, which describes the introduction of rabbits into Australia. Write a brief paragraph on the reasons why placental mammals such as rabbits that did not evolve in Australian habitats are showing greater fitness than native mammals.

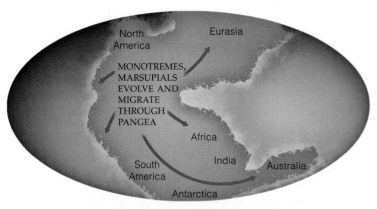

a About 150 million years ago, during the Jurassic

North America

Eurasia

MONOTREMES, MARSUPIALS EVOLVE AND MIGRATE THROUGH PANGEA

Africa

India

South America

Australia

Antarctica

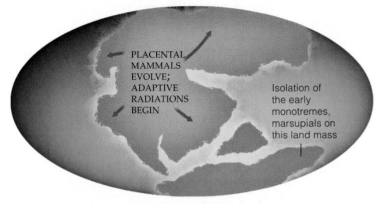

b Between 100 and 85 million years ago, during the Cretaceous

PLACENTAL MAMMALS EVOLVE; ADAPTIVE RADIATIONS BEGIN

Isolation of the early monotremes, marsupials on this land mass

c About 20 million years ago, during the Miocene

North America

ADAPTIVE RADIATIONS OF MORE EVOLVED PLACENTAL MAMMALS

Eurasia

South America

Continued isolation of early monotremes and marsupials

Africa

Extinctions of mammals

Antarctica

d About 5 million years ago, during the Pliocene

DISPERSAL OF HIGHLY EVOLVED PLACENTAL MAMMALS

South America

Extinctions of many marsupials and early placental mammals

Figure 24.31 Adaptive radiations of mammals.

Figure 24.32
Model for a long-term shift in the global climate in the Miocene on into the Pliocene.

ATLANTIC OCEAN

a   **b**

6. Before the Isthmus of Panama formed (Figure 24.32*a*), the salinity levels of ocean surface currents were much the same around the world. Circulation patterns kept Arctic waters warm. After the isthmus formed (Figure 24.32*b*), North Atlantic surface currents became saltier and heavier, so they sank before reaching the Arctic. With colder water in polar regions, the Arctic ice cap formed and ushered in a long-term trend toward a cooler, drier climate in Africa. Referring to Section 24.10, speculate on how the change may have been a factor in hominid evolution.

7. The first mammals spread their toes apart to support the body as they walked or ran on four legs. Primates still spread fingers or toes. Many also cup their fingers, as when monkeys lift food to the mouth. In ancient tree-dwelling primates, modifications to hand bones allowed them to wrap their fingers around objects (prehensile movements) and touch a thumb to the tip of a finger (opposable movements). Later, hands were freed of load-bearing functions, and the power grip and precision grip evolved:

power grip   precision grip

Write a paragraph on the roles these skeletal modifications may have played in the cultural evolution of early humans.

Selected Key Terms

Vertebrates

amniote egg 24.7
amphibian 24.6
behavioral flexibility 24.9
bird 24.8
bony fish 24.5
cartilaginous fish 24.5
chordate 24.1
convergent evolution 24.9
dentition 24.9
feather 24.8
filter feeder 24.2
fin 24.3
gill 24.3

gill slit 24.2
hagfish 24.4
jaw 24.3
lamprey 24.4
lancelet 24.2
lobe-finned fish 24.5
lung 24.3
mammal 24.9
nerve cord 24.1
notochord 24.1
ostracoderm 24.3
pharynx 24.1
placenta 24.9
placoderm 24.3
reptile 24.7
scale (fish) 24.5

swim bladder 24.5
tunicate 24.2
vertebra 24.3
vertebrate 24.1

Human Evolution

African emergence model 24.12
australopith 24.10
bipedalism 24.10
culture 24.10
hominid 24.10
human 24.11
multiregional model 24.12
Primates (order) 24.12

Readings

Gould, S. J. (general editor). 1993. *The Book of Life*. New York: Norton. Splendid easy-to-read essays, gorgeous illustrations.

Romer, A. S., and T. S. Parsons. 1986. *The Vertebrate Body*. Sixth edition. Philadelphia: Saunders. Still one of the best.

Tattersall, I. 1998. *Becoming Human: Human Evolution and Human Uniqueness*. Pennsylvania: Harvest Books.

On-Line readings at Student Guide for InfoTrac: www.brookscole.com/biology

BIODIVERSITY IN PERSPECTIVE

The Human Touch

In 1722, on Easter morning, a European explorer landed on a small volcanic island and found a few hundred skittish, hungry Polynesians living in caves. The island had no trees whatsoever, only dry, withered grasses and scorched, shrubby plants. There were about 200 massive stone statues near the coast and 700 unfinished, abandoned ones in inland quarries (Figure 25.1). Some weighed fifty tons. Without trees for wood and fibrous plants to make ropes, how were they erected? There were no wheeled carts or strong animals. How did the statues get from the quarries to the coast?

Two years later James Cook visited the island. When islanders paddled out to his ships, he noted that they didn't have the knowledge or materials to keep their canoes from leaking; they spent half the time bailing water. He saw only four canoes on the whole island. Nearly all of the statues had been tipped over, often onto spikes that shattered the faces on impact.

Later, researchers solved the mystery of the statues. Easter Island, as it came to be called, has a mere 165 square kilometers (64 square miles) of land. Voyagers from the Marquesas discovered this eastern outpost of Polynesia around A.D. 350, possibly after being blown off course by storms. The place was a paradise. Its fertile volcanic soil supported dense palm forests, hauhau trees, toromino shrubs, and lush grasses. The new arrivals built large, buoyant, ocean-going canoes of long, straight palms strengthened with rope made of fibers from hauhau trees. They used toromino wood as fuel to cook fishes and dolphins; they cleared forests to plant taro, bananas, sugarcane, and sweet potatoes.

By 1400, between 10,000 and 15,000 descendants were living on the island. Society had become highly structured, given the need to cultivate, harvest, and distribute food for so many. Crop yields had declined; precious soil nutrients had been carried off by each harvest and by ongoing, neglected erosion. Edible species vanished from the waters around the island, so fishermen had to build larger canoes to sail out ever farther, into the open and more dangerous sea.

Figure 25.1 On Easter Island, a few of the massive stone tributes to the gods. Apparently islanders erected the statues as a plea for divine intervention after their once-large population wiped out the biodiversity of their tropical paradise. The plea didn't have any effect whatsoever on reversing the losses on land and in the surrounding sea. The human population didn't recover, either.

Survival was at stake. Those in power appealed to the gods. They directed the community to carve divine images of unprecedented size and power, and to move the newly carved statues over miles of rough terrain to the coast. Jo Anne Van Tilburg recently demonstrated that they probably lashed the statues to canoe-shaped rigs, then rolled them along a horizontal "ladder" of greased logs. She got the idea after observing how existing islanders haul canoes from water onto land.

By now, islanders had all of the arable land under cultivation. They had eaten all the native birds, and no new birds came to nest. They were raising and eating rats, the descendants of hitchhikers on the first canoes. They coveted palm seeds, now a scarce delicacy.

Wars broke out over dwindling food and space. By about 1550 no one was venturing offshore to harpoon fishes or dolphins. They couldn't build canoes because the once-rich forests were gone. They had cut down all of the palms. Hauhau trees were extinct; islanders had used every last one as firewood for cooking.

The islanders turned to the only remaining source of animal protein. They started to hunt and eat one another.

Central authority crumbled, and gangs replaced it. As gang wars raged, those on the rampage burned the remaining grasses to destroy hideouts. The rapidly dwindling population retreated to caves and launched raids against perceived enemies. Winners ate the losers and tipped over the statues. Even if the survivors had wanted to leave the island, they no longer had a way to do so. What possibly could they have been thinking when they chopped down the last palm?

Easter Island initially sustained a society that rose to greatness amid abundant resources, then abruptly fell apart. In life's greater evolutionary story, its loss of biodiversity might not even warrant a footnote. After all, compare the loss to the many millions of species that appeared and became extinct in the distant past!

Yet there is a lesson here. The near-total destruction that 15,000 Polynesians wrought on biodiversity was confined to a small bit of land in the vastness of the Pacific Ocean. What are the requirements and whims of *6 billion* people doing to global biodiversity today? Is the loss of species on Easter Island an isolated case? Or should we take what happened there as a warning of a worldwide extinction crisis of our own making, one that is even now under way? Let's take a look.

Key Concepts

1. The current range of global biodiversity is largely an outcome of an overall pattern of abrupt extinction crises and slow recoveries. It takes tens of millions of years to reach the level of biodiversity that prevailed before such mass extinctions.

2. The loss of individual species as well as extinctions of major groups may have complicated causes, the understanding of which requires scientific analysis.

3. Species found only in a restricted geographic region and extremely vulnerable to extinction are classified as endangered species. Over the past four decades, rates of extinction have been rising through a combination of habitat loss and fragmentation, species introductions, overharvesting, and illegal trading in wildlife.

4. The rapidly growing human population is expected to reach 9 billion by 2050. Its demands for food, materials, and living space are threatening biodiversity around the world and are the basis of a new extinction crisis.

5. Conservation biology takes a three-pronged approach to preserving biodiversity. It includes the systematic survey of the full range of biodiversity, analysis of its evolutionary and ecological origins, and identification of methods to maintain and use biodiversity for the benefit of the human population.

6. Systematic surveys focus initially on identifying hot spots, which are limited areas where many species are in danger of extinction owing to human activities. At the next research level, broader or multiple hot spots are researched. Data are combined into a global picture of ecoregions where biodiversity is most threatened.

ON MASS EXTINCTIONS AND SLOW RECOVERIES

Based on many lines of evidence accumulated over the past few centuries, an estimated 99 percent of all species that have ever lived are extinct. Even so, the full range of biodiversity is greater now than it has ever been at any time in the past.

Reflect on the evolutionary stories of the preceding chapters in this unit. For at least the first 2 billion years of life's history, single-celled prokaryotes dominated the evolutionary stage. They did so until the Cambrian period, when atmospheric oxygen started to approach current levels. At that time, some of the single-celled eukaryotes began their genetic divergences and intricate species interactions, which led to the origin of diverse protistans, plants, fungi, and animals.

You saw how mass extinctions reduced biodiversity on land and in the seas. Each global episode spurred evolutionary change and radiations into newly vacated adaptive zones. However, recovery to the same level of biodiversity was exceedingly slow, requiring 20 to 100 million years. Figure 25.2a is a review of the pattern of five major extinctions and slow recoveries.

That pattern is only a composite of what happened to the major taxa. Lineages, remember, differ in their time of origin, the extent to which the member species diverged, and how long they persisted. If you consider ongoing survival and reproduction to be the measures of success, then each became a loser or a winner when environmental conditions changed in drastic or novel ways. To appreciate this point, reflect on Figure 25.2b, which shows the evolutionary history of representative lineages. Many such histories were combined to give us the overall pattern of Figure 25.2a.

Era	Period	MASS EXTINCTION UNDER WAY
CENOZOIC	QUATERNARY	With high population growth rates and cultural practices (e.g., agriculture, deforestation), humans become major agents of extinction.
	1 mya	
	TERTIARY	
	65	*MASS EXTINCTION*
MESOZOIC	CRETACEOUS	Slow recovery after Permian extinction, then adaptive radiations of some marine groups and plants and animals on land. Asteroid impact at K–T boundary, 85% of all species disappear from land and seas.
	138	
	JURASSIC	
	205	
	TRIASSIC	
	240	*MASS EXTINCTION*
PALEOZOIC	PERMIAN	Pangea forms; Land area exceeds ocean surface area for first time. Asteroid impact? Major glaciation, colossal lava outpourings, 90%–95% of all species lost.
	290	
	CARBONIFEROUS	
	360	*MASS EXTINCTION*
	DEVONIAN	More than 70% of marine groups lost. Reef builders, trilobites, jawless fishes, and placoderms severely affected. Meteorite impact, sea level decline, global cooling?
	410	
	SILURIAN	
	435	*MASS EXTINCTION*
	ORDOVICIAN	Second most devastating extinction in seas; nearly 100 families of marine invertebrates lost
	505	
	CAMBRIAN	
	570	*MASS EXTINCTION*
	(precambrian)	Massive glaciation. 79% of all species lost, including most marine microorganisms

a

b

Figure 25.2 (a) Review of the five greatest mass extinctions and subsequent slow recoveries in the past. (More mass extinctions occurred than are shown here.) Compare Figure 18.5, in Section 18.2. (b) Within the framework of this generalized graph, patterns of extinction and recoveries differed significantly for different major taxa, as these selected examples indicate.

in the western United States. Each **riparian zone** is a relatively narrow corridor of vegetation along a stream or river. Its plants afford a major line of defense against flood damage by sponging up water during spring runoffs and summer storms. Shade from the canopy of taller shrubs and trees helps conserve water during droughts. Riparian zones also offer food, shelter, and shade for wildlife, particularly in arid and semiarid regions. For example, in the western United States, 67 to 75 percent of the endemic species must spend all or part of their life cycle in riparian zones. They include more than 136 species of songbirds, some of which nest only in riparian vegetation.

Cattle raised in the American West provide beef for much of the human population. Compared with wild ungulates, cattle drink a lot more water, so they tend to congregate at rivers and streams. There they trample and feed until grasses and herbaceous shrubs are gone. It takes only a few head of cattle to destroy riparian vegetation. All but 10 percent of the riparian vegetation of Arizona and New Mexico has already disappeared, mainly into the stomachs of grazing cattle.

Restricted access and development of watering sites for cattle away from streams can help conserve riparian zones. So can the rotation of livestock and provision of supplemental feed at different grazing areas. Cattle ranchers resist implementing such measures. Putting in more fencing, for example, is costly.

Jack Turnell is one rancher who knows almost as much about riparian biodiversity as he knows about cattle. He runs cattle on a 32,000-hectare (80,000-acre) ranch south of Cody, Wyoming, and on 16,000 hectares of Forest Service land for which he has grazing rights. Ten years after he took over the ranch, he decided to raise cattle in an unconventional way by systematically rotating them away from riparian areas. Turnell also crossed Hereford and Angus cattle with a breed from France that does not consume as much water. He made most of his ranching decisions in consultation with specialists in range and wildlife management.

Willows and other plants are sustaining diversity in the restored zones. And Turnell's ranching approach is profitable; grasses maintained in the riparian zones help put more meat on his cattle. His may be only a small step toward sustainable ranching, but it appears to be a step in the right direction.

Throughout much of the world, the unprecedented rate of human population growth is the elephant in the room that no one wants to talk about.

Protecting biodiversity depends on finding ways for people to make a living from it without destroying it. Can such a balance be struck among so many billions of people?

SUMMARY   *Gold* indicates text section

1. Global biodiversity is greater now than ever, but the current rate of species losses is high enough to suggest that an extinction crisis is under way. *25.1*

2. The history of life indicates that, after global mass extinctions, 20 million to 100 million years pass before biodiversity recovers to the preceding level. *25.1*

  a. Different taxa have different evolutionary histories. Some were eliminated by major extinction events. Others passed through them relatively unscathed.

  b. Mass extinctions, and extinction of single species, may have obvious or hidden, complicated causes.

3. Human population size, expected to reach 9 billion within fifty years, is causing the sixth major extinction crisis. Human population growth is most rapid in areas with the richest, most vulnerable biodiversity. *25.2*

4. Habitat losses, habitat fragmentation, introduction of species into novel habitats, overharvesting, and illegal wildlife trading threaten endemic species. An endemic species originated in and is limited to one geographic region. An endangered species is defined as an endemic species extremely vulnerable to extinction. *25.2*

5. Habitat loss refers to physical reduction or chemical pollution of suitable places for species to live. Habitat fragmentation is carving a habitat into isolated patches. It puts species at risk, as by chopping up populations to sizes that cannot promote successful breeding. *25.2*

6. Island biogeography models help predict the number of current and future extinctions. A habitat island is a habitat in a sea of possibly destructive human activities, such as logging. Generally, destruction of 50 percent of an island habitat (or habitat island) will drive one-tenth of its species to extinction. Destruction of 90 percent will drive one-half to extinction. *25.2*

7. Indicator species are birds and other easily tracked species that can provide warning of changes in habitats and impending widespread loss of biodiversity. *25.2*

8. Humans are directly destroying coral reefs and other regions, as when commercial fishermen dynamite reefs to harvest fish. They may be destroying them indirectly, as by contributing to global warming (with concurrent rises in sea surface temperatures and sea levels). *25.3*

9. Conservation biology is a field of pure and applied research. Its goal is to conserve and use biodiversity in sustainable ways. The field encompasses: *25.5*

  a. A systematic survey at three ever more inclusive levels of the full range of biodiversity.

  b. Analysis of biodiversity's origins in evolutionary and ecological terms.

  c. Identification of methods that might maintain and use biodiversity for the good of the human population, which may otherwise destroy it.

10. Today the systematic survey of conservation biology is proceeding at three levels: *25.5*

a. Local hot spots (for example, an isolated valley) are identified and indicator species inventoried. The hot spots are habitats with many species in greatest danger of extinction because of human activities.

b. Major hot spots (or multiple ones) are inventoried. Research stations are set up across these broader areas to gather data by latitude and elevation.

c. Data gathered at the first two levels are combined with data on ecoregions; these are the most vulnerable, broad regions of land and seas throughout the world.

d. Data gathered for such maps will be refined into an increasingly detailed picture of global biodiversity.

11. Protecting biodiversity depends on finding ways for people to make a living from it without destroying it. To counter growing economic demands, biodiversity's future economic value must be determined. Methods by which local economies can exploit that biodiversity in sustainable ways must be developed. *25.6*

12. When people are given a choice between saving a species or themselves, that species will lose. Millions of people are making that choice as an outcome of their demands for food, shelter, and material goods. *25.6*

Review Questions

1. How many major mass extinctions have occurred, including the one that is now under way? 25.1

2. List four human activities that are major contributors to the current extinction crisis. Which threaten coral reefs? 25.2, 25.3

3. Define endangered, endemic, and indicator species. 25.2

4. Distinguish between island habitat and habitat island. 25.2

5. State the goal of conservation biology and briefly describe its three-pronged approach to achieving that goal. 25.5

6. Define hot spot. Why are conservation biologists focusing on hot spots rather than quickly completing a global survey? 25.5

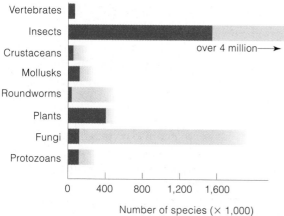

Figure 25.13 Current species diversity for a few major taxa. *Red* indicates the number of named species; *gold* indicates the estimated number of species not yet discovered.

1. Following mass extinctions in the past, recovery to the same level of biodiversity has taken many _____ of years.
   a. hundreds   b. millions   c. billions

2. Species may be driven to extinction by _____ .
   a. obvious factors   c. human activities
   b. complicated, hidden factors   d. all of the above

3. The goal(s) of conservation biology is (are) to _____ .
   a. conduct a three-level, systematic survey of all biodiversity
   b. analyze biodiversity's evolutionary and ecological origins
   c. identify ways to maintain and use biodiversity for people
   d. all of the above

4. Strip logging _____ .
   a. can sustain forests   c. destroys wild habitats
   b. is profitable   d. both b and c

Critical Thinking

1. Figure 25.13 compares the current estimated number of species for some major taxa. Given what you learned about taxa from chapters in this unit, which are most vulnerable in the current extinction crisis? Which are likely to pass through it with much of their biodiversity intact? List some reasons why (for example, compare the distribution of member species).

2. Visit or study a riparian zone near where you live. Imagine visiting it five years from now, then ten years after that. Given its location, what kinds of changes do you predict for it?

3. Mentally transport yourself to a tropical rain forest of South America. Imagine you don't have a job. There are no jobs in sight, not even in overcrowded cities some distance away. You have no money or contacts to get you anywhere else. Yet you are the sole supporter of your large family. Today a stranger approaches you. He tells you he will pay good money if you can capture alive a certain brilliantly feathered parrot in the forest. You know the parrot is rare; it is seldom seen. However, you have an idea of where it lives. What will you do?

4. In his review of this chapter, ecologist Robin Tyser offered these comments: Students might find the current biodiversity crisis too overwhelming to contemplate. But people are making a difference. For example, they helped reestablish viable bald eagle populations in the continental United States and wolves in northern Wisconsin. Daniel Janzen is working to recreate a dry forest ecosystem in Costa Rica. There is much good in the world. Do a literature search or computer research and report on one or two success stories you find inspiring.

Selected Key Terms

conservation biology 25.5	endangered species 25.2	hot spot 25.5
coral reef 25.3	habitat fragmentation 25.2	indicator species 25.2
ecoregion 25.5	habitat island 25.2	riparian zone 25.6
	habitat loss 25.2	strip logging 25.6

Readings

Cox, G. 1999. *Alien Species in North America and Hawaii: Impacts on Natural Ecosystems*. Washington, D.C.: Island Press.

Wilson, Edward O. 1992. *The Diversity of Life*. Cambridge, Massachusetts: Belknap Press. This slender book proves that Wilson is one of the most eloquent defenders of life's diversity.

On-Line readings at Student Guide for InfoTrac: www.brookscole.com/biology

V Plant Structure and Function

A flowering plant (Prunus) busily doing what it does best: producing flowers for the fine art of reproduction.

26

PLANT TISSUES

Plants Versus the Volcano

On a clear spring day in 1980, in a richly forested region of the Cascade Range of southwestern Washington, Mount Saint Helens erupted and 500 million metric tons of ashes were blown skyward. Within minutes, shock waves from the blast flattened or incinerated hundreds of thousands of mature trees growing near the northern flanks of the mountain.

Rivers of hot volcanic ash surged down the slopes at rates exceeding forty-four meters per second. They turned into nightmarish rivers of cementlike mud when intense heat from the blast melted and released more than 75 billion liters of water that had been locked up in the mountain's snowfields and glacial ice.

In one mind-numbing moment, about 40,500 hectares—100,000 acres—of magnificent forests dominated by hemlock and Douglas fir were transformed into barren sweeps of land (Figure 26.1a,b). In the aftermath of the violent eruption, we gained stunning insight into what the world must have looked like long ago, before the first plants started to colonize habitats on land.

Yet it did not take long for existing plants to move back into habitats that their ancestors had claimed. In less than a year's time, the seeds of a variety of flowering plants, including fireweed and blackberry, sprouted near the grayed trunks of fallen trees around Mount Saint Helens.

Even before ten years had passed, young willows and alders took hold near riverbanks, and low shrubs cloaked the land (Figure 26.1c). Their presence afforded pockets of shade, which favored the germination of seeds of the slower growing but ultimately dominant species, the hemlocks and Douglas fir (Figure 26.1d). In less than a century, the forest will be as it once was.

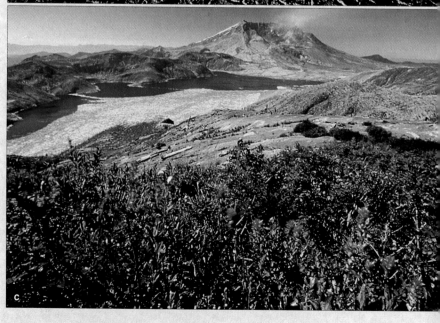

Figure 26.1 (**a,b**) Grim reminder of what the world would be like without plants: the aftermath of the violent eruption of Mount Saint Helens in 1980. Nothing remained of the vast forest that surrounded this Cascade volcano. (**c**) In less than a decade, however, seed-bearing vascular plants were making a comeback. (**d**) Twelve years after the volcanic eruption, seedlings of a dominant species, Douglas fir, were starting to reclaim the land.

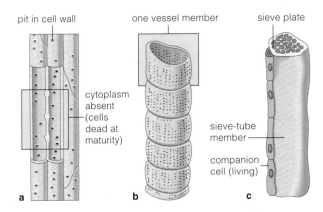

Figure 26.8 From xylem, portions of (**a**) tracheids and (**b**) a vessel. Pipelines made of such cells conduct water and dissolved ions. (**c**) One type of phloem cell. Long tubes of many cells conduct sugars and other organic compounds.

Figure 26.9 Light micrograph of a section through the upper surface of a kaffir lily leaf. The plant cuticle is made of secretions from epidermal cells. Inside the leaf are many photosynthetic parenchyma cells.

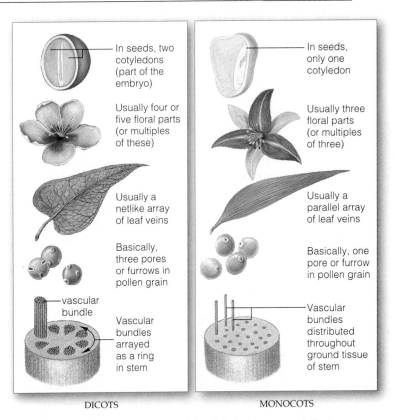

DICOTS   MONOCOTS

Figure 26.10 Comparison of the defining features of dicots and monocots. Both classes of flowering plants consist of the same simple and complex tissues, but their body parts show some differences in structural organization.

DERMAL TISSUES   A dermal tissue system, **epidermis**, covers surfaces of primary plant parts. In most plants, it is mainly a single layer of unspecialized cells. Waxes and the fatty substance cutin coat the outermost cell walls. The surface coating is a **cuticle**. A plant cuticle functions in restricting water loss and often in resisting attacks by certain microorganisms (Figure 26.9).

Stem and leaf epidermis contains many specialized cells. For instance, pairs of guard cells change shape in response to changing conditions. As they do, a gap—or **stoma** (plural, stomata)—closes or opens between them. The next chapter looks at how stomata work as control points for the movement of water vapor, oxygen, and carbon dioxide across the epidermis. **Periderm** replaces epidermis in stems and roots with secondary growth. Dead cork cells in this protective covering have a wall heavily impregnated with suberin, a fatty substance.

Dicots and Monocots—Same Tissues, Different Features

The **dicots** and **monocots**, recall, are the two classes of flowering plants (Section 22.8). Most trees and shrubs other than conifers—such as maples, elms, roses, cacti, peas, beans, lettuces, cotton, and carrots—are dicots.

Palm trees, lilies, orchids, ryegrass, bamboos, wheat, corn, sugarcane, and pineapples are familiar monocots.

Dicots and monocots are similar in structure and function, but they differ in some distinctive ways. For instance, dicot seeds have two cotyledons and monocot seeds have only one. Cotyledons are leaflike structures commonly known as seed leaves. They form in seeds as part of a plant embryo, and they store or absorb food for it. After the seed germinates, cotyledons wither and new leaves grow and start to make food. Figure 26.10 shows other differences between dicots and monocots.

Most of the plant body (ground tissue system) consists of parenchyma, collenchyma, and sclerenchyma. Each of these simple tissues is composed of only one type of cell.

Xylem and phloem are vascular tissues. In xylem, pipelines made of tracheids and vessel members conduct water and dissolved ions. In phloem, sieve-tube members interact with companion cells to distribute organic compounds.

Of two dermal tissues, epidermis covers the surfaces of the primary plant body. Periderm replaces epidermis on plant parts with extensive secondary growth.

Monocots and dicots consist of the same tissues, but each has some of the tissues organized in distinctive ways.

PRIMARY STRUCTURE OF SHOOTS

How Stems and Leaves Form

Next time you or a friend eats a bundle of bean sprouts or alfalfa sprouts, pull one aside to look at its structure. That seedling started forming while it was still inside a seed coat. It already has a primary root and shoot. In the primary shoot's tip, apical meristem and its descendant tissues are laying out orderly frameworks for the stem primary structure (Figure 26.11). Below the shoot tip, tissues become specialized as cells divide at different rates in prescribed directions, and differentiate in size, shape, and function. These tissues make up distinctive stem regions, leaves, and lateral (axillary) buds from which lateral shoots develop. Lateral shoots give rise to side branches and reproductive structures.

Briefly, as a typical shoot lengthens, bulges of tissue develop along the sides of apical meristem. Each bulge is one immature leaf (Figures 26.11 and 26.12). While growth continues, the stem lengthens between tier after tier of new leaves. Each part of the stem where one or more leaves are attached is a node. As shown in Figure 26.2, the stem region between two successive nodes is an internode. Buds develop in the upper angle where leaves attach to the stem. Each **bud** is an undeveloped shoot of mostly meristematic tissue, often protected by bud scales (modified leaves). As Chapter 28 describes, buds give rise to new stems, leaves, and flowers.

Internal Structure of Stems

While the primary plant body of a monocot or dicot is forming, the ground, vascular, and dermal tissues of its stems become organized in distinctive ways. Most often, primary xylem and phloem develop inside the same sheath of cells, as **vascular bundles**. The bundles are multistranded cords threading lengthwise through the ground tissue system of primary and lateral shoots.

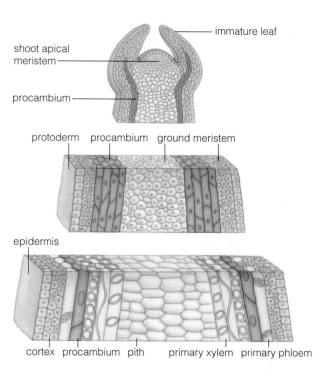

Figure 26.11 Successive stages in primary growth that started with activity at the shoot apical meristem of a typical dicot, then continued at the primary meristems derived from it. Notice the progressive differentiation of most of the tissue regions.

They commonly develop in two genetically dictated patterns. In most dicot stems, long bundles form a ring that divides the ground tissue into a cortex and pith (Figure 26.13a). A stem's **cortex** is the region between the vascular bundles and the epidermis. Its **pith** is the center of the stem, inside the ring of vascular bundles. Ground tissue of the plant's roots becomes similarly divided, into root cortex and pith. A different pattern is common inside the stems of most monocots and some

Figure 26.12 (**a**) Light micrograph of a *Coleus* shoot tip, cut longitudinally through its center. (**b**) Scanning electron micrograph of its surface. (**c**) New *Coleus* leaves.

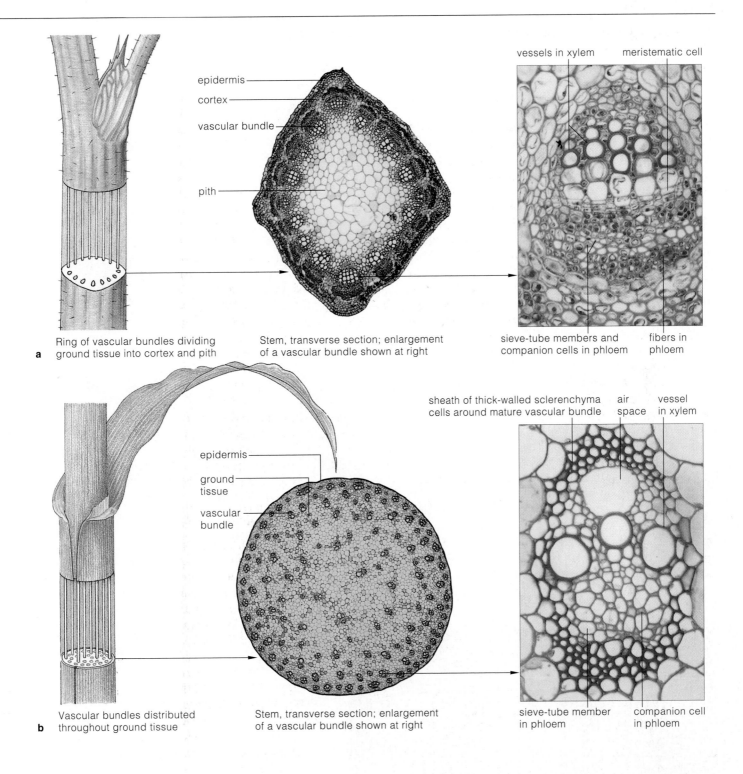

epidermis
cortex
vascular bundle
pith

vessels in xylem   **meristematic cell**

sieve-tube members and companion cells in phloem   **fibers in phloem**

Ring of vascular bundles dividing
a ground tissue into cortex and pith

Stem, transverse section; enlargement
of a vascular bundle shown at right

epidermis
ground tissue
vascular bundle

sheath of thick-walled sclerenchyma cells around mature vascular bundle   **air space**   **vessel in xylem**

sieve-tube member in phloem   **companion cell in phloem**

Vascular bundles distributed
b throughout ground tissue

Stem, transverse section; enlargement
of a vascular bundle shown at right

Figure 26.13 Internal organization of cells and tissues inside the stems from a dicot and a monocot. (**a**) Part of a stem from alfalfa (*Medicago*), a dicot. In many species of dicots and conifers, the vascular bundles develop in a more or less ringlike array in the ground tissue system, as shown here. The portion of the ground tissue between the ring and the surface of the stem is the cortex. The portion enclosed within the ring is the pith. (**b**) Part of a stem from corn (*Zea mays*), a monocot. In most monocots and some nonwoody dicots, vascular bundles are scattered through the ground tissue, as shown.

dicots. The long vascular bundles are scattered through the ground tissue (Figure 26.13*b*). How substances are conducted through such vascular systems is a topic of the next chapter.

The shoot apical meristem gives rise to the primary plant body, which develops a distinctive internal structure (as in the pattern in which its vascular bundles are arranged).

A CLOSER LOOK AT LEAVES

Similarities and Differences Among Leaves

Every **leaf** that forms is a metabolic factory equipped with many photosynthetic cells. Yet leaves vary greatly in size, shape, surface details, and internal structure. A duckweed leaf is no more than 1 millimeter (0.04 inch) across; the leaves of some water lilies are 2 meters (6.5 feet) across. Different leaves look like blades, spikes, cups, needles, feathers, tubes, and other structures. They differ greatly in color, odor, and edibility; many produce toxins. The leaves of birches and other species of *deciduous* plants wither and then drop from stems as winter approaches. The leaves of camellias and other *evergreen* plants also drop, but not all at the same time.

A typical leaf has a flat blade, as in Figures 26.14*a* and 26.15. It has a stalk (petiole) that attaches it to the stem. *Simple* leaves are undivided, but many are lobed. *Compound* leaves have blades divided into leaflets, all oriented in the same plane. Leaves of most monocots, such as ryegrass and corn, are flat surfaced like a knife blade. The blade's base encircles and sheaths the stem.

Leaves of most species are thin, with a high surface-to-volume ratio. Their flat surface grows and orients itself perpendicular to light. A leaf often projects from a stem in patterns that minimize shading of other leaves.

For example, a clover leaf stalk is attached to the stem at right angles to its neighbors (Figure 26.15*c*).

Such leaf adaptations intercept as much of a plant's energy source —sunlight—as possible. They also help oxygen diffuse out and carbon dioxide diffuse in. And when a leaf

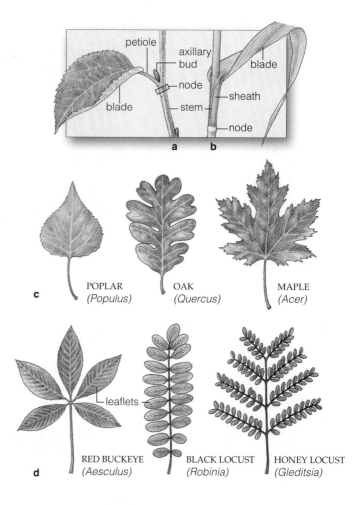

Figure 26.14 Common leaf forms of (**a**) dicots and (**b**) monocots. Examples of (**c**) simple leaves and (**d**) compound leaves.

is thick, you can assume it belongs to a succulent or some other plant of arid habitats and serves in water storage as well as photosynthesis. And still another example: Leaves of many desert plants orient themselves parallel with the sun's rays to reduce heat absorption to tolerable levels.

Leaf Fine Structure

In its fine structure, too, each leaf is adapted to intercept energy from the sun's rays and promote gas exchange. In addition, many have distinctive surface specializations.

Figure 26.15 (**a**) Decaying dicot leaf, with its netlike veins. (**b**) Parallel veins of a monocot leaf (*Agapanthus*). (**c**) Leaf orientation of a four-leaf clover (*Trifolium*).

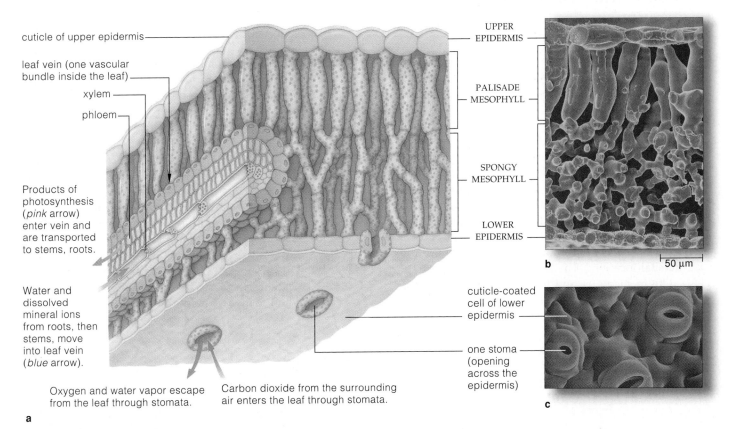

cuticle of upper epidermis

leaf vein (one vascular bundle inside the leaf)

xylem

phloem

Products of photosynthesis (*pink* arrow) enter vein and are transported to stems, roots.

Water and dissolved mineral ions from roots, then stems, move into leaf vein (*blue* arrow).

Oxygen and water vapor escape from the leaf through stomata.

Carbon dioxide from the surrounding air enters the leaf through stomata.

a

UPPER EPIDERMIS

PALISADE MESOPHYLL

SPONGY MESOPHYLL

LOWER EPIDERMIS

50 µm

b

cuticle-coated cell of lower epidermis

one stoma (opening across the epidermis)

c

Figure 26.16 (**a**) Diagram of leaf structure for many kinds of flowering plants. (**b**) Scanning electron micrograph of the tissue organization of a leaf from the kidney bean plant (*Phaseolus*). Notice the compact organization of the epidermal cells. (**c**) Stomata, each a tiny opening across the epidermis, appear when paired guard cells are in their plumped configuration. See also Figure 6.15.

LEAF EPIDERMIS Epidermis covers every leaf surface exposed to the air. It may be smooth, sticky, or slimy, with "hairs," scales, spikes, hooks, glands, and other surface specialties. The Chapter 27 introduction has two examples. A cuticle covers the sheetlike, compact array of epidermal cells; it restricts the loss of precious water (Figures 26.9 and 26.16). Most leaves have many more stomata on their lower surface. In arid or cold habitats, the stomata are often located at depressions in the leaf surface, together with thickly coated epidermal hairs. Both leaf adaptations help conserve water.

MESOPHYLL—PHOTOSYNTHETIC GROUND TISSUE As you read earlier, **mesophyll**, a type of parenchyma with photosynthetic cells and air spaces, extends through a leaf's interior (Figure 26.16). Most cell walls are exposed to air. Plasmodesma connect the cytoplasm of adjoining cells and allow rapid exchange of substances (Section 4.11). Carbon dioxide from the outside reaches cells by diffusing through stomata and the air spaces; oxygen diffuses in the opposite direction.

Leaves oriented perpendicular to the sun have two mesophyll layers. Attached to the upper epidermis is

palisade mesophyll—columnar parenchymal cells with more chloroplasts and more photosynthetic potential, compared to cells of the *spongy* mesophyll layer below them (Figure 26.16). Monocot leaves grow vertically and intercept light from all directions; as Figure 6.15 shows, their mesophyll is not organized as two layers.

VEINS—THE LEAF'S VASCULAR BUNDLES Leaf **veins** are vascular bundles, usually strengthened with fibers. Their continuous strands of xylem rapidly move water and dissolved nutrients to all mesophyll cells, and the continuous strands of phloem carry the photosynthetic products—especially sugars—away from them. In most dicots, the veins branch lacily into a number of minor veins embedded inside mesophyll. In most monocots, veins are more or less similar in length, and they run parallel with the leaf's long axis (Figure 26.15).

A leaf's structure is adapted for sunlight interception, gas exchange, and distribution of water, dissolved nutrients, and photosynthetic products. Leaves of each species have a characteristic size, shape, and often surface specializations.

PRIMARY STRUCTURE OF ROOTS

Taproot and Fibrous Root Systems

When a seed germinates, the first part to poke through the seed coat is a **primary root** (Figure 26.17). In nearly all dicot seedlings, it increases in diameter as it grows downward. Later on, **lateral roots** form in the primary root's tissues, perpendicular to its axis, and then erupt through epidermis. Youngest lateral roots are closest to the root tip. A **taproot system** is a primary root and its lateral branchings. Dandelions, carrots, and oak trees are examples of plans with this system (Figure 26.17c).

By comparison, the primary root of most monocots, such as grasses, is short-lived. In its place, adventitious roots arise from the stem, and then lateral roots branch from these. (*Adventitious* means the structures form at an unusual location.) The lateral roots are all more or less alike in diameter and length. Together, roots that form this way are a **fibrous root system** (Figure 26.17d).

Internal Structure of Roots

In Figure 26.17a, notice the meristems inside the tip of one root. Many cellular descendants of these meristems divide, enlarge, elongate, and become cells of primary tissue systems. Notice also the root cap, a dome-shaped mass of cells at the tip. Apical meristem produces the cap, which in turn protects the meristem.

Protoderm gives rise to root epidermis, the plant's absorptive interface with the soil. Some epidermal cells send out extensions called **root hairs**. Collectively, root hairs enormously increase the surface area available for taking up water and dissolved nutrients. Only first-time or foolish gardeners would yank a plant from the ground when transplanting it. Such yanking would tear off too much of the highly fragile absorptive surface.

Apical meristem also gives rise to the ground tissue system and to a **vascular cylinder**. A vascular cylinder

VASCULAR CYLINDER:
endodermis
pericycle
xylem
phloem

cortex

epidermis

fully grown root hair

Vessels have now matured; root hairs and the vascular cylinder are about to form.

Cells elongate; sieve tubes of phloem form and mature; xylem's vessel members start to form.

Most cells have stopped dividing.

Cells are dividing rapidly at apical and primary meristems.

quiescent center (no cell division)

a   root cap

b

c

d

Figure 26.17   Roots. (**a**) Generalized sketch of a primary root, showing the location of the zones where cells divide, then elongate, then differentiate. (**b**) Light micrograph of a root tip of corn (*Z. mays*), longitudinal section. The oldest root cells are farthest from apical meristem, which a root cap protects. The root cap cells secrete mucigel, a polysaccharide-rich slime that lubricates the root as ongoing cell divisions push it forward through soil. Mucigel ends up coating the root epidermis. It might enhance the uptake of dissolved mineral ions and mycorrhizae formation.

(**c**) Taproot system of the California poppy (*Eschscholzia californica*). (**d**) Fibrous root system of a grass plant.

Table 27.1 *Essential Elements and Plant Function*

MACRONUTRIENT	Some Functions	Some Deficiency Symptoms
Carbon Hydrogen Oxygen	Raw materials for second stage of photosynthesis	None; all are abundantly available (water, carbon dioxide are sources)
Nitrogen	Protein, nucleic acid, coenzyme, chlorophyll component	Stunted growth; light-green older leaves; older leaves yellow and die (these are symptoms that define a condition called chlorosis)
Potassium	Activation of enzymes; contributes to water–solute balances that influence osmosis*	Reduced growth; curled, mottled, or spotted older leaves; burned leaf edges; weakened plant
Calcium	Component in control of many cell functions; cementing of cell walls	Terminal buds wither, die; deformed leaves; poor root growth
Magnesium	Chlorophyll component; activation of enzymes	Chlorosis; drooped leaves
Phosphorus	Phospholipid, nucleic acid, ATP component	Purplish veins; stunted growth; fewer seeds, fruits
Sulfur	Component of most proteins, two vitamins	Light-green or yellowed leaves; reduced growth

MICRONUTRIENT	Some Functions	Some Deficiency Symptoms
Chlorine	Role in root and shoot growth and photolysis	Wilting; chlorosis; some leaves die
Iron	Roles in chlorophyll synthesis and in electron transport	Chlorosis; yellow and green striping in leaves of grass species
Boron	Roles in germination, flowering, fruiting, cell division, nitrogen metabolism	Terminal buds, lateral branches die; leaves thicken, curl, become brittle
Manganese	Chlorophyll synthesis; coenzyme action	Dark veins, but leaves whiten and fall off
Zinc	Role in forming auxin, chloroplasts, starch; enzyme component	Chlorosis; mottled or bronzed leaves; root abnormalities
Copper	Component of several enzymes	Chlorosis; dead spots in leaves; stunted growth
Molybdenum	Part of enzyme used in nitrogen metabolism	Pale green, rolled or cupped leaves

* All mineral elements contribute to water–solute balances, but potassium is notable because there is so much of it.

These usually are dissolved in soil water in ionic forms that reversibly bind with clay. Ions of calcium (Ca^{++}) and potassium (K^{+}) are examples. Plants easily exchange hydrogen ions for these weakly bound elements in clay.

Nine essential elements are *macro*nutrients. Normally they are required in amounts above 0.5 percent of the plant's dry weight (weighed after all the water has been

Figure 27.4 Erosion in an agricultural field. Water's erosive force forms gullies that channel runoff from the land. As gullies grow deeper and wider, erosion becomes more and more rapid. When topsoil is depleted, productivity declines and fertilizers usually must be trucked in to replace lost nutrients.

removed from the plant). The other elements listed are *micro*nutrients; they make up traces (usually a few parts per million) of the dry weight. Even the trace amounts are essential for normal growth.

Leaching and Erosion

Leaching refers to the removal of some of the nutrients in soil as water percolates through it. Leaching is most pronounced in sandy soils, which are not as good as clay at binding nutrients. It is not the same as **erosion**, which is the movement of land under the force of wind, running water, and ice (Figure 27.4). For example, each year, erosion from farmlands in the Mississippi River watershed puts about 25 billion metric tons of topsoil into the Gulf of Mexico. Whether by leaching or erosion, the loss of nutrients from soil is bad for plants and for all organisms that depend on plants for survival.

The mineral component of soil includes particles ranging from large-grained sand to silt and fine-grained clay. These particles, clay especially, reversibly bind water molecules and dissolved mineral ions and thereby make them more accessible for uptake by plant roots.

Soil also contains humus, which is a reservoir of organic material, rich in organic acids, in different stages of decay.

Most plants grow best in soils having equal proportions of sand, silt, and clay, as well as 10 to 20 percent humus.

Nutrients are essential elements. No other element can substitute for their direct or indirect roles in the metabolic activities that sustain growth and keep organisms alive.

HOW DO ROOTS ABSORB WATER AND MINERAL IONS?

In terms of energy outlays, mining the soil for mineral ions and water molecules clinging to clay particles is an expensive task. Plants spend considerable energy on building extensive root systems. Wherever the soil's texture and composition change, new roots must form to replace old ones and snake out in different regions. It isn't that roots "explore" soil for resources. Rather, the patches of soil where concentrations of water and mineral ions are greater stimulate outward growth.

Absorption Routes

Think back on the preceding chapter's discussion of a typical root's structure (Section 26.5). Water molecules in soil are only weakly bound to clay particles, so they readily cross the root epidermis and continue into the **vascular cylinder**, a column of vascular tissue. There, a cylindrical layer of cells, an **endodermis**, wraps around the column. A band of waxy deposits—the **Casparian strip**—is embedded in abutting endodermal cell walls (Figure 27.5). Water molecules can't penetrate it. They infiltrate only the unwaxed wall regions, pass through cells, then cross unwaxed wall regions on the opposite side. This is the only way water and solutes move into the vascular cylinder. Like all cells, endodermal cells have many transport proteins embedded in the plasma membrane. The proteins let some solutes but not others cross it (Section 5.7). *The transport proteins of endodermal cells are control points where plants adjust the quantity and types of solutes absorbed from soil water.*

Roots of many plants also have an **exodermis**, a cell layer just beneath their surface (Figure 27.5a). Walls of exodermal cells commonly have a Casparian strip that functions like the one next to a root vascular cylinder.

Specialized Absorptive Structures

ROOT HAIRS   Vascular plants require great amounts of water. Roots of a mature corn plant absorb more than three liters of water daily. They could not do so without **root hairs**. Recall, from the preceding chapter, that root hairs are slender extensions of specialized epidermal cells. They greatly increase the surface area available for absorption (Section 26.5 and Figure 27.6). When a plant is putting on primary growth, its system of roots may develop millions or billions of root hairs.

ROOT NODULES   Certain bacteria and fungi help many plants absorb dissolved nutrients and receive something in return. As you read in Chapter 22, any two-way flow of benefits between species is a symbiotic interaction we call **mutualism**. As an example, think about how nitrogen deficiency limits plant growth. There is an abundance of gaseous nitrogen ($N \equiv N$) in the air. But plants do not have the

exodermis
root hair
epidermis
newly forming vascular cylinder
cortex
Casparian strip (*gold*) within all the abutting walls of cells of the endodermis

a

conducting cell of primary phloem
conducting cell of primary xylem
endodermal cells

b Vascular cylinder

paths of water molecules
In root cortex, water molecules pass through and between walls of cells
endodermal cells
vascular cylinder

c Casparian strip (*gold*)

Casparian strip

d Two endodermal cells. For clarity, the cytoplasm and most of the cell wall (*blue*) are not shown. Water and solutes can only move into the vascular cylinder by passing through the cells and unwaxed parts of their walls, not in between the cells.

Figure 27.5   Control of the uptake of water and dissolved nutrients inside roots. (**a,b**) Roots of most flowering plants have an endodermis (a cell layer around the vascular cylinder) and an exodermis (a cell layer just beneath the epidermis). (**c**) Abutting walls of cells of both layers contain a waxy Casparian strip. The strip keeps water from moving indiscriminately *around* the cells and into the vascular column. It makes water move *through* the cells. (**d**) That way, transport proteins that span the plasma membrane of these cells can selectively control water and nutrient uptake.

Figure 27.6
Scanning electron micrograph of root hairs (extremely fine extensions of specialized root epidermal cells) of a young root.

root hairs   root   250 µm

Figure 27.8   Demonstration of the effect of root nodules on plant growth. At *left*, rows of soybean plants growing in nitrogen-poor soil. At *right*, growing in the same soil, plants in these rows were inoculated with *Rhizobium* cells and developed root nodules.

a Root nodule

b Infection thread of cellulose that symbiotic bacteria induce a plant to form after they infect a root hair

infection thread

root hair

c A root nodule developing through repeated divisions of infected plant cells and bacterial cells inside them

Figure 27.7   (**a**) Nutrient uptake at root nodules of legumes that are mutualists with nitrogen-fixing bacteria (*Rhizobium* and *Bradyrhizobium*). When infected by the bacteria, root hair cells form a thread of cellulose deposits. Bacteria use the thread as a highway to invade plant cells in the root cortex. (**b,c**) Infected plant cells and the bacterial cells inside them divide repeatedly. Together they form a swollen mass that becomes a root nodule. The bacteria start fixing nitrogen when membranes of plant cells surround them. The plant takes up some of the nitrogen; the bacteria take up some photosynthetic compounds.

metabolic means to engage in **nitrogen fixation**. With this process, certain enzymes break all three covalent bonds in gaseous nitrogen, then attach nitrogen atoms to organic compounds. To get high crop yields, farmers either apply nitrogen-rich fertilizers or encourage the growth of certain nitrogen-fixing bacteria that naturally inhabit soil. Such bacteria can convert gaseous nitrogen to forms that they—and plants—can use. In this respect, string beans, peas, alfalfa, clover, and other legumes have an advantage. Nitrogen-fixing bacteria reside as symbionts in their roots, in localized swellings called **root nodules** (Figures 27.7 and 27.8). The bacterial cells withdraw some photosynthetically produced organic compounds that were distributed from leaves to the roots. But in return they provide the plants with some of the nitrogen that they secured.

MYCORRHIZAE   Also think back on the **mycorrhizae** (singular, mycorrhiza). As described in Chapter 22, a mycorrhiza is a symbiotic interaction between a young root and a fungus. Hence the name, meaning "fungus-root." Fungal filaments—hyphae—form a velvety cover around roots or penetrate the root cells. Collectively, hyphae have a large surface area that absorbs mineral ions from a larger volume of soil than the roots can do. The fungus can absorb sugars and nitrogen-containing compounds from root cells. The root cells obtain some scarce minerals that the fungus is better able to absorb.

Gymnosperms and flowering plant roots control the uptake of water and dissolved nutrients at the vascular cylinder's endodermis and at a similar layer near the root surface.

Root hairs, root nodules, and mycorrhizae greatly enhance the uptake of water and dissolved nutrients.

HOW IS WATER TRANSPORTED THROUGH PLANTS?

Transpiration Defined

By now, you have a sense of how the distribution of water and dissolved mineral ions to all living cells is central to plant growth and functioning. Let's turn now to a model for how water actually moves from a plant's roots to its stems, then into leaves.

Plants, recall, use only a fraction of the water they absorb for growth and metabolism. Most of that water is lost mainly through the numerous stomata in leaves. Evaporation of water molecules from leaves, stems, and other plant parts is a process called **transpiration**.

Cohesion–Tension Theory of Water Transport

This brings up an interesting question. Assuming that plants lose most of the absorbed water from their leaves, how does water actually get *to* the leaves? What gets it to the uppermost leaves of plants, including redwoods and other trees that may be more than 100 meters tall?

In a plant's vascular tissues, water moves through a complex tissue called **xylem**. Recall, from Section 26.2, that the water-conducting cells of xylem are **tracheids** and **vessel members**. Figure 27.9 provides examples. These cells are dead at maturity, and only their lignin-impregnated walls are left. This means the conducting cells in xylem can't be actively pulling water "uphill."

Figure 27.10 *Facing page:* Diagram illustrating the cohesion–tension theory of water transport in vascular plants.

Some time ago, the botanist Henry Dixon came up with a useful way to explain water transport in plants. By his **cohesion–tension theory**, the water inside xylem is pulled upward by air's drying power, which creates continuous negative pressures—that is, tensions. These tensions extend downward from leaves to roots. Figure 27.10 illustrates Dixon's theory. Think about these three points as you review the illustration:

First, air's drying power causes transpiration: the evaporation of water from all plant parts exposed to air—but most notably at stomata. Transpiration puts the water confined in xylem's waterproofed conducting tubes into a state of tension. And that tension extends from veins inside leaves, down through the stems, and on into young roots where water is being absorbed.

Second, the unbroken, fluid columns of water show *cohesion*; they resist rupturing while they are pulled up under *tension*. (Here you may wish to reflect on Section 2.5.) The collective strength of all the hydrogen bonds between water molecules in the narrow, tubular xylem cells imparts this cohesion.

Third, for as long as molecules of water continue to escape from a plant, the continuous tension inside the

a Tracheids have unperforated, tapered end walls. In intact xylem, small pits in the walls of adjoining tracheids match up. The pits are too small to permit rapid flow, but they do confine air bubbles to individual tracheids in the system. Bubbles can obstruct water transport.

b Close-up of three of the adjoining vessel members that make up a vessel. The thick, finely perforated walls of these dead cells connect one after another to form vessels, another type of water-conducting tube in xylem. The walls of all these tubes are impregnated with lignin, which strengthens them and also makes them waterproof.

c Perforation plate at the end wall of one type of vessel member. The perforated ends permit water and air bubbles to flow unimpeded through the conducting tube. This may be a reason why natural selection favored conserving tracheids and vessel members in the same plants.

Figure 27.9 Scanning electron micrographs and sketches of a few types of tracheids and vessel members from xylem. These water-conducting tubes are made of the walls of cells that are dead at maturity. The cell walls still remain interconnected, and they form the tubes. Tracheids probably evolved before vessel members. Both occur in nearly all vascular plants.

mesophyll (photosynthetic cells)   vein   upper epidermis

stoma

1 The Driving Force of Evaporation Into Dry Air

Point 1: Transpiration is the evaporation of water molecules from aboveground plant parts, especially at stomata. The process puts the water in xylem in a state of tension that extends from roots to leaves.

vascular
xylem   cambium   phloem

growing cells also remove small amounts of water from xylem

2 Cohesion in Xylem of Roots, Stems, and Leaves

Water Uptake in Growth Regions

Point 2: The collective strength of hydrogen bonds among water molecules, which are confined within the narrow water-conducting tubes in xylem, imparts cohesion to the water. Hence the narrow columns of water in xylem resist rupturing under the continuous tension.

vascular cylinder   endodermis   cortex   water molecule   hair cell

3 Water Uptake From Soil by Roots

Point 3: As long as water molecules continue to escape by transpiration, that tension will drive the uptake of replacement water molecules from soil water.

xylem permits more molecules to be pulled upward from the roots, and therefore to replace them.

Hydrogen bonds are strong enough to hold water molecules together inside the xylem. However, they are not strong enough to prevent the water molecules from breaking away from one another during transpiration and then escaping from leaves, through stomata.

Transpiration is the process of evaporation from plant parts.

By the cohesion–tension theory of water transport, this process is the key source of tensions in water in xylem. The tensions extend from leaves to roots, and they allow columns of water molecules that are hydrogen-bonded to one another to be pulled upward through the plant body.

HOW DO STEMS AND LEAVES CONSERVE WATER?

At least 90 percent of the water that enters a leaf goes right on through it and evaporates into the surrounding air. Cells use only 2 percent of the water that stays in the leaf for photosynthesis, membrane functions, and other events. That tiny amount is vital. Plants wilt and water-dependent events are severely disrupted when water loss exceeds water uptake at roots for extended periods (Figure 27.11). Yet land plants are not entirely at the mercy of changes in the availability of soil water. They have a cuticle, and they have stomata.

The Water-Conserving Cuticle

Even mildly water-stressed plants would rapidly wilt and die without their **cuticle** (Figure 27.12). Epidermal cells secrete this translucent, water-impermeable layer, which coats cell wall regions exposed to the air. At the cuticle surface are deposits of waxes, which are water-insoluble lipids having long fatty-acid tails. The cuticle itself consists of waxes embedded in **cutin**, an insoluble lipid polymer. Beneath all the waxes, cellulose fibers weave through the cutin. A layer of polysaccharides (pectins) often helps bind the cuticle to the cell walls.

A cuticle does not bar the passage of light rays into photosynthetic parts of the plant. It does restrict water loss. It also restricts the *inward* diffusion of the carbon dioxide necessary for photosynthesis and the *outward* diffusion of oxygen by-products. Recall, from Section 6.6, that a buildup of oxygen in the air spaces inside a leaf has bad effects on the rate of photosynthesis.

Controlled Water Loss at Stomata

How do carbon dioxide and oxygen get past the cuticle-covered, water-conserving epidermis? They cross it at many **stomata** (singular, stoma). At all these openings, carbon dioxide enters and water evaporates from the plant. A pair of specialized cells, the **guard cells**, define each opening (Figure 27.13). As each swells with water, the force of internal fluid pressure distorts its wall. An osmotically induced, internal fluid pressure that builds up against cell walls is called **turgor pressure**. It makes

Figure 27.11 Osmosis and the wilting of a plant with soft green leaves. When such a plant is growing well, you can bet soil water is dilute (hypotonic) compared to fluids in the plant's living cells.

In response to solute concentration gradients, water osmotically moves into a plant cell. Internal fluid pressure builds up against the cell's wall (Section 5.9). Water also is squeezed out when this turgor pressure becomes great enough to counter the attractive force of cytoplasmic fluid. (Cytoplasm usually has more solutes compared to soil water.) Both forces have the potential to cause water to move directionally. The sum of the two opposing forces is the "water potential."

When soft plant parts are erect, as much water is moving into cells as is moving out. The constant pressure keeps cell walls plump. When the soil dries or gets too salty, water's concentration gradient reverses and the cells lose water. Osmotically induced shrinkage of cytoplasm in all of the young cells of soft plant parts results in wilting; the parts droop with the loss of turgor pressure.

A simple experiment demonstrates the wilting effect. Put 10 grams of table salt (NaCl) in 60 milliliters of water. Pour the salty solution into the soil around a tomato plant. The plant starts to collapse after 5 minutes. In less than 30 minutes, you see severe wilting.

Figure 27.12 Waxy cuticle on upper leaf epidermis. Water, carbon dioxide, and oxygen cross a cuticle mainly at stomata. In general, plants in deserts and other seasonally dry habitats have thick cuticles, which provide greater protection against water loss. Aquatic plants have thin cuticles or none at all.

cuticle

mesophyll inside leaf

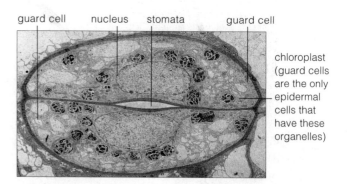

guard cell   nucleus   stomata   guard cell

chloroplast (guard cells are the only epidermal cells that have these organelles)

Figure 27.13 Guard cells in stem epidermis from a prickly pear (beavertail cactus). Cacti have stomata on stems and tiny leaves, if any. Their spines, modified nonphotosynthetic shoots, deter browsing animals while reducing the surface area from which water can evaporate.

Figure 27.14 Stomatal action. Whether a stoma is open or closed during a given interval depends on changes in the shape of the two guard cells that define this small opening across the cuticle-covered epidermis. (**a**) Stomatal opening. Water enters collapsed guard cells, which swell under turgor pressure and move apart and so form a stoma. (**b**) Stomatal closing. Water leaves swollen guard cells, which collapse against each other and close the stoma.

Figure 27.15 Experimental evidence that potassium ions accumulate in guard cells that are expanding with incoming water. Strips from leaf epidermis of a dayflower (*Commelina communis*) were immersed in solutions with dark-staining substances that bind preferentially with potassium ions. (**a**) In leaf samples with opened stomata, most of the ions became concentrated in guard cells. (**b**) In leaf samples with closed stomata, very little potassium entered guard cells; most was present in other epidermal cells.

guard cells bend in such a way that a gap (one stoma) forms between them. When two guard cells lose water and their internal fluid pressure drops, they collapse against each other and close the gap (Figure 27.14).

In most plants, stomata remain open during the day, when photosynthesis takes place. They do lose water—but they gain carbon dioxide for the reactions. Stomata stay closed at night. Then, plants conserve water, and precious carbon dioxide accumulates inside the leaf as the cells busily engage in aerobic respiration.

Figure 27.16 (**a**) Stomata at holly leaf surface. (**b**) What they look like when a holly plant is growing in industrialized regions. Gritty airborne pollutants clog its stomata and prevent much of the sun's rays from reaching photosynthetic cells in the leaf.

Guard cells are the only epidermal cells that contain chloroplasts. Whether a stoma opens or closes depends on how much water and carbon dioxide its guard cells hold. Photosynthesis starts after the sun comes up. As the morning progresses, carbon dioxide levels drop in cells—including guard cells. The decrease helps trigger active transport of potassium ions into the guard cells (Figure 27.15). So do blue wavelengths of light, which penetrate the atmosphere better as the sun arcs higher in the sky. Water follows potassium ions into the cells, by osmosis. That inward movement of water provides the fluid pressure to open the stoma.

Carbon dioxide levels in the cells rise when the sun sets and photosynthetic activity ceases. Potassium, then water, move out of guard cells. And so stomata close.

CAM plants, including most cacti, conserve water differently. They open stomata at night, when they fix carbon dioxide by the metabolic C4 pathway described in Section 6.6. The next day, when their stomata close, CAM plants use carbon dioxide in photosynthesis.

As this section makes clear, plant survival depends on stomata. Think about it when you are out and about on smog-shrouded days (Figure 27.16).

Water-dependent events in plants are severely disrupted when water loss exceeds uptake at roots for extended periods. Wilting is one observable outcome.

Transpiration and gas exchange occur mainly at stomata. These numerous small openings span the waxy cuticle, which covers all plant epidermal surfaces exposed to air.

Plants open and close stomata at different times to control water loss, carbon dioxide uptake, and oxygen disposal, all of which affect rates of photosynthesis and plant growth.

HOW ARE ORGANIC COMPOUNDS DISTRIBUTED THROUGH PLANTS?

Whereas xylem distributes water and minerals through the plant, the vascular tissue called **phloem** distributes organic products of photosynthesis. Like xylem, phloem consists of many conducting tubes, fibers, and strands of parenchyma cells, all of which extend through the plant body. Unlike xylem, phloem contains conducting tubes of a type called **sieve tubes**. These tubes consist of *living* cells, positioned side by side and end to end in vascular tissues. Figure 27.17 is an example of the cells, which are called sieve-tube members. Dissolved sugars and other organic compounds flow swiftly through the large pores in their abutting end walls.

Adjoining the long sieve tubes are **companion cells**. Like the sieve-tube members, these are differentiated parenchyma cells. As you will see, companion cells help load organic compounds into sieve tubes.

To understand what goes on inside phloem, think of what happens to sucrose and other organic products of photosynthesis. Leaf cells use some of the products for their own activities. The rest travels to roots, stems, buds, flowers, and fruits. In most cells, carbohydrates get stored as starch, in plastids. The proteins and fats, which the cells synthesize from carbohydrates and the amino acids distributed to them, often get stored in seeds. Avocados and some other fruits also accumulate fats.

Starch molecules are too large for transport across the plasma membrane of cells that produce them. Also, they are too insoluble for transport to other parts of the plant. Overall, proteins are too large and fats are too insoluble for transport from the storage sites. *However, the cells convert the storage forms of organic compounds to solutes of smaller size that are far more easily transported through phloem.*

For instance, the cells degrade starch to glucose monomers. When one of these monomers combines with fructose, the result is sucrose —an easily transportable sugar.

Simple experiments with aphids revealed that sucrose is the main carbohydrate that is transported inside phloem. These insects were anesthetized by exposure to high

Figure 27.18 A honeydew droplet exuding from the end of an aphid gut. The tubular mouthpart of the insect penetrated a conducting tube in phloem. The sugary fluid was under high pressure in phloem and was forced out through the gut's terminal opening.

levels of carbon dioxide as they were feeding on juices in conducting tubes of phloem. The body of the aphids was detached from their mouthparts—which were still embedded in the sieve tubes. For most of the plants studied, sucrose was the most abundant carbohydrate in the fluid that was being forced out of the tubes.

Translocation

Translocation is the technical name for the transport of sucrose and all other organic compounds through the phloem of a vascular plant. High pressure drives this process. Often the pressure is five times higher than in automobile tires. Aphids demonstrate the magnitude of it when they force their mouthparts into sieve tubes and feed upon the dissolved sugars. The high pressure can force fluid through an aphid gut and out the other end, as "honeydew" (Figure 27.18). Park a car under some trees being attacked by aphids, and it might get spattered by sticky honeydew droplets, thanks to the high fluid pressure in phloem.

Pressure Flow Theory

Phloem translocates the photosynthetic products along decreasing pressure and solute concentration gradients. The *source* of the flow is any region of the plant where organic compounds are being loaded into sieve tubes. Common sources are mesophylls—the photosynthetic tissues in leaves. The flow ends at a *sink*, which is any plant region where the products are being unloaded, used, or stored. Example: While flowers and fruits are growing during development, they are sink regions.

Why do organic compounds flow from a source to a sink? According to the **pressure flow theory**, internal pressure builds up at the source end of the sieve tube system and *pushes* the solute-rich solution on toward any sink, where solutes are being removed.

sieve plate

companion cell

sieve-tube member

Figure 27.17 Examples of sieve-tube members, living cells that interconnect to form a conducting tube in phloem. Living companion cells expend ATP energy to load organic compounds into adjacent sieve-tube members.

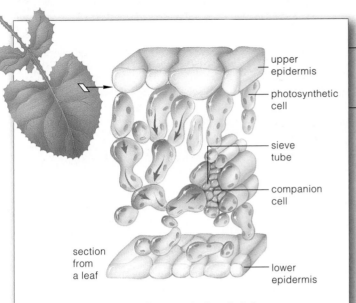

upper
epidermis

photosynthetic
cell

sieve
tube

companion
cell

section
from
a leaf

lower
epidermis

a *Loading at a source.* Photosynthetic cells in leaves are a common source of organic compounds that must be distributed through a plant. Small, soluble forms of these compounds move from the cells into phloem (a leaf vein).

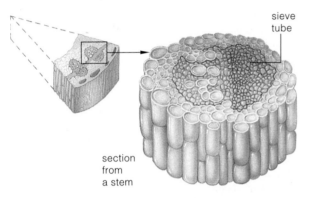

sieve
tube

section
from
a stem

b *Translocation along a distribution path.* Fluid pressure is greatest inside sieve tubes at the source. It pushes the solute-rich fluid to a sink, which is any region where cells are growing or storing food. There, the pressure is lower because cells are withdrawing solutes from the tubes.

sieve
tube

section
from
a root

c *Unloading at the sink.* Solutes are unloaded from sieve tubes into cells at the sink; water follows. Translocation continues as long as solute concentration gradients and a pressure gradient exist between the source and the sink.

Figure 27.19 Translocation of organic compounds in sow thistle (*Sonchus*). Review Figure 6.3 to get an idea of how translocation relates to photosynthesis in vascular plants.

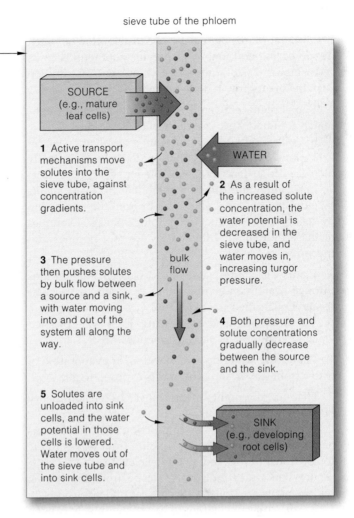

sieve tube of the phloem

SOURCE
(e.g., mature
leaf cells)

WATER

1 Active transport mechanisms move solutes into the sieve tube, against concentration gradients.

2 As a result of the increased solute concentration, the water potential is decreased in the sieve tube, and water moves in, increasing turgor pressure.

3 The pressure then pushes solutes by bulk flow between a source and a sink, with water moving into and out of the system all along the way.

bulk
flow

4 Both pressure and solute concentrations gradually decrease between the source and the sink.

5 Solutes are unloaded into sink cells, and the water potential in those cells is lowered. Water moves out of the sieve tube and into sink cells.

SINK
(e.g., developing
root cells)

Experimental evidence for the pressure flow theory was obtained from studies of sow thistle (*Sonchus*). Use Figure 27.19 to track what happens after sucrose moves from photosynthetic cells into small veins in the leaves of this plant. Companion cells in veins spend energy to load sucrose into adjacent sieve-tube members. As the sucrose concentration increases in the tubes, water also moves in, by osmosis. More internal fluid pressure is exerted against the sieve-tube walls. When the pressure increases, it pushes the sucrose-laden fluid out of the leaf, into the stem, and on to the sink.

Plants store organic compounds in the form of starch, fats, and proteins. They convert the storage forms to sucrose and other small units that are soluble and easily translocated.

Translocation is the distribution of organic compounds to different plant regions. It depends on concentration and pressure gradients in the sieve-tube system of phloem.

Gradients exist as long as companion cells load compounds into the sieve tubes at sources, such as mature leaves, and as long as compounds are removed at sinks, such as roots.

SUMMARY   *Gold* indicates text section

1. Vascular plants depend on the distribution of water, dissolved mineral ions, and organic compounds to its cells. Figure 27.20 summarizes how that distribution sustains plant growth and survival. *27.1–27.5*

2. Root systems efficiently take up water and nutrients, which commonly are scarce in soil. *CI, 27.1*

  a. Collectively, a vascular plant's root hairs (slender extensions of specialized root epidermal cells) greatly increase the surface area available for absorption. *27.2*

  b. Bacterial and fungal symbionts help many plants take up mineral ions. They benefit from the interaction by using some products of photosynthesis. Examples of these mutualistic interactions are root nodules and mycorrhizae. *27.2*

3. Plants distribute water and dissolved mineral ions through water-conducting tubes of xylem, a vascular tissue. Cells called tracheids and vessel members form the tubes but are dead at maturity. Their waterproofed walls interconnect as narrow pipelines. *27.3*

4. Plants lose water by transpiration, or evaporation of water from leaves and other parts exposed to air. *27.3*

5. Here are the points of the cohesion–tension theory of water transport in plants: *27.3*

  a. Inside xylem's water-conducting cells, continuous negative pressures (tensions) extend from leaves to the roots. Transpiration causes the tension.

  b. When water molecules escape from the leaves, replacements are pulled into the leaf under tension.

  c. The collective strength of numerous hydrogen bonds between water molecules imparts cohesion that allows water molecules to be pulled up as continuous fluid columns. The "cohesion" part of the theory refers to the capacity of the hydrogen bonds between water molecules to resist rupturing when put under tension.

6. Most plants have a waxy, water-impermeable cuticle covering their aboveground parts. They lose water and take up carbon dioxide at stomata. A pair of guard cells defines each of these microscopically small openings across the epidermis of leaves and stems. Guard cells are specialized parenchyma cells. *27.4*

7. Stomata open and close at different times. Control of their action balances water conservation with carbon dioxide uptake and the release of oxygen. *27.4*

  a. In many kinds of plants, stomata open during the day. Such plants lose water but take in carbon dioxide for photosynthesis. Stomata close at night; thus, water and the carbon dioxide released by aerobically respiring cells are conserved for the next day.

  b. CAM plants open stomata and fix carbon dioxide by the C4 pathway at night. During the day, they close their stomata and use the carbon fixed the night before for photosynthesis.

Figure 27.20 Summary of the interdependent processes that sustain the growth of vascular plants. All living cells in plants require oxygen, carbon, hydrogen, and at least thirteen mineral ions. They all produce ATP to drive metabolic activities.

8. Plants distribute organic compounds through sieve tubes of phloem, a vascular tissue. Translocation is the distribution of organic compounds in phloem. *27.5*

9. According to the pressure flow theory, translocation is driven by differences in solute concentrations and pressure between source and sink regions. A source is any site where organic compounds are being loaded into sieve tubes (e.g., mature leaves). A sink is any site where compounds are being unloaded from sieve tubes. Roots are examples. Solute concentration gradients and pressure gradients exist for as long as companion cells expend energy to load solutes into the sieve tubes and for as long as solutes are removed at the sink. *27.5*

Review Questions

1. Define soil, then distinguish between: *27.1*
   a. humus and loam
   b. leaching and erosion
   c. macronutrient and micronutrient (for plants)

2. Define nutrient. What are signs that a plant is deficient in one of the essential nutrients listed in Table 27.1? *27.1*

3. What is the function of the Casparian strip in roots? *27.2*

4. Using Dixon's model, explain how water moves from soil upward through tall plants. *27.3*

5. Describe the structure and function of a plant cuticle. *27.4*

6. Which type of ion influences stomatal action? *27.4*

7. Explain translocation according to the pressure flow theory described in this chapter. *27.5*

Self-Quiz   ANSWERS IN APPENDIX III

1. Carbon, hydrogen, oxygen, nitrogen, and potassium are examples of _____ for plants.
   a. macronutrients   d. essential elements
   b. micronutrients   e. both a and d
   c. trace elements

2. A _____ strip in endodermal cell walls forces water and solutes to move through root cells, not around them.
   a. cutin   b. lignin   c. Casparian   d. cellulose

3. The nutrition of some plants depends on a root–fungus association known as a _____ .
   a. root nodule   c. root hair
   b. mycorrhiza   d. root hypha

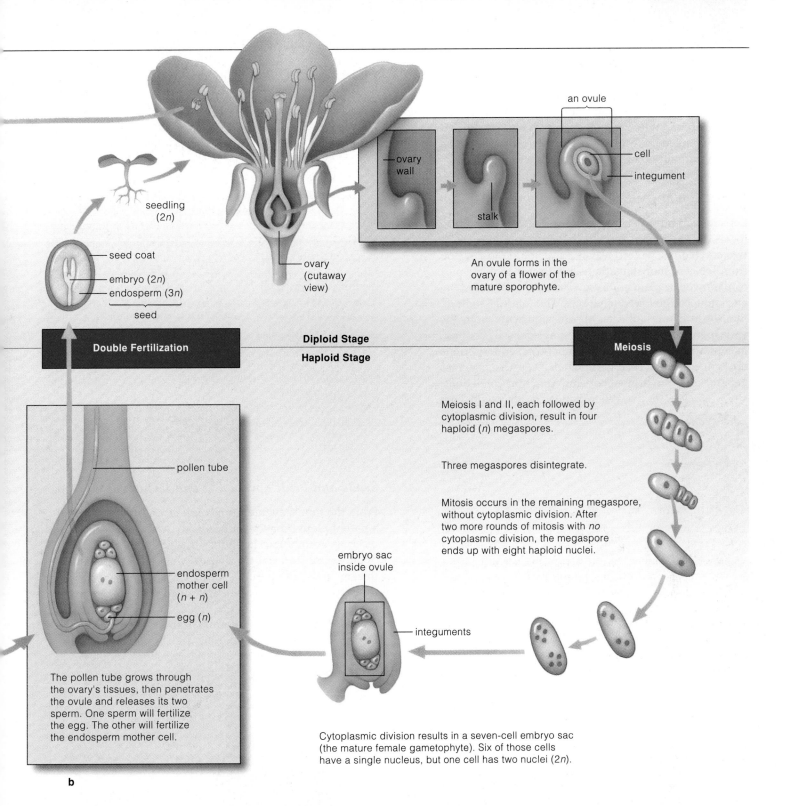

seedling
(2n)

seed coat

embryo (2n)

endosperm (3n)

seed

ovary
(cutaway
view)

an ovule

ovary
wall

cell

integument

stalk

An ovule forms in the
ovary of a flower of the
mature sporophyte.

Double Fertilization

Diploid Stage

Haploid Stage

Meiosis

Meiosis I and II, each followed by
cytoplasmic division, result in four
haploid (n) megaspores.

Three megaspores disintegrate.

Mitosis occurs in the remaining megaspore,
without cytoplasmic division. After
two more rounds of mitosis with *no*
cytoplasmic division, the megaspore
ends up with eight haploid nuclei.

pollen tube

endosperm
mother cell
(n + n)

egg (n)

embryo sac
inside ovule

integuments

The pollen tube grows through
the ovary's tissues, then penetrates
the ovule and releases its two
sperm. One sperm will fertilize
the egg. The other will fertilize
the endosperm mother cell.

Cytoplasmic division results in a seven-cell embryo sac
(the mature female gametophyte). Six of those cells
have a single nucleus, but one cell has two nuclei (2n).

b

toward the egg chamber and sexual destiny. When the
pollen tube reaches an ovule, it penetrates the embryo
sac, and its tip ruptures to release two sperm.

"Fertilization" generally means fusion of a sperm
nucleus with an egg nucleus. But **double fertilization**
occurs in flowering plants. In species having a diploid
chromosome number, one sperm nucleus fuses with an
egg nucleus to form a diploid (2n) zygote. Meanwhile,
the other sperm nucleus fuses with both nuclei of the

endosperm mother cell. The resulting cell has a triploid
(3n) nucleus and will give rise to the nutritive tissue.

In flowering plants, sperm cells form within pollen grains,
the male gametophytes. Ovules form inside ovaries, and
female gametophytes (with egg cells) develop within them.

After pollination and double fertilization, an embryo and
nutritive tissue form in the ovule, which becomes a seed.

FROM ZYGOTES TO SEEDS AND FRUITS

After fertilization, the newly formed zygote embarks on a course of mitotic cell divisions that lead to a mature embryo sporophyte. It develops as part of an ovule and is accompanied by formation of a **fruit**: a mature ovary, with or without other, neighboring tissues.

Formation of the Embryo Sporophyte

Let's look at how a shepherd's purse (*Capsella*) embryo forms. By the time it reaches the stage in Figure 28.7*e*, two **cotyledons**, or seed leaves, have started to develop from lobes of meristematic tissue. Cotyledons are part of all flowering plant embryos. Dicot embryos have two, and monocot embryos have one. The *Capsella* embryo, like those of most dicots, absorbs nutrients from the endosperm and stores them in its cotyledons. By contrast, in corn, wheat, and most other monocots, endosperm is not tapped until a seed germinates. Digestive enzymes get stockpiled in a monocot embryo's thin cotyledons. When activated, they will help transfer stored endosperm to the growing seedling.

nucleus

vacuole

a

Figure 28.7 Development of a dicot seed—shepherd's purse (*Capsella*). The micrographs are not to the same scale. (**a**) The single-celled zygote. (**b–d**) The early embryo is identified by the *yellow* boxes. The row of cells below each box transfers nutrients to the embryo sporophyte from the parent plant. The embryo in (**e**) is well developed, and the one in (**f**) is mature. (**g**) *Capsella* fruit.

Table 28.1 *Categories and Examples of Fruits*

SIMPLE FRUITS From one ovary of one flower.

Dry fruit

> *Dehiscent.* Fruit wall splits along definite seams to release seeds. Legume (e.g., pea, bean), poppy, larkspur, mustard
>
> *Indehiscent.* Fruit wall does not split on seams to release seeds. Acorn, grains (e.g., corn), sunflower, carrot, maple

Fleshy fruit

> *Berry.* Compound ovary, many seeds. Tomato, grape, banana
> Pepo. Ovary wall has hard rind. Cucumber, watermelon
> Hesperidium. Ovary wall has leathery rind. Orange, lemon
>
> *Drupe.* One or two seeds. Thin skin, part of flesh around a usually hard seed. Peach, cherry, apricot, almond, olive

AGGREGATE FRUITS

Many ovaries of one flower, all attached to the same receptacle. Many seeds. Raspberry, blackberry (not really "berries")

MULTIPLE FRUITS

Combined from ovaries of many flowers. Pineapple, fig

ACCESSORY FRUITS

Most tissues of the flesh are not derived from the ovary; e.g., mainly from the receptacle. Pome (apple, pear), strawberry

Seeds and Fruit Formation and Dispersal

From the time a zygote forms until an embryo matures, a parent plant transfers nutrients to ovule tissues. Food reserves accumulate in endosperm or cotyledons. The ovule separates from the ovary's wall; its integuments thicken and harden into a seed coat. The embryo, food reserves, and coat are a self-contained package, a **seed**, which is defined as a mature ovule.

As seeds form, other floral parts start to form fruits. Depending on the species, they are simple, aggregate, or multiple fruits (Table 28.1). When mature, simple dry fruits such as acorns are intact; others, such as pea pods,

seed coat

embryo's shoot tip

embryo's cotyledons (two)

embryo sporophyte

endosperm

embryo's root tip

mature embryo within ovule

A fruit (a mature ovary) cut open to show seeds (the mature ovules). The embryo in each seed is at stage (**f**).

b Upper part of the zygote gives rise to embryo

c Globular stage of embryo

d Heart-shaped stage of embryo

e

f

g

seed

fleshy fruit, from ovary tissue

remnants of sepals, petals,

expanding receptacle

ovary tissue
seed
enlarged receptacle

d

one of many individual fruits

receptacle

ovary

c

wing

seed (in carpel)

e

Figure 28.8 Fruits. (**a**) Peach, a drupe. (**b**) Pineapple (*Ananas*), a multiple fruit. Native Americans cultivated pineapples. To them, it was symbol of hospitality. They used the juice as the base for an alcoholic beverage and to poison arrow tips. Christopher Columbus thought the fruit looked like pine cones, hence the name. (**c**) Strawberry (*Fragaria*), a fragrant accessory fruit. (**d**) Fruit formation on an apple (*Malus*) tree. When petals drop, eggs are fertilized. (**e**) Winged seeds of maple (*Acer*).

unzip and so release seeds. Figure 28.8 shows a simple fleshy fruit (a peach) and a multiple fruit (a pineapple, with ovaries of many separate flowers grown together into a fleshy mass). A strawberry, an aggregate fruit, has many seeds on a fleshy, enlarged receptacle. An apple, an accessory fruit, is mainly an enlarged receptacle and calyx.

Fruits have a common function—*seed dispersal*. To gain insight into their structure, think about how they coevolved with dispersing agents—air or water currents and animals. Wind-dispersed maple fruits have fleshy winglike projections (Figure 28.8*e*). They may spin away from parent plants that outcompete seedlings for light, water, and nutrients. As other examples, air currents lift the plumelike "parachute" of dandelion fruits and also disperse orchid seeds as fine as dust particles.

Other fruits taxi to new locations on or in animals, including many insects, birds, and mammals. Some can adhere to feathers, feet, or fur with hooks, spines, hairs, and sticky surfaces. The fruits of cocklebur, bur clover, and bedstraw are like this. Embryo sporophytes inside the seed coats of fleshy fruits withstand the attacks by potent digestive enzymes inside an animal gut. Besides digesting the fruit's flesh, the enzymes digest some of the seed coat. The digested portions make it easier for

embryo sporophytes to break through their hard coat after seeds are expelled from the animal body and start to germinate. We return to this topic in Section 28.6.

Some water-dispersed fruits have heavy wax coats, and others, including sedges, have sacs of air that help them float. With their thick, water-repelling skin, fruits of coconut palms are adapted to travel across the open ocean. If they don't beach soon enough, saltwater may penetrate the skin and kill their embryos, but coconuts can travel hundreds of kilometers before that happens.

Humans are grand dispersing agents. Past explorers carried seeds all over the world. Cacao, oranges, corn, and many other plants were domesticated and currently reproduce in their new habitats, in far greater numbers than they did on their own. And in evolutionary terms, reproductive success is what life is all about.

A mature ovule, which encases an embryo sporophyte and food reserves inside a protective coat, is a seed.

A mature ovary, with or without additional floral parts that have become incorporated into it, is a fruit.

Seeds and fruits are structurally adapted for dispersal by air currents, water currents, and many kinds of animals.

ASEXUAL REPRODUCTION OF FLOWERING PLANTS

Asexual Reproduction in Nature

The features of sexual reproduction that we considered in earlier sections dominate flowering plant life cycles. Bear in mind, many species also reproduce asexually by modes of **vegetative growth**, as listed in Table 28.2. In essence, new roots and shoots grow right out from extensions or fragments of parent plants. Such asexual reproduction proceeds by way of mitosis, so that a new generation of offspring is genetically identical to the parent; it is a clone.

One "forest" of quaking aspen (*Populus tremuloides*) provides us with an impressive example of vegetative reproduction. The leaves of this flowering plant species tremble even in the slightest breeze, hence the name. Figure 28.9 is a panoramic view of the shoot systems of a single individual. The root system of the parent plant

Table 28.2 *Asexual Reproductive Modes of Flowering Plants*		
Mechanism	Examples	Characteristics
VEGETATIVE REPRODUCTION ON MODIFIED STEMS		
1. Runner	Strawberry	New plants arise at nodes along aboveground, horizontal stems.
2. Rhizome	Bermuda grass	New plants arise at nodes of underground horizontal stems.
3. Corm	Gladiolus	New plants arise from axillary buds on short, thick, vertical underground stems.
4. Tuber	Potato	New shoots arise from axillary buds on tubers, which are the enlarged tips of slender underground rhizomes.
5. Bulb	Onion, lily	New bulbs arise from axillary buds on short underground stems.
PARTHENOGENESIS		
	Orange, rose	An embryo develops without nuclear or cellular fusion (for example, from an unfertilized haploid egg or by developing adventitiously, from tissue surrounding the embryo sac).
VEGETATIVE PROPAGATION		
	Jade plant, African violet	A new plant develops from tissue or structure (a leaf, for instance) that drops from the parent plant or is separated from it.
TISSUE CULTURE PROPAGATION		
	Orchid, lily, wheat, rice, corn, tulip	A new plant is induced to arise from a parent plant cell that has not become irreversibly differentiated.

gave rise to adventitious shoots, which became separate shoot systems. Barring rare mutations, individuals at the north end of this vast clone are genetically identical to individuals at the south end. Water travels from roots near a lake all the way to shoot systems in much drier soil. Dissolved ions travel in the opposite direction.

As long as environmental conditions favor growth and regeneration, such clones are about as close as one can get to being immortal. No one knows how old the aspen clones are. The oldest known clone is a ring of creosote bushes (*Larrea divaricata*) growing in the Mojave Desert. It has been around for the past 11,700 years.

The reproductive possibilities are amazing. Watch strawberry plants send out horizontal, aboveground stems (runners), then watch the new roots and shoots develop at every other node. The oranges you eat may be descended from a single tree in southern California that reproduced by parthenogenesis. With this type of process, an embryo develops from an unfertilized egg.

Parthenogenesis can be stimulated when pollen has contacted a stigma even though a pollen tube has not grown through the style. Maybe certain hormones from the stigma or pollen grains diffuse to the unfertilized egg and trigger formation of an embryo. That embryo is $2n$ as a result of fusion of the products of mitotic cell division in the egg. A $2n$ cell outside the gametophyte also may be stimulated to develop into an embryo.

Induced Propagation

Most houseplants, woody ornamentals, and orchard trees are clones. People typically propagate them from cuttings or fragments of shoot systems. For example, with suitable encouragement, a severed African violet leaf may form a callus from which adventitious roots develop. A callus is one type of meristem. Remember, a meristem is a localized region of plant cells that retain the potential for mitotic cell division.

Or consider a twig or bud from one plant, grafted onto a different variety of some closely related species. Vintners in France, for instance, graft prized grapevines onto disease-resistant root stock from America.

Frederick Steward and his coworkers pioneered in **tissue culture propagation**. They cultured small bits of phloem from the differentiated roots of carrot plants (*Daucus carota*) in rotating flasks. They used a liquid growth medium that contained sucrose, minerals, and vitamins. The liquid also contained coconut milk, which Steward knew was rich in then-unidentified growth-inducing substances. As the flasks rotated, individual cells that were torn away from the tissue bits divided and formed multicelled clumps, which sometimes gave rise to new roots (Figure 28.10a). Steward's experiments were among the first to demonstrate that some cells of

Figure 28.9 A mere portion of Pando the Trembling Giant, so named by Michael Grant and his coworkers at the University of Colorado, who studied its genetic makeup. This "forest" in Utah is actually a single asexually reproducing male organism, of a type called quaking aspen (*Populus tremuloides*). Its root system extends through about 106 hectares (about 262 acres) of soil and functionally supports its 47,000 genetically identical shoots; that is, the trees. By one estimate, this clone weighs more than 5,915,000 metric tons.

Figure 28.10 (**a**) Cultured cells from a carrot plant. At the time this photograph was taken, roots and shoots of embryonic plants were already forming. (**b**) Young orchid plants, developed from cultured meristem and young leaf primordia. Orchids are one of the most highly prized cultivated plants. Before the meristems were cloned, they were difficult to hybridize. From the time seeds form, it can take seven years or more until a new plant bears flowers. In natural habitats, the seeds normally will not germinate unless they interact with a specific fungus.

specialized tissues still house the genetic instructions required to produce an entire individual.

Researchers now use shoot tips and other parts of individual plants for tissue culture propagations. The techniques prove useful when an advantageous mutant arises. Such a mutant may show resistance to a disease that is crippling to wild-type plants of the same species. Tissue culture propagation can result in hundreds, even thousands of identical plants from merely one mutant specimen. The techniques are already being employed in efforts to improve major food crops, including corn, wheat, rice, and soybeans. They also are being used to increase production of hybrid orchids, lilies, and other prized ornamental plants (Figure 28.10*b*).

Besides reproducing sexually, flowering plants engage in asexual reproduction, as by vegetative growth.

28.6

PATTERNS OF EARLY GROWTH AND DEVELOPMENT—AN OVERVIEW

Seed Germination

Let's first consider the overall patterns of growth and development for dicots and monocots, starting with the events that take place inside a seed. Figure 28.11 shows the embryo sporophyte inside a grain of corn. (A grain, remember, is a seed-containing dry fruit.) The growth of the embryo idles before or after the seed is dispersed from the parent plant. Later, if all goes well, the seed germinates. **Germination** is the process by which some immature stage in the life cycle of a species resumes growth after a period of arrested development.

Germination depends on environmental factors, such as soil temperature, moisture, and oxygen level, and the number of daylight hours. The factors vary with the seasons. For instance, mature seeds don't hold enough water for cell expansion or metabolism. In many land habitats, ample water is available on a seasonal basis, so seed germination coincides with the spring rains. By the process of imbibition, water molecules move into a seed, being attracted mainly to hydrophilic groups of proteins stored in endosperm or cotyledons. As more water moves in, the seed swells and its coat ruptures.

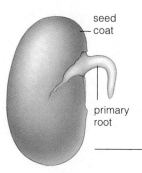

Figure 28.12 (**a,b**) Pattern of growth and development of a dicot, the common bean plant (*Phaseolus vulgaris*). When this seed germinates, the embryo in it resumes growth. A hypocotyl (a hook-shaped part of the shoot below the cotyledons) forces a channel through soil so the food-storing cotyledons are pulled up without being torn apart. At the surface, sunlight exposure makes the hook straighten. For several days, photosynthetic cells make food in the seedling's cotyledons. These wither and fall off. Foliage leaves take over photosynthesis. Flowers form in buds at nodes. (**c**) Cotyledons breaking through a seed coat.

a Bean seedling at the close of germination.

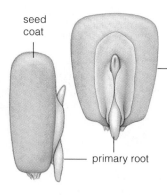

Figure 28.13 Pattern of growth and development of a monocot, using corn (*Zea mays*) as the example. (**a,b**) After germination of a corn grain, the coleoptile protects new leaves as the seedling grows through the soil. In corn seedlings, adventitious roots develop at the coleoptile's base. (**c**) Coleoptile and primary root of a corn seedling. (**d**) Coleoptile and first foliage leaf of two seedlings poking above the soil surface.

a Corn grain at the close of germination.

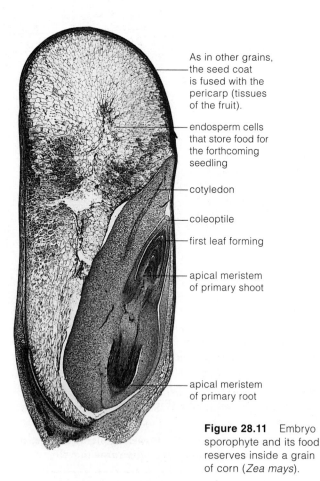

As in other grains, the seed coat is fused with the pericarp (tissues of the fruit).

endosperm cells that store food for the forthcoming seedling

cotyledon

coleoptile

first leaf forming

apical meristem of primary shoot

apical meristem of primary root

Figure 28.11 Embryo sporophyte and its food reserves inside a grain of corn (*Zea mays*).

Once the seed coat splits, more oxygen reaches the embryo, and aerobic respiration moves into high gear. The embryo's meristematic cells now divide rapidly. The root meristem is usually activated first. Its descendants divide, elongate, and give rise to the first, primary root of a seedling sporophyte. Germination is over when the seedling's primary root breaks through the seed coat.

Genetic Programs, Environmental Cues

Figures 28.12 and 28.13 show patterns of germination, growth, and development for dicots and monocots. The patterns have a heritable basis, being dictated by genes. All cells in a plant arise from the same cell (a zygote), so they generally inherit the same genes. But unequal divisions between daughter cells and their positions in one foliage leaf (this type is divided into three leaflets)

primary leaf

primary leaf

withered cotyledon

point at which cotyledons were attached

cotyledons (two)

hypocotyl

branch roots

primary root

root nodule

c

b Structures that form during early growth and development of a bean plant

coleoptile

branch root

primary root

first foliage leaf

first internode of stem

adventitious root

branch root

primary root

prop roots that form on corn seedlings and that afford additional support for the rapidly growing stem

c

first foliage leaf

coleoptile

d

b Structures that form during early growth and development of a corn plant

the developing plant lead to differences in metabolic equipment and output. Cell activities start to vary with the onset of selective gene expression (Chapter 14). For example, genes that govern the synthesis of growth-stimulating hormones are activated in some cells but not others. As you will see shortly, such events seal the developmental fate of different cell lineages.

Bear in mind, plants commonly adjust prescribed growth patterns in response to unusual environmental pressures. Suppose a seed germinates in a vacant lot and a heavy paper bag blows on top of it. The primary

shoot will quickly bend and grow out from under the bag, in the direction of sunlight impinging on it. Interactions among enzymes, hormones, and other gene products in its cells result in this growth response.

Plant growth involves cell divisions and cell enlargements. Plant development requires cell differentiation, as brought about by selective gene expression.

Interactions among genes, hormones, and the environment govern how each individual plant grows and develops.

EFFECTS OF PLANT HORMONES

Hormones are signaling molecules secreted by specific cells that travel to other, target cells and influence gene activity. Any cell having receptors for a hormone is its target. Hormones orchestrate growth and development. They respond to each other's signals. They respond to cues afforded by seasonal change, as when days grow longer and warmer in spring.

Before getting into hormonal effects on growth and development, be sure you know the difference between these two processes. In general, **growth** of a multicelled organism means its cells increase in number, size, and volume. **Development** is the emergence of specialized, morphologically different body parts. In other words, we must measure plant growth in *quantitative* terms, and plant development in *qualitative* terms.

As you know from Section 26.1, cell divisions and enlargements underlying growth occur at meristems. About half the resulting daughter cells do not increase in size but rather retain the capacity to divide. The rest enlarge, often by twenty times. Meristem descendants divide in prescribed planes and expand in particular directions (Figure 28.14). The eventual outcome will be plant parts having specific shapes and functions.

When young cells of an embryo sporophyte or a seedling take up water, turgor pressure (internal fluid pressure) increases against their primary wall and drives enlargement. Imagine blowing up a balloon. While it is soft, the balloon inflates easily. Similarly, a soft-walled cell under turgor pressure expands quickly. That wall thickens as polysaccharides are added to it, and more cytoplasm forms between it and a central vacuole.

Hormones help control the selective gene expression that underlies cell differentiation and the emergence of basic patterns of development. At least five different classes of hormones are known to have predictable and important effects on flowering plants, starting with the

a Growth proceeds in all directions

b Growth in longitudinal direction only

c Cellulose rings thicken the secondary wall of this tracheid, a conducting tube of xylem

Figure 28.14 How different cell shapes arise. In each cell, microtubules in the cytoplasm inherited from the parent cell are already oriented in prescribed patterns. The patterns govern how cellulose microfibrils will be oriented in the cell wall. With random orientation (**a**), a primary wall is elastic all over, so the cell can expand in all directions and assume a spherical shape. (**b**,**c**) When cellulose strands form in a transverse orientation, the cell can only lengthen.

germination of the new seed. As with animals, genetic instructions in a plant's DNA govern the synthesis of enzymes and other proteins necessary for metabolism, hence for all cell activities. *And hormonal action affects how and when each type of enzyme functions.*

The main plant hormones are gibberellins, auxins, cytokinins, abscisic acid, and ethylene (Table 28.3).

Gibberellins influence stem lengthening. They help buds and seeds break dormancy to resume growth in spring. In at least some cases they influence flowering.

Auxins play roles in stem lengthening and responses to gravity and light. They also make coleoptiles grow longer (Figure 28.15). A coleoptile is a thin sheath that keeps the primary shoot of some plants from shredding as growth pushes it up through soil. IAA (indoleacetic acid) is the most pervasive auxin in nature. Orchardists use auxins to thin flowers in spring so that their trees yield fewer but larger fruits. They also use auxins to prevent premature fruit drop, which cuts labor costs; all fruit in the orchard can be picked at the same time.

Some synthetic auxins serve as *herbicides*. At proper concentrations, these compounds kill some plants but not others (Section 3.7). 2,4-D controls broadleaf weeds that can outcompete cereal plants for nutrients. It does not seem to harm humans when properly handled. But mixing it with equal parts of a related compound, 2,4,5-T, yields *Agent Orange*. This herbicide was used to clear forested war zones during the Vietnam conflict. Later, laboratory experiments linked trace amounts of dioxin, a

Table 28.3 *Plant Hormones and Their Effects*

GIBBERELLINS. Promote stem elongation; help end dormancy of seeds and buds; contribute to flowering. Notable amounts in apical meristems of buds, roots, and leaves and in embryos.

AUXINS. Promote cell elongation in coleoptiles and stems; roles in phototropism and gravitropism. Notable amounts in bud and leaf apical meristems and in embryos in seeds.

CYTOKININS. Promote cell division and leaf expansion; retard leaf aging. Synthesized in roots and diffuses elsewhere.

ABSCISIC ACID. Promotes stomatal closure; promotes bud, seed dormancy. Notable amounts in leaves, stems, unripened fruit.

ETHYLENE. Promotes fruit ripening and abscission of leaves, flowers, fruits. Notable amounts in fruit, stems, leaves, roots, seeds.

Figure 28.25 (**a–c**) Results of experiments demonstrating that short-day plants flower by measuring the length of night, not daylength. Each horizontal bar signifies twenty-four hours. Daylight hours are coded *yellow*; night hours are *dark blue*. (**a**) Plants grown in a room where simulated nights were shorter than required for flowering didn't flower; they flowered when nights were longer (**b**). Plants exposed to a pulse of white light in the middle of the night didn't flower (**c**). Phytochrome is the receptor in short-day plants that responds to light. (**d**) A short pulse of red light at night will inhibit flowering. (**e**) Flowering did occur when 10 minutes of far-red light followed a pulse of red light.

a long day   short night

b short day   long night

c pulse of white light

d pulse of red light

e 10 minutes far-red light follows pulse of red light

flowers

b

Figure 28.26 Experiment involving flowering responses of (**a**) spinach, a long-day plant, and (**b**) chrysanthemum, a short-day plant. The plant at the left grew under short-day conditions. The plant at the right grew under long-day conditions.

Plants do not all flower in response to the same environmental cues. Poinsettias, chrysanthemums, and cockleburs are examples. They are *short-day* plants that flower in late summer or early fall, when daylength is shorter than a critical value (Figures 28.24 and 28.25). Spinach, irises, and other *long-day* plants form flowers in spring, when daylength exceeds some critical value (Figure 28.26). *Day-neutral* plants simply flower when they are mature enough to do so.

Spinach plants won't flower and form seeds unless they are exposed to ten hours of darkness (or fourteen hours of daylight) for two weeks. That's why you would not want to start, say, a spinach seed farm in the tropics (where this set of cues is not available). By contrast, chrysanthemum growers stall the flowering process by exposing plants to a flash of light at night to break up one long night into two "short" nights. That's why you can buy chrysanthemums in spring even though they naturally bloom in autumn. Cockleburs normally flower after only one night that's longer than eight and one-half hours. If artificial light interrupts their dark period for even a minute, they won't flower. And why didn't poinsettias planted by California's interstate highways put out flowers? Headlights from vehicles zipping past during the night blocked the flowering response.

Besides phytochrome, other kinds of hormones that are not yet identified probably influence flowering and other growth responses. Some of the hormones may be produced in leaves and transported to new buds. For example, if you trim all but one leaf from a cocklebur, and then cover the remaining leaf with black paper for eight and one-half hours, the plant will flower. If you cut off that one leaf right after the dark period is over, you will not see any cocklebur flowers.

Like other organisms, flowering plants have biological clocks, which are internal time-keeping mechanisms.

Phytochrome, a blue-green pigment, is part of a switching mechanism for phototrophic responses to light of red and far-red wavelengths. Its active form, Pfr, might interact with other hormones to control which kinds of enzymes are being produced in particular cells.

Different enzymes are required to complete flowering and other growth responses, which are influenced by daylength and other environmental cues.

LIFE CYCLES END, AND TURN AGAIN

Senescence

While leaves and fruits are growing, cells inside them produce auxin (IAA), which moves into stems. There it interacts with cytokinins and gibberellins to maintain growth. When autumn approaches and the length of daylight decreases, plants start to withdraw nutrients from leaves, stems, and roots, then distribute them to flowers, fruits, and seeds. Deciduous plants, which shed leaves when a growing season ends, channel nutrients to storage sites in twigs, stems, and roots before their leaves die and drop. The dropping of flowers, leaves, fruits, and other plant parts is known as **abscission**. An abscission zone is made of thin-walled parenchyma cells at the base of a petiole or some other part about to drop from the plant. Figure 28.27 shows one example.

Senescence is the sum total of processes that lead to death of a plant or some of its parts. The recurring cue is a decrease in daylength. But other factors, such as drought, wounds, and nutrient deficiencies, also bring it about. The cue causes a decline in IAA production in leaves and fruits. A different signal that the plant itself may produce stimulates abscission zone cells to make ethylene. Cells enlarge, deposit suberin in their walls, and produce enzymes that digest cellulose and pectin in middle lamellae. A middle lamella is the cementing layer between plant cell walls (Section 4.11). While cells continue to enlarge and their walls become digested, they separate from one another. Eventually the leaf or other plant part above the abscission zone drops away.

Interrupt the diversion of nutrients into flowers, seeds, or fruits, and you can prevent senescence inside a plant's leaves, stems, and roots. For example, if you remove each new flower or seed pod from a plant, its leaves and stems will remain vigorous and green much longer. Gardeners routinely remove flower buds from many kinds of plants to maintain vegetative growth.

Entering and Breaking Dormancy

As autumn approaches and days grow shorter, many perennials and biennials start to shut down growth. They do so even when temperatures are still mild, the sky is bright, and water is plentiful. When a plant stops growing under conditions that seem (to us) suitable for growth, it has entered a state of **dormancy**, in which its metabolic activities idle. Ordinarily, the plant's buds will not resume growth until there is a convergence of precise environmental cues in early spring.

Short days and long, cold nights are strong cues for dormancy. (So is dry, nitrogen-deficient soil.) Test this for yourself by interrupting the long dark period of, say, Douglas firs with a short period of red light. The plants will respond as if nights are shorter and days are

Figure 28.27 Abscission zone in maple (*Acer*). The section is through the base of a leaf petiole.

stem tissues   cells of abscission zone

abscission zone, where base of leaf joins a stem

Three Douglas fir plants exposed to different day-night cycles during three year-long experiments

Figure 28.28 Effect of the relative length of day and night on Douglas firs. The plant at left was exposed to 12-hour light/12-hour dark cycles for a year. Its buds became dormant; daylength was too short. The plant at right was exposed to 20-hour light/4-hour dark cycles; growth continued. The middle plant was exposed to 12-hour light/11-hour dark cycles with 1 hour of light in the middle of the dark period. The light interruption prevented bud dormancy; it caused Pfr formation at a sensitive time in the normal day–night cycle.

longer. They will continue to grow taller (Figure 28.28). In this experiment, conversion of Pr to Pfr by red light during the dark period prevents dormancy. In nature, buds might enter dormancy because less Pfr can form when daylength shortens in late summer.

A dormancy-breaking process is at work between fall and spring. The temperature may become milder, and rains and nutrients become available again. With the return of favorable conditions, the cycling of life begins anew. Seeds germinate; buds resume growth and give rise to new leaves, then to flowers. Depending on the species of plant, breaking dormancy probably involves gibberellins and abscisic acid, and it requires exposure to low winter temperatures at specific times.

Multiple cues from the environment influence hormonal secretions that stimulate or inhibit processes of growth and development during the life cycle of plants. The cues include changes in daylength, temperature, moisture, and nutrient availability.

SUMMARY   *Gold* indicates text section

1. Sexual reproduction is the main reproductive mode of all flowering plant life cycles. A diploid sporophyte (spore-producing plant body) alternates with haploid gametophytes (gamete-producing bodies). *28.1*

a. The sporophyte is a multicelled vegetative body with roots, stems, leaves, and, at some point, flowers. Air and water currents, as well as animals, coevolved with flowers as pollinating agents. Gametophytes form in male and female floral parts. *28.1*

b. Many flowering plants also reproduce asexually. They do so naturally, as with runners, rhizomes, and bulbs; and artificially, as by cuttings and grafting. *28.5*

2. Flowers typically have sepals, petals, one or more stamens (male reproductive parts), and carpels (female reproductive parts), most or all attached to a receptacle, the modified end of a floral shoot. *28.1–28.2*

a. Anthers of stamens contain pollen sacs in which cells divide by meiosis. A wall develops around each resulting haploid cell (a microspore), which gives rise to a sperm-bearing pollen grain (male gametophyte).

b. A carpel (or two or more fused carpels) has an ovary where eggs develop, fertilization takes place, and seeds mature. A stigma (sticky or hairy surface tissue) above the ovary captures pollen grains and promotes their germination. Ovules form on the ovary's inner wall. Each ovule is a female gametophyte with an egg cell, an endosperm mother cell, a surrounding tissue, and one or two protective layers called integuments.

3. At double fertilization, one sperm nucleus fuses with an egg nucleus and so forms a diploid zygote. The other sperm nucleus and both nuclei of another cell in the female gametophyte fuse to form a cell that will give rise to endosperm, a nutritive tissue. *28.3*

4. After fertilization, the endosperm forms, the ovule expands, the embryo sporophyte develops, and the integuments harden and often thicken. A fully matured ovule is a seed (its integuments have become the seed coat). While seeds form, the ovaries develop into fruits, which will help protect and disperse the seeds. *28.3*

5. After dispersal, seeds germinate—the embryo inside absorbs water, resumes growth, and breaks through the seed coat. The seedling increases in size; its tissues and organs develop. Fruits and seeds form. Older leaves drop. Plant hormones govern such patterns of growth and development. They adjust patterns in response to unpredictable and recurring environmental cues, such as seasonal changes in daylength. *28.4–28.11*

6. We know of five classes of plant hormones. Auxins and gibberellins promote stem elongation. Cytokinins promote cell division and leaf expansion, and retard leaf aging. Abscisic acid promotes bud and seed dormancy, and limits water loss by triggering stomatal closure. Ethylene promotes fruit ripening and abscission. *28.7*

7. Plant parts make tropic responses to light, gravity, and other environmental conditions. Hormones induce a difference in the rate and direction of growth on two sides of the part, which causes it to turn or move. *28.9*

8. A biological clock is a type of internal mechanism that measures time. It has a biochemical basis. *28.10*

a. In photoperiodism, plants respond to a change in the relative length of daylight and darkness, as occurs seasonally. The blue-green pigment phytochrome is the switching mechanism of the clock that promotes and inhibits germination, stem elongation, leaf expansion, stem branching, and flower, fruit, and seed formation.

b. Long-day plants flower during spring or summer, when there are more hours of daylight than darkness. Short-day plants flower when daylength is less. Day-neutral plants flower regardless of daylength.

9. Senescence is the sum of processes leading to the death of a plant or plant structure. Dormancy is a state in which a biennial or perennial plant stops growing even if conditions seem suitable for continued growth. A decrease in Pfr levels may trigger dormancy. *28.11*

Review Questions

1. Label the floral parts. Explain floral function by relating the parts to events in the life cycle of flowering plants. *28.1*

2. Distinguish between:
   a. Sporophyte, gametophyte *28.1*
   b. Megaspore, microspore *28.3*
   c. Pollination, fertilization *28.3*
   d. Pollen grain, pollen tube *28.1, 28.3*
   e. Ovule, female gametophyte *28.3*
   f. Seed, fruit *28.4*
   g. Senescence, dormancy *28.11*

3. Describe the steps by which a seven-cell, eight-nucleate embryo sac (a type of female gametophyte) forms. *28.3*

4. Define and describe one mode of vegetative growth. *28.5*

5. Name the three key factors that interact to dictate patterns of plant growth and development. *28.6*

6. List five types of plant hormones and briefly describe the known functions of each. *28.7*

7. Define plant tropism and give a specific example. *28.9*

8. What is phytochrome, and what is its role in flowering or some other process? *28.10*

Figure 28.29
Seeds inside the fruit of a papaya (*Carica papaya*).

Self-Quiz <small>ANSWERS IN APPENDIX III</small>

1. The flowers of many species coevolved with insects, birds, and other agents that function as _____ .

2. The _____ , which bears flowers, roots, stems, and leaves, dominates the life cycle of flowering plants.
   a. sporophyte
   b. gametophyte

3. A _____ is a vessel, the lower portion of which is an ovary in which eggs develop, fertilization occurs, and seeds mature.
   a. pollen sac   b. carpel   c. receptacle   d. sepal

4. After meiosis within pollen sacs, haploid _____ form.
   a. megaspores
   b. microspores
   c. stamens
   d. sporophytes

5. After meiosis in ovules, _____ megaspores form.
   a. two   b. four   c. six   d. eight

6. Cotyledons develop as part of all flowering plant _____ .
   a. seeds   b. fruits   c. embryos   d. ovaries

7. Seeds are mature _____ ; fruits are mature _____ .
   a. ovaries; ovules
   b. ovules; stamens
   c. ovules; ovaries
   d. stamens; ovaries

8. Which of the following statements is false?
   a. Auxins and gibberellins promote stem elongation.
   b. Cytokinins promote cell division but retard leaf aging.
   c. Abscisic acid promotes water loss and dormancy.
   d. Ethylene promotes fruit ripening and abscission.

9. Plant hormones _____ .
   a. interact with one another
   b. are influenced by environmental cues
   c. are active in plant embryos within seeds
   d. are active in adult plants
   e. all of the above

10. Light of _____ is the strongest stimulus for phototropism.
    a. red wavelengths
    b. far-red wavelengths
    c. green wavelengths
    d. blue wavelengths

11. The flowering process is a _____ response.
    a. phototropic
    b. gravitropic
    c. photoperiodic
    d. thigmotropic

12. Match the terms with the most suitable description.
    ____ double fertilization
    ____ ovule
    ____ mature female gametophyte
    ____ asexual reproduction
    ____ coevolution

    a. formation of zygote and first cell of endosperm
    b. outcome of two species interacting in close ecological fashion over geologic time
    c. contains a female gametophyte and has potential to be a seed
    d. an embryo sac, commonly with seven cells (one has two nuclei)
    e. mitotic cell division at bud or node produces a new plant

Critical Thinking

1. Observe flowers growing in the area where you live. Given the coevolution of flowering plants and their pollinators, name the kinds of pollinators your floral neighbors depend on.

2. Wanting to impress her many friends with her sophisticated botanical knowledge, Dixie Bee is preparing a plate of tropical fruits for a party and cuts open a papaya (*Carica papaya*) for the first time. Inside are great numbers of slime-covered seeds, surrounded by soft flesh and soft skin (Figure 28.29). Knowing that her friends will ask what sort of fruit this is, she panics, runs to her biology book, and opens it to Section 28.4. What does she find out? What will she tell them when they ask what kinds of agents disperse the seeds of such a fruit from the parent plant?

3. Before cherries, apples, peaches, and many other fruits ripen and the seeds inside them mature, their flesh is bitter or sour. Only later does it become tasty to animals that assist in seed dispersal. Develop a hypothesis of how this feature improves the odds for the plant's reproductive success.

4. Given what you know about the primary growth of plants (Section 26.1), propose an explanation of why plant hormones need not travel very far from the cells that secrete them.

5. Plant growth depends on photosynthesis, which depends on sunlight energy. How, then, can seedlings germinated in a dark room grow taller than seedlings that germinated in the sun?

6. *Solar tracking* refers to the observation that many plants are able to maintain the flat blades of their leaves at right angles to the sun throughout the day. Sunflowers are an example. This tropic response maximizes light harvest by leaves. Name one type of molecule that might be involved in the response.

7. Belgian scientists isolated a mutant of wall cress (*Arabidopsis thaliana*) that produces excess amounts of auxin. Predict what some of this mutant plant's phenotypic traits might be.

8. Cattle receive somatotropin, an animal hormone that makes them grow bigger (added weight means more profits). There is growing concern that such hormones may have unforeseen effects on beef-eating humans. Would you think plant hormones applied to crop plants can affect humans also? Why or why not?

Selected Key Terms

abscisic acid *28.7*
abscission *28.11*
auxin *28.7*
biological clock *28.10*
carpel *28.1*
circadian rhythm *28.10*
coevolution *CI*
cotyledon *28.4*
cytokinin *28.7*
development *28.7*
dormancy *28.11*
double fertilization *28.3*

endosperm *28.3*
ethylene *28.7*
flower *CI*
fruit *28.4*
gametophyte *28.1*
germination *28.6*
gibberellin *28.7*
gravitropism *28.9*
growth *28.7*
hormone *28.7*
megaspore *28.3*
microspore *28.3*
ovary *28.1*
ovule *28.3*
photoperiodism *28.10*

phototropism *28.9*
phytochrome *28.10*
pollen grain *28.1*
pollination *28.3*
pollinator *CI*
seed *28.4*
senescence *28.11*
sporophyte *28.1*
stamen *28.1*
statolith *28.9*
thigmotropism *28.9*
tissue culture propagation *28.5*
vegetative growth *28.5*

Readings

Grant, M. October 1993. "The Trembling Giant." *Discover* 14(10): 82–89.

Rost, T., M. Barbour, C. R. Stocking, and T. Murphy. 1998. *Plant Biology*. Belmont, California: Wadsworth.

On-Line readings at Student Guide for InfoTrac: www.brookscole.com/biology

VI Animal Structure and Function

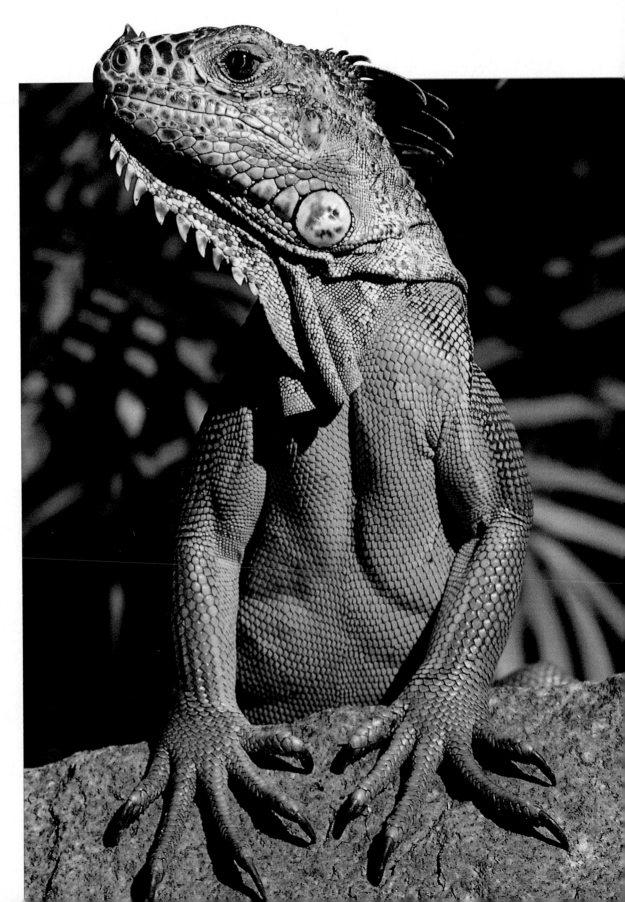

How many and what kinds of body parts does it take to function as a lizard in a tropical forest? Make a list of what comes to mind as you start reading Unit VI, then see how resplendent the list can become at the unit's end.

TISSUES, ORGAN SYSTEMS, AND HOMEOSTASIS

Meerkats, Humans, It's All the Same

After a cold night in Africa's Kalahari Desert, animals small enough to fit inside a coat pocket emerge stiffly from their burrows. These "meerkats" are a type of mongoose. They stand on their hind legs and face east, exposing their chilled bodies to the warm rays of the morning sun (Figure 29.1). Meerkats don't know it, but sunning behavior helps their enzymes. If their body's internal temperature were to fall below some tolerable range, the action of countless enzyme molecules in their cells would falter and metabolism would suffer.

Once meerkats warm up, they fan out from their burrows and search for food. Into the meerkat gut go insects and the occasional lizard. These are pummeled, dissolved, and digested into glucose and other nutritious bits small enough to move across the gut wall, into the blood, and on to the body's cells. Aerobic machinery in the cells cracks apart molecules of glucose and other organic compounds and releases energy. A respiratory system supplies the machinery with oxygen and takes away carbon dioxide leftovers.

All of this activity changes the composition and volume of the **internal environment**—blood and the interstitial fluid (tissue fluid) that bathes all living cells of any complex animal. Drastic changes in these fluids would kill the cells, but a urinary system works to keep

that from happening. Governing this system and the others are the nervous and endocrine systems. They interact as a central command post to mobilize the body as a whole for everything from simple housekeeping tasks to heart-thumping flights from predators.

And so meerkats start us thinking about this unit's topics: how the animal body is structurally put together (its *anatomy*) and the adaptations by which it functions in the environment (its *physiology*). This chapter is an overview of the animal tissues and organ systems we will be considering. It also introduces the key concept of **homeostasis**. With respect to animals, the word means stable operating conditions in the internal environment, as brought about by the coordinated activities of cells, tissues, organs, and organ systems.

Amazingly, all complex animals consist of only four basic types of tissues. These are epithelial, connective, muscle, and nervous tissues. A **tissue** is a group of cells and intercellular substances, all interacting in one or more tasks. As an example, muscle tissue takes part in contraction. An **organ** consists of different tissues that are organized in specific proportions and patterns. Thus every vertebrate heart has predictable proportions and arrangements of epithelial, connective, muscle, and nervous tissues. An **organ system** consists of two or more organs that are interacting physically, chemically, or both in a common task, as when interconnected arteries and other vessels transport blood through the body under the driving force of a beating heart.

Cells, tissues, organs, and organ systems split up the work, so to speak, in ways that contribute to survival of the body as a whole. This is sometimes referred to as a **division of labor**. By the end of this unit, you may have an abiding appreciation of the sheer magnitude of the division of labor among the body's separate parts and the extent to which their activities are integrated.

As you will see, whether you look at a flatworm or salmon, a meerkat or human being, the animal body is structurally and physiologically adapted to perform four overriding tasks:

1. *Maintain conditions in the internal environment within ranges that are most favorable for cell activities.*

2. *Acquire nutrients and other raw materials, distribute them through the body, and dispose of wastes.*

3. *Afford protection against injury or attack from viruses, bacteria, and other agents of disease.*

4. *Reproduce, then often help nourish and protect the new individuals during their early growth and development.*

Key Concepts

1. The cells of most animals interact at three levels of organization—in tissues, many of which are combined into organs, which are components of organ systems.

2. Most animals are constructed of four types of tissues: epithelial, connective, muscle, and nervous tissues.

3. Each animal cell engages in basic metabolic activities that ensure its own survival. At the same time, animal cells of a given tissue perform one or more activities that contribute to the survival of the animal as a whole.

4. The internal environment consists of all fluids that are not inside the body's cells—that is, blood and interstitial fluid.

5. The combined contributions of cells, tissues, organs, and organ systems help maintain stability in the internal environment, which is required for the survival of each individual cell. This concept helps us understand the functions of any organ or organ system.

6. Homeostasis is the formal name for stable operating conditions in the internal environment.

Figure 29.1 In the Kalahari Desert, gray meerkats (*Suricata suricatta*) face the sun's warming rays, as they do every morning. This simple behavior helps them maintain internal body temperature. How animals function in their environment is the focus of this unit.

EPITHELIAL TISSUE

General Characteristics

We commonly refer to an epithelial tissue as **epithelium** (plural, epithelia). This tissue has a free surface, which faces a body fluid or the outside environment. *Simple* epithelium, with only a single layer of cells, functions as a lining for body cavities, ducts, and tubes. *Stratified* epithelium, which has two or more cell layers, typically functions in protection, as it does in your skin. Figure 29.2 shows examples of this type of animal tissue.

All cells in epithelium are positioned close together, with little intervening material. They make, absorb, and secrete a variety of substances. As is the case for cells in nearly all animal tissues, specialized junctions serve as structural and functional links between them.

Cell-to-Cell Contacts

In epithelium and other tissues, we see three classes of cell junctions. **Tight junctions** prevent substances from leaking across the tissue. **Adhering junctions** are like spot welds; they cement neighboring cells together. **Gap junctions** are open channels between the cytoplasm of abutting cells. They facilitate the rapid transfer of ions and small molecules from one cell to its neighbors.

Consider tight junctions. The cells in some epithelia have parallel rows of proteins that form tight seals and fuse each cell with its neighbors. The junctions prevent most substances from leaking across the tissue's free surface. Substances reach tissues below only by passing

free surface of epithelium

simple squamous epithelium

basement membrane

a connective tissue

Figure 29.2 (a) Some basic characteristics of epithelium. Epithelium has a free surface exposed to some body fluid or to the outside environment. Between the opposite surface and the underlying connective tissue is a basement membrane. All epithelial cells rest on this type of membrane. The sketch below the athlete shows this arrangement for one type of epithelium that is a single layer of cells. The light micrograph below her arm is a section of the upper portion of stratified epithelium. Stratified epithelium consists of more than one layer of cells, which are flattened out near the free surface.

(b) Light micrographs and sketches of three simple epithelia, showing three common shapes of cells for this tissue.

b

TYPE: Simple squamous
DESCRIPTION: Friction-reducing slick, single layer of flattened cells
COMMON LOCATIONS: Lining of blood and lymph vessels, heart; air sacs of lungs; peritoneum
FUNCTION: Diffusion; filtration; secretion of lubricants

TYPE: Simple cuboidal
DESCRIPTION: Single layer of squarish cells
COMMON LOCATIONS: Ducts, secretory part of small glands; retina; kidney tubules; ovaries, testes; bronchioles
FUNCTIONS: Secretion, absorption

TYPE: Simple columnar
DESCRIPTION: Single layer of tall cells; free surface may have cilia, mucus-secreting glandular cells, microvilli
COMMON LOCATIONS: Glands, ducts; gut; parts of uterus; small bronchi
FUNCTION: Secretion; absorption; ciliated types move substances

Awareness of sounds and sights, of odors, of hunger and passion, fear and rage—all begin with the flow of information along communication lines of the nervous system. Do the lines remain silent until they receive outside signals, much as telephone lines wait to carry calls from all over the country? Absolutely not. Even before you were born, excitable cells called neurons became organized into gridworks in your newly forming tissues and started chattering among themselves. All through your life, in moments of danger or reflection, supreme excitement or sleep, their chattering has never stopped and will not stop until the time you die.

Three classes of neurons interact in all vertebrate nervous systems. **Sensory neurons** respond to specific stimuli and relay information about them to the spinal cord and brain. A **stimulus** is a specific form of energy, such as light. **Interneurons** are in the spinal cord and brain. They receive and process sensory input, then influence the activity of other neurons. **Motor neurons** relay information away from the brain and spinal cord to the body's effectors—muscles or glands—which carry out the specified responses (Figure 30.1b). More than half the volume of the nervous system consists of cells, collectively called neuroglia, that metabolically assist, protect, and structurally support the neurons.

Our initial focus in this chapter will be on the structure and function of neurons. Later on, we will consider how neurons interact in nervous systems.

STIMULUS
(input)

↓

RECEPTORS
(sensory neurons)

↓

INTEGRATORS
(interneurons of brain,
spinal cord)

↓

motor neurons

↓

EFFECTORS
(muscles, glands)

↓

RESPONSE
(output)

b

Key Concepts

1. Neurons are the basic units of communication in nearly all nervous systems. Collectively, many neurons interact to detect and integrate information about external and internal conditions, then select or control muscles and glands in ways that produce suitable responses.

2. Neurons are a type of excitable cell, which means a suitable stimulus can disturb the distribution of electric charge across their plasma membrane. Such disturbances are the basis of messages that travel from the input zone of a neuron to its output zone near a neighboring cell.

3. Information flow through nervous systems depends on the moment-by-moment integration of excitatory and inhibitory signals that act upon each neuron in a given pathway.

4. The simplest nervous systems are the nerve nets of radial animals, such as hydras and sea anemones. The nervous systems of most animals show pronounced cephalization and bilateral symmetry.

5. Vertebrate nervous systems are functionally divided into central and peripheral regions. The brain and spinal cord make up the central nervous system. Paired nerves that thread through the remainder of the body are the main components of the peripheral nervous system.

6. The somatic nerves of the peripheral nervous system deal with skeletal muscles. The autonomic nerves deal with the heart, lungs, and other soft internal organs that are generally enclosed in body cavities.

7. The vertebrate brain has three functional divisions called the hindbrain, midbrain, and forebrain. Its most ancient parts deal with reflex control of breathing, blood circulation, and other basic functions that are essential for staying alive.

NEURONS—THE COMMUNICATION SPECIALISTS

Functional Zones of a Neuron

Look at a neuron's nucleated cell body and you see it has cytoplasmic extensions that differ in number and length. Typically, the cell body and slender extensions called **dendrites** are *input* zones for incoming signals. At a nearby patch of plasma membrane called the *trigger* zone, the incoming signals are added up. From there, signals may flow to an **axon**, a slender but often longer extension that's a *conducting* zone. An axon propagates very rapidly from a trigger zone down to its branched endings. Axon endings are the neuron's *output* zones, where information is sent to other cells (Figure 30.2).

A Neuron at Rest, Then Moved to Action

In the body's communication lines, each neuron is only a single unit. It can only accept messages about stimuli at its input zones and swiftly convey signals about them to its output zones. In essence, each "signal" is a self-propagating disturbance to the distribution of electric charge across that neuron's plasma membrane.

When a neuron is not being bothered with incoming messages, it is maintaining a difference in electric charge across its plasma membrane. Its cytoplasmic fluid next to the membrane is negatively charged, compared with interstitial fluid just outside. We measure such charges

in millivolts. Think of the amount of energy inherent in the steady voltage difference across a neural membrane as the **resting membrane potential**. For many neurons, the amount is about −70 millivolts.

When a weak signal reaches a patch of membrane in the neuron's trigger zone, the voltage difference at the patch changes slightly, if at all. But a strong signal may trigger an **action potential**. We define this as a fleeting reversal in the steady voltage difference across the neural membrane. The cytoplasmic fluid inside the membrane becomes positive, relative to the fluid outside. As you will see, that reversal triggers a sequence of equivalent reversals, away from the neuron's trigger zone.

Gradients Required for Action Potentials

How does the neuron maintain readiness for an action potential? It does so by working with and against ion gradients across its plasma membrane.

First, every cell membrane has a lipid bilayer that blocks the passage of potassium ions (K^+), sodium ions (Na^+), and other charged substances. The bilayer allows a neuron to build up differences in ion concentrations across its plasma membrane. Second, such ions do flow from one side of the membrane to the other—but only through the interior of transport proteins that span the bilayer (Figure 30.3). And that flow is controlled.

dendrites

INPUT ZONE

cell body

TRIGGER ZONE

10 μm

dendrite

axon

cell body

a

dendrites

dendrite

cell body

axon

b

dendrites

dendrite

cell body

axon

c

CONDUCTING ZONE

axon

axon endings

d Scanning electron micrograph and diagram of a typical motor neuron

OUTPUT ZONE

Figure 30.2 Examples of neurons that differ in the number of cytoplasmic extensions of the cell body. (**a**,**b**) Certain types have one axon with dendritic branchings or one axon and one dendrite. Many sensory neurons are like this. (**c**) Other types have one axon and a profusion of dendrites; we find this kind in the mammalian brain. (**d**) Micrograph and sketch of the functional zones of a motor neuron.

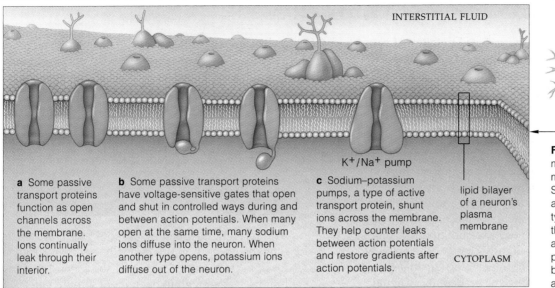

INTERSTITIAL FLUID

trigger zone

a Some passive transport proteins function as open channels across the membrane. Ions continually leak through their interior.

b Some passive transport proteins have voltage-sensitive gates that open and shut in controlled ways during and between action potentials. When many open at the same time, many sodium ions diffuse into the neuron. When another type opens, potassium ions diffuse out of the neuron.

c Sodium–potassium pumps, a type of active transport protein, shunt ions across the membrane. They help counter leaks between action potentials and restore gradients after action potentials.

K⁺/Na⁺ pump

lipid bilayer of a neuron's plasma membrane

CYTOPLASM

Figure 30.3 How ions move across the plasma membrane of a neuron. Spanning the lipid bilayer are channel proteins, a type of transport protein that passively moves ions across. Sodium–potassium pumps that also span the bilayer actively move ions across.

Suppose a motor neuron has 15 sodium ions inside the membrane for every 150 outside. Also suppose that it has 150 potassium ions inside for every 5 outside. You can depict each ion's concentration gradient in this way (from the large to the small letters):

outside

plasma membrane

inside

Gradients such as these determine the net direction in which sodium and potassium ions move. These ions diffuse through the interior of channel proteins, a type of transport protein in the membrane. Some channels never shut, so ions can leak through them all the time. But other transport proteins have molecular gates that open only after the neuron is adequately stimulated.

Suppose a motor neuron is at rest. Gated proteins in its membrane are closed, so sodium ions can't rush in. Some potassium leaks out through a few open channels and makes the cytoplasm a bit more negative, so some potassium is attracted back in. When the inward pull of electric charge balances the outward force of diffusion, potassium ions show no net movement. Concentration and electric gradients across the membrane now make the neuron ready to respond to stimulation.

Sodium–potassium pumps maintain the gradients and restore them after their reversal during an action potential. The pumps are transport proteins spanning the membrane (Section 5.7). When they get an energy boost from ATP, they actively transport potassium into the neuron and sodium out at the same time. Even in a resting neuron, a tiny fraction of the outward-leaking

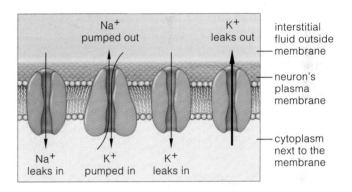

Na⁺ pumped out

K⁺ leaks out

interstitial fluid outside membrane

neuron's plasma membrane

Na⁺ leaks in

K⁺ pumped in

K⁺ leaks in

cytoplasm next to the membrane

Figure 30.4 Pumping and leaking processes that dictate the distribution of sodium and potassium ions across the plasma membrane of a neuron at rest. The arrow widths indicate the magnitude of the movements. Notice how the total inward and outward movements for each kind of ion are balanced.

potassium isn't attracted back in. And a tiny fraction of the sodium is leaking in through a few open channels (Figure 30.4). If the leaks went unattended, the crucial gradients would gradually disappear.

With this bit of information on the gradients across the neural membrane, we are ready to look at how an action potential arises at a trigger zone and propagates itself, undiminished, to an output zone.

An undisturbed neuron maintains ion concentration and electric gradients across its plasma membrane. We measure this as a steady voltage difference, and energy inherent in it is the resting membrane potential. An abrupt, short-lived reversal in that voltage difference is an action potential.

Sodium–potassium pumps restore and maintain the resting membrane potential in between action potentials.

TRIGGERING AND PROPAGATING ACTION POTENTIALS

Action potential propagation isn't hard to follow if you already know something about the gradients across the neural membrane. And so we now build on Section 30.1.

Approaching Threshold

When you weakly stimulate a neuron at its input zone, you disturb the ion balance across its membrane, but not much. Imagine putting a bit of pressure on the skin of a snoozing cat by gently tapping a toe on it. Tissues beneath the skin surface have receptor endings—input zones of sensory neurons. Patches of plasma membrane at these endings deform under pressure and let some ions flow across. The flow slightly changes the voltage difference across the membrane. In this case, pressure has produced a graded, local signal.

STIMULUS

trigger zone

influx, the cytoplasmic side of the membrane becomes less negative. This causes more gates to open and more sodium to enter. The ever increasing, inward flow of sodium is a case of **positive feedback**, whereby an event intensifies as a result of its own occurrence:

more Na⁺ ions flow into the neuron

more gated channels for Na⁺ open

neuron becomes more positive inside

At threshold, the opening of more sodium gates no longer depends on stimulus strength. Now the positive-feedback cycle is under way, and the inward-rushing sodium itself is enough to open more sodium gates.

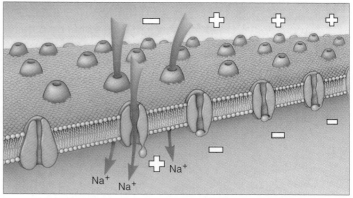

a Membrane at rest (inside negative with respect to the outside). An electrical disturbance (*red* arrow) spreads from an input zone to an adjacent trigger region of the membrane, which has a great number of gated sodium channels.

b A strong disturbance initiates an action potential. Sodium gates open. The sodium inflow decreases the negativity inside the neuron. The change causes more gates to open, and so on until threshold is reached and the voltage difference across the membrane reverses.

Figure 30.5 Propagation of an action potential along the axon of a motor neuron.

Graded means that signals arising at an input zone vary in magnitude. They are small to large, depending on the stimulus intensity or duration. *Local* means these signals do not spread far from the site of stimulation. Why? It takes certain kinds of ion channels to propagate a signal, and input zones simply don't have them.

When a stimulus is intense or long lasting, graded signals spread from the input zone into an adjoining trigger zone. This patch of membrane is richly endowed with voltage-sensitive gated channels for sodium ions. *And this is where a certain amount of change in the voltage difference across the plasma membrane triggers an action potential.* The amount is the neuron's threshold level.

The stimulus causes positively charged sodium ions to flow into the neuron, as in Figure 30.5. With the ion

An All-or-Nothing Spike

Figure 30.6 shows a recording of the voltage difference across the plasma membrane before, during, and after an action potential. Notice how the membrane potential spikes once threshold is reached. Every action potential in a neuron spikes to the same level above threshold as an *all-or-nothing* event. Once a positive-feedback cycle starts, nothing will stop the full spiking. If threshold is not reached, the disturbance to the plasma membrane will subside as soon as the stimulation stops.

Each spike lasts only for a millisecond or so. Why? At the patch where the charge reversed, gated sodium channels closed and shut off the sodium inflow. Also, about halfway into that reversal, potassium channels

Stellar nerve, with giant axon inside

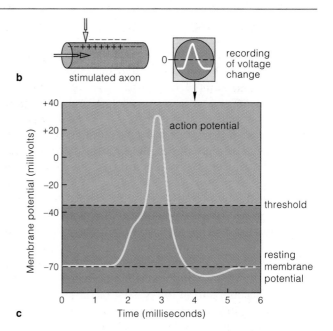

b stimulated axon

recording of voltage change

electrode outside axon

electrode inside axon → ++++++++++++++

no change in voltage registers on oscilloscope screen

a unstimulated axon

Figure 30.6 Action potentials. (**a**) When researchers started to study neural function, a squid (*Loligo*) provided them with evidence of action potential spiking. This squid's "giant" axons are large enough to slip electrodes inside. (**b**) The researchers put electrodes inside and outside the axon, then stimulated it. The electrodes detected the voltage change, which showed up as deflections in a beam of light across the screen of an oscilloscope connected to the electrodes. (**c**) A typical waveform (*yellow* line) for an action potential on an oscilloscope screen.

c

Membrane potential (millivolts) vs Time (milliseconds) graph showing action potential waveform with +40, +20, 0, −20, −40, −70 on y-axis and 0–6 on x-axis. Labels: action potential, threshold, resting membrane potential.

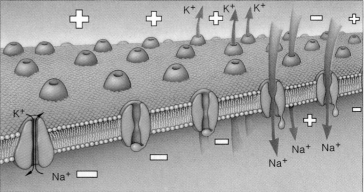

c With the reversal, sodium gates shut and potassium gates open (*pink* arrows). Potassium follows its gradient out of the neuron. Voltage is restored. The disturbance triggers an action potential at the adjacent site, and so on, away from the point of stimulation.

d Following each action potential, the inside of the plasma membrane becomes negative once again. However, the sodium and potassium concentration gradients are not yet fully restored. Active transport at sodium–potassium pumps restores them.

opened, so potassium ions flowed out. This restored the voltage difference at the patch but not the original ion gradients. Sodium–potassium pumps actively transport sodium back outside and potassium inside. Once this is done, most potassium gates close and sodium gates are in their original position—ready to be opened with the arrival of a suitable signal at the membrane patch.

The Direction of Propagation

During an action potential, the inward rush of sodium ions affects the charge distribution across the adjacent membrane patch, where an equivalent number of gated channels open. Gated channels open in the *next* patch, and the next, and so on. This positive feedback event is self-propagating and does not diminish in magnitude. You might be wondering: Do action potentials spread

back to the trigger zone? No. For a brief period after the inward rush of sodium, a membrane patch remains insensitive to stimulation. The transport proteins having the sodium gates are inactivated, so ions cannot move through them. This is one reason why action potentials do not spread back to the site where they were initiated but rather propagate themselves away from it.

Ions cross the neural membrane through transport proteins that serve as gated or open channels. At a suitably disturbed trigger zone, sodium gates open in an ever increasing way, and the inward-rushing sodium causes an action potential. Sodium–potassium pumps restore the original ion gradients.

Sodium gates across the membrane are briefly inactivated after an action potential, which is one reason why an action potential is self-propagating away from a trigger zone.

CHEMICAL SYNAPSES

When action potentials reach a neuron's output zone, they usually do not proceed farther than this. But their arrival may induce the neuron to release one or more **neurotransmitters**, which are signaling molecules that diffuse across chemical synapses. A **chemical synapse** is a narrow cleft between a neuron's output zone and an input zone of a neighboring cell (Figures 30.7 and 30.8). Some clefts are between two neurons. Others intervene between a neuron and a muscle cell or gland cell.

One of the two cells stores neurotransmitter inside certain vesicles. This *pre*synaptic cell has gated channels for calcium ions, which are more concentrated on the outside of the resting membrane. The gates open when an action potential arrives. As ions follow the gradient, into the cell, the flow induces synaptic vesicles to fuse with and become part of the plasma membrane, and so they dump neurotransmitter into the synaptic cleft.

Neurotransmitter molecules diffuse across the cleft. The *post*synaptic cell's membrane has protein receptors that bind to specific neurotransmitters. Binding changes the receptor shape, so a passage opens up through the protein's interior. Ions now cross the plasma membrane by diffusing through the passageway (Figure 30.8c).

How does the postsynaptic cell respond? It depends on the type and concentration of neurotransmitter, the types of receptors, and the number and responsiveness of gated channels in that cell's membrane. Such factors

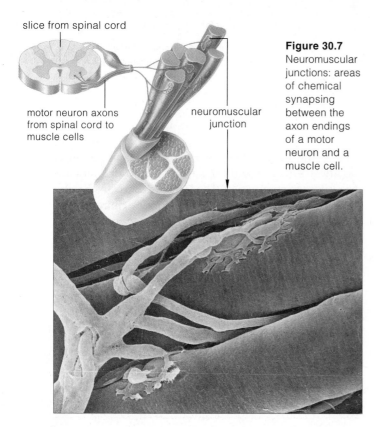

slice from spinal cord

motor neuron axons from spinal cord to muscle cells

neuromuscular junction

Figure 30.7 Neuromuscular junctions: areas of chemical synapsing between the axon endings of a motor neuron and a muscle cell.

Figure 30.8 *Facing page:* Example of a chemical synapse. (**a,b**) A thin cleft is all that separates a presynaptic cell from a postsynaptic cell. The arrival of an action potential at an axon ending triggers the release of molecules of neurotransmitter from the presynaptic cell. (**c**) The molecules diffuse across the cleft. They bind to receptors that are part of gated proteins of the postsynaptic cell membrane. The gates open. Ions flow in and trigger a graded potential at the membrane site.

influence whether a given neurotransmitter will have an *excitatory* effect and help drive the postsynaptic cell's membrane toward the threshold of an action potential. They influence whether it will have an *inhibitory* effect and drive the membrane away from threshold.

Consider the neurotransmitter **acetylcholine** (ACh). It has both excitatory and inhibitory effects on the brain, spinal cord, glands, and muscles. For example, it acts at chemical synapses between a motor neuron and muscle cell, as in Figure 30.7. ACh is released from the motor neuron, diffuses across the cleft, and binds to receptors on the muscle cell membrane. In this kind of cell it has excitatory effects; it can trigger action potentials, which in turn initiate muscle contraction (Section 33.7).

A Smorgasbord of Signals

Acetylcholine is only one of a veritable smorgasbord of signals that neurons deliver to target cells. For example, serotonin's targets in brain regions help control sensory perception, sleeping, body temperature, and emotions. Norepinephrine's targets influence emotions, dreaming, and waking up. Dopamine, too, works in brain regions that deal with emotions. GABA (gamma aminobutyric acid) is the most common inhibitory signal working in the brain. Valium and other antianxiety drugs work by enhancing GABA's effects.

Neuromodulators are signaling molecules that reduce or magnify a neurotransmitter's effects on neighboring or distant neurons. Examples: Substance P induces pain perception, and endorphins (natural painkillers) inhibit the release of substance P from sensory nerves. Also, neuromodulators may influence memory and learning, sexual activity, body temperature, and emotions.

Synaptic Integration

Anywhere from 1,000 to 10,000 lines of communication synapse on a typical neuron in your brain, which holds at least 100 billion neurons. As long as you are alive, those neurons hum with messages about doing what it takes to be a human. At any moment, a great number of excitatory and inhibitory signals may wash over input zones of a postsynaptic cell. Some drive its membrane

a

plasma membrane of an axon ending of presynaptic cell

synaptic vesicle

synaptic cleft

membrane receptor for neurotransmitter

plasma membrane of postsynaptic cell

b

presynaptic cell
synaptic cleft
postsynaptic cell

c

neurotransmitter molecule in synaptic cleft

ions (*red*) that affect membrane excitability

receptor for the neurotransmitter on gated channel protein in plasma membrane of postsynaptic cell

closer to threshold; others maintain the resting level or drive it away from threshold. Said another way, signals compete for control of the neuron's membrane.

All synaptic signals are graded potentials. The type we call an EPSP (for excitatory postsynaptic potential) has a *depolarizing* effect. This simply means it drives a membrane closer to threshold. An IPSP (for inhibitory postsynaptic potential) may have a *hyperpolarizing* effect and drive a membrane away from threshold, or it may help maintain the membrane at its resting level.

With **synaptic integration**, competing signals that reach an input zone of a neuron at the same time are summed. By this process, two or more signals arriving at a neuron may be dampened, suppressed, reinforced, or sent onward to other cells in the body. Figure 30.9 is a composite of an EPSP recording, an IPSP recording,

and a recording of their summation. Integration occurs when neurotransmitter molecules from more than one presynaptic cell reach a neuron's input zone at the same time. Integration also occurs after neurotransmitter is released swiftly and repeatedly from a presynaptic cell responding to a rapid series of action potentials.

How Is Neurotransmitter Removed From the Synaptic Cleft?

Information flow through the nervous system depends on the prompt, controlled removal of neurotransmitter molecules from synaptic clefts. Some of the molecules simply diffuse away from the cleft. Enzymes inside the cleft cleave others, as when acetylcholinesterase breaks apart ACh. Also, transport proteins in the membrane of presynaptic cells actively pump molecules back inside or into neighboring neuroglial cells.

What happens if neurotransmitter accumulates in the cleft? As one example, cocaine blocks the uptake of dopamine. Molecules of this neurotransmitter linger in synaptic clefts and keep on stimulating target cells. At first the abnormal stimulation invites euphoria (intense pleasure). Its later effects are disastrous (Section 30.11).

Figure 30.9 Example of synaptic integration. The *yellow* line shows how an EPSP of a certain magnitude would register on an oscilloscope screen if it were acting alone. The *purple* line shows the effect of an IPSP if *it* were acting alone. Suppose both signals arrive at a postsynaptic cell membrane at the same time. The *red* line shows their effect. In this case, when the two signals are integrated, threshold is not reached. So an action potential (the *white* dashed line) cannot be initiated in the target cell.

Neurotransmitters are signaling molecules that bridge a synaptic cleft. The cleft is a thin gap between two neurons or between a neuron and a muscle cell or gland cell.

Neurotransmitters have excitatory or inhibitory effects on different kinds of target cells. Synaptic integration is the moment-by-moment combining of excitatory signals and inhibitory signals acting on the same postsynaptic cell.

The summation process allows messages traveling through the nervous system to be reinforced or downplayed, sent onward or suppressed. The process is vital for normal body functioning.

PATHS OF INFORMATION FLOW

Blocks and Cables of Neurons

By synaptic integration, messages arriving at a neuron may be reinforced and sent on to its neighbors. What determines the direction in which a message travels? It depends on how neurons are organized in the body.

For example, your brain's staggering numbers of neurons engage in something like block parties. Regional blocks of hundreds or thousands of them get excitatory and inhibitory signals. They integrate signals entering their block, then send out new ones in response. In the regions of *divergent* circuits, the processes of neurons in one block fan out to form connections with other blocks. In regions of *convergent* circuits, signals from many are relayed to just a few. In regions of *reverberating* circuits, neurons synapse back on themselves and repeat signals like gossip that just won't go away. Such circuits make your eye muscles twitch rhythmically as you sleep.

In the cablelike **nerves**, long axons of many sensory neurons, motor neurons, or both permit long-distance communication between the brain and spinal cord and the rest of the body. Connective tissue bundles most of the axons in parallel array (Figure 30.10). Each long axon has a myelin sheath, which enhances the rate of action potential propagation. The sheath is a series of neuroglial cells (Schwann cells) wrapped like jelly rolls around the axon. It hampers ion movements across the

Figure 30.10 Structure of a nerve. Axons in the nerve are bundled together in wrappings of connective tissue.

the nerve's outer wrapping

blood vessels

nerve fascicle (a number of axons bundled inside connective tissue)

axon

myelin sheath

membrane. But between each cell is an exposed gap or node with voltage-sensitive, gated sodium channels in the neural membrane (Figure 30.11). Ion disturbances tend to flow until reaching a node, where they initiate a new action potential. Action potentials are propagated at 120 meters per second in large sheathed axons.

Multiple sclerosis is an autoimmune disorder. White blood cells wrongly attack the myelin of sheathed axons in the spinal cord. Chronic inflammation damages the sheaths, and then the axons they enclose. Some people have a genetic predisposition to the disorder, but viral infection may trigger it. The disruption in information flow leads to muscle weakening, pain, fatigue, uncontrolled movements, and other symptoms. The disorder affects more than a million people.

Reflex Arcs

Figure 30.12 is one example of the direction of information flow through a nervous system. It shows the roles of sensory and motor neurons of certain nerves in the stretch reflex. **Reflexes** are automatic movements made in response to stimuli. Sensory neurons synapse directly upon motor neurons in the simplest reflex arcs.

By the *stretch reflex*, a muscle contracts after gravity or some other load stretches it. Suppose

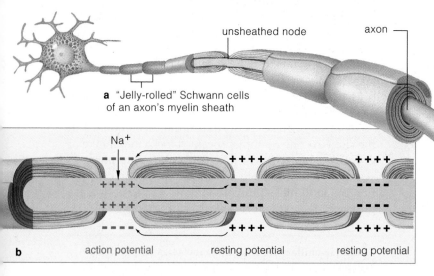

unsheathed node

axon

a "Jelly-rolled" Schwann cells of an axon's myelin sheath

Na⁺

action potential   resting potential   resting potential

b

K⁺

Na⁺

resting potential restored   action potential   resting potential

c

Figure 30.11 Action potential propagation in a sheathed neuron. (**a**) A myelin sheath (a series of cells wrapped like jelly rolls around an axon) blocks ion flow across the neural membrane. Ions cross only at unsheathed nodes between the jelly rolls. Each node contains dense arrays of gated sodium channels. (**b**) A disturbance caused by an action potential spreads along the axon. When it reaches a node, sodium gates open in the positive feedback cycle that triggers a new action potential. (**c**) The disturbance spreads swiftly to the next node, where it triggers a new action potential, and so on down the line.

receptor endings | peripheral axon | cell body | axon | axon ending | cell body | axon | cell body | axon | axon endings

dendrites

a SENSORY NEURON   INTERNEURON   dendrites   MOTOR NEURON

STIMULUS
Biceps stretches.

1 Fruit being loaded into a bowl puts weight on an arm muscle (biceps) and stretches it. Will the bowl drop? NO! Muscle spindles enclosed in the muscle sheath also are stretched.

2 Stretching stimulates sensory neurons that have receptor endings in the muscle spindle. The stimulation is strong enough to generate action potentials, which are propagated along the neurons toward the spinal cord.

3 Axon endings of the sensory neurons synapse on the input zone of a motor neuron in the spinal cord. Neurotransmitter released from the sensory neurons diffuses across the synaptic cleft and stimulates the motor neuron.

4 The stimulation is strong enough to generate action potentials, and these are swiftly propagated along the motor neuron's axon.

5 Axon endings of the motor neuron synapse with muscle cells in the stretched muscle.

6 Neurotransmitter released from the motor neuron's axon endings stimulates the plasma membrane of muscle cells.

RESPONSE
Biceps contracts.

7 Stimulation results in contraction of the stretched muscle. Continued stimulations and contractions hold the bowl steady.

muscle spindle   muscle cell

b

Figure 30.12 (**a**) General direction of information flow in nervous systems. Sensory neurons relay information *into* the spinal cord and brain, where they synapse with interneurons. Interneurons *within* the spinal cord and brain integrate signals. Many synapse with motor neurons, which carry signals *away* from the spinal cord and brain.

(**b**) Organization of nerves in a reflex arc that deals with muscle stretching. In a skeletal muscle, stretch-sensitive receptors of a sensory neuron are located in muscle spindles. The stretching generates action potentials, which reach axon endings in the spinal cord. These synapse with a motor neuron that carries signals to contract, from the spinal cord back to the stretched muscle.

you hold out a bowl and keep it stationary as someone puts some peaches into it. The peaches add weight to the bowl. Your hand starts to drop, which causes the biceps, a muscle in your arm, to stretch. Stretching activates the receptor endings of sensory organs called muscle spindles. The endings, enclosed in a sheath running parallel with the muscle, are input zones of sensory neurons. Inside the spinal cord, axons of the neurons synapse with motor neurons—the axons of which extend back to the stretched muscle (Figure 30.12). Action potentials that arrive at the axon endings trigger the release of ACh, which initiates contraction in the manner described in Chapter 33. As long as the receptor activity continues, motor neurons are excited even more, and the stimulation allows them to hold your hand's position against the force of gravity.

In nearly all reflex pathways, sensory neurons also interact with interneurons, which activate or suppress all motor neurons required for a coordinated response.

Vertebrates have interneurons organized in information-processing blocks. Cablelike nerves that have long axons of sensory neurons, motor neurons, or both connect the brain and spinal cord with the rest of the body.

Reflex arcs, in which sensory neurons synapse directly on motor neurons, are the simplest paths of information flow.

Skewed Information Flow

What happens if something disrupts information flow in nervous systems? Ask *Clostridium botulinum*, an anaerobic, endospore-forming bacterium that normally lives in soil. It can cause the disease *botulism*. Its endospores may be present in improperly preserved, canned, or stored food, and these resting structures can germinate in anyone who eats the food. One metabolic product of the bacterium is toxic. It enters neurons that synapse with muscle cells and blocks the release of acetylcholine (ACh). Without ACh, muscles cannot contract. They become more and more flaccid and paralyzed. Unless a poisoned individual receives antitoxins within ten days, breathing becomes impossible and the heart stops beating.

A related bacterium, *C. tetani*, lives in the gut of horses, cattle, and other grazing animals, even many people. Its endospores survive in soil, especially manure-rich ones. They persist for years if sunlight and oxygen do not reach them. They resist disinfectants, heat, and boiling water. When they enter the body through a deep puncture or a deep cut, they germinate in anaerobic, dead tissues. The bacteria do not spread from dead tissues. They produce a toxin that the blood or nerves deliver to the spinal cord and brain. In the spinal cord, the toxin acts on interneurons that help control motor neurons. It blocks the release of inhibitory neurotransmitters (GABA and glycine) and frees the motor neurons from normal inhibitory control. Symptoms of the disease *tetanus* are about to begin.

After four to ten days, the overstimulated muscles stiffen and go into spasm. They cannot be released from contraction, and prolonged, spastic paralysis follows. Fists and jaws may remain clenched (the disease is also called lockjaw). The back may arch severely and permanently. Once respiratory and cardiac muscles become paralyzed, death nearly always follows.

Vaccines were not available for soldiers of early wars, when dead cavalry horses and manure littered battlefields (Figure 30.13). Today, vaccines have all but eradicated tetanus in the United States.

Figure 30.13 Painting of a young victim of a contaminated battle wound, as he lay dying in a military hospital.

INVERTEBRATE NERVOUS SYSTEMS

We turn now from the messages sent through nervous systems to the systems themselves. At this point in the book, you know all animals except sponges have some type of nervous system in which nerve cells, such as neurons, are oriented relative to one another in signal-conducting and information-processing highways. At a minimum, the cells making up the communication lines receive information about changing conditions outside and inside the body, then they elicit suitable responses from muscle and gland cells.

Regarding the Nerve Net

To appreciate the diversity of nervous systems, start by recalling that animals first evolved in the seas. It is in the seas that we still find animals having the simplest nervous systems. They are the cnidarians, including sea anemones. These invertebrates have *radial* symmetry. Their body parts are arranged around a central axis, a bit like the spokes of a bike wheel (Section 23.1).

Radial animals rely on a **nerve net**, a loose mesh of nerve cells intimately associated with epithelial tissue (Figure 30.14). The nerve cells interact with sensory cells and contractile cells along reflex pathways in the same epithelial tissue. In reflex pathways, remember, sensory stimulation triggers simple, stereotyped movements.

In all cnidarians, one pathway dealing with feeding behavior extends from sensory receptors in the tentacles, along nerve cells, to contractile cells organized around the mouth (Figure 30.14*a*). In jellyfishes, other reflexes are the basis of movements necessary to swim (slowly) and to keep the soft body right-side up.

The nerve net itself extends through the animal's body, but information flow through it is not focused. It simply commands the body wall to slowly contract and expand or tentacles to move through water. Its owners will never dazzle you with bursts of speed or precision acrobatics. Either they are weak swimmers or they are sedentary types that spend most of their life attached to substrates. Yet in their watery world, the nerve net is adaptive. It is equally responsive to signals about food or danger that might come from any direction.

On the Importance of Having a Head

Flatworms are the simplest animals having a bilateral nervous system (Figure 30.14). *Bilateral* symmetry means having equivalent body parts on the left and right sides of the body's midsagittal plane. (Imagine sliding down a staircase banister that turns into a razor and you may never forget the location of the midsagittal plane.) Both sides have the same array of muscles that function in moving the body forward. Both have the same array of nerves to control the muscles, and so on.

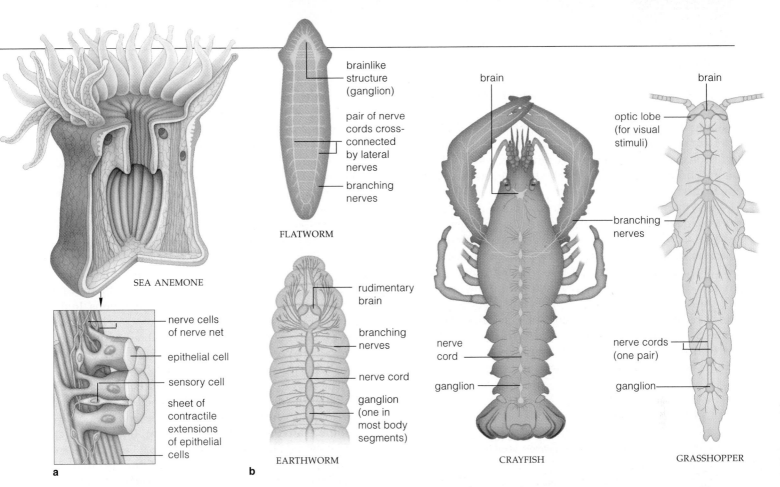

SEA ANEMONE

nerve cells
of nerve net

epithelial cell

sensory cell

sheet of
contractile
extensions
of epithelial
cells

a

FLATWORM

brainlike
structure
(ganglion)

pair of nerve
cords cross-
connected
by lateral
nerves

branching
nerves

EARTHWORM

rudimentary
brain

branching
nerves

nerve cord

ganglion
(one in
most body
segments)

b

CRAYFISH

brain

branching
nerves

nerve
cord

ganglion

GRASSHOPPER

brain

optic lobe
(for visual
stimuli)

nerve cords
(one pair)

ganglion

Figure 30.14 (**a**) Nerve net of a sea anemone, a cnidarian. Its nerve cells interact with sensory and contractile cells. Both cell types are embedded in epithelium between the outer epidermis and a jellylike midlayer (mesoglea) of the body wall. (**b**) Bilateral nervous systems of a few invertebrates. The sketches are not to scale relative to one another.

The flatworm's ladderlike nervous system has two cordlike nerves. The two run longitudinally through the body and have many side branches. In their head end, some flatworms have two ganglia. A ganglion (plural, ganglia) is a cluster of nerve cell bodies that is a local integrating center. Those in a flatworm head coordinate signals from paired sensory organs (two eyespots, for example), and provide a bit of control over the nerves.

Did bilateral nervous systems evolve from nerve nets? Maybe. In nearly all animals more complex than flatworms, we find local nerve nets—plexuses—such as the one in your intestinal wall. More intriguing, a self-feeding larval stage called a planula develops in some cnidarian life cycles. Like the flatworms, a planula has a flattened body and uses cilia to swim or crawl about:

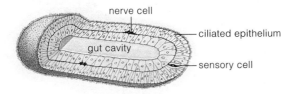

nerve cell

ciliated epithelium

gut cavity

sensory cell

Visualize a planula crawling on the Cambrian seafloor. A mutation blocks its metamorphosis into the adult yet doesn't stop reproductive organs from maturing. This does happen in some larvae (Sections 18.5 and 24.3). The planula kept on crawling but now had the capacity to reproduce. Its offspring inherited the mutant gene for forward mobility. They crawled longer and entered new parts of their habitat—food-rich, food-poor, and maybe dangerous. The mutation had survival value, because it led to more rapid, more effective responses to diverse stimuli. So a concentration of sensory cells in the body's leading end—not the trailing end—was favored.

Regardless of how it happened, cephalization (the formation of a head) and bilateral symmetry developed in most invertebarate lineages (Chapter 23). Patterns of cephalization and bilateral symmetry also are evident in the paired sensory structures, brain centers, nerves, and skeletal muscles of all vertebrates, including you.

All animals except sponges have a nervous system, in which nerve cells are oriented relative to one another in signal-conducting and information-processing highways.

Radial animals have a nerve net, a diffuse mesh of nerve cells that take part in simple reflex pathways involving sensory cells and contractile cells of an epithelial tissue.

Cephalized, bilateral animals have a nervous system with a brain or ganglia at the head end, as well as paired nerves and paired sensory structures.

VERTEBRATE NERVOUS SYSTEMS—AN OVERVIEW

Evolution of the Spinal Cord and Brain

Hundreds of millions of years ago, the earliest fishlike vertebrates were evolving (Section 24.3). A column of bony segments was taking over the functions of their notochord, a long rod of stiffened tissue that worked with their segmented muscles to bring about movement. Above the notochord, the hollow, tubular nerve cord was undergoing modification. It was the forerunner of the spinal cord and brain. These developments were the basis of different life-styles. Not long afterward, genetic divergences from lineages of filter-feeding, scavenging fishes gave rise to swift, jawed predators of the seas.

At first, simple reflex pathways prevailed. Sensory neurons synapsed directly with motor neurons, which directly signaled muscles to contract. No other neurons altered the flow of signals from reception of a stimulus to the response. In the changing world of fast-moving vertebrates, however, the types of predators and prey that were better equipped to sense the presence of food and danger had the competitive edge.

The senses of smell, hearing, and balance became keener among the vertebrates that invaded land. Bones and muscles evolved in ways that allowed specialized movements. Also, the brain became variably thickened with nervous tissue that could integrate rich sensory information and issue orders for complex responses.

The oldest regions of the vertebrate brain still deal with reflex coordination of respiration and other vital functions. But a great many interneurons now synapse on the sensory and motor neurons of ancient pathways and with one another in the newer brain regions. In the most complex vertebrates, interneurons receive, store, retrieve, and compare information about experiences. They weigh possible responses. And they give our own species the capacity to reason, remember, and learn.

The nerve cord persists in all vertebrate embryos. We call it the **neural tube**. While an embryo grows and develops, its neural tube expands. It becomes regionally modified into the brain and spinal cord (Figure 30.15a). A vertebral column encloses the spinal cord. Adjacent tissues give rise to nerves that thread through all body regions and connect with the spinal cord and brain.

The System's Functional Divisions

Figure 30.16 gives you some idea of the expansions of nervous tissue in the human nervous system. It also shows the major paired nerves of this bilateral system. What it cannot show are the 100 billion interneurons in the brain alone. To be sure, humans do have the most intricately wired nervous system in the animal world. Even so, you still find similar patterns of organization and information flow among other vertebrates.

FOREBRAIN. Receives, integrates sensory information from nose, eyes, and ears; in land-dwelling vertebrates, contains the highest integrating centers

MIDBRAIN. Coordinates reflex responses to sight, sounds

HINDBRAIN. Reflex control of respiration, blood circulation, other basic tasks; in complex vertebrates, coordination of sensory input, motor dexterity, and possibly mental dexterity

(start of spinal cord)

a Diagram of how the anterior end of the dorsal, hollow nerve cord of vertebrates expanded into functionally distinct regions, which also increased in complexity in certain lineages

olfactory lobe (part of forebrain)

forebrain

midbrain

hindbrain

b

FISH (shark)

AMPHIBIAN (frog)

REPTILE (alligator)

BIRD (goose)

MAMMAL (horse)

Figure 30.15 Evolutionary trend toward an expanded, more complex brain. The trend became apparent after morphological comparisons were made of the brains of some vertebrates. These dorsal views are not to the same scale.

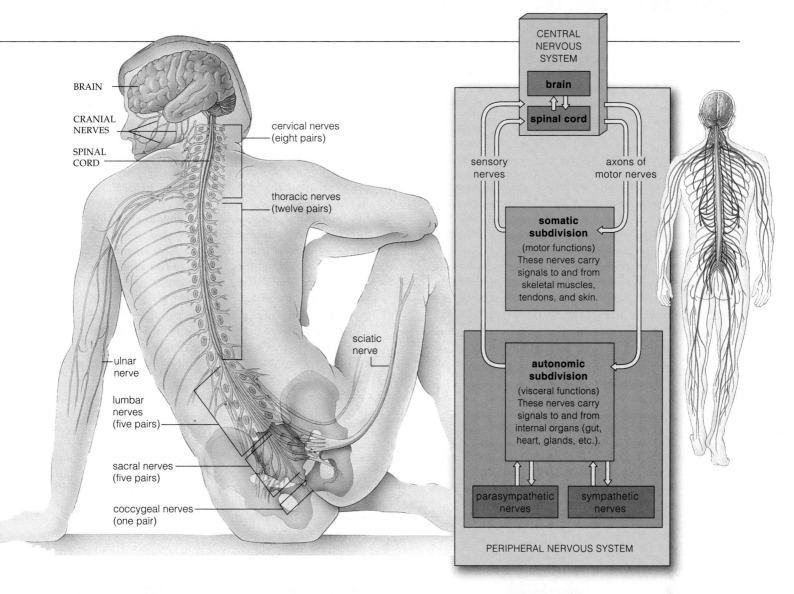

Figure 30.16 Sketch of the brain, spinal cord, and some major peripheral nerves of the human nervous system. The system also incorporates twelve pairs of cranial nerves that originate from the brain. Other vertebrates have a similar nervous system.

Investigators typically approach the complexity of the vertebrate nervous system by functionally dividing it into central and peripheral regions (Figure 30.17). All of the interneurons are confined to the **central nervous system**, which consists of the spinal cord and brain. The **peripheral nervous system** consists mainly of nerves that extend through the rest of the body. Nerves carry signals into and out of the central nervous system.

Inside the brain and spinal cord, the communication lines are called tracts, not nerves. The tracts making up *white* matter have axons with glistening white myelin sheaths; they function in rapid signal transmission. By contrast, *gray* matter consists of unmyelinated axons, dendrites, and cell bodies of neurons, plus neuroglial cells. Neuroglia, remember, protects or structurally and functionally supports neurons. It makes up more than half the volume of vertebrate nervous systems.

Figure 30.17 Functional divisions of the human nervous system. The central nervous system is color-coded *blue*, somatic nerves *green*, and autonomic nerves *red*. Sometimes nerves that carry sensory input to the central nervous system are called *afferent* (a word meaning "to bring to"). Nerves that carry motor output away from the central nervous system to muscles and glands are *efferent* ("to carry outward").

The coevolution of nervous, sensory, and motor systems made more complex life-styles possible among vertebrates.

The vertebrate nervous system has become so intricately wired that the structure and functions of its central and peripheral regions are described separately.

The central nervous system consists of the brain and spinal cord. The peripheral nervous system consists of nerves that thread through the rest of the body and carry signals into and out of the central region.

Myelinated axons make up the white matter of tracts inside the spinal cord and the brain. Neuroglia, unmyelinated axons, dendrites, and cell bodies of neurons make up the gray matter of those tracts.

THE MAJOR EXPRESSWAYS

Let's now take a look at the peripheral nervous system and the spinal cord. The two interconnect as the main expressways for information flow through the body.

Peripheral Nervous System

SOMATIC AND AUTONOMIC SUBDIVISIONS In humans, the peripheral nervous system includes thirty-one pairs of *spinal* nerves, which connect with the spinal cord. The system also includes twelve pairs of *cranial* nerves that connect directly with the brain.

Cranial and spinal nerves are further classified by function. Those carrying signals about moving the head, trunk, and limbs are **somatic nerves**. Their sensory axons deliver information from receptors in the skin, skeletal muscles, and tendons into the central nervous system. Their motor axons deliver the commands issued by the brain and spinal cord to the body's skeletal muscles.

By contrast, spinal and cranial nerves dealing with smooth muscle, cardiac (heart) muscle, and glands are **autonomic nerves**. They deal with the viscera—internal organs and structures. Autonomic nerves carry signals to and from these organs and structures.

PARASYMPATHETIC AND SYMPATHETIC NERVES Figure 30.18 shows the two categories of autonomic nerves. We call them parasympathetic and sympathetic. Normally they work antagonistically, with signals from one kind opposing signals from the other. However, both carry excitatory and inhibitory signals to the internal organs. Often their signals arrive at the same time at muscle or gland cells and compete for control over them. In such cases, synaptic integration at the cellular level leads to minor adjustments in the organ's level of activity.

Parasympathetic nerves dominate when the body is not receiving much outside stimulation. They tend to

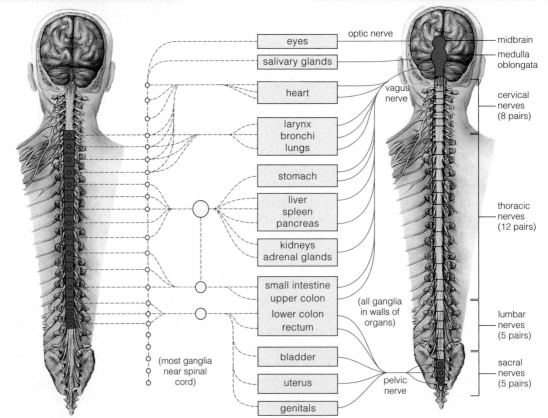

SYMPATHETIC OUTFLOW FROM THE SPINAL CORD

Examples of Responses
Heart rate increases
Pupils of eyes dilate (widen, let in more light)
Glandular secretions in airways to lungs decrease
Salivary gland secretions thicken
Stomach and intestinal movements slow down
Sphincters (rings of muscle) contract

PARASYMPATHETIC OUTFLOW FROM THE SPINAL CORD AND BRAIN

Examples of Responses
Heart rate decreases
Pupils of eyes constrict (keep more light out)
Glandular secretions in airways to lungs increase
Salivary gland secretions become dilute
Stomach and intestinal movements increase
Sphincters (rings of muscle) relax

Figure 30.18 Autonomic nervous system. This diagram shows the major sympathetic nerves and parasympathetic nerves leading from the central nervous system to some major organs. Remember, there are *pairs* of both kinds of nerves, servicing the right and left halves of the body. The ganglia are clusters of the cell bodies of neurons that have their axons bundled together in nerves.

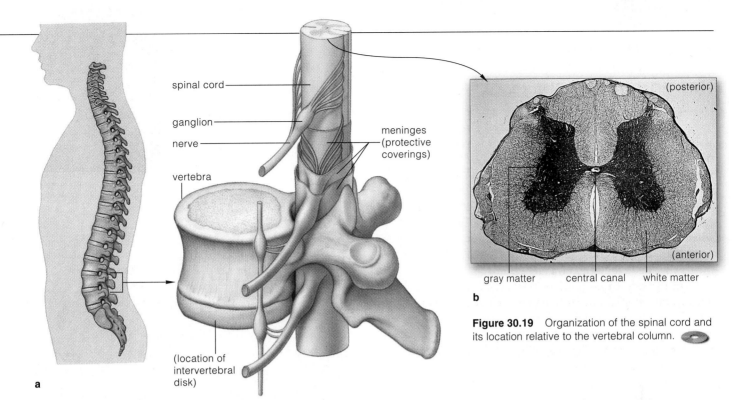

spinal cord

ganglion

nerve

vertebra

meninges (protective coverings)

(location of intervertebral disk)

a

(posterior)

gray matter   central canal   white matter

(anterior)

b

Figure 30.19 Organization of the spinal cord and its location relative to the vertebral column.

slow down the body overall and divert energy to basic "housekeeping" tasks, such as digestion.

Sympathetic nerves dominate in times of sharpened awareness, excitement, or danger. They tend to shelve housekeeping tasks. At the same time, they prepare the animal to fight or escape when threatened or to frolic, as in play or sexual behavior. Right now, sympathetic nerves are commanding your heart to beat a bit faster, and parasympathetic nerves are commanding it to beat a bit slower. Integration of the opposing signals adjusts the heart rate. If something scares or excites you, the parasympathetic input to the heart drops. Sympathetic signals trigger the release of epinephrine, which makes your heart beat faster. They also incite you to breathe faster and start to sweat. In this state of intense arousal, you are primed either to fight (or play) hard, or to get away fast. Hence the name *fight–flight response*.

Suppose that the stimulus for a fight–flight response ends. Sympathetic activity may decrease abruptly, and parasympathetic activity may rise suddenly. You might observe this "rebound effect" after someone has been instantly mobilized to rush onto a road to save a child from an oncoming car. The person might well faint as soon as the child has been swept out of danger.

The Spinal Cord

By definition, the **spinal cord** is a vital expressway for signals between the peripheral nervous system and the brain. Also, sensory and motor neurons make direct reflex connections in the spinal cord, as they do for the stretch reflex. The spinal cord threads through a canal made of bones of the vertebral column (Figure 30.19*a*).

The bones, and ligaments attached to them, protect the cord. So do meninges, three tubelike coverings around the spinal cord and brain. The coverings are tough but vulnerable. *Meningitis*, an often-fatal disease, is brought on by certain viral and bacterial infections. Symptoms include severe headaches, fever, a stiff neck, and nausea.

Signals swiftly travel up and down the spinal cord in bundles of myelinated axons. These axons, which make up the cord's white matter, surround the gray matter (Figure 30.19*b*). Gray matter, again, consists of dendrites and cell bodies of neurons, and neuroglial cells. It plays a key role in controlling reflexes for limb movements, as when you dance or wave your arms about, and for organ activity, such as bladder emptying.

Experiments tell us about such pathways. Examples: Between a frog's spinal cord and brain are circuits that command bent legs to straighten. Sever the circuits at the base of the brain, and the legs become paralyzed— but only for about a minute. Extensor reflex pathways in the spinal cord recover quickly and have the frog hopping about in no time. The recovery time is minimal after similar damage in humans and other primates— the vertebrates with the greatest cephalization.

Nerves of the peripheral nervous system connect the brain and spinal cord with the rest of the body.

The somatic nerves of the peripheral nervous system deal with skeletal muscle movements. Its autonomic nerves deal with smooth muscle, cardiac muscle, and glands.

The spinal cord is a vital expressway for signals between the brain and peripheral nerves. Some of its interneurons also exert direct control over certain reflex pathways.

THE VERTEBRATE BRAIN

Evolutionary High Points

The **brain**—the body's master control center—receives, integrates, stores, and retrieves sensory information. It also issues commands to adjust activities in response to stimuli. Skull bones, fluid, and membranes protect it. The brain has three functional divisions, and the **brain stem**—the most ancient nervous tissue with basic reflex centers—persists in all of them. The gray matter of that ancient tissue expanded and became far more complex when the senses of smell, hearing, and vision started to become more important among early vertebrates.

The medulla oblongata, cerebellum, and pons are parts of the **hindbrain** (Figure 30.20). Reflex centers for respiration, circulation, and other basic tasks are in the medulla oblongata, which coordinates motor responses with complex reflexes, such as coughing, and influences sleeping and arousal. The cerebellum deals with motor skills; it integrates sensory input from the eyes, ears, and muscle spindles with motor signals from the forebrain. Its expansion in humans might relate to language and some other forms of mental dexterity. Many bands of axons extend from both sides of the cerebellum to the pons (meaning bridge). The pons is a traffic center for information flow between the cerebellum and forebrain.

The **midbrain** coordinates reflex responses to sights and sounds. In fishes and amphibians, its roof of gray matter—the tectum—is a center for coordinating nearly all sensory input and initiating motor responses. (If you surgically remove a frog's highest brain center, its activities won't change much if its tectum is intact.) In most vertebrates (not mammals), the midbrain has a pair of optic lobes: centers for input from the eyes. In mammals, the eyes formed functional connections with the forebrain, and the tectum became a reflex center that only relays sensory signals to the forebrain.

What about the **forebrain**? For much of vertebrate history, chemical odors in aquatic habitats were the key clues to survival. An olfactory lobe dealing with odors from prey, predators, and mates was a major forebrain structure. So were paired outgrowths of the brain stem where olfactory input and responses to it could become integrated. Especially in lineages that invaded land, the outgrowths expanded and became the cerebrum. A thin layer of gray matter evolved into the **cerebral cortex**.

Another forebrain region, the thalamus, evolved as a coordinating center for sensory input and as a relay station to the cerebrum. Below this, the hypothalamus evolved into the premier center for homeostatic control of the internal environment. The hypothalamus became central to behaviors related to internal organs, such as thirst, hunger, and sex, and to emotional expression.

The brain isn't tidily subdivided into three regions. An ancient mesh of interneurons extends from the upper

Figure 30.20 (a) Human brain. This dorsal view shows how a longitudinal fissure separates the two cerebral hemispheres. (b) Sagittal view of the right cerebral hemisphere. Not visible is the reticular formation; it extends between the upper spinal cord and the cerebrum.

left cerebral hemisphere   right cerebral hemisphere

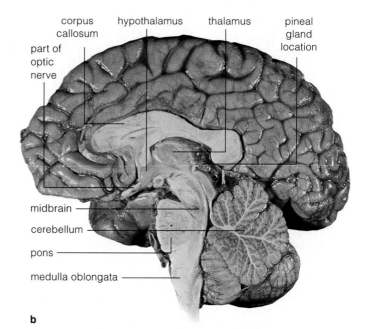

corpus callosum   hypothalamus   thalamus   pineal gland location

part of optic nerve

midbrain
cerebellum
pons
medulla oblongata

b

spinal cord through the brain stem, to the cerebrum's higher integrative centers. This network is the reticular formation. It persists as a low-level pathway to motor centers of the medulla oblongata and spinal cord. It also influences activities of the nervous system as a whole.

Functional Divisions of the Human Brain

An average-sized person's brain weighs 1,300 grams (3 pounds). Over half of it is neuroglia. Your "humanness" arises in the 100 billion interneurons of the cerebrum. A fissure divides the cerebrum into two hemispheres, each having frontal, occipital, temporal, and parietal lobes (Figure 30.21). The left hemisphere deals mostly with analytical skills, speech, and mathematics. It typically dominates the right hemisphere, which deals more with visual–spatial relationships and music. Each responds to sensory input mainly from the opposite side of the body. For example, pressure signals from the right arm travel to the left hemisphere. A band of nerve tracts (corpus callosum) coordinates activities of the two hemispheres.

What we comprehend, communicate, remember, and consciously act upon arises in the cerebral cortex. Here

SOMATIC SENSATIONS

Somatic sensations begin with receptors in the body's surface tissues, skeletal muscles, and walls of internal organs. The receptors are most highly developed in birds and mammals; amphibians have a few, and apparently fishes have none. Receptor inputs travel to the spinal cord and on into the **somatosensory cortex**, part of the gray matter of the cerebral hemispheres (Section 30.9). The interneurons in this brain region are organized like maps corresponding to particular parts of the body's surface. Map regions with the largest areas correspond to body parts that have the greatest sensory acuity and that require the most intricate control. Such body parts include the fingers, thumbs, and lips (Figure 31.4).

You and other mammals discern sensations of touch, pressure, cold, warmth, and pain. **Pain** is perception of injury to some body region. Sensations of *somatic* pain start with signals from pain receptors in skin, skeletal muscles, joints, and tendons. The sensations of *visceral* pain, associated with the internal organs, are related to excessive chemical stimulation, muscle spasms, muscle fatigue, inadequate blood flow to organs, distension of digestive tract regions, and other abnormal conditions.

Free nerve endings are the simplest receptors. These thinly myelinated or unmyelinated (naked) branched endings of sensory neurons are distributed in skin and internal tissues. Different types are mechanoreceptors, thermoreceptors, and pain receptors. All adapt slowly to stimulation. One subpopulation gives rise to a sense of prickling pain, as when you jab a finger with a pin. Another contributes to itching or warming sensations that are provoked by chemicals, including histamine. Two thermoreceptive types have peak sensitivities that are higher and lower than normal body temperature, respectively. One mechanoreceptive type coils around hair follicles and detects hair movements (Figure 31.5).

Capsules encase and shield some receptors near the body surface from all stimuli except pressure changes. One type, Meissner's corpuscle, adapts very slowly to low-frequency vibration. It is abundant in fingertips, lips, eyelids, nipples, and genitals. Another, the bulb of Krause, is a thermoreceptor that kicks in at 20°C (68°F) or lower. Below 10°C, it contributes to painful freezing sensations. Slowly adapting, encapsulated types called Ruffini endings respond to steady touching, pressure, and temperatures above 45°C (113°F).

The Pacinian corpuscle, also encapsulated, is widely distributed deep in the skin, where it responds to fine textures. It also resides near freely movable joints and in some internal organs. Onionlike layers of membrane alternating with fluid-filled spaces enclose this sensory ending (Figure 31.5). They help it detect rapid pressure changes associated with touch and vibrations.

Sensing limb movements and the body's position in space requires mechanoreceptors in skeletal muscle,

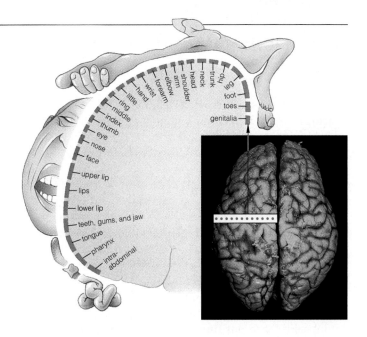

Figure 31.4 Differences in the representation of different body parts in the primary somatosensory cortex. This region is a strip of cerebral cortex, a little wider than 2.5 centimeters (about an inch), from the top of the head to above the ear.

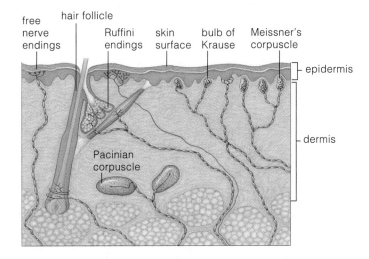

Figure 31.5 A sampling of receptors in human skin.

joints, tendons, ligaments, and skin. Examples include the stretch receptors of muscle spindles. As described in Section 30.4, these sensory organs run parallel with skeletal muscle cells. The degree of their responses to stimulation depends on how much and how fast the muscle is being stretched.

Diverse sensory receptors near the body surface and in internal organs detect touch, pressure, temperature, pain, motion, and positional changes of body parts. Their input travels through the spinal cord to the somatosensory cortex and other brain regions where somatic sensations arise.

SENSES OF HEARING AND BALANCE

Evolution of the Vertebrate Ear

Many arthropods and most vertebrates have a sense of **hearing**—sound perception. Sounds are vibrations, or wavelike forms of mechanical energy. You might wish to review Figure 31.6, which describes their properties.

Your sense of hearing evolved from structures that at one time had nothing at all to do with hearing. Here is the story, in brief: Most animals assess and respond to displacements from a natural, *equilibrium* position in which the body is balanced relative to gravity, velocity, acceleration, and other forces that act on its positions and movements. For instance, animals right themselves after tilting or turning upside-down. Most vertebrates have a pair of **inner ears** on both sides of their brain (Figure 31.7). Inner ears initially evolved as organs of equilibrium. Each has fluid-filled sacs and canals used to detect rotational, linear, and accelerated motions of the head. In all amphibians, birds, and mammals, the brain now integrates sensory input from the inner ears, eyes, skin, and joints to arrive at sensations of balance.

After vertebrates invaded the land, hearing became far more crucial than it had been in aquatic habitats,

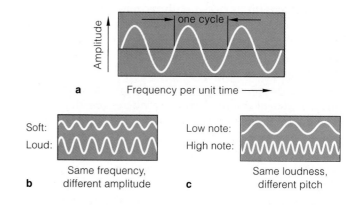

a

b Same frequency, different amplitude   c Same loudness, different pitch

Figure 31.6 Properties of sound—waves of compressed air. (**a**) Sounds differ in pressure, or *amplitude*, which we show as wave forms. We can measure amplitude in decibels. Every ten decibels is a tenfold increase in intensity above the faintest sound humans hear. (**b**) A sound's intensity, or loudness, depends on amplitude. (**c**) A sound's *frequency* is the number of wave cycles per second. Each cycle extends from the start of one wave peak to the start of the next. The more cycles per second, the higher the frequency and perceived *pitch*. Unlike a tuning fork's pure tone, most sounds are combinations of waves of different frequencies. Combinations of overtones give them their timbre (quality), a property that helps you recognize, say, voices of people you know on a phone.

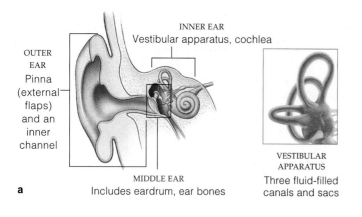

a

Figure 31.7 Sensory reception at the human ear. This sensory organ collects, amplifies, and sorts out sound waves (acoustical stimuli). The external flaps of the outer ear collect sound waves, which move through an auditory canal to the eardrum. The eardrum is also called the tympanic membrane.

b

c Sound reception in the inner ear. An eardrum vibrates with incoming sound waves. Middle ear bones pick up the vibrations. They amplify the stimulus by transmitting the force of pressure waves to a smaller surface, the oval window. The "window" is an elastic membrane over the entrance to the cochlea, part of the inner ear. Here, we "uncoiled" the cochlea for clarity.

The oval window bows in and out from the sound waves, thus producing waves of fluid pressure in two ducts (scala vestibuli and scala tympani). The waves reach another membrane, the round window, which bulges under the pressure and forces fluid to move back and forth in the inner ear.

Pressure waves are sorted out in the cochlear duct, the third duct inside the inner ear. One of its membranes (the basilar membrane) is narrow and stiff, but then it broadens and becomes more flexible deeper in the coil. At different points along its length, the basilar membrane vibrates more strongly to sounds of different frequencies.

a   b

Figure 31.8 Results of an experiment on the effect of intense sound on the inner ear. (**a**) From a guinea pig ear, three rows of hair cells projecting into the tectorial membrane in the organ of Corti. (**b**) Hair cells in that organ after exposure for twenty-four hours to noise levels comparable to extremely loud music.

To give you a sense of what "loud" is, a ticking watch measures 10 decibels (100 times louder than the threshold of hearing for humans). A normal conversation is about 60 decibels (about a million times louder), a food blender operating at high speed is about 90 decibels (a billion times louder), and a hard rock or heavy metal concert, about 120 decibels (a trillion times louder).

where sounds are muffled. The parts of the internal ear dealing with equilibrium didn't change much. But other parts evolved into structures that were able to receive and process even faint sounds traveling through air.

A **middle ear** evolved that amplified and sent air waves to the inner ear. In reptiles, a depression on each side of the head became an eardrum, a thin membrane that could vibrate swiftly in response to air waves. In existing crocodiles, birds, and mammals is an air-filled cavity and small bones that transmit vibrations to the inner ear. Those bones had structurally supported gill pouches and evolved as part of the jaw joint in ancient

fishes (Section 24.3). In other words, bones that once had roles in gas exchange and feeding in fishes became specialized for hearing in reptiles, birds, and mammals.

Among mammals, paired **external ears** that collect sound waves became well developed. Each external ear typically has a pinna, or sound-collecting, skin-covered and folded flaps of cartilage projecting from the side of the head. It also has a channel to the middle ear.

Figure 31.7 shows the external, middle, and inner ear of humans. Like other mammals (and birds), it has a cochlea, a part of the inner ear that receives and sorts out sound waves. Its acoustical receptors—**hair cells**—bend in response to pressure waves. This starts a flow of information along an auditory nerve that leads from the receptors to brain centers. Prolonged exposure to intense sounds permanently damages hair cells. These cells are not adapted to extremely loud music, roars of jet planes, and other recent developments (Figure 31.8).

A Closer Look at Inner Ear Functions

In each of your inner ears is a **vestibular apparatus** (Figure 31.7). When your head rotates, fluid in a canal corresponding to the direction of rotation is displaced the opposite way. The resulting fluid pressure activates an array of hair cells that bend in response to pressure. Deforming their plasma membrane may trigger action potentials, and so messages may reach adjacent sensory neurons. A different part of the vestibular apparatus is activated when you start and stop moving in a straight line. In *motion sickness*, a monotonous linear, angular, or vertical motion has overstimulated its hair cells.

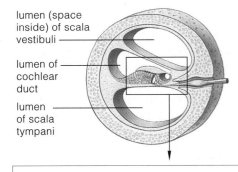

lumen (space inside) of scala vestibuli

lumen of cochlear duct

lumen of scala tympani

hair cell in organ of Corti

tectorial membrane

lumen of cochlear duct

basilar membrane

lumen of scala tympani

to auditory nerve

ORGAN OF CORTI
(organ of hearing that rests on top of basilar membrane)

d At the organ of Corti, 16,000 hair cells project into a jellylike flap (tectorial membrane) over the basilar membrane. When the flap vibrates suitably, hair cells bend, action potentials arise, and messages travel along an auditory nerve to the brain.

Organs of equilibrium help keep the body balanced in relation to gravity, velocity, acceleration, and other forces that influence its position and movement. Humans have such organs in the vestibular apparatus of each inner ear.

Hearing among land vertebrates involves structures that collect, amplify, and sort out sound waves traveling through air. In the inner ear, sound waves produce fluid pressure variations that hair cells transduce into action potentials.

SENSE OF VISION

All organisms are sensitive to light. Sunflowers track the sun as it arcs across the sky; single-celled amoebas abruptly stop moving when you shine a light on them. However, **vision** requires eyes and image formation in brain centers that receive and interpret signals about shapes, brightness, positions, and movements of visual stimuli. **Eyes** are sensory organs with a dense array of photoreceptors. Each photoreceptor contains pigments that can absorb photon energy, which can be converted into excitation energy in sensory neurons.

A Sampling of Invertebrate Eyes

Some eyeless invertebrates respond to light by way of photoreceptors embedded in their integument. Others have dense arrays of photoreceptors in eyelike patches or depressions in the integument. Vision evolved first among fast-moving predators that had to discriminate quickly among prey and other objects in their rapidly changing visual field. A **visual field** is the part of the outside world that an animal sees. Different groups of photoreceptors sample light intensity of different parts of the field. Most invertebrates don't have good vision; they are too small to have enough photoreceptors. For instance, a planarian's pigment cup only has about 200. Each square millimeter of an octopus eye has 70,000.

Image formation benefits from a **lens**, a transparent body that bends all light rays from a given point in the visual field so they converge onto photoreceptors. The round abalone lens bends light rays but not to the same focal point, so images are blurred (Figure 31.9*b*). Things improve with a **cornea**. This transparent cover directs light rays onto a lens, as in octopus eyes (Figure 31.9*d*).

Spiders have *simple* eyes, in which one lens services all of the photoreceptors. Crustaceans and insects have *compound* eyes. These eyes contain photosensitive units,

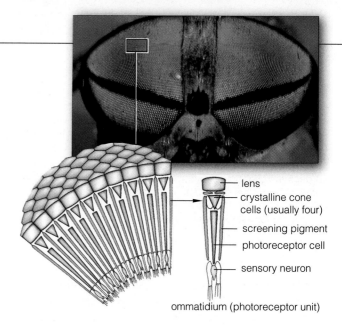

lens
crystalline cone cells (usually four)
screening pigment
photoreceptor cell
sensory neuron

ommatidium (photoreceptor unit)

Figure 31.10 Compound eyes of a deerfly. The lens of each photosensitive unit directs light onto a crystalline cone, which focuses light onto photoreceptor cells below it.

each with a lens and a bundle of photoreceptors (Figure 31.10). Some have thousands of units called ommatidia (singular, ommatidium). Each unit samples a small part of the visual field and contributes a tiny bit of sensory information that helps the brain form a visual mosaic.

Of all invertebrates, octopuses and squids have the most complex eyes. Like vertebrates, they have **camera eyes**, so named because the eyeball is structured like a camera. Inside it is a darkened chamber (Figure 31.9*d*). Light enters at a pupil, an opening in a contractile tissue ring called an iris (equivalent to a camera diaphragm). Behind the pupil, a lens focuses light on a **retina**, a tissue packed with photoreceptors (in the camera, onto light-sensitive film). Similarities in cephalopod and vertebrate eyes might be a result of convergent evolution.

Vertebrate Eyes

In most vertebrates, the eyeball has a three-layer wall. The *outer* layer has a sclera of dense, white fibers that protect most of the eyeball, and a cornea of transparent collagen fibers that cover the front of the eyeball.

The *middle* layer has a choroid, ciliary body, iris, and pupil. The choroid, a dark-pigmented and vascularized tissue, absorbs light the photoreceptors miss and stops it from scattering in the eye. Its iris is a pigmented ring suspended behind the cornea. The pupil at the center is an adjustable entrance for light. In bright light, circular muscles contract and shrink the diameter of the pupil. In dim light, radial muscles contract and enlarge it.

The *inner* layer, including the retina, is located at the back of the human eyeball (Figure 31.11). In some birds, it is located at the top of the eyeball (Figure 31.12).

The eye's interior incorporates a lens constructed of layers of transparent proteins. Bathing the lens is a clear

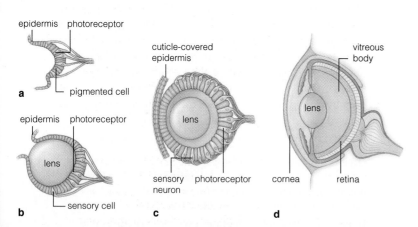

a
epidermis   photoreceptor
pigmented cell

b
epidermis   photoreceptor
lens
sensory cell

c
cuticle-covered epidermis
lens
sensory neuron   photoreceptor

d
vitreous body
lens
cornea   retina

Figure 31.9 Organization of invertebrate eyes, longitudinal section. There are far more photoreceptors than are shown. (**a**) Limpet ocellus, a shallow depression in epidermis that has light-sensitive receptors. (**b**) Abalone eye, with its spherical, transparent lens. (**c**) Snail eye. (**d**) Octopus eye.

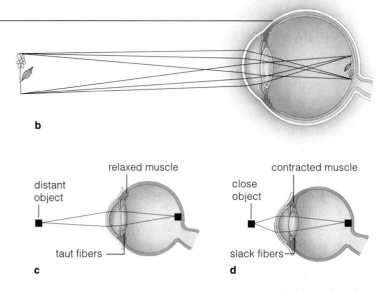

a

Figure 31.11 (**a**) Structure of the human eye.

(**b**) Pattern of retinal stimulation in the human eye. Light rays converge at the back of the eye and stimulate the retina in a distinct way. The cornea has a curved surface, so light rays coming from a given point in the visual field hit it at different angles and their trajectories change. (Rays of light bend at the boundaries between substances of different densities, and the bending sends them in new directions.) Because of their newly angled trajectories, light rays that converge at the back of the eyeball stimulate the retina in a pattern that is upside-down and reversed left to right, relative to the original source of the rays.

Focusing mechanisms use a ciliary muscle that encircles and attaches to a lens. (**c**) When it relaxes, the lens flattens; the focal point moves farther back and brings distant objects into focus. (**d**) When the muscle contracts, the lens bulges; the focal point moves closer and brings close objects into focus.

Figure 31.12 Here's looking at you! In owls and some other birds of prey, photoreceptors are more concentrated on top of the inner eyeball, not at back. These birds look down more than up when they fly and scan the ground for a meal. When on the ground, they cannot easily see objects above them unless they turn their head almost upside-down.

fluid, the aqueous humor. Behind the lens is a chamber filled with a jellylike substance, the vitreous body.

Light rays from sources at varying distances strike the cornea at different angles. But adjustments in lens position or shape usually focuses all of them onto the retina. This **visual accommodation** keeps the light rays from distant objects from being focused in front of the retina. Visual accommodation also prevents light rays from close objects from being focused behind it.

Fish and reptile eye muscles move the lens forward or back, like a camera's focusing device. Extending the distance between the lens and retina moves the focal point forward; shrinking the distance moves it back. The lens shape gets adjusted in birds and mammals. It flattens and moves the focal point back when a muscle encircling it relaxes. It bulges and moves the focal point forward when the muscle contracts (Figure 31.11*c,d*).

When an eyeball is not shaped quite right, the lens is positioned too close to or too far from the retina, so accommodation cannot bend the incoming rays of light to the same focal point. This happens with *astigmatism,* a condition in which the cornea has an uneven curvature.

Nearsightednesss (myopia) is common when the eyeball's horizontal axis is longer than the vertical. It is common when the ciliary muscle that adjusts the lens contracts too strongly and thus focuses light from distant objects in front of the retina, not onto it. *Farsightedness* is the opposite of myopia. The lens is "lazy" or the eyeball's vertical axis is longer than its horizontal axis. Either way, light from close objects gets focused behind the retina. Eyeglasses typically correct the visual disorders.

Vision requires eyes (sensory organs containing a dense array of photoreceptors) and a capacity for image formation in the brain, based on patterns of visual stimulation.

For most vertebrate eyes, the eyeball has three layers, and it encloses a lens, an aqueous humor, and a vitreous body.

A protective sclera and a light-focusing cornea make up the outer layer. The middle layer has a vascularized, pigmented choroid and other parts that admit and control incoming light. Photoreception occurs at the retina of the inner layer.

Adjustments in the positioning or shape of the lens focus incoming visual stimuli onto the retina.

VISUAL PERCEPTION

A sensory pathway from the retina to the brain detects, transmits, and also combines raw visual information. It leads to awareness of light and shadows, of colors, of near and distant objects in the outside world. The flow of information starts when light strikes dense arrays of **rod cells** and **cone cells**, two classes of photoreceptors (Figure 31.13). Rod cells detect very dim light. At night or in dark places, they contribute to coarse perception of movement by reporting changes in the intensity of light across the visual field. Cone cells detect bright light; they contribute to sharp vision and color perception in the daytime.

Part of a rod cell's membrane is folded into several hundred disks. Each disk is loaded with 10^8 molecules of rhodopsin, a visual pigment. The stacked membrane increases the density of pigments, hence the odds of intercepting packets of light energy—photons—of specific wavelengths. Action potentials that arise from absorption of even a few photons can lead to conscious awareness of objects in dimly lit surroundings.

Rhodopsin effectively absorbs photons of blue-to-green wavelengths (Section 6.2). Absorption changes the pigment's shape, which triggers a cascade of reactions that affect ion channels and pumps across the rod cell's plasma membrane. The voltage difference across the membrane shifts as gated sodium channels close, and a graded potential results. This graded potential reduces the ongoing release of a neurotransmitter that inhibits neighboring sensory neurons. Released from inhibition, the neighbors fire off signals about the visual stimulus that will travel to the brain.

The sense of color and of daytime vision starts with photon absorption by red, green, and blue cone cells, each having a distinct visual pigment. Here, too, photon absorption slows the release of a neurotransmitter that otherwise inhibits nearby sensory neurons. At the back of the eyeball is the fovea, a funnel-shaped depression near the retina's center. This retinal area has the highest density of photoreceptors and offers the greatest visual acuity. Its cone cells are the ones that discriminate most precisely between adjacent points in space.

Distinct layers of neurons are located above the rods and cones. Bipolar sensory neurons, then ganglion cells, accept visual information from these layers. Input from 125 million or so rods and cones converges on a mere 1 million ganglion cells. Then the axons of ganglion cells converge to form two optic nerves to the brain (Figure 31.13). But synaptic integration and visual processing start even before the information is sent to the brain.

Integration starts with the retina's organization. Its surface is divided into receptive fields: restricted areas that affect the activity of individual sensory neurons.

Figure 31.13
(**a**) Rods (*red*) and cones (*yellow*). (**b**) The sensory pathway from retinas to the brain.

For instance, each ganglion cell's field is a tiny circle. Some cells respond best to a spot of light with a dark ring in the field's center. Others respond to a fast change in light intensity, to a spot of one color, or to motion. Some respond to diagonal objects and ignore diffuse, uniform illumination. Lateral interactions among other players—horizontal cells and amacrine cells—dampen or enhance signals before the ganglion cells get them.

Remember, an individual's visual field is the part of the outside world that it actually sees. The right side of both retinas intercepts light from the *left* half of the visual field; the left side intercepts light from the *right* half. An optic nerve leading out from each eye delivers signals about a stimulus from the *left* visual field to the right cerebral hemisphere, and delivers signals from the *right* visual field to the left hemisphere (Figure 31.13).

Axons of the optic nerves end in a layered brain region, the lateral geniculate nucleus. Each layer has a map corresponding to the receptive fields, and it deals with one aspect of a stimulus—form, movement, depth, color, texture, and so on. After initial processing, signals rapidly and simultaneously reach different portions of the visual cortex. There, final integration organizes the electrical activity and produces the sensation of sight.

Eyes collect and integrate information on distance, shape, brightness, position, and movement of visual stimuli. Their sensory pathways start at receptive fields in the retina and proceed along optic nerves to the brain, where signals are processed and integrated in the visual cortex.

SENSES OF TASTE AND SMELL

We leave this chapter with two final examples of the special senses: taste and smell. Both are *chemical* senses. Their sensory pathways start at chemoreceptors, which are activated when they bind a chemical substance that is dissolved in the fluid bathing them. These receptors wear out and new ones replace them on an ongoing basis. In both pathways, sensory input travels from the receptors through the thalamus and on to the cerebral cortex, where perceptions of the stimulus take shape and undergo fine-tuning. The input also travels to the limbic system, which integrates it with emotional states and stored memories (Sections 30.9 and 30.10).

Different animals taste with their mouth, antennae, legs, tentacles, or fins, depending on where their **taste receptors** are. In your mouth, throat, and especially the upper surface of your tongue, these chemoreceptors are located in about 10,000 sensory organs called taste buds (Figure 31.14). In a taste bud, fluids in the mouth move through a pore to receptor cells. Thousands of perceived tastes are some combination of five primary sensations: sweet (as elicited by glucose and other simple sugars), sour (acids), salty (NaCl and other salts), bitter (plant toxins, including alkaloids), and *umami* (glutamate, the brothy or savory taste typical of aged cheese and meat).

Figure 31.15 Sensory pathway from the sensory endings of olfactory receptors in the nose to the cerebral cortex and limbic system. Receptor axons pass through holes in a bony plate between the nasal lining and the brain. In the earliest vertebrates, an olfactory bulb and olfactory lobe dominated the forebrain. In certain lineages, the olfactory bulb became less important when the cerebrum expanded greatly in size and functions, as Section 30.9 indicates.

Figure 31.14 Taste receptors inside a human tongue. Circular papillae ring epithelial tissue that contains taste buds. The filamentous papillae do not contribute to taste; their movements direct chemical-laden fluid past the receptors.

Olfactory receptors detect water-soluble or volatile (easily vaporized) substances. A human nose has about 5 million; a bloodhound nose has more than 200 million. The receptor axons lead into one of two olfactory bulbs. In these small brain structures, the axons synapse with groups of cells that sort out the components of a given scent and relay the information along an olfactory tract for further processing in the cerebrum (Figure 31.15).

Pheromones, signaling molecules secreted from the exocrine glands of one individual, influence the social behavior of other individuals of its species, including potential sex partners (Chapter 41). Bombykol is one of the sex pheromones. Olfactory receptors of a male silk moth hit by only one bombykol molecule per second fire action potentials. They help a male find a female in the dark, even if she is more than a kilometer upwind. Human pheromones may influence sexual preferences, menstrual cycles, and menopause. Receptors for them, and the olfactory pathway itself, are not yet identified.

The senses of taste and smell both start at chemoreceptors. Both involve sensory pathways that lead to processing regions in the cerebral cortex and in the limbic system.

SUMMARY   **Gold** indicates text section

1. A stimulus is a specific form of energy that the body detects by means of sensory receptors. A sensation is an awareness that stimulation has occurred. Perception is understanding what the sensation means. *CI*

2. Sensory receptors are endings of sensory neurons or specialized cells next to them. They respond to specific forms of energy, such as mechanical pressure and light. Animals can only respond to aspects of the internal or external environment when they have receptors that are sensitive to the energy of specific stimuli. *CI*

   a. Mechanoreceptors, including free nerve endings and hair cells, detect mechanical energy related to touch, pressure, and motion and changes in position.

   b. Thermoreceptors detect radiant energy (heat).

   c. Pain receptors (nociceptors) detect tissue damage.

   d. Chemoreceptors, such as olfactory receptors and taste receptors, detect chemical substances dissolved in the fluids bathing them.

   e. Osmoreceptors detect changes in water volume (hence solute concentrations) in the surrounding fluid.

   f. Photoreceptors detect light energy. The rods and cones of the retina in the human eye are examples.

3. A sensory system has receptors for specific stimuli and nerve pathways from those receptors to receiving and processing centers in the brain. The brain assesses a particular stimulus based on which nerve pathway is delivering signals, the frequency of signals traveling on individual axons of that pathway, and the number of axons that were recruited into action. *31.1*

4. Somatic sensations include touch, pressure, pain, temperature, and muscle sense. The receptors associated with these sensations are not localized in a single organ or tissue. Special senses include taste, smell, hearing, balance, and vision. Receptors associated with these senses typically reside in sensory organs, such as eyes, or some other particular body region. *31.2*

5. Organs of equilibrium, as in the vertebrate inner ear, detect gravity, velocity, acceleration, and other forces that affect body positions and movements. The sense of hearing (sound perception) in vertebrates requires components of the outer, middle, and inner ear that collect, amplify, and sort out sound waves. *31.3*

6. Vision requires eyes: sensory organs having a dense array of photoreceptors, as in a retina (Table 31.1). It requires a capacity for image formation in the brain, based on patterns of visual stimulation. The sense of vision and discrimination among objects first evolved among fast-moving, predatory animals. *31.4, 31.5*

7. The senses of taste and smell both involve sensory pathways from chemoreceptors to processing regions in the cerebral cortex and limbic system. *31.6*

Table 31.1 *Summary of Vertebrate Eye Components*		
THREE LAYERS FORMING WALL OF EYEBALL		
Outer layer	*Sclera.* Protects eyeball	
	Cornea. Focuses light	
Middle layer	*Choroid.* Blood vessels nutritionally support wall's cells; pigments prevent light scattering	
	Ciliary body. Its muscles control lens shape; its fine fibers hold lens in upright position	
	Iris. Adjustments here control incoming light	
	Pupil. Serves as entrance for light	
Inner layer	*Retina.* Absorbs and transduces light energy	
	Fovea. Increases visual acuity	
	Start of optic nerve. Carries signals to brain	
INTERIOR OF EYEBALL		
Lens	Focuses light onto photoreceptors	
Aqueous humor	Transmits light, maintains pressure	
Vitreous body	Transmits light, supports lens and eyeball	

Review Questions

1. Define stimulus, sensation, and perception. *CI*

2. Name six categories of sensory receptors and the type of stimulus energy that each kind detects. *CI*

3. What are the basic components of a sensory system? How does the brain assess the nature of a given stimulus? *31.1*

4. How do somatic sensations differ from special senses? *31.2*

5. What is pain? Describe one type of pain receptor. *31.2*

6. Which evolved first, a sense of balance or a sense of hearing? Do both involve the outer, middle, and inner ear? *31.3*

7. Label the component parts of the human ear: *31.3*

auditory nerve

8. How does vision differ from light sensitivity? What sensory organs and structures does vision require? *31.4, 31.5*

9. Label the component parts of the human eye: *31.4*

10. What are the five primary taste sensations? *31.6*

Figure 31.16
Flamingos at a large lake in Tanzania, East Africa.

what Tim sees

what Francesca sees

Self-Quiz   ANSWERS IN APPENDIX III

1. A _____ is a specific form of energy that is detected by a sensory receptor.

2. Conscious awareness of a stimulus is called a _____ .

3. _____ is understanding what particular sensations mean.

4. Each sensory system consists of _____ .
   a. nerve pathways from specific receptors to the brain
   b. sensory receptors
   c. brain regions that deal with sensory information
   d. all of the above

5. _____ is (are) a decrease in the response made to an ongoing stimulus.
   a. Perception   c. Visual accommodation
   b. Sensory adaptation   d. both b and c

6. _____ detect mechanical energy associated with changes in pressure, position, or acceleration.
   a. Chemoreceptors   c. Photoreceptors
   b. Mechanoreceptors   d. Thermoreceptors

7. Detecting light energy is the function of _____ .
   a. chemoreceptors   c. photoreceptors
   b. mechanoreceptors   d. thermoreceptors

8. Vision requires _____ .
   a. a tissue with dense arrays of photoreceptors
   b. eyes
   c. image-forming centers in the brain
   d. all of the above

9. The outer layer of the human eyeball includes the _____ .
   a. lens and choroid   c. retina
   b. sclera and cornea   d. both a and c

10. The inner layer of the human eyeball includes the _____ .
   a. lens and choroid   c. retina
   b. sclera and cornea   d. both a and c

11. Match each term with the most suitable description.
   _____ somatic senses   a. evolved first as organ of
   _____ umami   equilibrium
   _____ special senses   b. taste, smell, hearing, balance,
   _____ variations in   and vision
          intensity of   c. one of five primary taste
          stimulus   sensations
   _____ inner ear   d. encoded in the frequency and
                               number of action potentials
                           e. touch, pressure, temperature,
                               pain, and muscle sense

Critical Thinking

1. Wayne, on standby for the last flight from San Francisco to New York, was assigned the last available seat on the plane.

It was in the last row, where vibrations and noise from the engines are most pronounced. When Wayne got off the plane in New York, he was speaking very loudly and having trouble hearing what other people were saying. The next day, things were back to normal. Speculate on what happened to his sense of hearing during the flight.

2. Juanita made an appointment with her doctor because she was experiencing recurring episodes of dizziness. Her doctor immediately asked whether "dizziness" meant she had a sense of lightheadedness, as if she were going to faint, or whether it meant she had sensations of *vertigo*—that is, a feeling that she herself or objects near her were spinning around. Why do you suppose her doctor considered this clarification important early in his evaluation of her condition?

3. To Tim, a graduate student studying the social behavior of flamingos in Tanzania, the birds closest to him along the shore of a lake are in sharp focus, but the ones flying some distance away look fuzzy (Figure 31.16a). Francesca, another student, says the birds closest to her look fuzzy but those in the distance are in focus (Figure 31.16b).

Is Tim nearsighted or farsighted, and is the focal point of light rays from the flamingos falling in front of or behind his retinas? What about Francesca?

Selected Key Terms

camera eye *31.4*	olfactory	somatic sensation *31.1*
chemoreceptor *CI*	receptor *31.6*	somatosensory
cone cell *31.5*	osmoreceptor *CI*	cortex *31.2*
cornea *31.4*	pain *31.2*	special sense *31.1*
echolocation *CI*	pain receptor *CI*	stimulus *CI*
external ear *31.3*	perception *CI*	taste receptor *31.6*
eye *31.4*	pheromone *31.6*	thermoreceptor *CI*
hair cell *31.3*	photoreceptor *CI*	vestibular
hearing *31.4*	retina *31.4*	apparatus *31.3*
inner ear *31.3*	rod cell *31.5*	vision *31.4*
lens *31.4*	sensation *CI*	visual
mechanoreceptor *CI*	sensory	accommodation *31.4*
middle ear *31.3*	adaptation *31.1*	visual field *31.4*
	sensory system *CI*	

Readings

Delcomyn, F. 1998. *Foundations of Neurobiology*. New York: Freeman.

Romer, A., and T. Parsons. 1986. *The Vertebrate Body*. Sixth edition. New York: Saunders.

Sherwood, L. 1997. *Human Physiology*. Third edition. Belmont, California: Wadsworth.

Zeki, S. September 1992. "The Visual Image in Mind and Brain." *Scientific American* 267(3): 68–76.

On-Line readings at Student Guide for InfoTrac: www.brookscole.com/biology

ENDOCRINE CONTROL

Hormone Jamboree

In the 1960s, at her camp in a forest by the shore of Lake Tanganyika in Tanzania, the primatologist Jane Goodall let it be known that bananas were available. One of the first chimpanzees attracted to the delicious food was a female—Flo, as she came to be called (Figure 32.1). Flo brought along her offspring, an infant female and a juvenile male, and exhibited commendable parental behavior toward them.

Three years passed, and Goodall observed that Flo's earlier preoccupation with motherhood gave way to a preoccupation with sex. She also observed that male chimpanzees followed Flo to the camp by the lake and stayed for more than the bananas.

Sex, Goodall discovered, is *the* premier binding force in the social life of chimpanzees. These primates do not mate for life as eagles do, or wolves. Before the rainy season begins, mature females that are entering their fertile cycle become sexually active. Changing hormone levels drive behavioral and physical changes. Sexual organs, for instance, become swollen and vivid pink. The swellings are strong visual signals to males. They are the flags of sexual jamborees, of great gatherings of highly stimulated chimps in which any males present may have a turn at copulating with the same female.

The gathering of many flag-waving females draws together chimpanzees, which forage alone or in small family groups for most of the year. It reestablishes bonds that unite a rather fluid community.

Infants and juveniles now play with one another and with the adults. Their aggressive and submissive jostlings help map out future dominance hierarchies. Flo happened to be a high-ranking female in the social hierarchy. Through her sexual attractiveness and direct solicitations, she built strong alliances with many males.

Figure 32.1 (**a**) Primatologist Jane Goodall in Gombe National Park, near the shores of Lake Tanganyika, scouting for chimpanzees. (**b**) Flo and two of her offspring, all subjects of long-term field observations. Flo helped Goodall clarify the central role of sex—and of the hormones that orchestrate it—in the social life of these primate relatives of humans.

b

Through her high status and aggressive behavior, she helped her male offspring win confrontations with other young male chimps.

The hormone-induced swelling during the fertile period of female chimps lasts somewhere between ten and sixteen days. Yet females are fertile for only one to five days. Sex hormones induce swelling even after a female gets pregnant. We can hypothesize that sexual selection has favored prolonged swelling. Males groom a sexually attractive female more often, protect her, give her more food, and let her tag along to new foraging sites. The more that males accept a female, the higher she rises in the social hierarchy—and, evolutionarily speaking, the more her individual offspring benefit.

Through their effects, hormones help orchestrate the growth, development, and reproductive cycles of nearly all animals, from the invertebrate worms to chimpanzees and humans. They influence minute-by-minute and day-to-day metabolic functions. Through their interplays with one another and with the nervous system, hormones profoundly influence the physical appearance, well-being, and behavior of individuals. Behavior, too, affects whether individuals will survive, either on their own or as part of social groups.

This chapter focuses primarily on hormones—on their sources, targets, and interactions as well as the mechanisms involved in their secretion. If the details start to seem remote, remember that this is the stuff of life. Hormones underwrote Flo's appearance, behavior, and rise through the chimpanzee social hierarchy. Just imagine what they've been doing for you.

Key Concepts

1. For nearly all animals, hormones and other signaling molecules have central roles in integrating the activities of individual cells in ways that benefit the whole body.

2. Only the cells with molecular receptors for a specific hormone are its targets. Hormones operate by serving as signals for change in the activities of their targets.

3. Many types of hormones influence gene transcription and protein synthesis in target cells. Other types call for alterations in existing molecules and structures in cells.

4. Some hormones work by binding with and altering membrane characteristics, such as the permeability of a plasma membrane to a particular solute. Others interact with a cell's DNA, either directly or after first binding with a receptor in the cytoplasm.

5. Some hormones help the body adjust to short-term shifts in diet and levels of activity. Others help induce long-term adjustments in cell activities that bring about the body's growth, development, and reproduction.

6. Among vertebrates, the hypothalamus and pituitary gland interact in ways that coordinate the activities of a number of endocrine glands. Together, they exert control over many aspects of the body's functioning.

7. Hormones, neural signals, shifts in local chemical conditions, and environmental cues function as triggers for the secretion of particular hormones.

THE ENDOCRINE SYSTEM

Hormones and Other Signaling Molecules

Throughout their lives, cells must respond to changing conditions by taking up and releasing various chemical substances. In vertebrates, the responses of millions to many billions of cells must become integrated in ways that help keep the whole body alive and functioning.

Signaling molecules help integrate activities within and between cells, and between tissues. We call them hormones, neurotransmitters, local signaling molecules, and pheromones. Each acts on target cells. A "target" is any cell that has receptors for a signaling molecule and that may change its activities in response to it. Targets may or may not be adjacent to cells that send signals.

By definition, **animal hormones** are the secretory products of endocrine glands, endocrine cells, and some neurons that the bloodstream delivers to nonadjacent target cells. They are this chapter's focus.

Neurotransmitters, released from axon endings of neurons, act swiftly on target cells by diffusing across a tiny synaptic cleft between them. You read about them in Section 30.3. **Local signaling molecules** released by many types of cells alter conditions in local tissues. For instance, some prostaglandins restrict or enhance blood flow to certain tissues. They do this by causing smooth muscle cells in blood vessel walls to contract or relax.

Pheromones, nearly odorless secretions of certain exocrine glands, diffuse through water or air to targets outside the animal body. These hormone-like secretions act on cells of other individuals of the same species and help integrate social behavior. For example, female silk moths secrete bombykol as a sex attractant, and soldier termites secrete alarm signals when ants attack their colony (Chapter 41). In vertebrates, the vomeronasal organ is a common pheromone detector. Even humans have one, but we don't know much about its function. Human pheromones or other chemical signals do affect our sexual behavior (Section 31.6). Do they also work at the subconscious level to trigger those inexplicable impressions, such as instant good or bad "feelings" about somebody you have just met? Maybe.

Discovery of Hormones and Their Sources

The word "hormone" dates to the early 1900s. W. Bayliss and E. Starling were trying to find out what triggers the secretion of pancreatic juices when food is traveling through the canine gut. As they knew, acids are mixed with food in the stomach, and the pancreas secretes an alkaline solution after the acidic mixture is propelled forward into the small intestine. Was the nervous system or something else stimulating the pancreatic response?

To find an answer, Bayliss and Starling blocked the nerves—but not blood vessels—leading to a laboratory

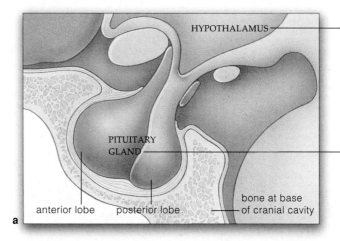

Figure 32.2 (**a**) A major neural–endocrine control center. The pituitary gland interacts closely with the hypothalamus, a part of the brain that also secretes some hormones. (**b**) *Facing page:* Overview of the key components of the human endocrine system and of the primary effects of their main hormonal secretions. The system also includes endocrine cells in many organs, such as the liver, kidneys, heart, small intestine, and skin.

animal's upper small intestine. Later, when acidic food entered the small intestine, the pancreas still secreted the alkaline solution. Even more telling, extracts of cells from the intestinal lining—a glandular epithelium—also induced the response. Glandular cells had to be the source of the pancreas-stimulating substance.

The substance came to be called secretin. Proof of its existence and mode of action confirmed a centuries-old idea: *The bloodstream picks up internal secretions that can influence the activities of organs inside the body.* Starling coined the word "hormone" for such internal glandular secretions (after *hormon*, meaning to set in motion).

Later on, researchers identified many other kinds of hormones and their sources. Figure 32.2 is a simplified picture of the locations of the following major sources of hormones for the human body. Bear in mind, most vertebrates have the same hormone sources:

Pituitary gland
Adrenal glands (*two*)
Pancreatic islets (*numerous cell clusters*)
Thyroid gland
Parathyroid glands (*in humans, four*)
Pineal gland
Thymus gland
Gonads (*two*)
Endocrine cells of the hypothalamus, stomach, small intestine, liver, kidneys, heart, placenta, skin, adipose tissue, and other organs

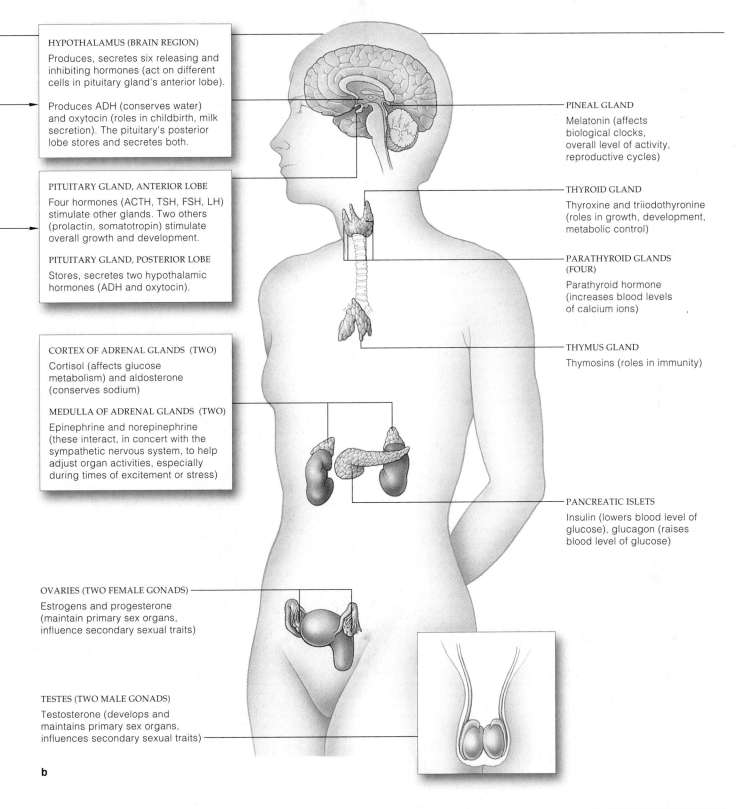

HYPOTHALAMUS (BRAIN REGION)

Produces, secretes six releasing and inhibiting hormones (act on different cells in pituitary gland's anterior lobe).

Produces ADH (conserves water) and oxytocin (roles in childbirth, milk secretion). The pituitary's posterior lobe stores and secretes both.

PITUITARY GLAND, ANTERIOR LOBE

Four hormones (ACTH, TSH, FSH, LH) stimulate other glands. Two others (prolactin, somatotropin) stimulate overall growth and development.

PITUITARY GLAND, POSTERIOR LOBE

Stores, secretes two hypothalamic hormones (ADH and oxytocin).

CORTEX OF ADRENAL GLANDS (TWO)

Cortisol (affects glucose metabolism) and aldosterone (conserves sodium)

MEDULLA OF ADRENAL GLANDS (TWO)

Epinephrine and norepinephrine (these interact, in concert with the sympathetic nervous system, to help adjust organ activities, especially during times of excitement or stress)

OVARIES (TWO FEMALE GONADS)

Estrogens and progesterone (maintain primary sex organs, influence secondary sexual traits)

TESTES (TWO MALE GONADS)

Testosterone (develops and maintains primary sex organs, influences secondary sexual traits)

PINEAL GLAND

Melatonin (affects biological clocks, overall level of activity, reproductive cycles)

THYROID GLAND

Thyroxine and triiodothyronine (roles in growth, development, metabolic control)

PARATHYROID GLANDS (FOUR)

Parathyroid hormone (increases blood levels of calcium ions)

THYMUS GLAND

Thymosins (roles in immunity)

PANCREATIC ISLETS

Insulin (lowers blood level of glucose), glucagon (raises blood level of glucose)

b

Collectively, the body's sources of hormones came to be called the **endocrine system**. The name implies there is a separate control system for the body, apart from the nervous system. (*Endon* means "within" and *krinein* is taken to mean "secrete.") Later, biochemical research and electron microscopy studies revealed that endocrine sources and the nervous system function in intricately connected ways, as you will see shortly.

Integration of cell activities depends on hormones and other signaling molecules. Each type of signaling molecule acts on target cells, which are any cells that have receptors for it and that may alter their activities in response to it.

Components of the endocrine system and certain neurons produce and secrete hormones, which the bloodstream takes up and distributes to nonadjacent target cells.

SIGNALING MECHANISMS

The Nature of Hormonal Action

Hormones and other signaling molecules interact with protein receptors of target cells. Their interactions have diverse effects on physiological processes. Some kinds of hormones induce a target cell to increase its uptake of glucose, calcium, or some other substance. Others stimulate or inhibit a target in ways that alter its rates of protein synthesis, modify the structure of proteins or other elements in the cytoplasm, or change cell shape.

To a large extent, responses to hormones depend on two factors. *First,* different hormones act on different mechanisms in target cells. *Second,* not all types of cells can respond to a given signal. For example, many have receptors for cortisol, so this hormone has widespread effects through the body. By contrast, only certain cells of small tubes in the kidneys have receptors for ADH, a hormone that initiates a highly directed response to a decrease in the volume of extracellular fluid.

With these points in mind, let's look briefly at the effects of two main categories of signaling molecules, the steroid and peptide hormones (Table 32.1).

Characteristics of Steroid Hormones

Steroid hormones are lipid-soluble molecules, derived from cholesterol, that are secreted by cells in adrenal glands and in primary reproductive organs, or gonads. Testosterone, one of the sex hormones, is an example.

Think of how testosterone affects secondary sexual traits associated with "maleness." These traits develop normally only if target cells have working receptors for testosterone. In *testicular feminization syndrome,* a genetic disorder, these receptors are defective. Genetically, the affected person is male (XY). He has functional testes that are secreting testosterone. But the target cells just can't respond to the hormone, so the secondary sexual traits that do develop are like those of females.

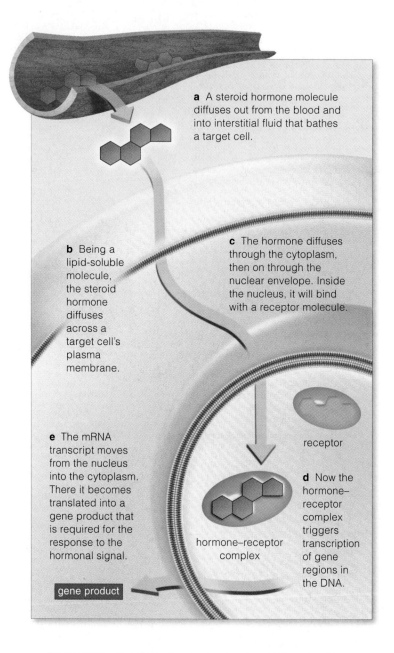

a A steroid hormone molecule diffuses out from the blood and into interstitial fluid that bathes a target cell.

b Being a lipid-soluble molecule, the steroid hormone diffuses across a target cell's plasma membrane.

c The hormone diffuses through the cytoplasm, then on through the nuclear envelope. Inside the nucleus, it will bind with a receptor molecule.

e The mRNA transcript moves from the nucleus into the cytoplasm. There it becomes translated into a gene product that is required for the response to the hormonal signal.

receptor

d Now the hormone–receptor complex triggers transcription of gene regions in the DNA.

hormone–receptor complex

gene product

Figure 32.3 Example of a mechanism by which a steroid hormone initiates changes in a target cell's activities.

Table 32.1 *Two Main Categories of Hormones*	
Type	Examples
Steroid and steroid-like hormones	Estrogens (feminizing effects), progestins (related to pregnancy), androgens (such as testosterone; masculinizing effects), cortisol, aldosterone. Thyroid hormones and vitamin D act like steroid hormones.
Peptide hormones	
Peptides	Glucagon, ADH, oxytocin, TRH
Proteins	Insulin, somatotropin, prolactin
Glycoproteins	FSH, LH, TSH

How does a steroid hormone operate? Being lipid soluble, it can diffuse directly across the lipid bilayer of a cell's plasma membrane (Figure 32.3). Once inside the cytoplasm, the hormone molecule commonly moves into the nucleus, where it binds to a receptor. In other cases, it binds with a receptor in the cytoplasm, then the hormone–receptor complex moves into the nucleus. The complex interacts with a specific region of DNA. Different complexes inhibit or stimulate transcription of gene regions into mRNA. Translation of such mRNA

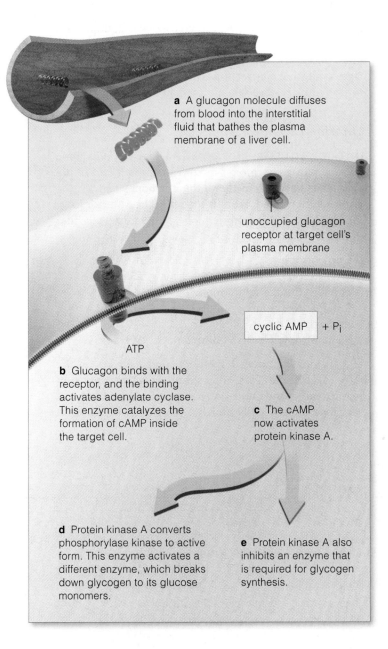

a A glucagon molecule diffuses from blood into the interstitial fluid that bathes the plasma membrane of a liver cell.

unoccupied glucagon receptor at target cell's plasma membrane

cyclic AMP $+ P_i$

ATP

b Glucagon binds with the receptor, and the binding activates adenylate cyclase. This enzyme catalyzes the formation of cAMP inside the target cell.

c The cAMP now activates protein kinase A.

d Protein kinase A converts phosphorylase kinase to active form. This enzyme activates a different enzyme, which breaks down glycogen to its glucose monomers.

e Protein kinase A also inhibits an enzyme that is required for glycogen synthesis.

Figure 32.4 Example of a mechanism by which a peptide hormone initiates changes in a target cell's activities. When the hormone glucagon binds at a receptor, it initiates reactions in the cell. In this case cyclic AMP, which is one type of second messenger, relays the signal into the cell interior.

vitamin D receptors. Thyroid hormones and vitamin D are not steroid hormones, but they behave like them.

Characteristics of Peptide Hormones

Traditionally, **peptide hormones** have been defined as water-soluble signaling molecules with 3 to 180 amino acids. Many peptides, polypeptides, and glycoproteins are in this category. When peptide hormones bind to receptors at the plasma membrane, specific membrane-bound enzyme systems become activated. These systems initiate reactions that lead to a cellular response.

Think of a liver cell. Glucagon, a peptide hormone, binds with a receptor that spans the plasma membrane and extends into the cytoplasm. Binding action triggers formation of a **second messenger**—a small molecule in the cytoplasm that relays signals from hormone–receptor complexes at the plasma membrane into the cell. cAMP (cyclic adenosine monophosphate) is this messenger.

An enzyme system at the plasma membrane starts the response to glucagon binding. Adenylate cyclase, an enzyme, is activated. It causes a cascade of reactions by converting ATP to many cyclic AMP molecules (Figure 32.4). These products serve as signals to convert many molecules of a protein kinase, another type of enzyme, to its active form. The kinases activate other enzymes, and so on until a final reaction converts the glycogen stored inside the cell to glucose. In short order, a huge number of molecules execute the hormonal response.

Or think of a muscle cell. It has receptors that bind insulin, a protein hormone. As one of its functions, an insulin–receptor complex induces glucose transporters to move through the cytoplasm and insert themselves into the plasma membrane. The transporters, a type of protein, help the muscle cell take up glucose faster. The signal also activates enzymes that help store glucose.

Bear in mind, there are other hormone categories, including the catecholamines such as epinephrine. Like glucagon, epinephrine binds with membrane receptors and triggers the release of a second messenger, cyclic AMP, that assists in the cellular response.

Hormones interact with receptors at the plasma membrane or inside the cytoplasm of target cells. Hormone binding directly influences protein synthesis in target cells.

Steroid hormones interact with a cell's DNA after entering the nucleus directly or after binding with a receptor in the cytoplasm. Some act by altering membrane properties.

Peptide hormones do not exert their effects by entering a cell. When they bind to a membrane receptor, the binding itself is a signal for enzyme-mediated, intracellular events. Often a second messenger inside the cytoplasm relays the signal into the cell interior.

---

transcripts results in enzymes and other proteins that can carry out a response to the hormonal signal.

Steroid hormones may exert effects in still another way. Some types have receptors on plasma membranes. They modify some function in a target cell by altering the properties of the membrane itself.

Part of a group of genes that specifies receptors for steroid hormones also specifies thyroid hormone and

THE HYPOTHALAMUS AND PITUITARY GLAND

From Section 30.9, you already know the **hypothalamus** is part of the forebrain's centers for homeostatic control of the internal environment, the viscera, and emotional states. Secretory neurons extend down into a slender stalk at its base, then into the lobed, pea-size **pituitary gland**. The hypothalamus and the pituitary interact as a major neural–endocrine control center.

The *posterior* lobe of the pituitary gland secretes two hormones that the hypothalamus produces. Its *anterior* lobe produces and secretes its own hormones. Most of these help control the release of hormones from other endocrine glands (Table 32.2). The pituitary of many vertebrates—*not* humans—has an intermediate lobe as well. In many cases, the third lobe secretes a hormone that governs reversible changes in skin or fur color.

Posterior Lobe Secretions

Figure 32.5*a* shows how hypothalamic neurons extend into the posterior lobe and end next to a capillary bed. The neurons produce antidiuretic hormone (ADH) and oxytocin, then store them in their axon endings. They secrete these hormones into interstitial fluid. Hormone molecules diffuse into capillaries, and the bloodstream distributes them through the body. ADH acts inside the kidneys. Its action adjusts the volume and composition of extracellular fluid in ways that help conserve water when necessary. Oxytocin has roles in reproduction. It triggers uterine contractions during childbirth and also induces milk release when offspring are being nursed.

Anterior Lobe Secretions

ANTERIOR PITUITARY HORMONES Other hypothalamic neurons release hormones next to a capillary bed in the stalk at its base. From there, the hormones flow into a *second* capillary bed, then diffuse into the anterior lobe. (Figure 32.6). Their anterior lobe targets are cells that secrete these hormones:

Corticotropin	ACTH
Thyrotropin	TSH
Follicle-stimulating hormone	FSH
Luteinizing hormone	LH
Prolactin	PRL
Somatotropin (growth hormone)	STH (or GH)

All of these hormones have widespread effects on the body. ACTH and TSH orchestrate the secretions from the adrenal glands and thyroid gland, respectively, as described shortly. FSH and LH have essential roles in

Table 32.2 *Hormones Released From the Mammalian Pituitary Gland*

Pituitary Lobe	Secretions	Designation	Main Targets	Primary Actions
POSTERIOR Nervous tissue (extension of hypothalamus)	Antidiuretic hormone	ADH	Kidneys	Induces water conservation as required during control of extracellular fluid volume and solute concentrations
	Oxytocin	OCT	Mammary glands	Induces milk movement into secretory ducts
			Uterus	Induces uterine contractions during childbirth
ANTERIOR Glandular tissue, mostly	Corticotropin	ACTH	Adrenal cortex	Stimulates release of cortisol, an adrenal steroid hormone
	Thyrotropin	TSH	Thyroid gland	Stimulates release of thyroid hormones
	Follicle-stimulating hormone	FSH	Ovaries, testes	In females, stimulates estrogen secretion, egg maturation; in males, helps stimulate sperm formation
	Luteinizing hormone	LH	Ovaries, testes	In females, stimulates progesterone secretion, ovulation, corpus luteum formation; in males, stimulates testosterone secretion, sperm release
	Prolactin	PRL	Mammary glands	Stimulates and sustains milk production
	Somatotropin (or growth hormone)	STH (GH)	Most cells	Promotes growth in young; induces protein synthesis, cell division; roles in glucose, protein metabolism in adults
INTERMEDIATE* Glandular tissue, mostly	Melanocyte-stimulating hormone	MSH	Pigmented cells in skin and other integuments	Induces color changes in response to external stimuli; affects some behaviors

* Present in most vertebrates (not humans). MSH is associated with the anterior lobe in humans.

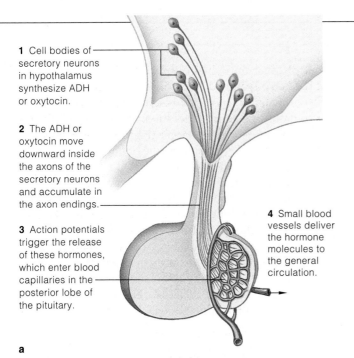

1 Cell bodies of secretory neurons in hypothalamus synthesize ADH or oxytocin.

2 The ADH or oxytocin move downward inside the axons of the secretory neurons and accumulate in the axon endings.

3 Action potentials trigger the release of these hormones, which enter blood capillaries in the posterior lobe of the pituitary.

4 Small blood vessels deliver the hormone molecules to the general circulation.

a

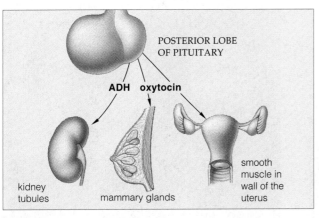

POSTERIOR LOBE OF PITUITARY

ADH **oxytocin**

kidney tubules

mammary glands

smooth muscle in wall of the uterus

b

Figure 32.5 (**a**) Functional links between the hypothalamus and the posterior lobe of the pituitary gland. (**b**) Main targets of the posterior lobe secretions.

1 Cell bodies of different secretory neurons in the hypothalamus secrete releasing and inhibiting hormones.

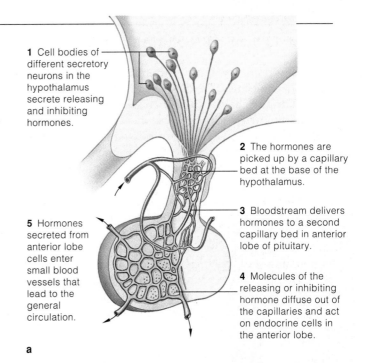

2 The hormones are picked up by a capillary bed at the base of the hypothalamus.

3 Bloodstream delivers hormones to a second capillary bed in anterior lobe of pituitary.

5 Hormones secreted from anterior lobe cells enter small blood vessels that lead to the general circulation.

4 Molecules of the releasing or inhibiting hormone diffuse out of the capillaries and act on endocrine cells in the anterior lobe.

a

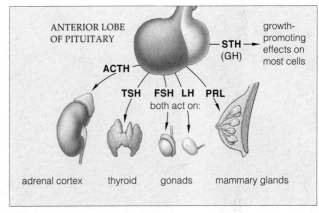

ANTERIOR LOBE OF PITUITARY

STH (GH) growth-promoting effects on most cells

ACTH

TSH **FSH** **LH** **PRL**
both act on:

adrenal cortex   thyroid   gonads   mammary glands

b

Figure 32.6 (**a**) Functional links between the hypothalamus and the anterior lobe of the pituitary gland. (**b**) Main targets of the anterior lobe secretions.

gamete formation and hormonal secretions required in sexual reproduction of animals, a topic of Chapter 39. Somatotropin affects metabolism in many tissues and triggers liver secretions that affect growth of bone and soft tissues (Table 32.2 and Figure 32.6b). Prolactin acts on different cell types. It's best known for stimulating and sustaining milk production in mammary glands, after other hormones prime the tissues. It also affects hormone production in the ovaries of some species.

ABOUT THE HYPOTHALAMIC TRIGGERS Most of the hypothalamic hormones acting in the anterior lobe of the pituitary are **releasers**. This means they stimulate secretion of hormones from target cells. For example, GnRH (gonadotropin-releasing hormone) brings about secretion of FSH and LH, which are gonadotropins. As

another example, TRH (thyrotropin-releasing hormone) stimulates secretion of thyrotropin. Other hypothalamic hormones acting in the anterior lobe are **inhibitors** of their targets. Somatostatin is one inhibitor. It is a signal to slow down somatotropin and thyrotropin secretion.

The hypothalamus and pituitary gland produce eight kinds of hormones and interact to control their secretion.

The posterior lobe of the pituitary stores and secretes two hypothalamic hormones, ADH and oxytocin, both of which target specific cell types.

The anterior lobe of the pituitary produces and secretes six hormones, ACTH, TSH, FSH, LH, PRL, and STH. These trigger release of other hormones from other glands, with a variety of effects throughout the body.

EXAMPLES OF ABNORMAL PITUITARY OUTPUT

The body does not churn out enormous quantities of hormone molecules. Two researchers, Roger Guilleman and Andrew Schally, realized this when they isolated the first known releasing hormone. In their four-year attempt to secure TRH, they dissected 500 metric tons of brains and 7 metric tons of hypothalamic tissue from sheep and ended up with only a single milligram of it.

Yet normal body functions depends on tiny but significant amounts of hormone secretions, which commonly are released in short bursts. The frequency of secretion must be controlled to block the underproduction or overproduction of a hormone. When something interferes with controls over hormonal output, the form and function of the body is altered, sometimes in abnormal ways.

For instance, *gigantism* results from overproduction of somatotropin. Affected adults are proportionally like average-size people but larger (Figure 32.7a,b). *Pituitary dwarfism* develops as an outcome of underproduction of somatotropin. Proportionally, affected adults are like an average person but much smaller (Figure 32.7b).

Sometimes somatotropin output becomes excessive in adults, when long bones have stopped lengthening. *Acromegaly* follows. In time, cartilage, bone, and other connective tissues in the jaws, feet, and hands thicken abnormally. So do epithelia of the skin, nose, eyelids, lips, and tongue. Figure 32.7c shows an affected female.

Another example: ADH secretion may dwindle or stop if the pituitary's posterior lobe is damaged, as by a

Figure 32.7 Examples of the outcome of abnormalities in the secretion of a hormone. (**a**) Manute Bol, an NBA center, is 7 feet 6–3/4 inches tall, an outcome of excessive secretion of somatotropin (STH) in childhood. (**b**) More examples of the effect of STH on body growth. The male at the center of this photograph is affected by gigantism, which resulted from excessive STH production in childhood. The male at the right is affected by pituitary dwarfism, an outcome of an underproduction of STH in childhood. The male at the left is of average height.

(**c**) Acromegaly, which resulted from excessive production of STH during adulthood. Before this female reached maturity, she was symptom-free.

blow to the head. This is one cause of *diabetes insipidus*. Symptoms include excessive excretion of dilute urine and life-threatening dehydration. Patients may respond to hormone replacement therapy involving injections or nasal spray applications of synthetic ADH.

Generally, endocrine glands release very small amounts of hormones at frequencies governed by control mechanisms. When controls fail, oversecretion or undersecretion follows and may cause alterations in the body's form or functions.

SOURCES AND EFFECTS OF OTHER HORMONES

Table 32.3 lists hormones from endocrine sources other than the pituitary. The rest of this chapter gives a few examples of their effects and of the controls over their output. As you read, keep the following points in mind.

First, hormones often interact with one another. One or more may oppose, add to, or prime target cells for another hormone's effects. *Second*, negative feedback mechanisms often control secretion. When a hormone's concentration increases or decreases in a body region, the change triggers events that respectively dampen or stimulate its further secretion. *Third*, a target cell may react differently at different times. Responses depend on the hormone's concentration and on the functional state of the cell's receptors. *Fourth*, environmental cues may be important mediators of hormonal secretion.

Hormone secretion and its effects depend on hormone interactions, feedback mechanisms, variations in the state of target cells, and sometimes environmental cues.

Table 32.3 *Hormone Sources Other Than the Mammalian Hypothalamus and Pituitary Gland*

Source	Secretion(s)	Main Targets	Primary Actions
ADRENAL CORTEX	Glucocorticoids (including cortisol)	Most cells	Promote breakdown of glycogen, fats, and proteins as energy sources; thus help raise blood level of glucose
	Mineralocorticoids (including aldosterone)	Kidney	Promote sodium reabsorption (sodium conservation); help control the body's salt–water balance
ADRENAL MEDULLA	Epinephrine (adrenaline)	Liver, muscle, adipose tissue	Raises blood level of sugar, fatty acids; increases heart rate and force of contraction
	Norepinephrine	Smooth muscle of blood vessels	Promotes constriction or dilation of certain blood vessels; thus helps control the flow of blood volume to different body regions
THYROID	Triiodothyronine, thyroxine	Most cells	Regulate metabolism; have roles in growth, development
	Calcitonin	Bone	Lowers calcium level in blood
PARATHYROIDS	Parathyroid hormone	Bone, kidney	Elevates calcium level in blood
GONADS			
Testes (in males)	Androgens (including testosterone)	General	Required in sperm formation, development of genitals, maintenance of sexual traits, growth, and development
Ovaries (in females)	Estrogens	General	Required for egg maturation and release; preparation of uterine lining for pregnancy and its maintenance in pregnancy; genital development; maintenance of sexual traits; growth, development
	Progesterone	Uterus, breasts	Prepares, maintains uterine lining for pregnancy; stimulates development of breast tissues
PANCREATIC ISLETS	Insulin	Liver, muscle, adipose tissue	Promotes cell uptake of glucose, thus lowers glucose level in blood
	Glucagon	Liver	Promotes glycogen breakdown, raises glucose level in blood
	Somatostatin	Insulin-secreting cells	Inhibits digestion of nutrients, hence their absorption from gut
THYMUS	Thymopoietin	T lymphocytes	Poorly understood regulatory effect on T lymphocytes
PINEAL	Melatonin	Gonads (indirectly)	Influences daily biorhythms, seasonal sexual activity
STOMACH, SMALL INTESTINE	Gastrin, secretin, etc.	Stomach, pancreas, gallbladder	Stimulate activities of stomach, pancreas, liver, gallbladder required for food digestion, absorption
LIVER	Somatomedins	Most cells	Stimulate cell growth and development
KIDNEYS	Erythropoietin	Bone marrow	Stimulates red blood cell production
	Angiotensin*	Adrenal cortex, arterioles	Helps control secretion of aldosterone (hence sodium reabsorption, and blood pressure)
	1,25-hydroxyvitamin D_6* (calcitriol)	Bone, gut	Enhances calcium reabsorption from bone and calcium absorption from gut
HEART	Atrial natriuretic hormone	Kidney, blood vessels	Increases sodium excretion; lowers blood pressure

* Kidneys produce *enzymes* that modify precursors of this substance, which enters the general circulation as an activated hormone.

FEEDBACK CONTROL OF HORMONAL SECRETIONS

What controls hormonal secretion? A look at a few of the endocrine glands listed in Table 32.3 will give you a sense of some of the control mechanisms. Briefly, the hypothalamus, pituitary, or both signal a gland to step up or slow down its secretory activity, so a hormone's concentration changes in the blood or elsewhere. With that change, a feedback mechanism trips into action.

By **negative feedback**, a rise in the concentration of a secreted hormone triggers events that *inhibit* further secretion. By **positive feedback**, a rising concentration of a secreted hormone triggers events that *stimulate* its further secretion.

Negative Feedback From the Adrenal Cortex

Humans have a pair of adrenal glands, one above each kidney (Figure 32.2b). Some cells of the **adrenal cortex**, an adrenal gland's outer layer, secrete glucocorticoids and other hormones. Glucocorticoids raise the blood level of glucose. One type, cortisol, acts when the body is so stressed that the glucose level declines below a set point that acts as an alarm. It starts a stress response, which a negative feedback mechanism can later cut off.

Figure 32.8 shows what goes on. The hypothalamus secretes the releasing hormone CRH when a too-low level of glucose in blood sets off the alarm. CRH makes the anterior pituitary secrete corticotropin (ACTH). In turn, ACTH causes cells of the adrenal cortex to secrete cortisol. In response to that cortisol, liver cells break down their stored glycogen, adipose cells break down fat, and skeletal muscle cells break down proteins. The breakdown products—glucose, fatty acids, and amino acids—enter blood. Except for brain cells, cells take up all three, not just glucose, as energy sources (Section 7.6). With more glucose and alternative energy sources available, the glucose level rises above the set point. If the threat to homeostasis ends, the hypothalamus and pituitary will issue signals to inhibit cortisol secretion.

Chronic stress, injury, or illness causes the nervous system to keep this response going. Why? Prolonged inflammation damages tissues, and cortisol suppresses the inflammation. Hence the use of cortisol-like drugs for asthma and other chronic inflammatory disorders.

Local Feedback in the Adrenal Medulla

Some neurons reside in the **adrenal medulla**, the adrenal gland's inner region. Their secretions, epinephrine and norepinephrine, are neurotransmitters in some contexts and hormones in others. Sympathetic nerves deliver hypothalamic signals to the adrenal medulla. Suppose they call for norepinephrine's release. Soon, hormone molecules collect in the synaptic cleft between target cells and axon endings from the nerve. A local negative feedback mechanism now operates at receptors on the axon endings. Excess amounts of norepinephrine bind to the receptors and shut down its further release.

When you are excited or stressed, epinephrine and norepinephrine help adjust blood circulation, and fat and carbohydrate metabolism. They boost heart rate, make arteriole diameters widen or narrow in different regions, and dilate airways to the lungs. These highly controlled events direct more of the total blood volume to heart and muscle cells that are to increase oxygen flow to cells demanding a lot of energy. These are the features of the *fight–flight response* (Section 30.8).

Cases of Skewed Feedback From the Thyroid

The human **thyroid gland** is located at the base of the neck in front of the trachea, or windpipe (Figures 32.2b and 32.9a–c). Thyroxine and triiodothyronine, its main hormones, have widespread effects. In their absence, many tissues cannot develop normally. Also, metabolic rates of warm-blooded animals depend on them. The importance of feedback control of their secretion comes into sharp focus in cases of abnormal thyroid output.

STIMULUS: Body is stressed; huge demand for glucose results in too-low glucose level in blood.

+ → HYPOTHALAMUS → **−**

CRH

ANTERIOR PITUITARY

ACTH

adrenal cortex

cortisol

adrenal cortex
adrenal medulla
adrenal gland
kidney

If stress ends, hypothalamus and pituitary will detect rise in blood glucose level and inhibit further cortisol secretion.

1. Blood glucose uptake inhibited in many tissues, especially muscles (but not the brain).

2. Proteins degraded in many tissues, especially muscles. Free amino acids converted to glucose, also used to synthesize or repair cell structures.

3. Fats in adipose tissue degraded to fatty acids, which are released to the blood as alternative energy sources (conserves blood glucose for brain).

More glucose as well as alternative sources of energy enter the blood.

Figure 32.8 Structure of the human adrenal gland. One gland rests on top of each kidney. The diagram shows a negative feedback loop that governs cortisol secretion from the adrenal cortex.

Figure 32.9 (**a**) Human thyroid gland, anterior view. (**b**) Posterior view, showing the location of four parathyroid glands. (**c**) Position of the thyroid in the body. (**d**) A mild case of goiter, displayed by Maria de Medici in the year 1625. A rounded neck was considered to be a sign of great beauty during the late Renaissance. It showed up regularly in parts of the world where iodine supplies were not sufficient for normal thyroid function.

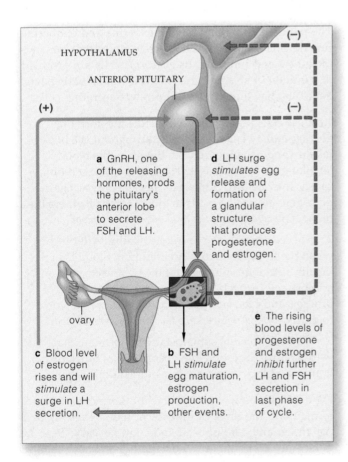

Figure 32.10 Feedback loops to the hypothalamus and the pituitary gland from the ovaries during the menstrual cycle, a recurring reproductive event. Positive feedback triggers egg release from an ovary. Negative feedback after its release prevents release of another egg until the cycle is over.

Example: Thyroid hormones cannot be synthesized without iodide, a form of iodine. Iodine-deficient diets cause one or both lobes of the thyroid gland to enlarge (Figure 32.9d). The enlargement, a *simple goiter*, occurs after low blood levels of thyroid hormones cause the anterior pituitary to secrete THS (thyroid-stimulating hormone). The thyroid attempts to make the hormones but cannot do so, which leads to continued secretion of TSH, and so on in a sustained, abnormal feedback loop. *Hypothyroidism* is the clinical name for low blood levels of thyroid hormones. Affected adults may be sluggish, overweight, intolerant of cold, dry-skinned, confused, and depressed. Simple goiter is no longer common in places where people use iodized salt.

Grave's disorder and other forms of *toxic goiter* result from excessive levels of thyroid hormones in blood, a condition we call *hyperthyroidism*. Symptoms include heat intolerance, irritability, anxiety, tremors, fatigue, difficulty sleeping, protruding eyes, and an irregular, pounding heartbeat. Some of the cases are autoimmune disorders; antibodies wrongly stimulate thyroid cells (Section 35.9). Other cases arise after the thyroid has become inflamed or develops nodules or tumors.

Feedback Control of the Gonads

Gonads are *primary* reproductive organs, which make and secrete gametes and certain sex hormones. Testes (singular, testis) in males and ovaries in females are examples. Testes secrete testosterone; ovaries secrete estrogens and progesterone. These hormones influence secondary sexual traits (as they did for those chimps described earlier), and feedback controls govern their secretion. Figure 32.10 is a preview of the feedback loops from ovaries to the hypothalamus and pituitary during the menstrual cycle, a key topic of Chapter 39.

Feedback mechanisms control secretions from endocrine glands. In many cases, feedback loops to the hypothalamus, pituitary, or both govern the secretory activity.

Negative feedback slows down the further secretion of a hormone. Positive feedback enhances its further secretion.

RESPONSES TO LOCAL CHEMICAL CHANGES

Hormones and neural signals are not the main controls over some endocrine glands or cells. Chemical changes in the local neighborhood stimulate or inhibit hormone secretions, too, as the following examples illustrate.

Secretions From Parathyroid Glands

Humans have four **parathyroid glands** on the posterior surface of the thyroid gland (Figure 32.9*b*). The glands secrete parathyroid hormone (PTH), the key regulator of calcium levels in blood. Calcium ions have roles in muscle contraction, enzyme action, blood clotting, and other activities. PTH is secreted when the calcium level is low. Its secretion slows when the calcium level rises. PTH acts on cells present in the skeleton and kidneys.

PTH induces living bone cells to secrete enzymes that digest bone tissue, and this releases calcium and other minerals into interstitial fluid. Calcium enters blood, then it enters nephrons in the kidneys, where PTH enhances its reabsorption. PTH also induces some kidney cells to secrete enzymes that act on blood-borne precursors of an active form of vitamin D_3, a hormone (Section 33.2). The activated form stimulates intestinal cells to absorb much more calcium from the lumen of the gut. In children who have vitamin D deficiency, not enough calcium and phosphorus are absorbed, so rapidly growing bones develop improperly. The resulting bone abnormality, *rickets*, is characterized by bowed legs, a malformed pelvis, and in many cases a malformed skull and rib cage (Figure 32.11).

Figure 32.11 A child who is affected by rickets.

Effects of Local Signaling Molecules

Many cells detect changes in the surrounding chemical environment and alter their activity, often in ways that counteract or amplify those changes. The cells secrete various local signaling molecules, the effects of which are confined to the immediate vicinity of change. Target cells take up most signaling molecules so rapidly that few enter the general circulation.

At least sixteen types of prostaglandins, each having a twenty-carbon fatty acid structure with a five-carbon ring, occur in many tissues. Production and secretion of these local signaling molecules rise as local chemical conditions change, with potent results. Some make rings of smooth muscle in arteriole walls constrict or dilate; this helps redirect blood flow through the body. Others have roles in inflammation, control of stomach acid and intestinal motility, and childbirth. They also function in overall endocrine control. They might have therapeutic use for asthma, ulcers, hypertension, and heart attacks.

Those growth factors you read about in Section 14.4 are local signaling molecules. They affect transcription of genes required in development. Example: Rita Levi-Montalcini discovered one nerve growth factor (NFG) that helps neurons survive and guides their direction of growth in an embryo. We know that in the presence of NFG, immature neurons can survive indefinitely in tissue culture. Without it, they die within a few days.

Secretions From Pancreatic Islets

The pancreas is a gland having exocrine and endocrine functions. Its *exocrine* cells secrete digestive enzymes into the small intestine. Its *endocrine* cells are grouped in 2 million or so small clusters called **pancreatic islets**. Each islet has three types of hormone-secreting cells:

1. *Alpha* cells in the pancreas secrete the hormone glucagon. Between meals, cells throughout the body take up and use glucose from blood, so the blood level of glucose decreases. Then, glucagon secretion causes liver cells to convert its storage polysaccharide glycogen and amino acids to glucose, which enters the blood. *Glucagon raises the level of glucose in blood.*

2. *Beta* cells secrete the hormone insulin. After meals, when the level of glucose circulating in blood is high, insulin stimulates glucose uptake by muscle cells and adipose cells especially. It promotes the synthesis of proteins and fats, and it inhibits protein conversion to glucose. *Insulin lowers the level of glucose in blood.*

3. *Delta* cells secrete somatostatin, a hormone that helps control digestion and absorption of nutrients. It also can block secretion of insulin and glucagon.

Figure 32.12 shows how pancreatic hormones interact to maintain the level of glucose in blood even though the times and amounts of food intake vary. Bear in mind, insulin is the only hormone that causes cells to take up and store glucose in forms that can be swiftly tapped when required. Its crucial role in carbohydrate, protein, and fat metabolism becomes clear when we consider people who cannot produce enough insulin or whose target cells cannot respond to it.

Figure 32.12 Some of the homeostatic controls over glucose metabolism.

Following a meal, glucose enters the blood faster than cells can take it up. The blood glucose level rises, and the chemical change stimulates pancreatic beta cells to secrete insulin. The main targets, liver and muscle cells, use glucose at once and store excess amounts as glycogen. In this way, insulin *lowers* the glucose level in blood.

Between meals, the blood glucose level decreases, and the chemical change stimulates pancreatic alpha cells to secrete glucagon. Target cells with receptors for this hormone convert glycogen back to glucose, which enters the blood. In this way, glucagon *raises* the glucose level in blood.

The nervous system helps control glucose metabolism. For example, as you read earlier, the hypothalamus orders the adrenal medulla to secrete glucocorticoids when the body is being stressed. As one effect, glycogen is broken down to glucose in the liver, and glycogen synthesis slows, in liver and muscle tissue especially.

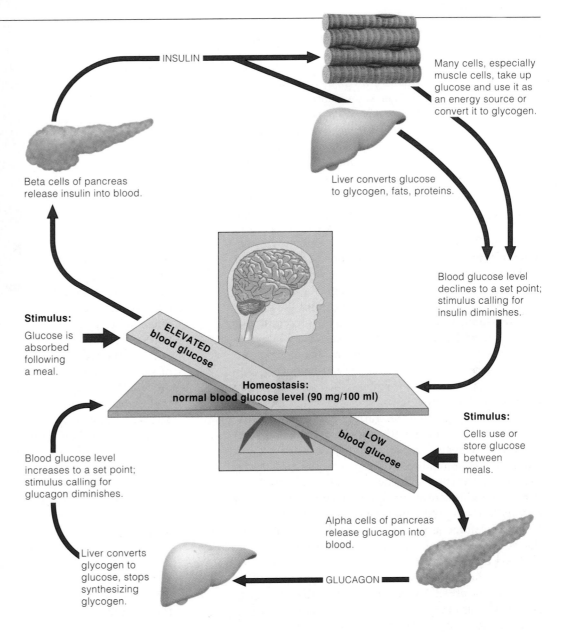

INSULIN

Many cells, especially muscle cells, take up glucose and use it as an energy source or convert it to glycogen.

Beta cells of pancreas release insulin into blood.

Liver converts glucose to glycogen, fats, proteins.

Blood glucose level declines to a set point; stimulus calling for insulin diminishes.

Stimulus:
Glucose is absorbed following a meal.

ELEVATED blood glucose

Homeostasis:
normal blood glucose level (90 mg/100 ml)

LOW blood glucose

Stimulus:
Cells use or store glucose between meals.

Blood glucose level increases to a set point; stimulus calling for glucagon diminishes.

Liver converts glycogen to glucose, stops synthesizing glycogen.

Alpha cells of pancreas release glucagon into blood.

GLUCAGON

For example, insulin deficiency may lead to *diabetes mellitus*. In this disorder, excess glucose accumulates in blood, then in urine. Urination becomes excessive, so the body's water–solute balance is disrupted. Affected people become dehydrated and thirsty, abnormally so. Without a steady supply of glucose, their body cells start depleting their own fats and proteins as sources of energy. Weight loss is one outcome. Another is that ketones accumulate in the blood and urine. Ketones are normal acidic products of fat breakdown. When they accumulate, they contribute to water losses and alter the body's acid–base balance. Such imbalances disrupt brain function. In extreme cases, death follows.

In *type 1 diabetes*, the body mounts an autoimmune response against its insulin-secreting beta cells. White blood cells mistakenly identify the beta cells as foreign and kill them. A combination of genetic susceptibility

and environmental factors gives rise to this disorder, which is less common than other forms of diabetes but more dangerous in the short term. Usually, symptoms first appear in childhood and adolescence (the disorder is also called juvenile-onset diabetes). Type 1 diabetic patients survive with insulin injections.

In *type 2 diabetes*, insulin levels are close to or above normal, but target cells cannot respond to it. Symptoms usually develop in middle age, for beta cells of affected people produce less insulin over time. Affected people lead normal lives by controlling diet and weight. Often prescription drugs enhance insulin action or secretion.

Secretions from some endocrine glands and endocrine cells are direct homeostatic responses to chemical change in the local neighborhood.

HORMONES AND THE ENVIRONMENT

This last section of the chapter invites you to reflect on a key point. An individual's growth, development, and reproduction begin with genes and hormones, and so does behavior. *But certain environmental factors commonly influence gene expression and hormonal secretion, and they do so in predictable ways.* Later chapters invite analysis of specific environmental effects on animals. For now, it is enough to consider the following examples.

Daylength and the Pineal Gland

Inside the brain is the **pineal gland**, a photosensitive organ (Figure 32.2). It secretes the hormone melatonin in the absence of light, so the blood level of melatonin varies from day to night and through the seasons. In many species, these variations help control the growth and development of gonads, the primary reproductive organs. How? Their melatonin is a piece of a **biological clock**, an internal timing mechanism. It helps control reproductive cycles and reproductive behavior.

Think of a hamster in winter, when there are more hours of darkness than in summer. The melatonin level in its blood is high, and this suppresses sexual activity. In summer, when daylength is longest, the blood level is low and hamster sex peaks. Similarly, in the fall and winter, melatonin indirectly suppresses the growth of a male white-throated sparrow's gonads (Figure 32.13*a*). When daylight increases in spring, stepped-up gonadal activity leads to production of hormones that influence singing behavior, as described in Section 41.1. With his distinctive song, the male sparrow defines his territory.

Does melatonin influence human behavior as well? Perhaps. Clinical observations and studies suggest that decreased melatonin levels may trigger **puberty**, the age at which reproductive organs and structures start to mature. For example, some patients who did not have a functional pineal gland entered puberty prematurely.

Figure 32.13 (**a**) Male white-throated sparrow, belting out a song that began, indirectly, with an environmentally induced decrease in melatonin secretion from the pineal gland. (**b**) Annie blanketing her winter blues.

Melatonin also acts on neurons that make the body temperature decline and make you drowsy after sunset, when light wanes. At sunrise, less melatonin is secreted and body temperature rises. In response, you wake up and become active. A biological clock governs the cycle of sleep and arousal. This one ticks in synchrony with daylength. We have indirect evidence of its action from night workers who cannot get to sleep in the morning. The same thing happens to travelers from the United States to Paris who go through four days of "jet lag." Two or three hours past midnight they sit up in bed, but two hours past noon they are ready for sleep.

Seasonal affective disorder (SAD) hits some people in winter. They become exceedingly depressed, binge on carbohydrates, and have an almost overwhelming need to sleep (Figure 32.13*b*). Such "winter blues" might develop when a biological clock is out of sync with the seasonally shorter daylengths. Clinically administered melatonin worsens the seasonal symptoms. Exposure to intense light, which shuts down pineal activity, often leads to dramatic improvement.

Thyroid Function and Frog Habitats

Since 1994, in natural habitats around the world, the number of deformed frogs has been skyrocketing. The reasons are not fully understood. In 1999, researchers decided to expose *Xenopus laevis* embryos to Minnesota and Vermont lakewater. Half the samples came from lakes where deformity rates were low. The other half came from "hot spots" with as many as twenty kinds of dissolved pesticides—and with high deformity rates.

When the frog embryos developed into tadpoles, the ones that had been raised in the hot-spot water had bent spines, malformed eyes, and a malformed mouth. Some did not metamorphose into adults (Figure 32.14). The other frog embryos of the sampling were normal.

Thyroid hormones orchestrate much of vertebrate development. For instance, a surge in thyroid hormone output stimulates a tadpole to metamorphose into an

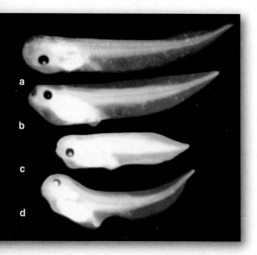

Figure 32.14 *Xenopus laevis* tadpoles showing possible environmental effects on the thyroid, hence on development. (**a**) Tadpole raised in water taken from a lake where there were few deformed frogs. (**b**–**d**) Three tadpoles raised in water taken from three "hot spot" lakes that had increasingly higher concentrations of dissolved chemical compounds, both natural and synthetic.

| absence of suitable stimulus | | suitable internal signals | suitable environmental cues |

```
absence of              suitable      suitable
suitable stimulus       internal      environmental
                        signals       cues
       ↓                       ↓          ↓
hormone-secreting              brain regions
neurons of X organ
produce, release               ↓
molt-inhibiting
hormone (MIH)           signals inhibit MIH
                        production in X organ;
       ↓                MIH level in blood drops
MIH stored in sinus            ↓
gland, then released
into blood              Y organ is free to
                        produce ecdysone
       ↓
MIH stops cells in Y           ↓
organ, an endocrine
gland located near      MOLTING STIMULATED
mouthparts or at base
of antennae, from
producing ecdysone

       ↓
MOLTING INHIBITED

a                       b
```

Figure 32.15 (**a,b**) Steps in the hormonal control of molting in crustaceans, including crabs (**c**). The steps differ a bit in insects, which do not use a molt-inhibiting hormone. Rather, stimulation of the insect brain makes certain neurons secrete ecdysiotropin. This hormone induces other neurons to make and secrete still another hormone, which targets ecdysone-producing cells in prothoracic glands. (**d**) This insect, one of the cicadas, is emerging from its old cuticle.

adult. Are conditions in hot-spot habitats interfering with thyroid functions? Maybe. Embryos that developed in hot-spot water displayed few symptoms, or none at all, when they were given extra thyroid hormones.

Other studies correlated some frog deformities with ultraviolet radiation and with infections by a parasitic protistan (Section 24.6). "Chemical soup" habitats that disrupt thyroid function are now added to the list.

Comparative Look at a Few Invertebrates

Although this chapter's focus has been on vertebrates, do not lose sight of the fact that all organisms produce signaling molecules of one sort or another. Let's look at hormonal control of **molting**, a periodic discarding and replacement of a hardened cuticle that otherwise would limit increases in body mass. Molting occurs in the life cycle of all insects, crustaceans, and other invertebrates with thick cuticles (Sections 23.11, 23.13, and 23.15).

Although details vary from group to group, molting is largely under the control of **ecdysone**. This steroid hormone is derived from cholesterol and is chemically related to many crucial vertebrate hormones. In insects and crustaceans, molting glands synthesize and store

ecdysone and then release it for distribution through the body at molting time. Hormone-secreting neurons in the brain seem to control its release. Apparently they respond to a combination of environmental cues, such as light and temperature, as well as internal signals.

Figure 32.15 provides examples of the control steps, which differ in crustaceans and insects. During premolt and molting periods, coordinated interactions among ecdysone and other hormones bring about structural and physiological changes. The interactions make the old cuticle separate from the epidermis and muscles. They induce changes that dissolve inner layers of the cuticle and recycle the remnants. The interactions also trigger changes in metabolism and in the composition and volume of the internal environment. They promote cell divisions, secretions, and pigment formation, all of which go into making the new cuticle. Simultaneously, hormonal interactions control heart rate, muscle action, color changes, and other physiological processes.

Environmental cues, such as changes in light intensity from day to night and seasonal changes in daylength, influence certain hormonal secretions.

SUMMARY

Gold indicates text section

1. The cells of complex animals continually exchange substances with the body's internal environment. Their withdrawals and secretions are integrated in ways that ensure cell survival through the whole body. *32.1*

2. Integration of cell activities requires the stimulatory or inhibitory effects of signaling molecules. *32.1, 32.2*

   a. Signaling molecules are chemical secretions from a cell that adjust the behavior of other, target cells.

   b. Any cell with molecular receptors for a signaling molecule is its target. Target cells may or may not be adjacent to the cell that sends the signal.

   c. There are different kinds of signaling molecules. Hormones as well as neurotransmitters, local signaling molecules, and pheromones are the main kinds.

3. In target cells, hormones influence gene activation, protein synthesis, and alterations in existing enzymes, membranes, and other cellular components. Hormones exert physiological effects by interacting with specific protein receptors at the plasma membrane or inside the cytoplasm of target cells. *32.2*

   a. Steroid hormones enter the target cell's nucleus directly, or after binding with an intracellular receptor, or by binding with plasma membrane receptors.

   b. Being water soluble, the protein hormones cannot enter target cells. They bind to membrane receptors at the cell surface. Their effect is exerted with the help of transport proteins as well as second messengers in the cytoplasm, some of which trigger the actual response.

4. The posterior lobe of the pituitary stores and secretes two hypothalamic hormones, ADH and oxytocin. ADH targets kidney cells, thus influencing extracellular fluid volume. Oxytocin acts on cells in mammary glands and the uterus to influence reproductive events. *32.3*

5. Hypothalamic hormones known as releasing and inhibiting hormones control secretions from a variety of cells of the anterior lobe of the pituitary gland. *32.3*

6. The anterior lobe makes and secretes six hormones: ACTH, TSH, FSH, LH, PRL, and STH. These trigger secretion from the adrenal cortex, the thyroid gland, gonads, and mammary glands. They elicit a variety of responses throughout the body. *32.3*

7. The vertebrate body has other sources of hormones (e.g., adrenal medulla; parathyroid, thymus, and pineal glands; the pancreatic islets; and endocrine cells in the stomach, small intestine, liver, and heart). *32.5*

8. Interactions among hormones, feedback mechanisms, the number and kind of target cell receptors, variations in the state of target cells, and often environmental cues influence secretion of a hormone and its effects. *32.6*

9. The secretion of local signaling molecules, such as prostaglandins, is a direct response to a change in the localized chemical environment. *32.7*

   a. In general, secretion of hormones such as insulin and parathyroid hormone can change rapidly when the extracellular concentration of some substance must be homeostatically controlled.

   b. Hormones such as somatotropin have prolonged, slow, and often irreversible effects, as on development.

10. Environmental cues, such as the change in sunlight intensity from day to night and the seasonal changes in daylength, influence some hormonal secretions. *32.8*

Review Questions

1. Name the endocrine glands typical of most vertebrates and state where each is located in the human body. *32.1*

2. Distinguish among hormones, neurotransmitters, local signaling molecules, and pheromones. *32.1*

3. A hormone molecule binds with a receptor on a plasma membrane. It does not enter the cell. Binding activates a second messenger in the cell, which triggers an amplified response to the hormonal signal. State whether the molecule is a steroid hormone or a peptide hormone. *32.2*

4. Which secretions of the posterior lobe of the pituitary gland have the targets indicated? (Fill in the blanks.) *32.3*

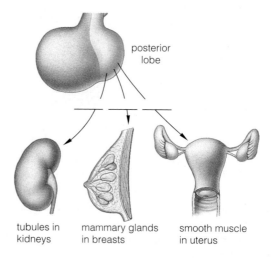

posterior lobe

tubules in kidneys   mammary glands in breasts   smooth muscle in uterus

5. Which secretions of the anterior lobe of the pituitary gland have the targets indicated? (Fill in the blanks.) *32.3*

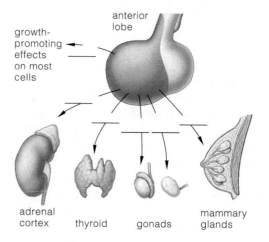

anterior lobe

growth-promoting effects on most cells

adrenal cortex   thyroid   gonads   mammary glands

1. _____ are molecules released from a signaling cell that have effects on target cells.
   a. Hormones
   b. Neurotransmitters
   c. Pheromones
   d. Local signaling molecules
   e. both a and b
   f. a through d

2. Hormones are products of _____ .
   a. endocrine glands
   b. some neurons
   c. exocrine cells
   d. a and b
   e. a and c
   f. a, b, and c

3. Second messengers include _____ .
   a. steroid hormones
   b. protein hormones
   c. cyclic AMP
   d. both a and b

4. ADH and oxytocin are hypothalamic hormones secreted from the _____ lobe of the pituitary gland.
   a. anterior
   b. posterior
   c. intermediate
   d. secondary

5. GnRH is a _____ secreted by hypothalamic neurons.
   a. releasing hormone
   b. inhibiting hormone
   c. corticotropin
   d. somatotropin

6. Which do *not* stimulate hormone secretions?
   a. neural signals
   b. local chemical changes
   c. hormonal signals
   d. environmental cues
   e. All of the above can stimulate secretion.

7. _____ lowers blood sugar levels; _____ raises it.
   a. Glucagon; insulin
   b. Insulin; glucagon
   c. Gastrin; insulin
   d. Gastrin; glucagon

8. The pituitary detects a rising hormone concentration in blood and inhibits the gland secreting the hormone. This is a _____ feedback loop.
   a. positive
   b. negative
   c. long-term
   d. b and c

9. Match the hormone source with the closest description.
   _____ adrenal medulla
   _____ thyroid gland
   _____ parathyroids
   _____ pancreatic islets
   _____ pineal gland
   _____ prostaglandin
   a. affected by daylength
   b. potent local effects
   c. raise blood calcium level
   d. epinephrine source
   e. insulin, glucagon
   f. hormones require iodide

Critical Thinking

1. The zebra offspring being nursed in Figure 32.16 is too young to nourish itself by eating grasses. Its source of nutrients is its mother's milk. Outline how secretions from the hypothalamus and the pituitary gland influence milk production and secretion.

2. In winter, with its fewer daylight hours than in the summer, Maxine became exceedingly depressed, craved carbohydrate-rich foods, and stopped exercising regularly. She put on a great deal of weight. Her doctor diagnosed her condition as *seasonal affective disorder* (SAD), the winter blues. Maxine was advised to purchase a cluster of intense, broad-spectrum lights and sit near them for at least an hour each day. The treatment quickly lifted the cloud of depression. Use your knowledge of the secretory activity of the pineal gland to explain why Maxine's symptoms appeared and why the prescribed therapy worked.

3. Marianne is affected by *type 1 insulin-dependent diabetes*. One day, after injecting herself with too much insulin, she starts to shake and feels confused. Her doctor recommends a glucagon injection. What caused her symptoms? How would an injection of glucagon help?

Figure 32.16 Female zebra nursing her offspring.

4. In one application of recombinant DNA technology, growth hormone—somatotropin—is now commercially available for treating pituitary dwarfism. Although it's illegal to do so, some athletes use somatotropin instead of anabolic steroids. (Refer to Section 33.9, *Critical Thinking* question 4.) Why? Somatotropin can't be detected by drug test procedures currently employed in sports medicine. Explain how athletes might believe this hormone can improve their performance.

5. *Osteoporosis* is a condition in which loss of calcium results in thin, brittle bones. Combined with other treatments, vitamin D_3 injections are sometimes recommended. Explain why.

Selected Key Terms

adrenal cortex *32.6*	neurotransmitter *32.1*
adrenal medulla *32.6*	pancreatic islet *32.7*
biological clock *32.8*	parathyroid gland *32.7*
ecdysone *32.8*	peptide hormone *32.2*
endocrine system *32.1*	pheromone *32.1*
gonad *32.6*	pineal gland *32.8*
hormone, animal *32.1*	pituitary gland *32.3*
hypothalamus *32.3*	positive feedback *32.6*
inhibitor (hypothalamic) *32.3*	puberty *32.8*
local signaling	releaser (hypothalamic) *32.3*
molecule *32.1*	second messenger *32.2*
molting *32.8*	steroid hormone *32.2*
negative feedback *32.6*	thyroid gland *32.6*

Readings

Goodall, J. 1986. *The Chimpanzees of Gombe.* Cambridge, Massachusetts: Belknap Press of Harvard University Press.

Goodman, H. 1994. *Basic Medical Endocrinology.* Second edition. New York: Raven Press.

Hadley, M. 1995. *Endocrinology.* Fourth edition. Englewood Cliffs, New Jersey: Prentice-Hall.

Raloff, J. 2 October 1999. "Thyroid Linked to Some Frog Defects." *Science News* 156:212.

Sherwood, L. 1997. *Human Physiology.* Third edition. Belmont, California: Wadsworth.

On-Line readings at Student Guide for InfoTrac:
www.brookscole.com/biology

33

PROTECTION, SUPPORT, AND MOVEMENT

Of Men, Women, and Polar Huskies

In 1989 Will Steger and his dog-sled team walked on ice for seven months, enduring temperatures of −113°F and a blizzard that lasted for more than seven weeks. They crossed Antarctica, all 6,023 kilometers (3,741 miles) of it. In 1995 this legendary polar explorer set out with four men, two women, and thirty-three sled dogs to cross 3,220 kilometers of the Arctic Ocean in one season. Ice blankets this northernmost ocean in winter and thins treacherously during the spring thaw. The sled dogs stayed with the team for two-thirds of the journey. Steger flew them out only when the team had to cross too much melting ice and switched to using canoes.

To Steger's mind, polar huskies were the heroes of the crossings, the members of the team that worked hardest and pulled all the weight (Figure 33.1). The traits of this mixed breed were modified through years of artificial selection among Canadian and Greenland huskies (bred for size and strength), Siberian huskies (bred for intelligence), and Alaskan racing dogs (bred for spirit and endurance).

A husky's leg bones are sturdy yet lightweight. Its forelegs move freely, thanks to a deep but not too broad rib cage. You won't see the massive muscles of its hind legs in sprinting greyhounds or cheetahs. They are the muscles of a load-pulling, long-distance runner. The husky has tough, calloused foot pads—cushions against sharp ice and frozen rock. Like many other mammals, it has a fur coat with underhair: a dense, soft, insulative layer that traps heat. Its coarser, longer, slightly oily guard hairs protect the underhair from wear and tear. On winter nights, a husky gets comfortable and covers its nose with its furry tail, oblivious of drifting snow.

Steger and his teammates, Victor Boyarsky, Julie Hanson, Martin Hignell, Paul Pregont, and Takako Takano, did not even approach the polar husky's stamina and built-in protection against the elements. Long before the polar crossings, the team followed a regimen of diet and exercise to put their arm and leg muscles in peak condition. Human legs are adapted for long-distance walking, not load-pulling motion.

Figure 33.1   In Ely, Minnesota, Will Steger and his polar huskies warming up for an Arctic crossing.

Figure 33.2 From left to right, overview of the human integumentary, muscle, and skeletal systems.

And our skin can't withstand bitter cold. Lacking the fur coat of mammals that evolved in polar climates, team members used clothes that could insulate and protect them from cold without restricting movement. From this perspective, it was human ingenuity that let humans keep company with polar huskies, which are supremely adapted for the challenges of life on ice.

With this chapter, we turn to three systems that give the body of most animals its superficial features, shape, and capacity to move. We call them the integumentary system, muscle system, and skeletal system. Figure 33.2 shows examples from a representative vertebrate.

Animals ranging from worms to humans have an outer covering, or **integument** (after the Latin *integer*, meaning to cover). Most are tough, pliable barriers against many environmental threats. Roundworms and insects, crabs, and other arthropods have a chitin-hardened cuticle (Chapter 23). Vertebrate **skin** is an integument that includes various structures derived from epidermal cells of its outer tissue layers.

Also, regardless of the species, movement of the body or parts of it requires contractile cells and some medium or structure against which contractile force is applied. As you will see, hydrostatic skeletons, exoskeletons, and endoskeletons receive the applied force among different animal groups.

Key Concepts

1. Nearly all animals have an integument, some type of skeleton, and muscles. An integument is the outer covering of the animal body. Vertebrate skin is a prime example.

2. Skin protects the body from abrasion, ultraviolet radiation, bacterial attack, and other environmental assaults. It also contributes to overall body functioning, as when it helps control moisture loss.

3. Three categories of skeletal systems are common in the animal kingdom. We call them hydrostatic skeletons, exoskeletons, and endoskeletons. Each has body fluids or structural elements, such as bones, against which a contractile force can be applied.

4. Bones are collagen-rich, mineralized organs. They function in movement, protection and support of soft organs, and mineral storage. Blood cells form only in certain bones. Ligaments or cartilage bridges the joints between bones, and tendons attach them to skeletal muscles.

5. Many responses to changes in external and internal conditions involve muscles that move the animal body or parts of it. In response to suitable stimulation, the cells of muscle tissue contract, or shorten.

6. Smooth muscle and cardiac muscle are responsible for the motions of internal organs. Skeletal muscle helps move the body's limbs and other structural elements and maintain their spatial positions.

7. In each muscle cell, many threadlike structures called myofibrils are divided into sarcomeres. The sarcomere is the basic unit of contraction. It has parallel arrays of actin and myosin filaments. ATP-driven interactions between the actin and myosin shorten the sarcomeres of a muscle and collectively account for its contraction.

VERTEBRATE SKIN

No garment comes close to having the qualities of skin. Skin can hold its shape after repeated stretchings and washings, block harmful rays from the sun, kill many bacteria on contact, hold in moisture, fix small cuts or burns, *and* last as long as its owner. Skin makes vitamin D, required for calcium metabolism. It helps adjust body temperature when the nervous system rapidly directs the flow of blood—which transports metabolic heat—to and from blood vessels in skin, the body's largest organ. And signals from sensory receptor endings in skin help the brain assess what's going on in the outside world.

Vertebrate skin consists of a dermis and epidermis. Like puff pastry, **epidermis** has sheetlike layers; it is a stratified epithelium. Many junctions structurally and functionally join its cells (Section 29.1). Ongoing, rapid

mitotic cell divisions push epidermal cells from deeper layers to the sheet's free surface. Because of wear and tear at the surface, along with pressure exerted by the growing cell mass, older cells are flattened and dead by the time they reach the outer layers. There, they are abraded off or flake away on an ongoing basis.

Keratinocytes, the most abundant epidermal cells, make keratin, a water-insoluble protein. Its **melanocytes** produce and donate melanin, a brown-black pigment, to the keratinocytes. Melanin blocks harmful ultraviolet radiation. Humans generally have the same number of melanocytes, but skin color varies owing to differences in the distribution and metabolic activity of these cells. For example, melanocytes in *albinos* cannot produce all of the enzymes necessary for melanin production. Pale skin contains little melanin, so the pigment hemoglobin inside red blood cells is not masked. Skin appears pink because hemoglobin's red color shows through thin-walled blood vessels and through the epidermis itself, all of which are transparent. Carotene, a yellow-orange pigment, also contributes to skin color.

The **dermis** is mainly a dense connective tissue with stretch-resisting elastin fibers and supportive collagen fibers. Blood and lymph vessels, and receptor endings of sensory nerves, thread through it. Below the dermis, a hypodermis anchors skin to underlying structures, yet lets it move a bit (Figure 33.3). Fats stored here help insulate the body and cushion some of its parts.

Human skin contains sweat glands, oil glands, and hair follicles (husklike cavities). Fluid secreted by sweat glands is 99 percent water, with dissolved salts, traces of ammonia, vitamin C, and other substances. Except on foot soles and the palms of hands, sebaceous glands (oil glands) lubricate and soften skin and hair. Their secretions kill many bacteria. *Acne*, an inflammation of skin, develops after bacteria infect oil gland ducts.

Beyond the typical array of epithelia and glands, structures derived from skin vary within and between vertebrates. For instance, birds have feathers, bills, and claws. Porcupines have quills. Plant-eating mammals have hooves and horns, you have nails, and so on.

Each **hair** is a flexible structure rooted in skin, with a shaft above the skin's surface (Section 3.5). Near its base, its mostly keratinized cells divide, are pushed upward, then flatten and die. Flattened cells of the shaft's outer layer overlap like roof shingles. If mechanically abused, these frizz out as "split ends." An average human scalp has about 100,000 hairs, but genes, nutrition, hormones, and stress influence hair growth and density.

outermost epidermal layer (all dead cells)

keratinized cells being flattened

rapidly dividing cells of epidermis

dermis

a

dead, flattened epidermal cells around a shaft of hair

hair

EPIDERMIS

DERMIS

HYPODERMIS

oil gland

hair follicle

blood vessels

sensory neuron

sweat gland

smooth muscle

b

Figure 33.3 (**a**) Micrograph of the thickened skin structure typical of human palms and soles. (**b**) Sketch of the skin structure typical of body parts with hair, oil glands, and sweat glands.

Like most animals, vertebrates have an integument, a protective covering that typically is tough yet pliable. Their integument, skin, consists of epidermis and dermis.

Sunlight and Skin

THE VITAMIN CONNECTION Sit outside in the sun and you give some of your skin's epidermal cells the chance to make cholecalciferol, a form of vitamin D. **Vitamin D** is a generic name for steroid-like compounds that help the body absorb calcium from food. Expose one type of skin cell to sunlight and it produces vitamin D from a precursor molecule related to cholesterol. The cells then secrete vitamin D, which the bloodstream takes up and transports to absorptive cells of the intestinal lining. This is a hormone-like action. It means that skin acts like an endocrine gland when exposed to sunlight.

As you know, the sun's rays also contain ultraviolet (UV) wavelengths, which are harmful. Dark skin is able to block UV radiation while still making vitamin D. It also blocks the breakdown of a B vitamin, **folate**. Among other things, folate happens to be essential for proper embryonic development, hence for reproduction.

Humans first evolved beneath the intense sun of the African savanna. Skin alone would have supplied our early ancestors with enough vitamin D. What happened when some populations moved out of the tropics, to higher latitudes? What happened when they moved into caves and bundled up in clothing? Dietary sources of vitamin D became more important. In addition, selection pressure for embryo-protecting dark skin probably eased up. Nina Jablonksi and George Chaplin compared satellite data on annual UV exposure levels with skin reflectance (the higher the reflectance, the lighter the skin). In more than fifty countries throughout the world, they matched up increasingly light skin with decreasing UV levels.

SUNTANS AND SHOE-LEATHER SKIN When exposed to UV radiation, the melanocytes in skin are stimulated to make melanin. Production accelerates slowly, then peaks about ten days after the initial exposure. Before then, unprotected skin can get mildly to severely burned.

Ongoing exposure produces the "tan" that so many light-skinned people covet (Figure 33.4). Dark-skinned people have better initial protection from UV radiation. Yet even in naturally dark skin, prolonged UV exposure causes elastin fibers in connective tissue of the dermis to clump together. Skin loses its resiliency as a result. In time, it starts looking like old shoe leather.

Skin will age anyway. As any person grows older, epidermal cells divide less often. Skin gets thinner and more vulnerable to injury; glandular secretions that once kept it soft and moistened dwindle. Collagen and elastin fibers in the dermis break down and become sparser, so skin loses its elasticity and wrinkles deepen. However, you can accelerate the process and look older faster in several ways, as by continually exposing your skin to dry winds or tobacco smoke—and by excessive tanning.

SUNLIGHT AND THE FRONT LINE OF DEFENSE Besides having cells that make melanin and keratin, skin also has cells that help protect the body against pathogens and against cancer. The two defending cell types of the epidermis are Langerhans and Granstein cells.

Langerhans cells are phagocytes that develop in bone marrow, then take up stations in skin. After engulfing virus particles or bacteria, they display molecular alarm signals at their plasma membrane. The signals mobilize the body's immune system. UV radiation damages these cells. This may be why sunburns often trigger *cold sores*, small, painful blisters that announce the recurrence of a *Herpes simplex* infection. Nearly everyone harbors the *H. simplex* virus. It remains hidden in the face, inside a ganglion. (A ganglion, recall, is a cluster of neuron cell bodies.) Sunburns and other stress factors can activate the virus. When that happens, virus particles move down the neurons to the axonal endings in skin. There they infect epithelial cells and cause eruptions.

When UV radiation damages Langerhans cells, it weakens one of your first lines of defense. It may activate proto-oncogenes and initiate the cancerous transformation of skin cells. (You read about this in the Chapter 14 introduction and Section 14.4.) Skin cancers grow rapidly and may spread to adjacent lymph nodes unless they are surgically removed.

In some as-yet undetermined way, **Granstein cells** interact with the white blood cells that can put the brakes on immune responses in the skin. By issuing suppressor signals, they help keep responses from spiraling out of control. Although their roles are not completely understood, researchers found that these cells are less vulnerable than Langerhans cells to the damaging effects of ultraviolet radiation.

Figure 33.4 Demonstration of how shoe-leather skin forms.

SKELETAL SYSTEMS

Animals move by activating, contracting, and relaxing muscle cells. But muscle cells alone can't move the body or parts of it. *Muscles require the presence of some medium or structural element against which the force of contraction is applied.* A skeletal system fulfills this requirement.

Three types of animal skeletons are common. With a **hydrostatic skeleton**, muscles work against an internal body fluid and redistribute it within a confined space. The confined fluid resists compression, like a waterbed. With an **exoskeleton**, rigid, *external* body parts such as a cuticle or a shell receive the applied force of muscle contraction. With an **endoskeleton**, *internal* body parts receive the applied force of muscle contraction.

We see hydrostatic skeletons among the soft-bodied invertebrates, such as a sea anemone with a soft body and saclike gut (Figure 33.5). When longitudinal muscles in its body wall are contracted (shortened) and radial ones are relaxed (lengthened), its body is squat. When longitudinal muscles relax and radial muscles contract (thus forcing fluid out of the gut), the body lengthens upright. An earthworm, too, has a hydrostatic skeleton. It uses the coelomic chambers of its segmented body (Section 23.10). Contracting and relaxing the fluid-filled segments in sequence moves the body forward. The body thrashes forward and back by alternating contractions of muscles on one side and then the other.

Recall how an arthropod has a hinged exoskeleton? Its hard parts are moved like levers by sets of muscles attached to them. Small contractions bring about large movements. This is especially true of a winged insect's cuticle. It extends over certain body segments and gaps between segments. The cuticle is pliable at the gaps. It

Figure 33.5 A sea anemone's hydrostatic skeleton. Contractile cells in the body wall run longitudinal to the main body axis and radially around the gut. *Left:* Radial cells are relaxed; longitudinal ones are contracted. This resting position is assumed at low tide, when currents don't deliver food. *Right:* The radial cells are contracted; longitudinal ones are relaxed. The body extends upward, to its feeding position.

resting position   feeding position

functions like a hinge when certain muscles raise either the wing or body parts to which the wings are attached (Sections 18.4 and 23.11).

Most vertebrates have a bony endoskeleton (Figures 33.6 and 33.7). In humans, it has 206 bones. Its pectoral girdle (at the shoulders), pelvic girdle (at the hips), and paired arms, hands, legs, and feet are the *appendicular* portion. Its pectoral girdles include slender collarbones and flat shoulder blades. Fall on an outstretched arm and you might dislocate a shoulder blade or fracture a collarbone, which is the most frequently broken bone.

The *axial* portion of the human skeleton has skull bones, twelve pairs of ribs, a breastbone, and twenty-six vertebrae. **Vertebrae** (singular, vertebra) are bony

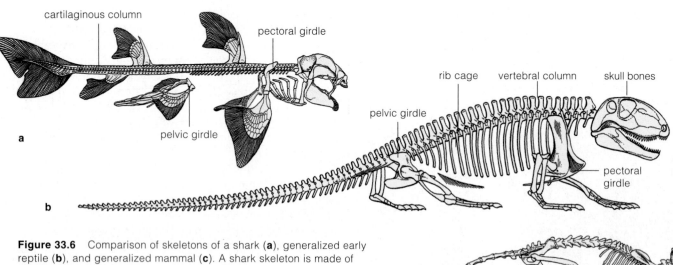

Figure 33.6 Comparison of skeletons of a shark (**a**), generalized early reptile (**b**), and generalized mammal (**c**). A shark skeleton is made of calcium-hardened cartilage. As in bony fishes, its pelvic and pectoral girdles serve as a stable base for fin motion. In four-legged vertebrates, these girdles transfer body weight to limbs. They have more surface area for attaching limb muscles and are tied closer to the body. Thus they offer better support and assist in the thrust of locomotion.

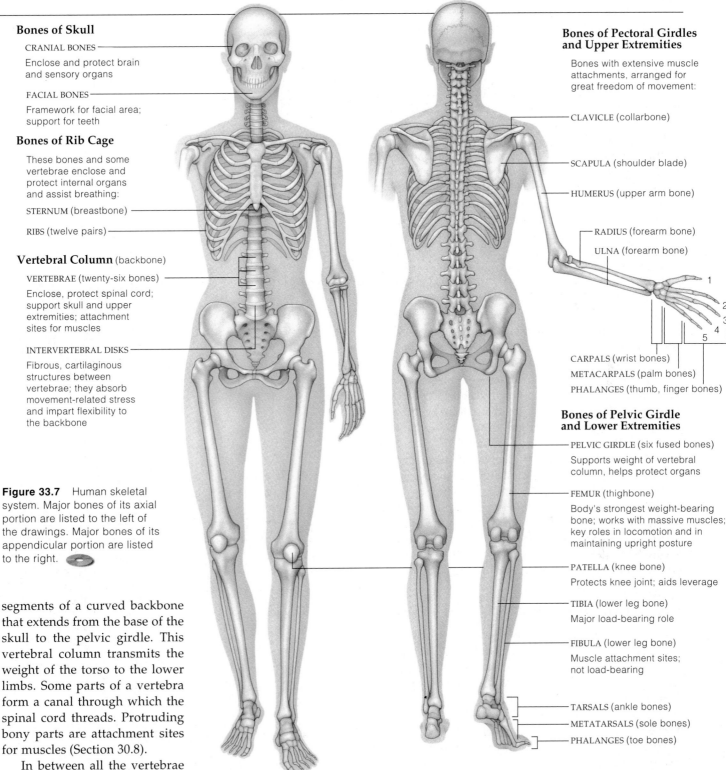

Bones of Skull

CRANIAL BONES

Enclose and protect brain and sensory organs

FACIAL BONES

Framework for facial area; support for teeth

Bones of Rib Cage

These bones and some vertebrae enclose and protect internal organs and assist breathing:

STERNUM (breastbone)

RIBS (twelve pairs)

Vertebral Column (backbone)

VERTEBRAE (twenty-six bones)

Enclose, protect spinal cord; support skull and upper extremities; attachment sites for muscles

INTERVERTEBRAL DISKS

Fibrous, cartilaginous structures between vertebrae; they absorb movement-related stress and impart flexibility to the backbone

Figure 33.7 Human skeletal system. Major bones of its axial portion are listed to the left of the drawings. Major bones of its appendicular portion are listed to the right.

Bones of Pectoral Girdles and Upper Extremities

Bones with extensive muscle attachments, arranged for great freedom of movement:

CLAVICLE (collarbone)

SCAPULA (shoulder blade)

HUMERUS (upper arm bone)

RADIUS (forearm bone)

ULNA (forearm bone)

CARPALS (wrist bones)

METACARPALS (palm bones)

PHALANGES (thumb, finger bones)

Bones of Pelvic Girdle and Lower Extremities

PELVIC GIRDLE (six fused bones)

Supports weight of vertebral column, helps protect organs

FEMUR (thighbone)

Body's strongest weight-bearing bone; works with massive muscles; key roles in locomotion and in maintaining upright posture

PATELLA (knee bone)

Protects knee joint; aids leverage

TIBIA (lower leg bone)

Major load-bearing role

FIBULA (lower leg bone)

Muscle attachment sites; not load-bearing

TARSALS (ankle bones)

METATARSALS (sole bones)

PHALANGES (toe bones)

segments of a curved backbone that extends from the base of the skull to the pelvic girdle. This vertebral column transmits the weight of the torso to the lower limbs. Some parts of a vertebra form a canal through which the spinal cord threads. Protruding bony parts are attachment sites for muscles (Section 30.8).

In between all the vertebrae are **intervertebral disks**. These are cartilaginous shock absorbers and flex points that permit some movement. Severe or rapid shock often forces a disk to slip out of place or rupture. *Herniated disks* are a painful, less-than-advantageous outcome of bipedalism. Remember, our early hominid ancestors walked about on four legs; later ones were walking upright before 4 million years ago. The change put a pronounced S-shaped curve in the backbone. So today,

the older we get, the longer the backbone continues to fight gravity in a compromised way, and the more back pain we may suffer.

Hydrostatic skeletons, exoskeletons, and endoskeletons all have structural elements or body fluids against which the force of contraction can be applied.

A CLOSER LOOK AT BONES AND JOINTS

Bone Structure and Function

By definition, **bones** are complex organs that function in movement, protection, support, mineral storage, and formation of blood cells (Table 33.1). Bones that support and anchor skeletal muscles help maintain or change the positions of body parts. Some form hard compartments that enclose and protect the brain, the lungs, and other internal organs. Bones are reservoirs for calcium and phosphorus ions. Continual deposits and withdrawals of ions from bone help maintain blood levels of calcium and phosphorus, thus supporting metabolic activities. Some bones (not all) are sites of blood cell formation.

Human bones range in size from middle ear bones smaller than lentils to clublike femurs, or thighbones (Figure 33.8). They have long, flat, short or cubelike, or irregular shapes. All consist of connective tissues (bone tissue especially) and epithelia. Bone tissue consists of **osteocytes**—mature bone cells—and collagen fibers in a calcium-hardened ground substance.

The shaft and both ends of thighbones have *compact* bone tissue, which resists mechanical shock. This tissue develops as Haversian systems: many thin, cylindrical, dense layers around interconnecting canals for blood vessels and nerves that service living bone cells. Also in the bone ends and shaft is *spongy* bone tissue, which imparts strength but doesn't weigh much. Abundant spaces make it look spongy, but its flattened parts are firm. **Red marrow**, a major site of blood cell formation, fills spaces in some bones, such as the breastbone. The cavities in most mature bones hold **yellow marrow**, a mostly fatty region that converts to red marrow and makes more red blood cells in times of severe blood loss.

Bone Formation and Remodeling

Bone tissue first forms in an embryo, where many bones are constructed upon cartilage models. **Osteoblasts** are the bone-forming cells. Using the models, they form and secrete organic substances, which become mineralized. Cartilage in the shaft breaks down, which opens up a marrow cavity. Once osteoblasts have been imprisoned by their own secretions, we call them osteocytes.

The total bone mass in healthy young adults doesn't change much, but minerals and osteocytes are removed and replaced on an ongoing basis. In **bone remodeling**, minerals are normally deposited and removed at the same time. The process adjusts bone strength and helps maintain blood levels of calcium and phosphorus. Two cell types interact in this process. Osteoblasts deposit bone, and **osteoclasts** secrete enzymes that digest bone's organic matrix. Thanks to these bone-degrading cells, liberated ions enter interstitial fluid, from which they can be reabsorbed by the bloodstream.

Figure 33.8 (**a**) Human thighbone structure, and (**b**) appearance of its spongy and compact bone tissue. Thin, dense layers of compact bone tissue form cylindrical, interconnected arrays around the canals for blood vessels and nerves. Each array is a Haversian system. The blood vessel inside services osteocytes, living bone cells in small tissue spaces connected by tunnels.

nutrient canal

location of yellow marrow

compact bone tissue

spongy bone tissue

a

space occupied by living bone cell

blood vessel

HAVERSIAN SYSTEM

55 µm

spongy bone tissue

compact bone tissue

blood vessel

outer layer of dense connective tissue

b

Table 33.1 *Functions of Bone*

1. *Movement.* Bones interact with skeletal muscle to change or maintain the position of the body and its parts.

2. *Support.* Bones support and anchor muscles.

3. *Protection.* Many bones form hard compartments that enclose and protect soft internal organs.

4. *Mineral storage.* Bones are a reservoir for calcium and phosphorus, the deposits and withdrawals of which help maintain ion concentrations in body fluids.

5. *Blood cell formation.* Only certain bones contain regions where blood cells are produced.

Figure 33.9 Bone affected by osteoporosis. (**a**) Section through normal bone tissue; mineral deposits continually replace withdrawals. (**b**) After the onset of osteoporosis, mineral replacements lag behind withdrawals. The tissue erodes; bones become hollow and brittle.

Bones and the Blood Level of Calcium

Calcium is central to neural function, contraction, and other vital activities, and its level in blood is one of the most tightly controlled aspects of metabolism. Bones and teeth store all but 1 percent or so of a body's calcium. Negative feedback loops to glands from the blood govern the rates of calcium release and uptake. The thyroid gland releases calcitonin when the calcium level rises. This hormone suppresses osteoclast action and slows calcium's release into the blood. Parathyroid glands release parathyroid hormone (PTH) when the calcium level falls. PTH promotes calcium release from bones and the kidneys, enhances osteoclast action, and helps activate vitamin D. Vitamin D stimulates calcium absorption from the small intestine (Section 33.2).

When a bone is stressed, as when it is subjected to compression, mineral deposition by osteoblasts exceeds withdrawals by osteoclasts. That's why bones of people who exercise rigorously are denser and stronger than bones of people who don't. In injured bones, mineral withdrawals predominate. Also, as a person ages, the backbone and other bones decline in mass. We call this condition *osteoporosis* (Figure 33.9). Decreased physical activity, declining activity of the bone-forming cells, loss of calcium, excessive protein intake, and sex hormone deficiencies contribute to the disorder.

Where Bones Meet—Skeletal Joints

Joints, areas of contact or near-contact between bones, have a distinct bridge of connective tissue. Very short connecting fibers bridge *fibrous* joints. Cartilage straps bridge *cartilaginous* joints. **Ligaments**, or long straps of dense connective tissue, bridge *synovial* joints.

vein — femur

artery — patella

— fat pad

— ligament

muscle — tibia

Figure 33.10 Human knee joint, longitudinal section.

Fibrous joints hold teeth in their sockets. They join the flat skull bones of a fetus. At childbirth, the loose connections allow bones to slide over each other a bit and prevent skull fractures. A newborn's skull still has fibrous joints and membranous areas called soft spots, or fontanels. By childhood, the fibrous tissue hardens and the skull bones are fused together as a single unit.

Cartilaginous joints bridge the vertebrae, ribs, and breastbone, and allow slight movements. Synovial joints move freely; ligaments stabilize them. The knee joint is an example (Figures 33.7 and 33.10). Cartilage cushions the abutting bones and helps absorb shocks. A flexible capsule of dense connective tissue surrounds the area of contact. Cells of a membrane that lines the interior of the capsule secrete a fluid that lubricates the joint.

Like other joints, knee joints are vulnerable to stress. They let you swing, bend, and turn the long bones below them. As you run, they absorb the force of your weight. Abruptly stretch or twist a knee joint too far and you *strain* it. Tear its ligaments or tendons and you *sprain* it. Move the wrong way and you may dislocate attached bones. In collision sports such as football, a blow to the knee often severs ligaments. These must be reattached within ten days. Why? Phagocytes in a lubricating fluid inside the joint usually clean up after everyday wear and tear. When presented with torn ligaments, these phagocytes indiscriminately turn the tissue to mush.

Joint inflammation and degenerative disorders are collectively called "arthritis." In *osteoarthritis*, cartilage at knees and other freely movable joints wears off as a person ages. Most often, the joints in the fingers, knees, hips, and vertebral column are affected. In *rheumatoid arthritis*, synovial membranes in joints become inflamed and thicken, cartilage degenerates, and bone deposits accumulate. These are the outcomes of an autoimmune response, whereby the body is abnormally triggered into attacking itself (Chapter 35). A bacterial or viral infection may trigger this degenerative disorder, which has a genetic basis. Rheumatoid arthritis can begin at any age, but symptoms usually emerge before age fifty.

Bones are collagen-rich, mineralized organs that function in movement, protection, support, storage of calcium and other minerals, and blood cell formation.

SKELETAL–MUSCULAR SYSTEMS

How Do Muscles and Bones Interact?

Skeletal muscles are the functional partners of bones. Each skeletal muscle contains bundles of hundreds to many thousands of muscle cells, which look like long, striped fibers. In muscle tissue, remember, muscle cells contract (shorten) in response to adequate stimulation. They lengthen in response to gravity and other loads. When you run, dance, breathe, squint, or scribble notes, many muscle cells are working to move your body or change the positions of some of its parts.

Connective tissue bundles muscle cells together and extends past them to form **tendons**. Each tendon, a cord or strap of dense connective tissue, attaches a muscle to bone (Figure 33.11). Most of these attachment sites are like a gearshift in a car. *They form a lever system in which a rigid rod is attached to a fixed point and can move about it.* Muscles connect to bones (rigid rods) near a joint (fixed point). When they contract, they transmit force to bones and so make them move. Tendons often rub against bones, as at the knees and wrists. They slide inside fluid-filled sheaths that help reduce friction.

Skeletal muscles interact with one another, also. Some work in pairs or groups to promote a movement. Others work in opposition; the action of one opposes

Figure 33.11 Tendon. A bursa is a flattened sac filled with synovial fluid. Bursae are located between tendons and bones (or some other structure). They help reduce friction during movements.

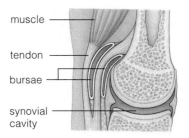

muscle
tendon
bursae
synovial cavity

or reverses the action of another. Figure 33.12 shows how opposing muscle groups work to move frog legs.

Also look at Figure 33.13. Extend your right arm forward, then place your left hand over the biceps in the upper right arm and slowly bend the elbow. Feel the biceps contract? Even when a biceps contracts only a bit, it causes a large motion of the bone connected to it. This is the case for most leverlike arrangements.

Bear in mind, only *skeletal* muscle is the functional partner of bone. As mentioned earlier, smooth muscle

c The first muscle group in the frog's upper hindlimb contracts again and draws the leg back toward the body.

b An opposing muscle group attached to the limb forcefully contracts and pulls the limb back. The contractile force, applied against the ground, propels the frog forward.

a A frog muscle attached to each upper hindlimb contracts and pulls the leg slightly forward relative to the body's main axis.

Figure 33.12 Frog demonstrating how a small decrease in the length of contracting muscles can produce large movements. Its leap depends on opposing muscle groups attached to the upper limb bone of each hind leg. (**a,c**) One muscle group pulls the limb forward a bit, toward the body's midline. (**b**) Another pulls the limb back and away from the body.

Triceps relaxes.

Biceps contracts at the same time, and pulls forelimb up.

Triceps contracts, pulls the forelimb down.

At the same time, biceps relaxes.

a

b

Figure 33.13 Two opposing muscle groups in a human arm. (**a**) When the triceps relaxes and its opposing partner (biceps) contracts, the elbow joint flexes and the forearm bends upward. (**b**) When the triceps contracts, the forearm straightens out.

PROPERTIES OF WHOLE MUSCLES

Muscle Tension and Muscle Fatigue

Whether a muscle actually shortens during cross-bridge formation depends on the external forces acting on it. Collectively, the cross-bridges exert **muscle tension**. By definition, this is a mechanical force that a contracting muscle exerts on an object, such as a bone. Opposing it is a load, either the weight of an object or gravity's pull on the muscle. Only when muscle tension exceeds the load does a stimulated muscle shorten.

Isotonically contracting muscles shorten and move a load (Figure 33.19). An *isometrically* contracting muscle develops tension but doesn't shorten. With *lengthening* contraction, a muscle lengthens when an external load is greater than its tension for the period of contraction. This happens to leg muscles as you walk down stairs.

A muscle's tension relates to the formation of cross-bridges in its cells and to the number of cells recruited into action. Consider a **motor unit**: a motor neuron and all muscle cells that form junctions with its endings. By stimulating that motor unit with an electrical impulse, we can induce an action potential and make a recording of an isometric contraction. It takes a few milliseconds for tension to increase, then it peaks and declines. This response is a **muscle twitch** (Figure 33.20*a*). Its duration depends on the load and on cell type. For example, fast-acting muscle cells rely on glycolysis—not efficient, but fast—and use up ATP faster than slow-acting cells do.

If another stimulus is applied before the response is finished, the muscle twitches again. **Tetanus** is a large contraction that results from the repeated stimulation of a motor unit, so twitches mechanically run together. (In *tetanus*, a disease by the same name, toxins disrupt muscle relaxation, as described in Section 30.5.) Figure 33.20*c* shows a recording of tetanic contraction.

Continuous, high-frequency stimulation that keeps a muscle in a state of tetanic contraction leads to *muscle fatigue*, or a decline in tension. After a few minutes of rest, a fatigued muscle will contract again in response to stimulation. The extent of recovery depends largely on how long and how often it was stimulated before. Muscles associated with brief, intense exercise (such as weight lifting) fatigue fast but recover fast. The muscles associated with prolonged, moderate exercise fatigue slowly but take longer to recover, often up to twenty-four hours. The molecular mechanisms causing muscle fatigue are unknown, but glycogen depletion is a factor.

Effects of Exercise and Aging

A muscle's properties depend on how often, how long, and how intensely it is put to use. With regular **exercise** (that is, increased levels of contractile activity), muscle cells do not increase in number. Rather, they increase in

a   **b**

contracted muscle can shorten

contracted muscle can't shorten

Figure 33.19 (**a**) Isotonic contraction. A load is less than a muscle's peak capacity to contract. Thus the muscle can contract, shorten, and lift the load. (**b**) Isometric contraction. A load exceeds a muscle's peak capacity. It contracts, but it can't shorten.

Figure 33.20 Three recordings of twitches in muscles artificially stimulated in three different ways. (**a**) A single twitch. (**b**) Summation of twitches following six stimulations per second. (**c**) Tetanic contraction following twenty stimulations per second.

size and metabolic activity, and become more resistant to fatigue. Think of *aerobic exercise*—not intense, but long in duration. Aerobic exercise increases the number of mitochondria in both fast and slow muscle cells, and increases the blood capillaries that service them. Such physiological changes improve endurance. By contrast, *strength training* (intense, short-duration exercise such as weight lifting) makes fast-acting muscle cells form more myofibrils and enzymes for glycolysis. Strong, bulging muscles may develop, although these muscles do not have much endurance. They fatigue rapidly.

Muscle tension declines in adult humans after age thirty or forty. Older people might exercise just as long and intensely as younger ones, but their muscles cannot adapt (change) in response to the same extent. Even so, some adaptation can be beneficial. Aerobic exercise can improve blood circulation. And, as it turns out, modest strength training slows the loss of muscle tissue that is an inevitable part of the aging process.

Properties of muscles vary with age and levels of activity.

33.9

Gold indicates text section

1. Most animals have an integumentary system, which covers the body's surface. Examples are insect cuticles and vertebrate skin. Skin protects against abrasion, UV radiation, dehydration, and many pathogens. It helps control body temperature; blood flow to skin dissipates heat. Its sensory receptors detect external stimuli. Skin exposed to sunlight has an endocrine function; it helps produce vitamin D, a hormone-like substance required for absorption of calcium from food. *CI, 33.1, 33.2*

2. Skin consists of two regions: an outer epidermis and underlying dermis, where rapid divisions replace cells continually being shed or abraded away. Keratinocytes (keratin producers) are the most abundant cells. Others are melanocytes (melanin producers), and Langerhans cells and Granstein cells (both types defend the body against pathogens and cancer cells). *33.1, 33.2*

3. Movement of the animal body or parts of it requires contractile cells and some medium or structure against which contractile force can be applied. *33.3*
   a. With hydrostatic skeletons, such as that of a sea anemone, body fluids accept the contractile force and, as a result, are redistributed within a confined space.
   b. With exoskeletons, such as that of an insect, rigid external body parts accept the force of contraction.
   c. With endoskeletons, rigid internal structures such as bones receive the applied force of contraction.

4. Bones are organs with osteocytes (living bone cells) in a mineralized, collagen-rich ground substance. They have roles in moving, protecting, and supporting body parts; storing minerals; and, in bones that contain red and yellow marrow, forming blood cells. *33.4*

5. Tendons attach skeletal muscles to bones. Skeletal muscles and bones interact as a system of levers, with rigid rods (bones) moving at fixed points (joints). Many muscles work together or in opposition to bring about movement or positional changes in body parts. *33.5*

6. A skeletal joint, a gap between bones, is bridged by connective tissues. Short fibers bridge fibrous joints. Cartilage bridges cartilaginous joints. Ligaments bridge synovial joints. *33.4*

7. Smooth, cardiac, and skeletal muscle cells contract (shorten) in response to adequate stimulation. Many soft internal organs incorporate smooth muscle. Only the heart incorporates cardiac muscle. Skeletal muscle is the functional partner of bones. *33.5*

8. Inside each skeletal muscle cell are many myofibrils arranged parallel with the long axis. These threadlike structures contain filaments of actin and myosin, also organized in parallel. Every myofibril is transversely divided into sarcomeres, the basic units of contraction. The parallel orientation of all parts directs the contractile force at a bone to be pulled in some direction. *33.6*

9. In response to stimulation from the nervous system, skeletal muscles shorten by decreases in the length of all of its sarcomeres. Key points for the sliding-filament model of muscle contraction follow: *33.6, 33.7*
   a. Action potentials cause the release of calcium ions from a membrane system (sarcoplasmic reticulum) that threads around the cell's myofibrils. Calcium diffuses into sarcomeres, and binds to and pulls aside accessory proteins on actin filaments to expose binding sites for myosin. Cross-bridges can now form.
   b. Each cross-bridge is a brief attachment between a myosin head and an actin binding site. Cross-bridges form during repeated, ATP-driven power strokes. The repeated strokes make actin filaments slide past myosin filaments, and collectively they shorten the sarcomere.

10. Muscle cells obtain the ATP for contraction in three ways: *Dephosphorylation of creatine phosphate*, a direct, fast pathway good for a few seconds of contraction. *Aerobic respiration*, a pathway that predominates during prolonged, moderate exercise. *Glycolysis*, which takes over if intense exercise exceeds the body's capacity to deliver oxygen to muscle cells. *33.7*

11. Muscle tension is a mechanical force caused by cross-bridge formation. The load (gravity or weight) is an opposing force. Stimulated muscles shorten if tension exceeds the load and lengthen if tension is less than the load. Exercise and aging affect muscle properties. *33.8*

Review Questions

1. List the functions of skin. Describe the distinctive features of its regions. Is the hypodermis part of skin? *33.1, 33.2*

2. Identify four cell types in vertebrate skin and briefly describe their functions. *33.1, 33.2*

3. Distinguish between:
   a. hydrostatic skeleton, exoskeleton, and endoskeleton *33.3*
   b. vertebra and intervertebral disk *33.3*
   c. ligament and tendon *33.4, 33.5*

4. What are the functions of bones? What is a joint? *33.4*

5. Which hormones help control calcium ion concentrations in blood? Name their general effects on bone tissue turnover. *33.4*

6. In what respect are tendons like a car's gearshift? *33.5*

7. Review Figure 33.15. On your own, sketch and label the fine structure of a muscle, down to one of its individual myofibrils. Identify the basic unit of contraction in the myofibrils. *33.6*

8. What role does calcium play in control of contraction? What role does ATP play, and by what routes does it form? *33.6, 33.7*

Self-Quiz   ANSWERS IN APPENDIX III

1. Which is *not* a function of skin?
   a. resist abrasion   c. initiate movement
   b. restrict dehydration   d. help control temperature

2. _____ are shock pads and flex points.
   a. Vertebrae   c. Marrow cavities
   b. Femurs   d. Intervertebral disks

Figure 33.21 (**a**) A demonstration of the curious invasions of the integument that have no immediately obvious adaptive value. (**b**) Extremely pumped-up muscles of a human male.

3. Blood cells form in _____ .
   a. red marrow
   b. all bones
   c. certain bones only
   d. a and c

4. The _____ is the basic unit of contraction.
   a. myofibril
   b. sarcomere
   c. muscle fiber
   d. myosin filament

5. Muscle contraction requires _____ .
   a. calcium ions
   b. ATP
   c. action potential arrival
   d. all of the above

6. ATP for muscle contraction can be formed by _____ .
   a. aerobic respiration
   b. glycolysis
   c. creatine phosphate breakdown
   d. all of the above

7. Match the M words with their defining feature.
   ____ muscle
   ____ muscle twitch
   ____ muscle tension
   ____ melanin
   ____ myosin
   ____ marrow
   ____ metacarpals
   ____ myofibrils
   ____ muscle fatigue

   a. actin's partner
   b. all in the hands
   c. blood cell production
   d. decline in tension
   e. brownish-black pigment
   f. motor unit response
   g. force exerted by cross-bridges
   h. bundles of muscle cells in a connective tissue sheath
   i. threadlike parts in muscle cell

Critical Thinking

1. The nose, lips, tongue, navel, and private parts are common targets for *body piercing*, cutting holes into the body so jewelry and other objects can be threaded through them (Figure 33.21*a*). *Tattooing*, or using permanent dyes to create patterns in skin, is another fad. Besides being painful, both skin invasions invite bacterial infections, chronic viral hepatitis, AIDS, and other nasty diseases if piercers and tattooers reuse unsterilized needles, dye, razors, gloves, swabs, or trays. Months may pass before problems arise, so the cause-and-effect connection isn't always obvious. If you dismiss the risks and think unregulated tissue invasions are okay, how could you make sure the equipment used is sterile?

2. For young women, the recommended daily allowance (RDA) of calcium is 800 milligrams/day. During pregnancy, the RDA is 1,200 milligrams/day. Why the increase? What might happen to a pregnant woman's bones without the larger amount?

3. Compared to most people, Joe and other long-distance runners have a greater number of muscle fibers with more mitochondria. Sprinters have muscle fibers that contain more of the enzymes necessary for glycolysis but fewer mitochondria. Think about how these two forms of exercise differ. Then explain why the muscle fibers differ between the two kinds of runners.

4. Extreme sexual dimorphism in body size involves differences in muscle mass (compare Section 16.9). This may be a measure of how much male mammals invest in fighting capacity. Maybe the cost of securing and using resources to grow and maintain a massive body is offset by reproductive rewards. Now think of the social rewards for millions of athletes who use anabolic steroids to increase muscle mass and strength (Figure 33.21*b*). *Anabolic steroids*, synthetic hormones, mimic testosterone. This sex hormone governs secondary sexual traits and also increases aggressive behavior, which is often associated with maleness.

By stimulating protein synthesis, anabolic steroids induce rapid gains in muscle mass and strength during weight-training and exercise programs. However, at high concentrations, they trigger a sharp decline in testosterone production, resulting in acne, baldness, shrinking testes, infertility, and maybe early heart disease in men. Even occasional use may damage kidneys or set the stage for cancers of the liver, testes, and prostate gland. Among women, anabolic steroids deepen the voice and produce pronounced facial hair. Breasts may shrink and the menstrual cycle may become irregular.

Not every steroid user develops severe side effects. More common are mental difficulties, called '*roid rage* or *body-builder's psychosis*. Users become irritable and increasingly aggressive. Some men have become uncontrollably aggressive, manic, and delusional. One steroid user accelerated his car to high speed and deliberately drove it into a tree.

Using natural selection theory as the basis for argument, would you say the advantages of anabolic steroid use outweigh the disadvantages? This is not a trick question.

Selected Key Terms

actin 33.6
action potential 33.7
bone 33.4
bone remodeling 33.4
creatine phosphate 33.6
cross-bridge formation 33.6
dermis 33.1
endoskeleton 33.3
epidermis 33.1
exercise 33.8
exoskeleton 33.3
folate 33.2
Granstein cell 33.2
hair 33.1

hydrostatic skeleton 33.3
integument CI
intervertebral disk 33.3
joint 33.4
keratinocyte 33.1
Langerhans cell 33.2
ligament 33.4
melanocyte 33.1
motor unit 33.8
muscle tension 33.8
muscle twitch 33.8
myosin 33.6
osteoblast 33.4
osteoclast 33.4

oxygen debt 33.6
red marrow 33.4
sarcomere 33.6
sarcoplasmic reticulum 33.7
skeletal muscle 33.5
skin CI
sliding-filament model 33.6
tendon 33.5
tetanus 33.8
vertebra (vertebrae) 33.3
vitamin D 33.2
yellow marrow 33.4

Readings

Sherwood, L. 1997. *Human Physiology*. Third edition. Belmont, California: Wadsworth.

Travis, J. 15 January 2000. "Boning Up." *Science News* 157: 41–43.

On-Line readings at Student Guide for InfoTrac:
www.brookscole.com/biology

CIRCULATION

Heartworks

For Dr. Augustus Waller, Jimmie the bulldog was no ordinary pooch. Connected to wires and soaked to his ankles in buckets of salty water, Jimmie was on a four-footed exploration of the heart (Figure 34.1a). Press your fingers to your chest a few inches left of center, between the fifth and sixth ribs, and feel the repetitive thumpings of your heart. The same rhythms intrigued Waller and other nineteenth-century physiologists. They wondered: Is there a pattern of electrical currents that corresponds to each heartbeat? Could they find out by devising a painless way to record currents at the body surface?

That's where Jimmie and the buckets of salty water came in. Saltwater happens to be an efficient conductor of electricity. In Waller's experiment, it picked up faint signals from Jimmie's beating heart through the skin of his legs and conducted them to a crude monitoring device. With that device, Waller made one of the first recordings of heart activity (Figure 34.1c). Today we

call such a recording an ECG. That's the abbreviation for electrocardiogram.

A graph of your heart's normal electrical activity would look very much the same. The pattern emerged a few weeks after you started growing, by mitotic cell divisions, from a fertilized egg in your mother. Early on, some embryonic cells differentiated into cardiac muscle cells and started to contract on their own. One small patch of the cells took the lead. It's been functioning as your heart's pacemaker ever since.

If all goes well, that patch of cardiac muscle cells will continue to contract as it should until the day you die. It is a natural pacemaker; it sets the baseline rate at which blood is pumped out of the heart, through blood vessels, then back to the heart. The rate, about seventy beats every minute, is moderate. But commands from the nervous and endocrine systems continually adjust it. When you run, skeletal muscle cells demand more blood-borne oxygen and glucose than they do when

JIMMIE   DR. WALLER

a   BUCKET   BUCKET   b

Figure 34.1   A bit of history in the making. (**a**) Jimmie the bulldog taking part in a painless experiment. (**b**) The physiologist Augustus Waller and his beloved pet bulldog sharing a moment in Waller's study after the experiment, which yielded one of the world's first electrocardiograms (**c**).

food, water intake   oxygen intake

DIGESTIVE SYSTEM   RESPIRATORY SYSTEM

elimination of carbon dioxide

nutrients, water, salts   oxygen   carbon dioxide

CIRCULATORY SYSTEM   URINARY SYSTEM

water, solutes

elimination of food residues   rapid transport to and from all living cells   elimination of excess water, salts, wastes

Figure 34.2 Diagram of the functional connections between the circulatory, respiratory, and digestive systems, which interact in transporting substances to and from all living cells in the animal body. Their integrated activities help maintain favorable operating conditions in the internal environment.

you sleep. At such times of intense exercise, your heart starts pounding more than twice as fast, and this helps deliver sufficient blood to muscle tissues.

We've come a long way from Waller and Jimmie in monitoring the heart. Sensors now detect faint signals characteristic of an impending heart attack. Internists use computers to analyze a patient's beating heart, and ultrasound probes to get images of it on video screens. Cardiologists routinely substitute battery-powered pacemakers for faltering natural ones.

With this chapter, we turn to the **circulatory system**, the means by which substances move rapidly to and from the interstitial fluid that bathes all living cells in nearly all animals. The system continually accepts the oxygen, nutrients, and other substances that an animal secures, through its respiratory and digestive systems, from the environment (Figure 34.2). Simultaneously it picks up carbon dioxide and other wastes from cells and gets them to the respiratory and urinary systems for disposal. Its operation is central to maintaining the operating conditions of the internal environment within a tolerable range. And that is the state we call homeostasis.

Key Concepts

1. All cells survive by exchanging substances with their surroundings. In most animals, substances rapidly move to and from cells by way of a closed circulatory system.

2. Blood, a fluid connective tissue, is the transport medium of circulatory systems. It transports oxygen, carbon dioxide, plasma proteins, vitamins, hormones, lipids, and many other solutes. Blood also transports metabolically generated heat.

3. Birds and mammals have a four-chambered, muscular heart that pumps blood through two separate circuits of blood vessels, both of which lead back to the heart. One circuit flows through the lungs, where it picks up oxygen that cells require, mainly for aerobic respiration. The other circuit delivers oxygenated blood to all body regions, from which it picks up carbon dioxide wastes from aerobic respiration.

4. Arteries are large-diameter, low-resistance vessels that rapidly transport oxygenated blood from the heart. Veins are large-diameter, low-resistance vessels that transport oxygen-poor blood to the heart and serve as blood volume reservoirs.

5. The flow volume through each organ is regulated at arterioles. In response to control signals, the diameter of these blood vessels widens in some parts of the body and narrows in others. The coordinated responses of arterioles throughout the body divert more of the flow volume to organs that are most active at a given time.

6. Small-diameter, thin-walled capillaries interconnect in capillary beds, which are diffusion zones. When a volume of blood spreads out through a capillary bed, it slows owing to the total cross-sectional area of all the capillaries, which is greater than that of arterioles. The slowdown allows time for exchanges between blood and interstitial fluid.

7. Pressure forces some fluid out of capillaries, into interstitial fluid. Drainage vessels of the lymphatic system return it to the blood. Organs of that system work to cleanse blood of infectious agents and other threats to the body.

CIRCULATORY SYSTEMS—AN OVERVIEW

General Characteristics

Imagine that an earthquake has closed off the highways around your neighborhood. Grocery trucks can't enter and waste-disposal trucks can't leave, so food supplies dwindle and garbage piles up. Cells would face similar predicaments if your body's highways were disrupted. The highways are part of a circulatory system, which swiftly moves substances to and from cells. The system helps maintain a favorable neighborhood, so to speak, and this is vital work. Your differentiated cells are just too specialized to fend for themselves. They interact to maintain the volume, composition, and temperature of **interstitial fluid**, a tissue fluid bathing them. **Blood**, a circulating connective tissue, interacts with interstitial fluid. Its continuous deliveries and pickups help keep conditions tolerable for enzymes and other molecules that carry out cell activities. Interstitial fluid and blood are the body's "internal environment."

Blood flows inside blood vessels, or tubes that differ in wall thickness and diameter. A muscular pump, the **heart**, generates pressure that keeps blood flowing. As is the case for many animals, you have a *closed* circulatory system; it confines blood inside continuously connected walls of a heart and blood vessels. This isn't the case in *open* circulatory systems of arthropods, such as insects, and most mollusks. In an open system, blood flows out from the vessels into sinuses, which are small spaces in body tissues. Blood mingles with the tissue fluids and then moves back to the heart through openings in the blood vessels or the heart wall (Figure 34.3*a*,*b*).

Think about the overall "design" of a closed system. Because the heart pumps incessantly, the *volume* of blood flowing through blood vessels has to equal the heart's output at any given time. The flow's *velocity* (speed) is highest in the large-diameter transport vessels. It slows in specific body regions, where there must be enough time for blood to exchange substances with cells. The required slowdown proceeds at **capillary beds**. There, blood spreads out through many small-diameter blood vessels called **capillaries**. During any interval, the same volume of blood is moving forward through the beds as elsewhere in the system. But it's doing so at a more leisurely pace because of the great total cross-sectional area of the blood capillaries. The simple analogy given in Figure 34.3*e* may help you grasp this concept.

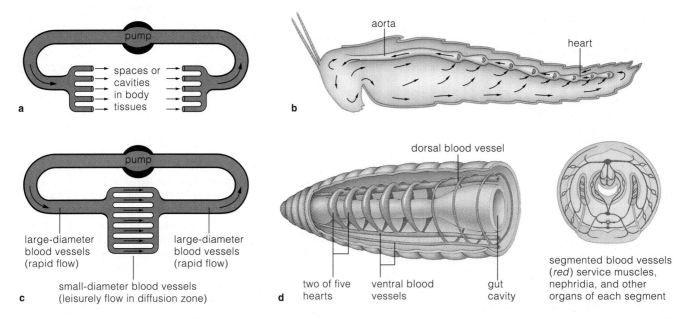

Figure 34.3 Open and closed circulatory systems. (**a**,**b**) In a grasshopper's open system, a heart (not like yours) pumps blood through a vessel, the aorta. Blood moves into tissue spaces and mingles with fluid bathing cells. It reenters the heart at openings in the heart wall. (**c**,**d**) An earthworm's closed system confines blood in several pairs of muscular hearts near the head end and inside blood vessels.

(**e**) Relationship between flow velocity and total cross-sectional area in any closed circulatory system. Visualize two fast rivers flowing into and out from a lake. The flow *rate* is the same in all three places; an identical volume of water moves from point *1* to point *3* during the same interval. Flow *velocity* decreases within the lake. Why? The volume spreads out through a larger cross-sectional area and moves forward a shorter distance during the same length of time.

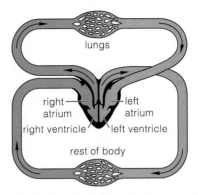

a In fishes, a two-chambered heart (atrium, ventricle) pumps blood in one circuit. Blood picks up oxygen in gills, delivers it to rest of body. Oxygen-poor blood flows back to heart.

b In amphibians, a heart pumps blood through two partially separate circuits. Blood flows to lungs, picks up oxygen, returns to heart. It mixes with oxygen-poor blood still in heart, flows to rest of body, returns to heart.

c In birds and mammals, the heart is fully partitioned into two halves. Blood circulates in two circuits: from the heart's right half to lungs and back, then from the heart's left half to oxygen-requiring tissues and back.

Figure 34.4 Comparison of the closed circulatory system of four vertebrate groups.

Evolution of Vertebrate Circulatory Systems

Humans and other existing vertebrates have a closed circulatory system, although the pump and plumbing for fishes, amphibians, birds, and mammals differ in their details. The differences evolved over hundreds of millions of years and arose during the move of some vertebrate lineages onto land. Recall from Section 27.3 that fishes, the first vertebrates, had gills. Gills, like all respiratory structures, have a thin, moist surface that oxygen and carbon dioxide can diffuse across. Later, in the ancestors of all land vertebrates, lungs evolved that supplemented gas exchange. Being *internally moistened* sacs, lungs had advantages for the move onto dry land. Also advantageous were concurrent modifications in circulatory systems, which pick up oxygen from lungs and deliver carbon dioxide wastes to them.

Consider this: In fishes, blood flows in *one* circuit (Figure 34.4*a*). Pressure generated by a two-chambered heart forces it from gill capillary beds into the largest artery, through capillary beds of organs, and back to the heart. Extensive gill capillaries offer so much resistance to flow that pressure drops considerably before blood enters the main artery. Blood delivery is enough for the activity level of most fishes. It would not be good enough for the more active life-styles of most land vertebrates.

As amphibians were evolving, their heart became partitioned into right and left halves—only partly so, but enough to pump blood through *two* partially separated circuits (Figure 34.4*b*). The flow separation continued in reptiles called crocodilians. It became complete in the two side-by-side pumps of birds and mammals (Figure 34.4*c*). Their heart's *right* half pumps oxygen-poor blood to lungs, where blood picks up oxygen and gives up carbon dioxide. The freshly oxygenated blood flows to the heart's left half. This route is the **pulmonary circuit**.

In the **systemic circuit**, the heart's *left* half pumps the freshly oxygenated blood to every tissue and organ where oxygen is used and carbon dioxide forms. Then the oxygen-poor blood flows to the heart's right half.

The double circuit is a rapid and efficient mode of blood delivery. It supports the high levels of activity typical of vertebrates whose ancestors evolved on land.

Links With the Lymphatic System

The heart's pumping puts pressure on blood flowing through the circulatory system. Partly because of this pressure, small amounts of water and a few proteins dissolved in blood are forced out of the capillaries and become part of interstitial fluid. Drainage vessels pick up excess interstitial fluid and reclaimable solutes and return them to the circulatory system. These vessels are part of the **lymphatic system**. Later, you will see how other parts of the system help cleanse fluid inside it.

Closed circulatory systems confine blood in one or more hearts and a network of blood vessels. In open systems, blood also intermingles with tissue fluids.

The closed system of vertebrates transports substances to and from interstitial fluid and is functionally connected with the lymphatic system.

Blood flows rapidly in large-diameter vessels between the vertebrate heart and capillary beds. At these beds, the flow velocity drops and exchanges are made between blood and interstitial fluid, and between interstitial fluid and cells.

In fishes, blood flows in one circuit from and back to the heart. In birds and mammals, blood flows in two circuits, through a heart partitioned as two side-by-side pumps. The double circuit supports the high levels of activity typical of vertebrates that evolved on land.

CHARACTERISTICS OF BLOOD

Functions of Blood

Blood is a connective tissue with multiple functions. It transports oxygen, nutrients, and other solutes to cells. It carries away metabolic wastes and secretions such as hormones. It helps stabilize internal pH and serves as a highway for phagocytic cells that fight infections and scavenge debris in tissues. In birds and mammals, blood helps equalize body temperature by moving excess heat from regions of high metabolic activity (such as skeletal muscles) to the skin, where heat can be dissipated.

Blood Volume and Composition

The volume of blood depends on body size and on the concentrations of water and solutes. Blood volume for average-size adult humans is about 6 to 8 percent of the total body weight. That amounts to about four or five quarts. As is the case for all vertebrates, human blood is a sticky fluid, thicker than water and slower flowing. Its components are the **plasma**, **red blood cells**, **white blood cells**, and **platelets**. Normally, plasma accounts for 50 to 60 percent of the total blood volume.

PLASMA   Prevent a blood sample in a test tube from clotting, and it separates into a red, cellular portion and the plasma, a straw-colored liquid, which floats on the cellular portion (Figure 34.5). Plasma is mostly water, and it serves as a transport medium for blood cells and platelets. Plasma also functions as a solvent for ions and molecules, including hundreds of different plasma proteins. Some of the plasma proteins transport lipids and fat-soluble vitamins through the body. Others have

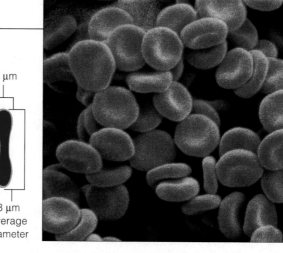

2 µm

8 µm
average
diameter

Figure 34.6 Size and shape of red blood cells.

roles in blood clotting or in defense against pathogens. Collectively, the concentrations of the plasma proteins influence the fluid volume of blood, for it affects the movement of water between the blood and interstitial fluid. Glucose and other simple sugars as well as lipids, amino acids, vitamins, and hormones are dissolved in plasma. So are oxygen, carbon dioxide, and nitrogen.

RED BLOOD CELLS   Erythrocytes, or red blood cells, are biconcave disks (Figure 34.6). They transport the oxygen used in aerobic respiration and carry away some carbon dioxide wastes. When oxygen first diffuses into blood, it binds to hemoglobin. It's this iron-containing pigment that makes red blood cells look red. Section 3.5 shows its structure. Oxygenated blood is bright red. Poorly oxygenated blood is darker red, but it appears blue in blood vessel walls near the body surface.

Stem cells in bone marrow give rise to red blood cells (Figure 34.7). Generally speaking, **stem cells** stay

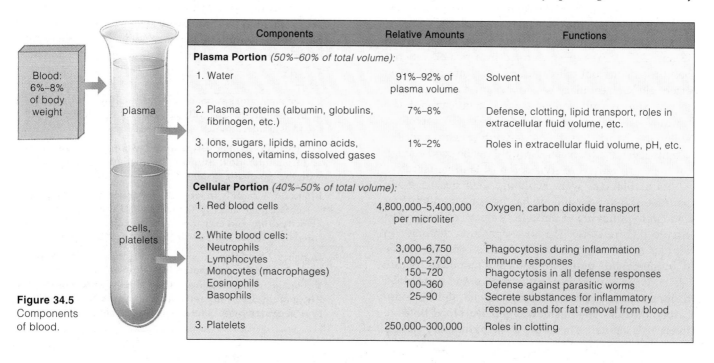

Figure 34.5
Components
of blood.

Blood:
6%–8%
of body
weight

plasma

cells,
platelets

Components	Relative Amounts	Functions
Plasma Portion (50%–60% of total volume):		
1. Water	91%–92% of plasma volume	Solvent
2. Plasma proteins (albumin, globulins, fibrinogen, etc.)	7%–8%	Defense, clotting, lipid transport, roles in extracellular fluid volume, etc.
3. Ions, sugars, lipids, amino acids, hormones, vitamins, dissolved gases	1%–2%	Roles in extracellular fluid volume, pH, etc.
Cellular Portion (40%–50% of total volume):		
1. Red blood cells	4,800,000–5,400,000 per microliter	Oxygen, carbon dioxide transport
2. White blood cells:		
Neutrophils	3,000–6,750	Phagocytosis during inflammation
Lymphocytes	1,000–2,700	Immune responses
Monocytes (macrophages)	150–720	Phagocytosis in all defense responses
Eosinophils	100–360	Defense against parasitic worms
Basophils	25–90	Secrete substances for inflammatory response and for fat removal from blood
3. Platelets	250,000–300,000	Roles in clotting

NK cells
eosinophils
neutrophils
basophils
mast cells
B lymphocytes (mature in bone marrow)
T lymphocytes (mature in thymus)

forerunners of the white blood cells (leukocytes)

stem cells that multiply and differentiate in bone marrow

?

monocytes (immature phagocytes)

red blood cells (erythrocytes)

dendritic cells

mature macrophages

megakaryocytes

platelets

Figure 34.7 Cellular components of blood. Growth factors, including erythropoietin and interleukin 3, stimulate growth and differentiation in the cell lineages. The next chapter looks at white blood cell functions. Section 34.10 describes the role of platelets in blood clotting.

unspecialized and retain their capacity for mitotic cell division. Their daughter cells divide, also, but only a portion go on to differentiate into specialized types.

Mature red blood cells no longer have their nucleus, nor do they require it. They have enough hemoglobin, enzymes, and other proteins to function for about 120 days. The body's own phagocytes engulf the oldest red blood cells and those already dead. Normally, ongoing replacements keep the cell count fairly stable.

A **cell count** is a measure of the number of cells of a given type in a microliter of blood. For example, the average number of red blood cells per cubic centimeter is 5.4 million in males and 4.8 million in females.

WHITE BLOOD CELLS Leukocytes, or white blood cells, arise from stem cells in bone marrow. They function in daily housekeeping and defense. Some types patrol tissues. They target or engulf damaged or dead cells and anything chemically recognized as foreign to the body. Many others are massed together in lymph nodes and the spleen, which are organs of the lymphatic system. There they divide to produce armies of cells that battle specific viruses, bacteria, and other threats to health. Figure 34.7 shows the main categories of white blood

cells, which differ in size, nuclear shape, and staining traits. Their numbers change, depending on whether an individual is active, healthy, or under siege. You'll read about them in the next chapter. For now, a few examples will give you a sense of what they do. Neutrophils are fast-acting phagocytes. Like basophils, they have roles in inflammatory responses to invasions and tissue damage. All immune responses against specific threats use two classes of lymphocytes: B and T cells. Macrophages and dendritic cells can call up immune responses. NK cells destroy virus-infected and cancerous body cells.

PLATELETS Some stem cells in bone marrow give rise to megakaryocytes. These "giant" cells shed cytoplasmic fragments wrapped in a bit of plasma membrane. The membrane-bound fragments are what we call platelets. Each platelet lasts only five to nine days, but hundreds of thousands are always circulating in blood. They can release substances that initiate blood clotting.

Vertebrate blood has major roles in transporting substances, defense, clotting, and maintaining the volume, composition, and temperature of the internal environment.

Blood Disorders

The body continually replaces blood cells, for good reason. Besides aging and dying off regularly, blood cells typically encounter a variety of pathogens that use them as places to complete their life cycle. Besides this, sometimes blood cells malfunction as a result of gene mutations.

RED BLOOD CELL DISORDERS Consider the **anemias**. These disorders result from too few red blood cells or deformed ones. Oxygen levels in blood cannot be kept high enough to support normal metabolism. Shortness of breath, fatigue, and chills follow. *Hemorrhagic* anemias result from sudden blood loss, as from a severe wound; *chronic* anemias result from ongoing but slight blood loss, as from an undiagnosed bleeding ulcer or hemorrhoids.

Some bacteria and protozoans replicate in blood cells. They escape by lysis and kill the cells, thus causing some *hemolytic* anemias. Insufficient iron in the diet results in *iron deficiency* anemia; red blood cells can't make enough normal hemoglobin without iron. B_{12} *deficiency* anemia is a potential hazard for strict vegetarians and alcoholics. Red blood cells form but cannot divide without vitamin B_{12}, which you can get from meats, poultry, and fish.

As described elsewhere in the book, a mutation that gives rise to an abnormal form of hemoglobin can result in *sickle-cell* anemia. Another mutation partially or fully blocks synthesis of the globin chains of hemoglobin; it causes *thalassemias*. Too few red blood cells form, and those that do are thin and fragile.

Polycythemias—symptoms of far too many red blood cells—make blood flow sluggish. Some bone marrow cancers lead to this condition. So can *blood doping*. Some athletes about to compete in strenuous events withdraw and store their own red blood cells, then reinject them a few days prior to competition. The withdrawal triggers red blood cell formation as the body attempts to replace "lost" cells, so when withdrawn cells are put back in the body, the cell count bumps up. The idea is to increase oxygen-carrying capacity and endurance. Blood doping leads to temporary high blood pressure and also lowers blood's viscosity. Some believe blood doping works, but others call it unethical. It is banned from the Olympics.

WHITE BLOOD CELL DISORDERS Possibly you've heard of *infectious mononucleosis*. An Epstein–Barr virus causes this highly contagious disease, which results from the formation of too many monocytes and lymphocytes. After a few weeks of fatigue, aches, low-grade fever, and a chronic sore throat, patients usually recover.

Recovery is chancy for *leukemias*. As you read in Chapter 11, these cancers suppress or impair the formation of white blood cells in bone marrow. Cancer cells can be killed by radiation therapy and chemotherapy, but side effects are usually severe. Remissions may last for months or years. A "remission" is a symptom-free period of a chronic illness. New gene therapies are leading to remissions for some leukemias, so far with only mild side effects.

BLOOD TRANSFUSION AND TYPING

Concerning Agglutination

Whenever blood volume or blood cell counts decline, automatic countermeasures kick in. If the volume were to decrease by more than 30 percent, circulatory shock would follow and could cause death.

Blood from donors can be transfused into patients affected by a blood disorder or severe blood loss. Such **blood transfusions** can't be hit-or-miss. Why not? *Red blood cells of a potential donor and recipient may not have the same kind of recognition proteins at their surface.* Some of the proteins are "self" markers. They identify cells as belonging to one's own body. If the donor's cells have the "wrong" marker, a recipient's immune system will recognize them as foreign, with serious consequences.

Figure 34.8 shows what happens when blood from incompatible donors and recipients intermingles. In a defense response called **agglutination**, proteins called antibodies that are circulating in the plasma act against

Figure 34.8 Light micrographs showing (**a**) an absence of agglutination in a mixture of two different yet compatible blood types and (**b**) agglutination in a mixture of incompatible types. (**c**) Responses in blood types A, B, AB, and O when mixed with samples of the same and different types.

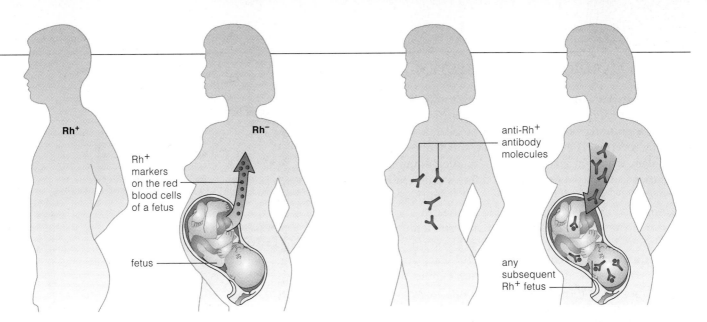

a A forthcoming child of an Rh⁻ woman and Rh⁺ man inherits the gene for the Rh⁺ marker. During childbirth, some of its cells bearing the marker may leak into the maternal bloodstream.

b The foreign marker stimulates antibody formation. If the same woman becomes pregnant again and if her second fetus (or any other) inherits the gene for the marker, her circulating anti-Rh⁺ antibodies will act against it.

Figure 34.9 Antibody production in response to Rh⁺ markers on red blood cells of a fetus.

the foreign cells and cause them to clump together. As the next chapter describes, antibodies bind to specific foreign markers and target any cell or particle bearing them for destruction by the immune system. Also, the large number of foreign cells transfused into a patient end up in numerous clumps that can clog small blood vessels and damage tissues. Without treatment, death may follow. The same thing can happen during certain pregnancies when antibodies diffuse from a mother's circulatory system into that of her unborn child.

Based on an understanding of cell surface markers and antibodies, scientists have devised ways to analyze the forms of self markers on a person's red blood cells. Blood typing is based on these analyses.

ABO Blood Typing

Molecular variations in one kind of self marker on red blood cells are analyzed with **ABO blood typing**. The genetic basis of such variation is described in Section 10.4. People with one form of the marker are said to have type A blood; those with another form have type B blood. Those with both forms of the marker on their red blood cells have type AB blood. Others do not have either form of the marker; they have type O blood.

If you are type A, your antibodies ignore A markers but will act against B markers. If you are type B, your antibodies will ignore B markers but will act against A markers. If you are type AB, your antibodies ignore both forms of the marker, so you can tolerate donations of type A, B, AB, or O blood. If you are type O, you have antibodies against both forms of the marker, so your options are limited to type O donations.

Rh Blood Typing

Rh blood typing is based on the presence or absence of the Rh marker (first identified in blood of *Rh*esus monkeys). If you are type Rh⁺, your blood cells bear this marker. If you are type Rh⁻, they do not. Usually people don't have antibodies against Rh markers. But an Rh⁻ recipient of transfused Rh⁺ blood will make antibodies against them, and these stay in the blood.

If an Rh⁻ woman becomes impregnated by an Rh⁺ man, there is a chance that the fetus will be Rh⁺. At childbirth, some fetal red blood cells typically leak into the woman's bloodstream. At that time her body will produce antibodies against Rh (Figure 34.9). And if she becomes pregnant again, Rh antibodies will enter the bloodstream of the new fetus. If its blood is type Rh⁺, her antibodies will cause fetal red blood cells to swell, rupture, and release hemoglobin.

Erythroblastosis fetalis is a severe outcome of mixing Rh⁺ and Rh⁻ types. Too many cells are killed, and the fetus dies. If the condition is diagnosed before birth, a newborn can survive when its blood is slowly replaced with transfusions that are free of Rh antibodies. Today, a known Rh⁻ woman can be treated right after her first pregnancy with an anti-Rh gamma globulin (RhoGam) that will protect her next fetus. The drug will inactivate any Rh⁺ fetal blood cells that are circulating through her bloodstream before she can become sensitized and start producing potentially dangerous antibodies.

To avoid symptoms of blood incompatibilities, red blood cells should be typed before transfusions or pregnancies.

HUMAN CARDIOVASCULAR SYSTEM

"Cardiovascular" comes from the Greek *kardia* (heart) and Latin *vasculum* (vessel). In a human cardiovascular system, a muscular heart pumps the blood into large-diameter **arteries**. The blood flows into small, muscular **arterioles**, which branch into even smaller diameter capillaries (introduced earlier). Blood flows continually from capillaries into small **venules**, and from there to large-diameter **veins** that return it to the heart.

As in most vertebrates, a partition separates a human heart into a double pump, which drives blood through two cardiovascular circuits (Figure 34.10). Each circuit has its own arteries, arterioles, capillaries, venules, and veins. The *pulmonary* circuit, a short loop, oxygenates blood. It leads from the heart's right half to capillary beds in both lungs, then returns to the heart's left half. The *systemic* circuit, a longer loop, starts at the heart's left half. Its main artery, the **aorta**, accepts oxygenated blood, which flows through arterioles, capillary beds in all regions, then veins that deliver the oxygen-poor blood to the heart's right half. Figure 34.11 identifies key blood vessels of both circuits and their functions.

For most of the systemic branchings, a given volume of blood flows through one capillary bed. The branch to the intestines is one of the exceptions. First blood picks up glucose and other absorbed substances from one bed, then moves through another capillary bed in the liver —a very important organ in nutrition. The second bed gives the liver time to process absorbed substances.

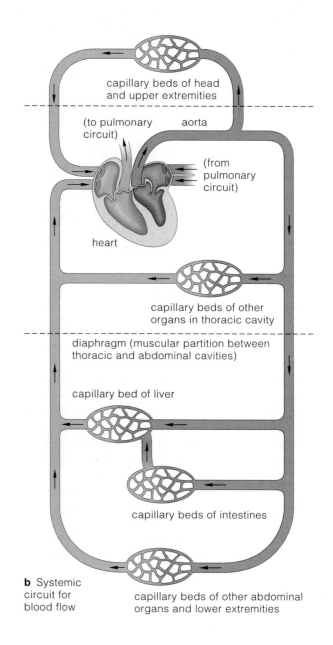

b Systemic circuit for blood flow

capillary beds of head and upper extremities

(to pulmonary circuit)   aorta

(from pulmonary circuit)

heart

capillary beds of other organs in thoracic cavity

diaphragm (muscular partition between thoracic and abdominal cavities)

capillary bed of liver

capillary beds of intestines

capillary beds of other abdominal organs and lower extremities

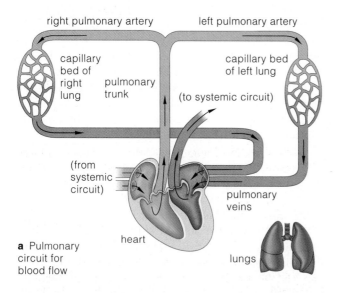

a Pulmonary circuit for blood flow

right pulmonary artery   left pulmonary artery

capillary bed of right lung   pulmonary trunk

capillary bed of left lung

(to systemic circuit)

(from systemic circuit)

pulmonary veins

heart   lungs

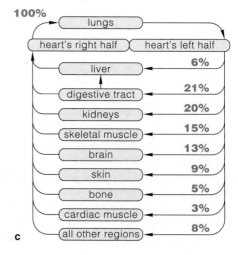

100%

lungs

heart's right half   heart's left half

liver   6%

digestive tract   21%

kidneys   20%

skeletal muscle   15%

brain   13%

skin   9%

bone   5%

cardiac muscle   3%

all other regions   8%

c

Figure 34.10 (**a**,**b**) Pulmonary and systemic circuits for blood flow through the human cardiovascular system. Blood vessels carrying oxygenated blood are color-coded *red*. Those carrying oxygen-poor blood are color-coded *blue*. (**c**) Distribution of the heart's output in a person at rest. The lungs receive all blood pumped out of the heart's right half. Organs serviced by the systemic circuit receive the portions indicated from the heart's left half. Control mechanisms adjust the flow distribution when required.

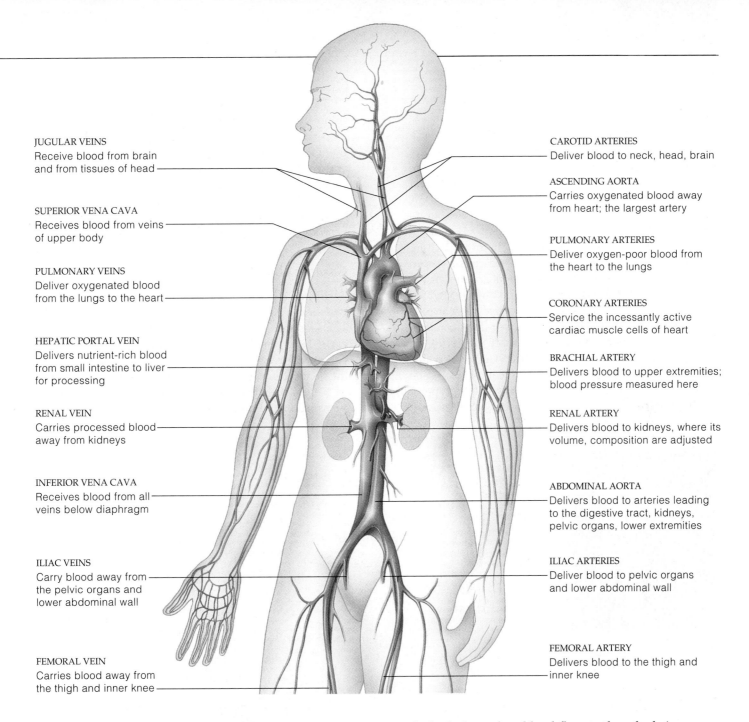

JUGULAR VEINS
Receive blood from brain and from tissues of head

SUPERIOR VENA CAVA
Receives blood from veins of upper body

PULMONARY VEINS
Deliver oxygenated blood from the lungs to the heart

HEPATIC PORTAL VEIN
Delivers nutrient-rich blood from small intestine to liver for processing

RENAL VEIN
Carries processed blood away from kidneys

INFERIOR VENA CAVA
Receives blood from all veins below diaphragm

ILIAC VEINS
Carry blood away from the pelvic organs and lower abdominal wall

FEMORAL VEIN
Carries blood away from the thigh and inner knee

CAROTID ARTERIES
Deliver blood to neck, head, brain

ASCENDING AORTA
Carries oxygenated blood away from heart; the largest artery

PULMONARY ARTERIES
Deliver oxygen-poor blood from the heart to the lungs

CORONARY ARTERIES
Service the incessantly active cardiac muscle cells of heart

BRACHIAL ARTERY
Delivers blood to upper extremities; blood pressure measured here

RENAL ARTERY
Delivers blood to kidneys, where its volume, composition are adjusted

ABDOMINAL AORTA
Delivers blood to arteries leading to the digestive tract, kidneys, pelvic organs, lower extremities

ILIAC ARTERIES
Deliver blood to pelvic organs and lower abdominal wall

FEMORAL ARTERY
Delivers blood to the thigh and inner knee

Figure 34.11 Location and functions of major blood vessels of the human cardiovascular system.

Figure 34.10*c* shows you how the two blood circuits distribute the heart's output to different organs. The percentages listed are for a person at rest. As you will see, change in levels of physical activity, plus shifting internal and external conditions, lead to adjustments in the flow distribution by the systemic route. Example: Blood flow to and from a skeletal muscle varies greatly, depending on what it's being called on to do at a given time. The same is true of skin. When the body gets too cold, blood flow—hence metabolic heat—gets diverted away from the skin's extensive capillary beds. When the body is too hot, blood flow to these beds increases and heat radiates away from the skin's surface. Only the brain is exempt; it cannot tolerate flow variations.

The human cardiovascular system consists of two separate circuits, pulmonary and systemic, for blood flow.

The pulmonary circuit is a short loop from the heart's right half, through both lungs, to the heart's left half. Oxygen-poor blood flowing into the circuit rapidly picks up oxygen and gives up carbon dioxide at capillary beds in the lungs.

The longer systemic circuit starts at the heart's left half and aorta. It delivers oxygen and accepts carbon dioxide at capillary beds of all metabolically active regions. Then its veins deliver oxygen-poor blood to the heart's right half.

THE HEART IS A LONELY PUMPER

Heart Structure

A human heart beats 2.5 billion times during a seventy-year life span, so you know it must be a durable pump. Its structure speaks of its durability. The pericardium, a double sac of tough connective tissue, protects and anchors the heart to nearby structures (*peri*, around). A fluid between its layers lubricates the heart during its perpetual twisting motions. The inner layer is actually part of the heart wall (Figure 34.12). The myocardium, the bulk of the wall, has cardiac muscle cells tethered to elastin and collagen fibers. The fibers are so densely crisscrossed, they serve as a "skeleton" against which contractile force can be applied. Cardiac muscle cells need so much oxygen, they have a separate, *coronary* circulation; coronary arteries branch off the aorta and lead to the heart's own capillary bed. The heart wall's glistening, inner layer is composed of connective tissue and endothelium. This sheet of single cells, a type of epithelium, occurs only in the heart and blood vessels.

Each half of the heart has two chambers: an atrium (plural, atria) and a ventricle. Blood flows into the atria, then the ventricles, then out through the great arteries: the aorta and pulmonary trunk. Between each atrium and ventricle is an AV valve (short for atrioventricular). Between each ventricle and the artery leading out from it is a semilunar valve. Both types are *one-way* valves. They are alternately forced open and then shut to help keep the blood moving in the forward direction only.

Cardiac Cycle

Each time the heart beats, its four chambers go through phases of systole (contraction) and diastole (relaxation). The sequence of contraction and relaxation is a **cardiac cycle**. First, relaxed atria fill with blood. Increasing fluid pressure forces both AV valves open. Blood flows into the ventricles, which fill when the atria contract (Figure 34.13). As the filled ventricles contract, the rising fluid pressure forces the AV valves shut. It rises so sharply

a

right lung left lung

1
2
3
4
5
6
7
8

diaphragm pericardium

rib cage

b

superior vena cava (from head, upper limbs)

arch of aorta

right semilunar valve (shown closed); to the pulmonary trunk

trunk of pulmonary arteries

left semilunar valve (shown closed); to aorta

right pulmonary veins (from lungs)

left pulmonary veins (from lungs)

right atrium

left atrium

right AV valve (shown open)

left AV valve (shown open)

right ventricle

left ventricle

(muscles that prevent valve from everting)

endothelium and underlying connective tissue

inferior vena cava (from trunk, legs)

septum (partition between heart's two halves)

heart's apex

myocardium

inner layer of pericardium

c

Figure 34.12   (a) Photograph of the human heart and (b) its location in the thoracic cavity. (c) Cutaway view of the heart, showing its wall and internal organization.

a Diastole (mid- to late). Ventricles fill, atria contract.

c Diastole (early). Both chambers relaxed.

b Ventricular systole (atria are still in diastole). Ventricles eject.

Figure 34.13 Blood flow during part of a cardiac cycle. Blood and heart movements generate a "lub-dup" sound at the chest wall. At each "lub," AV valves are closing as the ventricles contract. At each "dup," the semilunar valves are closing as ventricles relax.

— cardiac muscle cells

abutting plasma membranes of two cardiac muscle cells

adhering junction

gap junctions

a Abutting ends of cardiac muscle cells, longitudinal section. Tight junctions and adhering junctions hold them together. Gap junctions span the plasma membranes of both cells. They offer low electrical resistance and enhance a rapid spread of excitation from cell to cell.

Figure 34.14 (**a**) Cell junctions between two cardiac muscle cells. (**b**) Where noncontracting cardiac muscle cells that make up the cardiac conduction system are located.

SA node (cardiac pacemaker)

AV node

AV bundle

electrical bridge between atria and ventricles— the branchings of cardiac cells that conduct signals but don't contract

b

above the pressure in the great arteries that it forces the semilunar valves open—and blood leaves the heart. The ventricles relax, semilunar valves close, and the already filling atria are ready to repeat the cycle. And so, atrial contraction simply helps fill the ventricles. *Contraction of the ventricles is the driving force for blood circulation.*

Cardiac Muscle Contraction

Like skeletal muscle, cardiac muscle is striated (striped). In response to action potentials, the sarcomeres of its cells contract in the manner predicted by the sliding-filament model (Section 33.6). This tissue, too, requires ATP for contraction, which a great many mitochondria in the myocardium provide. But cardiac muscle cells are structurally unique. They are branching, short, and connected at their end regions. Gap junctions span both plasma membranes of adjoining cells. They let action potentials spread swiftly through cardiac tissue in waves of excitation that wash over the heart (Figure 34.14).

In addition, about 1 percent of the cardiac muscle cells *do not* contract. They function instead as a **cardiac conduction system**. About seventy times each minute, these specialized cells initiate and propagate waves of excitation that travel in a rhythmic, orderly sequence from atria to ventricles. Their synchronized excitation directs the heart's efficient pumping. Each wave starts

at the SA node, a cluster of cell bodies in the wall of the right atrium (Figure 34.14). It passes through the wall to another cluster of cell bodies, the AV node. This is the only electrical bridge between the atria and ventricles, which connective tissue insulates everywhere else. After the AV node, conducting cells are arranged as a bundle in the partition between the heart's two halves. These cells branch, and their branches deliver the excitatory wave up the ventricle walls. The ventricles contract in response with a twisting movement, upward from the heart's apex, that ejects blood into the great arteries.

The SA node fires action potentials faster than the rest of the system and serves as the **cardiac pacemaker**. Its spontaneous, rhythmic signals are the basis for the normal rate of heartbeat. The nervous system can only *adjust* the rate and strength of contractions dictated by the pacemaker. Even if all nerves leading to a heart are severed, the heart will keep on beating!

The heart's construction reflects its role as a durable pump. Under the spontaneous, rhythmic signals from the cardiac pacemaker, its branching, abutting cardiac muscle cells contract in synchrony, almost as if they were a single unit.

Although the heart has four chambers (each half has one atrium and one ventricle), contraction of the ventricles is the driving force for blood circulation away from the heart.

BLOOD PRESSURE IN THE CARDIOVASCULAR SYSTEM

As you have seen, blood flows to and from the human heart through arteries, arterioles, capillaries, venules, and then veins. Figure 34.15 shows the structure of these blood vessels. Arteries are the main transporters of oxygenated blood through the body. Arterioles are sites of control over the volume of blood flow through every organ. Blood capillaries and, to a lesser extent, venules are the diffusion zones of the cardiovascular system. Veins function as blood volume reservoirs and transporters of oxygen-poor blood back to the heart.

Two factors greatly influence the flow rate through each type of blood vessel. First, the flow rate is directly proportional to the pressure gradient between the start and end of the vessel. Second, the flow rate is inversely proportional to the vessel's resistance to flow.

Blood pressure is fluid pressure imparted to blood by heart contractions. In the pulmonary and systemic circuits, the *difference* in pressure between two points determines the flow rate. The beating heart establishes high pressure at the start of a circuit. Flowing blood rubs against the vessel walls, and the friction impedes flow. The major factor determining resistance to flow is blood vessel radius. Simply put, resistance increases as the tubes narrow down. Even a twofold decrease in radius increases the resistance sixteenfold.

Mainly because of frictional losses after the heart pumps out blood, pressure differs along the circuits. It is highest in contracting ventricles and still high at the start of arteries. It continues to drop until the relaxed atria, where blood pressure is lowest (Figure 34.16).

a ARTERY

b ARTERIOLE

c CAPILLARY

d VEIN

Figure 34.15 Structure of blood vessels. The basement membrane around each vessel's endothelium is a noncellular layer, rich with proteins and polysaccharides. Small venules resemble capillaries in structure and function.

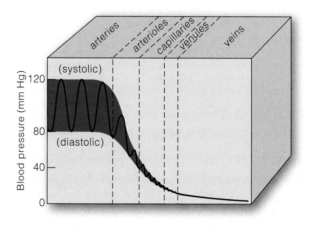

Figure 34.16 Plot of the decline in fluid pressure for a volume of blood moving through the systemic circuit.

Arterial Blood Pressure

With their large diameter and low resistance to flow, arteries are fast transporters of oxygenated blood. They also are pressure reservoirs that smooth out pulsations in pressure caused by each cardiac cycle. Their thick, muscular, elastic wall bulges from the large volume of blood that ventricular contraction forces into them. The wall recoils, and this forces the blood forward through the circuit while the heart is relaxing.

Suppose you want to measure your blood pressure daily. Typically, you'd take a reading from the brachial artery of an upper arm while resting (Figure 34.17). For one cardiac cycle, *systolic* pressure is the peak pressure that contracting ventricles exert against the artery wall. *Diastolic* pressure is the lowest arterial blood pressure of a cardiac cycle, reached when ventricles are relaxing. You can estimate the mean arterial pressure as *diastolic pressure + 1/3 pulse pressure*. (The difference between the highest and lowest readings is the pulse pressure.)

Figure 34.17 Measuring blood pressure.

A hollow cuff attached to a pressure gauge is wrapped around the upper arm. The cuff is inflated with air to a pressure above the highest pressure of the cardiac cycle—at systole, when ventricles contract. Above this pressure, you cannot hear sounds through a stethoscope positioned below the cuff and above the artery, for no blood is flowing through the vessel. Air in the cuff is slowly released, so some blood flows into the artery. The turbulent flow causes soft tapping sounds. When the tapping starts, the gauge's value is the systolic pressure, or about 120 mm mercury (Hg) in young adults at rest. This value means the measured pressure would force mercury to move upward 120 millimeters in a narrow glass column.

More air is released from the cuff. Just after the sounds grow dull and muffled, blood is flowing continuously, so the turbulence and tapping end. The silence corresponds to diastolic pressure at the end of a cardiac cycle, just before the heart pumps out blood. The reading is usually about 80 mm Hg. In this case, pulse pressure (the difference between the highest and lowest pressure readings) is 120 – 80, or 40 mm Hg. Assuming you're an adult in good health, this average resting value stays fairly constant over a few weeks or months.

How Can Arterioles Resist Flow?

Track a volume of blood through the systemic circuit and you notice the greatest pressure drop at arterioles (Figure 34.16), for they offer the greatest resistance to flow. The drop allows time for controls to divert more or less of the total volume of flow to different organs. Control proceeds at rings of smooth muscle in arteriole walls. Some control signals cause smooth muscle cells to relax, which results in **vasodilation**. The word means an enlargement, or dilation, of blood vessel diameter. Other signals cause contraction of the smooth muscle cells, which brings about a decrease in the blood vessel diameter, or **vasoconstriction**.

The nervous and endocrine systems can control the arteriole diameter. For example, sympathetic nerves end on smooth muscle cells of many arterioles. Increased activity in these nerves initiates action potentials that stimulate contraction and widespread vasoconstriction. Epinephrine and angiotensin are two of the signaling molecules that trigger changes in arteriole diameter.

Arteriole diameter also is adjusted when changes in metabolic activity shift the localized concentrations of substances in a tissue. Such local chemical changes are "selfish" in that they invite or divert the blood flow to meet the tissue's own metabolic needs. For example, when you run, the oxygen level falls in skeletal muscle tissue, and levels of carbon dioxide, potassium and hydrogen ions, and other substances rise. The changed conditions make arterioles in the vicinity dilate. Now more blood can flow through active muscles to deliver more raw materials and carry away cell products and metabolic wastes. When the muscles relax and demand fewer blood deliveries, the oxygen level increases and makes the arterioles constrict.

Controlling Mean Arterial Blood Pressure

Mean arterial pressure depends on cardiac output and on total resistance through the vascular system. Cardiac output is adjusted by controls over the rate and strength of heartbeats. The total resistance is adjusted mainly by coordinating vasoconstriction at arterioles. Of course, organs and tissues vary in their demands for blood's pickups and deliveries, so the balance between cardiac output and total resistance is juggled all the time.

A **baroreceptor reflex** is the key short-term control over arterial pressure. It starts as baroreceptors detect changes in arterial mean pressure and pulse pressure. The most vital of these mechanoreceptors are in carotid arteries (which deliver blood to the brain) and in the aortic arch (which delivers it to the rest of the body). Baroreceptors continually generate signals that reach a control center in the part of the hindbrain called the medulla oblongata. In response, the center adjusts its commands, which flow along sympathetic nerves and parasympathetic nerves to the heart and blood vessels. As an example, when the mean arterial pressure rises, the center commands the heart to beat more slowly and contract less forcefully.

Long-term controls over blood pressure are exerted at kidneys, which adjust the volume and composition of the blood. And that is a topic of another chapter.

Mean arterial pressure is an outcome of controls over the heart's output and over the total resistance to blood flow through the vascular system, as exerted mainly at arterioles.

FROM CAPILLARY BEDS BACK TO THE HEART

Capillary Function

THE NATURE OF THE EXCHANGES   Capillary beds, again, serve as diffusion zones for exchanges between blood and interstitial fluid. Living cells in any body tissue die quickly when deprived of these exchanges. Brain cells die within four minutes.

Between 10 billion and 40 billion capillaries service the human body. Collectively, these components of the circulatory system offer a tremendous surface area for gas exchanges. They extend into nearly all tissues. At least one threads as close as 0.001 centimeter to each living cell. Proximity to cells is vital, because shorter distances mean faster diffusion rates. Think about this: It would take years for oxygen to diffuse on its own from your lungs all the way down your legs. By then your toes, and the rest of you, would long be dead.

Red blood cells (8 micrometers wide) squeeze single file through capillaries, which are 3 to 7 micrometers wide and deform only a bit. The squeeze puts oxygen-transporting red blood cells, and substances dissolved in plasma, in direct contact with the exchange area—the capillary wall—or only a short distance away from it.

And remember, capillaries collectively present a greater cross-sectional area than the arterioles leading into them, so the flow velocity cannot be as great in the capillary beds. The slowdown means there is plenty of time for interstitial fluid to exchange substances with the 5 percent or so of the total volume of blood moving forward through these beds during a specified interval. Here you may wish to reflect on Figure 34.3e.

Each capillary is a single sheet of endothelial cells in the form of a tube (Figure 34.15c). The cells abut one another, but the "fit" differs in different body regions. In most cases, narrow, water-filled clefts occur between endothelial cells. In capillaries that service the brain, tight junctions join the cells and clefts are nonexistent; substances cannot "leak" between cells but must move through them. Thus the junctions are a functional part of the blood–brain barrier described in Section 30.9.

ARTERIOLE END OF CAPILLARY BED	
Outward-Directed Pressure:	
Hydrostatic pressure of blood in capillary:	35 mm Hg
Osmosis due to plasma proteins:	28 mm Hg
Inward-Directed Pressure:	
Hydrostatic pressure of interstitial fluid:	0
Osmosis due to interstitial proteins:	3 mm Hg
Net Ultrafiltration Pressure:	
(35 − 0) − (28 − 3) = 10 mm Hg	
ULTRAFILTRATION FAVORED	

VENULE END OF CAPILLARY BED	
Outward-Directed Pressure:	
Hydrostatic pressure of blood in capillary:	15 mm Hg
Osmosis due to plasma proteins:	28 mm Hg
Inward-Directed Pressure:	
Hydrostatic pressure of interstitial fluid:	0
Osmosis due to interstitial proteins:	3 mm Hg
Net Reabsorption Pressure:	
(15 − 0) − (28 − 3) = − 10 mm Hg	
REABSORPTION FAVORED	

Figure 34.18   Bulk flow in an idealized capillary bed. The fluid movement plays no significant role in diffusion of solutes. However, it is important in maintaining the distribution of extracellular fluid between blood and interstitial fluid.

The fluid movements across a capillary wall result from the two opposing effects of ultrafiltration and reabsorption.

At a capillary's arteriole end, the difference between blood pressure and interstitial fluid pressure forces some plasma, but very few plasma proteins, to leave the capillary. Plasma moves by bulk flow through clefts between endothelial cells of the capillary wall. Ultrafiltration is the bulk flow of fluid *out* of the capillary.

Reabsorption is an osmotic movement of some interstitial fluid *into* the capillary as a result of a difference in the water concentration between interstitial fluid and plasma. Plasma, with its dissolved protein components, has a greater solute concentration and therefore a lower water concentration.

Reabsorption near the end of a capillary bed tends to balance ultrafiltration at the beginning. Normally there is only a small *net* filtration of fluid, which the lymphatic system returns to the blood.

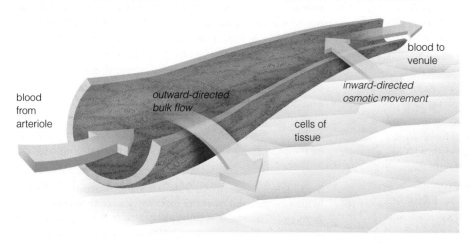

blood from arteriole

outward-directed bulk flow

inward-directed osmotic movement

cells of tissue

blood to venule

Self-Quiz   ANSWERS IN APPENDIX III

1. Cells directly exchange substances with _____ .
   a. blood vessels   c. interstitial fluid
   b. lymph vessels   d. both a and b

2. Which are *not* components of blood?
   a. plasma
   b. blood cells and platelets
   c. gases and other dissolved substances
   d. All of the above are components of blood.

3. The _____ produces red blood cells, which transport
   _____ and some _____ .
   a. liver; oxygen; mineral ions
   b. liver; oxygen; carbon dioxide
   c. bone marrow; oxygen; hormones
   d. bone marrow; oxygen; carbon dioxide

4. The _____ produces white blood cells, which function in
   _____ and _____ .
   a. liver; oxygen transport; defense
   b. lymph nodes; oxygen transport; pH stabilization
   c. bone marrow; housekeeping; defense
   d. bone marrow; pH stabilization; defense

5. In the pulmonary circuit, the heart's _____ half pumps
   blood to lungs, then _____ blood flows to the heart.
   a. right; oxygen-poor   c. right; oxygen-rich
   b. left; oxygen-poor   d. left; oxygen-rich

6. In the systemic circuit, the heart's _____ half pumps
   _____ blood to all body regions.
   a. right; oxygen-poor   c. right; oxygen-rich
   b. left; oxygen-poor   d. left; oxygen-rich

7. Blood pressure is high in _____ and lowest in _____ .
   a. arteries; veins   c. arteries; ventricles
   b. arteries; relaxed atria   d. arterioles; veins

8. _____ contraction drives blood circulation; and blood
   pressure is highest in contracting _____ .
   a. Atrial; ventricles   c. Ventricular; arteries
   b. Atrial; atria   d. Ventricular; ventricles

9. Which is *not* a function of the lymphatic system?
   a. delivers disease agents to disposal centers
   b. produces lymphocytes
   c. delivers oxygen to cells
   d. returns water and plasma proteins to blood

10. Match the type of blood vessel with its major function.
    _____ arteries   a. diffusion
    _____ arterioles   b. control of blood volume distribution
    _____ capillaries   c. transport, blood volume reservoirs
    _____ venules   d. overlap capillary function
    _____ veins   e. transport, pressure reservoirs

11. Match the components with their most suitable description.
    _____ capillary beds   a. two atria, two ventricles
    _____ lymph vascular system b. bathes body's living cells
    _____ human heart chambers   c. driving force for blood
    _____ blood   d. zones of diffusion
    _____ heart contractions   e. starts at capillary beds
    _____ interstitial fluid   f. fluid connective tissue

Critical Thinking

1. Shirelle, who is using a light microscope to examine a human tissue specimen, sees red blood cells moving single file through thin-walled tubes. She makes a photomicrograph (Figure 34.26). What type of blood vessel has she captured on film?

Figure 34.26   Light micrograph of a branching blood vessel.

2. In individuals who have weak or leaky valves in their veins, fluid pressure associated with the backflow of blood causes venous walls below the valves to bulge outward. In time, the walls become stretched and flabbily distorted, a condition called *varicose veins*. Some people are genetically predisposed to develop the condition, but the cumulative mechanical stress associated with prolonged standing, pregnancy, and aging can contribute to it. With chronic varicosity, the legs themselves become swollen. Speculate on why this swelling might happen. Also speculate on why veins close to the leg surface are more susceptible to varicosity than those deeper in the leg tissues.

3. Infection by *Streptococcus pyrogenes*, a hemolytic bacterium, may trigger an inflammation that ultimately damages valves in the heart. The disease symptoms of *rheumatic fever* follow. Explain how this disease must affect the heart's functioning and what kinds of symptoms would arise as a consequence.

Selected Key Terms

ABO blood typing 34.4
agglutination 34.4
anemia 34.3
aorta 34.5
arteriole 34.5
artery 34.5
baroreceptor reflex 34.7
blood 34.1
blood pressure 34.7
blood transfusion 34.4
capillary (blood) 34.1
capillary bed 34.1
cardiac conduction system 34.6
cardiac cycle 34.6
cardiac pacemaker 34.6
cell count 34.2
circulatory system CI
HDL 34.9
heart 34.1
hemostasis 34.10
interstitial fluid 34.1

LDL 34.9
lymph node 34.11
lymph vascular system 34.11
lymphatic system 34.1
plasma 34.2
platelet 34.2
pulmonary circuit 34.1
reabsorption 34.8
red blood cell 34.2
Rh blood typing 34.4
spleen 34.11
stem cell 34.2
systemic circuit 34.1
thymus gland 34.11
ultrafiltration 34.8
vasoconstriction 34.7
vasodilation 34.7
vein 34.5
venule 34.5
white blood cell 34.2

Readings

Baraga, M. 3 May 1996. "Finding New Drugs to Treat Stroke." *Science*, 274.

Little, R., and W. Little. 1989. *Physiology of the Heart and Circulation.* Fourth edition. Chicago: Year Book Medical.

Zamir, M. September–October 1996. "Secrets of the Heart." *The Sciences.* Effects of exercise on heart function.

On-Line readings at Student Guide for InfoTrac:
www.brookscole.com/biology

35

IMMUNITY

Russian Roulette, Immunological Style

Until about a century ago, smallpox epidemics swept repeatedly through the world's cities. Some outbreaks were so severe that only half of the stricken survived. Survivors had permanent scars on the face, neck, arms, and shoulders. Yet seldom did they contract the disease again; they were said to be "immune" to smallpox.

No one knew what caused smallpox, but the idea of acquiring immunity was dreadfully appealing. In twelfth-century China, people in good health gambled with deliberate infections. They sought out individuals with mild cases of smallpox, who displayed only mild scarring. They removed bits of crusts from the scars, ground them up, and inhaled the powder.

By the seventeenth century Mary Montagu, wife of the English ambassador to Turkey, was championing inoculation. She went so far as to poke bits of smallpox scabs into her children's skin. Others soaked threads in fluid from smallpox sores, then poked them into skin. Some people survived the chancy practices and they

acquired immunity to smallpox. Yet many developed raging infections. As if the odds weren't dangerous enough, the crude inoculation practices also made the body vulnerable to several other infectious diseases.

While this immunological version of Russian roulette was going on, Edward Jenner was growing up in the countryside of England. At the time, it was common knowledge that people who contracted cowpox never got smallpox. (Cowpox is a mild disease that can be transmitted from cattle to humans.) No one thought

Figure 35.1 (a) Statue honoring Edward Jenner's work to develop a vaccine against smallpox, one of the most dreaded diseases in human history. (b) Computer model for an antibody molecule, a type of weapon that Jenner's procedure mobilized. (c) White blood cell being attacked by HIV (*blue particles*), the virus that causes AIDS. Immunologists are working to develop effective weapons against this modern-day scourge.

much about this until 1796, when Jenner, by now a physician, injected material from a cowpox sore into the arm of a healthy boy. Six weeks later, after the reaction subsided, Jenner injected some material from *smallpox* sores into the boy (Figure 35.1*a*). He had hypothesized that the earlier injection might provoke immunity to smallpox. Fortunately he was right; the boy remained free of smallpox. Jenner had developed an effective immunization procedure against a specific pathogen.

The French mocked Jenner's procedure by calling it **vaccination**. The word literally means "encowment." Much later Louis Pasteur, an influential French chemist, developed similar immunization procedures for other diseases. In acknowledgment of Jenner's pioneering work, Pasteur also called his procedures vaccinations, and only then did the word become respectable.

By Pasteur's time, improved light microscopes had revealed bacteria, fungal spores, and other previously invisible forms of life. As Pasteur himself discovered, microorganisms abound in ordinary air. Did some cause diseases? Probably. Could they drift onto food or drink and spoil it? He demonstrated that they could and did.

Pasteur found a way to kill most of the suspected pathogens (disease agents) in food or beverages. As he and others knew, boiling killed the suspects. He also knew that boiling changes the properties of wine—or beer or milk, for that matter. He devised a way to heat beverages at a temperature low enough not to ruin them but high enough to kill most of the pathogens suspended in them. We still rely on his antimicrobial method, which was named pasteurization in his honor.

In the late 1870s Robert Koch, a German physician, linked a specific pathogen to a specific disease: anthrax. In one experiment, Koch injected blood from animals with anthrax into healthy animals. Later, the blood of the recipient animals teemed with cells of the bacterium *Bacillus anthracis*—and the animals developed anthrax. More convincing, injections of *B. anthracis* cells cultured outside of the animal body also caused the disease!

Thus, by the beginning of the twentieth century, the promise of understanding the basis of infectious disease and immunity loomed large—and the battles against those diseases were about to begin in earnest. Since that time, spectacular advances in microscopy, biochemistry, and molecular biology have increased our understanding of the body's defenses. Today we have greater insight into responses to tissue damage in general and immune responses to specific pathogens or tumor cells. The responses are the focus of this chapter.

Key Concepts

1. The vertebrate body has physical, chemical, and cellular defenses against pathogenic microorganisms, malignant tumor cells, and other agents capable of destroying tissues and the individual itself.

2. In the early stages of tissue invasion and damage, white blood cells and certain proteins in the plasma portion of blood escape from capillaries. They execute a rapid counterattack in response to a general alarm, not to the presence of a particular pathogen, so we call this a nonspecific inflammatory response. Phagocytic white blood cells ingest invading agents and clean up tissue debris. Plasma proteins promote phagocytosis, and some also destroy invaders directly.

3. If the invasion persists, certain white blood cells make immune responses. The cells can chemically recognize distinct configurations on molecules that are abnormal or foreign to the body, such as those on bacterial cells and viruses. If the foreign or abnormal molecule triggers an immune response, it is called an antigen.

4. In one type of immune response, activated B cells produce and secrete enormous quantities of antibodies. Antibodies are molecules that bind to a specific antigen, thus tagging it for destruction.

5. In another type of immune response, cytotoxic T cells directly destroy body cells already infected by a variety of intracellular pathogens as well as tumor cells.

THREE LINES OF DEFENSE

You continually cross paths with astoundingly diverse **pathogens**—the viruses, bacteria, fungi, protozoans, parasitic worms, and other agents that cause diseases. Having coevolved with most of them, you and other vertebrates have three lines of defense, so you need not lose sleep over this. Most pathogens cannot breach the body surface. If they do, they face chemical weapons and white blood cells, or leukocytes, that attack anything perceived as foreign. Other white blood cells zero in on specific targets. Table 35.1 lists the three lines of defense.

Surface Barriers to Invasion

Most often, pathogens can't get past the linings of body surfaces, including skin. Think of skin as a habitat of low moisture, low pH, and thick layers of dead cells. Normally harmless bacteria tolerate these conditions. Few pathogens can compete with dense populations of established types unless conditions change. Repeatedly enclose your toes in warm, damp shoes, for instance, and you may be inviting certain fungi to penetrate the sodden, weakened tissues and cause *athlete's foot.*

Similarly helpful bacteria live on the mucous lining of the gut or vagina. Think of *Lactobacillus* populations in the vagina. Lactate, their acidic fermentation product, helps maintain a low pH that most bacteria and fungi can't tolerate. Think of the branching, tubular airways into the lungs. When air rushes in, it flings airborne bacterial cells into the sticky, mucus-coated tube walls. That mucus contains protective substances, especially a type of **lysozyme**. Enzymes in this category digest cell walls of bacteria and thus invite their death. As a final touch, broomlike cilia lining the airways sweep away trapped and enzymatically destroyed pathogens.

There are more defenses. Tears, saliva, and gastric fluid contain lysozyme and other protective substances. Tears give eyes a sterile wash. Urine's flushing action and low pH help remove pathogens from the urinary tract. Diarrhea flushes pathogens from the gut, which is why stopping diarrhea can prolong infection. (Even so, severe cases lead to dehydration, especially in children. Oral hydration therapy uses a glucose–salt solution to reverse such dangerous losses of water and ions.)

Nonspecific and Specific Responses

All animals react to tissue damage. Even small aquatic invertebrates have phagocytic cells and antimicrobial substances, including lysozymes. But the more complex the animal, the more complex are the systems to defend it. Think back on the evolution of the circulatory and lymphatic systems of vertebrates that invaded the land (Section 34.1). As the circulation of body fluids became more efficient, so did the means to defend the body. Phagocytic cells and plasma proteins could be swiftly transported to tissues under attack and could intercept pathogens trickling along the vascular highways.

Vertebrates came to be equipped with sets of plasma proteins. Some proteins promote rapid clot formation after tissue damage; others destroy invading pathogens or target them for phagocytosis. In addition, exquisitely focused responses to specific dangers evolved.

In short, *internal defenses against a tremendous variety of pathogens are in place even before damage occurs.* Ready and waiting are white blood cells and plasma proteins. They make *nonspecific* responses to damaged tissue in general, not to a particular pathogen. Other white blood cells may recognize unique molecular configurations of a *specific* pathogen. The resulting "immune" response will run its course whether or not tissues are damaged.

Table 35.1	*The Vertebrate Body's Three Lines of Defense Against Pathogens*

BARRIERS AT BODY SURFACES (*nonspecific* targets)

Intact skin; mucous membranes at other body surfaces

Infection-fighting chemicals (e.g., lysozyme in tears, saliva)

Normally harmless bacterial inhabitants of skin and other body surfaces that can outcompete pathogenic visitors

Flushing effect of tears, saliva, urination, and diarrhea

NONSPECIFIC RESPONSES (*nonspecific* targets)

Inflammation:

1. Fast-acting white blood cells (neutrophils, eosinophils, and basophils)
2. Macrophages (also take part in immune responses)
3. Complement proteins, blood-clotting proteins, and other infection-fighting substances

Organs with pathogen-killing functions, such as lymph node

Some cytotoxic cells (e.g., NK cells) with a range of targets

IMMUNE RESPONSES (*specific* targets only)

T cells and B cells; dendritic cells, macrophages alert them

Communication signals (e.g., interleukins and interferons) and chemical weapons (e.g., antibodies, perforins)

Intact skin, mucous membranes, antimicrobial secretions, and other barriers at the body surface constitute the first line of defense against invasion and tissue damage.

Inflammation and other internal, nonspecific responses to invasion are the second line of defense.

Immune responses against specific invaders, as executed by armies of white blood cells and their chemical weapons, are the third line of defense.

COMPLEMENT PROTEINS

A set of plasma proteins has roles in nonspecific *and* specific defenses. Collectively, they are the **complement system**. About twenty kinds circulate in the blood, in inactive form. If even a few molecules of one kind are activated, they trigger cascading reactions that activate many molecules of another complement protein. Each molecule activates many molecules of another kind of protein at the next reaction step, and so on. Deploying huge numbers of molecules has the following effects.

Figure 35.2 Micrograph of the surface of a pathogenic cell. Membrane attack complexes made the pores.

pore

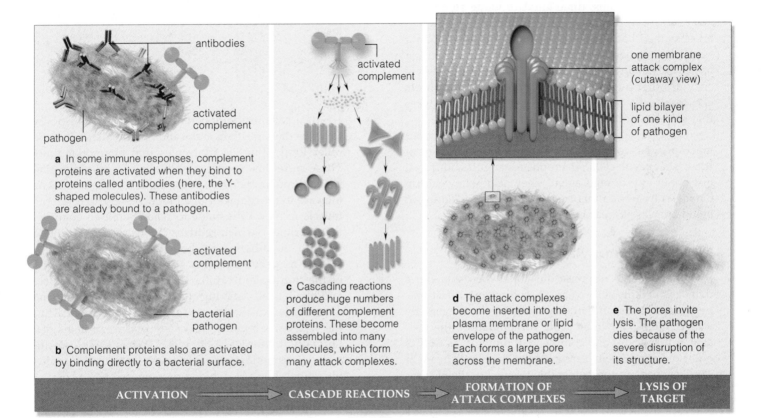

antibodies

activated complement

pathogen

activated complement

a In some immune responses, complement proteins are activated when they bind to proteins called antibodies (here, the Y-shaped molecules). These antibodies are already bound to a pathogen.

activated complement

bacterial pathogen

b Complement proteins also are activated by binding directly to a bacterial surface.

activated complement

c Cascading reactions produce huge numbers of different complement proteins. These become assembled into many molecules, which form many attack complexes.

one membrane attack complex (cutaway view)

lipid bilayer of one kind of pathogen

d The attack complexes become inserted into the plasma membrane or lipid envelope of the pathogen. Each forms a large pore across the membrane.

e The pores invite lysis. The pathogen dies because of the severe disruption of its structure.

ACTIVATION → CASCADE REACTIONS → FORMATION OF ATTACK COMPLEXES → LYSIS OF TARGET

Figure 35.3 Formation of membrane attack complexes. One reaction pathway starts when complement proteins bind to bacterial surfaces. Another operates in immune responses to specific invaders. Both pathways result in membrane pore complexes that induce lysis in pathogens.

Some complement proteins form attack complexes: molecular structures with an inner channel (Figures 35.2 and 35.3). The complexes are inserted into the plasma membrane of many pathogens and initiate lysis. They also are inserted into the wall of certain bacterial cells. **Lysis**, remember, is the gross structural disruption of a plasma membrane or cell wall; it results in cell death.

Certain activated proteins promote inflammation, a nonspecific defense response described next. By cascades of synthesis reactions, they form concentration gradients in irritated or damaged tissues. Phagocytic white blood cells follow the gradients to invasion sites. In addition, complement proteins bind to the surface of many types

of invaders. Together, they form a complement "coat" on the invaders that promotes phagocytosis.

In such ways, the complement proteins target many bacteria, parasitic protistans, and enveloped viruses.

The complement system, a set of about twenty kinds of plasma proteins circulating in blood, takes part in cascades of reactions that defend the body against many bacteria, some parasitic protistans, and enveloped viruses.

Complement proteins take part in nonspecific defenses. They also take part in immune responses, as when some are activated by antibody molecules bound to pathogens.

INFLAMMATION

The Roles of Phagocytes and Their Kin

Certain types of white blood cells take part in an initial response to a damaged tissue. White blood cells, recall, arise from stem cells in bone marrow. Many of the cells circulate in blood and lymph. A great many take up stations in lymph nodes as well as in the spleen, liver, kidneys, lungs, and brain. Here you may wish to scan Sections 34.2 and 34.11 once more; they introduce the components of blood and the lymphatic system.

Like SWAT teams, three kinds of white blood cells react swiftly to danger in general but are not adapted for sustained battles. **Neutrophils**, the most abundant kind, phagocytize bacteria. They ingest, kill, and digest bacterial cells into molecular bits. **Eosinophils** secrete enzymes that put holes in parasitic worms. **Basophils** and **mast cells** make and secrete chemical substances that help keep inflammation going after it starts.

Although slower to act, the white blood cells called **macrophages** are "big eaters." Figure 35.4 shows one of these phagocytic cells. Macrophages engulf and digest just about any foreign agent. They also help clean up damaged tissues. Immature macrophages circulating in blood are classified as monocytes.

The Inflammatory Response

An inflammatory response develops when something damages or kills cells of any tissue region. Infections, punctures, burns, and other assaults are the triggers. By a mechanism known as **acute inflammation**, the fast-acting phagocytes and complement proteins, as well as other plasma proteins, escape from the bloodstream at capillary beds in the vulnerable tissue. Localized signs that they have entered interstitial fluid and that acute inflammation is proceeding include redness, warmth, swelling, and pain, as listed in Table 35.2.

Table 35.2	*Localized Signs of Inflammation and Their Causes*
Redness	Arteriolar vasodilation increases blood flow to the affected tissue.
Warmth	Arteriolar vasodilation delivers more blood, carrying more metabolic heat, to the tissue.
Swelling	Chemical signals increase capillary permeability; plasma proteins escape and disrupt fluid balance across capillary walls; localized edema results.
Pain	Nociceptors (pain receptors) are stimulated by increased fluid pressure, local chemical signals.

Mast cells reside in connective tissues. In response to tissue damage, they make and release **histamine** and other local signaling molecules into interstitial fluid. These secretions trigger vasodilation of arterioles that snake through the tissue. Vasodilation, remember, is an increase in the diameter of a blood vessel when smooth muscle in its wall is made to relax. So blood engorges the arterioles—which reddens the tissue and warms it up with blood-borne metabolic heat.

The released histamine also increases permeability of the thin-walled capillaries in the tissue. It induces endothelial cells making up the capillary wall to pull apart farther at the narrow clefts between them. Thus the capillaries become "leaky" to plasma proteins that normally do not leave the blood. When some proteins leak out, osmotic pressure increases in the surrounding interstitial fluid. Also, blood pressure is higher because of the increased flow of blood to the tissue. As a result, ultrafiltration rises and reabsorption declines across the capillary wall. Localized edema is the outcome of the

Figure 35.4
(**a**) White blood cell squeezing out of a blood capillary at a cleft between endothelial cells. (**b**) Macrophage, caught in the act of engulfing a yeast cell.

a Bacteria invade a tissue and directly kill cells or release metabolic products that damage tissue.

b Mast cells in tissue release histamine, which then triggers arteriole vasodilation (hence redness and warmth) as well as increased capillary permeability.

c Fluid and plasma proteins leak out of capillaries; localized edema (tissue swelling) and pain result.

d Complement proteins attack bacteria. Clotting factors wall off inflamed area.

e Neutrophils, macrophages, and other phagocytes engulf invaders and debris. Macrophage secretions attract even more phagocytes, directly kill invaders, and call for fever and for T and B cell proliferation.

Figure 35.5 Acute inflammation in response to a bacterial invasion. The response involves delivering phagocytes and plasma proteins to the tissue. Collectively, these components of blood inactivate, destroy, or wall off the invaders, remove chemicals and cellular debris, and prepare the tissue for repair. These are their functions in all inflammatory responses.

shift in the fluid balance across the wall. (Here you may wish to review Section 34.8.) The tissue swells with fluid, and certain free nerve endings threading through it give rise to sensations of pain. Voluntary movements aggravate the pain, so an affected person tends to avoid movements and, in doing so, promotes tissue repair.

Within hours of the first physiological responses to the damage, neutrophils are squeezing across capillary walls. They swiftly go to work. Monocytes arrive later, differentiate into macrophages, and engage in sustained action (Figure 35.5). While macrophages are engulfing the invaders and debris, they secrete a number of local signaling molecules that act as chemical mediators.

Chemical mediators called *chemotaxins* attract more phagocytes. *Interleukins* stimulate the formation of B cell and T cell armies, as described shortly. *Lactoferrin* directly kills bacteria. *Endogenous pyrogen* triggers the release of certain prostaglandins, which have the effect of raising the set point on the hypothalamic thermostat that controls the body's core temperature. A **fever** is a core temperature that reaches the higher set point.

A fever of about 39°C (100°F) is not a bad thing. It increases body temperature to a level that is "too hot" for the functioning of most pathogens. It also promotes an increase in the host's defense activities. *Interleukin-1* induces drowsiness, which reduces the body's demands for energy. And so more energy may be diverted to the

tasks of defense and tissue repair. Macrophages take part in the cleanup and repair operations.

Besides complement proteins, the plasma proteins leaking into the tissue include clotting factors, which are activated by chemicals that the phagocytes secrete. Fibrin threads form, stick to exposed collagen fibers in the damaged tissue, and trap blood cells and platelets to form a clot (Section 34.10). By walling off inflamed areas, clots can prevent or delay the spread of invaders and toxic chemicals into surrounding tissues. After the inflammation subsides, anticlotting factors, which also escaped from the capillaries, dissolve the clots.

An inflammatory response develops in a local tissue when cells are damaged or killed, as by infection. It proceeds during both nonspecific and specific defenses of tissues.

Mast cells in damaged or invaded tissues secrete histamine, which causes arterioles to vasodilate and increases capillary permeability to fluid and to plasma proteins. The tissue warms and reddens because of the localized vasodilation. Edema results from the fluid imbalance across the capillary wall. The tissue swelling causes pain.

The response involves phagocytes such as macrophages, which engulf invaders and debris and secrete chemical mediators. It requires plasma proteins, such as complement proteins that target invaders for destruction, as well as clotting factors that wall off the inflamed tissue.

35.4

OVERVIEW OF THE IMMUNE SYSTEM

Defining Features

Sometimes physical barriers and inflammation are not enough to overwhelm an invader, so an infection may become well established. Then, the body calls upon its third line of defense—the B and T cells of the immune system. Four features define the **immune system**: self/nonself recognition, specificity, diversity, and memory. *Self/nonself recognition* and *specificity* mean that B and T cells ignore the body's own cells but attack specific foreign agents. *Diversity* means that, collectively, B and T cell receptors have the potential to respond to a billion specific threats. *Memory* means a portion of the B and T cells formed during a primary response to a foreign agent is set aside for a future battle with that agent.

Self/nonself recognition starts at unique molecular configurations that give every kind of cell, virus, and substance a unique identity. An individual's own cells bear self-marker configurations. That individual's B and T cells chemically recognize self markers and normally don't attack them. If they come across nonself markers on specific foreign agents or abnormal configurations on altered body cells, they divide again and again, and form huge populations. Any molecular configuration that incites formation of lymphocyte armies and is their target is an **antigen**. The most important antigens are proteins at the surface of pathogens and tumor cells.

All cells resulting from the divisions are sensitized to the same antigen. But some subpopulations become *effectors*: fully differentiated types that kill the enemy. Others become *memory* cells. Instead of joining the first response against a specific antigen, they enter a resting phase. But they remember that antigen. If it shows up again, they will join a larger, faster response to it.

In short, immune responses involve three events: recognition of antigen, repeated cell divisions that form huge populations of lymphocytes, and differentiation into subpopulations of effector and memory cells that have receptors for one kind of antigen.

The Key Defenders

Interactions among antigen-presenting cells, activated T helper cells, B cells, and T cytotoxic cells are central to all immune responses. Here, you can start thinking about how some of them process and display antigen, and how others recognize and attack its owners.

MHC markers are named after genes that code for them (the *Major Histocompatibility Complex*). They are among protein molecules projecting above the plasma membrane of body cells. When paired up with antigen, a cell's MHC markers represent a call to arms.

Any cell that can (1) process and display antigen in association with MHC markers and (2) activate T cells

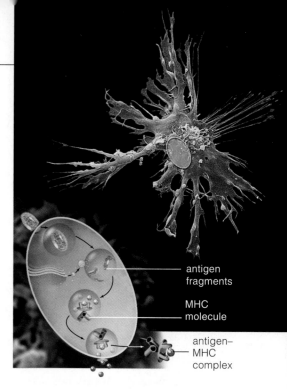

Figure 35.6
A dendritic cell —one of the major antigen-presenting cells. This class of cells patrol the blood, internal organs, and skin. (The Langerhans cells described in Section 33.2 are dendritic.) They ingest, process, and then display antigen with MHC molecules.

antigen fragments

MHC molecule

antigen–MHC complex

with a chemical signal is an **antigen-presenting cell**. Macrophages, B cells, and dendritic cells (Figure 35.6) are professionals in this enterprise. They ingest antigen, then the endocytic vesicle that forms around it fuses with a lysosome. The lysosome's potent enzymes cleave antigen molecules into fragments. Most of the fragments are exocytosed, but some associate with MHC markers and form **antigen–MHC complexes**. These move to the plasma membrane, where they are displayed.

Helper T cells can bind to antigen–MHC complexes. Binding stimulates them to produce and secrete signals that promote immune responses. The signals call for mitotic cell divisions of any B and T cell sensitive to the antigen, and for differentiation into subpopulations of effector and memory cells (Figure 35.7).

Effector **cytotoxic T cells** and **natural killer (NK) cells** carry out *cell-mediated* responses against infected body cells and tumor cells. All recognize antigen–MHC complexes, provided the MHC marker has a coreceptor called CD8 bound to it. They release chemical weapons that form pores in the plasma membrane of the targets.

Only antigen-sensitized **B cells** carry out *antibody-mediated* responses. They alone make and secrete many antigen-binding receptor molecules called **antibodies**. When macrophages, NK cells, or neutrophils contact a foreign agent with antibody bound to it, they destroy it.

The Main Targets

Any cell with antigen–MHC complexes at its surface turns on cytotoxic cells with receptors for that antigen. This includes body cells infected with an intracellular pathogen; they, too, display bits of antigen with MHC markers at their surface. Tumor cells and cells of organ transplants also are targets. Besides this, antibodies can directly inactivate antigen-bearing cells and toxins in

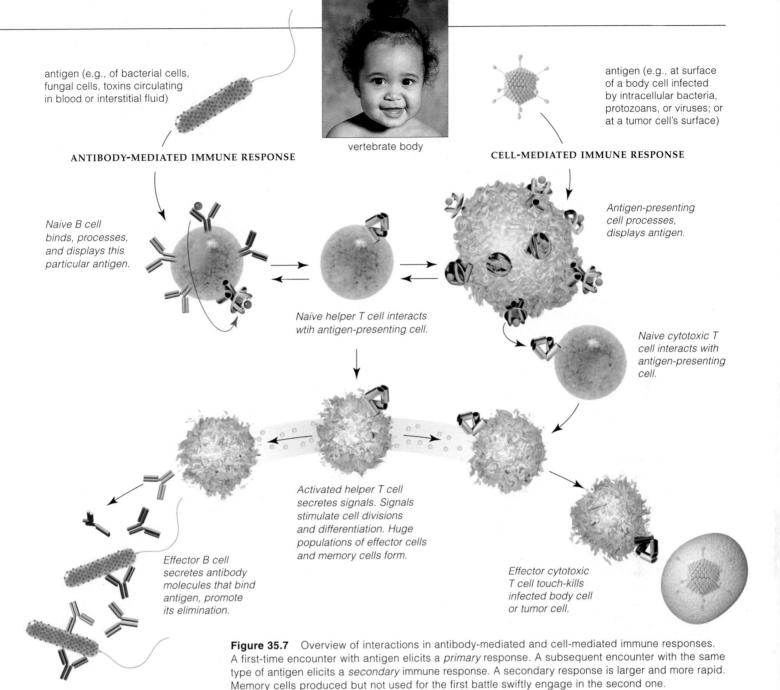

antigen (e.g., of bacterial cells, fungal cells, toxins circulating in blood or interstitial fluid)

vertebrate body

antigen (e.g., at surface of a body cell infected by intracellular bacteria, protozoans, or viruses; or at a tumor cell's surface)

ANTIBODY-MEDIATED IMMUNE RESPONSE

CELL-MEDIATED IMMUNE RESPONSE

Naive B cell binds, processes, and displays this particular antigen.

Antigen-presenting cell processes, displays antigen.

Naive helper T cell interacts wtih antigen-presenting cell.

Naive cytotoxic T cell interacts with antigen-presenting cell.

Activated helper T cell secretes signals. Signals stimulate cell divisions and differentiation. Huge populations of effector cells and memory cells form.

Effector B cell secretes antibody molecules that bind antigen, promote its elimination.

Effector cytotoxic T cell touch-kills infected body cell or tumor cell.

Figure 35.7 Overview of interactions in antibody-mediated and cell-mediated immune responses. A first-time encounter with antigen elicits a *primary* response. A subsequent encounter with the same type of antigen elicits a *secondary* immune response. A secondary response is larger and more rapid. Memory cells produced but not used for the first battle swiftly engage in the second one.

the blood or interstitial fluid. Antibodies also promote inflammation and activate complement proteins.

Control of Immune Responses

Antigen provokes an immune response—and removal of antigen stops it. For example, by the time the tide of battle turns, effector cells and their secretions already have killed most of the antigen-bearing agents in the body. With less antigen around to stimulate cells, the response declines, then stops. In addition, **suppressor T cells** release signals that help shut down both cell-mediated and antibody-mediated responses.

The immune system is defined by self/nonself recognition, specificity, diversity, and memory. Collectively, T and B cells bear receptors for millions of specific antigens: foreign molecular configurations. Recognition of antigen, not self markers on the body's own cells, triggers immune responses.

Huge T and B cell armies form by repeated mitotic cell divisions after antigen recognition. These differentiate into subpopulations of effector and memory cells, all of which are sensitized to that one kind of antigen.

Antigen-sensitized cytotoxic cells execute cell-mediated responses. Antigen-sensitized B cells execute antibody-mediated responses.

HOW LYMPHOCYTES FORM AND DO BATTLE

B and T Cell Formation

Your body is continually exposed to a mind-boggling array of antigens. However, antigen receptors of all of its B and T cell populations combined show staggering diversity—enough to recognize a billion or so different antigens. How does such diversity arise? *In each new T and B cell, gene sequences with instructions for building the antigen-binding parts of receptor molecules get shuffled, extensively and randomly, before they are expressed.* Figure 35.8 shows what happens before the antigen receptors called antibodies are synthesized. Antigen receptors are mostly proteins—polypeptide chains. While each chain is forming, its antigen-binding region folds up into a unique array of bumps and grooves having a certain distribution of charge. The only antigen in the world that will bind to that region will have complementary grooves, bumps, and distribution of charge.

As you read earlier, B cells arise from stem cells in bone marrow, and each acquires its unique antigen receptors before leaving it. As a B cell matures, it starts making many copies of a unique, typically Y-shaped antibody molecule. Each molecule moves to the plasma membrane, where the tail of the Y becomes embedded in the lipid bilayer and its two arms project above it. Soon, the cell bristles with antigen receptors—bound antibodies. It is ready to defend the body as a "naive" B cell, meaning it has not yet met up with the antigen it is genetically programmed to detect. Its membrane-bound antibodies will not recognize any antigen–MHC complexes. They will recognize antigen alone.

T cells also arise from stem cells in bone marrow. Then they migrate to the thymus gland, and there they mature and acquire unique antigen-binding receptors

Figure 35.9 Clonal selection of a B cell with the one antibody able to combine with a specific antigen. Only antigen-selected B cells and T cells give rise to a clonal population of immunologically identical cells.

called **TCRs** (short for *T-Cell Receptors*). Bristling with antigen receptors, naive T cells leave the thymus and enter the general circulation. Their TCRs will ignore free antigen; they will ignore unadorned MHC markers at the surface of the body's own cells. Their TCRs will recognize antigen–MHC complexes only.

And so each B cell or T cell bears only one type of antigen receptor. By the theory of *clonal selection*, of all cells in the body, antigen "chooses" (binds to) only the B or T cell that is displaying a receptor specific for it (Figure 35.9). Once activated, that cell will be the start of the repeated mitotic cell divisions that form a huge population of genetically identical cells—a clone—all with the same antigen receptor.

The clonal selection theory also explains how an individual acquires immunological memory of the first encounter with antigen. The term refers to the body's

Figure 35.8 Generation of antigen receptor diversity, using the antibody molecule as an example. Antibodies are proteins. Protein-building instructions are encoded in genes. In chromosomes with genes for antibodies, long DNA stretches contain different versions of the segments coding for variable regions of an antibody molecule. (Compare Figure 35.13a.)

Different V and J segments code for the variable region. A recombination event occurs in this region when each B cell is maturing. Any V segment may be joined to any of the J segments. Afterward, DNA intervening between them is excised. The new sequence is attached to a C (*Constant*) segment. This completes a rearranged antibody gene. That gene will be present in every descendant of the B cell.

a As B cell matures, different segments of antibody-coding genes recombine at random into a final gene sequence.

b The final sequence is transcribed into mRNA.

c Processing yields a mature mRNA transcript (e.g., introns excised, exons spliced).

d mRNa is translated into one of the polypeptide chains of an antibody molecule.

Figure 35.10 Immunological memory. Not all B and T cells are activated during a primary immune response to an antigen. A large number continue to circulate as memory cells, which become activated during a secondary immune response.

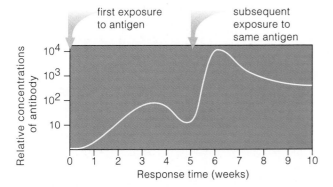

Figure 35.11 Example of the differences in magnitude and duration between a primary and a secondary immune response to the same antigen. The primary response peaked twenty-four days after it started. The secondary response peaked after seven days (the span shown between weeks five and six). Antibody concentration during the secondary response was 100 times greater, 10^4 compared to 10^2.

capacity to make a secondary immune response to any subsequent encounter with the same type of antigen that provoked the primary response (Figure 35.10).

Memory cells that form during a primary immune response don't engage in that battle. They circulate for years or for decades. Compared to a primary response, antigen is intercepted sooner by battalions that already formed and are patrolling tissues. Thus populations of effector cells form faster, so the infection is terminated before the individual gets sick. Far greater numbers of memory B and C cells form in a secondary response. Figure 35.11 gives an example. In evolutionary terms, advance preparations against further encounters with the same pathogen may bestow a survival advantage on the individual.

Figure 35.12 Lymph nodes, which contain organized arrays of antigen-presenting cells and lymphocytes. Compare Section 34.11).

Lymphocyte Battlegrounds

Where is antigen actually captured and dealt with? In the skin, mucous membranes, and internal organs, the macrophages and dendritic cells that form an early line of defense pick up antigen by means of phagocytosis or endocytosis. Then they migrate into regional lymph nodes (Section 34.11), where macrophages and dendritic cells present antigen to naive B and T cells.

Antigen in interstitial fluid enters the lymph vascular system. Because lymph drains into the blood, there is always a chance that antigen will be circulated throughout the body. However, lymph nodes trap most of it. Also, antigen that does manage to enter the blood filters through the spleen (Figure 35.12). Each day, more circulating lymphocytes move through the spleen than through all lymph nodes combined.

In lymph nodes, antigen first percolates through the region occupied by B cells, macrophages, and dendritic cells. There it is captured, processed, and presented to helper T cells, thereby activating them. Other dendritic cells that bear receptors for antibody and complement may activate B cells in the nodes. Deeper in the node, T cells start multiplying and maturing into effector and memory cells. So do B cells, which mature and start to secrete antibodies. Quantities of antibodies leave the node in lymph. So do enormous numbers of newly formed lymphocytes. At the same time, the antigen has been stimulating great numbers of lymphocytes to slip out of neighboring venules and enter the lymph node. Hence those swollen nodes you've probably felt from time to time under your jawline.

By recombination of segments drawn at random from the receptor-encoding regions of DNA, each B cell or T cell receives a gene sequence for just one of a billion possible kinds of antigen receptors.

Immunological specificity means the clonal descendants of an antigen-"chosen" cell bind only to the selecting antigen.

Immunological memory means B and T cells set aside during a primary response can make a faster and greater secondary response to another encounter with the same antigen.

Macrophages, dendritic cells, B cells, and T cells mount immune responses in lymphoid organs. They are organized for efficient interactions in the lymph nodes and spleen.

ANTIBODY-MEDIATED RESPONSE

B cells without helper T cells would be like a symphony orchestra without the conductor. (The same goes for cytotoxic T cells.) Activated helper T cells unleash both the antibody-mediated and cell-mediated responses to antigen. Their activation must be tightly controlled to put the immune fighters on suitable courses of action. Limiting how they recognize antigen is a key control. Helper T cells recognize antigen *only* when a fragment of it is associated with an MHC molecule at the surface of a B cell, macrophage, or dendritic cell. Also, after helper T cells interact with the antigen–MHC complex, they must be stimulated to divide by a chemical signal that the antigen-presenting cell releases. The resulting clone of helper T cells drives the immune responses.

The Roles of Antibodies

Let's now follow a naive B cell bristling with bound antibodies. All of its antibody molecules are alike; they are specific for only one antigen. Figure 35.13 shows a typical antibody's antigen-binding sites. When antigen binds with some of the molecules, it cross-links them together. The cross-linking activates receptor-mediated endocytosis, and so the antigen is moved into the cell (Section 5.8). There, antigen–MHC complexes form and move to the B cell surface, where they are displayed.

Suppose a responsive helper T cell meets up with a B cell's antigen–MHC complex. The two cells exchange a costimulatory signal, then disengage. When the B cell again encounters unprocessed antigen, its antibodies bind to it. The binding, in conjunction with interleukins secreted from nearby activated helper T cells, drives the B cell into mitosis. Its clonal descendants differentiate

into B effector and B memory cells (Figure 35.14). The effectors (also called plasma cells) produce and secrete huge numbers of antibodies. When freely circulating antibody binds to it, antigen gets tagged for destruction, as by phagocytes and complement activation.

Remember, the main targets of antibody-mediated responses are extracellular pathogens and toxins freely circulating in extracellular fluid. *Antibodies cannot bind to pathogens or toxins that are hidden inside a host cell.*

Classes of Immunoglobulins

B effector cells produce five classes of antibodies that are known as **immunoglobulins**, or **Igs**. These protein products of gene shufflings form when B cells mature and when they are engaged in immune responses. We abbreviate them as IgM, IgD, IgG, IgA, and IgE. Each has antigen-binding sites *and* other sites having special roles. When secreted, they assume these forms:

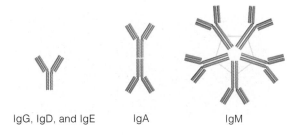

IgG, IgD, and IgE   IgA   IgM

IgM is the first one to be secreted during a primary response and the first one produced by newborns. IgM molecules aggregate into a structure having ten antigen binding sites. This makes it more efficient at binding

antigen-binding site

antigen-binding site

variable region of heavy chain

hinge region (flexible)

variable region of light chain

constant region of light chain

constant region of heavy chain (*bright green*), including the hinge region

a

antigen (*red*) on bacterial cell

binding site on one kind of antibody molecule for a specific antigen

b

antigen (*red*) on virus particle

binding site on a different kind of antibody molecule for a different antigen

c

Figure 35.13 Antibody structure. (**a**) Each antibody consists of four polypeptide chains, commonly bonded together in a Y-shaped configuration. (**b**,**c**) At the antigen-binding sites of the molecule, antigen fits into grooves and onto protrusions.

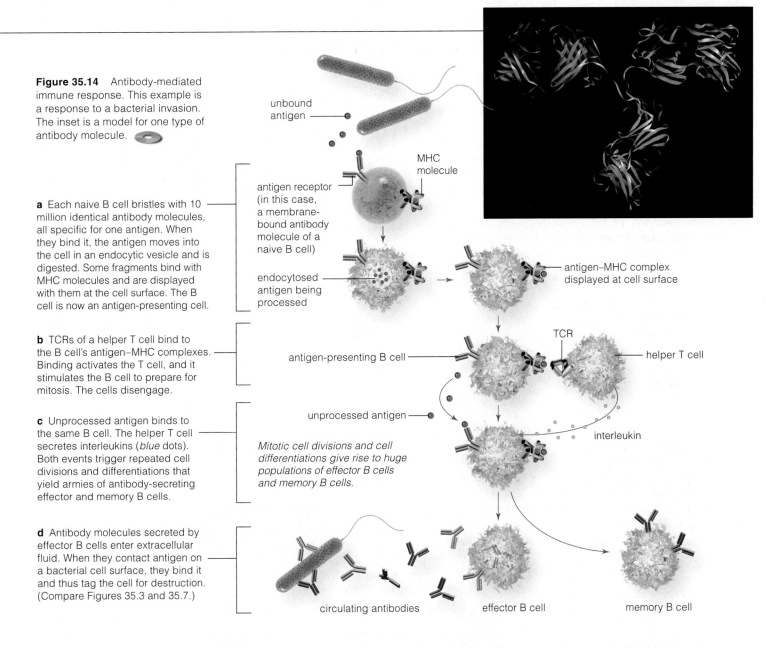

Figure 35.14 Antibody-mediated immune response. This example is a response to a bacterial invasion. The inset is a model for one type of antibody molecule.

unbound antigen

MHC molecule

antigen receptor (in this case, a membrane-bound antibody molecule of a naive B cell)

endocytosed antigen being processed

antigen–MHC complex displayed at cell surface

TCR

antigen-presenting B cell

helper T cell

unprocessed antigen

interleukin

Mitotic cell divisions and cell differentiations give rise to huge populations of effector B cells and memory B cells.

circulating antibodies

effector B cell

memory B cell

a Each naive B cell bristles with 10 million identical antibody molecules, all specific for one antigen. When they bind it, the antigen moves into the cell in an endocytic vesicle and is digested. Some fragments bind with MHC molecules and are displayed with them at the cell surface. The B cell is now an antigen-presenting cell.

b TCRs of a helper T cell bind to the B cell's antigen–MHC complexes. Binding activates the T cell, and it stimulates the B cell to prepare for mitosis. The cells disengage.

c Unprocessed antigen binds to the same B cell. The helper T cell secretes interleukins (*blue* dots). Both events trigger repeated cell divisions and differentiations that yield armies of antibody-secreting effector and memory B cells.

d Antibody molecules secreted by effector B cells enter extracellular fluid. When they contact antigen on a bacterial cell surface, they bind it and thus tag the cell for destruction. (Compare Figures 35.3 and 35.7.)

clumped targets, such as agglutinating red blood cells (Section 34.4) and dense aggregations of virus particles.

Along with IgM, *IgD* is the most common antibody bound at the plasma membrane of naive B cells. It may take part in the activation of helper T cells.

IgG makes up 80 percent or so of the blood-borne immunoglobulins. It's the most efficent one at turning on complement proteins, and it neutralizes many toxins. This long-lasting antibody easily crosses the placenta. It helps protect the developing fetus with the mother's acquired immunities. IgG secreted into early milk is also absorbed into a suckling newborn's bloodstream.

IgE triggers inflammation after attacks by parasitic worms and other pathogens. The tails of IgE antibodies lock on to basophils and mast cells, with their antigen receptors facing outward. The antigen binding induces basophils and mast cells to release histamine and thus promote inflammation. As you will see shortly, IgE also is involved in allergic reactions, including asthma, hay fever, and hives.

IgA is the main immunoglobulin in exocrine gland secretions, including tears, saliva, and breast milk. It also is in mucus that coats surfaces of the respiratory, digestive, and reproductive tracts—the areas that are vulnerable to most infectious agents. Like IgM, it can form larger structures that can bind larger antigens. Bacteria and viruses can't bind to the cells of mucous membranes when secreted IgA is bound to them. IgA is known to be effective in fighting the agents that cause salmonella, cholera, gonorrhea, influenza, and polio.

Antibodies secreted by B effector cells bind to antigens of extracellular pathogens or toxins and tag them for disposal, as by phagocytes and complement. Five antibody classes (immunoglobulins) offer defenses against diverse threats.

35.7

CELL-MEDIATED RESPONSE

If every antigen stayed out in the open in the internal environment, maybe the antibody-mediated response would be sufficient to dispose of them. But a number of pathogens evade antibodies. They hide in body cells, drain the life out of them, and often reproduce in them. They are exposed only briefly after they slip out of one cell and before they infect others. You've already read about some of the intracellular viruses, bacteria, fungi, protozoans, and sporozoans. You know about genetically altered body cells, such as tumor cells, that pose other threats. All are targets of cell-mediated responses.

Helper T cells and cytotoxic T cells, along with NK cells, macrophages, neutrophils, eosinophils, and some other types, are the executioners. Local concentrations of interleukins, interferons, and other signals get their attention and stimulate them to divide, differentiate, and attack. Interferons also call for increases in making MHC molecules, thus enhancing the capacity of some cells to display antigen. If antibodies bind with exposed antigen, they, too, will call in some of the defenders.

Cytotoxic T cells are so sensitive to antigen–MHC complexes and to altered configurations on body cells,

a A virus infects a macrophage. The host cell's pirated metabolic machinery makes viral proteins, which are antigenic. Enzymes of the macrophage cleave the antigens into fragments. Some fragments bind to a certain class of MHC molecules.

b Some antigen fragments are bound to class I MHC molecules. (These class I proteins occur on all nucleated cells and present antigen that originates inside infected cells.) Antigen–MHC complexes move to the cell surface and are displayed.

f The TCRs of a responsive cytotoxic T cell bind to the antigen–MHC complexes of the macrophage. Interleukins being secreted from the helper T cell stimulate the cytotoxic T cell to initiate cell divisions and differentiations. Large populations of effector and memory T cells form.

h An effector cytotoxic T cell contacts the infected cell and then touch-kills it. The effector cell releases perforins and toxic substances (*green* dots) onto the cell, which programs it for death.

i The effector disengages from the doomed cell and reconnoiters for more targets. Meanwhile, perforins make holes in the infected cell's plasma membrane. Toxins move into the cell, disrupt its organelles, and make the DNA disassemble. The cell dies.

c A different macrophage engulfs and digests antigen of the same type of virus. Antigen becomes bound to a different class of MHC molecules inside the macrophage.

d Antigen–MHC complexes move to the cell surface, where they are displayed.

e A responsive helper T cell binds to the antigen–MHC complexes of the antigen-presenting cell. This binding stimulates the cell to secrete interleukins (*yellow* dots). The interleukins stimulate the helper T cell to secrete different interleukins (the *blue* dots).

g The same virus infected a cell in the respiratory tract lining. Antigen synthesized by the host cell is processed. Some gets bound to MHC molecules. The complexes end up being displayed at the infected cell's surface.

Figure 35.15 Diagram of a T cell–mediated immune response.

Figure 35.16 Cytotoxic T cell caught in the act of touch-killing a tumor cell.

they don't need further signals to start dividing. Also, they display and secrete molecules used in touch-killing infected and altered body cells. Example: Perforins are protein molecules that form doughnut-shaped pores in a target's plasma membrane. The pores look similar to those shown in Figure 35.2. T cytotoxic cells persuade target cells to commit suicide by **apoptosis**, a process described in Section 14.4. The cell's cytoplasm dribbles out, its organelles are disrupted, and its DNA becomes fragmented. After making its lethal hit, the cytotoxic T cell disengages from the doomed cell, and it resumes reconnoitering (Figures 35.15 and 35.16).

Cytotoxic T cells also contribute to the rejection of tissue and organ transplants. Parts of MHC markers on donor cells are different enough from the recipient's to be recognized as antigens. Other parts are enough alike to call for the costimulatory signal. MHC typing and matching donors to recipients minimize the risk.

What about NK cells? Antigen–MHC complexes do not activate them, so other receptors must be involved. Some of their receptors block attacks on cells with MHC markers, but apparently the inhibition is lifted for cells with altered or deficient numbers of MHC molecules. However they become activated, NK cells are mobilized when infections by some viruses, intracellular bacteria, and tumor cells lead to rapidly rising concentrations of interferons and interleukins. Being the first immune cells called into battle, they buy time while others are being activated, dividing, and differentiating.

Intracellular pathogens and altered body cells are targets of cell-mediated immune responses.

Helper T cells and cytotoxic T cells are antigen-specific defenders. NK cells, macrophages, neutrophils, eosinophils, and other cells make nonspecific responses. All cytotoxic types touch-kill infected cells and tumor cells.

Cancer and Immunotherapy

Carcinomas, sarcomas, leukemia—these chilling words refer to malignant tumors in skin, bone, and other tissues. Such tumors arise when viral attack, irradiation, or chemicals alter genes and cells turn cancerous (Section 14.4). The transformed cells divide repeatedly. Unless they can be destroyed or surgically removed, they kill the individual.

Transformed cells often bear abnormal proteins and protein fragments bound to MHC. Immune responses to them may make a tumor regress but may not be enough to kill it. Also, some tumors "hide" by releasing copies of the abnormal proteins, which saturate antigen receptors. Hidden tumors that reach a critical mass may overwhelm the immune system. The idea behind *immunotherapy* is to enlist white blood cells to destroy this threat and others.

MONOCLONAL ANTIBODIES   Two decades ago, Cesar Milstein and Georges Kohler made limited amounts of antibodies that home in on tumor-specific antigens. They injected an antigen into mice, which made antibodies against it. Then they fused antibody-producing B cells from the mice with cells extracted from the B cell tumors. Clonal descendants of such hybrid cells made *monoclonal antibodies* (identical copies of antibodies). However, mice cannot produce useful amounts. Genetic engineers have now designed cows that can secrete antibodies into milk, but it may be difficult to isolate these molecules from bacteria or other pathogens that may be in the milk.

Today, genetically engineered plants can manufacture monoclonal antibodies. Cornfields might mass-produce them. Besides being cost-effective, *plantibodies* are safer than antibodies from cattle. (Few plant pathogens infect people.) The first plantibody used on human volunteers prevented infection by a bacterial agent of tooth decay.

MULTIPLYING THE TUMOR KILLERS   Lymphocytes often infiltrate tumors. Researchers have removed them from a tumor and exposed them to lymphokine, an interleukin. Large populations of tumor-infiltrating lymphocytes with enhanced killing abilities have formed. These *LAK* cells (short for *Lymphokine-Activated Killers*) appear to be somewhat effective when injected back into a patient.

THERAPEUTIC VACCINES   On the horizon are *therapeutic vaccines*—wake-up calls against specific tumor cells. One approach has been to genetically engineer antigen so it becomes obvious to the killer lymphocytes. Example: Remember those Langerhans cells (Section 33.2)? Like other dendritic cells, these phagocytes prowl on many pseudopods. Once activated, they swiftly migrate to the lymph nodes and sound the immune alarm. Dirk Schadendorf harvested dendritic cells from melanoma patients. He cultured them with ground-up tumor cells, then injected them at intervals into his patients' lymph nodes or skin. A year later, two patients had no trace of melanoma. In three others, tumors shrank more than half their size. Such vaccines may soon be available.

DEFENSES ENHANCED, MISDIRECTED, OR COMPROMISED

Immunization

Immunization refers to various processes that promote immunity against diseases. With *active* immunization, an antigen-containing preparation called a **vaccine** is taken orally or is injected into the body, as in Figure 35.17. An initial injection triggers a primary immune response. A subsequent injection—a booster—elicits a secondary response; more effector and memory cells form rapidly and provide long-lasting protection.

Many vaccines are manufactured from weakened or killed pathogens or use inactivated natural toxins. Still others are made of harmless viruses that have genes from three or more pathogens inserted into their DNA or RNA. After the vaccination, body cells use the novel genes to make antigens, and immunity is established.

Passive immunization protects individuals infected by agents of hepatitis B, diphtheria, tetanus, measles, and some other diseases. At-risk patients receive injections of a purified antibody. The best source is someone who has already made a lot of the required antibody. The effects don't last; a patient hasn't made antibodies or memory cells. But injections help counter the initial attack.

A vaccine may fail or have harmful effects. In a few reported cases, vaccines have caused immunological problems; a few have also resulted in neurological disorders. It's a good idea to assess the risks and benefits before agreeing to any immunization procedure.

Allergies

In many people, normally harmless substances provoke inflammation, excess mucus secretion, and sometimes immune responses. Such substances are **allergens**, and hypersensitivity to them is called an **allergy**. Common allergens are pollen, many drugs and foods, dust mites, fungal spores, insect venom, perfumes, and cosmetics.

Some people are genetically predisposed to develop allergies. Infections, emotional stress, or changes in air temperature also trigger reactions that otherwise might not occur. Upon exposure to an antigen, IgE is secreted and binds to mast cells. When it then binds antigen, the mast cells secrete prostaglandins, histamine, and other substances that fan inflammation. Copious amounts of mucus are secreted, and airways constrict. Stuffed-up sinuses, labored breathing, sneezing, and a drippy nose are symptoms of allergic responses in *asthma* and *hay fever* (Figure 35.18). Each year in the United States, more than 14.5 million people suffer asthma attacks.

In a few cases, the inflammatory reactions wash through the body in a life-threatening event called *anaphylactic shock*. Example: someone allergic to bee or wasp venom may die within minutes of a sting. Airways constrict massively and fluid escapes rapidly from grossly permeable capillaries. Blood pressure plummets; the circulatory system may collapse.

Antihistamines (anti-inflammatory drugs) often relieve symptoms. With desensitization programs, skin tests may identify offending allergens. Larger doses of them are given slowly. Each time, the body makes more circulating IgG, and B memory cells form. Thus IgG, not IgE, binds with the allergen.

Figure 35.17 From the Centers for Disease Control and Prevention, guidelines for immunizing children as of December 2000. Doctors routinely immunize infants and children. Low-cost or free vaccinations are available at many community clinics and health departments. You can look up more information on website http://aap.org/family/parents/immunize/htm.

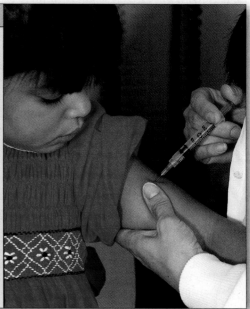

RECOMMENDED VACCINES	RECOMMENDED AGES
Hepatitis B	Birth–2 months
Hepatitis B booster	1–4 months
Hepatitis B booster	6–18 months
Hepatitis B assessment	11–12 years
DTP (Diphtheria; Tetanus; and Pertussis, or whooping cough)	2, 4, and 6 months
DTP booster	15–18 months
DTP booster	4–6 years
DT	11–16 years
HiB (*Haemophilus influenzae*)	2, 4, and 6 months
HiB booster	12–15 months
Polio	2 and 4 months
Polio booster	6–18 months
Polio booster	4–6 years
MMR (Measles, Mumps, Rubella)	12–15 months
MMR booster	4–6 years
MMR assessment	11–12 years
Varicella	12–18 months
Varicella assessment	11–12 years
Hepatitis A (in selected areas)	1–12 years

Review Questions

1. While jogging barefoot in the surf, your toes accidentally land on a jellyfish. Soon they are swollen, red, and warm to the touch. Using the diagram at right as a guide, describe events that result in these signs of inflammation. *35.3*

2. Distinguish between:
   a. neutrophil and macrophage *35.3*
   b. cytotoxic T cell and natural killer cell *35.4, 35.6*
   c. effector cell and memory cell *35.4*
   d. antigen and antibody *35.4*

3. Describe the events by which a macrophage turns into an antigen-presenting cell. *35.4*

4. Why is a vaccine to control AIDS so elusive? *35.10*

Self-Quiz (ANSWERS IN APPENDIX III)

1. _____ can deter many pathogens at body surfaces.
   a. Intact skin and mucous membranes
   b. Tears, saliva, gastric fluid
   c. Resident bacteria
   d. Urine flow
   e. all of the above

2. Complement proteins functions in defense. They _____ .
   a. neutralize toxins
   b. enhance resident bacteria
   c. promote inflammation
   d. form pore complexes that cause lysis of pathogens
   e. both a and b
   f. both c and d

3. Macrophages are derived from _____ .
   a. basophils
   b. monocytes
   c. neutrophils
   d. eosinophils

4. _____ are certain molecules that lymphocytes recognize as foreign and that elicit an immune response.
   a. Interleukins
   b. Antibodies
   c. Immunoglobulins
   d. Antigens
   e. Histamines

5. The most important antigens are _____ .
   a. nucleotides
   b. triglycerides
   c. steroids
   d. proteins

6. Immunological specificity is based on _____ .
   a. antigen receptor diversity
   b. recombinations of receptor genes
   c. mast cell proliferation
   d. both a and b
   e. all of the above

7. Antibody-mediated responses work best against _____ .
   a. intracellular pathogens
   b. extracellular pathogens
   c. extracellular toxins
   d. both b and c
   e. all of the above

8. Immunoglobulins _____ increase antimicrobial activity in mucus-coated surfaces of some organ systems.
   a. IgA   b. IgE   c. IgG   d. IgM   e. IgD

9. _____ would be a target of an effector cytotoxic T cell.
   a. Extracellular virus particles in blood
   b. A virus-infected body cell or tumor cell
   c. Parasitic flukes in the liver
   d. Bacterial cells in pus
   e. Pollen grains in nasal mucus

10. Match the immunity concepts.
   ____ inflammation
   ____ antibody secretion
   ____ fast-acting phagocyte
   ____ immunological memory
   ____ allergy
   ____ autoimmunity
   a. neutrophil
   b. effector B cell
   c. nonspecific response
   d. improper immune response to body's own components
   e. basis of secondary response
   f. hypersensitivity to allergen

Critical Thinking

1. As described in the chapter introduction, Edward Jenner lucked out. He performed a potentially harmful experiment on a boy who managed to survive it. What would happen if a would-be Jenner tried to do the same thing today?

2. Researchers have been attempting to develop a way to get the immune system to accept foreign tissue as "self." Speculate on some of the clinical applications of such a development.

3. Before each flu season, you get an influenza vaccination. This year you come down with "the flu" anyway. What do you suppose happened? (There are at least three explanations.)

4. Infection by a deadly strain of *Ebola* virus starts with high fever and flu-like aches, nausea, vomiting, and diarrhea. Blood vessels are destroyed. Blood seeps into tissues and out through the body's orifices. The liver and kidneys rapidly turn to mush. There is no vaccine. Suppose a patient is given a blood serum transfusion from another who survived the disease. Explain why a transfusion might increase chances of survival.

5. Ellen developed *chicken pox* when she was in kindergarten. Later in life, when her children developed chicken pox, she remained healthy even though she was exposed to countless virus particles daily. Explain why.

6. Write a short essay on how immune responses contribute to stability in the internal environment. (Compare Section 29.7.)

Selected Key Terms

allergen *35.9*
allergy *35.9*
antibody *35.4*
antigen *35.4*
antigen–MHC complex *35.4*
antigen-presenting cell *35.4*
apoptosis *35.7*
autoimmunity *35.9*
B lymphocyte (B cell) *35.4*
basophil *35.3*
complement system *35.2*
cytotoxic T cell *35.4*
eosinophil *35.3*
fever *35.3*
helper T cell (CD4 lymphocyte) *35.4*
histamine *35.3*
immune system *35.4*
immunization *35.9*
immunoglobulin (Ig) *35.6*
inflammation, acute *35.3*
lysis *35.2*
lysozyme *35.1*
macrophage *35.3*
mast cell *35.3*
MHC marker *35.4*
natural killer (NK) cell *35.4*
neutrophil *35.3*
pathogen *35.1*
suppressor T cell *35.4*
T lymphocyte (T cell) *35.4*
TCR *35.5*
vaccination *CI*
vaccine *35.9*

Readings

Goldsby, R., T. Kindt, and B. Osborne. 2000. *Kuby Immunology.* Fourth edition. New York: Freeman.

Pines, M. (editor). 1998. *Arousing the Fury of the Immune System.* Chevy Chase, Maryland: Howard Hughes Medical Institute.

Stine, G. 2001. *AIDS Update 2001.* Upper Saddle River, New Jersey: Prentice-Hall.

On-Line readings at Student Guide for InfoTrac: www.brookscole.com/biology

36

RESPIRATION

Conquering Chomolungma

To experienced climbers, Chomolungma may be the
ultimate challenge (Figure 36.1). The summit of this
Himalayan mountain, also known as Everest, is 9,700
meters (29,128 feet) above sea level. It is the highest
place on Earth. Iced-over vertical rock, driving winds,
blinding blizzards, and heart-stopping avalanches
await the mountain's challengers. So does the threat
that oxygen-poor air poses to the brain.

Most of us live at low elevations. Of the air we
breathe, one molecule in five is oxygen. When we travel
2,400 meters (about 8,000 feet) or more above sea level,
there is less air pressure pushing down from above. Gas
molecules are more spread out, and the breathing game
changes. We face *hypoxia*, or cellular oxygen deficiency.

Sensing the oxygen deficiency, the brain responds by
making us hyperventilate; it makes us breathe faster
and deeper than usual. Above 3,300 meters (10,000
feet), hyperventilating severely disrupts ion balances
in cerebrospinal fluid. It may lead to heart palpitations,
shortness of breath, headaches, nausea, and vomiting.
These are signs that the body's cells are, in a manner
of speaking, screaming for oxygen.

Living for months at high elevations helps climbers
adapt physiologically to the thinner air. For example,
mechanisms raise the red blood cell count, hence the
body's oxygen-carrying capacity. Professional climbers
know that, above 7,000 meters, oxygen scarcity and low
air pressure combine to make blood capillaries leaky.

Figure 36.1 A
climber inching
toward the summit
of Chomolungma,
where oxygen is
brutally scarce.

food, water intake   oxygen intake

DIGESTIVE SYSTEM   RESPIRATORY SYSTEM   elimination of carbon dioxide

nutrients, water, salts   oxygen   carbon dioxide

CIRCULATORY SYSTEM   URINARY SYSTEM

water, solutes

elimination of food residues   rapid transport to and from all living cells   elimination of excess water, salts, wastes

Figure 36.2 Interactions between the respiratory system and other organ systems, the foundation for homeostasis in complex animals.

More plasma escapes through clefts between endothelial cells of the capillary wall. Brain and lung tissues swell with excess fluid. Climbers will become comatose and die if this severe form of edema is not reversed. When prepared rescuers reach stricken climbers in time, they often zip them up in airtight bags. They pump bottled oxygen into the bag and remove carbon dioxide from it until the "air" inside is more like the air at 2,400 meters.

Few of us will ever find ourselves near the peak of Chomolungma, pushing our reliance on oxygen to the limits. Here in the lowlands, disease, smoking, and other environmental assaults push it in more ordinary ways, although the risks can be just as great.

The point is this: *Each animal has a body plan adapted to the oxygen levels of a particular habitat.* That body plan allows for **respiration**—movement of oxygen into the internal environment and movement of carbon dioxide out of it. Why is this crucial? Animals, remember, have high energy demands. Their cells require a great deal of oxygen, especially for aerobic respiration. And as this metabolic pathway releases energy from organic compounds, it produces carbon dioxide wastes.

This chapter samples a few **respiratory systems**, which function in the exchange of gases between the body and the environment. Together with other organ systems, they also contribute to homeostasis—that is, to maintaining internal operating conditions for all of the body's living cells (Figure 36.2).

Key Concepts

1. Of all organisms, multicelled animals require the most energy to drive their metabolic activities. At the cellular level, the energy comes mainly from aerobic respiration, an ATP-producing metabolic pathway that uses oxygen and produces carbon dioxide wastes.

2. By a physiological process called respiration, animals move oxygen into their internal environment and give up carbon dioxide to the external environment.

3. Oxygen diffuses into the animal body as a result of a pressure gradient. The pressure of this gas is higher in the outside air than it is in metabolically active tissues, where cells rapidly use oxygen. Carbon dioxide follows its own gradient in the opposite direction. Its pressure is higher in tissues, where it is a by-product of metabolism, than it is in the air.

4. In most respiratory systems, oxygen and carbon dioxide diffuse across a respiratory surface, such as the thin, moist respiratory membrane inside human lungs. Blood flowing through the body's circulatory system picks up oxygen and gives up carbon dioxide at this respiratory membrane.

5. Gas exchange is most efficient when the rate of air flow matches the rate of blood flow. The nervous system brings the rates into balance by controlling the rhythmic pattern and magnitude of breathing.

THE NATURE OF RESPIRATION

The Basis of Gas Exchange

A concentration gradient, recall, is a difference in the number of molecules (or ions) of a substance between two regions. Like other substances, oxygen and carbon dioxide tend to diffuse down a concentration gradient; each shows a net outward movement from the region where molecules of it are colliding more frequently (Section 5.6). The process of respiration is based on the tendency of these two gases to follow their respective concentration gradients—or, as we say for gases, to diffuse down **pressure gradients**—that form between the animal and its surroundings.

The gases don't exert the same pressure. Pump air into a flat tire near a beach in San Diego or Miami or anywhere else at sea level. You will fill it with about 78 percent nitrogen, 21 percent oxygen, 0.04 percent carbon dioxide, and 0.96 percent of some other gases. By mercury barometer measures, atmospheric pressure at sea level is about 760 mm Hg (Figure 36.3). Oxygen exerts just part of that total pressure on the tire wall. Its "partial" pressure is greater than carbon dioxide's. And so oxygen's **partial pressure**—its contribution to the total atmospheric pressure—is $760 \times 21/100$, or about 160 mm Hg. When similarly measured, carbon dioxide's partial pressure is about 0.3 mm Hg.

Gases enter and leave an animal body by crossing a **respiratory surface**, a thin layer of epithelium or some other body tissue. The surface must be kept moist at all times; gaseous molecules can't diffuse across it unless they are dissolved in fluid. What dictates the *number* of gas molecules that can move across a respiratory surface during any specified interval? According to **Fick's law**, the greater the surface area and the larger the partial pressure gradient, the faster the diffusion rate will be.

Which Factors Influence Gas Exchange?

SURFACE-TO-VOLUME RATIO   All animal body plans promote favorable rates of inward diffusion of oxygen and outward diffusion of carbon dioxide. For example, animals without respiratory organs are tiny, tubelike, or flattened, and gases diffuse directly across the body surface. These body plans meet the constraint imposed by the surface-to-volume ratio, as Section 4.1 describes. To get a sense of the ratio's effects, imagine a flatworm growing in all directions like an inflating balloon. The worm's surface area does not increase at the same rate as its volume. Once its girth exceeds even a millimeter, the diffusion distance between the body's surface and internal cells will be so great that the worm will die.

VENTILATION   Large-bodied, active animals have huge demands for gas exchange, more than diffusion alone

Figure 36.3   Atmospheric pressure as measured with a device called a mercury barometer. Part of the device is a glass tube in which the height of a column of mercury (Hg) increases or decreases, depending on air pressure outside the device. At sea level, the mercury rises to about 760 millimeters (29.91 inches) from the base of the tube. At this level, the pressure that the column of mercury exerts inside the tube is equal to the atmospheric pressure on the outside.

760 mm Hg

can satisfy. Diverse adaptations make the exchange rate more efficient. For example, a flap of tissue above many fish gills (respiratory organs) moves back and forth, thus stirring the surrounding water. The stirring brings more dissolved oxygen closer to the gills and moves carbon dioxide away. Among vertebrates, a circulatory system rapidly transports oxygen to cells and carbon dioxide to gills or lungs for disposal. As another example, you breathe in and out to ventilate your lungs.

TRANSPORT PIGMENTS   Rates of gas exchange get a boost with transport pigments—mainly **hemoglobin**—that help maintain the steep pressure gradients across the respiratory surface. For example, at the respiratory surface in a human lung, the oxygen concentration is high, and each hemoglobin molecule in blood weakly binds up to four oxygen molecules (Section 3.5.) The circulatory system swiftly transports hemoglobin away from the respiratory surface. In oxygen-poor tissues, oxygen follows its gradient and diffuses away from the hemoglobin, thereby helping maintain the gradient that entices oxygen into blood. **Myoglobin**, a similar but smaller pigment, has a higher oxygen-binding capacity. It's notably abundant in heart muscle cells and skeletal muscle cells that contract slowly and resist fatiguing.

As the chapter introduction states, respiration is a process by which animals move oxygen into the internal environment, especially for use in aerobic respiration, then rids the body of this pathway's carbon dioxide wastes.

Oxygen and carbon dioxide enter and leave the internal environment by diffusing across a moist respiratory surface. Like other atmospheric gases, they tend to move down their respective pressure gradients. Each type of gas exerts only part of the total pressure across a respiratory surface.

Gas exchange depends on steep partial pressure gradients between the outside and inside of the animal body. The greater the area of the respiratory surface and the larger the partial pressure gradient, the faster diffusion will proceed.

INVERTEBRATE RESPIRATION

Flatworms, earthworms, and many other invertebrates are not massive, and their life-styles do not depend on high metabolic rates (Figure 36.4a). Demands for gas exchange are simply met by **integumentary exchange**, in which gases diffuse directly across the body's surface covering (integument). This mode of respiration only works when the surface is moist, and invertebrates that rely exclusively on it are restricted to aquatic or damp habitats. (By contrast, amphibians and some other large animals use integumentary exchange also, but only as a supplement to other modes of respiration.)

Many invertebrates of aquatic habitats have moist, thin-walled respiratory organs called **gills**. Extensively folded gill walls have an increased respiratory surface area that enhances the rates of exchange between blood or some other body fluid and the surroundings. Figure 36.4b shows the much-folded gill of a sea hare (*Aplysia*). By supplementing integumentary exchange, the gill helps provide enough oxygen for this big mollusk. Some sea hares are 40 centimeters (nearly 16 inches) long.

Spiders and other invertebrates of dry habitats have a small, thick, or hardened integument that is not well

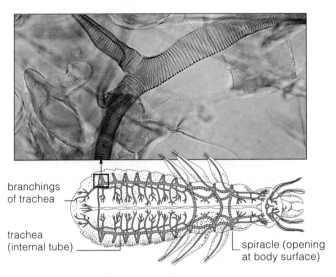

branchings of trachea

trachea (internal tube)

spiracle (opening at body surface)

Figure 36.5 General plan of insect tracheal systems. Chitin rings reinforce many of the branching tubes of the system.

endowed with blood vessels. Their integument holds in precious water, but it is not a good respiratory surface. Such animals have an *internal* respiratory surface. For instance, most spiders have internal respiratory organs with thin, folded walls that resemble book pages; hence the name, book lungs (Section 23.12). Like most insects, millipedes, and centipedes, some other spiders have a system of internal tubes for **tracheal respiration**.

Figure 36.5 shows the tracheal system of an insect. Small openings perforate the insect's integument. Each opening, a spiracle, is the start of a tube that branches within the body. Each of the last branchings dead-ends at its fluid-filled tip, where gases diffuse directly into tissues. The tube tips are especially profuse in muscle and other tissues with high oxygen demands.

We find hemoglobin or other respiratory pigments in many invertebrates, although they are rare in insects. Respiratory pigments increase the capacity of body fluids to transport oxygen, as they do in vertebrates. In species having a well-developed head, oxygenated blood tends to circulate first through the head end and then through the rest of the body.

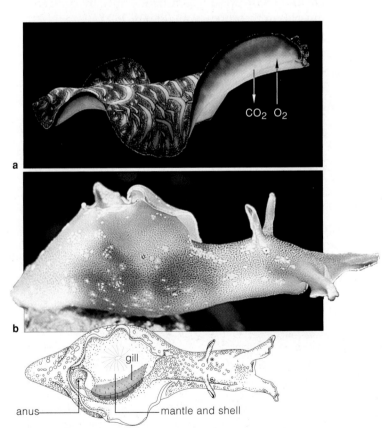

CO₂ O₂

a

b

gill

anus

mantle and shell

Figure 36.4 Invertebrates of aquatic habitats. (**a**) A flatworm, small enough to get along well without an oxygen-transporting circulatory system. Dissolved oxygen in such habitats reaches individual cells simply by diffusing across the body surface. (**b**) Gill of a sea hare (*Aplysia*), one of the gastropods.

Flatworms and some other invertebrates of aquatic or moist habitats do not have a massive body. By their simple mode of respiration, called integumentary exchange, oxygen and carbon dioxide diffuse directly across the body surface.

Most marine invertebrates and many freshwater types have gills of one sort or another. These respiratory organs have moist, thin, and often highly folded walls.

Most insects, millipedes, centipedes, and some spiders use tracheal respiration. Gases flow through open-ended tubes that start at the body surface and end directly in tissues.

VERTEBRATE RESPIRATION

Gills of Fishes and Amphibians

Many vertebrates use gills as respiratory organs. A few kinds of fish larvae and a few amphibians have *external* gills projecting into water. Adult fishes have a pair of *internal* gills: rows of slits or pockets at the back of the mouth that extend to the body's surface (Figure 36.6a). Whatever its form, each gill has a wall of moist, thin, vascularized epithelium.

In fishes, water flows into the mouth and pharynx, then over arrays of filaments in the gills (Figure 36.6b). Blood vessels thread through the respiratory surfaces in each filament. First water flows past a vessel leading to the rest of the body. The blood inside it has less oxygen than the water does, so oxygen diffuses into the blood. The same volume of water flows over a vessel leading into the gills. Although it already gave up some oxygen, it still holds more than the blood in this vessel does. So more oxygen diffuses into the filament. Any movement of two fluids in opposing directions is a **countercurrent flow**. With such a mechanism, a fish can extract about 80 to 90 percent of the oxygen dissolved in the water flowing past. That is more than the fish would get from a one-way flow mechanism at far less energy cost.

Lungs

Some fishes and all amphibians, birds, and mammals have a pair of **lungs**, internal respiratory surfaces in the shape of a cavity or sac. More than 450 million years ago, lungs started to develop in some fish lineages as outpouchings of the gut wall. They evolved rapidly by way of natural selection, possibly because they offered survival advantages. In oxygen-poor habitats, a pair of lungs afforded a larger surface area for gas exchange. Also, when some vertebrates entered dry habitats on land, lungs were better than gills (Section 24.6). Why? Gill filaments stick together and can't function unless water flows through them and keeps them moist.

Lungfishes of oxygen-poor habitats still have gills. They also use tiny lungs as a backup for gas exchange.

Amphibians never completed the transition to land. Salamanders especially use their skin as a respiratory surface for integumentary exchange. Frogs and toads rely more on small lungs for oxygen uptake, but most of their carbon dioxide wastes diffuse outward across the skin. Frogs also are heavy-duty breathers; they *force* air into their lungs, which they empty by contracting muscles in their body wall (Figure 36.7).

Water flows in through open mouth

FISH GILL

Water flows over gills, then out.

a

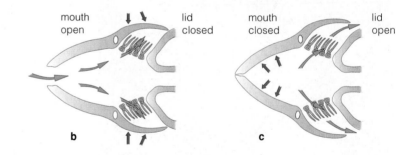

mouth open — lid closed

mouth closed — lid open

b

c

Figure 36.6 Example of fish gills. (**a**) One of a pair of gills. A bony lid, removed for this sketch, protects each gill.

(**b**) Gill ventilation. Opening the mouth and closing the lid pulls water across the gill. (**c**) Closing the mouth and opening the lid forces water to the outside.

(**d**) Gas exchange at a gill. Filaments in a gill have vascularized respiratory surfaces. The filament organization directs incoming water past the gas exchange surfaces. (**e**) A blood vessel delivers oxygen-poor blood from the body into each filament. Another delivers oxygenated blood into the body. Blood flows from one into the other and runs counter to the direction of water flow over respiratory surfaces. This countercurrent flow favors the movement of oxygen (down its partial pressure gradient) from water into the blood.

gill arch

gill filament

d A fish exhales water past its gill filaments, which are organized to direct the water past a gas exchange surface.

respiratory surface

direction of water flow

direction of blood flow

oxygenated blood back toward body

oxygen-poor blood from deep in body

e An efferent blood vessel (*blue*) delivers oxygen-poor blood to the filament. An afferent vessel (*red*) transports oxygenated blood back to the body. Blood in capillaries connecting the two vessels flows counter to the direction of water flowing over the gas exchange surface.

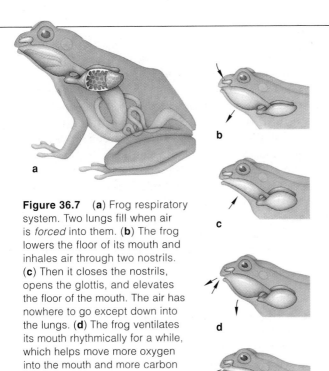

Figure 36.7 (**a**) Frog respiratory system. Two lungs fill when air is *forced* into them. (**b**) The frog lowers the floor of its mouth and inhales air through two nostrils. (**c**) Then it closes the nostrils, opens the glottis, and elevates the floor of the mouth. The air has nowhere to go except down into the lungs. (**d**) The frog ventilates its mouth rhythmically for a while, which helps move more oxygen into the mouth and more carbon dioxide out of it. (**e**) Muscles in the body wall above the lungs contract, the lungs elastically recoil, and air is forced out.

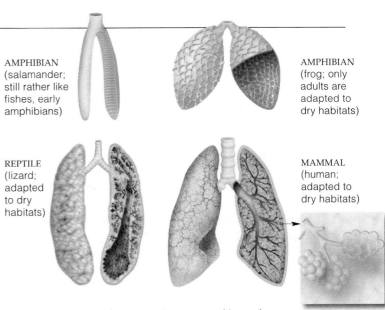

Figure 36.8 Generalized structure of lungs from four vertebrates, suggestive of an evolutionary trend from sacs for simple gas exchange to a larger, more complex respiratory surface area.

AMPHIBIAN (salamander; still rather like fishes, early amphibians)

AMPHIBIAN (frog; only adults are adapted to dry habitats)

REPTILE (lizard; adapted to dry habitats)

MAMMAL (human; adapted to dry habitats)

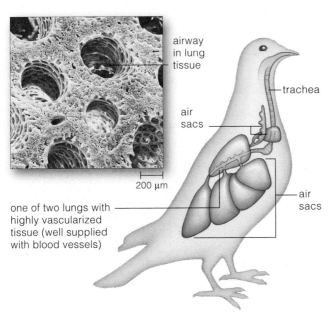

airway in lung tissue

trachea

air sacs

200 μm

one of two lungs with highly vascularized tissue (well supplied with blood vessels)

air sacs

Figure 36.9 A bird's respiratory system. Two small, inelastic lungs have a number of attached air sacs. When the bird inhales, it draws air into the sacs through tubes, open at both ends, that thread through vascularized lung tissue. This tissue is the respiratory surface, where gases are exchanged.

When the bird exhales, it forces air out of the sacs, through the tubes, and out of the trachea. Thus, air isn't just drawn into bird lungs. It is continuously drawn through them and across respiratory surfaces. This unique ventilating system supports the high metabolic rates that birds require for flight and other energy-intensive activities.

Frogs also use the lungs to make sounds. So do all mammals except whales. Sounds start near the entrance to a larynx. Part of a mucous membrane folds into this airway to the lungs. The folds are **vocal cords**; the gap between them is a **glottis**. Frogs force air back and forth through the glottis, and between the lungs and a pair of pouches on the floor of the mouth. Air flow makes the vocal cords vibrate. Like other vertebrates, frogs make different sounds by controlling the vibrations.

Paired lungs are the dominant respiratory organs in reptiles, birds, and mammals (Figure 36.8). Breathing moves air by bulk flow into and out of the lungs, where capillaries thread lacily around the respiratory surface. Concentration gradients for oxygen and carbon dioxide are so steep that both gases diffuse rapidly across the surface. Oxygen enters the capillaries and is circulated quickly through the body. In regions with low oxygen levels, it diffuses into interstitial fluid, then into cells. Carbon dioxide moves quickly in the other direction and is expelled from the lungs.

This mode of gas exchange is embellished a bit only in birds. As described in Figure 36.9, birds are unique in that air not only flows into and out from their lungs; it also flows *through* their lungs. With that exception in mind, we turn next to the human respiratory system, for its operating principles apply to most vertebrates.

A countercurrent flow mechanism in fish gills compensates for low oxygen levels in aquatic habitats. Internal air sacs—lungs—are more efficient in dry habitats on land.

Amphibians use integumentary exchange and force air into and out of small lungs. Ventilation of paired lungs is the major mode of respiration in reptiles, birds, and mammals.

HUMAN RESPIRATORY SYSTEM

What Are the System's Functions?

Getting oxygen from air and expelling carbon dioxide are the key functions of the human respiratory system. Breathing, a form of ventilation, alternately moves air into and out of its paired lungs. Each lung has about 300 million outpouchings. These tiny air sacs are called **alveoli** (singular, alveolus). Control mechanisms adjust breathing rates so that the inflow and outflow of air at alveoli match metabolic demands for gas exchange.

The system has other functions. Breathing is used in vocalizations, such as speech. It enhances the return of venous blood to the heart and helps dispose of excess heat and water. Breathing controls adjust the body's acid–base balance. Carbon dioxide, recall, helps form carbonic acid (H_2CO_3), which works with bicarbonate (HCO_3^-) as a buffer system. HCO_3^- accepts or releases hydrogen ions (H^+), depending on blood's pH. Deep, rapid breathing expels more carbon dioxide, so less carbonic acid forms and the body loses acid. Shallow, slow breathing has the opposite effect. Carbon dioxide builds up, more H_2CO_3 forms, so the body gains acid.

The respiratory system also has built-in mechanisms for dealing with airborne foreign agents and substances

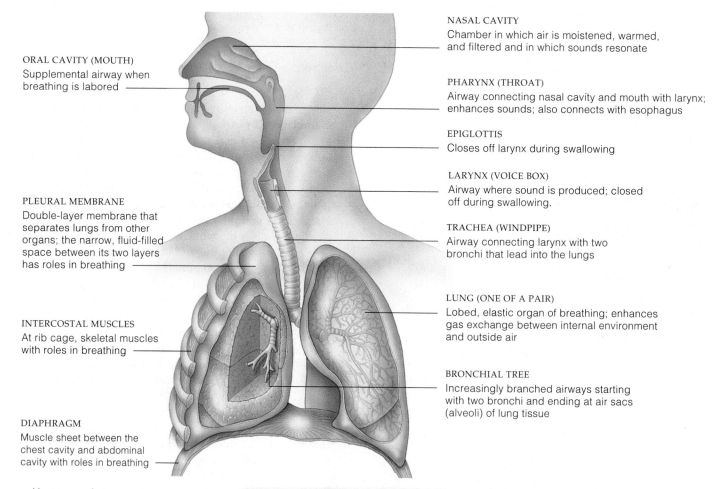

ORAL CAVITY (MOUTH)
Supplemental airway when breathing is labored

PLEURAL MEMBRANE
Double-layer membrane that separates lungs from other organs; the narrow, fluid-filled space between its two layers has roles in breathing

INTERCOSTAL MUSCLES
At rib cage, skeletal muscles with roles in breathing

DIAPHRAGM
Muscle sheet between the chest cavity and abdominal cavity with roles in breathing

NASAL CAVITY
Chamber in which air is moistened, warmed, and filtered and in which sounds resonate

PHARYNX (THROAT)
Airway connecting nasal cavity and mouth with larynx; enhances sounds; also connects with esophagus

EPIGLOTTIS
Closes off larynx during swallowing

LARYNX (VOICE BOX)
Airway where sound is produced; closed off during swallowing.

TRACHEA (WINDPIPE)
Airway connecting larynx with two bronchi that lead into the lungs

LUNG (ONE OF A PAIR)
Lobed, elastic organ of breathing; enhances gas exchange between internal environment and outside air

BRONCHIAL TREE
Increasingly branched airways starting with two bronchi and ending at air sacs (alveoli) of lung tissue

a Human respiratory system

Figure 36.10 **(a)** Components of the human respiratory system and their functions. Muscles, including the diaphragm, and parts of the axial skeleton have secondary roles in respiration. **(b,c)** Location of alveoli relative to bronchioles and to the pulmonary (lung) capillaries.

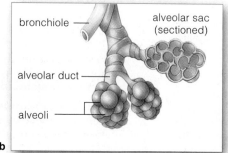

bronchiole

alveolar sac (sectioned)

alveolar duct

alveoli

b

alveolar sac

pulmonary capillary

c

inhaled with air. Finally, it removes and inactivates or otherwise modifies numerous blood-borne substances before they can circulate through the rest of the body.

The respiratory system's role in respiration ends at alveoli. There, the circulatory system takes over the task of moving gases. Oxygen and carbon dioxide diffuse across both the alveoli and the pulmonary capillaries next to them. (The Latin *pulmo* means lung.)

From Airways Into the Lungs

It will take at least 300 million breaths to get you to age seventy-five. You might find yourself going without food for a few hours or days. But stop breathing even for five minutes and normal brain function is over.

Take a deep breath. Now look at Figure 36.10 to get an idea of where the air will travel in your respiratory system. Unless you are out of breath and panting, air has entered two nasal cavities, not your mouth. There, secretions of mucus warm and moisten it. In the cavities, ciliated epithelium and hairs filter dust and particles from air, and olfactory receptors function in the sense of smell (Section 31.6). Now air is poised at the **pharynx**, or throat. The pharynx is the entrance to the **larynx**, an airway having two paired folds of mucous membrane. The lower pair are vocal cords (Figure 36.11). When you breathe, air is forced in and out through the glottis, the gap between these cords. As in frogs (Section 36.3), the air flow makes the vocal cords vibrate in ways that can be controlled to make different sounds.

Inside the folds are thick bands of elastic ligaments attached to cartilage. When the larynx muscles contract and relax, these ligaments tighten or slacken to change how much the folds are stretched. The nervous system coordinates the narrowing and widening of the glottis. For example, when it calls for increased tension in the larynx muscles, the gap between the vocal cords shrinks and you make high-pitched sounds and squeaks. The lips, teeth, tongue, and the soft roof above the tongue are enlisted to modify different sounds into patterns of vocalization, such as speech and song.

When vocal cords are inflamed as an outcome of an infection or irritation, swelling of their mucous lining interferes with their capacity to vibrate. If the swelling causes hoarseness, this condition is called *laryngitis*.

The **epiglottis** is a tissue flap at the entrance to the larynx. When it points upward, air can move into the **trachea**, or windpipe. The epiglottis points downward and shuts off the trachea's entrance when you swallow. Food and fluids enter a different tube (the esophagus) that connects the pharynx with the stomach.

The trachea branches into two airways, which lead into tissues of the lungs. Each airway is a **bronchus** (plural, bronchi). Its epithelial lining has a profusion of

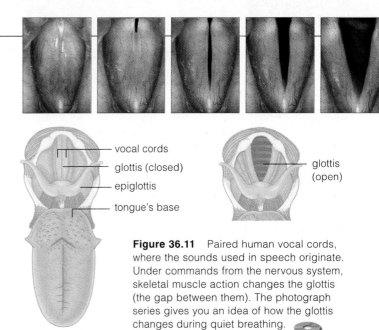

Figure 36.11 Paired human vocal cords, where the sounds used in speech originate. Under commands from the nervous system, skeletal muscle action changes the glottis (the gap between them). The photograph series gives you an idea of how the glottis changes during quiet breathing.

vocal cords
glottis (closed)
epiglottis
tongue's base
glottis (open)

cilia and mucus-secreting cells. The lining is a barrier to infection. Bacteria and airborne particles stick to the mucus, and cilia sweep the debris-laden mucus toward the mouth. The mucus is expelled or swallowed.

Human lungs are elastic, cone-shaped organs of gas exchange. They are located inside the rib cage to the left and right of the heart and above the diaphragm. The **diaphragm** is a muscular partition between the thoracic and abdominal cavities. The saclike pleural membrane lines the outer surface of the lungs and inner surface of the thoracic cavity's wall. The thoracic wall and lungs press its thin, opposing surfaces together.

A thin film of lubricating fluid cuts friction between the pleural membrane surfaces. In *pleurisy*, a respiratory ailment, the membrane becomes inflamed and swollen. Its two surfaces rub each other, so breathing is painful.

Inside each lung, air moves through finer and finer branchings of a "bronchial tree" (Figure 36.10*a*). These airways are **bronchioles**. Their finest branchings, the *respiratory* bronchioles, end in the cup-shaped alveoli. Alveoli usually are clustered as larger pouches called alveolar sacs, as in Figure 36.10*b,c*. Collectively, alveolar sacs offer a tremendous surface area for gas exchange with blood. If all the alveolar sacs were stretched out in one layer, they would cover a racquetball court floor!

Oxygen uptake and carbon dioxide removal are the major functions of the human respiratory system. Inside its pair of lungs, the circulatory system takes over the remaining tasks of respiration.

The respiratory system also has roles in moving venous blood to the heart, in vocalization, in adjusting the body's acid–base balance, in defense against harmful airborne agents or substances, in removing or modifying a number of blood-borne substances, and in the sense of smell.

CYCLIC REVERSALS IN AIR PRESSURE GRADIENTS

The Respiratory Cycle

There is a cyclic pattern to breathing, which ventilates the lungs. Each **respiratory cycle** consists of two actions: *inhalation* (a single breath of air drawn into the airways) and *exhalation* (a single breath out).

Look at Figure 36.12. *Inhalation always is an active, energy-requiring action.* When you are breathing quietly, inhalation is brought about by the contraction of the diaphragm and, to a lesser extent, external intercostal muscles. The outcome is an increase in the thoracic cavity volume. Breathe hard, and the volume increases more because neck muscles contract, thereby elevating the sternum and first two ribs attached to them.

During each respiratory cycle, the thoracic cavity's volume increases, then decreases. So pressure gradients between the air inside and outside the respiratory tract change. *Atmospheric* pressure, 760 mm Hg at sea level, is exerted by the combined weight of all the atmospheric gases on all airways. Before inhalation, *intrapulmonary* pressure (the pressure inside all alveoli) is also 760 mm Hg (Figure 36.12*d*).

A different gradient helps keep the lungs close to the thoracic cavity wall during the cycle even through exhalation, when lungs have a smaller volume than the thoracic cavity (Figure 36.12*e,f*). As the cavity expands, the lungs expand also, as a result of a pressure gradient across the lung wall.

In a person who is resting, the *intrapleural* pressure (inside the pleural sac) averages 756 mm Hg, which is less than atmospheric pressure. Intrapleural pressure is exerted outside the lungs (inside the thoracic cavity). As intrapleural pressure pushes inward on a lung wall, the intrapulmonary pressure is pushing outward. The difference between them—4 mm Hg—is large enough to make the lungs stretch and fill the thoracic cavity.

The cohesiveness of water molecules present in the intrapleural fluid (fluid in the pleural sac) also helps keep lungs close to the thoracic wall. By analogy, wet two small panes of glass and press them together. The panes easily slide back and forth but strongly resist being pulled apart. Similarly, intrapleural fluid "glues" lungs to the wall. As a result, when the thoracic cavity expands by inhalation, the lungs must expand, also.

Figure 36.12*b* shows what happens as you start to inhale. The diaphragm flattens and moves downward, and the rib cage is lifted upward and outward. As the thoracic cavity is expanding, the lungs expand along with it. At this time, the air pressure in all alveolar sacs combined is lower than the atmospheric pressure. Fresh air follows this pressure gradient and flows down into the airways, then into the alveoli.

Figure 36.12 (**a**) The location of lungs relative to the diaphragm. (**b,c**) Changes in the thoracic cavity's size during one respiratory cycle. The X-ray images show how the maximum inhalation possible changes the thoracic cavity volume. (**d–f**) Changes in the intrapulmonary pressure and lung volume in a respiratory cycle.

INWARD BULK FLOW OF AIR

OUTWARD BULK FLOW OF AIR

b Inhalation. The diaphragm contracts, moves down. External intercostal muscles contract and lift rib cage upward and outward. The lung volume expands.

c Exhalation. Diaphragm, external intercostal muscles return to resting positions. Rib cage moves down. Lungs recoil passively.

	d Before inhalation	**e** During inhalation (lungs expanded)	**f** During exhalation
Atmospheric pressure:	760	760	760
Intrapleural pressure:	756	754	756
Intrapulmonary pressure:	760	759	761

The second part of the respiratory cycle, exhalation, is passive when you are breathing quietly. The muscles that brought about inhalation relax and the lungs passively recoil, with no further energy outlays. The decrease in lung volume compresses air in the alveolar sacs. At this time, pressure inside the sacs is greater than atmospheric pressure. Air follows the gradient, flowing out from the lungs, as in Figure 36.12c.

Exhalation becomes active and energy-requiring only when you vigorously exercise and must expel more air. When exhalation is active, abdominal wall muscles contract. Abdominal pressure rises and exerts an upward-directed force on the diaphragm, which is thereby pushed upward. Also, the internal intercostal muscles contract, and so pull the thoracic wall inward and downward. Their action flattens the chest wall. In such ways, the dimensions of the thoracic cavity decrease. The lung volume decreases as well when elastic tissue of the lungs passively recoils.

Breathing at High Altitudes

From the chapter introduction, you already know that breathing becomes an agonizing challenge for climbers of Mount Everest and other cloud-piercing peaks. Even so, like llamas and other species that evolved at high elevations, millions of people live quite comfortably 4,800 meters (16,000 feet) or more above sea level.

Compared with people living at lower elevations, the permanent residents of high mountains have lungs with far more alveoli and blood vessels, which formed as they were growing up. Also, their heart developed larger ventricles, so it pumps larger volumes of blood. And more mitochondria formed in their muscle tissue.

Llamas have an advantage, too. Compared to human hemoglobin, llama hemoglobin has greater affinity for oxygen. It picks up oxygen more efficiently at the lower pressures characteristic of high altitudes (Figure 36.13).

Does this mean a healthy person who grew up by the sea can't pick up roots and move to the mountains? No. By mechanisms of **acclimatization**, people make long-lasting physiological and behavioral adaptations to a new habitat that differs from the one left behind. At high altitudes, gradual changes in the pattern and magnitude of breathing and heart output can supplant the initial and acute compensations someone makes to cellular oxygen deficiency (hypoxia).

Within a few days, the reduction in oxygen delivery stimulates kidney cells to secrete more **erythropoietin**, a hormone that induces stem cells in bone marrow to

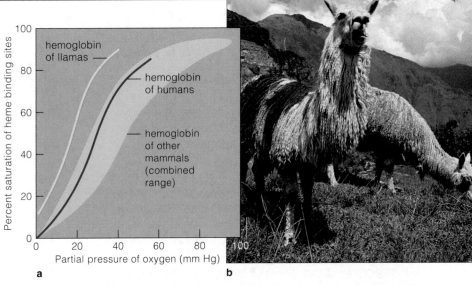

Figure 36.13 (**a**) Comparison of hemoglobin's binding and releasing capacity for humans, llamas, and other animals. (**b**) Llamas in the Peruvian Andes.

divide repeatedly and give rise to red blood cells. Each second in adults, 2 million to 3 million red blood cells form as replacements for ones that continually die off. Stepped-up erythropoietin secretion can increase that pace by six times during episodes of extreme stress. Increased numbers of circulating red blood cells raise the blood's oxygen-delivery capacity. When the oxygen concentration in the blood increases sufficiently, the kidneys slow down erythropoietin secretion.

Erythropoietin is the main hormonal mediator of red blood cell production. But men have a larger muscle mass and greater demands for oxygen. They depend as well on a sex hormone, testosterone, which stimulates increases in the basic rate of red blood cell production. That's why a blood sample from males normally has a greater percentage of red blood cells than an equivalent sample from females. But the compensatory increase in red blood cells comes at a cost. Having many more cells in the bloodstream increases resistance to blood flow, because the blood is more viscous, or "thicker." The heart must work harder to pump blood.

Breathing, which ventilates the lungs, has a cyclic pattern. The respiratory cycle consists of inhalation (one breath of air in) and exhalation (one breath of air out).

Inhalation is always an active, energy-requiring process involving contractions mainly of the diaphragm and the external intercostal muscles.

During quiet breathing, exhalation is a passive process. Muscles relax, the thoracic cavity volume decreases, and the lungs recoil elastically. Forceful exhalation is an active process that requires abdominal muscle contraction.

Breathing reverses pressure gradients between the lungs and the air outside the body.

GAS EXCHANGE AND TRANSPORT

Exchanges at the Respiratory Membrane

An alveolus is a cupped sheet of epithelial cells with a basement membrane. A pulmonary capillary consists of endothelial cells, and its basement membrane fuses with that of the alveolus. Together, the alveolar epithelium, capillary endothelium, and their basement membranes form a respiratory membrane (Figure 36.14). Gases can diffuse across this thin membrane very quickly.

Oxygen and carbon dioxide passively diffuse across the membrane in response to partial pressure gradients. An inward-directed gradient for oxygen is maintained because inhalations replenish oxygen and cells take it up an ongoing basis. An outward-directed gradient for carbon dioxide is maintained because cells continually produce carbon dioxide and exhalation removes it.

Oxygen Transport

Oxygen, like carbon dioxide, does not dissolve well in blood, so it can't be efficiently transported on its own. In vertebrates, as in other large animals, demands for oxygen are met with the help of hemoglobin molecules. These respiratory pigments are packed inside red blood cells. A hemoglobin molecule, recall, is a compact array of four polypeptide chains and four **heme groups**, each incorporating an iron atom that binds reversibly with oxygen. *Of all the oxygen inhaled into the human body, 98.5 percent of it is bound to heme groups of hemoglobin.*

Normally, inhaled air that reaches alveoli has plenty of oxygen. The opposite is true of blood in pulmonary capillaries. Thus, in the lungs, oxygen tends to diffuse into the plasma portion of blood. Then it diffuses into red blood cells where it binds with hemoglobin. In this way, it forms **oxyhemoglobin**, or HbO_2.

The amount of HbO_2 that forms in a given interval depends on oxygen's partial pressure. The higher the pressure, the greater will be the oxygen concentration. When plenty of oxygen molecules are available, they tend to randomly collide with the heme binding sites at a faster rate. The encounters continue until all four of the binding sites in hemoglobin are saturated.

HbO_2 molecules bind oxygen weakly and give it up where oxygen's partial pressure is lower than in the lungs. They give it up faster in tissues where blood is warmer, the pH is lower, and carbon dioxide's partial pressure is high. Such conditions prevail in contracting muscle and other metabolically active tissues.

Carbon Dioxide Transport

Carbon dioxide diffuses into blood capillaries in any tissue where its partial pressure is higher than it is in the blood flowing past. From there, three mechanisms transport it to the lungs. About 10 percent of the gas stays dissolved in blood. Another 30 percent binds with hemoglobin to form **carbamino hemoglobin** ($HbCO_2$). Most of it—60 percent—is transported in the form of bicarbonate (HCO_3^-). How do these HCO_3^- molecules form? First, carbonic acid forms when carbon dioxide combines with water in blood, but it quickly separates into bicarbonate and hydrogen ions (H^+):

$$CO_2 + H_2O \rightleftharpoons \underset{\text{CARBONIC ACID}}{H_2CO_3} \rightleftharpoons \underset{\text{BICARBONATE}}{HCO_3^-} + H^+$$

In blood plasma, that reaction converts only 1 of every 1,000 carbon dioxide molecules—which doesn't amount to much. It is a different story in red blood cells, which have the enzyme **carbonic anhydrase**. Enzyme action enhances the reaction rate 250 times! Most of the carbon dioxide not bound to hemoglobin becomes converted to carbonic acid this way. Conversion makes the blood level of carbon dioxide decline rapidly. Thus it helps maintain a gradient that promotes diffusion of carbon dioxide from interstitial fluid into the bloodstream.

What about the HCO_3^- that forms in the reactions? It tends to move out of red blood cells and into blood

a Surface view of capillaries associated with alveoli

pore for air flow between adjoining alveoli

Figure 36.14 Zooming in on the structure of the respiratory membrane inside the lungs.

red blood cell inside pulmonary capillary

air space inside alveolus

b Cutaway view of one of the alveoli and adjacent pulmonary capillaries

alveolar epithelium

capillary endothelium

fused basement membranes of both epithelial tissues

c Three components of the respiratory membrane

7. Hypoxia triggers increased _____ secretion.
   a. carbonic anydrase   c. erythropoietin
   b. carbon monoxide   d. myoglobin

8. After oxygen diffuses into pulmonary capillaries, it diffuses into _____ and binds with _____ .
   a. interstitial fluid; red blood cells
   b. interstitial fluid; carbon dioxide
   c. red blood cells; hemoglobin
   d. red blood cells; carbon dioxide

9. Oxyhemoglobin (HbO_2) gives up oxygen faster where the _____ than it is in the lungs.
   a. pH is lower   c. O_2 partial pressure is higher
   b. blood is cooler   d. CO_2 partial pressure is lower

10. Sixty percent of the carbon dioxide in human blood is in the form of _____ and thirty percent in the form of _____ .
    a. carbon dioxide; carbonic acid
    b. carbonic anhydrase; bicarbonate
    c. bicarbonate; carbamino hemoglobin
    d. carbamino hemoglobin; bicarbonate

11. Match the descriptions with their components in humans.
    ____ trachea   a. airway leading into a lung
    ____ pharynx   b. gap between vocal cords
    ____ alveolus   c. fine branch of bronchial tree
    ____ hemoglobin   d. windpipe
    ____ bronchus   e. respiratory pigment
    ____ bronchiole   f. site of gas exchange
    ____ glottis   g. throat

Critical Thinking

1. Some cigarette manufacturers are urging people to "smoke responsibly." What social and biological issues are involved here? For example, what about risks to a nonsmoking spouse or to children of a smoker? To nonsmoking patrons in restaurants? To the unborn child of a pregnant smoker? In your opinion, what behavior would constitute "responsible smoking"?

2. People sometimes poison themselves with carbon monoxide by building a charcoal fire in an enclosed area. Assuming help arrives in time, what would be the best treatment: (1) placing the victim outdoors in fresh air or (2) rapidly administering pure oxygen? Explain how you arrived at your answer.

3. When you swallow food, contracting muscles force the epiglottis down, to its closed position. This prevents food from going down the trachea and blocking gas flow. Yet each year, several thousand people choke to death after food enters the trachea and blocks air flow for as little as four or five minutes. The *Heimlich maneuver*—an emergency procedure only—often dislodges food stuck in the trachea (Figure 36.19).

   When correctly performed, the maneuver forcibly elevates the diaphragm, causing a sharp decrease in the thoracic cavity volume and an abrupt rise in alveolar pressure. Air forced up into the trachea by the increased pressure may be enough to dislodge the obstruction. Once the obstacle is dislodged, a doctor must see the person at once because an inexperienced rescuer can inadvertently cause internal injuries or crack a rib.

   Reflect now on our current social climate in which lawsuits abound. Would you risk performing the Heimlich maneuver to save a relative's life? A stranger's life? Why or why not?

4. Weddell seals can dive 600 meters and stay submerged for a half hour or more. Sperm whales can dive 2,250 meters and stay there well over an hour. If they cannot inhale oxygen and exhale carbon dioxide while under water, how do these and other air-breathing mammals routinely stay submerged for prolonged periods? Research and describe their adaptations for

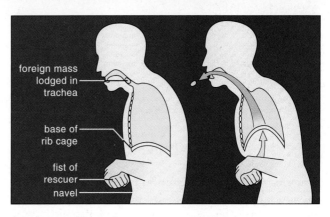

Figure 36.19 Heimlich maneuver to dislodge food stuck in the trachea. Stand behind the victim, make a fist with one hand, then position the fist, thumb-side in, against the victim's abdomen. The fist must be slightly above the navel and well below the rib cage. Now press the fist into the abdomen with a sudden upward thrust. Repeat the thrust several times if needed. The maneuver can be performed on someone who is standing, sitting, or lying down.

deep-sea diving. (*Hints:* These include increased reliance on myoglobin, selective distribution of oxygen to certain organs, and oxygen-collecting aspects of the respiratory system.)

5. Water pressure increases greatly with depth. Professional divers do not descend into very deep water without tanks of compressed air (air under pressure). Why the caution? Because of the increased pressure at depths, more gaseous nitrogen (N_2) than usual has become dissolved in body tissues. The outcome is *nitrogen narcosis*, or "raptures of the deep."

   At depths of 45 meters (150 feet), divers become euphoric and drowsy, as if they were tipsy with alcohol. Some have been known to offer their airtank mouthpiece to fishes. At lower depths, divers become clumsy and weak. Below 105 meters, they can slip into a coma. Gaseous nitrogen, a lipid-soluble gas, readily dissolves in the lipid bilayer of cell membranes. Given what you have learned about human body cells in preceding chapters, which type of cell is most likely to be a key player in nitrogen narcosis when excess nitrogen disrupts its membrane properties? (*Hint:* Reflect on the nature of the symptoms.)

Selected Key Terms

acclimatization *36.5*
alveolus (alveoli) *36.4*
bronchiole *36.4*
bronchus *36.4*
carbamino hemoglobin *36.6*
carbonic anhydrase *36.6*
countercurrent flow *36.3*
diaphragm *36.4*
epiglottis *36.4*
erythropoietin *36.5*
Fick's law *36.1*
gill (fish) *36.2*
glottis *36.3*
heme group *36.6*
hemoglobin *36.1*
integumentary exchange *36.2*
larynx *36.4*
lung *36.3*
myoglobin *36.1*
oxyhemoglobin *36.6*
partial pressure *36.1*
pharynx *36.4*
pressure gradient *36.1*
respiration *CI*
respiratory cycle *36.5*
respiratory surface *36.1*
respiratory system *CI*
trachea *36.4*
tracheal respiration *36.2*
vocal cord *36.3*

Readings

Gorman, J. 31 March 2001. "Breathing on the Edge." *Science News*, 159: 202–204.

Sherwood, L. 1997. *Human Physiology*. Second edition. Belmont, California: Wadsworth.

On-Line readings at Student Guide for InfoTrac:
www.brookscole.com/biology

DIGESTION AND HUMAN NUTRITION

Lose It—And It Finds Its Way Back

America's fixation on Beautiful People who have nary an ounce of extra fat on their utterly perfect selves is bad enough. After all, we might rationalize our own extra ounces with the truism that beauty is only skin deep. But we can't rationalize away the Fat Connection to atherosclerosis, heart attacks, strokes, colon cancer, and other dangerous ailments.

The proportion of body fat relative to total tissue mass is now supposed to be 18 to 24 percent for women less than thirty years old. For men, it is supposed to be no more than 12 to 18 percent. An estimated 34 million Americans don't even come close to the standards.

No matter how much we diet, the lost weight seems to find its way back. This physiological dilemma is an outcome of our evolutionary heritage. Like most other mammals, we have an abundance of fat-storing cells in adipose tissue. Collectively, the cells are an adaptation for survival, an energy warehouse that opens up when food is scarce. Once the fat-storing cells have formed, they are in the body to stay. Variations in food intake only affect how empty or full each cell gets.

Dieting opens the fat warehouse. The brain interprets this as *STARVATION!* and calls for a metabolic slowdown. In response, the body uses energy far more efficiently even for basic tasks, such as breathing and digesting food. It now takes less food to do the same things!

As you've probably heard, dieting is useless without long-term commitment to physical exercise. Why? Your skeletal muscles adapt to *STARVATION!* signals and burn less energy than before. Jog for four hours or play tennis for eight hours straight and you might lose a pound of fat. Meanwhile, your appetite surges. If and when you do stop dieting, your "starved" fat cells quickly refill. Unless you eat less and exercise moderately all through your life, you can't keep off extra weight (Figure 37.1).

What about gaining weight? A weight gain triggers increases in metabolism. You use 15 to 20 percent more energy than before until you lose the excess.

As if this were not enough, emotions affect weight gains and losses. As an extreme case, *anorexia nervosa* is a potentially fatal eating disorder based on a flawed assessment of body weight. Most anorexics dread being fat and hungry. Many starve themselves, overexercise, and have fears about growing up and of being sexually mature. They often have irrationally high expectations about their personal performance.

Another extreme is called *bulimia*, an out-of-control "oxlike appetite." A bulimic on an hour-long eating

Figure 37.1 A few organisms positively engaged in countering imbalances between caloric intake and energy output.

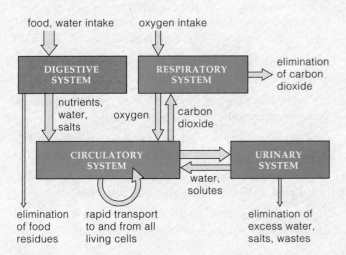

food, water intake   oxygen intake

DIGESTIVE SYSTEM → RESPIRATORY SYSTEM → elimination of carbon dioxide

nutrients, water, salts   oxygen   carbon dioxide

CIRCULATORY SYSTEM → URINARY SYSTEM

water, solutes

elimination of food residues   rapid transport to and from all living cells   elimination of excess water, salts, wastes

Figure 37.2 Functional links between the digestive, respiratory, circulatory, and urinary systems. These organ systems and others work together to supply cells with raw materials and eliminate wastes.

binge may take in 50,000 kilocalories' worth of food, then vomit or use laxatives to get rid of it. Some follow a binge–purge routine as an "easy" way to lose weight. Others may not even like to eat but purge themselves to relieve anger and frustration. The binge–purge routines range from once a month to a few times a day. Purgings damage the gut. Chronic vomiting brings up gastric fluid that erodes teeth to stubs. In severe cases, the stomach can rupture and the heart and kidneys fail.

Are severe eating disorders rare? No. In the United States, an estimated 7 million females and 1 million males are anorexic or bulimic. Most are in their teens and early twenties, but the number also is increasing among preadolescents. Each year, complications kill 5 to 6 percent of the weight-obsessed individuals.

With these sobering thoughts, we start our tour of **nutrition**. The word encompasses all those processes by which an animal ingests and digests food, then absorbs the released nutrients for later conversion to the body's own carbohydrates, lipids, proteins, and nucleic acids.

The nutritional processes proceed at the **digestive system**. This body cavity or tube mechanically and chemically reduces food to particles, then to molecules that are small enough to be absorbed into the internal environment. It also eliminates unabsorbed residues. Other organ systems, especially those shown in Figure 37.2, contribute to the nutritional processes.

Key Concepts

1. In complex animals, interactions among digestive, circulatory, respiratory, and urinary systems supply the body's living cells with raw materials, dispose of wastes, and maintain the volume and composition of extracellular fluid.

2. Most digestive systems include specialized regions for food transport, processing, and storage. Different regions mechanically break apart and chemically break down food, absorb breakdown products, and eliminate the unabsorbed residues.

3. "Nutrition" refers to all processes by which an animal ingests food, digests food, and then absorbs the released nutrients, which become converted to the body's own carbohydrates, lipids, proteins, and nucleic acids.

4. A subcategory of nutritional processes is the intake of foods that are good sources of vitamins, minerals, and a number of amino acids and fatty acids that the body itself cannot produce.

5. To maintain suitable body weight and overall health, energy intake must balance energy output by way of metabolic activity, regular physical exertion, and so on. Complex carbohydrates are the main source of dietary glucose. Typically, glucose is the body's main source of instant energy.

THE NATURE OF DIGESTIVE SYSTEMS

Incomplete and Complete Systems

Recall, from Chapter 23, that we do not see many organ systems until we get to the flatworms. Flatworms are among the invertebrates with an **incomplete digestive system**, a saclike, branching gut cavity having a single opening at the start of a tubular pharynx (Figure 37.3a). Food enters the sac, where it is partly digested, then is circulated to cells even as residues are sent out.

The saclike gut of flatworms, cnidarians, and a few other small invertebrates has a single opening for food intake *and* for waste disposal. The two-way traffic does not favor regional specialization. Other animals have a **complete digestive system**, a tube with an opening at one end for taking in food and an opening at the other end for eliminating unabsorbed residues (for example, mouth and anus). In between, the tube is subdivided into specialized regions, which function in the one-way transport, processing, and storage of raw materials.

Consider a frog's complete digestive system (Figure 37.3b). Between its mouth and the anus are a pharynx,

stomach, small intestine, and large intestine. A liver, gallbladder, and pancreas—all organs with accessory roles in digestion—connect with the small intestine. Birds, too, have a complete digestive system with a few unique regional specializations (Figure 37.3c).

Regardless of its complexity, a complete digestive system carries out five overall tasks:

1. **Mechanical processing and motility**: Movements that break up, mix, and propel food material.

2. **Secretion**: Release of digestive enzymes and other substances into the space inside the tube.

3. **Digestion**: Breakdown of food into particles, then into nutrient molecules small enough to be absorbed.

4. **Absorption**: Passage of digested nutrients and fluid across the tube wall and into body fluids.

5. **Elimination**: Expulsion of the undigested and unabsorbed residues from the end of the gut.

Correlations With Feeding Behavior

We can correlate any digestive system's specializations with feeding behavior. Consider the pigeon in Figure 37.3c. Its *food-gathering* region is a bill adapted to peck seeds off the ground. Seeds enter a mouth, a long tube (esophagus), and then a *food-processing* region, which is compactly centered in the body mass. The compactness helps balance the bird during flight. As in other seed eaters, part of the esophagus bulges out as a *crop*. This stretchable storage organ allows the bird to gulp a lot of food fast and make quick getaways from predators. Like all birds, a pigeon eats during the day and fills its crop before the sun sets. The first few hours after dark, it releases food to lessen the time of overnight fasting.

The first part of the pigeon stomach has a glandular lining, which secretes enzymes and other substances that function in digestion. The second part, the gizzard, is a muscular organ that grinds up food much as teeth and jaws do. A pigeon's intestines are proportionally shorter than those in ducks or ostriches. Such birds eat plant parts rich in tough, fibrous cellulose—which requires a much longer processing time than seeds do.

And what about the digestive system of pronghorn antelope (*Antilocapra americana*)? Fall through winter, on the mountain ridges from central Canada down into northern Mexico, pronghorn antelope browse on wild

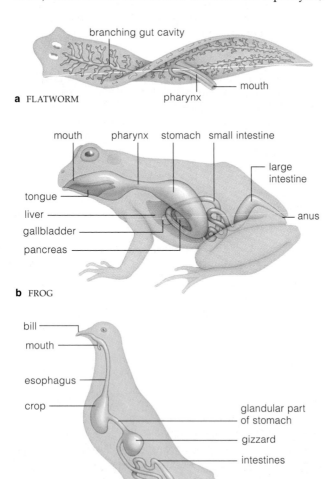

a FLATWORM

branching gut cavity

mouth

pharynx

b FROG

mouth · pharynx · stomach · small intestine

large intestine

tongue

liver

gallbladder

pancreas

anus

c BIRD (pigeon)

bill

mouth

esophagus

crop

glandular part of stomach

gizzard

intestines

cloaca (terminal opening; also has role in reproduction)

Figure 37.3 (**a**) Incomplete digestive system of a flatworm, which has a two-way traffic of food and undigested material through one opening. (**b,c**) Examples of a complete digestive system, a tube with specialized regions and an opening at each end.

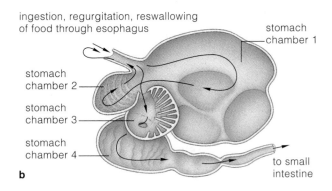

gumline

crown

crown

root

ANTELOPE
MOLAR

HUMAN
MOLAR

root

a

ingestion, regurgitation, reswallowing
of food through esophagus

stomach
chamber 1

stomach
chamber 2

stomach
chamber 3

stomach
chamber 4

to small
intestine

b

Figure 37.4 Some regional specializations of the complete digestive system of pronghorn antelope (*Antilocapra americana*).

(**a**) Comparison of antelope and human molars. (**b**) Multiple-chambered antelope stomach. The first chamber is a large pouch. The second is smaller, with a honeycombed inner surface (what butchers call "tripe"). In both chambers, food is mixed with fluid, kneaded, and exposed to fermentation activities of bacterial and protozoan symbionts. Some of the symbionts degrade cellulose; others synthesize organic compounds, fatty acids, and vitamins. The host uses a portion of these substances. Kneaded food is regurgitated into the mouth, rechewed, then swallowed again. It enters the third chamber, where it is pummeled once more before entering the last stomach chamber.

Figure 37.5 Exceptionally Big Mac of the snake world.

sage. Come spring, they move to open grasslands and deserts, there to browse on new growth (Figure 37.4).

Now think of your own cheek teeth, or molars, each having a flattened crown that is a grinding platform. The crown of an antelope's molars dwarfs yours (Figure 37.4*a*). Why the difference? You probably don't brush your mouth against dirt as you eat. An antelope does. Abrasive bits of soil enter its mouth along with tough plant material, so the crown wears out quickly. Natural selection has favored more crown to wear down.

Antelopes are **ruminants**, a type of hoofed mammal having multiple stomach chambers in which cellulose is slowly broken down. The breakdown gets under way inside the first two of four stomach sacs (Figure 37.4*b*). There, bacterial symbionts release digestive enzymes to access the nutrients in cellulose. As they do, their host regurgitates and rechews the contents of the first two sacs, then swallows again. (That's what "chewing cud" means.) Pummeling the plants more than once exposes more surface area to enzymes, which gives them more time to release nutrients that the antelope (as well as the bacteria) can use. The ruminant stomach system can steadily accept food during extended feeding times and slowly liberate nutrients when the animal rests.

Predatory and scavenging mammals have different specializations. They gorge on food when they can get it and might not eat again for some time (Figure 37.5). Part of their digestive system stores food, which they digest and absorb at a more leisurely pace. As you will

see, other organs with accessory roles in digestion help ensure that there will be an adequate distribution of nutrients between meals.

An incomplete digestive system is a saclike body cavity. Both food and residues enter and leave through the same opening. A complete digestive system is a tube that has two openings and regional specializations in between.

Complete digestive systems carry out the following tasks in controlled ways: movements that break up, mix, and propel food material; secretion of digestive enzymes and other substances; breakdown of food; absorption of nutrients; and expulsion of undigested and unabsorbed residues.

VISUAL OVERVIEW OF THE HUMAN DIGESTIVE SYSTEM

Humans have a complete digestive system, a tube with many specialized organs between two openings (Figure 37.6). If the tube of an adult were fully stretched out, it would extend 6.5 to 9 meters (21 to 30 feet). From start to finish, mucus-coated epithelium lines every surface facing the lumen. (A *lumen* is the space within a tube.) The thick, moist mucus helps protect the tube wall and promotes diffusion across its inner lining.

Ingested substances advance in one direction, from the mouth through the pharynx and esophagus, then the gastrointestinal tract, or gut. The human **gut** starts at the stomach and extends through the small intestine, large intestine (colon) and rectum, to the anus. Salivary glands and a gallbladder, liver, and pancreas function as accessory organs; they secrete a variety of substances into different regions of the tube.

Major Components:

MOUTH (ORAL CAVITY)

Entrance to system; food is moistened and chewed; polysaccharide digestion starts.

PHARYNX

Entrance to tubular part of system (and to respiratory system); moves food forward by contracting sequentially.

ESOPHAGUS

Muscular, saliva-moistened tube that moves food from pharynx to stomach.

STOMACH

Muscular sac; stretches to store food taken in faster than can be processed; gastric fluid mixes with food and kills many pathogens; protein digestion starts.

SMALL INTESTINE

First part (duodenum, C-shaped, about 10 inches long) receives secretions from liver, gallbladder, and pancreas.

In second part (jejunum, about 3 feet long), most nutrients are digested and absorbed.

Third part (ileum, 6–7 feet long) absorbs some nutrients; delivers unabsorbed material to large intestine.

LARGE INTESTINE (COLON)

Concentrates and stores undigested matter by absorbing mineral ions, water; about 5 feet long; divided into ascending, transverse, and descending portions.

RECTUM

Distension stimulates expulsion of feces.

ANUS

End of system; terminal opening through which feces are expelled.

Accessory Organs:

SALIVARY GLANDS

Glands (three main pairs, many minor ones) that secrete saliva, a fluid with polysaccharide-digesting enzymes, buffers, and mucus (which moistens and lubricates food).

LIVER

Secretes bile (for emulsifying fat); roles in carbohydrate, fat, and protein metabolism.

GALLBLADDER

Stores and concentrates bile that the liver secretes.

PANCREAS

Secretes enzymes that break down all major food molecules; secretes buffers against HCl from the stomach.

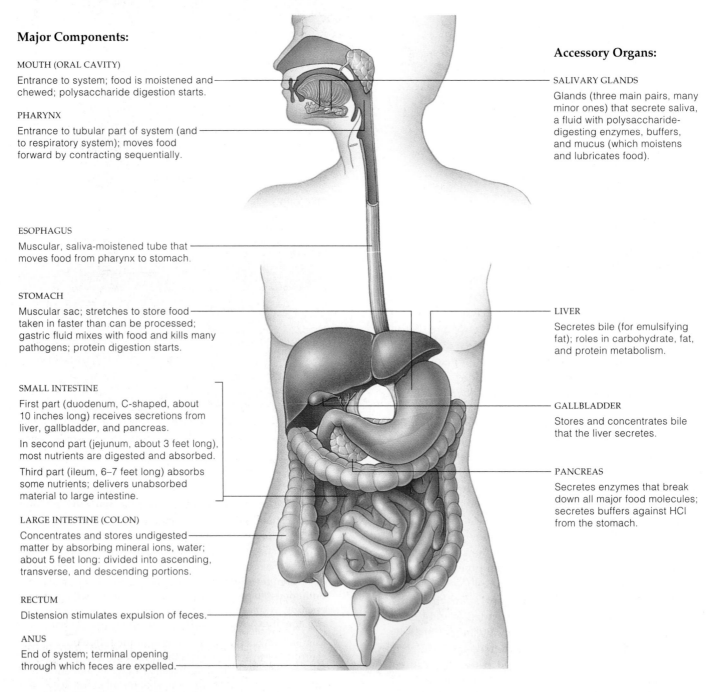

Figure 37.6 Overview of the component parts of the human digestive system and their specialized functions. Organs with accessory roles in digestion are also listed.

INTO THE MOUTH, DOWN THE TUBE

Food is chewed and polysaccharide breakdown begins in the oral cavity, or mouth. Thirty-two teeth typically project into an adult's mouth (Figure 37.7). Each **tooth** has an enamel coat (hardened calcium deposits), dentin (a thick, bonelike layer), and an inner pulp with nerves and tiny blood vessels. It is an engineering marvel, able to withstand years of chemical and mechanical stress. Chisel-shaped incisors shear off chunks of food. Cone-shaped canines tear food. Premolars and molars, with broad crowns and rounded cusps, grind and crush it.

Also in the mouth is a **tongue**, an organ consisting of membrane-covered skeletal muscles that function in positioning food in the mouth, swallowing, and speech. On that tongue's surface are many circular structures with taste buds embedded in their tissues (Figure 37.8). A taste bud contains sensory receptors that respond to chemical differences in dissolved substances. The brain uses this information to give rise to our sense of taste.

Chewing mixes food with **saliva**. This fluid contains an enzyme (salivary amylase), a buffer (bicarbonate, or HCO_3^-), mucins, and water. Salivary glands beneath and in back of the tongue produce and secrete saliva through ducts to the free surface of the mouth's lining. Salivary amylase breaks down starch. The HCO_3^- helps maintain the mouth's pH when you eat acidic foods. Modified proteins called mucins help form the mucus that binds food into a softened, lubricated ball (bolus).

Figure 37.8 Human tongue. Circular structures at its surface house many taste buds. Filamentous structures next to them help keep food moving in the mouth.

taste bud

sensory receptor

sensory nerve

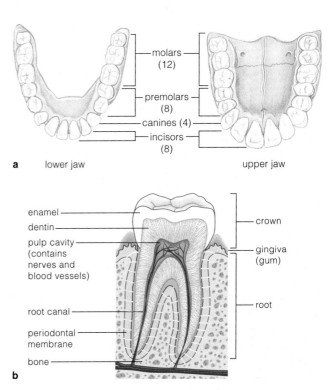

molars (12)

premolars (8)

canines (4)

incisors (8)

a   lower jaw   upper jaw

enamel

dentin

pulp cavity (contains nerves and blood vessels)

root canal

periodontal membrane

bone

crown

gingiva (gum)

root

b

Figure 37.7 (**a**) Number and arrangement of human teeth in the upper and lower jaw. (**b**) A molar's main regions: the crown and the root. Enamel capping the crown is made of calcium deposits; it is the hardest substance in the body.

Normally harmless bacteria live on and between teeth. Daily flossing, gentle brushing, and avoiding too many sweets help keep bacterial populations in check. Without such preventive measures, conditions favor infection that may lead to *caries* (tooth decay), *gingivitis* (inflamed gums), or both. Infections also can spread to the periodontal membrane that anchors teeth to the jawbone. In *periodontal disease*, infection slowly destroys the bone tissue around a tooth.

When you swallow, contractions of tongue muscles force boluses into the **pharynx**, the tubular entrance to the esophagus *and* trachea, an airway to the lungs. To block breathing as food leaves the pharynx, a flaplike valve (the epiglottis) and the vocal cords close off the trachea. That's why you normally don't choke on food (Sections 36.4 and 36.8, *Critical Thinking* question 3).

The tubular **esophagus** connects the pharynx with the stomach. Contractions of its muscular wall propel food past a sphincter, into the stomach. A **sphincter** is a ring of smooth muscles, the contractions of which close off a passageway or an opening to the body surface.

By the action of the mouth's teeth and tongue, food gets chewed, mixed with saliva, and bound into soft, lubricated balls that will be propelled through tubes to the stomach. Enzymes in saliva start the digestion of polysaccharides.

DIGESTION IN THE STOMACH AND SMALL INTESTINE

We arrive now at the premier food-processing organs, the stomach and small intestine. Both have layers of smooth muscles, the contractions of which break apart, mix, and directionally move food. The lumen of each organ receives digestive enzymes and other secretions that help break down nutrients into fragments and then molecules small enough to be absorbed. Table 37.1 lists the names and sources of the major digestive enzymes. Carbohydrate breakdown *starts* in the mouth. Protein breakdown *starts* in the stomach. Digestion of nearly all carbohydrates, lipids, proteins, and nucleic acids in food is *completed* in the small intestine.

The Stomach

The **stomach**, a muscular, stretchable sac (Figure 37.9a), has three important functions. First, it mixes and stores ingested food. Second, its secretions help dissolve and degrade the food, particularly proteins. Third, it helps control passage of food into the small intestine.

The stomach wall surface exposed to the lumen is lined with glandular epithelium. Each day, the lining's glandular cells secrete about two liters of hydrochloric acid (HCl), mucus, pepsinogens, and other substances that make up **gastric fluid**, or stomach fluid. Its potent acidity, in combination with strong contractions of the stomach wall, converts food into a thick, liquid mixture called **chyme**. Chyme's acidity kills many pathogens that enter the stomach with food. It also can cause *heartburn* when gastric fluid backs up into the esophagus.

Protein digestion starts as the high acidity alters the structure of proteins and exposes their peptide bonds. It also converts pepsinogens to pepsins: active enzymes that cleave the bonds. Protein fragments accumulate in the stomach's lumen. Meanwhile, glandular cells of the stomach lining secrete gastrin. This hormone stimulates cells of the lining that secrete HCl and pepsinogen.

Peptic ulcers form when the digestive enzymes and gastric fluid erode through the lining of the stomach or small intestine. Erosion follows when the body doesn't secrete enough mucus and buffers or secretes too much pepsin. Heredity, chronic stress, smoking, and excessive intake of alcohol or aspirin make things worse. At least 80 percent of these ulcers form after *Helicobacter pylori* infections (Section 20.1). Antibiotic therapy cures them.

The stomach empties by waves of contraction and relaxation, of a type called peristalsis. These waves mix chyme and gather force as they approach a sphincter at the stomach's base (Figure 37.9b). The arrival of a strong contraction closes the sphincter, which squeezes most of the chyme back. All of the chyme does move into the small intestine, but only in small amounts at a time.

The Small Intestine

The small intestine has three regions: the duodenum, jejunum, and ileum (Figure 37.6). On average, each day 9 liters of fluid enters the first region from the stomach and through the ducts of three organs: the **pancreas**, **liver**, and **gallbladder**. At least 95 percent of the fluid is absorbed across the intestine's epithelial lining.

Enzymes secreted by cells of the intestinal lining and pancreas digest food to monosaccharides (such as glucose), monoglycerides (each having a glycerol head and fatty acid tail), free fatty acids, free amino acids, nucleotides, and nucleotide bases. Example: Like the

Table 37.1 *Major Digestive Enzymes and Their Breakdown Products*

Enzyme	Source	Where Active	Substrate	Main Breakdown Products
CARBOHYDRATE DIGESTION				
Salivary amylase	Salivary glands	Mouth, stomach	Polysaccharides	Disaccharides
Pancreatic amylase	Pancreas	Small intestine	Polysaccharides	Disaccharides
Disaccharidases	Intestinal lining	Small intestine	Disaccharides	MONOSACCHARIDES* (such as glucose)
PROTEIN DIGESTION				
Pepsins	Stomach lining	Stomach	Proteins	Protein fragments
Trypsin and chymotrypsin	Pancreas	Small intestine	Proteins	Protein fragments
Carboxypeptidase	Pancreas	Small intestine	Protein fragments	AMINO ACIDS*
Aminopeptidase	Intestinal lining	Small intestine	Protein fragments	AMINO ACIDS*
FAT DIGESTION				
Lipase	Pancreas	Small intestine	Triglycerides	FREE FATTY ACIDS, MONOGLYCERIDES*
NUCLEIC ACID DIGESTION				
Pancreatic nucleases	Pancreas	Small intestine	DNA, RNA	NUCLEOTIDES*
Intestinal nucleases	Intestinal lining	Small intestine	Nucleotides	NUCLEOTIDE BASES, MONOSACCHARIDES*

* Breakdown products small enough to be absorbed into the internal environment.

Figure 37.9 (**a**) Stomach structure. (**b**) Peristaltic wave down the stomach. (**c**) Small intestine structure. The gut wall usually has an innermost mucosa: epithelium and an underlying connective tissue layer. Next is the submucosa, a connective tissue layer with blood and lymph vessels and a mesh of nerves that locally control digestion. Next are smooth muscle layers that differ in orientation, hence in the direction of contraction. The gut wall's outermost layer of connective tissue is the serosa.

pepsin secreted from cells in the stomach lining, two pancreatic enzymes (trypsin and chymotrypsin) digest proteins to peptides. Another cleaves peptides to free amino acids. The pancreas also secretes bicarbonate, a buffer that helps neutralize HCl from the stomach.

The Role of Bile in Fat Digestion

Besides enzymes, fat digestion requires **bile**. This fluid, which the liver secretes continually, contains bile salts, bile pigments, cholesterol, and lecithin, a phospholipid. One of the pigments is iron-rich bilirubin from degraded red blood cells. After the stomach empties, a sphincter closes the main bile duct from the liver. Bile backs up into the gallbladder, which stores and concentrates it.

By a process of **emulsification**, bile salts accelerate fat digestion. Most fats in food are triglycerides. Being insoluble in water, the triglycerides tend to cluster into large fat globules in chyme. But the intestinal wall has muscle layers (Figure 37.9c). As these move and agitate chyme, the fat globules break apart into small droplets that get coated with bile salts. Bile salts carry negative charges, so these coated droplets repel each other and stay separated. The suspension of fat droplets, formed by mechanical and chemical action, is the "emulsion."

Compared with fat globules, the emulsion droplets offer the fat-digesting enzymes a much greater surface area. And so the enzymes can break down triglycerides more rapidly to fatty acids and monoglycerides.

Controls Over Digestion

Homeostatic controls counter changes in the internal environment. By contrast, controls over digestion act *before* food is absorbed into the internal environment. The nervous system, endocrine system, and meshes of nerves (plexuses) in the gut wall exert the control.

When food enters the stomach, it distends the wall and stimulates mechanoreceptors. Signals travel along short reflex pathways to smooth muscles and glands (longer reflex pathways also signal the brain). As one response, gut wall muscles contract and glandular cells secrete enzyme-rich fluid into the lumen or hormones into blood. The response depends partly on the chyme volume and composition. A large meal activates more mechanoreceptors in the stomach wall, so contractions

become more forceful and the stomach empties faster. High acidity or a high fat content in the small intestine triggers hormonal secretions that lead to a slowdown in stomach emptying, so chyme is not moved forward faster than it can be processed. Fear, depression, and other emotional upsets also trigger slowdowns.

What are the gastrointestinal hormones? If chyme contains amino acids and peptides, cells of the stomach lining will secrete the gastrin that induces secretion of acid into the stomach. Secretin induces the pancreas to secrete bicarbonate. CCK (cholecystokinin) calls for the secretion of pancreatic enzymes and contraction of the gallbladder. When the small intestine holds glucose and fat, GIP (glucose insulinotropic peptide) causes cells to take up more glucose by stimulating insulin secretion.

Carbohydrate breakdown starts in the mouth, and protein breakdown starts in the stomach. In the small intestine, most large organic compounds are digested to molecules small enough to be absorbed into the internal environment.

Signals from the nervous system and a nerve plexus in the gut wall, along with endocrine secretions, control digestion.

ABSORPTION IN THE SMALL INTESTINE

Structure Speaks Volumes About Function

Most nutrient absorption occurs in the small intestine. (The stomach absorbs few substances, alcohol being one of them.) Figure 37.10 offers glimpses into the structure of the small intestine's wall. Focus first on the profuse folds of the mucosa that project into the lumen. Look closer and you see large numbers of tinier projections from each large fold. Look closer still and you see that epithelial cells at the surface of these tiny projections have a brushlike crown of great numbers of even tinier projections, all exposed to the intestinal lumen.

What is the significance of so much convolution? It has a highly favorable surface-to-volume ratio (Section 4.1). Collectively, all the projections from the intestinal mucosa ENORMOUSLY increase the surface area available to interact with chyme and absorb nutrients. Without that immense surface area, absorption would proceed hundreds of times more slowly, and that would not be enough to sustain human life.

Figure 37.10c,d shows **villi** (singular, villus), which are absorptive structures on mucosal folds. Each villus is about one millimeter long. Millions project from the mucosa. Their sheer density gives the mucosa a velvety appearance. Inside each villus is an arteriole, a venule, and a lymph vessel that function in moving substances to and from the general circulation (Figure 37.10e).

Most cells in the epithelial lining of a villus have **microvilli** (singular, microvillus), which are ultrafine, threadlike projections from their free surface. Each of these cells has 1,700 or so microvilli. Hence the name, "brush border" cell. Glandular cells in the lining make and secrete digestive enzymes (Table 37.1), Also, some phagocytic cells patrol and help protect the lining.

Figure 37.10 (**a**,**b**) Structure of the mammalian small intestine. Notice the circular folds of the intestinal mucosa, which are permanent. (**c**) Projecting from the free surface of each fold are many absorptive structures called villi. (**d**) Fine structure of one villus. Monosaccharides and most amino acids that cross the intestinal lining enter blood capillaries in the villus. Fats enter the lymph vessels. (**e**) Epithelial cell types at the free surface of a villus. The absorptive cell has a thick crown of microvilli facing the intestinal lumen.

a serosa (thin) connective tissue layer) — thin layer of longitudinal muscle — thick layer of circular muscle — submucosa — mucosa

b Extensive circular folds of mucosa

villi (many fingerlike, epithelium-covered projections from the mucosa)

connective tissue

vesicles

artery

vein

lymph vessel

c Villi on one of the folds, longitudinal section

epithelium

blood capillaries

lymph vessel

d One villus

microvilli at free surface of absorptive cells

cytoplasm

absorption   mucus secretion (goblet cell)   hormone secretion   phagocytosis, lysozyme secretion

e Spedialized cells making up the epithelium of a villus

a Digestion of carbohydrates to monosaccharides, and proteins to amino acids, is completed with enzymes secreted by the pancreas and by cells of the epithelial lining.

b Monosaccharides and amino acids are actively transported across the plasma membrane of the epithelial cells, then out of the same cells and into the internal environment.

c Emulsification: The constant movement of the intestinal wall breaks up fat globules into small emulsification droplets. Bile salts prevent the fat globules from re-forming. Pancreatic enzymes digest the droplets to fatty acids and monoglycerides.

d Micelles form as bile salts combine with digestion products and phospholipids. Products can readily slip into and out of the micelles.

e Concentration of monoglycerides and fatty acids in micelles enhances gradients that lead to diffusion of both substances across the lipid bilayer of the plasma membrane of cells making up the lining.

f Products of fat digestion reassemble into triglycerides in cells of the intestinal lining. These become coated with proteins, then are expelled (by exocytosis from the cells) into the internal environment.

Figure 37.11 Digestion and absorption in the small intestine.

What Are the Absorption Mechanisms?

Absorption, recall, is the passage of nutrients, water, salts, and vitamins into the internal environment. The intestinal wall's large absorptive surface facilitates the process, but so does the action of its smooth muscle. In **segmentation**, rings of circular muscle contract and relax repeatedly. This creates an oscillating (back and forth) movement that constantly mixes and forces the lumen contents against the wall's absorptive surface:

By the time the meal is halfway through the small intestine, most is broken apart and digested. Molecules of water cross the intestinal lining by osmosis. Mineral ions are selectively absorbed. Transport proteins in the plasma membrane of brush border cells actively shunt some breakdown products, including monosaccharides and amino acids, across the lining. By contrast, being

lipid-soluble, fatty acids and monoglycerides diffuse right across the plasma membrane's lipid bilayer. And bile salts help them do this (Figure 37.11).

By a process called **micelle formation**, the bile salts combine with fat digestion products and phospholipids to form tiny droplets (micelles). The product molecules in the micelles continuously exchange places with those dissolved in chyme. However, the micelles concentrate them next to the intestinal lining, and when they are concentrated enough, gradients promote their diffusion out of the micelles and into epithelial cells. Fatty acids and monoglycerides recombine inside the cells to form triglycerides. Then triglycerides combine with proteins into particles (chylomicrons) that leave the cells by way of exocytosis and enter the internal environment.

Once absorbed, the glucose and amino acids enter blood vessels directly. Triglycerides enter lymph vessels that eventually drain into blood vessels.

With its richly folded intestinal mucosa, millions of villi, and hundreds of millions of microvilli, the small intestine has a vast surface area for absorbing nutrients.

Substances pass through the brush border cells that line the free surface of each villus by active transport, osmosis, and diffusion across the lipid bilayer of plasma membranes.

DISPOSITION OF ABSORBED ORGANIC COMPOUNDS

Earlier in the book, in Section 7.6, you considered some of the mechanisms that govern organic metabolism—specifically, the disposition of glucose and other organic compounds in the body as a whole. You saw examples of the conversion pathways by which carbohydrates, fats, and proteins are broken apart to molecules that serve as intermediates in the ATP-producing pathway of aerobic respiration. Here, Figure 37.12 rounds out the picture by showing all the major routes by which organic compounds obtained from food can be shuffled and reshuffled in the body as a whole.

energy source. There is no net breakdown of protein in muscle tissue or any other tissue during this period.

Between meals, body cells tap their fat and glycogen stores. Adipose tissue cells dismantle fats to glycerol and fatty acids, and release these to the blood. Liver cells break apart glycogen and release glucose, which also enters blood. Most body cells take up fatty acids as well as glucose and use them for ATP production.

Bear in mind, the liver does more than store and convert organic compounds to different forms. Example: it helps maintain their concentrations in blood. It also

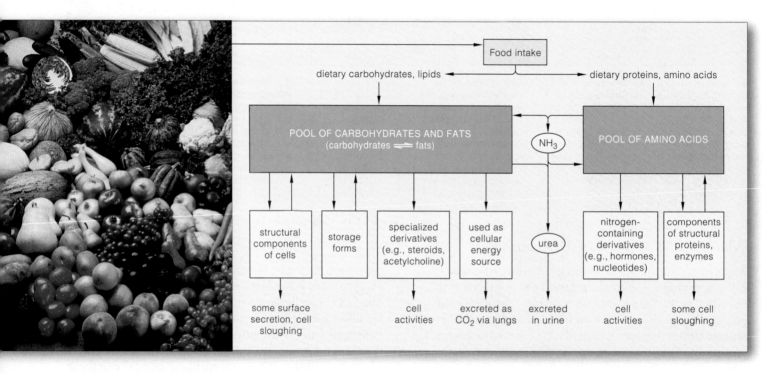

Figure 37.12 Summary of the major pathways of organic metabolism. Cells continually synthesize and tear down carbohydrates, fats, and proteins. Urea forms mainly in the liver. See also Sections 8.6 and 32.7.

All living cells in the body continually break apart most of their carbohydrates, lipids, and proteins, then use the breakdown products as energy sources or as building blocks. The nervous and endocrine systems interact to integrate this massive molecular turnover, but a treatment of how they do this would be beyond the scope of this book.

For now, simply reflect on a few key points about organic metabolism. When you eat, your body adds to its pools of materials. Excess amounts of carbohydrates and other organic compounds absorbed from the gut are transformed mostly into fats, which become stored in adipose tissue. Some are converted to glycogen in the liver and muscle tissue. *While* compounds are being absorbed and stored, most cells use glucose as the main

inactivates most hormone molecules, which are sent to the kidneys for excretion, in urine. The liver removes worn-out blood cells from the circulation and shuts down many potentially toxic compounds. For example, ammonia (NH_3), produced by amino acid breakdown, can be toxic in high concentrations. The liver converts NH_3 to urea, which is a less toxic waste product. Urea is excreted from the kidneys, as a component of urine.

Right after a meal, cells take up the glucose being absorbed from the gut and use it as a quick energy source. Excess amounts of glucose and other organic compounds become converted mainly to fats, which get stored in adipose tissue. Some of the excess is converted to glycogen, which gets stored mainly in the liver and in muscle tissue.

Between meals, cells tap fat reservoirs as their main energy source. Fats are converted to glycerol and fatty acids, both of which enter ATP-producing pathways.

THE LARGE INTESTINE

What happens to the material *not* absorbed in the small intestine? It moves into the large intestine, or **colon**. The colon concentrates and stores feces: a mixture of water, undigested and unabsorbed matter, and bacteria. The colon starts at a cup-shaped pouch (cecum). It ascends on the right side of the abdominal cavity and continues across to the other side. Then it descends and connects with a short tube, the rectum (Figures 37.6 and 37.13).

Colon Functioning

Material becomes concentrated as water moves across the colon's lining. Cells of the lining actively transport sodium ions out of the lumen. Ion concentrations fall, which increases the water concentration, so water also moves out of the lumen, by osmosis. These same cells secrete mucus that lubricates feces and helps keep them from mechanically damaging the colon wall. They also secrete bicarbonate, which buffers acidic fermentation products of ingested bacteria that managed to colonize the colon. The colonizers generally are harmless unless they breach the wall and enter the abdominal cavity.

Short, longitudinal bands of smooth muscle in the colon wall are gathered at their ends, like a series of full skirts nipped in at elastic waistbands. As they contract and relax, they move the lumen's contents back and forth against the colon wall's absorptive surface. Nerve plexuses largely control the motion, which is similar to segmentation in the small intestine but much slower.

Whereas the quick transit time through the small intestine does not favor the rapid population growth of ingested bacteria, the colon's slower motion favors it. The volume of cellulose fibers and other undigested material that cannot be decreased by absorption in the colon is called **bulk**. It adds to the volume of material and influences the transit time through the colon.

Following each new meal, secretion of gastrin and commands from the nervous system (along autonomic nerves) stimulate large portions of the ascending and transverse colon to contract together. Within seconds, the lumen's contents move as much as three-fourths of the colon's length, thus making way for incoming food.

The contents are stored in the last part of the colon until they distend the rectal wall enough to trigger a reflex action that leads to their expulsion. The nervous system controls defecation by stimulating or inhibiting the contraction of a muscle sphincter at the anus, the terminal opening of the gut.

Colon Malfunctioning

The normal frequency of defecation ranges from three times a day to once a week. Aging, emotional stress, a low-bulk diet, injury, or disease can result in delayed

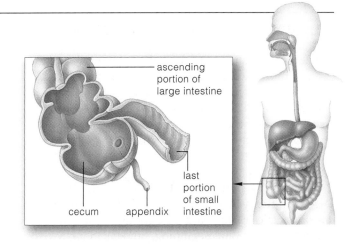

Figure 37.13 Cecum and appendix of the large intestine (colon).

defecation, called *constipation*. The longer the delay, the more water gets absorbed, so feces harden and dry out. The abdominal discomfort is accompanied by loss of appetite, headaches, and often nausea and depression.

Hard feces may become lodged inside the **appendix**, a narrow projection from the cecum (Figure 37.13). The appendix has no known digestive functions, but, like the ileum, it holds a concentration of bacteria-fighting white blood cells. Feces that obstruct the normal blood flow and mucus secretion to this projection can cause *appendicitis*, or an inflamed appendix. Unless removed, an inflamed appendix can rupture. Then bacteria, even otherwise harmless ones, would be free to enter the abdominal cavity and cause life-threatening infections.

The colon also is vulnerable to cancer, which occurs most often among the world's wealthiest and best-fed populations. Too many of these people skip meals, eat too much and too fast when they do sit down at the table, and generally give their gut erratic workouts. Their diet tends to be rich in sugar, cholesterol, and salt, and low in bulk. Too little bulk extends the transit time of feces through the colon. The longer that irritating, potentially carcinogenic material is in contact with the colon wall, the more damage it may cause. Symptoms of *colon cancer* include a change in bowel functioning, rectal bleeding, and blood in feces. Some people appear to be genetically predisposed to develop colon cancer, but a diet too low in fiber might also be a factor.

Colon cancer is almost nonexistent in rural India and Africa, where people can't afford to eat much more than fiber-rich whole grains. When these same people move to cities in the affluent nations and change their eating habits, the incidence of colon cancer increases.

The large intestine, or colon, functions in the absorption of water and mineral ions from the gut lumen. It also functions in the compaction of undigested residues into feces, for expulsion at the terminal opening of the gut.

HUMAN NUTRITIONAL REQUIREMENTS

You grow and maintain yourself by eating foods that are appropriate sources of energy and raw materials. Nutritionists measure food energy in **kilocalories**. A kilocalorie is 1,000 calories of heat energy, the amount needed to raise the temperature of 1 kilogram of water by 1°C. Raw materials in food will now be described.

New, Improved Food Pyramids

A few million years ago, our hominid ancestors dined mostly on fresh fruits and other fibrous plant material. From that nutritional beginning, humans in many parts of the world came to prefer low-fiber, high-fat foods. They still do even when they study **food pyramids**, or charts of well-balanced diets, from elementary school onward. Nutritionists revise the charts on an ongoing basis. Figure 37.14 is one version of daily portions for an average-sized adult male, expressed as percentages of energy intake. It suggests 55–60 percent complex carbohydrates, 15–20 percent proteins (less for females), and 20–25 percent fats. Alternative pyramids abound. Example: The *Mediterranean diet* pyramid has a large

base of grain products, too. But fruits and vegetables are the next largest groups, then legumes and nuts, then *olive oil* as a fat group, followed by cheese and yogurt. This pyramid limits the weekly intake of fish, poultry, eggs, simple sugars, and—at its tiny top—red meat. Olive oil is as much as 40 percent of the total energy intake. But it's monounsaturated and less likely than saturated fats to raise cholesterol levels. It also is a fine antioxidant, good at eliminating free radicals.

Carbohydrates

Starch and, to a lesser extent, glycogen should be the main carbohydrates in the human diet. These complex carbohydrates are easily degraded to glucose, the body's main energy source. Starch is abundant in fleshy fruits, cereal grains, and legumes, including beans and peas.

Foods rich in complex carbohydrates have another advantage: they typically are high in fiber. Section 37.7 touched upon the consequences of a low-fiber diet. As epidemiologist Denis Burkitt put it, when people pass small volumes of feces, you have large hospitals.

MILK, CHEESE YOGURT

Two servings/day (three/day for teenagers or pregnant, lactating, post-menopausal women)

1 cup nonfat or lowfat milk, fortified soy milk
1 cup yogurt or cottage cheese
1–1/2 ounces cheese

FRUITS

Two to four servings/day

1 medium apple, orange, banana, mango
1 wedge melon, pineapple
1 cup fresh berries
3/4 cup unsweetened fruit juice
1/2 cup unsweetened fruit
1/4 cup dried fruit

ADDED FATS AND SIMPLE SUGARS
Other foods have plenty. Restrict intake of butter, salad dressings, sweets, soft drinks, snacks

LEAN MEAT, POULTRY, FISH, LEGUMES, EGGS

Two to three servings/day

2–3 ounces trimmed, cooked, lean meat
2–3 ounces cooked poultry, no skin
2–3 ounces cooked fish
2 egg whites or 1 whole egg (limit yolks per week)
1/2 cup legumes (counts as 1/3 serving)
2 tablespoons peanut butter (counts as 1/3 serving)

VEGETABLES

Three to five servings/day

1/2 cup cooked or raw vegetables
1 cup dark green, leafy raw vegetables
3/4 cup vegetable juice

GRAIN PRODUCTS (BREADS, CEREALS, PASTA)

Six to eleven servings/day

1 biscuit, small roll, or muffin
1 slice bread
1/2 cup cooked cereal, pasta
1 ounce dry unsweetened cereal
1/2 cup rice (brown rice has more nutrients)
1/2 bagel, English muffin

Figure 37.14 One food pyramid, based on USDA guidelines. Good health requires daily selections among the alternatives from all groups *except* the "red-hot" tip of the pyramid. Added fats and simple sugars pile on calories but have few vitamins and minerals.

No Limiting Amino Acid	Low Lysine	Low Methionine, Cysteine	Low Tryptophan
Legumes: soybeans (e.g., tofu, soy milk) Grains: wheat germ Milk Cheeses (not cream) Yogurt Eggs (Meats)	Legumes: peanuts Grains: barley, rice, buckwheat, oats, corn, rye, wheat Nuts: almonds, cashews, coconut, walnuts, hazelnuts, pecans Seeds: pumpkin, sunflower	Legumes: beans (dried), black-eyed peas, garbanzos, lentils, lima beans, mung beans, peanuts Nuts: hazelnuts Fresh vegetables: asparagus, green peas, broccoli, mushrooms, parsley, potatoes, soybeans, Swiss chard	Legumes: beans (dried), garbanzos, lima beans, mung beans, peanuts Grains: cornmeal Nuts: almonds, English walnuts Fresh vegetables: corn, green peas, mushrooms, Swiss chard

Figure 37.15 Vegetarian diets. All eight essential amino acids, a small portion of the total protein intake, must be available at the same time, in certain amounts, if cells are to build their own proteins. Milk and eggs are among the *complete* proteins; they have high amounts of all eight in proportions that humans require.

Nearly all plant proteins are *incomplete*. Vegetarians must be careful to avoid protein deficiency. For instance, they might combine different foods from the three columns of incomplete proteins shown. Strict vegetarians (who also avoid dairy products and eggs) must take vitamins B$_{12}$ and B$_2$ (riboflavin) supplements. Animal protein is a luxury in most traditional societies. Their cuisines have good combinations of plant proteins, including rice/beans, chili/cornbread, tofu/rice, and lentils/wheat bread.

Unlike complex carbohydrates, the simple sugars don't offer fiber. Unlike whole foods, they don't offer vitamins and minerals. Yet each week, we in the United States typically consume up to two pounds of refined sugar (sucrose). Is this a far-fetched statistic? No. Look critically at the ingredients listed on the containers for dry cereals, frozen foods, soft drinks, and many other prepared foods. Sugars often are "hidden" in labels as corn syrup, corn sweeteners, dextrose, and so on.

Lipids

Your body can't function without fats and other lipids. For example, cell membranes require the phospholipid lecithin; and fats serve as energy reserves, cushion many internal organs, and act as insulation beneath the skin. Dietary fats also help you store fat-soluble vitamins. But the body can synthesize most of its own fats, including cholesterol, from proteins and carbohydrates. The ones it can't make are **essential fatty acids**, but whole foods offer plenty of these. Linoleic acid is an example. One teaspoon each day of olive oil, corn oil, or some other polyunsaturated fat in food provides enough of it.

On average, butter and other fats now make up 40 percent of the American diet. It should be 30 percent or less, according to most of the medical community.

Like other animal fats, butter is a saturated fat that tends to raise the level of cholesterol in the blood. As described in earlier chapters, cholesterol is used in the synthesis of bile acids and steroid hormones, and it is a required component of animal cell membranes. But too much cholesterol often interferes seriously with blood circulation (Section 34.9). Certain potato chips and other snack foods are made with edible but nondigestible oils such as *sucrose polyester*. They are aimed at people who crave fats enough to put up with the digestive upsets.

Proteins

Amino acid components of dietary proteins are needed for the body's own protein-building programs. Of the twenty common types, eight are **essential amino acids**.

That is, our body cells cannot synthesize them, so we must get enough of them from food. The eight amino acids are either methionine or cysteine (its metabolic equivalent), isoleucine, leucine, lysine, phenylalanine (or tyrosine), threonine, tryptophan, and valine.

Most animal proteins are *complete*: their amino acid ratios match human nutritional needs. Plant proteins are *incomplete;* they lack one or more essential amino acids. To get enough amino acids, vegetarians must eat certain combinations of different plants (Figure 37.15).

A measure called the net protein utilization (NPU) is used to compare proteins from different sources. NPU values range from 100 (all essential amino acids occur in ideal proportions) to 0 (one or more amino acids are absent; when eaten alone, the protein is not complete).

Protein-energy malnutrition (PEM), the world's most prevalent form of malnutrition, affects more than 500 million children today. Enzymes and other proteins are crucial for the body's structure and functions. PEM is especially worrisome for fetuses and infants. The brain grows and develops most rapidly early in life. Unless a mother-to-be takes in enough protein, especially just before and after birth, her child can expect irreversible mental retardation.

Ironically, people in developed countries often get too much protein, thus inviting acidosis, dehydration, diarrhea, fever, and excess ammonia and urea in blood. Diets high in purified protein may also promote calcium loss from bone. Fluid intake and kidney functions set the upper limit for safe protein intake. To calculate the most reasonable intake, use the guidelines in Section 37.10 to find the midpoint for your acceptable weight range. Assuming you are in good health, convert the pounds to kilograms (divide them by 2.2), and multiply the value by 0.8 (or by 0.9 for males up to 18 years old). This will give the energy intake in grams per day.

Individuals grow and maintain themselves in good health by eating certain amounts of complex carbohydrates, lipids, and proteins that provide them with adequate raw materials and energy, as measured in kilocalories.

VITAMINS AND MINERALS

Vitamins are *organic* substances that are essential for growth and survival, for no other substances can play their metabolic roles. Most plants are able to synthesize vitamins on their own. Most kinds of animals lost the ability to do so and must get vitamins from food.

At the minimum, human cells require the thirteen vitamins listed in Table 37.2. Each vitamin has specific metabolic functions. Many reactions use several types, and the absence of one affects the functions of others.

Minerals are *inorganic* substances that are essential for growth and survival; no other substances can play their metabolic roles. For instance, all of your cells need iron for their electron transport chains. Red blood cells don't function at all without iron in hemoglobin. And neurons stop functioning in the absence of sodium and potassium (Table 37.3).

People who are in good health get all the vitamins and minerals they need from a balanced diet of whole

Table 37.2 *Vitamins: Sources, Functions, and Effects of Deficiencies or Excesses**

Vitamin	Common Sources	Main Functions	Effects of Chronic Deficiency	Effects of Extreme Excess
FAT-SOLUBLE VITAMINS				
A	Its precursor comes from beta-carotene in yellow fruits, yellow or green leafy vegetables; also in fortified milk, egg yolk, fish liver	Used in synthesis of visual pigments, bone, teeth; maintains epithelia	Dry, scaly skin; lowered resistance to infections; night blindness; permanent blindness	Malformed fetuses; hair loss; changes in skin; liver and bone damage; bone pain
D	D_3 formed in skin and in fish liver oils, egg yolk, fortified milk; converted to active form elsewhere	Promotes bone growth and mineralization; enhances calcium absorption	Bone deformities (rickets) in children; bone softening in adults	Retarded growth; kidney damage; calcium deposits in soft tissues
E	Whole grains, dark green vegetables, vegetable oils	Counters effects of free radicals; helps maintain cell membranes; blocks breakdown of vitamins A and C in gut	Lysis of red blood cells; nerve damage	Muscle weakness, fatigue, headaches, nausea
K	Enterobacteria form most of it; also in green leafy vegetables, cabbage	Blood clotting; ATP formation via electron transport	Abnormal blood clotting; severe bleeding (hemorrhaging)	Anemia; liver damage and jaundice
WATER-SOLUBLE VITAMINS				
B_1 (thiamin)	Whole grains, green leafy vegetables, legumes, lean meats, eggs	Connective tissue formation; folate utilization; coenzyme action	Water retention in tissues; tingling sensations; heart changes; poor coordination	None reported from food; possible shock reaction from repeated injections
B_2 (riboflavin)	Whole grains, poultry, fish, egg white, milk	Coenzyme action	Skin lesions	None reported
Niacin	Green leafy vegetables, potatoes, peanuts, poultry, fish, pork, beef	Coenzyme action	Contributes to pellagra (damage to skin, gut, nervous system, etc.)	Skin flushing; possible liver damage
B_6	Spinach, tomatoes, potatoes, meats	Coenzyme in amino acid metabolism	Skin, muscle, and nerve damage; anemia	Impaired coordination; numbness in feet
Pantothenic acid	In many foods (meats, yeast, egg yolk especially)	Coenzyme in glucose metabolism, fatty acid and steroid synthesis	Fatigue, tingling in hands, headaches, nausea	None reported; may cause diarrhea occasionally
Folate (folic acid)	Dark green vegetables, whole grains, yeast, lean meats; enterobacteria produce some folate	Coenzyme in nucleic acid and amino acid metabolism	A type of anemia; inflamed tongue; diarrhea; impaired growth; mental disorders	Masks vitamin B_{12} deficiency
B_{12}	Poultry, fish, red meat, dairy foods (not butter)	Coenzyme in nucleic acid metabolism	A type of anemia; impaired nerve function	None reported
Biotin	Legumes, egg yolk; colon bacteria produce some	Coenzyme in fat, glycogen formation and in amino acid metabolism	Scaly skin (dermatitis), sore tongue, depression, anemia	None reported
C (ascorbic acid)	Fruits and vegetables, especially citrus, berries, cantaloupe, cabbage, broccoli, green pepper	Collagen synthesis; possibly inhibits effects of free radicals; structural role in bone, cartilage, and teeth; role in carbohydrate metabolism	Scurvy, poor wound healing, impaired immunity	Diarrhea, other digestive upsets; may alter results of some diagnostic tests

* Guidelines for appropriate daily intakes are being worked out by the Food and Drug Administration.

Table 37.3 *Major Minerals: Sources, Functions, and Effects of Deficiencies or Excesses***

Mineral	Common Sources	Main Functions	Signs of Severe Long-Term Deficiency	Signs of Extreme Excess
Calcium	Dairy products, dark green vegetables, dried legumes	Bone, tooth formation; blood clotting; neural and muscle action	Stunted growth; possibly diminished bone mass (osteoporosis)	Impaired absorption of other minerals; kidney stones in susceptible people
Chloride	Table salt (usually too much in diet)	HCl formation in stomach; contributes to body's acid–base balance; neural action	Muscle cramps; impaired growth; poor appetite	Contributes to high blood pressure in susceptible people
Copper	Nuts, legumes, seafood, drinking water	Used in synthesis of melanin, hemoglobin, and some transport chain components	Anemia, changes in bone and blood vessels	Nausea, liver damage
Fluorine	Fluoridated water, tea, seafood	Bone, tooth maintenance	Tooth decay	Digestive upsets; mottled teeth and deformed skeleton in chronic cases
Iodine	Marine fish, shellfish, iodized salt, dairy products	Thyroid hormone formation	Enlarged thyroid (goiter), with metabolic disorders	Goiter
Iron	Whole grains, green leafy vegetables, legumes, nuts, eggs, lean meat, molasses, dried fruit, shellfish	Formation of hemoglobin and cytochrome (transport chain component)	Iron-deficiency anemia, impaired immune function	Liver damage, shock, heart failure
Magnesium	Whole grains, legumes, nuts, dairy products	Coenzyme role in ATP–ADP cycle; roles in muscle, nerve function	Weak, sore muscles; impaired neural function	Impaired neural function
Phosphorus	Whole grains, poultry, red meat	Component of bone, teeth, nucleic acids, ATP, phospholipids	Muscular weakness; loss of minerals from bone	Impaired absorption of minerals into bone
Potassium	Diet provides ample amounts	Muscle and neural function; roles in protein synthesis and body's acid–base balance	Muscular weakness	Muscular weakness, paralysis, heart failure
Sodium	Table salt; diet provides ample to excessive amounts	Key role in body's salt–water balance; roles in muscle and neural function	Muscle cramps	High blood pressure in susceptible people
Sulfur	Proteins in diet	Component of body proteins	None reported	None likely
Zinc	Whole grains, legumes, nuts, meats, seafood	Component of digestive enzymes; roles in normal growth, wound healing, sperm formation, and taste and smell	Impaired growth, scaly skin, impaired immune function	Nausea, vomiting, diarrhea; impaired immune function and anemia

* The guidelines for appropriate daily intakes are being worked out by the Food and Drug Administration.

foods. Generally, vitamin and mineral supplements are necessary only for strict vegetarians, the elderly, and people suffering a chronic illness or taking medication that interferes with utilization of specific nutrients. As examples, vitamin K supplements aid calcium retention and lessen osteoporosis in elderly women. Vitamin C, vitamin E, and the beta-carotene precursor for vitamin A may lessen some aging effects and improve immune functioning by inactivating free radicals. A free radical, remember, is an atom or group of atoms that is highly reactive because it has an unpaired electron.

However, metabolism varies in its details from one person to the next, so no one should take massive doses of any vitamin or mineral supplement except under medical supervision. Also, excessive amounts of many

vitamins and minerals can harm everyone (Tables 37.2 and 37.3). Large doses of vitamins A and D are good examples. As is the case for all fat-soluble vitamins, excess amounts of these vitamins accumulate in tissues and interfere with normal metabolic activity. Similarly, sodium is present in plant and animal tissues and is a component of table salt. Sodium has roles in the body's salt–water balance, muscle activity, and nerve function. However, prolonged, excessive intake of sodium may contribute to high blood pressure in some people.

Severe shortages or self-prescribed, massive excesses of vitamins and minerals can disturb the delicate balances in body function that promote health.

Weighty Questions, Tantalizing Answers

Have you ever asked yourself *why* you want to weigh a certain amount? Are you merely fearful of "being fat"—that is, obese? By definition, **obesity** is an excess of fat in adipose tissues, most often caused by imbalances between caloric intake and energy output. One standard of what constitutes obesity is simply cultural, and it varies from one culture to the next. For example, one female student despaired over her plumpness until her graduate studies program in Africa, where many males found her to be one of the most desirable females on the planet. However, a lot depends on how healthy you want to be and how long you want to live. Thinner people generally live longer.

In the United States, obesity is now the second leading cause of death; 300,000 or so people die annually from preventable, weight-related conditions. Each year, the weight-related cases of type 2 diabetes, heart disease, hypertension, breast cancer, colon cancer, gout, gallstones, and osteoarthritis add up to a 100-billion-dollar drain on the economy. The future doesn't look any rosier: children today are 42 percent fatter than they were in 1980.

IN PURSUIT OF THE "IDEAL" WEIGHT   Figure 37.16 shows charts for estimating the "ideal" weight for adults. Another indicator of obesity-related health risk is the *body mass index* (BMI), as determined by this formula:

$$BMI = \frac{\text{weight (pounds)} \times 700}{\text{height (inches)}^2}$$

Shape Up America, a national initiative for promoting health, will quickly do the calculation for you. Their web site is http://www.shapeup.org. If your BMI value is 27 or higher, the health risk rises dramatically. Other factors that affect the risk are smoking habits, family history of heart disorders, use of sex hormones after menopause, and fat distribution. Fat stored above the belt, as with "beer bellies," definitely is not good.

Dieting alone can't lower a BMI value. Eat less, and the body slows its metabolic rate to conserve energy. So how do you function normally over the long term while maintaining acceptable weight? *You must balance caloric intake with energy output.* For most of us, this means eating controlled portions of low-calorie, nutritious foods *and* exercising regularly.

To figure out how many kilocalories you should take in daily to maintain a desired weight, multiply that weight (in pounds) by 10 if you are not active physically, by 15 if moderately active, and by 20 if highly active. From the value you get this way, subtract the following amount:

Age:	25–34	Subtract:	0
	35–44		100
	45–54		200
	55–64		300
	Over 65		400

For instance, if you want to weigh 120 pounds and are very active, 120 × 20 = 2,400 kilocalories. If you're thirty-five years old, then (2,400 − 100), or 2,300 kilocalories. Such calculations give a rough estimate of caloric intake. Other factors, including height, must also be considered. An active person 5 feet, 2 inches tall doesn't need as much energy as an active 6-footer who weighs just the same.

Figure 37.16   How to estimate the "ideal" weight for an adult. The values shown are consistent with a long-term Harvard study into the link between excessive weight and increased risk of cardiovascular disorders. Depending on certain factors (such as having a small, medium, or large skeletal frame), the "ideal" might vary by plus or minus 10 percent.

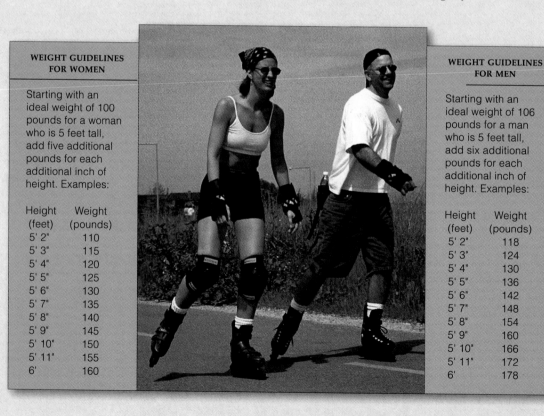

WEIGHT GUIDELINES FOR WOMEN

Starting with an ideal weight of 100 pounds for a woman who is 5 feet tall, add five additional pounds for each additional inch of height. Examples:

Height (feet)	Weight (pounds)
5' 2"	110
5' 3"	115
5' 4"	120
5' 5"	125
5' 6"	130
5' 7"	135
5' 8"	140
5' 9"	145
5' 10"	150
5' 11"	155
6'	160

WEIGHT GUIDELINES FOR MEN

Starting with an ideal weight of 106 pounds for a man who is 5 feet tall, add six additional pounds for each additional inch of height. Examples:

Height (feet)	Weight (pounds)
5' 2"	118
5' 3"	124
5' 4"	130
5' 5"	136
5' 6"	142
5' 7"	148
5' 8"	154
5' 9"	160
5' 10"	166
5' 11"	172
6'	178

MY GENES MADE ME DO IT   On the average, one out of three Americans has far more trouble keeping off excess weight than others do, and genes have a lot to do with it. Researchers suspected as much for a long time. Studies of identical twins offered clues. (*Identical twins* are born with identical genes.) As an outcome of family problems, some of the twins had been separated at birth and were raised apart, in different households. Yet, at adulthood, the separated twins had similar body weights! People, it seemed, were born with a "set point" for body fat. Could it be that whatever set point someone inherits is the one he or she will be stuck with for life?

Many experiments confirmed this suspicion. The first started in 1950 with a seriously obese mouse (Figure 37.17). In 1995, molecular geneticists identified one of its genes that affects the set point for body fat. They named it the *ob* gene (guess why). Which clues led to its discovery? The nucleotide sequence in the gene region of obese mice differs from that in normal mice. The gene is active in adipose tissue, nowhere else. Its biochemical message translates into a protein released from cells and picked up by the bloodstream; the gene specifies a hormone.

Jeffrey Friedman discovered the hormone in 1994 and called it **leptin**. A nearly identical hormone was isolated in humans. Leptin is just one of a number of factors that mediate the brain's commands to suppress or whip up appetite, but it's a crucial one. By assessing the blood levels of incoming hormonal signals, an appetite control center in the hypothalamus "decides" whether the body has taken in enough food to provide enough fat for the day. If so, hypothalamic commands go out to increase metabolic rates—and to stop eating.

If the *ob* gene is mutated, then its expression alone may be enough to disrupt the control center so a mouse's appetite skyrockets and metabolic furnaces burn less. In one experiment that supported this hypothesis, obese mice were injected with leptin. The mice quickly shed their excess weight.

Will genetic researchers go on to develop therapies for obesity in humans? Maybe. As you read in Chapter 32, however, hormonal control is tricky business. It now appears that, besides its role in controlling body weight, leptin is a selective inhibitor of bone formation, as mediated by the central nervous system. Thus the danger is that anti-obesity therapies will bring on osteoporosis.

a *1950*. Researchers at the Jackson Laboratories in Maine notice that one of their laboratory mice is extremely obese, with an uncontrollable appetite. Through crossbreeding of this apparent mutant individual with a normal mouse, they produce a strain of obese mice.

b *Late 1960s*. Douglas Coleman of the Jackson Laboratories surgically joins the bloodstreams of an obese mouse and a normal one. The obese mouse now loses weight. Coleman hypothesizes that a factor circulating in blood may be influencing its appetite, but he is not able to isolate it.

c *1994*. Late in the year, Jeffrey Friedman of Rockefeller University discovers a mutated form of what is now called the *ob* gene in obese mice. Through DNA cloning and gene sequencing, he defines the protein that the mutated gene encodes. The protein, now called leptin, is a hormone that influences the brain's commands to suppress appetite and increase metabolic rates.

d *1995*. Three different research teams develop and use genetically engineered bacteria to produce leptin, which, when injected in obese and normal mice, triggers significant weight loss, apparently without harmful side effects.

ob gene

protein product (leptin)

Figure 37.17 Chronology of research developments that revealed the identity of a key factor in the genetic basis of body weight.

SUMMARY

Gold indicates text section

1. For animals, *nutrition* refers to processes by which the body takes in, digests, absorbs, and uses food. *CI*

2. Most animals have a complete digestive system—a tube that has two openings (a mouth and an anus), and regional specializations between them. Mucus-coated epithelium lines and protects every exposed surface of the tube and facilitates diffusion across its wall. *37.1*

3. These activities proceed in digestive systems: *37.1*

    a. Mechanical processing and motility: Movements that break up, mix, and propel food material.

    b. Secretion: Release of digestive enzymes and other substances from the pancreas and liver, as well as from glandular epithelium, into the gut lumen.

    c. Digestion: Breakdown of food into particles, then into nutrient molecules small enough to be absorbed.

    d. Absorption: The diffusion or transport of digested organic compounds, fluid, and ions from the gut lumen into the internal environment.

    e. Elimination: The expulsion of undigested as well as unabsorbed residues at the end of the system.

4. The human digestive system has a mouth, pharynx, esophagus, stomach, small intestine, large intestine (colon), rectum, and anus. The salivary glands and the liver, gallbladder, and pancreas have accessory roles in the system's functions (Table 37.4). *37.2–37.5, 37.7*

5. Starch digestion starts in the mouth, and protein digestion starts in the stomach. Digestion is completed and most nutrients are absorbed in the small intestine. The pancreas secretes the main digestive enzymes. Bile from the liver assists in fat digestion. *37.3*

6. The nervous and endocrine systems, and meshes of neurons (plexuses) in the gut wall, interact to govern the digestive system. Many controls operate in response to the volume and composition of food in the gut. They cause changes in muscle activity and in secretion rates for hormones and enzymes. *37.4*

7. In absorption, cells of the intestinal lining actively transport glucose and most amino acids out of the gut lumen. Fatty acids and monoglycerides diffuse across the lipid bilayer of these cells. They are recombined in the cytoplasm as triglycerides, which then are released by exocytosis, into interstitial fluid. *37.5–37.6*

8. Nutritionists advise a daily food intake in certain proportions. Example: One diet for healthy adult males of average body weight recommends 55–60 percent complex carbohydrates, 15–20 percent proteins (less for females), 20–25 percent fats. A well-balanced diet of whole foods normally supplies all necessary vitamins and minerals. *37.8–37.9*

9. To maintain health and a given body weight, caloric (energy) intake must balance energy output. *37.10*

Table 37.4 *Summary of the Digestive System*

MOUTH (oral cavity)	Start of digestive system, where food is chewed, moistened; polysaccharide digestion begins
PHARYNX	Entrance to tubular parts of digestive and respiratory systems
ESOPHAGUS	Muscular tube, moistened by saliva, that moves food from pharynx to stomach
STOMACH	Sac where food mixes with gastric fluid and protein digestion begins; stretches to store food taken in faster than can be processed; gastric fluid destroys many microbes
SMALL INTESTINE	The first part (duodenum) receives secretions from the liver, gallbladder, and pancreas
	Most nutrients are digested, absorbed in second part (jejunum)
	Some nutrients absorbed in last part (ileum), which delivers unabsorbed material to colon
COLON (large intestine)	Concentrates and stores undigested matter (by absorbing mineral ions and water)
RECTUM	Distension triggers expulsion of feces
ANUS	Terminal opening of digestive system

Accessory Organs:

SALIVARY GLANDS	Glands (three main pairs, many minor ones) that secrete saliva, a fluid with polysaccharide-digesting enzymes, buffers, and mucus (which moistens and lubricates ingested food)
PANCREAS	Secretes enzymes that digest all major food molecules and buffers against HCl from stomach
LIVER	Secretes bile (used in fat emulsification); roles in carbohydrate, fat, and protein metabolism
GALLBLADDER	Stores and concentrates bile from the liver

Review Questions

1. Define the five key tasks carried out by a complete digestive system. Then correlate some organs of such a system with the feeding behavior of a particular kind of animal. *37.1*

2. Using the sketch to the right, list the organs and accessory organs shown and the main functions of each. *Table 37.4*

3. With respect to digestive systems, define segmentation. Does segmentation occur in the stomach? Does it occur in the small intestine, colon, or both? *37.5, 37.7*

4. Using the *black* lines shown in Figure 37.18 as a guide, name the types of breakdown products small enough to be absorbed across the small intestine's lining, into the internal environment. *37.5*

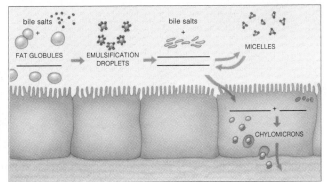

Figure 37.18 Fill in the blanks for substances that cross the lining of the small intestine.

Self-Quiz ANSWERS IN APPENDIX III

1. The _____ maintains the internal environment, supplies cells with raw materials, and disposes of metabolic wastes.
   a. digestive system
   b. circulatory system
   c. respiratory system
   d. urinary system
   e. interaction of all of the systems listed

2. Most digestive systems have regions for _____ food.
   a. transporting
   b. processing
   c. storing
   d. all of the above

3. _____ secretions do not assist in digestion and absorption.
   a. Salivary gland
   b. Thymus gland
   c. Liver
   d. Pancreas

4. Digestion is completed and most nutrients are absorbed in the _____ .
   a. mouth
   b. stomach
   c. small intestine
   d. colon

5. Bile has roles in _____ digestion and absorption.
   a. carbohydrate
   b. fat
   c. protein
   d. amino acid

6. Monosaccharides and most amino acids are absorbed _____ .
   a. by active transport
   b. by diffusion
   c. at lymph vessels
   d. as fat droplets

7. Most of our caloric intake should come from _____ .
   a. complex carbohydrates
   b. simple carbohydrates
   c. proteins
   d. lipids

8. On its own, the human body cannot produce all of the _____ it requires.
   a. vitamins and minerals
   b. fatty acids
   c. amino acids
   d. a through c
   e. a and c

9. Maintaining good health and normal body weight requires that _____ intake be balanced by _____ output.

10. Match the organ with its key digestive function(s).
   _____ gallbladder   a. secrete bile; many metabolic roles
   _____ stomach   b. digest, absorb most nutrients
   _____ colon   c. store, mix, dissolve food; start protein breakdown
   _____ pancreas
   _____ salivary   d. store, concentrate bile
          gland   e. concentrate undigested matter
   _____ small   f. secrete substances that moisten food,
          intestine   start polysaccharide breakdown
   _____ liver   g. secrete digestive enzymes and bicarbonate

Critical Thinking

1. A glassful of whole milk contains lactose, proteins, butterfat (mostly triglycerides), vitamins, and minerals. Explain what will happen to each component in your digestive tract.

2. As people age, the number of their body cells steadily decreases and energy needs decline. If you were planning an older person's diet, which foods would you emphasize, and why? Which ones would you deemphasize?

3. Using Section 37.10 as a reference, determine your ideal weight and then design a well-balanced program of diet and exercise that will help you achieve or maintain that weight.

4. Holiday meals often are larger than everyday ones and have a high fat content. After stuffing themselves at an early dinner on Thanksgiving Day, Richard and other members of his family feel uncomfortably full for the rest of the afternoon. Based on what you have learned about controls over digestion, propose a biochemical explanation for their discomfort.

5. Explain some of the ways in which your digestive system helps protect you against many pathogenic bacteria that may have contaminated the kinds of food you eat. (You may wish to refer to Section 35.1, also.)

6. Angelina is chronically anorexic. Clinical tests show that she has very low blood pressure and a below-average cardiac rate (bradycardia). The heart itself has become smaller. How did anorexia bring about these life-threatening conditions?

Selected Key Terms

appendix 37.7	liver 37.4
bile 37.4	micelle formation 37.5
bulk 37.7	microvillus (microvilli) 37.5
chyme 37.4	mineral 37.9
colon 37.7	nutrition CI
complete digestive system 37.1	ob gene 37.10
digestive system CI	obesity 37.10
emulsification 37.4	pancreas 37.4
esophagus 37.3	pharynx 37.3
essential amino acid 37.8	ruminant 37.1
essential fatty acid 37.8	saliva 37.3
food pyramid 37.8	segmentation 37.5
gallbladder 37.4	sphincter 37.3
gastric fluid 37.4	stomach 37.4
gut 37.2	tongue 37.3
incomplete digestive system 37.1	tooth 37.3
kilocalorie 37.8	villus (villi) 37.5
leptin 37.10	vitamin 37.9

Readings

Blaser, M. J. February 1996. "The Bacteria Behind Ulcers." *Scientific American*, 274.

Kassirer, J. 17 September 1998. *New England Journal of Medicine*. Editorial introduces reports on unproven and harmful uses of dietary supplements, some of which were contaminated.

Sherwood, L. 1997. *Human Physiology*. Third edition. Monterey, California: Brooks-Cole.

On-Line readings at Student Guide for InfoTrac:
www.brookscole.com/biology

38

THE INTERNAL ENVIRONMENT

Tale of the Desert Rat

Look closely at a fish or some other marine animal, and you'll find that its body cells are exquisitely adapted to life in a salty fluid. And yet, about 375 million years ago, some lineages of animals that evolved in the seas moved onto dry land. They were able to do so partly because they brought salty fluid along with them, as an *internal* environment for their cells. Even so, it was not a simple transition. The pioneers on land and their descendants faced intense sunlight, dry winds, more pronounced swings in temperature, water of dubious salt content, and sometimes no water at all.

How did the pioneers conserve or replace the water and salts lost as a result of everyday activities? How did they manage to stay comfortably warm when their surroundings became too cold or too hot? They must have done these things. Otherwise the composition, volume, and temperature of their internal environment would have spun out of control. In short, the question is this: *How did the land-dwelling descendants of marine animals maintain internal operating conditions and so prevent cellular anarchy?*

Observe any of their existing descendants and you get a sense of what some answers might be. Suppose you focus on a tiny mammal, a kangaroo rat living in an isolated desert of New Mexico (Figure 38.1). After a brief rainy season, the sun bakes the desert sand for months. The only obvious water is imported, sloshing about in the canteens of the occasional researcher and tourist. Yet with nary a sip of free water, a kangaroo rat routinely counters threats to its internal environment.

It waits out the heat of the day inside a burrow, then forages in the cool of night for dry seeds and maybe a succulent. It is not sluggish about this. It hops rapidly and far, searching for seeds and fleeing from coyotes and snakes. All that hopping requires ATP energy and water. Seeds, chockful of energy-rich carbohydrates, provide both. Metabolic reactions that release energy from carbohydrates and other organic compounds also yield water. Each day, such "metabolic water" makes up a whopping 90 percent of a kangaroo rat's total water intake. By comparison, metabolic water is only about 12 percent of the total water intake for your body.

	KANGAROO RAT	HUMAN
WATER GAIN (milliliters)		
by ingesting solids	6.0	850
by ingesting liquids	0.0	1,400
by metabolism	54.0	350
	60.0	2,600
WATER LOSS (milliliters)		
in urine	13.5	1,500
in feces	2.6	200
by evaporation	43.9	900
	60.0	2,600

Figure 38.1 Kangaroo rat, master of water conservation in a New Mexico desert. The chart will give you an idea of how kangaroo rats and humans gain and lose water. They do so in different ways. Even so, in both cases, *water losses balance out the gains* —just as they do in all animals. How this balancing happens, and why it absolutely must happen, is this chapter's initial focus.

food, water intake   oxygen intake

DIGESTIVE SYSTEM → RESPIRATORY SYSTEM → elimination of carbon dioxide

nutrients, water, salts   oxygen   carbon dioxide

CIRCULATORY SYSTEM ⇄ URINARY SYSTEM

water, solutes

elimination of food residues   rapid transport to and from all living cells   elimination of excess water, salts, wastes

Figure 38.2 Links between the urinary system and other organ systems that contribute to homeostasis, or stability in favorable operating conditions in the animal body.

A kangaroo rat resting in a cool burrow conserves and recycles water. As it inhales cool air into its warm lungs, water vapor condenses onto epithelium in its nose. Some of that water diffuses back into the body. Also, after a busy night of foraging, it empties its cheek pouches of seeds. As seeds leave its mouth, they soak up water dripping from its nose. When the kangaroo rat eats the dripped-upon seeds, it reclaims water.

A kangaroo rat can't lose water by perspiring; it has no sweat glands. It loses water when it urinates, but its two specialized kidneys don't let it piddle away much. Kidneys filter the blood's water and solutes, including dissolved salts. They continually adjust *how much water* and *which solutes* return to the blood or depart as urine.

Overall, kangaroo rats and all other animals take in enough water and solutes to replace daily losses (Figure 38.1). How they do so will be our initial focus. Later, we will consider how mammals withstand hot, cold, and often unpredictable temperature changes on land.

As a starting point, think back on the fluids inside most animals. **Interstitial fluid** fills the spaces between living cells and other components of tissues. Another fluid, **blood**, moves substances to and from all tissue regions by way of a circulatory system. Together, the interstitial fluid and blood are **extracellular fluid**. In many animals, a well-developed urinary system helps keep the composition and volume of extracellular fluid within tolerable ranges. As you'll see, the organ systems shown in Figure 38.2 interact with the urinary system in the performance of this homeostatic task.

Key Concepts

1. Animals are continually gaining and losing water and solutes, or dissolved substances. They continually produce metabolic wastes. Even with all of the inputs and outputs, the overall composition and volume of the extracellular fluid in the body remain relatively constant.

2. In humans as in other vertebrates, a urinary system is crucial to balancing the intake and output of water and solutes. This system filters water and solutes from blood on an ongoing basis. It reclaims some amount of both and eliminates the rest. Different amounts are reclaimed at different times, depending on what's required to maintain extracellular fluid.

3. Kidneys are blood-filtering organs, and the urinary system of vertebrates has a pair of them. Packed inside each kidney are a great number of nephrons.

4. At its receiving end, each nephron cups around a tuft of blood capillaries and accepts water and solutes from it. The nephron returns most of the filtrate to the blood by giving it up to capillaries that intertwine around the nephron's tubular parts.

5. Water and solutes not returned to the blood leave the body as a fluid called urine. During any interval, control mechanisms make adjustments that influence whether urine is concentrated or dilute. Two hormones, ADH and aldosterone, have key roles in the adjustments.

6. The internal body temperature of animals depends on a balancing of heat produced through metabolism, heat absorbed from the environment, and heat lost to the environment.

7. The internal body temperature is maintained within a favorable range through controls over metabolic activity and adaptations in body form and behavior.

URINARY SYSTEM OF MAMMALS

The Challenge—Shifts in Extracellular Fluid

Different solid foods and fluids intermittently enter a mammal's gut. Afterward, variable amounts of absorbed water, nutrients, and other substances move into the blood, then into interstitial fluid and on into cells. Such events could easily shift the volume and composition of extracellular fluid beyond tolerable limits. But the body makes compensatory adjustments that balance out the gains and losses. Within a given time frame, it takes in as much water and solutes as it gives up.

WATER GAINS AND LOSSES To start, think about how humans and other mammals *gain* water, mainly by two processes:

Absorption from the gut
Metabolism

Considerable water is absorbed from solids and liquids inside the gut. As you know, water also forms as a normal by-product of many metabolic reactions. In land mammals, how much water enters the gut in the first place depends on a thirst mechanism. When the body loses too much water, land mammals seek out streams, waterholes, and so on. We will be looking at the thirst mechanism later in the chapter.

Normally, the mammalian body *loses* water mainly by way of four physiological processes—especially the first process listed here:

Urinary excretion
Evaporation from lungs and the skin
Sweating, by mammals that sweat
Elimination, in feces

Urinary excretion affords the most control over water loss. The process eliminates excess water and solutes as urine. This fluid forms in a urinary system such as the one in Figure 38.3. Water evaporates from respiratory surfaces, too. Some species lose it in sweat. A mammal in good health tends to lose very little water from the gut; most is absorbed, not eliminated in feces.

SOLUTE GAINS AND LOSSES The mammalian body *gains* solutes mainly by the following four processes:

Absorption from the gut
Secretion from cells
Respiration
Metabolism

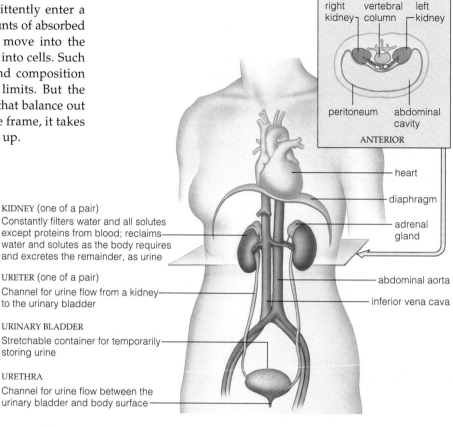

KIDNEY (one of a pair)
Constantly filters water and all solutes except proteins from blood; reclaims water and solutes as the body requires and excretes the remainder, as urine

URETER (one of a pair)
Channel for urine flow from a kidney to the urinary bladder

URINARY BLADDER
Stretchable container for temporarily storing urine

URETHRA
Channel for urine flow between the urinary bladder and body surface

Figure 38.3 Organs of the human urinary system and their functions. The two kidneys, two ureters, and urinary bladder are located *outside* the peritoneum, the membranous lining of the abdominal cavity. Compare Section 23.1.

For example, you absorb food, mineral ions, drugs, and food additives from the gut. Cell secretions and wastes, such as carbon dioxide, enter interstitial fluid and then blood. The respiratory system puts oxygen into blood; aerobically respiring cells put carbon dioxide into it.

Mammals typically *lose* solutes by three processes:

Urinary excretion
Respiration
Sweating, by some species

The urine of mammals includes various wastes formed by the breakdown of organic compounds. For example, ammonia forms as amino groups are split from amino acids. Then **urea**, a major waste, forms in the liver when two ammonia molecules join with carbon dioxide. Also in urine are uric acid (from nucleic acid breakdown), hemoglobin breakdown products (which give the urine much of its color), drugs, and food additives. Besides these solute losses, some mammals also lose mineral ions in sweat. And all mammals lose carbon dioxide, the most abundant waste, by way of respiration.

renal artery

kidney cortex

kidney medulla

renal pelvis

renal capsule

renal vein

ureter

a

b

orientation of nephrons relative to cortex and medulla

Bowman's capsule (*red*)

collecting duct

proximal tubule (*orange*)

distal tubule (*brown*)

loop of Henle (*yellow*)

c   **NEPHRON**

Bowman's capsule

glomerular capillaries (set of capillaries at start of nephron)

collecting duct

peritubular capillaries (around tubular parts of nephron)

d

Figure 38.4 (**a**) Human kidney and blood vessels servicing it. (**b**) Its functional units, called nephrons. Their size is greatly exaggerated here for clarity. (**c**) The nephron's functional zones. (**d**) Blood capillaries associated with nephrons. The start of a nephron, Bowman's capsule, balloons around a set of glomerular capillaries. A second set of blood vessels, the peritubular capillaries, threads around tubular parts of the nephron.

Components of the Urinary System

Mammals routinely counter shifts in the composition and volume of extracellular fluid, mainly with a **urinary system**. This consists of two kidneys, two ureters, one urinary bladder, and one urethra. **Kidneys** are a pair of bean-shaped organs, about as large as a fist in an adult human (Figures 38.3 and 38.4). Each has an outer capsule of connective tissue. Blood capillaries penetrate its two inner regions, the cortex and medulla.

Kidneys filter water, mineral ions, organic wastes, and other substances from blood. Then they adjust the filtrate's composition and return all but about 1 percent to the blood. The small portion of unreclaimed water and solutes is urine. By definition, **urine** is a fluid that rids the body of water and solutes that are in excess of the amounts required to maintain extracellular fluid.

Urine flows from each kidney into a **ureter**, a tubular channel between it and the **urinary bladder**, a muscular sac. Urine is briefly stored here before flowing into the **urethra**, a muscular tube that opens at the body surface.

Flow from the bladder (urination) is a reflex action. When the bladder is full, the sphincter around its neck opens, smooth muscle in its balloonlike wall contracts, and so urine is forced out through the urethra. Skeletal muscle surrounds the urethra. Its contraction, which is under voluntary control, prevents urination.

Nephrons—Functional Units of Kidneys

Each human kidney has more than a million **nephrons**, slender tubules packed inside lobes that extend from the kidney's cortex down through the medulla. *It is at the nephrons that water and solutes are filtered from blood and adjustments are made in the amounts to be reclaimed.*

A nephron starts as a **Bowman's capsule**, where its wall cups around a set of *glomerular* capillaries. The cup and capillaries interact as one blood-filtering unit, the **renal corpuscle** (Figure 38.4c). Next comes a **proximal tubule** (tubular region closest to the capsule), a hairpin-shaped **loop of Henle**, and a **distal tubule** (most distant from the capsule). A **collecting duct**, the nephron's last region, is part of a duct system leading to the kidney's central cavity (renal pelvis), then into a ureter.

Blood does not give up all of its water and solutes. The unfiltered part flows into capillaries threading lacily around the nephron's tubular parts. In these *peritubular* capillaries, the blood reclaims water and solutes. Then it flows into veins and back to the general circulation.

A urinary system counters unwanted shifts in the volume and composition of extracellular fluid. In its paired kidneys, water and solutes are filtered from blood. The body reclaims most of this, but the excess leaves the kidneys as urine.

URINE FORMATION

Urine forms in nephrons by three processes: filtration, tubular reabsorption, and tubular secretion. All three processes depend on properties of cells of the nephron wall. Cells in different wall regions differ in membrane transport mechanisms and in their permeability.

Blood pressure generated by the heart's contractions drives **filtration**—the forcing of water and all solutes except proteins out from glomerular capillaries into the fluid-filled space of Bowman's capsule. The protein-free filtrate moves into the proximal tubule (Figure 38.5).

Next, with **tubular reabsorption**, most of the water and solutes that moved down into the proximal tubule are returned to the peritubular capillaries (Figure 38.6). As you will see shortly, hormonal controls adjust the amounts conserved or excreted during a given interval.

Blood still inside peritubular capillaries holds excess ions (mainly H^+ and K^+), metabolites such as urea, and neurotransmitters, histamine, and maybe drugs or toxins. Such substances diffuse into the surrounding interstitial fluid. By **tubular secretion**, they now enter the nephron through proteins that function as channels and transporters across tubular portions of the nephron wall. Then they are excreted.

Factors Influencing Filtration

Each minute, about 1.5 liters (1–1/2 quarts) of blood flow through an adult's kidneys, and 120 milliliters of water and small solutes are filtered into nephrons. That's 180 liters of filtrate per day! A high filtration rate is possible mainly because the glomerular capillaries are 10 to 100 times more permeable to water and small solutes than other capillaries are. Also, blood pressure is high inside them. Why? Arterioles delivering blood to the nephron have a wider lumen and less flow resistance than those taking blood away. As a result, blood dams up inside the renal corpuscle, adding to the pressure in its capillaries.

Filtration rates also depend on flow volume to the kidneys. Neural, endocrine, and local controls maintain the flow even when blood pressure changes. Example: Dance until dawn and your nervous system commands an above-normal volume of blood to be diverted away from your kidneys and sent instead toward the energy-demanding cells in skeletal muscles and the heart. Such redistribution of blood volume involves coordinating the vasoconstriction and vasodilation of arterioles throughout the body.

blood vessel entering

blood vessel leaving

a Blood pumped from the heart travels to the renal artery, then into kidneys, where water and some solutes will be filtered from it. Most of the filtrate will return to the general circulation.

Bowman's capsule + glomerular capillaries = renal corpuscle

f Hormonal action adjusts the urine concentration. *ADH* promotes *water* reabsorption, so the urine is concentrated. When controls inhibit ADH secretion, urine is dilute.

Aldosterone promotes *sodium* reabsorption by stimulating sodium pumps. Because more sodium is reabsorbed, the urine has little sodium. When controls inhibit secretion, more sodium is excreted in urine.

b Filtration. At the start of the nephron, blood enters glomerular capillaries. Water and small solutes are filtered into Bowman's capsule.

COLLECTING DUCT

NEPHRON

d Tubular Secretion. Cells of the nephron's tubular wall regions secrete excess H^+ and a few other solutes into the fluid inside the nephron's lumen.

c Tubular Reabsorption. Water and many solutes cross the proximal tubule wall and enter interstitial fluid of the kidney cortex. Membrane transport proteins move most of the solutes across the wall. These materials then enter the peritubular capillaries.

e *After* its hairpin turn, the wall of the loop of Henle is impermeable to water. But its cells actively pump sodium and chloride ions out of the loop. Pumping makes the interstitial fluid saltier. As a result, even more water is drawn out of collecting ducts that also run through the medulla.

g Excretion. What happens to water and solutes that were not reabsorbed or that were secreted into the tubule? They flow through a collecting duct to the renal pelvis, then are eliminated from the body by way of the urinary tract.

Figure 38.5 Urine formation.

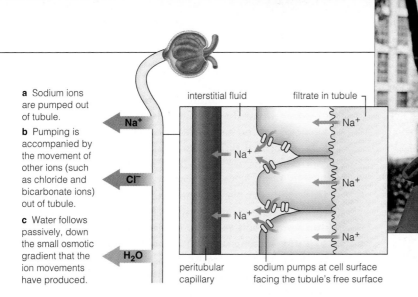

a Sodium ions are pumped out of tubule.

b Pumping is accompanied by the movement of other ions (such as chloride and bicarbonate ions) out of tubule.

c Water follows passively, down the small osmotic gradient that the ion movements have produced.

interstitial fluid filtrate in tubule

Na⁺

Cl⁻

H₂O

Na⁺

Na⁺

Na⁺

Na⁺

Na⁺

Na⁺

peritubular capillary

sodium pumps at cell surface facing the tubule's free surface

Figure 38.6 Membrane pumps in the nephron wall. Sodium and water are reabsorbed here in response to, say, gulping too much water or ingesting too much salt.

Also, the walls of arterioles leading to nephrons are sensitive to pressure. When blood pressure falls, the arterioles vasoconstrict, so less of the total blood volume is sent to the kidneys. When blood pressure rises, these arterioles vasodilate, so more blood flows in.

Reabsorption of Water and Sodium

REABSORPTION MECHANISM Kidneys precisely adjust how much water and sodium ions the body excretes or saves. Drink too much or too little water, wolf down salty potato chips, or lose too much sodium in sweat, and responses are quick. As filtrate enters the proximal tubules, cells actively transport some sodium out of it. Other ions follow sodium into interstitial fluid. Water leaves the filtrate, too, by osmosis. The proximal tubule wall is so highly permeable that about two-thirds of the filtrate's water is reabsorbed here (Figure 38.5c,d).

Interstitial fluid is saltiest around the hairpin turn of the loop of Henle. Water moves out of the filtrate by osmosis *before* the turn. The fluid left behind gets saltier until it matches interstitial fluid. The loop wall *after* the turn is impermeable to water, but sodium gets pumped out by active transport mechanisms. Interstitial fluid gets even saltier—so it attracts even more water out of the filtrate that is just entering the loop (Figure 38.5e).

Fluid arriving at a distal tubule is dilute. The stage is set for adjustments, which can lead to urine that is highly dilute, concentrated, or anywhere in between.

HORMONE-INDUCED ADJUSTMENTS When the volume of extracellular fluid drops, the hypothalamus calls for the release of antidiuretic hormone, or **ADH**, from the pituitary's posterior lobe (Section 32.3). ADH binds to receptors of distal tubules and collecting ducts, making their wall more permeable to water. More water gets reabsorbed so urine is more concentrated (Figure 38.5f). But less ADH is released when there is too much water. The wall of the tubes becomes less permeable, so less water is reabsorbed and the urine remains dilute.

Aldosterone promotes reabsorption of sodium. The extracellular fluid volume falls when too much sodium is lost. Sensory receptors in the heart and blood vessels detect the drop and stimulate renin-secreting gland cells in the wall of arterioles at glomeruli. The enzyme renin splits a plasma protein. A fragment is converted to the hormone **angiotensin II**. Among its many targets are aldosterone-secreting cells of the adrenal cortex, part of a gland on each kidney (Figure 38.3). The cells secrete more aldosterone, which stimulates cells of the distal tubules and collecting ducts to reabsorb sodium faster and excrete less sodium. Conversely, when there is too much sodium in the body, aldosterone secretion slows. Less sodium is reabsorbed, so more is excreted.

THIRST BEHAVIOR A **thirst center** in the hypothalamus induces water-seeking behavior. Osmoreceptors nearby warn of decreases in blood volume and rising blood solute levels. The signals activate the thirst center *and* ADH-secreting cells in the brain. While water intake is being encouraged, urinary output is reduced.

Angiotensin II also stimulates thirst by direct effects on brain cells controlling ADH secretion, so it helps to adjust the sodium balance *and* the water balance. Thirst also is initiated when free nerve endings detect dryness in the mouth, an early sign of dehydration.

Filtration, reabsorption, and secretion in the kidneys help maintain the composition and volume of extracellular fluid.

Blood filtration rates depend on high hydrostatic pressure (generated by heart contractions) and on neural, endocrine, and local control over how much blood flows into kidneys.

During tubular reabsorption, hormonal controls adjust the amounts of water and solutes in urine. ADH promotes water conservation and concentrated urine. Aldosterone promotes sodium conservation and dilute urine.

Certain excess solutes diffuse out of peritubular capillaries and into interstitial fluid. Then, by tubular secretion, they move into the nephron for subsequent excretion.

When Kidneys Break Down

By this point in the chapter, you probably have figured out that good health depends on nephron functioning. Whether by illness or accident, when the nephrons of both kidneys become damaged and no longer perform their regulatory and excretory functions, we call this **renal failure**. Chronic renal failure is irreversible.

Renal failure may occur after infectious agents reach the kidneys through the bloodstream or the urethra. It may occur after someone ingests lead, arsenic, pesticides, or other toxins. Ongoing high doses of aspirin and some other drugs also can bring it about. Abnormal retention of metabolic wastes, such as the by-products of protein breakdown, results in *uremic toxicity*. Atherosclerosis, heart failure, hemorrhage, and shock diminish blood flow and disrupt filtration pressure in the kidneys.

In *glomerulonephritis*, antibody–antigen complexes get trapped in glomeruli. Unless phagocytes remove them, the complexes keep activating complement and other agents that cause widespread inflammation and tissue damage. Fortunately, this type of disease is rare.

Kidney stones form when uric acid, calcium salts, and other wastes settle out of urine and collect in the renal pelvis. These hard deposits are usually passed in urine, but they can become lodged in the ureter or the urethra. When they disrupt urine flow, they must be medically or surgically removed to prevent renal failure.

About 13 million people in the United States alone suffer from renal failure. A *kidney dialysis machine* often restores the proper solute balances. Like the kidneys, it helps maintain extracellular fluid by selectively adjusting the solutes in blood. "Dialysis" refers to an exchange of substances across an artificial membrane interposed between two solutions that differ in composition.

In *hemodialysis*, the machine is connected to an artery or a vein. Then the patient's blood is pumped on through tubes that have been submerged in a warm saline bath. The bath's mix of salts, glucose, and other substances sets up the correct concentration gradients with blood. Then the blood flows back into the body. For kidney dialysis to have optimum effect, it must be performed three times a week. Each time, the procedure takes about four hours, because the patient's blood must circulate repeatedly to improve solute concentrations in her or his body.

In *peritoneal dialysis*, fluid of suitable composition is introduced into the patient's abdominal cavity, left in there for a specific length of time, then drained out. In this case, the cavity's lining itself, the peritoneum, acts as the membrane for dialysis.

Bear in mind, kidney dialysis is used as a temporary measure in reversible kidney disorders. In chronic cases, the procedure must be used for the rest of the patient's life or until a transplant operation provides her or him with a functional kidney. With treatment and controlled diets, many patients resume fairly normal activity.

THE BODY'S ACID–BASE BALANCE

In addition to maintaining the volume and composition of extracellular fluid, kidneys help keep it from getting too acidic or too basic (alkaline). The overall **acid–base balance** of that fluid is an outcome of controls over its concentrations of H^+ and other dissolved ions. *Metabolic acidosis* hints at the importance of the balancing acts. This condition results when the kidneys cannot secrete enough H^+ to keep pace with all the H^+ that is forming during metabolism. It is life-threatening.

Buffer systems, respiration, and urinary excretion all work in concert to provide control over the acid–base balance. A buffer system, remember, consists of weak acids or bases that can reversibly bind and release ions, thus helping to minimize shifts in pH (Section 2.6).

Normally, the extracellular pH of the human body should be maintained between 7.37 and 7.38. As you know, acids lower the pH and bases raise it. A variety of acidic and basic substances enter blood after being absorbed from the gut and as by-products of normal metabolism. Cell activities typically produce an excess of acids. These dissociate into H^+ and other fragments, and pH declines. The effect is minimized when excess hydrogen ions react with buffer molecules. An example is the *bicarbonate–carbon dioxide* buffer system:

$$H^+ + \underset{\text{BICARBONATE}}{HCO_3^-} \;\rightleftharpoons\; \underset{\text{CARBONIC ACID}}{H_2CO_3} \;\rightleftharpoons\; CO_2 + H_2O$$

In this case, the buffer system neutralizes excess H^+, and the carbon dioxide that forms during the reactions is exhaled from the lungs. Like other buffer systems in the body, however, this one has temporary effects only; it does not *eliminate* excess H^+. Only the urinary system can do so and thereby restore the buffers.

The same reactions proceed in reverse in cells of the nephron's tubular walls. HCO_3^- formed by the reverse reactions moves into interstitial fluid, then peritubular capillaries. Afterward, it enters the general circulation and buffers excess acid. The H^+ formed in the cells is secreted into the nephron and may join with HCO_3^-. The CO_2 that formed may be returned to blood, then exhaled. H^+ also may combine with phosphate ions or ammonia (NH_3), then leave the body in urine.

The kidneys work in concert with buffering systems that neutralize acids and with the respiratory system to help keep the extracellular fluid from becoming too acidic or too basic (alkaline).

A bicarbonate–carbon dioxide buffer system temporarily neutralizes excess hydrogen ions. The urinary system alone eliminates the excess ions and thus restores these buffers.

The bicarbonate–carbon dioxide buffer system is one of the key mechanisms that help maintain the acid–base balance.

ON FISH, FROGS, AND KANGAROO RATS

Now that you have a general sense of how your body maintains water and solute levels, consider what goes on in some other vertebrates, including that kangaroo rat hopping about at the start of the chapter.

Bony fishes and amphibians of freshwater habitats gain water and lose solutes (Figure 38.7*a*). Water moves into their internal environment by osmosis; they don't gain water by drinking it. Water diffuses across thin gill membranes in fishes and across the skin in adult amphibians. Excess water leaves as dilute urine, formed inside a pair of kidneys. In both groups of vertebrates, solute losses are balanced when more solutes come in with food and when gill or skin cells pump in sodium.

Body fluids of herring, snapper, and other marine fishes are about three times less salty than seawater. These fishes lose water by osmosis, then replace it by drinking more. They excrete ingested solutes against concentration gradients (Figure 38.7*b*). Fish kidneys do not have loops of Henle, so urine cannot ever become saltier than body fluids. Cells in the fish gills actively pump out most of the excess solutes in blood.

Figure 38.7*c* describes the water–solute balancing act in a salmon. This type of fish spends part of its life cycle in fresh water and another part in seawater.

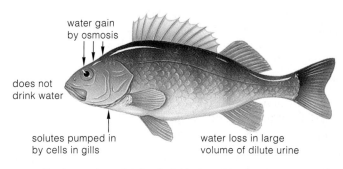

a Freshwater bony fish (body fluids far saltier than surroundings)

water gain by osmosis

does not drink water

solutes pumped in by cells in gills

water loss in large volume of dilute urine

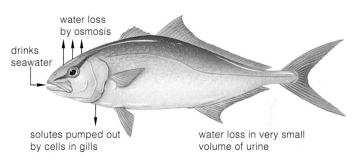

b Marine bony fish (body fluids less salty than surroundings)

water loss by osmosis

drinks seawater

solutes pumped out by cells in gills

water loss in very small volume of urine

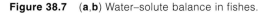

Figure 38.7 (**a**,**b**) Water–solute balance in fishes.

(**c**) Water–solute balance by salmon. This type of fish lives in saltwater *and* fresh water. Salmon hatch in streams and later move downstream to the open sea, where they feed and mature. They return to home streams to spawn.

For most salmon, salt tolerance is an outcome of changing hormone levels that appear to be triggered by increasing daylength in spring. Prolactin, a pituitary hormone, plays a key role in sodium retention in freshwater habitats. We know this because a freshwater fish that has its pituitary gland removed will die from sodium loss—but that fish will live if prolactin is administered to it.

A salmon's salt tolerance depends on cortisol. Secretion of this steroid hormone from the adrenal cortex correlates with rises in sodium excretion, sodium–potassium pumping by cells in the salmon's gills, and absorption of ions and water from the gut. In young salmon, cortisol secretion increases prior to the seaward movement—and so does salt tolerance.

salmon avoiding grizzly while maintaining water–solute balance

c

And what about that kangaroo rat! Proportionally, the loops of Henle of its nephrons are amazingly long, compared to yours. This means a great deal of sodium is pumped out from the nephron. Therefore, the solute concentration in the interstitial fluid around the loops becomes very high. The osmotic gradient between fluid in the loops of Henle and urine is *so* steep that nearly all the water that does reach the equally long collecting ducts gets reabsorbed. The kangaroo rats gives up only

a tiny volume of urine. And that urine is three to five times more concentrated than the most concentrated urine of humans.

The urinary systems of vertebrates differ in their details, such as the length of the nephron's loop of Henle. They are adapted to balance the body's gains in water and solutes with its losses of water and solutes in particular habitats.

MAINTAINING BODY TEMPERATURE

We turn now to another major aspect of the internal environment, its temperature. Many physiological and behavioral responses to change help maintain a **core temperature** within the tolerance range for the body's enzymes. *Core* means the body's internal temperature, as opposed to temperatures of tissues near its surface.

Figure 38.8 In hot desert sand at dusk, the J-shaped track of a sidewinder. As near as you can tell, how might this rattlesnake be gaining and losing heat?

Heat Gains and Heat Losses

Each metabolic reaction generates heat. If heat were to accumulate internally, the core temperature would rise. But a warm body loses heat to cool surroundings. Its core temperature stabilizes when the rate of heat loss balances the rate of metabolic heat production. The heat content of any complex animal depends on a balancing between heat gains and losses, as summarized here:

$$\begin{array}{ccccc} \text{CHANGE IN} & = & \text{HEAT} & + & \text{HEAT} & - & \text{HEAT} \\ \text{BODY HEAT} & & \text{PRODUCED} & & \text{GAINED} & & \text{LOST} \end{array}$$

Heat is gained and lost through exchanges at skin and other surfaces. Four processes—radiation, conduction, convection, and evaporation—drive the exchanges.

With **radiation**, the animal gains heat after being exposed to radiant energy—typically, sunlight—or to a surface warmer than its body surface temperature.

With **conduction**, an animal and an object in direct contact with it exchange heat in response to a thermal gradient between them. It loses heat when the object is cooler than its body, and it gains heat when that object is warmer than its body, as in Figure 38.8.

With **convection**, movement of air or water next to the body aids conductive heat loss. For example, as air heated by conduction moves away from a body, cooler air moves in to replace it. In turn, that air gains heat from the body and moves away, and so on. Heated air moves this way because it is less dense than cool air.

With **evaporation**, heat is lost as a liquid converts to a gas. Energy for the conversion comes from the heat content of the liquid (Section 2.5). Evaporation from a body surface has a cooling effect because the escaping water molecules carry away some energy with them.

Evaporation rates depend on humidity and on the rate of air movement. If air next to an animal is already saturated with water (when the local relative humidity is 100 percent), water will not evaporate. When air next to the body is hot and dry, evaporation may be the only means of countering metabolic heat production and the heat gains from radiation and convection.

Animals adjust heat losses and gains by changing their behavior and physiology, although some do this better than others. Like most animals, reptiles have low metabolic rates and poor body insulation; small ones especially absorb and gain heat quickly. They maintain the core temperature mostly by gaining environmental heat. Hence reptiles are among the **ectotherms**, which means "heat from outside."

When outside temperatures change, ectotherms use *behavioral* temperature regulation. Thus a lizard basks on warm rocks and gains heat by conduction. It keeps reorienting its body to expose the most surface area to the sun's infrared radiation. It loses heat after sunset. Before its metabolic rate slows, it hides under rocks or in crevices to conserve heat and avoid predators.

Most birds and mammals are **endotherms,** meaning "heat from within." They all have high metabolic rates; a foraging mouse uses up to thirty times more energy than a foraging lizard of the same weight. But that rate lets them stay active under a wide temperature range. Endotherms balance metabolism, heat gains and losses, and behavioral control of temperature. Their body form also is adapted to conserve or dissipate metabolic heat. Examples: Some mammals in cold habitats that have close relatives in warmer habitats have a more massive body. A large body mass, fur, fat layers, feathers, and clothing all help reduce heat loss (Figure 38.9).

Other birds and mammals are **heterotherms**. They are mainly endotherms but reduce metabolic rates by allowing the core temperature to drop at certain times. Activities idle, but the energy cost of maintaining body temperature is lowered. Example: Very high metabolic rates sustain tiny-bodied hummingbirds as they search for food during the day. Those rates plummet at night, when they can't feed; then, hummingbirds save energy by becoming almost as cool as their surroundings.

Warm, humid regions favor ectotherms, which need not spend much energy to maintain core temperatures. They can devote more energy to reproducing and other tasks. In tropical regions, reptiles exceed mammals in numbers and diversity. Endotherms have the edge in cool or cold regions. Thus arctic foxes, snowshoe hares, and some other endotherms do well in polar habitats, where you would never find a lizard.

Responses to Stressful Temperatures

The hypothalamus has control centers for maintaining the core temperature of mammals (Figure 29.14). They constantly get input from peripheral thermoreceptors in skin and from central thermoreceptors. Suppose the

Figure 38.9 How might the intrepid tourist relaxing on a deck chair of a ship off the coast of Antarctica be gaining and losing heat?

core temperature deviates from a set point. The centers integrate responses made by skeletal muscles, arterioles in skin and, in certain species, sweat glands. Negative feedback loops to the hypothalamus stop the responses when a suitable temperature returns.

RESPONSES TO COLD STRESS In mammals, cold stress brings on **peripheral vasoconstriction.** The diameter of arterioles constricts, so the blood's convective delivery of heat to body surfaces lessens. When your fingers are chilled, all but 1 percent of blood that would otherwise flow to your skin is diverted to other body regions.

Cold stress makes hairs and feathers "stand up." This **pilomotor response** creates a layer of still air next to skin and reduces convective and radiative heat loss. Behavioral changes also can minimize exposed surface areas and reduce heat loss, as when a cat curls up and when you hold both arms tightly against your body.

A **shivering response** to prolonged cold isn't useful for long. Rhythmic tremors begin as muscles contract ten to twenty times a second. Heat production rises by several times, but at a high energy cost. Prolonged or severe cold also leads to a hormonally induced rise in metabolic rates. This **nonshivering heat production** is greatest in *brown* adipose tissue. Hibernating animals, cold-acclimated animals, and human infants have this connective tissue. So do the ama of Japan and Korea who dive for shellfish six hours a day in frigid water. Adults who are not similarly adapted have little of it.

In *hypothermia*, core temperature is below normal. In humans, a drop of a few degrees alters brain function; more cooling leads to coma and death. Many mammals can recover from hypothermia. Frozen human cells die unless tissues thaw under medical supervision. Tissue destruction by localized freezing is called *frostbite*.

RESPONSES TO HEAT STRESS If a mammal gets too hot, the hypothalamus calls for **peripheral vasodilation**. It signals blood vessels in the skin to dilate. More blood flows to the skin, which can dissipate excess heat.

Evaporative heat loss is another response to heat stress. It occurs at moist respiratory surfaces and across skin. Animals that sweat lose some water this way. For instance, humans and some other mammals have sweat glands that release water and solutes through pores at the skin's surface. An average-sized adult human has 2–1/2 million or more sweat glands. For every liter of sweat produced, 600 kilocalories of heat energy are lost through evaporative heat loss.

Bear in mind, sweat that is dripping from skin does not dissipate heat. Outside temperatures must be high enough to cause water from sweat to *evaporate*. On hot, humid days, evaporation rates cannot match the rate of sweat secretion; air's high water content slows it down. During strenuous exercise, sweating may help balance heat production in skeletal muscle. Extreme sweating, as in a marathon race, forces the body to lose a vital salt—sodium chloride—as well as water. Such losses disrupt the composition and the volume of extracellular fluid. Runners collapse and faint with large losses.

What about mammals that sweat little or not at all? Some make behavioral responses, such as licking fur, or panting. "Panting" is shallow, rapid breathing that increases evaporative water loss from the respiratory tract. The body cools when water evaporates from the nasal cavity, mouth, and tongue (Section 29.7).

Sometimes peripheral blood flow and evaporative heat loss cannot counter heat stress, and *hyperthermia* results. In this condition, the core temperature rises above normal. For humans and other endotherms, even a few degrees above normal can be dangerous.

A *fever*, recall, is part of an inflammatory response to damaged tissues (Section 35.3). The hypothalamus resets the body's thermostat, which dictates what the core temperature is supposed to be. Mechanisms that raise metabolic heat production and cut heat loss are working; however, their operation maintains a higher temperature. A person feels chilled at a fever's onset. When the fever "breaks," peripheral vasodilation and sweating increase; the body is attempting to restore the normal core temperature and the person feels warm.

By bringing down a fever, anti-inflammatory drugs such as aspirin may actually prolong the healing time. Only when fevers approach dangerous levels should drugs be prescribed under medical supervision.

The internal, core temperature of an animal's body is being maintained when heat gains and heat losses are in balance.

Metabolic reactions generate heat inside the body. Radiation, conduction, and convection can move heat down thermal gradients that exist between the body and its surroundings. Evaporative heat loss carries heat away from the body.

Besides being morphologically adapted to their habitats, animals can make behavioral and physiological adjustments to environmental temperatures.

SUMMARY   *Gold* indicates text section

Control of Extracellular Fluid

1. The extracellular fluid in the animal body consists of certain types and amounts of substances dissolved in water. It is distributed as interstitial fluid (in tissue spaces) and blood. Its volume and composition are maintained when an animal's daily intake and output of water as well as solutes are in balance. The following processes maintain the balance in mammals: *CI, 38.1*

   a. Water is gained by absorption from the gut and by metabolism. It is lost by urinary excretion, evaporation from lungs and skin, sweating, and elimination of feces.

   b. Solutes are gained by absorption from the gut and by secretion, respiration, and metabolism. Solutes are lost by excretion, respiration, and sweating.

   c. Losses of water and solutes are controlled mainly by adjusting the volume and composition of urine.

2. The vertebrate urinary system has a pair of kidneys, a pair of ureters, a urinary bladder, and a urethra. *38.1*

3. Kidneys have many nephrons that filter blood and form urine. A nephron interacts with two sets of blood capillaries: glomerular and peritubular. *38.1, 38.2*

   a. The start of a nephron is cup-shaped (Bowman's capsule). It continues as three tubular regions (proximal tubule, loop of Henle, and distal tubule, which empties into a collecting duct).

   b. Together, Bowman's capsule and the set of highly permeable glomerular capillaries within it are a blood-filtering unit: a renal corpuscle. Blood pressure forces water and small solutes from capillaries, into the fluid-filled cup. Most of the filtrate gets reabsorbed at the nephron's tubular regions and is returned to the blood. A portion is excreted as urine.

4. Urine forms in the nephron by three processes: *38.2*

   a. Filtration of blood at the glomerulus, which puts water and small solutes into the nephron.

   b. Reabsorption. Water and solutes to be conserved leave the nephron's tubular parts and enter capillaries that thread around them. A small volume of water and solutes remains in the nephron.

   c. Secretion. A few substances can leave peritubular capillaries and enter the nephron, for disposal in urine.

5. Two hormones that act on cells of the wall of distal tubules and collecting ducts adjust urine concentration. ADH conserves water by enhancing reabsorption across the wall. In its absence, more water is excreted (urine is dilute). Aldosterone enhances sodium reabsorption. In its absence, sodium is excreted. Angiotensin II stimulates aldosterone secretion (sodium conservation) and thirst (water conservation by calling for more ADH). *38.2*

6. The urinary system acts in concert with buffers and with the respiratory system to maintain the acid–base balance of extracellular fluid. *38.4*

Control of Body Temperature

1. Maintaining an animal's core (internal) temperature depends on balancing metabolically produced heat and heat absorbed from and lost to the environment. *38.6*

2. Animals exchange heat with their environment by four processes: *38.6*

   a. Radiation. Emission from the body of infrared and other wavelengths. Radiant energy can be absorbed at the body surface, then converted to heat energy.

   b. Conduction. Direct transfer of heat energy from one object to another object in contact with it.

   c. Convection. Heat transfer by air or water currents; involves conduction and mass transfer of heat-bearing currents away from or toward the animal body.

   d. Evaporation. Conversion of liquid to a gas, driven by energy inherent in the heat content of the liquid. Some animals lose heat by evaporative water loss.

3. Core temperatures depend on metabolic rates and on anatomy, behavior, and physiology. *38.6*

   a. For ectotherms, core temperature depends more on heat exchange with the environment than on heat generated by metabolism.

   b. For endotherms, core temperature depends more on high metabolic rates and precise controls over heat produced and heat lost.

   c. For heterotherms, the core temperature fluctuates some of the time, and controls over heat balance come into play at other times. *38.6*

Review Questions

1. State the function of the urinary system in terms of gains and losses for the internal environment. State the components of the mammalian urinary system and their functions. *38.1*

2. Label the component parts of this kidney and nephron. *38.1*

3. Define filtration, tubular reabsorption, and secretion. How does urine formation help maintain an internal environment? *38.2*

4. Which hormone or hormones promote (a) water conservation, (b) sodium conservation, and (c) thirst behavior? *38.2*

5. Name and define the physical processes by which animals gain and lose heat. What are the main physiological responses to cold stress and to heat stress in mammals? *38.6*

1. In mammals, water intake depends on _____ .
   a. absorption from gut c. a thirst mechanism
   b. metabolism d. all of the above

2. In mammals, water is lost by way of the _____ .
   a. skin d. urinary system
   b. respiratory system e. c and d
   c. digestive system f. a through d

3. Water and small solutes enter nephrons during _____ .
   a. filtration c. tubular secretion
   b. tubular reabsorption d. both a and c

4. Kidneys return water and small solutes to blood by _____ .
   a. filtration c. tubular secretion
   b. tubular reabsorption d. both a and b

5. Certain excess solutes move out of peritubular capillaries that thread around the tubular parts of the nephron. These solutes are moved into the nephron during _____ .
   a. filtration c. tubular secretion
   b. tubular reabsorption d. both a and c

6. A nephron's reabsorption mechanism depends on _____ .
   a. osmosis across nephron wall
   b. active transport of sodium across nephron wall
   c. a steep solute concentration gradient
   d. all of the above

7. _____ promotes water conservation.
   a. ADH c. Low extracellular fluid volume
   b. Aldosterone d. both a and c

8. _____ enhances sodium reabsorption.
   a. ADH c. Low extracellular fluid volume
   b. Aldosterone d. both b and c

9. Match the term with the most suitable description.
   ____ renal corpuscle a. surrounded by saltiest fluid
   ____ distal tubule b. extra-long loops of Henle
   ____ loop of Henle c. involves buffer systems
   ____ acid–base balance d. blood-filtering unit
   ____ kangaroo rat e. ADH, aldosterone act here

10. Match the term with the most suitable description.
    ____ ectotherm a. heat transfer by air or water currents
    ____ endotherm b. body temperature fluctuates some of
    ____ evaporation   the time, is controlled at other times
    ____ heterotherm c. emission of radiant energy
    ____ radiation d. direct heat transfer between an object
    ____ conduction   and another object in contact with it
    ____ convection e. metabolism dictates core temperature
                      f. environment dictates core temperature
                      g. conversion of liquid to gas

Critical Thinking

1. Fatty tissue holds kidneys in place. Rarely, extremely rapid weight loss may cause the tissue to shrink and kidneys to slip from their normal position. If slippage puts a kink in one or both ureters and blocks urine flow, what may happen to the kidneys?

2. Drink one quart of water in one hour. What changes might you expect in your kidney function and in urine composition?

3. In 1912, the ocean liner *Titanic* left Europe on her maiden voyage to America—and a chunk of the leading edge of a glacier in Greenland broke away and floated out to sea. Late at night, off the Newfoundland coast, the iceberg and the *Titanic* made an ill-fated rendezvous (Figure 38.10). The *Titanic* was said to be unsinkable. Survival drills were neglected. There weren't enough

Figure 38.10 Sinking of the *Titanic*, based on eyewitness accounts.

lifeboats to hold even half the 2,200 passengers. The *Titanic* sank in about 2-1/2 hours. Rescue ships were on the scene in less than two hours, but 1,517 bodies were recovered from a calm sea. All the dead had on life jackets; none had drowned. They probably died from _____ . If so, how did their blood flow, metabolism, and skeletal muscle action change prior to death?

4. When iguanas have an infection, they rest for a prolonged period in the sun. Propose a hypothesis to explain why.

5. Out on a first date, Jon takes Geraldine's hand in a darkened theater. "*Aha!*" he thinks. "*Cold hands, warm heart!*" What does this tell us about the regulation of core temperature, let alone Jon?

Selected Key Terms

Salt–Water Balance
acid–base balance *38.4*
ADH *38.2*
aldosterone *38.2*
angiotensin II *38.2*
blood *CI*
Bowman's capsule *38.1*
collecting duct *38.1*
distal tubule *38.1*
extracellular fluid *CI*
filtration *38.2*
interstitial fluid *CI*
kidney *38.1*
loop of Henle *38.1*
nephron *38.1*
proximal tubule *38.1*

renal corpuscle *38.1*
renal failure *38.3*
thirst center *38.2*
tubular
   reabsorption *38.2*
tubular secretion *38.2*
urea *38.1*
ureter *38.1*
urethra *38.1*
urinary bladder *38.1*
urinary excretion *38.1*
urinary system *38.1*
urine *38.1*

Body Temperature 38.6
conduction
convection

core temperature
ectotherm
endotherm
evaporation
evaporative
   heat loss
heterotherm
nonshivering
   heat production
peripheral
   vasoconstriction
peripheral
   vasodilation
pilomotor response
radiation
shivering response

Readings

Sherwood, L. 1997. *Human Physiology*. Second edition. Monterey, California: Brooks-Cole.

Smith, H. 1961. *From Fish to Philosopher*. New York: Doubleday.

39

ANIMAL REPRODUCTION AND DEVELOPMENT

From Frog to Frog and Other Mysteries

With a quavering, low-pitched call that only a female of its kind could find seductive, a male frog proclaims the onset of warm spring rains, of ponds, of sex in the night. By August the summer sun will have parched the earth, and his pond dominion will be gone. But tonight is the hour of the frog!

Through the dark, a hormone-primed female moves toward the vocal male. They meet; they dally in the behaviorally prescribed ways of their species. He clamps his forelegs above her swollen abdomen and gives her a prolonged squeeze (Figure 39.1a). Out into the water streams a ribbon of hundreds of eggs, which the male blankets with a milky cloud of sperm. Soon afterward, fertilized eggs—**zygotes**—are suspended in the water.

For the leopard frog, *Rana pipiens*, a drama begins to unfold that has been reenacted each spring, with only minor variations, for many millions of years. Within a few hours after fertilization, each zygote divides into two cells, the two divide into four, then the four into eight. In less than twenty hours after fertilization, the mitotic cell divisions have produced a ball of cells no

larger than the zygote. It is an early embryonic stage, of a type known as a blastocyst.

The cells continue to divide, but now they start to interact by way of their surface structures and chemical secretions. At prescribed times, many change shape and migrate to prescribed positions. They all inherited the same genetic instructions from the zygote—yet now they start to differ in appearance and function!

Through their associations, the cells form layers of embryonic tissues and then embryonic organs. A pair of tissue regions at the embryo's surface interact with the tissues beneath them. Together they give rise to a pair of eyes. Within the embryo a heart is forming, and soon it starts an incessant, rhythmic beating. In less than a week, events have transformed the frog embryo into a swimming, algae-eating larva called a tadpole.

Several months pass. Legs form; the tail shortens and disappears. The mouth develops jaws that snap shut on insects and worms. Eventually the transformations lead to an adult frog. With luck the frog will avoid predators, disease, and other threats in the months ahead. In time

Figure 39.1 Reproduction and development of *Rana pipiens*, the leopard frog. (**a**,**b**) We zoom in on the life cycle as a male clasps a female in a reproductive behavior called amplexus. The female releases eggs into the water. The male releases sperm over the eggs. A zygote forms when an egg nucleus and a sperm nucleus fuse at fertilization. (**c**) Frog embryos suspended in the water. (**d**) A tadpole. (**e**) A transitional form between a tadpole and the young adult frog (**f**).

it may even call out quaveringly across a moonlit pond, and the life cycle will turn again.

Many years ago you, too, started a developmental journey when a zygote carved itself up. Three weeks into the journey, your embryonic body had the stamp of "vertebrate" on it. A mere five weeks after that, it was a recognizable human in the making!

With this chapter we turn to one of life's greatest dramas—the development of offspring in the image of sexually reproducing parents. The guiding question is this: *How does a single-celled zygote of a frog, human, or any other complex animal become transformed into all the specialized cells and structures of the adult form?* Some answers will become apparent as we move through a survey of basic principles, then through a case study of human reproduction and development.

DEVELOPING EMBRYO

Key Concepts

1. Sexual reproduction dominates the life cycles of nearly all animals. The separation into sexes requires specialized reproductive structures, hormonal control mechanisms, and forms of behavior. Having two distinct sexes affords a great selective advantage—variation in traits among the offspring. This advantage offsets the biological cost of the separation.

2. The life cycles of humans and many other animals proceed through six stages of embryonic development— gamete formation, fertilization, cleavage, gastrulation, organ formation, and growth and tissue specialization. Each stage builds on tissues and structures that formed during the stage preceding it.

3. In a developing embryo, the fate of each type of cell depends partly on cleavage, which distributes different regions of the fertilized egg's cytoplasm to different daughter cells. It also depends on interactions among cells of the embryo. These activities are the foundation for cell differentiation and morphogenesis.

4. With cell differentiation, each cell type selectively uses certain genes and synthesizes proteins not found in other types, and thereby becomes unique in structure and function. With morphogenesis, tissues and organs change in size, shape, and proportion. They also become organized relative to one another in prescribed patterns.

5. All complex, multicelled animals that show extensive cell differentiation undergo aging. Their cells gradually break down in structure and function, which leads to the decline of tissues, organs, and eventually the body.

6. Human males continually produce sperm from puberty onward. The sex hormones testosterone, LH, and FSH, are central to controlling male reproductive functions.

7. From puberty onward, human females are fertile on a cyclic basis. Each month during their reproductive years, an egg is released from one of a pair of ovaries, and the lining of the uterus is primed for a possible pregnancy. The hormones estrogen, progesterone, FSH, and LH are central to controlling the cyclic activity.

THE BEGINNING: REPRODUCTIVE MODES

Sexual Versus Asexual Reproduction

In earlier chapters, you learned about the cellular basis of **sexual reproduction**. Briefly, by this reproductive mode, meiosis and gamete formation typically proceed in two prospective parents. At fertilization, a gamete from one parent fuses with a gamete from the other to form the zygote, the first cell of the new individual. You also read about **asexual reproduction**, in which a single parent organism produces offspring by various means (but not by gamete formation). We now turn to a few structural, behavioral, and ecological aspects of the two reproductive modes.

Picture a scuba diver accidentally kicking a sponge. A tissue fragment breaks away from the sponge body, then grows and develops by mitotic cell divisions and cell differentiation into a new sponge. Or picture one of the flatworms undergoing transverse fission while it glides along through the water. First its body constricts below the midsection. The part behind the constriction grips a substrate and starts a tug-of-war with the part in front, and it splits off a few hours later. Both parts go their separate ways, regenerate what's missing, and thus become a whole worm. Only some species do this.

In such cases of *asexual* reproduction, all offspring are genetically the same as their individual parent, or nearly so. Phenotypically they are much the same, too. We can speculate that phenotypic uniformity is useful when each individual's gene-encoded traits are highly adapted to a limited and more or less consistent set of environmental conditions. Most variations introduced into the finely tuned gene package would not do much good, and often they could do harm.

Most animals live where opportunities, resources, and danger are highly variable. For the most part, such animals reproduce sexually, meaning female and male parents bestow different mixes of alleles on offspring. As you already learned from Section 9.1, the resulting variation in traits improves the odds that some of the offspring, at least, should survive and reproduce even if conditions change in the environment.

Costs and Benefits of Sexual Reproduction

Among animals, separation into sexes is not without cost. Some cells that can serve as gametes must be set aside and nurtured. Housing and delivering gametes close to or into a prospective mate require specialized reproductive structures. Mating often requires special forms of behavior, such as courtship, that can promote fertilization. Mating also requires built-in controls that can synchronize the timing of gamete formation, sexual readiness, even parental behavior in two individuals.

Figure 39.2 *Facing page:* Examples of where the embryos of invertebrates and vertebrates develop, how they are nourished, and how (if at all) parents protect them.

Snails (**a**) and spiders (**b**) are *oviparous*. They release eggs from which the young later hatch (*ovi–*, egg; *parous*, produce). Snails are hermaphroditic parents that leave eggs unprotected. Spider eggs develop in a silk egg sac that the female anchors to a substrate or carts around with her. The females often die soon after making the sac. But some guard the sac, then cart spiderlings about for a few days and even feed them.

(**c**) Birds also are oviparous. Their fertilized eggs have large yolk reserves, and they, too, develop and hatch outside the mother's body. Unlike snails, one or both parent birds expend considerable energy feeding and caring for the young.

Some fishes, all lizards, and many snakes are *ovoviviparous*. Their fertilized eggs develop inside the mother, then offspring are born live (*vivi–*, alive). Yolk reserves, not the mother's own tissues, sustain the eggs. As one example, the liveborn copperhead offspring in (**d**) are still in their egg sacs.

Most mammals are *viviparous*, meaning their young are born live. (**e, f**) In kangaroos and some other species, embryos are born in unfinished form. The young of this marsupial complete development inside a pouch on the mother's ventral surface. The juvenile stages (joeys) continue to draw nourishment from mammary glands in the mother's pouch. (**g**) By contrast, the human female retains the fertilized egg in her body. Maternal tissues nourish the developing individual until the time of its birth, in the manner described later in this chapter.

Just look at the question of *reproductive timing*. How do mature sperm in one individual become available exactly when the eggs mature in a different individual? Timing depends upon energy outlays for constructing, maintaining, and operating neural as well as hormonal control mechanisms in each parent. Also, parents must produce mature gametes in response to the same cues, such as a seasonal change in daylength, that mark the onset of the most appropriate time of reproduction for their species. Example: Male and female moose become sexually active only during late summer and early fall. The coordinated timing means that their offspring will be born the following spring—a time when the weather improves and food is plentiful for many months.

Locating and then actually recognizing a potential mate of the same species is another challenge. Different species invest energy when they synthesize signaling molecules—pheromones. They construct visual mating signals such as feathers of certain colors and patterns, and complex sensory receptors to detect the specific signals being sent. The males often expend astonishing amounts of energy to execute tricky courtship routines, as you will read in Chapter 41.

Ensuring the survival of offspring is costly (Figure 39.2). Many invertebrates, bony fishes, and frogs simply

liveborn snake inside egg sac

release eggs and motile sperm into the surrounding water, as in Figure 39.1*a*. If each adult were to produce only *one* sperm or *one* egg each season, the chances for fertilization would not be good. These species invest energy in making numerous gametes, often thousands of them. As another example, nearly all land animals rely on internal fertilization, the union of sperm and egg *within* the body of a female. They invest metabolic energy to construct elaborate reproductive organs, such as a penis (by which certain males deposit sperm in a female) and a uterus (a chamber inside the females of certain species where the embryo grows and develops).

Finally, animals set aside energy in forms that can *nourish the developing individual* until it has developed enough to feed itself. For instance, nearly all animal eggs contain **yolk**. This substance, rich in proteins and lipids, nourishes embryonic stages. The eggs of some species have much more yolk than others. Sea urchins make tiny eggs with little yolk, release large numbers of them, and thus limit the biochemical investment in each one. Within twenty-four hours, each fertilized egg has developed into a freely moving, self-feeding larva. Sea stars and other predators eat most of the eggs. So for sea urchins, bestowing as little metabolic energy as possible on each one of many individual gametes pays off, in terms of reproductive success.

By contrast, mother birds lay truly yolky eggs. Yolk nourishes the bird embryo through an extended period of development inside an eggshell that forms after the egg is fertilized. Your mother put tremendous demands on her body to protect and nourish you through nine

months of development inside her, starting from the time you were an egg with almost no yolk. After you implanted yourself into her uterus, physical exchanges with her tissues supported you through that extended pregnancy (Figure 39.2*g*).

As these few examples suggest, animals show great diversity in reproduction and development. However, as you will see in the sections to follow, some patterns are widespread throughout the animal kingdom, and they can serve as a framework for our reading.

Separation into male and female sexes requires special reproductive cells and structures, neural and hormonal control mechanisms, and forms of behavior. A selective advantage—variation in traits among offspring—offsets biological costs associated with the separation into sexes.

STAGES OF DEVELOPMENT—AN OVERVIEW

Embryos are a class of transitional forms between the fertilized egg and an adult. Although they all start out as a single cell, embryos of different species often look different as they grow and develop. For example, you do not look like a frog now, and you did not look like one when you and the frog were early embryos, either. However, despite such differences in appearance, it is possible to identify certain patterns in how embryos of nearly all animal species develop.

Figure 39.3 is an overview of the stages of animal development. During **gamete formation**, the first stage, eggs or sperm develop inside the reproductive organs of a parent body. **Fertilization**, the second stage, starts when sperm's plasma membrane fuses with the plasma membrane of an egg. Fertilization is over when the egg nucleus and sperm nucleus fuse and form a zygote.

The third stage, **cleavage**, is a program of mitotic cell divisions that divide the volume of egg cytoplasm into a number of **blastomeres**—smaller cells, each with its own nucleus. There is no growth during this stage. Cleavage only increases the number of cells; it does not change the original volume of the egg cytoplasm.

As cleavage draws to a close, the pace of mitotic cell division slackens. The embryo enters **gastrulation**. This fourth stage of animal development is a time of major cellular reorganization. The new cells become arranged into a gastrula with two or three primary tissues, often called germ layers. Cellular descendants of the primary tissues give rise to all tissues and organs of the adult:

1. **Ectoderm**. This is the *outermost* primary tissue layer, the one that forms first in the embryos of every animal. Ectoderm is the embryonic forerunner of the cell lineages that give rise to tissues of the nervous system and to the integument's outer layer.

2. **Endoderm**. Endoderm is the *innermost* primary tissue layer. It is the embryonic forerunner of the gut's inner lining and organs derived from the gut.

3. **Mesoderm**. This *intermediate* primary tissue layer gives rise to muscle, most of the skeleton; circulatory, reproductive, and excretory organs; and connective tissues of the gut and integument. It evolved hundreds of millions of years ago, and it was a pivotal step in the evolution of nearly all large, complex animals.

After the primary tissue layers form in embryos, they give rise to subpopulations of cells. This marks the onset of **organ formation**. The subpopulations become unique in structure and functions, and their descendants give rise to different kinds of tissues and organs.

During the last stage of animal development, **growth and tissue specialization**, organs increase in size, and they gradually assume specialized functions. This stage continues into adulthood.

Now take a look at Figure 39.4. Its photographs and sketches show several stages in the embryonic development of a typical animal, a frog. Take a moment to study this

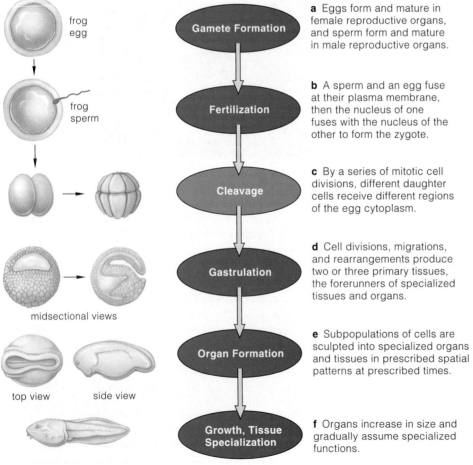

frog egg

frog sperm

midsectional views

top view   side view

Gamete Formation

a Eggs form and mature in female reproductive organs, and sperm form and mature in male reproductive organs.

Fertilization

b A sperm and an egg fuse at their plasma membrane, then the nucleus of one fuses with the nucleus of the other to form the zygote.

Cleavage

c By a series of mitotic cell divisions, different daughter cells receive different regions of the egg cytoplasm.

Gastrulation

d Cell divisions, migrations, and rearrangements produce two or three primary tissues, the forerunners of specialized tissues and organs.

Organ Formation

e Subpopulations of cells are sculpted into specialized organs and tissues in prescribed spatial patterns at prescribed times.

Growth, Tissue Specialization

f Organs increase in size and gradually assume specialized functions.

Figure 39.3 Overview of the stages of animal development. We use a few forms that appear during the frog life cycle as examples.

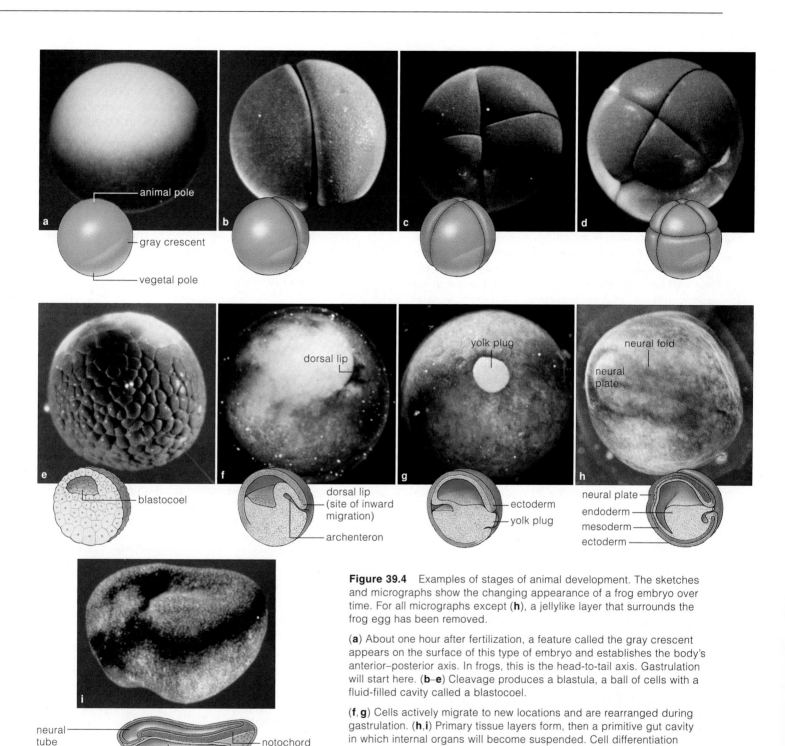

a animal pole
gray crescent
vegetal pole

b

c

d

e blastocoel

f dorsal lip
dorsal lip (site of inward migration)
archenteron

g yolk plug
ectoderm
yolk plug

h neural fold
neural plate
neural plate
endoderm
mesoderm
ectoderm

i neural tube
gut cavity
notochord

Figure 39.4 Examples of stages of animal development. The sketches and micrographs show the changing appearance of a frog embryo over time. For all micrographs except (**h**), a jellylike layer that surrounds the frog egg has been removed.

(**a**) About one hour after fertilization, a feature called the gray crescent appears on the surface of this type of embryo and establishes the body's anterior–posterior axis. In frogs, this is the head-to-tail axis. Gastrulation will start here. (**b–e**) Cleavage produces a blastula, a ball of cells with a fluid-filled cavity called a blastocoel.

(**f, g**) Cells actively migrate to new locations and are rearranged during gastrulation. (**h,i**) Primary tissue layers form, then a primitive gut cavity in which internal organs will become suspended. Cell differentiation proceeds, moving the embryo on its way to becoming a tadpole, as shown in Figure 39.1.

illustration, for it reinforces an important concept. The structures that form during one stage of development serve as the foundation for the stage that comes after it. Successful development depends on whether all of the structures develop according to normal patterns, in a prescribed sequence.

Animal development proceeds from gamete formation and fertilization through cleavage, gastrulation, then organ formation, and finally growth and tissue specialization.

Development cannot proceed properly unless each stage is successfully completed before the next begins.

EARLY MARCHING ORDERS

Information in the Egg Cytoplasm

Why don't you have an arm attached to your nose or toes growing from your navel? The patterning of body parts for all complex animals, including yourself, starts with messages built into the cytoplasm of immature eggs—or **oocytes**—even before sperm enter the picture. A **sperm**, remember, consists only of paternal DNA and a bit of equipment, including a tail, that helps it reach and penetrate an egg. Compared to a sperm, an oocyte is much larger and more complex (Section 9.5).

While an oocyte is maturing, its volume increases. Enzymes, mRNA transcripts, and other factors become stockpiled in different parts of the cytoplasm. Typically they will be activated, after fertilization, for use in the early rounds of DNA replication and cell division. Also present are tubulin molecules and factors that will help control the angle and timing of tubulin assembly into microtubules for mitotic spindles. How the cytoplasm gets cut depends on such factors and on the yolk in the cytoplasm. The amount of yolk and its distribution will influence how large the resulting blastomeres will be.

Such regionally localized aspects of the oocyte are "maternal messages." We find evidence of their effects as early as fertilization. As an example, a frog egg has pigment granules concentrated near one pole and yolk near the other. The egg's cortex (the plasma membrane and the cytoplasm just under it) undergoes structural reorganization when a sperm fertilizes the egg. Part of the cortex shifts toward the point of sperm entry, and it exposes a crescent-shaped area of yolky cytoplasm. This area of intermediate pigmentation, a gray crescent, forms near the frog egg's midsection. It establishes the body's anterior–posterior axis. Section 29.6 depicts the directional planes for this body axis and others.

In itself, a gray crescent is not evidence of regional differences in maternal messages. We get such evidence from experiments and by observing embryos in which localized cytoplasmic differences are obvious enough to be tracked during development (Figure 39.5). Thus, if you were to track the development of fertilized frog eggs, you would observe that the gray crescent is the site where gastrulation normally begins.

Cleavage—The Start of Multicellularity

Once an egg is fertilized, the zygote enters cleavage. Beneath its plasma membrane, its midsection has a ring of microfilaments made of the contractile protein actin. As the microfilaments slide past one another, the ring tightens and pinches in the cytoplasm. The cell surface above the shrinking ring is drawn inward, as a cleavage furrow (Section 9.5). It is this force of contraction that splits the cytoplasm into two blastomeres.

Simply by virtue of where they form, blastomeres end up with different maternal messages. This outcome

Figure 39.5 Examples of experiments that illustrate how the cytoplasm of a fertilized egg has localized differences that help determine the fate of cells in a developing embryo. The cortex of frog eggs contains granules of dark pigment concentrated near one pole. At fertilization, a portion of the granule-containing cortex shifts toward the point of sperm entry. The shift exposes lighter colored, yolky cytoplasm in a crescent-shaped gray area:

The first cleavage normally puts part of the gray crescent in both of the resulting blastomeres.

(**a**) For one experiment, the first two blastomeres that formed were physically separated from each other. Each blastomere still gave rise to a whole tadpole.

(**b**) For another experiment, a fertilized egg was manipulated so the cut through the first cleavage plane missed the gray crescent. Of two resulting blastomeres, only one received the gray crescent. It alone developed into a normal tadpole. The other, deprived of maternal messages present in the gray crescent, gave rise to a ball of undifferentiated cells.

gray crescent of frog zygote

first cleavage

Daughter cells (blastomeres) are separated experimentally.

A normal tadpole develops. A normal tadpole develops.

a EXPERIMENT 1

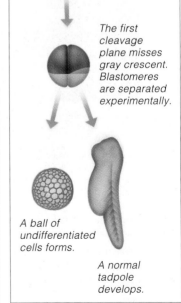

gray crescent of frog zygote

The first cleavage plane misses gray crescent. Blastomeres are separated experimentally.

A ball of undifferentiated cells forms.

A normal tadpole develops.

b EXPERIMENT 2

Figure 39.6 Comparison of the early cleavage planes for (**a**) sea urchins and (**b**) mammals. The early cuts of a fertilized egg are radial in sea urchins and rotational in mammals. (**c**) To get a sense of the impact of cleavage, consider Sabra and Nina, identical twins who started out life from the same zygote. The first two blastomeres formed at cleavage, the inner cell mass, or another early embryonic stage split. The split was the start of two genetically identical, look-alike individuals.

of cleavage, called **cytoplasmic localization**, helps seal the developmental fate of each cell's descendants. Its cytoplasm alone may have molecules of a protein that can activate, say, a gene coding for a certain hormone. And its descendants alone will make that hormone.

Cleavage Patterns

The simplest cleavage pattern has full cuts perpendicular to the mitotic spindle. The cuts parcel out a copy of the nucleus to each blastomere, so all the blastomeres are genetically equivalent. The cuts made during cleavage only divide the volume of cytoplasm into increasingly smaller cells; the blastomeres don't grow in size.

In most animals, a zygote's genes are silent during early cleavage; proteins and mRNAs stockpiled in the cytoplasm control how fast the cuts proceed and how the blastomeres become arranged. In mammals, certain genes must be activated first. Cleavage cannot even be completed without the protein products of those genes.

We can correlate cleavage patterns of major animal groups with the amount and distribution of yolk, which usually restricts cleavage. Cuts may go right through a nearly yolkless egg. Early cuts go only partway through an egg with yolk stockpiled at one end. That egg has polarity. Its *vegetal* pole is the yolk-rich end; its *animal* pole is closest to the nucleus. Frog eggs are like this.

Other heritable factors affect the cuts. Frog and sea urchin eggs undergo *radial* cleavage; cleavage furrows run horizontal and vertical to the animal–vegetal axis (Figure 39.6*a*). Successive cuts produce a **blastula**, an embryonic stage in which the blastomeres surround a fluid-filled cavity (a blastocoel). But a sea urchin egg is nearly yolkless, so the cuts produce horizontal rows of blastomeres. A frog egg's yolk impedes cuts near the vegetal pole (Figure 39.4*e*). More, smaller blastomeres form near its animal pole, and the blastocoel is offset. Between the time that 16 to 64 blastomeres form, any amphibian blastula looks like a mulberry. It's called a morula (after the Latin word for mulberry).

What about eggs of reptiles, birds, and most fishes? They undergo *incomplete* cleavage. The large volume of yolk restricts the early cuts to a small, caplike region near the animal pole. The result is two flattened layers of cells with a narrow cavity in between.

What about a mammal's egg? It undergoes *rotational* cleavage, with blastomeres dividing slowly at different times. The first cut is vertical through both poles. In the next cleavage, one blastomere is cut vertically and one horizontally (Figure 39.6*b*). The third cleavage yields a loose arrangement with space between eight cells. The cells undergo compaction; they abruptly huddle into a compact ball. Tight junctions stabilize the ones outside and seal off those inside. Gap junctions form between the inner cells and facilitate chemical communication among them. Descendants of the outer cells form a thin surface layer. Their secretions form a fluid-filled cavity in the ball, and the inner cells mass together against one side of it. We call this type of blastula a **blastocyst**. In all mammals, including humans, the inner cell mass gives rise to the embryo proper (Section 39.12).

Once in a while, the first two blastomeres, the inner cell mass, or even a later stage splits. Such a split may result in *identical twins*, which have the same genetic makeup and share a common placenta (Figure 39.6*c*). By contrast, *fraternal twins* arise from two oocytes that matured and then became fertilized during the same menstrual cycle. Each is serviced by its own placenta.

The egg cytoplasm contains maternal instructions in the form of regionally distributed enzymes and other proteins, mRNAs, cytoskeletal elements, yolk, and other factors.

Cleavage divides a zygote into blastomeres, each with a localized part of maternal messages inherent in the egg cytoplasm. We call this outcome cytoplasmic localization.

Differences in the amount and distribution of yolk and other heritable factors give rise to different patterns of cleavage. Those patterns contribute to differences in body plans among different animal groups.

HOW DO SPECIALIZED TISSUES AND ORGANS FORM?

Nearly all animals have a gut, with tissues and organs that function in digestion and absorption of nutrients. They have surface parts that protect internal parts and detect what is going on outside. In between, most have organs, such as those dealing with structural support, movement, and blood circulation. This three-layer body plan emerges as cleavage ends and gastrulation begins. The embryo's size increases little, if any. But cells start migrating to new, predictable locations to form primary tissue layers—ectoderm, endoderm, and mesoderm.

Figure 39.7 shows steps in the formation of a sea urchin gastrula. Figure 39.8 shows the emergence of a bird embryo's anterior–posterior axis. In all vertebrates, this axis defines where a **neural tube**, the forerunner of a brain and spinal cord, will form. All such organs now start to form by cell differentiation and morphogenesis.

Cell Differentiation

All cells of an embryo have the same number and kinds of genes (they all descend from the same zygote). They all activate the genes that specify proteins necessary for cell survival, such as histones and glucose-metabolizing enzymes. From gastrulation onward, certain groups of genes are used in some cells but not in others. When a cell selectively activates genes and synthesizes proteins not found in other cell types, we call this process **cell differentiation**. You read about the molecular basis of selective gene expression in Chapter 14. Basically, the process yields the proteins required for distinctive cell structures, products, and functions.

For instance, when your eye lenses developed, only some cells could activate genes for crystallins, a family of proteins that are assembled into transparent fibers for each lens. Long crystallin fibers formed in the cells,

a Blastula   **b** Cell migrations in early gastrula

Figure 39.7 (**a**,**b**) A few early stages of development of a sea urchin (*Lytechinus*), cross-section. After cleavage, the resulting blastula becomes transformed into a gastrula. (**c**) Scanning electron micrograph of cells at the gastrula surface cells. Some cells have begun their inward migration.

endoderm cells
c migrating inward

forcing them to lengthen and flatten. The differentiated cells provided each lens with unique optical properties. And crystallin-producing cells are only 1 of 150 or so differentiated cell types now present in your body.

As many experiments show, nearly all cells become differentiated with no loss of genetic information. For instance, John Gurdon stripped unfertilized eggs of an African clawed frog (*Xenopus laevis*) of their nucleus. He ruptured the plasma membrane of intestinal cells from tadpoles of the same species. He left the nucleus and much of the cytoplasm of these fully differentiated cells intact and inserted them into the enucleated eggs. Some eggs gave rise to a complete frog! The intestinal cells still had the same number and kinds of genes as

neural groove

eye
brain
heart
wing bud
neural tube
leg bud
tail

22 HOURS   29 HOURS   45 HOURS   55 HOURS   72 HOURS

168 HOURS (SEVEN DAYS OLD)

Figure 39.8 Onset of organ formation in a chick embryo during the first seven days of development. The heart begins to beat between thirty and thirty-six hours. You may have observed such embryos at the yolk surface of raw, fertilized eggs.

Figure 39.9 Examples of morphogenesis. (**a**) Cell migration. A nerve cell (*orange*) "climbs" through a developing brain to its final position, using a glial cell (*yellow*) as a highway. (**b**) How a neural tube forms. As gastrulation ends, ectoderm is a uniform sheet of cells. In some cells, microtubules lengthen. The elongating cells form a neural groove. In other cells, rings of microfilaments at one end constrict, so the cells become wedge shaped. Their part of the sheet folds over to form the tube.

the zygote; its nucleus still had all the genes required to make all cell types that make up a frog.

Embryonic cells of humans also retain the capacity to give rise to a whole individual. This is what happens after spontaneous separation of human blastomeres at the two-cell stage. The split does not result in two half-embryos, but rather in identical twins (Figure 39.6).

Morphogenesis

Morphogenesis refers to a program of orderly changes in an embryo's size, shape, and proportions, the result being specialized tissues and early organs. As part of the program, cells divide, grow, migrate, and change in size. Tissues expand and fold, and the cells in some of them die in controlled ways at prescribed locations.

Think of active cell migration. *Cells send out and use pseudopods that move them along prescribed routes.* When they reach their destination, they establish contact with cells already there. For instance, forerunners of neurons interconnect this way as a nervous system is forming.

How do cells "know" where to move? They respond to adhesive cues, as when migrating nerve cells stick to adhesion proteins on glial cells but not blood vessels (Figure 39.9a). They respond to chemical gradients, too. Their migrations may be coordinated by the synthesis, release, deposition, and removal of specific chemicals in the extracellular matrix. And adhesive cues tell cells when to stop. Cells will migrate to regions of strongest adhesion, but once there, further migration is impeded. Section 20.7 describes the chemical-induced migration of cells of the slime mold *Dictyostelium discoideum*.

Also, *whole sheets of cells expand and fold inward and outward.* Microtubules lengthen and microfilament rings constrict inside the cells. The assembly and disassembly of these components of the cytoskeleton are aspects of a controlled program of changes in cell shape.

Through such controlled, localized events, the size, shape, and proportions of body parts emerge. Example: Figure 39.9b shows what happens after three primary tissues form in embryos of amphibians, reptiles, birds, and mammals. At the embryo midline, ectodermal cells elongate and form a neural plate—the first sign that a region of ectoderm is on its way to becoming nervous tissue. Microtubules lengthen in some cells; other cells become wedge shaped as a microfilament ring contracts

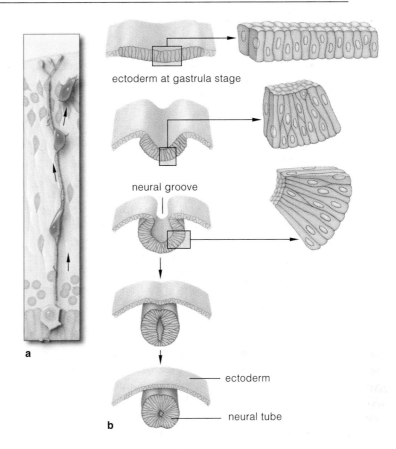

ectoderm at gastrula stage

neural groove

a

ectoderm

neural tube

b

and constricts them at one end. The changes in shape cause tissue flaps to fold over and meet at the midline, thus forming the neural tube (compare Section 30.7).

As a last example, *programmed cell death helps sculpt body parts.* Cells that function for a limited time only in embryos execute themselves. By this form of cell death, called apoptosis, molecular signals from certain cells activate weapons of self-destruction stockpiled inside other, target cells. Section 14.6 describes what happens to targets at the molecular level. For now, think about a human embryo's hand when it is still a paddle (Section 8.4). Cartilage models of the digits form inside it. Cells in thin tissue zones between digits have receptors for proteins, each a product of selective gene expression. When the proteins bind to their receptors, cells commit mass suicide, and digits separate. In rare cases, a gene mutation blocks apoptosis and the digits stay webbed.

In cell differentiation, a cell selectively uses certain genes and makes certain proteins, not found in other cell types, for distinctive cell structures, products, and functions.

Morphogenesis is a program of orderly changes in body size, shape, and proportions. It results in specialized tissues and organs at prescribed locations, at prescribed times.

Morphogenesis involves cell division, active cell migration, tissue growth and foldings, changes in cell size and shape, and programmed cell death by way of apoptosis.

PATTERN FORMATION

In developing embryos, differences between cells arise in spatially organized ways. Cells divide, differentiate, and live or die; they migrate and stick to like cells in tissues. Their cellular descendants fill in body details in ordered patterns. Cells of amphibian embryos make a head at one end and a tail at the other, and so on.

A cell locked in at a given place in the embryo has chemical memory; it selectively reads some genes but not others. It makes and secretes molecular signals that change the activities of its neighbors, and it responds to neighborhood signals. And its memory is passed on to all of its descendants, which go on to form tissues in places where we expect them to be. Now is the time of **embryonic induction**—the developmental fates of the embryonic cell lineages change when exposed to gene products from adjacent tissues. Through these changes, specialized tissues and organs emerge from clumps of embryonic cells in predictably ordered, spatial patterns. We call their emergence **pattern formation**.

We have experimental evidence of cell memory. For instance, researchers excised the dorsal lip of a normal axolotl embryo, then grafted it into a novel location in a different axolotl embryo (Figure 39.10). Gastrulation proceeded at the recipient's own dorsal lip *and at the graft*. Siamese twins—a double embryo having two sets of body parts—developed. Such studies showed that the dorsal lip is central to organizing this amphibian's main axis; how else would the two-tailed, two-headed tadpole in Figure 39.10 form? In fact, the dorsal lip was the first embryonic signaling center to be discovered.

a Dorsal lip excised from donor embryo is grafted to an abnormal site in another embryo.

b Graft induces another site of invagination.

c Gastrula develops into a "double" tadpole. Most of its tissues originated from the host embryo.

Figure 39.10 Experimental evidence that the dorsal lip of an embryonic axolotl (a type of amphibian) controls gastrulation. When transplanted to a different site in another axolotl embryo, it organized the formation of a second set of body parts.

Embryonic Signals in Pattern Formation

Cytoplasmic localization during cleavage ensures that different cells will send and receive different chemical signals. While the embryo forms, cells that started out as neighbors use the signals. So do cells that meet up by gastrulation and other morphogenetic movements.

For example, cells of the dorsal lip are the source of a morphogen. **Morphogens** are degradable molecules that diffuse as long-range signals from an embryonic signaling center. The concentration gradient helps cells assess their position and affects how they differentiate. The signal is strongest at the start of the gradient and weakens with distance. Cells in different body regions affected by the gradients receive different information, so they selectively express different genes as a result.

Other signals are short range, involving cell-to-cell contacts. Example: Changes in the expression of genes for cadherins, integrins, and other glycoproteins at the cell surface alter how cells associate with one another when the gastrula, then organs, form. When a cell starts to make a cytoskeletal protein, it may move or lengthen in some direction. If adhesion proteins are affected, a cell may cohere with neighbors, break free and migrate, or become segregated from cells of an adjoining tissue.

Also, cells respond differently to signals depending on when they receive them. Graft a bit of animal pole epithelium from an early gastrula over a rudimentary eye of a later embryo, and it gives rise to something like neural tube tissue. Leave it alone in vitro for a few hours before the graft, and it differentiates into a lens. Leave it even longer in vitro and it will not respond at all to the inductive signals from the eye tissue.

A Theory of Pattern Formation

Do the same classes of genes govern spatial patterning in all animals? It now appears so. After studies of many diverse animals, researchers have put together a theory of pattern formation. Here are its key points:

1. The formation of tissues and organs in ordered, spatial patterns starts with cytoplasmic localization. It unfolds through short-range cell-to-cell contacts, and through long-range signals—chemical gradients—that weaken with distance and thus provide different information to cells in different locations.

2. Morphogens and other inducer molecules diffuse through embryonic tissues, sequentially activating classes of **master genes**, such as the homeotic genes found in all major animal groups.

3. Master gene products and control elements interact to map out the overall body plan. Products appear as blocks of genes are activated and suppressed in cells along the embryo's anterior–posterior axis and dorsal–ventral axis. Other gene products interact to fill in the details of specific body parts.

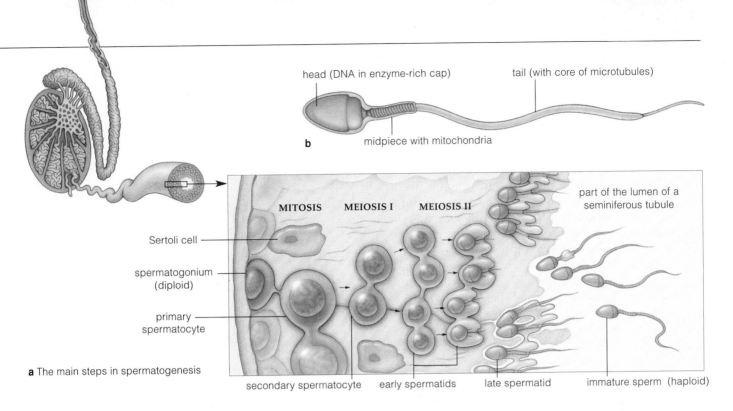

head (DNA in enzyme-rich cap)   tail (with core of microtubules)

b   midpiece with mitochondria

a The main steps in spermatogenesis

Sertoli cell

spermatogonium (diploid)

primary spermatocyte

MITOSIS   MEIOSIS I   MEIOSIS II

part of the lumen of a seminiferous tubule

secondary spermatocyte   early spermatids   late spermatid   immature sperm (haploid)

a Level of testosterone in blood decreases; the hypothalamus secretes GnRH, a releasing hormone.

(+)

HYPOTHALAMUS

(−)   (−)

f Elevated level of testosterone in blood inhibits secretion of GnRH.

g High sperm count induces Sertoli cells to secrete inhibin, which inhibits secretion of GnRH and LH.

(−)

ANTERIOR PITUITARY

b GnRH stimulates secretion of LH, FSH from anterior lobe of pituitary.

(+)

TESTIS (two)

c LH prompts Leydig cells in testes to produce and release testosterone.

d Sertoli cells bind FSH and testosterone, and function in spermatogenesis at puberty.

e Testosterone is used to stimulate formation and development of sperm.

Figure 39.15 Negative feedback loops to the hypothalamus and to the pituitary gland's anterior lobe from the testes. With these loops, excess levels of testosterone in blood slow down the mechanisms leading to its production. This helps maintain the testosterone level in amounts required for sperm formation.

Figure 39.14 (**a**) Sperm formation. The process starts with a spermatogonium, a type of diploid germ cell. Mitosis, meiosis, and incomplete cytoplasmic divisions result in a clone of immature haploid cells that differentiate and develop into mature sperm. (**b**) Structure of a mature sperm from a human male.

abbreviation stands for *Luteinizing Hormone*, the other for *Follicle-Stimulating Hormone*.) We now know their molecular structure is identical in males and females.

The hypothalamus controls secretion of LH, FSH, and testosterone, and thus controls sperm formation. When blood levels of testosterone and other factors are low, it secretes the releasing hormone GnRH (Figure 39.15). GnRH calls on the anterior lobe to step up the release of LH and FSH, which have targets in testes.

LH prods Leydig cells to secrete testosterone, which assists in stimulating the formation and development of sperm. Sertoli cells have receptors for FSH, which is required to jump-start spermatogenesis at puberty. We don't know whether FSH is also necessary for normal functioning of mature human testes.

Figure 39.15 also shows how feedback loops to the hypothalamus can cut back on testosterone secretion and sperm formation. An elevated testosterone level in blood slows down the release of GnRH. Also, when the sperm count is high, Sertoli cells release inhibin. This protein hormone acts on the hypothalamus and pituitary to cut back on the release of GnRH and FSH.

Sperm formation depends on the hormones LH, FSH, and testosterone. Negative feedback loops from the testes to the hypothalamus and pituitary gland control their secretion.

REPRODUCTIVE SYSTEM OF HUMAN FEMALES

The Reproductive Organs

We turn next to the reproductive system of the human female. Figure 39.16 shows its components, and Table 39.2 summarizes their functions. The female's *primary* reproductive organs, her pair of ovaries, produce eggs and secrete sex hormones. Her immature eggs are called oocytes. An oocyte released from an ovary moves into one of a pair of oviducts (also called Fallopian tubes). An oviduct is a channel into the **uterus**, a pear-shaped, hollow organ in which embryos grow and develop.

A thick layer of smooth muscle, the myometrium, makes up most of the uterine wall. As you will see, the **endometrium**, the inner lining of the wall, is vital for development. It consists of connective tissues, glands, and blood vessels. The narrowed-down portion of the uterus is the cervix. A vagina, a muscular tube, extends from the cervix to the surface of the body. The vagina receives sperm and functions as part of the birth canal.

At the body surface are external genitals (vulva), including organs for sexual stimulation. Outermost are a pair of fat-padded skin folds (the labia majora). They enclose a smaller pair of skin folds (the labia minora), which are highly vascularized but have no fatty tissue. The smaller folds partly enclose the clitoris, a sex organ highly sensitive to stimulation. At the body surface, the urethra's opening is positioned about midway between the clitoris and the vaginal opening.

Overview of the Menstrual Cycle

Females of most mammalian species follow an *estrous* cycle. They are fertile and in heat (sexually receptive to males) only at certain times of year. By contrast, female primates, including humans, follow a **menstrual cycle**. They become fertile intermittently, on a cyclic basis, and heat is not synchronized with fertile periods. Said another way, female primates of reproductive age can become pregnant only at certain times of year, but they may be receptive to sex at any time.

Briefly, an oocyte matures and escapes from an ovary during each menstrual cycle. Also during the cycle, the endometrium becomes primed to receive and nourish a forthcoming embryo *if* a sperm penetrates the oocyte and fertilization occurs. However, if that oocyte does not get fertilized, then about four to six tablespoons of blood-rich fluid from the uterus will start flowing out through the vaginal canal. Such a recurring blood flow is called menstruation. It means "there is no embryo at this time," and it marks the first day of a new cycle. The uterine nest is being sloughed off and is about to be constructed once again.

The events just sketched out proceed through three phases. The cycle starts with a *follicular* phase. This is

a

Figure 39.16 *Above and facing page:* (**a**) Position of the human female reproductive system relative to the pelvic girdle and urinary bladder. The midsagittal section in (**b**) shows the components of the system and lists their functions.

Table 39.2	*Female Reproductive Organs*
Ovaries	Oocyte production and maturation, sex hormone production
Oviducts	Ducts for conducting oocyte from ovary to uterus; fertilization normally occurs here
Uterus	Chamber in which new individual develops
Cervix	Secretion of mucus that enhances sperm movement into uterus and (after fertilization) reduces embryo's risk of bacterial infection
Vagina	Organ of sexual intercourse; birth canal

the time of menstruation, endometrial breakdown and rebuilding, and oocyte maturation. The next phase is limited to the release of an oocyte from the ovary. We call it **ovulation**. During the *luteal* phase of the cycle, an endocrine structure (corpus luteum) forms and the endometrium is primed for pregnancy (Table 39.3).

All three phases are governed by feedback loops to the hypothalamus and pituitary gland from the ovaries. FSH and LH promote cyclic changes in the ovaries. As you will see, FSH and LH also stimulate the ovaries to

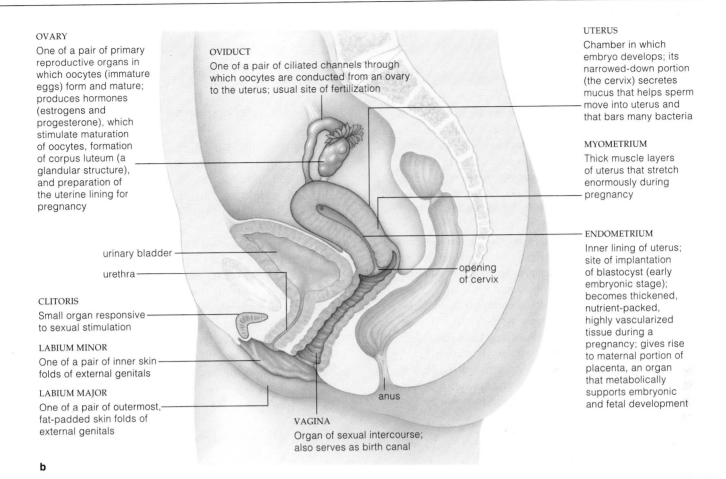

OVARY
One of a pair of primary reproductive organs in which oocytes (immature eggs) form and mature; produces hormones (estrogens and progesterone), which stimulate maturation of oocytes, formation of corpus luteum (a glandular structure), and preparation of the uterine lining for pregnancy

OVIDUCT
One of a pair of ciliated channels through which oocytes are conducted from an ovary to the uterus; usual site of fertilization

UTERUS
Chamber in which embryo develops; its narrowed-down portion (the cervix) secretes mucus that helps sperm move into uterus and that bars many bacteria

MYOMETRIUM
Thick muscle layers of uterus that stretch enormously during pregnancy

ENDOMETRIUM
Inner lining of uterus; site of implantation of blastocyst (early embryonic stage); becomes thickened, nutrient-packed, highly vascularized tissue during a pregnancy; gives rise to maternal portion of placenta, an organ that metabolically supports embryonic and fetal development

urinary bladder

urethra

opening of cervix

CLITORIS
Small organ responsive to sexual stimulation

LABIUM MINOR
One of a pair of inner skin folds of external genitals

LABIUM MAJOR
One of a pair of outermost, fat-padded skin folds of external genitals

anus

VAGINA
Organ of sexual intercourse; also serves as birth canal

b

Table 39.3	Events of the Menstrual Cycle	
Phase	Events	Days of Cycle*
Follicular phase	Menstruation; endometrium breaks down	1–5
	Follicle matures in ovary; endometrium rebuilds	6–13
Ovulation	Oocyte released from ovary	14
Luteal phase	Corpus luteum forms, secretes progesterone; the endometrium thickens and develops	15–28

* Assuming a 28-day cycle.

secrete sex hormones—**estrogens** and **progesterone**—to promote the cyclic changes in the endometrium.

A human female's menstrual cycles start between ages ten and sixteen. Each cycle lasts for about twenty-eight days, but this is only the average. It runs longer for some women and shorter for others. The menstrual cycles continue until a woman is in her late forties or early fifties, when her supply of eggs is dwindling and secretions of hormones slow down. This is the onset of *menopause*, the twilight of reproductive capacity.

You may have heard of *endometriosis*, a condition that arises after endometrial tissue abnormally spreads and grows outside the uterus. Endometrial scar tissue may form on ovaries or oviducts and lead to infertility. In the United States, 10 million women may be affected annually. Perhaps the condition arises when menstrual flow backs up through the oviducts and spills into the pelvic cavity. Or perhaps some embryonic cells became positioned in the wrong place before birth and were stimulated to grow when sex hormones became active at puberty. Whatever the case, estrogen still stimulates cells in the mislocated tissue. The resulting symptoms include pain during menstruation, sex, and urination.

Ovaries, the female primary reproductive organs, produce oocytes (immature eggs) and sex hormones. Endometrium lines the uterus, a chamber in which embryos develop.

Secretion of sex hormones (estrogens and progesterone) is coordinated on a cyclic basis through the reproductive years.

A menstrual cycle starts with menstruation, breakdown and rebuilding of the endometrium, and maturation of an oocyte. After an oocyte escapes from an ovary (ovulation), the cycle ends with the formation of a corpus luteum and an endometrium that is primed for pregnancy.

FEMALE REPRODUCTIVE FUNCTION

Cyclic Changes in the Ovary

Take a moment to review Section 10.5, the generalized picture of meiosis in an oocyte. A normal baby girl has about 2 million primary oocytes in her ovaries. By the time she is seven years old, only about 300,000 remain; her body resorbed the rest. Her primary oocytes have already started meiosis, but the nuclear division process was arrested in a genetically programmed way. Meiosis resumes in one oocyte at a time, starting with the first menstrual cycle. Only about 400 to 500 oocytes will be released during the reproductive years.

Figure 39.17 shows a primary oocyte near an ovary's surface. A cell layer (of granulosa cells) surrounds and nourishes it. We call a primary oocyte and the cell layer around it a follicle. As the menstrual cycle starts, the hypothalamus is secreting GnRH in amounts that cause the anterior pituitary to step up *its* secretion of FSH and LH. The blood level of these hormones rises, and the increase causes the follicle to grow (Figure 39.18).

The oocyte starts to increase in size, and more cell layers form around it. Deposits of glycoproteins collect between the layers and the oocyte, and they widen the distance between them. In time they make up the zona pellucida, a noncellular coating around the oocyte.

Outside the zona pellucida, FSH and LH stimulate cells to secrete estrogens. An estrogen-containing fluid accumulates in the follicle, and estrogen levels in blood start to increase. About eight to ten hours before being released from the ovary, the oocyte completes meiosis I. Then its cytoplasm divides, forming two cells.

One cell, the *secondary* oocyte, ends up with nearly all of the cytoplasm. The other cell is the first of three **polar bodies**. The meiotic parceling of chromosomes among four cells gives the secondary oocyte a haploid chromosome number—the precise number required for gametes and for sexual reproduction.

About halfway through each menstrual cycle, the pituitary gland detects the rise in estrogens in blood. It responds with a brief outpouring of LH. The surge in LH causes rapid vascular changes that make the follicle swell, and it makes enzymes digest the bulging follicle wall. The weakened wall ruptures. Fluid, together with the secondary oocyte, escapes (Figures 39.17 and 39.18).

Thus, the midcycle surge of LH triggers ovulation—the release of a secondary oocyte from the ovary.

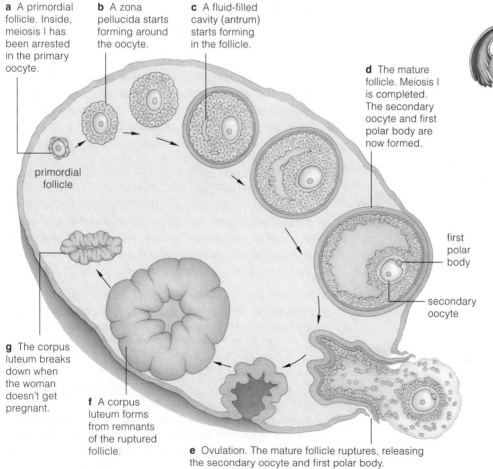

a A primordial follicle. Inside, meiosis I has been arrested in the primary oocyte.

b A zona pellucida starts forming around the oocyte.

c A fluid-filled cavity (antrum) starts forming in the follicle.

d The mature follicle. Meiosis I is completed. The secondary oocyte and first polar body are now formed.

primordial follicle

g The corpus luteum breaks down when the woman doesn't get pregnant.

f A corpus luteum forms from remnants of the ruptured follicle.

e Ovulation. The mature follicle ruptures, releasing the secondary oocyte and first polar body.

first polar body

secondary oocyte

oviduct

OVARY

uterus

vagina

Figure 39.17 Cyclic events in a human ovary, cross-section.

A follicle stays in the same place inside an ovary all through one menstrual cycle. It does not "move around" as in this diagram, which only shows the *sequence* in which events occur. In the cycle's first phase, a follicle grows and matures. At ovulation, the second phase, the mature follicle ruptures and releases a secondary oocyte. In the third phase, a corpus luteum forms from the follicle's remnants. If the woman doesn't get pregnant during the cycle, the corpus luteum self-destructs.

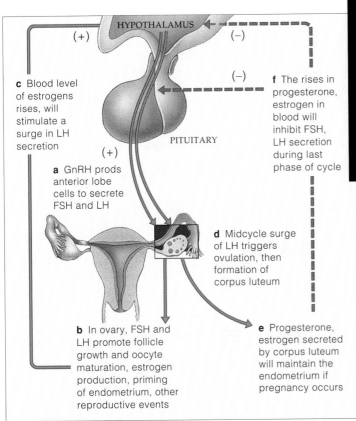

c Blood level of estrogens rises, will stimulate a surge in LH secretion

(+) **HYPOTHALAMUS** (−)

(−)

(+) **PITUITARY**

a GnRH prods anterior lobe cells to secrete FSH and LH

f The rises in progesterone, estrogen in blood will inhibit FSH, LH secretion during last phase of cycle

d Midcycle surge of LH triggers ovulation, then formation of corpus luteum

b In ovary, FSH and LH promote follicle growth and oocyte maturation, estrogen production, priming of endometrium, other reproductive events

e Progesterone, estrogen secreted by corpus luteum will maintain the endometrium if pregnancy occurs

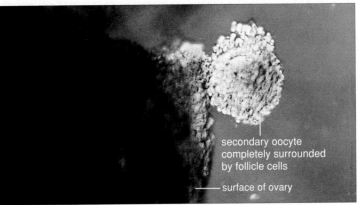

secondary oocyte completely surrounded by follicle cells

surface of ovary

Figure 39.18 Feedback control of hormonal secretion during a menstrual cycle. A positive feedback loop from an ovary to the hypothalamus causes a surge in LH secretion that triggers ovulation. The light micrograph shows a secondary oocyte being released from an ovary at this time. Afterward, negative feedback loops to the hypothalamus and pituitary inhibit FSH secretion. They block maturation of another follicle until the cycle is over.

Cyclic Changes in the Uterus

The estrogens released early in the menstrual cycle also help pave the way for a possible pregnancy. Estrogens stimulate growth of the endometrium and its glands. Just before the midcycle surge of LH, follicle cells start to secrete some progesterone as well as estrogens. Blood vessels grow quickly in the thickened endometrium. At ovulation, estrogens act on tissues around the cervical canal, which leads into the vagina. The cervix starts to secrete large quantities of a thin, clear mucus that is an ideal medium for sperm to swim through.

The midcycle surge of LH that triggers ovulation also triggers the formation of a **corpus luteum** in the ovary. This yellowish glandular structure forms by cell differentiation among granulosa cells left behind in the follicle. (Its name means yellow body.) Secretions from the corpus luteum influence the rest of the cycle.

The corpus luteum secretes progesterone and some estrogen. Progesterone prepares the reproductive tract for the arrival of a blastocyst, the type of blastula that forms from a fertilized mammalian egg (Section 39.3). For example, this hormone makes cervical mucus turn thick and sticky. The mucus may keep normal bacterial residents of the vagina out of the uterus. Progesterone also will maintain the endometrium during pregnancy.

A corpus luteum persists for about twelve days. All the while, the hypothalamus is calling for minimal FSH secretion, which stops other follicles from developing. If a blastocyst does not burrow into the endometrium, the corpus luteum will self-destruct in the last days of the cycle. It will secrete certain prostaglandins that can disrupt its own functioning.

After this, progesterone and estrogen levels in blood decline quickly, and so the endometrium starts to break down. Deprived of oxygen and nutrients, the lining's blood vessels constrict and tissues die. Blood escapes as weakened capillary walls start to rupture. Blood as well as the sloughed endometrial tissues make up the menstrual flow, which continues for three to six days. Then the cycle begins anew, with rising estrogen levels stimulating repair and growth of the endometrium.

By menopause, the supply of oocytes is dwindling, hormone secretions slow down, and in time menstrual cycles—and fertility—will be over. Eggs that are still to be released late in a woman's life are at some risk of alterations in chromosome number or structure when meiosis resumes. A newborn with Down syndrome, as described in Section 11.10, is one possible outcome.

During one menstrual cycle, FSH and LH stimulate growth of an ovarian follicle (a primary oocyte and the surrounding layer of cells). The first meiotic cell division results in one secondary oocyte and the first polar body.

A midcycle surge of LH triggers ovulation—the release of the secondary oocyte and the polar body from the ovary.

Early on, estrogens call for endometrial repair and growth. Then estrogens and progesterone prepare the endometrium and other parts of the reproductive tract for pregnancy.

VISUAL SUMMARY OF THE MENSTRUAL CYCLE

By now, you probably have come to the conclusion that the menstrual cycle is not a simple tune on a biological banjo. It's a full-blown hormonal symphony!

Before continuing with your reading, take a moment to review Figure 39.19. It correlates the cyclic events in the ovary and uterus with all the coordinated changes in hormone levels that bring about those events. The illustration may leave you with a better understanding of what goes on.

Figure 39.19 Changes in the ovary and uterus, as correlated with changing hormone levels during each turn of the menstrual cycle. *Green* arrows show which hormones dominate the cycle's first phase (when the follicle matures), and then the second phase (when the corpus luteum forms).

(a–c) FSH and LH secretions bring about the changes in ovarian structure and function. **(d,e)** Estrogen and progesterone secretions from the ovary stimulate the changes in the endometrium.

PREGNANCY HAPPENS

Sexual Intercourse

Suppose the secondary oocyte is on its way down an oviduct when a female and male are engaged in sexual intercourse, or coitus. The male sex act requires *erection*, whereby a normally limp penis stiffens and lengthens; and *ejaculation*, the forceful expulsion of semen into the urethra and out from the penis. As Figure 39.12 shows, the penis incorporates cylinders of spongy tissue. Many friction-activated sensory receptors are on its mushroom-shaped tip, the glans penis. In males *not* sexually aroused, large blood vessels that lead into the cylinders are vasoconstricted. In aroused males, they vasodilate and blood flows into the penis faster than it flows out, so blood collects in the spongy tissue. The resulting engorgement stiffens and lengthens the organ, this being helpful for penetration into the vaginal canal.

During coitus, pelvic thrusts stimulate the penis, the female's clitoris, and the vaginal wall. The mechanical stimulation induces involuntary contractions in the male reproductive tract. It forces sperm from the epididymes and the contents of seminal vesicles and the prostate gland into the urethra. The substances mix and form semen, which is ejaculated into the vagina. A sphincter closes the bladder's neck and prevents urination during ejaculation.

Emotional intensity, hard breathing, heart pounding, and the contraction of skeletal muscles in general accompany rhythmic throbbing of pelvic muscles. During *orgasm*, the end of the sex act, strong sensations of physical release, warmth, and relaxation dominate. Similar sensations typify female orgasm. It's a common misconception that a female cannot become pregnant if she doesn't reach orgasm. Don't believe it.

Fertilization

An ejaculation can put 150 million to 350 million sperm in the vagina. If they arrive a few days before or after ovulation or any time between, fertilization can occur. Less than thirty minutes after sperm arrive, contractions move them deep into the female's reproductive tract. Just a few hundred actually reach the upper part of the oviduct, where fertilization usually takes place.

Think back on the micrograph that opens Unit II, which depicts sperm around a secondary oocyte. When

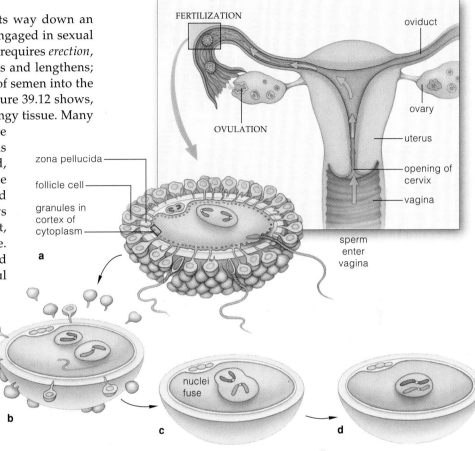

Figure 39.20 Fertilization. (**a**) Many sperm surround a secondary oocyte. Acrosomal enzymes clear a path through the zona pellucida. (**b**) When a sperm does penetrate the secondary oocyte, granules in the egg cortex release substances that make the zona pellucida impenetrable to other sperm. Penetration also stimulates meiosis II of the oocyte's nucleus. (**c**) The sperm's tail degenerates and its nucleus enlarges and fuses with the oocyte nucleus. (**d**) With fusion, fertilization is over. The zygote has formed.

sperm reach an oocyte, they release digestive enzymes that clear a path through the zona pellucida. Although many sperm might get this far, usually only one fuses with the oocyte (Figure 39.20). It degenerates inside the oocyte's cytoplasm until only its nucleus and centrioles remain. Penetration induces the secondary oocyte and the first polar body to finish meiosis II.

There are now three polar bodies and a mature egg, or **ovum** (plural, ova). The sperm nucleus and the egg nucleus fuse. Together, their chromosomes restore the diploid number for a brand new zygote.

The intense physiological events that accompany coitus have one function: to put sperm on a collision course with an egg. Fertilization is over with the fusion of a sperm nucleus and egg nucleus, which results in a diploid zygote.

FORMATION OF THE EARLY EMBRYO

Pregnancy lasts an average of thirty-eight weeks from the time of fertilization. It takes about two weeks for a blastocyst to form. The time span from the third to the end of the eighth week is the *embryonic* period, when the major organ systems form. When it ends, the new individual has distinctly human features and is called a **fetus**. In the *fetal* period from the start of the ninth week until birth, organs enlarge and become specialized.

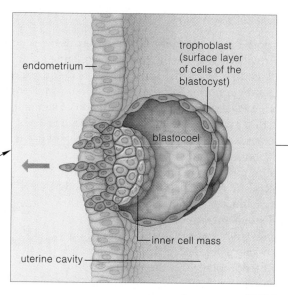

endometrium

trophoblast (surface layer of cells of the blastocyst)

blastocoel

inner cell mass

uterine cavity

d DAY 5. A blastocoel (a fluid-filled cavity) forms in the morula as a result of secretions from the surface cells. By the thirty-two-cell stage, cells of an inner cell mass are already differentiating. They will give rise to the embryo proper. This embryonic stage is called a blastocyst.

inner cell mass

c DAY 4. By 96 hours, there is a ball of sixteen to thirty-two cells that is shaped like a mulberry. It is a morula (after *morum*, Latin for mulberry). Cells of the surface layer will function in implantation and will give rise to a membrane, the chorion.

b DAY 3. After the third cleavage, the cells suddenly huddle together into a compacted ball, which becomes stabilized by numerous tight junctions among the outer cells. Gap junctions form among the interior cells and enhance intercellular communication.

a DAYS 1–2. Cleavage begins within 24 hours after fertilization. The first cleavage furrow extends between the two polar bodies. Subsequent cuts are rotational, so the resulting cells are not symmetrically arranged (compare Section 39.3). Until the eight-cell stage forms, the cells are loosely arranged, with considerable space between them.

e DAYS 6–7. Surface cells of the blastocyst attach to the endometrium and start to burrow into it. Implantation is under way.

actual size

We typically call the first three months of pregnancy the *first* trimester. A *second* trimester extends from the start of the fourth month to the end of the sixth. The *third* trimester extends from the seventh month until birth. Beginning with Figure 39.21, the next series of illustrations shows characteristic features of the new individual at progressive stages of development.

Cleavage and Implantation

Three to four days after fertilization, a zygote is already undergoing cleavage as it tumbles through the oviduct. Genes are being expressed; early divisions depend on their products. At the eight-cell stage, the cells huddle into a compact ball. By the fifth day, there is a surface layer of cells (the trophoblast), a cavity filled with their secretions (a blastocoel), and a cluster of interior cells (an inner cell mass). These are the defining features of a human blastocyst (Figure 39.21d).

Six or seven days after fertilization, **implantation** is under way. By this process, the blastocyst adheres to the uterine lining, some of its cells send out projections that invade the mother's tissues, and connections start to form that will metabolically support the developing embryo through the months ahead. While the invasion

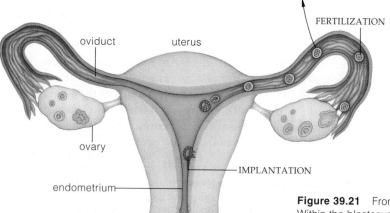

oviduct

uterus

FERTILIZATION

ovary

IMPLANTATION

endometrium

Figure 39.21 From fertilization through implantation. A blastocyst forms at cleavage. Within the blastocyst, the inner cell mass gives rise to the disk-shaped early embryo. Three extraembryonic membranes (the amnion, chorion, and yolk sac) start forming. A fourth membrane (allantois) forms after the blastocyst is implanted.

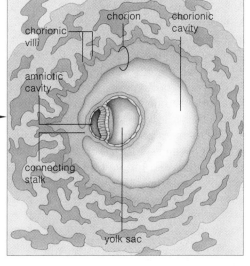

start of amniotic cavity   start of embryonic disk

start of yolk sac

blood-filled spaces

start of chorionic cavity

chorionic villi

chorion

chorionic cavity

amniotic cavity

connecting stalk

yolk sac

f DAYS 10–11. The yolk sac, embryonic disk, and amniotic cavity have started to form from parts of the blastocyst.

actual size

g DAY 12. Blood-filled spaces form in maternal tissue. The chorionic cavity starts to form.

actual size

h DAY 14. A connecting stalk has formed between the embryonic disk and chorion. Chorionic villi, which will be features of a placenta, start to form.

 actual size

is proceeding, the inner cell mass develops into two cell layers of a flattened and somewhat circular shape. The two layers make up the embryonic disk—and in short order they will give rise to the embryo proper.

Extraembryonic Membranes

As implantation progresses, membranes start to form outside the embryo. First a fluid-filled *amniotic* cavity opens up between the embryonic disk and part of the blastocyst's surface (Figure 39.21*f*). Then cells migrate around the wall of the cavity and form the **amnion**, a membrane that will enclose the embryo. Fluid inside the cavity will function as a buoyant cradle where the embryo can grow, move freely, and be protected from abrupt temperature changes and mechanical impacts.

While the amnion forms, other cells migrate around the inner wall of the blastocyst's first cavity. They form a lining that becomes a **yolk sac**. This extraembryonic membrane speaks of the evolutionary heritage of land vertebrates (Section 24.7). For most animals that make shelled eggs, the sac holds nutritive yolk. In humans, part of it becomes a site of blood cell formation. Part will give rise to germ cells, the forerunners of gametes.

Before a blastocyst is fully implanted, spaces open in maternal tissues and fill with blood seeping in from ruptured capillaries. Inside the blastocyst, a different cavity opens up around the amnion and yolk sac. Now fingerlike projections start to form on the cavity lining, which is the **chorion**. This new membrane will become part of the placenta, a spongy, blood-engorged tissue.

After the blastocyst is implanted, an outpouching of the yolk sac will become the third extraembryonic membrane—the **allantois**. The allantois has different functions in different animal groups. Among reptiles, birds, and some mammals, it serves in respiration and in storing metabolic wastes. In humans, blood vessels for the placenta and the urinary bladder form from it.

One more point should be made here. Cells of the blastocyst secrete a hormone, HCG (*Human Chorionic Gonadotropin*), which stimulates the corpus luteum to keep on secreting progesterone and estrogen. In this way, the blastocyst itself prevents menstrual flow and works to avoid being sloughed off until the placenta takes over the function of secreting HCG, about eleven weeks later. By the start of the third week, HCG can be detected in samples of the mother's blood or urine. At-home *pregnancy tests* have a treated "dip-stick" that changes color when HCG is present in urine.

A human blastocyst is composed of a surface layer of cells around a fluid-filled cavity (blastocoel) and an inner cell mass, which will give rise to the embryo proper.

Six or seven days after fertilization, the blastocyst implants itself in the endometrium. Now projections from its surface invade maternal tissues, and connections start to form that in time will metabolically support the developing embryo.

Some parts of the blastocyst give rise to an amnion, yolk sac, chorion, and allantois. These extraembryonic membranes serve different functions. Together they are vital for the structural and functional development of the embryo.

EMERGENCE OF THE VERTEBRATE BODY PLAN

By the time a woman has missed her first menstrual period, cleavage is completed and gastrulation is under way. Gastrulation, recall, is a stage when extensive cell divisions, migrations, and rearrangements produce the primary tissue layers (Sections 39.2 and 39.4).

By now, the embryonic disk is surrounded by the amnion and chorion except where a stalk connects it to the chorion's inner wall. Earlier, the inner cell mass had behaved *as if* it were still developing on the large ball of yolk that evolved in reptilian ancestors of mammals. (Thus it becomes a *flattened* two-layer embryonic disk, as in reptiles and birds.) Cell divisions and migrations of one layer formed the yolk sac lining. The other layer now starts to generate the embryo proper.

For example, by the eighteenth day, the embryonic disk has two neural folds, which will merge to form a neural tube (Figure 39.22b). Early in the third week, a sausage-shaped outpouching forms on the yolk sac. It is the allantois (Greek *allas*, meaning sausage). In humans, recall, it only helps form the urinary bladder and blood vessels for the placenta. Some mesoderm folds into a tube that becomes the notochord. A notochord is only a structural framework; bony segments of the vertebral column soon form around it. Toward the end of the third week, some mesoderm gives rise to the **somites**. These paired segments are the embryonic source of most bones, of skeletal muscles of the head and trunk, and of the dermis overlying these regions.

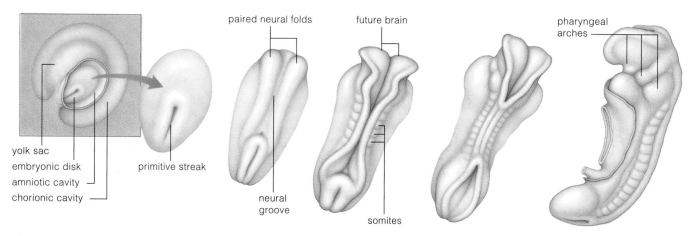

a DAY 15. A faint band appears around a depression along the axis of the embryonic disk. This is the primitive streak, and it marks the onset of gastrulation in vertebrate embryos.

b DAYS 18–23. Organs start to form through cell divisions, cell migrations, tissue folding, and other events of morphogenesis. Neural folds will merge to form the neural tube. Somites (bumps of mesoderm) appear near the embryo's dorsal surface. They will give rise to most of the skeleton's axial portion, skeletal muscles, and much of the dermis.

c DAYS 24–25. By now, some embryonic cells have given rise to pharyngeal arches. These will contribute to the formation of the face, neck, mouth, nasal cavities, larynx, and pharynx.

Figure 39.22 Hallmarks of the embryonic period of humans and other vertebrates. A primitive streak, then a notochord form. Neural folds, somites, and pharyngeal arches form later. (**a**,**b**) Dorsal views, of the embryo's back. (**c**) Side view. Compare the diagrams with the Figure 39.24 photographs.

Through cell divisions and migrations, the midline of that layer thickens faintly around a depression at its surface. This "primitive streak" lengthens and thickens further the next day. It marks the onset of gastrulation (Figure 39.22a). It also defines the anterior–posterior axis of the embryo and, in time, its bilateral symmetry. Endoderm and mesoderm form from cells migrating inward at this axis. Now pattern formation begins, as described in Section 39.5. Embryonic inductions and interactions among classes of master genes map out the basic body plan characteristic of all vertebrates. And specialized tissues and organs start forming in orderly sequences, in prescribed parts of the embryo.

Pharyngeal arches start to form that will contribute to the face, neck, mouth, nose, larynx, and pharynx. Tiny spaces open up in parts of the mesoderm. In time the spaces will interconnect as the coelomic cavity.

During the third week after fertilization—when a pregnant woman has missed her first menstrual period—the basic vertebrate body plan emerges in the new individual.

A primitive streak, neural tube, somites, and pharyngeal arches form during the embryonic period of all vertebrates. Formation of the primitive streak establishes the body's anterior–posterior axis and its bilateral symmetry.

WHY IS THE PLACENTA SO IMPORTANT?

Even before the beginning of the embryonic period, extraembryonic membranes have been collaborating with the uterus to sustain the embryo's rapid growth. By the third week, tiny fingerlike projections from the chorion have grown profusely into the maternal blood that has pooled in endometrial spaces. The projections —chorionic villi—enhance the exchange of substances between the mother and the new individual. They are functional components of the placenta.

The **placenta**, a blood-engorged organ, consists of endometrial tissue and extraembryonic membranes. At full term, a placenta will make up a fourth of the inner surface of the uterus (Figure 39.23).

a

Figure 39.23 (**a**) Relationship between fetal and maternal blood circulation in a full-term placenta. (**b**). Blood vessels extend from the fetus, through the umbilical cord, and into chorionic villi. Maternal blood spurts into spaces between the villi, but the two bloodstreams do not intermingle. Oxygen, carbon dioxide, and other small solutes diffuse across the placental membrane surface.

A placenta is the body's way of sustaining the new individual while allowing its blood vessels to develop apart from the mother's. Oxygen and nutrients diffuse out of the maternal blood vessels, across the placenta's blood-filled spaces, then into embryonic blood vessels. (The vessels converge in the umbilical cord, the lifeline between the placenta and the new individual.) Carbon dioxide and other wastes diffuse in the other direction. The mother's lungs and kidneys quickly dispose of them.

After the third month, the placenta secretes progesterone and estrogens—and so maintains the uterine lining.

The placenta is a blood-engorged organ of endometrial and extraembryonic membranes. It permits exchanges between the mother and the new individual without intermingling their bloodstreams. Thus it sustains the individual and allows its blood vessels to develop apart from the mother's.

EMERGENCE OF DISTINCTLY HUMAN FEATURES

WEEK 4

yolk sac
connecting stalk
embryo

WEEKS 5–6

a

forebrain
future lens
pharyngeal arches
developing heart
upper limb bud
somites
neural tube forming
lower limb bud
tail

actual length

b

head growth exceeds growth of other regions
retinal pigment
future external ear
upper limb differentiation (hand plates develop, then digital rays of future fingers; wrist, elbow start forming)
umbilical cord formation between weeks 4 and 8 (amnion expands, forms tube that encloses the connecting stalk and a duct for blood vessels)
foot plate

actual length

Figure 39.24 (**a**) Human embryo four weeks after fertilization. Like all vertebrates, it has a tail and pharyngeal arches. (**b**) Embryo at five to six weeks. (**c**) Embryo at the boundary between the embryonic and fetal periods. It now has human features. It floats in amniotic fluid. The chorion covers the amniotic sac but has been pulled aside here. (**d**) Fetus at sixteen weeks. Movements begin as nerves make functional connections with forming muscles. Legs kick, arms wave, fingers grasp, the mouth puckers. These reflex actions will be vital skills in the world outside the uterus.

When the fourth week of the embryonic period ends, the embryo is 500 times its starting size. The placenta sustained the growth spurt, but now the pace slows as details of organs fill in. Limbs form; toes and fingers are sculpted from embryonic paddles. The umbilical cord also forms, and the circulatory system becomes intricate. Growth of the all-important head now surpasses that of any other body region (Figure 39.24). The embryonic period ends after the eighth week is completed. No longer is the embryo merely "a vertebrate." By now, its features clearly define it as a human fetus.

During the second trimester, the fetus is moving its facial muscles. It frowns; it squints. It busily practices

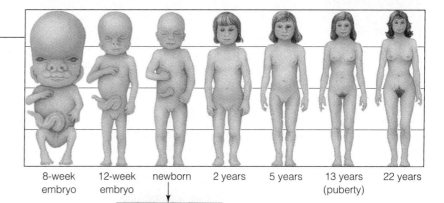

Figure 39.29 Observable, proportional changes in the human body during prenatal and postnatal growth. Changes in overall physical appearance are slow but noticeable until the teenage years. For example, the head becomes proportionally smaller, compared to the embryonic period. Legs become longer; the trunk gets shorter.

8-week embryo | 12-week embryo | newborn | 2 years | 5 years | 13 years (puberty) | 22 years

After he thawed the cells and placed them in a culture medium, they proceeded to complete the in-vitro cycle of fifty doublings and died on schedule.

No cell in a human body divides more than eighty or ninety times. You may well wonder: If an internal clock ticks off their life span, then how can *cancer* cells go on dividing? The answer provides us with a clue to why *normal* cells can't beat the clock.

As you know, cells duplicate their chromosomes before they divide. Capping the chromosome ends are repetitive DNA sequences known as telomeres. A bit of each telomere is lost during each nuclear division, and when only a nub remains, cells stop dividing and die. The exceptions are cancer cells and reproductive cells. Both make telomerase. This enzyme causes telomeres to lengthen. After somatic cells were experimentally exposed to telomerase, they went on dividing well past their normal life span, with no apparent bad effects.

THE CUMULATIVE ASSAULTS HYPOTHESIS A different hypothesis holds that aging is an outcome of ongoing damage at the molecular and cellular levels. By this hypothesis, damage accumulates in DNA as a result of ongoing environmental assaults and declines in DNA's self-repair mechanisms.

For example, free radicals, those rogue molecular fragments described in Chapter 5, bombard DNA and all other biological molecules. This includes the DNA of mitochondria—the power plants of eukaryotic cells. Structural changes in DNA endanger the synthesis of functional enzymes and other proteins necessary for normal life processes. And free radicals are implicated in many age-related problems, such as atherosclerosis, cataracts, and Alzheimer's disease.

Problems in DNA replication and repair operations also have been implicated in aging. Researchers have correlated *Werner's syndrome*, an aging disorder, with a mutation in the gene that specifies a helicase. Such enzymes unwind nucleotide strands from each other. The mutated helicase probably doesn't compromise the replication of DNA; affected people don't die at once. (In their thirties, they start aging abnormally fast and die before reaching fifty.) But a nonmutated form may be vital for repairs. People with Werner's syndrome

accumulate mutations at rates that are above normal. Sooner or later, damage may interfere with cell division. Like the skin cells of very old folks, skin cells of Werner's patients will not divide many times at all.

It may be that both hypotheses have merit. Aging may be an outcome of many interconnected processes in which genes, hormones, environmental assaults, and a decline in DNA repair mechanisms come into play. Consider how living cells of all tissues depend upon exchanges of materials with extracellular fluid. Also consider how collagen is a structural element of many connective tissues. If something shuts down or mutates collagen-encoding genes, the missing or altered gene product might disrupt the flow of oxygen, nutrients, hormones, and so forth to and from cells through every connective tissue. Repercussions from such a mutation would ripple through the body.

Similarly, if mutations result in altered self markers on the body's cells, do T cells of the immune system perceive them as foreign and go on the attack? If such autoimmune responses were to increase over time, they certainly would promote the greater vulnerability to disease and stress associated with old age.

In evolutionary terms, reproductive success means surviving long enough to produce offspring. Humans can reach sexual maturity *and* help their own children reach adulthood in thirty years. We don't *need* to live longer. But we among all animals have the capacity to think about it, and we generally have decided that we like life better than the alternative. Eventually, though, even the most tenacious clingers to life must confront their own mortality; and perhaps in time we all might learn to accept the inevitable with wisdom and grace.

The human life cycle flows naturally from the time of birth, growth, and development, to production of the individual's own offspring, and on through aging to the time of death.

Aging may result from many factors, including an internal clock that ticks off the life spans of all cells, tissues, and organs, and accumulated errors and damaged areas in DNA.

CONTROL OF HUMAN FERTILITY

Some Ethical Considerations

The transformation of a zygote into an adult of intricate detail raises profound questions. *When does development begin?* As you have seen, major developmental events unfold even before fertilization. *When does life begin?* During her lifetime, a woman can produce as many as 500 eggs, all of which are alive. With one ejaculation, a man can release a quarter of a billion sperm, which are alive. Even before sperm and egg merge by chance and establish the genetic makeup of a new individual, they are as much alive as any other form of life. It is scarcely tenable, then, to say that life begins at fertilization. *Life began more than 3 billion years ago; and each gamete, each zygote, and each sexually mature individual is but a fleeting stage in the continuation of that beginning.*

This greater perspective on life cannot diminish the meaning of conception. It is no small thing to entrust a new individual with the gift of life, wrapped in the unique evolutionary threads of our species and handed down through an immense sweep of time.

Yet how can we reconcile the marvel of individual birth with growing awareness of the astounding birth rate for the human species? Almost 15,000 newborns are entering the world every hour. By the time you go to bed tonight, there may be 360,470 more newborns on Earth than there were last night at that hour. In three months there may be 32,892,930 more—almost as many people as there are now in the entire state of California. *Within three months.*

Worldwide, human population growth is rapidly outstripping resources. Each year, many millions face the horrors of starvation. Living as we do on one of the most productive continents on Earth, few of us know what it means to give birth to a child, to give it the gift of life, and have no food to keep it alive.

And how can we reconcile the marvel of birth with the confusion that surrounds unwanted pregnancies? Even highly developed countries do not have adequate educational programs concerning fertility. And many people aren't inclined to exercise control. In 1998 nearly 500,000 babies were born to teenage mothers in the United States. Many parents actually encourage early boy–girl relationships without thinking through the risks of premarital intercourse and unplanned pregnancies. Advice is often condensed to a terse "Don't do it. But if you do it, be careful!" And each year, there are between 1.2 and 1.3 million abortions among all age groups.

The motivation to engage in sex has been evolving for more than 500 million years. And a few centuries of moral and ecological arguments for its suppression have not stopped all that many unwanted pregnancies. Besides this, complex social factors have contributed to a population growth rate that is out of control.

How will we reconcile our biological past and the need for a stabilized cultural present? Whether and how human fertility is to be controlled is one of the most volatile issues of our time. We will return to this issue in the next chapter, in the context of principles that govern the growth and stability of all populations. Here, we briefly consider some control options.

Birth Control Options

The most effective method of birth control is complete *abstinence*, no sexual intercourse whatsoever. Data show that it's unrealistic to expect many people to practice it.

A less reliable variation of abstinence is the *rhythm method*. The idea is to avoid intercourse in a woman's fertile period, starting a few days before ovulation and ending a few days after. A woman identifies and tracks her fertile period. She also keeps records of the length of her menstrual cycles, takes her temperature every morning when she wakes up, or both. (The body's core temperature rises by one-half to one degree just before a fertile period.) But ovulation may not be regular, and miscalculations are frequent. Also, sperm deposited in the vagina a few days before ovulation may survive until ovulation. The rhythm method *is* inexpensive; it costs nothing after you buy a thermometer. It does not require fittings and periodic medical checkups. But its practitioners run a risk of pregnancy (Figure 39.30).

Withdrawal, or removing the penis from the vagina before ejaculation, dates back at least to biblical times. But withdrawal requires very strong willpower, and the practice may fail anyway. Fluid released from the penis just before ejaculation may contain some sperm.

Douching (rinsing the vagina with a chemical right after intercourse) is next to useless. Sperm move past the cervix, and out of reach of a douche, ninety seconds after ejaculation.

Controlling fertility by surgical intervention is less chancy. In *vasectomy*, a physician makes a tiny incision in a man's scrotum, then severs and ties off each vas deferens. The operation takes only twenty minutes and requires only a local anesthetic. After the operation, sperm can't leave the testes, so they can't be present in semen. So far, there is no firm evidence that vasectomy disrupts hormonal functions or adversely affects sexual activity. Vasectomies can be reversed. But half of those who submit to surgery later develop antibodies against sperm and may not be able to regain fertility.

Females may opt for a *tubal ligation*. By this surgical intervention, oviducts are cauterized or cut and tied off, usually in a hospital. When the operation is performed correctly, tubal ligation is the most effective means of birth control. A few women have recurring pain in the pelvic area. The operation sometimes can be reversed.

EXTREMELY EFFECTIVE

Total abstinence	100%
Tubal ligation or vasectomy	99.6%
Hormonal implant (Norplant)	99%

HIGHLY EFFECTIVE

IUD + slow-release hormones	98%
IUD + spermicide	98%
Depo-Provera injection	96%
IUD alone	95%
High-quality latex condom + spermicide with nonoxynol-9	95%
Oral contraceptive (the Pill)	94%

EFFECTIVE

Cervical cap	89%
Latex condom alone	86%
Diaphragm + spermicide	84%
Billings or Sympto-Thermal Rhythm Method	84%
Vaginal sponge + spermicide	83%
Foam spermicide	82%

MODERATELY EFFECTIVE

Spermicide cream, jelly, suppository	75%
Rhythm method (daily temperature)	74%
Withdrawal	74%
Condom (cheap brand)	70%

UNRELIABLE

Douching	40%
Chance (no method)	10%

Figure 39.30 Comparison of the effectiveness of methods of contraception in the United States. The percentages also indicate the number of unplanned pregnancies per 100 couples who used only that method of birth control for a year. For example, "94% effectiveness" for oral contraceptives (the Pill) means that 6 of every 100 women will still become pregnant, on the average.

Less drastic methods of controlling fertility involve physical or chemical barriers that prevent sperm from entering the uterus and oviducts. *Spermicidal foam* and *spermicidal jelly* are toxic to sperm. They are transferred from an applicator into the vagina before intercourse. These products aren't always reliable unless used with another device, such as a diaphragm and a condom. *IUDs* are coils inserted into the uterus by a physician. They sometimes invite pelvic inflammatory disease.

A *diaphragm* is a flexible, dome-shaped device. It is inserted into the vagina and positioned over the cervix before intercourse. It is relatively effective when fitted initially by a doctor, used with foam or a jelly before each sexual contact, inserted correctly each time, and left in place only for a prescribed length of time.

Good brands of *condoms*—thin, tight-fitting sheaths worn over the penis during intercourse—are up to 95 percent effective if used with a spermicide. Only latex condoms offer protection against sexually transmitted diseases (Section 39.20). However, condoms often tear and leak, at which time they become absolutely useless.

The *birth control pill* is an oral contraceptive made of synthetic estrogens and progestins (progesterone-like hormones). Taken daily except for the last five days of a menstrual cycle, it suppresses oocyte maturation and ovulation. Often it corrects erratic menstrual cycles and reduces cramps, but in some women it causes nausea, weight gain, tissue swelling, and headaches. With more than 50 million women taking it, "the Pill" is the most-used fertility control method. It is 94 percent effective. Earlier formulations were linked to blood clots, high blood pressure, and maybe breast cancer. Lower doses of newer formulations might lessen the risk of breast and endometrial cancer. The possibility of a correlation between the Pill and breast cancer is still under study.

Progestin injections or implants block ovulation. A *Depo-Provera* injection works for three months and is 96 percent effective. *Norplant* (six rods implanted under skin) works for five years and is 99 percent effective. Both may cause sporadic, heavy bleeding, and doctors have some trouble surgically removing Norplant rods.

A pregnancy test won't register positive until after implantation. According to one view, a woman is not pregnant until that time, so *morning-after pills* intercept pregnancy. Actually, such pills work up to seventy-two hours following unprotected sex by interfering with the hormones that control events between ovulation and implantation. One is Preven, a kit with high doses of birth control pills that suppress ovulation and block corpus luteum function. RU-486, more commonly used as an abortion pill, also is used as a morning-after pill. It can block fertilization and progesterone-dependent implantation. It also may induce abortion when taken within seven weeks of implantation (progesterone also is vital to maintain pregnancy). It may trigger elevated blood pressure, blood clots, or breast cancer in some women. RU-486, now known as Mifeprex, is available in Europe. Although controversial, its use in the United States was granted FDA approval in 2000. By contrast, Preven had been used before as a contraceptive, and its repackaging as a morning-after pill has not been very controversial. It gained FDA approval in 1998.

Whether and how human fertility is to be controlled is a volatile issue. It centers on reconciling our biological past with seriously divergent views about our cultural present.

Sexually Transmitted Diseases

At some point in his or her life, at least one of every four individuals in the United States who engage in sexual intercourse will become infected by the pathogens that cause **sexually transmitted diseases**, or **STDs**. Tens of millions are already infected; STDs have reached epidemic proportions. Two-thirds are under age twenty-five; one-fourth are teenagers. Women and children are hardest hit. Worse yet, antibiotic-resistant strains of bacteria are on the rise, and some viral diseases simply cannot be cured.

Urban poverty, prostitution, intravenous drug abuse, and sex-for-drugs are fanning the epidemic. Yet STDs also are rampant in high schools and colleges, where students too often think, "It can't happen to me." They reject the idea that no sex—abstinence—is the only safe sex. In one poll of high school students, two-thirds of the respondents said they don't use condoms. More than 40 percent have two or more sex partners.

The economics of this health problem are staggering. In 1993, the Centers for Disease Control tallied the annual cost of treating the most prevalent STDs: *Herpes*, $759 million; gonorrhea, $1 billion; chlamydial infection, $2.4 billion; and pelvic inflammatory disease, $4.2 billion. This does not include the accelerating cost of treating patients with AIDS. In many developing countries, AIDS alone threatens to overwhelm health-care delivery systems and to unravel decades of economic progress.

The social consequences are sobering. Mothers bestow a chlamydial infection on 1 of every 20 newborns in the United States alone. They bestow type II *Herpes* virus on 1 in 10,000 newborns; one-half of the infected babies die, and one-fourth have severe neural defects. Every year, 1 million women develop pelvic inflammatory disease, and 100,000 to 150,000 become infertile.

AIDS   Someone can become infected by HIV, the human immunodeficiency virus, and not even know it. However, the infection marks the start of a titanic battle that the immune system almost certainly will lose (Section 35.10). At first there may be no outward symptoms. Five to ten years later, a set of chronic disorders develops. They are evidence of *AIDS* (*A*cquired *I*mmune *D*eficiency *S*yndrome). When the immune system finally is weakened, the stage is set for opportunistic infections. Normally harmless resident bacteria are the first to take advantage of the lowered resistance. Dangerous pathogens also take their toll. In time, the immune-compromised person simply is overwhelmed.

Most commonly, HIV spreads through anal, vaginal, and oral intercourse and by IV drug users. Most infections occur by transfer of blood, semen, urine, or vaginal

secretions between people. Cuts and abrasions on the penis, vagina, rectum, and oral membranes invite HIV into the internal environment of a new host.

Today there is no vaccine against HIV and no effective treatment for AIDS. If you get it, you die. *There is no cure*.

HIV was not identified until 1981, but it was present in some parts of Central Africa for at least several decades before that. In the 1970s and early 1980s, it spread to the United States and other developed countries, where most of those initially infected were male homosexuals. Today, many in the heterosexual population are infected or at risk.

Free or low-cost, confidential testing for HIV exposure is available at public health facilities and at many doctor's offices. People who are worried should know there may be a time lag between their first exposure and the first test to come out positive. It takes a few weeks to six months or more before detectable amounts of antibodies will form in response to infection. (The presence of antibodies in the blood indicates exposure to the virus.) Anyone who tests positive is capable of spreading the virus.

Public education programs attempt to stop the spread of HIV. Most health-care workers advocate safe sex, yet there is great confusion about what "safe" means. Many advocate the use of high-quality latex condoms *and* a spermicide that contains nonoxynol-9 to help prevent the transmission of the virus. Even then, there is a slight risk. As an unfortunate couple learned in 1997, no one should participate in open-mouthed kissing with someone who tests positive for HIV. Caressing is not risky *if* there are no lesions or cuts where body fluids that harbor the virus can enter the body. Pronounced lesions caused by other sexually transmitted diseases may increase susceptibility to HIV infection.

In sum, AIDS reached pandemic proportions mainly for three reasons. First, it took a while to discover that the virus can travel in semen, blood, and vaginal fluid and that *behavioral* controls can limit its spread. Second, it took time to develop ways to test symptom-free carriers, who infect others. Third, many still don't realize the medical, economic, and social consequences affect everyone.

Figure 39.31   Bacterial agents of gonorrhea.

GONORRHEA   During intercourse, *Neisseria gonorrhoeae* (Figure 39.31) can enter the body at mucous membranes of the urethra, cervix, and anal canal. This bacterium causes *gonorrhea*. An infected female may notice a slight vaginal discharge or burning sensation while she is urinating. If the infection spreads to her oviducts, it may induce severe cramps, fever, vomiting, and scarring, which may cause sterility. Males have more obvious symptoms. Within a week of infection, the penis discharges yellow pus. Urinating is more frequent and may be painful.

Figure 39.32
The bacterial agents of (**a**) syphilis and (**b**) chlamydia. (**c**) Genital warts on the external genitalia of a young woman.

This STD is rampant even though prompt treatment quickly cures it. Why? Women experience no troubling symptoms during early stages. Also, infection does not confer immunity to the bacterium, perhaps because there are sixteen or more different strains of it. Contrary to common belief, someone can get gonorrhea over and over again. Also, oral contraceptives, used so widely, promote infection by altering vaginal pH. Populations of resident bacteria decline—and *N. gonorrhoeae* is free to move in.

SYPHILIS *Syphilis*, a dangerous STD, is caused by the spirochete *Treponema pallidum* (Figure 39.32*a*). Sex with an infected partner puts the motile, spiral bacterium on the surface of genitals or into the cervix, vagina, or oral cavity. It can enter the body through tiny epidermal cuts. One to eight weeks later, new treponemes are twisting about in a flattened, painless chancre (local ulcer). This first chancre is a symptom of the primary stage of syphilis. By then, treponemes are in the blood. Treponemes can cross the placenta of an infected, pregnant woman. The result will be a miscarriage, stillbirth, or syphilitic newborn.

Chancres usually heal, but treponemes multiply in the spinal cord, brain, eyes, mucous membranes, bones, and joints. More chancres and a skin rash develop during this infectious, secondary stage of syphilis. Immune responses succeed in about 25 percent of the cases, so the symptoms do subside. Another 25 percent remain infected but are symptom free. In the rest, minor to significant lesions and scars appear in the skin and liver, bones, and other internal organs. Few treponemes form during this tertiary stage. But the host's immune system is hypersensitive to them. Chronic immune reactions may severely damage the brain and spinal cord and lead to general paralysis.

Probably because the symptoms are so alarming, more people seek early treatment for syphilis than for gonorrhea. Later stages require prolonged treatment.

PELVIC INFLAMMATORY DISEASE *Pelvic inflammatory disease* (PID) is one of the most serious complications of gonorrhea, chlamydial infections, and other STDs. It also can arise when normal bacterial residents of the vagina ascend to the pelvic region. Most commonly, the uterus, oviducts, and ovaries are affected. There is bleeding and vaginal discharge. Pain in the lower abdomen may be as severe as an acute appendicitis attack. The oviducts may become scarred. Pelvic inflammatory disease invites abnormal pregnancies as well as sterility.

GENITAL HERPES About 25 million Americans have *genital herpes*, caused by the type II *Herpes simplex* virus. Infection requires direct contact with active *Herpes* viruses or sores that contain them. Mucous membranes of the mouth and the genitals are highly susceptible to invasion. The symptoms are often mild or absent. Among infected women, small, painful blisters may appear on the vulva, cervix, urethra, or anal tissues. Among men, blisters form on the penis and anal tissues. Within three weeks, the virus enters latency, and the sores crust over and heal.

The virus is reactivated sporadically. Each time, it causes painful sores at or near the original infection site. Sexual intercourse, menstruation, emotional stress, or other infections trigger *Herpes* flare-ups. Acyclovir, an antiviral drug, decreases healing time and often decreases the pain.

CHLAMYDIAL INFECTION *Chlamydia trachomatis* (Figure 39.32*b*), a parasitic bacterium, spends part of its life cycle in genital and urinary tract cells. It causes several diseases, including *NGU* (chlamydial nongonococcal urethritis). Chlamydial infections are now the most common reported STDs in the United States; 4 million are infected annually. Rates are highest for individuals between ages fifteen and nineteen. *C. trachomatis* and *N. gonorrhoeae* can enter new hosts at the same time by vaginal, oral, or anal intercourse. Prompt doses of penicillin cure the gonorrhea—but not NGU, which requires both tetracycline and sulfonamide.

NGU leads to inflammation of the cervix and, in both sexes, the urethra. Infected people may notice a burning sensation while urinating. When they are symptom-free, they do not seek treatment and complications develop. In males, the prostate becomes swollen and inflamed. In females, the infection may spread into the uterus and the oviducts to cause pelvic inflammatory disease.

Symptom-free infected individuals are unwittingly spreading destructive chlamydial infections through all ethnic groups, among poor and affluent alike.

GENITAL WARTS More than sixty types of the human papillomaviruses (HPV) are known. A few cause *genital warts*, or benign, bumplike growths (Figure 39.32*c*). HPV infection of the genitals and the anus is one of the most prevalent STDs in this country. Usually type 16 HPV does not cause obvious warts, but it may cause precancerous sores and cancers of the cervix, vagina, vulva, penis, and anus. In one Seattle study, 22 percent of female college students who were examined tested positive for the virus.

To Seek or End Pregnancy

Because of sterility or infertility, about 15 percent of all couples in the United States cannot conceive. For example, hormonal imbalances in the female body may prevent ovulation. Or the male's sperm count may be so low that fertilization would be next to impossible without medical intervention.

In vitro fertilization means conception outside the body (literally "in glass" petri dishes or test tubes). It may occur if the couple can produce normal sperm and oocytes. First the woman receives injections of a hormone that prepares her ovaries for ovulation. Afterward, the preovulatory oocyte is removed from her with a suction device. In the meantime, sperm from the male are placed in a solution that simulates fluid in the oviducts. A few hours after the sperm and suctioned oocyte make contact in a petri dish, fertilization may occur. Twelve hours later, the zygote is transferred to a solution that can sustain it through the initial cleavages. Two to four days later, the resulting ball of cells is transferred to the female's uterus.

Each attempt at in vitro fertilization costs about 8,000 dollars, and most attempts aren't successful. In 1994, each "test-tube" baby cost the nation's health-care system about 60,000 to 100,000 dollars, on average. The childless couple may believe no cost is too great. But in an era of increased population growth and shrinking medical coverage, is the cost too great for society to bear? Court battles are being waged over this issue.

At the other extreme is **abortion**, the dislodging and removal of the blastocyst, embryo, or fetus from the uterus. At one time in the United States, abortions were forbidden by law unless the pregnancy endangered the mother's life. Later the Supreme Court ruled that the government does not have the right to forbid abortions during the early stages of pregnancy (typically up to five months). Before this ruling, there were dangerous, traumatic, and often fatal attempts to abort embryos, either by pregnant women themselves or by quacks.

From a clinical standpoint, vacuum suctioning and other methods make abortion painless for the woman, rapid, and free of complications when performed during the first trimester. Abortions performed in the second and third trimesters are extremely controversial unless the mother's life is threatened. Even so, for both medical and humanitarian reasons, the majority of people in this country generally agree that the preferred route to birth control is not through abortion. Rather, it is through sexually responsible behavior that prevents unwanted pregnancy from happening in the first place.

This biology textbook cannot offer you the "right" answer to a moral question because of the reasons given in Section 1.7. It *can* provide you with a serious, detailed description of how a new human individual develops. Your choice of the "right" answer to the question of the morality of abortion will be just that—your choice—and one that can be based on objective insights into the nature of life.

SUMMARY   Gold indicates text section

Principles of Reproduction and Development

1. For animals, sexual reproduction is the dominant reproductive mode. It requires specialized reproductive structures, control mechanisms, and forms of behavior that assist fertilization and support the offspring. *39.1*

2. Embryonic development typically proceeds through six stages: *39.2–39.3*

   a. Gamete formation, when oocytes (immature eggs) and sperm form in reproductive organs. Molecular and structural components are stockpiled and are localized in different parts of the oocyte. *39.2*

   b. Fertilization, from the time a sperm penetrates an egg to the fusion of sperm and egg nuclei that results in a zygote (fertilized egg). *39.2*

   c. Cleavage, when mitotic cell divisions transform a zygote into smaller cells (blastomeres). Cleavage does not increase the original volume of egg cytoplasm; it only increases the number of cells. It regionally divides the yolk, mRNAs, proteins, cytoskeletal elements, and other "maternal messages" among new blastomeres, an outcome called cytoplasmic localization. *39.2*

   d. Gastrulation, when primary tissue layers (germ layers) form. Cell divisions, cell migrations, and other events lead to the formation of endoderm, ectoderm, and (in most species) mesoderm. All tissues of the adult body develop from primary tissue layers. *39.3*

   e. The onset of organ formation. Different organs start developing by a tightly orchestrated program of cell differentiation and morphogenesis. *39.2*

   f. Growth and tissue specialization, when organs enlarge overall and acquire their specialized chemical and physical properties. The maturation of tissues and organs continues into post-embryonic stages. *39.2*

3. Embryos cannot develop properly unless each stage of development is successfully completed before the next stage begins. *39.2*

4. In cell differentiation, a cell selectively uses certain genes and synthesizes proteins not found in other cell types. The outcome is subpopulations of specialized lineages of cells that differ from one another in their structure, biochemistry, and functioning. *39.4*

5. Morphogenesis starts at gastrulation. It changes the embryo's size, shape, and proportions. *39.4*

6. Pattern formation means the emergence of the basic body plan and body parts in specific regions of the embryo, in orderly sequence. Cytoplasmic localization helps seal the fate of cells. Cell determination depends also on interactions among classes of master genes that specify certain products. These products are spatially organized and create chemical gradients. The gradients influence differentiation, hence the identity of each cell lineage in the embryo. *39.5*

Human Reproduction and Development

1. Humans have a pair of primary reproductive organs (sperm-producing testes in males, and egg-producing ovaries in females), accessory ducts, and glands. Testes and ovaries also produce sex hormones that govern reproductive functions and secondary traits. *39.6–39.8*

   a. The hormones LH, FSH, and testosterone control sperm formation. They are part of feedback loops from the testes to the hypothalamus and anterior lobe of the pituitary gland. *39.7*

   b. The hormones estrogen, progesterone, FSH, and LH control the maturation and release of oocytes from the ovary, as well as cyclic changes in the endometrium (the inner lining of the uterus). Their secretion is part of feedback loops from ovaries to the hypothalamus and to the anterior lobe of the pituitary gland. *39.8–39.10*

2. A menstrual cycle is a recurring cycle of intermittent fertility during the reproductive years of female humans and other primates. *39.9, 39.10*

   a. Follicular phase: One of many follicles matures inside an ovary. Each follicle is an oocyte and the cell layer surrounding it. Meanwhile, the endometrium is prepared for a possible pregnancy. It breaks down each cycle if pregnancy does not occur.

   b. Ovulation: A midcycle surge of the blood level of LH triggers ovulation, the release of a secondary oocyte from the ovary.

   c. Luteal phase: After ovulation the corpus luteum, a glandular structure, develops from the remnants of the follicle. It secretes progesterone and some estrogen that prime the endometrium for fertilization. If fertilization occurs, the corpus luteum is maintained.

3. After fertilization, a blastocyst forms by cleavage and becomes implanted in the endometrium. Three primary tissue layers form that give rise to all organs. They are ectoderm, endoderm, and mesoderm. *39.11, 39.12*

4. Four extraembryonic membranes form as embryos of humans and other vertebrates develop: *39.12, 39.13*

   a. The amnion becomes a fluid-filled sac around the embryo, which it protects from drying out, mechanical shock, and abrupt temperature changes.

   b. The yolk sac stores nutritive yolk in most shelled eggs. In humans, part of the sac becomes a major site of blood formation, and some of its cells give rise to germ cells. (Germ cells later give rise to sperm and eggs.)

   c. The chorion is a protective membrane enclosing the embryo and the other extraembryonic membranes. It also becomes a major component of the placenta.

   d. In humans, the blood vessels for the placenta arise from the allantois, as does the urinary bladder.

5. The placenta, a blood-engorged organ, is composed of endometrium and extraembryonic membranes. The placenta permits embryonic blood vessels to develop independently of the mother's but also allows oxygen, nutrients, and wastes to diffuse between them. *39.14*

Review Questions

1. What is the main benefit of sexual reproduction? What are some of its biological costs? *39.1*

2. Briefly state the key events of gamete formation, fertilization, then cleavage, gastrulation, and organ formation. At which stage is the frog embryo in the photograph at right? *39.2*

3. Does cleavage increase the volume of cytoplasm, the number of cells, or both, compared to the zygote? *39.2*

4. Give a brief example of cytoplasmic localization. *39.3*

5. Define cell differentiation and morphogenesis. *39.4*

6. Define pattern formation. Then explain the key points of the theory of pattern formation. *39.5*

7. Label the components of the human male reproductive system and state their functions: *39.6, 39.7*

8. Label the components of the human female reproductive system and state their functions: *39.8, 39.9*

9. Does sperm formation require mitosis, meiosis, or both? *39.7*

10. Study Figure 39.15 on sperm formation. Then sketch the feedback loops to the hypothalamus and anterior pituitary from the testes. Include the names of releasing hormone, hormones, and cells in the testes involved in these loops. *39.7*

11. Study Figure 39.18. Then, sketch the feedback loops to the hypothalamus and the anterior pituitary from the ovaries. These loops govern the menstrual cycle. Name the releasing hormone, hormones, and ovarian structures involved in the loops. *39.9*

12. State the embryonic source of the amnion, yolk sac, chorion, and allantois. State the role each extraembryonic membrane plays in the structure or functioning of an embryo. *39.12*

Self-Quiz ANSWERS IN APPENDIX III

1. During cleavage, the new blastomeres are allocated different regions of the fertilized egg's cytoplasm. This is called _____ .
   a. cytoplasmic localization   c. cell differentiation
   b. embryonic induction   d. morphogenesis

2. Primary tissue layers first appear _____ .
   a. in the egg cortex   c. in the gastrula
   b. during cleavage   d. in primary organs

3. In organisms ranging from fruit flies to humans, the body's organization relative to its long axis depends on _____ .
   a. morphogens   c. embryonic induction
   b. homeotic genes   d. all of the above

4. The _____ , a fluid-filled sac, surrounds and protects the embryo from mechanical shocks and keeps it from drying out.
   a. yolk sac   b. allantois   c. amnion   d. chorion

5. At full term, a placenta _____ .
   a. is composed of extraembryonic membranes
   b. directly connects maternal and fetal blood vessels
   c. keeps maternal and fetal blood vessels separated

6. Match the development stage with its description.
   _____ cleavage
   _____ gamete formation
   _____ organ formation
   _____ growth, tissue specialization
   _____ gastrulation
   _____ fertilization

   a. egg and sperm mature in parents
   b. sperm nucleus, egg nucleus fuse
   c. formation of primary tissue layers
   d. cytoplasmic localization, not gene interactions, dominate in most animals (not in mammals)
   e. organs, tissues increase in size, acquire specialized properties
   f. starts when primary tissue layers split into subpopulations of cells

---

Critical Thinking

1. In normal *Drosophila* larvae, a cluster of cells is the embryonic source of antennae on the head. When a larva has a certain mutant gene, it develops legs instead of antennae on its head, as in Figure 39.33. Would you say that the mutated gene is first expressed during cleavage or organ formation?

2. Before an amphibian egg enters cleavage, you divide it so only one daughter cell gets the gray crescent. Separate the two cells and only the one with the gray crescent forms an embryo with a long axis, notochord, nerve cord, and back muscles. The other forms a shapeless mass of immature gut cells and blood cells. Does cytoplasmic localization

Figure 39.33 Mutant fruit fly with two legs (not two antennae) on its head.

or embryonic induction play a greater role in these outcomes? Explain your answer.

3. Despite obvious differences in early stages of development, chick and human embryos resemble one another once early organ formation is under way. Using the theory of pattern formation, speculate on the role of embryonic induction in bringing about the similarities between these vertebrate lineages.

4. Infection by the rubella virus apparently has an inhibitory effect on mitosis. Serious birth defects result when a woman is infected during the first trimester of pregnancy, but not later. Review the developmental events that unfold during pregnancy and explain why this might be so.

5. Suppose you are an obstetrician advising a woman who just learned she is pregnant. What instructions would you provide concerning her diet and behavior during pregnancy?

6. In the United States, teenage pregnancies and STD infections are rampant. Suppose the office of the Surgeon General asks you to join a task force that will recommend practices to reduce the incidence of both. What practices might be most successful? What kind of enthusiasm or resistance might they provoke among the teenagers in your community? Among adults? Explain why.

7. By the traditional view, separation of the sexes has selective advantages. By a highly provocative, alternative view, males and females have been waging a "gender war," at the level of genes, for billions of years. Matt Ridley's *Genome* (1999, New York: HarperCollins) may challenge your thinking on this topic.

Selected Key Terms

Principles

asexual reproduction *39.1*
blastocyst *39.3*
blastomere *39.2*
blastula *39.3*
cell differentiation *39.4*
cleavage *39.2*
cytoplasmic localization *39.3*
ectoderm *39.2*
embryo *39.2*
embryonic induction *39.5*
endoderm *39.2*
fertilization *39.2*
gamete formation *39.2*
gastrulation *39.2*
growth, tissue specialization *39.2*
master gene *39.5*
mesoderm *39.2*
morphogen *39.5*
morphogenesis *39.4*
neural tube *39.4*
oocyte *39.3*
organ formation *39.2*
pattern formation *39.5*
sexual reproduction *39.1*
sperm *39.3*
yolk *39.1*
zygote *CI*

Human Case Study

abortion *39.21*
aging *39.18*
allantois *39.12*
amnion *39.12*
chorion *39.12*
corpus luteum *39.9*
endometrium *39.8*
estrogen *39.8*
fetus *39.12*
FSH *39.7*
implantation *39.12*
in vitro fertilization *39.21*
LH *39.7*
menstrual cycle *39.8*
ovulation *39.8*
ovum *39.11*
placenta *39.14*
polar body *39.9*
progesterone *39.8*
puberty *39.6*
seminiferous tubule *39.6*
somite *39.13*
STD *39.20*
testis (testes) *39.6*
testosterone *39.7*
uterus *39.8*
yolk sac *39.12*

Readings

Nusslein-Volhard, C. August 1996. "Gradients That Organize Embryonic Development." *Scientific American*, 54–61.

On-Line readings at Student Guide for InfoTrac:
www.brookscole.com/biology

been a symbol of the good life, drawing people from all over the world. But its schools are overcrowded, funds for welfare and other social services are shrinking, and sewage treatment plants are near capacity. Shortages of water are frequent and often severe. Statewide electric power shortages and rolling blackouts started in 2001. The shortages, together with ongoing salinization of cropland and air pollution, are eroding the agricultural and technological bases of this state. At this writing, the economic repercussions of California's slowdown are affecting individuals throughout the nation.

Claiming that population growth affects economic health, many governments restrict immigration. Only the United States, Canada, Australia, and a few other countries are now accepting large annual increases. Elsewhere, appalling living conditions, extremely harsh governmental policies, and civil strife have combined to promote the exodus of more people than better-off nations can easily absorb.

A Question of Resource Consumption

This chapter opened with a brief look at the conditions endured by most people in India, with its whopping 16 percent of the human population. By comparison, the United States has merely 4.7 percent. Yet which country is the most "overpopulated"—not in terms of actual numbers, but rather in terms of resource consumption and instigation of environmental damage?

The highly industrialized United States produces 21 percent of the world's goods and services. However, its people consume fifty times more goods and services than the average person in India. Also, its people use 25 percent of the world's processed minerals, and as much of the available, nonrenewable sources of energy. They generate at least 25 percent of all pollution and trash. By contrast, India produces about 1 percent of all goods and services. It uses 3 percent of the available minerals and nonrenewable energy resources. It only generates about 3 percent of the pollution and trash.

Extrapolating from these numbers, G. Tyler Miller estimated that it would take 12.9 billion impoverished individuals living in India to have as much impact on the environment as 258 million people in the United States. Think about this estimate when you also reflect on the biodiversity crisis and conservation programs, as sketched out in Chapter 25.

Differences in population growth among countries correlate with levels of economic development, hence with economic security (or lack of it) of individuals. Growth rates are low in preindustrial, industrial, and post-industrial stages and are greatest during the transition to industrialization.

SOCIAL IMPACT OF NO GROWTH

For us, as for all species, the biological implications of extremely rapid growth are staggering. Yet so are the social implications of what will happen when (and if) the human population decreases to the point of zero population growth—and stays there.

For instance, in a growing population, most people are in younger age brackets. If living conditions ensure constant growth over time, the age distribution should guarantee availability of a future work force. This has social implications. Why? *It takes a large work force to support the older age brackets.* In the United States, older, nonproductive people expect the federal government to provide them with subsidized medical care, low-cost housing, and many other social programs. However, as a result of improved medicine and hygiene, people in these brackets are living far longer than the elderly did when the nation's social security program was set up. The current cash benefits exceed the contributions that they made to the program when *they* were younger.

If the population ever does reach and maintain zero growth, a larger proportion of individuals will end up in the older age brackets. Even slower growth will pose problems. What happens when baby-boomers retire? Will all those nonproductive members continue to get goods and services if productive ones must carry more and more of the economic burden? This isn't an abstract question. Put it to yourself. Just how much economic hardship are you willing to bear for the sake of your parents? For your grandparents? And how much will your children be willing or able to bear for you?

We have arrived at a major turning point, not only in our biological evolution but in our cultural evolution as well. The decisions awaiting us are among the most difficult we will ever have to make—yet it is clear that they must be made, and soon.

All species face limits to growth. We might think we are different from the rest, for our special ability to undergo rapid cultural evolution has enabled us to postpone the action of most of the factors that limit growth. But the crucial word is *postpone*. No amount of cultural intervention can hold back the ultimate check of limited resources and a damaged environment.

We have sidestepped a number of the smaller laws of nature. In doing so we have become more vulnerable to those laws which cannot be repealed. Today, there may be only two options available. Either we make a global effort to limit population growth in accordance with environmental carrying capacity, or we wait until the environment does it for us.

In the final analysis, no amount of cultural intervention can repeal the ultimate laws governing population growth as imposed by the carrying capacity of the environment.

SUMMARY

1. A population is a group of individuals of the same species occupying a given area. It has a characteristic size, density, distribution, and age structure as well as characteristic ranges of heritable traits. *40.1*

2. The growth rate for a population during a specified interval may be determined by calculating the rates of birth, death, immigration, and emigration. To simplify calculations, we may put aside effects of immigration and emigration, and combine the birth and death rates into a variable *r* (net reproduction per individual per unit time). Then we may represent population growth (*G*) as $G = rN$, where *N* is the number of individuals during the interval specified. *40.3*

    a. In cases of exponential growth, the reproductive base of a population increases and its size expands by ever increasing increments during successive intervals. This trend plots out as a J-shaped growth curve.

    b. Any population will show exponential growth as long as its per capita birth rate stays even slightly above its per capita death rate.

    c. In logistic growth, a low-density population slowly increases in size, goes through a rapid growth phase, then levels off in size once carrying capacity is reached.

3. Carrying capacity is the name ecologists give to the maximum number of individuals in a population that can be sustained indefinitely by the resources available in their environment. *40.4*

4. The availability of sustainable resources as well as other factors that limit growth dictates population size during a specified interval. The limiting factors vary in their relative effects and vary over time, so population size also changes over time. *40.4*

5. Limiting factors such as competition for resources, disease, and predation are density-dependent. Density-independent factors, such as weather on the rampage, tend to raise the death rate or lower the birth rate more or less independently of population density. *40.4*

6. Patterns of reproduction, death, and migration vary over the life span of a species. Environmental variables help shape the life history (age-specific) patterns. *40.5*

7. The human population has now surpassed 6 billion. Its annual growth rate is below zero in a few developed countries and above 3 percent in some less developed countries. In 2000 the annual growth rate for the global human population was 1.26 percent. *40.7–40.8*

8. The human population's rapid growth in the past two centuries occurred through expansion into many diverse habitats and through agricultural, medical, and technological developments that raised the carrying capacity. Ultimately, we must confront the reality of the carrying capacity and limits to growth. *40.7–4.10*

Review Questions

1. Define population size, density, and distribution. Describe a typical population's age-structure categories. *40.1*

2. Define exponential growth. Be sure to state what goes on in the age category that is a foundation for its occurrence. *40.3*

3. Define carrying capacity. Describe its effect as evidenced by a logistic growth pattern. *40.4*

4. Give examples of the limiting factors that come into play when a population of mammals (for example, rabbits, deer, or humans) reaches very high density. *40.4, 40.5*

5. Define doubling time. In what year is the human population expected to surpass 9 billion? *40.3, 40.7*

6. How did earlier human populations expand steadily into new environments? How did they increase the carrying capacity in their habitats? Have we now avoided some limiting factors on population growth? Or is the avoidance an illusion? *40.7*

Self-Quiz   ANSWERS IN APPENDIX III

1. _____ is the study of how organisms interact with one another and with their physical and chemical environment.

2. A _____ is a group of individuals of the same species that occupy a certain area.

3. The rate at which a population grows or declines depends upon the rate of _____ .
    a. births   c. immigration   e. all of the above
    b. deaths   d. emigration

4. Populations grow exponentially when _____ .
    a. population size expands by ever increasing increments through successive time intervals
    b. size of low-density population increases slowly, then fast, then levels off once carrying capacity is reached
    c. Both a and b are characteristic of exponential growth.

5. For a given species, the maximum rate of increase per individual under ideal conditions is the _____ .
    a. biotic potential   c. environmental resistance
    b. carrying capacity   d. density control

6. Resource competition, disease, and predation are _____ controls on population growth rates.
    a. density-independent   c. age-specific
    b. population-sustaining   d. density-dependent

7. Which of the following factors does *not* affect sustainable population size?
    a. predation   c. resources   e. All of the above can
    b. competition   d. pollution   affect population size.

8. In 2000, the average annual growth rate for the human population at midyear was _____ percent.
    a. 0   b. 1.05   c. 1.26   d. 1.55   e. 2.7   f. 4.0

9. Match each term with its most suitable description.
    ____ carrying   a. maximum rate of increase per
         capacity   individual under ideal conditions
    ____ exponential   b. population growth plots out
         growth   as an S-shaped curve
    ____ biotic   c. the maximum number of
         potential   individuals sustainable by
    ____ limiting factor   an environment's resources
    ____ logistic growth   d. population growth plots out
                              as a J-shaped curve
                            e. short supply of any resource
                              essential to population growth

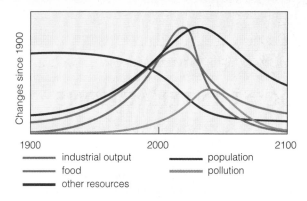

Figure 40.15 Computer-based projection of what may happen if human population size continues to skyrocket without dramatic policy changes and technological innovation. The assumptions used are that the population has already overshot the carrying capacity and current trends will continue unchanged.

Critical Thinking

1. If house cats that have not been neutered or spayed live up to their biotic potential, two can be the start of many kittens—12 the first year, 72 the second year, 429 the third, 2,574 the fourth, 15,416 the fifth, 92,332 the sixth, 553,019 the seventh, 3,312,280 the eighth, and 19,838,741 kittens the ninth year. Is this a case of logistic growth? Exponential growth? Irresponsible cat owners?

2. A third of the world population is below age fifteen. Describe the effect of this age distribution on the future growth rate of the human population. If you conclude that it will have severe impact, what sorts of humane recommendations would you make to encourage individuals of this age group to limit family size? What are some social, economic, and environmental factors that might keep them from following the recommendations?

3. Write a short essay about a population having one of the age structures shown below. Describe what may happen to younger and older groups when individuals move into new categories.

4. Figure 40.14 charts the legal immigration to the United States between 1820 and 1997. (The Immigration Reform and Control Act of 1986 accounted for the most recent dramatic increase; it granted legal status to undocumented immigrants who could prove they had lived in the country for years.) During the 1980s and 1990s, an economic downturn fanned resentment against newcomers. Many people now say legal immigration should be restricted to 300,000–450,000 annually and we should crack down on undocumented immigrants. Others argue that such a policy would diminish our reputation as a land of opportunity. They also say it would discriminate against legal immigrants during crackdowns on others of the same ethnic background. Do some research, then write an essay on the pros and cons of

both positions. (You may wish to start with web site http://www.census.gov.)

5. In his book *Environmental Science*, G. Tyler Miller points out that the projected increase in the human population to over 9 billion by the year 2050 raises serious questions. Will there be enough food, energy, water, and other resources to sustain that many people? Will governments be able to provide adequate education, housing, medical care, and other social services for all of them? Computer models suggest that the answers are no (Figure 40.15). Yet some people claim we can adapt politically and socially to a far more crowded world, assuming harvests improve through technological innovation, every inch of arable land is cultivated, and everybody eats only grain.

There are no easy answers to these questions. If you have not yet been doing so, start following the arguments in your local newspapers, in magazines, on television, and on the Web. This will allow you to become an informed participant in a global debate that surely will have impact on your future.

Selected Key Terms

age structure *40.1*	life history pattern *40.5*
biotic potential *40.3*	limiting factor *40.4*
capture–recapture method *40.2*	logistic growth *40.4*
carrying capacity *40.4*	migration *40.3*
cohort *40.5*	per capita *40.3*
demographic transition model *40.9*	population density *40.1*
demographics *40.1*	population distribution *40.1*
density-dependent control *40.4*	population size *40.1*
density-independent factor *40.4*	*r* (net reproduction per
doubling time *40.3*	individual per unit time) *40.3*
ecology *CI*	reproductive base *40.1*
emigration *40.3*	survivorship curve *40.5*
exponential growth *40.3*	total fertility rate *40.8*
immigration *40.3*	zero population growth *40.3*

Figure 40.14 Chart of legal immigration to the United States between 1820 and 1997.

Readings

Bloom, D. and D. Canning. 18 February 2000. "The Health and Wealth of Nations." *Science* 287:1207–1209. Argues that poor health is more than a consequence of low income; that it is also one of its fundamental causes.

Cohen, J. E. 1995. *How Many People Can the Earth Support?* New York: Norton. No pat answer to title's question.

Miller, G. T. 2002. *Living in the Environment.* Twelfth edition. Belmont, California: Brooks/Cole.

41

SOCIAL INTERACTIONS

Deck the Nest With Sprigs of Green Stuff

In 1890, bird fanciers imported more than a hundred starlings (*Sturnus vulgaris*) from Europe and released them in New York City's Central Park. In less than a century, descendants of the introduced species had expanded their geographic range from coast to coast. They are so good at gleaning food from the agricultural fields of North America that they have outmultiplied native birds. They also have evicted great numbers of them from scarce nesting sites in the cavities of trees.

Once a male and female commandeer a tree cavity, they gather dry grass and twigs and use them to build or rebuild a nest (Figure 41.1). Theirs is no ordinary nest. They *decorate* the nest bowl with sprigs, freshly plucked.

Why do starlings do this? Do the sprigs of greenery camouflage the nest from predators? Not likely. The nests are already concealed in tree cavities. Do sprigs serve as insulative material that helps keep forthcoming eggs warm? Actually, the still-moist, green plant parts would promote heat loss, not heat conservation. Well, do the green sprigs combat parasites? Think of how mites parasitize birds and infest nest cavities. In short order, even a few tiny mites can produce thousands of descendants. In large numbers, mites can suck enough blood from a nestling to weaken it. They interfere with the nestling's growth, development, and survival.

Larry Clark and Russell Mason decided to test the third hypothesis. As they knew, starlings don't weave just any green plant material into the nests. Starlings choosily favor the leaves of certain plants, such as wild carrot (Figure 41.1). So the two biologists built a set of experimental nests, some with freshly cut wild carrot leaves and some without. They removed the natural nests that pairs of starlings had constructed and had already started using. Half of the nesting pairs got replacement nests decorated with sprigs of wild carrot.

Replacement nests for the other pairs of starlings were sprigless.

Figure 41.2 shows the results. The number of mites in greenery-free nests was consistently greater than the number in nests decorated with greenery. At the end of one experiment, the sprig-free nests teemed with an average of 750,000 mites. Nests with sprigs contained a mere 8,000 mites.

Shoots of wild carrot happen to contain a highly aromatic steroid compound. The compound almost certainly repels herbivores and helps plants survive. By coincidence, the compound also prevents mites from maturing sexually—and so prevents mite population explosions in nests festooned with wild carrot.

Figure 41.1 A most excellent fumigator in nature—a European starling (*Sturnus vulgaris*). Starlings combat infestations of mites by decorating previously owned nests with fresh sprigs of wild carrot (*Daucus carota*), shown at left.

Figure 41.2 (a) Experimental results that point to the adaptive value of starling nest-decorating behavior. Nests designated *A* were kept free of fresh sprigs of wild carrot and other plants that contain aromatic compounds. The compounds prevent baby mites from developing into (b) adult mites. Other nests, designated *B*, had fresh sprigs tucked in every seven days. Head counts of mites (*Ornithonyssus sylviarum*) infesting the nests were made at the time the starling chicks left the nest, twenty-one days after this experiment was under way.

Far from being a trivial behavior, then, "decorating" a nest with aromatic greenery has adaptive value. It fumigates the nest and thus increases an individual's chance of producing healthy, surviving offspring.

And so starlings lead us into the fascinating world of research into behavior. As you will see, some of the studies focus on the adaptive value of behavioral traits to an individual's reproductive success. Others focus on the internal mechanisms that allow individuals to behave as they do.

As a starting point, recall that information encoded in genes directs the formation of tissues and organs that make up the animal body, including those of its nervous system. That system detects, processes, and integrates information about stimuli in the internal and external environments, then commands muscles and glands (also built according to genetic instructions) to make suitable responses. Other gene products called hormones contribute to the responses. Because genes specify all of the substances required for the structure and function of nervous and endocrine systems, *they are the heritable foundation for responses that animals make to stimuli.* We call these observable, coordinated responses to stimuli **animal behavior**.

Key Concepts

1. Animals make coordinated responses to stimuli. We call these responses animal behavior. The cooperative, interdependent relationships among individuals of a species are forms of social behavior.

2. Genes underlie the behavioral ability of individuals, as when instructions encoded in certain genes govern the development of the nervous and endocrine systems. These organ systems allow an animal to detect, process, and issue commands for behavioral responses to stimuli.

3. Most newly born or hatched animals have the neural wiring and motor systems necessary to perform some behaviors without having to learn them through actual experience. Sign stimuli, which are simple, well-defined environmental cues, trigger such instinctive behaviors.

4. For nearly all animal species, the nervous system has a capacity to process and retain information about specific experiences in the environment, then to use the information to vary or change behavioral responses. The outcomes are called learned behaviors.

5. Like other traits having a genetic basis, behavior has evolved by way of natural selection, which occurs when genetically different individuals in a population have different numbers of surviving offspring. Behavioral mechanisms that enhance the ability of individuals to pass on their genes to offspring have been favored.

6. Evolved modes of communication underlie social behavior. Communication signals, which are encoded in stimuli such as species-specific odors, body coloration and patterning, postures, and movements, hold clear meaning for both the sender and the receiver of signals.

7. Living in a social group has costs and benefits, as measured by an individual's ability to pass on its genes to offspring. Not every environment favors the evolution of such groups. In some cases, a solitary life-style allows an individual to leave more descendants than belonging to a large group would allow.

8. In altruism, an individual helps others in ways that require it to sacrifice its own reproductive success. The evolution of altruism requires special circumstances in which individuals can propagate their genes indirectly, by helping their relatives reproduce successfully.

BEHAVIOR'S HERITABLE BASIS

Genes and Behavior

An animal's nervous system, recall, is wired to detect, interpret, and issue commands for response to stimuli— that is, to specific aspects of the external and internal environments. One way or another, certain products of genes control each of the steps necessary to assemble and operate that system's receptors, nerve pathways, and brain. If we define animal behavior as coordinated responses to stimuli, then behavior starts with genes.

Stevan Arnold found evidence of the genetic basis of behavior by studying feeding preferences of coastal and inland populations of a snake species in California. Garter snakes near the coast prefer banana slugs (Figure 41.3*a*). Snakes living inland prefer tadpoles and small fishes. Offer them a banana slug and they ignore it.

In one set of experiments, Arnold offered captive newborn garter snakes a chunk of slug as the first meal. Coastal snake offspring usually ate the chunk and even flicked their tongue at cotton swabs drenched in essence of slug. (Snakes "smell" by tongue-flicking, which pulls chemical odors into the mouth.) However, inland snake offspring ignored the swabs and rarely ate slug meat. Here was an obvious difference between captive baby snakes that had no prior, direct experience with slugs. Arnold concluded the snakes were programmed before birth to accept or reject slugs; they certainly did not learn feeding preferences by taste trials. Maybe coastal and inland populations have allelic differences for the genes affecting how odor-detecting mechanisms form when a garter snake embryo is developing.

To test his hypothesis, Arnold crossbred coastal and inland snakes. If genes underlie the difference in food preferences, he figured that the hybrid offspring from

Figure 41.3 (**a**) Banana slug, food for (**b**) an adult garter snake of coastal California. (**c**) A newborn garter snake from a coastal population, tongue-flicking at a cotton swab drenched with tissue fluids from a banana slug.

snakes of both populations might make an *intermediate* response to slug chunks and odors. Results matched his prediction. Compared with a typical newborn inland snake, many baby snakes of mixed parentage tongue-flicked more often at the slug-scented cotton swabs— but less often than newborn coastal snakes did.

Hormones and Behavior

Gene products called hormones also guide behavior, including birdsong. As an example, each spring a male zebra finch (Figure 41.4) repeats a song with splendid consistency and clarity. In part, his singing arises from seasonal differences in how much melatonin is secreted from his pineal gland (Section 32.7). A high blood level of melatonin suppresses the growth and functions of a songbird's gonads. Suppression reverses when pineal gland photoreceptors absorb sunlight energy and issue signals that cause a slowdown in melatonin secretion.

Winter has fewer daylight hours than spring; there isn't enough light to slow melatonin secretion. Come

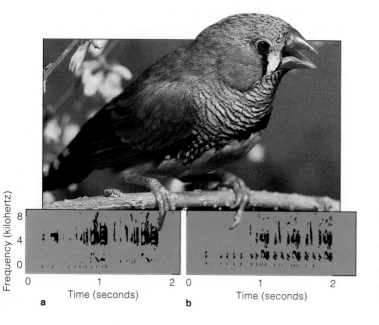

Figure 41.4 Hormones and the territorial song of zebra finches. (**a**) Sound spectrogram of the male's song. The spectrogram is a visual record of each note's pitch (that is, its frequency, measured in kilohertz). (**b**) Sound spectrogram of a female zebra finch that was experimentally converted to a singer. When she was a nestling, she received extra estrogen. At adulthood, she received an implant of testosterone—and started singing. Estrogen normally organizes the song system (certain parts of the brain) in developing songbird embryos. Testosterone activates the system in adult males.

Figure 41.5 Examples of instinctive responses that human offspring make to sign stimuli. (**a**) A close-up face of an adult triggers smiling behavior in young infants. (**b**) Somewhat older infants instinctively imitate adult facial expressions.

spring, daylength increases. Hormonal suppression is lifted, so the gonads increase in size and secrete more estrogen and testosterone. These sex hormones are the start of gender-related differences in singing behavior. As most songbird embryos develop, estrogen controls the formation of their **song system**, which consists of several brain structures that will control the muscles of a vocal organ. Male and female songbirds differ in the size and structural components of this system.

Unlike humans, females of songbird species are XY and males are XX. Certain Y chromosome genes specify products that block estrogen production. Before a *male* bird hatches, its gonads secrete estrogen which, at high levels, stimulates development of a masculinized brain. The estrogen gets converted to testosterone. The gonads enlarge and secrete more testosterone as the breeding season starts. By acting on cells in the song system, this hormone induces metabolic changes that will prepare the male to sing when suitably stimulated.

Instinctive Behavior Defined

With their tongue-flicking and strikes at prey, newborn garter snakes can start us thinking about **instinctive behavior**. This term means the behavior is performed without having been learned through actual experience in the environment. Instead, the nervous system of a newly born or hatched animal is prewired to recognize **sign stimuli**—one or two simple, well-defined cues in the environment that can trigger a suitable response. The snake's response is a stereotyped motor program. When the snake recognizes certain sign stimuli, a **fixed action pattern** follows. It is a program of coordinated muscle activity that runs to completion independently of feedback from the environment.

Another example: When human infants are two or three weeks old, they tend to smile instinctively when an adult's face comes close to their own (Figure 41.5a). Infants make the same response to an overly simplified stimulus—a flat, face-sized mask with two dark spots

Figure 41.6 Example of a social parasite. European cuckoos lay eggs in nests of other species. In response to an environmental cue, this cuckoo hatchling is executing a behavior that is innate. It has inherited the knowledge of what to do without having to learn it. Even before its eyes open, it responds to the spherical shape of the host's eggs and shoves them out of the nest. The foster parents keep on feeding the usurper.

where eyes would be on a human face. A mask with one "eye" won't do the trick. As Figure 41.5b suggests, older infants continue to show instinctive behavior.

And remember the cuckoo, a social parasite? Adult females lay their eggs in nests of other species. If newly hatched cuckoos eliminate the natural-born offspring, they will receive the undivided attention of their foster parents (Section 42.4). They are blind at birth, but when they contact an egg or another round object, they carry out a fixed action pattern. They maneuver the egg onto their back, then push it from the nest (Figure 41.6).

Animals clearly have a genetically based capacity to respond automatically to certain environmental cues. But they also have the means to process information about specific experiences, then use the information to vary or change their responses. As you will see next, the outcome is what we call learned behavior.

Genes underlie animal behavior—coordinated responses to stimuli. Certain gene products are essential in constructing and operating the nervous system, which governs behavior. Other gene products called hormones also influence the mechanisms required for particular forms of behavior.

Animals start out life neurally wired to recognize important cues and to make an instinctively suitable response, one that has not been learned through actual experience.

LEARNED BEHAVIOR

Animals process and integrate information gained from experiences, then use it to vary or change responses to stimuli. We call this **learned behavior**. For example, a young toad's nervous system commands it to flip its sticky tongue instinctively at any dark object crossing its field of vision. In the toad world, such objects usually are edible insects. What if a dark object is a bumblebee that stings the toad tongue? Thereafter, that toad will avoid all "bumblebee-sized black-and-yellow-banded objects that sting." **Imprinting**, another example, is a time-dependent form of learning triggered by exposure to sign stimuli, most often at a sensitive period when

Figure 41.7 No one can tell these imprinted baby geese that Konrad Lorenz is not Mother Goose. (Refer to Table 41.1.)

Table 41.1 *A Few Categories of Learned Behavior*

IMPRINTING This time-dependent form of learning involves exposure to sign stimuli, most often early in development. For instance, in response to a moving object and probably to certain sounds, baby geese imprint on the mother and follow her during a short, sensitive period after hatching. They are neurally wired to learn crucial information—the identity of the individual that will protect them in the months ahead. Usually that individual is the mother or father. Imprinting occurs among many animals.

Konrad Lorenz, an early ethologist, must have presented the baby geese in Figure 41.7 with sign stimuli that made them form an attachment to him. As another outcome of imprinting, birds also direct sexual attention to members of the species they had been sexually imprinted upon when young.

CLASSICAL CONDITIONING Ivan Pavlov's early experiments with dogs are an example of classical conditioning. Just before eating, dogs will salivate. Pavlov's dogs were conditioned to salivate even in the absence of food. They did so in response to the sound of a bell or a flash of light that they initially associated with getting food. In this case, the dogs learned to associate an automatic, unconditioned response with a novel stimulus that does not normally trigger the response.

OPERANT CONDITIONING An animal learns to associate a voluntary activity with its consequences, as when a toad learns to avoid stinging insects after its first attempt to eat them.

HABITUATION An animal learns by experience *not* to respond to a situation if the response has neither positive nor negative consequences. Thus pigeons and other birds living in cities learn not to flee from people who pose no threat to them.

SPATIAL (LATENT) LEARNING By inspecting its environment, an animal acquires a mental map of a particular region, often by learning the position of local landmarks. For example, blue jays can store information about the position of dozens, if not hundreds, of places where they have stashed food.

INSIGHT LEARNING An animal abruptly solves some problem without trial-and-error attempts at the solution. Chimpanzees often exhibit insight learning in captivity when they suddenly solve a novel problem that their captors devise for them. Some chimps abruptly stacked and stood on several boxes *and* used a stick to reach bananas suspended out of their reach.

---

an animal is young. The imprinting of baby geese is a classic example (Table 41.1 and Figure 41.7).

Don't fall into the trap of thinking that instinct is governed only by genes and learned behavior, only by the environment. Behavior arises from gene expression *and* experiences. In different habitats, songbirds use variations, or dialects, of their species song. As Peter Marler found out, male songbirds acquire the full song ten to fifty days after hatching by listening to other birds sing it. The nervous system of, say, a male white-crown sparrow is prewired to recognize that song; its learning mechanism is primed to select and respond to acoustical cues in the environment. But its rendition of the species song is influenced by what it actually hears during the sensitive period.

The male birds learn parts of a song by picking up cues when other males sing. Marler raised white-crown nestlings in soundproof chambers so they could not hear adult males. The captives sang when mature, but not with the detailed structure of a typical adult song. Marler exposed other isolated captives to recordings of white-crown sparrows *and* song sparrows. At maturity, the captives sang just the white-crown song. They even mimicked the dialects. Marler's experiments support the hypothesis that birdsong requires *a genetically based capacity to learn* from acoustical cues.

This isn't the whole story. In another experiment, Marler let young, hand-reared white-crowns interact with a "social tutor" of a different species, as opposed to listening to taped songs. The males tended to learn the tutor's song. Their primed learning mechanism had responded to *social experience* as well as acoustical cues.

In instinctive behavior, animals make complex, stereotyped responses to specific, often simple environmental cues. In learned behavior, responses may vary or change as a result of individual experiences in the environment.

Whether instinctive or learned, behavior develops through interactions between genes and environmental experiences.

THE ADAPTIVE VALUE OF BEHAVIOR

If genes directly or indirectly govern forms of behavior, then forms of behavior must be subject to evolution by natural selection. **Natural selection**, recall, is a result of differences in reproductive success among individuals that differ in heritable traits. Alleles underlying the most adaptive versions of a trait tend to increase in frequency in a population, whereas other alleles do not. (Alleles are different molecular forms of the same gene.) In time, genetic changes that yield greater reproductive success for individuals spread through the population.

By using the theory of evolution by natural selection as a point of departure, we should be able to develop and test possible explanations of why a behavior has endured. We should be able to discern how some actions bestow reproductive benefits that offset reproductive costs (disadvantages) associated with them. If a behavior is adaptive, it must promote the *individual's* production of offspring. Keep this thought in mind while you read through the following terms, which you will encounter repeatedly in the remainder of this chapter:

1. **Reproductive success**: The number of surviving offspring that the individual produces.

2. **Adaptive behavior**: Any behavior that promotes propagation of an individual's genes; its frequency is maintained or increases in successive generations.

3. **Social behavior**: Cooperative, interdependent relationships among individuals of the species.

4. **Selfish behavior**: Within a population, any form of behavior that increases an individual's chance to produce or protect offspring of its own regardless of the consequences for the group as a whole.

5. **Altruistic behavior**: Within a population, a self-sacrificing behavior. The individual behaves in a way that helps others but lessens or precludes its own chance of producing offspring.

When biologists speak of selfish or altruistic behavior, they don't mean an individual is consciously aware of what it's doing or of the behavior's reproductive goal. A lion doesn't have to know that eating zebras is good for reproductive success. Its nervous system just calls for HUNTING BEHAVIOR! when the lion is hungry and sees a zebra. This behavior persists in the population because genes responsible for it are persisting also.

As a case in point, Norwegian lemmings disperse from a population when density skyrockets and food is scarce. Many die during the exodus. Are they helping the species by committing suicide to get rid of "excess" individuals? Or do some just happen to die as a result of starvation, predation, or accidental drowning while scurrying to new locations where they may reproduce?

One of Gary Larson's instructive cartoons depicts a population of lemmings plunging over a cliff above water, presumably in the act of mass suicide. But one lemming has an inflated inner tube around its waist! If many lemmings were suicidally altruistic, then only the "selfish" ones would reproduce. Over time, genes underlying their altruism would disappear from the population. An alternative hypothesis is that individual lemmings disperse to less crowded places where each will have a better chance to survive and reproduce.

Another case: In northern forests, ravens scavenge carcasses of deer, elk, or moose, which are few and far between. When one of these large birds comes across a carcass—even in winter, when food is scarce—it often calls loudly and attracts a crowd of similarly hungry ravens. This calling behavior puzzled Bernd Heinrich. On the surface, it seemed to work against the caller's interests. Wouldn't a quiet raven eat more, increase its chance of surviving, and also leave more descendants compared to an "unselfish" vocalizer? If its behavior is an outcome of natural selection, then the biological cost in terms of lost calories and nutrients must be offset by some reproductive benefit for individual callers. What could be the benefit?

Maybe a lone bird picking at a carcass is vulnerable to predators that lie in wait for it. If that were so, then other ravens attracted by the calls could help keep an eye out for danger. But ravens are very big, agile birds that have very few known enemies. Heinrich stealthily watched ravens feeding alone and feeding in pairs at a carcass. He never saw a predator attack any of them.

Then he realized that territory may have something to do with it. A **territory** is an area that one or more individuals defend against competitors. After hauling a cow carcass into a Maine forest, Heinrich observed that single or paired ravens don't always vocally advertise food. Maybe the silent ravens were adults that had already staked out a large territory, which happened to include the spot where he put the carcass.

A pair of ravens would gain nothing by attracting others to their territory. But what if the Maine forest had been *subdivided* into territories and was defended by powerful adults? A wandering young bird would be lucky to eat at all in an aggressive pair's territory. But recruiting a gang of other, nonterritorial ravens might overwhelm the resident pair's defensive behavior.

As it turns out, only the wandering ravens advertise carcasses. Their calling behavior is basically selfish and adaptive. It gives them a shot at otherwise off-limits food and thereby promotes their reproductive success.

Behavioral biologists generally find it profitable to look for evidence of natural selection of the individual's traits rather than something that benefits the species as a whole.

COMMUNICATION SIGNALS

Competing for food, defending territory, alerting others to danger, advertising sexual readiness, forming bonds with a mate, caring for offspring—such *intraspecific* interactions depend on forms of communication among animals, as in the case of those vocalizing ravens.

The Nature of Communication Signals

Intraspecific interactions involve mixes of instinctive and learned behaviors by which individuals send and respond to information-laden cues, encoded in stimuli. These are **communication signals**, with unambiguous meaning for the species. They include specific odors, colors, patterning, sounds, postures, and movements.

Communication signals sent by one individual, the **signaler**, are meant to change the behavior of others of the species. Those responding are the **signal receivers**. A signal evolves or persists when it tends to help the reproductive success of the sender *and* receiver. When a signal proves disadvantageous to either party, then natural selection will favor individuals that don't send the signal or don't respond to it.

Remember **pheromones**? They are chemical signals between individuals of the same species. As you know, chemical odors from food and danger were the main stimuli that early animals had to deal with. As animals evolved, most started to rely on *signaling* pheromones to induce a receiver to respond swiftly. Some, including chemical alarm signals, call for aggressive or defensive behaviors. Others are sex attractants. Bombykol molecules released from female silk moths are like this (Section 32.1). The *priming* pheromones bring about generalized physiological responses. For instance, a volatile odor in the urine of certain male mice triggers and also enhances estrus in female mice.

Figure 41.8 A dog using a play bow to solicit a romp.

Acoustical signals also abound in nature, as when male songbirds sing to secure territory and attract a mate. At night, tungara frogs issue distinctive "whine-chuck" calls to attract females and warn rival males.

Some signals never vary. For example, zebra ears laid back flat against the head convey hostility, but ears pointing up convey its absence. Other signals convey the intensity of the signaler's message. A zebra with laid-back ears isn't too riled up when its mouth is just a bit open. When its mouth is gaping, watch out. That combination is a **composite signal**: a communication signal with information encoded in two cues (or more).

Signals often take on different meaning in different contexts. A lion emits a spine-tingling roar to keep in touch with its pride *or* to threaten rivals. Also, a signal can convey information about signals to follow. Dogs and wolves solicit play behavior with a play bow, as shown to the left in Figure 41.8. Because of the bow, subsequent behavioral patterns that a signal receiver would construe as aggressive, sexual, or exploratory in other contexts are construed as playful only.

Examples of Communication Displays

The play bow is a **communication display**—a pattern of behavior that is a social signal. Another common pattern, the **threat display**, unambiguously announces that a signaler is prepared to attack a signal receiver. When a rival for a receptive female confronts him, a dominant male baboon will role his eyes upward and "yawn" to expose his formidable canines (Figure 41.9). Often the rival backs down. The signaler benefits, for he retains access to the female without putting up a

Figure 41.9 *Above:* Exposed canines, part of a male baboon's threat display. *Below:* Part of a courtship display, with visual, tactile, and acoustical signals, that often precedes copulation. Here a male albatross spreads his wings, an information-laden cue for the female. See also Section 17.2.

Figure 41.10 Dances of the honeybees, as examples of tactile displays. (**a**) Honeybees that visit food sources close to their hive perform a *round* dance on the honeycomb. Worker bees that maintained contact with the forager through the dance go out and search for food near the hive.

(**b**) Bees trained to visit feeding stations over 100 meters from the hive do a *waggle* dance. During the dance, a bee makes a straight run and waggles its abdomen. (**c**) As Karl von Frisch discovered, a straight run's orientation varies, depending on the direction in which food is located. When he put a dish of honey on a direct line between the hive and the sun, foragers that located it returned to the hive and oriented their straight runs right up the honeycomb. When he put the honey at right angles to a line between the hive and the sun, the foragers made their straight runs at 90 degrees to vertical. Thus, a honeybee "recruited" into foraging can orient its flight *with respect to the sun and the hive*. It will waste less time and energy during its food-gathering expedition.

Waggle dancers also vary the dance speed to convey even more information about the distance of a food source. When a site is 150 meters away, the dance is executed much faster, with more waggles per straight run, compared with a dance about a food source that is, say, 500 meters away.

a

When bee moves straight down comb, recruits fly to source directly away from the sun.

When bee moves to right of vertical, recruits fly at 90° angle to right of the sun.

When bee moves straight up comb, recruits fly straight toward the sun.

b

c

fight. The signal receiver benefits because he avoids a serious beating, infected wounds, and possibly death.

Such displays are *ritualized*, with intended changes in the function of common behavior patterns. Normal movements may be exaggerated yet simplified; postures may be frozen. Body parts such as feathers, manes, and claws may be conspicuously enlarged, patterned, and colored. Ritualization is often developed to an amazing degree in **courtship displays** between potential mates. For instance, food-related displays are common among many birds. A male bird may emit calls while bowing low, as if to peck the ground for food. When a female comes running, he may spread his wings, as if to focus attention on the ground in front of him (Figure 41.9). Here, one behavior (searching for food) has changed to attract a female, the intent being copulation. Another example: Firefly courtship displays use bioluminescent flashes (Section 6.9, *Critical Thinking* question 5). A male firefly uses a light-generating organ to emit a bright, flashing signal. A few seconds later, a receptive female of his species may answer him with a flash. Both may flash back and forth until they meet up.

Similarly, in **tactile displays**, a signaler touches the receiver in ritualized ways. After discovering a source of pollen or nectar, a foraging honeybee returns to its colony, a hive, and performs a complex dance. It moves about in a circle, jostling in the dark with a crowd of workers. Other bees may follow and maintain physical contact with the dancer. Honeybees that do so acquire information about the general location, distance, and direction of pollen or nectar (Figure 41.10).

Illegitimate Signalers and Receivers

Sometimes the wrong parties intercept communication signals. Termites will respond with defensive behavior if they catch a whiff of a scent from an invading ant. That scent is meant to identify the ant as a member of an ant colony, and it elicits cooperative behavior from other ants. The scent does have evolved functions, but announcing an invasion of a termite colony is not one of them. A termite also might detect the scent and kill the ant. In that case, it is an **illegitimate receiver** of a signal meant for individuals of a different species.

We even see **illegitimate signalers**. Certain assassin bugs hook dead and drained bodies of termite prey on their dorsal surface to acquire the odor of the victims. By deceptively signaling that they "belong" to a termite colony, they can more easily hunt termite prey. Another illegitimate signaler is the female of certain predatory fireflies. If one notices a male firefly's flash, she flashes back. If she can lure him into attack range, he becomes her meal—an evolutionary cost of having an otherwise useful response to a come-hither signal.

A communication signal between individuals of the same species is an action having a net beneficial effect on both the signaler and receiver. Natural selection tends to favor communication signals that promote reproductive success.

MATES, PARENTS, AND INDIVIDUAL REPRODUCTIVE SUCCESS

For reasons we need not explore here, most people find mating and parenting behaviors of animals fascinating. How useful is selection theory in helping us interpret such behavior? Let's take a look.

Sexual Selection Theory and Mating Behavior

Competition among members of one sex for access to mates is common. So is choosiness in selecting a mate. Recall, from Section 16.9, that such activities are forms of a microevolutionary process called **sexual selection**. Sexual selection favors traits that give the individual a competitive edge in attracting and holding on to mates.

But *whose* reproductive success is it—the male's or female's? Males, recall, produce large numbers of tiny sperm, and females produce far larger but fewer eggs. For the male, success generally depends on how many eggs he fertilizes. For the female, it depends more on how many eggs she produces or how many offspring she can care for. In most cases, *the key factor influencing a female's sexual preference is the quality of the mate—not a quantity of partners*. Hangingflies, sage grouse, and bison offer examples of how most females dictate the rules of male competition (Figures 41.11 through 41.13). They also show how males employ tactics that might help them fertilize as many eggs as possible.

Female hangingflies (*Harpobittacus apicalis*) choose males that offer superior material goods. And so you may observe some males capturing and killing a moth or some other insect. After they do so, they release a sex pheromone to attract females to their "nuptial gift,"
as in Figure 41.11*a*. A female will select the male that displays the larger calorie-rich offering. She lets him mate, but only *after* she has been eating a gift for five minutes or so. Then she accepts sperm and holds them in a storage sac, but only as long as the food holds out. Before twenty minutes are up, she can break off the mating at any point. If she does, she might well mate with another male and accept his sperm. Doing so will dilute the reproductive success of her first partner.

In western regions of North and South Dakota, we find sage grouse (*Centrocercus urophasianus*) dispersed among stands of sagebrush. They never roam far from its protective cover during their spring courtship and summer nesting. Males make no attempt to secure and hold on to a large territory. During the breeding season, they congregate in a type of communal display ground called a **lek**. Each male stakes out a few square meters as his territory. The females are attracted to the lek—not to feed or nest, but rather to watch the males. With tail feathers erect and large neck pouches puffed out, a male emits booming calls and stamps about rather like a wind-up toy on his display ground (Figure 41.11*b*). Females tend to select and mate with one male only. Then they go off to nest by themselves in sagebrush, unassisted by a sexual partner. Many often choose the same male, so most of the males never mate.

Another example: The females of a wolf pack court males. Why? Unlike most male mammals, which don't invest as much as females do in raising offspring, male wolves help pups by giving them food, defending the feeding territory, and driving off infanticidal intruders.

Figure 41.11 (a) Male hangingfly dangling a moth as a nuptial gift for a future mate. Females of some hangingfly species choose sexual partners on the basis of the size of prey offered to them. (b) Male sage grouse. Its ornamental feathers are visual signals that function in a courtship display—a dance performed at a lek. Males vigorously defend a small patch of this communal mating area. The females (the smaller brown birds) observe the prancing males before choosing the one they will mate with.

AN EVOLUTIONARY VIEW OF HUMAN SOCIAL BEHAVIOR

If we can analyze the evolutionary basis of the behavior of termites, naked mole-rats, and other animals, would it not be rewarding to analyze such a basis of human behavior, too? Many resist the idea. Apparently, they believe that attempts to identify the adaptive value of a human trait are attempts to define its moral or social advantage. However, there is a clear difference between trying to explain something in terms of its evolutionary history and attempting to justify it. "Adaptive" does not mean "morally right." To biologists, it only means valuable with respect to the transmission of the individual's genes.

Consider the case of altruistic behavior that we call **adoption**, or the acceptance of offspring of other individuals as one's own. If we wish to study adoptive behavior, we may use evolutionary theory to come up with a hypothesis about it.

This does not mean that we are attempting to judge whether adoption is a moral behavior or even a socially desirable behavior. "Morality" and "desirability" are issues about which biologists have no more to say than anybody else.

Figure 41.20 Two adult emperor penguins competing to adopt an orphan, which penguins accept as substitute offspring.

Start with a premise that all adaptations have costs and benefits. One cost is that some adaptive behaviors may be *redirected* under rare or unusual circumstances. In many species, adults that lost offspring do adopt a substitute. Some cardinals have fed goldfish. A whale tried to lift a log out of water as if it were a distressed infant that needed help to reach the water's surface. Emperor penguins fight to adopt orphans (Figure 41.20).

As John Alcock suggests, such examples constitute a test of a hypothesis about adoptions by us. Human adoptive behavior occurs "by mistake" when otherwise adoptive parental behavior gets focused on unrelated offspring. Suppose it arises through a frustrated desire to bear children (hypothesis). If so, then we can expect that couples who lost an only child or are sterile should be especially prone to adopt strangers (prediction).

Now consider this. For most of their evolutionary history, humans have lived in small groups and, later, small villages. They probably had few opportunities or showed little inclination to adopt strangers. They were likely to accept a child of a deceased relative. But how would adoptive parents gain genetic representation in the next generation? According to theories of natural selection and indirect selection, individuals act in ways that promote genetic self-interest. So we might predict that parents with dependent, care-requiring children of their own are much less likely to adopt substitutes, compared with adults who have no children or who already raised children and are living by themselves. We could test the prediction by piecing together a large enough sampling of information, ideally from diverse existing human cultures, on a connection between childless conditions and adoption.

Indirect selection also favors adults who focus their parenting behavior on relatives and thereby perpetuate shared genes in an indirect way. We might even predict that such adults are far more likely to be related to an adopted child than we would expect by chance alone. Behaviorist Joan Silk tested the prediction in traditional societies. Her results showed that people will adopt related children far more often than nonrelated children.

Particularly in large, industrialized societies that have welfare agencies and other means of adoption assistance, individuals become parents of nonrelatives. Such societies are evolutionary environments. Parenting mechanisms did evolve in the past. Perhaps their redirection toward nonrelatives says more about our evolutionary history than it does about the transmission of one's genes.

The bottom line is this: Evolutionary hypotheses about the adaptive value of behavior lend themselves to testing. Through testing, we can gain understanding about the evolution of human behavior.

It is possible to test evolutionary hypotheses regarding the adaptive value of different forms of human behavior.

There might be greater acceptance of such testing when more people come to understand that adaptive behavior and socially desirable behavior are separate issues.

In biology, "adaptive" means only that a specified trait of an individual has proved beneficial in the transmission of genes responsible for the trait.

SUMMARY *Gold* indicates text section

1. Animal behavior (coordinated responses to stimuli) starts with genes that specify products required for the development and operation of nervous, endocrine, and skeletal–muscular systems. *CI, 41.1*

2. A behavior performed without having been learned by experience is instinctive, a prewired response to one or two simple, well-defined environmental cues (sign stimuli) that trigger a suitable response, such as a fixed action pattern. Experiences may lead to variations or changes in responses (learned behavior). Learning is a result of genetic and environmental inputs. *41.1, 41.2*

3. Behavior with a genetic basis is subject to evolution by natural selection. It evolved as a result of individual differences in reproductive success in past generations, when reproductive benefits of a behavior exceeded its reproductive costs (or disadvantages). *41.3*

4. Social groups require cooperative interdependency among individuals of a species. Sociality is promoted by communication signals, cues sent by one individual (a signaler) that can change the behavior of another of the same species (a signal receiver). *41.4*

5. Chemical, visual, acoustical, and tactile signals are components of communication displays. *41.4*

   a. Pheromones function in chemical communication. Signaling pheromones, such as sex attractants and alarm signals, can induce immediate change in the behavior of a receiver. Priming pheromones elicit a generalized physiological response by the receiver.

   b. Visual signals (observable actions or cues) are key components of courtship displays and threat displays.

   c. Acoustical signals (sounds with precise, species-specific information) include the mating calls of frogs.

   d. Tactile signals are specific forms of physical contact between a signaler and a receiver.

6. Members of the same species often create obstacles to one another's reproductive success, as by selective mate choice and by competing for access to mates. *41.5*

7. Social groups have costs, including competition for limited resources and greater exposure to contagious diseases and parasites. Benefits outweigh the costs in some cases, as when predation pressure is strong. *41.6*

8. In self-sacrificing (altruistic) behavior, individuals give up chances to reproduce while helping others of their group. In most social groups, individuals do not sacrifice reproductive chances to help others. *41.6*

   a. A dominance hierarchy is a social ranking in which some individuals give way to others of their group. Often, dominant individuals force subordinates to relinquish food or some other resource.

   b. When subordinates give in to dominant members of their group, they may receive compensatory benefits of group living, such as safety from predators. In some species, subordinates may reproduce if they live long enough and dominant ones slip in the hierarchy or die.

9. By one theory of indirect selection, genes associated with caring for relatives (not one's direct descendants but individuals that bear a fraction of the same genes) are favored in some cases. Indirect selection can lead to extreme altruism, as among workers of some species of social insects and naked mole-rats. The workers usually do not reproduce. Altruism helps reproducing relatives survive, so altruistic individuals pass on, "by proxy," genes underlying development of this behavior. *41.7*

10. Testing the adaptive value of human behavior starts with understanding that "adaptive" only means a trait is helpful in the transmission of an individual's genes; socially desirable behavior is a separate issue. *41.8*

Review Questions

1. Explain how genes and their products, including hormones, influence mechanisms required for forms of behavior. *41.1*

2. Define these terms: instinctive behavior, sign stimulus, fixed action pattern, and learned behavior. *41.1, 41.2*

3. Contrast altruistic behavior with selfish behavior. Why does either form of behavior persist in a population? *41.3*

4. Describe some characteristics of communication signals. Then give an example of a communication display. *41.4*

5. Describe some feeding behavior or mating behavior in terms of natural (individual) selection. *41.3, 41.5*

6. List some benefits and costs of sociality. *41.6, 41.7*

Self-Quiz   ANSWERS IN APPENDIX III

1. "Starlings minimize nest mites by festooning nests with wild carrot," said Clark and Mason. Their statement was _____ .
   a. an untested hypothesis   c. a test of a hypothesis
   b. a prediction   d. a proximate conclusion

2. Genes affect the behavior of individuals by _____ .
   a. influencing the development of nervous systems
   b. affecting the kinds of hormones in individuals
   c. governing the development of muscles and skeletons
   d. all of the above

3. Generally, living in a social group costs the individual, in terms of _____ .
   a. competition for food, other resources
   b. vulnerability to contagious diseases
   c. competition for mates
   d. a through c

4. A statement that overcrowding causes lemmings to disperse to areas that are more favorable for reproduction _____ .
   a. is consistent with Darwinian evolutionary theory
   b. is based on a theory of evolution by group selection
   c. is supported by the finding that most animals behave altruistically during their lives

5. The genetic similarity between an uncle and his nephew is _____ .
   a. the same as between a parent and its offspring
   b. greater than between two full siblings
   c. dependent on how many other nephews the uncle has
   d. less than that between a mother and her daughter

Figure 41.21 The puffed-up anterior segments of a caterpillar under siege.

Figure 41.23 Soldier termites guarding a break in a tunnel that leads from dead wood to a nest underground.

6. Match the terms with their most suitable description.

____ fixed action pattern	a. time-dependent form of learning requiring exposure to key stimulus
____ altruism	b. genes plus actual experience
____ basis of instinctive *and* learned behavior	c. stereotyped motor program that runs to completion independently of feedback from environment
____ imprinting	d. assisting another individual at one's own expense

Critical Thinking

1. A large caterpillar is crawling along a branch in a tropical forest. You poke it, and one of its responses is to let go of the branch and puff up anterior segments of its body (Figure 41.21). Propose an internal mechanism that may contribute to this defensive behavior. Also propose how the behavior may have adaptive value. How would you test your two hypotheses?

2. A cheetah scent-marks plants in its territory by releasing certain chemicals from exocrine glands. What evidence would you require to demonstrate that the cheetah's action is an evolved communication signal?

3. You observe mallard ducks in a small pond. Then you notice a rooster wading out to the ducks with amorous intent (Figure 41.22). What probably happened to the rooster early in life?

4. Develop an adaptive value hypothesis for this observation: Male lions kill offspring of females they acquire after chasing away previous pride holders that had mated with the females. How would you test your infanticide hypothesis?

5. What are the likely evolutionary costs and benefits to a male hangingfly of offering a nuptial gift to its potential mate? How might a female's choosiness depend on the costs and benefits?

6. In a eucalyptus forest in Queensland, Australia, you notice a narrow, brittle tube running along the trunk of a dead tree. You decide to chip a few fragments from it. Sunlight pours into the breached tube wall. Some small, nearly white insects bang their head against the wall, then scurry away. The head banging, an acoustical signal, makes vibrations that alert soldier termites. Soldiers run to the breach and make a defensive stand (Figure 41.23). Each has a swollen, eyeless, tapering head with a long,

pointed "nose." Each disturbed soldier termite shoots thin jets of silvery goo out of its nose! The silvery strands release volatile compounds that attract more soldier termites that battle the danger, which more typically is an invasion by ants.

Insects in the ruptured tunnel belong to a complex termite colony. Whereas soldiers protect the colony, the pale ones are workers. They build tunnels that lead to places where they safely gather fibers of wood. They carry fibers to an underground nest where termites live and cultivate an edible fungus. Fungal hyphae grow into the wood fibers and absorb nutrients from them. The termites eat some portions of the fungus. They chew the wood into small particles that they can swallow. Unlike most termite species, they don't produce enzymes that can digest cellulose in wood. Symbiotic microorganisms in their gut do this, and the symbionts and termites both absorb the released nutrients.

Soldier termites and worker termites are sterile; neither can reproduce. They engage in self-sacrificing behavior that helps others survive. A single queen and one or more kings serve as "parents" for the entire society.

Identify some of the communication signals and the manner in which individuals in the colony respond to them. Speculate on the adaptive value of the social behaviors just described. How, for instance, could being sterile help an individual's chances for genetic success?

Selected Key Terms

adaptive behavior *41.3*	fixed action pattern *41.1*	selfish behavior *41.3*
adoption *41.8*	illegitimate receiver *41.4*	selfish herd *41.6*
altruistic behavior *41.3*	illegitimate signaler *41.4*	sexual selection *41.5*
animal behavior *CI*	imprinting *41.2*	sign stimulus *41.1*
communication display *41.4*	indirect selection *41.7*	signal receiver *41.4*
	instinctive behavior *41.1*	signaler *41.4*
communication signal *41.4*	learned behavior *41.2*	social behavior *41.3*
	lek *41.5*	song system *41.1*
composite signal *41.4*	natural selection *41.3*	tactile display *41.4*
courtship display *41.4*	pheromone *41.4*	territory *41.3*
dominance hierarchy *41.6*	reproductive success *41.3*	threat display *41.4*

Readings

Alcock, J. 1998. *Animal Behavior: An Evolutionary Approach.* Sixth edition. Sunderland, Massachusetts: Sinauer.

Frisch, K. von. 1961. *The Dancing Bees.* New York: Harcourt Brace Jovanovich. A classic on the behavior of honeybees.

Wilson, E. O. 1975. *Sociobiology: The New Synthesis.* Cambridge, Massachusetts: Harvard University Press.

On-Line readings at Student Guide for InfoTrac:
www.brookscole.com/biology

Figure 41.22 Behaviorally confused rooster.

COMMUNITY INTERACTIONS

No Pigeon Is An Island

Flying through the rain forests of New Guinea is an extraordinary pigeon with cobalt blue feathers and lacy plumes on its head (Figure 42.1). It is about as big as a turkey, and it flaps so slowly and noisily that its flight sounds like an idling truck. Like eight species of smaller pigeons in the same forest, it perches on branches to eat fruit. How is it possible that *nine* species of large and small fruit-eating pigeons live in the space of the same forest? Wouldn't competition for food leave one species the winner? Actually, each species in that rain forest lives, grows, and reproduces in a characteristic way, as defined by its relationships with other organisms and with the surroundings.

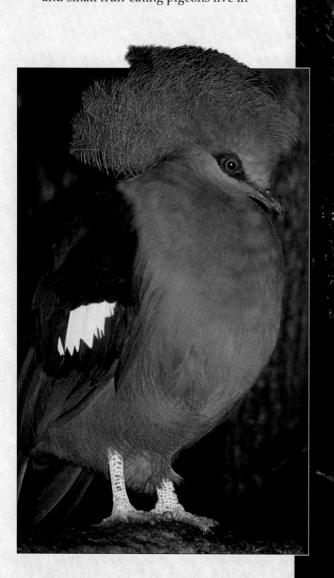

Figure 42.1 A cobalt blue, turkey-sized Victoria crowned pigeon, one of nine species of pigeons living in the same tropical rain forest of New Guinea. In this habitat, each species has its own niche.

Big pigeons perch on the sturdiest branches when they eat. And they eat big fruit. Smaller pigeons, with their smaller bills, can't peck open big fruit. They eat small fruit hanging from slender branches that aren't sturdy enough to support the weight of a turkey-sized pigeon. Species of trees in the forest differ with respect to fruit size and the diameter of fruit-bearing branches. They attract pigeons with different traits. In such ways, the nine pigeon species *partition* the fruit supply.

And how do individual trees benefit from enticing pigeons to eat? Seeds in their fruits have tough coats, which resist the action of digestive enzymes in a pigeon gut. During the time it takes for ingested seeds to travel through the gut, pigeons fly about and dispense seed-containing droppings in more than one place. In this way, they tend to disperse seeds some distance from the parent plants. Later, when seedlings grow, the odds are better that at least some of them won't have to compete with their parents for sunlight, water, and mineral ions. The seeds that drop close to home can't compete in any significant way with the resource-gathering capacity of mature trees, which already have well-developed, extensive roots and leafy crowns.

Within the same forest, leaf-eating, fruit-munching, and bud-nipping insects interact with other organisms and their surroundings in certain ways. So do nectar-drinking, flower-pollinating bats, birds, and insects. And so do great numbers of beetles, worms, and other invertebrates that busily extract energy from remains and wastes of other organisms on the forest floor. By their activities, they cycle nutrients back to the trees.

Like humans, then, no pigeon is an island, isolated from the rest of the living world. The nine species of New Guinea pigeons eat fruit of different sizes. They disperse seeds from different sorts of trees. Dispersal influences where new trees will grow and where the decomposers will flourish. Ultimately, tree distribution and decomposition activities influence how the entire forest community is organized.

Directly or indirectly, interactions among coexisting populations organize the community to which they belong. With this chapter, we turn to community interactions that influence all populations over time and in the space of their environment.

Key Concepts

1. A habitat is the type of place where individuals of a species normally live. A community is an association of all the populations of species that occupy the same habitat.

2. Every species in the community has its own niche, defined as the sum of all activities and relationships in which its individuals engage as they secure and use the resources required for their survival and reproduction.

3. The structure of a community starts with adaptive traits that give individuals of each species the capacity to respond to the physical and chemical features of their habitat, and to levels and patterns of resource availability over time.

4. Interactions among species influence the structure of the community. They include mutually beneficial interactions, competition, predation, and parasitism.

5. Community structure also depends on the geographic location and size of the habitat, the rates at which the member species arrive and disappear, and the history of physical disturbances to the habitat.

6. The first species to occupy a particular type of habitat are replaced by others, which are replaced by still others, and so on in an expected sequence. This process, known as primary succession, produces a climax community. A climax community is a stable, self-perpetuating array of species in balance with one another and with the environment.

7. Different stages of succession often prevail within the same habitat owing to local differences in the soil and other habitat conditions, recurring disturbances such as seasonal fires, and chance events.

WHICH FACTORS SHAPE COMMUNITY STRUCTURE?

Think of a clownfish darting above a coral reef, a maple tree on a Vermont hillside, or a mole making a burrow in soft earth. The type of place where you will normally observe a clownfish, maple, or mole is its **habitat**. The habitat of an organism is characterized by physical and chemical features, such as its temperature and salinity, and by the array of other species living in it. Directly or indirectly, the populations of all species in a habitat associate with one another as a **community**.

Five factors shape the structure of a community. *First*, interactions among climate and topography help dictate the habitat's temperatures, rainfall, soil types, and other conditions. *Second*, the kinds and amounts of food and other resources that become available through the year influence which species can live there. *Third*, individuals of each species have adaptive traits that allow them to survive and exploit specific resources in the habitat. *Fourth*, species in the habitat interact by competition, predation, and mutually helpful actions. *Fifth*, community structure is influenced by the overall pattern of population size (and the history of changes in its size), by arrivals and disappearances of species, and by physical disturbances to the habitat.

Together, these factors help determine the number of species at several "feeding levels," starting with the producers and continuing through levels of consumers. They influence population sizes. They also help dictate the overall number of species. For example, high solar radiation, warm temperatures, and high humidity in tropical habitats favor growth of many kinds of plants, which support many kinds of animals. Conditions in arctic habitats do not favor great numbers of species.

Chapters to follow deal with energy flow through feeding levels and with geographic factors influencing community structure. Here we begin with interactions among species, using the niche concept as our guide.

The Niche

If the organisms of a community all share the same habitat—that is, if they all have the same "address"—in what respects do they differ? Each kind is distinct in terms of its "profession" in the community, meaning the sum of the activities and relationships in which it engages to secure and use the resources necessary for its survival and reproduction. This is its **niche**.

For each species, a *fundamental* niche is the one that might prevail in the absence of competition and other factors that could constrain its acquisition and use of resources. However, as you will see, such constraining factors do come into play in communities. They tend to bring about a more constrained, *realized* niche that shifts in large and small ways over time, as individuals of the species respond to a mosaic of changes.

Categories of Species Interactions

Dozens to hundreds of species interact in diverse ways even in simple communities. In spite of this diversity, we can identify six categories of interactions that have different effects on population growth (Table 42.1).

Table 42.1 *Categories of Two-Species Interactions**

Type of Interaction	Direct Effect on Species 1	Direct Effect on Species 2
Neutral relationship	0	0
Commensalism	+	0
Mutualism	+	+
Interspecific competition	−	−
Predation	+	−
Parasitism	+	−

* 0 means no direct effect on population growth;
+ means positive effect; − means negative effect.

Each species has a neutral relationship with most species in its habitat. For example, Canadian lynx and grasses do not affect each other directly. Interactions with other species link them only indirectly. The lynx preys on and decreases the number of snowshoe hares that eat plants; plants fatten hares, the prey of lynx.

Commensalism directly helps one species but does not affect the other much, if at all. For instance, some birds use tree branches for roosting sites. The trees get nothing but are not harmed. With **mutualism**, benefits flow both ways between the interacting species. Don't think of this as cozy cooperation. Benefits flow from a two-way exploitation. With **interspecific competition**, disadvantages flow both ways between species. Finally, **predation** and **parasitism** are interactions that directly benefit one species (either the predator or the parasite) and directly hurt the other (the prey or host).

Commensalism, mutualism, and parasitism are cases of **symbiosis** ("living together"). For at least part of the life cycle, individuals of two or more species interact with neutral, positive, or negative effects on each other.

A habitat is the type of place where individuals of a species normally live. A community consists of all populations that live in the habitat.

Community structure arises from the habitat's physical and chemical features, resource availability over time, adaptive traits of its members, how the members interact, and the history of the habitat and its occupants. Species may have neutral, positive, or negative effects on one another.

A niche is the sum of all activities and relationships in which individuals of a species engage as they secure and use the resources necessary to survive and reproduce.

MUTUALISM

Mutualistic interactions in which positive benefits flow both ways abound in nature. When trees provide New Guinea pigeons with food and when the pigeons help disperse seeds from the trees to new germination sites, both function as mutualists. Many other kinds of plants and animals enter into such interactions. For example, most flowering plants, and the insects and birds, bats, and other animals that pollinate them, are mutualists. The introduction to Chapter 28 gives vivid examples.

of absorptive structures interact in two ways. Either the fungal hyphae penetrate root cells or they grow as a dense, velvety mat around them. The fungus is good at absorbing mineral ions from soil, and the plant comes to depend on taking some of them. In turn, the fungus withdraws some sugar molecules that the plant makes by photosynthesis. The fungus depends upon the plant for its reproductive success, too. When photosynthesis ceases, the fungus stops producing spores.

Figure 42.2 One mutualistic interaction on a rocky slope of Colorado's high desert.

(**a**) Different kinds of flowering plants of the genus *Yucca* are each pollinated exclusively by one species of yucca moth (**b**). This insect cannot complete its life cycle with any other plant.

The adult stage of the moth life cycle coincides with blossoming of yucca flowers. By using her specialized mouthparts, a female moth gathers sticky pollen and rolls it into a ball. Then she wings her way to another flower. She pierces the wall of the flower's ovary, where seeds form and develop, and lays her eggs inside. As she crawls out of the flower, she pushes a ball of pollen onto a pollen-receiving surface.

Pollen grains germinate and grow down through tissues of the ovary. They deliver sperm to the flower's eggs. The seeds develop after fertilization. Meanwhile, the moth eggs develop to the larval stage. (**c**) When larvae emerge, they eat a few seeds, then gnaw their way out of the ovary. The seeds that moth larvae do not eat give rise to new yucca plants.

Some forms of mutualism are *obligatory*. That is, the individuals of one species cannot grow and reproduce in the absence of intimate dependency with individuals of another species during the life cycle. This is the case for interactions between yucca plants and yucca moths. Each plant of this genus (*Yucca*) can be pollinated only by one species of the yucca moth genus (*Tegeticula*). In addition, the larval stages of the moth grow only in the yucca plant; they eat only yucca seeds (Figure 42.2).

We also find cases of obligatory mutualism between many fungi and plants. Mycorrhizae, remember, are intimate ecological interactions between fungal hyphae and young roots (Chapters 21 and 27). The two kinds

And what about the apparent endosymbiotic origin of eukaryotes? Long ago, phagocytic bacteria may have engulfed aerobic bacterial cells. These resisted digestion, tapped host nutrients, and reproduced independently. In time the hosts came to rely on ATP produced by the guests—which became mitochondria and chloroplasts. If those prokaryotic cells had not evolved as such close mutualists, you and all other eukaryotic species would not even be around today (Section 19.4).

In cases of mutualism, each of the participating species reaps benefits from the interaction.

COMPETITIVE INTERACTIONS

Where you find limited supplies of energy, nutrients, living space, and other natural resources, you observe organisms competing for a share of them. *Intraspecific* competition occurs between individuals of the same population or species. As you probably deduced from the preceding chapter, their interactions can be fierce. *Interspecific* competition occurs between populations of different species and usually isn't as intense. Why not? *Requirements of two species may be similar, but they never are as close as they are for individuals of the same species.*

Consider just two forms of competitive interactions. Sometimes individuals have equal access to a required but limited resource. If one is better than the other at exploiting it, competition can reduce the supply of that shared resource. (When you and a friend use straws to share a small milkshake, you might not get much at all if your friend uses a jumbo straw.) And sometimes individuals control access to a resource, letting others use only some of it or none at all, regardless of whether the resource is scarce or abundant. (For instance, even if you shared a ten-gallon milkshake, you still would get less if your friend pinched your straw.)

Competition abounds in nature. Chipmunks exclude other species of chipmunks from their habitats (Figure 42.3). A strangler fig tree wraps around other trees as a framework for its growth, and eventually it kills them.

From early spring until late summer, a male *broadtailed* hummingbird busily chases other males and females of its species from blossoms in his territory in the Rockies. In August, however, *rufous* hummingbirds of the Pacific Northwest migrate through the Rockies on their way to their wintering grounds in Mexico. Rufous males prove to be far more aggressive, hence stronger competitors for available food. As they follow the migratory route, they evict male broadtails from broadtail territories.

a
Relative population density — Time (days) 0 4 8 12 16 20 24

b
Relative population density — Time (days) 0 4 8 12 16 20 24

c
Relative population density — Time (days) 0 4 8 12 16 20 24

ALPINE CHIPMUNK LODGEPOLE CHIPMUNK YELLOW PINE CHIPMUNK LEAST CHIPMUNK

Figure 42.3 Example of competition in nature. On the eastern slopes of the Sierra Nevada, different chipmunk species occupy different habitats. The alpine habitat is at the highest elevation. Below it are the lodgepole pine, piñon pine, and then sagebrush habitats. The least chipmunk lives at the base of the mountains in sagebrush. Its adaptations would allow it to move up into the piñon pine habitat, but the aggressively competitive behavior of yellow pine chipmunks in that habitat won't let it. Food preferences keep the yellow pine chipmunk out of the sagebrush habitat.

Competitive Exclusion

Any two species differ in their adaptations for getting food and avoiding enemies. In some way, one usually competes more effectively for scarce resources. The two are less likely to coexist in the same habitat if they use resources in very similar ways. G. Gause demonstrated this by growing two species of *Paramecium* separately, then together (Figure 42.4a–c). Both species used the same food (bacteria) and competed intensely for it. As results from experimental tests show, two species that use identical resources can't coexist indefinitely. Many other experiments have supported this concept, which ecologists now call **competitive exclusion**.

Gause also studied two other *Paramecium* species that didn't overlap as much in their requirements. When grown together, one species tended to feed on bacteria suspended in culture tube liquid. The other ate yeast cells at the bottom of the test tube. Population growth rates slowed for both of the species—but the overlap in resource use wasn't enough for one species to fully exclude the other. The two continued to coexist.

Field experiments also reveal effects of competition. Robert Paine used removal experiments to document effects of a **keystone species**, a dominant species that dictates community structure by affecting abundances

Figure 42.4 Results of competitive exclusion between two protistan species that compete for the same food. (**a**) *Paramecium caudatum* and (**b**) *P. aurelia* were grown in separate culture tubes and established stable populations. The S-shaped growth curves in these graphs indicate stability. (**c**) Then the populations were grown together. *P. aurelia* (*red* curve) drove the other species toward extinction (*blue* curve in **c**). This experiment and others suggest that two species cannot coexist indefinitely in the same habitat *when they require identical resources.* If their requirements do not overlap much, one might influence the population growth rate of the other, but they may still coexist.

d Algal diversity in tidepools

e Algal diversity on rocks that become exposed at high tide

Figure 42.5 Effect of competition and predation on community structure. (**a**) Grazing periwinkles (*Littorina littorea*) affect the number of algal species in different ways in different marine habitats. (**b**) *Chondrus* and (**c**) *Enteromorpha*, two kinds of algae in their natural habitat. (**d**) In tidepools, periwinkles graze on the dominant alga (*Enteromorpha*); less competitive algae that might otherwise be overwhelmed have an advantage. (**e**) Algal diversity is lower on rocks exposed at low tide. There, periwinkles don't graze on *Chondrus* and other kinds of red algae, the dominant species.

of other species. All along North America's west coast, surf, tides, and storms repeatedly disturb the intertidal zones, and living space is scarce. Paine put a keystone species, the sea star *Pisaster ochraceus*, in control plots with its prey: mussels (*Mytilus*), barnacles, limpets, and chitons. He removed sea stars from experimental plots. Mussels took over those plots; they crowded out seven other invertebrate species. Mussels, the main prey of sea stars, are the strongest competitors in their absence. So sea star predation normally maintains the diversity of prey species because it blocks competitive exclusion by mussels. Remove all the sea stars, and the community shrinks from fifteen species to eight.

Similarly, Jane Lubchenco found that the periwinkle *Littorina littorea*, an alga-eating mollusk, promotes *or* limits diversity in different habitats. In tidepools, it eats dominant algal species, thereby helping many other, less competitive algae live. On rocks exposed at high tide only, *L. littorea* ignores tough, unpalatable species and favors the competitively weaker algal species it ignores in tidepools. And so *L. littorea* promotes algal diversity in tidepools—yet lowers it on rocks exposed at high tide (Figure 42.5).

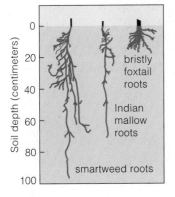

Figure 42.6 Partitioning of resources among three annual plants in a plowed, abandoned field. They differ in adaptations for securing soil water and mineral ions.

Resource Partitioning

Think back on the nine species of fruit-eating pigeons in the same New Guinea forest. They all use the same resource: fruit. Yet they overlap only slightly in their use of it, for each is adapted to eat fruits of a particular size. They are a fine example of **resource partitioning**—the *subdividing* of some category of similar resources that lets competing species coexist.

Another competitive situation arose among three species of annual plants in plowed, abandoned fields. Like other plants, all three require sunlight, water, and dissolved mineral ions. Yet each species is adapted to exploiting a different part of the habitat (Figure 42.6). Drought-tolerant foxtail grasses have a shallow, fibrous root system that quickly absorbs rainwater. They grow where moisture in soil varies from day to day. Mallow plants, with a taproot system, grow in deeper soil that is moist early in the growing season but drier later. Smartweed's taproot system branches in topsoil and soil below the roots of the other species. It grows where soil is continuously moist. In sum, species may coexist in the same habitat even when their niches do overlap.

In some competitive situations, all individuals have equal access to a resource that they all require, but some are better than others at exploiting it. In other competitive situations, some individuals control access to a shared resource.

The more that two species in the same habitat differ in their use of resources, the more likely they can coexist.

Two competing species also may coexist by sharing the same resource in different ways or at different times.

PREDATION AND PARASITISM

To keep things simple, let's use two broad definitions for interactions between consumers and their victims. **Predators** are animals that feed on living organisms—their **prey**—but do *not* take up residence on or in them. Prey may or may not die from the interaction. **Parasites** live in or on other living organisms—their **hosts**—and feed on specific host tissues for part of their life cycle. Hosts may or may not die from the interaction.

Many adaptations of predators (and parasites) and their victims arose by **coevolution**, the joint evolution of two or more species that exert selection pressure on each other as a result of close ecological interaction. Suppose a mutant prey organism has a new, heritable means of defense, which later spreads through the prey population. Some individual predators are better than others at countering the defense. They tend to eat more and have a better chance to survive and reproduce. In time the forms of traits *most* effective at overcoming the defense increase in frequency in the population. Now the better predators exert selection pressure that favors better prey defenses—and so on through time.

Dynamics of Predator–Prey Interactions

The outcome of predator–prey interactions in a specified interval depends partly on the carrying capacity of the prey population. **Carrying capacity**, remember, is the maximum number of individuals that the resources in a given environment can maintain indefinitely (Section 40.4). The reproductive rates of the predator and prey populations also influence the outcome. So do predator responses to increases in prey density.

When predation is keeping a prey population from exceeding the carrying capacity, both populations tend to coexist at fairly steady levels. More prey are around; predators reproduce fast and eat more prey. Population

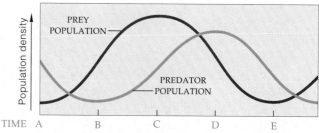

Figure 42.7 Idealized cycling of abundances of predator and prey. (For clarity, the diagram exaggerates predator density; predators are usually less common than prey during the cycle.) This pattern arises owing to time lags in predator responses to changes in abundance of prey. At time A, prey density is low; predators have a harder time getting food; their population is declining. In response to the decline, prey density rises. But predators don't start increasing until they start reproducing at time B. Both populations grow until the predator increase causes the prey population to decline (time C to E). Predators continue to increase and take out more prey. Having fewer prey leads to starving predators, so the predator population growth rate slows (starting at time D). At time E, a new cycle starts.

densities fluctuate when predators don't reproduce as fast as the prey, when they can eat only so many prey organisms in a given interval, and when the carrying capacity for the prey population is high.

Figure 42.7 shows an idealized pattern of the cyclic changes corresponding to time lags in the predator's response to changes in prey abundance. In nature, we sometimes observe such a correspondence—but other factors influence the pattern. Figure 42.8, for instance, shows a ten-year cycling in populations of the Canadian

Figure 42.8 Correspondence between abundances of Canadian lynx (*dashed* line) and snowshoe hares (*solid* line), based on counts of pelts sold by trappers to Hudson's Bay Company over ninety years. The photograph underscores the observation by Charles Krebs that predation causes mass paranoia among hares, which continually look over their shoulders during the declining phase of each cycle.

The graph is a good test of whether you will readily accept someone else's conclusions without questioning their basis in science. (Remember those sections in Chapter 1 on scientific methods?) What other factors may have impacted on the cycle? Did weather vary, with more severe winters imposing greater demand for hares (to keep lynx warmer) and higher death rates? Did the lynx compete with other predators, such as owls? Did predators turn to alternative prey during low points of the hare cycle? When fur prices rose in Europe, did trapping increase? When the supply of pelts outstripped the demand, did trapping decline?

Figure 42.9
Adult roundworms
(*Ascaris*), a type
of endoparasite,
packed inside
a small intestine
from a host pig.

lynx and showshoe hare. Charles Krebs led a long-term study in Alaska's Yukon to identify the basis of this cycling. For ten years, he tracked hare densities in 1-square-kilometer control plots and experimental plots that kept out mammalian predators (by electric fences), had more food, or had fertilizer to increase plant growth. The team placed radio collars on 1,000+ hares, lynx, squirrels, lynx, and coyotes. In predator-free plots, hare density doubled. In plots with extra food, it tripled. With fewer predators plus extra food, density increased elevenfold. But manipulations only delayed the cyclic declines; they didn't stop them. The fence could not keep out owls and other raptors. So predators got 83 percent of the collared hares, only 9 percent of which starved to death. Therefore, a simple predator–prey or plant–herbivore model can't explain the results. More variables are operating in a *three-level* interaction among plants, herbivores, and carnivores.

Dynamics of Parasite–Host Interactions

Parasites have pervasive influences on populations. By draining hosts of nutrients, they alter how much energy and how many nutrients the population is withdrawing from the habitat. Weakened hosts are more vulnerable to predators and less attractive to potential mates. Some infections result in sterility. Some may alter the ratio of host males to females. In such ways, parasitic infections lower the birth rate, raise the death rate, and influence intraspecific and interspecific competition.

The persistent drain of nutrients during a parasitic infection indirectly causes death if the weakened host just can't fight off secondary infections. Evolutionarily speaking, though, killing the host doesn't do much for the parasite's reproductive success. Infections of longer duration give the parasite more time to produce more offspring. So natural selection tends to favor parasite and host adaptations that promote a level of mutual tolerance and less-than-fatal effects. Death usually will result only when a parasite attacks a novel host, which has no coevolved defenses against it, or when there are too many individual parasites driving the infection.

*Ecto*parasites live on a host's surface; *endo*parasites live inside it. Some use one or more hosts throughout the life cycle. Others are free-living for part of it. Many use arthropods, including insects, to taxi to a new host. The **microparasites** among them include many bacteria, viruses, protozoans, and sporozoans (Chapter 20). Most reproduce rapidly. The larger, **macroparasites** include many flatworms, roundworms, and arthropods such as fleas and ticks (Chapter 23, and Figures 24.12 and 42.9).

Then there are parasitic plants. Nonphotosynthetic types withdraw nutrients and water from young roots of host plants. Species in the broomrape family do this to oak and beech tree roots. Other plants do so but still retain the capacity for photosynthesis. Mistletoe is like this; its extensions invade sapwood.

Not all parasites feed on tissues. **Social parasites** complete their life cycle by exploiting social behaviors of another species. For instance, a cowbird never builds a nest, incubates eggs, or cares for offspring. It removes an egg from another bird's nest and proceeds to lay a "replacement." Some birds hatch the alien egg and raise the hatchling. Usually, the young cowbird aggressively pushes the smaller, remaining rightful occupants out of the nest or demands and gets most of the food.

Many parasites are being commercially raised and selectively released as *biological controls*. They are touted as an alternative to chemical pesticides. But less than 20 percent of the existing selections qualify as effective defenses against pests. As outlined by C. Huffaker and C. Kennett, effective biological control agents display five attributes. They are well adapted to a host species and to their habitat. They are exceptionally efficient in searching for a host. Their rate of population growth is high relative to that of the host species. Their offspring are mobile enough to ensure adequate dispersal. And the lag time between their responses to changes in the numbers of the host population is minimal.

Releasing more than one kind of biological control agent in an area may trigger competition among them and eventually lessen their overall effectiveness. Also, a shotgun approach to biological control is risky. There is a possibility that the parasites will attack nontargeted species. In 1983, for example, F. Howarth reported that the populations of butterflies and moths native to the Hawaiian Islands are declining. Introductions of wasps that were supposed to serve as biological controls over something else are partly to blame. Later in the chapter, we will return to what can happen to native populations when a new species is introduced into a habitat.

Predators and their prey, as well as parasites and their hosts, interact in long-term, coevolutionary contests.

Predator and prey populations may coexist at fairly steady levels. Others may undergo recurring cycles of abundance and crashes, erratic cycles, or prey extinction.

Natural selection favors parasitic species that temper their demands in ways that ensure an adequate supply of hosts.

A Coevolutionary Arms Race

Populations of other species are part of any organism's environment, and the ones that interact as predators and prey exert continual selection pressure on each other. One must defend itself and the other must overcome the defenses. This is the basis of a coevolutionary arms race that has resulted in some truly amazing adaptations.

CAMOUFLAGE   Consider prey species that **camouflage** themselves; they can hide in the open. Such organisms have adaptations in form, patterning, color, and behavior that help them blend with their surroundings and escape detection. Figure 42.10 shows classic examples, including a desert plant (*Lithops*) that resembles a small rock. Only during a brief rainy season does *Lithops* flower. That is when other plants grow profusely and distract herbivores from *Lithops*—and when free water instead of juicy plant tissues is available to quench an animal's thirst.

WARNING COLORATION   Many prey species taste bad, are highly toxic, or inflict pain on attackers. Toxic types often have **warning coloration**, or conspicuous patterns and colors that predators learn to recognize as *AVOID ME!* signals. For example, maybe a young, inexperienced bird will spear a yellow-banded wasp or an orange-patterned monarch butterfly once. It will quickly learn to associate the distinctive colors and patterning with a painful sting or with the vomiting of foul-tasting butterfly toxins.

Truly dangerous or repugnant species make little or no attempt to conceal themselves. Skunks are like this. So are frogs of the genus *Dendrobates*; they are among the most vivid and most poisonous organisms (Section 29.1).

MIMICRY   Many prey organisms bear close resemblance to dangerous, unpalatable, or hard-to-catch species. Any close resemblance in form, behavior, or both between one species that serves as a *model* for deception and another that is its *mimic* is called **mimicry**. Figure 42.11 shows how some tasty but weaponless mimics physically resemble species that predators have learned to ignore. In "speed" mimicry, sluggish prey species look like swift-moving species that predators have given up trying to catch.

MOMENT-OF-TRUTH DEFENSES   When luck runs out, survival of prey organisms that are cornered or under attack may turn on a last-ditch trick. Suppose a leopard manages to run down a tasty baboon, as in Figure 41.17. By turning abruptly and displaying formidable canines, the baboon may startle and confuse this predator long enough for a getaway. Other cornered animals may spew irritating chemical repellents or toxins. Earwigs, skunks, and stink beetles produce disgusting odors. Some beetle species take aim and let loose with noxious sprays.

Similarly, many plants synthesize predator repellents. Tannins in the foliage and seeds of certain plants taste

Figure 42.10   A few prey organisms demonstrating the fine art of camouflage. (**a**) What bird??? When a predator approaches its nest, the least bittern stretches its neck (which is colored like the surrounding withered reeds), thrusts its beak upward, and sways gently like reeds in the wind. (**b**) An unappetizing bird dropping? No. This edible caterpillar's body coloration and the stiff body positions it assumes help camouflage it (hide it in the open) from predatory birds. (**c**) Find the plants (*Lithops*) hiding in the open from herbivores by their stonelike form, pattern, and coloring.

Figure 42.15 Part of the fence against 200–300 million rabbits that are destroying Australia's vegetation.

their path. Attempts to dig up or burn kudzu are futile. Grazing goats and herbicides help, but goats eat other plants, too, and herbicides contaminate water supplies. Kudzu could reach the Great Lakes by the year 2040.

On the bright side, a Japanese firm is constructing a kudzu farm and processing plant in Alabama. Asians use a starch extract from kudzu in drinks, herbal medicines, and candy. The idea is to export the starch to Asia, where the demand currently exceeds the supply. Also, kudzu may eventually help reduce logging operations. At the Georgia Institute of Technology, researchers report that kudzu might be one alternative source of paper.

THE RABBITS THAT ATE AUSTRALIA During the 1800s, British settlers in Australia just couldn't bond with the koalas and kangaroos, so they started to import familiar animals from their homeland. In 1859, in what would be the start of a wholesale disaster, a northern Australia landowner imported and then released two dozen wild European rabbits (*Oryctolagus cuniculus*). Good food and good sport hunting, that was the idea. An ideal rabbit habitat with no natural predators—that was the reality.

Six years later, the landowner had killed 20,000 rabbits and was besieged by 20,000 more. The rabbits displaced livestock, even kangaroos. Now Australia has 200 to 300 million hippityhopping through the southern half of the country. They overgraze perennial grasses in good times and strip bark from shrubs and trees during droughts. You know where they've been; they transform grasslands and shrublands into eroded deserts (Figure 42.15). They have

been shot and poisoned. Their warrens have been plowed under, fumigated, and dynamited. Even when all-out assaults reduced their population size by 70 percent, the rapidly reproducing imports made a comeback in less than a year. Did the construction of a 2,000-mile-long fence protect western Australia? No. Rabbits made it to the other side before workers finished the fence.

In 1951, government researchers introduced myxoma virus by way of mildly infected South American rabbits, its normal hosts. This virus causes *myxomatosis*. The disease has mild effects on South American rabbits that coevolved with the virus but nearly always had lethal effects on *O. cuniculus*. Biting insects, mainly mosquitoes and fleas, quickly transmit the virus from host to host. Having no coevolved defenses against the novel virus, the European rabbits died in droves. But, as you might expect, natural selection has since favored rapid growth of populations of *O. cuniculus* resistant to the virus.

In 1991, on an uninhabited island in Spencer Gulf, Australian researchers released a population of rabbits that they had injected with a calicivirus. The rabbits died quickly and relatively painlessly from blood clots in their lungs, heart, and kidneys. In 1995, the test virus escaped from the island, possibly on insect vectors. It has been killing 80 to 95 percent of the adult rabbits in Australian regions. At this writing, researchers are now questioning whether the calicivirus should be used on a widespread scale, whether it can jump boundaries and infect animals other than rabbits (such as humans), and what the long-term consequences will be.

Table 42.2 *Detrimental Effects of Some Species Introduced Into the United States*

Species Introduced	Origin	Mode of Introduction	Outcome
Water hyacinth	South America	Intentionally introduced (1884)	Clogged waterways; other plants shaded out
Dutch elm disease: *Ophiostoma ulmi* (fungus) Bark beetle (vector)	Europe	Accidental; on infected elm timber (1930) Accidental; on unbarked elm timber (1909)	Millions of mature elms destroyed
Chestnut blight fungus	Asia	Accidental; on nursery plants (1900)	Nearly all eastern American chestnuts killed
Argentine fire ant	Argentina	In coffee shipments from Brazil? (1891)	Crop damage; native ant communities; destroyed; ground-nesting birds killed
Japanese beetle	Japan	Accidental; on irises or azaleas (1911)	Over 250 plant species (e.g., citrus) defoliated
Sea lamprey	North Atlantic	Ship hulls, through canals (1860s, 1921)	Trout, whitefish destroyed in Great Lakes
European starling	Europe	Intentional release, New York City (1890)	Native songbirds displaced; crop damage; swine disease vector; huge flocks noisy, messy
House sparrow	England	Released intentionally (1853)	Crop damage; native songbirds displaced; transmission of some diseases

PATTERNS OF BIODIVERSITY

As you might well conclude from Chapter 25 and the preceding discussions, biodiversity differs greatly from one habitat to another. Also, as ecologists have found through many field investigations, biodiversity differs in significant ways between geographic regions.

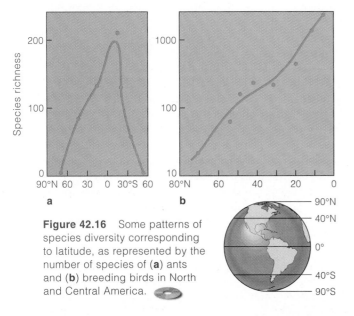

Figure 42.16 Some patterns of species diversity corresponding to latitude, as represented by the number of species of (**a**) ants and (**b**) breeding birds in North and Central America.

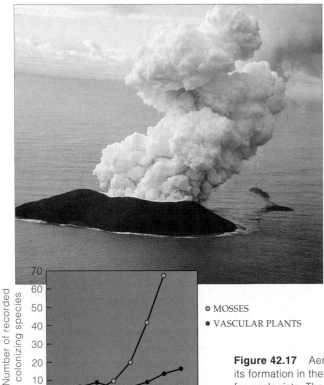

Figure 42.17 Aerial photograph of Surtsey, a volcanic island, at the time of its formation in the sea. Newly formed, isolated islands are natural laboratories for ecologists. The chart gives the number of colonizing species of mosses and vascular plants recorded on Surtsey between 1965 and 1973.

What Causes Mainland and Marine Patterns?

The most striking pattern of biodiversity corresponds to distance from the equator. For most groups of plants and animals, *the number of coexisting species on land and in the seas is greatest in the tropics, and it systematically declines from the equator to the poles*. Figure 42.16 shows two examples of this biodiversity pattern. What factors underlie it? Three are paramount.

First, as described in Chapter 44, tropical latitudes intercept more sunlight of consistently higher intensity, rainfall is heavier, and the growing season is longer. As one outcome, *resource availability tends to be greater and more reliable in the tropics than elsewhere*. All year long, different tree species in humid tropical forests put out new leaves, flowers, and fruit. Year in and year out, they support diverse herbivores, nectar foragers, and fruit eaters. Such specialization can't evolve to a comparable degree in temperate and arctic regions.

Second, *species diversity might be self-reinforcing*. The diversity of tree species in tropical forests is far greater than in comparable forests at higher latitudes. When more plant species compete and coexist, more herbivore species also evolve and coexist partly because no single herbivore can overcome all the chemical defenses of all the different plants. Typically, then, more predators and parasites evolve in response to a diversity of prey and hosts. The same applies to diversity on tropical reefs.

Third, *over geologic time, speciation rates in the tropics exceeded those of extinction rates*. Declines in biodiversity occurred, but mainly during mass extinctions. Global temperatures drop after such episodes. Species adapted to cool climates tend to survive, but those adapted to the temperatures of the tropics do not. Also, millions of species in tropical forests may disappear in the next decade, a topic addressed in Chapters 25 and 45.

What Causes Island Patterns?

Some islands are laboratories for biodiversity studies. For instance, in 1965, a volcanic eruption formed a new island southwest of Iceland. Within six months, bacteria, fungi, seeds, flies, and some seabirds were established on it. One vascular plant appeared after two years, and a moss two years after that (Figure 42.17). When soils improved, the number of plant species increased. All the colonists originated in Iceland. None originated on the island, which was named Surtsey. Yet the number

a

Number of species (percent of sample studied)

Distance from New Guinea (kilometers)

b

Number of species

less than 300 km

more than 300 km

Area (square kilometers)

c

Figure 42.18 Examples of island biodiversity, as correlated with (1) distance from a source of colonizing species and (2) the island's area.

(**a**) Distance effect. Biodiversity on islands of a specified size declines with increasing distance from the source of colonizing species. Each graph point denotes the number of species of land birds occupying lowland areas of South Pacific islands. The large island of New Guinea is the source of colonists for the smaller islands, each of which is at least 500 kilometers away. To correct for differences in island size (area), the number of species of birds on each of the distant islands is expressed as a percentage of the number of bird species on an island of equivalent size close to New Guinea.

(**b**) Area effect. Among islands similarly distant from their source of colonizing species, the larger ones support larger numbers of species. Graph points denote the number of bird species on subtropical and tropical islands. *Solid green* circles and the upper line are for islands less than 300 kilometers from the source of colonists. *Orange* triangles and the lower line denote islands more than 300 kilometers from the source areas. As you can see, this graph incorporates data on the distance effect as well as on the area effect.

(**c**) Wandering albatross, one travel agent for jump dispersals. Seabirds that island-hop long distances often have a few seeds stuck to their feathers. Seeds that successfully germinate in a new island community may give rise to a population of new immigrants.

of new species will not increase indefinitely. Why not? Studies of community patterns on islands around the world provide us with two models.

First, islands far from a source of potential colonists receive few colonizing species, and the few that arrive are adapted for long-distance dispersal (Figure 42.18). This is the **distance effect**. Second, larger islands tend to support more species than smaller islands at equivalent distances from source areas. This is the **area effect**. The larger islands tend to have more and varied habitats; most have more complex topography and extend higher above sea level. Thus they favor species diversity. Also, being bigger targets, they intercept more colonists.

Most importantly, extinctions suppress biodiversity on small islands. There, populations are small and far more vulnerable to storms, volcanic eruptions, disease, and random shifts in birth and death rates. As for any island, the number of species reflects a balance between

immigration rates for new species and extinction rates for established ones. Small islands that are distant from a source of colonists have low immigration rates and high extinction rates, so they support few species once the balance has been struck for their populations.

For most groups of organisms, the number of coexisting species is highest in the tropics and systematically declines from the equator to the poles.

A region's biodiversity depends on many factors, such as its climate, topographical variation, possibilities for dispersal, and evolutionary history—including extinctions and the time span available for speciation.

Disturbances to habitats tend to work against competitive exclusion and therefore to favor increases in biodiversity.

Island biodiversity is a balance between immigration rates for new species and extinction rates for established species.

SUMMARY   *Gold* indicates text section

1. A habitat is the type of place where individuals of a given species normally live—that is, their "address." A community consists of all populations of all the species that occupy the habitat and that directly or indirectly associate with one another. *42.1*

2. Each species has its own niche, or "profession," in the community. This is defined as the sum of activities and relationships in which its members engage as they secure and use the resources they require to survive and reproduce. *42.1*

3. Mutualism, commensalism, competition, predation, and parasitism are species interactions that directly or indirectly link populations in a community. *42.1–41.4*

4. Two species requiring the same limited resource tend to compete, as by exploiting it as fast or efficiently as possible or interfering with the other's use of it. *42.3*

 a. According to the competitive exclusion concept, when two (or more) species require identical resources, they cannot coexist indefinitely.

 b. Species are more likely to coexist if they differ in their use of resources. They may also coexist by using a shared resource in different ways or at different times.

5. Some predator and prey populations may coexist at stable levels if predators keep prey from overshooting the carrying capacity. Others show recurring or erratic cycles of abundance and crashes. Delays in predator responses to changes in prey density, and changes in the prey's food supplies, influence the cycles. *42.4*

6. Predators and their prey coevolve. After a novel, heritable trait appears and gives prey an edge in their coevolutionary contest, selection pressure works on the predator populations, which may evolve in response. The same occurs when a novel, heritable trait arises in a predator population. *42.4, 42.5*

 a. Evolved defenses by prey include threat displays, chemical weapons, mimicry, and camouflaging.

 b. Predators overcome defenses of prey by adaptive behavior (including stealth), camouflaging, and so on.

7. Parasites and hosts coevolve; the resistant hosts and only moderately harmful parasites are favored. *42.4*

8. By the classical model for ecological succession, a community develops in predictable sequence, from its pioneer species to an end array of species that persists over an entire region. *42.6*

9. A stable, self-perpetuating array of species that are in equilibrium with one another and with a particular environment is called a climax community. *42.6*

10. Similar climax stages may persist within the same region yet show variation as a result of environmental gradients and species interactions. *42.6*

11. Recurring, small-scale changes are an aspect of the internal dynamics of communities. Other changes, such as long-term shifts in climate and species introductions, may permanently alter community structure. *42.7, 42.8*

 a. Community structure reflects an uneasy balance of forces, including predation (as by keystone species) and competition, that have been operating over time.

 b. Community structure can change permanently by the introduction of species that one way or another expanded their geographic range. Dispersals may occur when a population expands its home range gradually. Jump dispersals rapidly put individuals into distant habitats. Extremely slow dispersals also have occurred over geologic time as a result of plate tectonics.

12. The number of species in a community depends on the size of a region, the colonization rate, disturbances, and extinction rates. It depends also on the levels and patterns of resource availability. *42.9*

13. Biodiversity tends to be highest in the tropics and to decline systematically toward polar regions, except during times of mass extinction. *42.9*

Review Questions

1. Distinguish between the habitat and the niche of a species. Why is it difficult to define "the human habitat"? *42.1*

2. Describe competitive exclusion. How might two species that compete for the same resource coexist? *42.3*

3. Define the difference between a predator and a parasite. *42.4*

4. Define primary and secondary succession. *42.6*

5. What is a climax community? How does the climax-pattern model help explain its structure? *42.6*

Self-Quiz   ANSWERS IN APPENDIX III

1. A habitat _____ .
 a. has distinguishing physical and chemical features
 b. is where individuals of a species normally live
 c. is occupied by various species
 d. both a and b
 e. a through c

2. A two-way flow of benefits in mutualistic interactions between species is an outcome of _____ .
 a. close cooperativeness   c. resource partitioning
 b. two-way exploitation   d. competitive coexistence

3. A niche _____ .
 a. is the sum of activities and relationships in a a community by which individuals of a species secure and use resources
 b. is unvarying for a given species
 c. shifts in large and small ways
 d. both a and b
 e. both a and c

4. Two species in one habitat can coexist when they _____ .
 a. differ in their use of resources
 b. share the same resource in different ways
 c. use the same resource at different times
 d. all of the above

Figure 42.19 (**a**) Flesh fly and (**b**) weevil (*Zygops*).

5. A predator population and prey population _____ .
   a. always coexist at relatively stable levels
   b. may undergo cyclic or irregular changes in density
   c. cannot coexist indefinitely in the same habitat
   d. both b and c

6. Parasites _____ .
   a. tend to kill their hosts   c. feed on host tissues
   b. can kill novel hosts   d. both b and c

7. In _____ , a disturbed site in a community recovers and moves again toward the climax state.
   a. the area effect   c. primary succession
   b. the distance effect   d. secondary succession

8. The biodiversity of a given region is an outcome of _____ .
   a. climate and topography   d. both a and b
   b. possibilities for dispersal e. a through c
   c. evolutionary history

9. Match the terms with the most suitable descriptions.
   ____ geographic   a. opportunistic colonizer of barren
        dispersal   or disturbed places
   ____ area effect   b. dominates community structure
   ____ pioneer   c. individuals leave home range, become
        species   successfully established elsewhere
   ____ climax   d. more biodiversity on large islands than
        community   small at same distance from source
   ____ keystone   e. stable, self-perpetuating array
        species   of species

Critical Thinking

1. Think of possible examples of competitive exclusion besides the ones used in this chapter, as by considering some of the animals and plants living in your own neighborhood.

2. Flesh flies (Figure 42.19*a*) have a gray and black body, red eyes, and red tail ends. They are fast fliers, and bird predators soon give up trying to catch them. Many sluggish insects, such as the weevil *Zygops rufitorquis*, resemble flesh flies (Figure 42.19*b*). This appears to be a case of _____ .

3. Frog populations all over the world declined dramatically during the 1990s. Individual frogs were being born with bizarre deformities, such as too few or too many legs (Sections 24.6 and 32.8). Should we view what is happening to frogs as a warning of drastically changing environments that will affect us, also? Are the same declines and deformities occurring in all aquatic habitats, or are there different problems in different regions? Use the library or a computer search engine to find information. (Try these key words: *frog deformities*.) Then write a brief list of possible answers to the preceding questions.

4. The water hyacinth (*Eichhornia crassipes*) is an aquatic plant native to South America. In the 1880s, someone fancied the plant's blue flowers and displayed it at an exposition in New Orleans. Other flower fanciers took home clippings and put them in ponds and streams. Unchecked by natural predators, the fast-growing hyacinths spread through nutrient-rich waters, displaced many native species, and choked rivers and canals

Figure 42.20 Water hyacinths choking a waterway in Florida.

(Figure 42.20). They have spread as far west as San Francisco. Do some research to see whether the United States now limits imports and, if so, how effective the restrictions have been.

5. Somewhere between predators and parasites are *parasitoids*, insect larvae that always kill what they eat. Peter Price studied the evolutionary effects of a parasitoid wasp that lays eggs on sawfly cocoons. Sawflies lay eggs in trees. Fertilized eggs give rise to fly larvae, which eat, grow, then drop to the forest floor. The larvae spin cocoons after burrowing into leaf litter. Some burrow deeper than others. The first adult sawflies emerge from cocoons that were the least deeply buried. Later in the season, more sawflies emerge from the more deeply buried ones.

These parasitoid wasps tend to lay eggs on the cocoons closest to the surface of leaf litter. They exert strong selection pressure on the sawfly population, for deep-burrowing sawfly individuals are more likely to escape detection. There are only so many flies near the surface, so the wasp able to locate cocoons deeper in the litter will be better at getting food for her larvae. Explain how the host species stays ahead in this coevolutionary contest.

Selected Key Terms

area effect (on island   keystone species *42.3*
  biodiversity) *42.9*   macroparasite *42.4*
camouflage *42.5*   microparasite *42.4*
carrying capacity *42.4*   mimicry *42.5*
climax-pattern model *42.6*   mutualism *42.1*
cimax community *42.6*   niche *42.1*
coevolution *42.4*   parasite *42.4*
commensalism *42.1*   parasitism *42.1*
community *42.1*   pioneer species *42.6*
competitive exclusion *42.3* predation *42.1*
distance effect (on island   predator *42.4*
  biodiversity) *42.9*   prey *42.4*
ecological succession *42.6* primary succession *42.6*
endangered species *42.8*   resource partitioning *42.3*
exotic species *42.8*   secondary succession *42.6*
geographic dispersal *42.7*   social parasite *42.4*
habitat *42.1*   symbiosis *42.1*
host *42.4*   warning coloration *42.5*
interspecific competition *42.1*

Readings

Krebs, C., et al. January 2000. "What Drives the 10-Year Cycle of Snowshoe Hares?" *Bioscience* 51(1): 25–35.

Smith, R. 1996. *Elements of Ecology*. Fifth edition. New York: HarperCollins.

43

ECOSYSTEMS

Crêpes for Breakfast, Pancake Ice for Dessert

Think of Antarctica and you think of ice. Mile-thick slabs of the stuff hide all but a small fraction of a vast continent whipped by fierce winds and kept frozen by murderously low temperatures, on the order of –100°F. Yet, on patches of exposed rocky soil and on nearby islands, bacteria, lichens, and mosses endure. During the breeding season, a variety of penguins and seals form great noisy congregations, then reproduce and raise offspring. They cruise offshore or venture out in the open ocean in pursuit of krill, fishes, and squids.

The first explorers of Antarctica called it "The Last Place on Earth." For all of its harshness, however, the beauty of this ecosystem fired the imagination of those first travelers and others who followed them.

In 1961, thirty-eight nations signed a treaty to set aside Antarctica as a reserve for scientific research. This was the start of scientific outposts—and of the trashing of Antarctica. At first, researchers discarded a few oil drums and old tires. Then prefabricated villages went up. Onto the ice and into the water went garbage, used equipment, sewage, and chemical wastes.

Antarctica even became a destination of cruise ships (Figure 43.1a). Every summer thousands of tourists, fortified by three sumptuous meals a day plus snacks, are ferried from ship to land across channels glistening with pancake ice. They trample the sparse vegetation and bob around the penguins, cameras clicking.

We do not know what tourism's long-term impact will be, but it pales when compared to other assaults. Nations looking for new sources of food started licking their chops over Antarctica's krill. Word got out about potentially rich deposits of uranium, oil, and gold. By

Figure 43.1 (a) A boatload of tourists crossing a channel of "pancake ice" off the coast of Antarctica. (b) On the shore of an island near Antarctica, tourists get close to nature—maybe too close.

1988, treaty nations were poised to authorize digs and drillings. But oil spills or commercial harvesting of krill could easily destroy the fragile ecosystem. For instance, tiny, shrimplike krill are the only food source for Adélie penguins and a key food for baleen whales and other marine animals which are, in turn, food for still others.

In 1991, the treaty nations thought about all of this and imposed a fifty-year ban on mineral exploration. Research stations started to bury or incinerate wastes, treat raw sewage, and take other pollution-curbing steps. Tour operators promised to supervise tourists. Krill are not being harvested on a massive scale. Yet.

Because Antarctica seems remote from the rest of the world, it is easier to see how species interconnect there. We can ask whether harm to one species or one habitat will lead to collapse of the whole and be fairly sure of the answer. What about places not as sharply defined? Are they as vulnerable to disturbance or more resilient? We won't know for sure until researchers gain deeper insights into the evolutionary and ecological histories of their species, including the suspected great numbers of species we haven't even discovered yet.

In the meantime, biodiversity is rapidly declining in ecosystems on land and in the seas. Competition from exotic species is one threat; the human propensity to overexploit species and destroy habitats is another. *Conservation biology* is an attempt to counter the threat by working to lower current extinction rates and design management programs that will help sustain the most vulnerable ecosystems. Chapter 25 introduced the challenges. To gain more insight into the magnitude of the task, start with a few principles of how ecosystems work. They are the topic of this chapter.

Key Concepts

1. An ecosystem is an association of organisms and their physical environment, interconnected by an ongoing flow of energy and a cycling of materials through it.

2. An ecosystem is an open system, with inputs, internal transfers, and outputs of energy and nutrients.

3. Energy flows in only one direction through every ecosystem. Most commonly, energy flow begins when photosynthetic autotrophs harness sunlight energy and convert it to forms that they and other organisms of the ecosystem can use. Autotrophs are primary producer organisms for the ecosystem.

4. Energy-rich organic compounds that the primary producers synthesize become incorporated in their body tissues. They are stored forms of energy, and they serve as the foundation for the ecosystem's food webs. Such webs consist of a number of interconnected food chains.

5. Each chain in a food web extends in a straight-line sequence from producers through various consumers, decomposers, and detritivores, to the highest consumer level.

6. Over time, most of the energy that entered a food web is lost to the environment, mainly in the form of metabolic heat. Nutrients are cycled within food webs, although some amounts are lost to the environment.

7. The water, carbon, nitrogen, phosphorus, and other substances required for primary productivity move through biogeochemical cycles that are global in scale. Ions or molecules of the substances move slowly from environmental reservoirs, then rapidly among organisms of food webs, then back to their reservoirs.

8. Human activities are disrupting the natural cycles of materials and are thereby endangering ecosystems.

b

THE NATURE OF ECOSYSTEMS

Overview of the Participants

Diverse natural systems abound on the Earth's surface. In climate, landforms, soil, vegetation, animal life, and other features, deserts differ from hardwood forests, which differ from tundra and prairies. In biodiversity and physical properties, the open seas differ from reefs, which differ from lakes. *Yet despite the differences, such systems are alike in many aspects of structure and function.*

With few exceptions, these systems run on sunlight captured by autotrophs (self-feeders). Plants and other photosynthesizers are the most common autotrophs. They convert sunlight energy to chemical energy and use it to produce organic compounds from inorganic raw materials. Because they secure energy directly from their environment, they are **primary producers** for the entire system (Figure 43.2).

All other organisms in the system are heterotrophs, not self-feeders. They extract energy from compounds that primary producers put together. **Consumers** feed on tissues of other organisms. Those consumers called *herbivores* eat only plants. By converting energy stored in plant tissues into animal tissues, they become food for *carnivores* and *parasites*, which feed on living animal tissues. Nonliving products and remains of producers and consumers feed **decomposers**. These heterotrophic fungi and bacteria practice extracellular digestion and absorption: they secrete enzymes that digest organic compounds in the surroundings, and their cells absorb some breakdown products. In this respect, the decomposers are distinct from detritivores. **Detritivores** are heterotrophs that ingest decomposing bits of organic matter, such as leaf litter. Earthworms are examples.

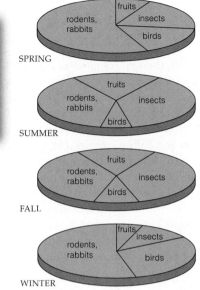

Figure 43.3 Red fox, an omnivore. Its diet shifts with seasonal changes in available food. In summer, it feeds mainly on fruits and insects. In spring and fall, it focuses mainly on rodents and rabbits.

Not all consumers fall into simple categories. Many are *omnivores*, which dine on assorted animals, plants, fungi, protistans, even bacteria. The red fox is like this (Figure 43.3). Also, a fox that comes across a dead bird may scavenge it. A *scavenger* is any animal that ingests nonliving plant and animal tissues all or some of the time. Vultures, termites, and many beetles do this.

What are the nutrients for the system? The primary producers take up small inorganic compounds—water and carbon dioxide—from the environment as sources for hydrogen, oxygen, and carbon. They also take up phosphorus, nitrogen, and other minerals. These are the raw materials for carbohydrates, lipids, proteins, and nucleic acids. Nutrients return to the environment as decomposers and detritivores work over the remains of organisms. And unless something removes nutrients from the system, as when minerals end up in a stream that flows away from a meadow, autotrophs typically take them up again.

What we have just described in broad outline is an ecosystem. An **ecosystem** is an array of organisms and their physical environment, interacting through a one-way flow of energy and a cycling of materials. It is an open system, unable to sustain itself. It runs on *energy inputs* (as from the sun) and often *nutrient inputs* (as from a creek delivering dissolved minerals into a lake). It has *energy outputs* and *nutrient outputs*. The energy cannot be recycled. Over time, most of the energy that autotrophs fixed is lost to the environment, mainly as metabolically generated heat. Nutrients are cycled, but some slip out. Most of this chapter explores ecosystem inputs, internal transfers, and outputs.

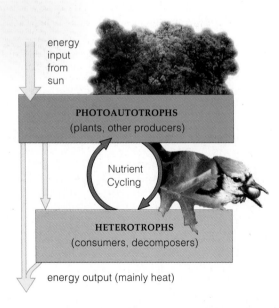

Figure 43.2
Simple model of ecosystems. Energy flows in one direction: into the ecosystem, through its living organisms, and then out from it. Its nutrients are cycled among autotrophs and heterotrophs. For this model, energy flow starts with autotrophs that can capture energy from the sun.

energy input from sun

PHOTOAUTOTROPHS
(plants, other producers)

Nutrient Cycling

HETEROTROPHS
(consumers, decomposers)

energy output (mainly heat)

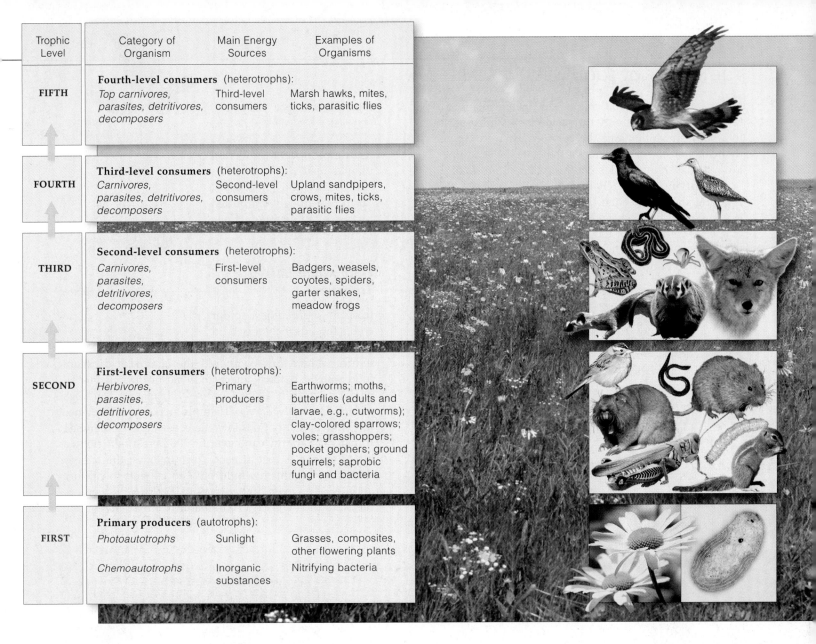

Trophic Level	Category of Organism	Main Energy Sources	Examples of Organisms
FIFTH	**Fourth-level consumers** (heterotrophs): *Top carnivores, parasites, detritivores, decomposers*	Third-level consumers	Marsh hawks, mites, ticks, parasitic flies
FOURTH	**Third-level consumers** (heterotrophs): *Carnivores, parasites, detritivores, decomposers*	Second-level consumers	Upland sandpipers, crows, mites, ticks, parasitic flies
THIRD	**Second-level consumers** (heterotrophs): *Carnivores, parasites, detritivores, decomposers*	First-level consumers	Badgers, weasels, coyotes, spiders, garter snakes, meadow frogs
SECOND	**First-level consumers** (heterotrophs): *Herbivores, parasites, detritivores, decomposers*	Primary producers	Earthworms; moths, butterflies (adults and larvae, e.g., cutworms); clay-colored sparrows; voles; grasshoppers; pocket gophers; ground squirrels; saprobic fungi and bacteria
FIRST	**Primary producers** (autotrophs): *Photoautotrophs*	Sunlight	Grasses, composites, other flowering plants
	Chemoautotrophs	Inorganic substances	Nitrifying bacteria

Figure 43.4 Simplified picture of trophic levels in tallgrass prairie, a type of grassland in parts of the American Midwest.

Structure of Ecosystems

We can classify all organisms of an ecosystem by their functional roles in a hierarchy of feeding relationships called **trophic levels** (*troph*, nourishment). "Who eats whom?" we may ask. When organism **B** eats organism **A**, energy gets transferred to **B** from **A**. All organisms at a given trophic level are the same number of transfer steps away from some energy input into an ecosystem.

Figure 43.4 shows organisms of a prairie ecosystem. Being closest to the energy input (sunlight), plants and other producers are at the first trophic level. They feed primary consumers, such as grasshoppers (herbivores) and earthworms (detritivores) at the next trophic level. Primary consumers feed carnivores and parasites at the third trophic level, and so on up the hierarchy.

At each trophic level, organisms interact with the same sets of predators, prey, or both. The omnivores feed at several levels, so they are partitioned among different levels or assigned one of their own.

A **food chain** is a straight-line sequence of steps by which energy stored in autotroph tissues passes on to higher trophic levels. We have a hard time finding such simple, isolated chains in nature. Why? Species usually belong to more than one food chain, especially species at a low trophic level. It is more accurate to visualize food chains *cross-connecting* with one another, as **food webs**. And that is the topic of the next section.

Producers, consumers, decomposers, detritivores, and their physical environment make up an ecosystem. A one-way energy flow and a cycling of materials interconnects them.

A food chain, a straight-line sequence of who eats whom in an ecosystem, starts with autotrophs. A food web is a network of cross-connected food chains.

THE NATURE OF FOOD WEBS

Make a sketch of a food chain—say, from a plant in a tallgrass prairie, to a cutworm that eats its juicy parts, to a baby snake that eats the cutworm, to a sandpiper that eats the snake, and a hawk that eats the sandpiper. Your sketch should look something like Figure 43.5.

Identifying a food chain is a simple way to start thinking about who eats whom in ecosystems, but this is just part of the picture. Most often, a complex array of species competes for food, particularly at the lower trophic levels. For instance, as the food web in Figure 43.6 shows, tallgrass

Figure 43.5 *Right:* Example of a simple food chain.

prairie producers (flowering plants) feed diverse insects and herbivorous mammals. Yet it is a hugely simplified picture of all the interactions in this ecosystem!

How Many Energy Transfers?

Imagine separating a spider web's threads. Ecologists did this for many food webs in nature. When they compared the chains of different food webs, a pattern emerged: *In most cases, energy that producers initially captured passes through no more than four or five trophic levels.* It simply makes no difference how much energy goes into the ecosystem. Remember, transfers of energy can never be 100 percent efficient; a bit of energy is lost at each step. In time, the energy that some organism would have to expend to capture another at an even higher trophic level would be more the amount of energy obtainable from it. Even rich

MARSH HAWK

UPLAND SANDPIPER

GARTER SNAKE

CUTWORM

PLANTS

HIGHER TROPHIC LEVELS

Complex array of carnivores, omnivores, and other consumers. Many feed at more than one trophic level all the time, seasonally, or when an opportunity presents itself

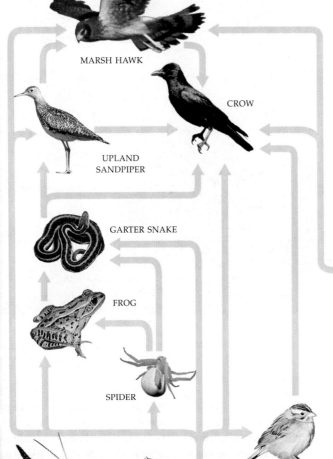

MARSH HAWK

UPLAND SANDPIPER

CROW

GARTER SNAKE

FROG

SPIDER

WEASEL

BADGER

COYOTE

SECOND TROPHIC LEVEL

Primary consumers (such as herbivores)

CLAY-COLORED SPARROW

PRAIRIE VOLE

POCKET GOPHER

GROUND SQUIRREL

EARTHWORMS, INSECTS (E.G., GRASSHOPPERS, CUTWORMS)

FIRST TROPHIC LEVEL

Primary producers

GRASSES, COMPOSITES

Figure 43.6 Small, highly simplified sampling of connecting interactions in a tallgrass prairie food web.

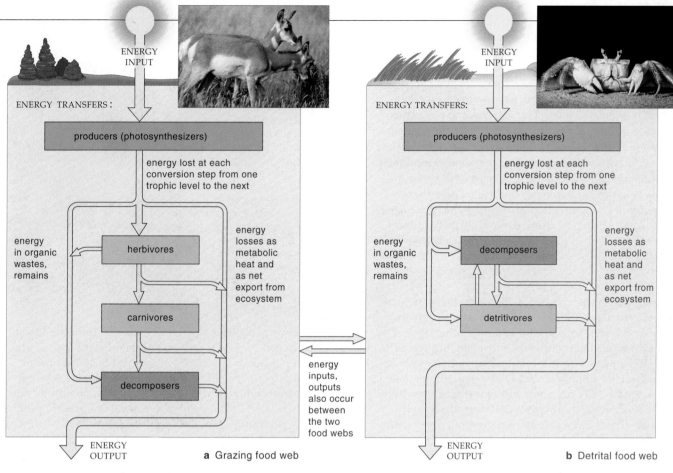

Figure 43.7 One-way flow of energy through (**a**) a grazing food web and (**b**) a detrital food web.

ecosystems that support complex food webs, including the one in Figure 43.6, don't have lengthy food chains.

Field studies and computer simulations of the food webs of marine, freshwater, and terrestrial ecosystems reveal more patterns. For instance, chains of food webs tend to be short where environmental conditions vary. Shifts in temperature, moisture, salinity, and pH are such variables. Chains are longer in stable environments, as in parts of the deep ocean. The most complex webs have the most herbivorous species, but their chains are shortest. You find such webs in grasslands. In contrast, the simplest food webs have more top carnivores.

Two Categories of Food Webs

In what direction does energy flow through ecosystems on land? Plants fix only a small part of the energy from the sun. They store half the energy in chemical bonds of new tissues but lose the rest as metabolic heat. Other organisms draw upon the energy that became stored in plant tissues, remains, and wastes. Then they, too, lose metabolic heat. Taken together, *all of these heat losses represent a one-way flow of energy out of the ecosystem.*

Energy from a primary source flows in one direction through two kinds of webs. In a **grazing food web**, it goes from photoautotrophs to herbivores, then through carnivores. By contrast, in a **detrital food web**, energy flows from photoautotrophs through detritivores and decomposers (Figure 43.7). Both webs cross-connect in nearly all ecosystems. For example, a connection forms when an opportunistic herring gull of a grazing food web gulps down a crab of a detrital food web.

The amount of energy moving through food webs differs from one ecosystem to the next and often varies with the seasons. In most cases, however, most of the net primary production moves through detrital food webs. You may doubt this. After all, when cattle graze heavily on pasture plants, about half of the net primary production enters a grazing food web. But cattle do not use all of the stored energy. A lot of undigested plant parts and feces become available for the decomposers and detritivores. Marshes are a similar case in point. In these aquatic ecosystems, most of the stored energy is not used until after parts of the marsh grasses die and become available for detrital food webs.

The loss of energy at each transfer in a food chain limits the number of trophic levels in ecosystems to four or five.

Tissues of living photosynthesizers are the basis of grazing food webs. Remains and wastes of photosynthesizers and consumers are the basis of detrital food webs.

Biological Magnification in Food Webs

ECOSYSTEM MODELING We turn now to a premise that opened this chapter—that disturbances to one part of an ecosystem often can have unexpected effects on other, seemingly unrelated parts. One approach to predicting unforeseen effects of disturbances is through **ecosystem modeling**. With this method, researchers identify crucial bits of data on different ecosystem components, and then use computer programs and models to combine the bits. Results help predict outcomes of the next disturbance.

A potential danger is that investigators may not have identified all of the key relationships in the ecosystem under study and incorporated them accurately into the computer model. The most crucial fact may be one that we do not yet know, as the following example makes clear.

DDT IN FOOD WEBS DDT, recall, is a synthetic organic pesticide (Section 3.7). This fairly stable hydrocarbon is nearly insoluble in water, so you might think that it would exert its effects only where applied. But winds carry DDT in vapor form; water moves fine particles of it. Being highly soluble in fats, DDT can accumulate in tissues. Hence DDT can show **biological magnification**. By this occurrence, a substance that degrades slowly or not at all gets more and more concentrated in tissues of organisms at higher trophic levels of a food web.

Most of the DDT that becomes concentrated in all the organisms that a consumer will eat during its lifetime will end up in the consumer's own tissues. DDT and modified forms of it disrupt metabolic activities and are toxic to many aquatic and terrestrial animals.

Figure 43.8 Peregrine falcon, a top carnivore of certain food webs. It almost became extinct as a result of biological magnification of DDT. A wildlife management program successfully increased its population sizes. Peregrine falcons were reintroduced into wild habitats. They also have adapted to cities. There, these raptors hunt pigeons, large populations of which are·a messy nuisance.

Several decades ago, DDT started to infiltrate food webs and act on diverse organisms in ways that no one had predicted. Where it was sprayed to control Dutch elm disease, songbirds died. In small streams flowing in forests where it was sprayed to kill budworm larvae, fish died. In fields sprayed to control one kind of pest, new pests moved in. *DDT was indiscriminately killing the natural predators that keep pest populations in check!*

Then side effects of biological magnification showed up in habitats far removed from where DDT had been applied—*and much later in time.* Most vulnerable were brown pelicans, bald eagles, peregrine falcons, and other top carnivores of some food webs (Figures 43.8 and 43.9).

Why? A DDT breakdown product interferes with physiological processes. As one outcome, bird eggs developed brittle shells; many chick embryos did not even hatch. Some species faced extinction.

In the United States, DDT has been banned since the 1970s except where necessary to protect public health. Many species hit hardest have recovered somewhat. Some birds still lay thin-shelled eggs; they pick up DDT at their winter ranges in Latin America. Even as late as 1990, a fishery near Los Angeles was closed. DDT from industrial waste discharges that had stopped twenty years before was still contaminating that ecosystem.

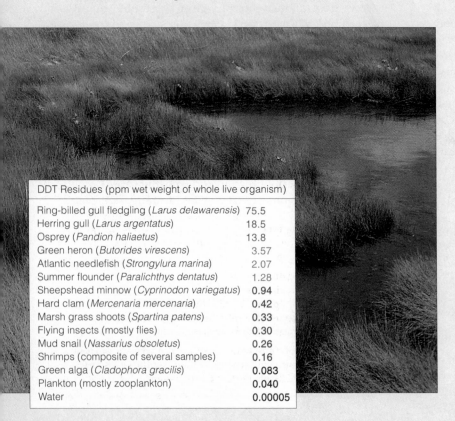

DDT Residues (ppm wet weight of whole live organism)	
Ring-billed gull fledgling (*Larus delawarensis*)	75.5
Herring gull (*Larus argentatus*)	18.5
Osprey (*Pandion haliaetus*)	13.8
Green heron (*Butorides virescens*)	3.57
Atlantic needlefish (*Strongylura marina*)	2.07
Summer flounder (*Paralichthys dentatus*)	1.28
Sheepshead minnow (*Cyprinodon variegatus*)	0.94
Hard clam (*Mercenaria mercenaria*)	0.42
Marsh grass shoots (*Spartina patens*)	0.33
Flying insects (mostly flies)	0.30
Mud snail (*Nassarius obsoletus*)	0.26
Shrimps (composite of several samples)	0.16
Green alga (*Cladophora gracilis*)	0.083
Plankton (mostly zooplankton)	0.040
Water	0.00005

Figure 43.9 Biological magnification in an estuary on the south shore of Long Island, New York, as reported in 1967 by George Woodwell, Charles Wurster, and Peter Isaacson. The researchers knew of broad correlations between the extent of DDT exposure and mortality. For instance, residues in birds known to have died from DDT poisoning were 30–295 ppm, and they were 1–26 ppm in several fish species. Some DDT concentrations measured during this study were below lethal thresholds but were still high enough to interfere with reproductive success.

SEDIMENTARY CYCLES

A Look at the Phosphorus Cycle

We continue our survey of biogeochemical cycling with one of the sedimentary cycles. In the **phosphorus cycle**, phosphorus is passed through food webs as it moves from land, to ocean sediments, then back to land (Figure 43.16). The largest reservoir is the Earth's crust, just as it is for other minerals.

Most phosphorus in rock formations is in the form of phosphates. Weathering and erosion put phosphates into streams and rivers, which move them to sediments in the seas. Phosphorus collects mainly on the submerged continental shelves. Along with other minerals, it forms insoluble deposits. Millions of years go by. Where movements of crustal plates uplift part of the seafloor, phosphates are exposed on drained land surfaces. Over geologic time, weathering releases phosphates from the exposed rocks, so the cycle's geochemical phase begins again.

The cycle's ecosystem phase is brief, compared with the long-term geochemical phase. All organisms need phosphorus for synthesizing phospholipids, NADPH, ATP, nucleic acids, and other compounds. Plants take up dissolved, ionized forms of phosphorus so fast and efficiently that they often reduce its concentrations in soil to extremely low levels. Herbivorous animals get phosphates by dining on plants; carnivores obtain it by dining on the herbivores. Both types of animals excrete phosphates as a waste in urine and feces. Decomposers in soil release phosphates. Plants take up the mineral and conribute to its rapid cycling in the ecosystem.

Eutrophication

Of all minerals, phosphorus is the main limiting factor in natural ecosystems. Soils hold little of it; sediments tie up most of it in aquatic ecosystems. Phosphorus doesn't have a gaseous phase, so ecosystems lose little of their scarce supply to the atmosphere. It helps that phosphorus resists leaching. Producers are adapted to respond efficiently to low phosphorus concentrations.

Phosphorus gets concentrated in eroded sediments, agricultural runoff, and outflows from industries and sewage treatments. Dissolved phosphorus that enters streams, rivers, lakes, and estuaries can promote dense algal blooms. Like plants, photosynthetic algae cannot grow without phosphorus, nitrogen, and other minerals. Freshwater ecosystems usually have plenty of nitrogen-fixing bacteria, so nitrogen is not a limiting factor on growth of algae. Phosphorus is. Decomposition of the remains from algal blooms depletes water of oxygen, which kills off the fish and other organisms.

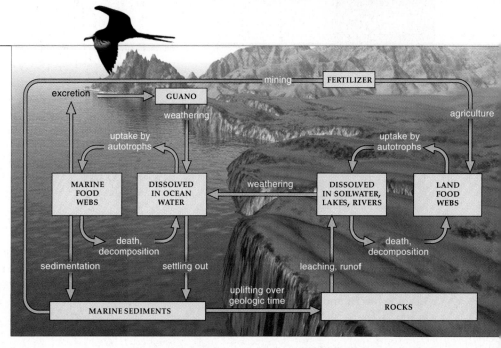

Figure 43.16 The phosphorus cycle. Most of the phosphorus moves in the form of phosphate ions (PO_4^{3-}).

Figure 43.17 Experiment demonstrating lake eutrophication in Ontario, Canada. Researchers stretched a plastic curtain across a channel between two basins of the same lake. They added phosphorus, carbon, and nitrogen to one basin (the basin in the *background*) and carbon and nitrogen to the other (the basin in the *foreground*). Within two months, the phosphorus-enriched basin showed signs of accelerated eutrophication; a dense algal bloom turned the water green.

Eutrophication is the name for nutrient enrichment of an ecosystem that is naturally low in nutrients. As you will be reading in Section 44.10, eutrophication is a natural process, but phosphorus inputs accelerate it. Field experiments, including the one shown in Figure 43.17, nicely demonstrate this outcome.

Sedimentary cycles, in combination with the hydrologic cycle, move phosphorus and most of the other mineral elements through terrestrial and aquatic ecosystems.

CARBON CYCLE

In an essential atmospheric cycle, carbon moves through the lower atmosphere and all food webs on its way to and from the ocean's water and sediments, and rocks. Its global movement is the **carbon cycle** (Figure 43.18). Sediments and rocks hold most of the carbon, followed by the ocean, then soil, the atmosphere, and biomass on land. It enters the atmosphere as cells engage in aerobic respiration, when fossil fuels burn, and when volcanoes erupt and so release the carbon from rocks that make up the Earth's crust. Most atmospheric carbon occurs in the form of carbon dioxide (CO_2). Carbon dissolved in the ocean is mainly in the forms of bicarbonate and carbonate.

When watching bubbles escape from a glass of carbonated soda left in the sun, you might wonder: Why doesn't all the CO_2 dissolved in the warm ocean surface waters escape to the atmosphere? Driven by winds and regional differences in water density, water makes a gigantic loop from the surface of the Pacific and Atlantic oceans to the seafloor of the Atlantic and Antarctic seafloors. There, the dissolved CO_2 moves into deep storage reservoirs before the seawater loops back up (Figure 43.19). This loop of ocean water is a key factor in carbon's distribution and global budget:

a

MAIN CARBON RESERVOIRS AND HOLDING STATIONS:

	p (10^{15} grams)
Sediments and rocks	
Ocean (dissolved forms)	39,700
Soil	1,500
Atmosphere	750
Biomass on land	715

ANNUAL FLUXES IN GLOBAL DISTRIBUTION OF CARBON:

From atmosphere to plants (carbon fixation)	120
From atmosphere to ocean	107
To atmosphere from ocean	105
To atmosphere from plants	60
To atmosphere from soil	60
To atmosphere from fossil fuel burning	5
To atmosphere from net destruction of plants	2
To ocean from runoff	0.4
Burial in ocean sediments	0.1

When photosynthetic autotrophs engage in carbon dioxide fixation, they lock up billions of metric tons of carbon atoms in organic compounds each year (Sections

Figure 43.18 (**a**) Global carbon cycle. The portion of the diagram on this page shows the movement of carbon through typical marine ecosystems. The *facing page* shows carbon's movement through ecosystems on land. The *gold* boxes indicate the main carbon reservoirs, the storage volumes of which are listed in the text.

Figure 43.19 Loop of ocean water that delivers carbon dioxide to its deep ocean reservoir. It sinks in the cold, salty North Atlantic and rises in the warmer Pacific.

The figure shows a diagram of the carbon cycle overlaid on a landscape photograph, with the following labeled boxes and arrows:

ATMOSPHERE (mainly carbon dioxide)

volcanic action

combustion of fossil fuels

photosynthesis

aerobic respiration

combustion of wood (for clearing land; or for fuel)

TERRESTRIAL ROCKS

weathering

LAND FOOD WEBS
producers, consumers, decomposers, detritivores

deforestation

SOIL WATER
(dissolved carbon)

death, burial, compaction over geologic time

PEAT, FOSSIL FUELS

leaching, runoff

6.8 and 43.4). Yet the average time that an ecosystem holds a given carbon atom varies greatly. For example, organic wastes and remains decompose so quickly in tropical rain forests that not much carbon accumulates at soil surfaces. In bogs, marshes, and other anaerobic sites, decomposers cannot degrade organic compounds to smaller bits, so carbon gradually accumulates in peat and other forms of compressed organic matter.

Also, in food webs of ancient aquatic ecosystems, carbon became incorporated in staggering numbers of shells and other hard body parts. Foramiferans and other shelled organisms died, sank through water, then were buried in sediments. The carbon remained buried for many millions of years in deep sediments until part of the seafloor was uplifted above the ocean surface by geologic forces of the sort described in Section 18.2. Remember those white cliffs of Dover (Section 20.8)?

As a final example, plants of the vast swamp forests of the Carboniferous also incorporated carbon atoms in their organic compounds. As you read in Section 22.4, those compounds were gradually converted into coal, petroleum, and gas reserves—which humans now tap as fossil fuels. Over the course of a few hundred years,

we have been extracting and burning fossil fuels that took millions of years to form. A major problem is that fossil fuels are nonrenewable resources (Chapter 45). Another problem is that they are the energy source of choice for a huge portion of the 6.1 billion people on Earth—mainly in the developed countries, which can afford them. At present, fossil fuel burning and other human activities are releasing more carbon into the atmosphere than can be cycled naturally to the ocean's storage reservoirs.

As you might deduce from the amounts listed on the preceding page, the ocean can remove only about 2 percent of the excess carbon entering the atmosphere. Most scientists now think the excess is amplifying the greenhouse effect, which means it may be contributing to global warming. The next section looks at this effect and some possible outcomes of modifications to it.

The ocean and atmosphere interact in the global cycling of carbon. Fossil fuel burning and other human activities appear to be contributing to imbalances in the global carbon budget.

From Greenhouse Gases to a Warmer Planet?

GREENHOUSE EFFECT   Atmospheric concentrations of gaseous molecules play a profound role in shaping the average temperature near the surface of the Earth. That temperature has enormous effects on the global climate.

Countless molecules of carbon dioxide, water, ozone, methane, nitrous oxide, and chlorofluorocarbons are key players in interactions that dictate the global temperature. Collectively, the gases act somewhat like a pane of glass in a greenhouse—hence their name, "greenhouse gases." Wavelengths of visible light can pass around them and reach the Earth's surface. But greenhouse gases impede the escape of longer, infrared wavelengths—heat—from the Earth into space. How? Gaseous molecules can absorb these wavelengths and radiate much of the absorbed energy back toward the Earth, as shown in Figure 43.20.

The constant radiation of heat energy from greenhouse gases toward Earth proceeds lockstep with the constant bombardment and absorption of wavelengths from the sun, and so heat builds up in the lower atmosphere. The **greenhouse effect** is the name for this warming action.

GLOBAL WARMING DEFINED   Without the action of greenhouse gases, the Earth's surface would be cold and lifeless. However, there can be too much of a good thing. Largely as an outcome of human activities, greenhouse gases are building to atmospheric levels that are higher than they were in the past. Figure 43.21 documents the increase. As many researchers suspect, greenhouse gases may be contributing to long-term higher temperatures at the Earth's surface, an effect called **global warming**. At present, the United States and some other developed countries are generating the most greenhouse gases.

EVIDENCE OF AN INTENSIFIED GREENHOUSE EFFECT
In the 1950s, researchers on Hawaii's highest volcano started measuring the atmospheric concentrations of different greenhouse gases. That remote site is almost free of local airborne contamination and is representative of conditions for the Northern Hemisphere. The Chapter 3

introduction outlined what they found. Briefly, in the Northern Hemisphere, carbon dioxide levels follow the annual cycles of plant growth. They decline in summer when photosynthesis rates are highest. Then they rise in winter, when photosynthesis declines but aerobic respiration is still proceeding.

The troughs and peaks along the graph line in Figure 43.22a show the annual lows and highs. For the first time, scientists saw the integrated effects of carbon balances for land and water ecosystems of an entire hemisphere. Notice the midline of the troughs and peaks in the cycle. *And notice that the concentration is steadily increasing.*

Current data from many satellites, weather stations and balloons, research ships at sea, and supercomputer programs suggest that we are past the point of being able to reverse some impacts of climate change. In 2001, the United Nation's Intergovernmental Panel on Climate Change reported that global warming and climate change are real. They concluded that human activities, including fossil fuel burning, are contributing factors. In the past decade—the warmest on record—hurricanes, floods, tornadoes, and droughts were intense. Alaska saw rare thunderstorms; Hilo, Hawaii, saw homes washed away by 27 inches of rainfall in 24 hours.

CAUSES AND EFFECTS: SOME PREDICTIONS   The lower atmosphere's temperature may spike higher by 2.5 to 10.4 degrees in this century. If that happens, then we might predict these outcomes:

Rising sea level. If the atmosphere warms by only 4°C (7°F), the sea level will rise by about 0.6 meter (2 feet). Why? The ocean surface temperature will increase, and water expands when heated. Also, glaciers and polar ice sheets will melt faster (*Critical Thinking* question 5 in Section 43.12). The volume of water released would flood low coastal regions. High tides and storm waves will exacerbate effects of a long-term rise in sea level. Low islands and coastal plains—which support many cities and agricultural lands—will be under water.

a The sun's rays penetrate the lower atmosphere, warm the Earth's surface.

b The surface radiates heat (infrared wavelengths) to the atmosphere. Some heat escapes into space. But greenhouse gases and water vapor absorb some infrared energy and radiate a portion of it back toward Earth.

c Increased concentrations of greenhouse gases trap more heat near Earth's surface. Sea surface temperature rises, more water evaporates into atmosphere. Earth's surface temperature rises.

Figure 43.20   The greenhouse effect, as executed mainly by carbon dioxide, water vapor, ozone, methane, nitrous oxide, and various chlorofluorocarbons present in the lower atmosphere.

Figure 43.21 Correlation of changes in atmospheric CO_2 levels (*white* graph line) with ice ages and interglacials. The 1985 mean temperature was the baseline for tracking global changes in temperature (*red* graph line). Arrows show start of the last ice age (*red*), last interglacial (*yellow*), and the industrial revolution (*white*).

Increased rainfall and flooding. Global warming will change regional patterns of precipitation. Much of the Northern Hemisphere will get more rain. Snow will melt faster in its great mountain ranges, so spring runoff and summer droughts will be more severe. The already bad flooding and mudslides in the tropics and in California will worsen. Interiors of the great continents will become drier, and deserts will expand.

Heat waves and wildfires. Heat waves will intensify in the Northern Hemisphere, and deaths from heat stroke will probably climb in cities ranging from Shanghai to Chicago. Hotter, drier summers will also invite more wildfires in Alaska, California, Florida, Brazil, and Asia.

Crop failures. Droughts will intensify in the United States, Mexico, Brazil, Africa, Europe, and Asia. Major crop losses will increase global malnutrition and famine.

Contagious disease. Warmer, wetter conditions along the coasts will favor cholera, malaria, and other serious diseases. For example, the West Nile virus, introduced into the United States in 1999, is already expanding its range in the eastern United States.

Water wars. By 2015, an estimated 3 billion people will not have enough fresh water. We will be returning to this prospect in Section 45.7.

Some argue that the global climate is too complex and unpredictable for us to accept data gathered so far. But the majority now agree there is cause for concern. We already see actions being taken. Chapter 45 outlines some of them, such as developing energy alternatives to fossil fuels. Are the actions none too soon? Future climate changes might turn out to be twice as violent as scientists were predicting just five years ago.

a Carbon dioxide (CO_2). Of all human activities, the burning of fossil fuels and deforestation contribute the most to increasing atmospheric levels.

b CFCs. Until restrictions were in place, CFCs were widely used in plastic foams, refrigerators, air conditioners, and industrial solvents.

c Methane (CH_4). Termite activity and anaerobic bacteria living in swamps, landfills, and the stomachs of cattle and other ruminants produce great amounts of methane as a by-product of metabolism.

d Nitrous oxide (N_2O). Denitrifying bacteria produce N_2O in metabolism. Also, fertilizers and animal wastes release enormous amounts; this is especially the case for large-scale livestock feedlots.

Figure 43.22 Recent increases in atmospheric concentrations of four greenhouse gases.

NITROGEN CYCLE

Since the time of life's origin, the atmosphere and ocean have contained nitrogen. This component of all proteins and nucleic acids moves in an atmospheric cycle called the **nitrogen cycle**. Gaseous nitrogen (N_2) makes up 80 percent or so of the atmosphere, the largest reservoir. Successively smaller reservoirs are seafloor sediments, ocean water, soil, land biomass, atmospheric nitrous oxide (N_2O), and the smallest, marine biomass.

The atoms of N_2 are joined by triple covalent bonds ($N \equiv N$). Few organisms have the metabolic means to break them. Only certain bacteria, volcanic action, and lightning convert N_2 into forms that enter food webs.

Of all nutrients required for plant growth, nitrogen often is the scarcest. Nearly all of the nitrogen in soils was put there by nitrogen-fixing organisms. Ecosystems lose it when other bacteria "unfix" the fixed nitrogen. Land ecosystems lose more by leaching, although this is the basis of nitrogen inputs into aquatic ecosystems such as streams, lakes, and the seas (Figure 43.23).

Cycling Processes

Let's follow nitrogen atoms through the portion of the nitrogen cycle that proceeds in an ecosystem on land. They are objects of processes called nitrogen fixation, assimilation and biosynthesis, decomposition, and then ammonification.

In **nitrogen fixation**, a few kinds of bacteria convert N_2 to ammonia (NH_3), which quickly dissolves in the cytoplasm, thus forming ammonium (NH_4^+). In aquatic ecosystems, *Anabaena, Nostoc,* and other cyanobacteria are nitrogen fixers. In many land ecosystems, *Rhizobium* and *Azotobacter* fix nitrogen. Collectively, these bacteria fix about 200 million metric tons of nitrogen each year!

Decomposition and **ammonification** are processes by which bacteria and fungi degrade nitrogenous wastes and remains of organisms. Decomposers use part of the released proteins and amino acids during metabolism. But most of the nitrogen remains in the decay products, in the form of ammonia or ammonium, which plants take up. Nitrifying bacteria also act on the ammonia or ammonium. In **nitrification**, they strip the compounds of electrons, and nitrite (NO_2^-) is the result. Other kinds of bacteria use the nitrite during their metabolism and produce nitrate (NO_3^-), which plants then take up.

Most plants growing in nitrogen-poor soil interact as mutualists with fungi. Their roots form mycorrhizae with fungal hyphae, the collective surface area of which absorbs a lot of mineral ions from soil. Clover, beans, peas, and other legumes are mutualists with nitrogen-

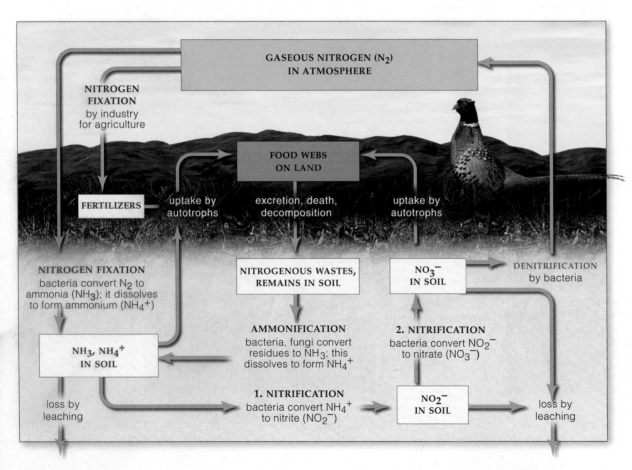

Figure 43.23 The nitrogen cycle in an ecosystem on land. Activities of nitrogen-fixing bacteria make nitrogen available to plants. Other bacterial species cycle nitrogen to plants when they convert organic wastes to ammonium and nitrates. The largest nitrogen reservoir by far is the atmosphere (mainly N_2), followed by ocean sediments, ocean water, and soil. Lesser amounts of nitrogen are in the form of nitrous oxide (N_2O) in the atmosphere and in the biomass of the seas.

fixing bacteria that form root nodules. Chapter 21 and Section 27.2 describe these interactions. How do plants use the nitrogen provided by bacterial activities? Their cells assimilate nitrogen and channel much of it into synthesizing amino acids, proteins, and nucleic acids. Plant tissues, remember, are the sole nitrogen source for animals, which directly or indirectly feed on plants.

Nitrogen Scarcity

Ammonium, nitrite, and nitrate that form during the nitrogen cycle are vulnerable to leaching and runoff. Leaching, recall, is the removal of some soil nutrients as water percolates through it (Section 27.1). Besides losses from leaching, some nitrogen is lost to the air by **denitrification**. By this process, certain bacteria convert nitrate or nitrite to N_2 and N_2O (nitrous oxide). Usually, most denitrifying bacteria rely on aerobic respiration. In waterlogged or poorly aerated soil, they use anaerobic pathways in which nitrate, nitrite, or N_2O (not oxygen) is the final electron acceptor. Section 7.5 describes this metabolic strategy. During the reactions, fixed nitrogen is converted to N_2, much of which escapes to the air.

Also, nitrogen fixation comes at high metabolic cost to plants that are mutualists with the nitrogen fixers. In exchange for nitrogen, plants give up sugars and other photosynthetic products that require large investments of ATP and NADPH. Such plants have a competitive edge in nitrogen-poor soil. In nitrogen-rich soil, species that don't pay the metabolic price often displace them.

Human Impact on the Nitrogen Cycle

Humans are having disproportionately bad effects on the nitrogen cycle, compared to other species. Where deforestation and grassland conversion for agriculture are rampant, losses of nitrogen from soil are huge. With each forest clearing and harvest, nitrogen stored in the plants is lost. Soil erosion and leaching take away more.

Many farmers counter nitrogen losses by rotating crops, as by alternating wheat with legumes. Rotation can help keep soil stable and productive. In developed countries, farmers also spread nitrogen-rich fertilizers. They even select new strains of crop plants that have a greater capacity to take up fertilizers from soil. With such practices, crop yields per hectare have doubled and even quadrupled over the past forty years.

But heavy fertilizer applications alter soil water and can harm plants. Remember, the surface charges of soil particles attract ions that weathering, irrigation, and fertilizers introduce to soils. By a pH-dependent process called **ion exchange**, ions dissociate from soil particles, and other ions dissolved in soil water replace them. Any change in the composition of soil water disrupts

Figure 43.24
Dead and dying spruce trees in a forest in Germany. They are cited as casualties of extremely high concentrations of nitrogen oxides and other forms of air pollution.

the number and kinds of ions in exchangeable form—which are the kinds that plants take up. Calcium and magnesium are the most abundant exchangeable ions. When nitrogen fertilizers enter soil, ammonium ions also assume importance. This is especially the case for ammonium sulfate $(NH_4)_2SO_4$, for all of its nitrogen is in the form of ammonium. Heavy applications of such a fertilizer greatly increase the acidity of soil water. Its many hydrogen ions compete for binding sites on soil particles. One outcome is that the levels of calcium and magnesium ions necessary for plant growth decline.

Humans introduce excess nitrogen into ecosystems in other ways, as when nitrogen-rich sewage typically enters rivers, lakes, and estuaries (Section 45.7). As another example, fossil fuel burning by power plants and vehicles having combustion engines releases NO_2 and other oxides of nitrogen into the atmosphere (Section 45.1). Winds carry the air pollutants to forests at high elevations and high northern latitudes, which have nitrogen-poor soils. Nitrogen is a limiting factor on tree growth, and the forest trees are adapted to storing and cycling it. With more nitrogen, they grow faster, well into the summer. But the late growth is vulnerable to autumn frost and dies during winter. In addition, their nitrogen-stimulated growth outstrips the availability of phosphorus and other nutrients, so the trees suffer the effects of nutrient deficiencies. Their leaves yellow and drop, so rates of photosynthesis decline. The decline has repercussions on the growth of all plant parts. The weakened trees are far more susceptible to disease and other environmental stresses (Figure 43.24).

The nitrogen cycle turns on processes of nitrogen fixation, assimilation and biosynthesis by all organisms, and on decomposition. Its natural cycling depends on the activity of nitrogen-fixing bacteria and on mycorrhizae.

Ecosystems lose nitrogen naturally by the activities of denitrifying bacteria, which convert nitrite or nitrate to forms that escape into the atmosphere.

Ecosystems often lose or gain too much nitrogen through human activities, in ways that compromise plant growth.

43.12

SUMMARY   *Gold* indicates text section

1. An ecosystem consists of an array of producers, consumers, detritivores, and decomposers along with their environment. It is an open system, with inputs, internal transfers, and outputs of energy and nutrients. There is a one-way flow of energy into and out from it and a cycling of materials among its organisms. *43.1*

2. Sunlight is the initial energy source for nearly all ecosystems. Photoautotrophs (the primary producers) convert that energy to other forms, such as chemical bond energy of ATP. They also assimilate nutrients that the ecosystem's heterotrophs require. *43.1, 43.2*

   a. Many heterotrophs are consumers. The herbivores feed upon algae and plants. Carnivores ingest animals. Parasites withdraw nutrients from the tissues of living hosts. Omnivores use a variety of food sources.

   b. Decomposers also are heterotrophs. Most (certain fungi and bacteria) obtain energy and nutrients from organic remains and wastes. Detritivores, such as crabs and earthworms, are heterotrophs that ingest bits of decomposing or dead organisms or organic products.

3. In ecosystems, feeding relationships are structured as a hierarchy of energy transfers (trophic levels). *43.2*

   a. Primary producers are at the first trophic level; consumers are at successively higher trophic levels.

   b. Herbivores are second-level consumers; carnivores and parasites are at higher levels. The detritivores and decomposers are assigned to all trophic levels above the first level. So are the parasites. Humans and other omnivores obtain energy from more than one source, so they cannot be assigned to a single trophic level.

4. A food chain is a straight-line sequence by which energy stored in autotroph tissues is passed on through higher trophic levels. Most cross-connect as food webs. In grazing food webs, energy is transferred from the primary producers to herbivores, then carnivores. In detrital food webs, energy is transferred from primary producers through detritivores and decomposers. *43.2*

5. The amount of useful energy that flows through successive levels of consumers declines at each energy transfer in a food web because of metabolic heat loss and shunting of food energy into organic wastes. This inefficiency typically limits the number of trophic levels in a food web to no more than four or five. *43.2*

   a. In environments that show variations in salinity, temperature, and other environmental conditions, food webs tend to have short chains. In stable environments, such as parts of the deep ocean, food chains are longer.

   b. The most complex food webs have short chains; the simplest have the most top carnivores.

6. Disturbance to one aspect of an ecosystem usually has unexpected effects on other, seemingly unrelated parts. To predict consequences of a disruption, as by computer modeling, researchers attempt to identify all the potential relationships and incorporate them into the model. *43.3*

7. Primary productivity is the rate at which primary producers capture and store a given amount of energy in a given interval. Net primary productivity is the rate at which energy is stored in producers in excess of the rate of its release by way of aerobic respiration. *43.4, 43.5*

8. Primary productivity and ecosystem structure require cycling of nutrients as well as energy inputs (Figure 43.25). Water and minerals pass on through the environment, then organisms, and then back to the environment in biogeochemical cycles.

   a. Water moves through a hydrologic cycle. In land ecosystems, plants stabilize soil and minimize nutrient loss during the cycle, as by runoff. *43.7*

   b. In a sedimentary cycle, a nutrient (phosphorus, for example) is in solid form, and the main reservoir is the Earth's crust. *43.8*

   c. In atmospheric cycles, a nutrient prevails mainly in gaseous form. Examples: carbon in carbon dioxide, nitrogen in gaseous nitrogen. *43.9, 43.11*

9. Sedimentary cycles and the hydrologic cycle interact and move mineral nutrients on a global scale. *43.7*

   a. Certain human activities are depleting minerals from ecosystems, as when weathered soils of tropical forests are cleared for agriculture.

   b. Some human activities are accelerating the process of eutrophication. They are adding nutrients to aquatic ecosystems that are naturally low in these nutrients and thereby promote destructive algal blooms.

10. In the carbon cycle, carbon moves through the ocean (its main reservoir), the atmosphere (as carbon dioxide, mostly), and ecosystems. Fossil fuel burning, logging, and the conversion of natural ecosystems for crops and grazing affect the global carbon budget. *43.9, 43.10*

11. Like water, ozone, CFCs, methane, and nitrous oxide, carbon dioxide is a greenhouse gas. Fossil fuel burning and other human activities that release carbon to the atmosphere may be adding to global warming. *43.10*

12. For land ecosystems, nitrogen is a limiting factor in the total net primary productivity. Gaseous nitrogen is abundant in the atmosphere. Nitrogen-fixing bacteria convert N_2 to ammonia and nitrates, which producers take up. Mycorrhizae and root nodules, two symbiotic interactions, enhance the nitrogen uptake. *43.11*

Producers trap, convert, and use or store some energy from the sun.

AUTOTROPHS (plants, other primary producers)

NUTRIENT CYCLING

HETEROTROPHS (consumers, decomposers, detritivores)

Energy is transferred from one organism to another. In time, all of the energy flows back to the environment.

ONE-WAY FLOW OF ENERGY

Figure 43.25 Summary of energy flow and nutrient cycling in ecosystems.

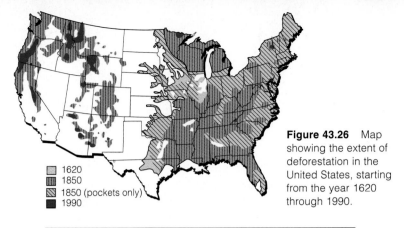

Figure 43.26 Map showing the extent of deforestation in the United States, starting from the year 1620 through 1990.

1620
1850
1850 (pockets only)
1990

Figure 43.27 Satellite image of an iceberg roughly the same size as Connecticut. This one started to break away from Antarctica in 2000 and may be the longest ever recorded.

Review Questions

1. Define an ecosystem in terms of inputs and outputs. *43.1*

2. Define primary producer, consumer, decomposer, and detritivore. Give an example of each. *43.1*

3. Define and describe trophic levels. Which class of organisms is farthest from the energy input into an ecosystem? *43.1, 43.2*

4. Define food chain and food web. How do grazing food webs differ from detrital food webs? *43.1, 43.2*

5. Distinguish between biomass pyramid and energy pyramid. What did the pyramids for Silver Springs reveal? *43.4, 43.5*

6. Define three types of biogeochemical cycles and give an example of each. *43.6–43.11*

7. Define nitrogen fixation. What is the difference between nitrification and denitrification? *43.11*

Self-Quiz   ANSWERS IN APPENDIX III

1. Ecosystems have _____ .
   a. energy inputs and outputs
   b. one trophic level
   c. nutrient cycling but not outputs
   d. a and b

2. Trophic levels are _____ .
   a. structured feeding relationships
   b. who eats whom in an ecosystem
   c. a hierarchy of energy transfers
   d. all of the above

3. Primary productivity on land is affected by _____ .
   a. photosynthesis and respiration by plants
   b. how many plants are neither eaten nor decomposed
   c. rainfall and temperature
   d. all of the above

4. Match the ecosystem terms with the suitable description.
   _____ producers   a. herbivores, carnivores, omnivores
   _____ consumers   b. feed on partly decomposed matter
   _____ decomposers   c. degrade organic remains, wastes
   _____ detritivores   d. photoautotrophs

Critical Thinking

1. Imagine and describe an extreme situation in which you are a participant in a food chain rather than a food web.

2. Marguerite is growing a vegetable garden in Maine. What are the variables that can affect its net primary production?

3. List as many of the agricultural products and manufactured goods as you can identify that you depend upon. Are any of them implicated in the amplified greenhouse effect?

4. Biologists Reed Noss, J. Michael Scott, and Edward LaRoe issued a report on *endangered ecosystems* in the United States. They categorize thirty natural, once-vast domains as in danger of vanishing (Figure 43.26). Agricultural conversions, urban growth, and other human activities account for the losses. More biodiversity was lost at the ecosystem level than was generally recognized. Do some research on a habitat that is part of such an ecosystem. Get a sense of its decline in biodiversity, of the kinds of organisms that once flourished there. Would you take part in efforts to set aside land for restoration? Or do you consider such efforts too intrusive on individual rights of property ownership?

5. *Polar ice shelves* are vast, thickened sheets of ice floating on seawater. Measurements taken between 1978 and the present indicate that Antarctic ice shelves are retreating. A chunk of the Larsen Ice Shelf the size of Rhode Island broke away from Antarctica in 1995 and drifted off into the open ocean. Figure 43.27 shows another gargantuan chunk, 37 kilometers wide and 295 kilometers long, that started to split away from the Ross Ice Shelf in 2000. Also, in 1996, seawater 200 meters below the Arctic ice cap was 1°C warmer than just five years ago. Do you suppose that the retreating and fragmentation of the polar ice sheets are evidence of a long-term trend in global warming? If not, what other factors might cause such impressive events?

Selected Key Terms

ammonification *43.11*	food chain *43.1*
biogeochemical cycle *43.6*	food web *43.1*
biological magnification *43.3*	global warming *43.10*
biomass pyramid *43.4*	grazing food web *43.2*
carbon cycle *43.9*	greenhouse effect *43.10*
consumer *43.1*	hydrologic cycle *43.7*
decomposer *43.1*	ion exchange *43.11*
decomposition *43.11*	nitrification *43.11*
denitrification *43.11*	nitrogen cycle *43.11*
detrital food web *43.2*	nitrogen fixation *43.11*
detritivore *43.1*	phosphorus cycle *43.8*
ecosystem *43.1*	primary producer *43.1*
ecosystem modeling *43.3*	primary productivity *43.4*
energy pyramid *43.4*	trophic level *43.1*
eutrophication *43.8*	watershed *43.7*

Readings

Brady, N., and R. Weil. 1996. *The Nature and Properties of Soils*. Eleventh edition. New Jersey: Prentice-Hall.

Krebs, C. 1994. *Ecology*. Fourth edition. New York: HarperCollins.

On-Line readings at Student Guide for InfoTrac:
www.brookscole.com/biology

THE BIOSPHERE

Does a Cactus Grow in Brooklyn?

Suppose you live in the American Southwest but find yourself touring an African desert. There you notice a flowering plant with spines, tiny leaves, and fleshy, columnlike stems—just like some cactus plants back home (Figure 44.1). Or suppose you live in the coastal hills of California and decide to tour the Mediterranean coast, the southern tip of Africa, or even central Chile. There you come across many-branched, tough-leafed woody plants—very much like the many-branched, tough-leafed chaparral plants back home.

In both cases, vast geographic and evolutionary distances separate the plants. Why, then, are they alike? The question intrigues you, so you decide to compare their locations on a global map. You see that American and African desert plants live about the same distance from the equator. Chaparral plants and their distant look-alikes grow along the western and southern coasts of continents between latitudes 30° and 40°. As Charles Darwin and other naturalists did long ago, you have just stumbled onto one of many predictable patterns in nature.

What causes such patterns? With this question, we turn to **biogeography**—the study of the distribution of organisms, past and present, and of diverse processes that underlie their distribution patterns. Think back on those "accidents of history," including the colossal breakup of Pangea more than 100 million years ago. When chunks of that supercontinent began to slowly drift apart, most species had no choice but to go along for the ride. The travelers were dispersed to different isolated locations. And there they diverged genetically from their parent populations. Among the changing populations on the fragment that became Australia were ancestors of eucalyptus trees and platypuses, of wombats and kangaroos.

Species also owe their distribution to topography, climate, and species interactions. With diligence, you might grow a cactus under artificial lights in a heated room in Brooklyn or another New York City borough. Plant that cactus outside, and it won't last one winter.

As this last example reminds us, we humans tinker with species distributions. Not all of our tinkering is as harmless as growing a cactus in Brooklyn. Think of our predatory effects on the world's fisheries and the effects of our pesticide battles with insect competitors for food. Earlier chapters gave you a general picture of predation, competition, and other species interactions. This chapter considers the physical forces shaping the biosphere itself. Use it as a foundation for addressing the impact of the human species on the biosphere, the topic of the chapter to follow.

Start with the definition of the **biosphere**: the sum total of all places in which organisms live. Organisms occupy the Earth's waters—the hydrosphere—which includes the ocean, ice caps, and other forms of liquid and frozen water. They occupy soils and sediments of the lithosphere, the Earth's outer, rocky layer. And they occupy the lower **atmosphere**, which consists of gases and airborne particles that envelop the Earth.

Figure 44.1 Life in the biosphere—a case of morphological convergence. (**a**) *Echinocerus*, a member of the cactus family (Cactaceae), grows in hot deserts of the American Southwest. (**b**) In deserts of southwestern Africa, we find *Euphorbia*, of the spurge family (Euphorbiaceae). Although the lineages appear to be closely related, they are geographically and evolutionarily distant. Long ago, plants in both lineages put similar structures, including their leaves, to similar uses in similar habitats. Their descendants ended up resembling each other.

Air in the upper atmosphere, seventeen kilometers or so above the Earth's surface, is too thin to support life.

In this vast biosphere you find ecosystems ranging from continent-straddling forests to a rainwater pool in a small cup-shaped array of leaves. As you will see, climate profoundly influences all ecosystems except for a few at hydrothermal vents on the ocean floor.

Climate means average weather conditions, such as temperature, humidity, wind speed, cloud cover, and rainfall, over time. Many factors contribute to it. The main ones are variations in the amount of solar radiation streaming in, the Earth's daily rotation and its path around the sun, the world distribution of continents and oceans, and land mass elevations. All of these factors interact to produce prevailing winds and ocean currents that influence the global patterns of climate. The physical and chemical development of sediments and soils also depend on climate.

Climate, together with the composition of sediments and soils, influences the growth of primary producers—and, through them, the distribution of ecosystems.

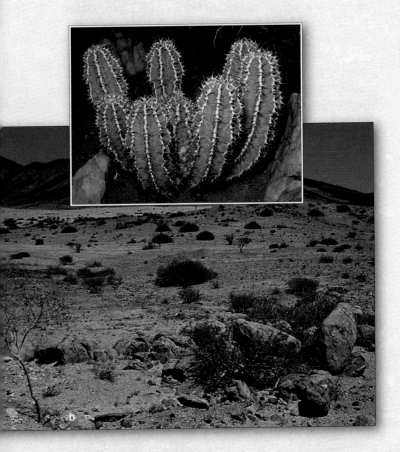

Key Concepts

1. Energy from the sun is the initial energy source for nearly all ecosystems on Earth. In addition, solar energy influences the global distribution of ecosystems. It does so by continually providing heat energy that warms the atmosphere and drives the Earth's weather systems.

2. Interactions among global air circulation patterns, ocean currents, and diverse topographic features result in regional variations in patterns of temperature and rainfall. The variations influence the composition of soils and sediments, the growth and distribution of primary producers, and, through them, the distribution of ecosystems.

3. A biome is a large, regional unit of land that can be characterized by the climax vegetation of the ecosystems within its boundaries. Deserts and broadleaf forests are examples. Their distribution corresponds roughly with regional variations in climate, topography, and soil type.

4. The water provinces cover more than 71 percent of the Earth's surface. The oceans hold all but 3 percent of the free-flowing water. Inland seas, estuaries, and diverse bodies of fresh water, including lakes, streams, and rivers, hold the rest.

5. Each freshwater and marine ecosystem has gradients in light availability, temperature, and dissolved gases. The gradients vary daily and with the changing seasons. Primary productivity, and the composition of species, depend on them.

AIR CIRCULATION PATTERNS AND REGIONAL CLIMATES

Each winter Pacific Grove, California, is host to great gatherings of tourists and monarch butterflies (Figure 44.2). Similarly, caribou, Canada geese, sea turtles, and whales undertake vast migrations, moving to and from overwintering grounds. Other animals stay put. Their adaptations in body form, physiology, and behavior help them endure seasonal change. In Canada and the United States, Japan and Italy, flowering plants leaf out, flower, bear fruit, then drop leaves. In the ocean, astronomical numbers of photosynthetic protistans and bacteria show seasonal bursts of primary productivity. Like nearly all organisms, they are exquisitely attuned to regional climates and seasonal change.

Climate starts with incoming rays from the sun. Of the total amount of solar radiation reaching the outer atmosphere, only about half gets through to the Earth's surface. Ozone (O_3) and oxygen (O_2) molecules in the upper atmosphere absorb most of the wavelengths of ultraviolet radiation. Such wavelengths are lethal for most forms of life. Absorption is greatest at altitudes between 17 and 27 kilometers above sea level, where molecules of ozone are the most concentrated (Figure 44.3). Hence the name "ozone layer." Clouds, particles of dust, and water vapor suspended in the atmosphere absorb other wavelengths or reflect them into space.

Radiation penetrating the atmosphere warms the Earth's surface, which gives up heat by radiation and evaporation. The lower atmosphere's molecules absorb some heat and reradiate part of it toward the Earth. The effect is a bit like heat retention in a greenhouse, which lets in the sun's rays while retaining heat that the plants and soil inside are giving up (Section 43.10). Why is this greenhouse effect important? *Heat energy from the sun warms the atmosphere and ultimately drives the Earth's great weather systems.*

The sun's effect varies from one latitude to the next. Its rays are more concentrated at the equator than the poles, so air is heated more at the equator (Figure 44.4). A global pattern of air circulation starts as warm air at the equator rises and spreads north and south. Under the moving air, the Earth rotates faster at the equator than it does at the poles, and this creates worldwide belts of prevailing east and west winds. The differences in solar heating at different latitudes and the modified patterns of air circulation help define the world's major **temperature zones**, which are shown in Figure 44.5a.

The differences also help define rainfall patterns at different latitudes. Warm air holds more moisture than cool air. Air warmed at the equator picks up moisture from the seas. It rises to cooler altitudes and gives up moisture as rain, which supports luxuriant growth of equatorial forests. Being drier now, the air moves away

Figure 44.2 Monarch butterflies. These migratory insects gather each winter in trees of California coastal regions and central Mexico. They travel hundreds of kilometers south to those places, which are cool and humid in winter. If they were to stay in their northern breeding grounds, monarchs would risk being killed by more severe climatic conditions.

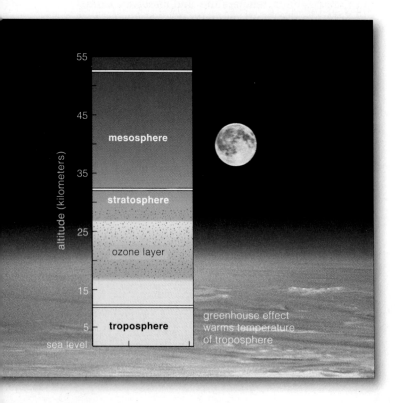

Figure 44.3 Earth's atmosphere. Most of the global air circulation proceeds in the troposphere, where temperature decreases rapidly with altitude. Most of the ultraviolet wavelengths in the sun's rays are absorbed in the upper atmosphere, mainly at the ozone layer. The ozone layer typically ranges between 17 and 27 kilometers above sea level.

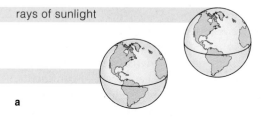

a

Figure 44.4 Global air circulation patterns brought about by three interrelated factors. *First*, the sun's rays are less spread out in equatorial regions than in polar regions (**a**). Warm equatorial air rises and spreads northward and southward to produce an initial pattern of air circulation (**b**).

Second, the nonuniform distribution of land creates air pressure variations. Land absorbs and gives up heat faster than the ocean. So depending on what's below them, some parcels of air rise or sink faster than others. Air pressure is lower where air rises, and it is greater where air sinks. The regional pressure differences create winds that disrupt the overall movement of air from the equator to the poles.

Third, the Earth's rotation and overall shape introduce easterly and westerly deflections in wind directions. With each full rotation, the surface turns faster beneath the air masses at the equator (where the Earth's diameter is greatest) and slower beneath air masses at the poles. Thus, a rising air mass cannot move "straight north" or "straight south." Its deflection is the source of prevailing east and west winds (**c**).

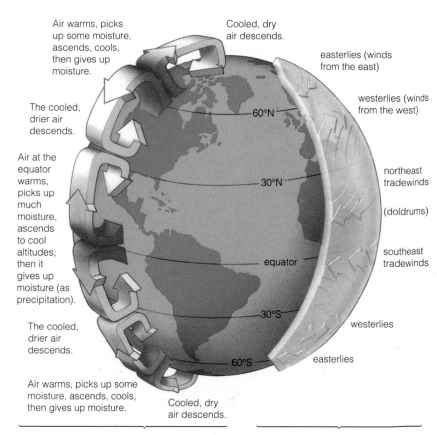

Air warms, picks up some moisture, ascends, cools, then gives up moisture.

The cooled, drier air descends.

Air at the equator warms, picks up much moisture, ascends to cool altitudes; then it gives up moisture (as precipitation).

The cooled, drier air descends.

Air warms, picks up some moisture, ascends, cools, then gives up moisture.

Cooled, dry air descends.

Cooled, dry air descends.

easterlies (winds from the east)

westerlies (winds from the west)

northeast tradewinds

(doldrums)

southeast tradewinds

westerlies

easterlies

60°N

30°N

equator

30°S

60°S

b Initial pattern of air circulation

c Deflections in the paths of air flow near the Earth's surface

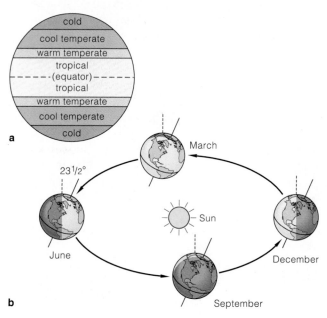

cold
cool temperate
warm temperate
tropical
(equator)
tropical
warm temperate
cool temperate
cold

a

23½°

March

June

Sun

December

September

b

Figure 44.5 (**a**) World temperature zones. (**b**) Annual variation in incoming solar radiation. The northern end of the Earth's fixed axis tilts toward the sun in June and away from it in December. Thus the equator's position relative to the day–night boundary of illumination varies over the year. Such variations in sunlight intensity and daylength cause seasonal temperature variations in the two hemispheres. Seasonal change is more pronounced with distance from the equator. It is greatest in the interior of continents, away from the ocean's moderating effects.

from the equator. It becomes warmer and drier when it descends at latitudes of about 30°. Deserts tend to form at these latitudes. Farther north and south, air again picks up some moisture and ascends to high altitudes. It creates another moist belt at latitudes of about 60°. Air descends in polar regions, where low temperatures and almost nonexistent precipitation give rise to cold, dry polar deserts.

The amount of solar radiation reaching the surface also varies seasonally because of the Earth's rotation around the sun (Figure 44.5*b*). This leads to seasonal changes in daylength, prevailing wind directions, and temperature. And so primary productivity alternately rises and falls on land and in the seas—and migrations shift many kinds of animals to new locations.

Latitudinal differences in the amount of solar radiation reaching the Earth produce global air circulation patterns. The Earth's rotation and its overall shape affect the patterns to produce latitudinal belts of temperature and rainfall.

Also, the Earth's annual rotation around the sun introduces seasonal changes in winds, temperatures, and rainfall.

Such factors influence the locations of different ecosystems.

THE OCEAN, LANDFORMS, AND REGIONAL CLIMATES

Ocean Currents and Their Effects

The **ocean**—a continuous body of water—covers more than 71 percent of the Earth. The uppermost 10 percent moves in currents that distribute nutrients in marine ecosystems and that influence regional climates. Solar heat and wind friction drive the surface currents.

Latitudinal and seasonal variation in solar heating warm and cool water on a vast scale. The volume of water expands when heated and shrinks when cooled. So sea level is about 8 centimeters (3 inches) higher at the equator than at the poles. The sheer volume of water in this "slope" is enough to get surface waters moving in response to gravity, usually toward the poles. Along the way, surface waters warm air parcels above them. At midlatitudes, about *10 million billion* calories of heat energy per second are transferred to the air!

Rapid mass flows of water—*currents*—arise mostly from the "tugging" of trade winds and westerlies. The Earth's rotation, the distribution of land masses, and the shapes of oceanic basins influence the direction and properties of the currents. Water circulates clockwise in the Northern Hemisphere and counterclockwise in the Southern Hemisphere, as shown in Figure 44.6.

Swift, deep, and narrow currents of nutrient-poor waters move parallel with the east coast of continents. As an example, 55 million cubic meters per second of warm water move northward as the Gulf Stream along the east coast of North America. Slower, shallow, and broad currents paralleling the west coast of continents move cold water toward the equator. As you will see, they move deep, nutrient-rich waters to the surface.

Why are Pacific Northwest coasts cool, foggy, and mild in summer? Winds nearing the coast give up heat to cold coastal water, which the California Current is moving toward the equator. Why are Baltimore and Boston muggy in summer? Air above the Gulf Stream gains heat and moisture, which southerly and easterly winds deliver to those cities. Why are winters milder in London and Edinburgh than in Ontario and central Canada, even though they are all at the same latitude? The North Atlantic Current picks up warm water from the Gulf Stream. As it flows past northwestern Europe, it gives up heat energy to the prevailing winds.

Figure 44.6 Generalized map of major climate zones correlated with surface currents and drifts of the world ocean. Warm surface currents move from the equator toward the poles. Water temperatures differ with latitude and depth. The differences contribute to regional differences in temperature and rainfall. In which direction does a given current flow? It depends on prevailing winds, the Earth's rotation, gravity, the shapes of ocean basins, and the positions of land masses.

warm surface current   cold surface current

dry   warm temperate   subpolar
tropical   cool temperate   polar (ice)
cold

a Winds carry moisture inland from Pacific Ocean

b Clouds and rain on windward side of mountain range

c Rainshadow on leeward side of mountain range

4,000/
3,000/ 85
2,000/ 85
1,800/ 125
1,000/ 25
1,000/ 85
moist habitats
15/ 25

Figure 44.7 Rain shadow effect, a reduction of rainfall on the side of high mountains away from prevailing winds. Plants adapted to arid or semiarid conditions grow in rain shadows. *Blue* numbers show average yearly precipitation in centimeters, as measured at different locations on both sides of the range. *White* numbers signify elevation in meters.

Regarding Rain Shadows and Monsoons

Mountains, valleys, and other aspects of topography influence regional climates. "Topography" refers to a region's physical features, such as elevation. Think of a warm air mass picking up moisture off California's western coast. It moves inland to the Sierra Nevada, a mountain range paralleling the coast. The air cools as it ascends to higher altitudes, then loses moisture as rain (Figure 44.7). Belts of vegetation at different elevations reflect the differences in temperature and moisture. At the western base of the range, we see grasslands with species adapted to semiarid conditions. Higher up we see deciduous and evergreen species adapted to more moisture and cooler air. Higher still, a subalpine belt supports just a few evergreen species that withstand a rigorously cold habitat. Above the subalpine belt, only low, nonwoody plants and dwarfed shrubs can grow.

Air warms and can hold more water after it flows over mountain crests. As it starts its descent, it retains moisture and draws more water out of plants and soil. The result is a **rain shadow**, a semiarid or arid region of sparse rainfall on the leeward side of high mountains. (*Leeward* is the direction not facing a wind; *windward* is the direction from which wind blows.) The Himalayas, Andes, Rockies, and other great mountain ranges cause impressive rain shadows. Similarly, on the windward side of Hawaii's high volcanic peaks are lush tropical forests; arid conditions prevail on the leeward side.

Monsoons are air circulation patterns that influence the continents north or south of warm oceans. The land heats intensely, so low-pressure air parcels form above it. The low pressure draws in moisture-laden air from the ocean, as from the Bay of Bengal into Bangladesh or from the Gulf of Mexico into deserts of the American Southwest. Fantastic summer thunderstorms are one result. Converging trade winds and the equatorial sun cause intense heating and rainfall in a zone called the doldrums. The air circulating north and south creates patterns of wet and dry seasons, as in eastern Africa.

Recurring sea breezes along coastlines are called "mini-monsoons." Here again, water and the adjacent land have different heat capacities. In the morning, the water's temperature does not rise as fast as the land's temperature. When warmed air above the land rises, cooler marine air moves in. After sunset, the land loses heat faster than water. Then, land breezes flow in the reverse direction (Figure 44.8).

cool air descends

land warmer than sea; breeze flows onshore

warm air ascends

a

warm air ascends

land cooler than sea; breeze flows offshore

cool air descends

b

Figure 44.8 Coastal breezes, afternoon (**a**) and night (**b**).

Surface ocean currents, in combination with global air circulation patterns, influence regional climates and help distribute nutrients in marine ecosystems.

Air circulation patterns, ocean currents, and landforms interact in ways that influence regional temperatures and moisture levels. Thus they also influence the distribution and dominant features of ecosystems.

REALMS OF BIODIVERSITY

Circulation patterns in the atmosphere and at the ocean surface, together with topography, give rise to regional differences in temperature and moisture. And these differences help explain why deserts, grasslands, forests, and tundra form in some places but not others. It helps us understand the distribution of ecosystems and their species, each adapted to regional conditions.

For example, the information provides clues to why many evolutionarily distant species look alike. Often they emerged through convergent evolution, as Section 18.4 describes. Briefly, their ancestors were not closely related but faced similar selection pressures in similar yet geographically distant places. By natural selection, their body plans underwent similar modifications and ended up looking alike. Ancestors of those cactus and spurge plants in Figure 44.1 both evolved in hot, dry

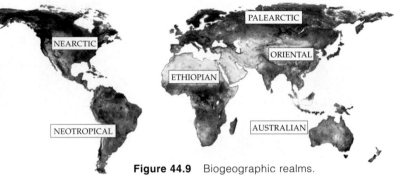

Figure 44.9 Biogeographic realms.

deserts, where water is scarce. Fleshy plant stems that have thick cuticles conserve precious water. And rows of sharp spines deter hungry, thirsty herbivores, which might otherwise chew on juicy plant parts.

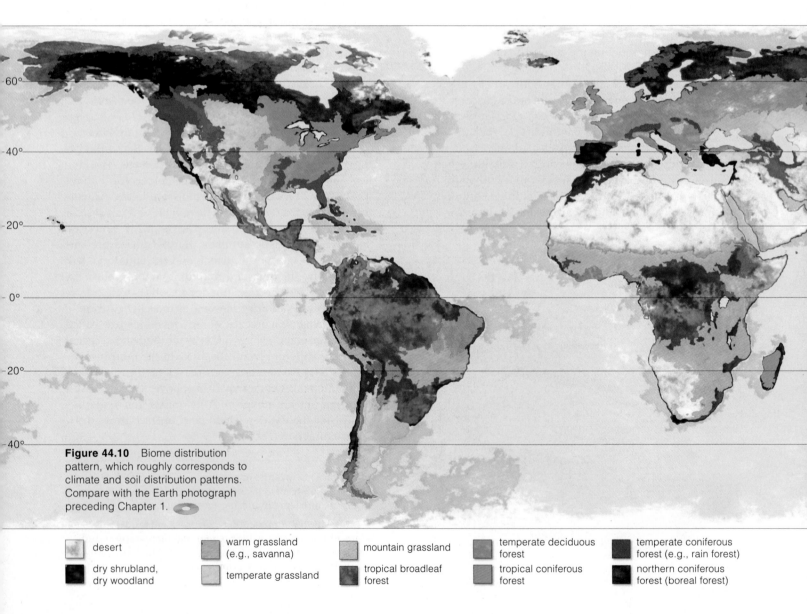

Figure 44.10 Biome distribution pattern, which roughly corresponds to climate and soil distribution patterns. Compare with the Earth photograph preceding Chapter 1.

- desert
- dry shrubland, dry woodland
- warm grassland (e.g., savanna)
- temperate grassland
- mountain grassland
- tropical broadleaf forest
- temperate deciduous forest
- tropical coniferous forest
- temperate coniferous forest (e.g., rain forest)
- northern coniferous forest (boreal forest)

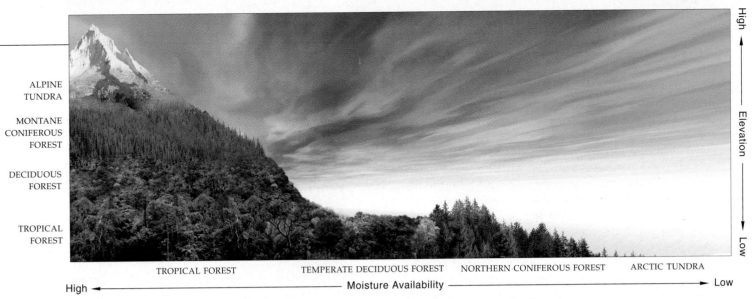

ALPINE
TUNDRA

MONTANE
CONIFEROUS
FOREST

DECIDUOUS
FOREST

TROPICAL
FOREST

High — Elevation — Low

TROPICAL FOREST   TEMPERATE DECIDUOUS FOREST   NORTHERN CONIFEROUS FOREST   ARCTIC TUNDRA

High ←——————— Moisture Availability ———————→ Low

Figure 44.11 For North America, changes in plant form along environmental gradients. Gradients in elevation and availability of water help dictate a biome's primary productivity. Other factors, such as mean annual air temperature and soil drainage, also are at work.

tropical dry forest

tundra

mountains, complex zonation

mangrove swamps

permanent ice cover

marine ecoregions

W. Sclater and, later, Alfred Wallace, bestowed the name **biogeographic realms** on six immense land areas, each with distinctive kinds and numbers of plants and animals (Figure 44.9). Biogeographic realms maintain their identity partly because of climate. They also tend to maintain a distinct identity when mountain ranges, oceans, and other physical barriers restrict gene flow, thus keeping the gene pools of each realm isolated.

A biogeographic realm may be divided into biomes. Each **biome** is a large region of land characterized by habitat conditions and community structure, including endemic species (which evolved nowhere else). Take a look at how Figure 44.10 correlates with environmental factors of the sort shown in Figure 44.11. It helps show that distinctive biomes prevail at certain latitudes and elevations. Biomes dominated by low-growing plants prevail in dry regions, at high elevations, and at high latitudes. Those dominated by tall, leafy plants prevail at tropical latitudes, temperate latitudes, and at lower elevations with warm temperatures and high rainfall. As you will see, soils also affect biome distribution.

Conservationists are working to locate, inventory, and protect "hot spots." These are portions of certain biomes that are the richest in biodiversity and the most vulnerable to species loss (Section 25.5). Twenty-four of the hot spots hold more than half of all land species.

In its Global 200 program, the World Wildlife Fund is targeting ecoregions as well as hot spots. **Ecoregions** are large areas representative of the globally important biomes *and* water provinces vulnerable to extinction. Figure 44.10 identifies the marine ecoregions.

The land and the seas contain realms of biodiversity. The unique identity of each biogeographic realm, biome, and ecoregion is an outcome of habitat conditions, the kinds and numbers of species, and their evolutionary history.

SOILS OF MAJOR BIOMES

Most regions of land have **soils**, which are mixtures of mineral particles and varying amounts of decomposing organic material (humus). As described in Section 27.1, the weathering of hard rocks produces coarse-grained gravel, then sand, silt, and finely grained clay. Water and air infiltrate spaces between particles. How much these particles compact together, and the proportions of each kind, differ within and between regions.

Soils have a layered structure, a *profile*, that reflects their stage of development. Figure 44.12 shows a few examples. Uppermost is topsoil, with the most humus and the most vulnerability to erosion. Topsoil may be less than a centimeter thick on steep slopes and more than a meter thick in grasslands. The lowest layer of soil consists of rocks in varying degrees of weathering.

O horizon: Pebbles, little organic matter

A horizon: Shallow, poor soil

B horizon: Leaching results in salinization (accumulated calcium, sodium)

C horizon: Rock fragments from uplands

DESERT SOIL

A horizon: Alkaline, deep, rich in humus

B horizon: Percolating water enriches with calcium carbonates

GRASSLAND SOIL

O horizon: Sparse litter

A–E horizons: Continually leached; iron, aluminum left behind impart red color to acidic soil

B horizon: Clays with silicates, other residues of chemical weathering

TROPICAL RAIN FOREST SOIL

O horizon: Scattered litter

A horizon: Rich in organic matter above humus layer unmixed with minerals

B horizon: Accumulated minerals leached from above

C horizon: Poorly weathered rocks

DECIDUOUS FOREST SOIL

O horizon: Well-defined, compacted mat of organic deposits resulting mainly from activity of fungal decomposers

A horizon: Acidic humus; most minerals leached out, silica retained

B horizon: Accumulated clays with oxides of iron and aluminum

CONIFEROUS FOREST SOIL

The growth of most plants suffers in poorly aerated, poorly draining soils. Remember, loam topsoils have the best mix of sand, silt, and clay for agriculture. They have enough coarse particles to promote drainage and enough fine particles to retain water-soluble mineral ions that serve as nutrients for plant growth. Gravelly or sandy soils promote rapid leaching, which depletes them of water and vital minerals. Clay soils with fine, closely packed particles are poorly aerated and do not drain well. Few plants grow in waterlogged clay soils.

As farmers know, soil affects primary productivity. They grow most crops in cleared, former grasslands. Many burn or clear-cut tropical forests for agriculture. But these biomes have very little topsoil above poorly

Figure 44.12 Soil profiles from a few representative biomes. Compare Section 27.1.

draining sublayers. Clearing exposes the topsoil, and heavy rains leach most of the nutrients (Section 25.6).

We turn now to the deserts, shrublands, woodlands, grasslands, various forests, and tundra. Bear in mind, none of these major biomes is uniform throughout. In each one, local climates, landforms, and other physical features favor patches of distinct communities.

Primary productivity depends on the soil profile and its proportions of sand, silt, clay, gravel, and humus.

DESERTS

Deserts form on land with less than ten centimeters or so of annual rainfall and high potential for evaporation. Such conditions prevail at latitudes of about 30° north and south. There we find great deserts of the American Southwest, of northern Chile, Australia, northern and southern Africa, and Arabia. Farther north are the high deserts of eastern Oregon, and Asia's vast Gobi and the Kyzyl-Kum east of the Caspian Sea. Rain shadows are the main reason these northern deserts are so arid.

after a rain. Annual and perennial species flower profusely but briefly after seasonal rains. Deep-rooted plants, including mesquite and cottonwood, commonly grow near the few streambeds that have a permanent underground water supply.

More than a third of the Earth's land surfaces is arid or semiarid, without enough rainfall to support crops. Crops in California's Imperial Valley and some other

Figure 44.13 Warm desert near Tucson, Arizona. Primary producers include creosote bushes, multistemmed ocotillos, tall saguaro cacti, and prickly pear cacti with rounded pads.

Deserts do not have lush vegetation. Rain falls in heavy, brief, infrequent pulses that swiftly erode the exposed topsoil. Humidity is so low that the sun's rays easily penetrate the air. They quickly heat the ground's surface, which radiates heat and cools quickly at night.

Although arid or semiarid conditions do not favor large, leafy plants, deserts show plenty of biodiversity. In a patch of Arizona's desert (as in Figure 44.13), you might find creosote and other deep-rooted, evergreen, woody shrubs, fleshy-stemmed, shallow-rooted cacti, tall saguaros, short prickly pears, and ocotillos, which drop leaves more than once a year and grow new ones

deserts need intensive soil management and irrigation. Without drainage programs, crop production declines from waterlogging and salt buildup. Many parts of the world are undergoing **desertification**. The word refers to conversion of grasslands and similarly productive biomes to desertlike wastelands with low biodiversity. We take a closer look at this trend in Section 45.6.

Where the potential for evaporation greatly exceeds sparse rainfall, deserts form. This condition prevails at latitudes 30° north and south and in rain shadows.

DRY SHRUBLANDS, DRY WOODLANDS, AND GRASSLANDS

Dry shrublands and **dry woodlands** prevail in western or southern coastal regions of the continents between latitudes 30° and 40°. These semiarid regions get a bit more rain than deserts, mostly during mild winters. Summers are long, hot, and arid. The dominant plants often have hardened, tough, evergreen leaves.

Dry shrublands get less than 25 to 60 centimeters of rain per year. We see them in California, South Africa, and Mediterranean regions, where they are known by such local names as fynbos and chaparral. California has 2.4 million hectares (6 million acres) of chaparral. In summer, lightning-sparked, wind-driven firestorms sweep through these biomes (Figure 44.14). The shrubs have highly flammable leaves and quickly burn to the ground. However, they are highly adapted to episodes of fire and soon resprout from their root crowns. Trees do not fare as well during firestorms. Shrubs actually "feed" the fires and have the competitive edge.

Dry woodlands dominate where annual rainfall is about 40 to 100 centimeters. The dominant trees can be tall, but they do not form a dense, continuous canopy. Eucalyptus woodlands of southwestern Australia and oak woodlands of California and Oregon are like this.

Grasslands cloak much of the interior of continents in zones between deserts and temperate forests. Warm temperatures prevail in the summer, and winters are exceedingly cold. The 25 to 100 centimeters of annual rainfall keep deserts from forming but aren't enough to support forests. Drought-tolerant primary producers survive strong winds, light and infrequent rainfall, and rapid evaporation. Dominant animals are grazing and burrowing species. Their grazing activities, combined with periodic fires, keep shrublands and forests from encroaching on the fringes of many grasslands.

North America's major grasslands are *shortgrass* and *tallgrass* prairie. Most form on flat or rolling land, as in Figure 44.15a. Roots of perennial plants extend profusely through the topsoil. In the 1930s, shortgrass prairie of the Great Plains was overgrazed and plowed under to grow wheat, which requires more water than this region sometimes gets. Strong winds, prolonged droughts, and unsuitable farming practices turned much of the prairie into the Dust Bowl (Section 45.6). John Steinbeck's *The Grapes of Wrath* and James Michener's *Centennial*, two historical novels, eloquently describe the consequences.

The tallgrass prairie once extended west from temperate deciduous forests (Figures 43.26 and 44.15b). A variety of legumes and composites, including daisies, thrived in the continent's interior, which had richer topsoil and slightly more frequent rainfall. Farmers converted nearly all of the original tallgrass prairie for agriculture, but certain areas are now being restored (Section 42.6).

Between the tropical forests and deserts of Africa, South America, and Australia are broad belts of grasslands with a smattering of shrubs or trees. These are the **savannas** (Figure 44.15c). Rainfall averages 90 to 150 centimeters a year, and prolonged seasonal droughts are common. Where rainfall is low, fast-growing grasses dominate. Acacia and other shrubs grow in regions that get a bit more moisture. Where the rainfall is higher,

Figure 44.14 California chaparral. The dominant plants are multibranched, woody, and typically only a few meters tall. Without periodic fires, they form a nearly impenetrable vegetation cover. The boxed inset shows a firestorm racing through a chaparral-choked canyon above Malibu.

Figure 44.15 (a) Rolling shortgrass prairie to the east of the Rocky Mountains. Once, bison were the dominant large herbivore of North American grasslands. At one time, the astoundingly abundant grasses supported 60 million of these hefty ungulates. Ungulates are hooved, plant-eating mammals.

(b) A rare patch of natural tallgrass prairie in eastern Kansas.

(c) The African savanna—a warm grassland with scattered stands of shrubs and trees. At present, more varieties and greater numbers of large ungulates live here than anywhere else. They include the migratory wildebeests shown in this photograph and giraffes, Cape buffalos, zebras, and impalas.

savannas grade into tropical woodlands with low trees and shrubs, and tall, coarse grasses. *Monsoon* grasslands form in southern Asia where heavy rains alternate with a dry season. Dense stands of tall, coarse grasses form, then die back and often burn during the dry season.

Plants of dry shrublands, dry woodlands, and grasslands are supremely adapted to surviving strong winds, grazing animals, and recurring episodes of drought and fire.

TROPICAL RAIN FORESTS AND OTHER BROADLEAF FORESTS

In forest biomes, tall trees grow close together, forming a fairly continuous canopy over a broad stretch of land. In general, there are three types of trees. Which type prevails in a region depends partly on distance from the equator. Evergreen broadleafs dominate between latitudes 20° north and south. Deciduous broadleafs are common at moist, temperate latitudes where winters are not severe. Evergreen conifers are common at high, colder latitudes and in mountains of temperate zones.

Evergreen broadleaf forests sweep across tropical zones of Africa, the East Indies and Malay Archipelago, Southeast Asia, South America, and Central America. Annual rainfall can exceed 200 centimeters and never is less than 130 centimeters. One biome, the **tropical rain forest**, depends on regular and heavy rainfall, an annual mean temperature of 25°C, and humidity of at least 80 percent (Figure 44.16). Evergreen trees of this highly productive forest produce new leaves and shed old ones all year long. But litter does not accumulate; decomposition and mineral cycling are notably rapid, given the hot, humid climate. The soils are weathered, humus-deficient, and poor nutrient reservoirs. We will take a closer look at these forests in the next chapter.

Leaving the tropical rain forests, we enter regions where temperatures remain mild but rainfall dwindles during part of the year. This is the start of **deciduous**

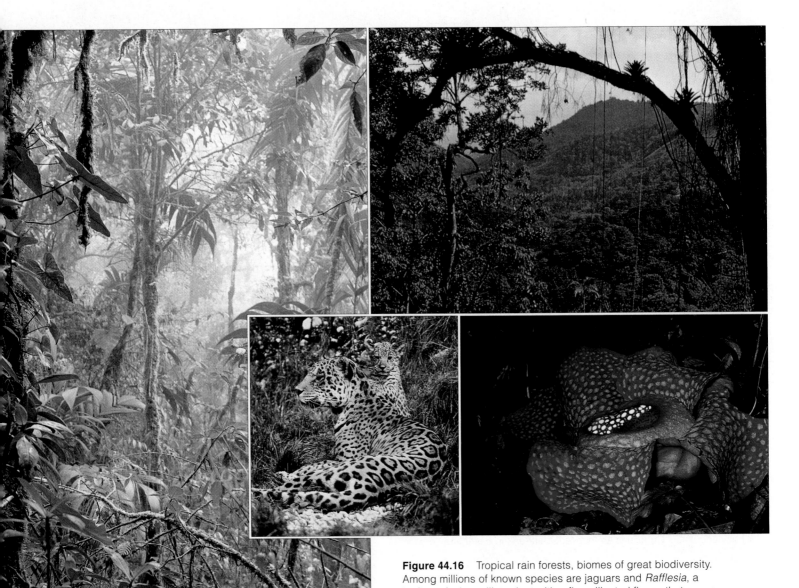

Figure 44.16 Tropical rain forests, biomes of great biodiversity. Among millions of known species are jaguars and *Rafflesia*, a leafless, foul-smelling plant with a fly-pollinated flower that grows three meters across. Bromeliads, orchids, and other epiphytes grow on trees, obtaining minerals from organic remains of leaves, insects, and other litter that have dissolved in tiny pockets of water.

SPRING

SUMMER

WINTER

AUTUMN

Figure 44.17 The changing appearance of part of a temperate deciduous forest south of Nashville, Tennessee, in spring, summer, autumn, and winter.

broadleaf forests. Trees of *tropical* deciduous forests drop some or all of their leaves in a pronounced dry season. *Monsoon* forests of India and southeastern Asia have such trees, also. Farther north, in the temperate zone, rainfall is even lower. Winters are cold; water is locked away as snow and ice. *Temperate* deciduous forests, such as those of the southeastern United States, prevail here (Figure 44.17). Decomposition is not as rapid as in the humid tropics, and many nutrients are conserved in accumulated litter on the forest floor.

Complex forests of ash, beech, birch, chestnut, elm, and deciduous oaks once stretched across northeastern North America, Europe, and East Asia. They declined drastically when farmers cleared the land. Pathogenic species introduced to North America killed nearly all chestnuts and many elms (Section 42.8). Now, maple and beech predominate in the northeast. Farther west, oak–hickory forests prevail, then oak woodlands that eventually grade into tallgrass prairie.

In forest biomes, conditions favor dense stands of tall trees that form a continuous canopy over a broad region.

CONIFEROUS FORESTS

Conifers (cone-bearing trees) are primary producers of **coniferous forests**. Most have thickly cuticled, needle-shaped leaves with recessed stomata—adaptations that help them conserve water during droughts and winter. They dominate the boreal forests, montane coniferous forests, temperate rain forests, and pine barrens.

Boreal forests stretch across northern Europe, Asia, and North America. Such forests also are called *taigas*, meaning "swamp forests." Most are in glaciated regions having cold lakes and streams (Figure 44.18*a*). It rains mostly in summer, and evaporation is low in the cool summer air. The cold, dry winters are more severe in eastern parts of these biomes than in the west, where

surrendered its rights to clear-cut one of the largest intact temperate rain forests outside of tropical regions. The tall trees, some of which are 800 years old, cloak British Columbia's Kitlope Valley.

Southern pine forests dominate the coastal plains of the south Atlantic and the Gulf states. Pine species are adapted to the dry, sandy, nutrient-poor soil and to natural fires or controlled burns. These fires do not get hot enough to damage the trees, and they open up the understory. Forests of pine, scrub oak, and wiregrass grow in New Jersey. Palmettos grow below pines and loblolly in the Deep South. During one recent Florida

ocean winds moderate the climate. Spruce and balsam fir dominate the boreal forests of North America. Pines, birches, and aspens take hold in the burned or logged areas. Where soil is poorly drained, highly acidic bogs dominated by peat mosses, shrubs, and stunted trees prevail. Boreal forests become much less dense to the north, where they grade into arctic tundra.

In the Northern Hemisphere, montane coniferous forests extend southward through the great mountain ranges. Spruces and firs dominate in the north and at higher elevations. They give way to firs and pines in the south and at lower elevations (Figures 3.1 and 44.18*b*). Some temperate lowlands support coniferous forests. One temperate rain forest that parallels the coast from Alaska on into northern California contains some of the world's tallest trees—Sitka spruce to the far north and redwoods to the south. Logging destroyed much of this biome. One bright note: In 1994, a timber corporation

Figure 44.18 (**a**) Spruce-dominated boreal forest. (**b**) Montane coniferous forest of Yosemite Valley in California's Sierra Nevada. (**c**) Firefighter retreating from flames in a forest of pines, oaks, and palmettos by Daytona Beach, Florida.

heat wave, fires destroyed more than 320,000 acres. Smaller burns had been prohibited in counties where housing and other developments encroached on these forests. Dense, tinder-dry understory had accumulated, and it fed the fires (Figure 44.18*c*).

Coniferous forests prevail in regions in which a cold, dry season alternates with a cool, rainy season.

$$-8 \quad -6 \quad -4 \quad -2 \quad 0 \quad 2 \quad 4 \quad 6 \quad 8$$
height (centimeters)

Figure 44.32 (**a**) Sea levels for the 1993–1994 winter between ENSOs. (**b**) Sea levels for the El Niño winter of 1997–1998. Values are based on measured increases in sea surface temperatures.

A CHOLERA CONNECTION During the 1997–1998 El Niño episode, 30,000 cases of cholera were reported in Peru, compared to only 60 from January to August in 1997. For some time, people knew that cholera was linked to water contaminated by *Vibrio cholerae* (Figure 44.33a). Epidemics develop after this bacterium infects humans, then enters water in feces expelled during the severe diarrhea that characterizes the disease.

What people did *not* know was where the bacterium lurked between cholera outbreaks. Year after year, no one could find it in humans or in water supplies. Then cholera would emerge simultaneously in places some distance apart—most often in coastal cities, where the urban poor draw water from rivers that enter the sea.

Rita Colwell, a microbiologist who later became the director of the National Science Foundation, suspected that humans did not harbor the pathogen between the outbreaks. Was there an environmental reservoir? Maybe, but nobody had detected any cells of *V. cholerae* in water samples subjected to standard culturing. Then Colwell had a flash of insight: What if no one could find the pathogen because it changes form and enters a dormant, sporelike stage between outbreaks?

During a cholera outbreak in Louisiana, she decided to use labeled antibodies that would bind to a certain protein at the bacterium's surface. Later, antibody tests in Bangladesh pinpointed the bacterium in fifty-one of fifty-two water samples. Standard culture methods had missed it in all but seven of the samples.

V. cholerae thrives in brackish water, rivers, estuaries, and the seas. Colwell knew that plankton (communities of mostly microscopic aquatic species) also thrive in these aquatic environments. She focused her search on the waters near Bangladesh, where outbreaks of cholera are endemic and seasonal. In time Colwell discovered a dormant stage of *V. cholerae* in copepods, marine animals that graze on algae and other members of phytoplankton (Figure 44.33b). The abundance of copepods—hence of *V. cholerae*—rise and fall with shifts in the abundance of phytoplankton, the "pastures of the seas."

Colwell already knew about seasonal variations in sea surface temperatures. Remember the saying, Chance favors the prepared mind? She discovered a correlation between seasonal temperature peaks in the Bay of Bengal with cholera cases in regional hospitals. Her correlation held for the 1990–1991 and 1997–1998 El Niño episodes (Figure 44.33c,d). Four to six weeks after the sea surface temperatures go up, so do the cases of cholera!

Together with her colleague Anwarul Huq, Colwell is working to identify other factors, such as shifts in salinity and nutrient content, that cause algal blooms, hence rises in copepod population size. Their goal is to integrate the data into a model that can be used to predict precisely where cholera outbreaks will occur and to give advance warning to filter drinking water. Meanwhile, Colwell is advising women in Bangladesh to use four layers of old sari cloth as filters, which remove 99 percent of *V. cholerae* cells from the water (Figure 44.33e). They can be rinsed in clean water, sun-dried, and used again and again.

Figure 44.33 (**a**) *Vibrio cholerae.* (**b**) Copepod. (**c**, **d**) Two satellite images correlating algal blooms with sea surface temperatures near South America during an El Niño episode. (**e**) In Bangladesh, Rita Colwell with unfiltered and filtered drinking water.

SUMMARY   *Gold* indicates text section

1. The biosphere encompasses the Earth's waters, the lower atmosphere, and the uppermost portions of its crust in which organisms live. Energy flows one way through the biosphere. Materials move through it on a grand scale to influence ecosystems everywhere. *CI*

2. The distribution of species through the biosphere is an outcome of the Earth's history, topography, climate, and interactions among species. *CI, 44.1–44.3*

    a. "Climate" refers to average weather conditions, including temperature, humidity, wind velocity, cloud cover, and rainfall over time. It results from differences in the amount of solar radiation reaching equatorial and polar regions, the Earth's daily rotation and its annual path around the sun, the distribution of continents and seas, and the elevation of land masses. *44.1*

    b. Interacting climatic factors underlie prevailing winds and ocean currents, which shape global weather patterns. Weather affects soil composition and water availability, which affects the growth and distribution of primary producers in ecosystems. *44.2*

3. The world's land masses are classified as six major biogeographic realms. Each is more or less isolated by oceans, mountain ranges, or desert barriers that tend to restrict gene flow between realms. Thus each tends to maintain a characteristic array of species. *44.3*

4. Biomes (a category of major ecosystems on land) are shaped by regional variations in climate, landforms, and soils. Dominant plants reflect prevailing conditions in the biomes called deserts, dry shrublands and dry woodlands, grasslands, broadleaf forests (e.g., tropical forests), coniferous forests, and tundra. *44.3–44.10*

5. Water provinces cover more than 71 percent of the Earth's surface. They include standing fresh water (such as lakes), running fresh water (such as streams), as well as the world oceans and seas. All aquatic ecosystems show gradients in light penetration, water temperature, salinity, and dissolved gases. These factors vary daily and seasonally, and affect primary productivity. *44.10*

6. Wetlands, intertidal zones, rocky and sandy shores, tropical reefs, and regions of the open ocean are major marine ecosystems. Photosynthetic activity is greatest in shallow coastal waters and in regions of upwelling: an upward movement of deep, cool ocean water that often carries nutrients to the surface. *44.11, 44.12*

Review Questions

1. Define biosphere. As part of your answer, also define atmosphere, lithosphere, and hydrosphere. *CI*

2. List the major interacting factors that influence climate. *CI*

3. List some of the ways in which air currents, ocean currents, or both may influence the region where you live. *44.1, 44.2*

4. Indicate where these biomes tend to be located in the world, and describe some of their defining features. *44.3–44.9*
    a. deserts   e. evergreen   h. alpine tundra
    b. dry shrublands   broadleaf forests   i. boreal forest
    c. dry woodlands f. deciduous forests j. montane
    d. grasslands   g. arctic tundra   coniferous forests

5. Define soils, then explain how the composition of regional soils affects ecosystem distribution. *44.4*

6. Describe the littoral, limnetic, and profundal zones of a large temperate lake in terms of seasonal primary productivity. *44.10*

7. Define the two major provinces of the world ocean. Is the open ocean devoid of life? *44.11*

8. Define and characterize the following types of ecosystems: mangrove wetland, estuary, and the intertidal zone. *44.12*

Self-Quiz   ANSWERS IN APPENDIX III

1. Solar radiation drives the distribution of weather systems and so influences _____ .
    a. temperature zones   c. seasonal variations
    b. rainfall distribution   d. all of the above

2. The _____ is a shield against ultraviolet wavelengths.
    a. upper atmosphere   c. ozone layer
    b. lower atmosphere   d. greenhouse effect

3. Regional variations in the global patterns of rainfall and temperature depend on _____ .
    a. global air circulation   c. topography
    b. ocean currents   d. all of the above

4. A rain shadow is a reduction in rainfall on the _____ of a mountain range.
    a. windward side   c. highest elevation
    b. leeward side   d. lowest elevation

5. Biogeographic realms are _____ .
    a. land and water provinces   c. divided into biomes
    b. six major land provinces   d. b and c

6. Biome distribution corresponds roughly with regional variations in _____ .
    a. climate   b. soils   c. topography   d. all of the above

7. Dominant plants of _____ are highly adapted to recurring episodes of lightning-sparked fires.
    a. dry shrublands   c. southern pine forests
    b. grasslands   d. all of the above

8. During _____ , deeper, often nutrient-rich water moves to the surface of a body of water.
    a. spring overturns   c. upwellings
    b. fall overturns   d. all of the above

9. Match the terms with the most suitable description.
    _____ boreal forest   a. high productivity; rapid cycling
    _____ permafrost   of nutrients (poor reservoirs)
    _____ chaparral   b. "swamp forest"
    _____ tropical   c. a type of dry shrubland
            rain forest   d. feature of arctic tundra

10. Match the terms with the most suitable description.
    _____ marine snow   a. deep, cool, often nutrient-rich
    _____ upwelling   ocean water moves upward
    _____ eutrophication   b. sediments, rocks of ocean bottom
    _____ estuary   c. partially enclosed mix of seawater
    _____ benthic   and fresh water
            province   d. nutrient enrichment of body
                                  of water; reduced transparency,
                                  phytoplankton blooms
                              e. basis of midocean food webs

OLIGOTROPHIC LAKE	EUTROPHIC LAKE
Deep, steeply banked	Shallow with broad littoral
Large deep-water volume relative to surface-water volume	Small deep-water volume relative to surface-water volume
Highly transparent	Limited transparency
Water blue or green	Water green to yellow- or brownish-green
Low nutrient content	High nutrient content
Oxygen abundant through all levels throughout year	Oxygen depleted in deep water during summer
Not much phytoplankton; green algae and diatoms dominant	Abundant, thick masses of phytoplankton; and cyanobacteria dominant
Abundant aerobic decomposers favored in profundal zone	Anaerobic decomposers
Low biomass in profundal	High biomass in profundal

b

Figure 44.34 (**a**) Crater Lake, Oregon, a collapsed volcanic cone filled with rainwater and melted snow. Like other volcanoes of the Cascade range, it started forming as a result of tectonic forces that prevailed at the dawn of the Cenozoic. (**b**) Major characteristics of oligotrophic and eutrophic lakes.

Critical Thinking

1. Kangaroos are furry, pouched mammals. Raccoons are furry placental mammals. Speculate on why Australia but not North America is the original home of kangaroos and why the reverse is true of raccoons. Use Section 24.9 and Section 24.13 *Critical Thinking* question 5 when devising an explanation.

2. Reflect on the world distribution of land masses and ocean water, then develop an explanation of why grassland biomes tend to form in the interior of continents, not along their coasts.

3. *Wetlands,* remember, are transitional zones between aquatic and terrestrial ecosystems in which plants are adapted to grow in periodically or permanently saturated soil. They include mangrove swamps and riparian zones (Section 25.6). Wetlands everywhere are being converted for agriculture, home building, and other human activities. Does the great ecological value of wetlands for the nation as a whole outweigh the rights of the private citizens who own many of the remaining wetlands in the United States? Should the private owners be forced to transfer title to the government? If so, who should decide the value of a parcel of land *and* pay for it? Or would such seizure of private property violate laws of the United States?

4. Observe conditions in a lake near your home or a place where you vacation. If lakes aren't part of your life, think about Oregon's Crater Lake instead (Figure 44.34*a*). Using the data in Figure 44.34*b* as a guide, would you conclude Crater Lake is oligotrophic or eutrophic? Will it remain so over time?

5. The United States Forest Service and Geophysical Dynamics Laboratory ran supercomputer programs to predict possible outcomes of a *global warming* trend. (Here you may wish to review Section 43.10.) According to their results, by the year 2030, the United States will experience recurring, devastating forest fires. Severe storms and more frequent rainfall in the western states will accelerate erosion, especially along the coasts and in steep foothills and mountain ranges.

Increases in the deposition of sediments will have adverse effects on rivers, streams, and estuaries. Dry shrublands and woodlands may replace hardwood forests in Minnesota, Iowa, and Wisconsin, and in parts of Missouri, Illinois, Indiana, and Michigan. The breadbasket of the American Midwest will shift north into Canada. Think about where you live now and where you plan to live in the future. How would such changes affect you? Should nations get serious about finding ways to counter global warming? Or do you believe that the scientists who are concerned about this are merely alarmist and that nothing bad will happen? Explain how you have reached your conclusion.

6. Think about the ocean, which covers the majority of the Earth's surface. Its deepest regions are remote from human populations, to say the least. Now think about a modern-day problem—where to dispose of nuclear wastes and other truly hazardous materials. You will be reading about this serious problem in the next chapter. For now, simply be aware that the United States government has stockpiles of very dangerous wastes and has nowhere to put them. Would it be feasible, let alone ethical, to use the deep ocean as a "burying ground" for them? Why or why not?

Selected Key Terms

atmosphere *CI*
biogeographic realm *44.3*
biogeography *CI*
biome *44.3*
biosphere *CI*
boreal forest *44.8*
climate *CI*
coniferous forest *44.8*
deciduous broadleaf forest *44.7*
desert *44.5*
desertification *44.5*
dry shrubland *44.6*
dry woodland *44.6*
ecoregion *44.3*

El Niño *44.12*
estuary *44.12*
eutrophication *44.10*
evergreen broadleaf forest *44.7*
fall overturn *44.10*
grassland *44.6*
hydrothermal vent *44.11*
intertidal zone *44.12*
lake *44.10*
mangrove wetland *44.12*
marine snow *44.11*
monsoon *44.2*
ocean *44.2*

permafrost *44.9*
rain shadow *44.2*
savanna *44.6*
soil *44.4*
southern pine forest *44.8*
spring overturn *44.10*
stream *44.10*
temperature zone *44.1*
tropical rain forest *44.7*
tundra *44.9*
ultraplankton *44.11*
upwelling *44.12*

Readings

Brown, J., and A. Gibson. 1983. *Biogeography*. St. Louis: Mosby.

Dold, C. February 1999. "The Cholera Lesson." *Discover*, 71–76.

Garrison, T. 1996. *Oceanography*. Second edition. Belmont, California: Wadsworth. Accessible; read its lyrical epilogue.

Olson, D. M., and E. Dinerstein, 1998. "The Global 200: A Representation Approach to Conserving the Earth's Most Biologically Valuable Ecoregions." *Conservation Biology* 12(3).

Smith, R. 1996. *Ecology and Field Biology*. Fifth edition. New York: HarperCollins.

On-Line readings at Student Guide for InfoTrac:
www.brookscole.com/biology

45

HUMAN IMPACT ON THE BIOSPHERE

An Indifference of Mythic Proportions

Of all the concepts introduced in the preceding chapter, the one that should be foremost in your mind is this: *The atmosphere, ocean, and land interact in ways that help dictate living conditions throughout the biosphere.* Driven by energy streaming in continually from the sun, these stupendous interactions give rise to globe-spanning temperatures and circulation patterns upon which life ultimately depends. This chapter turns to a related concept of equal importance. Simply put, *we have become major players in these interactions even before we fully comprehend how they work.*

To gain perspective on what is happening, think about something we take for granted—the air around us. The composition of the present-day atmosphere is a result of geologic and metabolic events—most notably, photosynthesis—that began billions of years ago. The first humans, recall, had evolved by 2.4 million years ago. Like us, they breathed oxygen from an atmosphere of ancient origins. Like us, they were protected from the sun's ultraviolet radiation by an ozone shield in the stratosphere. The size of their population was not much to speak of, and their effect on the biosphere was trivial. About 11,000 years ago, however, agriculture began in earnest, and it laid the foundation for huge increases in population size. Starting a few centuries ago, medical and industrial revolutions expanded that foundation for growth—and human population size skyrocketed in a mere blip of evolutionary time.

Today we are extracting huge amounts of energy and resources from the environment and giving back monumental amounts of wastes. As we do so, we are destabilizing ecosystems everywhere, even though the magnitude of change might not even be recognized when measured on the scale of a lifetime (Figure 45.1).

In a few developed countries, population growth has more or less stabilized, and the resource use per individual has dropped a bit. But resource utilization levels in those places are already high. In developing countries in Central America, Africa, and elsewhere, population sizes and resource consumption are rising fast, even though hundreds of millions of people are already malnourished or starving to death.

1900   1940   1954   1962

Many of the problems sketched out in this chapter are not going to disappear tomorrow. It will take many decades, even centuries, to reverse some trends that are already in motion—and not everyone is ready to make the effort. A few enlightened individuals in Michigan or Alberta or New South Wales can commit themselves to resource conservation and to minimizing pollution. But scattered attempts will not be enough. Individuals of all nations will unite to reverse global trends only when they perceive that the dangers of *not* doing so outweigh the personal benefits of ignoring them.

Does this seem pessimistic? Think of the exhaust fumes released to the air each time you drive a car or truck. Think of oil refineries, food-processing plants, and paper mills that supply you with goods and also release chemical wastes into the nation's waterways. Think of Mexico and other developing countries that produce cheap food by using an unskilled labor force and toxic pesticides. Unregulated pesticide applications poison the people who work the land, the land itself, and sometimes people who buy the exported produce.

Who changes behavior first? We have no answer. We can suggest, however, that a strained biosphere can rapidly impose an answer upon us. As an individual, you might choose to cherish or brood about or ignore any aspect of the world of life. Whatever your choice, the bottom line is that you and all other organisms are in this together. Our lives interconnect to degrees that we are only now starting to comprehend.

Figure 45.1 Tracking human population growth in one small part of the world, based on historical data and satellite imaging. *Red* denotes areas of dense human settlement in and around the San Francisco Bay Area and Sacramento since 1900.

Key Concepts

1. Human population growth has been skyrocketing ever since the mid-eighteenth century. At present, humans have the population size, the technology, and the cultural inclination to use energy and alter the environment at astonishing rates.

2. Pollutants are substances with which ecosystems have had no prior evolutionary experience, in terms of kinds or amounts, so adaptive mechanisms for dealing with them are not in place. In a more restrictive sense, pollutants are substances that accumulate in amounts that have adverse effects on human health, activities, and even survival itself.

3. Many pollutants, including the kinds that contribute to the formation of smog and acid rain, exert regionally harmful effects. The effects of other pollutants, including chlorofluorocarbons that attack the ozone layer in the stratosphere, are global in scale.

4. Conversion of marginally fertile lands for agriculture, highly mechanized deforestation, and other practices required to meet the demands of the growing human population are resulting in a loss of soil fertility and desertification. They also are causing a decline in the quality and quantity of one of the most crucial of all resources—fresh water.

5. Ultimately, the world of life runs on energy inputs from the sun that drive complex interactions among the atmosphere, ocean, and land. As a population, we are intervening in these interactions in ways that may have severe consequences in the near future.

6. We as a species must come to terms with principles of energy flow and principles of resource utilization that govern all systems of life on Earth.

AIR POLLUTION—PRIME EXAMPLES

Let's start this survey of the impact of human activities by defining pollution, a central problem. **Pollutants** are substances with which ecosystems have had no prior evolutionary experience, in terms of kinds or amounts, so adaptive mechanisms that might deal with them are not in place. From the human perspective, pollutants are substances that have accumulated enough to have adverse effects on our health, activities, or survival.

Air pollutants are premier examples. As Table 45.1 shows, they include carbon dioxide, oxides of nitrogen and sulfur, and chlorofluorocarbons. Also among them are photochemical oxidants formed when the sun's rays interact with certain chemicals. The United States alone releases 700,000 metric tons of air pollutants each day. Whether these remain concentrated at the source or are dispersed depends on local climate and topography.

Smog

When weather conditions trap a layer of cool, dense air under a layer of warm air, this is a **thermal inversion**. If the trapped air holds pollutants, winds can't disperse them, and they might accumulate to dangerous levels. Thermal inversions have been key factors in some of the worst air pollution disasters, because they intensify an atmospheric condition called smog (Figure 45.2).

Two types of smog, gray air and brown air, form in major cities. Where winters are cold and wet, **industrial smog** develops as a gray haze over industrialized cities that burn coal and other fossil fuels for manufacturing,

Table 45.1	*Major Classes of Air Pollutants*
Carbon oxides	Carbon monoxide (CO), carbon dioxide (CO_2)
Sulfur oxides	Sulfur dioxide (SO_2), sulfur trioxide (SO_3)
Nitrogen oxides	Nitric oxide (NO), nitrogen dioxide (NO_2), nitrous oxide (N_2O)
Volatile organic compounds	Methane (CH_4), benzene (C_6H_6), chlorofluorocarbons (CFCs)
Photochemical oxidants	Ozone (O_3), peroxyacyl nitrates (PANs), hydrogen peroxide (H_2O_2)
Suspended particles	Solids (dust, soot, asbestos, lead, etc.) and droplets (sulfuric acid, oils, pesticides, etc.)

heating, and generating electric power. Burning releases airborne pollutants, such as dust, smoke, soot, ashes, asbestos, oil, bits of lead and other heavy metals, and sulfur oxides. If not dispersed by winds and rain, the emissions may reach lethal concentrations. Industrial smog caused 4,000 deaths during a 1952 air pollution disaster in London. Before coal burning was restricted, New York, Pittsburgh, and Chicago also were gray-air cities. Today, most industrial smog forms in cities of China, India, and other developing countries as well as cities of coal-dependent countries of eastern Europe.

In warm climates, **photochemical smog** forms as a brown haze over large cities. It gets concentrated when the surrounding land forms a natural basin, as it does around Los Angeles and Mexico City (Figure 45.2). The key culprit is nitric oxide. It escapes from vehicles and reacts with oxygen in air to form nitrogen dioxide. Nitrogen dioxide exposed to sunlight reacts with hydrocarbons and forms photochemical oxidants. Most hydrocarbons are released from spilled or partly burned gasoline. Key oxidants are ozone and peroxyacyl nitrates, or **PANs**. Even traces of PANs sting eyes, irritate lungs, and damage crops.

Acid Deposition

Sulfur and nitrogen oxides are among the worst air pollutants. Coal-burning power plants, metal smelters, and factories emit most sulfur dioxides. Motor vehicles, power plants that burn gas and oil, and nitrogen-rich fertilizers produce nitrogen oxides. In dry weather, fine particles of oxides may briefly remain airborne, then fall as **dry acid deposition**. They form weak solutions of sulfuric and nitric acids when dissolved in atmospheric water. Winds may disperse them far from their source. If they fall to Earth in rain and snow, we call this wet acid deposition, or **acid rain**. The pH of normal rainwater is 5 or so. Acid rain can be 10 to 100 times more acidic, even as potent as lemon

Figure 45.2 (a) Normal pattern of air circulation in smog-forming regions. (b) Air pollutants trapped under a thermal inversion layer. (c) Mexico City on an otherwise bright, sunny morning. Topography, staggering numbers of people and vehicles, and industry combine to make its air among the world's smoggiest. Breathing this city's air is like smoking two packs of cigarettes a day.

cooler air
cool air
warm air
a

cool air
warm inversion layer
cool air
b

c

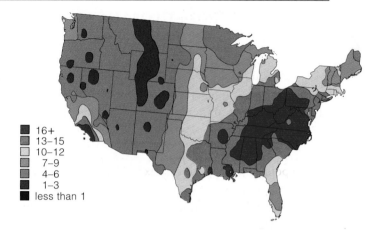

16+
13–15
10–12
7–9
4–6
1–3
less than 1

Figure 45.4 Regional differences in the concentrations of fine particles (2.5 μm or less) in micrograms per cubic meter of air, measured in 1992. Data are averages; they understate peak values in individual cities.

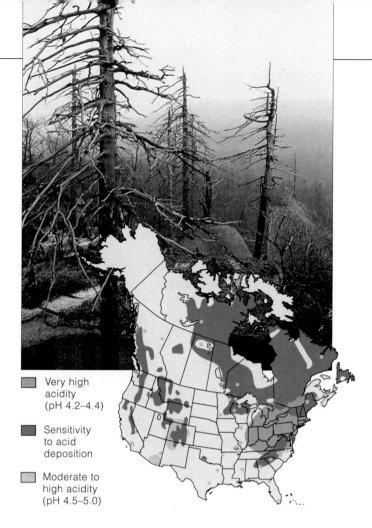

Very high acidity (pH 4.2–4.4)

Sensitivity to acid deposition

Moderate to high acidity (pH 4.5–5.0)

Figure 45.3 Map of average acidities of precipitation and soil sensitivities to acid deposition in 1984 for North American regions. The photograph, from Germany, is an example of how prolonged exposure to air pollutants is contributing to the rapid destruction of forests in much of the world. Weakened trees are more vulnerable to drought, disease, and insect pests.

juice! The deposited acids eat away at metals, marble buildings, rubber, plastics, nylon stockings, and other materials. They can disrupt the physiology of organisms, the chemistry of ecosystems, and biodiversity.

Depending on soil type and vegetation cover, some regions are far more sensitive than others to acid rain (Figure 45.3). Highly alkaline soil can neutralize acids before they enter the streams and lakes of watersheds. Water having a high carbonate content also neutralizes acids. However, many watersheds of northern Europe and southeastern Canada, and of regions throughout the United States have thin soils on top of solid granite. These soils cannot buffer much of the acidic inputs.

Rain in much of eastern North America is thirty to forty times more acidic than it was several decades ago. Crop yields are declining. Fish populations have already disappeared from more than 200 lakes in New York's Adirondack Mountains. By some estimates, fish will vanish from 48,000 lakes in Ontario, Canada, in the next two decades. Pollution from industrial regions

also is changing rainfall acidity enough to contribute to the decline of forest trees and mycorrhizae that support new growth (Chapter 21).

Researchers confirmed long ago that emissions from power plants, factories, and vehicles are the key sources of air pollutants. In 1995, researchers from the Harvard School of Public Health and Brigham Young University reported this finding from a comprehensive air quality study: Whack a year or so off your life span if you live in cities with the dirtiest air—especially air with fine particles of dust, soot, smoke, or acid droplets. Smaller particles are more easily inhaled and can damage lung tissue. Figure 45.4 summarizes air-quality data for the continental United States, gathered in 1992.

At one time the world's tallest smokestack, in the Canadian province of Ontario, accounted for 1 percent by weight of the annual worldwide emissions of sulfur dioxide. But Canada gets more acid deposition from industrialized regions of the midwestern United States than it sends across its southern border. Most of the air pollutants in Scandinavian countries, the Netherlands, Austria, and Switzerland arise in industrialized parts of western and eastern Europe. *Prevailing winds—hence air pollutants—do not stop at national boundaries.*

Pollutants are substances with which ecosystems have had no prior evolutionary experience, in terms of kinds or amounts, so adaptive mechanisms are not in place to deal with them.

Accumulated pollutants can harm organisms, as when they reach levels that adversely affect human health.

Smog formation and acid deposition are two examples of air pollution in specific regions, although prevailing winds often distribute pollutants beyond regional boundaries.

Smog forms mainly as a result of fossil fuel burning in urban and industrialized regions. Airborne acidic pollutants drift down as dry particles or as components of acid rain.

OZONE THINNING—GLOBAL LEGACY OF AIR POLLUTION

Now think about the ozone layer, almost twice as high above sea level as the top of Mount Everest, the highest place on Earth. Each September through mid-October, the layer becomes thinner at high latitudes. The seasonal **ozone thinning** is so pronounced, it was once called an "ozone hole." In 2000, ozone thinning over Antarctica was the most extensive ever recorded; it covered an area greater than North America (Figure 45.5). As in 1997 and 1999, the seasonal ozone loss at high northern latitudes was an alarming 60 percent.

Why is ozone reduction alarming? More ultraviolet radiation reaches the Earth and is triggering far more skin cancers, cataracts, and weakened immune systems. Such radiation also adversely affects photosynthesis. Substantial declines in the oxygen-releasing activity of phytoplankton alone could change the atmosphere's composition (Section 6.9).

CFCs (chlorofluorocarbons) are the major factors in ozone reduction. These compounds of carbon, fluorine, and chlorine are odorless and invisible. You find them in refrigerators and air conditioners (as the coolants), solvents, and plastic foams. CFCs slowly escape into air and resist breakdown. When a free CFC molecule absorbs ultraviolet light, it gives up one chlorine atom. If chlorine reacts with ozone, this yields oxygen and chlorine monoxide—which reacts with free oxygen to release another chlorine atom. Each released chlorine atom converts 10,000+ ozone molecules to oxygen!

Chlorine monoxide levels above Antarctica are 100 to 500 times higher than at midlatitudes. Why? High-altitude ice clouds form there during winter. Winds rotate around the South Pole for most of the winter, like a moving fence that keeps the ice clouds from spreading to other latitudes (Figure 45.5). This also happens on a lesser scale in the Arctic. Ice crystals are surfaces upon which chlorine compounds are swiftly degraded. When air warms in the spring, the chlorine is free to destroy ozone—hence the ozone thinning.

CFCs aren't the only ozone eaters. Methyl bromide, a fungicide, is nastier. It persists in the atmosphere only for a short time. Even so, it will account for about 15 percent of the ozone thinning in the years to come unless production stops.

Some scientists dismiss the threat of chemicals that contain chlorine and bromine. But the vast majority of those who have studied ecosystem modeling programs conclude that these chemicals pose long-term threats—not only to human health and certain crops but also to animal life in general. Substitutes are now available for most uses of CFCs; others are being developed.

Under international agreement, CFC production in developed countries has been phased out. By 2010, developing countries will phase out CFCs. And methyl bromide production will end by 2010. Recent computer models indicate the area of thinning should not get any larger. Assuming the international goals are met, it still will be 50 years before the ozone layer is restored to 1985 levels and another 100 to 200 years to total recovery, to pre-1950 levels. Meanwhile, you and all children and grandchildren of future generations will be living with the destructive effects of air pollutants. And if levels of greenhouse gases continue to rise, the stratosphere could cool enough to increase the size and duration of seasonal ozone depletions over the poles.

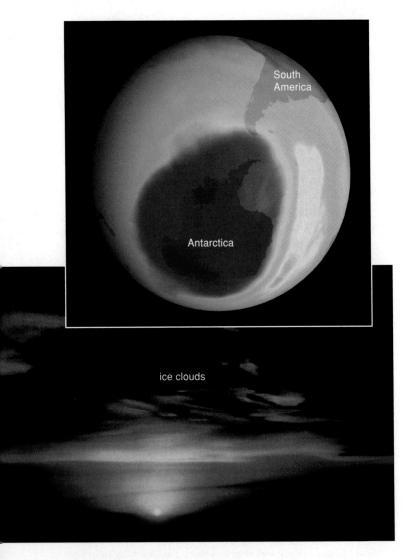

Figure 45.5 Seasonal ozone thinning above Antarctica in the spring of 2000, the largest yet recorded. Ice clouds above Antarctica have a role in the ozone thinning each spring.

Air pollution can have global repercussions, as when CFCs and other compounds contribute to a thinning of the ozone layer that shields life from the sun's ultraviolet radiation.

WHERE TO PUT SOLID WASTES? WHERE TO PRODUCE FOOD?

Oh Bury Me Not, In Our Own Garbage

In natural ecosystems, one organism's wastes serve as resources for others, so by-products of existence are cycled through the system. In developing countries, many resources are scarce. People conserve what they can and discard little. In the United States and some other developed countries, most people use something once, discard it, and buy another. Each year, millions of metric tons of solid wastes are dumped, burned, and buried. Paper products make up half the total volume, which also includes 50 billion nonreturnable cans and bottles. Every week in the United States, paper made from 500,000 trees ends up in Sunday newspapers. If every reader recycled merely one of ten newspapers, 25 million trees a year could be left alone. Recycling paper would reduce the airborne pollutants released during paper manufacturing by 95 percent and require 30 to 50 percent less energy than making new paper.

Our throwaway mentality is unique in the world of life. We are condoning a monumental accumulation of solid wastes in human ecosystems and natural ones. Bury the garbage in landfills? What happens when the space around cities runs out? Besides, landfills "leak" and eventually threaten groundwater supplies. Burn wastes in inefficient incinerators? These spew volumes of pollutants and ashes into the atmosphere.

Recycling is affordable and feasible. At the personal level, individuals can help initiate change by refusing to buy goods that are excessively wrapped, packaged in nondegradable containers, and designed for one-time use. They can participate in curbside recycling, by which they presort recyclable wastes. Such action may be more efficient than building huge resource recovery centers. These require so much trash to turn a profit that owners may end up encouraging the throwaway mentality. They also produce a toxic ash that must be disposed of in landfills, which eventually leak.

Converting Marginal Lands for Agriculture

The human population already uses nearly 21 percent of the Earth's land surfaces for cropland or for grazing. Another 28 percent is said to be potentially suitable for agriculture, but productivity would be so low that the conversion may not be worth the cost (Figure 45.6).

Asia and several other heavily populated regions now experience recurring, severe food shortages. Yet over 80 percent of their productive land is already under cultivation. Scientists have made valiant efforts to improve production on existing cropland. Under the banner of the **green revolution**, their research has been directed toward (1) improving the genetic character of crop plants for higher yields and (2) exporting modern

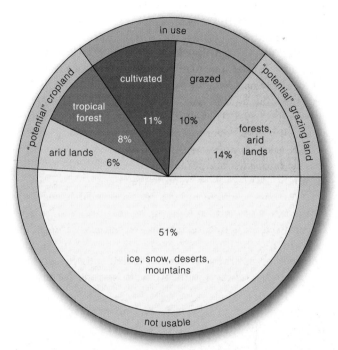

Figure 45.6 Classification of land with respect to its suitability for agriculture. Theoretically, clearing vast tracts of tropical forests and irrigating marginal lands could more than double the world's cropland. Doing so would destroy valuable forest resources, damage the environment, cause severe losses in biodiversity, and possibly cost more than it is worth.

agricultural practices and equipment to the developing countries. Many of these countries rely on *subsistence* agriculture, which runs on energy inputs from sunlight and human labor. They also rely heavily on *animal-assisted* agriculture, with energy inputs from oxen and other draft animals. By contrast, *mechanized* agriculture requires massive inputs of fertilizers, pesticides, and ample irrigation to sustain high-yield crops. It requires fossil fuel energy to drive farm machines. Crop yields are four times as high. But the modern practices use up 100 times more energy and minerals. Also, there are signs that limiting factors are coming into play to slow down further increases in crop yields.

Pressure for increased food production is greatest in parts of Central and South America, Asia, the Middle East, and Africa where human populations are rapidly expanding into marginal lands. Repercussions extend beyond national boundaries, as you will see next.

Our astounding population growth has impact on the Earth's land. We generate huge amounts of solid wastes but reuse or recycle very little. We are expanding into marginal lands for food production, a move that often has high energy costs and environmental consequences.

DEFORESTATION—AN ASSAULT ON FINITE RESOURCES

At one time, tropical forests cloaked regions that were, collectively, twice the size of Europe. For ten thousand years or more, they were home to an estimated 50 to 90 percent of all land-dwelling species. In less than four decades, human populations destroyed more than half of the forests, and most of their spectacular arrays of species may be lost. Each year, another 38 million acres are logged over. That's the equivalent of leveling thirty-four city blocks every minute. Logging extends beyond the tropics. Highly mechanized logging is proceeding in the once-vast temperate forests of the United States, Canada, Europe, Siberia, and elsewhere.

We have a name for the removal of all trees from large tracts of land for logging, agriculture, and grazing operations. It is **deforestation**. Why are we doing this? Paralleling the huge increases in the size of the human population are rapidly increasing demands for wood as fuel, lumber, and other forest products; and for grazing land and cropland. More and more people compete for dwindling resources. As you read in Chapter 25, they do so for economic profit, but also because alternative ways of life simply are not available to most people.

The world's great forests have profound influences on ecosystems. Like enormous sponges, the watersheds of forested regions absorb, hold, and gradually release water. By intervening in the downstream flow of water, they help control soil erosion, flooding, and a build-up of sediments that can clog rivers, lakes, and reservoirs. When the vegetation cover is stripped away, exposed soil becomes vulnerable to leaching of nutrients and to erosion, especially on steep slopes.

Deforestation is now greatest in Brazil, Indonesia, Colombia, and Mexico. If the clearing and destruction continue at present rates, only Brazil and Zaire will still have large tropical forests in 2010. By 2035, most of their forested regions will be cleared, also.

Figures 45.7 and 45.8 are close-up and panoramic views of what is happening in the Amazon basin of South America. Section 22.6 includes one example of heavy deforestation in North America.

In tropical regions, clearing the forests for agriculture sets the stage for long-term losses in productivity. The irony is that tropical forests are one of the worst places to grow crops or raise pasture animals. In intact forests, litter cannot accumulate, for the high temperatures and the heavy, frequent rainfall promote the rapid decomposition of organic wastes and remains. As fast as the decomposers release nutrients, the trees and other plants take them up. Deep, nutrient-rich topsoils simply cannot form.

Well before the advent of highly mechanized logging practices, people were practicing **shifting cultivation** (once referred to as slash-and-burn agriculture). They cut and burn trees,

Figure 45.7 Countries that are allowing the greatest destruction of tropical forests. *Red* denotes regions where 2,000 to 14,800 square kilometers are deforested each year. *Orange* denotes "moderate" deforestation, which encompasses areas of 100 to nearly 2,000 square kilometers.

ANDES

SMOKE
FROM
FIRES

Figure 45.8 The vast Amazon River basin of South America in September 1988. Smoke from fires that had been deliberately set during the dry season to clear tropical forests, pasturelands, and croplands completely obscured its features. The smoke extended to the Andes Mountains near the western horizon, about 650 miles (1,046 kilometers) away. That smoke cover was the largest that astronauts had ever observed. It extended almost 175 million square kilometers (1,044,000 square miles).

The smoke plume near the center of the photograph covered an area comparable to the immense smoke cover arising from the forest fire in Yellowstone National Park in that year. During the El Niño–induced drought of 1997, fires that had been set deliberately burned out of control. A smoke cover formed again.

Extensive deforestation is not confined to equatorial regions. For instance, in the past century, 2 million acres of redwood forests along the California coast were logged over. Most of the logging in temperate forests has occurred since 1950. Many of the logs are exported to lumber mills overseas, where wages are low.

then till the ashes into the soil. The nutrient-rich ashes can sustain crops for one to several seasons. Afterward, cleared plots are abandoned, for heavy leaching makes the soil infertile. When shifting cultivation is practiced on small, widely separated plots, a forest ecosystem isn't necessarily damaged extensively. But soil fertility plummets with population growth. Larger areas are cleared and plots are cleared again at shorter intervals.

Figure 45.9 Wangari Maathai, a Kenyan who organized the internationally acclaimed Green Belt Movement in 1977. The 50,000 members of this women's group are committed to establishing nurseries, raising seedlings, and planting a tree for each of the 27 million Kenyans. Together with half a million schoolchildren, they had planted more than 10 million trees by 1990. Their success inspired similar programs in more than a dozen countries in Africa. Dr. Maathai's efforts are not appreciated by her own government. Kenyan police have jailed her twice, and in 1992 they severely beat her because of her efforts on behalf of the environment.

Deforestation also affects the rates of evaporation, transpiration, runoff, and maybe regional patterns of rainfall. Trees release 50 to 80 percent of the water vapor above tropical forests. In logged-over regions, annual precipitation drops, and rain swiftly drains away from the exposed, nutrient-poor soil. The regions get hotter and drier—and soil fertility and moisture decline. In time sparse, dry grassland or desertlike conditions may prevail instead of the formerly rich, forested biomes.

Also, tropical forests absorb much of the sunlight reaching equatorial regions. Deforested land is shinier, so to speak, and reflects more incoming energy back into space. And the combined photosynthetic activity of so many trees affects the global cycling of carbon and oxygen. Extensive tree harvesting and burning release carbon stored in the tree biomass to the atmosphere, as carbon dioxide. In this way, deforestation might be a factor in the amplification of the greenhouse effect.

Conservation biologists are attempting to reverse the trend. As three examples, a coalition of 500 groups is dedicated to preserving Brazil's remaining tropical forests. In India, women have already built and installed 300,000 inexpensive, smokeless wood stoves. Over the past decade, the stoves saved more than 182,000 metric tons' worth of trees by reducing demands for fuelwood. In Kenya, women have planted millions of trees to hold soil in place and to provide fuelwood (Figure 45.9).

Once-vast forests helped sustain rapid increases in human population growth. Recent and highly mechanized modes of deforestation are rapidly depleting these finite resources.

You and the Tropical Rain Forest

Developing nations in Latin America, Southeast Asia, and Africa have the fastest-growing populations but not enough food, fuel, and lumber. Of necessity, they turn to their forests for growth-sustaining resources. Most of the forests may disappear within our lifetime. That possibility elicits the most outcries from concerned groups in highly developed nations—which happen to use most of the world's resources, including forest products. You read about economic aspects of this issue in earlier chapters. We invite you to reflect now on a few more aspects.

For purely ethical reasons, many people condemn the destruction of so much biodiversity. Tropical rain forests have the greatest variety and numbers of insects, and the world's largest ones. They are home to the most species of birds and to plants with the largest flowers. Living in the forest canopy and understory are monkeys, tapirs, and jaguars in South America and apes, okapi, and leopards in Africa. Massive vines twist around trees. Orchids, mosses, lichens, and other organisms grow on branches, absorbing minerals that rains deliver to them. Entire communities of microbes, insects, spiders, and amphibians live, breed, and die in small pools of water that collect in furled leaves.

Their disappearance will have repercussions through human life. Only a few strains of crop plants and livestock sustain most human populations, and they are vulnerable to ever evolving pathogens. Tissue-culture specialists and genetic engineers use genes of forest species to develop new or hybrid strains that can make our food base less vulnerable. Geneticists use them to develop more effective antibiotics and vaccines. Aspirin, the most widely used painkiller, is based on a chemical blueprint of an extract from tropical willow leaves. Many ornamental plants and spices and foods, including cocoa, cinnamon, and coffee, originated in the tropics. So did latex, gums, resins, dyes, waxes, and oils for tires, shoes, toothpaste, ice cream, shampoo, compact discs, condoms, and perfumes. Also think about this: Rampant burning of forests is releasing enough air pollutants to change the air we breathe and to help heat the planet during our lifetime.

And so conservation biologists rightly decry the mass extinction, the assaults on species diversity, the depletion of much of the world's genetic reservoir. Yet something else is going on here. Too many of us grow uneasy when we pass through obliterated forests in our own country. Is it because we are losing the comfort of our heritage—a connection with our evolutionary past? Many millions of years ago, our earliest primate ancestors moved into the trees of tropical forests. Through countless generations, their nervous and sensory systems evolved and became highly responsive to information-rich, arboreal worlds.

Does our neural wiring still resonate with rustling leaves, with shafts of light and mosaic shadows? Are we innately attuned to the forests of Eden—or have time and change buried recognition of home?

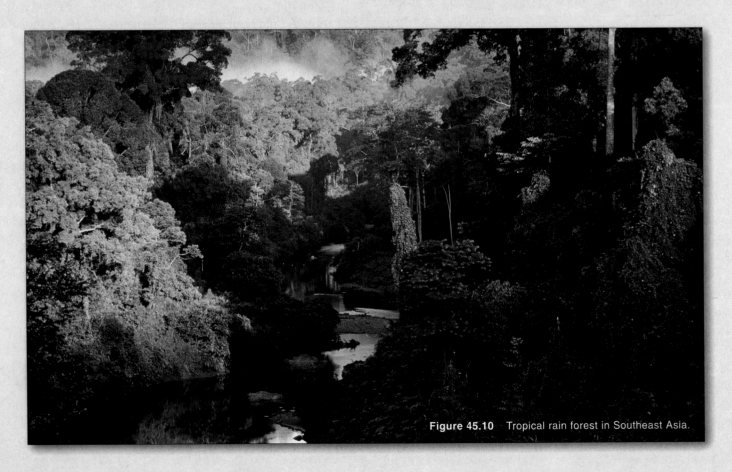

Figure 45.10 Tropical rain forest in Southeast Asia.

WHO TRADES GRASSLANDS FOR DESERTS?

Long-term shifts in climate often convert grasslands to deserts. Humans do, too. **Desertification** is the name for the conversion of large tracts of natural grasslands to a more desertlike condition. It applies also when the conversion of rain-fed or irrigated croplands causes a 10 percent or greater decline in agricultural productivity. During the past fifty years, 9 million square kilometers worldwide have become desertified.

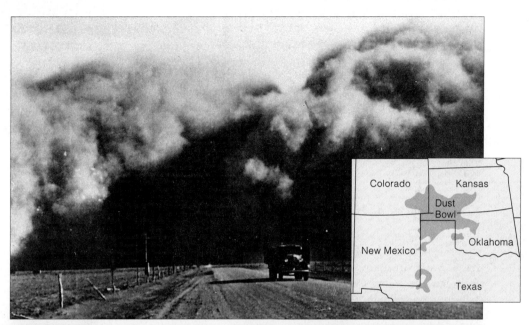

Figure 45.11 An awe-inspiring dust storm approaching Prowers County, Colorado, in 1934.

The Great Plains of the American Midwest are dry, windy grasslands that are subjected to pronounced, recurring droughts. Extensive conversion of these grasslands to agriculture began in the 1870s. Overgrazing left the ground bare across vast tracts. In May 1934, a cloud of topsoil that blew off the land blanketed the entire eastern portion of the United States, giving the Great Plains a dubious new name—the Dust Bowl. About 3.6 million hectares (9 million acres) of cropland were destroyed. Today, without large-scale irrigation and intensive conservation farming, desertlike conditions could prevail.

Figure 45.12 Desertification in the Sahel, a region of West Africa that forms a belt between the hot, dry Sahara Desert and tropical forests. This savanna country is undergoing rapid desertification as a result of overgrazing, overfarming, and prolonged drought.

Today, at least 200,000 square kilometers are being converted to deserts each year. Long droughts are accelerating the process, just as one did years ago in the Great Plains (Figure 45.11). At present, however, overgrazing of livestock on marginal lands is the source of most large-scale desertification.

In Africa, for instance, there are too many cattle in the wrong places. Cattle require more water than the native wild herbivores do, so they move back and forth between grazing areas and watering holes more often. As they do, they trample grasses and compact the soil surface (Figure 45.12). By contrast, gazelles and other wild herbivores get most (if not all) of their water from plants. They also are better at conserving water; they lose little in feces, compared to cattle.

In 1978 a biologist, David Holpcraft, began ranching antelopes, zebras, giraffes, ostriches, and other native herbivores. He also raised cattle as control groups to compare costs and meat yields on the same land. The native herds increased and provided tasty meat. Range conditions did not deteriorate; they improved. Vexing problems remained. Certain tribes in Africa have their own idea of what constitutes "good" meat, and some view cattle as the symbol of wealth in their society.

Without irrigation and conservation practices, grasslands converted for agriculture often end up as deserts.

A GLOBAL WATER CRISIS

The Earth has a tremendous supply of water, but most is just too salty for human consumption or agriculture. Imagine all that water in a bathtub. Withdraw all of the fresh, renewable portion (from lakes, rivers, reservoirs, groundwater, and other sources of surface water) and it would barely fill a teaspoon.

Why not consider **desalinization**—removal of salt from seawater? After all, there are unlimited supplies of seawater. Desalinization processes exist; they distill seawater or force it through membranes, by reverse osmosis. Expensive fuel energy drives these processes, so they may be feasible only in Saudi Arabia and a few other countries that have small population sizes, large reserves of energy, and lots of cash. Actually, in some situations they may be the only alternative to running out of drinkable water. That nearly happened in Santa Barbara and some other cities in California during one prolonged drought. Yet desalinization cannot solve the core problem. It may never be cost-effective for large-scale agriculture, and it makes mountains of salts.

Figure 45.13   Irrigated crops in the Sahara Desert, in Algeria.

Consequences of Heavy Irrigation

Large-scale agriculture accounts for nearly two-thirds of the human population's use of fresh water. In many cases, irrigation water from surface sources is piped into vast fields where water-demanding crops will not grow on their own. At its most extreme, irrigation has turned some hot deserts into lush gardens, although believing these can be maintained over the long term is a bit delusional (Figure 45.13).

Irrigation itself can change the land's suitability for agriculture. Concentrations of mineral salts commonly are high in piped-in water. In regions where soil drains poorly, evaporation may cause **salinization**: a buildup of salt in soil. Salinization can stunt the growth of crop plants, eventually kill them, and decrease yields.

Land that drains poorly also becomes waterlogged. When water accumulates underground, it slowly raises the **water table**, the upper limit at which the ground is fully saturated. When the water table rises close to the ground's surface, soil gets saturated with saline water, which can damage plant roots. Properly managing the water–soil system can reverse the salinization and the waterlogging. The economic cost of doing so is high.

Worse, groundwater overdrafts (the amount nature doesn't replenish) are high (Figure 45.14). For instance, overdrafts have depleted half of the Ogallala aquifer, which supplies irrigation water for 20 percent of the croplands in the United States. So much groundwater has been withdrawn for irrigation in some parts of the San Joaquin Valley in California that the surface of the water table has dropped by as much as six meters.

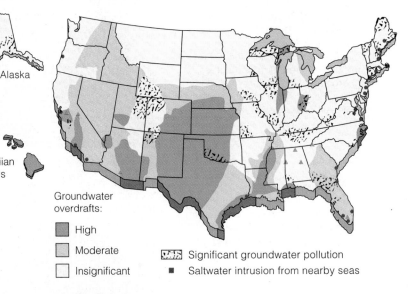

Groundwater overdrafts:

▪ High

▫ Moderate

☐ Insignificant

▦ Significant groundwater pollution

▪ Saltwater intrusion from nearby seas

Figure 45.14   Areas of aquifer depletion, saltwater intrusion, and groundwater contamination in the United States.

Water Pollution

Water pollution amplifies the problem of water scarcity. Inputs of sewage, animal wastes, and toxic chemicals from many power-generating plants and factories make water unfit to drink. Runoff from fields puts sediments and pesticides into water, along with phosphates and other plant nutrients that promote algal blooms.

Pollutants collect in lakes, rivers, and bays before reaching the oceans. Many cities throughout the world dump untreated sewage into coastal waters. The cities along rivers and harbors maintain shipping channels by dredging the polluted muck and barging it out to

Figure 45.19 City as ecosystem.

Critical Thinking

1. Kristen, a recent college graduate, is finding her idealism on a collision course with reality. She strongly believes people who live in the United States are obliged to make the world a level playing field for all human beings, with equality in resources, health care, education, economic security, and a pristine environment for all. Yet she also understands that the sheer size of the human population makes this impossible.

Kristen recently said she cannot be party to hard choices and actions that go against her ideals, and she just wants nature to solve the problem for us. Comment on this true story.

2. Investigate where the water for your own city comes from and where it has been. You may find the answer illuminating.

3. Make a list of advantages you personally enjoy as a member of an affluent, industrialized society. List some drawbacks. Do the benefits outweigh the costs? This is not a trick question.

4. It has been said that economic wars, more than military wars, will determine the winners and losers among nations in the near future. The economic growth of certain nations, including some in the former Soviet Union and the Far East, has had devastating impact on the environment. Elsewhere, efforts of conservation biologists to maintain standards of environmental protection add to the cost of goods produced and put practicing nations at serious disadvantage in this global competition.

Should the United States loosen environmental laws to help ensure its economic survival? Why or why not? Should it impose pressure on indifferent nations to encourage reduction in harmful practices? Why or why not?

5. To reduce the backlog of unnamed new species and secure funding, a group of German taxonomists is offering shoppers the opportunity, for $2,500 and up, to name a species (Biopat web site). Example: As a birthday present, investment banker Stan Lai's wife had a Bolivian yellow-striped brown toad named in her husband's honor (*Bufo stanlaii*, or "Stan Lai's toad").

Usually, a species name reflects its appearance or honors its discoverer. But not always; for example, one midge has the name *Dicrotendipes thanatogratus* (the "Grateful Dead" midge). The International Commission on Zoological Nomenclature in London denounces offering taxonomy to the public, which it believes will obscure science and hinder conservation efforts. What does Mrs. Lai think? "Now, every time I hear something about Bolivia, I'll pay attention. After all, I don't want to see our toad die."

Does inviting the public to pay their way into taxonomy undermine the science? Is it a good or bad way to raise public awareness of conservation biology?

6. Populations of every species utilize resources and produce wastes. Use Figure 43.19 as a starting point for a brief essay on the accumulation and uses of energy and materials, including wastes, in a human population of a large city. Contrast your description with the flow of energy and the cycling of materials that proceed in a natural population of some other organism.

7. Many psychologists believe that we spend much of our lives consciously or subconsciously searching for roots that might anchor us in bewildering or frightening times of change. Some philosophers have argued that each of us must find a mountain, river, backyard, or any other place that elicits a sense of sanctuary, of rooted connections with nature. Others believe that forming an emotional connection with nature is based on a romanticized or mystical view of it—and that time would be better spent developing a scientifically sound understanding of nature and the technological means to protect and sustain its resources.

Do you agree with one or the other point of view? Or do you suspect that the two are not mutually exclusive?

Selected Key Terms

acid rain *45.1*
chlorofluorocarbon (CFC) *45.2*
deforestation *45.4*
desalinization *45.7*
desertification *45.6*
dry acid deposition *45.1*
fossil fuel *45.8*
fusion power *45.9*
green revolution *45.3*
industrial smog *45.1*
meltdown *45.8*

ozone thinning *45.2*
PAN (peroxyacyl nitrate) *45.1*
photochemical smog *45.1*
pollutant *45.1*
salinization *45.7*
shifting cultivation *45.4*
solar–hydrogen energy *45.9*
thermal inversion *45.1*
wastewater treatment *45.7*
water table *45.7*
wind farm *45.9*

Readings

Frosh, R. September 1995. "The Industrial Ecology of the Twenty-First Century." *Scientific American* 273(3):178–181.

Kaiser, J. 18 February 2000. "Ecologists on a Mission to Save the World." *Science* 287:1188–1192. Addresses a growing debate over how far environmental scientists should go when interpreting their findings for policy makers.

Miller, G. T., Jr. 2001. *Living in the Environment.* Eighth edition. Belmont, California: Brooks/Cole.

APPENDIX I. CLASSIFICATION SCHEME

The classification scheme that follows is a composite of several that microbiologists, botanists, and zoologists use. Major groupings are agreed upon, more or less. There is not always agreement on what to name a particular grouping or where it may fit within the overall hierarchy. There are several reasons for the lack of total consensus.

First, the fossil record varies in its quality and in its completeness. Therefore, the phylogenetic relationship of one group to others is sometimes open to interpretation. Comparative studies at the molecular level are firming up the picture, but this work is still under way.

Second, ever since the time of Linnaeus, classification schemes have been based on the perceived morphological similarities and differences among organisms. Although some original interpretations are now open to question, we are so used to thinking about organisms in certain ways that reclassification often proceeds slowly.

For example, birds and reptiles traditionally have been placed in separate classes (Reptilia and Aves), yet there are many compelling arguments for grouping the lizards and snakes in one class and the crocodilians, dinosaurs, and birds in a separate class. Some favor six kingdoms but others favor three domains: archaebacteria, eubacteria, and eukaryota (alternatively, archaea, bacteria, and eukarya).

Third, researchers in microbiology, mycology, botany, zoology, and the other fields of biological inquiry have inherited a wealth of literature, based on classification schemes that were developed over time in each of those fields. Many see no good reason to give up the established terminology and thereby disrupt access to the past.

For instance, many microbiologists and botanists use *division*, and zoologists *phylum*, for taxa that are equivalent in the hierarchy of classification. Also, opinions are still polarized with respect to the kingdom Protista, certain members of which could just as easily be grouped with plants, fungi, or animals. Indeed, the term protozoan is a holdover from an earlier scheme in which some single-celled organisms were ranked as simple animals.

Given the problems, why do we even bother imposing artificial frameworks on the history of life? We do this for the same reason that a writer might decide to break up the history of civilization into several volumes, a number of chapters, and many paragraphs. Both efforts are attempts to impart obvious structure to what might otherwise be an overwhelming body of knowledge and to enhance the retrieval of information from it.

Finally, bear in mind that we include this classification scheme primarily for your reference purposes. Besides being open to revision, it also is by no means complete. It does not include the most recently discovered species, as from the mid-ocean. Many existing and extinct organisms of the so-called lesser phyla are not represented here. Our strategy is to focus mainly on organisms mentioned in the text. A few examples of organisms also are listed.

SUPERKINGDOM PROKARYOTA. Prokaryotes. Almost all microscopic species. DNA organized at nucleoid (a region of cytoplasm), not inside a membrane-bound nucleus. All are bacteria, either single cells or simple associations of cells. Autotrophs and heterotrophs. Table A on the following page lists representative types. Reproduce by prokaryotic fission, sometimes by budding and by bacterial conjugation.

The authoritative reference in bacteriology, *Bergey's Manual of Systematic Bacteriology,* calls this "a time of taxonomic transition." It groups bacteria mostly by numerical taxonomy (Section 20.1), not on phylogeny. The scheme presented here reflects strong evidence of evolutionary relationships for at least some bacterial groupings.

KINGDOM EUBACTERIA. Gram-negative, gram-positive forms. Peptidoglycan present in cell wall. Photosynthetic autotrophs, chemosynthetic autotrophs, and heterotrophs.

PHYLUM GRACILICUTES. Typical Gram-negative, thin wall. Autotrophs (photosynthetic and chemosynthetic) and heterotrophs. *Anabaena* and other cyanobacteria. *Escherichia, Pseudomonas, Neisseria, Myxococcus.*

PHYLUM FIRMICUTES. Typical Gram-positive, thick wall. Heterotrophs. *Bacillus, Staphylococcus, Streptococcus, Clostridium, Actinomycetes.*

PHYLUM TENERICUTES. Gram-negative, wall absent. Heterotrophs (saprobes, pathogens). *Mycoplasma.*

KINGDOM ARCHAEBACTERIA. Methanogens, extreme halophiles, extreme thermophiles. Evolutionarily closer to eukaryotic cells than to eubacteria. Strict anaerobes. Distinctive cell wall, membrane lipids, ribosomes, RNA sequences. *Methanobacterium, Halobacterium, Sulfolobus.*

SUPERKINGDOM EUKARYOTA. Eukaryotes. Both single-celled and multicelled species. Cells start out life with a nucleus (encloses the DNA) and usually other membrane-bound organelles. Chromosomes have many histones and other proteins attached.

KINGDOM PROTISTA. Diverse single-celled, colonial, and multicelled eukaryotic species. Existing species are unlike bacteria in characteristics and are most like the earliest, structurally simple eukaryotes. Autotrophs, heterotrophs, or both (Table 20.3). Reproduce sexually and asexually (by meiosis, mitosis, or both). Many related evolutionarily to plants, fungi, and possibly animals.

PHYLUM CHYTRIDIOMYCOTA. Chytrids. Heterotrophs; saprobic decomposers or parasites. *Chytridium.*

PHYLUM OOMYCOTA. Water molds. Heterotrophs. Decomposers, some parasites. *Saprolegnia, Phytophthora, Plasmopara.*

PHYLUM ACRASIOMYCOTA. Cellular slime molds. Heterotrophs with free-living, phagocytic amoeboid cells and spore-bearing stages. *Dictyostelium.*

PHYLUM MYXOMYCOTA. Plasmodial slime molds. Heterotrophs with free-living, phagocytic amoeboid cells and spore-bearing stages. Aggregate into streaming mass of cells that discard plasma membranes. *Physarum.*

Some Major Groups	Main Habitats	Characteristics	Representatives
EUBACTERIA			
Photoautotrophs:			
Cyanobacteria, green sulfur bacteria, and purple sulfur bacteria	Mostly lakes, ponds; some marine, terrestrial habitats	Photosynthetic; use sunlight energy, carbon dioxide; cyanobacteria use oxygen-producing noncyclic pathway; some also use cyclic route	*Anabaena, Nostoc, Rhodopseudomonas, Chloroflexus*
Photoheterotrophs:			
Purple nonsulfur and green nonsulfur bacteria	Anaerobic, organically rich muddy soils, and sediments of aquatic habitats	Use sunlight energy; organic compounds as electron donors; some purple nonsulfur may also grow chemotrophically	*Rhodospirillum, Chlorobium*
Chemoautotrophs:			
Nitrifying, sulfur-oxidizing, and iron-oxidizing bacteria	Soil; freshwater, marine habitats	Use carbon dioxide, inorganic compounds as electron donors; influence crop yields, cycling of nutrients in ecosystems	*Nitrosomonas, Nitrobacter, Thiobacillus*
Chemoheterotrophs:			
Spirochetes	Aquatic habitats; parasites of animals	Helically coiled, motile; free-living and parasitic species; some major pathogens	*Spirochaeta, Treponema*
Gram-negative aerobic rods and cocci	Soil, aquatic habitats; parasites of animals, plants	Some major pathogens; some fix nitrogen (e.g., *Rhizobium*)	*Pseudomonas, Neisseria, Rhizobium, Agrobacterium*
Gram-negative facultative anaerobic rods	Soil, plants, animal gut	Many major pathogens; one bioluminescent (*Photobacterium*)	*Salmonella, Escherichia, Proteus, Photobacterium*
Rickettsias and chlamydias	Host cells of animals	Intracellular parasites; many pathogens	*Rickettsia, Chlamydia*
Myxobacteria	Decaying organic material; bark of living trees	Gliding, rod-shaped; aggregation and collective migration of cells	*Myxococcus*
Gram-positive cocci	Soil; skin and mucous membranes of animals	Some major pathogens	*Staphylococcus, Streptococcus*
Endospore-forming rods and cocci	Soil; animal gut	Some major pathogens	*Bacillus, Clostridium*
Gram-positive nonsporulating rods	Fermenting plant, animal material; gut, vaginal tract	Some important in dairy industry, others major contaminators of milk, cheese	*Lactobacillus, Listeria*
Actinomycetes	Soil; some aquatic habitats	Include anaerobes and strict aerobes; major producers of antibiotics	*Actinomyces, Streptomyces*
ARCHAEBACTERIA (ARCHAEA)			
Methanogens	Anaerobic sediments of lakes, swamps; animal gut	Chemosynthetic; methane producers; used in sewage treatment facilities	*Methanobacterium*
Extreme halophiles	Brines (extremely salty water)	Heterotrophic; also, unique photosynthetic pigments (bacteriorhodopsin) form in some	*Halobacterium*
Extreme thermophiles	Acidic soil, hot springs, hydrothermal vents	Heterotrophic or chemosynthetic; use inorganic substances as electron donors	*Sulfolobus, Thermoplasma*

PHYLUM SARCODINA.   Amoeboid protozoans. Heterotrophs, free-living or endosymbiotic, some pathogens. Soft-bodied, with or without shell, locomotion by pseudopods. The rhizopods (naked amoebas, foraminiferans), *Amoeba proteus, Entomoeba*. Also actinopods (radiolarians, heliozoans).

PHYLUM CILIOPHORA.   Ciliated protozoans. Heterotrophs, predators or symbionts, some parasitic. All have cilia. Free-living, sessile, or motile. *Paramecium, Didinium*, hypotrichs.

PHYLUM MASTIGOPHORA.   Animal-like flagellated protozoans. Heterotrophs, free-living, many internal parasites. All with one to several flagella. *Trypanosoma, Trichomonas, Giardia*.

APICOMPLEXA.   Heterotrophs, sporozoite-forming parasites. Complex structures at head end. Most familiar members called sporozoans. *Cryptosporidium, Plasmodium, Toxoplasma*.

PHYLUM EUGLENOPHYTA.   Euglenoids. Mostly heterotrophs, some autotrophs (photosynthetic), some switch depending on environmental conditions. Most with one short, one long flagellum; red, green, or colorless. *Euglena, Peranema*.

PHYLUM PYRRHOPHYTA.   Dinoflagellates. Photosynthetic, mostly, but some heterotrophs. *Fiesteria, Gymnodinium breve*.

PHYLUM CHRYSOPHYTA.   Golden algae, yellow-green algae, diatoms. Photosynthetic. Some flagellated, others not. *Mischococcus, Synura, Vaucheria*.

PHYLUM RHODOPHYTA.   Red algae. Mostly photosynthetic, some parasitic. Nearly all marine, some in freshwater habitats. *Porphyra, Bonnemaisonia, Euchema*.

PHYLUM PHAEOPHYTA.   Brown algae. Photosynthetic, nearly all in temperate or marine waters. *Macrocystis, Fucus, Sargassum, Ectocarpus, Postelsia*.

PHYLUM CHLOROPHYTA.   Green algae. Mostly photosynthetic, some parasitic. Most freshwater, some marine or terrestrial. *Chlamydomonas, Spirogyra, Ulva, Volvox, Codium, Halimeda*.

PHYLUM ZYGOMYCOTA.   Zygomycetes. Zygosporangia (zygote inside thick wall) formed by sexual reproduction. Bread molds, related forms. *Rhizopus, Philobolus.*

PHYLUM ASCOMYCOTA.   Ascomycetes. Sac fungi. Sac-shaped cells form sexual spores (ascospores). Most yeasts and molds, morels, truffles. *Saccharomycetes, Morchella, Neurospora, Sarcoscypha, Claviceps, Ophiostoma, Candida, Aspergillus, Penicillium.*

PHYLUM BASIDIOMYCOTA.   Basidiomycetes. Club fungi. Most diverse group. Produce basidiospores inside club-shaped structures. Mushrooms, shelf fungi, stinkhorns. *Agaricus, Amanita, Craterellus, Gymnophilus, Puccinia, Ustilago.*

IMPERFECT FUNGI.   Sexual spores absent or undetected. The group has no formal taxonomic status. If better understood, a given species might be grouped with sac fungi or club fungi. *Arthobotrys, Histoplasma, Microsporum, Verticillium.*

LICHENS.   Mutualistic interactions between fungal species and a cyanobacterium, green alga, or both. *Lobaria, Usnea, Cladonia.*

PHYLUM RHYNIOPHYTA.   Earliest known vascular plants; muddy habitats. Extinct. *Cooksonia, Rhynia.*

PHYLUM PROGYMNOSPERMOPHYTA.   Progymnosperms. Ancestral to early seed-bearing plants; extinct. *Archaeopteris.*

PHYLUM PTERIDOSPERMOPHYTA.   Seed ferns. Fernlike gymnosperms; extinct. *Medullosa.*

PHYLUM CHAROPHYTA.   Stoneworts.

PHYLUM BRYOPHYTA.   Bryophytes: mosses, liverworts, hornworts. Seedless, nonvascular, haploid dominance. *Marchantia, Polytrichum, Sphagnum.*

PHYLUM PSILOPHYTA.   Whisk ferns. Seedless, vascular. No obvious roots, leaves on sporophyte. *Psilotum.*

PHYLUM LYCOPHYTA.   Lycophytes, club mosses. Seedless, vascular. Leaves, branching rhizomes, vascularized roots and stems. *Lepidodendron* (extinct), *Lycopodium, Selaginella.*

PHYLUM SPHENOPHYTA.   Horsetails. Seedless, vascular. Some sporophyte stems photosynthetic, others nonphotosynthetic, spore-producing. *Calamites* (extinct), *Equisetum.*

PHYLUM PTEROPHYTA.   Ferns. Largest group of seedless vascular plants (12,000 species), mainly tropical, temperate habitats. *Pteris, Trichomanes, Cyathea* (tree ferns), *Polystichum.*

PHYLUM CYCADOPHYTA.   Cycads. Gymnosperm group (vascular, bears "naked" seeds). Tropical, subtropical. Palm-shaped leaves, simple cones on male and female plants. *Zamia.*

PHYLUM GINKGOPHYTA.   Ginkgo (maidenhair tree). Type of gymnosperm. Seeds with fleshy outer layer. *Ginkgo.*

PHYLUM GNETOPHYTA.   Gnetophytes. Only gymnosperms with vessels in xylem and double fertilization (but endosperm does not form). *Ephedra, Welwitschia.*

PHYLUM CONIFEROPHYTA.   Conifers. Most common and familiar gymnosperms. Generally cone-bearing species with needle-like or scale-like leaves.
Family Pinaceae. Pines, firs, spruces, hemlock, larches, true cedars. *Abies, Cedrus, Pinus, Pseudotsuga.*
Family Cupressaceae. Junipers, cypresses. *Cupressus, Juniperus.*
Family Taxodiaceae. Bald cypress, redwoods, bigtree. *Metasequoia, Sequoia, Sequoiadendron, Taxodium.*
Family Taxaceae. Yews. *Taxus.*

PHYLUM ANTHOPHYTA.   Angiosperms (the flowering plants). Largest, most diverse group of vascular seed-bearing plants. Only organisms that produce flowers, fruits.

Class Dicotyledonae. Dicotyledons (dicots). Some families of some representative orders are listed:
Family Nymphaeaceae. Water lilies.
Family Papaveraceae. Poppies.
Family Brassicaceae. Mustards, cabbages, radishes.
Family Malvaceae. Mallows, cotton, okra, hibiscus.
Family Solanaceae. Potatoes, eggplant, petunias.
Family Salicaceae. Willows, poplars.
Family Rosaceae. Roses, apples, almonds, strawberries.
Family Fabaceae. Peas, beans, lupines, mesquite.
Family Cactaceae. Cacti.
Family Euphorbiaceae. Spurges, poinsettia.
Family Cucurbitaceae. Gourds, melons, cucumbers, squashes.
Family Apiaceae. Parsleys, carrots, poison hemlock.
Family Aceraceae. Maples.
Family Asteraceae. Composites. Chrysanthemums, sunflowers, lettuces, dandelions.

Class Monocotyledonae. Monocotyledons (monocots). Some families of several different orders are listed:
Family Liliaceae. Lilies, hyacinths, tulips, onions, garlic.
Family Iridaceae. Irises, gladioli, crocuses.
Family Orchidaceae. Orchids.
Family Arecaceae. Date palms, coconut palms.
Family Cyperaceae. Sedges.
Family Poaceae. Grasses, bamboos, corn, wheat, sugarcane.
Family Bromeliaceae. Bromeliads, pineapples, Spanish moss.

PHYLUM PLACOZOA.   Marine. Simplest known animal. Two cell layers, no mouth, no organs. *Trichoplax.*

PHYLUM MESOZOA.   Ciliated, wormlike parasites, about the same level of complexity as *Trichoplax.*

PHYLUM PORIFERA.   Sponges. No symmetry, tissues. *Euplectella.*

PHYLUM CNIDARIA.   Radial symmetry, tissues, nematocysts.
Class Hydrozoa. Hydrozoans. *Hydra, Obelia, Physalia, Prya.*
Class Scyphozoa. Jellyfishes. *Aurelia.*
Class Anthozoa. Sea anemones, corals. *Telesto.*

PHYLUM PLATYHELMINTHES.   Flatworms. Bilateral, cephalized; simplest animals with organ systems. Saclike gut.
Class Turbellaria. Triclads (planarians), polyclads. *Dugesia.*
Class Trematoda. Flukes. *Clonorchis, Schistosoma.*
Class Cestoda. Tapeworms. *Diphyllobothrium, Taenia.*

PHYLUM NEMERTEA.   Ribbon worms. *Tubulanus.*

PHYLUM NEMATODA.   Roundworms. *Ascaris, Caenorhabditis elegans, Necator* (hookworms), *Trichinella.*

PHYLUM ROTIFERA.   Rotifers. *Asplancha, Philodina.*

PHYLUM MOLLUSCA.   Mollusks.
Class Polyplacophora. Chitons. *Cryptochiton, Tonicella.*

Class Gastropoda. Snails (periwinkles, whelks, limpets, abalones, cowries, conches, nudibranchs, tree snails, garden snails), sea slugs, land slugs. *Aplysia, Ariolimax, Cypraea, Haliotis, Helix, Liguus, Limax, Littorina, Patella.*

Class Bivalvia. Clams, mussels, scallops, cockles, oysters, shipworms. *Ensis, Chlamys, Mytelus, Patinopectin.*

Class Cephalopoda. Squids, octopuses, cuttlefish, nautiluses. *Dosidiscus, Loligo, Nautilus, Octopus, Sepia.*

PHYLUM BRYOZOA.   Bryozoans (moss animals).

PHYLUM BRACHIOPODA.   Lampshells.

PHYLUM ANNELIDA.   Segmented worms.

Class Polychaeta. Mostly marine worms. *Eunice, Neanthes.*

Class Oligochaeta. Mostly freshwater and terrestrial worms, but many marine. *Lumbricus* (earthworms), *Tubifex.*

Class Hirudinea. Leeches. *Hirudo, Placobdella.*

PHYLUM TARDIGRADA.   Water bears.

PHYLUM ONYCHOPHORA.   Onychophorans. *Peripatus.*

PHYLUM ARTHROPODA.

Subphylum Trilobita. Trilobites; extinct.

Subphylum Chelicerata. Chelicerates. Horseshoe crabs, spiders, scorpions, ticks, mites.

Subphylum Crustacea. Shrimps, crayfishes, lobsters, crabs, barnacles, copepods, isopods (sowbugs).

Subphylum Uniramia.
Superclass Myriapoda. Centipedes, millipedes.
Superclass Insecta.
Order Ephemeroptera. Mayflies.
Order Odonata. Dragonflies, damselflies.
Order Orthoptera. Grasshoppers, crickets, katydids.
Order Dermaptera. Earwigs.
Order Blattodea. Cockroaches.
Order Mantodea. Mantids.
Order Isoptera. Termites.
Order Mallophaga. Biting lice.
Order Anoplura. Sucking lice.
Order Homoptera. Cicadas, aphids, leafhoppers, spittlebugs.
Order Hemiptera. Bugs.
Order Coleoptera. Beetles.
Order Diptera. Flies.
Order Mecoptera. Scorpion flies. *Harpobittacus.*
Order Siphonaptera. Fleas.
Order Lepidoptera. Butterflies, moths.
Order Hymenoptera. Wasps, bees, ants.

PHYLUM ECHINODERMATA.   Echinoderms.
Class Asteroidea. Sea stars. *Asterias.*
Class Ophiuroidea. Brittle stars.
Class Echinoidea. Sea urchins, heart urchins, sand dollars.
Class Holothuroidea. Sea cucumbers.
Class Crinoidea. Feather stars, sea lilies.
Class Concentricycloidea. Sea daisies.

PHYLUM HEMICHORDATA.   Acorn worms.

PHYLUM CHORDATA.   Chordates.

Subphylum Urochordata. Tunicates, related forms.

Subphylum Cephalochordata. Lancelets.

Subphylum Vertebrata. Vertebrates.
Class Agnatha. Jawless vertebrates (lampreys, hagfishes).
Class Placodermi. Jawed, heavily armored fishes; extinct.
Class Chondrichthyes. Cartilaginous fishes (sharks, rays, skates, chimaeras).
Class Osteichthyes. Bony fishes.
Subclass Dipnoi. Lungfishes.
Subclass Crossopterygii. Coelacanths, related forms.
Subclass Actinopterygii. Ray-finned fishes.
Order Acipenseriformes. Sturgeons, paddlefishes.
Order Salmoniformes. Salmon, trout.
Order Atheriniformes. Killifishes, guppies.
Order Gasterosteiformes. Seahorses.

Order Perciformes. Perches, wrasses, barracudas, tunas, freshwater bass, mackerels.
Order Lophiiformes. Angler fishes.
Class Amphibia. Mostly tetrapods; embryo in amnion.
Order Caudata. Salamanders.
Order Anura. Frogs, toads.
Order Apoda. Apodans (caecilians).
Class Reptilia. Skin with scales, embryo enclosed in amnion.
Subclass Anapsida. Turtles, tortoises.
Subclass Lepidosaura. *Sphenodon*, lizards, snakes.
Subclass Archosaura. Dinosaurs (extinct), crocodiles, alligators.
Class Aves. Birds. (In some of the more recent schemes, dinosaurs, crocodilians, and birds are grouped in the same category.)
Order Struthioniformes. Ostriches.
Order Sphenisciformes. Penguins.
Order Procellariiformes. Albatrosses, petrels.
Order Ciconiiformes. Herons, bitterns, storks, flamingoes.
Order Anseriformes. Swans, geese, ducks.
Order Falconiformes. Eagles, hawks, vultures, falcons.
Order Galliformes. Ptarmigan, turkeys, domestic fowl.
Order Columbiformes. Pigeons, doves.
Order Strigiformes. Owls.
Order Apodiformes. Swifts, hummingbirds.
Order Passeriformes. Sparrows, jays, finches, crows, robins, starlings, wrens.
Class Mammalia. Skin with hair; young nourished by milk-secreting glands of adult.
Subclass Prototheria. Egg-laying mammals (duckbilled platypus, spiny anteaters).
Subclass Metatheria. Pouched mammals or marsupials (opossums, kangaroos, wombats, Tasmanian devil).
Subclass Eutheria. Placental mammals.
Order Insectivora. Tree shrews, moles, hedgehogs.
Order Scandentia. Insectivorous tree shrews.
Order Chiroptera. Bats.
Order Primates.
Suborder Strepsirhini (prosimians). Lemurs, lorises.
Suborder Haplorhini (tarsioids and anthropoids).
Infraorder Tarsiiformes. Tarsiers.
Infraorder Platyrrhini (New World monkeys).
Family Cebidae. Spider monkeys, howler monkeys, capuchin.
Infraorder Catarrhini (Old World monkeys and hominoids).
Superfamily Cercopithecoidea. Baboons, macaques, langurs.
Superfamily Hominoidea. Apes and humans.
Family Hylobatidae. Gibbon.
Family Pongidae. Chimpanzees, gorillas, orangutans.
Family Hominidae. Existing and extinct human species (*Homo*) and australopiths.
Order Lagomorpha. Rabbits, hares, pikas.
Order Rodentia. Most gnawing animals (squirrels, rats, mice, guinea pigs, porcupines, beavers, etc.).
Order Cetacea. Whales, porpoises.
Order Carnivora. Carnivores.
Suborder Feloidea. Cats, mongooses, hyenas.
Suborder Canoidea. Dogs, weasels, skunks, otters, raccoons, pandas, bears.
Order Pinnipedia. Seals, walruses, sea lions.
Order Proboscidea. Elephants; mammoths (extinct).
Order Sirenia. Sea cows (manatees, dugongs).
Order Perissodactyla. Odd-toed ungulates (horses, tapirs, rhinos).
Order Artiodactyla. Even-toed ungulates (camels, deer, bison, sheep, goats, antelopes, giraffes, etc.).
Order Edentata. Anteaters, tree sloths, armadillos.
Order Tubulidentata. African aardvarks.

APPENDIX II. UNITS OF MEASURE

Metric-English Conversions

Length

English		Metric
inch	=	2.54 centimeters
foot	=	0.30 meter
yard	=	0.91 meter
mile (5,280 feet)	=	1.61 kilometer

To convert	multiply by	to obtain
inches	2.54	centimeters
feet	30.00	centimeters
centimeters	0.39	inches
millimeters	0.039	inches

Weight

English		Metric
grain	=	64.80 milligrams
ounce	=	28.35 grams
pound	=	453.60 grams
ton (short) (2,000 pounds)	=	0.91 metric ton

To convert	multiply by	to obtain
ounces	28.3	grams
pounds	453.6	grams
pounds	0.45	kilograms
grams	0.035	ounces
kilograms	2.2	pounds

Volume

English		Metric
cubic inch	=	16.39 cubic centimeters
cubic foot	=	0.03 cubic meter
cubic yard	=	0.765 cubic meters
ounce	=	0.03 liter
pint	=	0.47 liter
quart	=	0.95 liter
gallon	=	3.79 liters

To convert	multiply by	to obtain
fluid ounces	30.00	milliliters
quart	0.95	liters
milliliters	0.03	fluid ounces
liters	1.06	quarts

CHAPTER 1

1. Metabolism — 5
2. Homeostasis — 5
3. cell — 6
4. adaptive — 10
5. mutations — 10
6. d — 4
7. d — 4, 10
8. d — 10, 11
9. d — 13
10. c — 13
11. a — 13
12. c — 10
    e — 11
    d — 13
    b — 12
    a — 12

CHAPTER 2

1. b — 22
2. c — 27
3. e — 28, 29
4. e — 29
5. f — 30, 31
6. acid; base — 30
7. c — 21
   a — 31
   b — 25
   d — 29

CHAPTER 3

1. d — 36
2. e — 37
3. f — 38
4. b — 40
5. d — 42, 46
6. d — 45
7. d — 46
8. c — 42, 43
   e — 46
   b — 41
   d — 46
   a — 38

CHAPTER 4

1. c — 52–53
2. d — 52, 56
3. d — 56, 68
4. False; many — 68
   cells have an
   outer wall.
5. c — 52, 70
6. e — 64
   d — 65
   a — 52, 56, 61
   b — 62
   c — 62–63

CHAPTER 5

1. d — 79
2. a — 88–89
3. d — 85
4. b — 86–87
5. c — 81, 84
   g — 81, 85
   a — 81
   d — 81
   e — 81
   b — 82–83
   f — 81

CHAPTER 6

1. carbon dioxide, — 92
   sunlight energy
2. d — 94
3. b — 94
4. c — 94
5. d — 98
6. e — 100
7. c — 101
8. c — 101
9. d — 98
   e — 98–99
   a — 101, 102
   b — 101
   c — 98

CHAPTER 7

1. d — 110
2. c — 112
3. b — 110, 114
4. c — 111, 116
5. d — 118
6. c — 118
7. b — 118
8. d — 120
9. b — 112
   c — 118
   a — 114–115
   d — 116

CHAPTER 8

1. d — 128
2. b — 128
3. c — 128
4. a — 129
5. c — 130
6. c — 128
7. d — 130–131
8. d — 132
   b — 132
   c — 133
   a — 133

CHAPTER 9

1. d — 139, 140
2. a — 140
3. d — 140
4. b — 141
5. d — 141
6. c — 141
7. c — 141, 142
8. d — 141, 143
9. d — 140
   a — 140
   c — 141
   b — 141

CHAPTER 10

1. a — 155
2. b — 155
3. a — 155
4. b — 155
5. c — 156–157
6. a — 156–157
7. d — 158
8. b — 158
   d — 156
   a — 155
   c — 155

CHAPTER 11

1. c — 172
2. c — 177
3. d — 179
4. a — 179
5. e — 182–183
6. c — 184
7. d — 184
8. c — 172
   e — 183
   d — 184
   b — 182
   a — 170, 173
   f — 176

CHAPTER 12

1. c — 194
2. d — 194
3. d — 194–195
4. c — 195
5. a — 196–197
6. d — 197
7. d — 197
   b — 194–195
   c — 196
   a — 195

CHAPTER 13

1. c — 202
2. b — 202
3. c — 202
4. c — 204
5. a — 204
6. a — 204–205
7. e — 209
   c — 206
   a — 203
   f — 204
   d — 204
   g — 203
   b — 204

CHAPTER 14

1. d — 213
2. d — 213, 214
3. d — 214
4. a — 214
5. c — 214
6. a — 214
7. d — 213, 216
8. b — 216
9. d — 216
10. b — 219
11. b — 219
12. d — 217
    e — 216
    b — 218
    a — 214–215
    c — 217

CHAPTER 15

1. d — 222
2. c — 224
3. Plasmids — 224
4. a — 225
5. b — 225
6. a — 226
7. b — 227
8. d — 228
9. d — 227

   c — 230–231
   f — 223
   e — 229
   b — 232
   a — 233

CHAPTER 16

1. b — 239
2. d — 244–245
3. populations — 246
4. d — 246
5. c — 255
6. c — 249
7. b — 250
8. c — 253
9. c — 255
   d — 249
   a — 246–247
   b — 256

CHAPTER 17

1. d — 262–263
2. d — 262–263
3. c — 263
4. c — 268
5. b — 268
6. c — 268
   e — 268
   a — 268–269
   d — 269
   b — 269

CHAPTER 18

1. a — 281
2. c — 282–283
3. c — 286
4. c — 286
5. d — 286
6. e — 286
   b — 274
   f — 274–275
   c — 280
   a — 285
   d — 281

CHAPTER 19

1. e — 294–297, 302–307
2. b — 294–295
3. d — 294
4. c — 298
5. d — 298–299
6. d — 298
7. b — 298
   d — 299
   e — 302
   a — 304
   c — 307

CHAPTER 20

1. c — 314
2. c — 316
3. a — 316
4. d — 317
5. a — 318
6. a — 320–321
7. a — 322
8. f — 311, 323
9. a — 326
10. d — 326
11. d — 314

    e — 314
    b — 317
    c — 324
    a — 316

CHAPTER 21

1. b — 332
2. a — 334
3. a — 334
4. c — 334
5. c — 335
6. e — 336–337
   c — 336
   d — 334
   a — 334–335
   b — 337
   f — 337

CHAPTER 22

1. d — 342–343
2. c — 343
3. b — 344
4. c — 346
5. b — 343
6. c — 350
   e — 342
   g — 346
   h — 349
   f — 342, 344
   a — 342
   b — 342
   d — 351

CHAPTER 23

1. b — 358
2. c — 359
3. c — 362–363
4. c — 364–365
5. b — 364–367
6. b — 370–371
7. a — 372
8. d — 360
   f — 362
   c — 364
   g — 365
   e — 358, 381
   i — 368
   j — 370
   h — 372
   a — 378
   b — 357, 358

CHAPTER 24

1. d — 384
2. e — 384–385
3. d — 386–387
4. b — 386–387
5. b — 388
6. d — 390
7. f — 392
8. e — 392
9. d — 392, 395, 396
10. d — 394–395
11. e — 396
12. a — 400–401
13. d — 387
    e — 388
    c — 388–389
    b — 390
    g — 392

	f	394–395
	a	396

CHAPTER 25
1.	b	408
2.	d	409
3.	d	414–415
4.	a	416

CHAPTER 26
1.	a	423
2.	d	423
3.	a	424
4.	b	424
5.	c	424
6.	d	426
7.	c	429
8.	d	433
9.	b	423
	d	423
	e	424
	c	425
	f	430–431
	a	432–433

CHAPTER 27
1.	e	438–439
2.	c	440
3.	b	441
4.	a	441
5.	hydrogen bonds	442
6.	c	442
7.	d	442
8.	a	445
9.	d	445
10.	b	446
11.	c	444–445
	g	438
	e	446
	b	440
	d	442
	a	442
	f	446

CHAPTER 28
1.	pollinators	450
2.	a	452
3.	b	453
4.	b	454
5.	b	454
6.	c	456
7.	c	456
8.	c	462, 463
9.	d	462–463
10.	d	465
11.	c	466–467
12.	a	455
	c	456
	d	454–455
	e	458
	b	450

CHAPTER 29
1.	a	474
2.	b	476
3.	b	476
4.	b	477
5.	c	477
6.	c	478
7.	d	479
8.	d	480, 482
9.	c	482, 483
10.	a	482–483
11.	sensory receptors; integrator; effectors	482
12.	b	475
	h	475
	a	476
	c	482
	f	482
	g	483
	e	482
	d	482

CHAPTER 30
1.	d	490
2.	d	488–489
3.	a	492
4.	d	494
5.	False; also of brain	499
6.	b	492–493
	d	495
	c	490
	a	488
7.	e	500–501
	d	502
	b	502
	c	503
	a	502

CHAPTER 31
1.	stimulus	509
2.	sensation	508
3.	Perception	508
4.	d	509
5.	b	510
6.	b	509
7.	c	509
8.	d	514
9.	b	514, 518
10.	c	514, 518
11.	e	510, 511
	c	517
	b	510, 512–514
	d	510
	a	512

CHAPTER 32
1.	f	522
2.	d	522
3.	c	525
4.	b	526
5.	a	527
6.	e	529, 530, 532
7.	b	532
8.	b	530
9.	d	530
	f	530–531
	c	532
	e	532
	a	534
	b	532

CHAPTER 33
1.	c	540
2.	d	543
3.	d	544
4.	b	548
5.	d	549
6.	d	550
7.	h	548
	f	551
	g	551
	e	540
	a	548–549
	c	544
	h	543
	i	548
	d	551

CHAPTER 34
1.	c	556
2.	d	558
3.	d	558–559
4.	c	559
5.	c	557, 562
6.	d	557, 562
7.	b	566
8.	d	565, 566
9.	c	572–573
10.	e	566
	b	566–567
	a	566, 568
	d	566, 569
	c	566, 569
11.	d	556, 568
	e	572
	a	564
	f	556
	c	556
	b	556

CHAPTER 35
1.	e	578
2.	f	579
3.	b	580
4.	d	582
5.	d	582
6.	d	584
7.	d	586
8.	a	587
9.	b	588
10.	c	580–581
	b	586
	a	580
	e	584–585
	f	590
	d	591

CHAPTER 36
1.	a	599
2.	d	599, 606
3.	d	601
4.	b	602, 603
5.	c	602, 603
6.	d	604, 605
7.	c	605
8.	c	606
9.	c	606
10.	c	606
11.	d	603
	g	603
	f	602
	e	598
	a	603
	c	603
	b	603

CHAPTER 37
1.	e	613
2.	d	614
3.	b	617, 618
4.	c	618
5.	b	619
6.	a	621
7.	a	624
8.	d	625, 626
9.	caloric (or energy); energy	628
10.	d	619
	c	618
	e	623
	g	618–619
	f	617
	b	618, 620
	a	619

CHAPTER 38
1.	d	634
2.	f	634
3.	a	636
4.	b	636
5.	c	636
6.	d	637
7.	a	637
8.	b	637
9.	d	635
	e	637
	a	637
	c	638
	b	639
10.	f	640
	e	640
	g	640
	b	640
	c	640
	d	640
	a	640

CHAPTER 39
1.	a	650–651
2.	c	648, 652
3.	d	654
4.	c	667
5.	c	669
6.	c	650–651
	a	648
	f	648
	e	648
	c	648
	b	648

CHAPTER 40
1.	Ecology	685
2.	population	686
3.	e	688
4.	a	688–689
5.	a	689
6.	d	691
7.	e	690–691
8.	c	694
9.	c	690
	d	689
	a	689
	e	690
	b	690–691

CHAPTER 41
1.	a	702
2.	d	703
3.	b	713
4.	a	707
5.	d	714
6.	c	705
	d	707
	b	705, 706
	a	706

CHAPTER 42
1.	e	720
2.	b	720
3.	e	720
4.	d	722–723
5.	b	724–725
6.	d	724, 725
7.	d	728
8.	e	732–733
9.	c	730
	d	733
	a	728
	e	728
	b	722–723

CHAPTER 43
1.	a	738
2.	d	739
3.	d	743
4.	d	738
	a	738
	c	738
	b	738

CHAPTER 44
1.	d	758–759
2.	c	758
3.	d	760–761
4.	b	761
5.	d	763
6.	d	763
7.	d	766, 770
8.	d	772, 777
9.	b	770
	d	771
	c	766
	a	768
10.	e	774
	a	777
	d	773
	c	776
	b	774

CHAPTER 45
1.	c	783
2.	c	784
3.	b	784
4.	d	784–785
5.	c	786
6.	b	792
7.	c	792
8.	d	794
9.	d	794–795
10.	d	791
	b	788–789
	c	787
	a	796
	e	786

APPENDIX IV.   ANSWERS TO GENETICS PROBLEMS

CHAPTER 10

1. a. *AB*

  b. *AB, aB*

  c. *Ab, ab*

  d. *AB, Ab, aB, ab*

2. a. All of the offspring will be *AaBB*.

  b. 1/4 *AABB* (25% each genotype)
  1/4 *AABb*
  1/4 *AaBB*
  1/4 *AaBb*

  c. 1/4 *AaBb* (25% each genotype)
  1/4 *Aabb*
  1/4 *aaBb*
  1/4 *aabb*

  d. 1/16 *AABB* (6.25%)
  1/8 *AaBB* (12.5%)
  1/16 *aaBB* (6.25%)
  1/8 *AABb* (12.5%)
  1/4 *AaBb* (25%)
  1/8 *aaBb* (12.5%)
  1/16 *AAbb* (6.25%)
  1/8 *Aabb* (12.5%)
  1/16 *aabb* (6.25%)

3. Yellow is recessive. Because F_1 plants have a green phenotype and must be heterozygous, green must be dominant over the recessive yellow.

4. a. *ABC*

  b. *ABc, aBc*

  c. *ABC, aBc, ABc, aBc*

  d. *ABC, aBC, AbC, abC,*
  ABc, aBc, Abc, abc

5. Because all F_1 plants of this dihybrid cross had to be heterozygous for both genes, then 1/4 (25%) of the F_2 plants will be heterozygous for both genes.

6. a. The mother must be heterozygous for both genes. Both the father and their first child are homozygous recessive for both genes.

  b. The probability that their second child will not be a tongue roller and will have detached earlobes is 1/4 (25%).

7. a. Bill *AaEeSs*, and Marie *AAEESS*

  b. There is a 100% probability that all their children will have Bill's phenotype (flat feet, long eyelashes, and tendency to sneeze).

  c. Zero probability; no child of theirs will have high arches or short eyelashes, and none will be sneezy.

8. a. The mother must be heterozygous $I^A i$. The male with type B blood could have fathered the child if he were heterozygous $I^B i$.

  b. Genotype alone cannot prove the accused male is the father. Even if he happens to be heterozygous, *any* male who carries the *i* allele could be the father, including those heterozygous for type A blood ($I^A i$) or type B blood ($I^B i$) and those with type O blood (*ii*).

9. A mating between a mouse from a true-breeding, white-furred strain and a mouse from a true-breeding, brown-furred strain would provide you with the most direct evidence. Because true-breeding strains typically are homozygous for a trait being studied, all F_1 offspring from this mating should be heterozygous. Record the phenotype of each F_1 mouse, then let them mate with one another. Assuming only one gene locus is involved, these are possible outcomes for the F_2 offspring:

  a. All F_1 mice are brown, and their F_2 offspring segregate 3 brown : 1 white. *Conclusion*: Brown is dominant to white.

  b. All F_1 mice are white, and their F_2 offspring segregate 3 white : 1 brown. *Conclusion*: White is dominant to brown.

  c. All F_1 mice are tan, and the F_2 offspring segregate 1 brown : 2 tan : 1 white. *Conclusion*: The alleles at this locus show incomplete dominance.

10. You cannot guarantee that the puppies won't develop the disorder without more information about Dandelion's genotype. You could do so only if she is a heterozygous carrier, if the male is free of the alleles, and if the alleles are recessive.

11. Fred could use a testcross to find out if his pet's genotype is *WW* or *Ww*. He can let his black guinea pig mate with a white guinea pig having the genotype *ww*.

  If any F_1 offspring are white, then his pet's genotype is *Ww*. If the two parents are allowed to mate repeatedly and all the offspring of the matings are black, there is a high probability that his pet is *WW*. (If, say, ten offspring are all black, then the probability that the male is *WW* is about 99.9 percent. The greater the number of offspring, the more confident Fred can be of his conclusion.)

12. a. 1/2 red, 1/2 pink

  b. All pink

  c. 1/4 red, 1/2 pink, 1/4 white

  d. 1/2 pink, 1/2 white

13. 9/16 walnut comb
     3/16 rose comb
     3/16 pea comb
     1/16 single comb

14. Because both parents are heterozygotes (Hb^AHb^S), the following are the probabilities for each child:

   a. 1/4 Hb^SHb^S

   b. 1/4 Hb^AHb^A

   c. 1/2 Hb^AHb^S

15. A mating of two M^LM cats yields:

   1/4 homozygous dominant (MM)

   1/2 heterozygous (M^LM)

   1/4 homozygous recessive (M^LM^L)

Because M^LM^L is lethal, the probability that any one kitten among the survivors will be heterozygous is 2/3.

16. a. Both parents must be heterozygotes (Aa). Their children may be albino (aa) or unaffected (AA or Aa).

   b. All are aa.

   c. The albino father must be aa. They have an albino child, so the mother must be Aa. (If she were AA, they could not have an albino child.) The albino child is aa. The three unaffected children are Aa. There is a 50% chance that any child of theirs will be albino. The observed 3:1 ratio is not surprising, given the small number of offspring.

17. a. All of the offspring will have medium-red color corresponding to the genotype $A^1A^2B^1B^2$.

   b. All possible genotypes could appear in the following proportions:

1/16	$A^1A^2B^1B^1$	dark red
1/8	$A^1A^1B^1B^2$	medium-dark red
1/16	$A^1A^1B^2B^2$	medium red
1/8	$A^1A^2B^1B^1$	medium-dark red
1/4	$A^1A^2B^1B^2$	medium red
1/8	$A^1A^2B^2B^2$	light red
1/16	$A^2A^2B^1B^1$	medium red
1/8	$A^2A^2B^1B^2$	light red
1/16	$A^2A^2B^2B^2$	white

CHAPTER 11

1. a. Human males (XY) inherit their X chromosome only from their mother.

   b. In males, an X-linked allele will only be found on his one X chromosome. Males can produce two kinds of gametes: one kind with a Y chromosome free of the gene, and the other kind with an X chromosome bearing the X-linked allele.

   c. One. Each gamete of a woman who is homozygous for an X-linked allele will have an X chromosome that carries the allele.

   d. Two. If a female is heterozygous for an X linked allele, half of the gametes that she produces will contain one of the alleles and the other half will contain the other allele.

2. All of the offspring should be heterozygous for the gene and all should have long wings. However, because some have vestigial wings, the dominant allele might have mutated because of radiation.

3. Because Marfan syndrome is a case of autosomal dominant inheritance and because one parent bears the allele, the probability of any child inheriting the mutant allele is 50%.

4. Because the phenotype appeared in every generation shown in the diagram, this must be a pattern of autosomal dominant inheritance.

5. A daughter could develop this type of muscular dystrophy only if she were to inherit two X-linked recessive alleles—one from her father and one from her mother. However, if a male bears the allele on his X chromosome, it will be expressed, then he will develop the disorder, and most likely he will not father children because of his early death.

6. If no crossover occurs between the two genes, then half the chromosomes will carry alleles AB and half will carry alleles ab.

7. If the alleles are close together, it is highly unlikely that a crossover will separate them from each other during meiosis. The greater the distance between the two loci, the greater the probability that a crossover will separate them from each other.

8. a. Nondisjunction could occur in anaphase I or anaphase II of meiosis.

   b. As a result of a translocation, chromosome 21 (which is small) may become attached to the end of chromosome 14. Even though the chromosome number of the new individual would be 46, its somatic cells would contain the translocated chromosome 21, in addition to two normal chromosomes 21.

9. In the mother, a crossover between the two genes at meiosis generates an X chromosome that carries neither mutant allele.

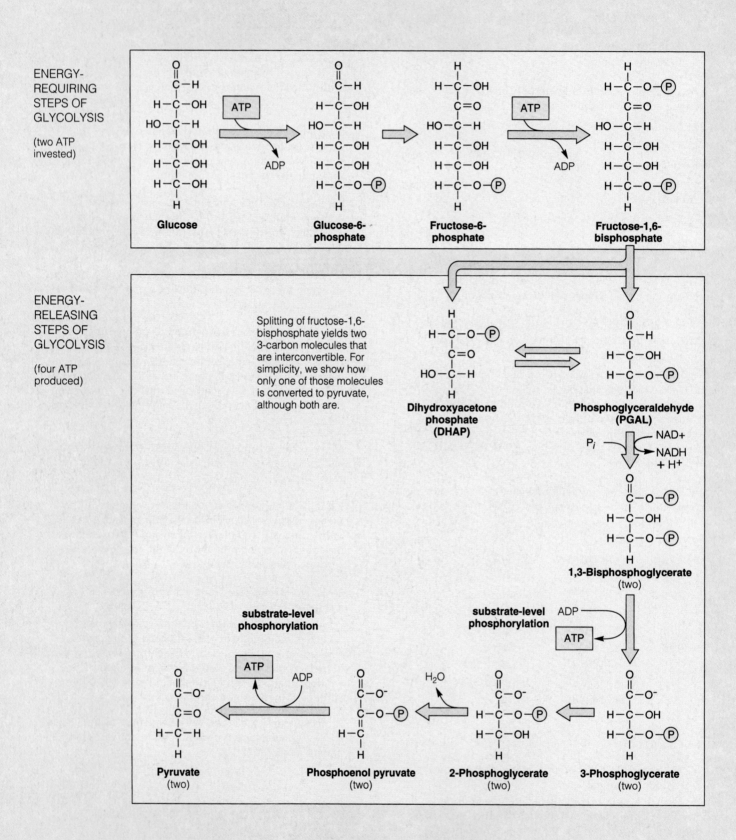

Figure A   Glycolysis, ending with two 3-carbon pyruvate molecules for each 6-carbon glucose entering the reactions. The *net* energy yield is two ATP molecules (two invested, four produced).

Step 1. Preparatory Conversions. COO^- group lost from pyruvate (as CO_2); hydrogen, electrons transferred to NAD^+, forming $NADH + H^+$. The 2-carbon acetyl fragment links with coenzyme A to form acetyl-CoA.

Step 2. Acetyl fragment transferred to oxaloacetate (the point of entry into the Krebs cycle), forming 6-carbon citrate.

Step 3. One H_2O lost then one H_2O added, converting citrate to its isomer, isocitrate. COO^- group lost from isocitrate (as CO_2). Hydrogen, electrons transferred from resulting compound to NAD^+, forming $NADH + H$.

Step 7. Oxaloacetate regenerated and hydrogen, electrons transferred to NAD^+, forming $NADH + H$.

Step 6. Electron transfers to FAD to form $FADH_2$.

Step 5. Substrate-level phosphorylation: Displacement of CoA group by phosphate and its transfer to GDP (forming GTP that donates phosphate group to ADP).

Step 4. COO^- group lost (as CO_2) from resulting compound; hydrogen, electron transfers to form another $NADH + H$. Resulting compound attached to CoA.

Pyruvate (from glycolysis)

Acetyl-CoA

Oxaloacetate

Citrate

Malate

Isocitrate

Fumarate

α-Ketoglutarate

Succinate

Succinyl CoA

KREBS CYCLE

Figure B Krebs cycle (also known as the citric acid cycle). *Red* identifies the carbon atoms entering the cyclic pathway by way of acetyl-CoA.

Figure C Calvin–Benson cycle of the light-independent reactions of photosynthesis.

Glossary

ABO blood typing Method of characterizing an individual's blood based on proteins (A, B, or their absence) at surface of red blood cells.

abortion Premature, spontaneous or induced expulsion of the embryo or fetus from uterus.

abscisic acid Plant hormone; induces stomatal closure, bud dormancy, and seed dormancy.

abscission (ab-SIH-zhun) Dropping of leaves, flowers, fruits, or other parts from a plant.

absorption Water and solute uptake by cell or multicelled organism, as when nutrients cross gut lining and enter internal environment.

accessory pigment Photosynthetic pigment; absorbs wavelengths missed by chlorophylls.

acclimatization Long-lasting physiological and behavioral adaptation to a new habitat.

acetylation Attachment of an acetyl group to a compound, such as DNA.

acetylcholine (ACh) Neurotransmitter; acts on brain, spinal cord, glands, and muscles.

acetyl-CoA (uh-SEET-ul) Coenzyme A attached to a two-carbon fragment from pyruvate, which it transfers to oxaloacetate for the Krebs cycle.

acid [L. acidus, sour] Any substance that, when dissolved in water, donates hydrogen ions.

acid–base balance State in which extracellular fluid is not too acidic or too basic, an outcome of controls over its concentrations of dissolved ions.

acidity Of a solution, an excess of hydrogen ions relative to hydroxyl ions.

acid rain Wet acid deposition; falling of rain (or snow) rich in sulfur and nitrogen oxides.

acoelomate (ay-SEE-luh-mate) Absence of a fluid-filled cavity between gut and body wall.

actin (AK-tin) Type of cytoskeletal protein.

action potential Abrupt, brief reversal in the resting membrane potential of excitable cells (e.g., neurons, muscle cells).

activation energy For each type of reaction, the minimum amount of collision energy that will drive reactants to an activated state.

active site Crevice in an enzyme molecule where a specific reaction is catalyzed (made to occur far faster than it would spontaneously).

active transport Pumping of a specific solute across a cell membrane against its concentration gradient through a transport protein's interior. Requires an energy boost, as from ATP.

adaptation [L. adaptare, to fit] Of evolution, being adapted (or becoming more adapted) to a set of environmental conditions. Of sensory neurons, decline or end of action potentials even if stimulus is maintained at constant strength.

adaptive radiation Macroevolutionary pattern; burst of genetic divergences from a lineage that give rise to many species, each using a novel resource or a new (or newly vacated) habitat.

adaptive trait Any aspect of form, function, or behavior that helps the individual survive and reproduce under prevailing conditions.

adaptive zone Any way of life available for organisms that are physically, ecologically, and evolutionarily equipped to live it.

adenine (AH-de-neen) A purine; a nitrogen-containing base in certain nucleotides.

ADH Antidiuretic hormone. Hypothalamic hormone; promotes water conservation.

adhering junction Junction where a mass of anchored proteins help adjoining cells adhere.

adipose tissue A connective tissue having an abundance of fat-storing cells.

ADP Adenosine diphosphate (ah-DEN-uh-seen die-FOSS-fate). A nucleotide coenzyme.

adrenal gland Endocrine gland. Its cortex secretes glucocorticoids, other hormones; its medulla secretes epinephrine, norepinephrine.

aerobic respiration (air-OH-bik) [Gk. aer, air, + bios, life] Oxygen-requiring pathway of ATP formation; from glycolysis, to Krebs cycle, to electron transport phosphorylation. Typical net energy yield: 36 ATP per glucose molecule.

African emergence model Holds that Homo sapiens originated in Africa, then replaced archaic Homo populations in other regions.

age structure Number of individuals in each age category for a population.

agglutination (ah-glue-tin-AY-shun) Defense response; antibodies cause antigen to clump together, which facilitates phagocytosis.

aging Of any multicelled organism showing extensive cell differentiation, the gradual and expected deterioration of the body over time.

AIDS Acquired immune deficiency syndrome. A set of chronic disorders following infection by the human immunodeficiency virus (HIV).

alcohol Organic compound that includes one or more hydroxyl groups (—OH); it dissolves readily in water. Sugars are examples.

alcoholic fermentation Anaerobic ATP-forming pathway. Pyruvate from glycolysis is degraded to acetaldehyde, which accepts electrons from NADH to form ethanol; NAD⁺ needed for the reactions is regenerated. Net yield: 2 ATP.

aldosterone (al-DOSS-tuh-rohn) Adrenal cortex hormone; helps control sodium reabsorption.

alga, plural **algae** Photosynthetic protistans, mainly; most are members of phytoplankton.

algal bloom Huge increases in aquatic algal population sizes after nutrient enrichments.

alkylating agent Any substance that transfers methyl or ethyl groups to a compound.

allantois (ah-LAN-twahz) [Gk. allas, sausage] Extraembryonic membrane used to exchange gases and store metabolic wastes of embryos of reptiles, birds, some mammals. In humans, it forms urinary bladder, placental blood vessels.

allele (uh-LEEL) One of two or more molecular forms of a gene that arise by mutation and code for different versions of the same trait.

allele frequency For a given gene locus, the relative abundances of each kind of allele among all the individuals of a population.

allergy Hypersensitivity to an allergen (any normally harmless substance that provokes inflammation and often immune responses).

allopatric speciation [Gk. allos, different, + L. patria, native land]. Speciation model. A physical barrier arises, separates populations or subpopulations of a species, ends gene flow, and favors divergences that end in speciation.

altruism (AL-true-iz-um) Behavior that lowers an individual's chance of reproductive success but helps others of its species.

alveolus (ahl-VEE-uh-lus), plural **alveoli** [L. alveus, small cavity] Cupped, thin-walled outpouching of respiratory bronchiole where lungs and blood exchange O_2 and CO_2.

amino acid (uh-MEE-no) Organic compound with an H atom, amino group, acid group, and R group, all covalently bonded to a carbon atom. Subunit of polypeptide chains.

ammonification (uh-moan-ih-fih-KAY-shun) Of nitrogen cycle, process by which soil fungi and bacteria break down nitrogenous wastes and remains to forms that plants can take up.

amnion (AM-nee-on) An extraembryonic membrane; boundary layer of a fluid-filled sac in which some vertebrate embryos develop.

amniote egg Egg that has extraembryonic membranes and often a shell. Contributed to successful invasion of land by vertebrates.

amphibian Vertebrate with a body plan and reproductive mode somewhere between fishes and reptiles and a mostly bony endoskeleton.

anaerobic electron transport (an-uh-ROW-bik) [Gk. an, without, + aer, air] Type of prokaryotic ATP-forming pathway. Uses electrons from an organic compound; an inorganic compound (not oxygen) is often the final electron acceptor. Variable but always small net energy yield.

anagenesis Speciation pattern; changes in allele frequencies and morphology accumulate within an unbranched line of descent.

analogous structures (ann-AL-uh-gus) [Gk. analogos, similar to one another] Body parts that once differed in evolutionarily distant lineages but converged in structure and function in response to similar environmental pressures.

anaphase (AN-uh-faze) Of mitosis, stage when sister chromatids of each chromosome move to opposite spindle poles. Of anaphase I (meiosis), each duplicated chromosomes and its homologue move to opposite poles. Of anaphase II (meiosis), sister chromatids of each chromosome move to opposite poles.

aneuploidy (AN-yoo-ploy-dee) Having one extra or one less chromosome relative to the parental chromosome number.

angiosperm (AN-gee-oh-spurm) [Gk. angeion, vessel, and spermia, seed] Flowering plant.

animal Multicelled, motile heterotroph that eats other organisms; has embryonic stages and usually tissues, organs, and organ systems.

Animalia Kingdom of animals.

annelid An invertebrate; a segmented worm (e.g., oligochaete, leech, or polychaete).

annual Plant that lasts one growing season.

anther [Gk. anthos, flower] Part of a stamen; pollen forms in it and is dispersed from it.

antibiotic Metabolic product of soil microbes that kills bacterial competitors for nutrients.

antibody [Gk. anti, against] Antigen-binding receptor. Only B cells make antibodies, then position them at their surface or secrete them.

anticodon Series of three nucleotide bases in tRNA; can base-pair with an mRNA codon.

antigen (AN-tih-jen) [Gk. anti, against, + genos, race, kind] Any molecular configuration that certain lymphocytes chemically recognize as nonself and that triggers an immune response.

antigen-presenting cell Cell that processes antigen and displays fragments of it with MHC molecules at the cell surface. Antigen–MHC complexes can stimulate immune responses.

aorta (ay-OR-tah) Of vertebrates, main artery of systemic circulation.

apical dominance Growth-inhibiting effect of a terminal bud on growth of lateral buds.

apical meristem (AY-pih-kul MARE-ih-stem) [L. *apex*, top, + Gk. *meristos*, divisible] Mass of self-perpetuating cells responsible for the primary growth at root tips and shoot tips.

apoptosis (APP-oh-TOE-sis) Programmed cell death of body cells finished with prescribed functions or altered, as by infection or cancer.

Archaebacteria Kingdom of prokaryotes more like eukaryotic cells than eubacteria; includes methanogens, halophiles, and thermophiles.

Archean Eon (3.9–2.5 bya) in which life arose.

archipelago Island chain some distance away from a continent.

area effect Idea that larger islands support more species than smaller ones at equivalent distances from sources of colonizer species.

arteriole (ar-TEER-ee-ole) Type of blood vessel between arteries and capillaries. Collectively, points of selective dilation and constriction.

artery Thick-walled, muscular, rapid-transport vessel that smooths out blood pressure pulses.

arthropod Invertebrate with hard exoskeleton, specialized segments, and jointed appendages.

artificial selection Selection of traits among a population under contrived conditions.

asexual reproduction Any reproductive mode by which offspring arise from a single parent and inherit the genes of that parent only.

atmosphere The volume of gases, airborne particles, and water vapor enveloping Earth.

atmospheric cycle A biogeochemical cycle in which an element occurs in a gaseous phase.

atom Smallest unit of an element that retains the element's properties.

ATP Adenosine triphosphate (ah-DEN-uh-seen try-FOSS-fate). Nucleotide with adenine, ribose, and three phosphate groups that drives most energy-requiring metabolic reactions.

australopith (OHSS-trah-low-pith) [L. *australis*, southern, + Gk. *pithekos*, ape] Early hominid.

autoimmunity Result of inappropriate attacks on normal body cells by lymphocytes.

automated DNA sequencing Recent method that determines the sequence of cloned DNA or PCR-amplified DNA in a few hours.

autonomic nervous system (auto-NOM-ik) All nerves from central nervous system to smooth muscle, cardiac muscle, and glands of viscera.

autosome Any chromosome of a type that is the same in males and females of the species.

autotroph (AH-toe-trofe) [Gk. *autos*, self, + *trophos*, feeder] Any organism that makes its own food with an environmental energy source (e.g., sunlight) and CO_2 as its carbon source.

auxin (AWK-sin) Plant hormone; induces stem lengthening and responses to gravity and light.

axon Neuron's conducting zone.

B cell B lymphocyte; only type of cell that produces and secretes antibodies.

bacterial conjugation Transfer of plasmid DNA from one prokaryotic cell to another.

bacteriophage (bak-TEER-ee-oh-fahj) Category of viruses that infect bacterial cells.

balanced polymorphism Form of selection in which two or more alleles for a trait are being maintained in a population over time.

bark All tissues external to vascular cambium.

Barr body Randomly condensed one of two X chromosomes in cells of female mammals.

basal body A centriole which, after giving rise to microtubules of a flagellum or cilium, remains attached to its base in the cytoplasm.

base Any substance that accepts hydrogen ions (H^+) when dissolved in water; OH^- forms after this. Also, an organic compound with a single or double ring structure containing nitrogen.

base-pair substitution One amino acid has replaced another during protein synthesis.

base sequence Sequential order of bases in a DNA or RNA strand.

behavior Response to external and internal stimuli based on sensory, neural, endocrine, and effector components. Has a genetic basis, can evolve, and can be modified by learning.

behavioral flexibility A capacity to expand on basic activities with novel forms of behavior and learning, especially among primates.

biennial (bi-EN-yul) Flowering plant that completes its life cycle in two growing seasons.

bilateral symmetry Body plan in which left and right halves generally are mirror images.

bile Liver secretion required for fat digestion.

binary fission Asexual reproductive mode; the body of protozoans and some other animals divides in two parts. *See* prokaryotic fission.

binomial system Of taxonomy, assigning a generic and a specific name to each species.

biodiversity [Gk. *bios*, life] All species living in a defined region, ranging from the smallest habitat to the biosphere, in a specified interval.

biogeochemical cycle Slow movement of an element from environmental reservoirs, through food webs, then back to the environment.

biogeography Scientific study of the world distribution of species.

biological clock Internal time-measuring mechanism; adjusts activities seasonally, daily, or both in response to environmental cues.

biological magnification Ever increasing concentration of a nondegradable or slowly degradable substance in body tissues as it is passed along food chains.

biological species concept Defines a species as one or more populations of individuals that are interbreeding under natural conditions, producing fertile offspring, and are isolated reproductively from other such populations. Applies to sexually reproducing species only.

biology The scientific study of life.

bioluminescence Production by an organism of fluorescent light.

biomass Combined weight of all organisms at a given trophic level in an ecosystem.

biome Large subdivision of a biogeographic realm, distinctive in its habitat conditions, community structure, and endemic species.

biosphere [Gk. *bios*, life, + *sphaira*, globe] All regions of the Earth's waters, crust, and atmosphere in which organisms live.

biosynthetic pathway Metabolic pathway by which organic compounds are synthesized.

biotic potential Of population growth for a given species, the maximum rate of increase per individual under ideal conditions.

bipedalism Habitually walking upright on two feet, as by ostriches and hominids.

bird Vertebrate with reptilian ancestry; only organism that produces feathers.

blastocyst (BLASS-tuh-sist) Type of blastula; a surface layer of blastomeres, a cavity filled with their secretions, and an inner cell mass.

blastomere One of the small, nucleated cells that form during cleavage of animal zygote.

blastula Early outcome of cleavage; a number of blastomeres enclose a fluid-filled cavity.

blood Fluid connective tissue of water, solutes, and formed elements (blood cells, platelets). Blood transports substances to and from cells, and helps maintain internal environment.

blood–brain barrier Mechanism that controls which solutes enter cerebrospinal fluid.

blood pressure Fluid pressure, generated by heart contractions, that circulates blood.

bone Vertebrate organ with mineral-hardened connective tissue (bone tissue); helps move body, protects other organs, stores minerals. Some (e.g., breastbone) produce blood cells.

bone remodeling Ongoing mineral deposits and withdrawals from bone that adjust bone strength and maintain levels of calcium and phosphorus in blood.

bony fish Type of fish with an endoskeleton, usually well developed, of bone.

boreal forest A swamp forest (or *taiga*) in glaciated regions with cold lakes and streams.

bottleneck Severe reduction in the size of a population, brought about by intense selection pressure or a natural calamity.

Bowman's capsule Part of a nephron; receives water and solutes being filtered from blood.

brain Of most nervous systems, integrating center that receives and processes sensory input and issues coordinated commands for responses by muscles and glands.

brain stem Most ancient nerve tissue in the vertebrate hindbrain, midbrain, and forebrain.

broadleaf forest Evergreen or deciduous forest of a tropical, subtropical, or temperate region in which tall trees form a continuous canopy (e.g., tropical or deciduous broadleaf forest).

bronchiole Finely branched airway; part of the bronchial tree inside the lung.

bronchus, plural **bronchi** (BRONG-cuss, BRONG-kee) [Gk. *bronchos*, windpipe] Tubular airway; branches from trachea and leads into lungs.

brown alga Mostly marine, photoautotrophic protistan, such as kelp, with chlorophylls a, c_1, and c_2, and carotenoids (e.g., fucoxanthin).

bryophyte Nonvascular land plant requiring free water for fertilization. Haploid dominance in life cycle. Some with cuticle and stomata. A moss, liverwort, or hornwort.

bud Undeveloped shoot, mainly meristematic tissue. Small, protective scales often cover it.

buffer system A weak acid and the base that forms when it dissolves in water. The two work as a pair to counter slight shifts in pH.

bulk Of the vertebrate gut, the volume of undigested material in the small intestine that cannot be decreased by absorption.

bulk flow In response to a pressure gradient, movement of more than one kind of molecule in the same direction in the same medium.

C3 plant Plant that uses three-carbon PGA as the first intermediate for carbon fixation.

C4 plant Plant that uses oxaloacetate (a four-carbon compound) as the first intermediate for carbon fixation. CO_2 is fixed twice, in two cell types; helps counter photorespiration.

Calvin–Benson cycle Light-independent cyclic reactions of photosynthesis that form sugars, using ATP energy, NADPH, and CO_2.

CAM plant Type of plant that conserves water by opening stomata only at night, when it fixes carbon dioxide by means of a C4 pathway.

camouflage Coloration, form, patterning, or behavior that helps predators or prey blend with the surroundings and escape detection.

cancer Malignant tumor; mass of altered cells that divide abnormally. Potentially lethal.

capillary, blood [L. *capillus*, hair] Blood vessel with thin endothelial wall and small diameter.

capillary bed Diffusion zone, consisting of great numbers of capillaries, where blood and interstitial fluid exchange substances.

carbamino hemoglobin $HbCO_2$, the form in which 30% of the CO_2 in blood is transported.

carbohydrate Molecule of carbon, hydrogen, and oxygen mostly in a 1:2:1 ratio. Main kinds are monosaccharides, oligosaccharides, and polysaccharides. They are structural materials, energy stores, and transportable energy forms.

carbon cycle An atmospheric cycle. Carbon moves from reservoirs (sediments, rocks, the ocean), through the atmosphere (mostly as CO_2), food webs, and back to the reservoirs.

carbon fixation First of the light-independent reactions. Rubisco, an enzyme, affixes carbon (from CO_2) to RuBP or to another compound for entry into the Calvin–Benson cycle.

carbonic anhydrase Red blood cell enzyme that converts CO_2 to bicarbonate, the form in which 60% of the CO_2 in blood is transported.

carcinogen (kar-SIN-uh-jen) Any substance or agent that can trigger cancer.

cardiac conduction system Specialized cardiac muscle cells that make the heart pump by way of their rhythmic, synchronized excitation.

cardiac cycle (KAR-dee-ak) [Gk. *kardia*, heart, + *kyklos*, circle] Sequence of muscle contraction and relaxation in one heartbeat.

cardiac muscle tissue A contractile tissue that is present only in the heart wall.

cardiac pacemaker Sinoatrial (SA) node; mass of self-excitatory cardiac muscle cells that are the basis of normal rate of heartbeat.

cardiovascular system Organ system that has blood, one or more hearts, and blood vessels.

carnivore Animal that devours other animals.

carnivorous plant Plant that traps, digests, and feeds on insects and other small animals.

carotenoid (kare-OTT-en-oyd) An accessory pigment of photosynthesis.

carpel (KAR-pul) Female reproductive part of a flower. The ovary, stigma, and often a style consist of one or more carpels.

carrying capacity The maximum number of individuals in a population (or species) that a given environment can sustain indefinitely.

cartilage Connective tissue with solid, pliable intercellular material that resists compression.

cartilaginous fish Jawed fish with a cartilage endoskeleton and prominent fins (e.g., sharks).

Casparian strip Narrow, waxy, impermeable band between walls of abutting cells making up root endodermis and exodermis.

catastrophism Idea that abrupt changes in the geologic or fossil record were divinely invoked.

cDNA DNA molecule copied from a mature mRNA transcript by reverse transcription.

cell [L. *cella*, small room] Smallest living unit; organized unit with a capacity to survive and reproduce on its own given DNA instructions, energy sources, and raw materials.

cell count The number of cells of a given type in one microliter of blood.

cell cycle Events by which a cell increases in mass, roughly doubles its cytoplasm, duplicates its DNA, and divides in two. Extends from the time a cell forms until it completes division.

cell differentiation Key development process. Cell lineages become specialized in structure, composition, and function by activating and suppressing part of the genome selectively.

cell junction Site where cells are interacting physically, functionally, or both.

cell plate In a dividing plant cell, a disklike structure that becomes a crosswall with new plasma membrane on both sides.

cell theory All organisms consist of one or more cells, the smallest units with a capacity for independent life that no longer spontaneously arise under existing conditions on Earth.

cell wall A semirigid, permeable structure external to the plasma membrane; helps many cells retain their shape and resist rupturing.

Cenozoic The present era (65 mya to present).

central nervous system Brain and spinal cord.

central vacuole Fluid-filled storage organelle of plant cell; its growth enhances cell surface area.

centriole (SEN-tree-ohl) Structure that gives rise to microtubules of cilia and flagella.

centromere (SEN-troh-meer) A constricted area of a chromosome that has attachment sites for spindle microtubules during nuclear division.

cephalization (SEF-ah-lah-ZAY-shun) [Gk. *kephalikos*, head] Over time, the concentration of sensory structures and nerve cells in a head.

cerebellum (ser-ah-BELL-um) Hindbrain region with reflex centers for maintaining posture and smoothing out limb movements.

cerebral cortex Thin surface layer of cerebral hemispheres; receives, integrates, and stores sensory information; coordinates responses.

cerebrospinal fluid Clear extracellular fluid, inside a system of canals and chambers, that bathes and protects the brain and spinal cord.

cerebrum (suh-REE-bruhm) Forebrain region that generally deals with olfactory input and motor responses. In mammals, it evolved into the most complex integrating center.

channel protein Type of transport protein, open or gated, that spans a cell membrane.

chemical bond A union between the electron structures of two or more atoms or ions.

chemical energy Potential energy of molecules.

chemical synapse (SIN-aps) Thin cleft between a presynaptic neuron and a postsynaptic cell. Neurotransmitter molecules diffuse across it.

chemiosmotic theory (KIM-ee-oz-MOT-ik) Idea that an electrochemical gradient drives ATP formation. H^+ accumulates in a membranous compartment, then ATP forms as it flows out through proteins (ATP synthases) in response to concentration and electric gradients.

chemoautotroph (KEE-moe-AH-toe-trofe) Any prokaryotic cell that synthesizes its own food using carbon dioxide as the carbon source and an inorganic substance as the energy source.

chemoreceptor Sensory receptor that detects ions or molecules dissolved in fluid bathing it.

chlorofluorocarbon (KLORE-oh-FLOOR-oh-car-bun) CFC; a compound of chlorine, fluorine, and carbon; is contributing to ozone thinning.

chlorophyll (KLOR-uh-fill) [Gk. *chloros*, green, + *phyllon*, leaf] Main photosynthetic pigment. Chlorophylls absorb all wavelengths of visible light but not much of green and yellow ones.

chloroplast (KLOR-uh-plast) The organelle of photosynthesis in plants and many protistans.

chordate Animal having a notochord, dorsal hollow nerve cord, pharynx, and gill slits in pharynx wall during at least part of life cycle.

chorion (CORE-ee-on) Type of extraembryonic membrane that becomes part of placenta. Villi form at its surface and facilitate exchanges of substances between the embryo and mother.

chromatid (CROW-mah-tid) Of a duplicated eukaryotic chromosome, one of two DNA molecules (with associated proteins) attached to each other at the centromere.

chromatin A cell's collection of DNA and all of the proteins associated with it.

chromosome (CROW-moe-some) [Gk. *chrōma*, color, + *soma*, body] Of eukaryotic cells, a DNA molecule, duplicated or unduplicated, with many associated proteins. Of prokaryotic cells, a circular DNA molecule.

chromosome number All chromosomes in a given type of cell. *See* haploidy; diploidy.

chrysophyte A category of photosynthetic protistans (e.g., golden algae, yellow-green algae, coccolithophores, diatoms).

cilium (SILL-ee-um), plural **cilia** Short motile or sensory structure of certain eukaryotic cells; its core is a 9 + 2 array of microtubules.

circadian rhythm (ser-KAYD-ee-un) [L. *circa*, about, + *dies*, day] Cycle of physiological events completed every twenty-four hours or so, independently of environmental change.

circulatory system Organ system that moves substances to and from cells, and often helps stabilize body temperature and pH. Typically consists of a heart, blood vessels, and blood.

cladogenesis [Gk. *clad-*, branch] Speciation pattern in which a lineage splits and isolated populations undergo genetic divergence.

cladogram Evolutionary tree diagram with branch points showing relative relationships. Groups closer together share a more recent common ancestor than those farther apart.

classification scheme A way of organizing and retrieving information about species.

cleavage Early stage of animal development. Mitotic cell divisions divide a fertilized egg into many smaller, nucleated cells; original volume of egg cytoplasm does not increase.

cleavage furrow Ringlike depression defining the cutting plane for a dividing animal cell.

cleavage reaction Enzyme action that splits a molecule in two or more parts (e.g., hydrolysis).

climate For a specified region, the prevailing weather conditions (e.g., temperature, cloud cover, wind speed, rainfall, and humidity).

climax community Array of species that has stabilized under prevailing habitat conditions.

climax pattern model Idea that environmental factors often vary in their effects across a large region, so stable communities other than the climax stage may also persist in that region.

cloaca (kloe-AY-kuh) Chamber or duct in last part of gut of some animals; roles in excretion, reproduction, and sometimes respiration.

cloning Making a genetically identical copy of DNA or of an organism.

cloning vector Any plasmid, viral DNA, or some other piece of DNA that researchers use to isolate and amplify DNA of interest.

club fungus Fungus having club-shaped cells that produce and bear spores.

cnidarian (nye-DAR-ee-un) Type of radial invertebrate at the tissue level of organization; the only nematocyst producer.

coal Nonrenewable energy source that formed over 280 million years ago from submerged, undecayed, and compacted plant remains.

codominance In heterozygotes, simultaneous expression of a pair of nonidentical alleles that specify different phenotypes.

codon One of 64 possible base triplets in an mRNA strand. A code word for an amino acid in a polypeptide chain; a few codons also act as START or STOP signals for translation.

coelom (SEE-lum) A peritoneum-lined cavity between the gut and body wall of most animals.

coenzyme Nucleotide that accepts electrons, H atoms that enzymes strip from substrates and transfers them to a different reaction site.

coevolution Joint evolution of two closely interacting species by changes in the selection pressures operating between the two.

cofactor Metal ion or coenzyme; it helps an enzyme catalyze a reaction or transfers atoms, electrons, or functional groups elsewhere.

cohesion Capacity to resist rupturing when placed under tension (stretched).

cohesion–tension theory Of plants, in response to transpiration, columns of water molecules are pulled up through xylem by the collective cohesive strength of their hydrogen bonds.

cohort Group of individuals from a population from the time of birth until the last one dies.

collenchyma (coll-ENG-kih-mah) A simple plant tissue that imparts flexible support during primary growth, as in lengthening stems.

colon (CO-lun) Large intestine.

commensalism Ecological interaction between two (or more) species in which one benefits directly and the other is affected little, if at all.

communication display Social signal, often ritualized with intended changes in functions of common patterns.

communication signal Social cue encoded in stimuli such as specific body coloration, body patterning, odors, sounds, and postures.

community All populations in a habitat. Also, a group of organisms with similar life-styles.

companion cell Specialized parenchyma cell; helps load conducting cells of phloem.

comparative morphology [Gk. *morph*, form] Scientific study of comparable body parts of adults or embryonic stages of major lineages.

competition, interspecific Type of ecological interaction in which individuals of different species compete for a share of resources.

competition, intraspecific Type of ecological interaction in which individuals of the same population compete for a share of resources.

competitive exclusion Theory that two or more species that require identical resources cannot coexist indefinitely.

complement system Set of proteins circulating in inactive form in vertebrate blood. Different kinds promote inflammation, induce lysis of pathogens, and stimulate phagocytes during nonspecific defenses and immune responses.

compound Molecule consisting of two or more elements in unvarying proportions (e.g., H_2O).

concentration gradient A difference in the number of molecules (or ions) of a substance between adjoining regions. Molecules collide constantly and careen outward to an adjoining region where they are less concentrated. All substances tend to diffuse down such gradients.

condensation reaction Covalent bonding of two molecules into a larger one; water often forms as a by-product.

conduction In temperature studies, exchange of heat between two touching objects owing to a thermal gradient between them.

cone cell In a vertebrate eye, a photoreceptor that responds to intense light and contributes to sharp daytime vision and color perception.

conifer A gymnosperm; most are evergreen woody trees or shrubs with thickly cuticled needle-like or scale-like leaves; adapted to conserve water in droughts and cold winters.

connective tissue Most abundant, pervasive type of animal tissue. Soft connective tissues —(1) loose, (2) dense, irregular, or (3) dense, regular—differ in their arrangement and number of fibroblasts and fibers, and in the amount of ground substance in which these are embedded. Specialized types include cartilage, bone tissue, adipose tissue, and blood.

conservation biology International systematic survey of the full range of biodiversity, analysis of its evolutionary and ecological origins, and identification of methods to maintain and use it for the benefit of the human population.

consumer [L. *consumere*, to take completely] A heterotroph that feeds on cells or tissues of other organisms (e.g., herbivores, carnivores).

continuous variation Of a population, a more or less continuous range of small differences in a given trait among its individuals.

contractile cell A cell that shortens in response to stimulation and returns to a resting position.

contractile vacuole (kun-TRAK-till VAK-you-ohl) [L. *contractus*, to draw together] Organelle in some protistans; takes up excess water and contracts to expel water through a pore.

control group A group used as a standard for comparison with an experimental group and, ideally, identical with it in all respects except for the one variable being studied.

convection Movement of air or water next to an object that aids conductive heat loss from it.

convergent evolution Dissimilar body parts in evolutionarily distant lineages become similar over time, as independent responses to similar kinds of environmental pressures.

coral reef A formation of accumulated hard parts of corals and other organisms in warm, clear waters between latitudes 25°N and 25°S.

core temperature Internal body temperature, as opposed to near its surface.

cork Periderm component; insulates, protects, and waterproofs woody stems and roots.

cork cambium Lateral meristem that replaces epidermis with cork on woody plant parts.

cornea Transparent cover over front of eye.

corpus callosum (CORE-pus ka-LOW-sum) Band of axons linking two cerebral hemispheres.

corpus luteum (CORE-pus LOO-tee-um) Of female mammals, a glandular structure that forms from cells of a ruptured ovarian follicle; it secretes progesterone and estrogen.

cortex A rindlike layer. In vascular plants, a ground tissue; supports parts and stores food.

cotyledon (KOT-uhl-EE-dun) Seed leaf. One develops in monocot seeds and has digestive roles; two develop in dicot seeds and store food for germination and early growth.

countercurrent flow Movement of two fluids in opposing directions, as in a fish gill.

courtship display Pattern of ritualized social behavior between potential mates. Commonly incorporates frozen postures, exaggerated yet simplified movements, and visual signals.

covalent bond (koe-VAY-lunt) [L. *con*, together, + *valere*, to be strong] Sharing of one or more electrons between atoms or groups of atoms. Nonpolar bonds share them equally. Polar bonds (slightly positive at one end, negative at the other) share them unequally.

creatine phosphate Organic compound that transfers phosphate to ADP in a rapid, short-term, ATP-generating pathway.

cross-bridge formation A reversible, ATP-driven interaction between a sarcomere's actin and myosin filaments; results in a short power stroke that is the basis of muscle contraction.

crossing over At prophase I of meiosis, an interaction in which nonsister chromatids of a pair of homologous chromosomes break at corresponding sites and exchange segments; genetic recombination is the result.

culture Sum of behavior patterns of a social group, passed between generations by way of learning and symbolic behavior.

cuticle (KEW-tih-kull) Body cover. Of plants, a transparent covering of waxes and cutin on outer epidermal cell walls. Of annelids, a thin, flexible coat. Of arthropods, a lightweight exoskeleton hardened with protein and chitin.

cutin Insoluble lipid polymer that functions as the ground substance for plant cuticles.

cycad Slow-growing gymnosperm; its lineage arose in the Permian. Resembles palm trees.

cyclic AMP (SIK-lik) A type of nucleotide (cyclic adenosine monophosphate) with roles in intercellular communication.

cyclic pathway of ATP formation Most ancient photosynthetic pathway. Membrane-bound photosystems give up electrons to transport systems that return them to photosystems. Electron flow across the membrane sets up H^+ gradients that drive ATP formation.

cyst Of many microorganisms, a resting stage with thick outer layers that typically forms under adverse conditions. Of skin, abnormal, fluid-filled sac without an external opening.

cytochrome (SIGH-toe-krome) Iron-containing protein molecule of electron transport systems.

cytokinin (SIGH-toe-KYE-nin) Type of plant hormone; stimulates cell division, promotes leaf expansion, and retards leaf aging.

cytomembrane system Organelles (ER, Golgi bodies) that modify, package, and distribute new protein and lipids; also various vesicles (e.g., lysosomes) that have diverse functions.

cytoplasm (SIGH-toe-plaz-um) All cell parts, particles, and semifluid substances between the plasma membrane and nucleus or nucleoid.

cytoplasmic division Cytokinesis; splitting of a parent cell into daughter cells.

cytoplasmic localization Parceling of a portion of maternal messages in the egg cytoplasm to each blastomere that forms during cleavage.

cytosine (SIGH-toe-seen) Pyrimidine; one of the nitrogen-containing bases in nucleotides.

cytoskeleton Dynamic internal framework that structurally supports eukaryotic cells and organizes and moves a cell and its structures.

decomposer [L. *dis–*, to pieces] Prokaryotic or fungal heterotroph; gets carbon and energy from products, remains, of organisms. Helps cycle nutrients to producers in ecosystems.

decomposition In general, processes by which prokaryotic cells and fungi degrade organic matter (e.g., nitrogenous wastes of organisms).

deductive logic Pattern of thinking; making inferences about specific consequences or predictions that must follow from a hypothesis.

deforestation Removal of all the trees from a large tract of land (e.g., Amazon River Basin).

degradative pathway A stepwise series of metabolic reactions that break down organic compounds to products having lower energy.

deletion At cytological level, loss of a segment from a chromosome. At molecular level, loss of one to a few base pairs from a DNA molecule.

demographic transition model Model that correlates changes in population growth with four stages of economic development.

demographics A population's vital statistics.

denaturation (deh-NAY-chur-AY-shun) Loss of a molecule's three-dimensional shape as weak bonds (e.g., hydrogen bonds) are disrupted.

dendrite (DEN-drite) [Gk. *dendron*, tree] Short, slender extension from cell body of a neuron; commonly a signal input zone.

dendritic cell Type of antigen-presenting cell.

denitrification (DEE-nite-rih-fih-KAY-shun) Conversion of nitrate or nitrite by certain soil bacteria to gaseous nitrogen (N_2) and a small amount of nitrous oxide (N_2O).

density-dependent control Factor that limits population growth by reducing the birth rate or increasing death and dispersal rates (e.g., predation, parasitism, disease, competition).

density-independent factor Factor that causes a population's death rate to rise independently of density (e.g., a severe storm or flood).

dentition (den-TIH-shun) Collectively, the type, size, and number of an animal's teeth.

derived trait A novel feature that evolved but once and is shared only by the descendants of the ancestral species in which it evolved.

dermal tissue system Tissues that cover and protect all exposed surfaces of a plant.

dermis Skin layer beneath the epidermis that consists primarily of dense connective tissue.

desalination Removing salt from seawater.

desert Biome that forms where the potential for evaporation greatly exceeds rainfall, and where soil is thin and vegetation sparse.

desertification (dez-urt-ih-fih-KAY-shun) The conversion of grassland or irrigated or rain-fed cropland to a desertlike condition.

detrital food web Network of food chains in which energy flows mainly from plants through arrays of detritivores and decomposers.

detritivore (dih-TRY-tih-vore) Heterotroph that feeds on decomposing particles of organic matter (e.g., crab, earthworm, roundworm).

deuterostome (DUE-ter-oh-stome) [Gk. *deuteros*, second, + *stoma*, mouth] Bilateral animal in which the anus forms from first indentation in the embryo (e.g., echinoderms, chordates).

development Of multicelled species, emergence of specialized, morphologically distinct body parts according to a genetic program.

diaphragm (DIE-uh-fram) [Gk. *diaphragma*, to partition] Muscular partition between thoracic and abdominal cavities with role in breathing. Also a contraceptive device inserted into the vagina to prevent sperm from entering uterus.

dicot (DIE-kot) [Gk. *di*, two, + *kotylēdōn*, cup-shaped vessel] Dicotyledon. Flowering plant generally characterized by embryos with two cotyledons; net-veined leaves; and floral parts arranged in fours, fives, or multiples of these.

diffusion Net movement of like molecules or ions down their concentration gradient.

digestive system Body sac or tube, often with specialized regions where food is ingested, digested, and absorbed, and where undigested residues are eliminated. Incomplete systems have one opening; complete systems have two.

dihybrid cross An intercross between two F_1 heterozygotes that are identical for two gene loci; the dihybrids are offspring of parents that bred true for different versions of two traits.

dinoflagellate Type of single-celled, flagellated, cellulose-plated protistan. Most are producers of marine phytoplankton; some cause red tides.

dinosaur One of a fabulous group of reptiles that originated in the Triassic and became the dominant land vertebrates for 125 million years.

diploidy (DIP-loyd-ee) Presence of two of each type of chromosome (i.e., pairs of homologous chromosomes) in a cell nucleus at interphase. *Compare* haploidy.

directional selection Mode of natural selection by which allele frequencies underlying a range of phenotypic variation shift in a consistent direction, in response to directional change or to new conditions in the environment.

disaccharide (die-SAK-uh-ride) [Gk. *di*, two, + *sakcharon*, sugar] A common oligosaccharide; two covalently bonded sugar monomers.

disease Outcome of infection when defenses aren't mobilized fast enough and a pathogen's activities interfere with normal body functions.

disruptive selection Mode of natural selection by which the different forms of a trait at both ends of the range of variation are favored and intermediate forms are selected against.

distal tubule Tubular part of nephron where water and sodium are selectively reabsorbed.

distance effect Idea that only species adapted for long-distance dispersal can be potential colonists of islands far from their home range.

diversity of life Sum of all variations in form, function, and behavior in all lineages, from the time of life's origin to the present.

division of labor Of a multicelled organism, the combined contributions of its cells, tissues, organs, organ systems—and even stages of development—to survival and reproduction.

DNA Deoxyribonucleic acid (dee-OX-ee-RYE-bow-new-CLAY-ik) Of cells and many viruses, the molecule of inheritance. H bonds join its two helically twisted nucleotide strands, one of which has instructions (in its base sequence) for synthesizing all of the enzymes and other proteins required to build and maintain cells.

DNA clone Many identical copies of DNA that was inserted into plasmids and later amplified.

DNA fingerprint DNA fragments inherited in a Mendelian pattern that give each individual a unique identity.

DNA ligase (LYE-gaze) Enzyme that seals new base-pairings during DNA replication.

DNA polymerase (poe-LIM-uh-raze) Enzyme of replication and repair that assembles a new strand of DNA on a parent DNA template.

DNA repair Enzyme-mediated process that fixes small-scale alterations in a DNA strand by restoring the original base sequence.

DNA replication Any process by which a cell duplicates its DNA molecules before dividing.

dominance hierarchy Social organization in which some individuals of the group have adopted a subordinate status to others.

dominant allele Of diploid cells, an allele that masks the phenotypic effect of any recessive allele paired with it.

dormancy [L. *dormire*, to sleep] A predictable time of metabolic inactivity for many spores, cysts, seeds, perennials, and some animals.

dosage compensation Any mechanism that balances gene expression between the sexes during critical early stages of development.

double fertilization Of flowering plants only, fusion of a sperm and egg nucleus, plus fusion of another sperm nucleus with nuclei of a cell that gives rise to a nutritive tissue (endosperm).

doubling time Time it takes for a population to double in size.

drug addiction Dependence on a drug, which assumes an "essential" biochemical role in the body following habituation and tolerance.

dry shrubland Biome that forms when annual rainfall is less than 25 to 60 centimeters; short, multibranched woody shrubs dominate.

dry woodland Biome that forms when annual rainfall is about 40 to 100 centimeters; it may have many tall trees but no dense canopy.

duplication Gene sequence repeated several to many hundreds or thousands of times. Even normal chromosomes have such sequences.

echinoderm Type of invertebrate with calcified spines, needles, or plates on body wall. Radial with some bilateral features (e.g., sea stars).

echolocation Use of echoes from self-generated ultrasounds as a navigational mechanism.

ecological succession Processes by which a community develops in sequence, from pioneer species to an end array of species that remain in equilibrium over some region.

ecology [Gk. *oikos*, home, + *logos*, reason] The scientific study of how organisms interact with one another and with their environment.

ecoregion Broad land or ocean region defined by climate, geography, and producer species.

ecosystem Array of organisms, together with their environment, interacting through a flow of energy and a cycling of materials.

ecosystem modeling Analytical method; uses computer programs and models to predict unforeseen effects of ecosystem disturbances.

ectoderm [Gk. *ecto*, outside, + *derma*, skin] The first-formed, outermost primary tissue layer of animal embryos; gives rise to nervous system tissues and integument's outer layer.

ectotherm Poorly insulated animal with low metabolic rates; maintains core temperature by absorbing environmental heat (e.g., from sun).

Ediacaran One of the species with a highly flattened body that arose in the precambrian.

effector Muscle (or gland) which, in response to neural commands, helps cause movement (or chemical change).

effector cell Differentiated cell of a lymphocyte subpopulation that, during immune responses, engages and destroys antigen-bearing agents.

egg Mature female gamete; an ovum.

El Niño Massive eastward movement of warm surface waters of the western equatorial Pacific; displaces cool water off South America. Recurs and disrupts climates throughout the world.

electromagnetic spectrum All wavelengths from radiant energy less than 10^{-5} nm long to radio waves more than 10 km long.

electron Negatively charged unit of matter, with particulate and wavelike properties, that occupies one of the orbitals around the atomic nucleus. Atoms gain, lose, or share electrons.

electron transfer A molecule donates one or more electrons to another, acceptor molecule.

electron transport phosphorylation (FOSS-for-ih-LAY-shun) Last stage of aerobic respiration; electrons from intermediates flow through a membrane transport system, to O_2. The flow sets up an electrochemical gradient that drives ATP formation at other membrane sites.

electron transport system Organized array of membrane-bound enzymes and cofactors that accept and donate electrons in series. It sets up an electrochemical gradient across the membrane that cells use to do work.

element Fundamental form of matter that has mass, occupies space, and cannot be broken apart into a different form of matter, at least by ordinary physical or chemical means.

embryo (EM-bree-oh) [Gk. *en*, in, + *bryein*, to swell] Of animals, a multicelled body formed by way of cleavage, gastrulation, and other early developmental events. Of plants, a young sporophyte until time of germination.

embryo sac The female gametophyte of flowering plants.

embryonic induction In a growing embryo, release of a gene product from one tissue that affects the development of an adjacent tissue.

emigration Permanent move by a resident out of a population.

emulsification Of chyme, a suspension of fat droplets coated with bile salts.

end product Substance present at the end of a metabolic pathway.

endangered species Endemic (native) species highly vulnerable to extinction.

endergonic reaction (en-dur-GONE-ik) Type of chemical reaction having a net gain in energy.

endocrine gland Ductless gland that secretes hormones, which the bloodstream distributes.

endocrine system Integrative system of cells, tissues, and organs, functionally linked to the nervous system, that exerts control by way of its hormones and other chemical secretions.

endocytosis (EN-doe-sigh-TOE-sis) Cell uptake of substances when part of plasma membrane forms a vesicle around them. Three routes are receptor-mediated endocytosis, bulk transport of extracellular fluid, and phagocytosis.

endoderm [Gk. *endon*, within, + *derma*, skin] Inner primary tissue layer of animal embryos; source of inner gut lining and derived organs.

endodermis Cell layer around root vascular cylinder; influences water and solute uptake.

endometrium (EN-doe-MEET-ree-um) [Gk. *metrios*, of the womb] Inner lining of uterus.

endoplasmic reticulum or **ER** (EN-doe-PLAZ-mik reh-TIK-yoo-lum) Organelle that starts at nucleus and curves through cytoplasm. New polypeptides get side chains in rough ER (has ribosomes on its cytoplasmic side); smooth ER (has no ribosomes) is a site of lipid synthesis.

endoskeleton [Gk. *endon*, within, + *skleros*, hard] Of chordates, an internal framework of cartilage, bone, or both; works with skeletal muscle to position, support, and move body.

endosperm (EN-doe-sperm) Nutritive tissue inside the seed of a flowering plant.

endospore Resting structure formed by some bacteria; encloses a duplicate of the bacterial chromosome and a portion of cytoplasm.

endosymbiosis Continuing physical contact between one species and another species that now lives and reproduces inside its body.

endotherm Animal with high metabolic rates, body form, and behavior to control heat gains and losses over a wide temperature range.

energy Capacity to do work.

energy carrier Molecule that delivers energy from one reaction site to another; mainly ATP.

energy pyramid Pyramidal diagram of an ecosystem's trophic structure; it shows energy losses at each transfer to another trophic level.

enhancer A short DNA base sequence that is a binding site for an activator protein.

entropy (EN-trow-pee) Measure of the degree of disorder in a system (how much energy has become disorganized, usually as dispersing heat, and is no longer available to do work). All systems need energy inputs to counter entropy.

enzyme (EN-zime) A type of protein or one of the few RNAs that catalyze reactions between substances, most often at functional groups.

eosinophil Fast-acting white blood cell; its enzyme secretions digest holes in parasitic worms during an inflammatory response.

epidermis Outermost tissue layer of plants and all animals above sponge level of organization.

epiglottis Flaplike structure between pharynx and larynx; its controlled positional changes direct air into trachea or food into esophagus.

epinephrine (ep-ih-NEF-rin) Adrenal hormone; raises blood levels of sugar and fatty acids, increases heart's rate and force of contraction.

epistasis (eh-PISS-tah-sis) Interaction among the products of two or more gene pairs.

epithelium (EP-ih-THEE-lee-um) Animal tissue that covers external surfaces and lines internal cavities and tubes. One surface is free and the other rests on a basement membrane.

equilibrium, dynamic [Gk. *aequus*, equal, + *libra*, balance] Point at which a reaction runs forward as fast as in reverse, so there is no net change in reactant or product concentrations.

erythrocyte (eh-RITH-row-site) [Gk. *erythros*, red, + *kytos*, vessel] Red blood cell.

erythropoietin Hormone secreted by kidney cells; induces stem cells in bone marrow to divide and give rise to red blood cells.

esophagus (ee-SOF-uh-gus) A muscular tube just after the pharynx in the digestive track of vertebrates and many invertebrates.

essential amino acid Any amino acid that an organism cannot synthesize for itself and must obtain from food.

essential fatty acid Any fatty acid an organism cannot synthesize and must obtain from food.

estrogen (ESS-trow-jen) Female sex hormone; helps oocytes mature, primes uterine lining for pregnancy, helps maintain secondary sexual traits, affects growth and development.

estuary (EST-you-ehr-ee) Partly enclosed coast region where seawater mixes with fresh water and runoff from land, as from rivers.

ethylene (ETH-il-een) Plant hormone; promotes fruit ripening and leaf, flower, fruit abscission.

Eubacteria Kingdom of all prokaryotic cells except archaebacteria.

euglenoid Single-celled, flagellated protistan; photosynthetic but with heterotrophic feeding apparatus (it had phagocytic ancestors).

eukaryotic cell (yoo-CARE-EE-oh-tic) [Gk. *eu*, good, + *karyon*, kernel] Cell having a nucleus and other membrane-bound organelles.

eutrophication Nutrient enrichment of a body of water; typically causes reduced transparency and a phytoplankton-dominated community.

evaporation [L. *e-*, out, + *vapor*, steam] The conversion of a substance from a liquid state to a gaseous state by an input of heat energy.

evaporative heat loss Response to heat stress (e.g., at respiratory surfaces and across skin).

evolution, biological [L. *evolutio*, unrolling] Genetic change in a line of descent. Outcome of microevolutionary events: gene mutation, natural selection, genetic drift, and gene flow.

evolutionary tree Treelike diagram; a branch point means divergence from a shared ancestor and branches signify separate lines of descent.

excretion Removal of excess water, solutes, wastes, and some harmful substances from body by way of a urinary system or glands.

exergonic reaction (EX-ur-GONE-ik) Chemical reaction that shows a net loss in energy.

exocrine gland (EK-suh-krin) [Gk. *es*, out of, + *krinein*, to separate] Glandular structure that secretes products, usually through ducts or tubes, to a free epithelial surface.

exocytosis (EK-so-sigh-TOE-sis) Release of a vesicle's contents at cell surface as it fuses with and becomes part of the plasma membrane.

exodermis Cylindrical sheet of cells inside the root epidermis of most flowering plants; helps control uptake of water and solutes.

exon One of the base sequences of an mRNA transcript that will become translated.

exoskeleton [Gk. *skleros*, hard, stiff] External skeleton (e.g., hardened cuticle of arthropods).

exotic species Species that left its home range and became established in a new community.

experiment, scientific Test that simplifies observation in nature or in the laboratory by manipulating and controlling the conditions under which the observations are made.

exponential growth (EX-po-NEN-shul) Pattern of population growth; the population size expands by ever increasing increments during successive intervals as its reproductive base enlarges; plots out as a J-shaped curve.

external ear Among mammals, one of a pair of sound-collecting flaps on the sides of the head.

extinction Irrevocable loss of a species.

extracellular digestion, absorption Mode of nutrition by which certain organisms grow in or on organic matter, digest it with secreted enzymes, then absorb breakdown products.

extracellular fluid Of most animals, all fluid not in cells; plasma (blood's liquid portion) plus interstitial fluid.

extracellular matrix The ground substance, fibrous proteins, and other materials between cells of animal tissues (e.g., cartilage).

extreme halophile Archaebacterium of unusually saline habitats (e.g., salt lakes).

extreme thermophile Archaebacterium of very hot aquatic habitats (e.g., hot springs).

eye Sensory organ that incorporates a dense array of photoreceptors.

FAD Flavin adenine dinucleotide, a type of nucleotide coenzyme; it transfers electrons and unbound protons (H^+) between reaction sites. At such times it is abbreviated $FADH_2$.

fall overturn Vertical mix of a body of water in fall. Its upper layer cools, gets more dense, sinks. Oxygenated surface water moves down, nutrients from bottom sediments move up.

family pedigree Chart of genetic relationship of family individuals through the generations.

fat Lipid with a glycerol head and one, two, or three fatty acid tails. Unsaturated tails have single covalent bonds in the carbon backbone; saturated tails have one or more double bonds.

fatty acid Molecule with a backbone of up to 36 carbon atoms, a carboxyl group ($-COO^-$ or $-COOH$) at one end, and hydrogen atoms at most or all of the other bonding sites.

feather Of birds, a lightweight structure used in flight, insulation, and behavioral displays.

feedback inhibition Mechanism by which a cellular change resulting from some activity shuts down the activity that brought it about.

fermentation [L. *fermentum*, yeast] Anaerobic pathway of ATP formation, from glycolysis to the transfer of electrons to an intermediate or a breakdown product. NAD^+ is regenerated. Net energy yield: 2 ATP per glucose molecule.

fern A seedless vascular plant of moist or wet habitats; has fronds often divided into leaflets.

fertilization [L. *fertilis*, to carry, to bear] The fusion of a sperm nucleus with the nucleus of an egg, thus forming a zygote.

fetus The stage of animal development after major organ systems formed in the embryo until the time of birth.

fever Any core temperature higher than the set point in the hypothalamic thermostat.

fibrous root system All lateral branchings of adventitious roots arising from a young stem.

Fick's law The larger the surface area and the greater the partial pressure gradient, the faster gases will diffuse across a respiratory surface.

filter feeder Animal that filters food from a current of water directed through a body part (e.g., through a sea squirt's pharynx).

filtration First step in forming urine; pressure of heart contractions filters blood (forces water and all solutes except proteins) into nephrons.

fin Of fishes generally, appendage that helps stabilize, propel, and guide body in water.

first law of thermodynamics The total amount of energy in the universe is constant. Energy cannot be created from nothing and existing energy cannot be destroyed.

fish An aquatic animal of the most ancient, diverse vertebrate lineage; one of the jawless, jawed cartilaginous, or jawed bony species.

fitness Increase in adaptation to environment, as brought about by genetic change.

fixation Loss of all but one kind of allele at a gene locus for all individuals in a population.

fixed action pattern Program of coordinated, stereotyped muscle activity that is completed independently of feedback from environment.

flagellum (fluh-JELL-um), plural **flagella** A motile structure of many free-living eukaryotic cells. Its core has a 9 + 2 array of microtubules. Typically longer than cilia.

flatworm One of the simplest invertebrates having organ systems; a turbellarian, fluke, or tapeworm. Some are notorious parasites.

flower Of angiosperms only, a reproductive structure with nonfertile parts (sepals, petals) and fertile parts (stamens, carpels) attached to a receptacle (modified base of a floral shoot).

fluid mosaic model A cell membrane consists of a lipid bilayer and proteins. Its lipids are mainly structural; they make it impermeable to water-soluble molecules, yet (by packing variations and movements), impart fluidity. Its diverse proteins perform most membrane functions (e.g., transport, signal reception).

food chain Straight-line sequence of steps by which energy stored in autotroph tissues passes on to higher trophic levels.

food pyramid Chart of a purportedly well-balanced diet; continually being refined.

food web Of ecosystems, cross-connecting food chains consisting of producers, consumers, and decomposers, detritivores, or both.

forebrain Most complex portion of vertebrate brain; includes cerebrum (and cerebral cortex), olfactory lobes, and hypothalamus.

forest Biome in which tall trees grow close enough to form a fairly continuous canopy.

fossil Recognizable, physical evidence of an organism that lived in the distant past.

fossil fuel Coal, petroleum, or natural gas; a nonrenewable energy source that formed long ago from remains of swamp forests.

fossilization How fossils form. An organism or evidence of it gets buried in sediments or volcanic ash; water and dissolved inorganic compounds infiltrate it; then chemical changes and pressure from accumulating sediments above transform it to stony hardness.

founder effect A form of bottlenecking. By chance alone, a few individuals that establish a new population have allele frequencies that differ from those of the original population.

free radical Any highly reactive molecular fragment having an unpaired electron.

fruit [L. after *frui*, to enjoy] Flowering plant's mature ovary, often with accessory structures.

FSH Follicle-stimulating hormone produced and secreted by the anterior lobe of pituitary gland; has reproductive roles in both sexes.

functional group An atom or a group of atoms that is covalently bonded to the carbon backbone of an organic compound and that influences its chemical behavior.

functional-group transfer Enzyme-mediated event in which a molecule donates one or more functional groups to another molecule.

Fungi Kingdom of fungi which includes major decomposers plus pathogens and parasites.

fungus Eukaryotic heterotroph; it secretes enzymes that digest food to bits that its cells absorb (extracellular digestion, absorption). The saprobes use nonliving organic matter; parasites feed on living organisms.

fusion power Release of energy from fuel (e.g., heated gas of H isotopes) that implodes when compressed to extremely high densities.

gallbladder Organ that stores bile secreted from liver; its duct connects to small intestine.

gamete (GAM-eet) [Gk. *gametēs*, husband, and *gametē*, wife] Haploid cell, formed by meiotic cell division of a germ cell; required for sexual reproduction. Eggs and sperm are examples.

gamete formation Formation of sex cells (e.g., sperm, eggs) in reproductive tissues or organs.

gametophyte (gam-EET-oh-fite) [Gk. *phyton*, plant] Haploid gamete-producing body that forms during plant life cycles.

ganglion (GANG-lee-on), plural **ganglia** Distinct cluster of cell bodies of neurons.

gap junction Cylindrical arrays of proteins in the plasma membrane that pair up as open channels for signals between adjoining cells.

gastric fluid Highly acidic mix of secretions from the stomach's glandular epithelium (HCl, mucus, pepsinogens, etc.) that act on food.

gastrulation (gas-tru-LAY-shun) Stage of animal development; reorganization of new cells into two or three primary tissue layers.

gel electrophoresis Laboratory technique used to distinguish among molecules. Applied electric field forces them to migrate through a viscous gel and distance themselves from one another by length, size, or electric charge.

gene [German *pangan*, after Gk. *pan*, all; *genes*, to be born] Unit of information for a heritable trait, passed from parents to offspring.

gene control One of the molecular mechanisms that govern when and how fast specific genes will be transcribed and translated, and whether gene products will be activated or inactivated.

gene flow Microevolutionary process; alleles enter and leave a population as an outcome of immigration and emigration, respectively.

gene frequency Abundance of an allele with respect to others at same locus in a population.

gene library Mixed collection of bacteria that house many different cloned DNA fragments.

gene locus A gene's chromosomal location.

gene mutation A small-scale change in the nucleotide sequence of a DNA molecule.

gene pair Two alleles at the same gene locus on a pair of homologous chromosomes.

gene pool All genotypes in a population.

gene therapy Generally, a transfer of one or more normal genes into an organism to correct or lessen adverse effects of a genetic disorder.

genetic abnormality A rare or less common version of a heritable trait.

genetic code [After L. *genesis*, to be born] The correspondence between nucleotide triplets in DNA (then mRNA) and specific sequences of amino acids in a polypeptide chain; the basic language of protein synthesis in cells.

genetic disease Illness in which expression of one or more genes increases susceptibility to infection or weakens immune response to it.

genetic disorder Any inherited condition that causes mild to severe medical problems.

genetic divergence Gradual accumulation of differences in gene pools of populations or subpopulations of a species after a geographic barrier arises and separates them; thereafter, microevolution occurs independently in each.

genetic drift Change in allele frequencies over the generations due to chance alone. Its effect is most pronounced in very small populations.

genetic engineering Deliberately altering the information content of DNA molecules.

genetic equilibrium In theory, a state in which a population is not evolving. These conditions are met: no mutation, very large size, isolation from others of the species, no natural selection (*all* members reproduce, by random mating).

genetic recombination Result of any process that puts new genetic information into a DNA molecule (e.g., by crossing over).

genome All the DNA in a haploid number of chromosomes for a given species.

genotype (JEEN-oh-type) Genetic constitution of an individual; a single gene pair or the sum total of an individual's genes.

genus, plural **genera** (JEEN-US, JEN-er-ah) [L. *genus*, race or origin] A grouping of species that are more closely related to one another in their morphology, ecology, and history than to other species at the same taxonomic level.

geographic dispersal An organism moves out of its home range and becomes established in a new community, as an exotic species.

geologic time scale Time scale for the Earth's history with major subdivisions corresponding to mass extinctions. Now radiometrically dated.

germ cell Animal cell of a lineage set aside for sexual reproduction; gives rise to gametes.

germination (jur-mih-NAY-shun) Of seeds and spores, resumption of growth after dormancy, dispersal from the parent organism, or both.

gibberellin (JIB-er-ELL-un) Plant hormone; promotes elongation of stems, helps seeds and buds break dormancy, and helps flowering.

gill Respiratory organ. Most gills have a thin, moist, vascularized layer for gas exchange.

gill slit Opening in a thin-walled pharynx. Gill slits serve in food-trapping and respiration; jaws evolved from certain gill slit supports.

ginkgo Deciduous gymnosperm; its ancestors were diverse in dinosaur times.

gland Secretory cell or structure derived from epithelium and often still connected to it.

global broiling Theory that an asteroid impact caused the K–T mass extinction by creating a colossal fireball, the debris from which raised global air temperature by thousands of degrees.

global warming A long-term increase in the temperature of the Earth's lower atmosphere.

glomerulus (glow-MARE-you-luss) [L. *glomus*, ball] Bowman's capsule, together with the glomeruli capillaries it cups around.

glottis Opening between two vocal cords.

glucagon (GLUE-kuh-gone) Hormone secreted by pancreas; stimulates cells to convert their stores of glycogen and amino acids to glucose.

glyceride (GLISS-er-eyed) Molecule having one, two, or three fatty acid tails attached to a glycerol backbone; one of the fats or oils.

glycerol (GLISS-er-ohl) [Gk. *glykys*, sweet, + L. *oleum*, oil] Three-carbon compound with three hydroxyl groups; component of fats and oils.

glycogen (GLY-kuh-jen) A highly branched polysaccharide made of glucose monomers; the main storage carbohydrate in animals.

glycolysis (gly-CALL-ih-sis) [Gk. *glykys*, sweet, + *lysis*, breaking apart] Breakdown of glucose or another organic compound to two pyruvate molecules. First stage of aerobic respiration, fermentation, and anaerobic electron transport. Oxygen has no role in glycolysis, which occurs in the cytoplasm of all cells. Two NADH form. Net yield: 2 ATP per glucose molecule.

glycoprotein Protein with linear or branched oligosaccharides covalently bonded to it; most surface proteins of animal cells and many proteins circulating in blood are glycoproteins.

gnetophyte Woody gymnosperm, a vine or shrub; unusual in having vessels in its xylem.

Golgi body (GOHL-gee) Organelle of lipid assembly, polypeptide chain modification, and packaging of both in vesicles for export or for transport to locations in cytoplasm.

gonad (GO-nad) Primary reproductive organ in which animal gametes are produced.

Gondwana Paleozoic supercontinent; with other land masses, it formed Pangea.

graded potential A local signal that alters the resting membrane potential at the input zone of a neuron or other excitable cell; magnitude varies, depending on the stimulus.

gradual model, speciation Idea that species arise by many small morphological changes that accumulate over great spans of time.

grassland Biome that has flat or rolling land, 25–100 centimeters of annual rainfall, warm summers, and a distinct array of grazers; fires recur and regenerate dominant plant species.

gravitropism (GRAV-ih-TROPE-izm) [L. *gravis*, heavy, + Gk. *trepein*, to turn] Directional plant growth in response to gravitational force.

gray matter Unmyelinated axons, dendrites, and cell bodies of neurons, plus neuroglial cells, in the brain and spinal cord.

grazing food web Network of food chains in which energy flows from plants to an array of herbivores, then to carnivores.

green alga Type of protistan biochemically, evolutionarily, and structurally most like plants (e.g., most are photoautotrophs with starch grains and chlorophylls *a*, *b* in chloroplasts).

green revolution In developing countries, use of improved crop strains, modern agricultural equipment and practices (e.g., heavy fertilizer and pesticide application) to raise crop yields.

greenhouse effect Atmospheric gases impede escape of infrared wavelengths (heat) from Earth's sun-warmed surface, absorb them, and radiate much of the heat back toward Earth.

ground meristem (MARE-ih-stem) [Gk. *meristos*, divisible] Primary meristem that gives rise to the ground tissue system of plants.

ground substance Intercellular material in a number of animal tissues; composed of cell secretions and other noncellular components.

ground tissue system Tissues, parenchyma especially, making up most of the plant body.

growth Of multicelled species, increases in the number, size, and volume of cells. Of bacterial populations, increases in the number of cells.

growth factor Protein that plays a role in the body's growth (e.g., by promoting mitosis).

growth ring One of the alternating bands of early and late wood; a "tree ring."

growth, tissue specialization Stage of animal development when organs enlarge and assume specialized functions; continues into adulthood.

guanine Nitrogen-containing base in one of four nucleotide monomers of DNA or RNA.

guard cell One of two adjoining cells defining a stoma, a type of opening where CO_2, O_2, and water vapor cross leaf or stem epidermis.

gut Generally, a sac or tube from which food is absorbed into internal environment. Also, a gastrointestinal tract (from stomach onward).

gymnosperm (JIM-noe-sperm) [Gk. *gymnos*, naked, + *sperma*, seed] Type of vascular plant in which seeds form on exposed surfaces of reproductive structures (e.g., on cone scales).

habitat [L. *habitare*, to live in] Place where an organism or species lives; characterized by its physical and chemical features and its species.

habitat island A region with endemic species (e.g., a forest or lake) located within a "sea" of habitats unsuitable for sustaining them.

habitat loss Reduction in suitable living space and habitat closure through chemical pollution.

hagfish Type of scavenging jawless fish that secretes slimy, smelly mucus.

hair A flexible structure rooted in skin, with a shaft above the skin's surface.

hair cell Mechanoreceptor that responds to being physically bent or tilted.

half-life The time it takes for half of a given quantity of any radioisotope to decay into a different, and less unstable, daughter isotope.

haploidy (HAP-loyd-ee) Presence of only half of the parental number of chromosomes in a spore or gamete, as brought about by meiosis.

hardwood Type of strong, dense wood with many vessels, tracheids, and fibers in xylem.

Hardy–Weinberg rule Allele frequencies stay the same over the generations when there is no mutation, the population is infinitely large and isolated from other populations of the species, mating is random, and all individuals are reproducing equally and randomly.

HDL A high-density lipoprotein in blood; it transports cholesterol to liver for metabolism.

hearing Perception of sound.

heart Muscular pump; its contractions keep blood circulating through the animal body.

heartwood A dry tissue at the core of aging stems and roots that no longer transports water and solutes. It helps the tree defy gravity and is a dumping ground for some metabolic wastes.

heat Thermal energy; a form of kinetic energy.

HeLa cell Cancer cell of a lineage established for research; now used in many laboratories.

helper T cell CD4 lymphocyte; has key roles in antibody- and cell-mediated immune responses. It induces responsive T and B cells to divide and differentiate rapidly into huge armies.

heme group Iron-containing functional group that reversibly binds oxygen; one associated with each of the four chains in hemoglobin.

hemoglobin (HEEM-oh-glow-bin) [Gk. *haima*, blood, + L. *globus*, ball] An iron-containing respiratory protein of red blood cells.

hemostasis (HEE-mow-STAY-sis) [Gk. *haima*, blood, + *stasis*, standing] Process that stops blood loss from damaged blood vessel by way of coagulation, blood vessel spasm, platelet plug formation, and other mechanisms.

herbivore [L. *herba*, grass, + *vovare*, to devour] Plant-eating animal (e.g., snail, deer, manatee).

hermaphrodite (her-MAH-froe-dyte) Individual having both male and female gonads.

heterocyst (HET-er-oh-sist) Cyanobacterial cell, self-modified, that synthesizes a nitrogen-fixing enzyme when nitrogen supplies dwindle.

heterotherm Generally, an endotherm that can lower metabolic rates; activities idle but it cuts energy cost of maintaining its core temperature.

heterotroph (HET-er-oh-trofe) [Gk. *heteros*, other, + *trophos*, feeder] Organism unable to make its own organic compounds; feeds on autotrophs, other heterotrophs, organic wastes.

heterozygous condition (HET-er-oh-ZYE-guss) [Gk. *zygoun*, join together] Having a pair of nonidentical alleles at a gene locus (that is, on a pair of homologous chromosomes).

higher taxon (plural, **taxa**) One of ever more inclusive groupings that reflect relationships among species. Family, order, class, phylum, and kingdom are examples.

hindbrain Medulla oblongata, cerebellum, and pons of vertebrate brain. Includes reflex centers for respiration, blood circulation, and other basic functions; also helps coordinate motor responses and many complex reflexes.

histamine Local signaling molecule that fans inflammation; makes arterioles vasodilate and capillaries more permeable (leaky).

histone Type of protein intimately associated with eukaryotic DNA and largely responsible for organization of eukaryotic chromosomes.

homeostasis (HOE-me-oh-STAY-sis) [Gk. *homo*, same, + *stasis*, standing] State in which physical and chemical aspects of internal environment (blood, interstitial fluid) are being maintained within ranges suitable for cell activities.

homeotic gene In all major animal groups, one of the master genes, the products of which interact with one another and with control elements to map out the overall body plan.

hominid [L. *homo*, man] All species on or near evolutionary road leading to modern humans.

hominoid Apes, humans, and recent ancestors.

homologous chromosome (huh-MOLL-uh-gus) [Gk. *homologia*, correspondence] Of cells with a diploid chromosome number, one of a pair of chromosomes identical in size, shape, and gene sequence, and that interact at meiosis. Nonidentical sex chromosomes (e.g., X and Y) also interact as homologues during meiosis.

homologous structures Of separate lineages, comparable body parts that show underlying similarity even when they may differ in size, shape, or function; outcome of morphological divergence from a shared ancestor.

homozygous condition (HOE-moe-ZYE-guss) For a specified trait, having a pair of identical dominant or recessive alleles at a locus.

hormone [Gk. *hormon*, stir up, set in motion] Signaling molecule secreted by one cell that stimulates or inhibits activities of any cell with receptors for it. Animal hormones are picked up and transported by the bloodstream.

horsetail Seedless vascular plant; has rhizomes, scale-shaped leaves, and hollow photosynthetic stems with silica-reinforced ribs.

host Living organism exploited by a parasite. A definitive host harbors the mature stage of a parasite's life cycle. One or more intermediate hosts harbor immature stages.

hot spot Limited area where human activities are driving many species to extinction.

human A primate of the species *Homo sapiens*.

human genome project Worldwide research project to sequence all 3.2 billion nucleotides in the DNA of human chromosomes.

humus Decomposing organic matter in soil.

hybrid offspring Of a genetic cross, offspring having a pair of nonidentical alleles for a trait.

hybrid zone Where adjoining populations are interbreeding and producing hybrid offspring.

hydrocarbon An organic compound that has only hydrogen bonded to a carbon backbone.

hydrogen bond Weak interaction between a small, highly electronegative atom and an H atom already taking part in a polar covalent bond in the same molecule or a different one.

hydrogen ion Free (or unbound) proton; one hydrogen atom that lost its electron and now bears a positive charge (H^+).

hydrologic cycle Biogeochemical cycle driven by solar energy; water is moved through the atmosphere, on or through land, to the ocean, and back to the atmosphere.

hydrolysis (high-DRAWL-ih-sis) [L. *hydro*, water, + Gk. *lysis*, loosening] Cleavage reaction that breaks covalent bonds and splits a molecule into two or more parts. H^+ and OH^- (from a water molecule) are often attached to the newly exposed bonding sites.

hydrophilic substance [Gk. *philos*, loving] A polar molecule or molecular region that easily dissolves in water (e.g., sugars).

hydrophobic substance [Gk. *phobos*, dreading] A nonpolar molecule or molecular region that strongly resists dissolving in water (e.g., oils).

hydrosphere All of the Earth's water.

hydrostatic pressure Pressure exerted by a volume of fluid against a wall, membrane, or some other structure that encloses the fluid.

hydrostatic skeleton A fluid-filled cavity or cell mass against which contractile cells act.

hydrothermal vent A steaming fissure in the deep ocean floor; has unique ecosystems.

hydroxide ion Ionized compound of one oxygen and one hydrogen atom (OH^-).

hypertonic solution A fluid having a greater solute concentration relative to another fluid.

hypha (HIGH-fuh), plural **hyphae** [Gk. *hyphe*, web] Fungal filament with chitin-reinforced walls; component of a mycelium.

hypodermis Subcutaneous layer with stored fat that helps insulate the body; also anchors skin yet allows it to move somewhat.

hypothalamus [Gk. *hypo*, under, + *thalamos*, inner chamber, or *tholos*, rotunda] Forebrain center of homeostatic control over the internal environment (e.g., salt–water balance, core temperature); influences hunger, thirst, sex, other viscera-related behaviors, and emotions.

hypothesis In science, a possible explanation of a phenomenon, one that has the potential to be proved false by experimental tests.

hypotonic solution A fluid that has a lower solute concentration relative to another fluid.

immigration One or more individuals moves from its population and takes up permanent residence in another population of its species.

immune system Interacting white blood cells that defend the vertebrate body through self / nonself recognition, specificity, and memory.

T and B cell antigen receptors ignore body's own cells yet collectively recognize a billion specific threats. Some B and T cells formed in a primary response are set aside as memory cells for future battles with the same antigen.

immunization Any process that promotes immunity against disease (e.g., vaccination).

immunoglobulin (Ig) One of five classes of antibodies, each with antigen-binding sites as well as other sites with specialized functions.

implantation Event in pregnancy. A blastocyst burrows into endometrium and establishes connections by which a mother will exchange substances with the embryo (and fetus) that develops from the blastocyst's inner cell mass.

imprinting Time-dependent form of learning, usually during a sensitive period for a young animal, triggered by exposure to sign stimuli.

in vitro fertilization Conception outside the body ("in glass" petri dishes or test tubes).

inbreeding Nonrandom mating among close relatives that share many identical alleles.

incomplete dominance Condition in which one allele of a pair is not fully dominant; a heterozygous phenotype somewhere between both homozygous phenotypes emerges.

independent assortment theory Mendelian theory that by the end of meiosis, each pair of homologous chromosomes (and linked genes on each one) are sorted before shipment to gametes independently of how the other pairs were sorted. Later modified to account for the disruptive effect of crossing over on linkages.

indicator species Any species that provides warning of changes in habitat and impending widespread loss of biodiversity.

indirect selection theory Idea that altruistic individuals can pass on their genes indirectly by helping relatives survive and reproduce.

induced-fit model When bound to an active site, a substrate alters an enzyme's shape, thus bringing about a more precise molecular fit between the two that promotes reactivity.

inductive logic Pattern of thinking; deriving a general statement from specific observations.

infection Invasion and multiplication of a pathogen in a host. Disease follows if defenses are not mobilized fast enough; the pathogen's activities interfere with normal body functions.

inflammation, acute Rapid response to tissue injury by phagocytes and certain proteins (e.g., histamine, complement, clotting factors). Signs include localized redness, heat, swelling, pain.

inheritance The transmission, from parents to offspring, of genes that specify structures and functions characteristic of the species.

inhibitor Generally, a substance able to bind with a specific molecule and interfere with its functioning (e.g., a hypothalamic hormone that suppresses anterior pituitary secretion).

insertion Insertion of one to a few bases into a DNA strand. Also, a movable attachment of muscle to bone.

instinctive behavior A behavior performed without having been learned by experience.

insulin Pancreatic hormone that lowers level of glucose in blood by causing cells to take up glucose; promotes protein and fat synthesis and inhibits protein conversion to glucose.

integration, neural [L. *integrare*, coordinate] Moment-by-moment summation of excitatory and inhibitory synapses acting on a neuron.

integrator A control center (e.g., brain) that receives, processes, and stores sensory input, issues commands for coordinated responses.

integument Of animals, protective body cover (e.g., skin). Of seed-bearing plants, one or more layers around an ovule; becomes a seed coat.

integumentary exchange (in-teg-you-MEN-tuh-ree) Respiration across a thin, moist, and often vascularized surface layer of animal body.

intermediate Substance that forms between the start and end of a metabolic pathway.

intermediate filament Cytoskeletal element; mechanically strengthens some animal cells.

internal environment Blood + interstitial fluid.

interneuron Neuron of brain or spinal cord.

internode In vascular plants, the stem region between two successive nodes.

interphase Of a cell cycle, interval between nuclear divisions when a cell increases in mass and roughly doubles the number of its cytoplasmic components. It also duplicates its chromosomes (replicates its DNA) during interphase, but *not* between meiosis I and II.

interstitial fluid (IN-ter-STISH-ul) [L. *interstitus*, to stand in the middle of something] The portion of extracellular fluid that occupies the spaces between animal cells and tissues.

intertidal zone Region between the low and high water marks of a rocky or sandy shore.

intervertebral disk Cartilaginous flex point and shock absorber between vertebrae.

intron A noncoding portion of a pre-mRNA transcript; excised before translation.

inversion Part of a chromosome that became oriented in reverse, with no molecular loss.

invertebrate Any animal without a backbone.

ion, negatively charged (EYE-on) An atom or a molecule that acquired an overall negative charge by gaining one or more electrons.

ion, positively charged Atom or molecule that acquired an overall positive charge by losing one or more electrons.

ion exchange The pH-dependent process by which ions dissociate from soil particles, and other ions dissolved in soil water replace them.

ionic bond Two ions being held together by the attraction of their opposite charge.

ionizing radiation High-energy wavelengths.

isotonic solution A fluid having the same solute concentration as a fluid against which it is being compared.

isotope (EYE-so-tope) One of two or more forms of an element's atoms that differ in the number of neutrons.

jaw Of chordates, either of two cartilaginous or bony parts bordering the mouth, typically bearing teeth used to secure or pummel food. Many invertebrates have similar opposing parts.

joint Area of contact between bones.

J-shaped curve Type of diagrammatic curve that emerges when unrestricted exponential growth of a population is plotted against time.

karyotype (CARE-ee-oh-type) Preparation of metaphase chromosomes sorted by length, centromere location, other defining features.

keratin Tough, water-insoluble protein made by vertebrate epidermal cells (keratinocytes).

key innovation A structural or functional modification to the body that by chance, gives a lineage opportunity to exploit aspects of the environment in more efficient or novel ways.

keystone species A species with a dominant role in shaping community structure.

kidney One of a pair of vertebrate organs that filter ions and other substances from blood; it controls amounts returned to help maintain the internal environment.

kilocalorie 1,000 calories of heat energy; the amount needed to raise the temperature of 1 kilogram of water by 1°C. Used as the unit of measure for the caloric content of foods.

kinase Enzyme (e.g., a protein kinase) that catalyzes a phosphate-group transfer.

kinetic energy Energy of motion.

Krebs cycle A stage of aerobic respiration, in mitochondria only; it (and a few preparatory steps) breaks down pyruvate to CO_2 and H_2O. Many coenzymes accept protons (H^+) and electrons from intermediates and deliver them to the next stage of reactions; 2 ATP form.

lactate fermentation An anaerobic pathway of ATP formation. Pyruvate from glycolysis is converted to three-carbon lactate, and NAD^+ is regenerated. Net energy yield: 2 ATP.

lake A body of standing fresh water in a basin formed by geologic events; characterized by light penetration, temperature, other features.

lamprey Type of predatory, jawless fish that has a cylindrical body, cartilaginous skeleton, and an oral disk with rasping parts.

lancelet A filter-feeding invertebrate chordate with a body tapered at both ends.

large intestine Colon.

larva, plural **larvae** Immature stage between an embryo and adult in many animal life cycles.

larynx (LARE-inks) Tubular airway leading to lungs. Contains vocal cords in some animals.

lateral root Outward branching from the first (primary) root of a taproot system.

LDL Low-density lipoprotein that transports cholesterol; cells take it up, but excess amounts contribute to atherosclerosis.

leaching Removal of some nutrients from soil as water percolates through it.

leaf Chlorophyll-rich plant part adapted for sunlight interception and photosynthesis.

learned behavior Lasting modification of a behavior as a result of experience or practice.

lek Communal display ground for courtship behavior by some animals, including birds.

lens A transparent body that bends light rays so that they converge onto photoreceptors.

leptin An appetite-mediating hormone.

lethal mutation Mutation with drastic effects on phenotype; usually causes death.

LH Luteinizing hormone. Anterior pituitary hormone; reproductive roles in males, females.

lichen (LY-kun) Symbiotic interaction between a fungus and photoautotroph.

life cycle A recurring pattern of genetically programmed events from the time individuals are produced until they themselves reproduce.

life history pattern Patterns of reproduction, survival, and life expectancy for a population.

ligament A strap of dense connective tissue that bridges a joint.

light-dependent reactions The first stage of photosynthesis. Sunlight energy is trapped and converted to chemical energy of ATP, NADPH, or both, depending on the pathway.

light-independent reactions Second stage of photosynthesis; sugar-building reactions that require phosphate-group transfers from ATP, electrons and H atoms from NADPH, and carbon from CO_2. The phosphorylated sugars from these reactions are then converted to end products (e.g., sucrose, cellulose, starch).

lignification Deposition of lignin in secondary wall of many plant cells. Stabilizes and protects wall components, lends strength, rigidity, and water-loss resistance to plant parts. Important factor in the evolution of vascular plants.

lignin Organic compound; three-carbon chain and oxygen atom linked to a six-carbon ring.

limbic system In cerebral hemisphere, centers that govern emotions; has roles in memory.

limiting factor Any essential resource which, in short supply, limits population growth.

lineage (LIN-ee-edge) Line of descent.

linkage group All genes on a chromosome.

lipid A mostly greasy or oily hydrocarbon; strongly resists dissolving in water but readily dissolves in nonpolar substances. All cells use lipids as storage forms of energy, structural materials as in membranes, and cell products.

lipid bilayer Phospholipids, mostly, arranged in two layers; the structural basis of all cell membranes. Hydrophobic tails are sandwiched between the hydrophilic heads; the heads are dissolved in intracellular or extracellular fluid.

liver In vertebrates and many invertebrates, a large gland that stores, converts, and helps maintain blood levels of organic compounds; inactivates most hormone molecules that have completed their tasks; inactivates compounds that can be toxic at high concentrations.

loam Soil with roughly the same proportions of sand, silt, and clay; best for plant growth.

lobe-finned fish Only bony fish with skeletal supporting elements inside a pair of thick, ventral fins; also has paired lungs. Lineage probably included ancestors of amphibians.

local signaling molecule A secretion by one cell type that alters chemical conditions only in localized tissue regions (e.g., prostaglandin).

logistic growth (low-JISS-tik) A population growth pattern. A low-density population slowly increases in size, enters a phase of rapid growth, then levels off in size once the carrying capacity has been reached.

loop of Henle Hairpin-shaped, tubular part of a nephron that reabsorbs water and solutes.

lung Internal, sac-shaped respiratory surface that originally evolved in oxygen-poor aquatic habitats. A few fishes and amphibians, birds, reptiles, and mammals have paired lungs.

lycophyte Seedless vascular plant; needs free water to complete life cycle. Most have leaves, roots, and stems (e.g., club mosses).

lymph (LIMF) [L. *lympha*, water] Tissue fluid that has entered vessels of lymphatic system.

lymph node Lymphoid organ packed with lymphocytes; major site for immune responses.

lymph vascular system Parts of lymphatic system; delivers excess tissue fluid, absorbed fats, and reclaimable solutes to blood.

lymphatic system Supplement to vertebrate circulatory system. Its vessels deliver fluid and solutes from interstitial fluid to blood; its lymphoid organs have roles in body defenses.

lysis [Gk. *lysis*, a loosening] Gross damage to a plasma membrane, cell wall, or both that lets the cytoplasm leak out; causes cell death.

lysosome (LYE-so-sohm) Important organelle of intracellular digestion.

lysozyme Infection-fighting enzyme present in mucous membranes (e.g., of mouth, vagina).

macroevolution Large-scale patterns, trends, and rates of change among higher taxa.

macrophage Phagocytic white blood cell; roles in nonspecific defense and immune responses.

malpighian tubule One of many small tubes that help land-dwelling insects dispose of toxic wastes without losing precious water.

mammal Vertebrate whose females nourish offspring with milk from mammary glands.

mangrove wetland A nutrient-rich ecosystem of tidal flats at tropical latitudes.

mantle Of mollusks, a tissue draped over the visceral mass. Of Earth, a zone of intermediate-density rocks beneath the crust.

marine snow Organic matter, drifting down from ultraplankton to midoceanic water; the basis of food webs and staggering biodiversity.

mass extinction Catastrophic event or phase in geologic time when entire families or other major groups are irrevocably lost.

mast cell Basophil-like cell; secretes histamine.

master gene One of the genes whose products govern the development of an embryo's body according to an inherited plan; all major groups have similar or identical master genes.

mechanoreceptor Sensory cell or nearby cell that detects mechanical energy (changes in pressure, position, or acceleration).

medulla oblongata Hindbrain region with reflex centers for basic tasks (e.g., respiration); coordinates motor responses with complex reflexes (e.g., coughing); also influences brain centers concerned with sleep and arousal.

medusa (meh-DOO-sah) [Gk. *Medousa*, one of three sisters in Greek mythology having snake-entwined hair] Of cnidarian life cycles, a free-swimming, bell-shaped stage, often with oral lobes and tentacles extending below the bell.

megaspore Haploid spore; forms by meiosis in the ovary of seed-bearing plants; one of its cellular descendants develops into an egg.

meiosis (my-OH-sis) [Gk. *meioun*, to diminish] Two-stage nuclear division process that halves the chromosome number of a parental germ cell nucleus, to the haploid number. Basis of gamete formation (and meiospore formation).

memory The capacity to store and retrieve information about past sensory experience.

memory cell B or T cell that formed during an immune response; it remains in a resting phase until a secondary immune response.

menopause (MEN-uh-pozz) [L. *mensis*, month, + *pausa*, stop] The time when the reproductive potential of human females draws to a close.

menstrual cycle A recurring cycle in adult human females. A secondary oocyte is released from an ovary and the uterine lining is primed for pregnancy, all under hormonal control.

meristem A localized region of a plant where undifferentiated cells give rise to cell lineages that form all of the mature tissues.

mesoderm (MEH-zoe-derm) [Gk. *mesos*, middle, + *derm*, skin] Primary tissue layer of all large, complex animals; gives rise to many internal organs and part of the integument.

mesophyll (MEH-zoe-fill) A photosynthetic parenchyma with an abundance of air spaces.

Mesozoic An era (240–65 mya) of spectacular expansion in the range of global diversity.

messenger RNA (mRNA) A single strand of ribonucleotides transcribed from DNA, then translated into a polypeptide chain. The only RNA encoding protein-building instructions.

metabolic pathway (MEH-tuh-BALL-ik) Orderly sequence of enzyme-mediated reactions by which cells maintain, increase, or decrease the concentrations of particular substances.

metabolism (meh-TAB-oh-lizm) [Gk. *meta*, change] All the controlled, enzyme-mediated chemical reactions by which cells acquire and use energy to synthesize, store, degrade, and eliminate substances in ways that contribute to growth, survival, and reproduction.

metamorphosis (me-tuh-MOR-foe-sis) [Gk. *meta*, change, + *morphe*, form] Major changes in body form of certain animals; hormonally controlled growth, tissue reorganization, and remodeling of body parts leads to adult form.

metaphase Of meiosis I, stage when all pairs of homologous chromosomes have become positioned at the spindle equator. Of mitosis or meiosis II, all the duplicated chromosomes are positioned at the spindle equator.

metastasis Abnormal migration of cancer cells, which may establish colonies in other tissues.

methanogen Anaerobic archaebacterium that produces methane gas as by-product.

methylation Attachment of a methyl group to an organic compound; a common gene control.

MHC marker Self-marker protein. Some kinds are on all body cells of the individual; others are unique to macrophages and lymphocytes.

micelle (my-SELL) Of fat digestion, tiny droplet of bile salts, fatty acids, and monoglycerides; role in fat absorption from small intestine.

microevolution Of a population, any change in allele frequencies resulting from mutation, genetic drift, gene flow, natural selection, or some combination of these.

microfilament [Gk. *mikros*, small, + L. *filum*, thread] Two thin, twisted polypeptide chains; cytoskeletal element that helps cell move, and helps produce and maintain cell shapes.

micrograph Photograph of an image that came into view with the aid of a microscope.

microorganism Organism, usually single celled, too small to be observed without a microscope.

microspore Walled haploid spore; becomes a pollen grain in gymnosperms or angiosperms.

microtubular spindle Bipolar array of many microtubules; forms for nuclear division and moves chromosomes apart in controlled ways.

microtubule (my-crow-TUBE-yool) Cylinder of tubulin subunits; cytoskeletal element with roles in cell shape, growth, and motion (e.g., part of cilia, flagella, and spindle apparatus).

microvillus (MY-crow-VILL-us) [L. *villus*, shaggy hair] Slender extension from free surface of certain cells; arrays of many microvilli greatly increase absorptive or secretory surface area.

midbrain Part of vertebrate brain with centers for coordinating reflex responses to visual and auditory input; also relays signals to forebrain.

migration Recurring round trip between two or more regions in response to environmental rhythms (e.g., seasonal change).

mimicry (MIM-ik-ree) Close resemblance in form, behavior, or both between one species (the mimic) and another (its model). Serves in

deception, as when an orchid mimics a female insect and so attracts males that pollinate it.

mineral Any element or inorganic compound that formed by natural geologic processes and is required for normal cell functioning.

mitochondrion (MY-toe-KON-dree-on) Double-membrane organelle of ATP formation. Only site of aerobic respiration's second and third stages. May have endosymbiotic origins.

mitosis (my-TOE-sis) [Gk. *mitos*, thread] Type of nuclear division that maintains the parental chromosome number for daughter cells. The basis of growth in size, tissue repair, and often asexual reproduction for eukaryotes.

mixture Two or more elements intermingled in proportions that can and usually do vary.

model Theoretical, detailed description or analogy that helps people visualize something that has not yet been directly observed.

molar Tooth with cusps that help crush, grind, and shear food; one of the cheek teeth.

molecular clock Model used to calculate the time of origin of one lineage relative to others; assumes that the lineage accumulates neutral mutations at predictable rates, measurable as a series of ticks back through time.

molecule Two or more atoms of the same or different elements joined by chemical bonds.

mollusk Only invertebrate with mantle draped over a soft, fleshy body; most have an external or internal shell. Diverse body plans and sizes (e.g., gastropods, bivalves, and cephalopods).

molting Periodic shedding of worn-out or too-small body structures. Permits some animals to grow in size or renew parts (e.g., shells, exoskeletons, hairs, feathers, and horns).

Monera In earlier schemes, one prokaryotic kingdom for archaebacteria and eubacteria.

monocot (MON-oh-kot) Monocotyledon; a flowering plant with one cotyledon in seeds, floral parts usually in threes (or multiples of three), and often parallel-veined leaves.

monohybrid cross Intercross between two F_1 heterozygotes that are identical for one gene locus; they are the offspring of two parents that breed true for different forms of a trait.

monomer Small molecule used as a subunit of polymers, such as sugar monomers of starch.

monosaccharide (MON-oh-SAK-ah-ride) [Gk. *monos*, alone, single, + *sakcharon*, sugar] One of the simple carbohydrates (e.g., glucose).

monsoon Air circulation pattern that moves moisture-laden air arising from warm oceans to continents north or south of them.

morphogen Type of inducer molecule that diffuses through embryonic tissues, activating master genes in sequence and so contributing to mapping out the overall body plan.

morphogenesis (MORE-foe-JEN-ih-sis) [Gk. *morphe*, form, + *genesis*, origin] Programmed, orderly changes in body size, proportion, and shape of an animal embryo through which all specialized tissues and early organs form.

morphological convergence Macroevolutionary pattern. In response to similar environmental pressures over time, evolutionarily distant lineages evolve in similar ways and end up being alike in appearance, functions, or both.

morphological divergence Macroevolutionary pattern. Genetically diverging lineages slowly undergo change from the body form of their common ancestor.

mosaic tissue effect Patches of body tissues in which one or the other of two X chromosomes is being expressed; outcome of X chromosome inactivation in all female mammals.

moss Most common kind of bryophyte.

motor neuron Neuron that relays signals from brain or spinal cord to muscle or gland cells.

motor protein Type of protein (e.g., myosin) attached to microfilaments and microtubules; used in cell movements (e.g., contraction).

motor unit A motor neuron and all muscle cells that form junctions with its endings.

multicelled organism Organism composed of many cells with coordinated metabolic activity; most show extensive cell differentiation into tissues, organs, and organ systems.

multiple allele system Three or more slightly different molecular forms of a gene that occur among individuals of a population.

multiregional model Holds that *Homo sapiens* evolved from *H. erectus* groups that had spread through much of the world by about 1 mya and evolved into regionally distinctive "races."

muscle fatigue Decline in tension of a muscle kept in a state of tetanic contraction as a result of continuous, high-frequency stimulation.

muscle tension Mechanical force exerted by a contracting muscle; resists opposing forces (e.g., gravity or weight of an object being lifted).

muscle tissue Tissue with arrays of cells able to contract under stimulation, then passively lengthen and return to their resting position.

muscle twitch Sequence of muscle contraction and relaxation in response to a brief stimulus.

mushroom Reproductive part of club fungi.

mutagen (MEW-tuh-jen) Environmental agent that can alter DNA's molecular structure.

mutation [L. *mutatus*, a change, + -*ion*, act, result, or process] Heritable change in DNA's molecular structure. Original source of all new alleles and, ultimately, the diversity of life.

mutation rate Of a gene locus, the probability that a spontaneous mutation will occur during or between DNA replication cycles.

mutualism [L. *mutuus*, reciprocal] Symbiotic interaction that benefits both participants.

mycelium (my-SEE-lee-um), plural **mycelia** [Gk. *mykes*, fungus, mushroom, + *helos*, callus] Mesh of tiny, branching filaments (hyphae); the food-absorbing portion of most fungi.

mycorrhiza (MY-coe-RIZE-uh) "Fungus-root." A form of mutualism between fungal hyphae and young plant roots. The plant gives up some carbohydrates and the fungus gives up some of its absorbed mineral ions.

myofibril (MY-oh-FY-brill) One of many striated threadlike structures inside a muscle cell.

myoglobin A pigment with a higher oxygen-binding capacity than hemoglobin; abundant in cardiac and skeletal muscle cells that can contract slowly and resist fatiguing.

myosin (MY-uh-sin) Motor protein with key roles in cell movements.

NAD+ Nicotinamide adenine dinucleotide. A nucleotide coenzyme; abbreviated NADH when carrying electrons and H+ to a reaction site.

NADP+ Nicotinamide adenine dinucleotide phosphate. A phosphorylated nucleotide coenzyme; abbreviated NADPH$_2$ when it is carrying electrons and H+ to a reaction site.

natural killer cell Cytotoxic lymphocyte that touch-kills tumor cells and virus-infected cells.

natural selection Microevolutionary process; the outcome of differences in survival and reproduction among individuals that differ in details of heritable traits.

negative control Use of regulatory proteins to trigger a slowdown in gene expression.

negative feedback mechanism A homeostatic mechanism by which a condition that changed as a result of some activity triggers a response that reverses the change.

nematocyst (NEM-at-uh-sist) [Gk. *nema*, thread + *kystis*, pouch] Cnidarian capsule that has a dischargeable, tube-shaped thread, sometimes barbed; releases a toxin or sticky substance.

nephridium (neh-FRID-ee-um), plural **nephridia** Unit that controls composition and volume of fluid in some invertebrates (e.g., earthworms).

nephron (NEFF-ron) [Gk. *nephros*, kidney] One of the kidney's urine-forming tubules; it filters water and solutes from blood, then selectively reabsorbs adjusted amounts of both.

nerve Sheathed, cordlike bundle of the axons of sensory neurons, motor neurons, or both.

nerve cell One of the cells that receive and integrate signals from sensory receptors, and guide responses to stimuli. *See also* Neuron.

nerve cord A prominent longitudinal nerve; most animals have one, two, or three. Chordate nervous systems arise from a dorsal nerve cord.

nerve net Simple nervous system in epidermis of cnidarians and some other invertebrates; a diffuse mesh of simple, branching nerve cells interacts with contractile and sensory cells.

nervous system Integrative organ system with nerve cells interacting in signal-conducting and information-processing pathways. Detects and processes stimuli, and elicits responses from effectors (e.g., muscles and glands).

nervous tissue Connective tissue composed of neurons and often neuroglia.

net population growth rate per individual (*r*) For population growth equations, a variable combining rates and death rates; assumes that both remain constant in specified interval.

neural tube The embryonic and evolutionary forerunner of brain and spinal cord.

neuroglia (NUR-oh-GLEE-uh) Collectively, cells that structurally and metabolically support neurons. They make up about half the volume of nervous tissue in vertebrates.

neuromuscular junction Synapse between a motor neuron's axon endings and a muscle cell.

neuron (NUR-on) Type of nerve cell; basic communication unit in most nervous systems.

neurotransmitter Any of a class of signaling molecules secreted by neurons. It acts on cells next to it, then is rapidly degraded or recycled.

neutral mutation A mutation with little or no effect on phenotype, so natural selection can't change its frequency in a population.

neutron Unit of matter, one or more of which occupies the atomic nucleus and has mass but no electric charge.

neutrophil Phagocytic white blood cell that acts fast against bacteria.

niche (NITCH) [L. *nidas*, nest] Sum total of all activities and relationships in which individuals of a species engage as they secure and use the resources required to survive and reproduce.

nitrification (nye-trih-fih-KAY-shun) Major nitrogen cycle process. Certain bacteria in soil break down ammonia or ammonium to nitrite, then other bacteria break down the nitrite to nitrate (a form that plants can take up).

nitrogen cycle Atmospheric cycle. Nitrogen moves from its largest reservoir (atmosphere), through the ocean, ocean sediments, soils, and food webs, then back to the atmosphere.

nitrogen fixation Major nitrogen cycle process. Certain bacteria convert gaseous nitrogen to ammonia, which dissolves in their cytoplasm to form ammonium (used in biosynthesis).

node Stem site where one or more leaves form.

noncyclic pathway of ATP formation (non-SIK-lik) [L. *non*, not, + Gk. *kylos*, circle] Light-dependent reactions of photosynthesis, using photolysis, two photosystems, two transport systems. H$_2$O molecules are split. Released electrons and H atoms are used in ATP and NADPH formation; the oxygen, a by-product, is the basis of Earth's oxygen-rich atmosphere.

nondisjunction Failure of sister chromatids or a pair of homologous chromosomes to separate during meiosis or mitosis. Daughter cells end up with too many or too few chromosomes.

non-ionizing radiation Wavelengths that boost electrons to a higher energy level. DNA easily absorbs one form, ultraviolet light.

notochord (KNOW-toe-kord) Of chordates, a rod of stiffened tissue (not cartilage or bone) that is a supporting structure for the body.

nuclear envelope Outermost portion of the cell nucleus; consists of a double membrane (two lipid bilayers and associated proteins).

nucleic acid (new-CLAY-ik) Single- or double-stranded chain of four kinds of nucleotides joined at their phosphate groups. Nucleic acids (e.g., DNA, RNA) differ in base sequences.

nucleic acid hybridization Any base-pairing between DNA or RNA from different sources.

nucleoid (NEW-KLEE-oid) Portion of bacterial cell interior in which the DNA is physically organized but not enclosed by a membrane.

nucleolus (new-KLEE-oh-lus) [L. *nucleolus*, tiny kernel] In a nondividing cell nucleus, a site for assembling protein and RNA subunits that will later join up as ribosomes in the cytoplasm.

nucleosome (NEW-klee-oh-sohm) A stretch of eukaryotic DNA looped twice around a spool of histone molecules; one of many units that give condensed chromosomes their structure.

nucleotide (NEW-klee-oh-tide) Small organic compound with deoxyribose (a five-carbon sugar), a nitrogenous base, and a phosphate group. Monomer for adenosine phosphates, nucleotide coenzymes, and nucleic acids.

nucleus (NEW-klee-us) [L. *nucleus*, a kernel] Of atoms, a central core of one or more protons and (in all but hydrogen atoms) neutrons. In a eukaryotic cell, the organelle that physically separates DNA from cytoplasmic machinery.

numerical taxonomy Study of the degree of relatedness between an unidentified organism and a known group through comparisons of traits. Used to classify prokaryotic species.

nutrient Element with a direct or indirect role in metabolism that no other element fulfills.

nutrition Processes of selectively ingesting, digesting, absorbing, and converting food into the body's own organic compounds.

nymph Immature, post-embryonic stage of some insect life cycles.

obesity Excess of fat in adipose tissue; caloric intake has exceeded the body's energy output.

ocean A continuous body of water that covers more than 71 percent of the Earth; its currents distribute nutrients in marine ecosystems and influence regional climates.

oligosaccharide (oh-LIG-oh-SAC-uh-RID) Short-chain carbohydrate of two or more covalently bonded sugar monomers (e.g., disaccharides).

omnivore [L. *omnis*, all, + *vovare*, to devour] Animal that eats at more than one trophic level.

oncogene (ON-koe-jeen) Any gene having the potential to induce cancerous transformation.

oocyte Type of immature egg.

oogenesis (oo-oh-JEN-uh-sis) Process by which a germ cell develops into a mature oocyte.

operator Very short base sequence between a promoter and bacterial genes; a binding site for a repressor that can block transcription.

operon Promoter–operator sequence serving more than one bacterial gene; part of a control that adjusts transcription rates up or down.

orbital One of the volumes of space around the atomic nucleus in which one or at most two electrons are likely to be at any instant.

organ Body structure with definite form and function that consists of more than one tissue.

organ formation Developmental stage in which primary tissue layers give rise to cell lineages unique in structure and function. Descendants of those lineages give rise to all the different tissues and organs of the adult.

organ system Organs interacting chemically, physically, or both in a common task.

organelle (or-guh-NELL) Membrane-bound sac or compartment in the cytoplasm having one or more specialized metabolic functions. Most eukaryotic cells have a profusion of them.

organic compound Molecule of one or more elements covalently bonded to some number of carbon atoms.

osmoreceptor Sensory receptor; detects shifts in water volume (solute concentration).

osmosis (oss-MOE-sis) [Gk. *osmos*, pushing] In response to a water concentration gradient, the diffusion of water between two regions that a selectively permeable membrane separates.

osmotic pressure Pressure that operates after hydrostatic pressure develops in an enclosed region (e.g., a cell); it counters water's inward diffusion (stops further rises in fluid volume).

osteoblast Cell that forms and deposits bone, thus storing minerals.

osteoclast Cell that secretes the enzymes that degrade bone tissue to release minerals.

ostracoderm One of the earliest jawless fishes; a bottom-dwelling filter feeder. Extinct.

ovary (OH-vuh-ree) In most animals, a female gonad. In flowering plants, the enlarged base of one or more carpels.

oviduct (OH-vih-dukt) One of a pair of ducts through which eggs travel from an ovary to the uterus. Formerly called a fallopian tube.

ovulation (OHV-you-LAY-shun) Release of a secondary oocyte from an ovary.

ovule (OHV-youl) [L. *ovum*, egg] Tissue mass in a plant ovary that becomes a seed; a female gametophyte with egg cell, nutritious tissue, and a jacket that will become a seed coat.

ovum (OH-vum) Mature secondary oocyte.

oxaloacetate (ox-AL-oh-ASS-ih-tate) A four-carbon compound with roles in metabolism (e.g., the point of entry into the Krebs cycle).

oxidation–reduction reaction A transfer of an electron (and often an unbound proton, or H^+) between atoms or molecules.

oxygen debt Lowered O_2 level in blood when muscle cells have used up more ATP than they have formed by aerobic respiration.

oxyhemoglobin Oxygen bound to hemoglobin (HbO_2) inside red blood cells.

ozone thinning A pronounced and seasonal thinning of the atmosphere's ozone layer.

pain Perception of injury to some body region.

Paleozoic Era (570 to 240 mya) consisting of the Cambrian, Ordovician, Silurian, Devonian, Carboniferous, and Permian.

pancreas (PAN-cree-us) Gland with roles in digestion and organic metabolism. Secretes enzymes and bicarbonate into small intestine; also secretes insulin and glucagon.

pancreatic islet Any of the 2 million or so clusters of endocrine cells of the pancreas.

Pangea Paleozoic supercontinent upon which the first terrestrial plants and animals evolved.

parapatric speciation Idea that neighboring populations can become distinct species while maintaining contact along a common border.

parasite [Gk. *para*, alongside, + *sitos*, food] Organism that lives in or on a host for at least part of its life cycle. It feeds on specific tissues and usually does not kill its host outright.

parasitism Symbiotic interaction in which a species that feeds on its tissues (a parasite) benefits and the other (its host) is harmed.

parasympathetic nerve (PARE-uh-SIM-pu-THET-ik) An autonomic nerve. Its signals slow down overall activities and divert energy to basic tasks; they also help make small adustments in internal organ activity by acting continually in opposition to sympathetic nerve signals.

parathyroid gland One of four glands, the secretions of which raise blood calcium levels.

parenchyma (par-ENG-kih-mah) Simple tissue that makes up the bulk of a plant; has roles in photosynthesis, storage, secretion, other tasks.

partial pressure Characteristic of a gas; its contribution to total atmospheric pressure.

passive transport Event in which a transport protein that spans a cell membrane passively permits a solute to diffuse through its interior. Also called facilitated diffusion.

pathogen (PATH-oh-jen) [Gk. *pathos*, suffering, + *genēs*, origin] Any virus, bacterium, fungus, protistan, or parasitic worm that can infect an organism, multiply in it, and cause disease.

pattern formation The ordered, sequential sculpting of embryonic cells into specialized animal tissues and organs; brought about by cytoplasmic localization and, later, inductive interactions among master genes, the products of which map out the basic body plan.

PCR Polymerase chain reaction. A method of enormously amplifying the quantity of DNA fragments cut by restriction enzymes.

peat bog Compressed, soggy, highly acidic mat of accumulated remains of peat mosses.

pedigree Diagram of the genetic connections among related individuals through successive generations; uses standardized symbols.

penis Male copulatory organ by which sperm is deposited in a female reproductive tract.

peptide hormone A hormone that binds to a membrane receptor, thus activating enzyme systems that alter a cell activity. A second messenger in the cell often relays its message.

perennial [L. *per-*, throughout, + *annus*, year] Plant that lives three or more growing seasons.

pericycle (PARE-ih-sigh-kul) [Gk. *peri-*, around, + *kyklos*, circle] One or more cell layers inside the endodermis; gives rise to lateral roots and contributes to secondary growth.

periderm Protective cover that replaces plant epidermis during extensive secondary growth.

peripheral nervous system (per-IF-ur-uhl) [Gk. *peripherein*, to carry around] All nerves leading into and out from the spinal cord and brain. Includes ganglia of those nerves.

peristalsis (pare-ih-STAL-sis) Recurring waves of contraction and relaxation of muscles in the wall of a tubular or saclike organ.

permafrost A water-impermeable, perpetually frozen layer, up to 500 meters thick, beneath the soil surface of the arctic tundra.

peroxisome A vesicle; its enzymes digest fatty acids and amino acids to hydrogen peroxide, which is then converted to harmless products.

PGA Phosphoglycerate (FOSS-foe-GLISS-er-ate) Important intermediate of glycolysis and of the Calvin–Benson cycle.

PGAL Phosphoglyceraldehyde. Intermediate of glycolysis and of the Calvin-Benson cycle.

pH scale Measure of the concentration of free hydrogen ions (H^+) in blood, water, and other solutions. pH 0 is the most acidic, 14 the most basic, and 7, neutral.

phagocytosis (FAG-oh-sigh-TOE-sis) [Gk. *phagein*, to eat, + *kytos*, hollow vessel] Of some cells, engulfment of an extracellular target by way of pseudopod formation and endocytosis.

pharynx (FARE-inks) Among invertebrates, a muscular tube to the gut. In some chordates, a gas exchange organ. In land vertebrates, a dual entrance to the esophagus and trachea.

phenotype (FEE-no-type) [Gk. *phainein*, to show + *typos*, image] Observable trait or traits of an individual that arise from gene interactions and gene–environment interactions.

pheromone (FARE-oh-moan) [Gk. *phero*, to carry, + *-mone*, as in hormone] Hormone-like, nearly odorless exocrine gland secretion. A signaling molecule between individuals of the same species that integrates social behavior.

phloem (FLOW-um) Plant vascular tissue. Live cells (sieve tubes) interconnect as conducting tubes for sugars and other solutes; companion cells help load solutes into the tubes.

phospholipid Organic compound that has a glycerol backbone, two fatty acid tails, and a hydrophilic head of two polar groups (one being phosphate). Phospholipids are the main structural component of cell membranes.

phosphorus cycle A sedimentary cycle; the movement of phosphorus (mainly phosphate ions) from land, through food webs, to ocean sediments, then back to land.

phosphorylation (FOSS-for-ih-LAY-shun) Means of activating molecules. An enzyme attaches inorganic phosphate to a molecule or mediates a phosphate-group transfer, as from ATP.

photoautotroph Photosynthetic autotroph; any organism that synthesizes its own organic compounds using CO_2 for carbon atoms and sunlight for energy. Nearly all plants, some protistans, and a few bacteria do this.

photolysis (foe-TALL-ih-sis) [Gk. *photos*, light, +-*lysis*, breaking apart] Reaction sequence in which photon energy splits water molecules. The released electrons and hydrogen are used in the noncyclic pathway of photosynthesis; the oxygen is released as a by-product.

photon One unit of energy of visible light.

photoperiodism Biological response to change in relative lengths of daylight and darkness.

photoreceptor Light-sensitive sensory cell.

photosynthesis Trapping of sunlight energy, followed by its conversion to chemical energy (ATP, NADPH, or both) and then synthesis of sugar phosphates, which become converted into sucrose, cellulose, starch, and other end products. The main pathway by which energy and carbon enter the web of life.

photosystem Cluster of many light-trapping pigments in a photosynthetic membrane.

phototropism [Gk. *photos*, light, + *trope*, a turning, direction] Change in the direction of cell movement or growth in response to light (e.g., as in a stem bending toward light).

phycobilin (FIE-koe-BY-lin) Type of accessory pigment; notably abundant in red algae and in cyanobacteria.

phylogeny Evolutionary relationships among species, starting with an ancestral form and including branches leading to its descendants.

phytochrome A light-sensitive pigment. Its controlled activation and inactivation affect plant hormone activities that govern leaf expansion, stem branching, stem lengthening, and often seed germination and flowering.

phytoplankton (FIE-toe-PLANK-tun) [Gk. *phyton*, plant, + *planktos*, wandering] Aquatic community of floating or weakly swimming photoautotrophs (e.g., "pastures of the seas").

pigment Any light-absorbing molecule.

pineal gland Light-sensitive endocrine gland; its melatonin secretions affect biological clocks, overall activity, and reproductive cycles.

pioneer species Any opportunistic colonizer of barren or disturbed habitats. Adapted for rapid growth and dispersal.

pith Of most dicot stems, ground tissue inside the ring of vascular bundles.

pituitary gland Endocrine gland that, with the hypothalamus, controls many physiological functions, including activity of many other endocrine glands. Its posterior lobe stores and secretes hypothalamic hormones. Its anterior lobe produces and secretes its own hormones.

placenta (plah-SEN-tuh) Blood-engorged organ of pregnant female placental mammals; made of endometrial tissue and extraembryonic membranes. Permits exchanges between the mother and fetus without intermingling their bloodstreams; sustains the new individual and allows its blood vessels to develop separately.

placoderm One of the earliest fishes with jaws and paired fins; extinct.

placozoan The simplest known animal; tiny, asymmetrically soft-bodied, two cell layers.

plankton [Gk. *planktos*, wandering] Of aquatic habitats, a community of suspended or weakly swimming organisms, mostly microscopic.

plant Generally, a multicelled photoautotroph with well-developed root and shoot systems;

photosynthetic cells that include starch grains as well as chlorophylls *a* and *b*; and cellulose, pectin, and other polysaccharides in cell walls.

Plantae Kingdom of plants.

planula Larva, usually with ciliated epidermis. Swims or creeps; may resemble first animals.

plasma (PLAZ-muh) Liquid portion of blood; mainly water in which ions, proteins, sugars, gases, and other substances are dissolved.

plasma membrane Outermost cell membrane; structural and functional boundary between cytoplasm and the fluid outside the cell.

plasmid A small, circular molecule of extra bacterial DNA that carries a few genes and is replicated independently of the chromosome.

plasmodesma (PLAZ-moe-DEZ-muh), plural **plasmodesmata** Plant cell junction; a channel that connects the cytoplasm of adjoining cells.

plate tectonics Theory that great slabs (plates) of the Earth's outer layer float on a hot, plastic mantle. All plates are slowly moving and have rafted continents to new positions over time.

platelet (PLAYT-let) Megakaryocyte fragment that releases substances used in clot formation.

pleiotropy (PLEE-oh-troe-pee) [Gk. *pleon*, more, + *trope*, direction] Positive or negative effects on two or more traits owing to expression of alleles at a single gene locus. Effects may or may not emerge at the same time.

polar body One of four cells that has formed by the meiotic cell division of an oocyte but that does not become the ovum.

pollen grain [L. *pollen*, fine dust] Immature or mature, sperm-bearing male gametophyte of gymnosperms and angiosperms.

pollen sac Chamber inside a flower's anther in which pollen grains develop.

pollen tube Sperm-carrying tube that grows from a germinated pollen grain, through carpel tissues to the egg inside an ovule.

pollination Arrival of a pollen grain on the landing platform (stigma) of a flower's carpel.

pollinator Agent (wind, water, an animal) that puts pollen from male on female reproductive parts of flowers of the same plant species.

pollutant Natural or synthetic substance with which an ecosystem has no prior evolutionary experience, in terms of kinds or amounts; it accumulates to disruptive or harmful levels.

polymer (POH-lih-mur) [Gk. *polus*, many, + *meris*, part] Large molecule of three to millions of monomers of the same or different kinds.

polymerase (puh-LIM-ur-aze) Type of enzyme that catalyzes a polymerization reaction, as in DNA replication and repair.

polymorphism (poly-MORE-fizz-um) [Gk. *polus*, many, + *morphe*, form] The persistence of two or more qualitatively different forms of a trait (morphs) in a population.

polyp (POH-lip) Vase-shaped, sedentary stage of cnidarian life cycles.

polypeptide chain An organic compound of three or more amino acids joined by peptide bonds; atoms of its backbone have this pattern: —N—C—C—N—C—C—. All proteins are composed of one or more polypeptide chains.

polyploidy (POL-ee-PLOYD-ee) Having three or more of each type of chromosome in the nucleus of a eukaryotic cell at interphase.

polysaccharide [Gk. *polus*, many, + *sakcharon*, sugar] Straight or branched chain of many covalently linked sugar units of the same or

different kinds. In nature, the most common types are cellulose, starch, and glycogen.

polytene chromosome Insect chromosome consisting of many in-parallel copies of DNA.

pons A hindbrain traffic center for signals between cerebellum and forebrain centers.

population All individuals of the same species that are occupying a specified area.

population density Count of individuals of a population in a habitat (by area or volume).

population distribution Dispersal pattern for individuals of a population through a habitat.

population size The number of individuals that make up the gene pool of a population.

positive control Use of regulatory proteins to promote gene expression.

positive feedback mechanism A homeostatic control mechanism. It sets in motion a chain of events that intensifies change from an original condition; after a limited time, intensification reverses the change.

potential energy A stationary object's capacity to do work owing to its position in space or to the arrangement of its parts.

predation Ecological interaction in which a predator feeds on a prey organism.

predator [L. *prehendere*, to grasp, seize] A heterotroph that eats other living organisms (its prey), does not live in or on them (as parasites do), and may or may not kill them.

prediction Statement about what you should observe in nature if you were to go looking for a particular phenomenon; the if–then process.

pressure flow theory Oganic compounds flow through phloem in response to pressure and concentration and pressure gradients between sources (e.g., leaves) and sinks (e.g., growing parts where they are being used or stored).

pressure gradient A difference in pressure between two adjoining regions.

prey Organism that another organism (e.g., a predator) captures as a food source.

primary growth Plant growth originating at root tips and shoot tips.

primary immune response Defensive actions by white blood cells and their secretions, as elicited by first-time recognition of antigen. Includes antibody- and cell-mediated responses.

primary producer Type of autotroph that secures energy directly from the environment and stores some in its tissues.

primary productivity, gross Of ecosystems, the rate at which primary producers capture and store a given amount of energy in their cells and tissues during a specified interval.

primary productivity, net Of ecosystems, the rate of energy storage in primary producer cells and tissues in excess of rate of aerobic respiration during a specified interval.

primary root First root of a young seedling.

primary wall A thin, flexible plant cell wall of cellulose, polysaccharides, and glycoproteins; allows growing cells to divide or change shape.

primate Mammalian lineage dating from the Eocene; includes prosimians, tarsioids, and anthropoids (monkeys, apes, and humans).

primer Short nucleotide sequence designed to base-pair with any complementary DNA sequence; acts as a START tag for replication.

prion Small infectious protein that causes rare, fatal degenerative diseases of nervous system.

probability The chance that each outcome of a given event will occur is proportional to the number of ways the outcome can be reached.

probe Very short stretch of DNA labeled with a radioisotope; it is designed to base-pair with part of a gene in a DNA sample being studied.

procambium (pro-KAM-bee-um) A meristem that gives rise to primary vascular tissues.

producer Autotroph (self-feeder); it nourishes itself using sources of energy and carbon from the physical environment. Photoautotrophs and chemoautotrophs are examples.

progesterone (pro-JESS-tuh-rown) Of female mammals, a sex hormone secreted by ovaries and the corpus luteum.

proglottid One of many tapeworm body units that bud in sequence, behind the scolex.

prokaryotic cell (pro-CARE-EE-oh-tic) [L. *pro*, before, + Gk. *karyon*, kernel] Archaebacterium or eubacterium; single-celled organism, most often walled; lacks the profusion of membrane-bound organelles observed in eukaryotic cells.

prokaryotic fission Cell division mechanism by which prokaryotic cells alone reproduce.

promoter Short stretch of DNA to which RNA polymerase can bind and start transcription.

prophase Of mitosis, a stage when duplicated chromosomes start to condense, microtubules form a spindle, and the nuclear envelope starts to break up. Duplicated pairs of centrioles (if present) are moved to opposite spindle poles.

prophase I First stage of meiosis I. Each duplicated chromosome starts to condense. It pairs with its homologue; nonsister chromatids usually undergo crossing over and are attached to spindle. One pair of duplicated centrioles (if present) is moved to the opposite spindle pole.

prophase II First stage of meiosis II. Spindle microtubules attach to each chromosome and move them to spindle's equator. A centriole pair (if present) is already at each spindle pole.

protein Organic compound of one or more polypeptide chains folded and twisted into a globular or fibrous shape, overall.

Protista Kingdom of protistans. Chytrids; water molds; slime molds; protozoans; sporozoans; euglenoids; chrysophytes; dinoflagellates; and red, brown, and green algae are major groups.

protistan (pro-TISS-tun) [Gk. *protistos*, primal] Photoautotroph or heterotroph (or both) unlike bacteria; some like earliest eukaryotic cells. Has a nucleus, larger ribosomes, mitochondria, ER, Golgi bodies, chromosomes with numerous proteins, and cytoskeletal microtubules. Range in size from microscopic algae to giant kelps.

proto-cell Hypothetic cell-like stage between chemical evolution and the first living cell.

proton Positively charged particle; one or more in the nucleus of each atom. An unbound (free) proton is called a hydrogen ion (H^+).

protostome (PRO-toe-stome) [Gk. *proto*, first, + *stoma*, mouth] Lineage of coelomate, bilateral animals that includes mollusks, annelids, and arthropods. The first indentation to form in protostome embryos becomes the mouth.

protozoan Type of protistan that may be like the single-celled heterotrophs that gave rise to animals. Amoeboid, animal-like, and ciliated protozoans are major categories.

proximal tubule Nephron's tubular portion extending from Bowman's capsule.

pseudopod A dynamically extending lobe of cytoplasm used for motility or engulfment.

puberty Post-embryonic stage; gametes start to mature and secondary sexual traits emerge.

pulmonary circuit Vertebrate cardiovascular route; O_2-poor blood flows from the heart to lungs, gets oxygenated, flows back to heart.

punctuation model, speciation Idea that most morphological changes occur in a brief span when populations start to diverge; speciation is rapid, and the daughter species change little for the next 2–6 million years or so.

Punnett-square method Construction of a simple diagram as a way to predict probable outcomes of a genetic cross.

pupa, plural, **pupae** An immature, post-embryonic stage of many insect life cycles.

purine A nucleotide base with a double ring structure (e.g., adenine or guanine).

pyrimidine (pih-RIM-ih-deen) A nucleotide base with a single ring structure (e.g., cytosine or thymine).

pyruvate (PIE-roo-vate) Organic compound with a backbone of three carbon atoms. Two molecules form as end products of glycolysis.

r Variable in population growth equations that signifies net population growth rate. Birth and death rates are assumed to remain constant and are combined into this one variable.

radial symmetry Animal body plan having four or more roughly equivalent parts around a central axis (e.g., sea anemone).

radiation Radiant energy such as sunlight or a surface warmer than body surface temperature; exposure to it can lead to a heat gain in the animal body.

radioisotope Unstable atom (uneven number of protons and neutrons). It spontaneously emits particles and energy, and so decays into a different atom over a predictable time span.

radiometric dating Method of measuring the proportions of (1) a radioisotope in a mineral trapped long ago in newly formed rock and (2) a daughter isotope that formed from it by radioactive decay in the same rock. Used to assign absolute dates to fossil-containing rocks and to the geologic time scale.

rain shadow A reduction in rainfall on the leeward side of a high mountain range that results in arid or semiarid conditions.

reabsorption In kidneys, diffusion or active transport of water and reclaimable solutes from nephrons into peritubular capillaries; amounts are controlled by ADH and aldosterone. At a capillary bed, osmotic movement of interstitial fluid into a capillary when water concentration between plasma and interstitial fluid differ.

reaction center The only molecule (a special chlorophyll *a*) that can pass electrons out of a photosystem, to a nearby acceptor molecule.

rearrangement, molecular Conversion of one organic compound to another through changes in its internal bonds.

receptor, molecular Protein at which some extracellular substance (e.g., a hormone) docks.

receptor, sensory Sensory cell or specialized cell adjacent to it that can detect a stimulus.

recessive allele [L. *recedere*, to recede] Allele whose expression in heterozygotes is fully or partially masked by expression of its partner. Fully expressed in homozygous recessives.

reciprocal cross A paired cross. In the first cross, one parent displays the trait of interest. In the second, the other parent displays it.

recombinant DNA Any DNA molecule that incorporates one or more nonparental base sequences. Outcome of microbial gene transfer in nature or of recombinant DNA technology.

recombinant DNA technology Procedures by which DNA molecules from different species are isolated, cut up, spliced together, and then enormously amplified to useful quantities.

recombination Any enzyme-mediated reaction that inserts one DNA sequence into another. Generalized recombination uses any pair of homologous sequences between chromosomes as substrates, as occurs at crossing over. Site-specific recombination uses a short stretch of homology between bacterial and viral DNA. A different reaction inserts transposons at new, random sites in bacterial or eukaryotic genomes; no homology is required.

red alga Type of photoautotrophic protistan; most are multicelled and aquatic; an abundance of phycobilins masks their chlorophyll *a*.

red blood cell Erythrocyte. Cell that functions in the rapid transport of oxygen in blood.

red marrow Site of blood cell formation in the spongy tissue of many bones.

reflex [L. *reflectere*, to bend back] Stereotyped, simple movement in response to stimuli. In the simplest reflex arcs, it results from sensory neurons synapsing directly on motor neurons.

refractory period Brief interval following an action potential when a small patch of neural membrane is insensitive to stimulation.

regulatory protein Component of mechanisms that control transcription, translation, and gene products by interacting with DNA, RNA, new polypeptide chains, or proteins (e.g., enzymes).

releaser Hypothalamic hormone; enhances or slows secretions from target cells in anterior lobe of pituitary gland.

renal corpuscle Bowman's capsule plus the glomerular capillaries it cups around.

repressor Protein that binds with an operator on bacterial DNA to block transcription.

reproduction Any process by which a parental cell or organism produces offspring. Among eukaryotes, asexual modes (e.g., binary fission, budding, vegetative propagation) and sexual modes. Prokaryotes use prokaryotic fission only. Viruses cannot reproduce themselves; host organisms execute their replication cycle.

reproductive base The number of actually and potentially reproducing individuals in a population.

reproductive isolating mechanism Heritable feature of body form, functioning, or behavior that prevents interbreeding between two or more genetically divergent populations.

reproductive success Production of viable offspring by the individual.

reptile Carnivorous species belonging to the first vertebrate lineage to escape dependency on free water by way of internal fertilization, efficient kidneys, amniote eggs, and other adaptations. Examples are dinosaurs (extinct), crocodilians, snakes, lizards, and tuataras.

resource partitioning Of two or more species that compete for the same resource, a sharing of the resource in different ways or at different times, which permits them to coexist.

respiration [L. *respirare*, to breathe] For all animals, exchange of environmental oxygen with CO_2 from cells (e.g., by integumentary exchange or by way of a respiratory system).

respiratory cycle One inhalation, exhalation.

respiratory surface Thin, moist epithelium that functions in gas exchange.

respiratory system Type of organ system that exchanges gases between the body and the environment and contributes to homeostasis.

resting membrane potential Of a neuron and other excitable cells, a voltage difference across the plasma membrane that holds steady in the absence of outside stimulation.

restoration ecology Attempts to reestablish biodiversity in ecosystems severely altered by mining, agriculture, and other disturbances.

restriction enzyme One of a class of bacterial enzymes that can cut apart foreign DNA that infects a cell, as by viral attack. Important tool of recombinant DNA technology.

reticular formation A mesh of interneurons extending from the upper spinal cord, through the brain stem, and into the cerebral cortex; it is a low-level pathway of information flow.

retina Dense array of photoreceptors in an eye.

reverse transcription Synthesis of DNA on an RNA template by reverse transcriptase, a viral enzyme. Basis of RNA virus replication cycle and of cDNA synthesis in the laboratory.

Rh blood typing Method of characterizing red blood cells by a certain membrane surface protein. Rh⁺ cells have it; Rh⁻ cells do not.

rhizoid Simple rootlike absorptive structure of some fungi and nonvascular plants.

rhizome Short, branched, mainly horizontal underground stem of some plant species.

ribosomal RNA (rRNA) Type of RNA that combines with proteins to form ribosomes, on which polypeptide chains are assembled.

ribosome Two subunits of rRNA and proteins briefly joined together as a structure on which mRNA is translated into polypeptide chains.

riparian zone Narrow corridor of vegetation along a stream or river.

RNA Ribonucleic acid. Any of a class of single-stranded nucleic acids that function in transcribing and translating the genetic instructions encoded in DNA into proteins.

RNA polymerase Enzyme that catalyzes the assembly of RNA strands on DNA templates.

RNA world One model for prebiotic evolution in which RNA was the template for protein synthesis before the evolution of DNA.

rod cell Vertebrate photoreceptor sensitive to very dim light. Contributes to the coarse perception of movement across visual field.

root Plant part, typically belowground, that absorbs water and dissolved minerals, anchors aboveground parts, and often stores food.

root hair Thin extension of a young, specialized root epidermal. Increases root surface area for absorbing water and minerals.

root nodule Localized swelling on a root of certain legumes and other plants. Develops when nitrogen-fixing bacteria infect the plant, multiply, and become mutualists with it.

roundworm Cylindrical, bilateral, cephalized animal with a false coelom and a complete digestive system. Most cycle nutrients in communities; many are parasites.

rubisco RuBP carboxylase; an enzyme that catalyzes attachment of the carbon atom from CO_2 to RuBP and so starts the Calvin–Benson cycle of the light-independent reactions.

RuBP Ribulose bisphosphate; five-carbon organic compound; used for carbon fixation in the Calvin–Benson cycle, which regenerates it.

ruminant Hoofed, herbivorous mammal with multiple stomach chambers.

sac fungus Fungus that forms sexual spores in sac-shaped cells; e.g., truffles, morels.

salinization Salt buildup in soil through poor drainage, evaporation, and heavy irrigation.

saliva Glandular secretion that mixes with food and starts starch breakdown in mouth.

salt Compound that releases ions other than H^+ and OH^- in solution.

sampling error Use of a sample or subset of a population, an event, or some other aspect of nature for an experimental group that is not large enough to be representative of the whole.

saprobe Heterotroph that obtains energy and carbon from nonliving organic matter and so causes its decay (e.g., many fungal species).

sapwood Of older stems and roots, secondary growth in between the vascular cambium and heartwood; wet, usually pale, not as strong.

sarcomere (SAR-koe-meer) One of many basic units of contraction, defined by Z lines, that subdivide a muscle cell. Shortens in response to ATP-driven interactions between its parallel arrays of actin and myosin components.

sarcoplasmic reticulum (sar-koe-PLAZ-mik reh-TIK-you-lum) System of membrane-bound chambers; threads around muscle cells and takes up, stores, and releases the calcium ions necessary for contraction.

savanna Broad belt of warm grassland with a smattering of shrubs and trees.

scale Of fishes, small, bony plates that protect the body surface without weighing it down.

sclerenchyma (skler-ENG-kih-mah) Simple plant tissue that supports mature plant parts and commonly protects seeds. Most of its cells have thick, lignin-impregnated walls.

second law of thermodynamics A law of nature stating that the spontaneous direction of energy flow is from organized forms to less organized forms; with each conversion, some energy is randomly dispersed in a form (usually heat) not as useful for doing work.

second messenger Molecule within a cell that mediates a hormonal signal.

secondary sexual trait A trait associated with maleness or femaleness but with no direct role in reproduction (e.g., distribution of body hair and body fat). The primary sexual trait is the presence of male or female gonads.

secondary wall Of older plant cells no longer growing but in need of structural support, a wall on the inner surface of the primary wall. Contains lignin in older cells of woody plants.

secretion Release of a substance from cells or glands to the external environment, to the lumen of some organ, or to interstitial fluid.

sedimentary cycle Biogeochemical cycle. An element having no gaseous phase moves from land, through food webs, to the seafloor, then returns to land through long-term uplifting.

seed Mature ovule with an embryo sporophyte inside and integuments that form a seed coat.

seed bank A safe storage facility where genes of diverse plant lineages are being preserved.

segmentation Of animal body plans, a series of units that may or may not be similar to one another in appearance. Of tubular organs, an oscillating movement produced by rings of circular muscle in the tube wall.

segregation, theory of [L. *se-*, apart, + *grex*, herd] Mendelian theory. Sexually reproducing organisms inherit pairs of genes (on pairs of homologous chromosomes), the two genes of each pair are separated from each other at meiosis, and they end up in separate gametes.

selective gene expression Control of which gene products a cell makes or activates during a specified interval. Depends on the cell type, adjustments to changing chemical conditions, external signals, and built-in control systems.

selective permeability Of a cell membrane, a capacity to let some substances but not others cross at certain sites, at certain times owing to its bilayer structure and its transport proteins.

selfish behavior An individual protects or increases its own chance to produce offspring regardless of consequences to its social group.

semen (SEE-mun) The sperm-bearing fluid expelled from a penis during male orgasm.

semiconservative replication [Gk. *hēmi*, half, + L. *conservare*, to keep] Mechanism of DNA duplication. A DNA double helix unzips and a complementary strand is assembled on exposed bases of each strand. Each conserved strand and its new partner wind up together as a double helix; a half-old, half-new molecule.

seminiferous tubule One of the small coiled tubes in testes where sperm start forming.

senescence (sen-ESS-cents) [L. *senescere*, to grow old] Processes leading to the natural death of an organism or to parts of it (e.g., leaves).

sensation Conscious awareness of a stimulus.

sensory adaptation Decrease in response to a stimulus maintained at constant strength.

sensory neuron Type of neuron that detects a stimulus and relays information about it toward an integrating center (e.g., a brain).

sensory system The "front door" of a nervous system; it detects external and internal stimuli and relays information to integrating centers that issue commands for responses.

sex chromosome A chromosome with genes that affect sexual traits. Depending on the species, somatic cells have one or two sex chromosomes of the same or different type (e.g., in mammals, XX females, XY males).

sexual dimorphism Occurrence of female and male phenotypes among the individuals of a sexually reproducing species.

sexual reproduction Production of offspring by meiosis, gamete formation, and fertilization.

sexual selection A microevolutionary process; a type of selection that favors a trait giving an individual a competitive edge in attracting or keeping a mate (favors reproductive success).

shell model Model of electron distribution in which all orbitals available to electrons of atoms occupy a nested series of shells.

shifting cultivation Cutting and burning trees in a plot of land, followed by tilling ashes into soil. Once called slash-and-burn agriculture.

shoot system Aboveground plant parts (e.g., stems, leaves, and flowers).

sieve-tube member One of the cells that join together as phloem's sugar-conducting tubes.

sign stimulus Simple environmental cue that triggers a response to a stimulus, which the nervous system is prewired to recognize.

sink Any part of a plant in which cells are storing or using food (e.g., roots).

sister chromatid Of a duplicated chromosome, one of two DNA molecules (and associated proteins) attached at the centromere until they are separated from each other at mitosis or meiosis; each is then a separate chromosome.

six-kingdom classification scheme A recent phylogenetic scheme that groups all organisms into the kingdoms Eubacteria, Archaebacteria, Protista, Fungi, Plantae, and Animalia.

skeletal muscle Organ with many muscle cells bundled inside a sheath of connective tissue that extends past the muscle, as tendons.

skeletal muscle tissue Striated contractile tissue that is the functional partner of bone.

skin Of vertebrates, the integument plus structures derived from epidermal cells.

sliding filament model Model for how muscles contract, based on the ATP-driven interactions between myosin and actin filaments in each sarcomere of each muscle cell. The sarcomere shortens as actin filaments projecting from its two sides are made to slide toward its center.

small intestine Part of the vertebrate digestive system in which digestion is completed and most dietary nutrients are absorbed.

smog, industrial Polluted, gray-colored air in industrialized cities during cold, wet winters.

smog, photochemical Polluted, brown, smelly air that forms in warm weather in large cities that have many gas-burning vehicles.

smooth muscle tissue Nonstriated contractile tissue found in soft internal organs.

social behavior Diverse interactions among individuals of a species, which display, send, and respond to shared forms of communication that have genetic and learned components.

sodium–potassium pump Type of membrane transport protein that, when activated by ATP, selectively transports potassium ions across a membrane against its concentration gradient, and passively allows sodium ions to cross in the opposite direction.

softwood Wood with tracheids but no vessels or fibers. Weaker, less dense than hardwood.

soil Mixture of mineral particles of variable sizes and decomposing organic material; air and water occupy spaces between particles.

solute (SOL-yoot) [L. *solvere*, to loosen] Any substance dissolved in a solution.

solvent Any fluid (e.g., water) in which one or more substances are dissolved.

somatic cell (so-MAT-ik) [Gk. *somā*, body] Any body cell that is not a germ cell. (Germ cells are the forerunners of gametes.)

somatic nervous system Nerves leading from a central nervous system to skeletal muscles.

somatic sensation Perception of touch, pain, pressure, temperature, motion, or positional changes of body parts.

somite One of many paired segments in a vertebrate embryo that give rise to most bones, skeletal muscles of head and trunk, and dermis.

source Any plant part where photosynthetic cells are making organic compounds.

special sense Vision, hearing, olfaction, or another sensation that arises from a particular location, such as the eyes, ears, or nose.

speciation (spee-see-AY-shun) The formation of a daughter species from a population or subpopulation of a parent species by way of microevolutionary processes. Routes vary in their details and duration.

species (SPEE-sheez) [L. *species*, a kind] One kind of organism. Of sexually reproducing organisms, one or more natural populations in which individuals are interbreeding and are reproductively isolated from other such groups.

specific name A name that, combined with a genus name, designates one species alone.

sperm [Gk. *sperma*, seed] Mature male gamete.

spermatogenesis (sper-MAT-oh-JEN-ih-sis) Formation of mature sperm from a germ cell.

sphere of hydration A clustering of water molecules around the molecules or ions of a substance placed in water owing to positive and negative interactions among them.

sphincter A ring of smooth muscles that can alternatively contract and relax to close off and open a passageway to the body surface.

spinal cord Part of central nervous system, in a canal inside the vertebral column; site of direct reflex connections between sensory and motor neurons; has tracts to and from the brain.

spindle apparatus Dynamic, temporary array of microtubules that moves chromosomes in precise directions during mitosis or meiosis.

spleen A lymphoid organ that is a filtering station for blood, a reservoir of red blood cells, and a reservoir of macrophages.

sponge Animal with no body symmetry, and no tissues, and phagocytic collar cells in its body wall. Lineage dates from precambrian.

spore A reproductive or resting structure of one or a few cells, often walled or coated; used for resisting harsh conditions, dispersal, or both. May be nonsexual or sexual, as formed by way of meiosis. Some bacteria as well as sporozoans, fungi, and plants form spores.

sporophyte [Gk. *phyton*, plant] Vegetative body that produces spore-bearing structures; grows by mitotic cell divisions from a plant zygote.

sporozoan Type of parasitic protistan. Forms a motile infective stage inside specific host cells; some form cysts. One end of the body has a complex of structures used to penetrate hosts.

spring overturn Of many bodies of water (e.g., large lakes in temperate zones) in spring, a downward movement of oxygenated surface water and upward movement of nutrient-rich water from below; fans primary productivity.

S-shaped curve Type of diagrammatic curve that emerges when logistic population growth is plotted against time.

stabilizing selection Mode of natural selection by which intermediate phenotypes in the range of variation are favored and extremes at both ends are eliminated.

stamen (STAY-mun) A male reproductive part of a flower (e.g., anther with stalked filament).

start codon Base triplet in mRNA that serves as the START signal for translation.

statolith A gravity-sensing mechanism based on clusters of particles.

STD One of the sexually transmitted diseases.

stem cell Self-perpetuating, undifferentiated animal cell. Some of its daughter cells are self-perpetuating; others become specialized (e.g., erythrocytes from stem cells in bone marrow).

steroid hormone Lipid-soluble hormone made from cholesterol that acts on a target cell DNA.

sterol (STAIR-all) Lipid with a rigid backbone of four fused carbon rings (e.g., cholesterol). Sterols differ in the number, position, and type of their functional groups.

stigma Sticky or hairy surface tissue on upper part of a carpel (or fused carpels) that captures pollen grains and favors their germination.

stimulus [L. *stimulus*, goad] A specific form of energy (e.g., pressure, light, and heat) that activates a sensory receptor able to detect it.

stoma (STOW-muh), plural **stomata** [Gk. *stoma*, mouth] A gap between two guard cells in leaf or stem epidermis. Opens or closes to control CO_2 movement into a plant and H_2O and O_2 out of it. Stomata help plants conserve water.

stomach Muscular, stretchable sac that mixes and stores ingested food, helps break it apart mechanically and chemically, and controls its expulsion (e.g., into the small intestine).

stop codon Base triplet in mRNA; STOP signal in translation that blocks further additions of amino acids to a new polypeptide chain.

strain One of two organisms with differences that are too minor to classify it as a separate species (e.g., *Escherichia coli* strain 018:K1:H).

stratification Stacked layers of sedimentary rock, built up by gradual deposition of volcanic ash, silt, and other materials over time.

stream A flowing-water ecosystem that starts out as a freshwater spring or seep.

strip logging A way to minimize downside of deforestation; only a very narrow corridor that parallels contours of sloped land is cleared; the upper part is used as a road to haul away logs, then is reseeded from intact forest above it.

strobilus Of many nonflowering plants, a cluster of spore-producing structures.

stroma [Gk. *strōma*, bed] A semifluid matrix between the thylakoid membrane system and two outer membranes of a chloroplast; a zone where sucrose, starch, cellulose, and other end products of photosynthesis are assembled.

stromatolite Fossilized mats of shallow-water microbial communities, mainly cyanobacteria, from Archean to Precambrian. Cell secretions blocked UV radiation but trapped sediments, and new mats grew on old ones; some are half a mile thick and hundreds of miles across.

substrate Reactant or precursor for a specific enzyme-mediated metabolic reaction.

substrate-level phosphorylation The direct, enzyme-mediated transfer of a phosphate group from a substrate to a molecule, as when an intermediate of glycolysis is made to give up a phosphate group to ADP, forming ATP.

succession, primary (suk-SESH-un) [L. *succedere*, to follow after] Ecological pattern by which a community develops in orderly progression, from the time that pioneer species colonize a barren habitat to the climax community.

succession, secondary Ecological pattern by which a disturbed area of a community recovers and moves back toward the climax state.

surface-to-volume ratio Mathematical relation in which volume increases with the cube of the diameter, but surface area increases only with the square. The volume of a cell expanding in diameter only would increase faster than the surface area of the plasma membrane servicing it; a major constraint on cell size and shape.

survivorship curve Plot of age-specific survival of a group of individuals in the environment, from the time of birth until the last one dies.

swim bladder Adjustable flotation device that helps many fishes maintain neutral buoyancy in water; its volume changes as it exchanges gases with blood.

symbiosis (sim-by-OH-sis) [Gk. *sym*, together, + *bios*, life, mode of life] Individuals of one species live near, in, or on those of another species for at least part of life cycle (e.g., in commensalism, mutualism, and parasitism).

sympathetic nerve An autonomic nerve; deals mainly with increasing overall body activities at times of heightened awareness, excitement, or danger; works continually in opposition with parasympathetic nerves to make minor adjustments in internal organ activities.

sympatric speciation [Gk. *sym*, together, + *patria*, native land] A speciation event within the home range of an existing species, in the absence of a physical barrier. Such species may form instantaneously, as by polyploidy.

synaptic integration (sin-AP-tik) Moment-by-moment combining of all excitatory and inhibitory signals arriving at the trigger zone of a neuron or some other excitable cell.

syndrome A set of symptoms that may not individually be a telling clue but collectively characterize a genetic disorder or disease.

systemic circuit (sis-TEM-ik) Of vertebrates, a cardiovascular route in which oxygen-enriched blood flows from the heart through the rest of the body (where it gives up oxygen and takes up carbon dioxide), then back to the heart.

T cell T lymphocyte; a type of white blood cell vital to immune responses (e.g., helper T cells and cytotoxic T cells).

tandem repeat One of many short sequences of DNA, occurring one after the other, in a chromosome. Used in DNA fingerprinting.

taproot system A primary root together with all of its lateral branchings.

target cell Any cell with molecular receptors that can bind a particular signaling molecule.

taxonomy Field of biology that deals with identifying, naming, and classifying species.

TCR Antigen-binding receptor of T cells.

tectum Midbrain's roof. Coordinating center for most sensory inputs and initiating motor responses in fishes and amphibians. In most vertebrates (not mammals), a reflex center that relays sensory input to forebrain.

telophase (TEE-low-faze) Of meiosis I, a stage when one member of each pair of homologous chromosomes has arrived at a spindle pole. Of mitosis and of meiosis II, the stage when chromosomes decondense into threadlike structures and two daughter nuclei form.

temperature A measure of the kinetic energy of ions or molecules in a specified region.

temperature zone Globe-spanning bands of temperature defined by latitude (e.g., cool temperate, warm temperate, equatorial).

tendon A cord or strap of dense connective tissue that attaches a muscle to bone.

territory An area that an animal is defending against competitors for mates, food, water, living space, and other resources.

test A means to determine the accuracy of a prediction, as by conducting experimental or observational tests and by developing models. Scientific tests are conducted under controlled conditions in nature or the laboratory.

testcross Experimental cross to determine whether an individual of unknown genotype that shows dominance for a trait is either homozygous dominant or heterozygous.

testis, plural **testes** A type of primary male reproductive organ (gonad); it produces male gametes and sex hormones.

testosterone (tess-TOSS-tuh-rown) A type of sex hormone necessary for the development and functioning of the male reproductive system of vertebrates.

tetanus (TET-uh-nuss) Of a muscle, a large contraction; repeated stimulation of a motor unit causes muscle twitches to run together. In a disease by the same name, a toxin prevents muscles from being released from contraction.

thalamus (THAL-uh-muss) A forebrain region that is a coordinating center for sensory input and a relay station for signals to the cerebrum.

theory, scientific An explanation of the cause or causes of range of related phenomena. It has been rigorously tested but is still open to tests, revision, and tentative acceptance or rejection.

thermal inversion A layer of dense, cool air trapped beneath a layer of warm air; can hold air pollutants close to the ground.

thermoreceptor Sensory cell or specialized cell next to it that detects radiant energy (heat).

thigmotropism (thig-MOE-truh-pizm) [Gk. *thigm*, touch] Orientation of the direction of growth in response to physical contact with a solid object (e.g., a vine curls around a post).

threshold Of excitable cells (e.g., a neuron or muscle cell), the minimum amount of change in the resting membrane potential that causes an action potential.

thylakoid Internal portion of a chloroplast's membrane system, often folded into flattened sacs, that forms a single compartment. Light-trapping pigments, and enzymes used to form ATP, NADPH, or both, are embedded in it.

thymine A nitrogen-containing base; one of the nucleotides in DNA (not in RNA).

thymus gland Lymphoid organ; its hormones influence T cells, which form in bone marrow but migrate to and differentiate in the thymus.

thyroid gland Endocrine gland; its hormones influence overall growth, development, and metabolic rates of warm-blooded animals.

tight junction Cell junction where strands of fibrous proteins oriented in parallel with a tissue's free surface collectively block leaks between the adjoining cells.

tissue Of multicelled organisms, a group of cells and intercellular substances that function together in one or more specialized tasks.

tissue culture propagation Inducing a new organism to grow from a cell of a parent tissue.

tonicity (toe-NISS-ih-TEE) The relative solute concentrations of two fluids (e.g., cytoplasmic fluid relative to extracellular fluid).

topsoil Uppermost soil layer, the one most essential for plant growth.

total fertility rate (TFR) The average number of children being born to women of a specified population during their reproductive years.

touch-killing Mechanism by which cytotoxic T cells directly release perforins and toxins onto a target cell and cause its destruction.

toxin Normal metabolic product of a species that can hurt or kill a different species.

trace element Any element that represents less than 0.01 percent of body weight.

tracer Substance with an attached radioisotope that researchers can track after delivering it into a cell, body, ecosystem, or some other system. Its emissions are detected as it moves through a pathway or reaches a destination.

trachea (TRAY-kee-uh), plural **tracheae** An air-conducting tube of respiratory systems. Of land vertebrates, the windpipe through which air passes between the larynx and bronchi.

tracheal respiration Of certain invertebrates (e.g., insects), respiration by way of finely branching tubes that start at openings in the integument and dead-end in body tissues.

tracheid (TRAY-kid) One of two types of cells in xylem that conduct water and mineral ions.

tract A cordlike bundle of axons of sensory neurons, motor neurons, or both inside the brain or spinal cord. Comparable to a nerve.

transcription [L. *trans*, across, + *scribere*, to write] First stage of protein synthesis. An RNA strand is assembled on exposed bases of an unwound strand of a DNA double helix. The transcript has a complementary base sequence.

transfer RNA (tRNA) An RNA that binds with and delivers amino acids to a ribosome and that pairs with an mRNA codon during the translation stage of protein synthesis.

translation Stage of protein synthesis when an mRNA's base sequence becomes converted to the amino acid sequence of a new polypeptide chain by rRNA, tRNA, and mRNA interactions.

translocation Of cells, movement of a stretch of DNA to a new chromosomal location with no molecular loss. Of vascular plants, distribution of organic compounds by way of phloem.

transpiration Evaporative water loss from a plant's aboveground parts, leaves especially.

transport protein Type of membrane protein used in active or passive transport of water-soluble substances across the lipid bilayer of a cell membrane. The solutes move across the membrane through the protein's interior.

transposon A stretch of DNA that can move spontaneously and at random to a different location in a genome; may inactivate genes and cause changes in phenotype.

triglyceride (neutral fat) A type of lipid that has three fatty acid tails attached to a glycerol backbone. Triglycerides are the body's most abundant lipids and its richest energy source.

trophic level (TROE-fik) All organisms the same number of transfer steps away from the energy input into an ecosystem.

tropical rain forest A biome characterized by regular, heavy rainfall, an annual mean temperature of 25°C, humidity of 80+ percent, and stunning but threatened biodiversity.

true breeding lineage Of sexually reproducing species, a lineage in which only one version of a trait appears over the generations in all parents and their offspring.

tumor Tissue mass with cells dividing at an abnormally high rate. If benign, cells stay in the home tissue; if malignant, they metastasize.

tundra Biome with poor drainage, extremely low temperatures, short growing season, poor decomposition; arctic tundra is a treeless plain between the polar ice cap and belts of northern boreal forests; alpine tundra occurs at high elevations in mountains throughout the world.

tunicate A baglike, filter-feeding urochordate.

turgor pressure (TUR-gore) Type of internal fluid pressure; it acts against a cell wall when water moves into the cell by osmosis.

ultrafiltration Bulk flow of a small amount of protein-free plasma from a blood capillary when the outward-directed effect of blood pressure exceeds the inward-directed osmotic movement of interstitial fluid.

ultraplankton Photosynthetic bacteria, no more than 2μ wide, that may account for 70 percent of the ocean's primary productivity.

uniformity theory Theory that Earth's surface changes in slow, uniformly repetitive ways except for catastrophes that occur annually, such as large earthquakes and floods. Helped change Darwin's view of evolution. Has since been replaced by plate tectonics theory.

upwelling An upward movement of deep, nutrient-rich water along coasts; it replaces surface waters that move away from shore when the prevailing wind direction shifts.

uracil (YUR-uh-sill) Nitrogen-containing base of a nucleotide in RNA but not DNA. Like thymine, uracil can base-pair with adenine.

urea Waste product formed in the liver from ammonia (derived from protein breakdown) and CO_2; excreted in urine.

ureter One of a pair of tubes that conduct urine from the kidneys to the urinary bladder.

urethra Urine-conducting tube from urinary bladder to an opening at the body's surface.

urinary excretion Mechanism by which excess water and solutes are removed from the body through the urinary system.

urinary system Organ system that adjusts the volume and composition of blood, and thereby helps maintain extracellular fluid.

urine Fluid consisting of any excess water, wastes, and solutes; it forms in kidneys by filtration, reabsorption, and secretion.

uterus (YOU-tur-us) [L. *uterus*, womb] Of a female placental mammal, a muscular, pear-shaped organ in which embryos are contained and nurtured during pregnancy.

vaccination Immunization procedure against a specific pathogen.

vaccine An antigen-containing preparation designed to increase immunity to diseases.

vagina Of female mammalian reproductive system, organ that receives sperm, forms part of birth canal, and channels menstrual flow.

variable Of an experimental test, a specific aspect of an object or event that may differ over time and among individuals. A single variable is directly manipulated in an attempt to support or disprove a prediction.

vascular bundle Array of primary xylem and phloem in multistranded, sheathed cords that thread lengthwise in the ground tissue system.

vascular cambium A lateral meristem that increases stem or root diameter.

vascular cylinder Arrangement of vascular tissues as a central cylinder in roots.

vascular plant Plant with xylem, phloem, and usually well-developed roots, stems, and leaves.

vascular tissue system Xylem and phloem, the conducting tissues that distribute water and solutes through a vascular plant.

vasoconstriction The diameter of one or more blood vessels decreases in response to stimuli.

vasodilation The diameter of one or more blood vessels increases in response to stimuli.

vegetative growth A new plant grows from an extension or fragment of its parent.

vein Of a cardiovascular system, any of the large-diameter vessels that lead back to the heart. Of leaves, one of the vascular bundles that thread through photosynthetic tissues.

venule A small blood vessel that serves as a transitional conducting tube between a small-diameter capillary and a larger diameter vein.

vertebra, plural **vertebrae** One of a series of hard bones that serve as a backbone and that protect the spinal cord.

vertebrate Animal having a backbone.

vesicle (VESS-ih-kul) [L. *vesicula*, little bladder] One of various small, membrane-bound sacs in cytoplasm that function in the transport, storage, or digestion of substances.

vessel member Type of cell in xylem; dead at maturity, but its wall becomes part of a water-conducting pipeline (a vessel).

vestibular apparatus Organ of equilibrium.

vestigial (ves-TIDJ-ul) Applies to a small body part, tissue, or organ abnormally developed or degenerated and unable to function like its normal counterpart (e.g., vestigial wings of mutant fruit flies; "tail bones" of humans).

villus (VIL-us), plural **villi** Any of several fingerlike absorptive structures projecting from the free surface of an epithelium.

viroid Infectious particle of short, tightly folded strands or circles of RNA.

virus A noncellular infectious agent made of DNA or RNA, a protein coat and, in some, an outer lipid envelope; it can be replicated only after its genetic material enters a host cell and subverts the host's metabolic machinery.

viscera All soft organs inside an animal body (e.g., heart, lungs, and stomach).

vision Perception of visual stimuli. Light is focused on a dense layer of photoreceptors that sample details of a light stimulus, which are sent to the brain for image formation.

visual accommodation Adjustments in a lens position or shape that focus light on a retina.

visual field The part of the outside world that an animal sees.

visual signal An observable action or cue that functions as a communication signal.

vitamin Any of more than a dozen organic substances that an organism requires in small amounts for metabolism but that it generally cannot synthesize for itself.

vocal cord Of certain animals, one of the thickened, muscular folds of the larynx that help produce sound waves for vocalization.

warning coloration Of many toxic species and their mimics, avoidance signals (strong colors and patterns) that predators learn to recognize.

water potential Sum of two opposing forces (osmosis and turgor pressure) that can cause a directional movement of water into or out of an enclosed volume (e.g., a walled cell).

water table Upper limit at which ground in a specified region is fully saturated with water.

water–vascular system Of sea stars and sea urchins, a system of many tube feet that are deployed in synchrony for smooth locomotion.

watershed Any specified region in which all precipitation drains into one stream or river.

wavelength A wavelike form of energy in motion. The horizontal distance between the crests of every two successive waves.

wax Organic compound; long-chain fatty acids packed together and attached to long-chain alcohols or carbon rings. Waxes have a firm consistency and repel water.

white blood cell A leukocyte (e.g., eosinophil, neutrophil, macrophage, T or B cell) that, with chemical mediators, counters tissue invasion and tissue damage. Different kinds take part in nonspecific and specific defense responses.

white matter Part of brain and spinal cord; its axons, in glistening white sheaths, specialize in rapid signal transmission.

wild-type allele Allele that occurs normally or with the greatest frequency at a given gene locus among individuals of a population.

wing A body part that functions in flight, as among birds, bats, and many insects.

X chromosome A type of sex chromosome. An XX mammalian embryo becomes female; an XY pairing causes it to develop into a male.

X chromosome inactivation One of the two X chromosomes in somatic cells of mammalian females condenses, which inactivates most of its genes. A dosage compensation mechanism.

X-linked gene A gene on an X chromosome.

X-linked recessive inheritance Recessive condition in which the responsible, mutated gene is on the X chromosome.

xylem (ZYE-lum) [Gk. *xylon*, wood] Of vascular plants, a complex tissue that conducts water and solutes through pipelines of interconnected walls of cells, which are dead at maturity.

Y chromosome Distinctive chromosome in males or females of many species, but not both (e.g., human males XY; human females, XX).

yellow marrow A fatty tissue in the cavities of most mature bones that produces red blood cells when blood loss from the body is severe.

Y-linked gene Gene on a Y chromosome.

yolk Protein- and lipid-rich substance that nourishes embryos in animal eggs.

yolk sac Extraembryonic membrane. In most shelled eggs, it holds nutritive yolk; in humans part becomes a site of blood cell formation and some cells give rise to forerunners of gametes.

zero population growth Of a population, no overall increase or decrease during a specified interval; population size is stabilized.

zooplankton A community of suspended or weakly swimming heterotrophs of aquatic habitats. Most species are microscopic.

zygomycetes Parasitic or saprobic fungus that forms a thick-walled sexual spore (a diploid zygote) in a thin cover (the zygosporangium).

zygote (ZYE-goat) First cell of a new individual, formed by fusion of a sperm nucleus with egg nucleus at fertilization; a fertilized egg.

CREDITS AND ACKNOWLEDGMENTS

This page is an extension of the copyright page.

We thank the following authors, agents, and publishers for permission to use the material listed. We also thank graphics artists for the final rendering of original illustrations developed by Cecie Starr and, more recently, Lisa Starr, in collaboration with general advisers, researchers, and content reviewers.

TABLE OF CONTENTS: iv (left to right) Derrell Fowler; Gary Head; model by Dr. David B. Goodin, The Scripps Research Institute / v Larry West/FPG / vi Dr. Pascale Madaule, France / vii (left) From "Multicolor Spectral Karyotyping of Human Chromosomes," by E. Schrock, T. Ried, et al, *Science*, 26 July 1966, 273:495; (right) William E. Ferguson / viii (left) Christopher Ralling; (right) Painting, Charles Knight (No. 2425), Courtesy Department of Library Services/American Museum of Natural History / ix (left to right) K. G. Murty/Visuals Unlimited; David M. Phillips/Visuals Unlimited; © Sherry K. Pittam; Edward S. Ross / x (left to right) Courtesy of Department of Library Services/American Museum of Natural History (Neg. K10273); P. J. Bryant, University of California, Irvine/BPS; Jean Paul Tibbles; Gary Head / xi David Cavagnaro/Peter Arnold; David M. Phillips/Visuals Unlimited; John Lotter Gurling/Tom Stack & Associates; David Macdonald / xii (left to right) Eric A. Newman; Lennart Nilsson © Boehringer Ingelheim International GmBH; Kenneth Garrett/National Geographic Image Collection; Prof. P. Motta/Dept. of Anatomy/University "La Sapienza," Rome/Science Photo Library/Photo Researchers / xiii (left to right) Courtesy The New York Academy of Medicine Library; Lisa Starr with photograph © 2001 PhotoDisc, Inc.; NSIBC/SPL/Photo Researchers; Dr. Douglas Coleman, The Jackson Library / xiv (left to right) Gunter Ziesler/Bruce Coleman; Ed Reschke; John H. Gerard; From Lennart Nilsson, *A Child Is Born*, © 1966, 1977 Dell Publishing Company, Inc., / xv (left to right) Nigel Cook/Dayton Beach News Journal/Corbis Sygma; Pieter Johnson; Stanley Sessions/Hartwick College; Alan and Sandy Carey / xvi Frans Lanting/Minden Pictures

INTRODUCTION NASA Goddard Space Flight Center

CHAPTER 1 1.1 John McColgan, Bureau of Land Management, Alaska Fire Service / 1.2 Lisa Starr / 1.3 Photograph, (all) Jack de Coningh / 1.4 © Y. Arthrus-Bertrand/Peter Arnold / 1.5 Art, Gary Head; Photographs, (left) Paul DeGreve/FPG; (right) Norman Meyers/Bruce Coleman / 1.6 Art, Lisa Starr; Photographs, (far left) Walt Anderson/Visuals Unlimited; (b) Gregory Dimijian/Photo Researchers; (c) Alan Weaving/Ardea, London / 1.7 (a) R. Robinson/Visuals Unlimited; (b) Tony Brain/SPL/Photo Researchers; (c) M. Abbey/Visuals Unlimited; (d) Edward S. Ross; (e) Gary Head; (f) Edward S. Ross; (g, h) Pat and Tom Leeson/Photo Researchers / 1.8 J. A. Bishop and L. M. Cook / 1.9 Courtesy Derrell Fowler, Tecumseh, OK / 1.10 Art, Gary Head / 1.11 Photographs, (all) Gary Head / 1.12 CNRI/SPL/Photo Researchers / 1.13 Art, Gary Head and Preface, Inc.; Photographs, Dr. Douglas Coleman, The Jackson Laboratory / 1.14 James Carmichael Jr./NHPA / page 19 © Cabisco/Visuals Unlimited

CHAPTER 2 2.1 Gary Head for Norman Terry, University of California, Berkeley / 2.2 Jack Carey / 2.3 Lisa Starr / 2.4 (Thyroid scans) Hank Morgan/Rainbow, computer-enhanced by Lisa Starr; Photograph, Gary Head / 2.5 (a) © John Griffin/MediChrome; (b, c) Art, Raychel Ciemma; (d) Dr. Harry T. Chugani, M.D., UCLA School of Medicine / 2.6 Micrograph, Maris and Cramer / 2.7, 2.8 Lisa Starr / 2.9 Lisa Starr and Gary Head / 2.10 (a, b) Art, Lisa Starr; Micrograph, © Bruce Iverson / page 26 Lisa Starr / 2.11 Vandystadt/Photo Researchers / 2.12 Lisa Starr / 2.13 Photograph, Steve Lissau/Rainbow; Art, Raychel Ciemma / 2.14 © Kennan Ward/Corbis Stock Market; Art, Raychel Ciemma / 2.15 (a) H. Eisenbeiss/Frank Lane Picture Agency; (b) Lisa Starr / 2.16 Lisa Starr / 2.17 Lisa Starr, using photographs © 2000 Photo Disc, Inc. / 2.18 Michael Grecco/Picture Group

CHAPTER 3 3.1 (left) NASA; (right) Dave Schiefelbein / page 35 © Ron Sanford/Michael Agliolo/Photo Researchers / 3.2 Art, Gary Head / page 36 Art, Precision Graphics and Lisa Starr / 3.3 Art, Lisa Starr; Photograph, Tim Davis/Photo Researchers / 3.4, 3.5 Lisa Starr / 3.6 Art and Photograph, Lisa Starr / 3.7 (a, b) Lisa Starr; (c) Photograph, David Scharf/Peter Arnold / 3.8 Art, Precision Graphics / 3.9 Clem Haagner/Ardea, London / 3.10 Art, Precision Graphics; (c) Micrograph,

Lewis L. Lainey; (d) Larry Lefever/Grant Heilman Photography; (e) Kenneth Lorenzen / page 41 Lisa Starr / 3.11 (a) Ron Davis/Shooting Star; (b) Frank Trapper/Corbis Sygma / 3.12–3.15 Art, Precision Graphics / 3.16, 3.17 Lisa Starr / 3.18 (a) CNRI/SPL/Photo Researchers; Art, (b) Robert Demarest; (c) Lisa Starr / page 46 Art, Gary Head / 3.19 Art, Precision Graphics / 3.20 Lisa Starr / 3.21 Photographs, (a) © Inga Spence/Tom Stack & Associates; (inset) Gary Head; Art, (b–d) Gary Head and Lisa Starr

CHAPTER 4 4.1 (a) © Bettmann/Corbis; (b) Armed Forces Institute of Pathology; (c) National Library of Medicine; (e) Francis A. Countway Library of Medicine / 4.2 (a) Manfred Kage/Bruce Coleman; (b) George J. Wilder/ Visuals Unlimited / 4.3–4.5 Art, Raychel Ciemma / page 54 Leica Microsystems, Inc., Deerfield, IL. / 4.6 Art, Raychel Ciemma; Micrograph, Driscoll, Youngquist, and Baldeschwieler/CalTech/Science Source/Photo Researchers / 4.7 (a–d) Jeremy Pickett–Heaps, School of Botany, University of Melbourne / 4.8 Art, Raychel Ciemma and Precision Graphics / 4.9 Art, Raychel Ciemma and Preface, Inc.; Micrograph, M. C. Ledbetter, Brookhaven National Laboratory / 4.10 Art, Raychel Ciemma and Preface, Inc.; Micrograph G. L. Decker / 4.11 Micrographs, Stephen Wolfe / 4.12 (a-left) Don W. Fawcett/Visuals Unlimited; (a-right) A. C. Faberge, *Cell and Tissue Research*, 151: 403–415, 1974; (b) Art, Lisa Starr / 4.13 Art, Raychel Ciemma and Precision Graphics / 4.14 Art, Raychel Ciemma; Micrographs (a, b) Don W. Fawcett/Visuals Unlimited / 4.15 Art, (left) Raychel Ciemma; (right) Robert Demarest after a model by J. Kephart; Micrograph, Gary Grimes / 4.16 Micrograph, Keith R. Porter; Art, Raychel Ciemma / 4.17 Art, (above) Lisa Starr; Micrograph, L. K. Shumway / 4.18 Courtesy Dr. Vincenzo Cirulli, Laboratory of Developmental Biology, The Whittier Institute for Diabetes, University of California-San Diego, La Jolla, California / 4.19 Lisa Starr / 4.20 Photographs (a) © Lennart Nilsson; Art, Precision Graphics after Stephen Wolfe, *Molecular and Cellular Biology*, Wadsworth, 1993 / 4.21 Ronald Hoham, Dept. of Biology, Colgate University / 4.22 Photographs, (a) Walter Hodge/Peter Arnold; (c) courtesy Burlington Mills; Micrograph, Biophoto Associates/ Photo Researchers; Art, Raychel Ciemma / 4.23 Microscopy, (a) George S. Ellmore; (b) Ed Reschke / 4.24 Art, Raychel Ciemma and Lisa Starr / 4.25 (a) Art, Lisa Starr; (b) Courtesy Dr. G. Cohen–Bazire; (c) K. G. Murti/Visuals Unlimited; (d) R. Calentine / Visuals Unlimited; (e) Gary Gaard, Arthur Kelman / 4.28 © Abraham Menashe

CHAPTER 5 5.1 (a, b) Models by Dr. David B. Goodin, The Scripps Research Institute / 5.2 Photographs, Gary Head; Art, Raychel Ciemma / 5.3 Evan Cerasoli / 5.4 Photographs, (above) NASA; (below) Manfred Kage/Peter Arnold / 5.5 (a–d) Lisa Starr, using photographs © 1997, 1998 Jet Propulsion Laboratory/California Institute of Technology and NASA, 1972 / 5.6 Art, Raychel Ciemma and Nadine Sokol / 5.7 Lisa Starr / 5.8 Art, Raychel Ciemma / 5.9 Lisa Starr / 5.10 Art, Gary Head / 5.11 Art, Precision Graphics / 5.12 Lisa Starr, using photographs © 1997, 1998 Jet Propulsion Laboratory, California Institute of Technology and NASA, 1972 / 5.13 (a, b) Thomas A. Steitz; (c) Lisa Starr / 5.14, 5.15 Lisa Starr / 5.16 Art, Precision Graphics; (b) © Douglas Faulkner/Sally Faulkner Collection / 5.17, 5.18, 5.19 Lisa Starr / 5.20 (a, b) Art, Raychel Ciemma; (c) M. Abbey/Visuals Unlimited / 5.21 Art, Precision Graphics / 5.22 Art, Raychel Ciemma; Micrographs, M. Sheetz, R. Painter, and S. Singer, *Journal of Cell Biology*, 70:193 (1976) by permission of The Rockefeller University Press / 5.23 Lisa Starr / 5.24, 5.25 Lisa Starr / 5.26 Frieder Sauer/Bruce Coleman

CHAPTER 6 6.1 Art, Raychel Ciemma; Micrograph, Carolina Biological Supply Company / 6.2 Art, Lisa Starr; Photograph, Wernher Krutein/PhotoVault / 6.3 Lisa Starr / pages 93, 94, 95 Lisa Starr / 6.4 Art, Gary Head and Lisa Starr with photograph from David Neal Parks / 6.5 (a) © Barker-Blakenship/FPG; (b) Art, Precision Graphics / 6.6 Lisa Starr after Stephen L. Wolfe, *Molecular and Cellular Biology*, Wadsworth / 6.7 Larry West/FPG / 6.8–6.11 Lisa Starr / 6.12 E. R. Degginger / 6.13, 6.14 Lisa Starr / 6.15 Art, Lisa Starr; Photographs, (a) © Bill Beatty/Visuals Unlimited; (b) © 2001 PhotoDisc, Inc.; (c, both) © 2001 PhotoDisc, Inc.; Micrographs, (above) © Bruce Iverson, computer-enhanced by Lisa Starr; (below) © Ken Wagner/Visuals Unlimited, computer-enhanced by Lisa

Starr / page 103 Steve Chamberlain, Syracuse University / 6.16 NASA / 6.17–6.19 Lisa Starr / 6.20 (a) Herve Chaumeton/Agence Nature; (b) Douglas Faulkner/Sally Faulkner Collection

CHAPTER 7 7.1 Stephen Dalton/Photo Researchers, computer-enhanced by Lisa Starr / 7.2 Art, Gary Head and Raychel Ciemma / 7.3 Photographs, (a-left) Gary Head; (a-right) Ed Reschke; (b) Paolo Fioratti; Art, (c) Lisa Starr / page 112 Lisa Starr / 7.4 (left) Art, Raychel Ciemma and Lisa Starr; (right) Lisa Starr after Ralph Taggart / 7.5 Art, (a) Raychel Ciemma; (b, c) Lisa Starr; Micrograph, Keith R. Porter / 7.6 Photograph, © 2001 PhotoDisc, Inc.; Art on photo, Raychel Ciemma; Art, Lisa Starr / 7.7, 7.8, 7.9 Lisa Starr / 7.10 (a) Lisa Starr; Photographs, (b) Adrian Warren/Ardea, London; (c) David M. Phillips/Visuals Unlimited / 7.11 William Grenfell/Visuals Unlimited / 7.12 Art, Lisa Starr; Photograph, Gary Head / page 122 (left) Art, Gary Head; (right) Photograph, R. Llewellyn/SuperStock / page 123 Lisa Starr / page 125 © Francis Leroy, Biocosmos/Science Photo Library/Photo Researchers

CHAPTER 8 8.1 (left and right, center) Chris Huss; (right, above and below) Tony Dawson / 8.2 Art, Gary Head / 8.3 Art, Lisa Starr and Raychel Ciemma; Micrographs (a) C. J. Harrison et al, *Cytogenetics and Cell Genetics*, 35: 21–27 © 1983 S. Karger, A. G. Basel; (c) B. Hamkalo; (d) O. L. Miller, Jr., Steve L. McKnight / 8.4 Art, Raychel Ciemma and Gary Head / 8.5 Micrographs, Andrew S. Bajer, University of Oregon / 8.6 Art, Gary Head / 8.7 Art, Raychel Ciemma and Gary Head; Photographs, © Ed Reschke / 8.8 Art, Lisa Starr; Micrograph, © R. Calentine/Visuals Unlimited / 8.9 Art, Raychel Ciemma; Micrograph, © D. M. Phillips/Visuals Unlimited / 8.10 (a–c, e) Lennart Nilsson from *A Child Is Born* © 1966, 1967 Dell Publishing Company, Inc.; (d) Lennart Nilsson from *Behold Man*, © 1974 by Albert Bonniers Förlag and Little, Brown & Company, Boston / 8.11 Photograph, (above) Courtesy of The Family of Henrietta Lacks; Micrograph, Dr. Pascal Madaule, France

CHAPTER 9 9.1 (a) Jane Burton/Bruce Coleman; (b) Dan Kline/Visuals Unlimited / 9.2 Art, Raychel Ciemma / 9.3 CNRI/SPL/Photo Researchers / 9.4 Art, Raychel Ciemma; Micrographs, with thanks to the John Innes Foundation Trustees, computer-enhanced by Gary Head / 9.5, 9.6 Art, Raychel Ciemma / 9.7 Art, Preface, Inc.; Photograph, (fly) David Maitland/Seaphot Limited / 9.8 Lisa Starr / 9.9 Art, Lisa Starr; Micrograph, David M. Phillips/Visuals Unlimited / 9.10 Art, Raychel Ciemma / 9.11 Art, Precision Graphics / 9.12 © Richard Corman/Corbis Outline

CHAPTER 10 10.1 (left) ZumaPress; (center) Focus on Sports; (right-above) The Moravian Museum, Brno; (right-below) Fabian/Corbis Sygma / 10.2 Art, Jennifer Wardrip; Photograph, Jean M. Labat/Ardea, London / 10.3 Lisa Starr / 10.4 Art, Precision Graphics / 10.5 Art, Raychel Ciemma and Precision Graphics / 10.6 Art, Precision Graphics / 10.7, 10.8, 10.9, page 158 Art, Raychel Ciemma and Precision Graphics / 10.10 Photographs, William E. Ferguson / 10.11 Preface, Inc. / 10.12 Art, Preface, Inc.; Micrographs, Stanley Flegler/Visuals Unlimited / 10.13 Art, Preface, Inc.; Photographs, (a, b) Michael Stuckey/Comstock, Inc., (c) Bosco Broyer, photograph by Gary Head / 10.14 David Hosking / 10.15 Ted Somes / 10.16 (above to below) Frank Cezus/FPG; Frank Cezus/FPG; © 2001 PhotoDisc, Inc.; Ted Beaudin/FPG; Stan Sholik/FPG / 10.17 (a) Art, Precision Graphics; Photograph, Daniel Fairbanks, Brigham Young University / 10.18 Art, D. & V. Hennings; Photograph, Jane Burton/Bruce Coleman / 10.19 (above) William E. Ferguson; (below) Eric Crichton/Bruce Coleman / 10.20 Photograph, © Pamela Harper/Harper Horticultural Slide Library; Art, Lisa Starr from Prof. Otto Wilhelm Thomé, Flora von Deutschland Österreich und der Schweiz. 1885, Gera, Germany / 10.21 Evan Cerasoli / 10.22 Leslie Faltheisek/Clacritter Manx / 10.23 Joe McDonald/ Visuals Unlimited

CHAPTER 11 11.1 From "Multicolor Spectral Karyotyping of Human Chromosomes," by E. Schrock, T. Ried, et al, *Science*, 26, July 1966, 273:495. Used by permission of E. Schrock and T. Reid and the American Association for the Advancement of Science, computer-enhanced by Lisa Starr / 11.2 Paul A. Thiessen / 11.3 Art, Raychel Ciemma and Preface, Inc. / 11.4 Art, Raychel Ciemma; Photograph, Charles D. Winters/Photo Researchers; Micrograph,

Omikron/Photo Researchers / **11.5** Art, Precision Graphics and Gary Head; Photographs (left) © 2001 EyeWire; (right) © 2001 PhotoDisc, Inc. / **11.6** (a) Photograph, From Lennart Nilsson, *A Child Is Born*, © 1966, 1977 Dell Publishing Company, Inc.; (b) Robert Demarest after Patten, Carlson & others; (c) Robert Demarest with permission from M. Cummings, *Human Heredity: Principles and Issues*, p. 126 of third Ed. © 1994 by Brooks/Cole. All rights reserved. / **11.7** Art, Raychel Ciemma and Preface Inc.; Photograph, Carolina Biological Supply Company / **11.8** Art, Raychel Ciemma / **11.9** Art, (a, b) Precision Graphics; Photograph, (b) Dr. Victor A. McKusick; (c) Steve Uzzell / **page 180** Art, Precision Graphics / **11.10** Lisa Starr / **11.11** (a) Lisa Starr; (b) Painting, Giraudon/Art Resource, New York / **11.12** (a) Lisa Starr; Photograph, (b) © Bettmann/Corbis; Art, (b) After V. A. McKusick, *Human Genetics*, 2nd Ed., © 1969. Reprinted by permission of Prentice-Hall, Inc., Englewood Cliffs, NJ / **11.13** C. J. Harrison / **11.14** Eddie Adams/AP Photo / **11.15** (a, b) Courtesy G. H. Valentine / **11.16** From "Multicolor Spectral Karyotyping of Human Chromosomes," by E. Schrock, T. Ried, et al, *Science*, 26, July 1966, 273:496. Used by permission of E. Schrock and T. Reid and the American Association for the Advancement of Science / **11.17** Art, Raychel Ciemma / **11.18** (a) Micrograph, Cytogenetics Laboratory, The University of California, San Francisco; Photograph, (a) courtesy of Peninsula Association for Retarded Children and Adults, San Mateo Special Olympics, Burlingame, CA; (b) after Collman and Stoller, *American Journal of Public Health*, 52, 1962; Photograph, (b) by permission of Carole Lafrate / **11.19** Art, Raychel Ciemma / **11.20** Art, Lisa Starr; Photographs, courtesy of Lennart Nilsson from *A Child Is Born*, © 1966, 1977 Dell Publishing Company, Inc. / **11.21** Photograph, (a) Fran Heyl Associates © Jacques Cohen, computer-enhanced by © Pix Elation; Art, (b) Raychel Ciemma / **11.22** Carolina Biological Supply Company / **page 189** © Mitchell Gerber/Corbis

CHAPTER 12 **12.1** A. C. Barrington Brown, © 1968 J. D. Watson / **12.2** Model by Dr. David B. Goodin, The Scripps Research Institute / **12.3** Art, Raychel Ciemma / **12.4** Art, (a, b) Raychel Ciemma; Photograph, (c) Lee D. Simon/Science Source/Photo Researchers / **12.5** Art, Raychel Ciemma / **12.6** Micrograph, Biophoto Associates/SPL/Photo Researchers; Art, Precision Graphics / **12.7** Lisa Starr, incorporating model by Dr. David B. Goodin, The Scripps Research Institute / **12.8** Courtesy of the Cold Springs Harbor Laboratory Archives / **12.9, 12.10** Art, Precision Graphics / **12.11** Photographs, (a) From Daniel Fairbanks; (b) PA News Photo Library

CHAPTER 13 **13.1** (above) Dennis Hallinan/FPG; (below) © Bob Evan/Peter Arnold / **13.2, 13.3** Art, Precision Graphics / **13.4** (a) Art, Palay/Beaubois; (b–e) Lisa Starr / **13.5** Art, Preface, Inc. / **13.6, 13.7** Art, Precision Graphics and Gary Head / **13.8** (a) tRNA model by Dr. David B. Goodin, The Scripps Research Institute; (b, c) Lisa Starr / **13.9** (a) Dr. Thomas A. Steitz; (b) Lisa Starr / **13.10** Lisa Starr / **13.11** (a) Lisa Starr; (b) Precision Graphics / **13.12** Lisa Starr / **13.13** (left) Nik Kleinberg; (right) Peter Starlinger / **13.14** Lisa Starr / **13.15** Courtesy of the National Neurofibromatosis Foundation

CHAPTER 14 **14.1** Photographs, (a) Ken Greer/Visuals Unlimited; (b) Biophoto Associates/Science Source/Photo Researchers; (c) James Stevenson/SPL/Photo Researchers; (right) Gary Head / **page 212** Lisa Starr / **14.2** Art, Raychel Ciemma / **14.3** Brian Matthews, University of Oregon / **14.4** Art, Palay/Beaubois and Hans & Cassady / **14.5** Art, (b) Raychel Ciemma / **14.6** Jack Carey and Lisa Starr / **14.7** Art, Raychel Ciemma; Micrograph, W. Beerman / **14.8** Frank B. Salisbury / **14.9** (a) Lennart Nilsson © Boehringer Ingelheim International GmbH; (b, c) Art, Betsy Palay/Atemis / **14.10** Art, Betsy Palay/Atemis / **14.11** Slim Films / **page 221** Lisa Starr

CHAPTER 15 **15.1** Lewis L. Lainey / **15.2** Science VU/Visuals Unlimited / **15.3** (a) Stanley N. Cohen/Science Source/Photo Researchers; (b) Dr. Huntington Potter and Dr. David Dressler / **page 224** Art, Preface, Inc. / **15.4** Lisa Starr with photograph © 2000 PhotoDisc, Inc. / **15.5, 15.6** Art, Gary Head / **15.7** (above) Damon Biotech, Inc.; (below) Cellmark Diagnostics, Abingdon, UK / **15.8, 15.9** Lisa Starr / **15.10** Stephen Wolfe, *Molecular and Cellular Biology* / **15.11** Art, Lisa Starr; Photograph, Herve Chaumeton/Agence Nature / **15.12** Keith V. Wood / **15.13** Photographs, (a) Dr. Vincent Chiang, School of Forestry and Wood Products, Michigan Technology University; (b) Courtesy Calgene LLC / **15.14**

R. Brinster, R. E. Hammer, School of Veterinary Medicine, University of Pennsylvania / **15.15** © Adrian Arbib/Still Pictures / **page 237** S. Stammers/SPL/Photo Researchers

CHAPTER 16 **16.1** Elliot Erwitt/Magnum Photos, Inc. / **16.2** Photographs, (a) Jen & Des Bartlett/Bruce Coleman; (b) Kenneth W. Fink/Photo Researchers; (c) Dave Watts/A.N.T. Photo Library / **16.3** Art, Raychel Ciemma / **16.4** Photographs, (a) Courtesy George P. Darwin, Darwin Museum, Down House; (b) Heather Angel, computer-enhanced by Lisa Starr; (c) Christopher Ralling; (e) Dieter & Mary Plage/Survival Anglia; Art, (d-above) Leonard Morgan; (d-below) Precision Graphics / **16.5** © Philip Boyer/Photo Researchers; (b) Field Museum of Natural History, Chicago. Artist, Charles R. Knight (Neg. CK21T) / **16.6** Photographs, (a) Heather Angel; (b) David Cavagnaro; (c) Dr. P. Evans/Bruce Coleman; (d) Alan Root/Bruce Coleman / **16.7** Courtesy of Down House and The Royal College of Surgeons of England / **page 246** Courtesy Derrell Fowler, Tecumseh, OK / **16.8** Alan Solem / **16.9** Art, Raychel Ciemma and Gary Head / **16.10** Art, Precision Graphics / **16.11** J. A. Bishop and L. M. Cook / **16.12** Art, Precision Graphics / **16.13** Photographs, (a, b) Warren Abrahamson; (c) Forest W. Buchanan/Visuals Unlimited; (d) Kenneth McCrea and Warren Abrahamson; computer-enhanced by Gary Head / **16.14** Art, Precision Graphics / **16.15** Photographs, (a, b) Thomas Bates Smith; (c) art, Preface, Inc. / **16.16** Bruce Beehler / **16.17** Art, Precision Graphics after Ayala and others / **16.18** (above) David Neal Parks; (below) W. Carter Johnson / **16.19** Art, Precision Graphics after computer models developed by Jerry Coyne / **16.20** Art, Raychel Ciemma; Photograph, David Cavagnaro / **16.21** Kjell Sandved/Visuals Unlimited / **page 259** Art, Precision Graphics / **16.22** Art, Precision Graphics using NIH data

CHAPTER 17 **17.1** (a) Photograph, Gary Head with snail courtesy of Larry Reed; (b) Art, adapted from R. K. Selander and D. W. Kaufman, *Evolution*, 29(3), December 31, 1975 / **17.2** Art, Jennifer Wardrip / **17.3** Art, Precision Graphics after F. Ayala and J. Valentine, *Evolving*, Benjamin-Cummings, 1979 / **17.4** Photographs, (a) G. Ziesler/ZEFA; (b) Alvin E. Staffan/Photo Researchers; (c) John Alcock/Arizona State University / **17.5** Photographs, (a, all) Fred McConnaughey/Photo Researchers; (b) Patrice Geisel/Visuals Unlimited; (right) Tom Van Sant/The Geosphere Project, Santa Monica, CA / **17.6** Art, Preface, Inc. / **17.7** Art, Raychel Ciemma / **17.8** Photograph, (above) Policy Programme of WWF Cameroon/www.wwfcameroon.org, computer-enhanced by Lisa Starr; (below) Tom Van Sant/The Geosphere Project, Santa Monica, CA / **17.9** Lisa Starr, after W. Jensen and F. B. Salisbury, *Botany: An Ecological Approach*, Wadsworth, 1972, and others / **17.10** Photographs, (left) © H. Clarke, VIREO/Academy of Natural Sciences; (right) Robert C. Simpson/Nature Stock; Art, Preface, Inc. / **17.11** Art, Precision Graphics / **17.12** Photographs (above) Jack Dermid; (below) Leonard Lee Rue III/FPG; Art, Precision Graphics after P. Dodson, *Evolution: Process and Products*, 3rd Ed., PWS / **17.13** Jen & Des Bartlett/Bruce Coleman

CHAPTER 18 **18.1, 18.2** Lisa Starr, with use of NASA satellite photographs / **18.3** Vatican Museums / **18.4** Photographs, (a) Donald Baird, Princeton Museum of Natural History; (b) A. Feduccia, *The Age of Birds*, Harvard University Press, 1980; (c) H. P. Banks; (d) Jonathan Blair / **18.5** (Globes), Lisa Starr; Chart, Gary Head / **18.6** Art, Preface, Inc. / **18.7** Photograph, (a) © 2001 PhotoDisc, Inc.; Art, (b, c) Lisa Starr / **18.8** Dan Peha/Index Stock Imagery / **18.9** Art, Leonard Morgan / **18.10** Lisa Starr, after A. M. Ziegler, C. R. Scotese, and S. F. Barrett, "Mesozoic and Cenozoic Paleogeographic Maps," and J. Krohn and J. Sundermann (Eds.), *Tidal Friction and the Earth's Rotation II*, Springer–Verlag, 1983; Photograph, Martin Land/Photo Researchers / **18.11** Art, Raychel Ciemma / **18.12** Art, (a) Lisa Starr and Raychel Ciemma, art reference, Natural History Collection, Royal BC Museum; Photographs, (a) Stephen Dalton/Photo Researchers, computer-enhanced by Lisa Starr; (b) Frans Lanting/Minden Pictures, computer-enhanced by Lisa Starr; (c) J. Scott Altenbach, University of New Mexico, computer-enhanced by Lisa Starr / **18.13** Art, Precision Graphics and Gary Head, after E. Guerrant, *Evolution*, 36:699–712; Photographs, Gary Head / **18.14** From T. Storer et al., *General Zoology*, sixth Ed., McGraw-Hill, 1979, reproduced by permission of McGraw-Hill, Inc. / **18.15** Art, Raychel Ciemma / **18.16** Art, Precision Graphics / **18.17** Photographs, (left to right) Kjell B. Sandved/Visuals Unlimited; Jeffrey Sylvester/FPG; Thomas D. Mangelsen/Images of Nature / **18.18** (left to right) Larry Lefever/Grant Heilman Photography;

R.I.M. Campbell/Bruce Coleman; Runk & Schoenberger/Grant Heilman Photography; Bruce Coleman / **18.21** Photograph, P. Morris/Ardea London; Art, Raychel Ciemma / **page 291** © 1990 Arthur M. Green

CHAPTER 19 **19.1** Jeff Hester and Paul Scowen/Arizona State University and NASA / **19.2** William K. Hartmann / **19.3** (a) Chesley Bonestell; (b) Art, Raychel Ciemma / **19.4** Art, Precision Graphics / **19.5** Microscopics, (a) Sidney W. Fox; (b) W. Hargreaves and D. Deamer / **19.7** Art, Raychel Ciemma and Precision Graphics / **19.8** Bill Bachmann/Photo Researchers / **19.9** Photographs, (a) Stanley M. Awramik; (b–f) Andrew H. Knoll, Harvard University / **19.10** Micrograph, P. L. Walne and J. H. Arnott, *Planta*, 77: 325–354, 1967 / **19.11** Art, Raychel Ciemma / **19.12** Robert K. Trench / **19.13** Photographs, (a, b) Neville Pledge/South Australian Museum; (c, d) Dr. Chip Clark / **19.14** (a, b) Illustrations by Zdenek Burian, © Jiri Hochman and Martin Hochman; (c) Patricia G. Gensel; (d) © John Barber / **19.15** Painting by Megan Rohn courtesy of David Dilcher; Art, Gary Head; Photographs, (from below clockwise) Runk & Schoenberger/Grant Heilman Photography; Robert and Linda Mitchell; Ed Reschke; Lee Casebere / **19.16** (a) © John Sibbick; (b, c) © Karen Carr/www.karencarr.com / **19.17** NASA Galileo Imaging Team / **19.18** Art, Raychel Ciemma / **19.19** Paintings, (a, b) © Ely Kish; (c) Field Museum of Natural History, Chicago. Artist, Charles R. Knight (Neg. CK8T) / **Globes, page 302–304, 307** Lisa Starr / **19.20** Lisa Starr

CHAPTER 20 **20.1** (a, b) Tony Brain and David Parker/SPL/Photo Researchers; (c) Lee D. Simon/Photo Researchers / **20.2** (a) © Oliver Meckes/Photo Researchers; (b) Courtesy Allan W. H. Bé and David A. Caron; (c) Lewis Trusty/Animals Animals / **20.3** (a) Lisa Starr; (b) CNRI/SPL/Photo Researchers / **20.4** (a) L. J. LeBeau, University of Illinois Hospital/BPS; (b, c) Courtesy of K. Amako and A.Umeda, Kyshu University, Japan / **20.5** © P. Hawtin, University of Southampton/SPL/Photo Researchers / **20.6** (a) Jack Jones, *Archives of Microbiology*, 136: 254–261, 1983. Reprinted by permission of Springer–Verlag, computer-enhanced by Lisa Starr; (b) © 2000 PhotoDisc, Inc.; (c) NASA; (d) © Alan L. Detrick, Science Source/Photo Researchers / **20.7** (a) Dr. Jeremy Burgess/Science Photo Library/Photo Researchers; (b) Tony Brain/SPL/Photo Researchers; (c) P. W. Johnson and J. MeN. Sieburth, Univ. Rhode Island/BPS / **20.8** Photographs, (above) Edward S. Ross; (center) Courtesy of Wadsworth Center, NY; (deer ticks) Courtesy Pfizer Central Research; Art, Preface, Inc. / **20.9, 20.10** Art, Raychel Ciemma / **20.11** Art, Raychel Ciemma after Stephen L. Wolfe and others / **20.12** (a) K. G. Murti/Visuals Unlimited; (b) Kenneth M. Corbett / **20.13** Art, Palay/Beaubois and Precision Graphics / **20.14** Art, Raychel Ciemma / **20.15** (a) Heather Angel; (b) W. Merrill / **20.16** Art, (a) Leonard Morgan; Photographs, (b) M. Claviez, G. Gerish, and R. Guggenheim; (c) London Scientific Films; (d–f) Carolina Biological Supply Company; (g) Courtesy Robert R. Kay from R. R. Kay, et al., *Development*, 1989 Supplement, pp. 81–90, © The Company of Biologists Ltd., 1989 / **20.17** Edward S. Ross / **20.18** Photographs, (a) M. Abbe/Visuals Unlimited; (b) G. Shih and R. Kessel/Visuals Unlimited; (c) John Clegg/Ardea, London; (e) Neal Ericsson / **20.19** Micrograph, (a) Sidney L. Tamm; Art, (b, c) Raychel Ciemma, redrawn from V. & J. Pearse and M. & R. Buchsbaum, *Living Invertebrates*, The Boxwood Press, 1987. Used by permission / **20.20** Micrographs, (a) Dr. Stan Erlandsen, University of Minnesota; (b) David M. Phillips/Visuals Unlimited / **20.21** (b) Lisa Starr, after Prescott, *Microbiology*, third Ed. / **20.22** Lauren and Homer, Photography by Gary Head / **20.23** Art, Leonard Morgan; Micrograph, Steven L'Hernault / **Table 20.2** Courtesy Saul Tzipori, Div. of Infectious Diseases, Tufts University School of Veterinary Medicine / **20.24** Art, (a-left) Raychel Ciemma; (a-right) Copyright © Richard Triemer, Rutgers; Micrographs, (b) Greta Fryxell, University of Texas, Austin; (c) Emiliania huxleyi photograph by Vita Pariente, scanning electron micrograph taken on a Jeol T330A instrument at the Texas A&M University Electron Microscopy Center; (d) S. Carty, Heidelberg College / **20.25** Photographs, (a-both, b-right) North Carolina State University, Aquatic Botany Lab; (b-left) C. C. Lockwood / **20.26** Art, Raychel Ciemma / **20.27** (a) Art from T. Garrison, *Oceanography: An Invitation to Marine Science*, Brooks/Cole, 1993; Photograph, (b) J. R. Waaland, University of Washington/BPS / **20.28** Photographs, (a) Manfred Kage/Peter Arnold; (b) Ronald W. Hoham, Dept. of Biology, Colgate University; (c) © Linda Sims/Visuals Unlimited; (d) Brian Parker/Tom

Stack & Associates; Art, (d) Raychel Ciemma / **20.29** Art, Raychel Ciemma; Photograph, D. J. Patterson/Seaphot Limited/Planet Earth Pictures / **20.30** Carolina Biological Supply Company

CHAPTER 21 **21.1** © Sherry K. Pittam / **21.2** (a–f) Robert C. Simpson/Nature Stock / **21.3** (a) Robert C. Simpson/ Nature Stock; (b) Jane Burton/Bruce Coleman; (c) Thomas J. Duffy / **21.4** Art, Raychel Ciemma; Micrograph, Garry T. Cole, University of Texas, Austin/BPS / **21.5** (a) Dr. P. Marazzi/SPL/Photo Researchers; (b) Eric Crichton/ Bruce Coleman / **21.6** Photographs, (a) Dr. John D. Cunningham/Visuals Unlimited; (b) © Michael Wood/ mykoweb.com; (c) © North Carolina State University, Department of Plant Pathology; (d) © Fred Stevens/ mykoweb.com; (e) Dennis Kunkel Microscopy, Inc.; (f) Garry T. Cole, University of Texas, Austin/BPS / **21.7** Raychel Ciemma; Micrographs, Ed Reschke / **21.8** (a) After Raven, Evert, and Eichhorn, *Biology of Plants*, 4th Ed., Worth Publishers, New York, 1986; Photograph, (b) © Mark E. Gibson/Visuals Unlimited / **21.9** (a) © 1990 Gary Braasch; (b) F. B. Reeves / **21.10** John Hodgin

CHAPTER 22 **22.1** (a) Pat and Tom Leeson/Photo Researchers; (b) Craig Wood/Visuals Unlimited; (c) Edward S. Ross; (d) © Greg Allikas/ www.orchidworks.com / **22.2** Art, Raychel Ciemma after E. O. Dodson and P. Dodson, *Evolution Process and Product*, third Ed., p. 401, PWS / **22.3** Art, Precision Graphics / **22.4** (a) Field Museum of Natural History, Chicago (Neg. 7500C); (b) Brian Parker/Tom Stack & Associates / **22.5** Photograph, Jane Burton/Bruce Coleman; Art, Raychel Ciemma / **22.6** (a) Fred Bavendam/Peter Arnold; (b) Dr. John D. Cunningham/ Visuals Unlimited / **22.8** (a) Kingsley R. Stern; (b) Ed Reschke/Peter Arnold; (c) William Ferguson; (d) W. H. Hodge; (e) Kratz/ZEFA / **22.9** Art, Raychel Ciemma; Photographs, (inset) A. & E. Bomford/Ardea, London; (below) Lee Casebere / **22.10** Art, Raychel Ciemma; Photographs, (left) Edward S. Ross; (right-above) Robert and Linda Mitchell Photography; (right-below) R. J. Erwin/Photo Researchers / **22.11** © 1994 Robert Glenn Ketchum / **22.12** (a) © Jeff Gnass Photography; (b) Joyce Photographics/Photo Researchers; (c) Kingsley R. Stern / **22.13** (a) Robert and Linda Mitchell; (b) Edward S. Ross; (c) William Ferguson / **22.14** Art, (a) Jennifer Wardrip; Photographs, (b) M. P. L. Fogden/Bruce Coleman; (c) L. Mellichamp/Visuals Unlimited; (d) Peter F. Zika/Visuals Unlimited; (e) Heather Angel / **22.15** Art, Raychel Ciemma / **22.16** (a) George Loun/Visuals Unlimited; (b) Earl Roberge/Photo Researchers; (c) John Mason/ Ardea, London; (d) Dick Davis/Photo Researchers; (e) Rein/ZEFA / **22.17** Bob Cerasoli / **22.18** (left and center) © 1989, 1991 Clinton Webb; (right) Gary Head

CHAPTER 23 **23.1** Courtesy of Department of Library Services, American Museum of Natural History (Neg. K10273) / **23.2** Art, Gary Head; Photograph, Lisa Starr / **23.3** Art, Leonard Morgan / **23.4** Art, Raychel Ciemma / **23.5** (left) After Laszlo Meszoly in L. Margulis, *Early Life*, Jones and Bartlett, 1982 / **23.6** Bruce Hall / **23.7** (a–c) Art, Raychel Ciemma; (d-left) art, Precision Graphics after Bayer and Owre, *The Free-Living Lower Invertebrates*, © 1968 Macmillan; Photographs, (d-right) Don W. Fawcett/Visuals Unlimited; (e) Marty Snyderman/Planet Earth Pictures / **23.8** Art, Raychel Ciemma after Eugene Kozloff / **23.9** (a, b) Art, Raychel Ciemma; Photographs, (c) Courtesy of Dr. William H. Hammer; (d-both) Kim Taylor/Bruce Coleman; (e-both) F. S. Westmorland/Tom Stack & Associates / **23.10** (a) Art, Precision Graphics after T. Storer et al., *General Zoology*, sixth Ed., © 1979 McGraw-Hill; Photographs, (b) Andrew Mounter/Seaphot Limited/Planet Earth Pictures; (c) Christian DellaCorte / **23.11** Art, Raychel Ciemma; Photograph, Kim Taylor/Bruce Coleman / **23.12** (a) Robert and Linda Mitchell Photography; (b) Cath Ellis, University of Hull/SPL/Photo Researchers / **23.13** Art, Lisa Starr; Micrograph, J. Sulston, MRC Laboratory of Molecular Biology / **23.14** Art, Raychel Ciemma / **23.15** Art, Raychel Ciemma; Photograph, Carolina Biological Supply Company / **23.16** (a) © L. Jensen/Visuals Unlimited; (b) Dianora Niccolini / **page 367** Art, Precision Graphics / **23.17** Photographs, (a) Gary Head; (b) Alex Kerstitch; (c) Jeff Foott/Tom Stack & Associates; (d) Frank Park/A.N.T. Photo Library; (e) Bob Cranston; Art, (f) Palay/Beaubois / **23.18** (a, b) Art, Raychel Ciemma / **23.19** J. A. L. Cooke/Oxford Scientific Films / **23.20** (a) © Cabisco/Visuals Unlimited; (b) Jon Kenfield/Bruce Coleman; (c) Art, Precision Graphics after Eugene Kozloff, *Invertebrates* © 1990 by Saunders College Publishing.

Reproduced by permission of the publisher; (d) Art, Precision Graphics, adapted from Rasmussen, "Ophelia," Vol. 11, in Eugene Kozloff, *Invertebrates*, 1990 / **23.21** Art, Raychel Ciemma / **23.22** Jane Burton/Bruce Coleman / **23.23** (a) Redrawn from *Living Invertebrates*, V. & J. Pearse/M. &. R. Buchsbaum, The Boxwood Press, 1987. Used by permission; Photographs, (b-above) Jane Burton/ Bruce Coleman; (b-below) Angelo Giampiccolo/FPG; (c) P. J. Bryant, University of California, Irvine/BPS; (d) Ken Lucas/Seaphot Limited/Planet Earth Pictures / **23.24** B. Backes, Arizona State University / **23.25** Photographs, (a) Herve Chaumeton/Agence Nature; (b) Frans Lanting/ Bruce Coleman; (c) Fred Bavendam/Peter Arnold; (d) Agence Nature; Art, Precision Graphics, after D. H. Milne, *Marine Life and the Sea*, Wadsworth, 1995 / **23.26** Art, Precision Graphics / **23.27** (a) Z. Leszczynski/Animals Animals; (b) Steve Martin/Tom Stack & Associates / **23.28** Art, Raychel Ciemma / **23.29** Art, Precision Graphics. From G. Pasteur, "Jean Henri Fabre," *Scientific American*, July 1994. © 1994 by Scientific American. All rights reserved / **23.30** Photographs, (a–h) Edward S. Ross; (i) David Maitland/Seaphot Limited/Planet Earth Pictures; (j) C. P. Hickman, Jr. / **23.31** (b) Marlin E. Rice, Iowa State University; (a, c) John Obermeyer, Purdue University, Dept. of Entomology / **23.32** (a) Chris Huss/The Wildlife Collection; (b) John Mason/Ardea, London; (c) Kjell B. Sandved; (d) Ian Took/Biofotos / **23.33** (a, b) Art, L. Calver; (c, d) Photographs, Herve Chaumeton/Agence Nature / **page 379** Inset, Jane Burton/Bruce Coleman / **23.34** Art, Raychel Ciemma / **23.35** Walter Deas/Seaphot Limited/ Planet Earth Pictures / **23.36** John H. Gerard / **23.37** (a) J. Solliday/BPS; (b) Herve Chaumeton/Agence Nature

CHAPTER 24 **24.1** (a) Jean Phillipe Varin/Jacana/Photo Researchers; (b) Tom McHugh/Photo Researchers / **24.2** Art, Gary Head / **24.3** (a) Rick M. Harbo; (b, c) Redrawn from *Living Invertebrates*, V. & J. Pearse and M. & R. Buchsbaum, The Boxwood Press, 1987. Used by permission / **24.4** (a) Art, Raychel Ciemma; (b) Runk & Schoenberger/Grant Heilman Photography / **24.5** Art, D. & V. Hennings and Precision Graphics / **24.6** (a–c) Art, Raychel Ciemma, adapted from A. S. Romer and T. S. Parsons, *The Vertebrate Body*, sixth Ed., Saunders College Publishing, 1986; (d) Photograph, Al Giddings/Images Unlimited / **24.7** (a, b) After D. H. Milne, *Marine Life and the Sea*, Wadsworth, 1995; (c) Photograph, Heather Angel / **24.8** (a) Erwin Christian/ZEFA; (b) Allan Power/Bruce Coleman; (c) Tom McHugh/Photo Researchers / **24.9** (a) Bill Wood/Bruce Coleman; (b) Art, Raychel Ciemma; (c) Robert and Linda Mitchell Photography; (d) Patrice Ceisel/ © 1986 John G. Shedd Aquarium; (e) © Norbert Wu/Peter Arnold; (f) From T. Garrison, *Oceanography: An Invitation to Marine Science*, Wadsworth, 1993 / **24.10** © Marianne Collins; (b, c) art, Laszlo Meszoly and D. & V. Hennings / **24.11** (a) Stephen Dalton/ Photo Researchers; (b) John Serraro/Visuals Unlimited; (c) Jerry W. Nagel; (d) Art, Leonard Morgan/Adapted from A. S. Romer and T. S. Parsons, *The Vertebrate Body*, sixth Ed., Saunders College Publishing, 1986; (e) Juan M. Renjifo/Animals Animals / **24.12** (a) Pieter Johnson; (b) Stanley Sessions, Hartwick College / **24.13** Art, Raychel Ciemma / **24.14** © 1989 D. Braginetz; (b) Z. Leszczynski/Animals Animals / **24.15** (a, c) Art, Raychel Ciemma; Photographs, (a) D. Kaleth/Image Bank; (b) Andrew Dennis/A.N.T. Photo Library; (c) Stephen Dalton/Photo Researchers; (d) Heather Angel / **24.16** Art, Lisa Starr; Photograph, J. L. G. Grande/Bruce Coleman / **24.17** (a) Gerard Lacz/A.N.T. Photo Library; (b) Rajesh Bedi; (c) Thomas D. Mangelsen/Images of Nature / **24.18** (a, b) Art, Raychel Ciemma / **24.19** Photographs, (a) Sandy Roessler/FPG; (b) Leonard Lee Rue III/FPG; (c) © Kevin Schafer/Tom Stack & Associates; (d) Art, Raychel Ciemma after M. Weiss and A. Mann, *Human Biology and Behavior*, 5th Ed., Harper Collins, 1990 / **24.20** Art, Raychel Ciemma / **24.21** Photographs, (a) D. & V. Blagden/A.N.T. Photo Library; (b) courtesy The Zoological Society of San Diego; The San Diego Zoo; (c) Clem Haagner/Ardea, London; (d) Douglas Faulkner/ Photo Researchers; (e) Christopher Crowley / **24.22** Photographs, (a) Larry Burrows/Aspect Picture Library; (b) Bruce Coleman; (c) Tom McHugh/Photo Researchers; (d) Courtesy of Dr. Takeshi Furuichi, Biology, Meiji–Gakuin University-Yokohama; (e) Time Inc. 1965/Larry Burrows Collection / **24.23** Lisa Starr / **24.24** (a) Dr. Donald Johanson, Institute of Human Origins; (b) Louise M. Robbins; (c, d) Kenneth Garrett/National Geographic Image Collection / **24.25** (a) Jean Paul Tibbles; (b, c) John Reader © 1981 / **24.26, 24.27, 24.28** Art, Raychel Ciemma / **24.29** Art, Preface, Inc. / **24.30** Lisa Starr, with photograph from NASA / **24.31** Lisa Starr / **24.32** Art, Precision Graphics after National Geographic,

February 1997, page 82 / **page 405** Art, D. & V. Hennings

CHAPTER 25 **25.1** Tom Till/© Stone / **25.2** Art, Gary Head / **25.3** (a) Painting by Charles Knight, Department of Library Services/American Museum of Natural History (2425); (b) Mansell Collection/Time, Inc. / **25.4** Lisa Starr / **25.5** Steve Hillebrand, U.S. Fish & Wildlife Service / **25.6** (a) Courtesy of Dan Stanford; (b) Lisa Starr, based on 1940–1984 reports from International Whaling Commission / **25.7** (a) C. B. & D. W. Frith/Bruce Coleman; (b) Douglas Faulkner/Photo Researchers; (c) Douglas Faulkner/ Sally Faulkner Collection / **25.8** (a) From T. Garrison, *Oceanography: An Invitation to Marine Science*, third Ed., Brooks/Cole, 2000. All rights reserved; (b), (above row, left to right) Peter Scoones/Planet Earth Pictures; Jim Doran; (insets) Douglas Faulkner/Sally Faulkner Collection; (center row, left to right) Jeff Rotman; Alex Kerstitch; Douglas Faulkner/Sally Faulkner Collection; (below row, all) Douglas Faulkner/Sally Faulkner Collection / **25.9** Eric Hartmann/Magnum Photos / **25.10** Lisa Starr / **25.11** Lisa Starr, with photographs © 2000 PhotoDisc, Inc. / **25.12** Photographs, Bureau of Land Management / **25.13** Lisa Starr / **page 419** © Gary Head

CHAPTER 26 **26.1** (a) R. Barrick/USGS; (b, c) © 1980 Gary Braasch; (d) Don Johnson/Photo Nats, Inc. / **26.2, 26.4** Art, Raychel Ciemma / **26.5** Micrography, James D. Mauseth, *Plant Anatomy*, Benjamin-Cummings, 1988 / **26.6** (all) Biophoto Associates / **26.7** (a) Jan Robert Factor/Photo Researchers; (b) D. E. Akin and I. L. Risgby, Richard B. Russel Agricultural Research Center, Agricultural Research Service, U.S. Department of Agriculture, Athens, Georgia; (c) Kingsley R. Stern / **26.8** Art, Precision Graphics / **26.9** George S. Ellmore / **26.10** Art, D. & V. Hennings / **26.11** Art, Raychel Ciemma / **26.12** (a) Robert and Linda Mitchell Photography; (b) Roland R. Dute; (c) Gary Head / **26.13** Art, D. & V. Hennings; Micrographs, (a-left) Ray F. Evert; (a-right) James W. Perry; (b-left) Carolina Biological Supply Company; (b-right) James W. Perry / **26.14** Art, D. & V. Hennings / **26.15** (a, above and below) Heather Angel; (b) Gary Head; (c) © 2001 PhotoDisc, Inc. / **26.16** (a) Art, Raychel Ciemma; (b) C. E. Jeffree, et al., *Planta*, 172(1):20–37, 1987. Reprinted by permission of C. E. Jeffree and Springer–Verlag; (c) Dr. Jeremy Burgess/SPL/Photo Researchers / **26.17** (a) Art after Salisbury and Ross, *Plant Physiology*, fourth Ed., Wadsworth; (b) Micrograph, John Limbaugh/Ripon Microslides, Inc.; (c, d) Art, Raychel Ciemma / **26.18** Micrographs, (a-both) Chuck Brown; (b) Carolina Biological Supply Company / **26.19** Photographs, John Limbaugh/Ripon Microslides; Sketch after T. Rost et al., *Botany: A Brief Introduction to Plant Biology*, second Ed., © 1984, John Wiley & Sons / **26.20** Art, (a) Raychel Ciemma, from Rost et al, *Plant Biology*, Wadsworth; (b) Art, Raychel Ciemma / **26.21** Lisa Starr / **26.22** H. A. Core, W. A. Cote, and A. C. Day, *Wood Structure and Identification*, second Ed., Syracuse University Press, 1979 / **26.23** Art, Raychel Ciemma / **26.24** Photographs, Edward S. Ross / **page 435** Art, Lisa Starr; Photograph, © Stephen J. Krasemann/National Audubon Society Collection/Photo Researchers

CHAPTER 27 **27.1** (a, b) Robert and Linda Mitchell Photography; (c) © John N. A. Lott, *Scanning Electron Microscope Study of Green Plants*, St. Louis: C. V. Mosby Company, 1976; (d) Robert C. Simpson/Nature Stock / **27.2** David Cavagnaro/Peter Arnold / **27.3** William Ferguson / **27.4** U.S. Dept. of Agriculture / **27.5** Art, Leonard Morgan; Micrograph, (b) Chuck Brown / **27.6** Jean Paul Reve / **27.7** Photographs, (a) Adrian P. Davies/ Bruce Coleman; (c) Mark E. Dudley and Sharon R. Long; (a, b) Art, Jennifer Wardrip / **27.8** NifTAL Project, Univ. of Hawaii, Maui / **27.9** Micrographs, (a) W. A. Cote of N. C. Brown Center for Ultra-Structure Studies, computer-enhanced by Lisa Starr; (b, c) H. A. Core, W. A. Cote and A. C. Day, *Wood Structure and Identification*, 2nd Ed., Syracuse University Press, 1979 / **27.10** Art, Raychel Ciemma / **27.11** Frank B. Salisbury / **27.12** George S. Ellmore / **27.13** W. Thomson, *American Journal of Botany*, 57(3):316, 1970 / **27.14** Art, Palay/Beaubois / **27.15** T. A. Mansfield / **27.16** Photograph, (left) John Lawlor/ FPG; Scanning micrographs, (a, b) Dr. Jeremy Burgess/ SPL/Photo Researchers / **27.17** Art, Hans & Cassady, Inc. / **27.18** Martin Zimmerman, *Science*, 1961, 133:73–79, © AAAS / **27.19** Art, (left) Palay/Beaubois; (right) Precision Graphics / **27.21** James T. Brock

CHAPTER 28 **28.1** (left) Courtesy Merlin D. Tuttle, Bat Conservation International; (right) John Alcock/Arizona

Precision Graphics / **38.3** Art, Robert Demarest / **38.4** Photograph, (a-left) © Ralph T. Hutchings; Art, Robert Demarest / **38.5** Art, (above) Robert Demarest; (below) Precision Graphics / **38.6** Art, Precision Graphics; Photograph, Gary Head / **38.7** (a, b) From T. Garrison, *Oceanography: An Invitation to Marine Science*, Brooks/Cole, 1993. All rights reserved; (c) Photograph, © Thomas D. Mangelsen/Images of Nature / **38.8** © Bob McKeever/ Tom Stack & Associates / **38.9** Colin Monteath, Hedgehog House, New Zealand / **38.10** © Bettmann/Corbis

CHAPTER 39 **39.1** Art, Raychel Ciemma; Photographs, (a) Hans Pfletschinger; (c) John H. Gerard; (d, e) © David M. Dennis/Tom Stack & Associates; (f) John Shaw/Tom Stack & Associates / **39.2** (a) Frieder Sauer/ Bruce Coleman (b) © Matjaz Kuntner; (c) Wisniewski/ZEFA; (d) L. L. Rue III; (e) Carolina Biological Supply Company; (f) Fred McKinney/FPG; (g) Gary Head / **39.3** Art, Palay/ Beaubois and Precision Graphics / **39.4** Art, L. Calver; Micrographs, Carolina Biological Supply Company / **39.5** Art, Robert Demarest / **39.6** Art, Precision Graphics; Photograph, Gary Head / **39.7** Art, after V. E. Foe and B. M. Alberts, *Journal of Cell Science*, 1983, 61:32, © The Company of Biologists; Micrograph, J. B. Morrill / **39.8** (Photographic series) Carolina Biological Supply Company; (right photograph) Peter Parks/Oxford Scientific Films/Animals Animals / **39.9** Art, (a) Lisa Starr; (b) Raychel Ciemma after B. Burnside, *Developmental Biology*, 1971, 26:416–441. Used by permission of Academic Press / **39.10** Art, (a, b) Lisa Starr; Photographs, (c) Prof. Jonathon Slack / **39.11** Art, Palay/Beaubois and Preface, Inc. / **39.12** Art, Raychel Ciemma, modified by Lisa Starr / **39.13** Art, (a) Raychel Ciemma; Micrograph, (b) Ed Reschke / **39.14** Art, Raychel Ciemma / **39.15** Art, Precision Graphics / **39.16** Art, Raychel Ciemma / **39.17** Art, (above) Robert Demarest; (below) Raychel Ciemma/ **39.18** Art, Precision Graphics; Photograph, Lennart Nilsson from *A Child Is Born*, © 1966, 1977 Dell Publishing Company, Inc. / **39.19** K. Somerville, Robert Demarest, and Preface, Inc. / **39.20–39.23** Art, Raychel Ciemma / **39.24**, Raychel Ciemma; Photographs, Lennart Nilsson, *A Child Is Born*, © 1966, 1977 Dell Publishing Company, Inc. / **39.25** Art, Raychel Ciemma, modified from K. L. Moore, *The Developing Human: Clinically Oriented Embryology*, fourth Ed., Philadelphia: W. B. Saunders Co., 1988 / **39.26** James W. Hanson, M.D. / **39.27** Art, Robert Demarest / **39.28** Art, Raychel Ciemma / **39.29** Art, Raychel Ciemma, adapted from L. B. Arey, *Developmental Anatomy*, Philadelphia, W. B. Saunders Co., 1965; Photograph, Lisa Starr / **39.30** Art, Preface Inc. / **39.31** CNRI/SPL/Photo Researchers / **39.32** (a) Dr. John D. Cunningham/Visuals Unlimited; (b) David M. Phillips/Visuals Unlimited; (c) © Kenneth Greer/Visuals Unlimited / **39.33** Carolina Biological Supply Co. / **page 683** Alan and Sandy Carey

CHAPTER 40 **40.1** (a) Gary Head; (b) Antoinette Jongen/ FPG / **40.2** Art, Precision Graphics / **40.3** (a) E. R. Degginger; (inset) Jeff Foott Productions/Bruce Coleman; (b) © Paul Lally/Stock, Boston / **40.4** Photograph, © Jeff Lepore/Photo Researchers; Art, Precision Graphics / **40.5** (a) Art, Precision Graphics; (b) Micrograph, Stanley Flegler/Visuals Unlimited / **40.6** Art, Precision Graphics / **40.7** Art, Precision Graphics; Photograph, E. Vetter/ZEFA / **40.8** Art, Precision Graphics; Photographs, (a) Jonathan Scott/Planet Earth Pictures; (b) Hal H. Harrison/Photo Researchers; (c) Fred Bavendam/Peter Arnold / **page 693** David Reznick/ University of California-Riverside; computer-enhanced by Lisa Starr / **40.9** Art, Precision Graphics; Photograph, NASA / **40.10** Art, Gary Head after G. T. Miller, Jr., *Environmental Science*, 6th Ed., Brooks/Cole, 1997. All rights reserved. / **40.11** Art, Precision Graphics and Preface, Inc.; Photograph, United Nations / **40.12** Data from Population Reference Bureau after G. T. Miller, Jr., *Living in the Environment*, 8th Ed., Brooks/Cole, 1993. All rights reserved / **40.13** Art, Precision Graphics; Photograph, United Nations / **40.14, 40.15** After G. T. Miller, Jr., *Environmental Science*, sixth Ed., Brooks/Cole, 1997. All rights reserved

CHAPTER 41 **41.1** (left) John Bova/Photo Researchers; (right) Robert Maier/Animals Animals / **41.2** (a) From L. Clark, *Parasitology Today*, 6 (11), Elsevier Trends Journals, 1990, Cambridge, U.K.; (b) Photograph, Jack Clark/ Comstock, Inc. / **41.3** (a) Eugene Kozlowf; (b, c) Stevan Arnold, Inc. / **41.4** Photograph, Hans Reinhard/Bruce Coleman; Spectogram, from G. Pohl-Apel and R. Sussinka, *Journal for Ornithologie*, 123:211–214 / **41.5** (a) Evan Cerasoli; (b) From A. N. Meltzoff and M. K. Moore, "Imitation of Facial and Manual Gestures by Human Neonate," *Science*, 1977, 198:75–78. ©1977 by the AAAS /

41.6 Eric Hosking / **41.7** Nina Leen in *Animal Behavior*, Life Nature Library / **page 707** (left) © 2001 PhotoDisc, Inc.; (right) © Ed Reschke / **41.8** Mark Bekoff / **41.9** (above) Edward S. Ross; (below) E. Mickleburgh/Ardea, London / **41.10** Art, D. & V. Hennings; Photograph, Steven Dalton/ Photo Researchers / **41.11** (a) John Alcock/Arizona State University ; (b) Ray Richardson/ Animals Animals / **41.12** Patricia Caulfield / **41.13** Michael Francis/The Wildlife Collection / **41.14** Frank Lane Agency/Bruce Coleman / **41.15** John Alcock/ Arizona State University / **41.16** Fred Bruemmer / **41.17** (a) John Dominis, *Life Magazine*, © Time, Inc.; (b) A. E. Zuckerman/Tom Stack & Associates / **41.18** (a, b) Kenneth Lorenzen / **41.19** Gregory D. Dimijian/Photo Researchers / **41.20** Yvon Le Maho / **41.21** Lincoln P. Brower / **41.22** F. Schutz / **41.23** John Alcock/Arizona State University

CHAPTER 42 **42.1** (left) Edward S. Ross; (right) Donna Hutchins / **42.2** (a, c) Harlo H. Hadow; (b) Bob and Miriam Francis/Tom Stack & Associates / **42.3** Art, Precision Graphics; Photograph, Clara Calhoun/Bruce Coleman / **42.4** Art, after G. Gause, 1934 / **42.5** Photographs, (a, c) Jane Burton/Bruce Coleman; (b) Heather Angel; Art, (d, e) Precision Graphics from Jane Lubchenco, *American Naturalist*, 112:23–29. © 1978 by The University of Chicago Press / **42.6** Art, after N. Weland and F. Bazazz, *Ecology*, 56: 681–688. © 1975 Ecological Society of America / **42.7** Art, Precision Graphics / **42.8** Art, Precision Graphics; Photograph, Ed Cesar/Photo Researchers / **42.9** © C. James Webb/ Phototake USA / **42.10** (a) James H. Carmichael; (b) Edward S. Ross; (c) W. M. Laetsch / **42.11** Photographs, (all) Edward S. Ross / **42.12** (a, b) Thomas Eisner, Cornell University; (c) Douglas Faulkner/Sally Faulkner Collection; (d) Edward S. Ross/Entymology, 1972, 104:1003–1016 / **42.13** (a–e) Roger K. Burnard; (f) E. R. Degginger; (g) Roger A. Powell / **42.14** Angelina Lax/Photo Researchers / **42.15** (left-inset) John Carnemolla/Australian Picture Library/Westlight; (right) Peter Bird/Australian Picture Library/Westlight / **42.16** Art, Gary Head after W. Dansgaard, et al, *Nature*, 364:218–220, 15 July 1993; D. Raymond et al., *Science*, 259:926–933, February 1993; W. Post, *American Scientist*, 78:310–326, July–August 1990 / **42.17** Art, after S. Fridriksson, *Evolution of Life on a Volcanic Island*, Butterworth: London, 1975; Photograph, Dr. Harold Simon/Tom Stack & Associates / **42.18** (a, b) Art, Gary Head after J. M. Diamond, *Proceedings of the National Academy of Sciences*, 69:3199–3201, 1972; Photographs, (a) © PhotoDisc, Inc.; (c) Frans Lanting/Minden Pictures, computer modified by Lisa Starr/ **42.19** (a, b) Edward S. Ross / **42.20** Heather Angel/Biofotos

CHAPTER 43 **43.1** (a) Wolfgang Kaehler; (b) Frans Lanting/ Minden Pictures / **43.2** Art, Gary Head and Lisa Starr / **43.3** Photograph, Alan and Sandy Carey; Art, Gary Head after R. L. Smith, *Ecology and Field Biology*, fifth Ed. / **43.4, 43.5, 43.6** Lisa Starr with photographs, tallgrass prairie, © Ann B. Swengel/Visuals Unlimited; bacterium, Stanley W. Watson, *International Journal of Systematic Bacteriology*, 21:254–270, 1971; cutworm, © Nigel Cattlin/Holt Studios International/Photo Researchers; gopher and prairie vole, © Tom McHugh/Photo Researchers; sparrow, © Rod Planck/Photo Researchers; yellow spider, Michael Jeffords; frog, John H. Gerard; garter snake, Michael Jeffords; weasel, courtesy Biology Department, Loyola Marymount University; plover, © O. S. Pettingill Jr./Photo Researchers; crow, © Ed Reschke; hawk, © J. Lichter/Photo Researchers; all others © 2000 PhotoDisc, Inc. / **43.7** Art, Precision Graphics and Gary Head; Photographs, (a) D. Robert Franz; (b) Frans Lanting/Bruce Coleman / **43.8** Hans Reinhard/Bruce Coleman / **43.9** Gary Head / **43.10** Gene C. Feldman and Compton J. Tucker/NASA, Goddard Space Flight Center / **43.11, 43.12** Art, Precision Graphics / **43.13** Art, Gary Head; Photograph, Gerry Ellis/The Wildlife Collection / **43.14** Lisa Starr / **43.15** Art, Gary Head / **43.16** Art, Lisa Starr; photograph, © 2000 PhotoDisc, Inc. / **43.17** D. W. Schindler, *Science*, 184:897–899 / **43.18** Art, Lisa Starr after Paul Hertz; Photograph, © 2000 PhotoDisc, Inc. / **43.19** Lisa Starr with NASA photograph / **43.20** Lisa Starr, with photographs from NASA JSC Digital Image Collection / **43.21** Lisa Starr, compilation of historical records from Vostok and Law Dome ice core samples, Barnola, Raynaud, Lorius, Laboratoire do Glaciologie et de Geophysique de l'Environnement; N.I. Barkov, Arctic and Antarctic Research Institute, and Atmospheric CO_2 concentrations from the South Pole; Keeling and Whorf, Scripps Institute of Oceanography; Jones, Osborn, and Briffa, Climatic Research Unit, University of East Anglia; Parker, Hadley Centre for Climate Prediction and Research; Vostok ice core

data set, Chapellaz and Jouzel, NOAA/NGDC Paleoclimatology Program / **43.22** (a) Lisa Starr, compilation of data from Mauna Loa Observatory, Keeling and Whorf, Scripps Institute of Oceanography; photographs, © 2000 PhotoDisc, Inc.; (b) Lisa Starr, compilation of data from Prinn et al, CDIAC, Oak Ridge National Laboratory; World Resources Institute; Khalil and Rasmussen, Oregon Graduate Institute of Science and Technology, CDIAC DB-1010; Leifer and Chan, CDIAC DB-1019; photographs, © 2000 PhotoDisc, Inc. and EyeWire, Inc.; (c, d) Lisa Starr, compilation of data from World Resources Institute; Law Dome ice core samples, Etheridge, Pearman, and Fraser, Commonwealth Scientific and Industrial Research Organisation; Prinn et al, CDIAC, Oak Ridge National Laboratory; Leifer and Chan, CDIAC DB-1019; photograph, © 2000 PhotoDisc, Inc. / **43.23** Art, Lisa Starr, with Gary Head after P. Hertz; Photograph, © 2000 PhotoDisc, Inc. / **43.24** USDA Forest Service / **43.26** After R. Noss, J. M. Scott, and E. LaRoe / **43.27** Courtesy of Matthew Lazzara, Antarctic Meteorological Research Center, University of Wisconsin–Madison

CHAPTER 44 **44.1** (a-above) Edward S. Ross; (a-below) David Noble/FPG; (b-above) Edward S. Ross; (b-below) Richard Coomber/Planet Earth Pictures / **44.2** Edward S. Ross / **44.3** Art, Preface Inc.; Photograph, NASA / **44.4** (b, c) Art, L. Calver / **44.5** Art, Precision Graphics / **44.6** Lisa Starr, using NASA photograph / **44.7, 44.8** Lisa Starr, after The World Wildlife Fund and other sources, and using NASA photograph / **44.11** Lisa Starr, with photographs © 2000 PhotoDisc, Inc. / **44.12** Art, D. & V. Hennings after Whittaker, Bland and Tilman / **44.13** Harlo H. Hadow / **44.14** (above) Dr. John D. Cunningham/Visuals Unlimited; (inset) AP/Wide World Photos; (below) Jack Wilburn/Animals Animals / **44.15** (a) Kenneth W. Fink/Ardea, London; (b) Ray Wagner/Save the Tall Grass Prairie, Inc.; (c) Jonathan Scott/Planet Earth Pictures / **44.16** (left) © 1991 Gary Braasch; (right) Thase Daniel; (below-left) Adolf Schmidecker/FPG; (below-right) Edward S. Ross / **44.17** (all) Thomas E. Hemmerly / **44.18** (a) Dennis Brokaw; (b) Jack Carey; (c) Nigel Cook/Dayton Beach News Journal/Corbis Sygma / **44.19** (a, b) Dr. Peter Kuhry,Arctic Center, University of Lapland-Finland; (c) Roger N. Clark. All rights reserved / **44.20** D. W. MacManiman / **44.21** Art, Precision Graphics / **44.22** Art, Precision Graphics, modified after E. S. Deevy, Jr., *Scientific American*, October 1951 / **44.23** E. F. Benfield, Virginia Tech / **44.24** Art, L. K. Townsend / **44.25** (a) Ulli Seer/© Stone; (b) McCutcheon/ZEFA; (c) after T. Garrison; (d) J. Frederick Grassle, Woods Hole Institution of Oceanography; (e) Robert Hessler, Woods Hole Institution of Oceanography / **44.26** (a) Courtesy of Everglades National Park; (b) © Beth Davidow/Visuals Unlimited / **44.27** Photograph, E. R. Degginger; Art, D. & V. Hennings / **44.28** Courtesy J. L. Sumich, *Biology of Marine Life*, fifth Ed., W. C. Brown, 1992 / **44.29–44.31** Lisa Starr / **44.32** NOAA/Laboratory for Satellite Altimetry / **44.33** (a) © Dennis Kunkel/ Phototake, NYC; (b) © Roland Birke/OKAPIA/Photo Researchers; (c, d) Courtesy Rita Colwell and A. Huq/UMBI and Byron Wood, Brace Lotitz/NASA; (e) Raghu Rai/Magnum Photos / **44.34** Jack Carey

CHAPTER 45 **45.1** Photograph, Gary Head; Maps, U. S. Geological Survey / **45.2** (a, b) Art, Lisa Starr; (c) Photograph, © United Nations / **45.3** Photograph, Heather Angel; Art, after G. T. Miller, Jr., *Environmental Science: An Introduction*, Brooks/Cole, 1986. All rights reserved, and the Environmental Protection Agency / **45.4** From Center for Air Pollution Impact and Trend Analysis (CAPITA), Washington University, St. Louis / **45.5** (above) NASA Goddard Space Flight Center, Science Visualization Studios; (below) National Science Foundation / **45.6** Data from G. T. Miller, Jr. / **45.7** Photograph, R. Bieregaard/Photo Researchers; Art after G. T. Miller, Jr., *Living in the Environment*, eighth Ed., Brooks/Cole, 1993. All rights reserved / **45.8** NASA / **45.9** William Campbell/Time magazine / **45.10** Gerry Ellis/The Wildlife Collection / **45.11** USDA Soil Conservation Service/Thomas G. Meier / **45.12** Agency for International Development / **45.13** Dr. Charles Henneghien/Bruce Coleman / **45.14** Water Resources Council / **45.15** Ocean Arks International / **45.16** After G. T. Miller, Jr., *Living in the Environment*, tenth Ed., Brooks/Cole. All rights reserved / **45.17** Photograph, (a) AP/Wide World Photos; (b) Art, Precision Graphics after, M. H. Dickerson, "ARAC: Modeling an Ill Wind," in *Energy and Technology Review*, August 1987. Used by permission of University of California, Lawrence Livermore National Laboratory and U.S. Dept. of Energy / **45.18** (a, b) Alex MacLean/Landslides / **page 797** J. McLoughlin/ FPG / **45.19** © 1983 Billy Grimes

A Abalone eye, 514, 514i
Abdominal aorta, 563i, 634i
Abdominal cavity, 359, 480i, 511i
Abortion, 186, 187, 676, 677, 680
Abrahamson, W., 252i
Abscisic acid, 462t, 463, 469
Abscission, 468, 468i
Absorption
    from gut, 614, 620–621, 620i, 621i, 630
    by mycorrhiza, 332, 338, 441
    by plants, 439, 440–441, 440i, 441i
Absorption spectrum, 96, 97i, 103
Abstinence, sexual, 676, 677i
Accessory fruit, 456t, 457, 457i
Accessory pigment, 96, 97, 327
Accessory protein, 67, 550
Acclimatization, 605
Accommodation, visual, 514
Acer, 428i, 433, 456i, 457i
Acetabularia, 328i
Acetyl-CoA, 117i, 120, 121i, 124
    Krebs cycle and, 114, 114i, 115i, 117i
Acetylaldehyde, 118, 119i
Acetylation, 214
ACh (acetylcholine), 492, 493, 495, 496,
    504, 550, 622i
Achillea millefolium, 166i
Achondroplasia, 179t, 180, 180i
Achoo syndrome, 168, 179t
Acid, defined, 30, 32, 32t
Acid deposition, 784–785, 785i, 794
Acid rain, 30i, 31, 31i, 784
Acid stomach, 30
Acid-base balance, 30, 602, 638
Acid–base reaction, 83
Acidic solution, defined, 30, 30i, 31, 31i
Acidosis, 31, 625, 638
Acne, 540
Acoelomate animal, 359i, 364–365
Acorn, 255, 255i, 456, 456t
Acoustical signal, 716
Acquired characteristics, Lamarck, 242
Acquired immunodeficiency syndrome
    (*See* AIDS)
Acrasiomycota, 321
Acromegaly, 528, 528i
Acrosome, 658
ACTH, 523i, 526, 526t, 527i
Actin, 43, 66, 66i, 478, 548, 548i, 549i, 550
Actinomycetes, 314
Action potential, 488, 490, 503, 506, 510
    in muscle cell, 550–551, 550i, 551
    nature of, 490–494, 490i–494i
    sensory system and, 516
Activation energy, 82–83, 82i
Activator protein, 214
Active site, defined, 83
Active transport, 86–87, 87i, 90, 90t
    by Na$^+$/K$^+$ pump, 489, 489i, 491i
    at nephron, 636i, 637
    in plants, 445
    at synaptic cleft, 493
Acute inflammation, 580, 581i
Acyclovir, 679
ADA, 591i
Adaptive radiation
    birds, 244, 245i, 265, 265i, 277i, 394
    early animals, 276, 277i, 299, 302, 304
    fishes, 277i, 302, 304, 368, 387, 402
    in Hawaiian Archipelago, 265, 265i
    hominoids, 399, 399i
    insects, 277i, 302, 304
    invertebrates, 302, 304
    mammals, 269, 269i, 276, 276i, 307,
        396, 403, 404, 405i
    nature of, 268–269, 277, 406, 406i
    plants, 276i, 303, 304, 304i, 343, 347
    protistans, 277i, 301
    reptiles, 277i, 304, 392, 394
Adaptive trait, defined, 10–11
Adaptive zone, 268–269, 302, 399
Adenine (A), 46, 46i, 78, 194, 194i, 195,
    195i, 201, 208, 295
Adenosine, 49i
Adenosine deaminase, 591i
Adenosine triphosphate (*See* ATP)

Adenovirus, 317i
Adenylate cyclase, 525
ADH (antidiuretic hormone), 523i, 524t,
    526, 526t, 527i, 528, 636i, 637
Adhering junction, 69, 474–475, 475i, 565i
Adhesion protein, 53, 53i, 653
Adipose tissue, 476t, 477, 477i, 570, 674
    brown, 641
    hormones and, 522, 529t, 530, 532
    metabolism and, 41, 120, 530, 530i,
        612, 622, 628, 629
Adoption behavior, 714
ADP, in ADP/ATP cycle, 79i
Adrenal gland, 500i, 523i, 524, 526
    cortex, 526t, 527i, 529t, 530, 530i, 637
    medulla, 529t, 530, 530i, 533i
Adrenaline (*See* Epinephrine)
Adult, defined, 361
Adventitious root, 430, 458, 461i
Aerobic exercise, 551
Aerobic respiration, defined, 90, 123
    in animals, 384, 549, 549i, 597, 598
    carbon cycle, 748–749, 748i, 750
    in muscle cells, 549, 549i
    origin of, 277i, 299
    overview, 110–111, 111i
    photosynthesis and, 92, 93i, 99, 103,
        110, 122, 296, 445, 448i, 473
    in plants, 110, 431, 432, 436, 445, 473
    in prokaryotes, 298i, 301, 314, 753
    reactions, 109, 110–117, 110i–117i, 123i
Afferent, defined, 499i
Aflatoxin, 47
African blood lily cell, 131i
African clawed frog, 652
African emergence model, 402, 402i
African finch, 253, 253i, 254
African savanna, 7i, 767i
African sleeping sickness, 311i, 323, 323i
African violet, 458i
Africanized bee, 108–109, 108i
Afterbaby blues, 674
Afterbirth, 673i
Agapanthus leaf, 428i
Agar, 327
Agaricus brunnescens, 334, 335
Agave, 353
Age spots, 74, 75i
Age structure, 686, 697i
Agent Orange, 462
Agglutination, 560–561, 560i
Aggregate fruit, 456, 456t, 457
Aging
    Caenorhabditis elegans, 75, 75i
    Down's syndrome and, 185, 185i
    human, 74–75, 75i, 545, 545i, 541, 551,
        575, 628, 631, 674–675
    in life history patterns, 692, 692t
    molecular basis, 74–75, 674–675
    of plants, 432–433
    progeria and, 179t, 182, 182i
Agnathan (Agnatha), 386i, 387
Agriculture, 765
    biological controls for, 251
    crop diseases, 230, 336
    deforestation and (*See* Deforestation)
    in deserts, 765, 791, 792, 792i
    genetic engineering and, 222–223,
        230, 236
    history of, 694–695, 782
    impact on biodiversity, 410, 787i, 788
    irrigation and, 792–793
    land for, 764, 766, 780, 783, 786, 787i,
        788, 791, 791i, 792, 794
    nutrients for, 20, 425, 425i, 747, 753
    pesticides and, 47, 47i, 236, 250–251,
        376i, 414, 742
    population growth and, 694–695, 782
    red tides and, 327
    runoff, 20, 20i, 104, 327, 439i, 776
    soils and, 20, 438, 764
    subsistence vs. mechanized, 788–799
Agrobacterium tumefaciens, 230–231, 231i
AIDS, 230, 325, 325t, 591, 592–593, 592i
Air circulation, global, 405, 405i,
    758–759, 759i

Air pollution, defined, 784
    acids in, 31, 31i, 784–785, 785i
    lichens and, 250, 332, 338
    ozone thinning and, 786
    particulates in, 785, 785i
    stomata and, 445, 445i
    sources, 784–785, 784t, 785i, 786, 794
Air pressure, 598, 598i, 604–605, 759,
    761, 778
Air sac, bird, 601, 601i
Aix sponsa, 37i
Ajellomyces capsulatus, 336
Alachlor (Lasso), 47i
Alagnak River, Alaska, 126, 126i
Alanine, 204i
Alar, 47i
Albatross, 260, 263i, 708i, 733i
Albinism, 162, 163i, 169, 169i, 186, 540
    koala, 397i
Albumin, 45, 232, 394i, 558i
Alcock, J., 714
Alcohol, 36, 63
    at blood–brain barrier, 503, 505
    effects of addiction, 505, 560
    pregnancy and, 673
Alcoholic fermentation, 118–119, 119i, 123
Aldehyde group, 36i
Alder, 420, 728i
Aldosterone, 524t, 529t, 636i, 637
Alexander the Great, gene flow and, 402
Alfalfa, 47, 427i
Alga (algae), 107, 107i, 320, 330t
    blue-green, 314 (*See* Cyanobacterium)
    brown, 311i, 320, 326, 328, 328i, 330t
    characteristics, 92, 94, 320, 326
    chrysophytes, 326–327, 326i
    community structure and, 723, 723i
    golden, 311i, 326
    green, 55i, 97, 1007, 107i, 320, 328–329,
        328i, 329i, 330t, 340i, 742i, 772
    origin of, 298i, 299, 300i
    red, 97, 107, 107i, 320, 326, 327, 327i,
        330i, 412
    reef-forming, 412
    single-celled, 326–329, 326i, 342, 342i
    yellow-green, 326
Algal bloom, 104, 326–327, 327i, 330,
    331i, 747, 747i, 779, 779i
Algin, 328
Alkaline solution, defined, 30
Alkaloid, 336, 517
Alkalosis, 31
Alkylating agent, defined, 209
All-or-nothing spike, 490–491, 491i
Allantois, 394i, 667, 668
Allele, defined, 140, 153, 155, 167, 246
    lethal, 169
    wild-type, defined, 172
Allele frequency, defined, 246, 247, 248
Allergen, 590
Allergic rhinitis, 453
Allergy, 587, 590, 594t
    pesticides and, 47
    smoking and, 609i
Alligator, 392–393, 498i
Allopatric speciation, 261, 264–265, 270
All-or-nothing spike, 490, 491i
Allosteric control, of enzymes, 84, 84i
Aloe vera, 353
Alpha cell, pancreas, 532, 533i
Alpine tundra, 763i, 771
Altruism, 703, 707, 714, 715, 716
Aluminosilicate, 438
Aluminum, 21i
Alvarez, L., 306
Alvarez, W., 306
Alveolar sac, 602i, 603–607, 606i, 607i
Alveolus, 602, 602i, 603, 606, 606i
Alzheimer's disease, 232i, 504, 675
Ama (Asian divers), 641
Amacrine cell, 516, 516i
Amanita, 334i
Amazon basin deforestation, 788, 789i
Amino acid
    abiotic synthesis, 294–295
    absorption from gut, 621, 621i

essential, 625, 625i
    metabolism, 63, 120, 124, 618, 618t
    in protein synthesis, 42–45, 42i–45i,
        201, 204–207, 204i–206i
    sequence comparisons, 284, 290
    structure, 42, 42i
Amino group, 36, 36i, 42, 42i
Aminopeptidase, 618t
Amish, Old Order, 257
Ammonia, 120, 294, 622, 634, 752
Ammonification, 752, 752i
Ammonite, 305i
Ammonium, 752
Amnesia, 504
Amniocentesis, 186, 186i, 187
Amnion, 174i, 394i, 667, 673, 681
Amniote egg, 392, 392i, 394i
Amniotic fluid, 593
Amoeba, 66, 87, 150i, 321, 321i, 322, 515
Amoeba proteus, 322i
Amoebic dysentery, 322
Amphetamine, 505
Amphibian (Amphibia)
    biodiversity through time, 408i
    characteristics, 384, 390–391, 403, 490i,
        502, 511, 512, 557, 557i, 600–601, 601i
    evolution, 302, 308i, 390, 390i, 498i, 557
    existing groups, 358t, 390–391, 391i, 726
Amplexus, 644i
Amplitude, sound, 512i
Amygdala, 504
Amylase, salivary, 617, 618t
Amyloplast, 65
Amylose, 39i
Anabaena, 314, 752
Anabolic pathway, defined, 81, 81i
Anabolic steroid, 553
Anaerobic electron transport, 109, 110,
    110i, 119, 123
Anaerobic metabolism, 110, 118–120,
    298, 300i, 314, 329t, 753
Anagenesis, 268
Analgesic, 505
Analogous structure, 281
Ananas, 457i
Anaphase I, meiosis, 141, 142i, 148t
Anaphase II, meiosis, 141, 142i, 143i, 148i
Anaphase, mitosis, 130, 130i, 131, 131i,
    133, 133i, 136, 149i
Anaphylactic shock, 590
Androgen, 524t, 529t
Anemia, 179t, 187, 560, 626t, 627t
Aneuploidy, 184, 188
Angel Island, human effect on, 684
Angina pectoris, 571
Angiography, 571
Angioplasty, 571
Angiosperm (*See also* Flowering plant),
    341, 342, 343, 351–352, 422, 453
    evolution, 302, 302i, 304, 304i, 342i,
        343, 343i, 348–349
    life cycles, 342–343, 342i, 348, 351, 352i
Angiotensin, 529t, 567, 637
Angraecum sesquipedale, 451
Anhidrotic ectodermal dysplasia, 179t,
    216, 216i, 485
Animal (Animalia)
    body plans, 648–655
    classification, 8, 8i, 9, 287, 358, 358t
    defining characteristics, 358–359, 359i
    evolution, 276, 276i, 298i, 299ff, 356ff.
    evolutionary trends, summary, 380i
    generalized life cycle, 146, 146i
    major groups, 356i, 358ff, 382ff.
    origin of, 99, 298i, 300–301, 302, 360
    "primitive," 356
    reproduction, 138–139, 138i, 146, 644ff.
Animal cell (*See also specific types*)
    characteristics, 56, 57i, 59i, 66, 66i,
        72t, 134i, 135, 135i, 331t
    human embryonic, 648ff.
Animal pole, 649i, 651
Animal-like flagellate, 311i, 322, 323, 323i
Annelid (Annelida) (*See also* Earthworm)
    characteristics, 358t, 370–371, 370i
    evolution, 357i, 367, 372, 380i

Annual (plant), 422, 432
Annual ring, 432i, 433, 433i
Anopheles, 324
Anorexia nervosa, 612
Ant, 709, 731t, 732i
Antacid, 30
Antarctica, 278, 736–737, 736i, 737i, 755, 755i, 786, 786i
Anteater, 397, 397i, 397t
Antelope, 614–615, 615i
Antenna, 370i, 372, 374, 374i, 376, 376i
Anterior (directional term), 359, 480i, 481i
Anterior pituitary, 523i, 526–527, 526t, 527i, 530, 530i, 531, 531i
    human reproduction and, 659, 662, 663i, 664i
Anther, 452i, 453, 454i, 469
Anthocyanin, 97, 97i, 452
Anthozoan, 362, 363
Anthrax, 577
Anthropoid, 398
Anti-acne drug, pregnancy and, 673
Antianxiety drug, 492
Antibiotic, defined, 11, 251
    bacteriophage vs., 14–15, 14i
    pregnancy and, 790
    producers of, 314, 332, 337
Antibiotic resistance, 11, 14–15, 14i, 74–75, 75i, 251, 315, 325
Antibody, 573i, 578t, 579i, 582–583
    actions, 561, 561i, 586, 587i
    monoclonal, 589
    structure, 48t, 576i, 584–587, 584i, 586i
Antibody-mediated immune response, 582, 583, 583i, 586–587, 587i, 588
Anticodon, defined, 204–207, 205i, 206i
Antidiuretic hormone, 526, 526t, 527i, 636i, 637 (*See also* ADH)
Antigen, defined, 582
    receptors, diversity of, 584, 584i
Antigen–MHC complex, 582, 582i, 583i, 584, 586, 587i, 588, 588i, 589
Antigen-presenting cell, 582, 582i, 583i, 586, 587i, 590
Antihistamine, 590
Anti-inflammatory drug, 590
Antilocapra americana, 614–615, 615i
Antioxidant, 74, 624
Antitrypsin, 608
Antrum, 662i
Anus, 616, 616i, 623, 630t
Aorta, 556i, 562, 563i, 564i, 567, 634i
    grasshopper, 385i
Aortic arch, 564i, 567
Aortic body, 607
Ape, 136i, 183, 398–399, 398i
Aphid, 138i, 139, 446, 446i
Apical dominance, 463
Apical meristem, 423, 423i, 430, 460i, 462t
Aplysia, 599, 599i
Apnea, and apneustic center, 607
Apoptosis, 219, 219i, 589, 653
Appendicitis, 623
Appendicular skeleton, human, 542, 543i
Appendix, 573, 573i, 623, 623i
Appetite control, 629
Apple (*Malus*), 456t, 457, 457i
Apple scab, 337i
Aqueous humor, 511t, 515, 515i
Aquifer depletion, 792, 792i
Arabidopsis thaliana, 355, 470
Arachnid, 373
Archaea (*See* Archaebacterium)
Archaeanthus linnenbergeri, 304i, 305i
Archaebacterium (Archaebacteria)
    characteristics, 8, 8i, 9, 52, 70–71, 72t, 104, 287, 312–313, 314, 314i, 330
    classification, 8, 286, 291, 312, 315
    major groups, 314, 314i, 315
    origin, 287, 298–299, 298i–299i, 312
Archaeopteryx, 288, 288i, 395
Archean, 276i, 298
Archenteron, 649i
Archipelago, defined, 265
    allopatric speciation on, 265, 265i
    Galápagos Islands, 242i, 243i, 244, 245i, 265, 265i
    Hawaiian Islands, 265, 265i
Arctic fox, 397i

Arctic tundra, 771, 771i
Ardipithecus ramidus, 399, 399i, 401i
Area effect, biodiversity, 733, 733i
Arginine, 204i
Aristotle, 240
Armadillo, 244, 244i
Armillaria ostoyae, 335
Arnold, S., 704
Arrhythmia, 571, 571i
Arsenic poisoning, 47, 638
Arteriole, 529t, 530, 555, 562, 566, 566i, 567, 574, 580, 620
Arteriosclerosis, defined, 570
Artery, 41i, 555, 562, 566, 566i, 570i, 574
Arthritis, 545
Arthropod (Arthropoda), 357i, 358t
    adaptations underlying success, 372
    ancestry, 357i, 367
    blood circulation, 556, 556i
    major groups, 372–377, 542
Artificial selection, defined, 10
    bees, 108–109
    crop plant, 223, 223i, 239
    dogs, 238–239, 238i, 538
    genetic engineering vs., 223, 239
    natural selection vs., 10–11
    pigeons, 10–11, 10i, 245
    yeast, 337
    zebroids, 271, 271i
Ascaris, 725i
Ascomycota, 334
Ascorbic acid (*See* Vitamin C)
Ascospore, 336, 337
Asexual reproduction, defined, 150
    advantages, 150, 646
    animals, 128, 139, 358, 364–366, 646
    by binary fission, 322, 323
    by fragmentation, 361, 646
    fungi, 334, 336i, 337, 337i
    by gemmules, 361
    by longitudinal fission, 322
    by mitotic cell division (*See* Mitosis)
    by parthenogenesis, 458, 458i
    plants, 128, 139, 451, 458–459, 458t
    by prokaryotic fission, 128, 312t, 316, 316i, 331t
    protistan, 320, 322–325, 329, 329i
    sexual vs., 128, 138, 140, 148, 646
    by tissue culture propagation, 458, 458t
    by transverse fission, 364, 646
    Trichoplax, 360
    by vegetative growth, 458–459, 458t
Asparagine, 204i
Aspartame, 186
Aspartate, 204i
Aspen, 231i, 458, 459i
Aspergillus, 336, 337
Aspirin
    fevers and, 638, 641
    renal failure and, 790
Ass (*Asinus*), 270
Assassin bug, 709
Association area, cerebral cortex, 503
Asteroid, defined, 306, 306i
    impacts, 269, 277i, 279, 302–306, 306i
Asthma, 532, 587, 590
Astigmatism, 515
Atherosclerosis, 570, 638, 675
Atherosclerotic plaque, 222, 222i, 570i, 571
Athlete
    anabolic steroid use by, 553
    blood doping by, 560
    cardiac rates, 571
Athlete's foot, 336i, 578
Atmosphere (air)
    circulation patterns, 405, 405i, 756, 757, 758i
    composition, 596, 598
    global broiling and, 306
    origin and evolution, 99, 122, 293, 294, 298i, 299, 301, 340
    pressure, and respiration, 598, 604–605
Atmospheric cycle, 745, 746, 752
Atoll, 412i
Atom, 4, 6i, 21, 22, 22i, 32, 32t
Atomic nucleus, 8, 22
Atomic number, 22, 22t
Atomic weight, 22

ATP (adenosine triphosphate)
    abiotic synthesis, 295
    active transport and, 86–87, 87i, 489
    roles, 5, 46, 48t, 64, 75, 78–79, 79i, 90t
    structure, 46, 78, 79i, 81i
ATP formation
    chemiosmotic theory, 116i
    by fermentation, 118–119, 118i, 119i
    by glycolysis, 111, 111i, 112, 112i, 113i
    in mitochondria, 64, 110, 114–117, 114i–117i
    in muscle cells, 549, 549i
    in photosynthesis, 65, 99i, 100
ATP synthase, 100, 100i, 114i, 116, 117i
Atrial fibrillation, 571
Atrial natriuretic hormone, 529t
Atrioventricular valve (*See* AV valve)
Atrium, heart, 564, 564i, 565
Atrazine, 47i
Auditory canal, 512i
Auditory nerve, 512i
Australia, human population, 698, 699
Australian sawfly caterpillar, 712, 712i
Australopith, 399
Australopithecus afarensis, 399, 399i
A. africanus, 399, 399i, 401i
A. anamensis, 399, 399i, 401i
A. boisei, 399, 399i
A. garhi, 399, 399i
A. robustus, 399, 399i, 401i
Autoimmune response
    aging and, 675
    in Grave's disorder, 531, 591, 703
    in multiple sclerosis, 494, 507, 591
    in rheumatoid arthritis, 545, 591
    in toxic goiter, 531
    in type 1 diabetes, 533
Automated DNA sequencing, 228, 228i, 232–233
Autonomic nerve, 487, 499i, 500, 500i, 506
Autosomal inheritance, 179t, 180, 180i, 211
Autosome, defined, 172, 184
Autotroph, defined, 92, 123, 739i
    ecosystem roles, 104, 123, 738
    pioneers on land, 340
Auxin, 462–463, 462t, 463i, 464, 464i, 469
AV bundle, 565i
AV node, 565, 565i
AV valve, 564, 564i, 565i
Aves (*See* Bird)
Axial skeleton, human, 542–543, 543i
Axillary bud (*See* Lateral bud)
Axolotl, Mexican, 389, 654, 654i
Axon, 488, 488i
Azospirillum, 314
Azotobacter, 752
AZT (azidothymidine), 593

B B$_{12}$ deficiency, 560
B cell, 559, 559i, 578t, 581–584, 583i, 594t
B lymphocyte (*See* B cell)
Baboon, 708–709, 708i, 713i
Baby-boomer, 696
Bacillus, 313, 313i
Bacillus, 310i, 313, 313i, 577
Backbone (*See* Vertebral column)
Bacterial conjugation, 316, 316i
Bacterial flagellum, 8i, 70, 70i, 72t, 313
Bacteriophage, 55i, 192, 193i, 317
    antibiotics vs., 14–15, 14i
    in DNA studies, 192–193, 193i
    multiplication cycles, 318, 318i
Bacterium (*See also* Archaebacterium, Eubacterium, Prokaryotic cell)
    beneficial, 311, 312, 615, 615i
    characteristics, 8–9, 8i, 312, 312i, 312t, 314–315, 330
    DNA, 214–215, 214i, 215i, 224–225, 224i, 229, 229i, 230
    enteric, 70i, 212–213, 312, 314
    growth, 310–311, 316
    metabolism, 92, 103, 110, 118, 298, 298i, 311, 312–314, 441, 441i, 774
    in ocean, 103, 104, 774
    pathogenic, 11, 70, 70i, 311, 313i, 315, 325t, 571, 575
    reproduction, 310, 316, 316i
Bahama woodstar, 451i
Baking soda, 30, 30i

Balance, sense of, 512, 513
Balanced polymorphism, 254–255
Balancing selection, 254–255
Bald eagle, 410i, 742
Ballerina, skeletal muscle, 548, 548i
Balling, R., 19
Balloon angioplasty, 571
Baltimore oriole, 267, 267i
Bamboo, 425
Banana, 456t
Banana slug, 704, 704i
Bangladesh, 761, 779, 779i
Barbiturate, 505
Bark, 432, 433i, 729
Bark beetle, 731t
Barkley, C., 152i
Barnacle, 374, 374i
Barombi Mbo, Lake, 266i
Baroreceptor reflex, 567
Barr body, 216, 216i
Barr, M., 216
Barrier reef, 412, 412i
Basal body, 67, 67i
Basal cell carcinoma, 212i
Base
    H$^+$-accepting substance, 30, 32, 32t
    nucleotide, 46, 194, 194i, 202, 202i, 618
Base pairing
    DNA, 195–197, 195i, 197i
    with primers, 258, 258i, 260
    with probes, 261, 261i
    in recombinant DNA, 256, 256i
    RNA–DNA, 228, 228i, 230–231
Base sequence, 195, 201, 204i, 235
    evolution and, 284, 284i
Basement membrane, 474i
Base-pair substitution, 208, 208i
Basic (alkaline) solution, 30
Basidiomycete, 334, 334i, 338
Basidiospore, 335
Basilar membrane, 512i, 513i
Basophil, 558i, 559, 559i, 578t, 580, 594t
Bat, 269i, 275i, 307, 450, 450i
    echolocation by, 508, 508i
    wing, 280i, 281, 281i
Bateson, W., 163
Bathyl zone, 774i
Bayliss, W., 522
bcl-2 suppressor gene, *219*
Beagle, voyage of, 242, 242i, 243i
Bean (*Phaseolus*), 454i, 460, 460i, 466i
Bear, 285i, 286, 410i
Becquerel, H., 22
Bee, 8i, 9i, 108–109, 377i, 451, 709, 709i, 714, 714i, 726i (*See also* Honeybee)
Beeswax, 41, 41i
Beetle, 7i, 377i, 450–451, 726, 726i, 731t
    genetic drift experiment with, 256, 256i
Behavior
    altruistic, 703, 707, 714, 715, 716
    bacterial, 315
    body temperature and, 640–641
    brain and, 500–505, 501i
    cannibalistic, 713
    cooperative, 407, 714i
    courtship, 263, 263i, 708i, 709, 710, 710i
    defensive, 712i, 726–727
    environment and, 706, 706t
    evolution of, 246, 707
    feeding, 496, 612–615, 704, 704i
    genetic basis for, 246, 704
    hormones and, 704–705, 704i
    human, 398, 502–505, 705, 705i, 715
    imprinted, 706, 706i, 706t
    instinctive, 705, 705i, 716
    learned, 396, 398, 703, 706, 706t
    mating, 646, 708–711, 710i, 711i
    parental, 395, 711, 711i, 715, 715i
    pheromone and, 517, 522, 708, 716
    play, 501, 708, 708i
    predator-avoidance, 712, 712i
    sexual, 520–521, 520i, 524, 644, 644i, 665
    social, 703, 707, 711–713, 712i, 713i, 715
    territorial, 707
    thirst, 637
    threat, 708, 708i, 713i
Behavioral flexibility, defined, 396
Behavioral isolation, 263, 263i

Bell-shaped curve, constructing, 165
   examples, 165i, 250i, 252i, 253i, 259i
Bengal tiger, wildlife trade and, 410i
Benthic province, 774, 774i
Benzene, 784t
Bermuda grass, 102, 102i, 458t
Berry, 456t
Beta cell, pancreas, 532, 533i
Beta-carotene, 97i, 626t
Bicarbonate (HCO₃)
   in acid–base balance, 31, 638
   breathing and, 602, 606–607
   in carbon cycle, 748i
   in digestive tract, 616, 618, 622
Bicarbonate-carbon dioxide buffer
   system, 31, 638
Biceps, 546, 546i, 547i, 548
Biennial (plant), 422
Big bang theory, 292–293, 308
Big laughing mushroom, 332i
Bighorn sheep, 8i, 9i, 410i
Bilateral symmetry, 358i, 359, 496
Bile, 616i, 621
Bile salt, 41, 621, 621i
Bilirubin, 619
Bill, bird, 244–245, 245i, 252i, 253, 253i
Bill, platypus, 382–383, 382i
Binary fission, 322, 323
Binomial system, species, 286, 289
Biochemistry, comparative, 284–285
Biodiversity (See also Diversity of life)
   analysis of, 407, 415
   basis of, 195, 306–307, 308t
   conservation biology and, 737
   defined, 407
   economic growth vs., 415, 416–417
   island, 732–733, 732i, 733i
   of major taxa, 418i
   patterns, 732–733, 732i, 733i,
      757–758, 757i
   perspective on, 408–409, 797
   realms, 774, 762–763, 762i, 763i
   restoration ecology and, 20, 729
   summary diagrams, 276i, 408i
   threats to, 410–411
   through time, 408–409, 408i, 417
   tropical forests and, 768i, 790
Bioethics
   abortion, 186, 187, 676, 677, 680
   biodiversity, 406–407, 410–417
   cloning, 198, 199, 232
   competition in science, 196
   creating life in test tubes, 236
   deforestation, 790
   fertility control, 676
   gene therapy, 233
   genetically engineered food, 236, 236i
   human cloning, 198, 232, 233
   human impact on environment, 797
   human population growth, 684–685,
      694–697, 699, 700, 782
   and limits of science, 15
   in vitro fertilization, 680
Biogeochemical cycle, 745–753, 745i–753i
Biogeographic realm, 762i, 763, 780
Biogeography, 240, 273, 278–279, 757ff.
Biological clock, 466–467, 469, 523i,
   534, 674–675
Biological control, 251, 725
Biological magnification, 742, 742i
Biological organization, 6, 6i, 16
Biological therapy, disease, 14–15, 14i
Biology, defined, 3
Bioluminescence, 107, 231i, 326, 709
Biomass pyramid, 734
Biome, 757, 762i, 763, 763i, 764i,
   765–771, 765i–771i, 780
Bioremediation, 20
Biosphere, 6, 6i, 756–757, 757i–758i, 780
   autotroph impact on, 104
   human impact on, 750–751, 753, 757
Biosynthesis, in cell cycle, 130–131, 218
Biosynthetic pathway, defined, 81, 81i
Biotech barnyard, 232
Biotic potential, 689
Biotin, 626t
Bipedalism, 398, 399, 399i
Bird (Aves), 41, 240i, 241i, 394–395, 395i,
   403, 515i, 614, 614i, 652i, 726i

behavior, 119, 119i, 263, 702–703,705,
   705i, 707, 708i, 710, 711, 711i, 725, 726i
   biodiversity, 732–733, 732i, 733i
   blood circulation, 557, 557i
   evolution, 288, 394–395, 498i
   life history pattern, 692, 692i
   migratory, 394i, 758, 776i
   as pollinators, 450, 451i
   reproduction, 646i, 647i
   respiration, 601, 601i
Bird of paradise, 254i
Bird song, 534, 534i, 704–705, 704i, 706
Birth control, 676–677, 677i
Birth, human, 259, 259i, 673
Bison, 711i, 767i
Biston betularia, 10i, 250, 251i, 251t
Bivalve, 368, 369i
Black bear, 285i
Black bread mold, 337, 337i
Black duck, 260
Black locust leaf, 428i
Black Sea, 272–273, 272i, 273i
Black widow, 373i
Blackberry, 456t
Blade, leaf, 428, 428i
Blastocoel, 649i, 651, 666, 666i
Blastocyst, 651, 663, 666, 666i
Blastomere, 648, 650, 650i
Blastula, 649i, 651, 652i, 663
Blending theory, 154
Blood, 476t, 477, 477i, 484, 555–558, 558i,
   559i, 562i, 569, 607, 638
   at crime scenes, 227
   filtration, 568i, 569, 580, 636–637, 636i
   loss, and yellow marrow, 544
   pH, 30i, 31, 638
Blood–brain barrier, 503, 568
Blood cell, 544, 544t, 559, 573i, 605, 667
Blood circulation (See also Cardiovascular
   System, Circulatory system)
   control of, 500i, 567
   coronary, 564–565, 565i, 571, 571i
   exercise and, 549, 551, 567
   fetal, 669, 669i
   flow distribution, 567, 568, 568i, 569
   flow resistance, 567
   flow velocity, 556, 557, 568, 569
   flow volume, 555, 556, 562
   gas exchange by, 597, 598, 600, 600i, 606
   hormones and, 530, 567
   invertebrate, 383, 384, 368i, 370i
   lymphatic system and, 572
   nature of, 556–557, 556i, 566–567
   pulmonary, 557, 562, 562i, 574
   systemic, 557, 562, 562i, 563, 566i, 574
   vertebrate, 386, 388, 392, 395, 557, 557i
Blood clotting, 82, 230, 581, 558i, 559, 627t
   in tissue defense, 570i, 571, 572
Blood doping, 560
Blood fluke, 354t, 364, 365, 366, 366i
Blood pressure, 529t, 566–567, 566i, 567i,
   568–569, 568i, 570
   controls over, 570
   in vascular disorders, 367, 560, 567, 574
Blood transfusion, 160, 560
Blood typing, 160, 160i, 560i, 561
Blood vessel (See also specific types)
   invertebrate, 370, 371, 556, 556i
   vertebrate, 556, 563, 566i, 575, 592, 595
Blue jay, 255, 255i
Blue offspring, 179t, 259
Blue whale, overharvesting, 411, 411i
Bluegill sunfish, 712
Blue-green alga (See Cyanobacterium)
Blue-headed wrasse, 264i
Blue-spotted reef ray, 388i
BMI, 628
Boa, 393
Body axis, vertebrate
   directional terms, 480i–481i, 648
   genes governing, 664, 668
Body build, climate and, 401, 401i, 640
Body-building, 124, 553
Body cavity, 358–359, 359i, 480, 480i
Body mass index (BMI), 628
Body piercing, 553, 553i
Body temperature, 640, 641, 656
   fever and, 581, 587, 641
Bog, 344, 344–345, 345i, 771

Bol, M., 528i
Bolus, 617
Bombykol, 517, 522, 708
Bond, C., 236
Bond, chemical, defined, 21
Bone, 527, 543i, 544–555, 544t
   bird, 395, 395i
   disorders, 528, 532, 532i, 545, 545i
   embryonic source, human, 668, 668i
   hormones and, 527, 529t, 532
Bone formation and remodeling, 544
Bone marrow, 544, 559, 573i, 580, 591i
Bone tissue, 476t, 477, 477i, 544, 544i, 545
Bonobo, 398i
Bony fish, 237i, 408i, 388–389, 389i
Booby, blue-footed, 242i
Book lung, 373, 373i, 599
Booster shot, 590
Boreal forest, 762i, 770, 770i
Borrelia burgdorferi, 315, 315i
Bottleneck, 256–257, 257i, 268
Botulism, 118, 315, 496
Bovine spongiform encephalopathy,
   232, 319
Bowman's capsule, 635, 635i, 636i
Boyarsky, V., 538
Brachial artery, 563i
Bradycardia, 571, 571i
Bradyrhizobium, 441i
Brain
   blood flow to, 562i, 563i, 567, 568
   invertebrate, 371, 497, 497i
   metabolic demands by, 20, 121
   sensory pathways to, 494, 495i, 508ff.
   vertebrate, 487, 498–499, 498i, 499i,
      502–503, 502i, 503i, 506t
Brain, human, 117i, 120, 502–503, 504
   development, 625, 672i
   drugs and, 486, 492, 504–505, 504t, 505i
   evolution, 396, 398–400, 498, 502
   number of neurons, 492, 498
   PET scans, 23i, 503i
   structure, 494, 502–503, 502i, 503i,
      506t, 509i, 511, 511i
Brain stem, 502
Breaching whale, 775i
Bread wheat, polyploidy of, 266, 267i
Breast cancer, 137, 628, 657, 674, 677
Breastbone (sternum), 542, 543i, 544
   bird, 395, 395i
Breathing, 602–605, 604i, 610, 611
   bird, 601, 601i
   disorders, 596, 603, 607, 608–609, 611
   frog, 601, 601i
   at high altitudes, 596–597, 605, 605i
Brinster, R., 232
Bristlecone pine, 350, 350i
Britten, R., 283
Brittle star, 378, 378i
Broadleaf forest, 768–769, 768i, 769i
Bronchial tree, 602i
Bronchiole, 602i, 603, 608
Bronchitis, 608
Bronchus, 500i, 603, 608i
Broomrape, 725
Brown adipose tissue, 641
Brown alga, 311i, 320, 326, 328, 328i
   commercial uses, 328
Brown bear, 285i
Brown pelican, 742
Brown, R., 51
Brown recluse, 381, 381i
Bryophyte, 342, 342, 342i, 344i, 345i
   evolution, 342–343, 342i, 343i
BSE, 232, 319
"Bubble boy," 233
Bubonic plague, 681, 694i
Buckeye (Aesculus) leaf, 428i
Bud, plant, 426, 426i, 428i, 432, 432i, 458t
Bud scale, 426i
Budding
   prokaryotes, 316, 322, 331t
   protozoan, 322
   viral, 318, 319i, 592i
Buffer, 36, 607, 616i, 617, 619
Buffer system, 31, 602, 638
Buffon, G. L. Leclerc de, 241
Bug (insect order), 374, 376i, 377i
Bulb, 458t

Bulb of Kraus, 511, 511i
Bulbourethral gland, 656t, 657, 657i, 658i
Bulimia, 612–613
Bulk, 623
Bulk flow, 88, 447i, 568i, 569, 601, 604i
Bull, J., 14, 15
Bulldog, 239, 554, 554i
Bullock's oriole, 267, 267i
Bult, C., 287
Bundle-sheath cell, 102, 102i
Bunkley-Williams, L., 412
Buoyancy, 362, 372, 388
Burgess Shale species, 302i, 356, 356i
Burkholder, J., 327
Burkitt, D., 624
Burn, skin, 479, 540
Bursa, 546i
Butorides virescens, 742i
Butter, 40, 48t
Buttercup, 431i, 726
Butterfly, 691, 725, 726, 758, 758i
   allele frequency examples, 248i–253i
Byssus, 200–201
BYU biology students, 165i

C C3 plant, 102, 102i
C4 plant, 102–103, 102i, 448
Cabbage plant, gibberellin and, 463i
Cacao, 353i, 457
Cactus, 102i, 103, 431, 444i, 445, 757i,
   758i, 765, 765i
Cadherin, 654
Caecilian, 391, 391i
Caenorhabditis elegans, 75, 75i, 365, 365i
Caffeine, 503, 504
Caiman, 396i
Calamites, 344
Calcitonin, 529t, 545
Calcitriol, 529t
Calcium, 21i, 22t, 25i
   absorption, vitamin D and, 530, 541
   atherosclerotic plaque and, 571
   bone and, 530, 544t, 545
   for contraction, 550–551, 551i
   dietary, 627t
   hormones and, 523i, 524, 529t, 532, 545
   loss from watershed, 746, 746i, 753
   membrane crossing by, 86
   metabolism, 532, 541, 544–545
   plant growth and, 439i, 466
   at synaptic cleft, 492
Calcium carbonate, 360, 368, 372, 378
Calcium pump, 86
Calcivirus, 731
Calico cat, 216, 216i
California Current, 760, 760i
California human population, 698–699
California poppy, 430i
Callus, 458
Caloric intake, human, 612, 612i, 626–627
Calvaria major, 409i
Calvin, M., 23
Calvin–Benson cycle, 101, 101i, 103, 105
Calyx, 438, 452, 452i
CAM plant, 102i, 103, 445, 448
Camarhynchus pallidus, 245i
Cambrian, 276i–277i, 302, 302i, 308i
   animals, 302, 302i, 356, 356i, 386
Camel, 397i
Camera eye, 514
Camouflage, 10, 250, 397i, 702, 726, 727i
cAMP, 48t, 215, 525, 525i
Campbell, L., 212, 212i
Camptodactyly, 164, 179t
Canada, human population, 697i
Canada goose, 124, 394i
Canadian lynx, 720, 742i, 725
Cancer, defined, 212
   bone marrow, 170, 560
   breast, 137, 628, 657, 674, 677
   cervical, 137, 679
   characteristics, 136, 218–219, 589
   Chernobyl and, 795i
   colon, 623, 628
   endometrial, 677
   immunotherapy for, 589
   Kaposi's sarcoma, 592, 592i
   leukemias, 560, 795i
   lung, 608–609

metastasized, 218, 218i
prostate, 657
respiratory tract, 589, 609i
skin, 212, 212i, 541, 589, 784
smoking and, 608–609, 609i
stomach, 313i
telomeres and, 675
testicular, 657
therapies for, 23, 137, 589
vaginal, 679
vulva, 679
Candida albicans, 336i, 337
Canine (tooth), 396i, 617, 617i
Cannabis, 505, 609
Cannibalism, 407, 713
CAP, 215
Cape buffalo, 767i
Capillary, blood, 555, 556, 562, 566, 566i
of gills, 557, 600i
glomerular, 635–637, 635i, 636i
during inflammation, 580, 581i
peritubular, 635–637, 635i, 636i
pulmonary, 602, 606
Capillary, lymph, 572, 572i, 573
Capillary bed, defined, 556
at blood–brain barrier, 503
function, 556, 568–569, 568i
human cardiovascular system, 562, 562i
leakage at high altitudes, 596–597
lymph vascular system and, 572–573
of pituitary, 526, 527i
Caplan, A., 236
Capsella, 456, 456i
Capsule, prokaryotic, 70, 70i, 312i, 313
Capture–recapture method, 687
Carbamino hemoglobin, 606
Carbohydrate, 35, 48t, 121i, 618t, 624
classes, 38–39
digestion, 618–619, 618t, 621i
in human nutrition, 613, 624–625
metabolism, 108ff., 120, 120i, 622, 622i
storage, in plants, 101, 446
Carbon, 20, 21i, 22t, 25, 25i, 34–36
central role in life, 34–35, 92, 122
global reservoirs, 748, 749i
numbering, ring structure, 38i
Carbon-14, 277i
Carbon cycle, 748–749, 748i–749i
Carbon dioxide
from aerobic respiration, 110–115
atmospheric, 35, 598, 751i, 789
carbon cycle and, 748, 748i
in integumentary exchange, 599, 599i
membrane crossing by, 83, 83i
in photosynthesis, 92ff., 436, 438, 444
transport in blood, 597, 600ff.
Carbon fixation, defined, 101
global reservoirs, 748–749, 749i
photorespiration and, 102–103, 102i
reactions, 101, 102, 102i
Carbon monoxide, 294, 784t
in cigarette smoke, 570
poisoning, 611
Carbonate (carbonic acid), 30, 31, 602, 748
Carbonic acid–bicarbonate buffer
system, 31, 602, 638
Carbonic anhydrase, 606
Carboniferous, 276i–277i, 302, 303, 343i,
344, 344i, 348, 387, 392, 749
Carbonyl compound, 36i
Carboxyl group, 36, 36i, 40, 40i
Carboxypeptidase, 618t
Carcinogen, 209, 212–213, 609, 793
Carcinoma, 212i, 589
Cardiac conduction system, 565, 585i
Cardiac cycle, 564, 565i, 566, 567i
Cardiac (heart) muscle, 478, 478i, 547,
548, 562i, 564, 565, 565i, 598
Cardiac pacemaker, 554, 565, 565i, 571
Cardiovascular system
diagram, vertebrates, 557i
disorders, 570–571
human, 562–563, 562i, 563i, 566–567
Carica papaya, 470
Carnivore, 9, 738, 739, 739i, 747
Carnivorous plant, 436, 436i, 437i
Carotene, 540
Carotenoid, 65, 97, 97i, 314i, 326, 328, 452
Carotid artery, 563i, 567

Carotid body, 607
Carpal, 543i
Carpel, 154i, 452, 452i, 453, 469
Carrageenan, 327
Carrot, 456t, 458, 459i, 702, 702i
Carrying capacity, 685, 690, 691i, 695,
700, 724
Carson, R., 414, 414i
Cartilage, 180, 476–477, 476i, 476t, 477i,
484, 543, 543i, 545
Cartilaginous fish, 388, 388i
Cartilaginous joint, 545
Cascading reactions, example, 579, 579i
Casparian strip, 440
Caspian tern, 711i
Cat, 85, 85i, 166, 169i, 216i, 257i,
307i, 324i
Catabolism, defined, 81, 81i
Catalase, 74, 74i
Cataract, 675
Catastrophism, 273
Catecholamine, 525
Caterpillar, 4, 4i, 712, 712i
Cattle, 319, 416–417, 416i, 791
cloned, 198, 198i
genetically engineered, 232
effect on riparian zones, 418–419, 418i
methane formation in, 314i, 751i
Cause-and-effect assumption, 12, 16, 750
CCK, 619
CD4 lymphocyte (helper T cell), 592
CD8, 582
cDNA, 225, 225i, 229, 232–233
Cecum, 623, 623i
Celery, 47
Cell (*See also specific types*)
basic features, 52–53, 52i, 53i
defined, 4, 5, 6, 6i, 50, 52
earliest, 296–299, 298i, 299i
early observations of, 50–51, 50i
excitable, 477, 487, 550
movements, 66–67, 67i
number in human body, 126
prokaryotic vs. eukaryotic, 72t, 128, 331t
shape, 28, 52, 54, 66, 72t, 462, 462i, 524
size, 52, 54, 55i
structure, 52–53, 66–70, 66i–70i
Cell communication
in apoptosis, 219, 219i, 589, 653
cardiac conduction system, 565, 565i
at chemical synapses, 492–493, 493i
in contraction, 550, 550i
in embryonic induction, , 654–655, 655i
at gap junctions, 69, 474–475, 475i,
478, 565, 565i, 651, 666i
by hormonal signals, 216–217, 522,
524–525, 524i, 525i, 529–533
in immune responses, 582–589,
583i–588i, 592–593
in inflammatory response, 580–581, 581i
in morphogenesis, 653, 653i, 668i, 680
neural pathways of, 487, 488, 494–495,
495i, 496
at plasmodesmata, 68i, 429
Cell count, human blood, 559, 560, 592
Cell cycle, 127, 130–131, 130i
cancer and, 130, 136, 218–219
controls over, 130, 131, 218
Cell death (apoptosis), 219, 589, 653
Cell differentiation, defined, 213, 645, 680
in animals, 652–653, 652i, 653i, 655i
Dictyostelium, 321i
in frog, 652–653
in human development, 213, 653, 666i
nature of, 198, 213, 216–217, 220
in plants, 426i, 462i
reversing, for cloning, 198
Cell division
animal, 126, 128, 217, 250–251, 526t,
535, 650, 651, 653–655
bacterial, 14, 128, 128t, 316, 316i
cancer and, 136, 137, 170, 218
fungal, 333, 335i, 337i
incomplete, sperm formation, 786, 787i
meiotic (*See* Meiosis)
at meristems, 501, 502, 504, 508i, 511,
511i, 540, 548
mitotic (*See* Mitosis)
nature of, 126–127, 128t 548, 548i, 630

plant, 128, 342, 452, 456, 458, 549,
549t, 460–463
prokaryotic vs. eukaryotic, 14i, 128,
150, 158, 331t
protistan, 372, 372i, 379, 379i
summary of mechanisms, 158t, 331t
Cell from Hell, 326i, 327, 327i
Cell junction
animal, 69, 69i, 474–475, 475i, 478, 651
cardiac muscle, 478, 565, 565i
plants, 68i, 429
Cell membrane (*See also* Plasma
membrane), 38–41, 41i, 52–53, 52i, 70
origin of, 297, 297i
solute crossings, 85–87, 85i–89i
Cell migration, 652i, 653, 653i, 667
Cell plate formation, 134, 134i
Cell theory, 51, 71
Cell wall, 58i, 68–69, 68i, 72t, 89, 89i
bacterial (prokaryotic), 70, 70i, 312i,
313, 313i, 314
fungal, 39, 56, 68
plant, 38, 39i, 56, 57i, 59i, 68, 68i, 424,
424i, 425, 429, 432,440, 442i, 462i, 468
prokaryotic vs. eukaryotic, 313, 330, 331t
protistan, 56, 68, 68i, 327, 328
Cell-mediated immune response, 582–
583, 583i, 588–589, 588i
Cellular work, nature of, 76
Cellulose, 38, 39i, 48t, 68, 615, 615i, 623
Cenozoic, 269i, 276i–277i, 304i, 307,
307i, 343i, 781i
Census, defined, 678
Centimeter, defined, 55i
Central nervous system, 499, 499i, 506
Central vacuole, 57i, 58i, 65i, 72t, 89i
Centriole, 57i, 67, 67i, 132, 132i
Centrocercus urophasianus, 710
Centromere, 128, 133, 136
Centuroides sculpturatus, 373i
Cephalization, 359, 497
Cephalochordate, 384
Cephalopod, 368, 368i, 369, 369i, 514
Cephalothorax, 374i
Ceratium, 68i
Cerebellum, 502, 502i, 506t
Cerebral cortex, 392, 502–504, 503i, 506t,
511i, 517, 517i
Cerebral hemisphere, 502, 502i, 516
Cerebrospinal fluid, 503, 503i, 593
Cerebrum, 502, 502i, 506t
Certhidea olivacea, 245i
Cervical cancer, 137, 679
Cervical cap, 677i
Cervical nerve, 499i
Cervix, 660, 660t, 661i, 663, 665i, 673, 673i
Cesium 137, 20
CFC, 750, 751i, 784t, 786
CFTR protein, 232
Chagas disease, 323
Chain of Being, 240, 242
Chambered nautilus, 302, 413i
Chamberlain, S., 103
Chameleon, 393, 479i, 726
Chancre, defined, 679
Channel protein, defined, 53, 53i
Chaparral, 766, 766i
Chaplin, G., 541
Chargaff, E., 194
Chase, M., 192, 193i
Checkpoint gene, 218
Cheese, 337
Cheetah, 257, 257i, 717
Chelicerate, 372
Chemical bond, defined, 21
Chemical burn, 30
Chemical energy, 76, 85
Chemical equation, defined, 24i
Chemical equilibrium, 80i
Chemical synapse, 492–493, 493i
Chemical work, 76, 78i
Chemiosmotic theory, 100, 100i
Chemoautotroph, 104, 312, 314, 739i, 775
Chemoheterotroph, 312–313, 314
Chemoreceptor, 509, 517
Chemotaxin, 581
Chen, J., 153i
Chernobyl, 20, 795, 795i

Cherry (*Prunus*), 41i, 452i, 454i, 456t
Chestnut blight, 336, 731t, 769
Chiang, V., 231i
Chiasma, 144i
Chick, 652i
Chicken, 163i, 280i
Childbirth, 259, 259i, 483, 526, 526t, 532,
673, 673i
Chimaera, 388, 388i
Chimpanzee, 150i, 283, 283i, 398i, 399i,
401i, 410i, 520, 521i, 713
China, human population, 696–697, 697i
Chipmunk, 722, 722i
Chironomus, 217i
Chiton, 368, 368i, 369
Chlamydia trachomatis, 679
Chlamydial infection, 678, 679, 679i
Chlamydomonas, 55i, 328i, 329, 329i
Chloride, 627t
Chlorine, 21, 21i, 22t, 25i, 26i, 439t, 793
Chlorine monoxide, 786
Chlorofluorocarbon, 750, 751i, 784t, 786
Chlorophyll, 65, 96–97, 97i, 98, 105, 326
structure, 296, 296i
Chlorophyta, 328
Chloroplast, 55i, 58i, 65, 65i, 67, 71, 72t
algal, 326–329, 326i
origin of, 65, 299i, 300–301, 308
structure, 19i, 57i, 94, 94i, 95, 105
Chlorosis, 439t
Chocolate, 353i
Cholecalciferol, 541
Cholecystokinin, 619
Cholera, 587, 695, 751, 779
Cholesterol, 48t, 52–53, 53i, 524, 619, 625
dietary, 179t, 222, 222i, 570–571, 570i
structure, 41, 41i
Chomolungma, 596, 596i
Chondrichthyes, 388
Chondrus, 723i
Chordate, 358t, 383, 383i, 384, 403
Chorion, 394i, 666i, 667, 667i, 669, 681
Chorionic villi sampling, 186–187
Chorionic villus, 667i, 669
Choroid, 511t, 514, 515i
Chriacus, 307i
Chromatid, defined, 128, 128i, 141
Chromatin, 60i, 60t, 61
Chromoplast, defined, 65
Chromosome, 127, 129i
abnormalities, 170, 170i, 172, 179t,
182–185, 182i–185i
defined, 60t, 61, 65
homologous, 140, 141i, 142i, 144–145
human, 129, 141i, 145, 173, 173i
metaphase, 129i, 170, 173, 173i, 179t
sex, 172 (*See also* X chromosome;
Y chromosome)
structure, overview, 128–129, 129i
Chromosome number, 128, 136, 140
changes in, 150i, 179t, 184–185, 280
diploid vs. haploid, 129, 140, 143i, 150
flowering plant, 266, 267i
meiosis and, 140–141, 146–147, 150i
mitosis and, 128–129, 131i
Chronic myelogenous leukemia, 170, 170i
Chrysanthemum, 467, 467i
Chrysophyte, 320, 326, 327, 330, 330t
Chylomicron, 621, 621i
Chyme, defined, 618
Chymotrypsin, 618t, 619
Chytrid, 320, 330
Cicada, 535i
Cichlid, 266, 266i
Ciliary body, 511t, 514
Ciliary muscle, 515i
Ciliophora, 322
Cilium, 67, 67i, 72t, 322–323, 323i
Circadian rhythm, 466, 506t
Circulatory system (*See also* Blood
circulation), 480i, 481, 555–557, 556i,
557i, 562–563, 562i, 566–567, 648
closed, 371, 556, 557, 557i, 574
invertebrate examples, 369, 370i, 556i
lymphatic system and, 557, 572–573,
572i, 578, 580, 620–621
open, 373, 556, 556i
vertebrate examples, 557, 557i

Cirrhosis, 505
Cladogenesis, 268
Cladogram, 286–287
Cladophora gracilis, 742i
Clam, 358i, 369i, 742i
Clark, L., 702
Classification scheme, defined, 286
  (*See also* Appendix I)
  five-kingdom, 287
  based on numerical taxonomy, 312
  phylogenetic, 286
  six-kingdom, 8, 286–287, 287i
  three-domain, 287, 287i
  two-kingdom, 8, 286
Clavaria, 333i
Claviceps purpurea, 336
Clavicle, 543i
Clay, 294, 296, 438
Cleavage, animal development, 648
  frog, 648i, 649i, 650–651, 650i
  human, 666–667, 666i, 680
  mammal, 651, 651i
  mechanism, 134i, 135
  patterns, 367, 651, 654–655, 654i, 655i
  sea urchin, 651, 651i
Cleavage, metabolic reaction, 37
Cleavage furrow, 135, 135i
Cleft lip, 186, 189
Climate
  biodiversity and, 762–763, 762i, 763i
  characteristics, 757, 758–759, 759i, 780
  in history of life, 272, 272i, 274, 276i, 278, 279, 303ff., 399, 797
  ocean currents and, 760, 760i
  primary productivity and, 34–35
Climate zone, 760i
Climax community, 728, 728i
Climax-pattern model, 728
Clitoris, 660, 661i
Clonal selection model, 584–585, 584i
Clone, defined, 140
Cloning
  of DNA, 224–225
  of mammals, 198, 198i, 232
Cloning vector, defined, 225
Closed circulatory system, 556, 556i, 557, 557i, 574
Clostridium botulinum, 315, 496
C. tetani, 315, 496
Clotting (*See* Blood clotting)
Club fungus, 334–335, 334i, 335i, 338
CML, 170, 170i
Cnidarian, 358i, 362–363, 496, 497i
Coagulation, 572
Coal, 303, 344, 344i, 794
Coat color, 85, 85i, 162, 162i, 166, 166i, 526
Cobalt 60, 23
Cobra, 393i
Cobra lily, 437i
Cocaine, 486, 504, 505–506, 506i, 673
Coccolithophore, 326, 326i
Coccus, 313, 313i
Coccyx, 241, 241i
Cochlea, 512i, 513, 513i
Codeine addiction, 505
Codium, 107, 107i
Codominance, allelic, 160
Codon, defined, 204, 204i
Coelacanth, 389, 389i
Coelom, defined, 359, 359i
Coelomate animal, 356i, 359i, 367, 383ff.
Coenzyme, 46, 46t, 81, 626t, 627t
  in glycolysis, 111–112
  in Krebs cycle, 111, 114–115, 114i, 115i, 120, 121i
  in photosynthesis, 79, 92, 93i, 94, 94i, 98
Coenzyme A, 114, 115i, 120, 121i
Coevolution, defined, 450
  parasite/host, 724, 725, 725i
  pathogen/immune system, 325
  pollinator/plant, 436–437, 436i
  predator/prey, 724–725, 724i
Cofactor, 81, 90t
Coffee, 30i, 504
Cohesion, defined, 29, 29i, 443i
Cohesion-tension theory, 442, 443i, 448
Cohort, defined, 692
Coitus, 665

Colchicine, 66, 131, 173, 184
Colchicum, 131, 173
Cold sore, 318, 541
Cold stress, responses to, 641
Coleman, D., 629i
Coleoptera, 377i
Coleoptile, 460i, 461i, 462, 465i
Coleus, 426i
Collagen, 45, 48t, 247, 476, 514, 540, 544, 564, 626t
Collar cell, sponge, 360, 361i
Collarbone, 543i
Collecting duct, nephron, 635, 635i, 636i
Collenchyma, 422, 424, 424i, 434, 434t
Colobus monkey, 397t
Colon (*See* Intestine, large)
Colonizer species, 728–729, 728i, 733, 733i
Color blindness, 177, 179i, 181
Color perception, 516
Columbus, C., 457i
Colwell, R., 779, 779i
Coma, 31
Comet, 304, 306
Commelina communis, 445i
Commensalism, 720, 720t
Communication display, 708–709
Communication signal, 517, 522, 708–709
Community, defined, 6, 6i, 720
Community structure, 719, 720, 720t
  instability vs. stability, 729–730
  keystone species and, 722–723, 723i
  successional models, 728–729, 728i, 729i
Companion cell, 424, 425i, 427i, 446, 446i
Comparative embryology, 282–283, 283i
Comparative morphology, 240–241, 241i, 280–281, 280i, 281i
Competition, biochemical, 296
Competition, biological
  coexistence and, 718–719, 723, 723i
  interspecific, 718–719, 720, 720t, 722–723, 722i–723i, 725, 729
  intraspecific, 244, 249, 685, 686, 722, 712–713, 725
  sexual, 708i, 709, 710, 710i
Competitive exclusion, 722, 722i
Complement, 578t, 579, 579i, 581i, 587
Complete digestive system, 614, 616
Complex carbohydrate, 38, 39i, 48t, 613, 624–625
Complex signal, 708
Complex tissue, plant, 424, 424i, 434, 434t
Composite (flower), 8i
Composite signal, 708
Compound, chemical, defined, 25, 32t
Compound eye, 514, 514i
Compound leaf, 428, 428i
Compound light microscope, 54, 55i
Concentration gradient, defined, 80, 80i
Condensation reaction, 37, 37i, 38i, 295
Condom, 677, 677i
Conduction, defined, 640
Conduction zone, axon, 488, 488i
Cone, 348i, 349, 350, 350i
Cone cell (receptor), 516, 516i
Conidiospore, 336i
Conifer, 34, 34i, 350, 350i, 354t, 355i
  evolution, 303i, 304i, 332, 343i, 349
  representative life cycle, 348i
Coniferous forest, 34, 34i, 349, 349i, 762i, 763i, 764i, 770
Conjugation
  bacterial, 313, 316, 316i
  ciliate, 323
  Spirogyra, 329i
Connective tissue, 476–477, 476i, 476t, 484, 540, 545, 546, 564, 626t, 648
  disorders, 179t, 189
Conservation biology, 407, 414, 417, 737
  hot spots, 414–415, 415i, 763
  human activities and, 416–417
  restoration efforts, 20, 729, 755, 789
Constipation, 623
Consumer, defined, 7, 7i, 738, 739i
Contagious disease, nature of, 694
Continental drift, 278–279, 730
Continental shelf, 774i
Continuous variation, 164–165, 246
Contraceptive, 676–677, 677i
Contractile vacuole, 91, 91i, 322, 323i, 326, 326i

Contraction, 362
  ACh (acetylcholine) and, 495, 495i
  ATP and, 549, 549i
  basis of, 548–599, 548i, 549i, 551, 551i
  calcium and, 543, 550
  cardiac muscle, 478, 564–565, 565i, 571
  skeletal muscle, 548–551, 549i, 550i, 551i
  skeletal systems and, 542
  smooth muscle, 478, 539, 546–547, 617, 618, 619i, 620, 622
Control group, defined, 13
  examples, 12i, 14i, 16, 217i, 231, 463i, 465i, 468i, 702, 704, 723, 725
Convection, defined, 640
Convergence, morphological, 281, 281i, 289, 757i–758i
Convergent circuit, brain, 494
Convergent evolution, 396, 397t, 514
Cooksonia, 275i, 346i
Copepod, 374i, 375, 772, 779, 779i
Copernicus, N., 15
Copper, 21i, 439t, 627t
Copperhead, 646i
Coprolite, 274
Coral, 358t, 363, 363i, 413i
Coral bleaching, 412
Coral reef, 302–303, 356, 412, 412i, 772
Corbin, K., 267
Core temperature, defined, 640
Cork, 50, 50i, 425, 432, 433i, 434t
Cork cambium, 423, 432, 433i, 434t
Corm, 458t
Corn (*Zea mays*), 150i, 230, 286i
  ancestral species, 230
  carbon fixation by, 102, 102i
  chloroplast, 65, 65i
  development, 427i, 430i, 431i, 460i
  gene activity in, 176, 209i, 217, 217i
  rootworm-damaged, 377i
Corn leaf blight, 230
Cornea, 511t, 514, 515, 515i
Corolla, 452, 453i
Coronary artery, 41i, 563i, 571
Coronary bypass, 571, 571i
Coronary circulation, 564
Corpus callosum, 502, 502i
Corpus luteum, 526t, 660, 661i, 661t, 662i, 663, 663i, 664i, 667, 673, 677
Cortex
  cell, 66, 650
  plant matter, 426, 426i, 427i, 430i, 431i
Corticotropin, 526, 526t, 530
Corticotropin releasing hormone, 674
Cortisol, 523i, 524, 526t, 529i, 530, 530i, 639i, 674
Cotton, 231, 231i, 424i
Cottonwood, 765
Cotyledon, 422i, 425, 425i, 456, 456i, 460i
Countercurrent flow, 600, 600i
Courtship, 263, 263i, 708i, 709, 710, 710i
Covalent bonding, 26–27, 27i, 32, 36, 76
Cowbird, 725
Cowpox, 576–577
Crab, 372, 373, 374–375, 374i, 375i, 535i, 741, 741i, 776, 777
Crabgrass, 102i
Crack (cocaine), 486, 505, 505i
Crain, R., 466
Cranial bone, 543i
Cranial cavity, 480i
Cranial nerve, 499i, 500
Crank ("poor man's cocaine"), 505
Crassulacean acid metabolism, 103
Crater Lake, Oregon, 781i
Craterellus, 333i
Crayfish, 497i
Creatine, 549
Creatine phosphate, 549i
Creosote bush, 458, 686, 687i, 765, 765i
Cretaceous, 276–277i, 304, 304i, 306, 306i, 309i, 343i, 393, 396, 405i
Creutzfeldt–Jakob disease, 319
CRH, 530, 530i, 674
Cri-du-chat syndrome, 179t, 183, 183i
Crick, F., 190, 190i, 195, 196
Crinoid, 378
Crocodile (crocodilian), 392–393, 392i, 395, 513
Crop, bird, 614, 614i

Cropland (*See* Agriculture)
Cross-bridge formation, 548–549, 549i, 550–551, 550i
Crossing over, 142i, 144–147, 144i, 171, 172, 172i, 177, 177i, 246
Crosslinking protein, 66
Crown, tooth, 615, 615i, 617, 617i
Cruise, T., 152i
Crustacean, 358i, 372, 374–375, 374i, 418i, 497i, 514, 535i, 775i
Cryphonectria parasitica, 336
Cryptosporidium, 324, 325i
Crystallin, 652
Cuckoo, 705, 705i
Cucumber, 456t
Cucurbitacin, 377i
Culture, defined, 398
Curie, M., 22
Cuticle
  animal, 365, 371, 372, 374
  plant, 41, 41i, 68, 69i, 342, 344, 354t, 425, 425i, 444–445, 444i, 445i
Cutin, 41, 444
Cuttlefish, 369i
Cyanobacterium, 314, 315i, 332, 338, 430, 752, 772, 773, 781i
Cyanophora paradoxa, 301i
Cycad, 303, 303i, 304i, 305i, 343i, 350, 350i
Cyclic AMP (cyclic adenosine monophosphate), 215, 525, 525i
Cyprinodon variegatus, 742i
Cyst, 322, 324
Cysteine, 204i, 625, 625i
Cystic fibrosis, 73, 73i, 179t, 232
Cytochrome, 84, 266, 284, 284i, 627t
Cytokeratin, 53
Cytokinesis (*See* Cytoplasmic division)
Cytokinin, 462t, 463, 469
Cytomembrane system, 62–63, 62i, 63i, 71
  polypeptide chains in, 206–207, 206i
Cytoplasm, defined, 51, 52, 60i
Cytoplasmic division (cytokinesis), 126–127, 128t, 134–135, 134i
Cytoplasmic localization, 651, 654, 655
Cytoplasmic streaming, 67
Cytosine, 46, 46i, 194, 194i, 201, 208, 209
Cytoskeleton, 56t, 57i, 65–66, 66i, 70, 72t
Cytotoxic cell, 582, 583i, 588–589, 594t

D 2,4-D, 462
da Vinci, Leonardo, 274
Darlingtonia californica, 437i
Darwin, C., 239, 242i
  arriving at natural selection theory, 3, 10–11, 154, 244–245, 249, 261
  on phototropism, 464
  voyage on *Beagle*, 242–243, 242i
Daucus carota, 702i
Dayflower, guard cells, 445i
Daylength
  pineal gland and, 534
  plant growth and, 466–467
Day-neutral plant, 467, 469
ddI, 593
DDT, 47, 47i, 742, 742i
Deamer, D., 297
Decibel, defined, 512i, 513
Deciduous forest, 762i–764i, 768–769, 769i
Deciduous plant, defined, 428
Decomposer, 7, 332, 738, 739i, 747
  ecosystem roles, 7, 7i, 8i, 9, 35, 752, 768, 769, 772
Deductive logic, defined, 12
Deer, 684, 687
Deer tick, 315i
Deerfly eye, 514i
Defecation, 623
Definitive host, 364
Deforestation, 349, 349i, 746, 746i, 749i, 751i, 753, 788–790, 788i, 789i
Degradative pathway, defined, 81
Dehiscent fruit, 456i
Dehydration, 89, 528, 533, 578, 625, 637
Deinonychus, 305i
Delbrück, M., 192
Deletion
  base-pair, 208, 208i
  chromosomal, 183, 183i
Delphinium decorum, 282, 282i

D. nudicaule, 282, 282i
Delta cell, pancreas, 532
Deltoid, 547i
Demographic transition model, 698, 698i
Demographics, defined, 686
Denaturation, defined, 45
Dendrite, defined, 488, 488i
Dendritic cell, 578t, 582, 582i, 594t
Dendrobates, 726
Dendrocopus pubescens, 252i
Denitrification, 753
Dense irregular connective tissue, 476, 476i, 476t, 484
Dense regular connective tissue, 476, 476i, 476t, 484
Dentin, 617, 617i
Dentition, 396
Deoxyribonucleic acid (*See* DNA)
Deoxyribose, 38, 46i, 194, 194i
Depolarization, membrane, 493
Depo-Provera injection, 677, 677i
Depressant, 505
Derived trait, defined, 286
Dermal tissue, plant, 421–423, 422i, 425
Dermaptera, 377i
Dermatitis, 626t
Dermis, 511i, 540, 540i
DeRouin, T., 15
Desalinization, 792
Desert, 632, 632i, 762i, 764i, 765, 765i
Desertification, 731, 765, 789, 791, 791i
Designer organ, 479
Desmid, 328
Detrital food web, 741, 741i
Detritivore, 738, 739, 739i, 741, 741i
Detritus, 371, 776
Deuterostome, 367
Development, defined, 4, 460
  control of, 213, 216–217, 219, 650–655
  evolutionary constraints on, 655
  growth vs., 460
Development, animal, 356
  cell differentiation in, 213, 652–653
  chick, 652i
  cleavage patterns in, 367, 651, 651i
  contraints on, 646–647, 655
  division of labor and, 372, 473
  frog, 644–645, 644i, 648i, 649i, 650–651
  insect, 655, 655i
  mammalian, 647, 651i, 653
  master genes for, 654–655
  morphogenesis in, 645, 653, 653i, 680
  morphological convergence in, 282
  pattern formation in, 654–655
  primary tissue layers, 358, 367, 367i
  sea urchin, 651, 651i
  stages, overview, 648–649, 648i
Development, human
  drug abuse and, 672–673
  embryonic, 135i, 175i, 554, 666–673, 66i–673i
  fetal, 669–673, 670i–672i
  hormones and, 523i, 529t, 532
  postnatal, 674–675, 675i
Development, plant
  controls over, 213, 217, 217i, 462ff.
  nature of, 421, 422–423, 462
  primary growth in, 423, 423i., 426–431, 426i–431i, 440, 446, 450, 460–461, 460i
  secondary growth in, 423, 423i, 432–433, 432i–433i
Devonian, 276i–277i, 303, 303i, 309i, 343i
Diabetes insipidus, 528
Diabetes mellitus, 479, 533, 570, 628
Diabetic, making insulin for, 230
Diabrotica virgifera, 377i
Dialysis, kidney, 638
Diaphorase, 259
Diaphragm, birth control device, 677
Diaphragm, muscle sheet, 359, 562i, 564i, 602i, 603, 604, 604i
Diarrhea, 578
Diastole, cardiac cycle, 565i, 566, 566i, 567i
Diatom, 326, 326i, 687, 772
Dickensonia, 302i
Dicot (Dicotyledonae)
  characteristics, 354t, 425i, 428i, 429
  classification, 342, 351, 354t
  development, 456i, 460i, 461i

growth, 426, 426i
  life cycle, 452, 454i
  monocot vs., 425, 425i, 427i
  primary root, 430, 431, 431i
Dictyostelium discoideum, 321i, 653
Dieting, 612, 628
Diffusion, defined, 75, 80
  at capillary bed, 568–569
  gas exchange and, 598, 606–607, 606i
  nature of, 80–89, 80i
Digestion, defined, 614, 630
  carbohydrate, 613, 618, 619, 618t, 621i, 622, 622i, 624
  by carnivorous plants, 436, 436i, 437i
  enzymes of, examples, 62, 85, 618t, 627t
  extracellular, 320, 332, 334
  fat (lipid), 41, 618t, 619, 621i
  intracellular, 56t, 63, 72t, 582
  protein, 613, 618–622, 618t, 621i, 622i
Digestive system, 613, 562i, 563, 613
  complete, 365, 614, 616
  earthworm, 371, 371i
  feeding behavior and, 614–615
  flatworm, 364–365, 364i
  human, 481, 481i, 616ff., 616i, 630t
  incomplete, 614, 614i
  ruminant, 615, 615i
Digitalin, 353
Dihybrid cross, 158, 159i, 167
Dimetrodon, 303i
Dinoflagellate, 68i, 326, 326i, 327, 330, 330t, 363, 412
Dinosaur, 277i, 288, 304–306, 305i, 308, 309i, 392, 392i, 395
Dionaea muscipula, 436, 436i
Dioxin, 462–463
Diphtheria, 590, 590i
Diploidy, 129, 140, 141i, 150, 150i
Diptera, 377i
Directional selection, 250–251, 250i, 258
Disaccharidase, 618t
Disaccharide, 38, 38i, 48t, 618t
Disease, defined, 179
  nature of, 323–325, 325t, 576–577
  in plants, 230, 319, 320, 320i, 377i
  population growth and, 694–695
Dismorphia, 726i
Dispersal, geographic, 260, 730–731, 756
Disruptive selection, 253, 253i, 258
Dissolved substance, defined, 29
Distal (directional term), 481i
Distal tubule, nephron, 635, 635i, 636i
Distalless gene, 655
Distance effect, biodiversity, 733
Disulfide bridge, 45i, 49
Divergent circuit, brain, 494
Diversity of life (*See also* Biodiversity)
  molecular basis, 4, 8, 10, 122, 190, 195
  nature of, 3, 8–9, 11, 16, 16t, 408, 418i
Division of labor, 372, 473
Dixon, H., 442
DNA, 48t
  bacterial, 70, 70i, 72t, 214–215, 214i, 215i, 312i, 312t, 313, 316, 316i
  basis of life, 4, 5, 8, 122, 190, 195, 297
  cloning, 140, 198, 225, 232
  discovery of, 190
  eukaryotic, 60, 61, 72t
  functions, 43i, 46, 72t, 192–193, 297
  hormones and, 524–525, 524i
  human, 224, 227–229
  hybridization, 225, 225i, 229, 284, 402i
  mammalian, 224, 232
  mitochondrial, 266, 301, 402i
  structure, 27, 27i, 46, 46i, 55i, 190, 191, 191i, 194–195, 195i, 199, 297
DNA amplification, 226, 226i, 229
DNA fingerprinting, 227, 227i, 714
DNA ligase, 197, 224
DNA polymerase, 226, 226i, 228
DNA repair, 197, 197i, 212
DNA replication
  in cell cycle, 130, 130i, 218
  nature of, 196, 196–197, 197i, 199
  transcription vs., 201, 202–203, 220
DNA sequencing, 227, 227i, 228, 228i, 232–233, 235
Dodo, 409, 409i
Dog, 238–239, 538, 538i, 554, 554i

Dolichohippus, 271
Dolly (the sheep), 198, 198i, 232
Dolomedes, 17, 17i
Dominance, allelic, 160, 160i, 155, 211
Dominance hierarchy, 520, 521, 712–713
Dopamine, 492, 504, 505
Dormancy, plant, 34, 468
Dorsal lip, 654, 654i
Dorsal surface, defined, 359, 480i
Dosage compensation, 216, 266
Dosidiscus, 368i
Double fertilization, 352i, 455, 455i, 469
Double helix, 195i, 203i (*See also* DNA)
Doubling time, population growth, 689
Douching as birth control, 676, 677i
Douglas fir, 420i, 421i, 468i
Douglas, M., 151, 151i
Dover, C. L. van, 103
Dover, England, 322i, 326
Down's syndrome, 179t, 184, 185i
Downing, 777
Downy mildew, 320, 320i
Downy woodpecker, 252, 252i
Dragonfly, 377i
Dromaeosaurus, 288i
Drosophila melanogaster
  chromosomes, 150i, 176, 176i
  development, 176, 176i, 654i, 655i
  experiments, 176, 176i, 189, 189i, 655, 684, 684i
  species, Hawaiian Islands, 265
Drought, 751, 791, 791i, 778
  stomatal response to, 102–103
Drug, defined, 504
  addiction, 504, 504t
  effects on brain, 504–505, 505i
  pregnancy and, 672–673, 672i
Drupe, 456t, 457i
Dry shrubland, 762i, 766–767
Dry woodland, 762i, 766–767
Dryas, 728i
Duchenne muscular dystrophy, 221
Duck, 37i
Duck-billed platypus, 382–383, 382i
Duckweed, 217
Dung beetle, 7i
Duodenum, 616i, 619i
Duplication, chromosome, 182, 182i
Dust Bowl, 766, 791i
Dutch elm disease, 731t, 742
Dwarfism, 179t, 221, 528, 528i, 537
Dynein, 67, 67i, 133
Dysentery, amoebic, 322

E
Eagle Nebula, 292i
Ear, human, 512, 512i, 513, 513i, 670i
Eardrum, 512i
Earlobe trait, 152, 152i
Early wood, 433
Earth, view from space, 1i
  crust, 21i, 278–279, 279i, 294
  origin and evolution, 276i, 292ff., 308i
Earthworm, 150i, 358t, 370i, 371i, 542, 556i
Earwig, 377i
Easter Island, human impact on, 406–407
Eating disorder, examples, 612–613
Ebola, 595
Ecdysiotropin, 535i
Ecdysone, 217, 217i, 535
ECG, 554, 554i, 571, 571i
Echinocerus, 757i
Echinoderm, 358t, 378–379, 386, 738i–739i
Echolocation, 508, 509i
Ecological pyramid, 734
Ecology, defined, 685
Economic growth
  biodiversity vs., 414–415
  demographics and, 698–699, 698i
Ecoregion, 415, 415i, 763
Ecosystem, defined, 6, 6i, 738
  conservation of, 20, 407, 414–417, 737
  endangered, 755
  energy flow through, 6–7, 737, 738i, 739–741, 739i–741i, 743–744, 744i, 754
  hydrothermal vent, 103, 774–775, 775i
  kelp bed as, 311, 311i, 328
  nutrient availability in, 738, 745–753, 745i–753i, 754, 775i, 776

participants, 738–739, 738i, 739i, 754
  water pollution and, 230, 773, 776, 792–793, 792i, 793i
Ecosystem modeling, 742
Ecstasy (MDMA), 505
Ectocarpus, 328
Ectoderm, 358, 481, 648, 649i
Ectoparasite, 725
Ectotherm, 640
Edema, 569, 580–581, 581i, 597
Ediacaran, 302, 302i
Eel, 389i
Effector cell, immunity, 582ff.
Effector, nervous system, 482, 482i, 487, 487i, 506
Efferent, defined, 499i
EGF, 218
Egg (*See also* Oocyte), 4–5
  amniote, 392, 394i
  amphibian, 390
  bird, 394–395, 394i, 646i, 647, 651i, 742
  Drosophila, 655i
  energy investments in, 646i, 647
  frog, 55i, 644, 645i, 649i, 650, 650i, 651
  human, 125i, 212, 660–663, 665, 665i, 676
  invertebrate, 4i, 7i, 361, 365–367, 365i, 366i, 373–376, 374i, 646i, 647, 651, 651i
  mammal, 382, 396, 397i651i, 651i
  plant, 347i., 450ff.
  reptile, 392, 392i, 393, 646i
Egg sac (case), spider, 373, 646i
Egg white, 30, 30i, 394i
Eichhornia crassipes, 735
Eicosanoid, 41
Einkorn, 267i
Ejaculation, penis, 665
Ejaculatory duct, 656, 656t, 657i
El Niño, 777, 778–779
El Niño Southern Oscillation, 778, 779i
Elastin, 69, 540
Electric gradient, defined, 80
Electrocardiogram, 554, 554i, 571, 571i
Electrochemical work, defined, 76
Electromagnetic spectrum, 96, 96i
Electron, 4, 5, 6i, 32t
  characteristics, 21–25, 24i, 25i, 32, 37, 75
Electron microscope, 54, 55i
Electron transfer, defined, 37, 75
Electron transport phosphorylation, 111, 111i, 116–117, 116i, 117i
Electron transport system, 79, 98, 100i, 123
  of bacterial plasma membrane, 119
  of chloroplast, 99i, 100i
  of mitochondria, 111, 114i, 116, 116i
Electronegative atom, 27
Element, 20–21, 22, 22t, 32, 32t, 293
  essential, for plants, 438–439, 439t
Elephant, 7i, 281, 307
Elephantiasis, 367, 569
Elimination, digestive tract, 614, 630
Ellis-van Creveld syndrome, 178i, 257
Elm, 731t, 769
Elodea, 19i, 99i
Embolus, 571
Embryo, defined, 186, 648
Embryo sac, flowering plant, 455, 455i
Embryonic disk, human, 667, 667i, 668
Embryonic induction, 654–655, 654i, 655i
Emigration, 255, 688
Emphysema, 608, 608i
Emu, 240i, 241i
Emulsification, 619, 621i
Enamel, tooth, 617, 617i
End product, defined, 81, 90t
Endangered species, 257, 410, 410i, 730
Endemic disease, 325, 410
Endemic species, 763
Endergonic reaction, 78, 79i
Endler, J., 693
Endocrine gland, defined, 475 (*See also* specific types; Hormone, animal)
Endocrine system, 480i, 522i, 523ff.
Endocytosis, 62i, 87, 87i, 90, 569, 585
Endoderm, 358, 648, 649i, 652i
Endodermis, 430i, 431, 440, 440i
Endogenous pyrogen, 581
Endometriosis, 661
Endometrium, 660, 661i, 661t, 663, 663i, 664i, 666i

Endoparasite, 725, 725i
Endoplasmic reticulum (ER), 56t, 57i, 58i, 59i, 62, 62i, 63i, 72t
  possible origin of, 300, 301i
Endorphin, 492, 505
Endoskeleton, 386, 539, 542, 552
Endosperm, 351, 351i, 352i, 454, 456, 456i
Endospore, 315
Endosymbiosis, 293, 299i, 326
  in origin of eukaryotes, 300–301, 721
Endothelium, 564, 564i, 568, 568i
Endotherm, 640
Energy carrier, 81
Energy, defined, 5, 7, 76
  forms of, 76, 78i, 90
  for human population, 794–796, 794i
  levels in atom, 24, 24i
  nature of conversions, 76–79, 76i–78i
  reserves in human body, 120
Energy flow, 744
  ecosystem, 6–7, 737, 738i, 739–741, 739i–741i, 743–744, 744i, 754
  major pathways, 92, 154, 738i, 739, 739i
Energy pyramid, 734, 744i, 745, 745i
Energy sources
  for cellular work, 38, 40–41, 530, 530i, 548, 622i
  for ecosystems, 738ff.
  for human population, 794–796, 794i, 795i, 796i
  at hydrothermal vents, 103
  sunlight, 7, 7i, 75, 77, 77i, 92, 96, 738ff.
Englemann, T., 92, 96
Enhancer (base sequence), 214, 220, 283
Enkephalin, 505
Enriquez, C., 325
ENSO, 778, 778i, 779i
Entamoeba histolytica, 322
Enterobacterium, 626t
Enterobius vermicularis, 366
Enteromorpha, 723i
Entropy, 77
Environment
  enzyme activity and, 84–85,
  evolution and, 11, 99, 103, 106, 209, 242, 244–245, 246ff., 292ff.
  gene expression and, 166, 166i, 706
  hormone action and, 534–535, 534i, 535i
  internal (*See* Internal environment)
  plant responses to, 340, 465–468
  population growth and, 694, 699
Environmental Protection Agency, 47i
Enzyme (*See also specific types*)
  activity, control of, 82, 84, 84i, 214, 218
  breakdown, aging and, 74–75
  function, 4, 5, 33, 46, 75, 78, 81, 82–85, 82i–84i, 90, 638
  of metabolic pathways, 81, 84, 84i
  models, 81, 83i
  pH and, 85, 85i, 638
  structure, 43, 83, 83i
  temperature and, 84–85, 85i, 264
Eocene, 276i–277i, 307, 307i
Eosinophil, 558i, 559i, 578t, 580, 594t
EPA, 47i
Ephedra, 350, 350i
Epidemic, defined, 325
Epidermal growth factor (EGF), 218
Epidermis, animal, 362, 362i, 479, 511i, 540, 540i
Epidermis, plant, 34, 422i, 423, 423i, 425, 425i, 427i, 429, 434, 434t
Epidermophyton floccosum, 336i
Epididymis, 656, 656t, 657i, 658i
Epiglottis, 602i, 603, 603i, 611, 617
Epilepsy, 507
Epinephrine, 523i, 525, 529t, 530, 567
Epiphyte, 347, 768i
Epistasis, 162
Epithelium, 66i, 472–473, 472i, 484, 540
  cnidarian, 362, 362i
  glandular, 474, 474i, 475
  intestinal, 620, 620i, 621
EPSP, 493
Epstein-Barr virus, 560
Equilibrium
  chemical, 80i
  genetic, 246–248, 249i, 258
  sense of, 511, 512, 513

Equisetum, 150i, 346, 346i
ER (*See* Endoplasmic reticulum)
Ergotism, 336
Erosion, 439, 753
Erythroblastosis fetalis, 561
Erythrocyte (*See* Red blood cell)
Erythropoietin, 529t, 559i, 605
Escherichia coli
  beneficial strains, 314, 316
  biological therapy and, 14i, 15
  gene control in, 214–215, 215i
  genome, 229
  harmful strains, 14i, 70i, 199, 315, 325
  plasmids, 36
  size/structure, 15i, 55i, 70i, 312i, 313i
  uses in research, 192, 193i, 224t, 230
Esophagus, 603, 616, 616i, 617, 630t
EST, 233
Estrogen
  breast cancer and, 674
  endocrine roles, 523i, 526t, 529t
  in menstrual cycle, 531, 531i, 661–663, 661i, 664i
  sexual traits and, 174i, 704i, 705
  structure, 37i, 41, 524t
Estrous cycle, 660
Estuary, 742i, 776, 776i
Ethanol, 118, 119i
Ethylene, 462t, 463, 465i
Eubacterium (Eubacteria)
  characteristics, 8, 8i, 9, 70, 72t, 312–315
  classification, 8–9, 287, 287i, 312, 315i
  origins, 298–299, 312, 312i
  representative groups, 298i, 312–313
Eucalyptus, 766
Eucheuma, 327
Eugenic engineering, 233
Euglena, 300i, 326, 326i
Euglenoid (Euglenophyta), 326, 330, 330t
Eukaryotic cell (*See also specific types*)
  characteristics, 9, 50ff., 300i, 320
  defining features, 52, 56, 56t, 57i, 72t
  origins, 293, 298–301, 298i–301i, 320, 721
  prokaryotic vs., 72t, 320, 331t
Eupenicillium, 336i
Euphorbiaceae, 757i, 758i
Euplectella, 361i
Eurosta solidaginis, 252, 252i
Eurytoma gigantea, 252i
Eutherian, 396
Eutrophication, 747, 747i, 773
Eutrophy, 773, 781i
Evaporation, defined, 29, 640
Evaporative heat loss, 641
Evaporative water loss
  animal, 29, 634, 640
  plant, 29, 29i, 68, 442–443, 443i
Evergreen, 428
Evergreen broadleaf forest, 768, 768i
Evolution, biological (*See also specific organisms*; Natural selection)
  branching vs. unbranched, 268, 268i
  convergent, 281, 281i, 514
  cultural vs., 400–401
  defined, 3, 10, 13, 16, 239, 241
  endosymbiosis in, 293, 299–301, 326, 721
  eukaryotic, 298i, 299ff., 320
  history of, 240–241, 258, 259, 272, 276i, 292ff., 408–409, 340i, 356, 360
  joint (coevolution), 325, 436–437, 436i, 450, 724–725, 724i, 725i
  mechanisms of, 238, 246ff., 262ff., 283
  molecular basis, 10–11, 16t, 120, 195, 247
  of pathogens, 230, 325
  perspective on, 120, 408–409, 797
  prokaryotic, 298–300, 298i, 300i, 312
  role of transposons, 283, 283i
  sexual reproduction and, 140, 246–248, 254, 646, 682
Evolution, molecular
  of aerobic respiration, 98, 120, 298–299
  of carbon-fixing pathways, 103
  of chlorophyll, 103, 296
  of DNA, 284, 296–297
  of enzymes, 103, 264, 295, 296
  of metabolic pathways, 120, 296, 320
  in origin of life, 292–297, 297i
  of photosynthesis, 98, 103, 122, 298, 320
Evolutionary tree diagram, 270, 357

examples, 269i, 285i, 343i, 357i, 383i
  how to construct, 268, 268i, 285, 285i
  of major taxa, 276i, 285i, 298i, 408i
Evolutionary trend
  animals, 358, 380i
  plants, 342–343, 343i, 308i
  vertebrates, 386–387
Excitatory postsynaptic potential, 493
Excretion, 475, 636i
Exercise, 551
  body-building, 124, 553
  bone density and, 545
  cardiovascular effect, 555i, 567, 569–571
  dieting and, 612
  heat production during, 641
Exergonic reaction, 78, 79i
Exhalation, 604, 604i, 605i
Exocrine cell, defined, 532
Exocrine gland, defined, 475, 475i, 517
Exocytosis, 62i, 87, 87i, 90, 569
Exodermis, plant, 440, 440i
Exon, 203, 203i, 225
Exoskeleton, 372, 542, 552
Exotic species, 730–731, 737 (*See also* Species introduction)
Expansion mutation, 181
Experiment, defined, 16
Experiment, examples
  abiotic synthesis of membranes, 297
  abiotic synthesis of organic compounds, 294–295, 295i
  aphids and phloem, 446
  artificial-sponge colonization, 730
  auxin effects, 462i, 464, 465i
  bacterial transformation, 192
  bee hybridization, 108–109
  biological therapy, 14–15, 14i
  bioluminescence, 231
  bird song, 704i
  caterpillar defensive behavior, 712
  circadian rhythm in plants, 467i
  cloning mammals, 198, 198i, 232
  competitive exclusion, 722, 722i
  computer simulation of origin of biological molecules, 297
  cytoplasmic localization, 650i
  dihybrid cross, 158–159, 159i
  DNA function 192, 193i, 199
  DNA replication, 199
  dormancy, fir seedlings, 468i
  Drosophila, 176, 176i, 189, 189i, 655, 684, 684i
  embryonic induction, 654, 654i
  Englemann's, 92, 92i
  eutrophication, 747, 747i
  flowering response, 467, 467i
  foolish seedling effect, 463, 463i
  frog neural function, 501
  gene therapy, 222
  gene transfers, 230–231, 231i, 233
  genetic engineering, 231, 231i, 232, 232i
  gibberellin effects, 463, 463i
  heart function, activity, Jimmy, 554, 554i
  Helmont's, 107
  keystone species, 722–723, 723i
  learning, bird song, 706
  mechanically stressed plant, 465, 465i
  Mendel's, 154ff.
  minimal organisms, 236
  monohybrid cross, 156, 157i
  natural selection, guppies, 693
  nature's, 223
  nesting behavior, 702–703, 703i
  nitrogen and plant growth, 441i
  obese mice, 629, 629i
  pancreatic secretion, 522
  photorespiration, tomato plants, 103
  plant hormones, 463i
  predator/prey interactions, 725
  pressure flow theory, 446
  raven calling behavior, 707
  receptive fields, visual stimuli, 514–516
  salmon salt tolerance, 639i
  self-replicating systems, 296–297
  snake feeding behavior, 704, 704i
  sound and inner ear, 513
  starling nest building, 702–703, 703i
  stomatal action, 445i
  testcross, 157

tissue culture propagation, 230, 458–459, 459i
  tonicity and plant wilting, 444i
  tonicity and red blood cells, 88i, 89
  wastewater treatment, 791, 791i
  watershed disturbance, 746, 746i
Experimental design, 12i, 13
Experimental group, 12i, 13, 14i, 16, 217i
Exponential growth, 685, 688i, 689
Expressed sequence tag, 233
External ear, 513
External oblique, 547i
Extinction, defined, 269, 270, 270t, 276 (*See also* Mass extinction)
Extracellular digestion, 320, 332, 334
  carnivorous plant, 436, 436i
Extracellular fluid, 482, 503, 503i, 568i, 633, 634–639, 642
Extreme halophile, 298i, 314, 314i
Extreme thermophile, 298i, 314, 314i
Eye, 510, 511t, 514–515, 514i, 515i
  human, 510, 511i, 652, 655
Eyespot, 326, 385i

F F_1 and F_2 generation, defined, 155, 314i
Facial bone, 543i
Fact, memory storage of, 504
"Facts change," 18
FAD, 48t, 111, 114–116, 114i, 115i, 117i
FADH$_2$, 111, 111i, 114–116, 114i–117i, 121i
Falcon, peregrine, 401i, 742, 742i
Fall overturn, lake, 772
Fallopian tube, 660 (*See also* Oviduct)
Familial cholesterolemia, 179t, 222
Family (taxon), 8, 286
Family planning, 696–697
Family tree (*See* Evolutionary tree)
Farmland, 439, 773 (*See also* Agriculture)
Farsightedness, 515
Fat, 40, 48t, 120 (*See also* Triglyceride)
  dietary, 612
  digestion, 41, 618t, 619, 621i
  metabolism, 110, 121i, 612, 622, 622t
  saturated vs. unsaturated, 40, 40i, 624
Fatty acid, 40, 40i, 48t, 121i
  essential, 625
  metabolism, 63, 120, 121i, 618–621, 618t, 621i, 622
Feather, 394, 395i
Feather star, 378i
Feces, 623, 747
Feedback mechanism, 484, 530–531, 530i, 531, 536, 659, 659i, 660–661, 663i
  enzymes and, 84, 84i
  homeostasis and, 482–483, 482i, 483i
  negative, 482, 482i, 483i
  positive, 483, 490
Feedforward mechanism, 797
Feline infectious peritonitis, 257
Femoral artery, 563i
Femoral vein, 563i
Femur, 543i, 545i (*See also* Thighbone)
Fermentation pathways, 118–119, 119i, 123, 120, 298
Fern, 303i, 304i, 305i, 342, 343i, 346, 354t
  life cycle, 342i, 347, 347i
Fertility, human, 676–677
Fertilization, 140, 146, 146i, 648, 648i, 680
  double, 352i, 455, 455i, 469
  human, 665, 665i
  internal, 647
  plant, 454, 454i
  in vitro, 187, 680
Fertilizer, 20, 47, 104, 747i, 751i, 753, 787
Fetal alcohol syndrome, 673, 673i
Fetoscopy, 187, 187i
Fetus, defined, 666
Fever, 581, 587, 641
Fiber
  connective tissue, 476–477, 476i
  plant, 424, 424i, 427i, 433, 434t
Fibrillin, 189
Fibrin, 581, 572i
Fibrinogen, 558i, 572, 572i
Fibroblast, 476
Fibrous joint, 545
Fibrous protein, defined, 45, 48t
Fibrous root system, 430, 430i
Fibula, 543i

Fig, 47, 456t, 722
Fight–flight response, 501, 530
Filament, of stamen 452i, 453
Filter-feeder, 374i, 384, 385, 387, 776
Filtration, blood, 636–637, 636i
Fin, 386, 388–390, 389i, 390i
Finch, 244–245, 245i, 253, 253i, 704, 704i
Fire ant, 731t
Fire, forest, 729, 766, 766i, 770, 770i
Firefly, 107, 231i, 709
First law of thermodynamics, 76, 77
Fish (See also specific types)
  adaptations, 389i, 498i, 513, 515, 557,
    557i, 598, 600, 600i, 639, 639i, 646i, 647i
  aquarium, 326, 326i
  bony, 388–389
  cartilaginous, 388, 388i
  evolution, 237i, 264, 264i, 266, 266i,
    276i, 303, 386–387, 386i, 498i
  jawed, 302, 387, 388–390, 388i, 389i
  jawless, 302, 387, 387i
  lobe-finned, 303, 389, 389i, 390, 390i
Fish kill, 20, 326–327, 326i
Fishing, commercial, 411, 412, 777
Fitness, defined, 249
Fixation, allelic, 256, 256i
Fixed action pattern, 705
Flagellum
  bacterial, 70, 70i, 72t, 312i, 313, 313i
  eukaryotic, 67, 67i, 320
  protistan, 320, 323, 323i, 326, 326i
  sperm (tail), 146, 658, 659i
Flame cell, 364, 364i
Flatworm (Platyhelminthes), 358t,
  364–365, 366, 497, 497i, 614i
Flavin adenine dinucleotide (See FAD)
Flavoprotein, 465
Flax, 68i, 424i
Flea, 377i, 691
Flemming, W., 170, 171
Flesh fly, 735, 735i
Flight, 394i, 395, 395i
Flood, biblical, 272–273, 274i
Flounder, 742i
Flower, 351, 351i, 352i, 450, 452, 452i
Flowering plant (angiosperm)
  asexual reproduction, 451, 458–459, 458t
  characteristics, 351, 354t, 425i, 452–453
  classification, 342, 354t
  development, 456, 456i, 462–4667, 467i
  evolution, 342, 343i, 354t, 450–451
  life cycles, 342–343, 342i, 351, 352i,
    452, 452i, 454i–455i, 468
  number of species, 422
  polyploidy, 266i, 267i
  primary growth, 421, 423, 423i
  secondary growth, 421, 423, 425,
    432–433, 432i, 433i
  sexual reproduction, 451, 454–455
  tissues, 421, 422–423, 422–425, 434t
Flu, "24-hour," 325 (See also Influenza)
Flu shot, 317
Fluid mosaic model, 52–53, 53i
Fluke (Trematoda), 358t, 365, 366, 366i
Fluorescent light, 98
Fluorine, 627t
Fly (See specific types)
Flying squirrel, 397t
Folate (folic acid), 541, 626t, 672
Follicle
  hair, 511i, 540i
  ovarian, 662, 662i, 663, 664i
Follicle-stimulating hormone, 526, 526t
Follicular phase, 660, 661t, 664i, 681
Fontanel, 545
Food chain, 739, 740, 740i, 754
Food poisoning, 325
Food production (See Agriculture)
Food pyramid, 624, 624i
Food web, 737, 739, 740–741, 740i, 741i,
  742, 754, 771, 774, 775
Foot
  chimpanzee, 398i
  human, 398, 540, 543i
  husky, 538
  molluscan, 368i, 369
  platypus, 832–833, 832i
Foot and mouth disease, 331
Foraminiferan, 311i, 322, 322i, 749

Forebrain, 498i, 502, 506t, 526, 670i
Formaldehyde, 294, 295, 296, 296i
Fossil, defined, 274
Fossil, examples, 241, 274, 275i, 283i
  Archaeopteryx, 288, 288i
  bat, 275i
  bony fish, 237i
  Cooksonia, 275i
  Dickensonia, 302i
  early eukaryotic cell, 300i
  early humans (Homo), 302i
  early prokaryotic cell, 300i
  Ediacaran, 302i
  first cells, 300i
  Glossopteris, 279i
  hominids, 399i
  ichthyosaur, 275i
  in situ, before analysis, 275i
  protozoan, 322
  Psilophyton, 303i
  Spriggina, 302i
  stromatolite, 298i
  trilobite, 302i
Fossil fuel, 35, 794
  carbon cycle and, 748, 749
  origin of, 304, 305, 344
  pollution by, 31, 31i, 35, 753, 784
  population growth and, 694–695, 698
Fossil record, 239, 240, 242i, 268, 269,
  273, 289, 312, 328
  animals, 244, 278, 280, 302, 302i, 303i,
    305i, 307i, 356, 356i, 360, 390i, 392
  earliest cells, 298–300
  hominids, 399, 400, 400i
  interpreting, 241, 242, 272, 274,
    276–277, 277i, 278–279, 288
  "missing links" in, 288
  plants and fungi, 278, 278i, 303i, 334,
    344, 440
Fossilization, 274–275, 277i, 312i, 288
Founder effect, 257, 257i, 271
Fovea, 511t, 515i, 516
Fox, 397i, 682, 738, 738i
Fox, S., 295, 297
Foxglove (Digitalis pupurea), 353
Foxtail, 723, 723i
Fragile X syndrome, 179t, 181, 181i
Fragmentation, reproduction by, 361, 458t
Frameshift mutation, 208, 208i, 209
Franklin, R., 194, 196, 196i
Free radical, 74, 75, 209, 624, 626t, 627, 675
Frequency, sound, 512i
Freshwater province, 772–773, 772i, 773i
Friedman, J., 629, 629i
Frigate bird, 395
Frisch, K. von, 709i
Frog, 55i, 60, 150i, 390–391, 391i, 498i,
  601i, 614, 614i, 726
  deformities, 391i, 534–535, 534i
  development, 534–535, 534i, 644–645,
    644i, 651, 735
Frontal lobe, 502, 503i
Frontal plane (directional term), 480i, 481i
Frostbite, 641
Fructose, 38, 38i, 657
Fruit, 351, 451, 456, 456–457, 456i, 457i
Fruit fly, 150i, 176, 176i, 189, 189i, 265,
  377i (See also Drosophila)
FSH, 523i, 524i, 526, 526t, 527, 527i
  in menstral cycle, 531i, 660–661, 662,
    663, 663i, 664i
  in spermatogenesis, 658–659, 659i
Fucoxanthin, 326, 328
Functional group, defined, 36, 36i, 37
Fundamental niche, 720
Fungicide, 47, 786
Fungus (Fungi), 333i, 334–335, 418i
  beneficial, 332, 334, 337, 338, 338i
  characteristics, 8, 8i, 9, 11, 39, 72t,
    332i–335i, 334–337
  classification, 8, 8i, 9, 287, 334, 337
  evolution, 298i, 334
  harmful, 334i, 336, 336i
Fusion power, 796

G G1 and G2, interphase, 130, 130i
GABA, 492, 496, 507
Galactose, 38, 180
Galactosemia, 179t, 180

Galápagos Islands, 242i, 244, 245i, 265
Galápagos Rift, 774
Galilei, G., 15, 50
Gall, 252, 252i
Gallbladder, 616, 616i, 630t
Gallstone, 628
Gamete, 139, 140, 140i, 143i, 680
  formation 146, 146i, 147i, 648, 648i
    658–659, 659i
Gametic mortality, 263
Gametophyte, 146i, 342, 342i
  algal, 327i, 328i, 329i
  bryophyte, 344, 345i
  flowering plant, 452–455, 452i, 454i, 458
  seed plant, 348–349, 352i
  seedless vascular plant, 346
Gamma aminobutyric acid, 492, 496, 507
Gamma globulin, 561
Gamma ray, 96i, 133
Ganglion, 371, 497, 541
Ganglion cell, retinal, 516, 516i
Gap gene, 655i
Gap junction, 69, 69i, 474–475, 475i, 478,
  565, 565i, 651, 666i
Garter snake, 704, 704i
Gas exchange, via respiration, 597
  in birds, 601, 601i
  at gills, 386–387, 557, 598, 599, 600, 600i
  across integument, 599, 599i
  in lungs, 389, 390, 606–607, 606i
Gastric fluid, 30, 30i, 69, 85, 475, 578,
  613, 616i, 618
Gastrin, 529t, 618, 619
Gastrocnemius, 547i
Gastrodermis, 362, 362i
Gastrointestinal tract, defined, 616
Gastropod, 368, 368i, 599, 599i
Gastrula, 648, 652i, 654i
Gastrulation, 648, 648i, 649i, 668, 668i, 680
Gated protein, defined, 53, 53i
Gause, G., 722
Gecko, 393
Gel electrophoresis, 227, 227i, 228, 228i
Gemmule, 361
"Gender war," 682
Gene, defined, 140, 153, 155, 167, 172, 188
  interactions, 162–165, 654–655, 655i, 246
  linked, 176, 176i, 177, 177i
  locus, 155, 155i, 176–177
  mutations (See Mutation)
Gene expression
  control of, 128, 213–218, 215i–217i
  environment in, 166, 166i, 217, 246, 704
  hormones and (See Hormone)
  in plants, 460–463
  selective, 128, 217, 220, 652, 653
  and unexpected phenotypes, 162, 246
Gene flow, 247, 252, 255, 255i, 258, 270t
  barriers to, 260, 262, 264–265
Gene library, 229, 229i
Gene locus, defined, 155, 155i
Gene pool, 246–247
Gene sequencing, 227, 227i, 228, 228i,
  232–233, 235
Gene therapy, 222–223, 233, 591i
Gene transfer, 231–236, 231i, 316i
Genetic code, 204, 204i, 290
  mitochondrial, 301
Genetic counseling, 186
Genetic disorder, nature of, 179, 179t
Genetic divergence, 262, 262i, 264, 270t
  in adaptive radiations, 268–269
Genetic drift, 247, 256–257, 258, 261, 261i
Genetic engineering, 223, 230–234, 231i,
  234, 236, 790
Genetic equilibrium, 246–248, 249i, 258
Genetic recombination, , 172, 172i, 197,
  223, 233, 246
Genetic screening, 186
Genital warts, 137, 679, 679i
Genome, defined, 224, 234
  automated DNA sequencing of, 228
  Caenorhabditis elegans, 365
  Escherichia coli, 229
  Haemophilus influenzae, 233
  human, 224, 227–229, 232–233, 236
  mapping of, 232–233, 402
Genomic library, 229
Genotype, defined, 155

Genus (taxon), 8, 286
Geoduck, 369
Geographic dispersal, 260–261, 730
Geographic isolation, 264–265, 264i, 265i
Geologic time scale, 276–277, 276i–277i,
  302–307, 308i–309i, 343i
Geospiza conirostris, 245i
G. scandens, 245i
Geringer, F., 182, 182i
Germ cell, defined, 128, 139, 481
German measles, 672
Germination, defined, 460, 469
  seed, 454, 460–461, 460i, 461i
Gey, G., 136
Gey, M., 136
GH (See Somatotropin)
Giardia lamblia, 323, 323i
Gibberella fujikuroi, 463
Gibberellin, 462, 462t, 463, 469
Gibbon, 183i, 398i, 401i
Gigantism, 528, 528i
Gill
  animal, 386, 386i, 557, 598, 599, 600, 600i
  mushroom, 335, 335i
Gill slit, 384–388, 386i–387i
Gingivitis, 617i
Ginkgo, 303, 304i, 342, 343i, 350, 350i, 354t
GIP, 619
Gizzard, 614, 614i
Glaciation, 277, 302, 728i, 746i, 772i
Gladiolus, 458t
Gland, defined, 475 (See also specific types)
Gland cell, 62, 475, 620
Glandular epithelium, 475, 475i, 522, 618
Glial cell, 653, 653i
Global broiling hypothesis, 306
Global warming, 35, 750–751, 751i, 781
Globin, 44, 44i
Globular protein, defined, 45, 48t
Globulin, 558i
Glomerulonephritis, 638
Glomerulus, 635, 635i, 636i, 638
Glossopteris, 278, 279i
Glottis, 601, 603, 603i
Glucagon, 120, 523i, 524i, 525, 525i, 529t,
  532, 533i
Glucocorticoid, 529t, 530, 533i
Glucose, 38, 38i, 39i, 48t, 78–79, 120
  abiotic synthesis, 295
  membrane crossing by, 85, 86, 120,
    530, 618t, 621
Glucose–6–phosphate, 110i, 120, 121i
Glucose formation
  photorespiration vs., 102–103, 105
  by photosynthesis, 92, 93i, 94–95, 95i,
    101–103, 101i
Glucose insulinotropic peptide, 619
Glucose metabolism, 38–39, 110–120
  hormones and, 523i, 525, 529t, 530,
    532–533, 533i
Glucose transporter, 530
Glutamate (glutamic acid), 42i, 204,
  204i, 208i, 517
Glutamine, 204i
Gluteus maximus, 547i
Glyceride, 48t
Glycerol, 40, 40i, 120, 121i
Glycine, 42i, 204i, 496
Glycocalyx, 313
Glycogen, 39, 48t, 120, 529i, 530, 533i, 551
Glycolate, 103
Glycolysis, defined, 109, 110, 113i, 123
  in aerobic respiration, 110–112, 111i,
    112i, 115, 117i, 120, 121i, 548
  early life and, 110, 115, 117i, 118
  in fermentation, 118, 118i, 119i, 122
  reactions, 112, 112i
Glycoprotein, 45, 524t, 654, 662
Glyptodont, 244, 244i
Gnetophyte, 342, 343i, 350, 354t
GnRH, 527, 659, 659i
  in menstrual cycle, 531i, 662, 663i, 664i
Goiter, 531, 627t
Goldenrod, 252, 252i
Golgi body, 56t, 57i, 59i, 62, 62i, 63i, 72t
Gonad, 363, 523i, 524, 527i, 529t, 531, 534
Gonadotropin, 527
Gonadotropin-releasing hormone, 527
Gondwana, 278, 279, 302, 302i, 308i, 309i

Gonorrhea, 11, 587, 678–679, 678i
Goodall, J., 520, 520i
Goose, 124, 394i, 498i, 687i, 706, 706i
Gorilla, 128, 398i, 401i, 410i
Gout, 628
Graded potential, 490, 493, 510, 516
Gradual model, speciation, 268
Grand Canyon, Arizona, 278i
Granstein cell, 541
Grant, M., 459i
Granulosa cell, 662, 662i, 663
Granum (grana), 65, 65i
Grape, 118, 119i, 320i, 456t
Grass, 20i, 58i, 453i (See also specific types)
Grasshopper, 497i, 556i
Grassland, 6i, 762i, 764i, 765–767, 791
  emergence during Eocene, 307, 307i
Grateful Dead midge, 799
Graves' disorder, 531, 591
Graves, J., 264
Gravitropism, 464, 464i
Gray crescent, 649i, 650, 650i
Gray matter, 499, 501, 502, 511
Grazing food web, 741, 741i
Great Barrier Reef, 412
Great Lakes, North America, 387
Green alga, 55i, 97, 1007, 107i, 320, 328–
  329, 328i, 329i, 330t, 340i, 742i, 772
Green Belt Movement, 789i
Green revolution, 787
Greenhouse effect, 749, 750–751, 750i,
  758, 789, 794
Greenhouse gas, 750, 784t, 786
Griffith, F., 192
Grip, in human evolution, 405, 405i
Grizzly, 410i
Gross primary productivity, 734
Ground meristem, 423, 423i, 426i
Ground tissue, plant, 421, 422, 422i, 423,
  423i, 425, 429
Groundwater, 746i, 787, 792, 792i
Growth, nature of
  animal, 523i, 529t, 532, 648, 648i
  plant, 462–463, 462t, 463i
Growth factor, 218, 559i
Growth hormone (See Somatotropin)
Growth ring, 433
Guanine (G), 46, 46i, 194, 194i, 201
Guar, 198
Guard cell, 425, 444–445, 444i, 445i, 448
Guilleman, R., 528
Gulf Stream, 760, 760i
Gull, 713, 742i
Guppy, 693, 693i
Gurdon, J., 652
Gut, defined, 359, 616
Gymnophilus, 333i
Gymnosperm, 341, 342, 350, 354t
  evolution, 277i, 303i, 304, 304i, 342i,
  343, 343i, 348
  life cycle, 342–343, 348–349, 348i

Habitat, defined, 686, 720, 734
Habitat island, 411
Habitat loss, 410–411, 417
Habituation, 504, 504t, 706t
Hadal zone, 774i
Haemophilus influenzae, 224, 228, 233, 590i
Hagfish, 387, 387i
Hair, 41, 42, 42i, 45, 45i, 538, 540, 540i
  color, genetic basis, 162, 162i
Hair cell (mechanoreceptor), 513
Hair follicle, 511i, 540i
Half-life, defined, 277, 277i
Halimeda, 329
Hallucinogen, 505
Halobacterium, 314i
Halophyte, 776
Hamilton, W., 714
Hammer, R., 232
Hand, evolution, 398–401, 405, 405i
  components, human skeleton, 543i
Hangingfly, 710, 710i
Hanson, J., 538
Haploidy, defined, 129, 141, 143i
Hardwood, 433, 433i
Hardy–Weinberg rule, 248–249, 258
Harpobittacus apicalis, 710, 710i
Hartshorn, G., 416

Haversian system, 544, 544i
Hawaiian Archipelago, 265, 265i
Hawkmoth, 451
Hay fever, 453, 454, 587, 590
Hayes, M., 182, 182i
Hayflick, L., 674–675
HbS mutation, 208, 254–255
HCG, 667
HDL, 222, 571
Hearing, 511, 512, 512i, 513, 513i
Heart, defined, 556
  human, 359, 529t, 556, 562–569, 562i,
  564i, 565i
  invertebrate, 368i, 369, 371, 371i, 373
Heart attack, 222, 232, 532, 570
Heart disease, 570, 608, 609i, 628
Heartbeat, 564–565, 571i, 605, 607
Heartburn, 618
Heartwood, 432–433, 433i
Heat (thermal) energy, defined, 76
Heat stress, mammalian, 641
Hectare, defined, 687
Heimlich maneuver, 611, 611i
HeLa cell, 136, 136i
Helianthus, 424i
Helicase, 675
Helicobacter pylori, 313i, 618
Heliozoan, 322
Helium, 22, 25, 25i, 293
Helix aspersa, 260–261, 260i
Helmont, J. B. van, 107
Helper T cell, 582, 583i, 586–589, 592, 594t
Heme group, 44, 44i, 485, 606
Hemichordate, 386
Hemiptera, 377i
Hemodialysis, 638
Hemoglobin, 48t, 558–559, 570, 627t
  blue form, 259
  functions, 44–45, 598, 599, 605i, 606, 607
  genetically engineered, 230, 231, 239
  HbS, 161, 161i, 208, 254–255
  llama, 605, 605i
  porphyria and, 485
  structure, 44–45, 44i
Hemophilia, 81i, 177, 179t, 181, 187, 189
Hemostasis, 572, 572i
Henslow, J., 242
Hepatic portal vein, 563i
Hepatitis, 325t, 590, 590i
Herbaceous plant, defined, 432
Herbicide, 47, 47i, 234, 462–463
Herbivore, 9, 738, 739, 747
Heritable trait, defined, 11, 16
Hermaphrodite, 364
Herniated disk, 543
Heroin, 505
Heron, green, 742i
Herpes, 592, 678, 679
Herpes simplex, 318, 541
Hershey, A., 192, 193i
Heterocephalus glaber, 714
Heterocyst, 314, 315i
Heterotherm, 640
Heterotroph, 92, 738, 739i
Heterozygous condition, 155, 156i, 167
Hexokinase, 83i
Hickman, C., 360
Hickory wood, 433i
High-density lipoprotein, 222, 571
Hignell, M., 538
Hindbrain, 498i, 502, 506t, 567
Hippocampus, 504
Hippopotomus, 307
Hippotrigris, 271
Hirudo medicinalis, 370i
Hives, 587
Hognose snake, 392i
Holley, R., 224
Holpcraft, D., 791
Homarus americanus, 286
Homeostasis, defined, 5, 472, 482
  and internal environment, 482, 555,
  555i, 556, 632, 633i

mechanisms for, 482–483, 528
  organ system roles in, 555, 555i, 597,
  597i, 613i, 619, 633i
Homeotic gene, 654, 655i
Hominid, 398, 399, 399i, 401i, 403
Hominoid, 399, 401i
Homo erectus, 400–401, 400i, 401i, 402, 403
H. habilis, 400, 400i, 401i, 403, 695
H. neanderthalensis, 400i, 401i
H. rudolfensis, 400i, 401i
H. sapiens, 150i, 400i, 401–403, 401i, 402i
Homologous chromosome, 140, 142i, 144,
  144i, 150i, 172, 176–177
Homologous recombination, 233
Homologous structure, 280, 280i, 281i, 286
Homozygous condition, 155, 156i, 166–
  169, 176i, 181, 186, 256
Honey locust, leaf, 428i
Honeybee, 709, 709i (See also Bee)
Honeycomb, 41, 41i, 714
Honeycreeper, 265, 265i
Hood, L., 232
Hooke, R., 50
Hookworm, 358t
Horizontal cell, retina, 516, 516i
Hormone, animal
  control over genes, 214, 216–217, 704
  environment and, 534–535
  functions, 521–536, 658–659
  mechanism of action, 524–525, 524i, 525i
  sources, 523i, 526, 526t, 529, 529t
  types, 37, 41, 523i
Hormone, plant, 462–463, 469, 462t
Horse, 271, 271i, 280–281, 307, 307i, 498i
Horseshoe crab, 373i
Horsetail, 342, 343i, 346–347, 346i, 354t
Host, defined, 724
  coevolution with pathogen, 325
  coevolution with parasite, 725
  definitive vs. intermediate, 364
Hot spot, 414–415, 534, 534i, 763
Housefly, 286i
Howarth, F., 725
Huffaker, C., 725
Human
  chromosome number, 128, 141, 141i,
  150i, 174, 174i, 178, 178i, 183, 283, 283i
  development (See Development, human)
  evolution, 277i, 383, 398–405, 401i
  genome, 224, 227–229, 232–233, 236
  homologous structures, 280i, 281
  organ systems, overview, 480–481, 480i
  taxonomic classification, 286i
Human chorionic gonadotropin, 667
Human immunodeficiency virus (See HIV)
Human impact on environment, 736–737,
  742i, 750–751, 753, 773, 776, 776i
Human papillomavirus (HPV), 137, 679
Human population
  biotic potential, 689
  growth, 406, 406i, 684–685, 694–700,
  694i–700i, 782–783, 782i–783i
Humboldt Current, 760i, 777, 778
Humerus, 543i
Humidity, 640
Hummingbird, 451i, 640, 722
Humus, 438, 764
Hunchback of Notre Dame, 179
Huntington disorder, 178i, 179t, 180, 181
Huq, A., 779
Husky, 483, 538, 538i
Hybrid offspring, defined, 155, 223
Hybrid zone, 267, 267i
Hydra, 362i
Hydrangea macrophylla, 166i
Hydrocarbon, 36, 784
Hydrochloric acid, 30, 30i, 31, 618
Hydrogen atom, 21i, 22, 22t, 24–26, 24i, 25i
Hydrogen bonding, 27, 27i, 32
  in DNA, 46, 46i, 191, 195i, 196
  in proteins, 44–45, 44i
  in transpiration, 442–443, 443i
  in water, 28i, 29, 443
Hydrogen cyanide, 294
Hydrogen ion (H+), 21, 30
  in aerobic respiration, 111ff.
  pH and, 30i, 31, 638
  in photosynthesis, 98ff.
Hydrogen peroxide, 63, 74, 784t

Hydrogen sulfide, 103, 119
Hydrologic cycle, 745, 746, 746i
Hydrolysis, 37, 37i
Hydrophilic substance, defined, 28, 32t
Hydrophobic substance, defined, 28, 32t
Hydrosphere, 746, 757
Hydrostatic pressure, 89, 568i
Hydrostatic skeleton, 363, 371, 542, 552
Hydrothermal vent, 103, 295, 314,
  774
Hydroxide ion, 30
Hydroxyl group, 36, 36i
Hydrozoan, 358t, 362, 362i
Hygrophorus, 333i
Hymenoptera, 377i
Hyperpolarization, membrane, 493
Hypertension, 532, 570, 628
Hyperthermia, 641
Hyperthyroidism, 531
Hypertonic solution, 88i, 89, 89i
Hyperventilation, 596
Hypha, fungal, 334, 335, 335i, 338
Hypnotic, effect on brain, 505
Hypocotyl, 460i, 461i
Hypodermis, 540, 540i
Hypothalamus, 502, 502i, 506t, 522i, 526,
  527i, 536, 629, 664i, 659, 659i
  blood–brain barrier and, 503
  body temperature and, 637, 641
  hormones of, 523i, 526, 527i, 530i
Hypothermia, 641
Hypothesis, nature of, 12, 13, 16
Hypothyroidism, 531
Hypotonic solution, 88i, 89, 89i
Hypoxia, 596

IAA (indoleacetic acid), 462, 468
Ice, structure of, 28i, 29
Ice age, 273, 277i, 306
Ice cloud, Antarctica, 786, 786i
Iceberg, from Ross Ice Shelf, 755i
ICE-like protease, 219, 219i
Ichthyosaur, 275i, 305i
Ichthyostega, 390i
IgA, 586, 587
IgD, 586, 587
IgE, 586, 587, 590
IgG, 586, 587, 590
IgM, 586–587
Iguana, 393
Ileum, 616i
Iliac artery, 563i
Iliac vein, 563i
Illegitimate receiver/signaler, 709
Immigration, 255, 688, 700
Immune deficiency, 591, 591i
Immune response, 578, 578t, 585, 594
  antibody-mediated, 582–583, 583i,
  586–587, 587i
  cell-mediated, 582, 583i, 588–589, 588i
  complement in, 578t, 579, 579i, 587
  control of, 583
  primary, 583i, 585, 585i, 590
  secondary, 583i, 585i, 590
Immune system, 578, 578t, 582–583
  defining characteristics, 582, 584–585
Immunization, 590, 590i
Immunoglobulin, 586–587
Imperfect flower, 453
Implantation, 666–667, 666i
Imprinting, 706, 706i, 706t
In vitro fertilization, 187, 198, 680
Inbreeding, 257
Incest, 257
Incisor, 396i, 617, 617i
Incomplete digestive system, 614, 614i
Incomplete dominance, 160, 160i
Incomplete metamorphosis, 376i
Indehiscent fruit, 456t
Independent assortment, 158–159, 167
  exceptions to, 172, 177
India, human population, 684–685, 684i,
  697i, 789
Indian pipe (Monotropa uniflora), 351, 351i
Indicator species, 411, 417
Indinavir, AIDS and, 593
Indirect selection theory, 714, 715
Indoleacetic acid (IAA), 462, 468
Indricotherium, 307i

Induced-fit model, 83, 83i
Inductive logic, defined, 12, 18
Industrial smog, 784
Infection, defined, 325
Infectious disease, 311, 317, 319, 320,
    322–325, 325t, 576–577
Inferior (directional term), 481i
Inferior vena cava, 563i, 564i, 634i
Inflammation, 532, 559, 578t, 579,
    580–581, 580i, 581i, 587, 590
Inflammatory response, 581i
Influenza, 317, 317i, 325t, 587
Infrared radiation, 96i
Inhalation, defined, 604, 604i
Inheritance, defined, 4, 155
Inheritance pattern
    autosomal dominant, 179t, 180, 180i, 189
    autosomal recessive, 179t, 180, 180i
    sex determination, 172, 174, 174i
    X-linked recessive, 179t, 181, 181i, 189
Inhibin, 659
Inhibiting hormone, 523i, 527
Inhibitory postsynaptic potential, 493
Inner ear, 512, 512i
Inorganic compound, defined, 36
Input zone, axon, 488, 488i, 506
Insect, 358t, 376–377, 514, 599i
    diversity through time, 418i
    mimicry, 726
    molting, 372, 372i, 535, 535i
    parasitoid, 735
Insecticide, 47, 353 (See also Pesticide)
Insight learning, 706t
Instinctive behavior, 705, 716
Insulin, 5, 43i, 48t, 120, 230, 523i, 524t,
    525, 529t, 532–533
Integrator, 482, 482i
Integrin, 654
Integument, 454, 455i, 539, 648
Integumentary exchange, 599
Integumentary system, defined, 480i
Intercostal muscle, 602i, 604, 604i
Interferon, 230, 578t
Interleukin, 559i, 578t, 581, 586, 587i, 588i
Intermediate, defined, 81, 90t
Intermediate filament, 66, 66i
Intermediate host, 364
Internal environment, 5, 31, 472, 473
    blood circulation and, 556
    salt–water balance in, 633, 634
    temperature of, 637, 640–641
Interneuron, 495, 495i, 498, 506, 511
Internode, 422i, 426i
Interphase (in cell cycle), 130, 131i, 132i
Interspecific competition, 720, 720t
Interstitial fluid, 482, 556, 568i, 569, 633
Intertidal zone, 774i, 776–777, 777i
Intestinal nuclease, 618t
Intestine, 562, 562i
    large 500i, 616, 616i, 623, 623i, 630t
    neural control of, 500i, 511i
    parasites of, 365, 366, 367
    small, 500i, 573, 573i, 616i, 618–619,
        618t, 620–621, 620i, 621i, 630t
Intraspecific competition, 708–709,
    712–713, 722
Intron, 203, 203i, 218, 225, 318
Inversion, chromosome, 182, 182i
Inversion, thermal, 784, 784i
Invertebrate (See also specific types)
    acoelomate, 357i, 359i, 364–367
    bilateral, 358, 358i
    blood circulation, 368i, 370, 370i, 371,
        371i, 373, 556, 556i
    characteristics, 357–359, 380i
    coelomate, 357i, 359i, 367ff.
    digestion, 614, 614i, 360, 361i, 362, 364,
        364i, 365, 371i, 379, 379i, 384
    endocrine function, 521, 535, 535i
    evolution, 276i, 302–303, 303i,
        356–357, 357, 358, 380i
    major groups, 358ff., 358t, 380
    nervous systems, 364i, 368, 369, 372,
        496, 496i
    origin of, 302, 356, 357, 360
    photoreception, 514i, 515
    pseudocoelomate (false coelom), 359i,
        365

radial, 358, 358i, 362–363
respiration, 372, 599, 599i
solute–water balance, 371, 371i, 376
skeletal–muscular systems, 369, 370,
    370i, 372, 372i, 374, 378, 378i, 542, 542i
Iodide, 531
Iodine, 21i, 22t, 23, 531, 627i, 795i
Ion, defined, 21, 26, 27, 32t
    electric gradients and, 86
Ionic bonding, 26, 26i, 27
Ionization, 21, 26, 26i
Ionizing radiation, 209
IPSP, 493
Iris, 164, 164i, 434i, 511t, 514, 515i
Iron, 22t, 103, 294, 439t
    in human body, 21i, 560, 606, 627t
Irrigation, 792, 792i
Isaacson, P., 742i
Island biogeography, 411, 417, 732–733,
    732i, 733i
Isocitrate, 115i
Isoleucine, 204i, 205, 625
Isometric contraction, 551, 551i
Isotonic contraction, 551, 551i
Isotonic solution, 88i, 89
Isotope, defined, 21, 22, 32, 32t
Isthmus of Panama, 264, 264i
IUD, 677, 677i
Ixodes, 315i

Jablonski, N., 541
Jade plant, 458t
Jaguar, 768i
Japanese beetle, 731t
Jaundice, 626i
Jaw, evolution of, 386, 386i
Jawed fish, 302, 303, 386i, 388–389, 388i
Jawless fish, 302, 386i, 387
Jay, blue, 255, 255i
Jejunum, 616i
Jellyfish, 358t, 362, 362i
Jenner, E., 576–577, 576i
Johnson, P., 391i
Joint, skeletal, 545, 545i
J-shaped growth curve, 688i, 689, 689i
    human population, 695i
Jugular vein, 563i
Jump dispersal, 730
Jurassic, 276i–277i, 304, 304i, 305i, 343i,
    396, 404, 405i
Juvenile, defined, 372

K (See Carrying capacity)
Kalat, J., 486
Kangaroo, 646i, 647i
Kangaroo rat, 632–633, 632i, 639
Kaposi's sarcoma, 592, 592i
Karyotype, 170, 170i, 173, 173i, 185i, 188
Kaposi, J., 486
Kelp bed, 311, 311i, 328
Kennett, C., 725
Kenya, 789, 789i
Keratin, 45, 48t, 49
Keratinocyte, 219, 540
Ketone, 36i, 533
Kettlewell, H. B., 250
Key innovation, defined, 269
Keystone species, 722–723
Khorana, G., 204
Kidman, N., 42i
Kidney, 500i, 635, 635i, 637i
    failure (renal), 638
    function, 526, 526t, 527i, 529t, 562i, 53i,
        625, 634i, 635
Kidney cell, 52i, 63, 117i, 532
Kidney dialysis, 638
Kidney stone, 638
"Killer bee," 108–109, 108i
Killifish, 693
Kilocalorie, defined, 76, 624, 628
Kinase, 218, 525
Kinesin, 133
Kinetic energy, defined, 76
Kingdom (taxon), 8, 286
Kingfisher, 111i
Klinefelter syndrome, 179t, 185
Knee joint, 543i
Koala, 397i, 397t
Koch, R., 577
Kohler, G., 589

Komodo dragon, 409i
Komodo monitor lizard, 393
Koshland, D., 83
Krebs cycle, 111, 111i, 114–115, 115i,
    121i, 123, 124
Krebs, H., 114
Krill, 737
K–T asteroid impact, 305, 306, 306i
Kubicek, M., 136
Kudzu, 730–731, 730i
Kurosawa, E., 463
Kuru, 319

Labium majora, 660, 661i
Labium minora, 660, 661i
Labor (See Childbirth)
Labrador retriever, 162, 162i
Lacks, H., 136, 136i
Lactate (lactic acid), 118, 578
Lactate fermentation, 118, 118i, 123
Lactation, 674, 674i
Lactobacillus, 118, 314, 578
Lactoferrin, 581
Lactose, 38, 180
Lactose operon, 215, 215i
LAK cell, 589
Lake, characteristics, 772–773, 772i
Lake Baikal, 279i, 775
Lamarck, J., 242
Lamella, middle, 58i, 68, 134i
Laminaria, 328
Lamprey, 387, 387i, 731t
Lancelet, 358t, 384–385, 385i, 403
Landfill, 787
Langerhans cell, 541, 582i, 589
Lanugo, 671
Larkspur, 282, 282i, 456t
LaRoe, E., 755
Larrea divaricata, 458, 686
Larsen Ice Shelf, calving from, 755, 755i
Larus, 742i
Larva, defined, 4–5, 361, 363
Laryngitis, 603
Larynx, 500i, 602i, 603, 609i
Late blight, 320, 690i
Late wood, 433, 433i
Lateral bud, 422i, 432i
Lateral geniculate nucleus, 516, 516i
Lateral meristem, 421, 423, 432
Lateral root, 422i, 430, 431, 431i
Latimeria, 389i
Latissimus dorsi, 547i
LDL, 222, 571
Leaching, 439, 753
Lead, 638, 784, 784t
Leaf, 428–429, 428i, 429i, 444–445, 445i
    color, 96, 97i
    dicot vs. monocot, 425i, 428, 428i
Leaf blight, 230
Leakey, M., 400
Learning (See Behavior, learned)
Lecithin, 619, 625
Leclerc de Buffon, Georges-Louis, 241
Leech, 358t, 370, 370i
Leeuwenhoek, A. van, 50, 50i
Leeward, defined, 761
Legume, 314, 441, 752–753
Lek, 710, 710i
Lemna, 217
Lemon, 30, 30i, 456t
Lemur, 397t
Lens, 511t, 514, 515, 515i, 670i
Lenski, R., 271
Lenticel, 432
Leptin, 629, 629i
Lepidoptera, 377i
Leptin, 629, 629i
Lethal mutation, 247
Leucine, 204i, 208i, 625
Leukemia, 170, 170i, 179t, 560, 795i
Leukocyte, 559i
Levi-Montalcini, R., 532
Levin, B., 14, 15
Leydig cell, 658i, 659, 659i
LH, 523i, 524t, 526, 526t, 527, 527i,
    658–659, 659i
    in menstrual cycle, 531i, 660–661, 662,
        663, 663i, 664i
Lichen, 250, 251i, 251t, 291i, 332, 332i,
    333, 338, 338i, 339, 728

Life
    characteristics, overview, 4–9, 11, 16t
    difficulty defining, 2, 3
    diversity of (See Biodiversity)
    emergent properties, 4–8, 16, 16t, 36, 122
    origin, 21, 103, 276i, 292–298, 298i, 775
    perspective on, 122, 408–409, 671, 797
    unity of (See Unity of life)
Life cycle, examples
    amphibian, 390, 391
    animal, generalized, 146, 146i
    arthropod, 372
    black bread mold (Rhizopus), 337, 337i
    blood fluke (Schistosoma), 366, 366i
    bryophyte, 345i
    club fungus, 355, 355i
    crab, 375i
    dicot (Prunus), 454–455, 454i, 468
    dinoflagellate (Pfeisteria), 326i, 327
    fern, 347i
    frog (Rana), 644–645, 644i
    green alga (Chlamydomonas), 329i
    gymnosperm (Pinus), 348i, 349
    hydrozoan (Obelia), 363i
    insects, 376
    monocot (Lilium), 352i
    moss (Polytrichum), 345i
    plant, 146, 146i, 342–343, 342i, 422
    red alga (Porphyra), 327i
    reptile, 392, 393
    slime mold (Dictyostelium), 321i
    sporozoan (Plasmodium), 324i
    tapeworm (Taenia), 366, 366i
    virus (See Multiplication cycle)
Life history pattern, 692, 693i
Life table, 692, 692t
Ligament, 539, 545, 545i
Ligase, DNA, 197
Light (See also Sunlight)
    in bioluminescence, 107, 230i, 709
    fluorescent, defined, 98
    photosynthesis and (See Photosynthesis)
    phototropism and, 464–465, 465i
    visible, 96, 96i
    vision and, 510, 514–515, 514i, 515i
Light microscope, 54, 55i
Light-dependent reactions, 94, 95i, 96,
    98–99, 105, 105i
Light-independent reactions, 94, 95i,
    101–103, 101i, 105, 105i
Lignin, 68, 231i, 342, 424
Lily (Lilium), 129, 351, 352i
Limbic system, 503, 504, 506t, 517
Limiting factor, 685, 690, 700
Limnetic zone, 772, 772i
Limpet, 138–139, 138i, 514i
Lindlow, S., 234
Lineage, evolutionary, 275
Lineage, true-breeding, 155
Linkage group, defined, 176, 177i, 188
Linneaus, C., 8
Linolenic acid, 40i, 625
Lion, 5i, 6i, 713
Lipase, 618t
Lipid, 35, 40–41, 48t, 62, 62i (See also Fat)
    dietary, 625
    origin of, 297, 297i
Lipid bilayer
    organization, 52–53, 52i, 53i, 122, 488
    origin of, 297, 297i
    of nuclear envelope, 60–61, 61i
    solute crossings of, 85–87, 85i–87i, 90i
Lipoprotein, 45, 222, 571
Lithops, 726, 726i
Lithosphere, 757
Littoral zone, 772, 772i, 777, 777i
Littorina littorea, 723, 723i
Liver, 500i, 505
    blood flow through, 562, 562i, 563i
    functions, 529t, 616, 616i, 622, 630t
    in glucose metabolism, 122, 533i, 622i
Liver cell, 38, 59i, 62, 64, 117i, 525, 532
Livestock, 751i, 791 (See also Cattle)
Lizard, 393, 393i, 646i, 647i
Llama, 605, 605i
Loam, 438, 764
Lobaria, 332, 332i
Lobe-finned fish, 389, 389i, 390, 390i
Lobster, 286, 374, 374i

Local signaling molecule, defined, 522
Lockjaw, 496
Logging, 416, 416i, 788, 789i
Logic, defined, 12, 16
Logistic growth, 690–691, 690
Loligo, 368, 491i
Long-day plant, 467, 467i, 469
Loop of Henle, 635, 635i, 636i, 637, 639
Loose connective tissue, 476, 476i, 476t
Lorenz, C., 706i, 706t
Louse (order Mallophaga), 377i
Low-density lipoprotein, 222, 571
LSD, 505
Lubchenco, J., 723
Luciferase, 107, 231i
Lucy, 399i
Lumbar nerve, 499i
Lumbricus terrestris, 150i
Lumen, defined, 616
Lung, 557, 600–601
    circulatory system and, 562, 562i, 563i
    disorders, 608–609, 608i
    evolution, 388, 600, 601
    human, 500i, 602–603, 602i
Lungfish, 389, 600
Luria, S., 192
Luteal phase, 660, 661i, 664i, 681
Luteinizing hormone (*See* LH)
Lycopersicon, 422i
Lycophyte, 342, 343i, 346, 346i, 354t
Lycopodium, 346i
Lyell, C., 242–243
Lyme disease, 315, 315i
Lymph capillary, 572, 572i, 573
Lymph node, 367, 559, 572i, 573, 573i, 578i, 585, 585i
Lymph vascular system, 572–573, 572i
Lymph vessel, 572, 573, 573i, 620, 620i
Lymphatic duct, 573i
Lymphatic system, 481i, 557, 559, 568i, 572–573, 572i, 573i, 574
Lymphocyte, 529t, 558i, 560, 572i, 573, 594t (*See also* White blood cell)
Lymphoid organ, 573
Lymphoid tissue, 573
Lymphokine, 589
Lynx, Canadian, 724–725, 724i
Lyon, M., 216
Lysine, 42i, 204i, 625, 625i
Lysis, 318, 579
Lysogenic pathway, 318, 318i
Lysosome, 57i, 59i, 63, 72t, 87, 124, 582
Lysozyme, 578
Lystrosaurus, 278, 304, 305i
Lytic pathway, 318, 318i

M Maathai, W., 789i
MacArthur, R., 411
Macaw, Amazon, 410i
Macrocystis, 311i, 328, 328i
Macroevolution, 270t, 273, 276, 289
Macronucleus, ciliate, 323, 323i
Macronutrient, plant, 439
Macrophage, 558i, 559, 559i, 578i, 580, 580i, 581, 581i, 582, 594t
Mad cow disease, 232, 319
Magicicada septendecim, 263, 263i
Magicicada tredecim, 263, 263i
Magnesium, 21, 21i, 22t, 439t, 627t, 753
Magpie, 394i
Maiasaura, 392, 392i
Maize, 414
Malaria, 254–255, 254i, 255i, 324, 324i, 325t, 751
Malathion, 47i
Malfroy, B., 75
Malignant melanoma, 212, 212i
Mallophaga, 377i
Mallow, 723, 723i
Malnutrition, 625, 694
Malpighian tubule, 376
Malthus, T., 244
Maltose, 38
Mammal (Mammalia), 397, 403, 542i, 557, 557i, 601i
    evolution, 269i, 307, 396, 404–405, 498i
Mammary gland, 526t, 527, 527i, 674, 674i
Manganese, 103, 439t

Mangrove wetland, 776, 776i
Mantle, molluscan, 368, 368i
Manx cat, 169, 169i
Maple (*Acer*), 428i, 433, 456t, 457i
Marfan syndrome, 179t
Margarita Teresa, 180i
Margulis, L., 300–301
Marijuana, 505, 609
Marine snow, 774
Marler, P., 706
Marsh, 741
Marsupial, 396, 397, 397i, 646i
Mason, R., 702
Mass extinction, defined, 269
    Cenozoic, 307
    Cretaceous (K–T), 305, 306, 306i
    Devonian–Carboniferous, 303, 308i
    humans as agents of, 307, 408
    Mesozoic, 276, 304–306, 305i
    nature of, 276, 408–409, 408i
    Ordovician–Silurian, 302
    Permian, 303
Mass number, 22, 22t
Mast cell, 559i, 580, 581i, 594t
Master gene, 654–655
Mastigophora, 323
Maternal messages in egg, 650
Mating behavior, 646, 708–711, 710i, 711i
Mayr, E., 262, 271
McCain, G., 18
McClintock, B., 208–209, 209i
MDMA (Ecstasy), 505
Measles, 325t, 590, 590i, 592
Mechanical work, 75, 76, 78i
Mechanoreceptor, 509, 510, 511, 619
Medulla oblongata, 502, 502i, 504, 506t, 567, 607
Medusa, 362–363, 362i, 363i
Meerkat, 472, 472i, 473i
Megakaryocyte, 559
Megaspore, 348i, 349, 352i, 451, 454, 454i
Meiosis, 128t
    chromosome number and, 139, 140–141, 150, 150i
    mitosis vs., 140, 141, 148, 148i–149i
    in oocyte formation, 146–147, 147i, 662, 662i, 665, 665i
    in sperm formation, 146, 177i, 658, 659i
    in spore formation, 143i, 146, 146i,
    stages of, 142i–143i, 144–145, 150
Meissner's corpuscle, 511, 511i
Melanin, 85i, 162, 162i–164i, 164, 166, 166i, 216i, 540, 541, 627t (*See also* Albinism)
Melanocyte, 540, 541
Melanocyte-stimulating hormone, 526t
Melanoma, malignant, 212, 212i, 589
Melatonin, 523i, 529t, 534, 534i, 704
Melosh, H. J., 306
Melov, S., 75
Meltdown, 794, 795i
Membrane attack complex, 579, 579i
Membrane pump, 86
Memnoliella, 336
Memory, 503, 504, 506t
Memory cell, 582, 583i, 585, 585i
Mendel, G., 152–153, 152i
    experimental approach, 154–159, 246
    on independent assortment, 158i, 159
    on segregation, 157
Meninges, 501, 501i
Meningitis, 325, 501
Menopause, 661, 663
Menstrual cycle, 660–663, 660i–664i, 681
Menstruation, 660
Mental retardation, 179t
Mercenaria mercenaria, 742i
Mercury barometer, 598i
Meristem, 421, 423, 423i, 426, 426i, 430, 430i, 434, 460i
Merozoite, 324, 324i
Mesoderm, 358, 481, 648, 649i
Mesoglea, 362–363, 362i
Mesophyll, 102, 102i, 424, 429, 434
Mesosphere, 758i
Mesozoic, 276i–277i, 304, 305i, 343i
Mesozoic era, 304–305
Mesquite, 765
Messenger RNA (*See* mRNA)
Metabolic acidosis, 638

Metabolic pathway, defined, 75, 81, 90
    evolution of, 298–299, 308
"Metabolic water," 632
Metabolism, defined, 5, 75, 90, 123
    adipose tissue in, 41, 120, 530, 622, 629
    ATP energy for, 76–79, 79i, 92, 110
    cell membranes and, 85–87, 85i
    concentration gradients and, 80–82, 85
    electron transfers in, 79, 79i
    enzymes and, 82–85
    evolution of, 296–297, 296i, 300–301
    nutrients for, 438–439, 613ff.
Metacarpal, 543i
Metamorphosis, 372, 376, 376i, 385i
Metaphase chromosome, 129i
Metaphase I, meiosis, 142i, 145, 145i, 148t
Metaphase II, meiosis, 143i, 149i
Metaphase, mitosis, 131i, 132–133, 132i, 133i, 136, 149i
Metastasis, 218, 218i
Metatarsal, 543i
Methane, 294, 314, 314i, 750, 751i, 784t
Methanococcus jannaschii, 287, 314i
Methanogen, 314, 314i
Methanosarcina, 8i
Methionine, 42i, 204, 204i, 625, 625i
Methyl bromide, 786
Methyl group, 36i
Methylation, 214
Mexico, human population, 696, 697i
Mexico City, 784i
Meyerowitz, E., 355
MHC, 582, 582i, 586
Micelle, 621, 621i
Michener, J., 766
Micrograph, 54, 55i, 58i
Micrometer, defined, 55i
Micrasterias, 328, 328i
Microevolution, 239, 247, 270, 270t, 710
Microfilament, 57i, 66–67, 66i, 134i, 135
Micrograph, 54, 55i, 58i
Micrometer, defined, 55i
Micronucleus, ciliate, 323, 323i
Micronutrient, plant, 439
Microorganism, defined, 310
Microparasite, 725
Microscope, 50, 50i, 54, 55i
Microspore, 348i, 349, 352i, 451, 454, 454i
Microtubule, 57i, 66, 66i, 67, 67i, 131i, 132, 133, 134i
Microvillus, 361i, 620
Microwave, 33, 96i
Midbrain, 498i, 502, 502i, 506t
Middle ear, 512i, 513
Middle lamella, 58i, 68, 68i, 134i, 468
Midge, 217i
Midsagittal plane, 480i, 481i, 496
Miescher, J. F., 190
Mifeprex, 677
Migration, 394i, 688, 758, 758i, 776
Mildew, downy, 320, 320i
Milk, 38, 526, 526t, 527, 587, 593, 674
Milk of magnesia, 30, 30i
Miller, G. T., 699, 796
Miller, S., 294–295, 295i
Millimeter, defined, 55i
Millipede, 375, 375i
Milstein, C., 589
Mimic, 726
Mineral, dietary, 626–627, 627t
Mineralocorticoid, 529t
Minnow, sheepshead, 742i
Miocene, 276i–277i, 307, 307i, 309i, 399
Miscarriage, defined, 184
Mistletoe, 351, 351i
Mite, 702
Mitochondrion, 55i, 56i, 57i–59i
    aerobic respiration and, 110, 110i, 111i, 114, 114i, 115i, 116i
    DNA, 266, 402i
    function, 64, 71, 72t
    in muscle cells, 553, 605
    origin of, 64, 299i, 300–301, 308
    structure, 64, 64i
Mitosis, 128, 128t, 129, 136
    cell cycle and, 130, 130i
    meiosis vs., 140, 141, 148, 148t–149i
    stages, 130–131, 131i, 132–133, 132i–133i
Mixture, defined, 25, 32t
Model, defined, 12
Modification enzyme, 224, 225

Molar (tooth), 396i, 615, 615i, 617, 617i
Mold, 320–321, 320i, 321i
Mole (Molar), defined, 33
Molecular clock, 284–285, 289
Molecule, defined, 4, 6i, 25, 32t
Mollusk (Mollusca), 358i, 368–369, 418i
Molting, 372, 372i, 535, 535i
Molybdenum, 439t
Monarch butterfly, 691, 726, 758, 758i
Moneran, 287 (*See also* Prokaryotic cell)
Mongoose, 472
Monkey, 401i
Monoclonal antibody, 589
Monocot (Monocotyledonae), 351
    characteristics, 354t, 428, 429
    development, 426–427, 456i, 460i, 461i
    dicot vs., 425, 425i, 427i
    life cycle, 352i
    primary root, 430, 431, 431i
Monocyte, 558i, 559i, 560, 571, 581
Monoglyceride, 618, 621
Monohybrid cross, 156, 156i, 167
Monomer, defined, 37
Mononucleosis, infectious, 560
Monosaccharide, 38, 38i, 48t, 618, 618t
Monosomy, 184
Monotropa uniflora, 351i
Monotreme, 396
Monsoon, 761
Monsoon forest, 768
Monsoon grassland, 767
Montagu, M., 576
Montane coniferous forest, 34, 34i, 763i, 770, 770i
Moore-Walker, N., 15
Moorhead, P., 674–675
Moray eel, 389i
Morchella deliciousa, 336i
Morgan, T., 176, 176i
Morning-after pill, 677
Morphogen, 654
Morphogenesis, 653i, 645, 653, 653i, 668i, 680
Morphological convergence, 281, 281i, 289, 757i–758i
Morphological divergence, 280–281, 280i, 289
Morphological trait, defined, 246
Morphology, comparative, 240–241, 241i, 280–281, 280i, 281i
Morula, 651, 666i
Mosaic tissue effect, 179t, 216, 216i
Mosasaur, 409i
Mosquito, 324
Moss, 340i, 344–345, 345i
Moth, 4–5, 4i, 10, 10i, 250, 251i, 377i, 517, 522, 721, 721i, 725
Motion sickness, 513
Motor cortex, 503i
Motor neuron, 479i, 495, 495i, 506, 550, 550i, 551
Motor protein, 66, 67, 124i, 132i, 548
Motor unit, 551
Moufet, T., 240
Mount Saint Helens, 420, 420i
Mountain building, 307
Mouse, 688, 726, 726i
    laboratory, 14–15, 192, 232, 629, 629i
Mouth, 517, 602i, 616, 616i, 617, 618i, 630t
mRNA, 46, 202, 203, 203i, 204, 204i, 205–207, 210, 213–216, 650
MSH, 526t
Mucigel, 430i
Mucin, 617
Mucosa, 619i
Mucus, 608, 616, 616i, 617, 623
Mucus-secreting cell, 608i, 620i
Mule, 223, 263
Multicelled organism, 6, 6i
Multiple allele system, 160
Multiple fruit, 456t, 457
Multiple sclerosis, 494, 507, 591
Multiplication cycle, viral, 318, 318i, 319i
Multiregional model, 402
Mumps, 590i
Muscle, 476, 542 (*See also specific types*)
    contraction (*See* Contraction)
    flight, 394i, 395, 395i
    segmented, 384, 385i, 478, 478i

Muscle fatigue, 551
Muscle tension, 551
Muscle tissue, 478, 484
Muscle twitch, 551, 551i
Muscular dystrophy, 179t, 189, 221
Muscular system, human, 480i, 481
Mushroom, 9, 334–335, 334i, 335i
Musk oxen, 712i
Mussel, 200, 200i
Mustard, 456t
Mutagen, defined, 209
Mutation, gene, defined, 10, 201, 208
    Drosophila, 176i, 189, 189i
    effects on development, 282–283
    effects on protein synthesis, 208–209
    genetic disorders from, 179, 180–183
    genetic divergence and, 261–264,
        288–289
    lethal, 169, 247
    nature of, 10, 16t, 201, 208, 258, 270t
    neutral, 247, 284–285
    rates, 209, 227, 247
    repair enzymes and, 197, 212
    sources (mutagens), 189, 206–207
    types, 181, 208–209, 208i, 209i, 210, 247
    and variation, 10–11, 160, 165, 166,
        201, 247, 250, 252, 255, 261
Mutualism, 332, 440, 720, 720t, 721, 721i
Mycelium, 320, 333, 334
Mycobacterium tuberculosis, 259
Mycorrhiza, 332, 338i, 339, 351, 441, 721
Myelin sheath, 494, 494i
Myocardium, 564, 564i
Myofibril, 548, 548i, 550
Myoglobin, 598
Myometrium, 660, 661i
Myopia, 515
Myosin, 478, 548, 548i, 549i
Mytilus, 200–201, 200i, 723
Myxobacterium, 315
Myxococcus xanthus, 315
Myxoma virus, 731
Myxomatosis, 731
Myxomycota, 321

N 9 + 2 array, 67, 67i
NaCl (*See* Sodium chloride)
NAD+, 46, 48t, 111, 112, 114–118, 114i–
    118i, 115i, 117i
NADH, 111–118, 111i–118i, 117i, 121i
NADPH, 98, 99i
Naive cell, immune system, 584
Naked mole rat, 714, 714i
Nanometer, defined, 55i
Nasal cavity, 602i, 603
Nassarius obsoletus, 742i
Natural killer cell (*See* NK cell)
Natural selection, 11, 13, 154, 270t
    in animal evolution, 282–283, 302, 359,
        392, 398, 598, 615, 707–709, 714, 715
    artificial selection vs., 10, 245
    balancing, 254–255, 255i
    basic premises, 10–11, 249
    Darwin and, 10, 16, 154 , 239, 244–245,
        249, 261, 288
    defined, 11, 154, 249, 258, 270
    development and, 282–283
    directional, 250–251, 250i
    disruptive, 253, 253i
    in genetic divergence, 262–265
    in human evolution, 283, 400, 402, 403
    life history patterns and, 692–693
    meiosis and, 140
    mutation and, 10–11, 247, 285
    of pathogens, 230, 325
    pesticide resurgence and, 250–251
    of photosynthetic pigments, 107
    in plant evolution, 100, 342, 442i, 450
    of rubisco, 100
    sexual, 254, 254i, 266, 394i, 521, 709
    speciation vs., 261
    stabilizing, 252, 252i, 261
Nautiloid, 302, 303i
Navel, 673
Neandertal, 401, 401i
Nearsightedness, 515
Nebula, 292i
Nectar guide, 8i, 9i, 451, 452
Needlefish, Atlantic, 742i

Negative control system, 214–215, 215i
Negative feedback, 84, 84i, 482, 482i, 483i,
    530–531, 530i, 531
Neisseria gonorrhoeae, 678
Nematocyst, 362, 362i
Nematode, 358t (*See* Roundworm)
Nephridium, 371, 371i
Nephron, 635, 635i, 637i
Neritic zone, 774, 774i
Nerve, 494, 494i, 506
Nerve cell (*See* Neuron)
Nerve cord, 371, 384, 385i, 498
Nerve growth factor, 532
Nerve net, 362, 496, 506
Nervous system, defined, 480i, 498
    abnormal development, human, 672
    earthworm, 371, 371i
    embryonic source, 481, 648, 650–653
    evolution, 486–487, 496–497, 498–499,
        498i, 499i
    flatworm, 364–365, 364i
    functioning, 488–491, 489i, 490i, 506
    invertebrate, 487, 496–497, 497i
    organization, 487, 495i, 499, 499i,
        500–501, 500i
    vertebrate, 487, 498–499, 498i, 499i,
        502–503, 502i, 503i, 506
Nervous tissue, defined, 479, 484
Net primary productivity, 734
Net protein utilization, 625
Neural fold, 668, 668i
Neural groove, 653i
Neural plate, 649i, 653
Neural tube, 498, 503, 649i, 652, 653i,
    668, 668i, 670i
Neurofibromatosis, 179t, 211, 211i
Neuroglia, 479, 494, 499
Neuromodulator, 492
Neuron (*See also specific types*), 130
    function, 479, 487–491, 490, 491, 550, 551
    pesticide effects on, 47
    structure, 479, 488, 488i
    types of, 487, 488, 488i
Neurospora crassa, 337
Neurospora sitophila, 337
Neurotransmitter, 492, 492i, 496, 504,
    506, 516, 522
Neutral mutation, 247, 284–285
Neutron, 4, 6i, 21, 22, 32t
Neutrophil, 558i, 559, 559i, 578t, 580,
    581, 581i, 594t
NF1 gene, 211
Niacin, 626t
Niche, 719, 720, 734
Nickel, 294
Nicotinamide adenine dinucleotide
    (*See* NAD+)
Nicotine, 503, 504, 570 (*See also* Smoking)
Night blindness, 626t
Nirenberg, M., 204
Nisbet, E., 103
Nitric acid, 30
Nitric oxide, 784, 784t
Nitrification, 752, 752i
Nitrobacter, 301i
Nitrogen, 20, 21i, 22, 22t, 25, 25i, 26, 752i
    algal blooms and, 327
    plants and, 439i, 440–441, 441i, 745, 747
Nitrogen cycle, 104, 119, 314, 332,
    752–753, 753i
Nitrogen dioxide, 784
Nitrogen fixation, 441, 441i, 752–753, 752i
Nitrogen gas, 611
Nitrogen narcosis, 611
Nitrous oxide, 750, 751i, 753
NK cell, 559, 559i, 578t, 582, 589, 594t
Nocardia otitidus caviarum, 224t
Nociceptor, 509, 580t
Node, 422i, 426i, 428i
Nondisjunction, defined, 184–185, 184i
Nonionizing radiation, 209, 212
Nonpolar covalent bond, 26–27, 32
Nonrenewable energy source, 794, 794i
Nonshivering heat production, 641
Norepinephrine, 492, 504, 505, 523i, 530
Nori, 327
Norplant, 677, 677i
Nose, 511i, 517, 517i (*See also* Nasal cavity)
Noss, R., 755

Nostoc, 70i, 71i, 752
Notochord, 384, 385i, 498, 649i, 668
Nuclear division (*See* Meiosis; Mitosis)
Nuclear energy, 794–795, 794i
Nuclear envelope, 59i, 60, 60i, 60t, 61i,
    132, 132i, 298i, 300, 301i
Nuclear pore, 61, 61i
Nucleic acid, defined, 4, 47, 48t (*See also*
    DNA; RNA)
    hybridization, 225, 225i, 229, 284, 402i
Nucleoid, 52, 70
Nucleolus, 57i, 60i, 60t, 61, 72t
Nucleoplasm, 57i, 60i, 60t
Nucleotide, 46, 46i, 194, 194i, 195, 195i,
    201, 202, 204, 618, 618t, 622i
Nucleus, atomic, 8, 22
Nucleus, cell, 51, 52, 56t, 57i–59i, 60–61
    origin of, 300, 301i
    structure, 57i–59i, 60i, 60t
Nudibranch, 369
Numerical taxonomy, 312
Nutrient, defined, 438
    absorption in human body, 620–621,
        620i, 621i
    cycling, 7i, 314, 737, 745–753, 745i–753i
    ecosystem, 738, 745–754, 745i–753i, 775i
Nutrition, 613, 624–627, 624i, 625i, 626t,
    627t, 630
    plant, 439–439, 439i, 448, 753
Nymph, 376, 376i
Nymphaea, 351i

O Oak (*Quercus*), 8, 428i, 766
Ob gene, 629, 629i
Obelia, 363, 363i
Obesity, 570, 628, 629, 674
Observational test, defined, 12
Occipital lobe, 502, 503i
Ocean, 746i, 760
    currents, 760–761, 760i, 777, 778, 779i
    provinces, 757, 774–775, 774i–775i
Oceanic zone, 774, 774i
Ocotillo, 765, 765i
OCT (*See* Oxytocin)
Octopus, 138, 358t, 368, 514, 514i
Odonata, 377i
Oil, 28, 40, 48t, 737 (*See also* Fossil fuel)
Oil gland, 540, 540i
Oil spill, 230, 794
Okazaki, R., 197i
Oleic acid, 40i
Olfactory bulb, 517i
Olfactory lobe, 498, 506t
Olfactory receptor, 517, 517i
Oligocene, 276i–277i, 307
Oligosaccharide, 38, 45, 48t, 53, 53i
Oligotrophy, 773, 781i
Olive, 456t
Olive oil, 624
Ommatidium, 514
Omnivore, 738, 738i, 739
On the Origin of Species, 245
Oncogene, 218
Onion, 458t
Oocyte, 146, 147i, 650, 660, 661i, 662, 663i
Oogenesis, 146, 147i, 648, 648i
Oogonium, 147i
Oomycota, 320
Open circulatory system, 556, 556i
Operant conditioning, 706t
Operator, 214, 215, 215i
Operon, 214, 215
Ophiostoma ulmi, 731t
Optic disk, 515i
Optic nerve, 500i, 502i, 511t, 515i, 516
Oral cavity, 617 (*See also* Mouth)
Orange tree, 456t, 458, 458t
Orangutan, 398i
Orbital, electron, 24–25, 24i, 25, 25i, 32
Orchid, 102i, 286i, 340i, 351, 458t, 459i
Order (taxon), 8, 286
Ordovician, 276i–277i, 343i
Organ, defined, 6i, 364, 472
Organ formation, 648, 648i, 652–653, 680
Organ of Corti, 513i
Organ system, 6i, 364, 472–473, 480
    human, 480–489, 480i, 481i, 484, 666

Organ transplant, 589
Organelle, 6i, 56, 56t, 300, 301, 301i
Organic compound, 35, 36, 37, 48
    distribution in plants, 446–447, 448
    origin of, 294–295, 295i
Organophosphate, 47i
Orgasm, 665
Oriole, 267, 267i
Orr, A., 266
Oryctolagus cuniculus, 731
Osmoreceptor, 509, 511
Osmosis, 88–89, 89i, 90, 568, 568i, 569
Osprey, 742i
Osteichthyes, 388–389
Osteoarthritis, 545, 628
Osteoblast, 544
Osteoclast, 544, 545
Osteocyte, 544, 544i
Osteoporosis, 537, 545, 545i, 627t, 629
Ostracoderm, 387
Ostrich, 240i, 394
Outer ear, 512, 512i
Output zone, axon, 506
Oval window, 512i
Ovary, human, 660, 660t, 661i, 662,
    662i, 663i, 664i, 665i
    hormones and, 523i, 526t, 529t, 531, 531i
Ovary, plant, 452i, 453, 457
Overgrazing, 791, 791i
Ovibos moschatus, 712i
Oviduct, 660, 660t, 661i, 662
Oviparity, 646i, 647i
Ovoviviparity, 646i, 647i
Ovulation, 526t, 660, 661t, 662, 662i,
    663, 663i, 677, 681
Ovule, 351i, 352i, 451, 452i, 454i, 455i, 457
Ovum, 146, 147i, 665
Owl, 515i
Oxaloacetate, 102i, 114, 114i, 115i
Oxidation–reduction, defined, 79
Oxidized substance, 79
Oxygen, 20, 21i, 22t, 25i, 26, 27, 28 (*See also*
    Aerobic respiration, Photosynthesis)
    membrane crossing by, 83, 569
Oxygen debt, 548
Oxyhemoglobin, 606
Oxytocin, 483, 524t, 526t, 527i, 674
Ozone, 750, 758, 784, 784t
Ozone "hole," 786
Ozone layer, 758, 758i, 782, 783, 786, 786i
Ozone thinning, 786, 786i

P Pacemaker, 554, 565, 565i, 571
Pacinian corpuscle, 511
Pain, 509, 511, 518, 580t
Paine, R., 271, 722–723
Pair-rule gene, 655i
Paleocene, 276i–277i, 307, 307i
Paleozoic, 276i–277i, 302–303, 343i, 373
Palisade mesophyll, 429, 429i
Palmiter, R., 232
Palp, invertebrate, 369i
Pan troglodytes, 150i
Pancreas, 60i, 62, 500i, 532–533, 609i,
    616, 616i, 618t, 630t
Pancreatic amylase, 618t
Pancreatic islet, 523i, 529t, 532–533
Pancreatic nuclease, 618t
Panda, 285i
Pandemic, defined, 325
Pandion haliaetus, 742i
Pando the Trembling Giant, 459i
Pangea, 278, 303, 303i, 304, 307, 309i, 757
Panting, 641
Pantothenic acid, 626t
Pap smear, 137
Papilla, 517i
Papillomavirus, 137, 679
Paradisaea raggiana, 254i
Paralichthys dentatus, 742i
Paramecium, 91, 91i, 322, 323i, 360, 722
Parapatric speciation, 261, 267i, 270
Parapod, 370, 370i
Parasite, 9, 333, 334, 336, 739, 739i
    defined, 339, 364, 724, 738
    interactions with host, 725
    types of, 334, 336, 705, 705i, 725
Parasitism, 720, 720t
Parasitoid, 735

Parasympathetic nerve, 499i, 500–501, 500i, 506
Parathyroid gland, 523i, 529t, 532
Parathyroid hormone, 523i, 532, 545
Parenchyma, 422, 424, 424i, 425i, 429, 432, 434, 434t
Parental behavior, 395, 711, 711i, 715, 715i
Parietal lobe, 502, 503i
Parkinson's disease, 504, 505
Parthenogenesis, 458, 458t
Partial pressure, in respiration, 598, 606
Passive transport, 86, 86i, 90
Pasteur, L., 577
Pasteurization, 577
Patella, 543i, 545i
Pathogen, defined, 70, 310, 577
    examples, 11, 70, 70i, 310, 315, 320–325, 571, 575, 576–577
    immune response to, 580–581, 581i, 587
Pattern formation, 654–655, 655i
Pauling, L., 190, 195
Pavlov, I., 706i
PCR, 226, 226i
Pea, garden, 128, 150i, 154ff., 154i
Peach, 456i, 457, 457i
Pear, 424i, 456t
Peat, 344–345, 345i, 771
Pectin, 424
Pectoral girdle, 542, 542i, 543i
Pectoralis major, 547i
Pedigree, 178, 178i, 181i, 188
Pedipalp, 373, 373i
Pelagic province, 774, 774i
Pellagra, 626t
Pellicle, 326, 326i
Pelvic cavity, 480i
Pelvic girdle, 241, 241i, 542, 542i, 543i
Pelvic inflammatory disease, 677, 678, 679
Pelvic nerve, 500i
Penguin, 40i, 280, 280i, 713i, 715, 715i, 737, 737i
Penicillin, 251
Penicillium, 336
Penis, 647, 656i, 657i, 658i
Pepo, 456t
Pepsin, 85, 618, 618t, 619
Pepsinogen, 618
Peptic ulcer, 618
Peptide bond, 42, 43i
Peptide group, 44, 44i
Peptide hormone, 524t, 525, 525i
Peptidoglycan, 313, 314
Peradectes, 307i
Perception, defined, 508, 518
Peregrine falcon, 410i, 742, 742i
Perennial (plant), 422
Perfect flower, 453
Perforin, 578i, 588i, 589
Pericardium, 564, 564i
Pericarp, 460i
Pericycle, 430i, 431
Periderm, 423, 425, 432, 433i, 434, 434t
Periodontal disease, 617i
Peripheral nervous system, 499, 499i, 500–501, 500i, 506
Peristalsis, 619i
Peritoneal dialysis, 638
Peritoneum, 359
Peritubular capillary, 635, 635i, 636, 636i
Periwinkle, 723, 723i
Permafrost, 771
Permanent wave, 42, 42i, 45i
Permeability, selective, 85, 85i
Permian, 276i–277i, 302, 303i, 309i, 343i
Peroxisome, 63
Peroxyacyl nitrate, 784, 784t
Pertussis, 590i
Pest resurgence, 251
Pesticide, 47, 47i, 414, 638, 742, 784t, 787
    alternatives, 725
Pesticide resistance, 250–251, 377i
PET scan, 23, 23i, 503i, 505i
Petal, 452, 452i, 457i
Petiole, 428, 428i
Pfiesteria piscicida, 326i, 327
PGA, 101, 101i
PGAL, 101, 101i, 112, 113i, 117i, 120, 121i
pH, 45, 578, 602, 607, 617, 638, 753
    effect on enzymes, 85, 85i

pH scale, 21, 30, 30i, 32
Phaeophyta, 328
Phagocyte, 541, 545, 559, 580, 581i, 585
Phagocytosis, 87, 90, 558i, 585
Phalanges, 543i
Phanerozoic, 276i–277i
Pharmaceutical, 231
Pharyngeal arch, 668, 668i, 670i
Pharynx, 364, 364i, 384–387, 385i, 600, 600i
    human, 602i, 603, 614, 616, 616i, 617, 630t
Phaseolus vulgaris, 460i
Pheasant, 394i
Phenotype, 155, 162–165, 162i–165i, 166, 166i, 167, 171, 246, 247
Phenylalanine, 42i, 204i, 625
Phenylketonuria, 179t, 186
Phenylpyruvate, 186
Pheromone, 517, 522, 708, 716
Philadelphia chromosome, 170, 170i, 171i
Phleum pratense, 58i
Phloem, 342, 422, 424i, 434, 434t, 458
    components, 423, 446, 446i
    function, 342, 424, 437, 446–447, 447i
    primary, 426, 426i, 427i, 429, 429i, 431i, 431i, 447i
    secondary, 432–433, 432i, 433i
Phloem conducting cell (sieve-tube member), 423i, 437i, 446, 446i, 447i
Phosphate group, 36i
Phosphate ion, 747, 747i
Phosphate-group transfer, 78–79, 112, 115i
Phosphatidylcholine, 48t
Phosphoglyceraldehyde (See PGAL)
Phosphoglycerate (See PGA)
Phospholipid, 41, 41i, 48t, 52–53, 52i
Phosphorus, 20, 21i, 22t, 23, 544, 627t
    as plant nutrient, 439t, 747, 747i
Phosphorus cycle, 104, 747, 747i
Phosphorylation, 78–79, 112, 113i
Photoautotroph, 92, 96, 103, 312, 314, 739i
Photochemical oxidant, 784, 784t
Photochemical smog, 784
Photoheterotroph, prokaryotic, 312
Photolysis, 98, 100, 100i
Photon, 96, 98, 98i, 103, 516
Photoperiodism, 466–467, 466i, 469
Photoreceptor, 103, 509, 510, 515, 515i, 516
Photorespiration, 102, 103, 105
Photosynthesis, 92, 94–95, 94i, 102i
    aerobic respiration vs., 93, 93i
    cyclic vs. noncyclic, 98–99, 99i
    end products, 94–95, 94i, 101
    equation, 95i, 105
    evolution, 98, 103, 122, 277i, 296–299
    leaf adaptations for, 429, 429i
    reactions, 79, 81, 90, 92i, 94, 94i, 96–97, 98–103, 98i–102i
    studies of, 23, 92, 92i
Photosystem, defined, 97, 98, 98i
Photosystems I and II, 97i, 98, 99i
Phototropism, 464–465, 465i
Photovoltaic cell, 796, 796i
Phycobilin, 97, 107, 327
Phycoerythrin, 97i
Phylogenetic reconstruction, 273
Phylogeny, defined, 286
Phylum (taxon), 8, 286
Physalia, 363, 363i
Physarum, 321, 321i
Phytochrome, 217, 217i, 466, 466i
Phytophthora infestans, 320
Phytoplankton, 326, 734, 772, 773, 774, 775i, 776, 777, 779, 781i, 786
Phytosterol, 53
PID (See Pelvic inflammatory disease)
Pigeon, 10–11, 10i, 11i
    community structure, 718–719, 718i
Pigment, 49, 452, 465
    coat color, 85i, 162, 162i, 166, 166i, 216i
    floral, 317i, 450, 452
    photosynthetic, 65, 72i, 96, 97i, 103, 326–328, 341
    phytochrome, 217, 466, 466i, 467
    in phototropism, 465
    skin, 74, 540 (See also Melanin)
    visual, 511t, 514i, 515, 516, 516i
Pike-cichlid, 693
Pilobolus, 339, 339i

Pilomotor response, 641
Pilus, 70, 70i, 312i, 313, 316
Pine forest, 770
Pine tree, 433, 433i
Pineal gland, 502i, 506t, 523i, 529t, 534
Pineapple, 102i, 456t, 457, 457i
Pinna, 513
Pinus longaeva, 350i
P. ponderosa, 348i
Pinworm, 358i, 366
Pioneer species, 340, 728, 728i
Pisaster ochraceus, 723
Pistil, 453
Pisum sativum, 150i, 154i, 155, 156i
Pitch, sound, 512i
Pith, 426, 426i, 427i, 431i
Pitman, W., 272
Pituitary dwarfism, 528, 528i, 537
Pituitary gland, 506t, 522i, 526, 527i, 528, 530i, 536, 659i, 663i, 664i
PKU, 186
Placenta, 174i, 396, 396i, 587, 667, 667i, 668, 669, 673, 673i, 681
Placoderm, 302, 386i
Placozoa, 358t, 360
Plague, 691
Planarian, 364i, 514
Plankton, defined, 322, 742i
Plant (See also specific types)
    biodiversity through time, 418i
    classification, 8, 8i, 9, 287, 341, 351, 354t
    diseases, 230, 319, 320, 320i, 377i
    evolution, 340, 340i, 342–343, 342i, 343i, 354t, 450–451
    growth, 421, 430–433, 432i, 460–465, 460i, 461i, 464i, 465i
    human uses of, 353, 353i, 355, 355i
    life cycles (See Life cycle, examples)
    nonphotosynthetic, 351, 351i
    number of known species, 351, 354t
    nutrition 436ff. (See also Photosynthesis)
    reproduction (See Reproduction, plant)
    responses to environment, 465–468
    specimen cuts, defined, 423i
    sugar transport, 446–447, 446i, 447i, 448
    toxins, nature of, 46
Plant hormone, 462ff., 462t, 469
Plant physiology, defined, 436
Planula, 363, 363i, 497
Plaque, atherosclerotic, 41i, 222i
Plasma, 477, 477i, 482, 558, 558i, 568i, 574
Plasma membrane, defined, 52
    components, 51–53, 53i, 70, 97, 119, 301, 312, 314, 316, 487i, 488i, 541, 582–589
    cytoskeletal supports, 53i, 66, 134i, 135
    function, 51–54, 56i, 57i, 70, 72t
    muscle cell, 478, 550, 550i, 565, 565i
    neuron, 477, 487–492, 487i, 488i, 510
    origin and evolution of, 297, 300, 301
    selective permeability, 52, 85, 85i
    solutes across, 75, 78i, 82, 85–87, 85i–87i, 91i, 440, 446, 475, 488–492, 516, 569
    structure, 51–53, 53i, 70
    vesicles across, 64i, 65, 87, 87i, 582, 582i
    water across, 85, 85i, 88–89, 89i, 91i, 48i
Plasma protein, 558, 558i, 568i, 569, 578, 579, 581, 581i
Plasmid, 224–225, 224i, 230–231, 315, 316
Plasmodesma, 68i, 69, 429
Plasmodium, 255, 321, 324, 324i
Plasmolysis, 89, 89i
Plasmopara viticola, 320, 320i
Plastid, 65, 464, 464i
Plate tectonics theory, 278–279, 503i
Platelet, 558i, 559, 571
Platyhelminthes, 358t
Platypus, 382–383, 382i, 397, 397t
Pleiotropy, 161, 161i
Pleistocene, 276i–277i, 307i
Pleural membrane, 602i, 603
Pleurisy, 603
Plexus, 497, 630
Pliocene, 276i–277i, 405i
Plutonium, 23, 795, 795i
Pneumatophore, 776
Pneumocystis carinii, 592
Pneumonia, 251, 325, 325i
Pneumotaxic center, 607
Poinsettia, 467

Polar bear, 28i, 727
Polar body, 144, 146, 147, 147i, 662, 662i
Polar covalent bond, 27, 32
Polar husky, 538, 538i
Polarity, 122
    animal embryos, 651, 655i
    water molecule, 28, 28i, 29, 32
Poliovirus, 55i, 587, 590i
Pollen grain, 52i, 343, 348i, 349, 351, 352i, 450, 451, 453, 453i
Pollen sac, 352i
Pollen tube, 352i, 454i
Pollination, 346, 349, 351, 352i, 454, 454i
Pollinator, 351, 351i, 450–451, 450i, 451i
Pollutant, 783, 784, 784t, 785i, 798
Pollution, 771i
    air, 18, 31i, 250, 332, 338, 445, 445i, 773, 784–786, 784t, 785i, 786, 794
    water, 230, 412, 449, 449i, 534–535, 535i, 742, 773, 776, 785, 776, 792–793
Polychaete, 358t, 370i
Polycythemia, 560
Polydactyly, 178i, 179t
Polymer, defined, 37
Polymerase chain reaction, 226, 226i
Polymorphism, 246, 254, 258
Polyp, cnidarian, 362, 362i, 363i
Polypeptide chain, 43, 43i, 61, 62, 62i, 201, 202, 204–207, 204i, 206i
Polyploidy, 184, 188, 266, 267i
Polysaccharide, 38, 48t, 617, 618i
Polysome, defined, 207
Polytene chromosome, 217
Polytrichum, 345i
Pome, 456t
Pons, 502, 502i, 506t, 607
Pool, in stream, 773, 773i
Pope John Paul II, 259
Poplar leaf, 428i
Poppy, 430i, 456t
Population, defined, 6, 6i, 239, 246, 261
    characteristics, 246–247, 262–263, 685, 686, 687, 691–693
    human (See Human population)
    genetic equilibrium in, 246–248
    size, 686–689, 688i
    as unit of evolution, 10–11, 246, 258
Population growth
    carrying capacity and, 685, 690, 691i, 700, 724
    exponential, 685, 688i, 689
    limits on, 685, 690–691, 691i, 694–695
    logistic, 690–691, 690i, 691i
    prokaryotic (bacterial), 310, 316, 689
    reproductive base for, 686, 689
    species interactions and, 720, 720t, 724
    zero, 688, 696–697, 697i, 699
Populus tremuloides, 458, 459i
Porifera, 358t, 360–361, 360i, 361i
Porphyra, 327, 327i
Porphyria, 485
Porphyrin, 296, 296i
Porpoise, 280, 280i
Portuguese man-of-war, 363, 363i
Positive control system, 214, 215
Positive feedback, 483, 490, 530
Positron-Emission Tomography, 23
Postelsia palmaeformis, 723
Posterior (directional term), 359, 480i, 481i
Postpartum depression, 674
Potassium, 20, 21i, 22t, 627t
    neural function and, 488–491, 489i, 491i
    in plants, 439t, 445, 445i
Potato, 47, 234, 320, 458i, 690i
Potential energy, 76, 78, 78i, 90
Prairie, 729, 739, 739i, 740i, 766, 767i
Praya dubia, 774
Praying mantis, 727i
Precursor, defined, 81
Predation, 614, 720, 720t, 724–725
Predator, 10, 39i, 64, 302, 305, 352, 362ff., 386ff., 498, 502, 614, 615, 724–725, 734
    microbial, 64, 298, 301, 310, 311, 315, 320, 321, 323, 330t
    sexual, human, 504
Predator avoidance, 386–387, 474i, 712–713, 712i, 713i, 724–727, 725i, 726i
Prediction, defined, 12, 16
Prefrontal cortex, 503, 503i, 504

Pregnancy, human, 680
   embryonic period, 666–671, 666i–667i
   events leading to, 665
   fetal period, 670–673, 670i–672i
   nutritional considerations, 625, 672–673
   prenatal diagnosis, 186–187
   risks during, 672–673, 672i
Pregnancy test, 667, 677
Pregont, P., 538
Preimplantation diagnosis, 187
Premolar, 396i, 617, 617i
Premotor cortex, 503
Prenatal diagnosis, 186–187, 186i, 187i
Pressure
   air, 598, 598i, 604–605, 759, 761, 778
   hydrostatic, 89, 568i
   negative, in transpiration, 442
   osmotic, 89, 568–569, 568i
   sense of, 511
   in translocation, 446–447, 447i
   turgor, 444, 444i, 445i, 462, 462i
Pressure flow theory, 446–447, 448
Pressure gradient, 80, 598, 604–605, 607i
Preven, 677
Prey, predation and, 724–725, 734
Price, P., 735
Primary cell wall, 68, 68i, 462i
Primary growth, plant, 421, 423, 423i, 434
Primary producer, 738ff., 739i (See also Autotroph; Producer)
Primary productivity, 734, 734i, 738–739, 772–774, 775i, 776
Primary root, 422i, 460i
Primary structure (amino acid), 43, 44, 45
Primary succession, 728, 728i
Primary tissue layer, 481, 648, 649i, 652i
Primate, 183, 283, 283i, 284i, 383, 398i
   evolution, 398–402, 401i, 403, 405, 405i
Primer, 226, 226i
Priming pheromone, 708
Primitive streak, 668, 668i
Prion, 319
PRL, 526, 527i (See also Prolactin)
Probability, 157
Probe, 229, 229i
Procambium, 423, 423i, 426i
Producer, defined, 5, 34, 310, 326, 328, 738
   ecosystem roles, 7, 7i, 9, 34, 96i, 102, 720, 737ff., 738i–753i
Profundal zone, 772, 772i, 781i
Progeria, 179t, 182, 182i
Progesterone, 523i, 526t, 529t, 531, 531i, 661, 661i, 661t, 663, 663i, 664i
Progestin, 524t, 677
Proglottid, 365, 366i
Prokaryotic cell (See also Bacterium)
   characteristics, 52, 70–71, 298, 298i–299i, 300i 312–313, 312t
   eukaryotic vs., 8, 72t, 331t
   structure, 70, 70i, 312–313, 312i, 313i
Prokaryotic fission, 128, 128i, 316, 316i
Prolactin (PRL), 217, 523i, 524t, 526, 526t, 527, 639i, 674
Proline, 42i, 204i, 208i
Prometaphase, 132, 136
Promoter, 202, 203i, 214, 215i, 220, 283
Pronghorn antelope, 614–615, 615i
Prop root, 461i, 776, 776i
Prophase I, meiosis, 142i, 144i, 148t, 172i
Prophase II, meiosis, 143i, 149i
Prophase, mitosis, 131i, 132, 132i, 133, 136
Prosimian, 398, 401i
Prostaglandin, 41, 524t, 532, 581, 657, 663
Prostate cancer, 657
Prostate gland, 656t, 657, 657i, 658i
Protein (See also specific types)
   chromosomal, 60, 60t, 61, 128–129, 129i, 142, 214, 214i
   function, 35–37, 38, 201, 206
   in human diet, 120, 121i, 625, 625i
   membrane, 52–53, 52i, 53i, 60, 218
   metabolism, 120, 121i, 618t, 619
   regulatory, nature of, 213–216
   structure, 36, 42–45, 43i–45i, 48i, 190, 204i, 208–209
Protein-energy malnutrition, 625
Protein hormone, 524, 524t, 525, 536

Protein kinase, 218
Protein synthesis, 42–44, 43i, 46, 201–207 (See also Transcription; Translation)
   summary diagram, 210i
Protein-mediated transport, 86–87, 86i, 87i
Proteinoid, 295
Proterozoic, 276i–277i, 299, 300
Protistan (Protista), 299i, 300, 300i, 301
   characteristics, 8i, 72t, 87, 104, 110, 310, 310i, 320i–329i
   classification, 8, 9, 287, 311, 320
   evolution, 298i, 300–301, 320
   groups, 320–324, 326–329, 330, 330t
Proto-cell, 297
Protoderm, 423, 423i, 426i
Proton, 4, 6i, 21, 22, 24, 32t
Protonephridium, 364, 364i
Protostome, 367
Protozoan, 322–323, 322i, 323i, 325t, 330
   biodiversity through time, 418i
Proximal (directional term), 481i
Proximal tubule, 635, 635i, 636, 636i, 637
Prunus, 419i
Pseudocoel, 359
Pseudocoelomate (false coelom), 359i, 365
Pseudomonad, 315
Pseudomonas, 70i, 71i, 325
Pseudopod, 66, 66i, 322
Psilophyta, 346
Psilophyton, 303i
Psilotum, 346i
Psychedelic, 505
Pterosaur, 280i
PTH, 523i, 532, 545
Puberty, 534, 656
Pueraria lobata, 730, 730i
Pulmonary artery, 562i, 563i, 564i, 607i
Pulmonary capillary, 602i, 606
Pulmonary circuit, 557, 562, 562i
Pulmonary vein, 562i, 563i, 564i, 607i
Pulse pressure, 566–567, 567i
Pumpkin, 21i
Punctuation model, speciation, 268
Punnett, R., 163
Punnett-square method, 157, 157i
Purine, 194
Pyloric sphincter, 619i
Pyracantha, 119i
Pyrenestes ostrinus, 253, 253i
Pyrimidine, 194
Pyrrhophyta, 326
Pyruvate, 110, 112, 115i, 121i
Python, 241i, 393, 508, 508i, 615i

Q Quadrat, defined, 687
Quadriceps femoris, 547i
Quaking aspen, 459i
Qualitative vs. quantitative variation, 246
Quaternary period, 276i–277i
Queen Victoria, 181, 181i
Quercus, 8 (See also Oak)

R R group, protein, 42, 44
Rabbit, 166, 166i, 731, 731i
Raccoon, 285i, 396i
Radial cleavage, 367, 651, 651i
Radial section, plant, 423i
Radial symmetry, 358–359, 358i, 496
Radiation (See also specific types), 96, 640
Radiation poisoning, 137, 795i
Radiation therapy, 23
Radioactive decay, 22, 277, 277i
Radioactive fallout, 795t
Radioisotope, 22, 23, 32t, 94, 193i, 277, 277i, 795, 795i
Radiolarian, 322, 322i
Radiometric dating, 276i, 277, 277i
Radium, 23, 226
Radius (bone), 543i
Radula, 368, 368i
Rafflesia, 768i
Ragweed, 453, 453i, 591i
Rain shadow, 761, 761i
Rainfall, patterns of, 18, 789
Ramaria, 334i
Rana pipiens, 150i, 644, 644i
Ranching, conservationist, 416–417

Ranunculus, 726
Raphus cucullatus, 409, 409i
"Raptures of the deep," 611
Raspberry, 456t
Rattlesnake, 163, 163i, 393i
Raup, D., 269
Raven, 707
Ray, cartilaginous fish, 388, 388i
Ray, in wood, 432, 433
Reabsorption, 568i, 569, 580, 636–637, 636i
Reactant, defined, 81
Reaction center, photosystem, 98, 98i
Realized niche, 720
Rearrangement, bond, 37
Receptacle (flower), 452, 452i, 457i
Receptor, molecular, 5, 43, 53, 53i, 63, 74, 87, 219, 222, 317, 467i, 571, 653
   bacterial, 315
   lymphocyte, 582, 582i, 583–589, 584i–589i, 591
Receptor, sensory, 5, 181, 326, 382, 383, 506–509, 508i, 511i, 513–515, 514i, 518, 567, 602, 607, 617, 619, 637, 640, 646
   neural function and, 488, 490, 492, 492i, 495–496, 495i, 502, 503, 504
   skin, 482, 482i, 511, 511i, 540, 664
Recognition protein, 53, 53i, 219
Recombinant DNA, 222ff., 224i, 225i, 235
Recombination, genetic, 172, 171i, 172, 172i, 197, 223, 233, 246
Recombination, homologous, 233
Rectum, 500i, 616, 616i, 623, 630t
Rectus abdominus, 547i
Recycling, solid wastes, 787
Red alga (Rhodophyta), 97, 107, 107i, 320, 326, 327, 327i, 330t, 412
Red blood cell, 130, 573, 529t, 544
   disorders, 161, 161i
   formation, 544, 544t, 559, 573i, 605, 667
   function, 161, 558–559, 558i, 574
   structure, 4, 55i, 161i, 558, 555i, 568
   tonicity, 88i, 89
Red marrow, 544
Red tide, 326 (See also Algal bloom)
Redwood, 8i, 9i, 433
Reef, coral, 363, 412, 412i, 413i
Reeve, H. K., 714
Reflex, 494, 495i, 506t
Regulatory protein, 214–215, 217
Reindeer, 691i
Relaxin, 673
Release factor, 206
Releasing hormone, 523i, 527
Remission, defined, 560
Renal artery, 563i, 635i, 636i
Renal capsule, 635i
Renal corpuscle, 635, 636i
Renal failure, 638
Renal pelvis, 635i, 638
Renal vein, 563i, 635i
Renewable energy source, 794, 794i, 796i
Renin, 637
Replication (See DNA replication)
Repressor protein, 214–215, 214i, 215i
Reproduction, 4, 126 (See also Asexual reproduction; Sexual reproduction)
Reproductive base, 686, 689
Reproductive isolating mechanism, 262–265, 263i–265i
Reproductive success, 692–693, 707
   of female vs. male, 710–711, 683
Reproductive system, defined, 481i
   flatworm, 364–365, 364i
   human female, 523i, 660–667, 660i–667i
   human male, 523i, 656–659, 656i–659i
   origin of tissues, 481, 648
Reptile (Reptilia), 392, 393, 393i, 403, 513, 515, 542i, 601i, 640
   evolution, 302, 302i, 304–305, 386–387, 392, 498i, 498i, 553, 557, 557i
Resource, environmental
   competition for, 722–723, 722i, 723i
   consumption, human, 696–697, 698–699, 788, 792–793, 794–795, 794i
Resource partitioning, 723, 723i
Respiration, 597, 598 (See also Gas exchange; Respiratory system)
   invertebrate, 599, 599i, 610
   vertebrate, 600–601, 600i, 601i

Respiratory cycle, 604–605, 604i, 605i
Respiratory membrane, 606, 606i
Respiratory surface, 598
Respiratory system, defined, 481i, 602–603
   human, 602–603, 602i, 610
   smoking and, 570, 608–609, 609i
Resting membrane potential, 488
Restoration ecology, 729
Restriction enzyme, 224, 224t, 225
Restriction fragment length polymorphism, 227
Reticular formation, 502
Retina, 511t, 514, 515i, 516, 670i
Retrovirus, 318–319, 592
Reverse transcriptase, 225, 225i, 592
Reynolds, W., 283
Reznick, D., 693
RFLP, 227
Rh blood typing, 561, 561i
Rhea, 240i, 241i
Rheumatic fever, 575
Rheumatoid arthritis, 545, 591
Rhinoceros, 307, 410i
Rhizobium, 314, 441i, 752
Rhizome, 346–347, 458t
Rhizophora, 776
Rhizopus stolonifer, 337, 337i
Rhodophyta, 327
Rhodopsin, 103, 516
RhoGam, 561
Rhyniophyte, 343i
Rhythm method, 676, 677i
Rib bone, 543i
Rib cage, 542i, 543i
Ribeiroia, 391i
Riboflavin, 626t
Ribonucleic acid (See RNA)
Ribose, 38, 295
Ribosomal RNA (See rRNA)
Ribosome, 52, 56t, 60–62, 62i, 63i, 72t
   bacterial (prokaryotic), 70, 70i
   eukaryotic, 57i–59i, 62, 63i, 204, 205, 205i, 206i, 207
Ribulose bisphosphate, 101, 102i
Rickets, 532, 532i, 626t
Riffle, stream, 773, 773i
Riparian zone, 416–417, 416i
Ritonavir, 593
RNA, 4, 5, 46, 46i, 48t, 61, 72t, 201, 202
RNA polymerase, 202, 215, 215i, 217
RNA virus, 317–319, 317i, 318, 325
RNA world, 296, 296i, 297, 297i
Robin, 119i
Rocky Mountain spotted fever, 373
Rod cell, eye, 516, 516i
Rodinia, 299, 308i
Root canal, 617i
Root cap, 422i, 430, 430i, 464
Root, defined, 422
   adventitious, 430, 458, 460i
   function, 422, 423, 431, 433, 437, 440–443, 440i–443i, 448, 464, 464i
   hormone effects on, 458, 460–464
   lateral, 430, 430i
   primary, 423, 423i, 426i 430–431, 430i, 431i, 460, 460i, 464
   symbionts, 333–338, 338i, 441, 441i, tissue components, 421, 423–425, 423i
   woody, 423, 423i, 432–433, 432i, 433i
Root hair, 422i, 430, 430i, 440, 441i
Root nodule, 440–441, 441i
Root system, 342, 421, 430, 430i, 458i
Rootworm, 377i
Ross Ice Shelf, calving from, 755
Rotifer, 358t
Rough endoplasmic reticulum, 57i, 62, 62i, 207
Round window, 512i
Roundworm, 358t, 365, 366, 367i, 569
rRNA, 202, 204–205, 206–207, 206i
RU-486, 677
Rubella, 590i, 672, 682
Rubisco, 101, 102, 105, 217, 217i
RuBP, 101, 102i
Ruffini ending, 511, 511i
Ruminant, 615, 615i
Run, stream, 773, 773i
Runner, plant, 458, 458t
Runoff, 104, 438, 439i, 449, 449i, 776

Ryan, W., 272
Rye, 336, 351, 353i, 431

S S interval, interphase, 130, 130i
SA node, 565, 565i
Sac fungus, 334, 337, 337i
Saccharomyces cerevisiae, 118, 119i, 337
Saccharomyces ellipsoideus, 118
Sacral nerve, 499i
Sagittaria sagittifolia, 262i
Saguaro, 410i, 450i, 765i
Sahel, desertification in, 791i
Salamander, 391, 391i
Salinization, 792, 792i
Salisbury, F., 465i
Saliva, 217, 217i, 578, 587, 593, 617
Salivary amylase, 617, 618t
Salivary gland, 500i, 616, 616i, 618t, 630t
Salix, 431i
Salmon, 126, 126i, 127i, 639, 639i
Salmonella, 316, 325, 331, 587
Salt, defined, 26i, 31, 32t
Salt marsh, 776i
Salt-water balance, 627, 627t, 633, 634, 636–637, 639
Saltwater intrusion, United States, 792i
Sampling error, 13, 13i, 16, 256
Sand dollar, 378
Sap, 88, 433
Saprobe, 333, 334, 339
Saprolegnia, 320, 320i
Sapwood, 433, 433i
Sarcodina, 322
Sarcomere, 548, 548i, 549i, 565
Sarcoplasmic reticulum, 550, 550i
Sarcoscypha coccinia, 336i
Sarcosoma, 333i
Sartorius, 547i
Satter, R., 466
Saturated fat, 40, 40i, 624, 625
Savanna, 7i, 762i, 766, 767i, 791i
Scala tympani, 512i, 513i
Scala vestibuli, 512i
Scale, fish, 388
Scallop, 368i
Scanning electron microscope, 54, 55i
Scanning tunneling microscope, 55i
Scapula, 543i
Scarlet fever, 251
Scavenger, 9, 738
Scenedesmusi, 55i
Schadendorf, D., 589
Schally, A., 528
Schistosoma japonicum, 366, 366i
Schistosome, 365
Schistosomiasis, 366
Schleiden, M., 51
Schliewen, U., 266
Schoener, A., 730
Schroder, R., 377i
Schwann cell, 494, 494i
Schwann, T., 51
Sciatic nerve, 499i
SCID, 591i
Scientific approach (methods), 12–13
Scientific theory, nature of, 13, 18
Sciuridae, 397t
Sclera, 511t, 514, 515i
Sclereid, 424, 424i, 434t
Sclerenchyma, 422, 424, 424i, 427i, 434t
Scolex, 365, 365i
Scorpion, 373, 373i
Scorpionfish, 726i
Scott, J. M., 755
Scrapie, 319
Scrotum, 656, 656i
Scutigera, 375
Scyphozoan (*See* Jellyfish)
Sea anemone, 358t, 362, 362i, 497i, 542
Sea cucumber, 378i
Sea horse, 389i
Sea lamprey (*See also* Lamprey), 731t
Sea lettuce, 328i
Sea squirt, 384, 385i
Sea star, 358t, 379, 379i, 723
Sea urchin, 130, 147i, 358t, 378i, 647, 651, 651i, 652i
Sea wasp, 362i
Seafloor spreading, 278, 279i

Seal, diving adaptation of, 611
Seasonal affective disorder, 534
Seawater, pH, 30, 30i (See also Ocean)
Seaweed, 9
Sebaceous gland, 540 (*See also* Oil gland)
Second law of thermodynamics, 76–77
Secondary cell wall, 68, 68i
Secondary growth, plant, 421, 423, 432–433, 432i, 434
Secondary sexual trait, 523, 523i, 529t, 656
Secondary succession, 728
Secondhand smoke, 607, 608
Secretin, 522, 529t, 619
Secretion, defined, 475, 614, 630
Sedimentary cycle, 745, 747
Sedimentary rock, 274–275, 277
Seed, 343, 348i, 350, 350i–352i, 451, 456i
   dispersal, 255, 255i, 456–457, 719
   germination, 460–461, 460i, 461i
Seed bank, global, 230
Seed coat, 426, 438, 439, 455i, 456, 457, 460, 460i
Seed plant, 342, 348–353 (*See also* Angiosperm; Gymnosperm)
Seedless vascular plant, 342, 346–347, 346i, 347i
Seedling 426, 430, 456, 457, 461i, 465i
Segal, E., 18
Segment polarity gene, 655i
Segmentation, 359, 370–372, 384, 621
Segmented worm (*See* Annelid)
Segregation theory, 156–157, 167, 171, 176i
Seizure, 505, 507
Selaginella, 346
Selander, R., 261
Selective permeability, membrane, 85, 85i
Selenium, 20, 20i
Self vs. nonself, 53, 560, 582
Selfish behavior, 707
Selfish herd, 712
Semen, 593, 656–657, 665
Semilunar valve, 564, 564i, 565
Seminal vesicle, 656i, 657i, 658
Seminiferous tubule, 656, 658, 658i
Senescence, 468, 469
Sensation, defined, 508, 509, 518
Sensory adaptation, 510
Sensory neuron, 486, 486i, 490, 492ff., 495i, 506, 510, 510i, 516, 540
Sensory pathway, nature of, 510, 510i
Sensory receptor (*See* Receptor, sensory)
Sensory system, 490–491, 490i, 491i, 508–509, 511
Sepal, 452, 452i, 457i
Sequoia, 729, 729i
Serine, 204i
Serosa, 619i
Serotonin, 492, 505
Serratus anterior, 547i
Sertoli cell, 658, 659, 659i
Sessions, S., 391i
Setae, 370
Severe combined immunodeficiency, 591i
Sewage sludge, 793i
Sex chromosome, defined, 141i, 172
   abnormalities, 184–185
Sex determination, 172, 174, 174i
Sex-linked gene, 176
Sexual behavior, 520–521, 524, 665
Sexual dimorphism, 254, 254i, 553
Sexual intercourse, 483 (*See also* Coitus)
Sexual reproduction, 150, 645
   asexual vs., 28, 140, 646
   chromosome number and, 140
   costs and benefits, 646–647
   defining characteristics, 128, 138–140
   frog, 644–645, 645i
   fungi, 334, 336i, 337, 337i
   human, 645, 647, 656, 665, 656i–665i, 662, 676–680, 682
   invertebrates, 138–139, 138i, 139i, 356, 360, 361, 363, 345, 366i, 372, 376, 646–647
   meiosis and (*See* Meiosis)
   in plants, 336, 451, 454–455, 454i, 455i
   protistans, 322, 329, 329i
   selective advantages, 140, 144, 148, 645, 646
Sexual selection, 254, 254i, 258, 710–711

Sexually transmitted disease, 678–679
Shaft, feather, 395i
Shark, 386i, 388, 388i, 498i, 542i
Shark Bay, Australia, 298i
Sheath, leaf, 428i
Shell model, atoms, 24–25, 24i, 25i
Shepherd's purse, 456, 456i
Shifting cultivation, 788–789
Shivering response, 641
Shoemaker-Levy 9 impact, 304, 306
Shoot, defined, 422
Shoot system, defined, 342
Short-day plant, 467, 467i, 469
Shortgrass prairie, 766, 767i
Shoulder blade, 543i
Shrimp, 103, 742i
Shrubland, 762i, 766–767
Sickle-cell anemia, 161, 161i, 179t, 187, 208, 208i, 560
   malaria and, 254–255, 254i, 255i
SIDS, 607
Sieve plate, 425i
Sieve-tube member, 424, 425i, 427i, 434t, 446, 447i
Sign stimulus, 705
Silica, 322, 326, 347, 360
Silicon, 21i
Silk gland, 373, 373i
Silk, J., 715
Silkworm moth, 4, 4i, 517, 522, 708
Sillén-Tullberg, B., 712
Silt, 438, 449, 449i
Silurian, 277i, 302, 303i, 308, 343i, 378
Silver, L., 198i
Silver Star Mountain, 34i
Simple fruit, 456t
Simple sugar, 36, 38, 48t
Simple tissue, plant, 424, 424i, 434, 434t
Simpson, G., 269
Siphonaptera, 377i
Siphonophore, 774
Sister chromatid, 128, 128i, 141
Six-kingdom scheme, 286i, 287
Skeletal muscle, 117i, 478, 478i
   blood flow and, 562i, 563
   bone and, 539, 544, 544t, 546, 546i
   circulation and, 554, 558, 563, 567, 569, 569i
   contraction, 548–551, 549i, 550i, 551i
   human, 447i, 547, 598
   in respiration, 601i, 602i, 604–605, 604i
Skeletal-muscular system, 546–547, 547i
Skeletal system, 539, 542–545, 542i, 543i
Skeleton, 395i
   bird, 395i
   cartilaginous, 387, 388
   exoskeleton, 372
   human, 542–543, 543i, 648
   hydrostatic, 363, 371, 542, 542i
Skin, 45, 526i, 539–541, 540i, 552, 578t, 585
   blood flow and, 562i, 563
   pigments, 162–163, 162i, 163i, 526
   receptors, in 490, 510, 510i, 511, 511i
Skull bone, 283, 283i, 399, 399i, 545
Slash-and-burn agriculture, 788
Sliding-filament model, 548, 549i, 565
Slime mold, 320–321, 321i, 330, 653
Slug, 300, 358i, 368i
Small intestine (*See* Intestine, small)
Smallpox, 576, 576i
Smartweed, 723, 723i
Smell, sense of, 503, 506t, 517
Smilodon, 307, 307i
Smith, H., 224, 233
Smith, H. W., 14
Smith, T., 253
Smithies, O., 233
Smog, 784, 784i, 798
Smoking, 570, 608–609, 609i, 673
Smooth endoplasmic reticulum, 57i, 62, 62i, 63i
Smooth muscle, 478, 478i, 546–547, 569, 571, 573, 618, 619i
Snail, 65, 86, 246i, 247i, 260–261, 260i, 358t, 368i, 514i, 646i, 647i
Snake, 241i, 392i, 393, 393i, 615i, 646i
   behavior, 704, 704i
   thermos, 508, 508i
Snapdragon, 160, 160i

Sneezing, 179t
Snowshoe hare, 724–725, 724i
Social behavior, 707, 715
Social parasite, 705, 705i
Sodium, 21, 21i, 22t, 25i, 26, 26i, 529t
   dietary, 627, 627t
   neural function and, 488–491, 489i, 491i
   in salt-water balance, 635, 635i, 637i
Sodium chloride, 26, 26i, 29, 31
Sodium hydroxide (NaOH), 30i
Sodium-potassium pump, 86, 489–491, 489i–491i, 506
Soft spot (skull), 545
Softwood, 433, 433i
Soil, 438–439, 438i, 728, 764, 764, 764i
   pollution, 753, 785
Soil profile, 764, 764i
Solar–hydrogen energy, 796, 796i
Solar tracking, 470
Solid waste, 787
Solidago altissima, 252, 252i
Solute, defined, 29, 32t
Solvent, defined, 29
Somatic cell, 128, 150i, 481
Somatic nerve, 487, 499i, 500, 506
Somatic sensation, 509, 510, 511
Somatomedin, 529t
Somatosensory cortex, 503, 503i, 511, 511i
Somatostatin, 527, 529t, 532
Somatotropin, 217, 221, 230, 232, 232i, 470, 523i, 524t, 525, 526t, 527, 528, 528i, 537
Somite, 668, 668i, 670i
Sorus, 347, 347i
Sound production, 601, 603, 603i
Sound, properties of, 512, 512i
Sow thistle (*Sonchus*), 447, 447i
Spanish flu, 2–3
Spartina, 20i, 742i, 776i
Special sense, defined, 509, 510
Speciation, defined, 261, 264, 270
   isolating mechanisms for, 262–263
   patterns, 262–263, 269i
Species, defined, 8, 261, 262, 270
   endemic, 763, 730
   exotic, 730–731, 737
   interactions, 720–725, 721i–724i, 734
   keystone, 722–723
   Mayr's concept, 262
   names for sale, 799
   naming, 8, 286
   pioneer, 340, 728, 728i
Species introduction, 387, 725, 730–731, 730i, 731i, 731i, 737
Spectral karyotyping, 170, 170i, 183i
Sperm, animal, 650
Sperm, human, 50i, 67i, 125i, 146, 627t
   formation, 656, 656t, 658–659, 658i, 659i
Spermaticide, 676–677, 677i
Spermatid, 146, 147i, 658
Spermatocyte, 146, 147i, 658, 659i
Spermatogenesis, 146, 147i, 658–659, 659i
Spermatogonium, 147i, 658, 659i
Sphagnum, 344–345, 345i
Sphenodon, 393i
Sphere of hydration, 29, 29i
Sphincter, defined, 617, 619i
Spicule, 360
Spider, 17, 17i, 514, 599, 646i, 647i, 686
   characteristics, 358t, 373, 373i, 381, 381i
Spinach, 467, 467i
Spinal cavity, 480i
Spinal cord, 498–499, 498i, 499i, 501, 501i, 506t
Spinal meningitis, 325, 501
Spinal nerve, 500
Spindle apparatus, 130–131, 132, 133
Spiracle, 599, 599i
Spiral cleavage, 367, 367i
Spirillum, 313
Spirochete, 315i
Spirogyra, 92i, 329i
Spleen, 500i, 559, 573, 573i, 585, 585i
Sponge (Porifera), 358t, 360–361, 360i, 361i
Spongy mesophyll, 429, 429i
Sporadic disease, defined, 325
Spore, defined, 146
   bacterial, 315, 315i
   formation by meiosis, 143i

fungal, 334–337, 339, 339i
plant, 342–343, 348, 352
slime mold, 321, 321i
Sporophyte, 342, 342i, 344, 345i, 346, 452, 452i, 456
embryo, 451, 456i, 460–462, 460i
Sporozoan, 320, 324, 324i, 325t
Sporozoite, 324, 324i, 325i
Sprain, defined, 545
Spriggina, 302i
Spring overturn, 772
Squamous cell carcinoma, 212i
Squid, 358t, 368, 368i, 491i, 514
SRY gene, 174i
S-shaped growth curve, 690–691, 690i
St. John's wort, 434i
Stabilizing selection, 252, 252i, 258, 261
Stachybotrys, 336
Stamen, 154i, 155, 452, 452i, 453
Staph infection, 11
Staphylococcus, 313i, 325
Starch, 95i, 101i, 617, 624
Starch grain, 38, 39i, 48t, 65, 101
Starfish (*See* Sea star)
Starling, E., 522
Starling, European, 702–703, 702i, 731t
Statolith, 464, 464i
STD, 678–679, 678i, 679i
Stearic acid, 40i
Steger, W., 538, 538i
Steinbeck, J., 766
Stellar nerve, 491i
Stem, 423, 426–427, 427i, 444
dicot vs. monocot, 426–427, 427i
Stem cell, animal, 559, 559i, 580
Sterility, 179t, 678, 679
Sternum, 543i (*See also* Breastbone)
Steroid hormone, 41, 524–525, 524i, 524t, 526t, 535, 536, 622i
Sterol, 41, 48t, 52
Steward, F., 230
STH, 525, 527i, 528i
Stice, S., 232
Stigma, 452i, 453, 454i
Stimulant, brain, 504
Stimulus, 5, 482, 482i, 483, 509, 518, 705
Stinkbug, 377i
Stinkhorn fungus, 8i
Stipe, 328, 328i
Stoma (stomata), 102, 102i, 342, 425, 429i, 444–445, 444i, 445i, 448, 770
in C4 vs. C3 pathways, 102, 102i
Stomach, 379, 388, 500i, 529i, 547i, 614–618, 616i, 618t
human, 359, 367, 613, 618, 619i, 630t
ruminant, 314i, 615, 615i
Stomach acid, 30, 30i, 532
Strain, defined, 545
Stratification, 274–275
Stratosphere, 758i
Strawberry plant, 234, 456t, 457, 457i, 458
Strength training, 551
Streptococcus, 316
S. pneumoniae, 192, 192i, 325
S. pyrogenes, 575
Streptomycin, 11, 15, 251, 673
Stress response, 530, 530i
Stress electrocardiogram, 571
Stretch receptor, 495i, 510, 511
Stretch reflex, 494–495, 495i
Strip logging, 416, 416i
Strobilus, 346, 346i
Stroke (brain damage), 570
Stroma, 65, 65i, 94, 95i
Stromatolite, 298, 298i, 299
Strongylura marina, 742i
Sturnus vulgaris, 702–703, 702i, 703i
Style, flower, 452i, 453
Suberin, 425, 432, 468
Submucosa, 619i, 620i
Subsistence agriculture, 787
Subspecies, defined, 267
Substance P, 492
Substrate, defined, 81, 90t
Succession, ecological, 728–729, 728i, 729i
Succinyl-CoA, 115i
Succulent, 102i
Sucrose, 38, 38i, 48t, 95i, 101, 101i, 446i
Sucrose polyester, 625

Sudden infant death syndrome, 607
Sugar, simple, 36, 38, 48t
Sugarcane, 38, 102i
Sulfate-reducing species, 119
Sulfide, 774–775
Sulfonamide, 679
Sulfur, 20, 21i, 22t, 103, 439t, 627t
oxides, 31i, 784, 784t, 785, 794
Sulfuric acid, 30, 784t
Sunburn, 541
Sunfish, 712
Sunflower, 20, 424i, 456t, 515
Sunlight, 77i, 96, 96i
climate and, 757, 758–759, 759i
input into ecosystems, 739, 739i, 741i, 743, 744, 744i, 759, 796, 796i
seasonal depression and, 534
skin cancer and, 212, 212i, 541
Suntan, 541
Supercontinent, 278–279, 299, 302, 303, 308i–309i
Superior (directional term), 481i
Superior vena cava, 563i, 564i
Supernatural, defined, 12
Superoxide dismutase, 74, 74i
Surface recognition protein, 53, 53i, 219
Surface-to-volume ratio, 54, 54i, 598, 620
Suricata suricatta, 472, 472i, 473
Survivorship curve, 692–693, 693i
Swallowing, 603, 617
Swamp, 438, 776, 776i
Sweat, 29, 641
Sweat gland, 179t, 216, 216i, 540, 540i, 641
Swim bladder, 388
Swimmeret, 374i
Symbiosis, 332, 338
Symmetry, body, 358–359
Sympathetic nerve, 499i, 500, 500i, 506, 530, 567
Sympatric speciation, 261, 266, 267i, 270
Synapsid, 396
Synaptic cleft, 492, 493i
Synaptic integration, 492–493, 493i, 500, 506, 516
Syndrome, 179
Synovial cavity, 546i
Synovial joint, 545
Syphilis, 679, 679i
System, defined, 76i
Systemic circuit, 557, 562, 562i, 566i, 607i
Systole, cardiac cycle, 565i
Systolic pressure, 566, 566i, 567i
Szuts, E., 103

T

2,4,5-T, 462–463
T cell (T lymphocyte), 559, 559i, 573, 578t, 581, 582, 583, 583i, 584
T tubule, 550i
Tachycardia, 571i
Tachyglossus, 397i
Tactile display, 709
Tactile signal, 716
Tadpole, 63, 534–535, 534i, 644i, 645i
Taenia saginata, 366i
Taiga, 770
Takano, T., 538
Tallgrass prairie, 739, 739i, 740i, 766, 767i
Tandem repeat, 227, 235
Tangential section, plant, 423i
Tannin, 726
Tapeworm, 358t, 365, 365i, 366i
Taproot system, 430, 430i
Tarsal, 543i
Tarsier, 398i
Tarsoid, 398
Taste bud, 517, 517i, 617
Taste, sense of, 517
Taxol, 66, 137
Taxonomy, 273, 286, 286i, 312
Taxus brevifolia, 66, 137
Tea, harvesting leaves, 353i
Tears, as body defense, 578, 587, 593
Tectorial membrane, 513i
Tectum, 502, 506i
Tegeticula, 721
Teleost, 389, 389i
Telophase I, meiosis, 142i, 148t
Telophase II, meiosis, 143i, 149i
Telophase, mitosis, 131i, 133, 133i, 136, 149i

Temnodontosaurus, 305i
Temperate forest, 755, 762i, 768, 768i
Temperature, body
blood flow and, 563
control of, 473, 473i, 482–483, 483i, 558, 632, 640–641, 641i, 642
enzymes and, 84–85, 85i, 641
rhythm method using, 676, 677i
scrotum and, 656
Temperature, defined, 29
Temperature, environmental
animal responses to, 472–473, 473i, 632
climate zones and, 760–761, 760i, 761i
greenhouse effect and, 749, 750–751, 750i, 758, 789, 794
plant dormancy and, 468
sea surface, 774, 775, 775i
Temperature zone, 758, 759i
Temporal lobe, 502, 503i
Tendon, 539, 546, 546i, 552
Tendril, 465, 465i
Tension, defined, 551
Teratogen, 672i
Terminal bud, 422i
Termite, 8i, 314, 522, 709, 717, 738, 751i
Tern, Caspian, 711, 711i
Territory, defined, 707
Terry, N., 20
Tertiary period, 276i–277i, 309i
Test, nature of, 12, 16
Test, observational (*See also* Experiment)
bacterial transformation, 192
biological magnification, DDT, 742
frog deformities, 534, 534i
inheritance patterns, 152ff.
light at hydrothermal vents, 103
of plate tectonics, 278–279, 279i
premature puberty, 534
raven calling behavior, 707
rhythmic leaf movements, 466i
root nodule effects on growth, 441i
tropisms in plants, 465i, 466i
Vibrio cholerae reservoir, 779
wastewater treatment, 793, 793i
wavelengths for photosynthesis, 92
XYY and criminal behavior, 185
Testcross, 157
Testicular cancer, 657
Testicular feminization syndrome, 524
Testis, 523i, 526t, 529t, 531, 656, 656i, 657, 657i, 658i
Testosterone, 37i, 523i, 524, 526t, 529t, 553
bird song, 704i, 705
spermatogenesis and, 658–659, 659i
Tetanus
disease, 118, 325, 325t, 551, 590, 590i
muscle, 496, 551, 551i
Tetany, 31, 551
Tethys Sea, 303, 303i, 309i
Tetracycline, 673, 679
Thalamus, 502, 502i, 506t, 517
Thalassemia, 560
Thalassoma, 264i
Thalidomide, 672–673
Thigh bone, 543i, 544, 544i (*See also* Femur)
Theobroma cacao, 353i
Theory, scientific, 3, 13, 15, 16
Therapeutic vaccine, 589
Therapsid, 396, 503i
Thermal energy, 76
Thermal inversion, 784, 784i
Thermocline, 772, 772i
Thermodynamics, 76, 77, 90
Thermophile, 314, 314i
Thermoreceptor, 508, 508i, 511, 518, 640
Thermus aquaticus, 224, 224t, 226
Theropod, 288
Thiamin, 626t
Thigmotropism, 465, 465i
Thirst behavior, 637
Thoracic cavity, 359, 480i, 564i
Thoracic duct, 573i
Thoracic nerve, 499i
Threat display, 708, 708i, 713i
Threonine, 204i, 208i, 625
Threshold, 493, 493i
Throat, 517
Thrombin, 82
Thrombus, 571

THS, 524t, 526, 526t, 527, 531
Thykaloid, defined, 94
Thylakoid membrane, 65, 65i, 94, 94i, 95i, 97i, 100i
Thymine (T), 46, 194, 194i, 201, 209, 297
Thymine dimer, 212, 212i
Thymosin, 523i, 529t
Thymus gland, 523i, 573, 573i, 584, 585i
Thyroid gland, 23, 23i, 523i, 526, 526t, 527i, 529t, 530–531, 545, 795i
Thyroid hormone, 524t, 525, 526t, 534–535, 591, 627t
Thyroid-stimulating hormone (TSH), 524t, 526, 526t, 527, 531
Thyrotropin, 526, 526t, 527
Thyrotropin-releasing hormone, 527
Thyroxine, 523i, 529t, 530
Tibia, 543i, 545i
Tibialis anterior, 547i
Tick, 39i, 315, 315i, 373
Tight junction, 69, 69i, 474–475, 475i, 503, 565i, 568, 651, 666i
TIGR Assembler, 233
Tilburg, J. A. van, 407
Tilia sp., 102i
Tissue, 6i, 472, 473
animal, 474–480, 476i, 478i, 484, 652–653
plant, 421, 422–423, 422–425, 434t
Tissue culture propagation, 458–459, 479
Titanic, 643
Toad, 286, 390–391, 391i
Tobacco mosaic virus, 55i
Tobacco plant, 219, 231 (*See also* Smoking)
Tobin, E., 217
Todd, J., 793
Tolerance, drug, 504, 504t
Tomato plant (*Lycopersicon*), 422i, 456t
Tongue, 511i, 517, 617
Tongue-rolling trait, 168, 168i
Tonicity, 88, 88i, 89
Tonsil, 573, 573i, 585i
Tool use, 400, 400i
Tooth
antelope, 615, 615i
human, 614, 615, 615i, 617
mammalian, 396, 396i, 545, 613
Tooth decay, 617
Topography, 761
Topsoil, 438, 439, 764, 791i
Tortoise, 393i
Total fertility rate, 696–697
Touch, sense of, 511
Toxoplasma, 324, 324i
Toxoplasmosis, 324
TPA, 232
Trace element, defined, 21
Trace fossil, 274
Tracer, 23, 32t, 94, 193i
Trachea, 602i, 603
Tracheal respiration, 599, 599i
Tracheid, 424, 434t, 442, 442i, 462i
Transcription, 201–203, 202i, 220, 592, 593
control of, 213, 214–218, 215i
eukaryotic, 216–218
prokaryotic, 214–215, 215i
Transfer RNA (*See* tRNA)
Transfusion, blood, 160
Translation of mRNA, 43, 43i, 201, 204, 204i, 206–207, 206i–207i
Translocation, 170, 170i, 182, 182i, 437, 446, 447i
Transmission electron microscope, 54, 55i
Transpiration, 437, 442–443, 443i, 448
Transport protein, 53, 53i, 79, 79i, 81, 86, 86i, 90, 90t
at blood–brain barrier, 503
neuron, 489–491, 489i–491i, 506
Transposon, 208–209, 209i, 283
Transverse fission, 364
Transverse plane, 480i, 481i
Transverse section, plant, 423i
Trapezius, 547i
Travisano, M., 271
Tree ring, 433, 433i
Tree shrew, 398i
Trematode, 391i
Treponema pallidum, 679
TRH, 524t, 527, 528
Triassic, 276i–277i, 304, 309i, 343i

Tribolium, 256
Triceps brachii, 546i, 547i
Triceratops, 304
Trichinella spiralis, 367, 367i
Trichomonad, 8i
Trichomonas vaginalis, 323, 323i
Trichomoniasis, 323i
Trichoplax adhaerens, 360, 360i
Trigger zone, axon, 488, 488i, 506
Triglyceride, 40–41, 40i
  metabolism, 120, 618t, 619, 621
Triiodothyronine, 523i, 529t, 530
Trilobite, 302, 302i, 372
Trisomy 21, 184
Triticum, 267i, 353i
tRNA, 202, 204–205, 205–207, 205i, 206i
Trophic level, 739
Trophoblast, 666, 666i
Tropical forest, 728–729, 732, 762i, 763i,
  764, 764i, 768, 768i, 788, 788i, 790
Tropism, plant, 464–465, 464i, 465i, 469
Tropomyosin, 550
Troponin, 550
Troposphere, 758i
True-breeding lineage, defined, 155
Truffle, 337
Trypanosoma, 311i, 323, 323i
Trypsin, 85, 618t, 619
Tryptophan, 84, 84i, 204i, 625, 625i
Tsetse fly, 323
TSH, 523i, 524t, 526, 526t, 527i
Tuatara, 393, 393i
Tubal ligation, 676–677, 677i
Tube foot, echinoderm, 379, 379i
Tube worm, 775i
Tuber, 458t
Tuberculosis, 11, 251, 259, 325, 325t
Tubular reabsorption, 636, 636i
Tubular secretion, 636, 636i
Tubulin, 66, 66i, 132
Tuco-tuco, 397t
Tumor, 212, 218i, 219, 589
Tundra, 763i, 771
Tunicate, 358, 384, 403
Turbellarian, 358t, 364
Turgor pressure, 444–445, 444i, 462
Turkey, 18, 119
Turnell, J., 417
Turner syndrome, 179t, 185
Turtle, 393
Twin, human, 651, 651i
Two-kingdom scheme, 8, 386
Tylosaurus, 409i
Tympanic membrane, 512i
Typhoid fever, 11, 320
Tyrosine, 42i, 162, 204i, 625
Tyrosine kinase, 218

U Ulcer, 313i, 532
Ulna, 543i
Ulnar nerve, 499i
Ultrafiltration, 568i, 569, 580
Ultraplankton, 774
Ultrasaur, 304
Ultrasound, 508
Ultraviolet radiation, 96, 96i, 97, 209,
  212, 212i, 540, 541, 758i, 786
Ulva, 328i
Umami, 517
Umbilical cord, 670, 670i, 673i
Ungulate, 767i
Uniformity theory, 243, 278
Uniramian, 372
United States
  aquifer depletions, 792, 792i
  endangered ecosystems, 742i, 755, 755i
  human population 692t, 696, 696i,
    697i, 698, 699, 701, 701i
Unity of life
  in ancestral traits, 274, 280–283
  molecular basis, 190, 195
  nature of, 3, 8, 9, 11, 16t, 270t
Universe, 292–293
Unsaturated fat, 40, 40i, 624, 625
Upwelling, 777, 777i
Uracil (U), 46, 202, 202i, 297
Uranium, 737
Urea, 622, 634
Uremic toxicity, 638

Ureter, 634i, 635, 635i
Urethra, 634i, 635, 656, 657i, 658i, 660, 661i
Uric acid, 634, 638
Urinary bladder, 500i, 609i, 634i, 635, 657i,
  661i, 667, 668
Urinary excretion, 634, 636i
Urinary system, 481, 481i, 633, 633i,
  634i, 635, 635i
Urination, 635
Urine, 533, 578, 593, 622, 634–637, 747
Urochordata, 384
Ursus americanus, 286
Ursus arctos, 286
Ursus maritimus, 286
Uterus, 137, 500i, 526t, 527i, 529t, 647,
  660, 660t, 661i, 662i, 663, 665i

V Vaccination, 577, 590, 590i
Vaccine, 192, 324, 590, 590i, 593, 790
Vacuole, 57i, 58i, 65, 72t
Vagina, 578, 657, 660, 660t, 661i, 662i, 665i
Vagus nerve, 500i
Valine, 42i, 204i, 208i, 625
Valproic acid, 507
Vampire stories, porphyria and, 485
Vancomycin, 251
Variable, defined, 13, 15, 16
Variation in traits
  continuous, 164–165, 165i, 246–247, 246i
  mutation and (*See* Mutation)
  natural selection and 10–11, 244–245,
    250–251
  qualitative vs. quantitative, 246
Varicose vein, 575
Vas deferens, 656, 656i, 657i, 658i, 676
Vascular bundle, 425i, 426, 427i, 429, 434
Vascular cambium, 423, 432–433, 433i
Vascular cylinder, 430–431, 430i,
  431i, 440, 440i
Vascular plant, 304i, 342, 346–353, 354t
Vascular tissue, 421, 422, 422i, 423 (*See
  also* Phloem; Xylem)
  brown alga, 328
  bryophyte, 344
  dicot vs. monocot, 426–427, 427i
  functions, 423, 424–425, 434
  leaf, 428i, 429, 429i
Vasectomy, 676
Vasoconstriction, 567, 641
Vasodilation, 567, 580, 580t, 581i, 641
Vegetal pole, 649i, 651
Vegetarian diet, 560, 625, 625i, 627
Vegetative growth, 458, 458t
Vegetative propagation, 458t
Vein, 555, 562, 566, 566i
  plant, 428i, 429i
Velociraptor, 304
Vena cava, 563i, 634i
Ventner, C., 232–233
Ventral surface, defined, 359, 480i
Ventricle
  brain, 503i
  heart, 564, 564i
Ventricular fibrillation, 571, 571i
Venturia inaequalis, 336i
Venule, 562, 569, 574, 620
Venus flytrap, 436, 436i
Vertebra, 386, 542–543
Vertebral column, 498, 501, 501i, 542i, 543i
Vertebrate (Vertebrata), defined, 358, 384
  characteristics, 384, 403, 511t,
    514–515, 557i, 714
  development, 282–283, 668, 668i
  diversity of, 418i
  evolution, 303, 383, 383i, 386–387,
    386i, 494, 517i, 557, 578
  major groups, 360t, 382ff.
Vertical zonation, 777, 777i
Vertigo, 519
Vesicle
  examples, 64i, 65, 87, 87i, 582, 582i
  types of, 56t, 62i, 63
Vessel member, 424, 425i, 427i, 434t, 442
Vestibular apparatus, 512i, 513
Villus, small intestine, 620, 620i
Vine, 465
Vibrio cholerae, 695, 779, 779i
Villus, small intestine, 620, 620i
Vine, 465
Virchow, R., 51

Viroid, 319
Virus (*See also* Bacteriophage)
  characteristics, 310i, 311, 317, 317i, 330
  DNA, 317, 317i, 318i, 319i
  enveloped, 317, 317i, 592, 593i
  infectious disease and, 325, 325t, 751
  multiplication, 318–319, 318i, 319i
  mutations in, 230
  in recombinant DNA work, 225, 230,
    233, 234
  RNA, 317, 317i, 318–319, 592, 593i
Viscera, defined, 511
Visible light, properties, 96, 96i
Vision, 511, 514–515, 514i, 515i
Visual accommodation, 515
Visual cortex, 503i, 516i
Visual field, 514
Vitamin, human diet, 626, 626t
  A, 627
  B, 541, 560, 672
  C, 38, 74, 609i, 627, 672
  D, 41, 222, 524t, 525, 532, 537, 540,
    541, 545, 627
  E, 74, 627
  K, 314, 627
Vitreous body, 515i
Viviparous, 646i, 647i
Vocal cord, 601, 603, 603i
Volatile substance, defined, 517
Volcanic eruption, 420, 420i, 732i, 748
Volvox, 77i, 328i, 329, 360
Vomeronasal organ, 522
Vomiting, 503, 505, 613
Vulva, 660, 661i

W Walker, N. M., 15
Wall cress, 470
Wallace, A., 239, 245, 245i
Waller, A., 551i, 554
Walrus, 269i
Warning coloration, 726
Wasp, 725, 726, 726i
Wastewater treatment, 793, 793i
Water (*See also* Ocean)
  early Earth's, 294, 308
  fresh, availability, 792, 792i
  metabolic, 632, 632i
  properties, 28–30, 30i, 32, 122, 442–443
  roles in life, 28, 124, 295i, 436ff., 632ff.
Water budget, global, 746i
Water conservation vs. loss
  human, 633ff., 632i
  kangaroo rat, 632i, 639
  plant, 102, 437, 444–445, 444i, 445i, 448
Water hyacinth, 731t, 735, 735i
Water mold, 320, 320i, 330
Water molecule, properties of, 25, 27,
  27i, 28i, 88–89, 122
Water pollution, 230, 773, 776, 792–793
Water province, 757, 772–780, 772i, 777i
Water strider, 29i
Water table, defined, 792
Water transport, plant, 442–443, 443i, 448
Watermelon, 456i
Watershed, 746, 788
Water-vascular system, 379, 379i
Watson, J., 190, 190i, 195, 196
Wavelength, defined, 54, 96, 96i
Wax, 34, 41, 48t
Weather systems, 758–759, 759i, 778–779
Wegener, A., 278
Weight, human, 612–613, 628–629, 628i
Weight training, 124, 553
Weismann, A., 170
Welwitschia, 350, 350i
Went, Fritz, 464
Werner's syndrome, 675
West Nile virus, 751
Wetland, 20, 20i, 776–777, 776i, 781
Wetness, sense of, 508
Wexler, N., 178i
Whale, 269i, 411, 411i, 611, 737, 775i
Wheat, 266, 267i, 284i, 353i
Whisk fern, 342, 346, 346i, 354t
White blood cell (lymphocyte)
  body locations, 559, 572i, 573i
  characteristics, 53, 558i, 574, 580, 580i
  disorders, 179t, 533, 560
  roles in immunity, 578ff., 594t

  roles in inflammation, 532, 559, 578t,
    579, 580–581, 580t, 581i
  types, 87, 559i, 578t, 594t
White matter, 499
White, T., 399
Whittaker, R., 287
Whooping cough, 325, 590i
Wildebeest, 767i
Wildlife trade, 410i
Wild-type allele, defined, 172
Wilkins, M., 194, 196
Williams, Ernest, 412
Willow tree root, 431i
Wilmut, I., 198
Wilson, E. O., 410, 411
Wilting, 439t, 444, 444i
Wind energy, 796, 796i
Windpipe (trachea), 23, 23i, 602i, 603
Windward, defined, 761
Wing, 281, 281i, 394i, 395i
Winter blues, 534
Withdrawal, as birth control, 676, 677i
Wobble effect, 205
Woese, C., 287
Wolf, 397t, 710–711, 711i
Wolf, Tasmanian, 397t
Wood, 421, 432–433, 432i, 434
Wood duck, 37i
Woodland, 762i, 766–767
Woodpecker, downy, 252, 252i
Woodwell, G., 742i
World Wildlife Fund, 415i
Wrasse, 264i
Wuchereria bancrofti, 367, 367i
Wurster, C., 742i

X X chromosome, 172, 174, 174i, 177
X chromosome inactivation, 216, 216i
Xanthophyll, 49
Xenopus laevis, 534, 534i, 652
Xist gene, 216, 220
X-linked anhidrotic dysplasia, 216, 216i
X-linked gene, 176, 176i
X-linked recessive inheritance, 179t,
  181, 181i
X-ray, 96i, 133, 209
X-ray diffraction image, 194, 196
Xylem, defined, 342, 344, 347
  characteristics, 424, 424i, 433, 434, 434t
  components, 424, 425i, 442–443,
    442i, 448
  functions, 342, 424, 432, 442–443, 443i
  in plant evolution, 342
  primary, 426, 426i, 427i, 428, 429i,
    431, 431i
  secondary, 432–433, 432i
XYY condition, 179t, 185

Y Y chromosome, 172, 174
Yarrow, 166, 166i
Yeast, 118, 284i, 337
Yellow marrow, 544, 544i
Yellow-green alga, 326
Yellowjacket, 726i
Yellowstone National Park, 789i
Yersinia pestis, 691
Yew, western, 66, 137
Y-linked gene, 176
Yolk, 646i, 647, 650, 51, 652i, 667, 668
Yolk plug, 649i
Yolk sac, 394i, 666i, 667, 667i, 668, 681
Yucca, 721, 721i

Z Z band, 548, 548i
Zea diploperennis, 414
Z. mays, 65i, 102i, 150i, 176, 209i
Zebra, 6i, 271, 537i
Zebra finch, 704, 704i
Zebroids, 271, 271i
Zero population growth, 688
Zinc, 21i, 439t, 627t
Zona pellucida, 662, 662i, 665, 665i
Zooplankton, 734, 742i, 772
Zygomycete), 334, 336–337, 338
Zygosporangium, 337
Zygote, defined, 146i, 648, 665i
  flowering plant, 456, 456i
  human, 644, 644i, 665, 665i, 666
  X0 (X-linked disorder), 184

Index of Applications

Abortion, 186, 187, 676, 677, 680
Abstinence, sexual, 676, 677i
Acclimatization, 605
Achondroplasia, 179t, 180, 180i
Achoo syndrome, 168, 179t
Acid deposition, 784–785, 785i, 794
Acid rain, 30i, 31, 31i, 784
Acid stomach, 30
Acidosis, 31, 625, 638
Acne, 540
Acromegaly, 528, 528i
Acute inflammation, 580, 581i
Acyclovir, 679
ADA deficiency, 591i
Adoption behavior, human, 714
Aerobic exercise, 551
Aflatoxin poisoning, 47
African emergence model, 402, 402i
African sleeping sickness, 311i, 323
Afterbaby blues, 674
Afterbirth, 673i
Agar, uses of, 327
Age spots, 74, 75i
Agent Orange, 462
Aging
 Down's syndrome and, 185, 185i
 human, 74–75, 75i, 545, 545i, 541,
  551, 575, 628, 631, 674–675
 progeria and, 179t, 182, 182i
Agriculture, 765
 biological controls for, 251
 crop diseases, 230, 336
 deforestation and (See
  Deforestation)
 in deserts, 765, 791, 792, 792i
 genetic engineering and,
  222–223, 230, 236
 history of, 694–695, 782
 impact on biodiversity, 410,
  787i, 788
 irrigation and, 792–793
 land for, 764, 766, 780, 783, 786,
  787i, 788, 791, 791i, 792, 794
 nutrients for, 425, 425i, 747, 753
 pesticides and, 47, 47i, 236,
  250–251, 376i, 414, 742
 population growth and,
  694–695, 782
 red tides and, 327
 runoff, 20, 20i, 104, 327, 439i, 776
 soils and, 20, 438, 764
 subsistence vs. mechanized,
  788–799
AIDS, 230, 325, 325t, 591, 592–593,
 592t
Air pollution, defined, 784
 acids in, 31, 31i, 784–785, 785i
 lichens and, 250, 332, 338
 ozone thinning and, 786
 particulates in, 785, 785i
 rainfall patterns and, 18
 sources, 784–785, 784t, 785i,
  786, 794
 stomata and, 445, 445i
Alachlor (Lasso), 47i
Alar, 47i
Albinism, 162, 163i, 169, 169i,
 186, 540
Alcohol, as beverage, 63
 at blood–brain barrier, 503, 505
 effects of addiction, 505, 560

pregnancy and, 673
Alexander the Great, gene flow and,
 402
Algal bloom, 104, 326–327, 327i, 330,
 331, 747, 747i, 779, 779i
Algin, uses of, 328
Alkalosis, 31
Allergen, 590
Allergic rhinitis, 453
Allergy, 587, 590, 594t
 pesticides and, 47
 smoking and, 609i
Alzheimer's disease, 232i, 504, 675
Ama (Asian divers), 641
Amanita poisoning, 334i
Amazon basin deforestation, 788,
 789i
Amish, Old Order, 257
Amnesia, 504
Amniocentesis, 186, 186i, 187
Amoebic dysentery, 322
Amphetamine, 505
Anabolic steroid, 553
Analgesic, 505
Anaphylactic shock, 590
Anemia, 179t, 187, 560, 626t, 627t
Anesthetic, 503
Angel Island, human effects on, 684,
 684i
Angina pectoris, 571
Angiography, 571
Angioplasty, 571
Angiotensin, 529t, 637
Anhidrotic ectodermal dysplasia,
 179t, 216, 216i, 485
Anorexia nervosa, 612
Antacid, 30
Antarctica, trashing of, 736–737
Anthrax, 577
Anti-acne drug, avoiding use
 during pregnancy, 673
Antianxiety drug, 492
Antibiotic
 pregnancy and, 790
 superbugs and, 11, 14
Antibiotic resistance, 11, 14–15,
 14i, 74–75, 75i, 251, 315
 day care centers and, 325
Antihistamine, 590
Anti-inflammatory drug, 590
Antioxidant, 74, 624
Apnea, 607
Appetite control, 629
Apple scab, 337i
Aquifer depletion, 792, 792i
Arctic tundra, vulnerability
 to development, 771i
Arrhythmia, 571, 571i
Arsenic poisoning, 47, 638
Arthritis, 545
Asbestos, 784, 784t
Aspirin
 fevers and, 638, 641
 renal failure and, 790
Asthma, 532, 587, 590
Astigmatism, 515
Atherosclerosis, 570, 638, 675
Atherosclerotic plaque, 222,
 222i, 570i, 571
Athlete
 anabolic steroid use by, 553

blood doping by, 560
 cardiac rates, 571
Athlete's foot, 336i, 578
Atrazine, 47i
Atrial fibrillation, 571
Australia, human population,
 698, 699
Autoimmune response
 aging and, 675
 in Grave's disorder, 531, 591, 703
 in multiple sclerosis, 494, 507, 591
 in rheumatoid arthritis, 545, 591
 in toxic goiter, 531
 in type 1 diabetes, 533
AZT (azidothymidine), 593

B_{12} deficiency, 560
Baby-boomer population, 696, 697i
Bacterium
 beneficial, 311, 312, 615, 615i
 pathogenic, 11, 70, 70i, 311, 313i,
  315, 325t, 571, 575
Baking soda, 30, 30i
Ballerina, skeletal muscle, 548, 548i
Balloon angioplasty, 571
Bangladesh, cholera in, 779, 779i
Barbiturate, 505
Bark beetle, as disease vector, 731t
Basal cell carcinoma, 212i
Behavior, human, 398, 502–505,
 705, 715
Bengal tiger, wildlife trade and, 410i
Benzene, as air pollutant, 784t
Biblical account of great flood, 272,
 274, 247i
Biodiversity
 analysis of, 407, 415
 conservation biology and, 737
 economic growth vs., 415, 416–417
 perspective on, 408–409, 797
 restoration ecology and, 20, 729
 threats to, 410–411
Bioethics
 abortion, 186, 187, 676, 677, 680
 cloning, 198, 199, 232
 competition in science, 13, 191,
  194, 196
 creating life in test tubes, 236
 deforestation, 790
 fertility control, 676
 gene therapy, 233
 genetically engineered food,
  236, 236i
 human cloning, 198, 232, 233
 human impact on biosphere,
  782ff., 797
 limits of science, 15
 repercussions, human population
  growth, 684–685, 694–699, 700,
  782, 799
 in vitro fertilization, 680
Biological therapy versus antibiotics,
 14–15, 14i
Bioremediation of contaminated
 soil, 20, 20i
Biosphere, human impact on,
 750–751, 753, 757, 782ff.
Biotech barnyards, 232
Birth control, 676–677, 677i
Birth, human, 259, 259i, 673

Blood, at crime scenes, 227
Blood circulation
 exercise and, 549, 551, 567
 fetal, 669, 669i
Blood clotting, 570i, 571, 572
Blood doping, 560
Blood fluke infection, 366, 366i
Blood pressure
 in cardiovascular disorders, 367,
  560, 567, 574
Blood transfusion, 160, 560
Blood typing, 160, 160i, 560i, 561
Blue offspring, 179t, 259
Blue whale, overharvesting, 411, 411i
BMI (body mass index), 628
Body build, climate effects on,
 401, 401i, 640
Body-building, 124, 553
Body piercing, 553, 553i
Bone disorders, 528, 532, 532i,
 545, 545i
Booster shot, vaccines, 590
Botulism, 118, 315, 496
Bovine spongiform encephalopathy,
 232, 319
Bradycardia, 571, 571i
Brain
 drug abuse and, 486, 492, 504–
  505, 504t, 505i
Breast cancer, 137, 628, 657,
 674, 677
Breathing
 disorders, 596, 603, 607–609, 611
 at high altitudes, 596–597, 605, 605i
Bronchitis, 608
Brown alga, commercial uses, 328
Brown pelican, DDT and, 742
Brown recluse bite, 381
BSE, 232, 319
"Bubble boys," gene therapy for, 233
Bubonic plague, 681, 694i
Bulimia, 612–613
Bulk, in diet, 623
Bulldog breeding, 239
Burn, skin, 479, 540
BYU biology students, 165i

Caffeine, 503, 504
Calcium
 atherosclerotic plaque and, 571
 human evolution and, 530, 541
 loss from watersheds, 746, 753
California human population,
 698–699
Caloric intake, human, 612,
 612i, 626–627
Camptodactyly, 164, 179t
Canada
 acid deposition and, 785
 human population, 697i
Cancer, 136, 212, 218–219, 589
 bone marrow, 170, 560
 breast, 137, 628, 657, 674, 677
 cervical, 137, 679
 Chernobyl and, 795i
 colon, 623, 628
 endometrial, 677
 Kaposi's sarcoma, 592, 592i
 leukemias, 560, 795i
 lung, 608–609

metastasized, 218, 218i
prostate, 657
respiratory tract, 589, 609i
skin, 212, 212i, 541, 589, 784
smoking and, 608–609, 609i
stomach, 313i
testicular, 657
therapies for, 23, 137, 589
vaginal, 679
vulva, 679
Candida infections, 336i, 337
Cannibalism, Easter Island, 407
Capillary beds, leakage at high
  altitudes, 596–597
Carbohydrates, in human nutrition,
  613, 624–625
Carbon monoxide
  as air pollutant, 784t
  in cigarette smoke, 570
  poisoning, 611
Carcinogen, 209, 212–213, 609, 793
Carcinoma, 212i, 589
Cardiac pacemaker, artificial, 554
Cardiovascular disorders, 367, 560,
  567, 570–571, 574
Carson, R., attacks on, 414
Cataract, of eye, 675
Cattle, 319, 416–417, 416i, 791
  cloned, 198, 198i
  effects on riparian zones, 418–419,
    418i
  genetically engineered, 232
Cell counts, human blood, 559,
  560, 592
Cell cycle and cancer, 130, 136,
  218–219
Cell differentiation
  reversing, for cloning, 198
Cell from Hell, 326i, 327, 327i
Cervical cancer, 137, 679
Cervical cap, 677i
CFC, 750, 751i, 784t, 786
CFTR production, for cystic
  fibrosis, 232
Chagas disease, 323
Chancre, 679
Chemical burn, 30
Chernobyl, 20, 795, 795i
Chestnut blight, 336, 731t, 769
Childbirth, 259, 259i, 483, 526,
  526t, 532, 673, 673i
China, human population growth,
  696–697, 697i
Chlamydial infection, 678, 679, 679i
Chlorine monoxide, 786
Chlorofluorocarbon, 750, 751i,
  784t, 786
Chlorosis, 439t
Chocolate, 353i
Cholera, 587, 695, 751, 779
Cholesterol, 179t, 222, 222i, 570–571,
  570i
Chorionic villi sampling, 186–187
Chromosome abnormalities, 170,
  170i, 172, 179i, 182–185, 182i–185i
Chronic myelogenous leukemia,
  170, 170i
Cirrhosis, 505
Cleft lip, 186, 189
Cloning, 198, 198i, 232
CML, 170, 170i
Coal, 303, 344, 344i, 794
Cocaine, 486, 504, 505–506, 506i, 673
Codeine addiction, 505
Coffee, 30i, 504

Coitus, 665
Cold sore, 318, 541
Cold stress, responses to, 641
Collarbone, fractures, 542
Color blindness, 177, 179t, 181
Condom, 677, 677i
Connective tissue disorders, 179t,
  189
Conservation biology, 407, 414,
  417, 737
  hot spots, 414–415, 415i, 763
  human activities and, 416–417
  restoration efforts, 20, 729, 755, 789
Constipation, 623
Contagious disease, nature of, 694
Contraceptive, 676–677, 677i
Coral bleaching, 412
Corn, rootworm-damaged, 377i
Corn leaf blight, 230
Coronary bypass, 571, 571i
Cortisol, in stress responses, 523i,
  524, 526t, 529t, 530, 530i, 639i, 674
Cotton, uses, 231, 231i, 424i
Cowpox, 576–577
Crack (cocaine), 486, 505, 505i
Crank ("poor man's cocaine"), 505
Creutzfeldt–Jakob disease, 319
Cri-du-chat syndrome, 179t, 183,
  183i
Cystic fibrosis, 73, 73i, 179t, 232

2,4-D poisoning, 462
ddI, and AIDS, 593
DDT, 47, 47i, 742, 742i
Deforestation, 349, 349i, 746, 746i,
  749i, 751i, 753, 788–790, 788i, 789i
Dehydration, 89, 528, 533, 578,
  625, 637
Demographic transition model,
  698, 698i
Demographics, 686
Denaturation, defined, 45
Depo-Provera injection, 677, 677i
Depressant, 505
Dermatitis, 626t
Desalinization, 792
Desertification, 731, 765, 789, 791,
  791i
Designer organ, 479
Development, human
  drug abuse and, 672–673
  embryonic, 135i, 175i, 554,
    666–673, 666i–673i
  fetal, 669–673, 670i–672i
  postnatal, 674–675, 675i
Diabetes insipidus, 528
Diabetes mellitus, 479, 533, 570, 628
Diabetic, insulin requirements, 230
Dialysis, kidney, 638
Diaphragm, birth control device, 677
Diarrheas, 325t, 578
Dieting, 612, 628
Digestion, human
  carbohydrate, 613, 618, 619, 618t,
    621i, 622, 622i, 624
  fat (lipid), 41, 618t, 619, 621i
  protein, 613, 618–622, 618t, 621i,
    622i
Digitalin, 353
Dioxin, 462–463
Diphtheria, 590, 590i
Disease, 323–325, 325t, 576–577
  population growth and, 694–695
DNA cloning, 140, 198, 225, 232

human, 224, 227–229
DNA fingerprinting, 227, 227i
Dodo, humans and, 409i
Dolly (the sheep), 198, 198i, 232
Douching as birth control, 676, 677i
Down's syndrome, 179t, 184, 185i
Downy mildew, 320, 320i
Droughts, 751, 791, 791i
  El Niño and, 778
Drug, defined, 504
  addiction, 504, 504t
  effects on brain, 504–505, 505i
  pregnancy and, 672–673, 672i
Duchenne muscular dystrophy, 221
Dust Bowl, 766, 791i
Dutch elm disease, 731t, 742
Dwarfism, 179t, 221, 528, 528i, 537
Dysentery, amoebic, 322

Earlobe trait, 152, 152i
Easter Island, human impact on,
  406–407
Eating disorder, examples, 612–613
Ebola infection, 595
ECG, 554, 554i, 571, 571i
Economic growth
  biodiversity vs., 414–415
  demographics and, 698–699, 698i
Ecosystem
  conservation, 20, 407, 414–417, 737
  endangered, 755
  water pollution and, 230, 773, 776,
    792–793, 792i, 793i
Ecosystem modeling, 742
Ecstasy (MDMA), 505
Edema, 569, 580–581, 581i, 597
Eicosanoid, 41
Ejaculation, 665
El Niño, 777, 778–779, 779i
Elephantiasis, 367, 569
Ellis-van Creveld syndrome, 178i,
  257
Embolus, 571
Emphysema, 608, 608i
Endangered species, 257, 410, 410i,
  730
Endemic disease, 325, 410
Endometriosis, 661
Endothelium, 564, 564i, 568, 568i
Energy
  human population, 794–796,
    794i–796i
  reserves in human body, 120
Enkephalin, 505
ENSO, 778, 778i, 779i
Enzyme breakdown, aging and,
  74–75
Ephedra, 350, 350i
Epidemic, defined, 325
Epilepsy, 507
Epstein-Barr viral infection, 560
Ergotism, 336
Erosion, 439, 753
Erythroblastosis fetalis, 561
Escherichia coli
  beneficial strains, 314, 316
  biological therapy and, 14i, 15
  harmful strains, 14i, 70i, 199, 315,
    325
Estrogen, breast cancer and, 674
Eugenic engineering, 233
Eutrophication, human-induced,
  747, 773
Exercise, 551

body-building, 124, 553
bone density and, 545
cardiovascular effects, 555i, 567,
  569, 570, 571
dieting and, 612
heat production during, 641
Exotic species, 730–731, 737

"Facts change," 18
Familial cholesterolemia, 179t, 222
Family planning, 696–697
Farmland, 439, 773 (*See also*
  Agriculture)
Farsightedness, 515
Fat, in diet, 40, 612
Feline infectious peritonitis, 257
Fertility, human, 676–677
Fertilization, human, 665, 665i
Fertilizer, 20, 47, 104, 747i, 751i,
  753, 787
Fetal alcohol syndrome, 673, 673i
Fetoscopy, 187, 187i
Fever, 581, 587, 641
Fire ant, 731t
Fire, forest, 766, 766i, 770, 770i
Fish kill, 20, 326–327, 326i
Fishing, commercial, 411, 412
Flatworm infections, 364–366
Flood, biblical, 272–273, 274i
Flu, "24-hour," 325 (*See also*
  Influenza)
Flu shots, 317
Fluke infection, 366, 366i
Food poisoning, 325
Food production (*See* Agriculture)
Foot and mouth disease, 331
Fossil fuel, 35, 794
  carbon cycle and, 748, 749
  origin of, 304, 305, 344
  pollution by, 31, 31i, 35, 753, 784
  population growth and, 694–695,
    698
Fragile X syndrome, 179t, 181, 181i
Franklin, R., attack on, 194, 196, 196i
Free radicals, aging and, 74, 75,
  627, 675
Frog deformities, 391i, 534–535, 534i
Frostbite, 641
Fungicide, 47, 786
Fungus
  beneficial, 332, 334, 337, 338, 338i
  harmful, 334i, 336, 336i
Fusion power, 796

Galactosemia, 179t, 180
Gallstone, 628
Gastroenteritis, 325t
"Gender war," 682
Gene therapy, 222–223, 233, 591i
Genetic counseling, 186
Genetic disorder, nature of, 179, 179t
Genetic engineering, 223, 230–234,
  231i, 234, 236, 790
Genetic screening, 186
Genital warts, 137, 679, 679i
Genome, human, 224, 227,
  232–233, 236
German measles, 672
Giardia infection, 323
Gigantism, 528, 528i
Gingivitis, 617i
Global warming, 35, 750–751,
  751i, 781

Glomerulonephritis, 638
Goiter, 531, 627t
Gonorrhea, 11, 587, 678–679, 678i
Gout, 628
Grateful Dead midge, 799
Graves' disorder, 531, 591
Great Lakes, lampreys in, 387
Green Belt Movement, 789i
Green revolution, 787
Greenhouse gases, 750, 784t, 786
Groundwater depletion, 792, 792i

Habitat islands, 411
Habitat losses, 410–411, 417
Habituation, to drugs, 504, 504t
Haemophilus infection, 224, 228, 233, 590i
Hair, human, 41, 42, 42i, 45, 45i, 162, 162i, 538, 540, 540i
Hallucinogen, 505
Hand, evolution, 398–401, 405, 405i
Hardwood, uses, 433i
Hay fever, 453, 454, 587, 590
HDL, 222, 571
Hearing, 511, 512, 512i, 513, 513i
Heart, human, 359, 529t, 556, 562–569, 562i, 564i, 565i
Heart attack, 222, 232, 532, 570
Heart disease, 570, 608, 609i, 628
Heartburn, 618
Heat stress, 641
Heimlich maneuver, 611, 611i
HeLa cells, 136, 136i
Hemodialysis, 638
Hemoglobin
   blue form, 259
   HbS, 161, 161i, 208, 254–255
Hemophilia, 81i, 177, 179t, 181, 187, 189
Hepatitis, 325t, 590, 590i
Herbicide, 47, 47i, 234, 462–463
Herniated disk, 543
Heroin, 505
Herpes infections, 318, 541, 592, 678, 679
High-density lipoprotein, 222, 571
Hirudo medicinalis, blood-letting, 370i
Histoplasmosis, 336
HIV infection, 317i, 318–319, 576i, 592–593, 678
Hives, 587
Hookworm infection, 358t
Hot spots, environmental, 414–415, 534, 534i, 763
HPV infections, 137, 679
Human
   evolution, 277i, 383, 398–405, 401i
   genome, 224, 227–229, 232–233, 236
   organ systems, overview, 480–481, 480i
   population growth, 407, 417
Human impact on environment, 736–737, 742i, 750–751, 753, 773, 776, 776i
Human papillomavirus infections, 137, 679
Human population
   biotic potential, 689
   growth, 406, 406i, 684–685, 694–700, 694i–700i, 782–783, 782i–783i
Hunchback of Notre Dame, 179
Huntington disorder, 178i, 179t, 180, 181
Hypertension, 532, 570, 628

Hyperthermia, 641
Hyperthyroidism, 531
Hyperventilation, 596
Hypnotic, effect on brain, 505
Hypothermia, 641
Hypothyroidism, 531
Hypoxia, 596

Ice age, humans and, 273
Immigration, human, 701
Immune deficiency, 591, 591i
Immunization, 590, 590i
In vitro fertilization, 187, 198, 680
Incest, 257
India, human population, 684–685, 684i, 697i, 789
Indicator species, 411, 417
Indinavir, AIDS and, 593
Industrial smog, 784
Infectious diseases, 311, 317, 319, 320, 322–325, 576–577
   deadliest, 325t
Influenza, 317, 317i, 587
Inheritance patterns, human
   autosomal dominant, 179t, 180, 180i, 189
   autosomal recessive, 179t, 180, 180i
   sex determination, 172, 174, 174i
   X-linked recessive, 179t, 181, 181i, 189
Insecticides, 47, 47i, 353
Intestinal parasites, 365, 366, 367
Ionizing radiation, 209
Irrigation, 792, 792i
Isometric exercises, 551, 551i
Isotonic exercises, 551, 551i
IUD, 677, 677i

Japanese beetle importation, 731t
Jaundice, 626t
Jellyfish sting, lethal, 362i

Kaposi's sarcoma, 592, 592i
Kidney dialysis, 638
Kidney failure, 638
Kidney stone, 638
"Killer bees," 108–109, 108i
Klinefelter syndrome, 179t, 185
Knee joint, vulnerability, 543
Kudzu, 730–731, 730i
Kuru, 319

Lactation, 674, 674i
Landfill, 787
Larsen Ice Shelf breakup, 755, 755i
Laryngitis, 603
Late blight, 320, 690i
LDL, 222, 571
Leaf blight, 230
Leech, medicinal use, 370i
Leukemia, 170, 170i, 179t, 560, 795i
Life table, U.S. population, 692, 692t
Livestock, 751i, 791 (*See also* Cattle)
Lockjaw, 496
Logging, 416, 416i, 788, 789i
Low-density lipoprotein, 222, 571
LSD, 505
Lung disorders, 325t, 608–609, 608i
Lyme disease, 315, 315i
Lysergic acid diethylamide, 505

Mad cow disease, 232, 319
Malaria, 254–255, 254i, 255i, 324, 324i, 325t, 751
Malignant melanoma, 212, 212i
Malnutrition, 625, 694
Marfan syndrome, 179t
Marijuana, 505, 609
Mass extinction, humans as agents of, 307, 408
MDMA (Ecstasy), 505
Measles, 325t, 590, 590i, 592
Melanoma, malignant, 212, 212i, 589
Meltdown, 794, 795i
Meningitis, 325, 501
Menopause, 661, 663
Menstruation, 660
Metastasis, 218, 218i
Methyl bromide, air pollutant, 786
Mexico, human population, 696, 697i
Mexico City smog, 784i
Mifeprex, 677
Milk of magnesia, 30, 30i
Minerals, in diet, 626–627, 627t
Miscarriage, 184
Monoclonal antibody, 589
Mononucleosis, infectious, 560
Monsoon, 761
Morning-after pill, 677
Motion sickness, 513
Multiple sclerosis, 494, 507, 591
Multiregional model, and human "races," 402
Mumps, 590i
Muscle fatigue, 551
Muscular dystrophy, 179t, 189, 221
Myopia, 515
Myxomatosis, 731

Neanderthal disappearance, 401
Nearsightedness, 515
Nervous system development, teratogens and, 672, 672i
Neurofibromatosis, 179t, 211, 211i
Night blindness, 626t
Nitrogen narcosis, 611
Nitrogen oxides, as air pollutants, 784, 784t
Nonionizing radiation, 209, 212
Nonrenewable energy, 794, 794i
Nonshivering heat production, 641
Nori, 327
Norplant, 677, 677i
Nuclear energy, 794–795, 794i
Nutrition, human, 613, 624–627, 624i, 625i, 626t, 627t, 630

Obesity, 570, 628, 629, 674
Ogallala aquifer, 792
Oil, 28, 40, 48t, 737 (*See also* Fossil fuel)
Oil spill, 230, 794
Olive oil, in human diet, 624
Oncogenes, cancer and, 218
Organ transplant, 589
Organophosphate, 47i
Orgasm, 665
Osteoarthritis, 545, 628
Osteoporosis, 537, 545, 545i, 627t, 629
Overgrazing, 791, 791i
Ozone "hole" (ozone thinning), 786, 786i

Pain, 509, 511, 518, 580t
Pandemic, nature of, 325
Pando the Trembling Giant, 459i
Pap smear, 137
Papillomavirus infection, 137, 679
Parkinson's disease, 504, 505
Pellagra, 626t
Pelvic inflammatory disease, 677, 678, 679
Penicillin resistance, 251
Peptic ulcer, 618
Peregrine falcon, endangered species, 410i, 742, 742i
Periodontal disease, 617i
Peritoneal dialysis, 638
Permanent wave, 42, 42i, 45i
Peroxyacyl nitrates (PANs), 784, 784t
Pertussin, 590i
Pest resurgence, 251
Pesticides, 47, 47i, 638, 742, 784t, 787
   alternatives to, 725
   effects on birds, 414
   resistance to, 250–251, 377i
PET scans, 23, 23i, 503i
   crack cocaine and, 505i
Pharmaceuticals, genetic engineering of, 231
Phenylketonuria, 179t, 186
Philadelphia chromosome, 170, 170i
Photochemical oxidant, 784, 784t
Photochemical smog, 784
Photovoltaic cell, 796, 796i
PID, 679
Pituitary dwarfism, 528, 528i, 537
PKU, 186
Plague, 691
Plant, 437, 444–445, 444i, 445i
   diseases, 230, 319, 320, 320i, 377i
   human uses of, 353, 355
Plasmodium infection, 255, 321, 324, 324i
Pleurisy, 603
Pneumocystis carinii infection, 592
Pneumonia, 251, 325, 325i
Poliovirus, 55i, 587, 590i
Pollutant, 783, 784, 784t, 785i, 798
Pollution, 771i
   air, 18, 31i, 784–785, 784t, 785i, 786, 794
   coral reefs and, 412
   effects on lichens, 332, 338
   effects on stomata, 445i
   water, 230, 534, 534i, 773, 785
Polycythemia, 560
Polydactyly, 178i, 179t
Pope John Paul II, 259
Population growth, human
   Easter Island, 406, 406i
   economy and, 698–699, 698i
   environment and, 782ff.
   limits on, 690–691, 691i, 694–695
   zero, 688, 696–697, 697i, 699
Porphyria, 485
Postpartum depression, 674
Potato famine, 320
Pregnancy, human, 680
   embryonic period, 666, 666–671, 666i, 666i–667i
   events leading to, 665
   fetal period, 670–673, 670i–672i
   nutritional considerations, 672–673
   prenatal diagnosis, 186–187
   risks during, 672–673, 672i
   unwanted, 676

Pregnancy test, 667, 677
Preimplantation diagnosis, 187
Prenatal diagnosis, 186–187
Preven, 677
Prions, 319
Progeria, 179t, 182, 182i
Progestin, 524t, 677
Prostate cancer, 657
Protein-energy malnutrition, 625
Protozoan infections, 322–323, 330
Psychedelic, 505
Puberty, 534, 656
Pulse pressure, 566, 567i

Queen Victoria, family inheritance pattern, 181, 181i

Radiation poisoning, 137, 795i
Radiation therapy, 23
Radioactive fallout, 795i
Ragweed, allergies and, 453, 591i
Rainfall, human-influenced, 18
Ranching, conservationist approach to, 416–417
"Raptures of the deep,"611
Recombinant DNA, and human affairs, 223, 224, 225i, 235, 258
Recycling solid wastes, 787
Red blood cell disorders, 161, 161i
Red tide, 326
Remission, from cancer, 560
Renal failure, 638
Renewable energy, 794, 794i, 796i
Respiratory system
   human, 602–603, 602i, 610
   smoking and, 570, 608–609, 609i
Restoration ecology, 729
Rh blood typing, 561, 561i
Rheumatic fever, 575
Rheumatoid arthritis, 545, 591
RhoGam, 561
Rhythm method, birth control, 676, 677i
Rickets, 532, 532i, 626t
Riparian zone, restoring, 416–417, 416i
Ritonavir, 593
Rocky Mountain spotted fever, 373
Root canal, 617i
Rootworm, pesticides and, 377i
Ross Ice Shelf breakup, 755, 755i
Roundworm infection, 366, 367i,
RU-486, 677
Rubella, 590i, 672, 682

SAD, 534
Sahel, desertification in, 791i
Salinization, 792, 792i
Scarlet fever, 251
Schistosoma infections, 365, 366, 366i
Schistosomiasis, 366
SCIDs, 591i
Scorpion sting, 373
Scrapie, 319
Sea wasp stings, 362i
Seasonal affective disorder, 534
Secondhand smoke, 607, 608
Seed bank, global, 230
Seizure, 505, 507
Selenium contamination, 20, 20i

Sequoia forest fires, 729
Severe combined immunodeficiency, 591, 591i
Sewage sludge, 793i
Sex chromosome, human abnormalities, 184–185
Sex determination, human, 172, 174, 174i
Sexual intercourse, 483
Sexually transmitted diseases, 678–679, 678i, 679i
Shifting cultivation, 788–789
Shivering response, 641
Shoemaker-Levy 9, 304, 306
Sickle-cell anemia, 161, 161i, 179t, 187, 208, 208i, 560
   and malaria, 254–255, 254i, 255i
SIDS, 607
Skeleton, human, 542–543, 543i
Skin, human, 45, 526t, 539, 540–541, 540i, 552, 578t, 585
   pigments, 162–163, 162i, 163i, 526
Slash-and-burn agriculture, 788
Smallpox, 576, 576i
Smog, 784, 784i, 798
Smoking, 570, 608–609, 609i, 673
Sneezing, 179t
Soft spot (skull), 545
Softwood, uses of, 433i
Soil, 438–439, 438i, 764, 764, 764i
   pollution, 753, 785
Solar-hydrogen energy, 796, 796i
Solid wastes, recycling and, 787
Spanish flu, 2–3
Species introductions, 387, 725, 730–731, 730i, 731i, 731t
Species names for sale, 799
Spermaticide, 676–677, 677i
Spinal meningitis, 325, 501
Sporadic disease, nature of, 325
Sporozoan infections, 320, 324, 324i, 325t
Sprain, 545
Squamous cell carcinoma, 212i
Staph infection, 11
Staphylococcus, 313i, 325
Starling, as exotic species, 702–703
STDs, 678–679, 678i, 679i
Sterility, human, 179t, 678, 679
Stimulant, effect on brain, 504
Stomach acid, human, 30, 30i, 532
Strain, muscle, 545
Strength training, 551
Streptococcus infections, 316, 325
Stress electrocardiogram, 571
Stress response, 530, 530i
Strip logging, 416, 416i
Stroke (brain damage), 570
Subsistence agriculture, 787
Substance P, 492
Sucrose polyester, 625
Sudden infant death syndrome, 607
Sulfonamide, 679
Sulfur oxide pollutants, 31i, 784, 784t, 785, 794
Sunburn, 541
Sunlight, 77i, 96, 96i
   climate and, 757, 758–759, 759i
   seasonal depression and, 534
   skin cancer and, 212, 212i, 541
Suntan, 541
Supernatural, defined, 12
Sustainable development, vs. biodiveristy, 415

Swallowing, 603, 611, 611i, 617
Sweating, 29, 641
Syndrome, defined, 179
Syphilis, 679, 679i

2,4,5-T, 462–463
Tachycardia, 571, 571i
Taenia saginata infection, 366i
Tapeworm infections, 358t, 365, 365i, 366i
Taxol, cancer and, 66, 137
Tea leaves, harvesting, 353i
Temperature, body
   rhythm method, 676, 677i
   scrotum and, 656
Temperature, environmental
   greenhouse effect and, 749, 750–751, 750i, 758, 789, 794
Temperature zone, 758, 759i
Testicular cancer, 657
Testicular feminization syndrome, 179t, 524
Tetanus (disease), 118, 551, 325t, 590
Tetracycline, 673, 679
Thalassemia, 560
Thalidomide, 672–673
Therapeutic vaccine, 589
Thermal inversion, 784, 784i
Thrombus, 571
Tick, disease vector, 373
Titanic, 643
Tolerance, drug, 504, 504t
Tongue-rolling trait, 168, 168i
Tooth decay, 617
Tourism, Antarctica, 737, 737i
Toxoplasma infection, 324, 324i
TPA, 232
Transfusion, blood, 160
Trichinella infection, 367, 367i
Trichomonad infections, 323, 323i
Trichomoniasis, 323i
Trisomy 21, 184
Tropical forest, deforestation, 728–729, 732, 762i, 763i, 764, 764i, 768, 768i, 788, 788i, 790
Trusting turkey, 18
Trypanosoma infections, 311i, 323, 323i
Tubal ligation, 676–677, 677i
Tuberculosis, 11, 251, 259, 325, 325t
Tumor, 212, 218i, 219, 589
Tundra, vulnerability, 771i
Turner syndrome, 179t, 185
Twins, human 651, 651i
Typhoid fever, 11, 320

Ulcer, 313i, 532
Ultraviolet radiation
   harmful effects, 209, 212, 212i
   ozone layer and, 786
Umami, 517
United States, human population 692t, 696, 696i, 697i, 698, 699, 701, 701i
Upwelling, fishing and, 777, 778
Uremic toxicity, 638
Urinary bladder, cancer, 609i

Vaccination, 577, 590, 590i
Vaccines, 192, 324, 590, 590i, 593, 790

Valproic acid, 507
Vampire stories, porphyria and, 485
Vancomycin, 251
Varicose veins, 575
Vas deferens, 656, 656t, 657i, 658i, 676
Vasectomy, 676
Vegetarian diets, 560, 625, 625i, 627
Ventricular fibrillation, 571, 571i
Vertigo, 519
Visual accommodation, 515
Visual cortex, 503i, 516i
Vitamins, human diet, 626, 626t
   A, 627
   B, 541, 560, 672
   C, 38, 74, 609i, 627, 672
   D, 41, 222, 524t, 525, 532, 537, 540, 541, 545, 627
   E, 74, 627
   K, 314, 627
Vomiting, 503, 505, 613

Wastewater treatment, 793, 793i
Water pollution, 230, 773, 776, 792–793
Water tables, declining, 792
Water wars, 793
Watershed clearing, city water supplies and, 746, 788
Weight, human, 612–613, 628–629, 628i
Weight training, 124, 553
Werner's syndrome, 675
West Nile virus, 751
Wetlands, human impact on, 20, 20i, 776–777, 776i, 781
Wheat, domestication, 266, 267i, 284i, 353i
White blood cell disorders, 179t, 533, 560
Whooping cough, 325, 590i
Wildlife trade, illegal, 410, 410i
Wind energy, 796, 796i
Winter blues , 534
Withdrawal, as birth control, 676, 677i
World Wildlife Fund ecoregion designations, 415i
Wrasse, 264i
Wuchereria bancrofti infection, 367, 367i

X-linked anhidrotic dysplasia, 179t, 216, 216i
X-linked recessive inheritance, human, 179t, 181, 181i
X-rays, cumulative effects, 96i, 133, 209
XYY condition, 179t, 185

Yeast infections, 337
Yersinia pestis infections, 691

Zero growth implications, human population, 688